Indoor Air Quality Engineering

Indoor Air Quality Engineering

Environmental Health and Control of Indoor Pollutants

Robert Jennings Heinsohn
John M. Cimbala
The Pennsylvania State University
University Park, Pennsylvania, U.S.A.

CRC Press
Taylor & Francis Group
Boca Raton London New York

CRC Press is an imprint of the
Taylor & Francis Group, an informa business

Portions of this text are reprinted from *Sources and Control of Air Pollution* by Heinsohn/Kabel, ©1999 Reprinted by permission of Prentice-Hall, Inc., Upper Saddle River, NJ.

Reprinted 2009 by CRC Press

Library of Congress Cataloging-in-Publication Data
A catalog record for this book is available from the Library of Congress.

ISBN: 0-8247-4061-0

This book is printed on acid-free paper.

Headquarters
Marcel Dekker, Inc.
270 Madison Avenue, New York, NY 10016
tel: 212-696-9000; fax: 212-685-4540

Eastern Hemisphere Distribution
Marcel Dekker AG
Hutgasse 4, Postfach 812, CH-4001 Basel, Switzerland
tel: 41-61-260-6300; fax: 41-61-260-6333

World Wide Web
http://www.dekker.com

The publisher offers discounts on this book when ordered in bulk quantities. For more information, write to Special Sales/Professional Marketing at the headquarters address above.

Current printing (last digit):
10 9 8 7 6 5 4 3 2 1

Preface

Background

Indoor air quality (IAQ) is the field of applied science concerned with controlling the quality of air inside buildings and other enclosures (automobiles, tunnels, etc.) to ensure healthy conditions for workers and the public, and a clean environment for the manufacture of products. While indoor air pollution is of primary concern, other factors such as noise, temperature, and odors also contribute to the quality of indoor air. To claim that all airborne contaminants can be eliminated from the indoor environment is naive and unachievable. More to the point and within the realm of achievement is the goal of controlling contaminant *exposure* within prescribed limits. To accomplish this goal, one must be able to describe the movement of particulate and gaseous contaminants in quantitative terms that take into account:

- the spatial and temporal rate at which contaminants are generated and emitted
- the velocity field of air in the indoor environment
- the spatial relationship between source and workers, and the openings through which air is withdrawn or added
- exposure limits (time-concentration relationships) that define unhealthy conditions

There is a sizable body of information on indoor air quality control. Many books have been written in which the conclusions and recommendations, while useful, fail to address the subject with the organization, scientific foundation, and analytical and mathematical rigor expected of today's engineers. For example, engineers are often asked to predict the performance of ventilation systems to ensure that they satisfy government or more stringent company standards, prior to constructing and testing the entire system. The design of ventilation systems requires making decisions in several stages:

(1) identify the contaminants and understand their effects on health
(2) select the maximum exposure limits that will be used as design criteria and standards to judge the system's performance
(3) design the "hood" and select the volumetric flow rates of exhaust (captured) air, recirculated air, and make-up air
(4) design the duct system, select the fans, and compute the operating costs
(5) select the gas cleaning device to remove contaminants from the captured air before discharge to the outside environment or recirculation to the indoor environment
(6) conduct laboratory experiments of the ventilation system, test the full scale system in the field, and sample the air in the vicinity of the source and workers to ensure that item (2) above is satisfied

Presently the body of knowledge on these subjects consists of general design guidelines that lack precision and detail. For example, present practices provide no way to estimate the mass concentration, c(x,y,z,t), at locations where workers are apt to be stationed. Thus, when a process produces contaminants, it is difficult to determine whether installation of a control system is necessary. Furthermore, if a control system is installed, it is difficult to determine whether it reduces concentrations to safe levels. It is thus prudent that we develop the ability to make such predictions and include them as integral steps in the design of products, systems, or processes. No longer can we afford to examine health and safety considerations *after* a design has been completed, and then to modify the design to satisfy these considerations by costly cycles of testing and retrofit.

Occupational health and safety, and indoor air pollution control in particular, is of interest to three types of students:

(a) engineering students wishing to incorporate indoor air pollution control in their studies
(b) engineering students planning careers in air pollution control
(c) students whose baccalaureate degrees are not in engineering but who are enrolled in graduate programs of industrial hygiene, environmental health engineering, etc., and who seek a rigorous presentation based on mathematics and principles of fluid mechanics.

Objective and Goals

The objective of this book is to establish fundamental quantitative relationships that describe the generation of particulate and gaseous contaminants, and that govern their movement in the indoor environment. Once established, these relationships can become the tools with which designers predict contaminant exposure. Individuals who design and operate systems to ensure indoor air quality function as engineers irrespective of their formal education. This book is therefore addressed to engineers and non-engineers alike, who are engaged in controlling indoor air pollution. This book is written as:

- a textbook for university senior-level and graduate-level courses in mechanical, civil, chemical, environmental, architectural, mining, agricultural, and biomedical engineering, and in industrial hygiene
- a source book for professionals enrolled in continuing education courses, workshops, and symposia
- a reference book for working professionals

It is assumed that readers are familiar with the fundamentals of calculus, vector notation, thermodynamics, fluid mechanics, mass transfer, and heat transfer. A background in numerical computation is not required, although many of the in-chapter illustrated examples require such computations.

The primary goal of this book is to organize existing information in ways compatible with other topics in baccalaureate engineering disciplines. A secondary goal is to contribute to the body of knowledge itself in order to advance our understanding of controlling contaminants in the indoor air environment. Specifically, this book seeks to instruct in:

- medical, economic, regulatory, and professional aspects of indoor air quality control
- contemporary engineering practices
- bodies of knowledge available in other disciplines that can be used to predict contaminant concentrations
- developments in computational fluid dynamics (CFD), which is rapidly becoming a useful tool for indoor air quality engineering

This book seeks to bridge the disciplines of engineering and industrial hygiene. Engineering knowledge contains the fundamental relationships needed to describe the movement of contaminants in the indoor environment; industrial hygiene contains knowledge that relates time and concentration to conditions injurious to health. Both disciplines are needed to create quantitative relationships that become criteria upon which products, systems, and processes can be designed.

Computational Tools

Computers make it possible to predict concentration fields in ways not possible in the past. As computer hardware and software improve, so also will our ability to analyze more complex flow fields. Computer techniques are integrated into this book. The three primary computer programs used are ***Microsoft Excel*** (spreadsheet), ***Mathcad*** (mathematical calculation software), and ***Fluent*** (computational fluid dynamics software). Much of what is done with these programs can also be performed with BASIC, FORTRAN, or C programs, if preferred. Several Excel and Mathcad files have been placed on this book's website as examples for use by the readers. Readers are encouraged to download and modify the files, in order to help them learn how to use the software.

The text is supplemented by a website at **www.mne.psu.edu/cimbala/** (click on Authoring, and then on the Heinsohn/Cimbala link), which contains the following material:

- Excel and Mathcad example files, including those used to solve most of the example problems in this book
- links to websites of relevance to this book
- additional material not included in this book
- computer program for designing a ventilation system

A unique feature of this book is the inclusion of CFD to predict the velocity and pollutant concentrations at arbitrary points inside a building or room containing one or more sources of pollution, one or more exhaust ducts, and one or more inlet fresh air ducts. While CFD is typically taught only in advanced courses in fluid mechanics, the availability of commercial CFD programs (such as Fluent) makes it possible for undergraduates to acquire proficiency heretofore the domain of graduate students.

Organization

This book is organized into ten chapters beginning with fundamental concepts of risk, contaminant concentration, and the respiratory system; and ending with the elements of future computational methods that are used to predict contaminant concentrations in the vicinity of processes equipped with industrial ventilation systems. Chapters 1 and 2 establish a quantitative relationship between air contaminants and risks to health. Chapter 3 defines criteria engineers must consider in the design of ventilation systems. Chapter 4 provides ways to estimate the rate at which contaminants are generated and emitted by industrial processes. Chapter 5 shows how the assumption of "well mixed" can be used to predict contaminant concentrations in rooms, containers, and tunnels. Chapter 6 summarizes contemporary practices used in the field of industrial ventilation. Chapter 7 develops the fundamental equations of irrotational flow, which can be used to approximate the velocity field of air and contaminants upstream of inlets. Chapter 8 presents the fundamental equations describing the movement of particles in air. Chapter 9 describes the physical processes governing the performance of devices designed to remove liquid and solid particles from a process gas stream. Chapter 10 describes the fundamentals concepts of CFD and illustrates its use to model the performance of local ventilation systems and air pollution control devices.

Each chapter contains homework problems suitable for use by instructors. Throughout the text accessible references in the professional literature are cited, a complete list of which appears at the end of the book; readers should become familiar with some of these references. Important terms in each chapter are bolded and italicized, and are listed in the Index. At the beginning of this book is a table of nomenclature that defines terms in the operational dimensions of mass (M), length (L), time (t), amount of matter (N), force (F), temperature (T), electrical charge (C), voltage (V), and energy (Q).

For convenience, operational dimensions are used rather than only primary dimensions (M, L, t, N, etc.).

The Appendices contain data needed in the homework problems. In addition they are organized to illustrate the technical data engineers need in professional practice. Instructors should place the following source materials on reserve in the library and create homework problems or class projects that require students to use this material:

- OSHA General Industry Standards
- ACGIH Industrial Ventilation, A Manual of Recommended Practice
- ASHRAE Handbooks

It is important that students become familiar with these original materials, not merely abridgments of them. Instructors should locate websites and source books of thermophysical data in the reference section of the library and assign homework that requires students to locate data not provided by the text, as this will be required of students after graduation.

Acknowledgments

The authors would like to acknowledge two people who contributed substantially to the editing of this book. Professor William P. Bahnfleth, Ph.D., Penn State Department of Architectural Engineering, helped to edit all the chapters, and contributed to the portions of this book that concern room ventilation and ASHRAE standards. Dr. Keith S. Karn, Ph.D., Xerox Corporation, and Adjunct Assistant Professor, University of Rochester, helped to edit the book in its early stages, providing us with some very valuable suggestions that have made this book more readable and introduced readers to the importance and practical applications of various topics in the book.

Finally, special thanks must go to our families, especially our wives, Anne Heinsohn and Suzanne Cimbala, for putting up with the many long hours when their husbands' faces were unseen – their heads being buried in either a book or a computer screen.

Robert Jennings Heinsohn
John M. Cimbala

Contents

Nomenclature

English symbols

symbol	description	dimensions*
a	radius of airway	L
a	speed of sound (acoustic velocity)	L/t
a, b	one-half major and minor diameter of an elliptical opening	L
a, b, c	nondimensional constants	-
a_1, a_2, a_3, a_4	nondimensional constants	-
a_j	aggregate risk per person in a population	-
a_p	surface area of bed packing per volume of bed	L^{-1}
A	area, airway or spray chamber cross-sectional area	L^2
A	coefficient in standard ordinary differential equation	-
A	annual cost	$
A	nondimensional constant or parameter	-
A	width of finite rectangular opening	L
A′	open cross-sectional area within a filter	L^2
A_0	duct cross-sectional area at section 0, frontal area of a filter	L^2
A_b	cross-sectional area of branch section of a wye, or of flowing bulk materials	L^2
A_{bed}	cross-sectional area of a transverse packed bed scrubber	L^2
A_B	cross-sectional area of blowing inlet	L^2
A_c	cross-sectional area of a duct, or of the common (outlet) section of a wye	L^2
A_f	total filtration area in a baghouse, face area of a filter	L^2
A_{face}	face area of an inlet or hood	L^2
A_i	inlet area, constant	L^2
A_i	nondimensional constant	-
A_j	aggregate risk from pollutant j	-
A_j	cross-sectional area of a push jet	L^2
A_L	effective air leakage area for calculation of infiltration rate	L^2
$A_{packing}$	surface area of an element of packing material in a transverse packed bed scrubber	L^2
$A_{packing,total}$	total surface area of all packing elements in a transverse packed bed scrubber	L^2
A_r	aspect ratio	-
A_s	body surface area, surface area, surface area of enclosure over which adsorption occurs, cross-sectional area of straight (inlet) section of a wye, surface area of hot surface, total collection surface area in an ESP	L^2
A_S	cross-sectional area of suction inlet	L^2

symbol	description	dimensions*
A_u	cross-sectional area of plume at elevation (Z)	L^2
A_x, A_y, A_z	components of the area	L^2
b	radius to outer electrode in an ESP	L
b, b_1, b_2	nondimensional constants	-
B	coefficient in standard ordinary differential equation, nondimensional constant or nondimensional parameter	-
B	constant defined by the one-hit equation	1/t
B	height of finite rectangular opening	L
B_0	nondimensional constant	-
c	speed of light	L/t
c	radius of circular inlet	L
c, c_j	mass concentration, mass concentration of species j	M/L^3
$\overline{c}$	average mass concentration	M/L^3
c(0), c(∞)	initial and final mass concentration	M/L^3
$c(D_p)$	local mass concentration of particles of diameter D_p	M/L^3
$c(D_{p,j})$	mass concentration of particles in the jth interval or class	M/L^3
$\overline{c}(D_p)$	average dust mass concentration in an enclosure as a function of particle diameter	M/L^3
c(i+1)	mass concentration determined in iteration number (i+1)	M/L^3
c_a	ambient mass concentration	M/L^3
c_{bulk}	bulk mass concentration	M/L^3
c_D	particle drag coefficient	-
c_f, c_L, c_H	final mass concentration, mass concentration at L or H	M/L^3
c_i, c_0	inlet, free-stream, or far-field mass concentration	M/L^3
c_{molar}, $c_{molar,j}$	molar concentration, molar concentration of species j	N/L^3
$\overline{c_{molar,j}}$	partial molar concentration of molecular species j	N/L^3
$c_{molar,L}$	average or total molar density of the liquid phase	N/L^3
$c_{molecular,j}$	molecular concentration (number of molecules per unit volume) of species j	$1/L^3$
c_{number}	number concentration (number of particles per unit volume)	L^{-3}
c_{oz}	mass concentration in occupied zone	M/L^3
c_r	pollutant mass concentration in the recirculation eddy	M/L^3
c_s	mass concentration in supply air	M/L^3
c_{ss}	steady-state mass concentration	M/L^3
c_v	specific heat at constant volume	Q/MT
c_p	specific heat at constant pressure	Q/MT
C	lung capacitance	L^5/F
C	Bernoulli constant	L^2/t^2
C	slip factor (Cunningham correction factor)	-
C_0	local loss coefficient of a minor loss (fitting, inlet, etc.)	-
C_b	local loss coefficient for branch section of a wye	-
C_f, C_{dc}	nondimensional constants associated with the pressure drop across filter material and dust cake respectively	-

symbol	description	dimensions*
C_{fu}	cost per unit of fuel	-
C_h	local loss coefficient for hood entry	-
C_H	head coefficient for a fan	-
C_L, C_R	capacitance of left and right lung	L^5/F
C_n	constant used to describe curvilinear flow	-
C_P	power coefficient for a fan	-
C_Q	capacity coefficient for a fan	-
C_s	stack coefficient for calculation of infiltration rate	$L^2t^{-2}T^{-1}$
C_s	local loss coefficient for a slot, a screen, or the main section of a wye	-
C_w	wind coefficient for calculation of infiltration rate	-
d	airway diameter, distance upwind of inlet face, distance between source and hood	L
d_n	collision diameter	L
d_p	depth of a pool of liquid	L
D	molecular diffusion coefficient, particle diffusion coefficient	L^2/t
D'	virtual diffusion coefficient	L^2/t
D	nondimensional constant, dilution factor, dilution ratio	-
D	diameter (of tank, tunnel, circular inlet, stack exit, etc.)	L
$\vec{D}$	directional vector	-
D_0	diameter of a spill	L
D_1, D_2, D_d, D_e	characteristic diameters of a standard Lapple cyclone	L
D_c	diameter of a collecting particle, collecting drop, or cloud; characteristic diameter of packing elements in a transverse packed bed scrubber	L
D_e	exit diameter of a duct	L
$D_{e,p}$	equivalent volume diameter of a particle	L
D_f	characteristic diameter of a single filter fiber	L
D_{ja}	diffusion coefficient of a contaminant (j) in air (a)	L^2/t
D_{jw}	diffusion coefficient of a contaminant (j) in water (w)	L^2/t
D_{min}	dose rate	M/t
D_p	diameter of a particle or drop	L
$D_{packing}$	average overall packing diameter of the irregularly shaped packing elements in a transverse packed bed scrubber	L
$D_{p,15.9}$	particle diameter at which cumulative distribution is 15.9%	L
$D_{p,50}$	median particle diameter (cumulative distribution is 50%)	L
$D_{p,84.1}$	particle diameter at which cumulative distribution is 84.1%	L
$D_{p,act}$, $D_{p,aero}$	actual and aerodynamic diameters of a particle	L
$D_{p,am}$	arithmetic mean particle diameter	L
$D_{p,cut}$	cut diameter (diameter of a particle whose collection efficiency in a cyclone is 50%)	L
$D_{p,gm}$	geometric mean particle diameter	L
$D_{p,j}$	mean particle diameter in the jth interval or class	L

symbol	description	dimensions*
$D_{p,m}$	mass median particle diameter	L
$D_{p,min}$	diameter of the most penetrating particle in a filter	L
$D_{p,p}$	equivalent projected area diameter of a particle	L
$D_{packing}$	average diameter of an element of packing	L
D_s	diameter of suction (exhaust) duct, duct diameter in straight section of a wye	L
$D_{s,p}$	equivalent surface area diameter of a particle	L
D_t	total dose	M
D_y	hydraulic diameter	L
DD_h	annual heating degree days	tT
e	effectiveness coefficient	-
E	photon energy	Q
E	nondimensional constant, parameter used in design of a push-pull ventilation system	-
E_∞	electric field intensity acting on a charged particle	V/L
E(t)	exposure as a function of time	Nt/L^3
E_c	electric field intensity at the outer edge of a corona	V/L
EF_c	automobile emission factor	M/L
E_{max}	maximum evaporation rate	Q/t
E_r	radial component of the electric field intensity	V/L
E_{req}	required evaporation rate	Q/t
En	nondimensional parameter to measure collective hazard	-
f	frequency, breathing frequency	t^{-1}
f	Darcy friction factor, dimensionless filling factor, fraction of contaminant remaining (Bohr model), probability density function (PDF)	-
f(R)	nondimensional function of R	-
f_d	solids fraction of dust cake	-
f_f	fit factor, fraction of a plume in the recirculation eddy, solids fraction of filter	-
f_i	volumetric flow rate factor for j^{th} component, Q_i/Q	-
f_j	mass fraction of molecular species j	-
f_{LK}	leak fraction	-
F_n	normal force	F
f_r	fraction of a plume captured by a recirculation eddy	-
f_{su}	plume surge frequency	t^{-1}
f_w	powder moisture content	-
F	buoyant flux parameter	L^4/t^3
F	distance canopy hood overhangs open vessel	L
F	future worth	$
F	nondimensional constant, bioavailability	-
F_{drag}	magnitude of the drag force	F
F_{el}	force on a charged particle due to electric field	F

symbol	description	dimensions*
F_s	dimensionless factor used in definition of ECL	-
FEV_1	forced expiratory volume in one second	L^3
$g(\alpha)$	particle removal rate by grinding	-
$g(D_p)_j$	mass fraction of particles in size range or class j	-
g	acceleration of gravity	L/t^2
G	nondimensional constant	-
h	height of liquid in a tank, distance between actual source and inlet face of canopy	L
h	heat transfer coefficient (also called film coefficient)	$Q/(tL^2T)$
$\hat{h}$	specific enthalpy (enthalpy per unit mass)	Q/M
h_0	height of dividing streamline at x = 0	L
h_∞	height of dividing streamline at x = infinity	L
h_c	convective heat transfer coefficient (film coefficient)	$Q/(tL^2T)$
$\hat{h}_{\text{dry air}}$	specific enthalpy of dry air	Q/M
$\hat{h}_{fg}$	specific enthalpy of vaporization	Q/M
$\hat{h}_g$	specific enthalpy of a saturated vapor	Q/M
$\hat{h}_{rp}$	specific enthalpy change associated with combustion	Q/M
h_{Lf}	major head loss	L
h_{LM}	minor head loss	L
h_{LT}	total head loss	L
h_s	geometric height of a stack above the height of a building	L
$\hat{h}_w$	specific enthalpy of water vapor	Q/M
H	Henry's law constant	FL/N
H	distance between liquid surface and face of canopy hood, filter thickness, height of a building, duct, enclosure, suction inlet baffle for push-pull system, or ESP collecting electrode, drop height of bulk materials	L
H'	alternative Henry's law constant	FL^{-2}
H''	nondimensional Henry's law constant	-
H_c	height of roof leading edge recirculation cavity	L
H_{dw}	downwash distance	L
H_r	height of a building eddy	L
i	annual rate of interest	-
$\hat{i}, \hat{j}, \hat{k}$	unit vectors in the x, y and z directions	-
I	intensity of an odor sensation (units vary)	-
I	sound intensity	Q/tL^2
I	electrical current in an ESP	C/t
I_0	reference value of sound intensity	Q/tL^2
I_i	incidence of occupational injury and illness	-

symbol	description	dimensions*
I_{lw}	Incidence of lost workdays due to occupational injury or illness	-
J	nondimensional constant, Jacobian, number of intervals or classes in a grouped data set	-
J_j	molar flux of molecular species j per unit area	N/L^2t
k	Boltzmann's constant = 1.38×10^{-23} J/(molecule K)	Q/T
k	constant of proportionality, ratio of specific heats (c_p/c_v)	-
k	thermal conductivity	Q/(tLT)
k	tunnel wall loss parameter	t^{-1}
k	turbulent kinetic energy per unit mass, k-ε turbulence model	L^2/t^2
k	volumetric flow rate per unit length, Q/(πL) for potential flows	L^2/t
k, k_1, k_2, etc.	nondimensional constants	-
$k_0(T,\Phi)$	mass transfer coefficient at a given temperature and relative humidity	L^3/t
k_{ab}	absorption coefficient	L^3/t
k_b	solubility of a contaminant in the blood	-
k_c	convective mass transfer coefficient	L/t
k_G	individual mass transfer coefficient for gas phase	$NF^{-1}t^{-1}$
k_L	individual mass transfer coefficient for liquid phase	L/t
$k_{m,j}$	mass transfer coefficient of molecular species j in the layer	L/t
k_r	first-order rate constant for removal of any contaminant	L^3/t
$k_{s,j}$	solubility constant of molecular species j in a layer	-
k_t	thermal conductivity of airway wall	Q/MtT
k_w	wall-loss coefficient or deposition velocity	L/t
K	dispersion coefficient	L
K	mass transfer coefficient, overall mass transfer coefficient	L/t
K	eddy diffusivity	L^2/t
K, K_1, K_2, etc.	nondimensional constants	-
K_0, K_L	nondimensional constants	-
K_f	residual drag of a filter	FtL^{-3}
K_{dc}	dust cake specific resistance	t^{-1}
K_G	overall mass transfer coefficient for gas phase	$NF^{-1}t^{-1}$
K_L	overall mass transfer coefficient for liquid phase	L/t
l	length of a fluid pathline from the inlet to a point in a room	L
l	penetration, pulvation, or stopping distance	L
L	characteristic length, long dimension, length of a building in the windward direction, length of ESP collecting electrode, length of straight-through cyclone, length of tunnel, length of duct, length of slot perpendicular to x-y plane, length of open vessel, longitudinal length of an airway, flange width, height of spray chamber	L
L	nondimensional constant, dust to displaced powder ratio (mg dust/kg displaced powder)	-

symbol	description	dimensions*
L'	hypothetical height	L
L_{dc}, $L_{dc}(t)$	dust cake thickness, dust cake thickness as a function of time	L
L/d	airway aspect ratio	-
L_1, L_2, L_3	characteristic lengths of a standard Lapple cyclone	L
L_c	length of roof leading edge recirculation cavity, critical length	L
L_e	length of airway needed for fully developed flow, equivalent length of duct	L
L_f	length of filter fiber per unit volume of filter $(4f_f/\pi D_f^2)$	L^{-2}
L_i	exposure limit for molecular species i	M/L^3
L_L	larger of H (building height) or W (building width)	L
L_p, L_I, L_w	sound level defined in terms of pressure, intensity and power	-
L_r	loading rate (number of containers per unit time)	t^{-1}
L_S	smaller of H (building height) or W (building width)	L
L_z, L_y	distance between flanking planes and center of rectangular inlet	L
m	mixing factor, safety factor, coefficient or exponent	-
m, m_j	mass, mass of molecular species j, mass of an ion	M
$\dot{m}$	mass flow rate	M/t
$m(D_p)$	mass distribution function	L^{-1}
m_a	mass of air	M
m_b	mass (of body, bulk materials in a vessel)	M
m_{bb}	body burden	M
$\dot{m}_b$, $\dot{m}_d$	mass flow rate of bulk materials, of pollutant discharge	M/t
$\dot{m}_{evap,t}$	total evaporation mass flow rate	M/t
m_g	mass of inert gas blanket	M
$\dot{m}_g$	pollutant generation expressed as a mass flow rate	M/t
$\dot{m}_j$	mass flow rate, emission rate, or evaporation rate of molecular species j	M/t
m_p	average mass of particle matter exhaled per cigarette	M
m_{puff}	mass released in a puff	M
m_s	mass of organic solvent	M
m_t	total mass of a mixture, total mass of particles	M
m_t'	metal removal rate by grinding	M/t
$\dot{M}$	metabolic rate	Q/t
M	nondimensional constant, moisture content (%)	-
M, M_j	molecular weight, molecular weight of species j	M/N
$M(D_p)$	cumulative mass distribution function	-
$\dot{M}_b$	basal metabolic rate	Q/t
M_j	jet momentum flux	ML/t^2
M_u	plume momentum flux	ML/t^2

symbol	description	dimensions*
n	nondimensional constant, exponent in Steven's law for odors, number of cigarettes per room, number of encounters between particle and falling drop, exponent for curvilinear flow, normalized free area of a screen	-
n, n_j	number of mols, number of mols of species j	N
$n(D_p)$	number distribution function	L^{-1}
n_c	traffic count, automobiles/km	L^{-1}
n_c	number of collectors per volume of chamber	L^{-3}
n_j	number of particles in class j, of diameter $D_{p,j}$	-
n_t	total mols of a mixture, total number of mols	N
n_t	total number of particles	-
N	number of persons in a population	-
N	number of room air changes per unit time (N = Q/V)	t^{-1}
N	number of line sinks	-
N	number of ions/volume	L^{-3}
$N(D_p)$	cumulative number distribution function	-
N_b	number of bags in a baghouse	-
N_D	diffusion parameter	-
N_e	number of turns in a cyclone	-
N_i	total number of injuries and illness	-
N_{iw}	total number of lost workdays	-
N_j	molar flux of molecular species j	N/L^2 t
N_j	cumulative distribution for particles less than $D_{p,max,j}$	-
N_S	specific speed (a dimensionless coefficient) for a fan	-
N_w	number mols of water per unit area and time	N/L^2t
O_{oxy}	oxygen consumption rate	L^3/t
p, q	nondimensional constants	-
P	penetration of a filter ($P = 1 - \eta$)	-
P	perimeter	L
P	present worth	$
P	pressure (static or overall)	F/L^2
P_0	background probability	-
P_0	inlet partial pressure, far-field pressure, sound reference pressure	F/L^2
P_∞	far-field pressure	F/L^2
P^*	pressure at choked flow	F/L^2
P′	airway perimeter	L
P_a	partial pressure in alveolar region, atmospheric pressure	F/L^2
P_{am}	log mean partial pressure difference	F/L^2
P_b, P_{blood}	partial pressure of a species in the blood	F/L^2
P_c	critical pressure	F/L^2
$P_{c,sat}$	saturation pressure of molecular species c	F/L^2
P_d	excess risk probability due to an atmospheric pollutant	-

symbol	description	dimensions*
$P_{i,g}$, $P_{i,b}$	partial pressure of molecular species i in gas or blood	F/L^2
P_j	partial pressure of species j	F/L^2
P_j^*	hypothetical (star state) partial pressure of species j	F/L^2
P_n	population	-
P_r	partial pressure in arterial blood	F/L^2
P_{sat}	saturation pressure (also called vapor pressure)	F/L^2
P_t	total pressure	F/L^2
P_v	partial pressure in venous blood	F/L^2
P_v	vapor pressure (also called saturation pressure)	F/L^2
$P_v(T)$	vapor pressure (saturation pressure) at temperature T	F/L^2
$P_{v,j}$	vapor pressure of molecular species j	F/L^2
P_y	partial pressure at an arbitrary location (y) in a capillary	F/L^2
q	turbulent fluctuation velocity scale (square root of turbulent kinetic energy per unit mass), q-ω turbulence model	L/t
q_e	charge on an electron, coulombs/ion	C
q_{fu}	available energy per unit of fuel	Q/M
q_l, q_b	total rate of heat transfer from the liquid and blood	Q/t
q_m, q_e	make-up and exhaust volumetric flow rate per volume of tunnel	t^{-1}
q_p	electrical charge per particle	C
q_{pd}	diffusion charge per particle	C
q_{pf}	field charge per particle	C
$q_{pf,s}$	saturation field charge on a particle	C
q_v	space charge density	C/L^3
Q	sound directivity	-
Q	inhalation rate, volumetric flow rate, total volumetric flow rate, volumetric flow rate into an inlet (sink strength)	L^3/t
$\dot{Q}$	rate of heat transferred into body	Q/t
Q/L	volumetric flow rate per unit length of slot	L^2/t
Q_1, Q_2	volumetric flow rates for push-pull ventilation system	L^3/t
Q_a	volumetric flow rate into (or out of) alveolar region	L^3/t
Q_a, Q_s	volumetric flow rates of air and scrubbing fluid	L^3/t
Q_a, Q_s, Q_d	volumetric flow rate of: ambient air, supply air, dilution air	L^3/t
Q_b	blood volumetric flow rate, volumetric flow rate of bled air, volumetric flow rate in branch	L^3/t
Q_B	volumetric flow rate of blowing air	L^3/t
$\dot{Q}_c$	rate of convection heat transfer	Q/t
$\dot{Q}_{conv}$	rate of heat transfer from body by convection	Q/t
Q_d	volumetric flow rate of displaced air (out)	L^3/t
Q_e, Q_m, Q_r	volumetric flow rate of: exhaust air, make-up air, recirculated air	L^3/t
$\dot{Q}_{evap}$	rate of energy transfer from body by evaporation	Q/t

symbol	description	dimensions*
Q_H	volumetric flow rate at hood inlet	L^3/t
Q_i	induced air volumetric flow rate (in)	L^3/t
$Q_{infiltration}$	volumetric flow rate due to infiltration into a building	L^3/t
Q_j	volumetric flow rate of species j, jth volumetric flow rate	L^3/t
Q_J	volumetric flow rate of air curtain jet	L^3/t
Q_L, Q_R	volumetric flow rates in left and right lung	L^3/t
$\dot{Q}_p$	heat transfer rate into a particle or drop	Q/t
$\dot{Q}_{rad}$	rate of radiant heat transfer from body	Q/t
Q_s	volumetric flow rate in straight section of a wye, volumetric flow rate of secondary air (in), volumetric flow rate of suction air	L^3/t
Q_{su}	plume surge volumetric flow rate	L^3/t
Q_t	minute respiratory rate	L^3/t
Q_u	volumetric flow rate of buoyant plume	L^3/t
Q_w	volumetric flow rate of withdrawn air (out)	L^3/t
r	radius, radial distance	L
r_0	radial distance to stagnation point	L
$r_1(t)$, $r_2(T)$	functions	-
r_1, r_2	inner and outer radii	L
r_{avg}	radius at which the local velocity is equal to U_{avg}	L
r_c	corona radius, characteristic radius of packing elements in a transverse packed bed scrubber	L
r_e	excess lifetime cancer risk	-
r_{plume}	plume radius	L
$r_{u,j}$	unit risk factor for molecular species j	-
r_w	radius of the corona wire in an ESP	L
R	nondimensional constant, airway resistance, ratio of v_S to v_c, diameter ratio (D_p/D_f), reflux ratio (Q_s/Q_a)	-
R	radius of disk	L
R or R_{gas}, R_j	specific ideal gas constant, R for molecular species j	Q/MT
R′	nondimensional constant	-
R_c	radius of a collecting spherical particle	L
R_{dc}	dust cake electrical resistance	Vt/C
R_L, R_G	resistance of the liquid and gas layer	-
R_L, R_R	resistance of left and right lung	Ft/L^5
R_p	response	varies
R_{pf}	ratio of particle diameter to filter fiber diameter	-
R_u	universal ideal gas constant	Q/NT
R_{vp}	ventilation-perfusion ratio (Q_a/Q_b)	-
R_{vpm}	modified ventilation-perfusion ratio [$Q_a/(Q_b k_b)$]	-
s	specific entropy	Q/MT
s	contaminant generation rate per volume of tunnel	M/L^3 t

symbol	description	dimensions*
s	source strength per unit area	$ML^{-2}t^{-1}$
s	slot width (s = 2w), stretch string distance, distance from elemental area to point in space, distance separating collecting plates in an ESP	L
s_1, s_2	constants	-
S	capillary surface area, cumulative airway surface area	L^2
S	source, source strength, contaminant emission rate	M/t
S	filter drag	FtL^{-3}
S′	net formaldehyde mass emission rate	M/t
S_B	width of blowing jet in push-pull system or Aaberg inlet	L
S_d	evaporation rate associated with displaced air	M/t
S_e	evaporation rate from an open vessel	M/t
S_f	evaporation rate due to filling	M/t
S_i	odor strength during period i	-
$S_{j,vol}$	rate of production of species j per unit volume in gas phase	$ML^{-3}t^{-1}$
S_s	evaporation rate from a spill	M/t
S_S	width of suction inlet in push-pull system or Aaberg inlet	L
t	time, lifetime	t
t	duct thickness	L
$t_{1/2}$	half-life (for first-order systems)	t
t′	time to empty a leaking vessel, characteristic time	t
t*	dimensionless time	-
t_{age}	local mean age	t
t_b	average length of time a cigarette burns	t
t_c	critical time, cleaning cycle time (for baghouses)	t
t_j	duration of exposure to sound level j	t
$t_{j,permitted}$	permitted duration of exposure to sound level j	t
t_{Lv}	Lagrangian turbulence time scale	t
t_m	arithmetic mean transient time	t
t_N	average residence time, $t_N = V/Q = 1/N$	t
$t_{room,avg}$	room mean age	t
t_s	time to achieve steady state concentration	t
t_{su}	duration of plume surge	t
t_{vc}	time to expire volume of air equal to vital capacity	t
t_{work}	total number of worker-hours during a specified period	t
T	temperature	T
T_0	ambient temperature	T
T_a'	dimensionless (normalized) air temperature	-
T_a, T_m, T_s	temperature of air or ambient, mucous, and skin	T
$T_{alveolar}$	temperature of air in alveolar region	T
T_b	boiling temperature	T
T_b, T_{blood}	blood temperature	T

symbol	description	dimensions*
T_c	body core temperature, critical temperature, temperature of plume during charging	T
T_{db}, T_{wb}, T_G	dry bulb, wet bulb, and globe temperature	T
T_f	flash point temperature	T
T_s	source temperature	T
T_w	wall temperature	T
$\hat{u}$	specific internal energy	Q/M
$\hat{u}_f$	specific internal energy of a saturated liquid	Q/M
U	velocity (actually speed, the magnitude of velocity), wind speed, air or gas speed (magnitude of air or gas velocity)	L/t
U	stored energy in body	Q
$\vec{U}$	fluid velocity, velocity of the carrier gas	L/t
$\|\vec{U}\|$	speed of air, speed of air along a pathline	L/t
U*	friction velocity	L/t
U*, v*	dimensionless gas and particle velocities	-
U_0	average velocity, characteristic velocity, centerline velocity, far-field velocity, wind velocity (Note: "velocity" is more properly "speed", the magnitude of velocity), air-to-cloth ratio for filters	L/t
U_1, U_2	gas velocities (actually speeds)	L/t
U_{10}	wind speed at an elevation of 10 m	L/t
U(0), U(x)	air speed at entrance and a location x inside a tunnel	L/t
U(r)	radial velocity (actually speed) in front of an inlet	L/t
U_a, U_f or U_{face}	air velocity, face velocity (actually speeds)	L/t
U_{avg}	average speed	L/t
U_{face}	magnitude of velocity at face	L/t
U_H	wind speed at height of building	L/t
U_J	magnitude of jet velocity, stack exit velocity	L/t
U_m	molar average gas speed	L/t
U_{max}	maximum (centerline) speed of plume	L/t
U_r, U_θ	radial and tangential gas velocity components	L/t
U_{rms}	root mean square speed	L/t
U_u	magnitude of plume mass-averaged velocity at elevation Z	L/t
U_x, U_y	components of gas velocity vector in x and y directions	L/t
v	specific volume	L^3/M
v	magnitude of particle velocity (particle speed)	L/t
$\vec{v}$	particle velocity	L/t
v_0	magnitude of velocity of room air current, i.e. draft	L/t
v_b	magnitude of velocity at face of blowing opening	L/t
v_B	magnitude of velocity of blowing jet	L/t
v_c	speed of automobiles in a tunnel, capture velocity, magnitude of cloud settling velocity, speed of a collecting particle	L/t

symbol	description	dimensions*
$v_{c,j}$	specific volume of molecular species j at its critical state	L^3/M
v_g	magnitude of velocity of evaporating vapor from open vessel	L/t
v_p	magnitude of particle velocity (particle speed)	L/t
v_r , v_θ	radial and tangential components of particle velocity	L/t
v_r, v_{rel}	magnitude of the relative velocity $\left\|\vec{v}_p - \vec{v}_c\right\|$	L/t
$\vec{v}_r$	relative velocity of a particle w.r.t. the fluid $\left(\vec{v}_r = \vec{v} - \vec{U}\right)$	L/t
v_{rx}, v_{ry}, v_{rz}	x, y, and z components of relative velocity	L/t
v_S	magnitude of velocity at face of suction opening	L/t
$v_s(t)$	instantaneous settling speed of a contaminant particle	L/t
v_s, $v_s(t)$	settling speed, instantaneous settling speed	L/t
v_t	terminal (gravimetric, steady-state) settling velocity (actually speed) of a particle (also called the fall velocity)	L/t
$v_{t,c}$	terminal velocity (actually speed) of collecting particle	L/t
v_x v_y	x and y components of particle velocity	L/t
V	total or overall volume, volume, lung volume	L^3
V	voltage	V
V_0	initial value of lung volume	L^3
V_0	voltage at the outer edge of the corona in an ESP	V
V_a	instantaneous volume of air in an enclosure	L^3
$V_a(t)$	volume of alveolar region as a function of time	L^3
V_b	pulmonary capillary blood volume, volume of the body, instantaneous volume of bulk materials in an enclosure	L^3
V_{bed}	total bed volume of a transverse packed bed scrubber	L^3
V_c	voltage at edge of corona in an ESP	V
V_{cross}	magnitude of cross-flow velocity (cross-flow speed)	L/t
V_d	anatomic dead space (volume of conducting airways), volume of dilution air	L^3
V_{dc}	voltage drop across a dust cake in an ESP	V
V_e	volume of an enclosure	L^3
V_f	final volume of inflated lung	L^3
V_{fr}	lung functional residual capacity	L^3
V_g	volume of metal removed by grinding	L^3
V_h	hood volume	L^3
V_j	volume of molecular species j	L^3
V_p	particle volume	L^3
V_r	lung residual volume	L^3
V_{store}	volume in which a contaminant is stored in the body	L^3
V_t	tidal volume	L^3
V_{vc}	vital capacity volume	L^3
w	slot half-width (for a rectangular slot)	L
w	drift velocity (actually speed), precipitation parameter	L/t
w(t)	dust cake loading/area of filter	M/L^2

symbol	description	dimensions*
W	flange width, short dimension, width of a building in cross wind direction, width of a duct or open vessel, distance between jet and exhaust opening for air curtain	L
$\dot{W}$	rate at which work is done by the body	Q/t
W, W_0	acoustical power, reference value of acoustical power	Q/t
W_r	width of building eddy	L
x	distance perpendicular to opening face, distance downwind from roof lip	L
x′	distance, point in a tunnel when stagnation conditions occur, U = 0	L
x, y, z	spatial coordinates or distances	L
x_c	distance to center of roof eddy	L
x_j	mol fraction of molecular species j in the liquid phase	-
x_j*	hypothetical mol fraction of species j in the liquid phase	-
x_r	downwind length of building eddy	L
y_{am}	log mean mol fraction	-
y_j	mol fraction of molecular species j in gas phase	-
z	vertical distance	L
z′	smaller of W or H	L
z*	distance between puncture and base of a spherical vessel	L
z_1, z_2	vertical elevation of points 1 and 2	L
Z	airway generation number	-
Z	distance between hood inlet and virtual source	L
Z, Z_c	compressibility factor, Z at the critical point	-
Z_I	height of roof leading edge recirculation cavity	L
Z_{II}	height of the turbulent shear zone	L
Z_{III}	height of roof wake	L

*The dimensions are defined as:

C = electrical charge
F = force
L = length
M = mass
N = mols
Q = energy
t = time
T = temperature
V = voltage

Greek symbols

symbol	description	dimensions*
α	angle, fraction of material absorbed by the body	-
α	angular displacement, turning angle	-
α_1, α_2, α_n	nondimensional constants	-
β	nondimensional constant, corner angle	-
β	isothermal expansion coefficient	T^{-1}
γ	ratio of specific heats, also called the isentropic coefficient	-
Γ	activity coefficient	-
δ or Δ	difference, e.g. δt = time difference	-
δ_m	concentration (mass) boundary layer thickness	L
δ_T	thermal (energy) boundary layer thickness	L
δ_u	velocity (momentum) boundary layer thickness	L
δP	pressure drop (or rise)	F/L^2
δP_c	pressure drop across dust cake upstream of a filter	F/L^2
δP_d	overall pressure drop across a dust-caked filter	F/L^2
δP_f	pressure drop across a filter	F/L^2
δP_{fan}	pressure rise across a fan	F/L^2
δt	increment of time, time step	t
ΔV	$V_f - V_0$	L^3
ε	turbulent dissipation rate per unit mass in k-ε turbulence model	L^2/t^3
ε	bed porosity of a transverse packed bed scrubber, porosity of a filter	-
ε_0	permitivity of free space	C^2/ML^2
$\vec{\zeta}$	vorticity vector	t^{-1}
η	efficiency (of air cleaner, filter, or fan)	-
$\eta(D_p)$	fractional or grade efficiency of an air cleaner	-
η_a	aspiration efficiency	-
η_d	single drop collection efficiency	-
η_f	single fiber collection efficiency	-
η_I, η_R, η_D, η_{DR}	single fiber collection efficiencies of impaction, interception, diffusion, combination of diffusion and interception	-
η_u	uptake absorption efficiency, or uptake efficiency	-
θ	angle, angular displacement	-
θ_c	critical angle in curvilinear flow	-
θ_τ	total angular displacement	-
κ	dielectric constant	-
λ	wavelength, mean free path of carrier gas	L
Λ	metal removal parameter	$L^3/(tF)$
μ	dynamic viscosity	Ft/L^2, $ML^{-1}t^{-1}$
ν	kinematic viscosity (μ/ρ)	L^2/t

symbol	description	dimensions*
ν	radiation frequency	t^{-1}
Π	dimensionless parameter in Buckingham Pi Analysis	-
ρ, ρ_m	density (usually of air or carrier gas), metal density	M/L^3
$\overline{\rho}$	bulk density of a gas mixture	M/L^3
ρ_a, ρ_p, ρ_w	density of air, density of a particle, density of water	M/L^3
ρ_b	density of bulk materials	M/L^3
ρ_c, $\overline{\rho}_c$	density and bulk density of the collecting liquid	M/L^3
ρ_{dc}	dust cake resistivity	$VtLC^{-1}$
$\overline{\rho}_i$	partial density of molecular species i	M/L^3
ρ_0	ambient air density at temperature T_0	M/L^3
σ	dispersion coefficient, standard deviation of particle diameters for particle distributions	L
σ_g	geometric standard deviation	-
τ	first-order time constant, lung time constant (τ= RC), time period	t
τ, τ_a	magnitude of shear stress, magnitude of air shear stress	F/L^2
τ_{ij}	viscous stress tensor in conservation of momentum equation	F/L^2
τ_p	particle relaxation time	t
ϕ	velocity potential	L^2/t
Φ	relative humidity, packing area ratio = $\pi D_{packing}{}^2/(4A_{packing})$	-
φ	association factor	-
X (chi)	particle dynamic shape factor	-
ψ	stream function	L^2/t
ω	specific dissipation rate per unit mass, q-ω turbulence model	t^{-1}
ω	fan RPM	t^{-1}
ω	humidity ratio	-
$\vec{\Omega}$, Ω	angular velocity vector, magnitude of angular velocity	t^{-1}

*The dimensions are defined as:

C = electrical charge	M = mass	t = time
F = force	N = mols	T = temperature
L = length	Q = energy	V = voltage

Subscripts

subscript	**description**
$(\ )_\theta$	θ (tangential) coordinate or component in the θ direction
$(\ )_\phi$	component in angular direction
$(\ )_0$	value in far field or surroundings (ambient air or water bath), initial value, inlet value
$(\ )_\infty$	property evaluated in the far-field
$(\ )_1$, $(\ )_2$	properties at points 1 and 2
$(\ )_a$ or $(\ )_{air}$	air property, conditions in the air
$(\ )_{actual}$	actual thermodynamic properties
$(\ )_{ambient}$	property of ambient conditions
$(\ )_{avg}$ or $(\ )_{ave}$	average value of a property
$(\ )_b$	property of the burned state, property of a building, property of bled fluid, bulk property, property of blood
$(\ )_B$	property of blowing air
$(\ )_{blood}$	property of blood
$(\ )_{bb}$	property of the body burden
$(\ )_c$	body core value, critical value, property of a collected fluid or of a common section, property at (thermodynamic) critical point, cloud property, property of a collecting particle
$(\ )_{cor}$	corrected value
$(\ )_{cut}$	cut property for a cyclone (property at the cut diameter)
$(\ )_d$	property of the dividing streamline, property of a drop
$(\ )_{da}$, $(\ )_{DA}$	property of dry air
$(\ )_{dc}$	property of dust cake
$(\ )_{design}$	design property
$(\ )_e$	equivalent property
$(\ )_e$	exit condition
$(\ )_E$	property of the exhaust from a room (when discussing ventilation systems)
$(\ )_f$	saturated liquid value
$(\ )_f$	final condition, filter property, streaming flow property
$(\ )_{face}$	space between respirator face-piece and face, property of an inlet face
$(\ )_g$	property of an inert gas blanket, geometric property
$(\ )_G$	property of a gas
$(\ )_H$	property at length H
$(\ )_{hood}$	hood property
$(\ )_i$	value at the air-liquid interface, value at i^{th} iteration, initial condition, inlet condition
$(\ )_{in}$, $(\ )_{inlet}$	inlet condition or property at an inlet
$(\ )_{iso}$	isolated property
$(\ )_j$	property of molecular species j or of a push jet
$(\ )_L$	property of a leak or of a liquid, condition at distance L
$(\ )_m$	property of mucus or of make-up air
$(\ )_{max}$	maximum value

subscript	description
$(\ )_{mixture}$	conditions of a mixture
$(\ )_{out}$	outlet or outflow condition
$(\ )_{overall}$	overall property
$(\ )_{oz}$	property of an occupied zone
$(\ )_{p}$	property of a particle
$(\ )_{P}$, $(\ )_{Q}$, $(\ )_{R}$	properties at point P, Q and R
$(\ )_{p,j}$	particle in size range or class j for a particle distribution
$(\ )_{packing}$	property of the packing material in a transverse packed bed scrubber
$(\ )_{r}$	property of recirculated air, property of r coordinate (radial component)
$(\ )_{ref}$	reference value
$(\ )_{room}$	property pertaining to the entire room
$(\ )_{s}$	property of supply, property of suction air, property of a surface, scrubbing liquid property
$(\ )_{sat}$	saturated property
$(\ )_{sf}$	streaming flow property
$(\ )_{slot}$	slot property
$(\ )_{ss}$	steady-state property
$(\ )_{STP}$	property at standard temperature and pressure (298.15 K, 101.3 kPa)
$(\ )_{t}$, $(\ )_{total}$	total or stagnation property
$(\ )_{t,c}$	terminal settling velocity of collecting particle
$(\ )_{u}$	property of a plume
$(\ )_{w}$	water property, property pertaining to adsorption of contaminants on walls, property of walls, property of withdrawn air, property relevant to inlet slot of width 2w
$(\ )_{water}$	water property
$(\ )_{x}$, $(\ )_{y}$, $(\ )_{z}$	property of x, y, or z coordinates, cartesian components

Abbreviations and acronyms

abbreviation	description
A/P	ratio of annualized cost to present worth
AAQS	ambient air quality standards
ACFM	actual cubic feet per minute (ft^3/min); volumetric flow rate at actual pressure and temperature
ACGIH	American Conference of Governmental and Industrial Hygienists
AD	alveolar duct
AIChE	American Institute of Chemical Engineers
AIHA	American Industrial Hygiene Association
ALJ	Administrative Law Judge
AMCA	Air Moving and Conditioning Association
AMTIC	ambient monitoring technical information center
ANSI	American National Standards Institute

abbreviation	description
AP-42	compilation of EPA emission factors
APCS	air pollution control system
AR	airway resistance
AS	alveolar sac
ASHRAE	American Society of Heating, Refrigerating and Air-Conditioning Engineers
ASME	American Society of Mechanical Engineers
ASTM	American Society for Testing and Materials
AWMA	Air & Waste Management Association (formerly Air Pollution Control Association, APCA)
BAL	broncho-alveolar lavage
BEI	biological exposure index
BEP	best efficiency point
BG	blast gate
BH	baghouse
Bi	Biot number
BL	bronchiole
BLS	Bureau of Labor Statistics
BR	bronchi
BRI	building related illness
BTU	British thermal unit
C	ceiling TLV (TLV-C) or PEL (PEL-C) which is never to be exceeded
CAAA	Clean Air Act Amendments of 1990
CAS	Chemical Abstract Service
CFC	Chlorofluorohydrocarbons
CFD	computational fluid dynamics
CFM	cubic feet per minute (ft^3/min)
CFR	Code of Federal Regulations
CHIEF	Clearinghouse for Inventories and Emission Factors (maintained by EPA)
CIS	Chemical Information Service
CMA	Chemical Manufacturers Association
CO	Compliance Officer (OSHA), also carbon monoxide
CQI	Continuous Quality Improvement
CRF	capital recovery factor
dB	decibel
DNA	deoxyribonucleic acid
DOAS	dedicated outdoor air system
DOE	Department of Energy
ECL	exposure control limit
EF	emission factor
EMS	environmental management system, Environmental Management Standards
EMTIC	emission measurement technical information center
EOM	extractable (particulate) organic matter
EPA	Environmental Protection Agency

abbreviation	description
ESP	electrostatic precipitator
ETS	environmental tobacco smoke
F/A	ratio of future worth to annualized cost
F/P	ratio of future worth to present worth
FCF	fixed cost factor
FDA	Food and Drug Administration
FPM	ft/min
FR	Federal Register
FTC	Federal Trade Commission
FVC	forced vital capacity
GATT	General Agreement on Tariffs and Trade
GPM	gallons per minute
HAP	hazardous air pollutant
HC	hydrocarbon
HCHO	formaldehyde
HE	heavy exercise conditions
HEPA	high efficiency particulate air (pertains to air filters)
HERP	human exposure dose/rodent potency dose
HEW	Health, Education, and Welfare (department of US government)
HHV	higher heating value
HP	horsepower
HSI	heat stress index
HVAC	heating, ventilating, and air conditioning
HVLV	high velocity - low volume
IAQ	indoor air quality
ICF	indirect cost factor
ID	inside diameter or inner diameter
IEC	International Electrochemical Commission
ISB	isothermal saturation boundary
ISO	International Organization for Standardization
Kn	Knudsen number
Ku	Kuwabara hydrodynamic factor (also called Kuwabara number)
Le	Lewis number
LE	light exercise conditions
LEL	lower explosion limit
LES	large eddy simulation
LVHV	low velocity - high volume
MAC	maximum acceptable concentration
MCI	methylisocyanate
MDR	multi-drug-resistant, a virulent strain of tuberculosis
ME	medium exercise conditions
MIA	minimum identifiable odor
MSDS	material safety data sheet

abbreviation	description
MSHA	Mine Safety and Health Administration
NAFTA	North American Free Trade Agreement
NFPA	National Fire Protection Association
NIH	National Institutes of Health
NIOSH	National Institute for Occupational Safety and Health
NIST	National Institute of Standards and Technology
NOEL	no observable effect level
NSC	National Safety Council
Nu	Nusselt number
ODE	ordinary differential equation
OSHA	Occupational Safety and Health Administration
OSHRC	Occupational Safety and Health Review Commission
OTA	Office of Technology Assessment
PAN	peroxyacetylnitrate
PDF	probability distribution function
PDR	Petition for Discretionary Review
Pe	Peclet number
PEL	permissible exposure limit
PERC	perchloroethylene ("perk")
PMA	Petition for Modification of Abatement (OSHA)
PMN	Premanufacture Notice
PPB	parts per billion (mol fraction)
PPM	parts per million (mol fraction)
Pr	Prandtl number
QLFT	qualitative fit-test
QNFT	quantitative fit-test
R	rest conditions
RBL	respiratory bronchiole
RD_{50}	reduced respiratory rate
Re	Reynolds number
REL	recommended exposure level
RF	radio frequency
RMS	root mean square
RPM	revolutions per minute
RQ	respiratory quotient
RTECS	Registry of Toxic Effects of Chemical Substances
SAE	Society of Automotive Engineers
SBS	sick building syndrome
Sc	Schmidt number
SCBA	self-contained breathing apparatus
SCFM	standard cubic feet per minute (ft^3/min) (hypothetical volumetric flow rate that would exist at STP conditions)
SCUBA	self-contained underwater breathing apparatus

abbreviation	**description**
SG	specific gravity
Sh	Sherwood number
SMR	standard mortality ratio
SP	static pressure
STEL	short-term exposure limit
Stk	Stokes number
STP	standard temperature (25 °C, 298.15 K) and pressure (101.325 kPa)
TAC	total annual cost
TAG	US Technical Advisory Group
TB	tuberculosis
TBL	tracheobronchial region
TCC	total capital cost
TCE	trichloroethylene
TD_{50}	daily dose rate to halve the percent of tumor-free animals
TDC	total direct cost
TDI	toluene diisocyanate
TFC	total fixed cost
TIC	total indirect cost
TL	terminal bronchiole
TLD	top level domain (same as top level extension) for web site names
TLV	threshold limit value
TRR	total revenue requirements
TSCA	Toxic Substances Control Act
TSP	total suspended particles or total suspended particulates
TTN	EPA's Technology Transfer Network
TTNBBS	EPA's Technology Transfer Network Bulletin Board System
TTS	temporary threshold shift
TVC	total variable cost
TWA	time-weighted average
TWA-TLV	time-weighted average threshold limit value
UEL	upper explosion limit
ULPA	ultra low penetration air (pertains to air filters)
UV	ultraviolet radiation
VOC	volatile organic compound
VP	velocity pressure
Wo	Womersley number

1

Introduction

In this chapter you will learn:

- how to anticipate and rank risks quantitatively
- strategies to control workplace contaminants
- generic classes of ventilation systems
- how to compute mass and volumetric flow rates and pollutant concentrations
- how to compute properties of ideal mixtures of gases and liquids
- how to solve first-order ordinary differential equations, both analytically and numerically
- about government regulations, agencies, and standards which are relevant to indoor air quality

It is probably safe to say that nothing in life is taken more for granted than the air we breathe. Due to some important pieces of legislation in the past two decades, *outdoor* air in the developed nations is generally cleaner and safer now than it was thirty years ago. Environmental catastrophes such as occurred in Bhopal, India in 1984 (Elsom, 1987) are rare and should be defined as ***industrial accidents***, not air pollution. ***Air pollution*** pertains to pollutant concentrations resulting from normal industrial operations that may be amplified from time to time by unusual meteorological conditions. Deaths related directly to air pollution occur rarely and involve far fewer deaths than occur because of accidents in the home. Two notorious air pollution episodes resulting in the loss of life were:

- Donora, PA - in 1948: Twenty people died when SO_2 and particle emissions accumulated in the Monongahela river valley during a 5 day period.
- London - in 1952: Several thousand people are believed to have died from large concentrations of SO_2 that accumulated over a period of approximately one week.

On the other hand, *indoor* air is not always as clean and safe, especially in buildings where hazardous chemicals are used or where combustion processes occur. Much attention has been given in the news to the effects of radon, asbestos, second-hand smoke, and more recently anthrax and other substances used by bioterrorists. ***Indoor air pollution*** describes the generation and transport of pollutants inside a variety of interior environments in which people live and work, such as

- industrial workplaces
- office buildings, schools, hospitals, and other public buildings
- tunnels
- mines
- private homes and apartment buildings

Pollutants (both vapors and particles) that unfortunately enter the indoor environment can often be captured by judiciously located inlets, ducts, and exhaust fans called ***local ventilation systems***, also called ***hoods***. Some common household examples include hoods above kitchen stoves, dryer vents, furnace flues, etc. in which captured air is exhausted to the outdoors. In industrial settings, environmental regulations prohibit the release of certain contaminants into the outdoor air, and such pollutants must be removed from the captured air by devices called ***air pollution control systems*** (APCSs). The primary goal of indoor air pollution engineering is to design ventilation systems and APCSs which effectively remove hazardous chemicals and particles from the air breathed by individuals in interior environments.

1.1 Indoor Air Pollution and Risks

This chapter begins by examining the broad range of workplace risks. The terms hazard and risk are often used as synonyms, but there are important differences. ***Hazard*** describes a situation that has the potential for producing an undesirable event, without expressing the likelihood that it will occur. ***Risk*** expresses not only the potential for producing an undesirable event, but the probability that it will occur. Similarly, exposure to materials considered ***hazardous*** jeopardize health, while exposure to materials considered ***toxic*** or ***noxious*** are known to cause illness or death. Materials that are not injurious to health but are nonetheless unpleasant or repulsive should be accurately called ***obnoxious***. ***Occupational health and safety*** pertains to risks in the workplace; indoor air pollution is only one of many such risks. While this book concentrates on indoor air pollution, other aspects of occupational health and safety such as noise, explosion, and heat stress are briefly discussed in later chapters.

To appreciate the risks associated with human endeavors, Starr (1969) suggested that it is useful to categorize risks as voluntary or involuntary. ***Voluntary risks*** are assumed by individuals of their own free will. Examples include risks associated with recreational sports and flying private aircraft. ***Involuntary risks*** are imposed on individuals because of circumstances beyond their control. Examples include risks associated with using elevators in tall buildings and flying in commercial aircraft. The issue becomes cloudy concerning private automobiles because individuals voluntarily drive their own automobiles yet are at risk from accidents caused by others. Figure 1.1 shows how voluntary and involuntary risks vary with the ***benefit*** people believe the activity provides. For a first approximation the risk is computed as the statistical probability of fatality per hour of exposure associated with the activity. The benefit derived from each activity is converted into a dollar equivalent as a measure of its value to the individual. For voluntary activities the amount of money spent on the activity by the average individual is assumed to be proportional to its benefit. In the case of involuntary activities, the contribution of the activity to the individual's annual income is assumed to be proportional to its benefit. While the approximations used by Starr are crude, it is apparent that the large difference between voluntary and involuntary risks suggests that individuals accept higher risks in voluntary activities but expect lower risk in involuntary activities. Other than self-employed people and some exotic occupations, risks associated with most employment are considered involuntary, and the public expects risks to be no larger than those associated with naturally occurring events. *Risks associated with air pollution are involuntary risks.*

An interesting comparison showing the complexity of voluntary and involuntary risks concerns the use of wood-burning stoves and the political action to enact nonsmoking ordinances. The concentrations of carbon monoxide, ***total suspended particles*** (TSP), and several polycyclic aromatic hydrocarbons (including benzo[a]pyrene) in homes using wood-burning stoves (Traynor et al., 1987) are comparable to the concentrations of these materials in public places in which smoking is allowed (Repace and Lowery, 1986; Sterling et al., 1982). Indoor pollution from wood-burning stoves is a voluntary risk to the homeowner while tobacco smoke in public buildings is an involuntary risk to the nonsmoker. While the risks to health may be comparable, many people object to the involuntary risk yet accept with equanimity the voluntary risk. Some observers would consider this hypocrisy, others would consider it the freedom of choice. The issue is even more complex, because in some regions of the United States, up to one half the outdoor TSP is due to wood-burning stoves. The irony is often lost that the voluntary risk to homeowners using wood-burning stoves is an *involuntary* risk to their neighbors due to outdoor air pollution. Many indoor and outdoor pollutants are related, and in many cases indoor pollutant concentrations exceed outdoor concentrations (Brickus et al., 1998).

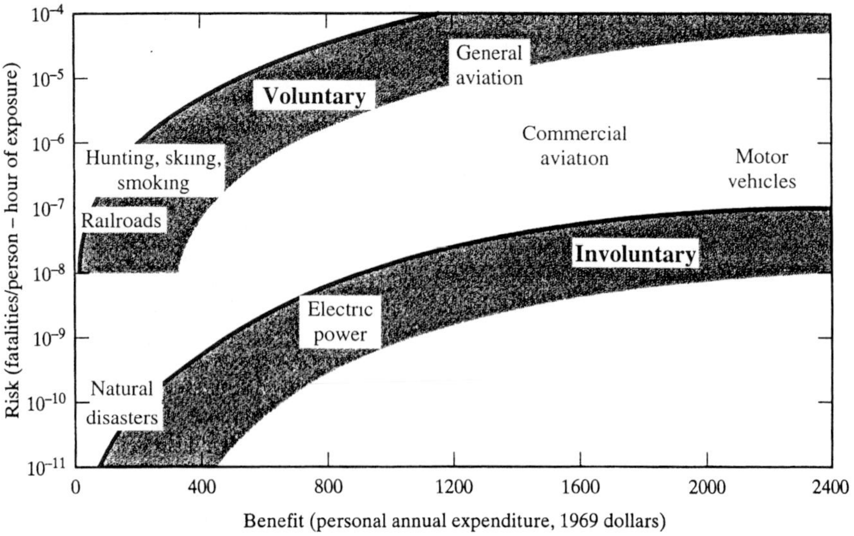

Figure 1.1 Risk vs. benefit for various kinds of voluntary and involuntary activities; for comparison, the risk of fatality by disease for the entire U.S. population is approximately 10^{-6} (abstracted from Starr, 1969).

Data on occupational injuries and deaths are sufficiently accurate and comprehensive to provide guidance for prevention. Data about occupational illness and exposure to indoor air pollutants are far less accurate or comprehensive. When considering the health effects of air pollution, it is necessary to have benchmarks for comparing environmental disease and injury statistics to those of the general population. Table 1.1 shows the number of people in the United States that died in the year 1997 (National Safety Council, 2000) from a variety of causes. (The total US population is approximately 280 million as of the 2000 census.)

To put information about environmental health hazards into perspective, consider government *estimates* that up to 3,500 people *may* die per year from radon or secondary tobacco smoke. It is helpful to realize that Table 1.1 indicates that that this is less than one fourth of the *actual* deaths due to falls, and less than one twelfth of the *actual* number of deaths due to motor-vehicle accidents. Reality rests in what actually occurs; estimates are speculation. Table 1.2 lists the number of deaths and nonfatal injuries due to various kinds of accidents involving vehicles. These numbers are useful as benchmarks for comparison to the number of deaths from other activities. For example, nearly 6,000 pedestrians are killed each year by automobiles, and almost 19,000 people are killed in vehicle-vehicle accidents, while only about 700 people die each year from air travel.

The ***Bureau of Labor Statistics*** (BLS) of the US Department of Labor compiles the most reliable data on occupation-related deaths, although it only counts about three-fourths of the nation's workforce. Adjusted to include the entire nation's workforce, the ***Congressional Office of Technology Assessment*** (OTA) found that approximately 5,000 deaths occur annually from occupational injuries, or about 25 deaths per each working day. In 1912 the work force was only half as large as now, yet an estimated 18,000 to 21,000 workers lost their lives that year in work-related incidents. Table 1.3 shows the number of deaths and disabling injuries from several industries in 1999. On the basis of number of deaths, construction was the highest, followed by transportation and agriculture. Mining was the lowest. However, on the basis of the on-the-job death *rates* (deaths/100,000 workers), mining and agriculture, followed by construction, were the riskiest occupational fields. Additional BLS data show that about half the fatal occupational injuries involve motor vehicles, off-road industrial vehicles, and

Table 1.1 Deaths (all ages) in the United States in 1997 (from National Safety Council, 2000).

Deaths due to diseases: (2,168,220 total deaths due to diseases)

total	cause of death	subtotal
726,974	heart disease	
539,577	cancer (approx. 150,000 lung cancer, 85% related to smoking)	
364,506	other diseases not listed here	
159,791	stroke (estimated)	
109,029	chronic obstructive pulmonary disease	
86,449	pneumonia and influenza	
62,636	diabetes mellitus	
25,331	nephritis and nephrosis	
25,183	chronic liver disease, cirrhosis	
22,401	septicemia	
16,735	arteriosclerosis	
16,516	AIDS (human immunodeficiency virus infection)	
13,092	certain conditions originating in prenatal period	

Accidental deaths: (95,644 total, 5,100 of which were in the workplace)

total	cause of death	subtotal
45,798	transportation accidents	
	motor-vehicle	43,458
	air and space transport	734
	water transport	758
	railway	527
	other road vehicle	220
	vehicle accidents not elsewhere classifiable	101
15,447	falls	
9,587	poisonings by solids and liquids	
	drugs, medicaments, and biologicals	9,099
	alcohol	342
	other solids and liquids	135
	foodstuffs and poisonous plants	11
5,629	other accidental deaths not listed here	
3,561	drowning (excluding water transport drowning)	
3,490	fire and flames	
3,043	complications, misadventures of surgical, medical care	
2,180	inhalation and ingestion of objects other than food	
1,316	natural and environmental factors	
	excessive cold	501
	hunger, thirst, exposure, and neglect	224
	excessive heat	182
	cataclysmic storms, and floods resulting from storms	136
	other injury caused by animal (19 of these from dog bites)	102
	poisoning by toxic reaction to venomous animals and plants	68
	lightning	58
	other natural and environmental factors	25
	cataclysmic earth surface movements and eruptions	20
1,145	mechanical suffocation	

total	cause of death	subtotal
1,095	inhalation and ingestion of food	
1,055	machinery	
981	firearm missile	
576	poisonings by gases and vapors	
	non-motor-vehicle exhaust gas	251
	motor-vehicle exhaust gas	208
	other gases and vapors	117
488	electric current	
149	explosive material	
104	cutting or piercing instruments or objects	
0	radiation	

Deaths other than due to accident or disease: (54,047 total)

total	cause of death	subtotal
30,535	suicide	
	firearms	17,566
	hanging, strangulation, and suffocation	5,413
	motor-vehicle exhaust gas	1,367
	jumping from high places	600
	cutting and piercing instruments	499
	other suicides	1,780
	poisoning by solid and liquid substances	3,310
19,491	homicide	
3,657	undetermined whether accidentally or purposely inflicted	
355	legal intervention	
9	operations of war	

falls. During the 1980's, approximately 17 workers a day died in workplace accidents and the death rate was 8.9 deaths per 100,000 workers; by 1989 it had dropped 37% to 5.6 deaths per 100,000 workers. By 1999, the death rate for all industry was under 4%. Death from motor vehicle accidents is the leading cause (approximately 23%), followed by machinery (13%), homicide (12%), falls (10%), and falling objects (7%). A startling statistic is that the third largest cause of death is homicide. Surprisingly, murder associated with robbery makes taxicabs drivers (27 deaths per 100,000 workers) and convenience store clerks (1,600 total homicides) risky occupations. The leading (28%) disabling, *nonfatal* injuries are over-extensions (largely to the back).

Table 1.2 Motor vehicle deaths and nonfatal injuries in the USA in 1999 (from National Safety Council, 2000).

type of vehicle accident	deaths (to nearest 100)	nonfatal injuries (to nearest 1000)
vehicle-vehicle	18,800	1,580,000
vehicle-fixed object	11,100	330,000
vehicle-pedestrian	5,800	95,000
noncollision accident (jackknife, etc.)	4,300	110,000
vehicle-pedalcycle	900	70,000
vehicle-train	300	2,000
other collision (vehicle-animal, vehicle-buggy, etc.)	100	13,000
total	**41,300**	**2,200,000**

Table 1.3 Worker deaths and injuries in the United States in 1999 with percent change from 1998; "nil" implies below 0.5% (from National Safety Council, 2000).

industry	workers[a] (to nearest 1,000)	deaths[a]	death rates[b]	disabling injuries (to nearest 10,000)
construction	8,479,000	1,190 (+ 5)	14.0 (- 1)	400,000
transportation	7,948,000	850 (+ 3)	10.7 (nil)	370,000
agriculture[c]	3,348,000	770 (- 4)	22.5 (nil)	150,000
services[c]	46,766,000	640 (+ 2)	1.4 (nil)	900,000
manufacturing	19,993,000	600 (- 4)	3.0 (nil)	670,000
government	20,118,000	470 (+ 1)	2.3 (- 4)	580,000
trade[c]	27,473,000	450 (+ 2)	1.6 (nil)	710,000
mining[c]	562,000	130 (- 8)	23.1 (nil)	20,000
all industry	**134,688,000**	**5,100 (nil)**	**3.78 (nil)**	**3,800,000**

Footnotes:

(a) Workers and death rates include persons 16 years or older; deaths include all ages.

(b) Deaths per 100,000 workers in each group.

(c) Agriculture includes forestry and fishing (see National Safety council). Mining includes quarrying, oil and gas extraction. Trade includes wholesale and retail trade. Services includes finances, insurance, and real estate.

Risks to maintenance workers are particularly high. While safe procedures are generally adopted for scheduled maintenance, workers also make emergency repairs following equipment breakdowns. To accomplish these repairs, safety controls and established procedures are sometimes circumvented. Because of the adverse conditions under which these repairs are made and pressure from superiors to resume production, careless practices are apt to occur.

Several methods are used to cite accident and loss statistics. Two commonly used parameters are: (a) fatality rates and (b) OSHA incidence rates. ***Fatality rate*** is the ratio of the number of deaths that occur in a certain activity divided by the total number of people engaged in that activity for a specified period of time. ***OSHA incidence rates*** (National Safety Council, 2000) are based on 100 worker-years, where a ***worker-year*** consists of 2000 hours (50 weeks per year, 40 hours per week). Thus 100 worker-years consist of an exposure of 200,000 worker-hours. OSHA incidence rates can be based on occupational illness and injury or on lost workdays.

- ***Occupational injuries*** are lacerations, fractures, amputations, etc., resulting from a single workplace accident.
- ***Occupational illnesses*** are acute or chronic disorders, other than occupational injuries, caused by factors encountered in the workplace.
- ***Lost workdays*** are the number of days (consecutive or nonconsecutive) in which a worker is prevented from working because of an occupational illness or injury. Lost workdays do not include the day of injury or the onset of illness, but include days in which workers:
  - are unable to go to work
  - are assigned temporary jobs
  - spend less than full time at their permanent jobs
  - cannot perform all the duties associated with their permanent jobs

OSHA incidence rate (I_i) is defined as the reported number of occupational injuries or illnesses per 200,000 worker-hours. Thus, for a given number (N_i) of reported injuries or illnesses during some measured number of worker-hours (t_{work}), the OSHA incidence rate (I_i) is defined by

$$I_i = \frac{N_i}{t_{work}} = \frac{N_i}{t_{work}}\left(\frac{200,000 \text{ worker-hr}}{200,000 \text{ worker-hr}}\right) \quad (1\text{-}1)$$

where the ratio in parentheses is obviously equal to 1, but is inserted in Eq. (1-1) to properly convert the units of I_i to the number of occupational injuries or illnesses per 200,000 worker-hours. For example, suppose 2 occupational injuries and 3 occupational illnesses are reported at a furniture manufacturing plant during 10,000 worker-hours. From Eq. (1-1), the OSHA incidence rate can be calculated as follows:

$$I_i = \frac{(2 \text{ injuries} + 3 \text{ illnesses})}{10,000 \text{ worker-hr}}\left(\frac{200,000 \text{ worker-hr}}{200,000 \text{ worker-hr}}\right) = 100\frac{\text{injuries or illnesses}}{200,000 \text{ worker-hr}}$$

Similarly, the ***OSHA lost workday rate*** (I_{iw}) is defined as the reported number of lost workdays per 200,000 worker-hours, and is defined by

$$I_{iw} = \frac{N_{iw}}{t_{work}} = \frac{N_{iw}}{t_{work}}\left(\frac{200,000 \text{ worker-hr}}{200,000 \text{ worker-hr}}\right) \quad (1\text{-}2)$$

where N_{iw} is the total number of reported lost workdays during t_{work} worker-hours.

Example 1.1 – Estimating Future Occupational Illness and Injury Rates

Given: In 1985 the Bureau of Labor Statistics reported that workers employed in the oil and gas extraction industry had experienced 143.8 injuries or illnesses per 200,000 worker-hours, and 5.30 lost workdays per 200,000 worker-hours. The industry employed 50,000 workers in 1985, each of whom worked on the average 2,000 hr/yr. Suppose economic conditions change, and by the year 2005, it is estimated that:

(a) the number of workers will increase by about 25.%
(b) each worker will work about 20.% more hours (overtime)

To do: Estimate the OSHA incidence rate (I_i) and lost workday rate (I_{iw}) in the year 2005. Also estimate the total number of expected injuries or illnesses, and the total number of lost workdays in the oil and gas extraction industry in 2005.

Solution: The incidence rates reflect the hazards associated with a particular occupation, not the number of workers in that occupation or how many hours they work. Consequently, unless the workplace has become more (or less) hazardous, the OSHA incidence rate (I_i) and lost workday rate (I_{iw}) should remain the same since these values are normalized on the basis of the number of worker-hours, which is independent of the number of workers and number of hours worked per year. Hence,

$$I_i(2005) = I_i(1985) = 143.8\frac{\text{injuries or illnesses}}{200,000 \text{ worker-hr}}$$

and

$$I_{iw}(2005) = I_{iw}(1985) = 5.30\frac{\text{lost workdays}}{200,000 \text{ worker-hr}}$$

During the year 2005, the total number of injuries and illnesses and the total number of lost workdays will increase owing to the greater number of workers who are working longer hours. In 2005 the total workforce will be (1.25)(50,000 workers), or 62,500 workers, each of whom will work (1.20)(2000 hr) = 2400 hr. The total number of worker-hours in the year will thus be

$$t_{work} = (62,500 \text{ workers})(2,400 \text{ hr}) = 150,000,000 \text{ worker-hr}$$

Using Eqs. (1-1) and (1-2), the following can be expected in 2005:

$$N_i(2005) = I_I(2005) \cdot t_{work} = \left(143.8 \frac{\text{injuries or illnesses}}{200{,}000 \text{ worker-hr}}\right)(150{,}000{,}000 \text{ worker-hr})$$

$$= 107{,}850 \cong 110{,}000 \text{ injuries or illnesses}$$

and

$$N_{IW}(2005) = I_{iw}(2005) \cdot t_{work} = \left(5.30 \frac{\text{lost workdays}}{200{,}000 \text{ worker-hr}}\right)(150{,}000{,}000 \text{ worker-hr})$$

$$= 3{,}975 \cong 4{,}000 \text{ lost workdays}$$

Discussion: The final numbers have been rounded to two significant digits, because these estimates can be no more precise than the least precise values used in the calculations, in this case the estimates of 25.% and 20.%, which are precise only to two significant digits. Significant digits will be discussed in more detail later in the chapter.

In 1982 the ***National Institute for Occupational Safety and Health*** (NIOSH) ranked occupational diseases and injuries in the following order of importance, based on the frequency of their occurrence:

(1) occupational lung disease (e.g. lung cancer, pneumoconiosis, and occupational asthma)
(2) musculoskeletal injuries (e.g. back injury, carpal tunnel syndrome, arthritis, and vibration white finger disease)
(3) occupational cancers (other than lung cancer)
(4) traumatic deaths, amputations, fractures, and eye losses
(5) cardiovascular diseases (e.g. myocardial infarction, stroke, and hypertension)
(6) reproductive problems
(7) neurotoxic illnesses
(8) noise-induced hearing loss
(9) dermatological problems (e.g. dermatitis, burns, contusions, and lacerations)
(10) psychological disorders

Of these, all but (2) and (4) can possibly be linked to indoor air quality. Occupational illness is difficult to define for three reasons. First of all, many occupational diseases are indistinguishable from non-occupational diseases. Secondly, the relationship between a disease and an occupational environment is not always recognized, and may be complicated by causes not related to occupation. Lastly, many diseases have long latency periods and occur after occupational exposure has ceased or the worker has changed jobs or retired. The last reason is the most troubling factor inhibiting accurate assessments of risk. Serious diseases such as respiratory and neurological disorders and cancers are not generally captured in BLS records of work-related illness; but after considerable debate, it is generally agreed (US Office of Technology Assessment, 1985) that approximately 5 percent of cancer deaths are related to occupational exposures.

Data on occupational illness obscure three important facts: occupational illness is preventable, workers in some industries have disproportionate risks, and once an occupational disease is identified controls can be adopted to reduce risks. Improvement of workplace health and safety proceeds in three steps:

(1) identify the hazard and its causative agents
(2) select preventive strategies
(3) acquire the authority to implement the necessary changes

Steps (1) and (2) are technical and require specialists. Step (3) is administrative and involves managers, employers, and elected officials. Depending on the severity of the hazard, step (3) can be taken *before* the hazards are fully identified and controls developed.

1.2 Management and Assessment of Risk

Health and safety are moral sentiments, and like other sentiments such as peace, freedom, and happiness, are *not* absolutes nor moral imperatives, however desirable they may be to some segments of society. They are, in fact, measured intangibly by the absence of their undesirable consequences. Nevertheless, they are of such evident importance that they must be assessed and managed. ***Risk management*** is the goal of personal and government policy, while ***risk assessment*** is an activity that estimates the spectrum and frequency of accidents and other negative events. Consider the automobile. If an environmental impact statement were prepared today it would reveal highly probable risks accounting for over 40,000 deaths per year. Do such risks warrant banishing automobiles? Of course not. Human intervention (either personal or governmental) manages the risks and dictates public policy. "How safe is safe enough?" is a rhetorical question that depends more on the management of risk than it does on quantitative statements derived from risk assessment. In the final analysis, public acceptance of risk depends more on the public's confidence to manage risks than on informed use of the quantitative estimates of the consequences, probabilities, and magnitudes of undesirable consequences.

Entrepreneurs strive to compress the time between invention and commercial development to a few years rather than decades as in years past. Governmental regulatory procedures exist for pre-market testing of new materials and devices, yet the number of new products is too large for an orderly testing of all of them. Our political system recognizes this fact and tempers its regulatory responsibility with the realization that the nation's future economy depends on sustaining a preeminent role in the development of new products, and maintaining or reassuming a position of leadership in process technology (Thurow, 1987). The dilemma has been phrased aptly (Nichols and Zeckhauser, 1985) as "*Where* should we spend *whose* money to undertake *what* programs to save *which* lives with *what* probability?"

A decision to develop new products is often cast in terms of the ***cost-benefit ratio***. If the ratio is small, conventional wisdom argues that the venture should proceed. All things being equal, it is financially prudent to pursue the venture that has the lower ratio. A low cost-benefit ratio is not a basic moral norm that is intrinsically "good", because it sidesteps the fundamental question, "Who benefits ... who pays?" In political terms, which political constituency derives the benefit and which political constituency bears the costs? Most large societal issues, such as pollution, poverty, etc., are issues where the two political constituencies are not the same (Wilson and Crouch, 1987; Ames et al., 1987; Slovic, 1987; Russell and Gruber 1987; Lave, 1987; Okrent, 1987; Dewees, 1987).

If an individual plans to buy a personal computer with his or her own funds, the person buying it also profits from its use. Thus if two machines have the same capabilities, buying the one with the lowest cost-benefit ratio is the obvious choice. On the other hand, consider the citizens of a small community troubled by a proposed municipal waste incinerator. Defining costs and benefits is a dilemma. The costs and benefits for the community whose waste will be treated are easier to calculate than the costs and benefits borne by the community in which the facility will be located. Citizens generating the waste benefit, but those in the community where the facility will be located bear the burden of additional truck traffic and anxiety about stack emissions. Possible benefits to the smaller community may be employment and tax revenue. The requirement that operating permits be secured is designed to assure the smaller community that it will not be subjected to unhealthy conditions, but such assurance may not remove its citizens' anxiety, particularly when they believe their political influence is insignificant compared to that of the community whose wastes they will treat.

To illustrate the difficulty in assessing risk, consider the creation of new materials. The age of metals gave way to the age of petrochemicals, which today is giving way to the age of composite materials, semiconductors, microelectronics, nanotechnology, and biotechnology. These emerging industries are quite different from past industries in several significant ways. First, unusual chemical compositions and manufacturing processes are created before the health and environmental implications are fully understood. Second, the pace of these developments is faster than the pace at which health tests can be conducted. Third, new products are developed by a diverse group of small manufacturers that are difficult to identify or monitor. Lastly, only small amounts of these new materials are actually used in final products. The building blocks of advanced products are metals, polymers, ceramics, semiconductors, and biochemicals which are assembled sometimes molecule by molecule using unique processes that may not be automated and may indeed be labor intensive. Some steps in the process involve exotic hazardous materials such as carcinogenic organics, highly toxic gases, and submicron particles (particles whose characteristic dimension is less than one micron, $D_p < 1$ μm) to whose surface highly active organics have been added.

Indoor air pollution involves both familiar pollutants and new materials (gases, vapors, or particles). The deleterious effects of familiar pollutants are well known. Whether new materials produce benign or deleterious effects is generally not known, although experts may anticipate these effects. New materials may not even be subject to regulations. Nevertheless, professional responsibility requires engineers to assess risk in one way or another. The majority of new materials is not invented by Fortune 500 companies but by small entrepreneurial firms that are labor intensive. The number of exposed people is small but the exposure is likely to be a large percentage of their workforce. Often, these firms are the least likely to have personnel whose sole occupational duty is health and safety. Thus there are no specialists to call upon as can be called upon in Fortune 500 companies. Lastly the volatility of these small firms and the mobility of their employees prevent the accuracy of defining the exposed population, which in turn inhibits the accuracy of epidemiological studies.

The greatest barrier to maintaining a safe and healthy environment is the inadequacy of data to estimate the health risk of new materials. For the most part, new materials are not such obvious hazards as pesticides, herbicides, or explosives. In 1984 there were approximately 60,000 chemicals and 2,000,000 mixtures in commercial use. Each year more than 1000 new chemical compounds are synthesized. Most of these materials do not reach the public as end products, but are used as intermediary materials in the production of finished products. These intermediate chemicals are often called ***chemicals of commerce***. Actions presently followed by regulatory agencies are based on the following assumptions:

- Every perceived risk cannot be evaluated at the same time.
- Plans must be developed to determine which materials pose the largest potential risk.
- The actual risk of materials having the largest potential risk will be determined first.

Evidence of risk involves several components. ***Acute*** (short-term) hazards are the easiest to diagnose because symptoms are immediate and sometimes dramatic. Toxins such as carcinogens, teratogens, neuropathogens, and mutagens produce the following ***chronic*** (long-term) effects and are more difficult to assess:

- disorders of the pulmonary, cardiovascular, and immune systems
- disorders of the skeletal system, blood, and bone marrow
- disorders of the skin and mucous membrane

- hypersensitivity

Assessing risk requires knowledge of the actual amount of material used and the percent discharged to the indoor environment. Merely listing materials that are used can create a false sense of risk. To begin with, an engineer should conduct an ***environmental audit*** to identify the amount of material procured over a period of time. Then he or she must estimate the amount of the material that remains in the product, plus the amount that is removed as waste. The difference between these is assumed to be the amount which escapes to the indoor and outdoor environment. Such audits are mass balances for the process, and can identify the material(s) posing a risk.

Responding to the ***Toxic Substances Control Act*** (TSCA), the EPA catalogued more than 56,000 manufactured or imported substances used in manufacture. The list is called the ***TSCA Inventory***. The TSCA Inventory excludes classes of materials regulated under other federal statutes such as for 8627 food additives, 1815 prescription and nonprescription drugs, 3410 cosmetic ingredients, and 3350 pesticides.

Identification of properties of chemicals that require special handling can be a complicated matter. Lists of chemicals are published by many governmental, trade, and professional agencies. The following are five familiar lists and the agency responsible for each:

- ***Environmental Protection Agency*** (EPA): Title III, Clean Air Act Amendments of 1990 (CAAA) - *189 Hazardous Air Pollutants* (HAPs)
- ***National Institute For Occupational Safety And Health*** (NIOSH) - *Registry of Toxic Effects of Chemical Substances* (RTECS)
- ***Occupational Safety and Health Administration*** (OSHA) - *List of Toxic And Hazardous Substances*
- ***National Institute For Occupational Safety and Health*** (NIOSH) - *Pocket Guide To Chemical Hazards*
- ***American Conference Of Governmental and Industrial Hygienists*** (ACGIH) - *Threshold Limit Values For Chemical Substances and Physical Agents in the Workroom Environment with Intended Changes* (published and updated yearly)

RTECS, also called ***The Registry***, is published and updated by NIOSH every few years in compliance with the ***1970 Occupational Safety and Health Act***. The Registry is the most comprehensive source to consult. It contains toxicity data extracted from the scientific and professional literature for a fraction of the approximately 65,000 chemical compounds listed. The toxicity data should not be considered as definitions of values for describing safe doses for human exposure. Table 1.4, taken from a study of a representative group of chemicals, shows the limited knowledge about the toxic properties of chemicals listed in RTECS (National Research Council, 1984). Most alarming is the fact that there is no toxicological information available for over 75% of all of the chemicals in the three "chemicals in commerce" production categories that represent the majority (both in number and volume) of chemicals used in the US.

The TSCA also requires NIH (National Institutes of Health) and EPA to create a computer database and search programs for chemicals listed in RTECS. The database is called the ***Chemical Information Service*** (CIS) and is an on-line service that enables users to retrieve toxicological data for chemicals identified by their RTECS numbers. In addition, CIS enables users to search for chemicals with specific toxicological properties, structures, dose, etc. The American Chemical Society created a registry that uniquely identifies specific compounds by a ***Chemical Abstract Service*** (CAS) number. The CAS inventory contains many substances posing no hazard to health and is considerably larger than RTECS. Many professional journals require authors to list the CAS numbers for all chemicals included in their article. Chemicals listed in the RTECS and CIS also cite the CAS numbers.

Table 1.4 Estimated mean percentage of materials for which health hazards are known; A – complete health hazard assessment possible; B – partial health hazard assessment possible; C – minimal toxicity information available; D – some toxicity information available, but below minimal; and E – no toxicity information available (from National Research Council,1984).

toxicity category	**size**	**A**	**B**	**C**	**D**	**E**
pesticides and inert ingredients of pesticide formulations	3,350	10	24	2	26	38
cosmetic ingredients	3,410	2	14	10	18	56
drugs and excipients used in drug formulations	1,815	18	18	3	36	25
food additives	8,627	5	14	1	34	46
chemicals in commerce: at least 1 million pounds per year	12,860	0	11	11	0	78
chemicals in commerce: less than 1 million pounds per year	13,911	0	12	12	0	76
chemicals in commerce: production unknown or inaccessible	21,752	0	10	8	0	82

Other compilations of this information can be found in engineering handbooks (Sax's "Dangerous Properties of Industrial Materials", 1979; Prugh's "Kirk-Othmer Encyclopedia of Chemical Technology", 1982; Verschueren's "Handbook of Environmental Data", 1983; and Budavari's "The Merck Index", 1996). Lastly, states such as Massachusetts and California publish a "Right-To-Know" list of substances. There is considerable information about the hazards and thermophysical properties of materials available on the Internet (see Section 1.12). It is not the shortage but the plethora of data that causes confusion.

Two procedures required by the federal government help to identify hazardous substances and to handle them properly. OSHA requires that whenever industrial chemicals are transferred between buyer and seller, a document called a ***material safety data sheet*** (MSDS) specifying the following properties accompany the transfer:

- manufacturer's name and chemical synonym
- hazardous ingredients (pigments, catalysts, solvents, additives, vehicle)
- physical data (boiling temperature, vapor pressure, solubility evaporation rate, percent volatile material)
- fire and explosion data (flash point, flammability limits, ignition temperature, fire fighting procedures)
- health hazard (threshold limit value, effects of overexposure, first aid procedures)
- spill and leak procedures, waste disposal methods

While the MSDS may not have all the data an engineer wants, it serves to alert engineers of possible hazards. Companies are obliged to file MSDSs and to make them available to workers; thus a "paper trail" can be followed to trace materials from creation to destruction. MSDSs for most substances can now be obtained free of charge via the Internet.

The second procedure to guide the public about hazardous new materials is called the ***Premanufacture Notice*** (PMN). TSCA legislation requires that EPA evaluate new substances and methods used to manufacture them, determine potential release points, estimate potential exposures, and determine whether it will be necessary to specify procedures to minimize exposure. Companies planning to manufacture or import chemicals not on the TSCA Inventory are required to notify the EPA with a PMN at least 90 days prior to action. The document requires the formula, chemical structure, use, and details about the production process so that points of release can be anticipated and estimates prepared of the emissions and human exposure. Other data required are the physical and chemical properties of the substance (vapor pressure, solubility in water or solvents, normal melting and boiling temperatures, particle size if it is a powder, Henry's law constant, pH, flammability, volatilization from water) and any toxicological data that may be available. New substances are screened by EPA and assigned a risk category that obliges the company to control release (general ventilation, protective clothing, respirators, glove boxes, etc.) or test the compound for its toxic or environmental effects. Toxicity is assessed by examining the toxicity of substances of comparable molecular structure and physical properties. Both MSDS and PMN procedures were created to minimize environmental exposures, prevent chronic exposures of people living near manufacturing plants, and reduce acute exposure of people affected by transportation accidents.

Key elements in risk assessment are the fields of epidemiology and toxicology. ***Epidemiology*** is the branch of medicine that investigates the cause of disease; ***toxicology*** is the study of the adverse effects of chemical agents on biologic systems (Casarett and Doull's Toxicology, 1986). Epidemiology begins with the symptoms human subjects display and relates them to certain causative factors that can be shown to be statistically significant. In a sense, epidemiology looks backwards to what has already happened. Thus it is inherently limited in its ability to assess the effects of new materials and new manufacturing processes. There are several limitations to epidemiology in environmental health:

- Information is often lacking about the duration and concentration of the exposure.
- There is frequently a long time between exposure and symptoms of the disease. Induction periods of 5 to 50 years are common in lung disease and cancer.
- Workers change occupations and/or employer, and it is thus difficult to define exposure.
- The number of individuals in the sample is often small, which complicates the determination of biologically significant elevations in risk.
- Multiple exposures to several chemicals in complex industrial settings make it difficult to determine causative agents.
- Agents affecting health (smoking, alcohol, drugs etc.) to which individuals are exposed are not identified fully, or at all.

Toxicological studies require considerable time and involve large subject populations. Because the subjects are often animals with different physiology, extrapolating the results to humans who might be subjected to low doses is fraught with debate and controversy. Occupational exposures that produce chronic effects require tests over the animal's lifetime. Occupational exposures related to pregnancy, offspring, and fertility of offspring require tests over several generations of animals. Lastly there are very few tests for neurological damage.

Risk assessment is a formal field of study in science and technology. An area where it has received considerable attention has been the design and operation of nuclear reactors. There exists a considerable body of knowledge on the theory of risk, and the subject has been treated mathematically with great sophistication (Brown, 1991). Applying the theory to occupational health and safety is hampered by an inability to define the relationships between the sources of accident and disease and an inability to express the occurrence of each as a probability. When safety and disease can be explained in probabilistic terms, the theory of risk analysis can be tapped.

1.2.1 Perspectives on Risk

In addition to an assessment of risk in quantitative terms, it is also necessary to place risks in *perspective,* i.e. to view risks in relation to one another (Ames et al., 1987; Lave, 1987; Okrent, 1987; Russell and Gruber, 1987; Slovic, 1987; Wilson and Crouch, 1987; and Gough, 1991), and to personal activities in which citizens engage both voluntarily and involuntarily. If perspective is lost, risk data are apt to acquire a false concreteness in which priorities are apt to be set foolishly. We live in an era when ***perceptions***, including those about risk, become so important they are apt to be taken as reality (Cox and Strickland, 1988). If perceptions were based on fact and presented dispassionately, there might be little need to worry. Unfortunately this is not the case. We also live at a time when ***propaganda*** is ubiquitous (e.g. advertising, partisan politics, collegiate athletics, etc.). Propaganda (Ellul, 1965 and Orwell, 1968) is not a lie or even a half-truth; rather propaganda is information disseminated to provoke a certain response by arousing our emotions rather than by engaging our minds. Perceptions manipulated by skillful propaganda impede formulation of sound public policies.

The Founding Fathers believed that democracy, along with the politics to sustain it, was created to safeguard space in which individuals can grow. The Constitution was devised to promote freedom and enterprise, and to encourage individuals to take full responsibility for their lives. In recent times, democracy has become diluted by the belief that the opportunity to pursue happiness is the right to possess it (as if equal opportunity implies equal outcome!). Thus the goal of politics has become looking after all citizens from cradle to grave, regardless of what they have done to deserve it. The goal of democracy has been redirected to creating a risk-free society with the government as the insurer of last resort. Accordingly, politics becomes a rescue service to preserve safety, health, and comfort.

An obsession with health is profoundly unhealthy; an obsession with safety is profoundly unsafe. If we believe that the state is to cushion us from misfortune, to compensate every loss and make up for every suffering, then we automatically relinquish control over our lives and narrow drastically the sphere of human action. Regulations of a mind-numbing complexity and number will govern private and corporate activity, consumer products, and employment, with the aim of ensuring that citizens can amble through a risk-free world, picking their pleasures from shelves loaded with packaged and sanitized products. If this is allowed to occur, we will suffer an enormous diminution in the value of human existence because risk will be removed from the heart of it. It is only by staking one's life that one fully possesses it (Scruton, 1999).

Individuals overreact to risks of low probability but large visibility (Viscusi, 1992). For example, terrorist bombing of commercial airliners is abominable and threatens political stability. News of aircraft accidents dominate the nation's attention from time to time. On the average, more people in the US are killed by lightening or drown in their bathtubs than are killed in all types of commercial airline accidents. Risks arising from acts of commission draw the public's attention whereas acts of omission are hardly reported. When the bearers of risk do not share in the cost of reducing the risk, extravagant remedies are apt to be adopted. Overreaction to small risks often impedes the technical progress that historically improves health. A small portion of the public opposes water chlorination but ignores its widely accepted benefit. Man-made carcinogens produce a strong public reaction whereas natural carcinogens (such as hydrazines in mushrooms or aflatoxin in peanut butter) are largely ignored.

For several decades, Americans have become increasingly preoccupied with self-centered behavior. There are many manifestations, e.g. fetishes of health, diet, and personal appearance, and an insatiable pursuit of pleasure, recreational activity, and carnal gratification. Longevity is presumed to define the good life. Forgotten is that living, as distinct from surviving, acquires value from the risks and sacrifices that benefit the general good, but may shorten life. It is the meaning and purpose of the life that is led, not longevity, which defines the moral dimension. We live in a secular age in which "caring for the environment" has replaced spiritual concerns. For these individuals, environmentalism is a secular religion; others consider this neo-paganism. Since environmental conditions can affect health, environmentalism reinforces our preoccupation with health. As evidence of our preoccupation with health, consider how much health care costs have grown recently. Over a period of two decades, health care expenditures, including workers' compensation, rose from approximately 2 percent of the gross national product in 1970 to about 11 percent in 1990, a factor of more than five. If there had been a commensurate improvement in health and safety, these expenditures would be seen as worthwhile. Unfortunately, health and safety have not improved by a factor of five. The reasons for the precipitous rise in costs are complex, but the public can ill-afford indifference to the burden these costs are to business and ultimately to the public.

The art of politics begins with persuasion, and if persuasión succeeds, programs can be conceived to achieve specific goals. Figure 1.2 shows the reinforcing incentives in environmentalism that inhibit political debate. These are listed and discussed below.

- Individuals are sensitive to the risks from involuntary activities (but ignore the higher risks of voluntary activities). They are anxious about health and can be aroused easily about the involuntary risk posed by environmental issues. Involuntary risks imbue people with the notion that they are victims and that their claims possess moral superiority.
- The media competes fiercely for an audience since the financial success of media organizations depends on the size of their audience. Environmental issues are cast as emergencies that are inherently newsworthy and chosen to sustain the public's anxiety.
- Elected lawmakers pass legislation and appropriate money for emotional issues that address the concerns of their constituents and enhance their reelection. Protecting the environment and health is a safe issue to support. There is little incentive for lawmakers to be candid about the full cost of such legislation.
- Governmental agencies need to sustain the public's anxiety about environmental risks because this anxiety secures their budget and mandate to manage programs passed by lawmakers. Professionals in government can hardly be expected to question publicly the need and direction of programs they are asked to implement.
- Researchers and academics have incentive to persuade the public that additional studies are needed because without acquiring money from governmental agencies, they would be unable to sustain their research programs.
- Environmental activists need to sustain the public's anxiety about involuntary environmental risks and urge the expenditure of pubic funds for additional study. Environmental lawyers derive their livelihood from litigation, and environmental organizations derive their operating funds from contributions sustained by the argument that an emergency exists.
- While some industries suffer from tighter pollution controls, others benefit because new economic opportunities are created for products consumers will be obliged to buy. Consider the following examples:

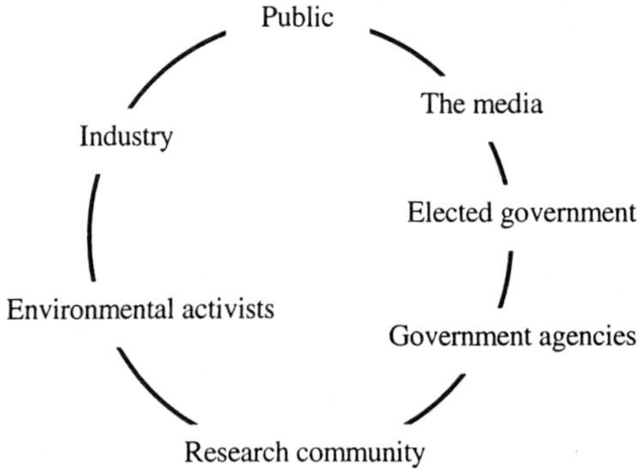

Figure 1.2 Illustration of how reinforcing incentives affect environmental policy; incentives of different political constituencies overlap to reinforce each other, and there is a conspicuous absence of conflicting incentives to restore balance.

- Banning chlorofluorohydrocarbons (CFCs) reduces profits from patented CFCs but creates a demand for CFC substitutes.
- Assisting the government in writing regulations enables companies to craft regulations that benefit them at the expense of their competitors.
- Reducing CO_2 emissions improves the sale of natural gas (CH_4) since burning CH_4 produces less CO_2 per BTU than burning coal and petroleum.
- Credits and allowances to reduce CO_2 emissions generate new wealth, because these arrangements have pecuniary value that can be bought and sold.

Explanations that expenditures to reduce emissions will be borne by the public through higher costs lack emotional appeal and adversely affect an industry's public image. Government actions that adversely affect one company's products become business opportunities for competing companies to devise new products. Thus among those who set policy, appropriate funds, and spend money, the incentives above reinforce each other; there are few incentives that compete with one another. We have one-dimensional incentives that pander to the public's anxiety about health and impending global decline. What is missing is *perspective*, the small voice within that asks, "Is the problem really this serious? Where are all the sick people? Where are all the dead bodies?" Or from the words of Henny Penny in the nursery rhyme, "Is the sky really falling?"

The restoration of sound environmental policies can be achieved by thoughtful rational debate when the public recognizes the counterproductive nature of some present policies; however, the process is slow. Counterbalance is achieved rapidly when policies are recognized as failures that do not achieve intended results. Such failures are revealed best by satirists, comedians, actors, artists, writers, etc. - those outside the arena of public debate who can portray the issue in oblique if not comic terms that enable the public to recognize the failure of a policy. To achieve perspective, citizens need to hear and see alternative points of view regarding pollution prevention and control. This is often difficult owing to competing, conflicting, and entwined interests. ***Competing interests*** are apt to exaggerate, advance half-truths, and misrepresent opposing interests to further their position. Examples of ***conflicting interests*** include:

- industries that seek to minimize expenditures for pollution control to remain competitive
- individuals who pursue personal freedom without governmental interference
- government agencies that implement air pollution legislation passed by elected officials.

Each of the above may operate at cross purposes to the other. ***Entwined interests*** occur when the goals of various independent elements of society reinforce one another and create a new and powerful constituency. Regardless of conflicting and entwined interests, the objective of pollution prevention and control is first and foremost to improve public health and safety. Policies must be measured in terms of the cost to the public and benefit derived by the public (Lehr, 1992). In the final analysis, one has to answer with candor, *are the benefits derived commensurate with the costs incurred*?

To acquire perspectives about risk, one must first understand the units with which data are presented (e.g. to ask "where are the zeros?" or if the values are small, "where is the decimal point?"). For example, if an air contaminant is described as having a mol fraction of 0.000001, or is described as being present at 1 PPM (part per million) or 1,000 PPB (parts per billion), layman are apt to respond quite differently unless they know that these concentrations are identical. The location of zeros enables one to either excite or lull without logical persuasion; for example, 0.000001 sounds small while 1,000 PPB sounds large. Presenting data in this way perverts logical debate, but unfortunately it is often employed. Lastly, even the units themselves tax comprehension, for example, trichloroethylene (TCE) in drinking water at a concentration of 1 PPB may sound ominous but such a ratio is equivalent to 1/6 of an aspirin tablet (325 mg) dissolved in a railroad tank car filled with water (16,000 gal).

Table 1.5 Risks of fatality from various activities (from Wilson and Crouch,1987; Lehr, 1992).

action	annual risk	uncertainty
cancer related to cigarettes (1 pack/day)	3.6×10^{-3}	factor of 3
all cancers	2.8×10^{-3}	10%
mountaineering (mountaineers)	6×10^{-4}	50%
motor vehicle accident (total)	2.4×10^{-4}	10%
police killed line of duty (total)	2.2×10^{-4}	20%
electrocution	1.1×10^{-4}	5%
motor vehicle accident (pedestrian)	4.2×10^{-5}	10%
alcohol, light drinker	2×10^{-5}	factor of 10
4 tablespoons peanut butter/day	8×10^{-6}	factor of 3
death due to home falls	3.5×10^{-6}	unknown
drinking water with EPA limit of chloroform	6×10^{-7}	factor of 10
drinking water with EPA limit of trichloroethylene	2×10^{-9}	factor of 10

Shown in Table 1.5 are ***risks*** of fatality (annual fatality rates) from various activities. For example, the risk of electrocution is 1.1×10^{-4}, or 0.011%, which means that approximately one out of every 9100 individuals dies by electrocution each year. One must be careful in using Table 1.5 because of the types of people affected. For the most part, police killed in the line of duty are healthy adults, mountaineers are healthy young people, and those killed in falls at home are elderly. Nonetheless, differences of several orders of magnitude exist between fatality rates due to cancer from smoking cigarettes, eating peanut butter, and drinking water containing chloroform or trichloroethylene.

An example where perspective has been debated is the removal of asbestos from schoolrooms. The probability of children contracting lung cancer (mesothelioma) is estimated (Lave, 1987) to be five per million lifetimes, less than 1/5,000 the chance of death faced by other events in children's lives. The risk to a building occupant for a ten-year exposure is less than one-fiftieth (1/50) the risk of a highway fatality resulting from commuting by car to and from the building (Dewees, 1987). In addition, improper removal of asbestos poses major risks to workers, their children, and to the population as a whole (Mossman et al., 1990). Consequently, it has been decided to leave the asbestos in many buildings until the building undergoes major renovation or demolition, since removal may not reduce already low concentrations, and a much larger improvement in public health could be bought by spending the money on other programs.

With regard to the ***carcinogenicity*** of certain chemical agents, a technique to improve our perspective has been proposed by Ames et al. (1987) and Gold et al. (1992). While animal cancer tests cannot be used to predict human risk with absolute certainty, such tests can be used to produce an index for setting priorities that reflect the carcinogenic hazard potential of certain chemical agents. A comparison of hazards from carcinogens ingested by humans should reflect the vastly different potency these carcinogens produce in humans. The usual measure of ***potency*** (TD_{50}) is the daily dose (in milligrams of intake per kilogram of body weight) at which fifty percent of any species will be tumor-free at the end of a standard lifetime. The lower the value of TD_{50}, the more hazardous is the chemical. Data from which a TD_{50} can be calculated for humans are rarely available. To arrive at an index, Ames et al. defined a term that expressed the human daily lifetime dose (in milligrams per kilogram of body weight) as a percentage of the rodent TD_{50} dose (also in milligrams per kilogram) for each carcinogen. Ames called the index the ***human exposure dose/rodent potency dose*** (***HERP***),

$$\text{HERP (\%)} = \frac{\text{human dose (mg/kg per day) over a lifetime}}{\text{rodent } TD_{50} \text{ (mg/kg per day) over a lifetime}} \times 100\% \tag{1-3}$$

In other words, a HERP of 100% means that a human intakes the *same* dose per unit of body mass as that which causes half of the tested rodents to develop tumors. The rodent TD_{50} values can be taken from a database of 975 chemicals, and typical human exposures can be estimated from data reported in the professional literature. Shown in Table 1.6 are selected HERP values taken from Ames et al. (1987). For example, the TD_{50} for exposure of mice to chloroform is 90 mg/kg. Thus consuming 83 μg of chloroform per day over the lifetime of a 70 kg human leads to a HERP of

$$\text{HERP} = \frac{\dfrac{0.083\ \text{mg}}{70\ \text{kg}}}{90\dfrac{\text{mg}}{\text{kg}}} \times 100\% = 0.0013\%$$

In Table 1.6, the HERP value for chloroform at this level of exposure (83 mg/kg) is listed as 0.001%, which agrees with the above calculation to a precision of one significant digit (usually about all one can expect for these kinds of data). Since low values of TD_{50} imply high potency, and high exposures increase the hazard, higher HERP values suggest greater possible hazard. Contaminated wells pose a

Table 1.6 HERP index (abstracted from Ames, Magaw, and Gold, 1987).

daily human exposure	dose per day per 70-kg person	HERP (%)
basil (1 g of dried leaf)	estragole (3.8 mg)	0.1
beer (12 oz)	ethyl alcohol (18 mL)	2.8
chlorinated tap water (1 L)	chloroform (83 μg)	0.001
comfrey herb tea (1 cup)	symphytine (38 μg)	0.03
contaminated well water (1 L) (Silicon Valley, CA)	trichloroethylene (2800 μg)	0.004
contaminated well water (1 L) (Woburn, Mass.)	trichloroethylene (267 μg)	0.0004
	chloroform (12 μg)	0.0002
	tetrachloroethylene (21 μg)	0.0003
conventional home (14 hr/day)	formaldehyde (598 μg)	0.6
	benzene (155 μg)	0.004
cooked bacon (100 g)	dimethylnitrosamine (0.3 μg)	0.003
	diethylnitrosamine (0.1 μg)	0.006
diet cola (12 oz)	saccharin (95 mg)	0.06
dried squid, broiled in gas oven (54 g)	dimethylnitrosamine (7.9 μg)	0.06
EDB daily intake (high exposure agricultural worker)	ethylene dibromide (150 mg)	140.0
mobile home air (14 hr/day)	formaldehyde (2.2 mg)	2.1
mushroom, one raw (15 g)	mix of hydrazines	0.1
mustard, brown (5 g)	allyl isothiocynanate (4.6 μg)	0.07
peanut butter (32 g)	aflatoxin (0.064 μg)	0.03
pesticide residue on food (daily diet, average intake)	PCB - polychlorinated biphenyls (0.2 μg)	0.0002
	DDE - principal metabolite of DDT (2.2 μg)	0.0003
	ethylene dibromide (0.42 μg)	0.0004
sake (250 mL)	urethane (43 μg)	0.003
swimming pool (1 hr, child)	chloroform (250 μg)	0.008
wine (250 mL)	ethyl alcohol (30 mL)	4.7
worker average daily intake	formaldehyde (6.1 mg)	5.8

considerably smaller risk than does diet cola, wine, beer, or several natural foods. Chlorine added to water kills bacteria but also interacts with organic matter to produce chloroform; nevertheless the amount of chloroform in tap water results in a lower HERP index than do common soft drinks and natural foods. Pesticide residues on food that once caused a great deal of anxiety are seen to result in a HERP of no particular concern. With respect to ethylene dibromide, the exposure of agricultural workers is huge compared to exposures associated with residues on foods. Plants produce their own toxins to combat a variety of insects and fungi; unfortunately some of these possess significant HERP. The aflatoxin in peanut butter is 2 PPB which corresponds to a HERP of 0.03% for a single daily peanut butter sandwich. Of course most people wouldn't eat a peanut butter sandwich every day of their lives, so the hazard from peanut butter is less than that implied by the HERP. The cancer risk in foods is nil compared to their nutritional value that sustains the body's immune system to combat cancer and other diseases.

It would be a mistake to use HERP data as absolute estimates of human hazard because of the uncertainty of applying rodent cancer tests to humans. At low dose rates, human susceptibility may differ systematically from rodent susceptibility, and the shapes of the dose-response curves are not known. In addition, as pointed out by Ames et al. (1987), there is significant discordance in the data, even between rats and mice, which are certainly more physiologically similar to each other than to humans. In fact, approximately 25% of the tested chemicals were carcinogenic to mice but not to rats, or vice-versa. In such cases, the HERP data are those of the more sensitive species, which may overestimate the potential hazard. In the final analysis the HERP index is not a scale of human risks but only a tool for estimating *relative* risks and to help set priorities. No society is risk-free, and to believe that life today is more risky than in the past is doubtful and certainly unproved.

Example 1.2 - Comparison of Risks using the HERP Index

Given: Formaldehyde is a known carcinogen, and formaldehyde vapors are given off by various materials present in our homes. Ray and Tara live in a conventional home. Tara is a stay-at-home mom, averaging 20 hours per day inside the house. Ray is concerned about the formaldehyde vapors to which Tara and their children are exposed. Tara jokes that by drinking a can of beer every evening, Ray is probably exposed to more of a cancer risk than she is by living in their home.

To do: Based on HERP data, estimate how many days (at 20 hours per day) of Tara's exposure to formaldehyde inside their conventional home would result in a cancer risk equivalent to Ray's cancer risk from the ethyl alcohol in one can of beer per day.

Solution: Let N_{days} represent the number of 20-hour days Tara stays in the house. Let N_{cans} represent the number of cans of beer Ray drinks each day. The HERP values make it possible to compare these two activities quantitatively. From Table 1.6,

- formaldehyde (14 hr per day in a conventional home, 598 μg of formaldehyde) HERP = 0.6%
- ethyl alcohol (1 can of beer per day, 18 mL of ethyl alcohol) HERP = 2.8%

Equating the risks associated with these two activities,

$$\text{risk in home} = N_{days} 20 \frac{\text{hr}}{\text{day}} \frac{0.6\%}{14 \text{ hr}} = N_{cans} \frac{2.8\%}{1 \text{ can}} = \text{risk of beer}$$

Setting $N_{cans} = 1$, one can solve for N_{days},

$$N_{days} = \frac{(1 \text{ can}) \frac{2.8\%}{1 \text{ can}}}{20 \frac{\text{hr}}{\text{day}} \frac{0.6\%}{14 \text{ hr}}} = 3.27 \text{ days} \cong 3. \text{ days}$$

Discussion: Tara is correct. The ethyl alcohol in beer is somewhat more hazardous than is the exposure to formaldehyde vapors in a conventional home. Expressed quantitatively, the ethyl alcohol in a can of beer presents a cancer risk equivalent to the formaldehyde exposure from breathing the air in a conventional home for approximately three 20-hour days. Aside from the risk to cancer, common sense indicates that intoxication from the ethyl alcohol in beer is a far more serious issue than is its chance to produce cancer. The final answer is reported to only one significant digit, since more than that cannot be justified. It is implicitly assumed in the above analysis that Ray and Tara are the same weight - a ratio of their weights can be inserted into the problem if that is not the case. The children may or may not have a greater risk, since although they are smaller than either of their parents, they also inhale less air - it is not obvious whether their dose of inhaled formaldehyde vapor per unit body weight is greater or less than that of their mother. Besides the uncertainty of applying rodent cancer tests to humans, as discussed above, there is also uncertainty about whether these carcinogens affect males, females, and children equivalently. Nevertheless, the HERP values indicate that the cancer risk from either of these activities is small; Tara and her children would be exposed to greater risk by leaving their home for a drive!

1.3 Liability

In the previous section, risk was discussed in terms of injury to individuals. There is another kind of risk of which engineers concerned with indoor air pollution must be aware, and that is the risk incurred in designing products whose performance (or lack thereof) may cause injury to others. In the event of such injury, the engineer may be liable, i.e. legally bound to make good on losses or damages incurred by the other party. ***Liability*** is a legal concept described by the theory of torts. A ***tort*** is a wrongful or injurious act for which civil action is brought. While a full discussion of the theory of torts is beyond the scope of this text, engineers should be aware of two classes of product liability - negligence and strict liability. To prevail in a product liability suit, the plaintiff (person initiating the suit) must prove all of the following:

- product contained a defect
- product left the manufacturer containing the defect
- defect was a significant factor causing the injury (***proximate cause***)
- compensable damage arose because of the injury

Defects arise from two types of errors: production errors and design errors. ***Production errors*** pertain to the manufacturer alone and occur when the manufacturer delivers a product which does not meet the manufacturer's standards. To prevail in a damage suit based on production errors, the plaintiff must show both of the following:

- the manufacturer failed to satisfy its own standards
- such a failure led to the injury

Design errors (or defects) involve events external to the manufacturer, and suits based on design errors must pass one of the following tests:

- ***negligence***, an inquiry that tests the defendant's conduct
- ***strict liability***, an inquiry that weighs utility against risk
- ***express warrantee and misrepresentation***, an inquiry that examines the product's performance against explicit claims made for the product

Negligence is the failure to use a reasonable amount of care, such that the result is injury or damage to another. The negligence standard concentrates on whether the engineer was careful, prudently trained, and properly supervised. Common law negligence exists if the plaintiff can prove "the violation of a statute which is intended to protect the class of persons (to which the plaintiff belongs) against the risk, is the type of harm which has in fact occurred" (West's Handbook Series, Rothstein, 1983). Liability due to negligence requires the plaintiff to show that there is a causal relationship between the violation and the injury.

Of the three tests mentioned above, ***strict liability*** is the most elusive. Strict liability weighs utility against risk. All products have utility, but they also have attendant risks. The issue is to determine the balance between the two. If utility outweighs risks, it is not "defective", that is to say it is not unreasonably dangerous. Strict liability turns on the issue of whether the product is ***unreasonably dangerous*** (or reasonably safe). If there are reasonable things the manufacturer did not do to make the product safer, the product could be ruled unreasonably dangerous. Strict liability does not depend on whether a product can cause injury, but only that the risks inherent in its use are those that a reasonable user ought to understand and avoid based on the following:

- usefulness and desirability of the product
- availability of other safer products that meet the same needs
- likelihood of injury and its probable seriousness
- obviousness of the danger
- common knowledge and normal public expectation of danger
- injury resulting from careful use that is in accord with written instructions and warnings
- ability to eliminate danger without seriously impairing the product's usefulness or making it unduly expensive

The courts have ruled that manufacturers must be as knowledgeable about their products as are experts in the field. The manufacturer must know all the requirements, standards, and codes that have been imposed by statute, issued by government agencies, published by technical and industry associations, and even those things known as ***good engineering practice*** that change with the state of the art and the passage of time. The standards to provide the necessary information are supposed to provide guidance, but unfortunately they sometimes are contradictory, omit specific details, and may appear so innocuous as to be useless. Responsible engineers do not design products that are intentionally defective, but they are nonetheless liable for technical inadequacies of which they are not aware, but are known by competing companies or by experts in the field, and if such inadequacies cause injury to persons or property and could have been avoided.

1.4 Indoor Air Pollution Control Strategy

There are three main control strategies to prevent chemical or biological toxins from contaminating air in the indoor environment:

- ***administrative controls***
- ***engineering controls***
- ***personal protective devices***

It must be emphasized that these strategies should be followed *in sequential order*, since administrative measures are superior and preferable to engineering controls, which in turn are superior to personal protective devices. The economics of each strategy must be considered as well. In the final analysis, combinations of methods may be the most effective for individual industries, and in many instances it may not even be necessary to install an industrial ventilation system. Figure 1.3 illustrates how all three strategies could (or should) be applied to a process in which dry ingredients are transferred to a vessel; the picture illustrates poor practice. An administrative control would be to transfer dry ingredients through permanent ductwork; this is by far the most effective way to prevent the conditions seen in the figure. If permanent ductwork is not possible, engineering controls involving a local ventilation system should be affixed to the vessel entrance to capture the dry ingredients as they enter the vessel. Only as a last resort should workers be fitted with personal protection equipment.

Figure 1.3 Example of an unsatisfactory process in which dry ingredients are transferred to a vessel (courtesy of Waukesha Cherry-Burrel, Inc.).

These three main control strategies are subdivided and explained in more detail below.

1.4.1 Administrative Controls

- ***Work practices*** are decisions made by management that specify the use of certain tools and procedures. Examples of work practices are procedures and tools used by workers in high-risk occupations, such as cleaning toxic waste sites and working with high-voltage electrical equipment or radioactive materials.
- ***Labeling and warning*** systems are written instructions attached to product containers indicating their hazards and methods to be followed in their use. Examples include labels on product containers and a material safety data sheet (MSDS) accompanying the transfer of each product.
- ***Education*** involves conveying information about the hazards associated with a product or process that may not be immediately apparent, and which require some degree of instruction. An example is the training and certification of workers who use agricultural chemicals.
- ***Waste disposal practices*** are followed when discarding hazardous materials to prevent them from inadvertently entering the workplace or contaminating municipal sanitation workers. An example is the need to drum certain chemicals for removal by certified waste management firms rather than discarding them into municipal sewers or solid waste sites.
- ***Environmental monitoring*** involves the analysis of air samples taken in the workplace to ensure that the concentration of certain compounds is within acceptable standards. Monitoring enables one to anticipate hazards and to prevent ill health. An example is monitoring the airborne lead concentration in plants that manufacture glass frit, i.e. particles of glass and lead used to manufacture porcelain-coated bathtubs, sinks, etc.
- ***Assignment scheduling*** is regulating the time a worker is exposed to high-risk conditions so that the time-averaged exposure is within acceptable limits. Examples of this practice are in the glass and metals industry where workers rotate between high-temperature and moderate-temperature assignments throughout the workday.

- ***Medical surveillance*** involves a medical examination of workers to detect unhealthy medical symptoms or hypersensitivity. An example is monitoring the lead in the blood of workers employed in the manufacture of glass frit.
- ***Housekeeping*** involves an array of obvious procedures for cleanup and tidy-up. For example, removing empty containers from the workplace from which vapors could escape, or removing inadvertent discharges (spills, drips, etc.) of materials by vacuum rather than by hand brooms.
- ***Dust suppression*** involves practices to minimize dust generated from stockpiles, bag-dumping, conveyor transfer points, drilling, etc. Examples include the use of wetting agents, windbreaks, or enclosures.
- ***Maintenance*** is the scheduled inspection, repair, and replacement of components of a process to prevent failures that would result in emission of contaminants.
- ***Sanitation*** is the application of hygienic practices to reduce the inhalation or ingestion of hazardous materials. For example, sanitation involves removing work clothes before entering eating areas, and properly cleaning work clothes so as not to expose those doing the cleaning.
- ***Management*** refers to the existence of organizational structures that provide managers the authority to affect change.

1.4.2 Engineering Controls

- ***Elimination*** is the total removal of the source of contamination. For example, replacing a solvent-based coating with a water-based coating eliminates emission of hydrocarbons as the coating dries.
- ***Substitution*** is the replacement of one toxic substance by a substance of lesser toxicity, for example, replacing benzene with toluene, replacing asbestos fibers with glass fibers, or replacing sand by steel shot for abrasive blasting of castings.
- ***Isolation*** requires placing an impervious covering over the source of contamination, or locating materials in an isolated location in a plant. An example is storing toxic materials in a designated closed room outside of the main building containing workers.
- ***Enclosure*** involves surrounding the process by a physical barrier or enclosure, thereby substantially reducing contaminants escaping to the workplace. Examples include using hopper valves to discharge a hopper into a closed vessel equipped with a bin vent filter instead of merely dumping the hopper's contents into an open vessel, or using a shake-out room where castings are removed from their molds.
- ***Process change*** involves installing a new machine or process that reduces the emission of a contaminant. Examples include replacing atmospheric cooking vessels with vessels that cook under vacuum so that any leakage would be *into* (not out of) the vessel, or redesigning dry cleaning equipment to eliminate the need for individuals to physically transfer garments between machines.
- ***Product change*** is a desirable option in instances where changing the product can reduce workplace emissions. For example, a machine component that had to be strengthened by welding a supporting brace could be redesigned as a single casting or forging that requires no welding. Similar advantages can be gained by eliminating gluing or soldering.
- ***Industrial ventilation*** is the installation of a configured inlet (***hood***) and an air mover to withdraw air surrounding the process, capture contaminants, and prevent their transfer to the workplace.

1.4.3 Personal Protective Devices

A variety of ***respirators*** covering parts of the face can remove particle, gas, and vapor contaminants from air inhaled by workers. Simple ***face masks***, as shown in Figure 1.4, remove nuisance dust, but cannot remove harmful vapors.

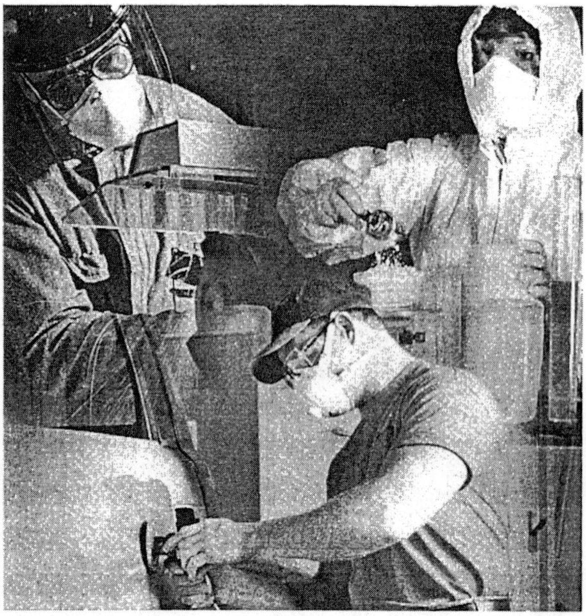

Figure 1.4 Disposable respirators for protection from dust, fumes, particulates, and nuisance odors (courtesy of Mine Safety Appliances International).

Full-face masks covering the entire face (mouth, nose, and eyes) by tight-fitting seals remove contaminants by cartridge adsorbers mounted in the mask, and protect the eyes as well (Figure 1.5a). ***Half-masks*** covering the mouth and nose remove gases and vapors by adsorbers in screw-type holders mounted in the mask (Figure 1.5b).

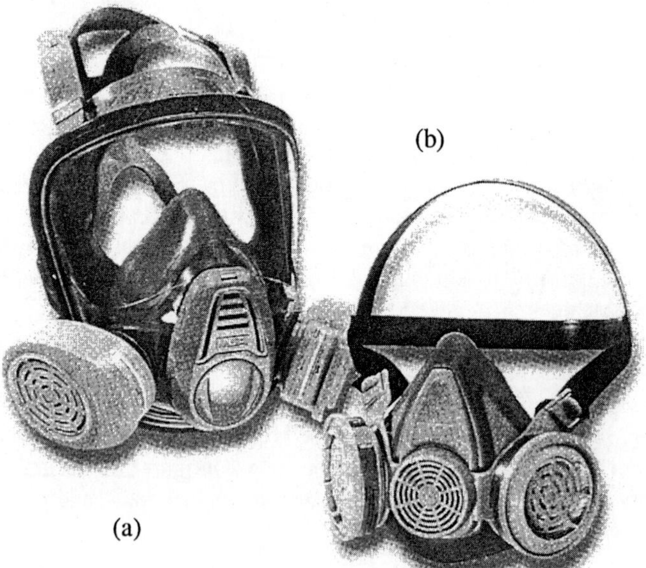

Figure 1.5 (a) Full-face, and (b) half-mask respirators (courtesy of Mine Safety Appliances International).

Lastly, a ***self-contained breathing apparatus*** (SCBA), consisting of a full-face respirator with portable air supply, is available for use in highly contaminated environments. (Readers may be more familiar with the acronym SCUBA, in which the letter U is added to indicate underwater use.) An example is shown in Figure 1.6. The air is filtered through air purifiers mounted on the worker's belt, a fan/motor blows air into the helmet, and a collar reduces leakage around the neck area. Alternatively, air can be supplied to a full-face mask through hoses connected to an external air supply. SCBA equipment is needed for applications such as welding and sand blasting, and when caustic vapors are involved. Shown in Figure 1.7 is a device in which cleaned room air is blown down between the face and face shield and out around the neck. The air is moved by a fan which is powered by a battery worn around the worker's waist.

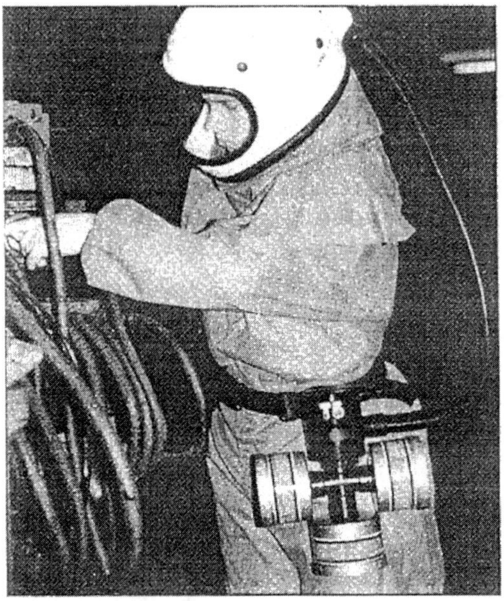

Figure 1.6 Full-face, positive pressure, helmet-type air purifier with collar and belt-mounted filter (courtesy of The St. George Company Ltd.).

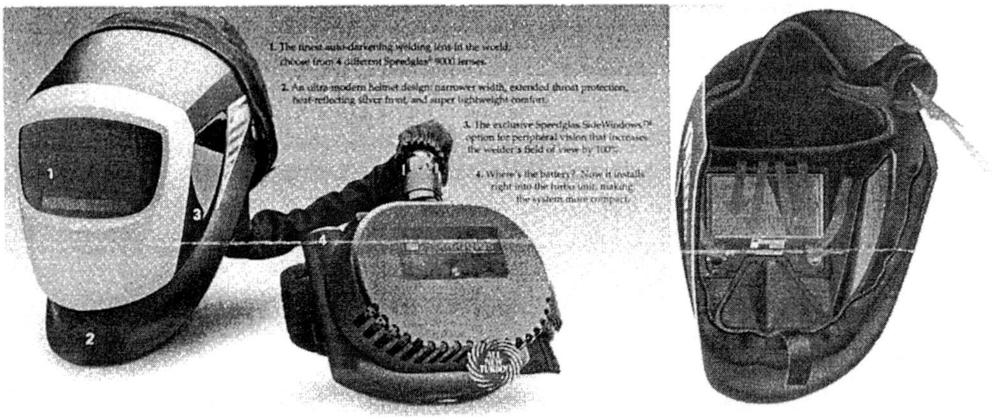

Figure 1.7 Powered air purifying respirator, with fan and battery pack for mounting on the worker's waist; air is supplied to the rear of the helmet, and a stream of filtered air passes over the worker's face (from Hornell Speedglas Inc.).

In an industrial setting, the success of a respirator program depends on the effectiveness of managerial procedures that oversee replacement of the cartridges, maintain the respirators, and ensure that they fit workers properly. Securing a tight seal between the mask and the face is extremely important. The contour of men's and women's faces varies; the contour is also different for individuals of different ethnic backgrounds. Beards and corrective glasses require special practices. Many people suffer from headaches induced by the tight elastic bands across the temple, or experience skin irritation from the seal. The effectiveness of respirators is easy to appreciate; the inconvenience in wearing them is often underestimated if not ignored. Lastly, any manager who contemplates prescribing respirators should wear one for an extended period of time before requiring others to do so!

OSHA requires individuals who are asked by their employers to wear respirators to satisfy a ***fit-test*** (29 CFR 1910.134). Fit-testing is required prior to using respirators on the job, and at least annually thereafter. In addition, fit-testing must be performed whenever there is a change in the employee's physical condition that could affect the respirator's fit, e.g. dental changes, or significant change in body weight. There are two basic types of fit-test: qualitative and quantitative. The ***qualitative fit-test*** (QLFT) depends on the subject's voluntary (or involuntary) response to a test agent. The response involves smell, taste, or irritation; i.e. does the individual smell, taste, or get irritated by the test agent? If the individual responds, the respirator fails the fit-test. The ***quantitative fit-test*** (QNFT) measures the ratio of the concentration of a test agent in the ambient environment ($c_{ambient}$) to the concentration in the space between the respirator face-piece and the individual's face (c_{face}). This ratio is called the ***fit factor***, (f_f),

$$\boxed{f_f = \frac{c_{ambient}}{c_{face}}} \tag{1-4}$$

Different classes of respirators have different fit-factors. A respirator with a fit-factor of 100 means that the ratio of the concentration in the ambient environment ($c_{ambient}$) is 100 times larger than the concentration inside the respirator, $c_{ambient} = 100\ c_{face}$.

The qualitative fit-test (QLFT) is a pass/fail test that does not actually measure leakage or concentrations. Test agents include saccharin (sweet), bitrex (a commercial aerosol with a bitter taste), isoamyl acetate (a detectable odor at 1 PPM), and stannic chloride (odorous smoke). If the individual does not respond to the test agent, the respirator passes the QLFT test; it can be assumed that the fit-factor is 100 or greater. Other details of fit-tests include:

- The individual must be instructed as to how to place the respirator over the face and adjust the straps.
- The respirator must be worn for at least 5 minutes prior to conducting the test.
- The individual must wear the respirator with whatever other devices are normally worn (glasses, contact lenses, hearing aides, etc.)
- Facial hair must be shaved so that there is no hair between the respirator's seal and the skin.

1.5 Fundamental Calculations

In the field of indoor air quality, fundamental calculations are performed involving molecular weights, ideal gases, mass and volumetric flow rates, concentrations, densities, and mol fractions. Engineers also need to be aware of the accuracy and precision of their measurements and calculations. The following sections provide a brief introduction to each of these topics.

1.5.1 Accuracy, Precision, and Significant Digits

Numbers are the currency of engineering, and it is imperative that numbers be used properly if "correct answers" are to be achieved with maximum possible certainty. Three principles govern the proper use of numbers: accuracy, precision, and significant digits.

- ***Accuracy error*** (*inaccuracy*) is defined as the reading minus the true value.
- ***Precision error*** is defined as the reading minus the average of readings.
- ***Significant digits*** are digits which are relevant and meaningful.

An instrument can be very precise without being very accurate, and vice-versa. For example, suppose the true value of temperature is 25.0 °C. Two thermometers A and B take five temperature readings each:

- Thermometer A: 25.5, 25.7, 25.5, 25.6, and 25.6 °C. Average of all readings = 25.6 °C.
- Thermometer B: 26.3, 24.5, 23.9, 26.8, and 23.5 °C. Average of all readings = 25.0 °C.

Clearly, thermometer A is more precise, since none of the readings differs by more than 0.1 degrees from the average. However, the average is 25.6 °C, which is 0.6 degrees greater than the true temperature, which indicates significant ***bias error***, also called ***constant error***. On the other hand, thermometer B is not very precise, since its readings swing wildly from the average; but its overall average is closer to the true value. Hence, thermometer B is more accurate than thermometer A, at least for this set of readings, even though it is less precise.

The difference between accuracy and precision can be illustrated effectively by analogy to shooting a gun at a target, as sketched in Figure 1.8. Shooter A is very precise, but not very accurate, while shooter B has better overall accuracy, but less precision.

Many engineers do not pay proper attention to the number of significant digits. The least significant numeral in a number implies the precision of the measurement or calculation. For example, a result written as 1.23 implies that the result is precise to within one digit in the second decimal place, i.e. the number is somewhere between 1.22 and 1.24. Expressing this number with any more digits is misleading. The number of significant digits is most easily evaluated when the number is written in exponential notation; the number of significant digits can then simply be counted, including zeroes. Some examples are shown in Table 1.7.

When performing calculations or manipulations of several parameters, the final result is generally only as precise as the least precise parameter in the problem. For example, suppose A and B are multiplied to obtain C. If A = 2.3601 (five significant digits), and B = 0.34 (two significant digits), then C = 0.80 (only two digits are significant in the final result). Note that most students are tempted to write C = 0.802434, with six significant digits, since that is what is displayed on a calculator after multiplying these two numbers. Let's analyze this simple example carefully. Suppose the exact value of B is 0.33501, which is read by the instrument as 0.34. Also suppose A is exactly 2.3601, as

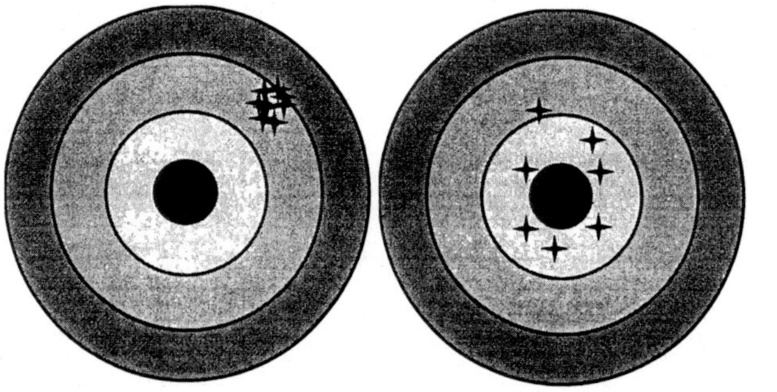

Figure 1.8 Illustration of accuracy versus precision. Shooter A is more precise, but less accurate, while shooter B is more accurate, but less precise.

Table 1.7 Significant digits.

number	exponential notation	number of significant digits
1.23	1.23×10^{1}	3
123,000	1.23×10^{5}	3
0.00123	1.23×10^{-3}	3
40,300	4.03×10^{4}	3
40,300.	4.0300×10^{4}	5
0.005600	5.600×10^{-3}	4
0.0056	5.6×10^{-3}	2
0.006	$6. \times 10^{-3}$	1

measured by a more accurate instrument. In this case, C = A times B = 0.79066 to five significant digits. Note that our first answer, C = 0.80 is off by one digit in the second decimal place. Likewise, if B is 0.34499, and is read by the instrument as 0.34, the product of A and B would be 0.81421 to five significant digits. Our original answer of 0.80 is again off by one digit in the second decimal place. The main point here is that 0.80 (to two significant digits) is the best one can expect from this multiplication since, to begin with, one of the values had only two significant digits. Another way of looking at this is to say that beyond the first two digits in the answer, the rest of the digits are meaningless or not significant. For example, if one reports what the calculator displays, 2.3601 times 0.34 equals 0.802434, the last four digits are *meaningless*. As shown above, the final result may lie between 0.79 and 0.81 - any digits beyond the two significant digits are not only meaningless, but *misleading*, since they imply to the reader more precision than is really there.

When writing intermediate results in a computation, it is advisable to keep one or two "extra" digits to avoid round-off errors; however, the final result should be written with the number of significant digits taken into consideration. The reader must also keep in mind that a certain number of significant digits in the result does not necessarily imply the same number of digits of overall accuracy. Bias error in one of the readings may, for example, significantly reduce the overall accuracy of the result, perhaps even rendering the last significant digit meaningless, and reducing the overall number of reliable digits by one. Finally, when the number of significant digits is unknown, the accepted engineering standard is three significant digits.

1.5.2 Mols, Molecular Weights, and Ideal Gases

A ***mol*** (sometimes gmol, g-mol, or mole, not to be confused with the rodent) denotes an *amount of matter*. Specifically one mol is 6.0251×10^{23} molecules of a substance, a standard number of molecules known as ***Avogadro's number***. Strictly speaking, mol does not have dimensions of mass; rather, mol is a primary dimension in and of itself, i.e. the amount of matter. Note that some authors, however, treat mol as a unit of mass. The number of mols of a substance is denoted by the letter n. ***Molecular weight*** (M) is defined as the number of grams (g) per mol (n) of a substance. The mass, m, of a substance is equal to the number of mols times the molecular weight of the substance,

$$\boxed{m = nM} \tag{1-5}$$

For elements, M is obtained from standard periodic charts (see also Appendix A.19). For example, the molecular weight of nitrogen is $M_{nitrogen}$ = 14.0067 g/mol. For molecules, M is the sum of M for each element comprising the molecule. Nitrogen in its gaseous state occurs as a diatomic molecule, N_2; thus, $M_{gaseous\ nitrogen}$ = 2(14.0067 g/mol) = 28.0134 g/mol. Air is a mixture of gases, but since air is made up predominantly of nitrogen gas, the molecular weight of air is very close to that of nitrogen,

$$M_{air} = 28.97 \frac{g}{mol}$$

In S.I. units, the kilogram (kg) is preferred over the gram; thus the ***kilogram-mol*** (kmol, sometimes kg-mol or kg-mole) is often used instead of the mol. By definition, a kmol is defined as 1000 mol, or 6.0251 x 10^{26} molecules of the substance. The molecular weight of air in terms of kg and kmol is then

$$M_{air} = \left(\frac{28.97\ g}{mol}\right)\left(\frac{1000\ mol}{kmol}\right)\left(\frac{kg}{1000\ g}\right) = 28.97\frac{kg}{kmol}$$

In English units, the pound-mass (lbm) is the standard unit of mass. In order to use the same molecular weights as those listed on the periodic chart, the ***pound-mol***, (lbmol, sometimes lb-mol, lbm-mol, or lbm-mole) is defined. The molecular weight of elemental nitrogen, for example, in English units is $M_{nitrogen}$ = 14.0067 lbm/lbmol, and the molecular weight of air is

$$M_{air} = 28.97\frac{lbm}{lbmol}$$

The ***ideal gas law*** for a single species of a gas is

$$\boxed{PV = nR_uT} \qquad (1\text{-}6)$$

where P is the absolute pressure of the gas, V is the volume occupied by the gas, n is the number of mols of the gas, R_u is the ***universal ideal gas constant***, and T is the absolute temperature of the gas. The ideal (perfect) gas law, Eq. (1-6), can be solved for the ***density*** (ρ), since density is defined as mass per unit volume,

$$\boxed{\rho = \frac{m}{V} = \frac{nM}{V} = \frac{PM}{R_uT}} \qquad (1\text{-}7)$$

where Eq. (1-5) has also been used. The universal ideal gas constant (R_u) is, as its name implies, *universal*, i.e. the *same* regardless of the gas being considered. In S.I. units (Joule, J, the unit of energy and Kelvin, K, the unit of temperature), the universal ideal gas constant is

$$R_u = 8.3143\frac{kJ}{kmol \cdot K} = 8314.3\frac{J}{kmol \cdot K}$$

Since R_u is written in terms of kmol, n must represent the number of kmols of the substance in the ideal gas law, Eq. (1-6). In English units (foot-pound, ft-lbf, the unit of energy and Rankine, R, the unit of temperature),

$$R_u = 1545.4\frac{ft \cdot lbf}{lbmol \cdot R}$$

Since R_u is given in terms of lbmol in the English system of units, the number of mols (n), must represent the number of lbmols of the substance in the ideal gas law, Eq. (1-6).

A more *general* formulation of the ideal gas law can be written for cases in which the gas pressure is too high to behave as an ideal gas,

$$\boxed{PV = \frac{ZmR_uT}{M} = ZnR_uT} \qquad (1\text{-}8)$$

where Z is the ***compressibility factor***. The density (ρ) is then

$$\boxed{\rho = \frac{m}{V} = \frac{PM}{ZR_uT}} \qquad (1\text{-}9)$$

If the total pressure (P) is well below the ***critical pressure*** (P_c), i.e. if $P << P_c$, the compressibility factor (Z) is equal to 1, and Eq. (1-9) reduces to Eq. (1-7). In air pollution control systems, the effects of Z can nearly always be ignored.

The ***specific ideal gas constant*** (R, sometimes R_{gas}) is *not* universal - its value depends on the specific gas being considered. R is defined as the universal gas constant divided by the molecular weight of the substance,

$$\boxed{R = \frac{R_u}{M}} \tag{1-10}$$

The dimensions of R are not the same as those of R_u, since molecular weight is a not a dimensionless quantity, although some authors treat it as such. The ideal gas law in terms of R is

$$\boxed{PV = mRT \quad \text{or} \quad P = \rho RT} \tag{1-11}$$

where P is the absolute pressure of the gas, V is the volume occupied by the gas, m is the mass of the gas, and T is the absolute temperature of the gas. For air in S.I. units,

$$R_{air} = \frac{R_u}{M} = \frac{8.3143 \frac{kJ}{kmol \cdot K}}{28.97 \frac{kg}{kmol}} = 0.2870 \frac{kJ}{kg \cdot K} = 287.0 \frac{J}{kg \cdot K}$$

For air in English units,

$$R_{air} = \frac{R_u}{M} = \frac{1545.4 \frac{ft \cdot lbf}{lbmol \cdot R}}{28.97 \frac{lbm}{lbmol}} = 53.34 \frac{ft \cdot lbf}{lbm \cdot R}$$

As a check, one can convert from S. I. to English units,

$$R_{air} = \left(0.2870 \frac{kJ}{kg \cdot K}\right)\left(\frac{1\ Btu}{1.055\ kJ}\right)\left(\frac{5\ K}{9\ R}\right)\left(\frac{778.17\ ft \cdot lbf}{Btu}\right)\left(\frac{0.4536\ kg}{lbm}\right) = 53.35 \frac{ft \cdot lbf}{lbm \cdot R}$$

The disagreement in the last digit is due to round-off errors in the conversion factors.

1.5.3 Mass Balance

In steady flow, the total mass of material entering a device must equal the total mass leaving the device, irrespective of whatever chemical reactions occur within the device. Practitioners must be able to perform elementary mass balances as shown in Example 1.3.

Example 1.3 - Carbon Monoxide from a Ventless Heater Burning Natural Gas

Given: Your elderly parents wish to provide supplementary heating to their living room, and agree to install a ventless fireplace that burns natural gas (CH_4). The fireplace uses ceramic artificial logs and embers to resemble burning wood, and is advertised as being 99% efficient; but the efficiency is never defined. You interpret the claim to mean that 99% of the natural gas burns to CO_2 and H_2O. At the end of the manufacturer's literature (and in small print) is the statement that the fireplace uses 10% excess air and that the CO to CO_2 ratio in the exhaust products is 0.0526% by volume. The fireplace is rated to consume natural gas at a rate of 20. g/min.

To do: Advise your parents if the addition of CO to the living room represents an inside air pollution problem.

Solution: The first step in the process is to compute the mass emission rate of CO. Details about the hazards of CO inhalation can be found in Chapter 2. How to relate the CO mass emission rate (g/s) to CO concentrations in the living room is an issue discussed in Chapter 5. If combustion were "ideal", there would be no excess oxygen and all the carbon and hydrogen would be burned to CO_2 and H_2O:

$$CH_4 + 2O_2 = CO_2 + 2H_2O$$

Actual combustion is not ideal, but requires some ***excess air***, and other products of combustion are formed; the ***stoichiometric equation*** for CH_4 combustion becomes

$$CH_4 + a(O_2 + 3.76N_2) = b(CO_2) + d(CO) + e(O_2) + f(N_2) + g(H_2O)$$

where the value 3.76 comes from the fact that normal atmospheric air contains 3.76 times more mols of molecular nitrogen than molecular oxygen. Under ideal conditions the value of coefficient a would be 2 (i.e. 2 mols). Since the actual combustion uses 10% excess air, the value of a is

$$a = 2(1.10) = 2.20.$$

While all the coefficients, a, b, d, e, f, and g are not needed to solve the problem, they will be computed for completeness. Constructing a mass balance for each of the atoms in the stoichiometric equation,

- carbon (C): $1 = b + d$
- hydrogen (H): $4 = 2g$, thus $g = 2$
- oxygen (O): $2a = 2b + d + 2e + g$
- nitrogen (N): $2(3.76)a = 2f$, thus $f = 3.76a = 8.272$

Because $(CO)/(CO_2) = 0.0526\%$ by volume, $d/b = 0.000526$. Thus, $1 = b + 0.000526b = 1.000526b$. Consequently, $b = 0.99947$, and $d = 0.0005257$. Returning to the oxygen balance,

$$2(2.20) = 2\,(0.99947) + 0.0005257 + 2e + 2$$

Solving for e yields

$$e = 0.20026$$

The mass flow of CO can be computed from the following:

$$\dot{m}_{CO} = \dot{m}_{CH_4}\frac{n_{CO}}{n_{CH_4}}\frac{M_{CO}}{M_{CH_4}} = 20.\frac{g}{min}\left(\frac{0.0005257 \text{ mol CO}}{1 \text{ mol } CH_4}\;\frac{28.00\frac{g}{\text{mol CO}}}{16.00\frac{g}{\text{mol } CH_4}}\right) = 0.0184\frac{g}{min}$$

Since there are only two significant digits in some of the available data, the final mass flow rate of CO is estimated to be 0.018 g/min.

Discussion: As discussed above, more than two significant digits were included in intermediate calculations to avoid round-off errors. This CO emission rate seems very small but it can cause CO poisoning if the room is small and the room's ventilation is inadequate. See Chapter 5 for details on how to compute the CO concentration in a room as a function of room volume, ventilation rate, source strength, etc.

1.5.4 Mass and Volumetric Flow Rate

As a general proposition, fluid flow should be described in terms of mass flow rate since there is no confusion about what kg/s means. However, owing to ease of measurement, volumetric flow rates have been used in engineering for quite some time. For liquids, the volumetric flow rate is not confusing since the density of a liquid is nearly constant. Gas flows are a different matter, however, and two terms are widely used to describe volumetric flow rates. In engineering units the two terms are ***standard cubic feet per minute*** (SCFM) and ***actual cubic feet per minute*** (ACFM). The unit of time, minutes, is sometimes replaced by seconds or hours and poses no confusion. A relationship between the two volumetric flow rate designations needs to be established. The mass flow rate of gas through a duct is given by

$$\boxed{\dot{m} = \int_A \rho U dA} \qquad (1\text{-}12)$$

where U is the velocity perpendicular to the differential area dA and ρ is the density of the gas at the actual temperature and pressure (T and P). If the density (ρ) does not vary across area A, Eq. (1-12) becomes

$$\dot{m} = \rho \int_A U dA = \rho Q \tag{1-13}$$

where Q (sometimes called Q_{actual}) is the ***actual volumetric flow rate*** at the actual temperature and pressure (T and P) of the gas,

$$Q = Q_{actual} = \int_A U dA \tag{1-14}$$

Typical units for Q are ACFM (actual cubic feet per minute) or actual cubic meters per second (m^3/s).

It is common practice in air pollution to speak of a ***standard volumetric flow rate*** in units such as SCFM (standard cubic feet per minute) which is a *hypothetical* volumetric flow rate equal to what the actual volumetric flow rate would be at ***standard temperature and pressure*** (STP), defined as

$$\begin{aligned} &\text{Standard pressure} = 1 \text{ atm} = 101.325 \text{ kPa} = 14.696 \text{ psia} \\ &\text{Standard temperature} = 25\ ^\circ\text{C} = 298.15 \text{ K} = 77\ ^\circ\text{F} = 536.67 \text{ R} \end{aligned} \tag{1-15}$$

Thus the standard volumetric flow rate is a hypothetical, or corrected, or reference volumetric flow rate; it is not the actual volumetric flow rate. Since the mass flow rate (ρQ) is independent of temperature and pressure,

$$Q_{STP} = Q_{actual}\left(\frac{\rho_{actual}}{\rho_{STP}}\right) = Q_{actual}\left(\frac{P_{actual}}{P_{STP}}\right)\left(\frac{T_{STP}}{T_{actual}}\right) \tag{1-16}$$

In the commonly used engineering units of cubic feet per minute, the above becomes

$$Q\ (\text{SCFM}) = Q\ (\text{ACFM})\left(\frac{P_{actual}}{P_{STP}}\right)\left(\frac{T_{STP}}{T_{actual}}\right) \tag{1-17}$$

There are times when engineers are casual and speak of the volumetric flow rate in the ambiguous unit CFM (ft^3/min); this leaves listeners in a quandary since they do not know what temperature and pressure are implied. If the actual temperature and pressure are close to the standard values, the ambiguity is inconsequential. Engineers should always explicitly indicate actual or standard volumetric flow rate. When using SI units the confusion seldom arises because the modifying terms "standard" and "actual" are generally not used. In SI units a volumetric flow rate is understood to be the actual value even though the modifying word "actual" is not used. Whenever possible, one should use the *mass* flow rate in appropriate units to avoid confusion.

- The term ACFM is incomplete without specifying the actual (absolute) pressure and the actual (absolute) temperature where the measurement is made. Thus one should always state that the volumetric flow rate is Q (ACFM) at pressure P and temperature T.
- The term SCFM has a unique meaning and there is no need to specify the actual pressure and temperature. If the user needs to know the actual fluid velocity, however, it will be necessary to specify the actual pressure and temperature of interest and then use Eq. (1-17) to convert to ACFM.

Example 1.4 - The Flow of Air Inside a Duct

Given: Air passes through a 10.0-inch (ID) duct at 240. °F and at a gage pressure of 80.5 psig. The average velocity through the duct (U) is equal to 25. ft/s.

To do: Compute the mass flow rate in lbm/min, and the volumetric flow rate in ACFM and SCFM.

Solution: The air density can be computed from the ideal gas law, Eq. (1-7),

$$\rho = \frac{PM}{R_u T} = \frac{\left((80.5+14.696)\dfrac{\text{lbf}}{\text{in}^2}\right)\left(\dfrac{144\ \text{in}^2}{\text{ft}^2}\right)\left(28.97\dfrac{\text{lbm}}{\text{lbmol}}\right)}{\left(1545.4\dfrac{\text{ft}\cdot\text{lbf}}{\text{lbmol}\cdot\text{R}}\right)(240.+459.67)\ \text{R}} = 0.3673\frac{\text{lbm}}{\text{ft}^3}$$

The mass flow rate is given by the integrated form of Eq. (1-12),

$$\dot{m} = \rho A U = \rho\frac{\pi D^2}{4}U = 0.3673\frac{\text{lbm}}{\text{ft}^3}\frac{\pi(10.0\ \text{in})^2}{4}25.\frac{\text{ft}}{\text{s}}\left(\frac{\text{ft}^2}{144\ \text{in}^2}\right) = 5.008\frac{\text{lbm}}{\text{s}}\left(\frac{60\ \text{s}}{\text{min}}\right) = 300.5\frac{\text{lbm}}{\text{min}}$$

The volumetric flow rate in ACFM is obtained from Eq. (1-13),

$$Q\ (\text{ACFM}) = \frac{\dot{m}}{\rho} = \frac{300.5\dfrac{\text{lbm}}{\text{min}}}{0.3673\dfrac{\text{lbm}}{\text{ft}^3}} = 818.1\ \text{ACFM at 80.5 psig and 240. }^\circ\text{F}$$

The volumetric flow rate in units of SCFM is given by Eq. (1-17),

$$Q\ (\text{SCFM}) = Q\ (\text{ACFM})\left(\frac{P_{\text{actual}}}{P_{\text{STP}}}\right)\left(\frac{T_{\text{STP}}}{T_{\text{actual}}}\right) = 818.1\frac{\text{ft}^3}{\text{min}}\left(\frac{(80.5+14.696)\dfrac{\text{lbf}}{\text{in}^2}}{14.696\dfrac{\text{lbf}}{\text{in}^2}}\right)\left(\frac{536.67\ \text{R}}{(240.+459.67)\ \text{R}}\right)$$

$$= 4065.\ \frac{\text{ft}^3}{\text{min}} = 4065.\ \text{SCFM}$$

Discussion: Units should always be included when working out solutions, as in this example, to avoid unit errors. Q (ACFM) could also have been found simply from Q = UA. The intermediate values above are written to four significant digits. However, the speed of the flow is given to only two significant digits. Hence, the final results for mass flow rate, ACFM, and SCFM should be more properly given as 300 lbm/min, 820 ACFM, and 4100 SCFM, respectively, all to two significant digits.

1.5.5 Contaminant Concentration and Mixtures of Ideal Gases

Air contaminants may be particles or gases. ***Particles*** may be solid and/or liquid. ***Gaseous contaminants*** may be truly gases or they may be potentially condensable vapors. Both contaminant gases and vapors that obey the ideal gas law will be called simply "gas" throughout this text. ***Fume*** is the generic name for contaminants generated by exothermic processes, often metallurgical processes. A fume is a mixture of gases, vapors, and particles in varying amounts. The particles are tiny (characteristic dimensions are generally considerably less than one micrometer), which for the most part are ***condensates*** of vapors produced by the exothermic process. Fumes may also contain small solid particles that were created as such.

The mass of a contaminant per unit volume of carrier gas is called the ***mass concentration*** (c). If more than one type of contaminant is being considered, a subscript (typically i, j, or k) is added (c_i, c_j, or c_k); alternatively, the contaminant name or its abbreviation or chemical formula may be used as the subscript (e.g. c_{mercury} or c_{Hg}). A variety of units are used for the mass concentration, but milligrams per cubic meter (mg/m^3) is common and will be used in this text unless the powers of ten suggest more convenient units. Many authors use the term "concentration" by itself; readers must be careful to understand what units are implied, since some users will be thinking of the mass per unit volume of carrier gas while others will be thinking of the number of mols, the number of molecules, or the number of particles per unit volume of carrier gas. Careful attention to the units will signify which definition is being used. There is no consensus on the word "concentration" because theoretical and experimental advantages accompany different conventions. To clarify the issue, users should demand that the units of concentration be specified.

The mass concentration of *gaseous* contaminants has the same meaning as that for particles,

$$\boxed{c_j = \frac{m_j}{V}} \tag{1-18}$$

where c_j is the mass concentration of species j, m_j is its mass, and V is the total volume occupied by the gas mixture. If the mass of each species is known, ***mass fraction*** (f_j) can also be defined:

$$\boxed{f_j = \frac{m_j}{m_t}} \tag{1-19}$$

where m_t is the total mass of the mixture. If the molecular weight of the contaminant is known, the concentration can alternately be given as a mol fraction. The ***mol fraction*** (y_j) of gaseous species j is

$$\boxed{y_j = \frac{n_j}{n_t}} \tag{1-20}$$

where n_j is the number of mols of the contaminant and n_t is the total number of mols in the gas mixture. The mol fraction expressed as a percent (%) is also called the ***percent by volume***. Note that since n_t is the sum of each n_j in the mixture,

$$\boxed{\sum_j y_j = \sum_j \frac{n_j}{n_t} = \frac{\sum_j n_j}{n_t} = \frac{n_t}{n_t} = 1} \tag{1-21}$$

By definition of the mol fraction (y_j) in Eq. (1-20), a value of $y_j = 10^{-6}$ would be one mol (or part) of species j per million total mols (or parts). This ratio is called ***parts per million*** (PPM), i.e. 1 PPM ≡ (1 part) / (10^6 parts), and

$$\boxed{PPM_j = 10^6 y_j} \tag{1-22}$$

Some authors use the unit PPMV to emphasize that the ratio is based on *volume*; this convention is not adopted in this text. The unit PPM is also used by some authors to mean the mass (in grams) of a pollutant per million grams of surrounding medium. To eliminate confusion, the abbreviation PPM can be defined with modifying words,

- ***PPM based on volume*** = ratio of mols of pollutant to 10^6 total mols
- ***PPM based on mass*** = ratio of the mass (in g) of pollutant to 10^6 g of total material

Because the concentration of air pollutants is often expressed as a mol fraction (y_j), it is the convention in this text to consistently define the abbreviation PPM only on the basis of volume. The words "on the basis of volume" will be omitted for the sake of brevity. In those occasions when a ratio based on mass is appropriate, the abbreviation PPM will *never* be used. Rather the ratio of the grams of pollutant per million grams of medium (or μg of pollutant per gram of medium) will be stated explicitly. For the very low concentrations often associated with the term parts per million, the ratio of the number of mols of species j per million mols of carrier gas is essentially equal to the PPM. For example, if 2 mols of a contaminant are added to one million moles of air, the mol fraction is equal to $2/(2 + 10^6) = 1.999996 \times 10^{-6}$, which is nearly identical to $2/10^6$ (2 PPM).

For a ***mixture of ideal gases***, the ideal gas law, Eq. (1-6), is valid for the jth molecular species, provided that the ***partial pressure*** (P_j) for that species is used in place of the total pressure (P),

$$P_jV = \left(\frac{m_j}{M_j}\right)R_uT = n_jR_uT \tag{1-23}$$

The partial pressure of species j is defined as the pressure that would result if that species were the only species in the volume (V). Eq. (1-23) applies to each species of gas in the ideal gas mixture. Suppose there are a total of N gas species in the mixture. The sum of all the partial pressures in the mixture must equal the total pressure (P) of the gas mixture, i.e. ***Dalton's law of additive pressures***:

$$P = \sum_{j=1}^{N} P_j = \sum_j P_j = \sum P_j \tag{1-24}$$

Note that for brevity, the upper limit (N) on the summation may be omitted as in Eq. (1-24), and in the material to follow. It is understood that the summation is from j = 1 to N. When it is obvious that the summation is over j, the lower limit may also be omitted for brevity, as in the far right-hand-hand side of Eq. (1-24). A different form of the ideal gas law can be written instead in terms of the ***partial volume*** (V_j) of species j, where the pressure is the total pressure of the gas mixture,

$$PV_j = n_jR_uT \tag{1-25}$$

The partial volume of species j is defined as the volume that species j would take up at pressure P if it were the only species. Eq. (1-25) applies to each species of gas in the ideal gas mixture. The sum of all the partial volumes in the mixture must equal the total volume (V) of the gas mixture, i.e. ***Amagat's law of additive volumes***:

$$V = \sum_j V_j \tag{1-26}$$

In terms of partial pressures and partial volumes, Eq. (1-20) becomes

$$y_j \equiv \frac{n_j}{n_t} = \frac{P_j}{P} = \frac{V_j}{V} \tag{1-27}$$

P and V without subscripts are the ***total pressure*** and ***total volume*** of the gas mixture; these are sometimes denoted with the subscript t, i.e. P_t and V_t. For indoor air pollution applications, the gas mixture is air with a small amount of contaminant, and the total pressure (P) is just the local (absolute) ambient pressure in the room. Often the molecular weight of the gas mixture, called the ***total molecular weight*** (M_t) is needed. M_t, sometimes referred to as the ***average molecular weight*** (M_{avg}), can be calculated from the following:

$$M_t = M_{avg} = \frac{m_t}{n_t} = \sum_j \left(y_jM_j\right) \tag{1-28}$$

where m_t and n_t are the total mass of the gas mixture and the total number of mols of the gas mixture respectively.

Combining Eqs. (1-18) and (1-23), the mass concentration (c_j) of gaseous species j per total volume of carrier gas, and the mol fraction (y_j) of species j are related as follows:

$$c_j = \frac{m_j}{V} = \frac{P_jM_j}{R_uT} = \left(\frac{P_j}{P}\right)\left(\frac{PM_j}{R_uT}\right) = y_j\frac{PM_j}{R_uT} \tag{1-29}$$

where Eq. (1-27) has also been applied. At standard temperature and pressure (STP), Eq. (1-15), the mass concentration in the units (mg/m^3) reduces to

$$c_j\left(\frac{mg}{m^3}\right)_{STP} \cong \frac{y_j M_j}{24.5\times10^{-6}} = \frac{PPM_j M_j}{24.5} \tag{1-30}$$

where y_j is the mol fraction, PPM_j x 10^{-6} is its equivalent, and 24.5 is a conversion factor. All the units in Eq. (1-30) have been accounted for already, so molecular weight M_j can be plugged in without units. If the actual pressure P and/or actual temperature T are *not* STP values, the mass concentration is then

$$c_j\left(\frac{mg}{m^3}\right)_{actual} \cong \frac{y_j M_j}{24.5\times10^{-6}}\left[\frac{P_{actual}}{P_{STP}}\right]\left[\frac{T_{STP}}{T_{actual}}\right] = \frac{PPM_j M_j}{24.5}\left[\frac{P_{actual}}{P_{STP}}\right]\left[\frac{T_{STP}}{T_{actual}}\right] \tag{1-31}$$

The reader is cautioned that in Eq. (1-31), P_{STP} and the actual pressure must be *absolute* pressures, and must be in the same units. Similarly, the two temperatures must be *absolute*, and must be expressed in compatible units. Since the mol fraction is defined as a ratio of numbers of mols in Eq. (1-27), the mol fraction is independent of temperature and pressure. Similarly independent is composition expressed in parts per million,

$$PPM_{j,\ STP} = PPM_{j,\ actual} \tag{1-32}$$

In air pollution engineering, the mass flow rate of a pollutant ($\dot{m}_j$) is often of greater concern than is the mass flow rate of the entire gas mixture ($\dot{m}$). To find $\dot{m}_j$, one simply needs to multiply the actual mass concentration (c_j, mass/volume) by the actual volume flow rate (Q, volume/time),

$$\dot{m}_j = c_j Q \tag{1-33}$$

Molar concentration ($c_{molar,j}$) is defined as the number of mols of contaminant j per unit volume of carrier gas, typically in units of gram-mol per cubic meter (mol/m^3), or kilogram-mol per cubic meter (kmol/m^3).

$$c_{molar,j} = \frac{n_j}{V} \tag{1-34}$$

In some texts, molar concentration and mass concentration are given the same symbol (c); the units indicate which one is being used. In this text, c by itself will always mean mass concentration to avoid confusion. The reader can easily verify that molar concentration and mass concentration are related by

$$c_{molar,j} = \frac{c_j}{M_j} \tag{1-35}$$

where M_j is the molecular weight of species j.

Mass and molar concentrations are sufficient to describe vaporous contaminants. However, as will be discussed in more detail in Chapter 8, these concentrations are insufficient to fully describe particulate contamination in air. In such cases, it is often convenient to define the ***number concentration***, $c_{number,j}$, which depends not only on mass concentration of species j, but also on the size distribution of the contaminant.

$$c_{number,j} = \frac{\text{number of particles of species j}}{V} \tag{1-36}$$

Finally, there are times when the concentration is so small that it is inconvenient to use either mass concentration or molar concentration. At extremely small concentrations, ***molecular concentration*** (number of molecules per cubic meter) can be used. Examples of such small concentrations are the atmospheric concentrations of hydroxyl or oxygen free radicals. In these cases, the units of molecular concentration of species j (using the symbol $c_{molecular,j}$) are molecules/m^3. By making use of Avogadro's number (6.0251 x 10^{23} molecules per mol), the mol fraction can be expressed as the ratio of molecules of contaminant to total molecules, rather than as mols of contaminant to total mols. At STP, the number of air molecules per cubic meter is 2.457 x 10^{25}. In terms of molecular concentration, an equation similar to Eq. (1-31) can be written,

$$c_{molecular,j}\left(\frac{\text{molecules}}{m^3}\right)_{actual} \cong 2.457\text{x}10^{25}\, y_j \left[\frac{P_{actual}}{P_{STP}}\right]\left[\frac{T_{STP}}{T_{actual}}\right] = 2.457\text{x}10^{19}\, PPM_j \left[\frac{P_{actual}}{P_{STP}}\right]\left[\frac{T_{STP}}{T_{actual}}\right] \quad (1\text{-}37)$$

Example 1.5 – Describing the Flow of Air and Pollutant in a Duct

Given: A duct transports air that contains gaseous hydrogen cyanide (HCN, M_{HCN} = 27.0). The HCN concentration is 100. PPM, i.e. $y_{HCN} = 10^{-4}$. The gas mixture travels through a duct of diameter (D) equal to 10.0 cm, with an average velocity of 2.50 m/s. The temperature (T) and pressure (P) of the gas mixture inside the duct are 500. K and 2.20 atm.

To do: Compute the following:

(a) the actual volumetric flow rate (Q) of the gas mixture expressed in units of m^3/s and ACFM
(b) the total mass flow rate of the gas mixture (kg/s)
(c) the total volumetric flow rate of the gas mixture expressed as SCFM
(d) the mass concentration (mg/m^3) of HCN inside the duct
(e) the HCN mass flow rate (mg/s)

Solution:

(a) The actual volumetric flow rate can be found from Eq. (1-14),

$$Q_{actual} = \int_A U dA = \frac{\pi}{4}D^2 U = \frac{\pi}{4}\left(0.100\ m^2\right)2.50\frac{m}{s} = 0.01963\frac{m^3}{s}\left(\frac{ft}{0.3048\ m}\right)^3\left(\frac{60\ s}{min}\right) = 41.6\ \text{ACFM}$$

Since $y_{HCN} = 10^{-4}$, $y_{air} = 1 - 10^{-4} = 0.9999$ from Eq. (1-21). The molecular weight of the gas mixture, i.e. the total molecular weight (M_t) is obtained from Eq. (1-28),

$$M_{mixture} = M_t = \sum_j y_j M_j = y_{HCN}M_{HCN} + y_{air}M_{air}$$

$$= \left(10^{-4}\right)27.0\frac{kg}{kmol} + \left(0.9999\right)28.97\frac{kg}{kmol} \cong 28.97\frac{kg}{kmol}$$

Since the HCN mol fraction is small (10^{-4}), the gas mixture behaves as if it were only air. The specific ideal gas constant of the air-contaminant mixture is given by

$$R_{mixture} = \frac{R_u}{M_{mixture}} = \frac{8.3143\frac{kJ}{kmol\cdot K}}{28.97\frac{kg}{kmol}} = 0.2870\frac{kJ}{kg\cdot K}$$

The density of the gaseous mixture can be found from the ideal gas equation, Eq. (1-11), written in terms of the specific ideal gas constant,

$$\rho = \frac{P}{R_{mixture}T} = \frac{2.20\ atm}{0.2870\frac{kJ}{kg\cdot K}(500.\ K)}\left(101.3\frac{kPa}{atm}\right)\left(\frac{kJ}{kN\cdot m}\right)\left(\frac{kN}{m^2\cdot kPa}\right) = 1.55\frac{kg}{m^3}$$

(b) The total mass flow rate is calculated from Eq. (1-13):

$$\dot{m} = \rho Q = \left(1.55\ \frac{kg}{m^3}\right)\left(0.01963\frac{m^3}{s}\right) = 0.0304\frac{kg}{s}$$

(c) The total volumetric flow rate in standard cubic feet per minute (SCFM) is found from Eq. (1-17),

$$Q\ (SCFM) = \left(41.60\ \frac{ft^3}{min}\right)\left(\frac{2.20\ atm}{1\ atm}\right)\left(\frac{298.15\ K}{500.\ K}\right) = 54.6\ SCFM$$

(d) The actual mass concentration of HCN (M = 27.0) is obtained from Eq. (1-31):

$$c_{HCN,\ actual}\left(\frac{mg}{m^3}\right) = \left[\frac{(100.\ PPM)(27.0)}{24.5}\right]\left(\frac{2.20\ atm}{1\ atm}\right)\left(\frac{298.15\ K}{500.\ K}\right) = 145.\frac{mg}{m^3}$$

(e) Finally, the HCN mass flow rate is calculated from Eq. (1-33):

$$\dot{m}_{HCN} = \left(c_{HCN,\ actual}\right)\left(Q_{total,\ actual}\right) = \left(145.\frac{mg}{m^3}\right)\left(0.01963\frac{m^3}{s}\right) = 2.85\frac{mg}{s}$$

Example 1.6 - Pollutant Concentration and Emission Rate

Given: A vent from a printing press discharges 2.0 ACFM of air containing phenol vapor at 200. °C, 90. kPa. A continuous emissions monitor shows that the discharge contains 4.5 PPM of phenol vapor. Appendix A.20 indicates that this concentration is between the low and high odor thresholds; workers may be able to detect the medicinal, sweet odor of phenol. The environmental compliance officer of the company is concerned about discharge to the atmosphere.

To do: Compute the actual phenol concentration in the exhaust stream (in units of mg/m^3) and the emission rate of phenol to the atmosphere (in units of mg/hr).

Solution: The molecular weight of phenol is 94.1. From Eq. (1-31), the actual concentration of phenol is

$$c_{phenol,\ actual}\left(\frac{mg}{m^3}\right) = \left[\frac{(4.5\ PPM)(94.1)}{24.5}\right]\left(\frac{90.\ kPa}{101.3\ kPa}\right)\left(\frac{298.15\ K}{473.15.\ K}\right) = 9.67\ \frac{mg}{m^3} \cong 9.7\ \frac{mg}{m^3}$$

The emission rate of phenol is

$$\dot{m}_{phenol} = c_{phenol,\ actual}Q_{actual} = 9.67\frac{mg}{m^3}2.0\frac{ft^3}{min}\left(\frac{60\ min}{ht}\right)\left(\frac{0.3048\ m}{ft}\right)^3 = 32.9\frac{mg}{hr} \cong 33.\frac{mg}{hr}$$

1.5.6 Mixtures of Ideal Liquids

An ***ideal solution*** is one in which no energy is generated (or consumed) when two or more individual species are mixed. Thus the total volume (V_t) of a mixture containing N distinct molecular species is equal to the sum of the volumes of the individual components.

$$\boxed{V_t = \sum_j^N V_j = \sum_j V_j = \sum V_j} \tag{1-38}$$

Note again that for brevity, the upper (and sometimes lower) limit on the summation is omitted in the material to follow. The total mass (m_t) of the mixture and the total number of mols (n_t) of the mixture can be expressed as

$$m_t = \sum_j m_j \tag{1-39}$$

and

$$n_t = \sum_j n_j = \sum_j \frac{m_j}{M_j} \tag{1-40}$$

The ***mol fraction*** (x_j) of species j in the liquid phase is

$$x_j = \frac{n_j}{n_t} \tag{1-41}$$

The ***average molecular weight*** (M_{avg}) of the liquid mixture can be defined by

$$M_{avg} = \frac{m_t}{n_t} = \frac{\sum_j (n_j M_j)}{n_t} = \sum_j \left(\frac{n_j}{n_t} M_j\right) = \sum_j (x_j M_j) \tag{1-42}$$

In this text the letter "y" always refers to a mol fraction in the gaseous phase, and the letter "x" refers to a mol fraction in the liquid phase. Readers should be aware that this convention is not necessarily followed throughout engineering.

The ***average density*** (ρ_{avg}), also called the ***total density*** (ρ_t), is defined by dividing the total mass (m_t) by the total volume (V_t),

$$\rho_{avg} = \rho_t = \frac{m_t}{V_t} = \sum_j \frac{m_j}{V_t} = \sum_j \bar{\rho}_j \tag{1-43}$$

where $\bar{\rho}_j$ is defined as the ***partial density***, which can be thought of as the density species j would have if it *alone* occupied the entire volume. The volume (V_j) occupied by each species is

$$V_j = \frac{m_j}{\rho_j} \tag{1-44}$$

The total volume (V_t) of an a ideal mixture is

$$V_t = \sum_j V_j = \sum_j \frac{m_j}{\rho_j} \tag{1-45}$$

The reciprocal of the average density ($1/\rho_{avg}$) can be expressed as

$$\frac{1}{\rho_{avg}} = \frac{V_t}{m_t} = \sum_j \left(\frac{m_j}{m_t}\frac{1}{\rho_j}\right) = \sum_j \frac{f_j}{\rho_j} \tag{1-46}$$

where f_j is the ***mass fraction*** of species j, which has the same definition for liquid mixtures as for gas mixtures, Eq. (1-19).

The ***average molar concentration*** ($c_{molar,avg}$) is equal to the total number of mols (n_t) divided by the total volume (V_t). The average molar concentration is also called the ***total molar concentration*** ($c_{molar,t}$), and is given by

$$c_{molar,avg} = c_{molar,t} = \frac{n_t}{V_t} = \sum \frac{n_j}{V_t} = \sum \left(\frac{m_j}{M_j V_t} \right) = \sum \left(\frac{m_j}{V_t} \frac{1}{M_j} \right) \tag{1-47}$$

Or, in terms of the partial density,

$$c_{molar,avg} = c_{molar,t} = \sum \frac{\overline{\rho_j}}{M_j} = \sum \overline{c_{molar,j}} \tag{1-48}$$

The term $\overline{c_{molar,j}}$ is defined as the ***partial molar concentration*** which can be interpreted as the concentration species j would have if it *alone* occupied the entire volume. Equations (1-43), (1-46), (1-47), and (1-48) suggest methods for calculating the total density and total concentration from pure component data. For ideal solutions these averaging methods are rigorous, i.e. $\rho_{avg} = \rho_t$ and $c_{molar,avg} = c_{molar,t}$. One may choose to calculate ρ_{avg} and $c_{molar,avg}$ this way for *non-ideal* solutions, but he or she should not expect the averaged values to equal the true total values.

For a single species (no subscript j), the molar concentration is related to the density by

$$c_{molar} = \frac{n}{V} = \frac{m}{M} \frac{1}{V} = \frac{m}{V} \frac{1}{M} = \frac{\rho}{M} \tag{1-49}$$

Written for a mixture, Eq. (1-49) provides an alternative definition to Eq. (1-42) for the average molecular weight (M_{avg}) or its inverse,

$$M_{avg} = \frac{\rho_t}{c_{molar,avg}} = \frac{\rho_t}{c_{molar,t}} = \frac{\dfrac{m_t}{V_t}}{\sum \left(\dfrac{m_j}{M_j} \dfrac{1}{V_t} \right)} \tag{1-50}$$

$$\frac{1}{M_{avg}} = \sum \left(\frac{m_j}{m_t} \frac{1}{M_j} \right) = \sum \frac{f_j}{M_j} \tag{1-51}$$

The mass fraction (f_j) of species j can be manipulated to yield

$$f_j = \frac{m_j}{m_t} = \frac{m_j}{\sum m_j} = \frac{\dfrac{m_j}{V_t}}{\sum \dfrac{m_j}{V_t}} = \frac{\overline{\rho_j}}{\rho_t} \tag{1-52}$$

The mol fraction in the liquid phase (x_j) can also be re-written in terms of partial molar concentration:

$$x_j = \frac{n_j}{n_t} = \frac{n_j}{\sum n_j} = \frac{\dfrac{n_j}{V_t}}{\sum \dfrac{n_j}{V_t}} = \frac{\overline{c_{molar,j}}}{c_{molar,t}} \tag{1-53}$$

Mol fraction and mass fraction are related by

$$f_j = \frac{m_j}{m_t} = \frac{\dfrac{m_j}{M_j}}{\dfrac{m_t}{M_{avg}} \dfrac{M_{avg}}{M_j}} = \frac{n_j}{n_t} \frac{M_j}{M_{avg}} = x_j \frac{M_j}{M_{avg}} \tag{1-54}$$

Example 1.7 - Composition of a Solvent Mixture

Given: The label on a gallon (3.788×10^{-3} m^3) of a commercial product states the mol fraction of its contents, i.e. water, ethyl benzene, butyl alcohol, and diacetone alcohol. The mol fraction (x_J), molecular weight (M_J), and density (ρ_J) of the pure species are shown in the table below:

species	x_j	M_j (kg/kmol)	ρ_j (kg/m^3)
water	0.553	18.0	1,000
ethyl benzene	0.255	106.2	870
butyl alcohol	0.057	74.1	810
diacetone alcohol	0.135	116.2	940
total	**1.000**		

To do: Compute the mass (m_J), mass fraction (f_J), and number of mols (n_j) of each species. Also compute the average molecular weight (M_{avg}) and average density (ρ_{avg}) of the liquid mixture.

Solution: From Eq. (1-42),

$$M_{avg} = \sum_j \left[x_j M_J\right] = \left[(0.553)(18.0) + (0.255)(106.2) + (0.057)(74.1) + (0.135)(116.2)\right]\frac{kg}{kmol}$$

$$= 57.0\frac{kg}{kmol}$$

The mass fraction f_J can be found from Eq. (1-54). The average density can be found from Eq. (1-46),

$$\frac{1}{\rho_{avg}} = \sum_J \frac{f_J}{\rho_j} = \frac{0.175}{1000} + \frac{0.475}{870} + \frac{0.080}{810} + \frac{0.275}{940} = 11.06\times10^{-4}\frac{m^3}{kg}$$

$$\rho_{avg} = 904.\frac{kg}{m^3}$$

Thus the mass of a gallon of the liquid mixture is

$$m_t = \rho_{avg}V = \left(904.\frac{kg}{m^3}\right)\left(3.788\times10^{-3}\ m^3\right) = 3.42\ kg$$

The mass of each species (m_J) can be found from the definition of mass fraction, Eq. (1-19),

$$m_j = f_J m_t$$

and the number of mols of each species (n_J) can be found from Eq. (1-5), solved for n_J,

$$n_j = \frac{m_j}{M_J}$$

Results are summarized in the table below:

species	f_j	m_j (kg)	n_j (kmol)
water	0.175	0.597	0.0332
ethyl benzene	0.475	1.62	0.0153
butyl alcohol	0.0741	0.253	0.00342
diacetone alcohol	0.275	0.941	0.00810
totals		**3.41**	**0.0600**

The total number of mols (n_t) can be calculated another way, i.e. from Eq. (1-42):

$$n_t = \frac{m_t}{M_{avg}} = \frac{3.42\ \text{kg}}{57.0\ \frac{\text{kg}}{\text{kmol}}} = 0.0600\ \text{kmol}$$

which agrees (to three digits of precision) to that obtained by summation in the above table. Finally, the total molar concentration ($c_{molar,t}$) is equal to

$$c_{molar,t} = \frac{\rho_{avg}}{M_{avg}} = \frac{904.\ \frac{\text{kg}}{\text{m}^3}}{57.0\frac{\text{kg}}{\text{kmol}}} = 15.9\frac{\text{kmol}}{\text{m}^3}$$

Alternately, it can be found directly from fundamental principles using Eq. (1-34)

$$c_{molar,t} = \frac{n_t}{V} = \frac{0.0600\ \text{kmol}}{0.003788\ \text{m}^3} = 15.8\frac{\text{kmol}}{\text{m}^3}$$

Discussion: The small discrepancy is due to round-off error. The total mass from summation in the table agrees (except for small round-off error) with that calculated above from the average density.

1.5.7 Humidity and Moist Air

Air in the atmosphere or in a building always has some moisture or humidity because water vapor molecules are mixed with air molecules. Because of its relationship to thermal comfort, the amount of humidity in the air is usually expressed in terms of the ***relative humidity*** (Φ), defined as

$$\boxed{\Phi = \frac{P_{water}}{\left(P_v\right)_{water}}} \qquad (1\text{-}55)$$

where P_{water} is the actual partial pressure of the water vapor in the air-water vapor mixture, and P_v is the ***vapor pressure*** (also called the ***saturation pressure***) of water at temperature T (see Appendix A.17). Note that in some texts, the symbol used for relative humidity is RH instead of Φ. Relative humidity can be expressed either as a fraction or as a percentage. For example, on a very hot summer day when the temperature is 35. °C and the local atmospheric pressure is 99.6 kPa, suppose the relative humidity is reported by a weather service to be 90.%. The partial pressure of the water vapor in the atmosphere can be computed as follows: First, the vapor pressure (P_v) of water at a given temperature can be found in any thermodynamics textbook, where it is usually called the saturation pressure (P_{sat}) instead of P_v. At 35 °C, P_v = 5.628 kPa. A relative humidity of 90.% means that Φ = 0.90, and the partial pressure of water vapor in the atmosphere can be found from Eq. (1-55),

$$P_{water} = \left(P_v\right)_{water} \Phi = (5.628\ \text{kPa})\ 0.90 = 5.065\ \text{kPa} \cong 5.1\ \text{kPa}$$

In addition, the mol fraction of water vapor in the air can be calculated using Eq. (1-27),

$$y_{water} = \frac{P_{water}}{P} = \frac{5.065\ \text{kPa}}{99.6\ \text{kPa}} = 0.0509 \cong 0.051$$

In other words, this highly humid (and highly uncomfortable) air actually contains only about 5.1% water vapor! Finally, Eq. (1-22) can be used to express the mol fraction of water vapor in units of parts per million, PPM,

$$\text{PPM}_{water} = y_{water} \times 10^6 = 0.0509\times10^6\ \text{PPM} = 50{,}900\ \text{PPM} \cong 51{,}000\ \text{PPM}$$

Heating, ventilating, and air conditioning (HVAC) engineers also make use of *absolute* humidity when designing HVAC systems. The most widely used measure of air moisture content is the ***humidity ratio*** (ω), which is defined as the ratio of water vapor mass to dry air mass. ***Dry air*** exists when all water vapor and contaminants have been removed from the air. Humidification and dehumidification processes add and remove moisture from the air, but dry air mass is conserved. Dry air thus serves as a convenient reference state, even though it may not actually exist anywhere in a building's ventilation system. When designing heating and cooling systems, HVAC engineers multiply the dry air mass flow rate by the humidity ratio to obtain the mass flow rate of water vapor. In addition, properties on psychrometric charts are defined in terms of dry air and humidity ratio. After a bit of algebra, one can convert the mol fraction of water vapor in air to humidity ratio,

$$\boxed{\omega = \frac{m_{water}}{m_{air}} = \frac{n_{water}}{n_{air}}\frac{M_{water}}{M_{air}} = \frac{y_{water}}{\left(1 - y_{water}\right)}\frac{M_{water}}{M_{air}}} \tag{1-56}$$

For the example given above, (35. °C, 99.6 kPa, and 90.% relative humidity), the humidity ratio is

$$\omega = \frac{y_{water}}{\left(1 - y_{water}\right)}\frac{M_{water}}{M_{air}} = \frac{0.0509}{\left(1 - 0.0509\right)}\frac{18.02}{28.97} = 0.0333 \cong 0.033$$

For air containing water vapor, the ***density of moist air*** can be estimated from various empirical relationships, such as the one given by Jennings (1988):

$$\boxed{\rho_{moist\ air} = \frac{PM_{air}}{ZR_u T}\left[1 - \frac{M_{water}}{M_{air}}\Phi f \frac{\left(P_v\right)_{water}}{P}\right]} \tag{1-57}$$

where M_{air} and M_{water} are the molecular weights of air and water respectively, and Φ is the relative humidity, used here as a fraction rather than as a percentage. The constant f in Eq. (1-57) is equal to 1.004 for air over a wide range of temperatures and pressures (Jennings, 1988). The compressibility factor (Z) is also very close to 1.

Most people believe that very humid air is "thicker" or more dense. Is this really true? The following example compares the density of moist air for different levels of relative humidity.

Example 1.8 - Density of Moist Air
Given: The density of moist air, Eq. (1-57).

To do: Plot the density of moist air at one standard atmosphere (1 atm or 14.696 psia) as a function of temperature in the range 40 °F < T < 100 °F for relative humidity ranging from 0 to 100 % ($0 < \Phi < 1$).

Solution: Vapor pressure of water is a function of temperature. An accurate estimate of its value for the temperature range 40-100 °F can be expressed as an analytical function of temperature using the ***Clausius-Clapeyron equation*** (see Chap 4). Alternatively, one can use tabulated values of vapor pressure (see Appendix A.17), and generate an empirical equation using curve-fitting software. The authors used the latter approach, with the cubic spline function in Mathcad; the Mathcad program to solve this problem can be found in the textbook's web site; the resulting plot is shown in Figure E1.8.

Discussion: Residents of the western mountain states feel oppressed when they experience the humid air of the east coast. To them, the air feels "heavy"; the above plot shows that quite the opposite is true! At any temperature, the density of moist air *decreases* as the water vapor concentration (and relative humidity) increases. As evident from the plot, the effect is more pronounced at high temperatures. The dry air of the west coast feels cooler than the humid air on the east coast, even at the same temperature. The reason is that in dry air, water can more easily evaporate from the body, offering more efficient cooling. The cooling mechanisms of the body are discussed in further detail in Chapter 3.

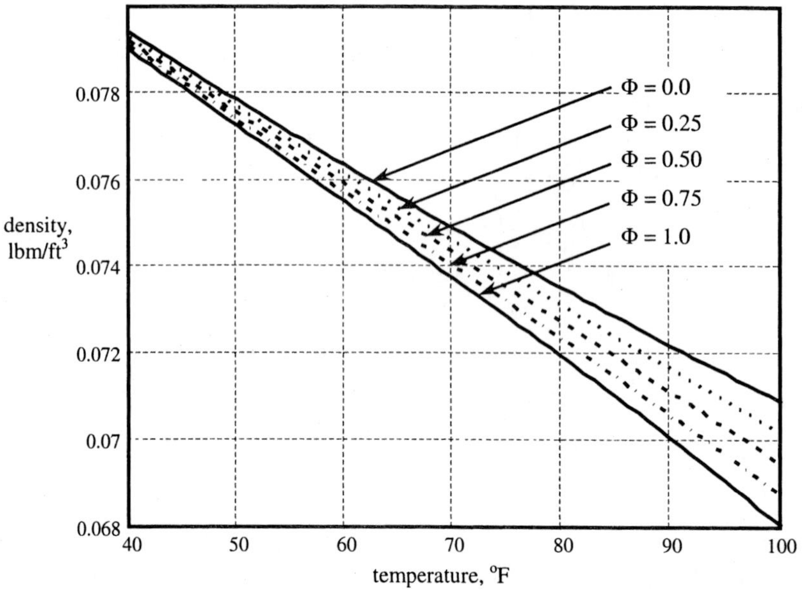

Figure E1.8 Density of moist air as a function of temperature for various values of relative humidity.

1.5.8 Solution of First-Order Ordinary Differential Equations

Many problems in indoor air pollution control, as well as in several other disciplines, require solution of a first-order ordinary differential equation (ODE). Suppose $y = y(t)$ is a parameter which is a function of time (t). Let $y(0)$ denote the value of y at time $t = 0$. It will be assumed that at times $t > 0$, some sudden change in the process causes y to grow (or decay). For example, suppose y is the mass concentration or molar concentration of a ***volatile organic compound*** (VOC) or other contaminant in a room. At $t = 0$, a source (S) of the VOC is turned on; the concentration of VOC in the room will begin to rise, eventually reaching some steady-state value. The time rate of change of parameter y in this example can be modeled mathematically by a first-order ODE, which in general can be written in standard mathematical form as

$$\boxed{a_1 \frac{dy}{dt} + a_0 y = b} \tag{1-58}$$

where a_0, a_1, and b are coefficients specific to the problem at hand, which may or may not be constants. After dividing by a_1 and rearranging, Eq. (1-58) can be re-written in a form that will be referred to in this text as the ***standard form***,

$$\boxed{\frac{dy}{dt} = B - Ay} \tag{1-59}$$

where

$$\boxed{A = \frac{a_0}{a_1} \qquad B = \frac{b}{a_1}} \tag{1-60}$$

If A and B are constants with respect to time, Eq. (1-59) can be solved analytically by separation of variables and integration. The initial condition at $t = 0$ is $y = y(0)$; at some arbitrary time t, $y = y(t)$. The solution of Eq. (1-59) is thus

$$\boxed{\frac{y_{ss}-y(t)}{y_{ss}-y(0)}=\exp(-At)} \qquad (1\text{-}61)$$

or, solving for y(t),

$$\boxed{y(t)=y_{ss}-(y_{ss}-y(0))\exp(-At)} \qquad (1\text{-}62)$$

where y_{ss} is the ***steady-state*** value, the value of y after a very long time has elapsed. y_{ss} is found by setting the left hand side of Eq. (1-59) to zero,

$$\boxed{y_{ss}=\frac{B}{A}} \qquad (1\text{-}63)$$

An alternate and equivalent form of Eq. (1-61) is

$$\boxed{\frac{y(t)-y(0)}{y_{ss}-y(0)}=1-\exp\left(-\frac{t}{\tau}\right)} \qquad (1\text{-}64)$$

where the ***first-order time constant*** (τ) has been introduced,

$$\boxed{\tau=\frac{1}{A}=\frac{a_1}{a_0}} \qquad (1\text{-}65)$$

It is left as an exercise for the reader to show that Eqs. (1-64) and (1-61) are indeed equivalent. The time constant (τ) has the same units as t (typically seconds, minutes, or hours), and represents the time at which the change in y, i.e. y(t) - y(0), grows (or decays) to approximately 63.2% of its final change, y_{ss} - y(0). This can be seen by letting t = τ in Eq. (1-64):

$$\frac{y(\tau)-y(0)}{y_{ss}-y(0)}=1-\exp(-1)$$

or

$$y(\tau)-y(0)=\left[1-\exp(-1)\right]\left[y_{ss}-y(0)\right]\approx 0.632\left[y_{ss}-y(0)\right]$$

An alternate time scale called the ***half-life*** ($t_{1/2}$) can be introduced as the time required for y(t) - y(0) to grow (or decay) to half (50%) of its final change y_{ss} - y(0). Using Eq. (1-64) again, with $t = t_{1/2}$,

$$\frac{y(t_{1/2})-y(0)}{y_{ss}-y(0)}=1-\exp\left(-\frac{t_{1/2}}{\tau}\right)=\frac{1}{2}$$

which yields

$$\boxed{t_{1/2}=-\ln\left(\frac{1}{2}\right)\tau \approx 0.693\tau} \qquad (1\text{-}66)$$

Finally, it is useful to solve for the time (t) at which y(t) reaches a certain value. t can be found in terms of parameter A, the time constant (τ), or the half-life ($t_{1/2}$): Solving for t in Eq. (1-61), and using Eqs. (1-65) and (1-66),

$$\boxed{t=-\frac{1}{A}\ln\left[\frac{y_{ss}-y(t)}{y_{ss}-y(0)}\right]=-\tau\ln\left[\frac{y_{ss}-y(t)}{y_{ss}-y(0)}\right]=\frac{t_{1/2}}{\ln(1/2)}\ln\left[\frac{y_{ss}-y(t)}{y_{ss}-y(0)}\right]} \qquad (1\text{-}67)$$

Throughout this text, readers will find numerous examples in which a differential equation can be written in the standard form of Eq. (1-59). In some examples, a spatial coordinate (x, y, or z) replaces time as the independent variable. In such a case, the above solutions are still valid, except that the initial and final (steady-state) conditions are replaced by appropriate ***boundary conditions***, and the first-order time constant is instead called a ***critical length***.

The analytical solutions given above pertain to the special case in which parameters A and B in Eq. (1-59) are constants. If instead they are functions of time, numerical schemes such as the ***Runge-Kutta*** marching technique (see Appendix A.12) can be applied to solve Eq. (1-59) numerically. This is accomplished most easily with mathematical computer programs such as Mathcad, but can also be programmed into spreadsheets such as Excel.

Example 1.9 – Runge-Kutta Numerical Solution

Given: Consider a first-order ODE in the form of Eq. (1-58) with A = 3.82, B = 10, and y(0) = 7.6.

To do:

(a) For t = 0 to 1, use the Runge-Kutta method to calculate y(t). Compare to the analytical solution.
(b) Repeat the calculations for the case in which A is not constant, but is a function of time, namely

$$A(t) = 3.82 + 2.0\sin(10\pi t)$$

Solution: The exact, analytical solution for Part (a) is given by Eq. (1-64), and is plotted in Figure E1.9. The authors solved the equation using both Mathcad and Excel, and the files can be obtained from the textbook's web site; the results for Part (a) and Part (b) are also plotted in Figure E1.9.

Discussion: The agreement in Part (a) between the Runge-Kutta numerical calculations and the exact analytical solution is excellent (in fact the two curves are indistinguishable in Figure E1.9). When A is a function of time as in Part (b), an analytical solution has not been attempted; the numerical solution shows oscillations, but has the same general decaying trend as in Part (a).

1.6 Government Regulations

Knowledge that certain occupations adversely affect workers' health dates from the time of Hippocrates, circa 400 BC. Diseases of the lung related to occupation were recognized ages ago (e.g. ***potters' rot*** of ceramic workers, ***grinders' rot*** of workers sharpening steel tools, ***coal miners' asthma***, etc.). Over 500 years ago the German physician Ellenborg studied the adverse effects of various toxic substances. Of particular importance in the field of industrial medicine is Bernardino Ramazzini, who in 1713, wrote "Diseases of Workers" that carefully documented diseases of working places.

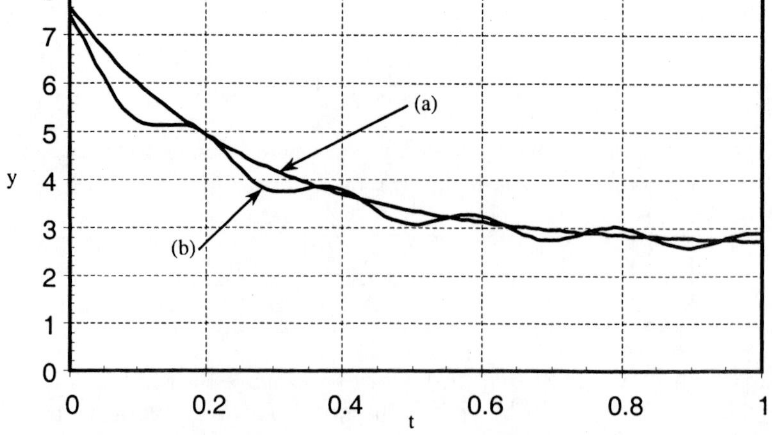

Figure E1.9 Solution of the ODE of Example 1.9; (a) A and B are constants, (b) A is a function of time, B is constant; Runge-Kutta solution (a) is compared to exact analytical solution.

Industrial disease and accident were once considered inevitable by-products of manufacturing. When the industrial revolution brought rapid prosperity to Britain, disease and accidents related to working conditions also rose. In 1802, the ***Health and Morals of Apprentices Act*** and the ***Factory Act of 1833*** were passed in Parliament describing acceptable working conditions for children. In 1842, Edwin Chadwick's report, "The Sanitary Condition of the Labouring Population" advanced the cause of public sanitation. Public support for child labor laws was aroused through the novels of Charles Dickens.

Some of the first occupational safety and health laws in this country involved seamen who manned US ships. In 1708 a sailor's hospital was established in Boston financed by a 6-cent per month pay deduction. In 1780 the ***Marine Hospital Service Bill*** was proposed in the House of Representatives and in 1798 the marine hospital was established from which the US Public Health Service evolved. (This is the reason why Public Health Officers and the Surgeon General wear Navy-like uniforms.) In the Congress, reports were prepared documenting occupational health hazards in 1837 and 1841 that dealt with industry in general. Some of the first industrial safety laws were passed in Massachusetts and concerned steam engines (1852), factory inspection (1867), and mechanical guards for textile spinning machinery (1877). In 1869 Massachusetts established the first State Bureau of Labor Statistics, and in 1881 the American Public Health Association was founded. By 1890, twenty-one (21) states had passed laws similar to those in Massachusetts. In 1908 the ***Workman's Compensation Act*** was passed concerning certain federal employees, and in 1911 several states began passing Workman's Compensation Acts for workers within their states.

In 1865 a bill was submitted to the Congress to create a Federal Mining Bureau but it wasn't until 1891 that the first legislation was passed. In 1893 the legislation was extended to safety equipment on railroads. In 1902 the Public Health Service (PHS) was established by Congress and in 1910 the Bureau of Mines was established in the Department of Interior. In 1916 Congress enacted the first federal child labor law but it was later ruled unconstitutional, and it wasn't until 1938 that a child labor law passed. In 1912 New York established a Division of Industrial Hygiene. By 1921, forty-six states had passed some form of worker compensation laws to relieve the hardship of injured workers and their families. In the mid 20th century the physician Alice Hamilton correlated occupational illness with exposure to toxic industrial materials. In 1943 her book, "Exploring the Dangerous Trades," established the foundation of modern occupational medicine. Programs proposed by Hamilton were slow to be adopted. On one hand it was hard for people to understand the long period between exposure and evidence of disease, and on the other hand there was skepticism that disease was related to working conditions. Initially occupational medicine programs were directed toward treatment of occupational injuries. The realization that prevention was more important and economical than treatment was slow to be adopted.

During the depression many laws were passed concerning collective bargaining, but little attention was given to worker health and safety. In 1936 the ***Walsh-Healey Public Contracts Act*** was the first major federal legislative attempt to enforce occupational safety and health standards. The act set mild standards that working conditions must not be unsanitary, hazardous, or dangerous to health and safety, but it contained little provision for enforcement. Thus as late as 1969 fewer than 3000 of the estimated 75,000 companies covered by the act were inspected, which resulted in only 34 formal complaints and 32 formal hearings. In 1947 the ***Taft-Hartley Act*** contained a provision that permitted workers to walk off the job if it was "abnormally dangerous". In 1948 President Truman called the first Presidential Conference on Industrial Safety.

In mine safety, Congress was more inclined to take action because conspicuous tragic mine accidents received considerable publicity and aroused the public's desire to take corrective actions. In

1951, 119 miners were killed in West Frankfort, Illinois, and in 1952 the ***Coal Mine Safety Act*** was passed. Six years later similar provisions were extended in the ***Maritime Safety Act***. Throughout the 1960s specialized acts were passed containing limited safety statutes. The first significant legislation in job safety and health occurred in 1966 with passage of the ***Metal and Nonmetallic Mine Safety Act***.

In 1968 President Johnson proposed the nation's first comprehensive occupational safety and health program. Several acts were passed in rapid succession:

- ***Coal Mine Health and Safety Act*** in 1969
- ***Construction Safety Act*** in 1969
- ***Federal Railway Safety Act*** in 1970

By 1970 the incidence of occupational disease was recognized as a serious financial loss. In addition to the personal tragedy of illness and death, over $1.5 billion was wasted annually in lost wages at a time with the GNP was only $8 billion. The number of disabling injuries per million workers was 20% higher in 1970 than it was in 1958. Lastly occupational health expertise was incapable of coping with rapidly advancing technology involving new materials of unknown toxicity and new processes that posed unique risks. During the late 1960s and early 1970s, the country was beset with civil and racial unrest and an unpopular war in Vietnam. The nation was in an introspective mood that fostered the passage of legislation to advance the cause of occupational safety and health as well as protection of the environment.

The 1970 the ***Occupational Safety and Health Act*** was designed "... to assure so far as possible every working man and woman in the nation safe and healthful working conditions and to preserve our human resources." The act created two federal agencies known today as OSHA and NIOSH. The legislative history of the 1970 bill greatly influenced the content and implementation of the act. The Democratic bills were introduced in the House of Representatives in January 1969 by Representative J. O'Hara, and in the Senate in May 1969 by Senator H. Williams. Republican President Nixon's proposal was introduced in August 1969. The major difference between the Democratic and Nixon bills concerned who should be responsible for setting and enforcing safety and health standards. The Democratic bill placed the responsibility with the Department of Labor while the Nixon bill vested authority in a new body, the ***National Occupational Safety and Health Board***. Organized labor supported the Democratic bill and industry favored the Nixon bill. Congressional debate was strongly partisan and ended with House and Senate Committee reports containing minority (Republican and dissenting) reports. Eventually both the House and Senate passed differing versions of the bill that went to a Conference Committee for resolution. In December of 1970 both the House and Senate approved the Conference Report compromise bill. Unlike the legislation creating the Environmental Protection Agency (EPA) where the agency was empowered to both set and enforce standards, the OSHA Act assigned standards and enforcement to two separate agencies.

The 1970 act created the Occupational Safety and Health Administration (OSHA) within the Department of Labor with powers to:

- promulgate standards to protect health and safety
- enter and inspect industrial workplaces
- issue citations for violations and recommend penalties
- supervise grants to states for occupational safety and health programs
- conduct short-term training of OSHA employees
- keep and maintain records

In addition, the 1970 act created the National Institute of Occupational Safety and Health (NIOSH) within the Department of Health and Human Services and granted it the authority to:

- develop standards for occupational safety and health
- conduct and support research into the cause, occurrence, incidence, etc., of occupational diseases and injuries
- develop educational programs to train qualified professionals

The numerous and sometimes hasty compromises of the 1970 legislation resulted in vague, redundant, and paradoxical clauses requiring judicial interpretation. In spite of its controversy, the legislation was one of the most important pieces of protective legislation coming out of President Johnson's Great Society program, and Congress has not passed any substantive amendments to the Act. However, there have been actions taken to limit OSHA enforcement powers attached to appropriation bills. The OSHA Act applies to all workers in every state except those workers over whom other state or federal agencies have statutory authority (for example mine and railroad workers). Among other requirements, every employer must comply with two provisions:

- employers must keep the workplace free from recognized hazards that can cause death or serious harm
- employers must comply with OSHA standards

Occupational health and safety standards are promulgated and enforced as follows:

- The Secretary of Labor was authorized to adopt national consensus standards without lengthy rulemaking procedures. This authority ended in 1973 and was superseded by procedures to modify or issue new statutes.
- The Secretary of Labor may also issue emergency temporary standards if it is determined that employees are subjected to grave danger or to agents that are known to be toxic or physically harmful. Such standards are effective upon publication in the Federal Registrar.
- Enforcement of the 1970 act was vested in OSHA, located in the Department of Labor.

All enforcement functions lay with OSHA but decisions may be appealed to higher levels of authority. Every state is covered by the Act but enforcement functions can be transferred to the state government if a state plan is approved by OSHA. At the time of this writing, only about half the states have sought and obtained such authority. The following are elements in the enforcement and appeal process:

- OSHA Compliance Officers (CO)
- OSHA Area Director
- Petition for Modification of Abatement (PMA)
- Occupational Safety and Health Review Commission (OSHRC)
- Administrative Law Judge (ALJ)
- Petition for Discretionary Review (PDR)
- US Court of Appeals

OSHA is empowered to inspect an industry covered by the act, issue citations, and award penalties. The OSHA ***Compliance Officer*** (CO) is authorized to inspect any part of an industrial facility without announcement. At the close of the visit the CO is required to conduct a closing conference during which safety and health conditions are to be discussed and possible violations identified. Following the filing of the CO's report, the OSHA Area Director decides whether to issue a citation, determines the penalty, and sets the date for the violation to be corrected. The citation must be in writing, describe the violation, and cite the relevant standards and regulations. Civil penalties are progressive, reflecting the level of risk, the employer's history of compliance, and the size of the employer:

- De Minimis Notice $0
- Nonserious $0 - $1,000
- Serious $1 - $1,000
- Repeated $0 - $10,000

- Willful $0 - $10,000
- Failure to Abate Notice $0 - $1,000 per day

There exist criminal sanctions for willful violations of the Act that result in the death of one of or more employees. These sanctions have been used.

The employer or employee (or union representative) has 15 days to file a notice to contest a violation. The Secretary of Labor must forward any contested violation to the ***Occupational Safety and Health Review Commission*** (OSHRC). The Commission is a quasi-judicial, independent administrative agency composed of three Presidential-appointed Commissioners who serve on staggered six-year terms. In cases before the OSHRC, the Secretary of Labor is referred to as the complainant, and has the burden of proving the violation; the employer is usually called the respondent. The hearing is presided over by an ***Administrative Law Judge*** (ALJ) of the Commission. After the hearing, the ALJ makes a decision to affirm the citation, modify or dismiss the citation, modify the violation penalty, or modify the abatement date. The ruling of the ALJ is transmitted to the OSHRC. Either party may file a ***Petition for Discretionary Review*** (PDR), asking that the ALJ's decision be reviewed. Even without a PDR, any Commission member may request a review of any or all of the ALJ's decision. Under these conditions, the OSHRC reconsiders the evidence and may issue a new decision. If, however, no member of the OSHRC requests a review within 30 days, the decision of the ALJ is final. Any party adversely affected by the final order of the OSHRC may file a petition for review with the US Court of Appeals for the circuit in which the citation was issued or in the US Court of Appeals for the District of Columbia. Such appeals must be filed within 60 days.

OSHA rules affecting health and safety change periodically as new information becomes available. Because of ambiguous language in the OSHA Act and the discretion given to OSHA to choose working conditions to investigate, the results of litigation and judicial review often become the basis for policy (Ricci and Molton, 1981). As a result of the 1980 Supreme Court ruling in *Industrial Union versus the American Petroleum Institute,* the court required OSHA to demonstrate that a proposed standard "is reasonably necessary and appropriate to remedy a significant risk of material health impairment". Unfortunately the significance of risk is shrouded by scientific uncertainty and issues that evolve from the political process. A critical question to be faced is the extent to which costs should be weighed against benefits. Unfortunately court rulings have been ambiguous. In 1978 the Fifth Circuit Court held that the benefits of the benzene standard must bear a "reasonable relationship" to its cost, but in 1979 the D. C. Circuit Court upheld OSHA's cotton dust standard in which OSHA did not need to balance costs and benefits.

It is important for engineers to understand the federal process called ***rule making***. The legislation creating OSHA defines its purpose and scope, establishes the governmental agency that has authority to enact and implement rules, and sometimes specifies certain issues to be regulated. Regulations are promulgated in a manner shown in Figure 1.9 and generally take considerable time since all interested parties must have the opportunity to make public comments. Newly adopted regulations and their amendments, presidential proclamations, executive orders, federal agency documents having applicability and legal effect, and other federal agency documents of public interest are published in the ***Federal Resister***. The Federal Register is published daily by the National Archives Administration, Washington DC 20408 to provide a uniform method to disseminate regulations and legal notices to the public. To cite any information in the Federal Register, one simply uses the volume and page number. For example *51FR 1234* means *volume 51, page 1234*. To keep track of all regulations relevant to an issue such as occupational health and safety, an engineer would need to monitor every issue of the Federal Register and extract the relevant information. The task is formidable but it is also unnecessary since it is accomplished in the ***Code of Federal Regulations*** (CFR).

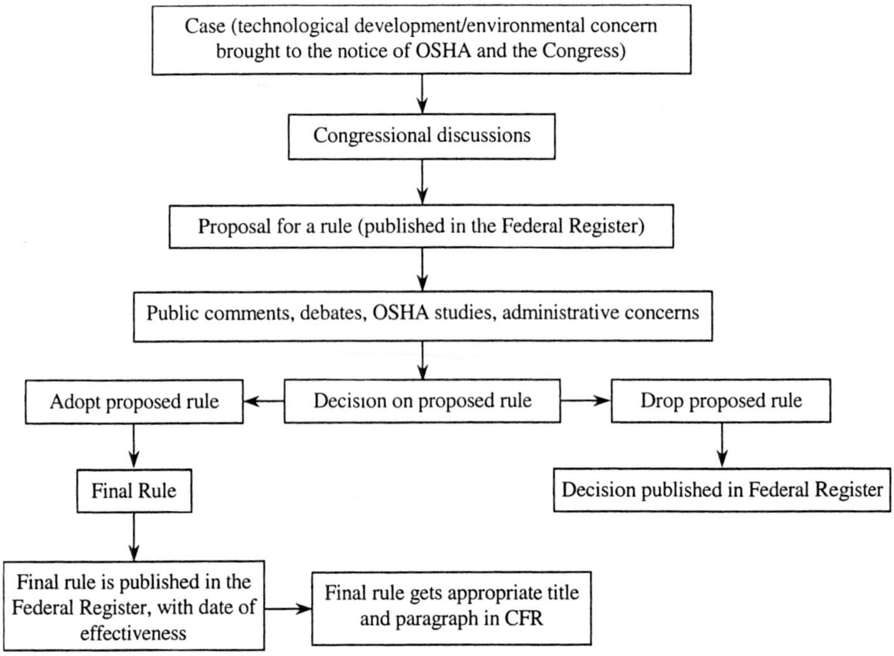

Figure 1.9 The process of promulgating OSHA regulations.

The Code of Federal Regulations is a systematic compilation of all general and permanent rules and their amendments published by the Federal Register pertaining to certain issues or agencies. The code is divided into 50 titles representing broad areas that are subjected to federal regulation. Each title is divided into chapters that usually bear the name of the issuing agency, and each chapter is subdivided into parts covering specific regulatory areas. For example,

- Title 29 - Labor, is composed of 8 volumes; each volume contains parts grouped in ways to facilitate their use. E.g. Chapter XVII, Parts 1900 to 1910, revised as of the current date, represents all regulations pertaining to the Occupational Safety and Health Administration, codified under this title as of the current date.
- Title 40 - Protection of Environment, is composed of 14 volumes; each volume contains parts grouped in ways to facilitate their use. For example, Parts 1 to 51, Part 52, and Parts 53 to 60, revised as of the current date, represent all regulations relating to the Environmental Protection Agency codified under this title as of the current date.

The CFR is kept up to date by the individual issuance of the Federal Register. These two publications must be used together to determine the latest version of any rule. A CFR is revised usually just once a year; a reference in it may be listed as "Reserved" which means that the reference will be, or is planned to be, incorporated in a future CFR. Each Federal Register lists the title number and parts of the CFR to which it makes reference. For example, a typical reference in the Federal Register could be 40CFR 270.70(a), which should be interpreted as

- Title 40 of the Code of Federal Regulations
- Part 270
- Paragraph 70(a)

Oftentimes, when items of importance to the protection of the environment are being discussed, Title 40 is assumed, and the Federal register cites a reference simply as 266.34(b); this requires the reader to seek Part 266, Paragraph 34(b) of Title 40 of the Code of Federal Regulations.

1.7 Standards

It is human nature for organizations engaged in common pursuits to seek uniform standards about aspects of their work so that they can converse with each other and the public in ways that enhance understanding, establish quality, ensure performance, and eliminate unnecessary costs. For example, it is logical that nuts and bolts (fasteners) used in everyday commerce have dimensions, thread designs, and male and female parts that are compatible. Rarely is there need for fasteners with unique dimensions, materials, and physical properties. In order to expand standardization to more complex items used in a world economy, one can appreciate the need for national and international organizations to establish useful standards of design and performance.

American National Standards Institute (ANSI) - ANSI was founded in 1918 by five engineering societies and three governmental agencies. ANSI is a private nonprofit organization supported by trade associations and professional societies. ANSI promotes the use of US standards internationally, and advocates US policies and technical positions. The primary goal of ANSI is to enhance the global competitiveness of US business by promoting consensual standards and their integrity. ANSI does not develop standards, rather it facilitates the development of standards by seeking consensus among qualified groups. ANSI is the sole US representative and dues-paying member of the two major non-treaty international standards organizations, the ***International Organization for Standardization*** (ISO) and the ***International Electrotechnical Commission*** (IEC). ANSI was one of the founding members of the ISO.

International Organization for Standardization (ISO) - While the standards of professional societies (ASME, SAE, etc.) and those of trade associations (API, CMA, etc.) and ANSI are quantitative criteria that products must satisfy, the ISO is a set of managerial standards and procedures embracing the concept of ***continuous quality improvement*** (CQI). CQI consists of procedures embedded in an organization that examine how it performs its work and changes work practices to improve the quality of the product or service. The process is continual and an integral part of the company's activity. CQI is the creation of W. Edwards Deming (1986) that transformed the world's industry in the last decades of the 20th century. Output is measured against quantitative measures of quality, and the process must demonstrate that quality continually improves.

The (ISO) was founded in 1947, and is a non-governmental international organization. It is important to recognize that the ISO has no legal enforcement authority. The ISO promotes the international harmonization of manufacturing, products, and communications. Enforcement is achieved through consensus and voluntary compliance aided by economic arbitration in a free-trade market. More than 120 countries belong to the ISO. The US is a full voting member and is represented by ANSI. The ISO produces internationally harmonized standards through a structure of Technical Committees that are divided into subcommittees and divided once again into Technical Working Groups. Standards germane to indoor and outdoor air pollution are grouped as ISO 14000 standards. The US Technical Advisory Group (TAG) to the Technical Committee has the responsibility for drafting the ISO 14000 standards. By 2004 American industries need to certify that they comply with ISO 14000 series International Environmental Management Standards in order to sell products in the European market.

The ISO 14000 approach is based on third-party registration and certified auditors working with accredited registrars. The standards provide guidance and a framework for international operations, products, and services. The standards do not dictate performance requirements of specific operational procedures, but rather require an organization to develop and abide by industry-based, free-market business principles. The ISO 14000 approach is based on the belief that is possible to include environmental considerations within "enlightened self-interest" in the operation, products, and services of the free-market economy. The ISO 14000 process is dynamic, requiring revisions based on consumer and stakeholder input that reflect a long-term commitment to minimize the adverse impact on the environment and improve occupational health and safety. It is important to recognize that ISO 14000 helps an organization establish and meet its own policy goals through setting objectives and targets, organizational structures that ensure accountability, management controls, and review functions. ISO 14000 does not set requirements for environmental compliance nor does it establish quantitative requirements for specific levels of pollution prevention or performance. These are issues within the jurisdiction of local, state, and federal agencies. The EMS specification document calls for environmental policies that include a commitment to both compliance with local environmental regulations and steps to prevent pollution.

About 10% of world trade is related to environmental and occupational health and safety issues. Industries in the developed countries incur costs for environmental improvements, and are at a disadvantage to industries in less developed countries that have not implemented policies for environmental pollution control and occupational health and safety. The disparity in environmental policy manifests itself in the worldwide economy by affecting costs and distorting trade. Under the ***General Agreement on Tariffs and Trade*** (GATT) it is illegal for the US to take unilateral action against countries to redress international trade disputes. The ***North American Free Trade Agreement*** (NAFTA) has adopted a similar approach. Thus the US supports the ISO to address environmental problems and the global recognition of trade-related environmental issues.

It is imperative that industries whose products and services impact the environment and occupational health and safety comply with ISO standards. In 1992 the ISO charged Technical Committee 207 with the task of developing international environmental management standards. ISO 14000 certification requires an organization to have an ***environmental management system*** (EMS) that establishes environmental policies, and demonstrates how the organization meets (or exceeds) regulatory requirements, how the organization's objectives are monitored, and how attainment is measured. The ISO 14000 ***Environmental Management Standards*** (also EMS, unfortunately) are a series of voluntary standards and guidelines which include the following:

- ***ISO 14001*** & ***14004*** - *environmental management system* (EMS): core document upon which all other standards are based, identifies environmental policies and procedures about how they are monitored and recorded
- ***ISO 14010*** to ***14012*** - *environmental auditing*: describes steps for auditing
- ***ISO 14020*** to ***14025*** - *labeling standards*
- ***ISO 14031*** - *performance evaluation*: describes the process for selecting environmental indicators and how performance is evaluated
- ***ISO 14040 to 14043*** - *life-cycle analysis*: cradle-to-grave analysis of the impact of production processes
- ***ISO 14060*** - *environmental aspects in product standards*: manual for product developers that defines factors to be taken into consideration for new products

Registration under ISO requires an organization to evaluate the environmental impact of its activities, products, and services. Registration begins by establishing an operational EMS, selecting an independent accrediting third-party registrar, submitting an application to the registrar for certification, selecting a lead auditor, preparing written auditors' reports, reviewing the policies and procedures in the company's EMS manual, and verifying procedures for continuous improvement.

American Society of Heating, Refrigerating and Air-Conditioning Engineers (ASHRAE) – ASHRAE is an international technical society founded in 1984 with membership in 2001 of over 50,000 in chapters world-wide. The stated purpose of the society is to advance "the arts and sciences of heating, ventilation, air conditioning and refrigeration for the public's benefit through research, standards writing, continuing education and publications." ASHRAE standards address a wide range of subjects related to the design and operation of air conditioning systems. Many are ANSI-certified and are referenced in building codes or used as models for standards developed by others. Three major ASHRAE standards widely used in design address various aspects of indoor air pollution:

- ***Standard 55-1992*** - *Thermal Environmental Conditions for Human Occupancy*, which specifies criteria of air temperature, humidity, and movement consistent with occupant comfort
- ***Standard 15-1994*** - *Safety Code for Mechanical Refrigeration*, which defines requirements for mechanical equipment rooms based on occupancy type, refrigeration system type, and the quantity and classification of refrigerant. The classification scheme for refrigerants by toxicity and flammability is itself established by another ASHRAE standard
- ***Standard 62-2001*** - *Ventilation for Acceptable Indoor Air Quality*, which establishes prescriptive requirements for ventilation air quantities as well as a compliance path based on control of specific air pollutants.

Other important standards define the performance of components, for example those used to remove solid contaminants from air streams:

- ***Standard 52.1-1992*** - *Gravimetric and Dust Spot Procedures for Testing Air Cleaning Devices Used in General Ventilation for Removing Particulate Matter*
- ***Standard 52.2-1999*** - *Method of Testing General Ventilation Air Cleaning Devices for Removal Efficiency by Particle Size.*

Although they are not, *per se*, standards, the four volumes of the ASHRAE Handbook series are widely viewed as establishing a standard of practice for the design of space-conditioning and refrigeration systems.

While there is no single internationally recognized occupational health and safety management system, there is a growing demand to use existing institutions for this purpose. A variety of nationally recognized standards and guidelines are commonly used. In the US, the standards of the ***American Industrial Hygiene Association*** are used. In Great Britain, the ***British Standards Institute*** is used.

1.8 Contributions from Professionals

Not all improvements in health arise from government regulations. Two important groups of professionals have also contributed handsomely to improving health:

- ***Home economists***: preparation and processing of food and hygiene in the home
- ***Sanitary engineers***: treatment of drinking water and sanitary waste water

1.8.1 Home Economics

Evolving in the later part of the19th century was a body of knowledge called ***home economics*** taught to young women in high school. With the acceptance of science as a method to achieve progress came the expectation that science could be put to use to improve the health of everyday people. Women learned nutrition and ways to improve the preparation and preservation of food, and personal hygiene reduced the incidence of disease throughout the nation. These improvements came about because home economics became a part of the required high school curricula. These improvements were implemented without fanfare or publicity, and proceeded slowly but resolutely in every public school in the nation. Several women initiated this movement, but chief amongst them was the protagonist Ellen Swallow Richards (1842-1911). Richards was educated as a chemist and spent her career at MIT contributing to the body of knowledge called home economics. What young women learned had enormous influence because it became implemented in peoples' homes, and affected everyone in the US, rich and poor alike. Because this material was required to be taught in public schools, the knowledge was applied in all homes throughout the land. To appreciate the significance of this body of knowledge, one should consider what exists today when home economics has all but been eliminated in today's schools, and men and women alike are indifferent to the preparation of food. Food is sometimes poorly prepared, improperly stored, allowed to stand unrefrigerated before or after serving, or not properly cleaned (especially poultry). Thus, it is no surprise that food poisoning occurs.

1.8.2 Sanitary Engineers

While empirical knowledge relating disease to drinking water was understood, it was Pasteur's discovery in the 19th century that living microbes caused infection which firmly established the need for what was later called ***sanitary engineering***. Principal accomplishments in the 19th century were steps to prevent wastewater from contaminating drinking water by using separate underground conduits for each. In the early 20th century it was discovered that chlorine and ozone were effective agents to purify drinking water. Simultaneously, primary, secondary, and tertiary treatment methods were adopted before wastewater could be discharged to the environment. When potable water was dispensed, and wastewater discharged within the same room (bathroom, kitchen, etc.), precautions had to be taken to prevent sewer gases and microbes from entering buildings through open drainpipes connected to underground sewer conduits. It wasn't until the later part of the 19th century that the simple ***water-sealed trap*** became a common and required item on all fixtures discharging wastewater. Similarly all pipes discharging wastewater were required to be equipped with a ***vent*** that prevented sewer gases from entering buildings.

1.9 Components of Industrial Ventilation Systems

The goal of an industrial ventilation system is to prevent contaminants from entering air in the indoor environment. There are several ways to accomplish this goal. Figure 1.10 shows the basic components of an industrial ventilation system.

All industrial ventilation systems contain some if not all these components:

- Source - Contaminant gases, vapors, and particles are generated by industrial activities and emitted into the air. The mass emission rate and physical properties of the contaminants have to be described quantitatively.
- Hood - The configured inlet used to capture contaminants.
- Ducts and fittings - A system of branching ducts connects the hoods to an exhaust fan.
- Air cleaning device - It may be necessary to remove contaminants from the collected air. If the contaminant concentration in the collected air exceeds environmental standards or if the collected air is recirculated, an ***air-cleaning device*** (also called an air pollution control system, or APCS) will be needed.
- Air mover (exhaust fan) - Air is drawn into a hood by an air mover which is generally a fan. The fan may be upstream or downstream of the air-cleaning device.
- Stack - A stack exhausts air to the atmosphere and prevents it from reentering the workplace.
- Recirculation - All or a portion of the collected air may be recirculated to the workplace. Recirculated air must be cleaned.
- Make-up air - Outside air added to the workplace is called make-up air. The temperature and humidity of the make-up air may be controlled.
- Exhaust air - It may be necessary to exhaust a portion of the room air.

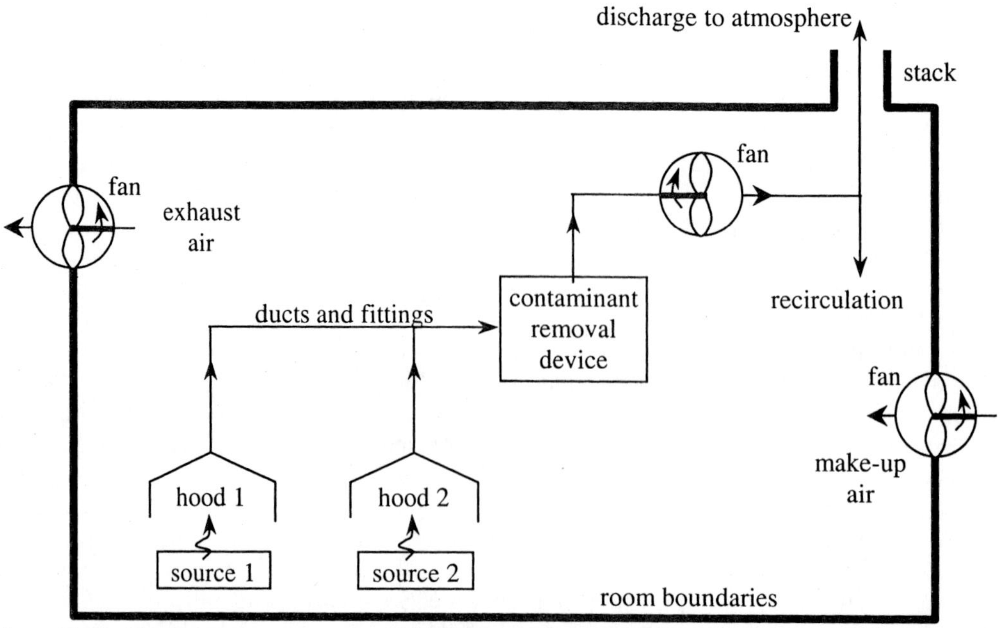

Figure 1.10 Typical components of an industrial ventilation system, with exhaust air, make-up air, a contaminant removal device, and recirculation.

1.10 Classification of Ventilation Systems

When selecting a ventilation system, one should never overlook the most obvious corrective steps (see Section 1.4) - modify the process to generate smaller amounts of contaminant and take steps to prevent contaminants from entering the indoor environment. In so doing, a ventilation system may not even be needed. Industrial practices are not sacrosanct; operational costs and contaminant exposure can often be reduced simultaneously by modifying the process or choosing different raw materials. For example, questions such as the following should be asked before any design is begun:

- Can streams of air discharged by machinery be directed elsewhere? Can the exhaust from pneumatic equipment be directed elsewhere?
- Can a liquid of lower volatility be used? Can its temperature be lowered to reduce the emission of vapor?
- Can the type of lubricant or its rate of flow be changed to reduce the production of mist from metal cutting and metal punching machines?

Once one has eliminated all opportunities to reduce the generation of contaminants, a ventilation system may have to be designed. To a great extent the choice of a ventilation system is dictated by how the emissions are generated. If there are many sources of emission and they are distributed throughout a process, such as dust generated in a quarry or vapors generated from freshly painted surfaces, they are called ***fugitive emissions*** or ***area-source emissions***. If the emissions are generated from well-defined points in a process, they are called ***point-source or line-source emissions*** and are the easiest to control.

In the field of indoor air pollution control, ***hood*** is the generic word given to a uniquely configured air inlet through which contaminated air is withdrawn. The word denotes the function to be performed rather than any particular geometrical configuration. Table 1.8 summarizes the two general classes of ventilation systems - ***general ventilation*** and ***local ventilation*** - and subdivisions within local ventilation systems.

Table 1.8 Classification of ventilation systems.

General ventilation

Withdrawal of air from the entire workplace and its replacement with a selected mixture of make-up fresh air and a cleaned portion of the air withdrawn from the workplace. (See Chapter 5.)

Local ventilation

Withdrawal of air and contaminants from a region close to the point of generation. (See Chapter 6.)

- Enclosure - A housing that virtually encloses the source. *Examples*: laboratory fume hood, glove box, spray booth, fumigation booth, grinding booth
- Receiving hood - A housing that collects contaminants because of their intrinsic motion. Receiving hoods move minimal amounts of air. *Examples*: canopy hood over a hot source, pedestal grinders, kitchen range hoods
- Exterior hood - A uniquely configured air inlet placed in close proximity of a source of contamination. Exterior hoods are also called ***capture hoods***. *Examples*: down-draft grinding benches, welding snorkels, lateral exhausters and push-pull exhausters (with and without side panels), high velocity-low volume systems, side-draft exhausters

The reader should be cautioned that the classifications are qualitative and not universally adhered to. Local ventilation systems with inlets in close proximity of a source that withdraw small amounts of air at high velocity are called ***high velocity - low volume*** (HVLV) systems. Figure 1.11 is an example of the effectiveness of a HVLV system to capture particles generated by sanding and grinding (Topmiller and Hampl, 1993). Figure 1.12 shows how a local ventilation system is incorporated within a sander. Figure 1.13 shows a welding gun integrated with a HVLV fume extracting local ventilation system that extracts welding fume (Cornu, 1993). The phrase ***close capture hood*** is also used. Figure 1.14 is an example of a close capture hood designed for a wood planing mill.

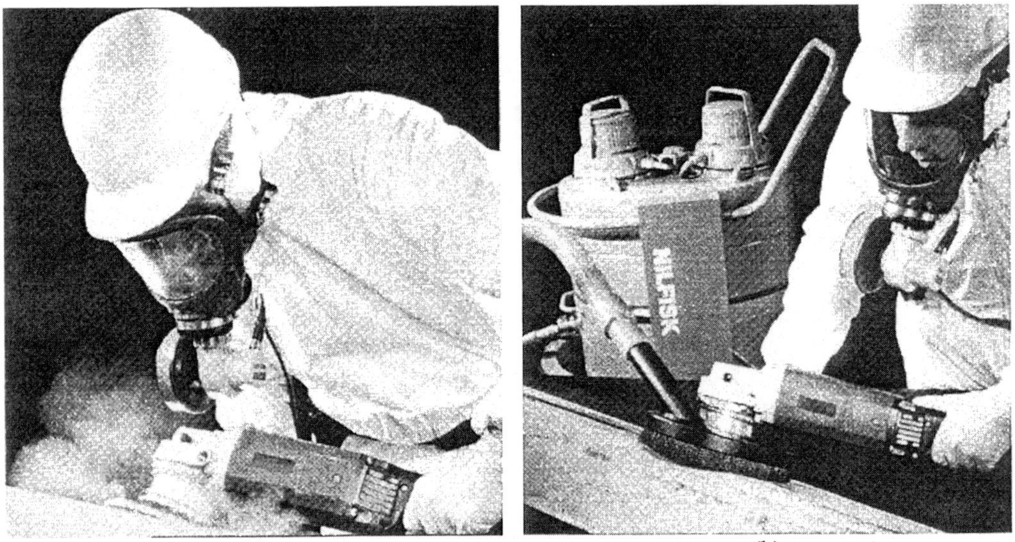

(a) (b)

Figure 1.11 Example of local ventilation system effectiveness: (a) uncontrolled surface grinding; (b) grinding with attached HVLV system (courtesy of Nilfisk of America).

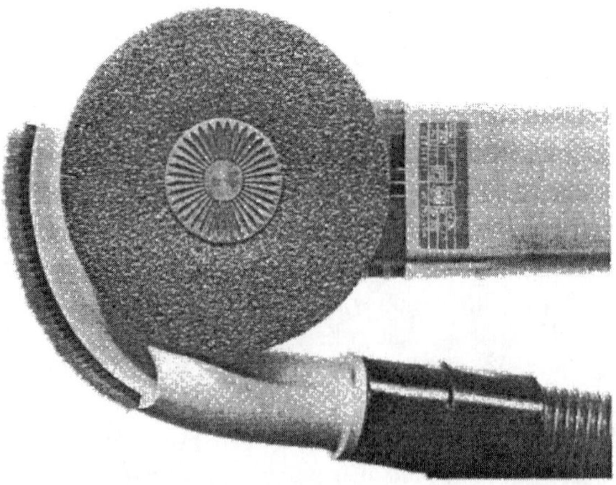

Figure 1.12 Surface sander with HVLV local ventilation system (from Fein Power Tools, Inc.).

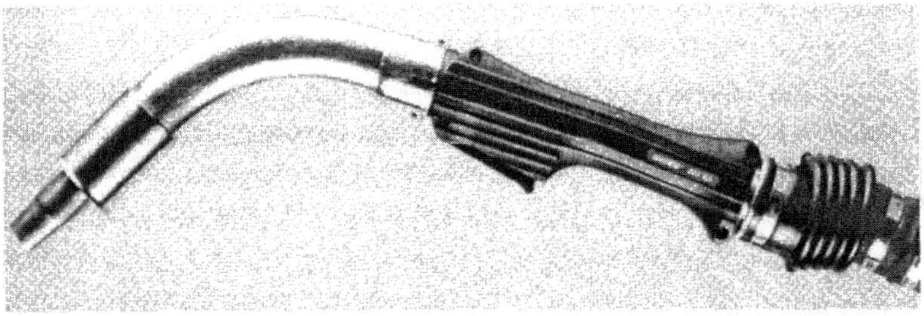

Figure 1.13 High velocity - low volume (HVLV) fume extracting welding gun; air is drawn in from the lower left of the photograph (courtesy of The Lincoln Electric Company).

Figure 1.14 Close capture hood (close-fitting canopy hood) for a wood planing mill. (courtesy of Donaldson Company, Inc.).

Inlets that are farther from the source and remove large quantities of air at low velocity are called ***low velocity-high volume*** (LVHV). Examples of these are canopy hoods shown in Figure 1.15 and ***side-draft hoods*** shown in Figure 1.16. There is inherent confusion as to what "V" represents in the acronyms HVLV and LVHV, but the reader should be prepared to encounter these acronyms.

Figure 1.15 Low velocity-high volume (LVHV) canopy hood in a foundry (from Envirex Co., 1978).

Figure 1.16 Side-draft hoods to capture fumes produced by pouring molten metal in sand molds (courtesy of CMI-Schneible Company).

As a general proposition, general ventilation should be avoided as a method of controlling airborne toxic substances. It is more economical to address each source and install appropriate local ventilation systems than to allow the contaminants to enter the entire indoor environment and be forced to remove and replace the entire air mass in the work space. One may wish to employ general ventilation to control temperature, humidity, and odor, but controlling industrial pollutants can be achieved more economically by local rather than general ventilation. In the microelectronics and biotechnology industries there is a need for ultra pure working environments, i.e. ***clean rooms***, and general ventilation is used. Figure 1.17 shows the details of a clean room. Even in these cases, capturing contaminants at the point of generation remains the preferred and most economical approach. While the workers' health is not the primary reason clean rooms are needed, the movement of air and contaminants follows the same principles used in industrial ventilation.

If it is possible to totally enclose the process, leaving the worker outside the enclosure, it should be done without hesitation. Such an ***enclosure*** will produce the maximum control with the minimum amount of exhaust air. A few examples where this may be possible are electric arc furnaces (Figure 1.18), ***laboratory fume hoods*** (Figure 1.19), and machines that fill vessels with powders and granular material (Figure 1.20).

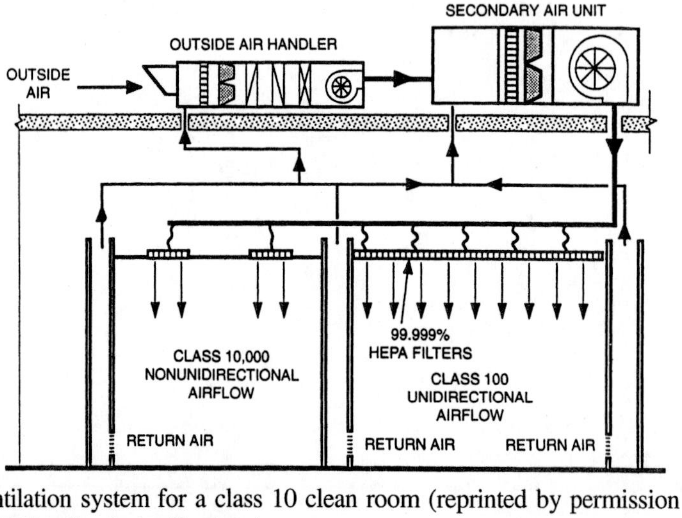

Figure 1.17 Ventilation system for a class 10 clean room (reprinted by permission from ASHRAE 1995 Handbook-Applications).

(a) 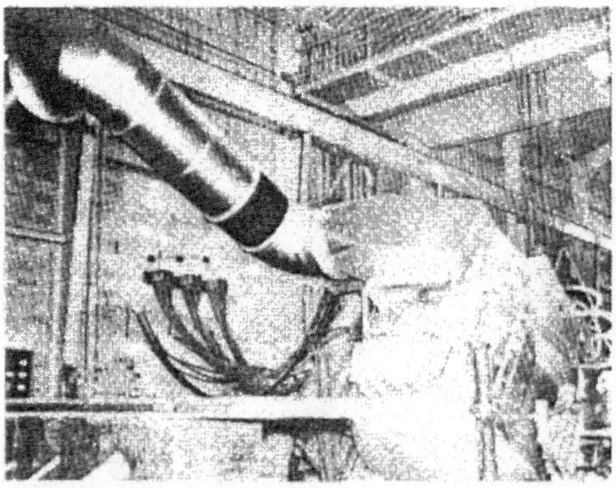 (b) 

Figure 1.18 Close-fitting hood for an electric arc furnace: (a) in place, and (b) swung open for charging (from Danielson, 1973).

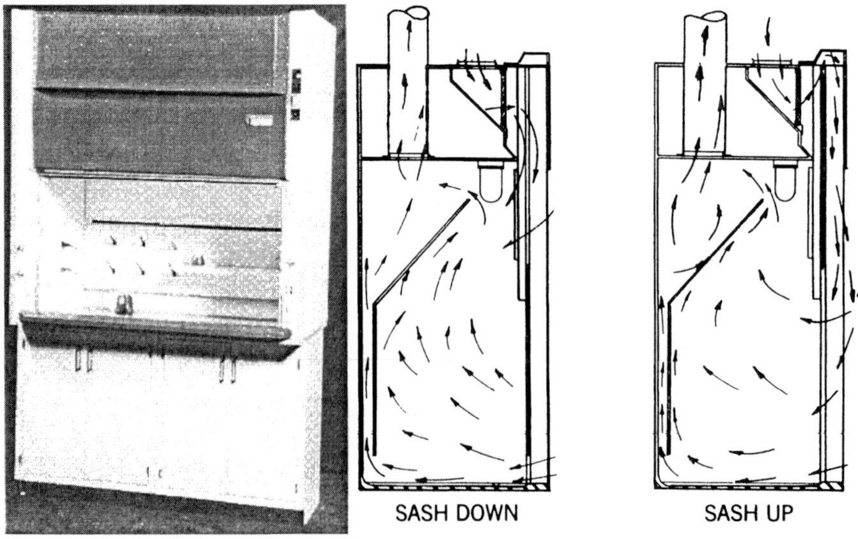

Figure 1.19 Uniflow laboratory fume hood with make-up air introduced as an air curtain to reduce withdrawal of conditioned room air (courtesy of Hemco Corp.).

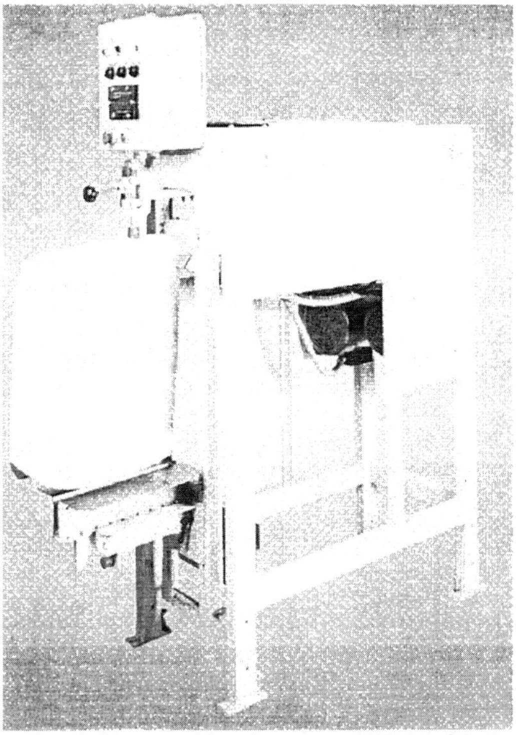

Figure 1.20 Bag filling and weighing machine (courtesy of Thiele Technologies, Barry-Wehmiller Inernational Resources).

If the operator cannot be excluded from the enclosure, a ***booth*** surrounding the operator wearing personal protective apparatus is the next most effective method to prevent contaminants from traveling to other parts of the workplace. Great care must be taken to ensure that contaminant concentrations in the operator's breathing zone do not exceed safe values. Examples of industrial operations where booths are attractive include grinding booths for portable hand-held sanders (a surface sander is shown in Figure 1.12), paint spraying booths, and welding booths. Figure 1.21 shows the differences between the spray cycle and cure cycle in an automobile painting booth. Figure 1.22 show a comparison of wood sanding in the open workplace and inside a booth. The booth in this example utilizes a system which captures and cleans dusty air, and then recirculates cleaned air back into the booth, thereby reducing the withdrawal of conditioned room air. Figure 1.23 shows a "water-wall" paint spray booth that captures oversprayed paint particles with a column of cleaning water.

Many industrial operations involve equipment and the movement of people and material that makes it impossible to enclose the source of contamination. In these cases ***exterior hoods*** should be used. Examples of such operations are ***lateral exhausters*** (Figure 1.24) for open vessels used for electroplating, galvanizing, pickling, and etching; ***bag dumping stations*** (Figure 1.25); ***snorkels*** for welding (Figure 1.26); and long-wall automatic miners used in coal mines.

(a)

(b)

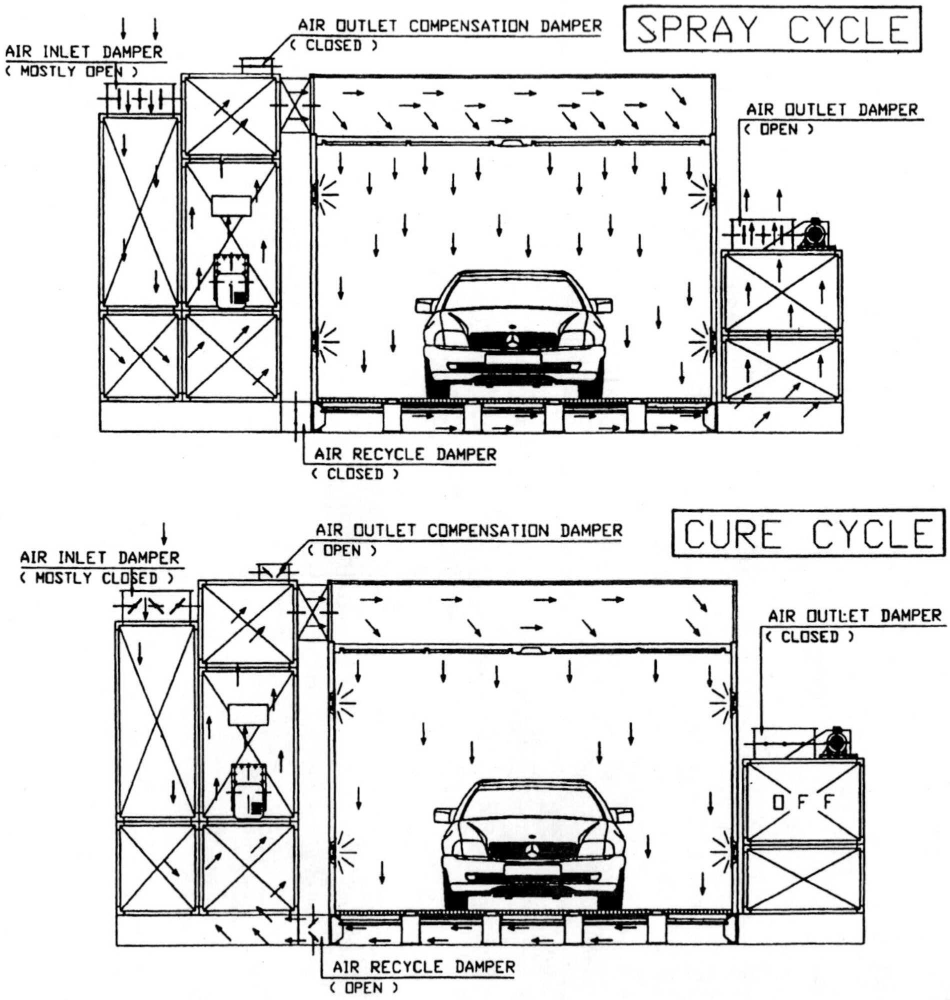

Figure 1.21 Enclosure for painting automobile parts or repainting automobiles: (a) spray cycle, and (b) cure cycle; flow enters from the top and exits through the floor (courtesy of American Paint Booths, Inc., SAICO Paint and Cure).

(a)

(b)

Figure 1.22 Wood sanding (a) in the open work place, and (b) inside a booth with recirculated air that captures particles and reduces the withdrawal of conditioned room air (courtesy of Donaldson Co., Inc., Torit Products).

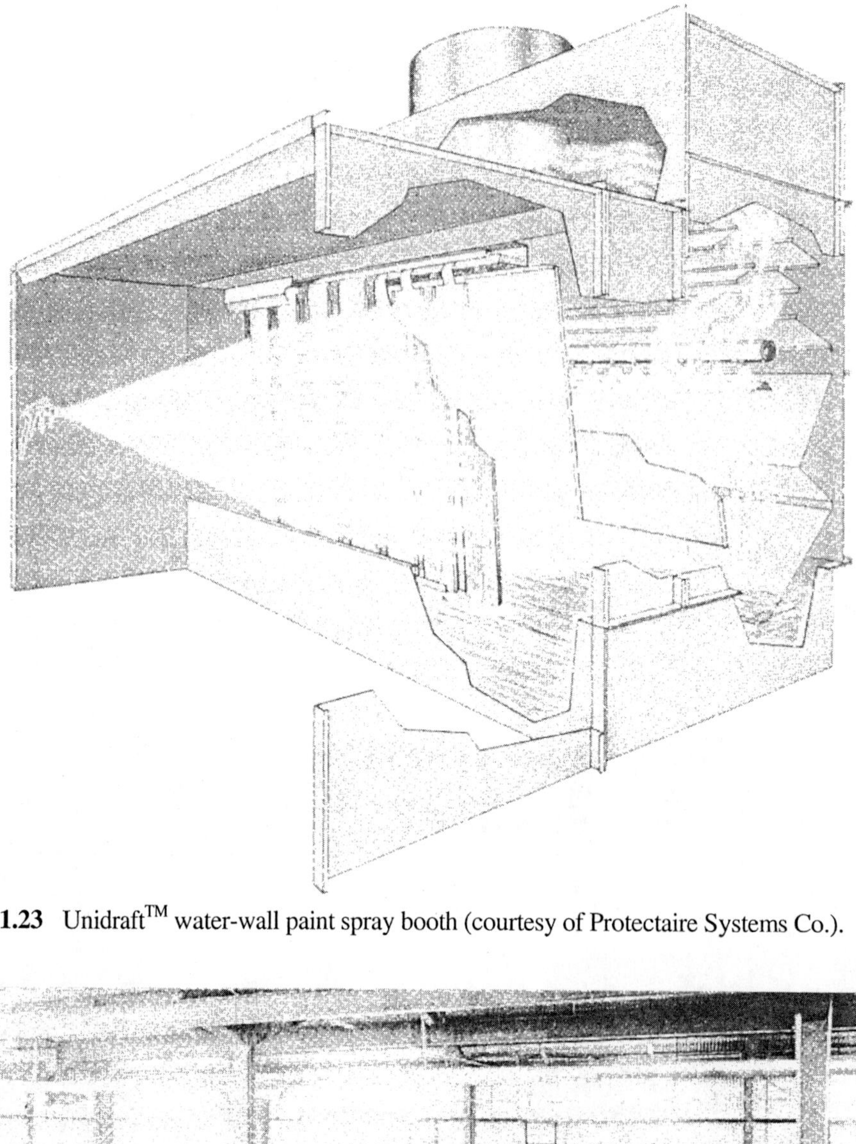

Figure 1.23 Unidraft™ water-wall paint spray booth (courtesy of Protectaire Systems Co.).

Figure 1.24 Example of open vessels with lateral exhausters on each side (from Danielson, 1973).

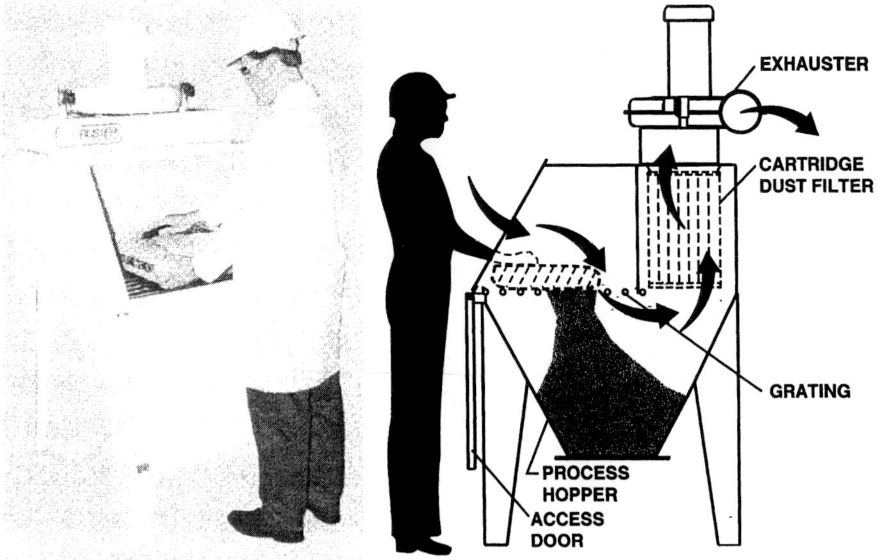

Figure 1.25 Manual bag dumping unit; the task is to dump the contents of the bag into a hopper; escaping dust is captured by an exhauster (courtesy of Dynamic Air, Inc.)

Figure 1.26 Snorkel fume extractor (courtesy of The Lincoln Electric Company).

The air flow required for exterior hoods is larger than for enclosures, and one must be careful that contaminant concentrations in the worker's breathing zone do not exceed safe values. ***Canopy hoods*** are located directly above the process (Figure 1.15) and try to capitalize on buoyancy. ***Restaurant canopy hoods,*** (Figure 1.27) can also employ a novel push-pull concept. Side-draft hoods have faces oriented diagonally to the process (Figure 1.16). While lateral exhausters withdraw air at right angles to the process plume (Figure 1.24), ***down-draft benches*** (Figure 1.28) withdraw air in the downward direction.

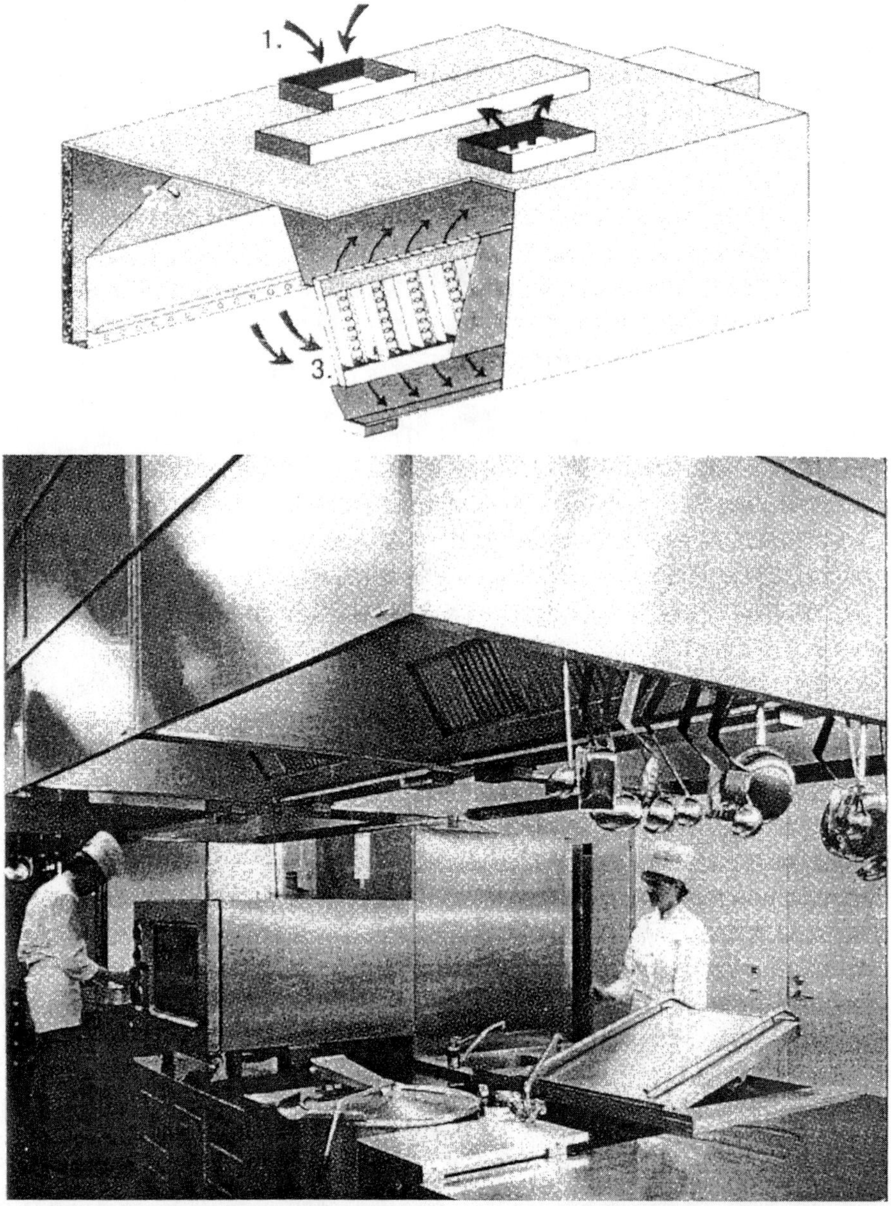

Figure 1.27 Restaurant canopy hood incorporating the push-pull concept, which maximizes capture and minimizes the amount of withdrawn air (courtesy of Halton Co.).

If the contaminant is a distinct stream of heavy particles or a buoyant plume, ***receiving hoods*** may be applicable. Examples where receiving hoods are effective include side-draft hoods to remove plumes from furnaces (Figure 1.16), wood working machines, and the ***swarf*** from pedestal grinders and ***swing grinders*** (Figure 1.29). Even in these cases, great care must be taken to ensure that the receiving hoods are located and sized to collect the contaminants, since room drafts may deflect buoyant plumes or small particles before they enter the receiving hood. In many cases receiving hoods are used in combination with exterior hoods. Dust generated from a wood planer can be captured by a close-fitting canopy hood (Figure 1.14).

Push-pull hoods (Figure 1.27, see also Chapter 6) represent a class of hoods whose full potential and limitations have not been fully assessed. Applications where push-pull systems are attractive include large open vessels. The blowing jet increases the size of the region of influence in front of the inlet through which the contaminants are withdrawn. The blowing jet is a jet pump that induces room air into the jet as the composite flow sweeps over the surface of the liquid and enters the exhaust inlet. Extreme care must be taken in the design of push-pull systems to ensure that the blowing jet does not inadvertently function as a mixer and increase the transfer of contaminant in the workplace. A jet with the wrong orientation or an exhaust inlet that cannot accommodate the total flow may increase the concentration in the general workplace.

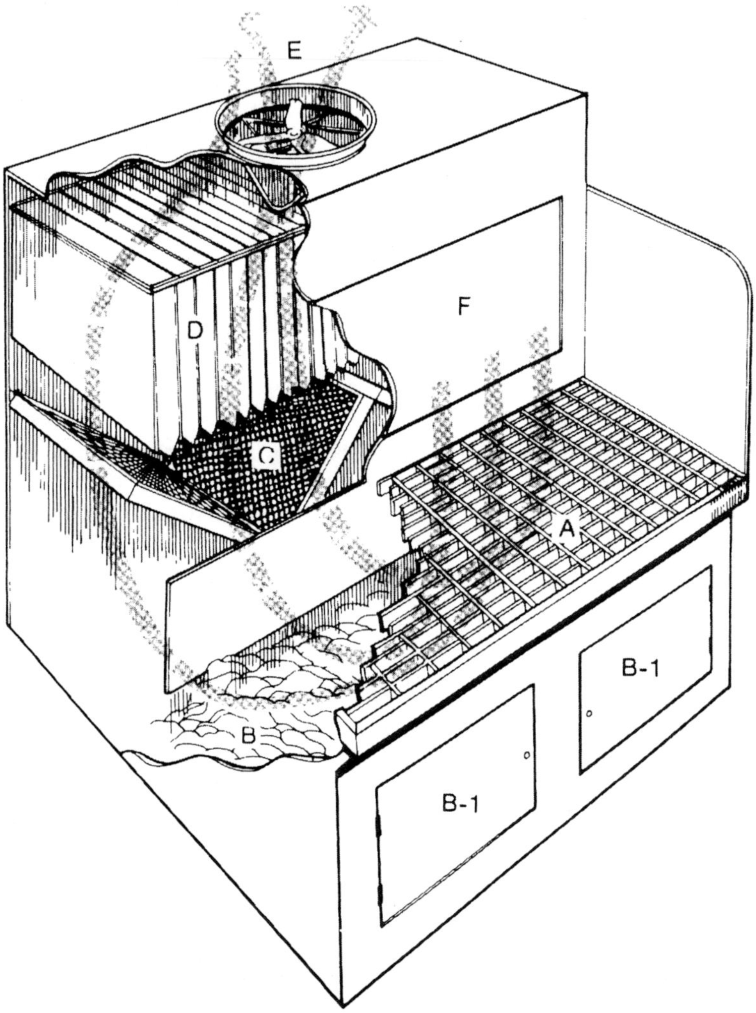

Figure 1.28 Downdraft grinding bench. Air travels downward through an open steel grating (A). Large particles fall out on the front floor (B), and are removed through out doors (B-1). Air then passes through reusable metallic viscous pre-filters (C), removing most of the particles from the air stream. High efficiency filters (D) remove remaining dust before returning air back into the work area through an axial flow fan (E). Access panels (F) allow for filter maintenance (courtesy of Wolverine Equipment Division).

Figure 1.29 Receiving hood for a swing grinder; arrow indicates face of receiving hood (from Envirex Co., 1978).

1.11 Deficiencies in Present Knowledge

It is prudent that the performance of industrial ventilation systems and APCSs be demonstrated prior to commitment of funds for their construction and installation. Major design changes performed after a system is built are ineffective and expensive substitutes for a priori analysis. There are six major deficiencies in our understanding of how to control indoor air pollution:

1. inability to predict the concentration at arbitrary points (x,y,z) in the room
2. inability to predict the performance of a conventional design operating at conditions different from those for which it was designed, i.e. ***off-design performance***
3. inability to predict the effects of drafts and wakes in the workplace
4. inability to generalize acceptable designs
5. inability to predict the behavior of new or unusual contaminant sources
6. unwillingness to abandon the concept of capture velocity

All these deficiencies arise from one fundamental shortcoming – the inability to describe the motion of contaminants from their source of origin to an arbitrary point in space (x,y,z) at an arbitrary time t.

1. Predicting the concentration Contemporary texts, handbooks, and manuals provide useful descriptive information but limited quantitative information about the design of ventilation systems for specific industrial operations (e.g. booths for swing grinders, pedestal grinders, laboratory hoods, etc.). Until the advent of ***computational fluid dynamics*** (CFD), designers had no way to estimate the contaminant concentration at selected points in the vicinity of the pollutant source, the worker, and heat sources. Thus designers could not assure themselves that the design would satisfy OSHA standards or more stringent company standards. In the 21st century such an inability is unacceptable professionally and unnecessary technically, now that CFD can be used to predict c(x,y,z,t); CFD is discussed in detail in Chapter 10.
2. Off-design performance Seldom is a ventilation system constructed or operated as planned. Practical considerations or administrative decisions often require the system's dimensions or the volumetric flow rate to change or be reconfigured. There is presently no way designers

can predict the consequences of these changes, and techniques to accomplish this are urgently needed if one is to know that the new system still satisfies OSHA requirements.

3. Drafts and wakes Theoretical consideration of ventilation systems is frequently based on the assumption that surrounding the process is a quiescent body of air. Nothing could be further from the truth. Every workplace contains equipment and workers performing tasks that produce ***drafts***, room air currents of unique and unpredictable character, and ***heat sources*** that create buoyant plumes. ***Spurious room air currents*** are produced from the following:

    - motion of workers and machinery in the vicinity of the source of contamination
    - aerodynamic wakes around workers and equipment
    - plumes of hot air rising from the surfaces of hot machinery or vapors rising from volatile liquids
    - jets of air from cooling fans within the machinery, doors and windows, workers' personal fans, or heaters and air jets from building heating or cooling systems

    Other factors affecting the motion of contaminants are walls, partitions, machinery surfaces, jigs, and fixtures. In some cases, impediments may be eliminated, but for the most part they have to be incorporated in the design of the ventilation system from the beginning. A vexing circumstance for engineers is trying to correct unwanted room air currents produced by machinery or new production methods put in place after the ventilation system has been installed, and about which the design engineer may not have been consulted.
4. Generalization Even if a specific ventilation system satisfies OSHA requirements, there is presently no way to scale such a design geometrically so that one can be assured that a larger or smaller design will also be successful. In addition, there is no way to know how much the volumetric flow rate should be increased or decreased.
5. New or unusual contaminant sources The professional literature contains a great deal of information on designs that have been shown to be successful. Unfortunately, it is rare that one is dealing with an industrial process that is identical to one for which the ventilation system is already published. At best it might be possible to select a ventilation system for a similar, but not identical industrial operation. Sometimes even this will not be possible and one will be forced to design a ventilation system from scratch. In either case, designers presently have difficulty estimating performance.
6. Capture velocity For many years hoods have been designed using the notion of ***capture velocity***. The ACGIH defines capture velocity as "the minimum hood-induced air velocity necessary to capture and convey the contaminant into the hood" (ACGIH, 1988). The use of this concept is regrettable. There is no way to postulate values of capture velocities for particles produced by different industrial operations. To specify such values is equivalent to assuming the contaminant's trajectory. One may wish to do so qualitatively, but quantitatively it cannot be done with the precision needed in design. As will be seen in Chapter 8, the viscosity of air reduces the relative velocity between particle and air to virtually zero in a very short time. The time it takes to do so and the distance the particle travels depend on the particle's size, initial velocity, and the carrier gas velocity. For gaseous contaminants the situation is vastly different. Thus it is illogical and fundamentally inconsistent with the principles of fluid mechanics to tabulate capture velocities. The suggestion that a moving air stream somehow captures contaminants only when it has a velocity equal to or larger than a capture velocity is bogus, and should be abandoned.

1.12 Professional Literature

A systematic summary of the professional literature on indoor air pollution and industrial ventilation has been prepared by Olander (1993). The literature often lacks the precision, rigor, and application of fundamental engineering principles that are required in a competitive world economy. If professionals are going to understand the motion of contaminants and are going to design collection systems skillfully, they will need to use more sophisticated analytical and experimental techniques than they have in the past. If students in mechanical, civil, chemical, architectural, environmental, and biomedical engineering are going to be attracted to the field, the level of technical discourse must be commensurate with what they have learned and used in their disciplines. If working professionals are going to be able to adopt technological developments from other fields of engineering, it is necessary that they acquire the sophistication used by professionals in these fields. It is our belief that indoor air pollution control, and industrial ventilation in particular, will witness a burst of activity in the near future when engineers in the fields of fluid mechanics and heat and mass transfer awake to the interesting and important problems that need to be solved.

It is not our intention to write an exhaustive review of all literature on indoor air pollution or the literature on fluid mechanics, numerical computation, or aerosol dynamics that pertains to industrial ventilation. Rather, what follows is a brief summary of several important books and sources of information with which readers should become familiar. Some of these books are old and qualitative in nature, but they are classics in the field. Anyone wishing to pursue careers in the field will find in these books rich sources of practical information. Experimental and analytical techniques have changed over the years, but the knowledge that was sought and the applications to which it was to be applied have changed little. Proceeding without such information is to risk reinventing the wheel!

Hemeon's book, *Plant and Process Ventilation* (1963), contains a great deal of practical qualitative information needed by individuals entering the field. Hemeon does not model ventilation systems analytically with the sophistication used in engineering today, but this eminently readable book contains many empirical equations that engineers can use to prepare initial designs of hood geometry and estimates of ventilation rates. The 3rd edition of Baturin's *Fundamentals of Industrial Ventilation* (1963) is largely qualitative but it is a thorough technical book on the subject. The material is essential for the working professional and contains numerous analyses and experimental data not found in the US literature of the time, e.g.

- velocity fields for air curtains
- velocity fields for unusual inlets and outlets
- air changes for natural ventilation in buildings
- velocity fields in buildings containing heat sources

It is believed that with advances in the field of ***computational fluid mechanics*** (CFD), there will be decreasing need for the material in Baturin and Hemeon because computational models can provide reliable guidance to designers more quickly and easily than in the past. Nevertheless, readers should review Baturin and Hemeon to acquire physical insight about the movement and control of contaminants.

The American Conference of Governmental and Industrial Hygienists (ACGIH) is a professional organization that disseminates information on methods to maintain a healthy environment in the workplace and methods to measure the concentration of contaminants. The ***Committee On Industrial Ventilation*** within ACGIH reviews developments in industrial ventilation, and every few years (since 1951) publishes a new edition of *Industrial Ventilation: A Manual of Recommended Practice*, often called simply the "***ventilation manual***". The ventilation manual contains sketches of specific ventilation systems for nearly 150 different industrial operations, discussions of how to design a duct system, how to select the ventilation flow rate, the proper fans, and appropriate gas cleaning devices, and how to test ventilation systems. The ventilation manual does not base its recommendations on rigorous engineering principles, but it is nevertheless an invaluable reference book for those responsible for the design and operation of ventilation systems. The ventilation manual provides valuable guidance to prepare the preliminary design of a ventilation system and to estimate the volumetric flow rate. The ventilation manual is the most widely used publication in the field, and engineers should be conversant with it. In the field of mine ventilation, the *Manual of Mine Ventilation Design Practices* (Bossard et al., 1983) provides a similar function.

The American Society of Heating, Refrigeration, and Air-Conditioning Engineers (ASHRAE), is an engineering professional society that addresses the issue of industrial ventilation. Of invaluable use to engineers are four ASHRAE Handbooks. Each year ASHRAE updates and republishes one volume of its four-volume set of handbooks. Several chapters in the ASHRAE HVAC Applications Handbook are devoted to the control of pollution in the indoor environment. The material is written for engineers and is more quantitative and precise than what is contained in ACGIH literature. The handbooks presume familiarity with subjects included in accredited engineering Bachelor of Science degree programs.

The ***Center For Chemical Process Safety*** (1990) is a directorate within the ***American Institute of Chemical Engineers*** and was established to create and disseminate knowledge on health, safety, and loss prevention. Many educators believe that embedding health and safety within the required courses in the college student's major is the most effective way to prepare engineers. Accordingly the Center developed 90 student problems and an instructor's guide in 1990 for undergraduate thermal sciences courses in all engineering disciplines, not just chemical engineering. The chemical engineering profession has taken explicit steps to educate students about plant safety and loss prevention. The book by Crowl and Louvar, *Chemical Process Safety: Fundamentals with Applications* (1990) presents the subject in quantitative terms to enable engineers to incorporate safety, health, and loss prevention as integral parts of the design. Their text serves as a model on how industrial ventilation can be incorporated in the design process.

The text by Alden and Kane, *Design of Industrial Ventilation Systems* (1982), has been used for many years and provides a quantitative basis to ACGIH procedures. The book by Goodfellow, *Advanced Design of Ventilation Systems for Contaminant Control* (1986), contains the best and most complete description of industrial ventilation practices that is available, and is indispensable to practicing design engineers. McDermott's book, *Handbook of Ventilation for Contaminant Control* (1976), is titled a handbook, but it is largely a summary of material that can be found in the ASHRAE and ACGIH handbooks. The book by Constance, *Controlling In-Plant Airborne Contaminants* (1983), is similar to McDermott (1976) in content and style. The text by Burgess, Ellenbecker, and Treitman, *Ventilation For Control of the Work Environment* (1989), contains both theory and practical applications for the design of industrial ventilation systems. This is an excellent book for those responsible for the specification, design, installation, and maintenance of industrial ventilation systems. It is written to supplement the ACGIH ventilation manual. Major articles are published in the leading journals (see below) on a regular basis. The article by Goodfellow and Smith (1982) summarizes the major publications in the field, and that by Billings and Vanderslice (1982) updates the science and technology in the field. In addition, some texts on industrial hygiene contain sections on industrial ventilation, e.g. the book entitled *The Industrial Environment - It's Evaluation and Control* by the US Department of Health, Education and Welfare (HEW) (1973), *Engineering Design for the Control of Workplace Hazards* by Wadden and Scheff (1987), and *Industrial Hygiene Engineering, Recognition, Measurement, Evaluation and Control* by Talty (1988).

The book *Industrial Ventilation and Air Conditioning* by Hayashi et al. (1985) is a detailed description of design procedures adopted by the Japanese Labor Department based on several decades of experimental research by T. Hayashi. The procedures are presently requirements for Japanese industry. The design procedures are similar to ACGIH procedures but incorporate ways to include cross drafts, round and rectangular sources (of different aspect ratios), buoyant plumes, etc. in more quantitative terms than ACGIH. The procedures contain considerably more information and computational opportunities than do ACGIH procedures. Hayashi's concepts are based on what is called ***cooperation theory***. The theory rests on the idea that contaminant control is governed by a "pull" flow that may be aided by a "push" flow. With or without the push flow, air is induced from the surrounding environment and drawn over the source of contamination and into the pull receiver. The essential element in good design, i.e. "cooperation", is to induce the minimum, essential amount of air using as little pull and push as possible. The concept is curiously described by Hayashi in terms of harmony (Siu, 1957), reminiscent of the way Herrigel (1953) writes of Zen in the art of archery.

The textbooks by Crawford, *Air Pollution Control Theory* (1976); Heinsohn, *Industrial Ventilation* (1991); and Heinsohn and Kabel, *Sources and Control of Air Pollution* (1999) are thorough expositions on the design, operation and performance of air pollution control systems. Analytical developments begin with engineering principles and proceed logically based on the assumptions that are made. The texts are excellent reference books for engineering professionals engaged in the design and operation of industrial ventilation systems. The handbook by Calvert and Englund, *Handbook of Air Pollution Technology* (1984) presents (but does not develop) the essential analytical relationships describing the operation of air pollution control systems. Other textbooks by Cooper and Alley (1986), Licht (1980), Flagan and Seinfeld (1988), and Seinfeld (1986) discuss the overall issue of air pollution, including control systems.

A major activity in architectural engineering is to describe the velocity field within (and between) rooms in buildings so as to achieve high entrainment of room air in jets of fresh air and simultaneously produce low air speeds in occupied zones. In Europe, Japan, and the US there is considerable interest in using CFD to model the transport of air contaminants and water vapor to ensure health and comfort. Students of industrial ventilation should acquaint themselves with the publications of Murakami and Kato (1989), Kurabuchi et al. (1989), Baker et al. (1989), and Nielsen (1989) because the results, arguments, modeling techniques, and experimental techniques are directly applicable to industrial ventilation.

The following is a brief list of textbooks and handbooks on environmental pollution. Listed also are materials from the related fields of industrial hygiene, industrial ventilation, and toxicology, since the principles by which contaminants are controlled and their toxic effects are the same.

Ecology

- Williamson, S. J. (1973)
- Elsom, D. M. (1992)

Health effects and industrial hygiene

- Cralley and Cralley, "Patty's Industrial Hygiene and Toxicology" (1979)

Atmospheric chemistry and physics

- Seinfeld, J. H. and Pandis, S. N. (1998)
- Finlayson-Pitts, B. J. and Pitts, J. N. Jr. (1986)
- Flagan, R. C. and Seinfeld, J. H. (1988)

Aerosols

- Hidy, G. M. and Brock, J. R. (1970)

- Friedlander, S. K. (1977)
- Hinds, W. C. (1982)
- Willeke, K. and Baron, P. A. (1994)

Air pollution control

- Crawford, M. (1976)
- Licht, W. (1980)
- Calvert, S. and Englund, H. M. (1984)
- Cooper, C. D. and Alley, F. C. (1986)
- Bunicore, A. J. and Davis, W. T. (1992)
- Theodore, L. and Bunicore, A. J. (1994)
- Boubel, R. W., Fox, D. L., Turner, B., and Stern, A. C. (1994)
- deNevers, N. (1995)
- Mycok, J. C., McKenna, J. D. and Theodore, L. (1995)
- Wark, K., Warner, C. F., and Davis, W. T. (1998)
- Heinsohn, R. J. and Kabel, R. L. (1999)

Control of contaminants in the indoor environment

- Burton, D. J. (1984)
- Wadden, R. A. and Scheff, P. A. (1987)
- Talty, J. T. (1988)
- Burgess, W. A., Ellenbecker, M. J. and Treitman, R. D. (1989)
- Heinsohn, R. J. (1991)
- Hays, S. M., Gobbell, R. V., and Ganick, N. R. (1994)
- Godish, T. (1995, 1998)
- Moffat (1997)
- Spengler, J. D., Samet, J. M., and McCarthy, J. F. (2000)
- American Conference of Governmental and Industrial Hygienists (ACGIH) Industrial Ventilation: A Manual of Recommended Practice (use current edition)

Chemical engineering

- Bisio, A. and Kabel, R. L. (1985)
- Fogler, H. S. (1992)
- Smith, J. M. and Van Ness, H. C. (1975)
- McCabe, W. L., Smith, J. C. and Harriott, P. (1993)

Professional Societies - The following publications of professional societies contain material relevant to environmental pollution control:

- Aerosol Science and Technology
- American Industrial Hygiene Association Journal (AIHA)
- Annals of Biomedical Engineering
- Annals of Occupational Hygiene
- Applied Industrial Hygiene
- Atmospheric Environment
- Environmental Science and Technology (ACS)
- Journal of Aerosol Science
- Journal of the Air & Waste Management Association (A&WMA)

- Transactions of the American Society of Heating, Refrigerating and Air-Conditioning Engineers (ASHRAE)
- Transactions of the American Society of Mechanical Engineers (ASME)

Governmental Agencies

The following agencies of the US government have specific missions devoted to the control of contaminants and publish technical reports expressly related to environmental pollution control. Their web sites should be consulted to keep abreast of changes in regulations and standards.

- Department of Energy (DOE)
- Environmental Protection Agency (EPA)
- Mine Safety and Health Administration (MSHA)
- National Institute for Occupational Safety and Health (NIOSH)
- Occupational Safety and Health Administration (OSHA)

Symposia, Conferences, and Professional Meetings

Various national and international symposia, conferences, and workshops are held every year. These meetings usually prove to be more valuable than one anticipates. Chance meetings with people in the hall, at dinner, etc. inevitably prove to be stimulating. Learning about research findings and new engineering practices is acquired more rapidly than by trying to keep up with the professional literature. The proceedings of many conferences are published but may not appear in university or corporate libraries. International meetings are of particular importance because Asians and Europeans follow practices that in many respects are superior to those used in the US.

Lastly engineers should subscribe to trade journals and magazines in environmental pollution control, some of which can be received by engineers working in the field at no cost. These publications contain primarily advertisements about new equipment, products, and services, and articles about general topics in the field. Subscribing to these publications is an excellent way for educators to keep abreast of new equipment and products and convey these to their students. For engineers in industry the publications are essential.

Internet

In the pursuit of their work, readers will need information that may not be in their personal library. It is to their advantage to perfect the art of browsing ***electronic bulletins*** and the ***Internet*** (Tencer, 1994) for this information. Since new sites are created on a frequent basis and other sites disappear almost as fast, a printed list of web addresses will have only transient value. Therefore, the authors have chosen instead to provide links on the present book's web site, so that they can be updated as necessary. At the time of this writing the web address, technically called a ***universal resource locator*** (URL), of most of the corporate, government, and professional organizations are readily constructed by www. followed by either an acronym or abbreviation, followed by the appropriate suffix (.com, .gov, etc.). These suffixes are formally called ***top level extensions*** or ***top level domains*** (TLDs); some examples are provided below:

- **Corporate** (commercial) - Use TLD .com or the more recent .biz; e.g. the URL for Chemical Information Services, Inc. is www.chemicalinfo.com.
- **U.S. Government** - Use TLD .gov; e.g. the National Safety Council's URL is www.nsc.gov.
- **Professional organizations** - Use TLD .org; e.g. the URL for the American Congress of Government and Industrial Hygienists is www.acgih.org.

Finding a web site is not always this straightforward; if such attempts do not work, users can easily locate a site by using an Internet search engine.

The following is a partial list of government agencies, professional organizations, and companies which the authors have included on the present book's web site; these can serve as jumping-off points for topics discussed in the text.

- ***American Congress of Government and Industrial Hygienists*** (ACGIH) - Member-based organization of professionals that advances worker health and safety through education and the development and dissemination of scientific and technical knowledge
- ***Chemical Abstract Service*** (CAS) - The world's largest and most comprehensive databases of chemical information
- ***Chemical Information Services, Inc.*** - Company which delivers information to the chemical and pharmaceutical industries
- ***Code of Federal Regulations*** (CFR) - National Archives and Records Administration
    - CFR Title 40 - Protection of the environment
    - CFR Title 29, Parts 1900 to 1910 - Labor
- ***Environmental Protection Agency*** (EPA) - Government agency to protect human health and to safeguard the natural environment.
    - Technology Transfer Network (TTN) - EPA's technical web site for air pollution information and bulletin board service (TTNBBS)
    - Clean Air Technology Center (CATC) - EPA's resource for emerging and existing air pollution prevention and control technologies
    - ClearingHouse for Inventories and Emission Factors (CHIEF) - EPA's source for the latest information on emission inventories and emission factors
- ***National Institute for Occupational Safety and Health*** (NIOSH) - Division of the Centers for Disease Control and Prevention (CDC)
    - Registry of Toxic Effects and Chemical Substances (RTECS) - Database of toxicological information compiled, maintained, and updated by NIOSH
- ***National Institute of Standards and Technology*** (NIST) - Non-regulatory federal agency within the Commerce Department's Technology Administration that promotes economic growth by working with industry to develop and apply technology, measurements, and standards
- ***National Safety Council*** (NSC) - Non-profit public service organization focusing on occupational health and safety, traffic safety, first-aid and home safety, and environmental health
- ***Occupational Safety and Health Administration*** (OSHA) - Division of the U.S. Department of Labor

Readers are urged to browse the present book's web site for other information as well, such as additional links, Mathcad and Excel solutions to chapter example problems, and additional material and class notes for some of the chapters.

1.13 Closure

Before one can design indoor air quality control devices or understand how they work, he or she needs to appreciate the concepts of hazard and risk; this introductory chapter therefore contains both a qualitative and quantitative discussion of risk, including its management and assessment. This is followed by a qualitative introduction to indoor air pollution control strategy. A thorough review is then given of the fundamentals of ideal gas mixtures, contaminant concentrations, humidity, and the solution of first-order ordinary differential equations. Professionals working in the field of indoor air quality need to become familiar with the various government agencies, standards, and regulations relevant to the field; the most important of these have been identified in this chapter. Several photographs and schematic diagrams of air pollution control devices are shown so that readers are introduced to the unique terminology associated with indoor air pollution control. This is followed by a list of deficiencies in our understanding of how to design and control indoor air pollution. Finally, readers who desire to delve more deeply into the subject are provided with a list of recommended books and other literature.

Chapter 1 - Problems

1. Use the Internet to answer the following questions. For each case, write the URL(s) from which the answers were obtained. Note: Answers are given for some cases, but since many of the parameters are subject to change, the answers may not be current.

(a) What are the motor vehicle death rates per 10^6 vehicle-miles in recent years?
(b) What is the title of Part 85, 40 CFR? [Control of pollution from motor vehicles and motor vehicle engines]
(c) What is the specific web address of the AP-42 emission factors for the organic chemical process industry contained in the CHIEF bulletin board?
(d) What is the title of Part 1910.146 App B, CFR 29? [Procedures for atmospheric testing in confined spaces]
(e) Describe the circumstances under which emergency temporary standards are set.
(f) Locate the NIOSH "Pocket Guide to Chemical Hazards". What does it contain and can it be downloaded?
(g) What is the address and telephone number of ACGIH?
(h) What is the ***permissible exposure limit*** (PEL) for ***hydrazine*** in PPM and mg/m^3? [1 PPM, 1.3 mg/m^3]
(i) Find the following information about styrene:

- formula [C_8H_8]
- molecular weight [104.15]
- normal boiling temperature (°C) [145.2 °C]
- Is it flammable, and if so what is the lower explosion limit? [yes it is flammable]
- specific gravity of the liquid [0.906]
- CAS number [100-42-5]
- vapor pressure
- solubility in water [0.032g/100 mL]
- Henry's law constant
- PEL [100 PPM, 420 mg/m^3]
- critical pressure and temperature [580 psia, 640.8 K]
- health effects [active oxidant that attacks mucus membrane, eyes, throat, nose]

(j) Find the specific heat at constant pressure (c_p) for methanol (CH_3OH) at 800 K. [19.07 cal/mol K]

2. An article in a professional journal lists the following chemicals and their CAS numbers: pyridine CAS 110-86-1; methyl chloride CAS 74-87-3; methylene chloride CAS 75-09-2; and acrolein CAS 107-02-8. For each of these compounds find the following:

- RTECS number
- permissible exposure limit (PEL)
- chemical formula
- vapor pressure at 25 °C
- phase at 25 °C, 1 atm
- lower explosion limit (LEL)
- health effects
- recommended personal protection equipment and sanitation

3. You plan to conduct an experiment that requires use of the following chemicals: dimethylamine, trimethylamine, and pentaborane. Consult one of the chemical registries and find the following for each compound:

- CAS and RTECS numbers
- permissible exposure limit (PEL)
- chemical formula
- vapor pressure at 25 °C
- phase at 25 °C and one atmosphere
- lower explosion limit (LEL)
- health effects
- recommended personal protection equipment and sanitation

4. Visit the Supply Room in your institution and ask the attendant to show you the MSDS for materials your institution uses that you know to be hazardous.

5. Many students have had jobs or summer jobs in which they encountered what they believe were hazardous working conditions. If you have had such an experience, use it for this homework problem. If not, visit a local auto body repair shop, automotive repair garage, construction site, commercial printer, or dry cleaning establishment, or take a public tour of a company, or walk through any place where industrial activity is going on. Identify specific industrial activities you believe are occupational hazards. Choose the most egregious example and discuss the applicability of *each* of the separate control strategies within the three main categories: administrative controls, engineering controls, and personal protective devices.

6. A smoker typically inhales one cigarette during an elapsed time of 20. seconds at a volumetric flow rate of 5.0 L/min. The concentration of smoke particles in the inhaled air is 10^{15} particles/m^3. Consider a nonsmoker in a room with smokers in which the smoke concentration is 0.30 mg/m^3. Assume the nonsmoker inhales air at a volumetric flow rate of 4.0 L/min. How long would the nonsmoker have to remain in the room to receive a dose of smoke equivalent to smoking one cigarette. Assume that smoke particles are spherical with a uniform diameter (D_p) of 0.1 micrometers and density 800. kg/m^3. Note that all the above cigarette data are accurate to only one significant digit.

7. Adding chlorine to drinking water kills microscopic disease-causing organisms but it also reacts with organic matter and produces small amounts of chloroform. The issue is more serious with treatment of surface water supplies because ground water contains negligible amounts of plant or animal matter. The US average chloroform concentration in tap water is 83. micrograms per L. Chlorinating swimming pools also produces chloroform, and while as a rule children don't drink the water, they ingest some of it and breath air above the water surface containing chloroform, receiving on average a total dose of 250 micrograms per hour of exposure. Assess the risk of playing in a pool compared with drinking tap water. Specifically, how many 10-oz glasses of tap water would a child have to drink to receive a chloroform dose equivalent to one hour's play in a pool? (For details, see Aggazzotti G, Fantuzzi G, and Predieri G, "Plasma Chloroform Concentrations in Swimmers Using Indoor Swimming Pools" Archives of Environmental Health, Vol. 45, No 3, May/June 1990, pp 175-179, for a discussion of chloroform exposure and indoor swimming pools.)

8. It was recently announced that the drinking water in your town contained 100 PPB (on a molar basis) of chloroform ($CHCl_3$, M = 119. kg/kmol). (The EPA maximum allowable concentration, by the way, is 100 PPB). At a public meeting several people are alarmed at the (involuntary) risk of tumors associated with drinking tap water. You assert that there is more (voluntary) risk from the saccharin in one 12-oz can of diet cola per day than all the tap water one could drink in a day! The audience is incredulous and hoots you down.

(a) Write an equation for the dose (i.e. mass) of chloroform in a glass of water, specifically $m_{chloroform}$ as a function of the mol fraction ($x_{chloroform}$) of chloroform in the water, the density (ρ) of water, the volume (V) of a glass of water, and the molecular weights of chloroform and water ($M_{chloroform}$ and M_{water} respectively).

(b) Plug in the numbers to calculate how many micrograms of chloroform are in one 12 fluid ounce glass of water. Show all your work, including units and unit conversions. Assume the density of water is 1000 kg/m^3.

(c) Using HERP data (Table 1.6) estimate how many 12-oz glasses of tap water per day one would need to drink in order to produce a cancer risk equivalent to drinking one 12-oz can of diet cola per day. Show all your work.

9. Using HERP data, how many sandwiches containing 32 grams of peanut butter produce risks equivalent to formaldehyde (HCHO) exposure inside a mobile home for 14 hours? Repeat for a conventional home.

10. Several years ago, a professor had a research project in which the chemical boron trifluoride was to be used. In order to protect the health of his graduate students, he needed to find the risks involved with using this chemical. A good and concise source of data about a chemical is its MSDS; for most chemicals MSDSs are now available on the web.

(a) Search the web to locate the ***NIOSH Pocket Guide to Chemical Hazards***. From there, print out the MSDS for boron trifluoride.

(b) What is the chemical formula for this chemical?

(c) What are the CAS number and the RTECS number for this chemical?

(d) What is the molecular weight of this chemical?

(e) What is the OSHA PEL (permissible exposure limit) of this chemical in parts per million?

(f) Perform the necessary conversions (show all your work, including all units) to convert the PEL in parts per million to mass concentration (c) in mg/m^3. Compare your result with that given on the MSDS. Hint: The molecular weight of air is 28.97. Assume the lab is at standard T and P.

(g) The laboratory dimensions are 20. x 12. x 9.5 ft. Suppose a small leak of the chemical occurs. Calculate the maximum mass (in mg) of this chemical which can be leaked without exceeding the PEL. Assume well-mixed conditions in the room and no ventilation.

11. A national firm recycles approximately 300,000 automobile batteries per year. Teams of 6 or 8 workers remove connecting wires, posts, and battery covers. The liquid electrolyte is then poured through grates to a tank below. The nickel-cadmium plates are removed, sorted into bins, and sold as scrap for the manufacture of stainless steel. Workers wear full face respirators, hard hats, and protective clothing. They also need to bend and lift batteries weighing approximately 50 pounds each. Recommend ways to minimize back, hand, and respiratory hazards. Also recommend ways to keep electrolyte vapors escaping from the tank beneath the grate from traveling to other parts of the plant or to the atmosphere. Include in your remarks all relevant administrative and engineering controls.

12. The combustion of natural gas produces fewer pollutants than does the combustion of gasoline. Politicians and journalists jump to the conclusion that natural gas should be used as an automotive fuel, particularly in cities that use automobile fleets (taxies, buses, trucks, etc.) with regular routes that return to a central location each day for maintenance and refueling. No one performs elementary calculations regarding the size and weight of the vessel needed to store high-pressure natural gas. An

1800-pound, 4-door sedan typically has a 20-gallon molded plastic fuel tank weighing approximately 5 lbm located in the rear of the car and a passenger compartment that is 5 m^3. Assume the following:

- higher heating value of gasoline - 48,256 kJ/kg
- higher heating value of natural gas - 55,496 kJ/kg
- density of gasoline - 900 kg/m^3
- critical pressure and temperature of natural gas (methane) 4.64 MPa, 191.1 K

Estimate the volume and weight of the following two natural gas tanks that store the same amount of energy as 20 gallons of gasoline:

(a) A high-pressure steel tank which stores the natural gas at an initial pressure of 200 atm (20,000 kPa). Assume that a high-pressure tanks weighs ten times the mass of gas it contains.
(b) A light-weight fiberglass tank at 5 atm (500 kPa). Assume that a fiberglass tank weighs 0.3 lbm for each cubic foot of gas stored at 5 atm.

Compare these values to the size of the passenger compartment and mass of a 4-door sedan. List procedures you think need to be taken to cope with:

- filling of the tanks with high pressure natural gas,
- natural gas leaks for the vehicles stored in a garage over night,
- gas leaks, fire, and explosion attendant to using natural gas as a fuel in a vehicle traveling in cities and on open roads.

For more details, see "Running on Methane" Mechanical Engineering, Vol. 112, No 5, pp 66-71, May 1990.

13. It is important to test all calculations against one's intuition and determine if the answer makes sense; a good engineer should have good intuition. Air speed is one quantity about which many students have little appreciation, i.e. they have little ability to discriminate 2 m/s from 20 m/s. One way to acquire such an appreciation is to experience air passing over the skin at 2 and 20 m/s.

MPH	ft/sec	m/s
5	7.33	2.23
10	14.67	4.47
15	22.00	6.71
20	29.33	8.94
25	36.67	11.18

To correlate air speed with the sensation produced by moving air, every student is urged to ask a friend to drive an automobile at the above speeds and to place his/her hand (or head) through an open window, and experience the sensation of air on the skin.

14. Locate CFR 29, Parts 1900 to 1910, in the library or on the Internet and answer the following:

(a) What is the title of 1910.94(c)?
(b) What is the title of 1910.107?
(c) Review Sections 1910.94(c) and 1910.107, answer the following question, and cite the specific passage that justifies your answer: "Can exhaust air from a paint spray booth be mixed with make-up (fresh) air for introduction into the spray booth?"
(d) Quote the full reference from which you can find the PEL of acetone.
(e) Quote the full reference for noise exposure.
(f) Quote the full reference that specifies the ventilation air requirements for welding in confined spaces.

15. Two years ago your firm received a citation for a "serious violation" that resulted in a penalty of \$1000 for benzyl chloride vapor concentrations that exceeded the OSHA PEL (1 PPM). A month ago an OSHA Compliance Officer returned for an inspection, made measurements, and found violations. A few days ago you were informed that a penalty of \$10,000 has been assigned for a "repeated violation". You believe the inspection was done improperly, that the penalty is unwarranted and you wish to appeal the case. Outline the procedure to appeal the case as far as the US Court of Appeals in the District of Columbia.

16. Consider for a moment different activities you engage in on a regular basis. Identify 3-5 voluntary and 3-5 involuntary risks associated with these activities. On the basis of exposure over a year, compare the personal hazard these risks present. Be as quantitative as possible.

17. People who smoke outdoors are sometimes criticized for polluting the air. Compare the amount of pollution created by smoking to that generated by automobiles. Consider only CO for the moment. Assume that all autos satisfy the 1990 Clean Air Act amendments and produce no more than 9 grams of CO per mile. The CO generated by cigarettes is given in Appendix A.7. If smokers consume on the average 20 cigarettes (1 pack) per day, and on the average autos accumulate 10,000 miles per year, estimate how many smokers produce the amount of CO generated by *one* automobile in a year.

18. A thermometer behaves as a first-order dynamic system with a time constant of 5.2 seconds. The thermometer is sitting in ice water (0.0 °C). It is suddenly placed into boiling water (100. °C). How long will it take for the thermometer to register (a) 50 °C? (b) 90 °C? (c) 99.9 °C?

19. Consider a room of volume V in which a source (S) of carbon monoxide (CO) is suddenly turned on. The volume flow rate of fresh air entering the room through the ventilation system is Q. (An equal amount of contaminated air is removed.) A first-order ordinary differential equation can be derived for the mass concentration (c_{CO}) of carbon monoxide in the room, if one assumes that the air in the room is well-mixed,

$$V\frac{dc_{CO}}{dt} = S - Qc_{CO}$$

Assume that the air in the room has no CO until the source is turned on.

(a) Re-write the above equation in standard form, i.e. in the form of Eq. (1-59). Solve for the constants A and B in that equation. (Give your answer in terms of parameters V, S, and Q.)
(b) What is the steady-state mass concentration of CO in the room? (Give your answer in terms of parameters V, S, and Q.)
(c) If the room volume were twice as large, but all other parameters remain the same, how does your answer to Part (b) change? Explain.

20. Begin with the ideal gas law for species j, Eq. (1-23).

(a) Re-write the ideal gas law in terms of molar concentration of species j ($c_{molar,j}$), partial pressure of species j (P_j), volume (V), temperature (T), and universal ideal gas constant (R_u). This form of the ideal gas law is often used by air pollution engineers.
(b) Re-write the ideal gas law in terms of mass concentration (c_j) instead. (Assume the molecular weight of the species is M_j.)

21. A combustion chamber contains octane (C_8H_{18}) and air. Prior to combustion, the molar oxygen-fuel ratio is 20:1, which means that there is one mol of octane molecules for every 20 mols of oxygen molecules (O_2). The air contains 3.76 mols of N_2 for every mol of O_2. It contains only trace amounts of argon, carbon dioxide, etc. The mixture is burned to produce *only* the following:

- carbon dioxide (CO_2)
- water vapor (H_2O)
- oxygen (O_2)
- nitrogen (N_2)

Assume complete combustion.

(a) Write the chemical equation for this combustion process in the following form:

$$\boxed{C_8H_{18} + a(O_2 + 3.76N_2) = b(CO_2) + c(H_2O) + d(O_2) + e(N_2)}$$

Find constants a, b, c, d, and e.

(b) Note that the ratio given in the problem statement is an oxygen-fuel ratio, not an air-fuel ratio. Calculate the molar air-fuel ratio in the combustion chamber before combustion occurs.

(c) Calculate the mass air-fuel ratio (AF), assuming $M_{air} = 28.97$ g/mol.

2

The Respiratory System

In this chapter you will learn:

- names and functions of components of the respiratory system and how pollutants affect these functions
- how to model the motion of air in the respiratory system
- how to model gas exchange in the respiratory system
- primary occupational lung diseases
- dose and risk relationships

The ***respiratory system*** (or ***pulmonary system***) of the human body is, in engineering terms, a marvelous design, like a fine-tuned machine. The respiratory system is intimately linked to, and works alongside, the cardiovascular system to supply life-sustaining oxygen to every cell in the body. So entwined are the heart and lungs that physiologists often refer to the respiratory and circulatory systems as one, i.e. the ***cardiopulmonary system***. As will be discussed in this chapter, there are components of the respiratory system which condition the incoming air (humidity and temperature), conduct that air through an intricate network of branching ducts, transfer oxygen *to* the bloodstream, extract waste gases *from* the bloodstream, and send waste gases back out of the body, recovering heat and humidity along the way! There is even a self-cleaning mechanism, designed to trap and expel inhaled particles. Unfortunately, this fine-tuned machine can also efficiently transfer *undesirable* gases into the bloodstream. In fact, the large surface area of the lung, its airways, and the thin membrane separating air space and capillaries make the lung the primary organ, not only for oxygen exchange, but also for absorption of toxins. In addition, numerous diseases can attack the respiratory system, reducing its efficiency and performance. The purpose of this chapter is to describe the respiratory system and explain how air contaminants (both gaseous and particulate) affect its function and may lead to lung disease.

Table 2.1 lists common industrial toxicants and the lung diseases they produce. Engineers interested in air pollution and industrial hygiene should learn the fundamental concepts of physiology and toxicology as they pertain to the respiratory system. It is recommended that engineers become familiar with the following books: Cralley and Cralley (1979), Guyton (1986), and Klaassen et al. (1986). Several chapters in research monographs (Maroni, et al., 1995; Slutsky et al., 1985; Paiva, 1985; Engle, 1985) and physiology texts used by medical students (West, 1974; Levitzky, 1986; Slonim and Hamilton, 1987; Murray, 1986; and Ganong, 1997) are particularly suited for readers of the present book, since they describe the motion of gas in the lung in terms congenial to engineers. The chapter on Indoor Environmental Health in the ASHRAE Fundamentals Handbook (2001) also presents material in ways engineers can appreciate.

Table 2.1 Selected industrial toxicants producing lung disease through inhalation (abstracted from Klaassen et al., 1986).

toxicant	chemical composition	occupational source	common name of disease	site of action	acute effect	chronic effect
aluminum	Al_2O_3	manufacturing of abrasives, smelting	aluminosis	upper airways, alveolar, interstitium	cough, shortness of breath	interstitial fibrosis
ammonia	NH_3	ammonia production, manufacturing of fertilizers, chemical production, explosives		upper airway	immediate upper and lower respiratory tract irritation, edema	chronic bronchitis
arsenic	As_2O_3, AsH_3 (arsine), $Pb_3(AsO_4)_2$	manufacturing of pesticides, pigments, glass, alloys		upper airways	bronchitis	lung cancer, bronchitis, laryngitis
asbestos	fibrous silicates (Mg, Ca, and others)	mining, construction, shipbuilding, manufacturing of asbestos-containing materials	asbestosis	parenchyma		pulmonary fibrosis, pleural calcification, lung cancer, pleural mesothelioma
beryllium	Be, $Be_2Al_2(SiO_3)_4$	ore extraction, manufacturing of alloys, ceramics	berylliosis	alveoli	severe pulmonary edema, pneumonia	pulmonary fibrosis, progressive dyspnea, interstitial granulomatosis, corpulmonale
chlorine	Cl_2	manufacturing of pulp and paper, plastics, chlorinated chemicals		upper airways	cough, hemoptysis, dyspnea, tracheo-bronchitis, broncho-pneumonia	
chromium (IV)	Na_2CrO_4 and other chromate salts	production of Cr compounds, paint pigments, reduction of chromite ore		nasopharynx, upper airways	nasal irritation, bronchitis	lung tumors and cancers

toxicant	chemical composition	occupational source	common name of disease	site of action	acute effect	chronic effect
coal dust	coal plus SiO_2 and other minerals	coal mining	pneumoconiosis	lung parenchyma, lymph nodes, hilus		pulmonary fibrosis
coke oven emissions	polycyclic hydrocarbons, SO_x, NO_x, and particulate mixtures of heavy metals	coke production		upper airways		tracheobronchial cancers
hydrogen fluoride	HF	manufacturing of chemicals, photographic film, solvents, plastics		upper airways	respiratory irritation, hemorrhagic pulmonary edema	
iron oxides	Fe_2O_3	welding, foundry work, steel manufacturing, hematite mining, jewelry making	siderotic lung disease: silver finisher's lung, hematite miner's lung, arc welder's lung	silver finishers: pulmonary vessels and alveolar walls; hematite miners: upper lobes, bronchi and alveoli; arc welders: bronchi	cough	silver finishers: subpleural and perivascular aggregations of macrophages; hematite miners: diffuse fibrosis-like pneumoconiosis; arc welders: bronchitis
kaolin	$Al_2O_3 \cdot 2SiO_2 \cdot 2H_2O$ plus crystalline SiO_2	pottery making	kaolinosis	lung parenchyma, lymph nodes, hilus		pulmonary fibrosis
nickel	NiCO (nickel carbonyl), Ni, Ni_2S_3 (nickel subsulfide), NiO	nickel ore extraction, nickel smelting, electronic electroplating, fossil fuel		parenchyma (NiCO), nasal mucosa (Ni_2S_3), bronchi (NiO)	pulmonary edema delayed by two days (NiCO)	squamous cell carcinoma of nasal cavity and lung
oxides of nitrogen	NO, NO_2, HNO_3	welding, silo filling, explosives manufacturing		terminal respiratory bronchi and alveoli	pulmonary congestion and edema	emphysema

Table 2.1 Continued

toxicant	chemical composition	occupational source	common name of disease	site of action	acute effect	chronic effect
ozone	O_3	welding, bleaching flour, deodorizing		terminal respiratory bronchi and alveoli	pulmonary edema	emphysema
perchoro-ethylene	C_2Cl_4	dry cleaning, metal degreasing, grain fumigating			pulmonary edema	
phosgene	$COCl_3$	production of plastics, pesticides, chemicals		alveoli	edema, pulmonary edema	bronchitis
silica	SiO_2	mining, stone cutting, construction, farming, quarrying	silicosis, pneumoconiosis	lung parenchyma, lymph nodes, hilus		pulmonary fibrosis
sulfur dioxide	SO_2	manufacturing of chemicals, refrigeration, bleaching, fumigation		upper airways	broncho-constriction, cough, tightness in chest	
talc	$Mg_3Si_4O_{10}(OH)_2$	rubber industry, cosmetics	talcosis	lung parenchyma, lymph nodes		pulmonary fibrosis
tin	SnO_2	mining, processing of tin	stanosis	bronchioles and pleura		widespread mottling of x-ray without clinical signs
toluene 2.4-diisocyanate (TDI)	$CH_3C_6H_3(NCO)_2$	manufacturing of plastics (polyurethane)		upper airways	acute bronchitis, bronchospasm, pulmonary edema	
xylene	$C_6H_4(CH_3)_2$	manufacturing of resins, paints, varnishes, other chemicals, general solvent for adhesives		lower airways	pulmonary edema	

2.1 Physiology

Normal quiet breathing is accomplished with the diaphragm. The lungs expand and contract by downward and upward movement of the diaphragm and by expansion and contraction of the rib cage. Expansion (or contraction) of the lungs causes the gas pressure within the lungs to become negative (or positive) with respect to atmospheric pressure. The pressure difference for normal breathing is approximately plus and minus 1 mm of mercury. Figure 2.1 shows the three regions of the respiratory system:

- nasopharyngeal
- tracheobronchial
- pulmonary

and the elements within each region. Each of these three regions is discussed below in some detail.

2.1.1 Nasopharyngeal Region

The ***nasopharyngeal region*** lies between the nostrils and larynx; it is also called the ***upper airways*** or the ***extrathoracic airways***. There are large convoluted surfaces in the nose called ***nasal turbinates*** containing blood vessels. The nasal turbinates are covered by a mucous membrane that transfers energy and water to fresh air during inhalation and removes them from exhaled air. Nasal turbinates are the primary sites for absorption of water soluble contaminants in air. Under quiet breathing most of the air passes through these nasal airways. Under vigorous exercise the mouth becomes the primary airway for inhalation and exhalation. The ***switching point*** is on average a

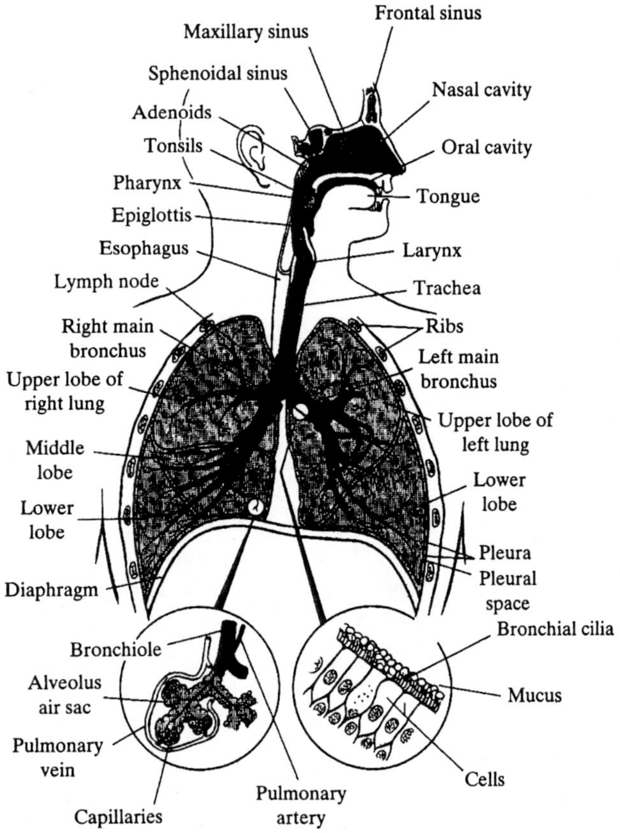

Figure 2.1 Components of the respiratory system (courtesy of the American Lung Association).

respiratory rate (also called minute volume) of 34.5 L/min. The upper airway volume is approximately 50 mL. The flow path between the lips and glottis is approximately 17 cm long and that from the nose to glottis is 22 cm long. The nasal passages are lined with cellular tissue and mucous glands (***vascular mucous epithelium***). Downstream of the ***pharynx*** is the ***larynx*** which constricts the flow of air producing a ***vena contracta*** and vibrating the vocal cords, enabling humans to produce sounds. The maximum cross-sectional area of the adult trachea is approximately 2.5 cm^2. The cross-sectional area and average gas velocity within the larynx increase with volumetric flow rate. Typical values are shown in Table 2.2. Instabilities and vortices are produced in the air stream passing through the apertures of the larynx. Deposition of particles in the larynx is a common source of malignant tumors (Martonen et al., 1993). During inspiration, air passing through the folds and vocal cords forms a high-velocity air stream (***laryngeal jet***) that affects the velocity of air and particle size distribution entering the tracheobronchial tree.

2.1.2 Tracheobronchial Region

The ***tracheobronchial region*** (TBL), as shown in Figure 2.2, consists of the ***trachea, bronchi***, and ***terminal bronchiole***. Typical values of the diameter and length of the adult trachea are 1.8 cm and 12.0 cm respectively. The typical diameter and length of the right and left primary bronchi are 1.2 cm and 4.8 cm respectively.

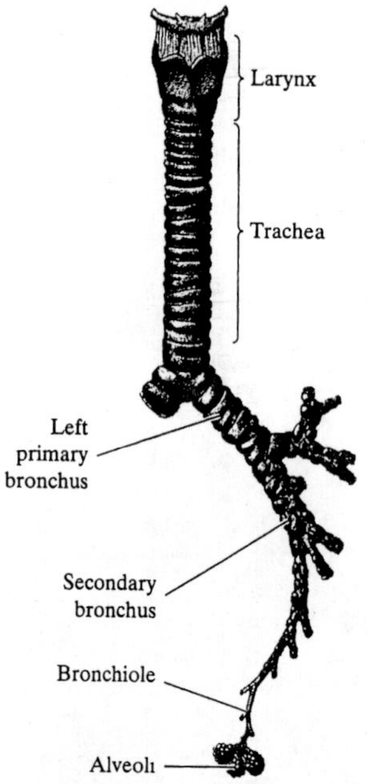

Figure 2.2 Schematic diagram of branching airways. The trachea divides into two primary bronchi, one to each lung. Each bronchus divides at least 20 times, and finally terminates in a cluster of alveoli (from Heinsohn & Kabel, 1999).

Table 2.2 Cross-sectional area and average gas velocity within the larynx as a function of volumetric flow rate.

Q_a (L/min)	A (cm^2)	U_a(m/s)
15.	0.88	1.7
30.	1.4	2.1
60.	2.4	2.5

The angle separating the right and left bronchi is about 70 degrees. The tracheobronchial region is also called the ***conducting airway***. The passageways from the nose to the terminal bronchiole are lined with ciliated epithelium and coated with a thin layer of ***mucus*** from mucus-secreting cells. This surface is called the ***mucosal layer*** or ***mucous membrane***. Mucus production is about 10 mL per 24 hours. In humans, the thickness of the mucus layer is from 5 μm to 10 μm. ***Cilia*** are hair-like organs protruding 3 μm to 4 μm into the mucus layer above the surface of the cell, as seen in Figures 2.3 and 2.4.

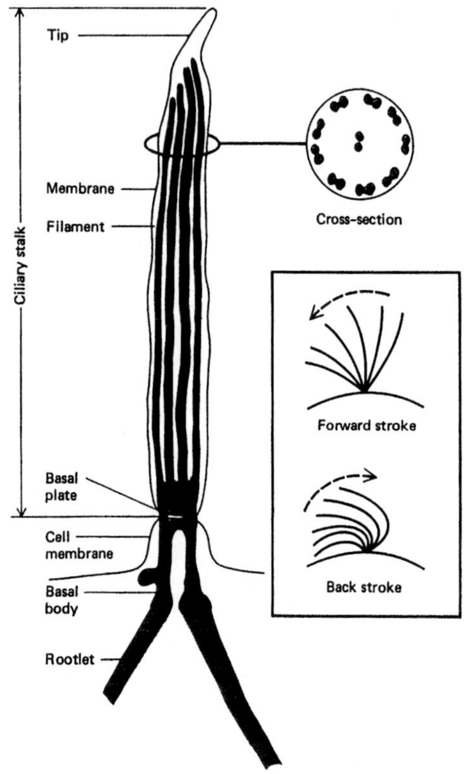

Figure 2.3 Structure and function of the cilium (adapted from Guyton, 1986).

Figure 2.4 Scanning electron micrograph of a portion of a bronchial passage showing cilia interspersed with mucus-secreting cells whose surfaces are covered with microcilli (from Heinsohn & Kabel, 1999).

As many as 200 cilia per cell protrude from special epithelial cells lining air passages. ***Mucociliary escalation*** is the whip-like movement of cilia occurring in the nasal cavity and respiratory tract. Cilia move forward (away from the lung) with a sudden rapid stroke, bending sharply where they are attached to the cell surface, displacing the mucus. They then move backward (toward the lung) slowly, without appreciably displacing any mucus. This whip-like beating movement propels mucus toward the pharynx at a net rate of approximately one cm/min. Cilia beat at a frequency of 10 to 20 Hz. Once in the pharynx, the mucus passes through the digestive system. The mechanism of cilia and mucus that transfers material from the lung to the digestive tract is called ***lung clearance***.

The trachea is called the ***first generation respiratory passage***. The trachea divides into the right and left bronchi at a point called the ***carnia***. The right and left main bronchi are called ***second generation respiratory passages***. Each division of bronchi thereafter is an additional generation. There are between 20 and 25 generations before air finally reaches alveoli. The final few generations are less than 1 mm to 1.5 mm in diameter and are called ***bronchiole***. All the intermediate passageways between the trachea and bronchiole are called ***bronchi***. To keep the trachea from collapsing, multiple rings of cartilage surround 83% of the circumference. In the bronchi, smaller amounts of cartilage provide partial rigidity. The bronchiole have no rigidity and expand and contract along with alveoli. The walls of the trachea and bronchi are composed of cartilage and smooth muscle. Bronchiole walls are entirely smooth muscle. The upper and conducting airways are relatively rigid and have a volume of around 150 mL. The upper conducting airways are called ***anatomic dead space*** because very little gas exchange occurs in these airways.

Under ***quiet breathing*** (10 to 15 breaths/minute) the pressure drop between alveoli and trachea is approximately 1 mm of mercury. The bronchi and trachea are very sensitive to touch, light, particles, and certain gases or vapors. The larynx and carnia are particularly sensitive. Nerve cells in these passageways initiate a series of involuntary actions that produce coughing. ***Coughing*** begins with inspiring about 2.5 liters of air followed by the spontaneous closing of the ***epiglottis*** and tightening of the vocal cords that trap air in the lungs. The abdominal muscles then contract, forcing the diaphragm to contract the lung, collapse the bronchi, constrict the trachea and increase the air pressure in the lungs to values as high as 100 mm Hg. The epiglottis and vocal cords suddenly open and air under pressure is expired. The rapidly moving air carries foreign matter upward from the bronchi and trachea. ***Sneezing*** is another involuntary reflex similar to coughing, except that the ***uvula*** is depressed and large amounts of air pass rapidly through the nose and mouth.

2.1.3 Pulmonary Region

The ***pulmonary region*** is also called the ***respiratory airspace***. The pulmonary region consists of tiny sacs called ***alveoli*** (Figure 2.5) clustered in groups and interconnected by openings called ***alveolar ducts*** (ADs). There are 300 to 500 million alveoli in the adult human. Alveoli are thin-walled polyhedral pouches whose characteristic width is 250 μm to 350 μm and have at least one side open to either a ***respiratory bronchiole*** (RBL) or an ***alveolar duct***. Each terminal bronchiole supplies air for a segment of the lung called an ***acinus***. The function of the alveoli is to provide a surface for the exchange of oxygen, carbon dioxide, and volatile metabolites between air and blood in the capillaries. The mean total alveoli surface area of a healthy man is about 143 m^2 (Murray, 1986). The alveoli are served by a labyrinth of venous and arterial capillaries approximately 8 μm in diameter. Approximately 85 to 95 percent of the alveolar surface is served by capillaries; hence, the surface area of the alveolar capillary interface is approximately 122 to 136 m^2. For comparison, this is more than 70 times the surface area of a man's skin, and 66% of the area of a singles tennis court (195.7 m^2). The mean thickness of the tissue between alveoli (***alveoli membrane***) is about 9 μm. The air-side of alveoli consists of squamous ***epithelial cells*** and rounded ***septal cells***. Mobile phagocytic cells and macrophage lie on the inner surface of the alveolus. ***Macrophage*** are white blood cells 7 μm to 10 μm in diameter that metabolize inhaled particulate material, spores, bacteria, etc. The process is called ***phagocytosis***. The metabolizing cells are also called ***phagocytes***. Infectious microorganisms, except those causing chronic bacterial and fungal infections such as tuberculosis and some viral diseases, are usually consumed by macrophage.

The branching airway is an ordered structure in which each airway divides into two airways of similar, if not identical, geometry. The ***Weibel symmetric model*** (Weibel, 1963), shown in Figure 2.6, is widely used to describe the structure of the conducting airways in terms of ***airway generation*** (Z). The tracheobronchial region consists of generations 0-16. Generations 0-3 contain cartilage and are called ***bronchi*** (BR). Generations 4-16 contain no cartilage and are called ***bronchiole*** (BL). Generation 16 is called the ***terminal bronchiole*** (TBL). The ***pulmonary region*** consists of generations 17-23 which are subdivided into 3 generations of ***partially alveolated respiratory bronchiole***, three generations of ***fully alveolated alveolar ducts***, and the dead-ended or terminal ***alveolar sacs*** (AS).

Figure 2.7 shows airway cross-sectional area (A), surface area (S), and local Reynolds number in the airways as functions of both distance (y) from the nasal tip and Weibel airway generation (Z). Once inside the lung, the airflow decelerates and the flow path decreases rapidly. While the Reynolds numbers are low, one should not assume that the flow is fully developed since bronchial bifurcations produce regions of recirculation and unsteadiness. Fully developed flow in a duct does not occur until duct length (L) exceeds the duct's ***entrance length*** (L_e), i.e. not until $L/L_e >$ 1, which occurs only in airway generations beyond Z = 6 or 7, in the bronchioles. The fluid mechanics of flow through the bronchioles will be discussed in more detail later.

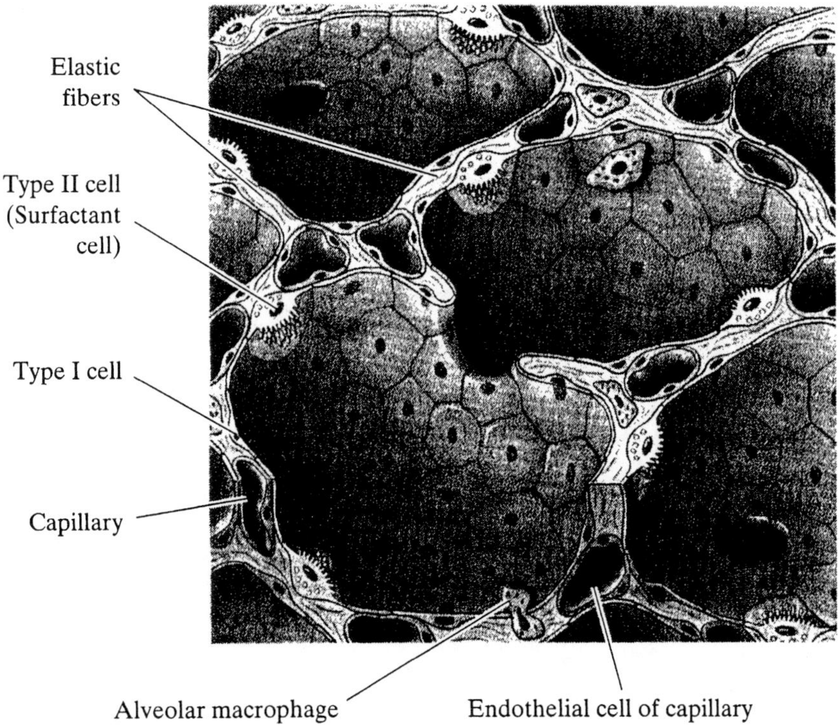

Figure 2.5 Alveoli in the pulmonary system; the alveoli are composed of type I cells for gas exchange and type II cells that synthesize surfactant; macrophage lying on the alveolar membrane ingest foreign material that reach the alveoli (from Heinsohn & Kabel, 1999).

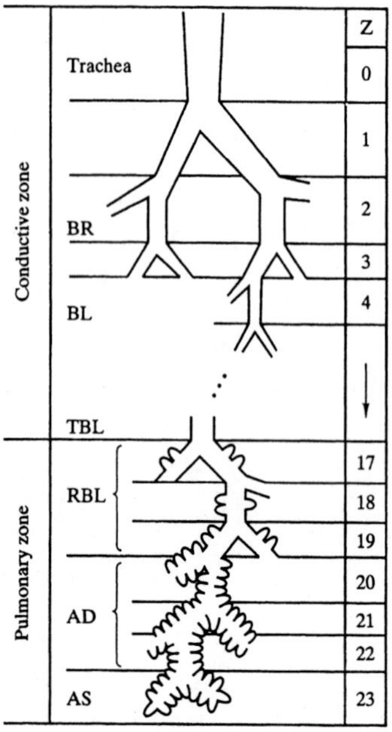

Figure 2.6 Systematic Weibel model; airway generation (Z), bronchi (BR), bronchioles (BL), terminal bronchiole (TBL), partially alveoliated respiratory bronchioles (RBL), fully alveoliated ducts (AD), and terminal alveolar sacs (AS) (redrawn from Ultman, 1985).

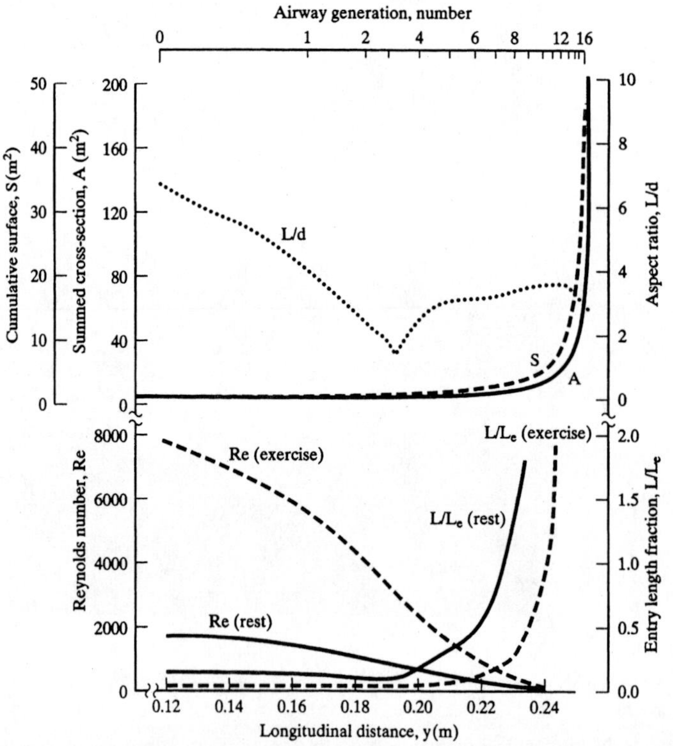

Figure 2.7 Geometric and aerodynamic characteristics of a symmetric Weibel model for two ventilation flow rates; rest (0.4 L/s) and exercise (1.6 L/s); L_e is the length needed for the flow to become fully developed (redrawn from Ultman, 1988).

The volume of the lungs changes and the honeycomb-alveoli structure expands and contracts as air is inhaled and exhaled. During rest, alveolar surface area expands by about 20% from beginning to end of a typical inhalation. This area expands by nearly 40% during moderate exercise (Ultman, 1989). Some debate exists about the existence and role of a thin (0.5 to 1 μm) liquid layer on the air side of alveoli. This aqueous layer is called a ***surfactant*** and is a complex mixture of proteins, predominately dipalmitoyl lecithin. The surfactant forms a liquid monolayer on the alveolus-air interface capable of lowering the normal surface tension (72 N/m) to nearly zero (Longo et al., 1993). To accomplish this reduction, the surfactant must adhere to the interface and be capable of maintaining a coherent, tightly packed monolayer that will not collapse even under high compression accompanying expiration. The collapse of the monolayer results in its expulsion from alveoli and an increase in surface tension. Deficiency and inactivation of the surfactant are contributing factors to a number of pulmonary diseases.

Arteries carry blood containing nutrients to tissue, and ***veins*** carry blood containing waste products away from tissue. Each nutrient artery entering an organ divides six to eight times after which the remaining vessels are called ***arterioles***. Arterioles are approximately 40 μm in diameter. Arterioles subdivide two to five times and become ***capillaries*** with diameters of 8 to 9 μm. The body contains nearly 10 billion capillaries with a total surface area of 500 to 700 m^2. Nearly all cells are within 20 to 30 μm from a capillary. Blood flows through capillaries in an intermittent fashion. On the return leg, capillaries recombine to form ***venules*** which in turn recombine to form veins. ***Perfusion*** is the phrase used to describe the flow of blood through the blood vessels in the lung. There is a large pressure drop associated with the flow of blood in arteries and arterioles. Thus the walls of these vessels are considerably stronger and thicker than those of veins and venules. Capillary walls are composed of unicellular endothelial cells surrounded by basement membrane. The thickness of the wall is about 0.5 μm.

Materials are transferred through the capillary membrane by a variety of processes broadly called ***diffusion***. Cells in the capillary wall are composed of materials in which different substances (nutrients or contaminants) are soluble. If the substance, such as oxygen or carbon dioxide, is soluble in ***lipids*** (materials within cells that are soluble in fat solvents but not water), it will diffuse through the cells in the capillary membrane containing lipids. Water-soluble but lipid-insoluble substances such as sodium and chloride ions, glucose, etc. pass through the membrane through slit-pores (6 to 7 nm wide). The phrase ***permeability*** is used to describe the diffusing capacity of different substances to pass through the capillary membrane. The concentration difference is the driving potential.

Approximately one sixth of body tissue is the space between cells called ***interstitium***. Fluid in this space, and on the outside of capillaries, is called ***interstitial fluid***. Figure 2.8 shows the elements of the interstitium. Interstitial fluid is a plasma derived from capillaries. Solid material is composed of two major elements, collagen fiber bundles and proteoglycan filaments. ***Collagen*** is a protein and major constituent in the intercellular connective tissue of meats that is not readily digested by most enzymes. Collagen fiber bundles have large tensile strength and provide tensile strength to the tissue. ***Proteoglycan filaments*** are small coiled protein molecules that form a "mat." The combination of proteoglycan filaments and interstitial fluid is called ***tissue gel***. After passing through the capillary wall, materials diffuse through the tissue gel, molecule by molecule, to cells receiving the material. Although virtually all fluid is bound in some way, small rivulets of "free fluid" are present. ***Edema*** is the name for a physical disorder that occurs when the free fluid accumulates in pockets and the rivulets expand enormously.

Not all material can be transferred to the blood by diffusion, e.g. insoluble material, indigestible bacteria, and dust particles. The ***lymphatic system*** is an additional route by which these materials are transferred from the interstitial space to the blood. ***Lymphatic fluid***, also called simply ***lymph***, is derived from interstitial fluid but is richer in proteins. Lymph is a yellowish, coagulable, plasma-like liquid containing white blood cells. Small vessels called ***lymphatic capillaries*** carry proteins and large particles of foreign matter from interstitial spaces. Lymphatic capillaries branch and merge as do veins and arteries. They merge at ***lymph nodes*** where additional chemical processes occur. Approximately 100 mL of lymph flow through the thoracic duct of a resting individual. An estimated additional 20 mL of lymph flow through other channels. Lymph is believed to be pumped by movement of muscles and by the numerous flaps that pass material to larger collecting lymphatic vessels. About one tenth of the fluids from the arteries passes through the lymphatic system instead of returning through venous capillaries. Many large inhaled particles find their way into the lymphatic system where they may pose a hazard to health. In particular, substances with large molecular weights that cannot diffuse may enter lymphatic capillaries. Figure 2.9 illustrates how the unique structure of lymphatic capillaries enables them to accept very large molecules and foreign particles. Anchoring filaments of the lymphatic capillary are attached to the connective tissue between cells (interstitial space). Adjacent ends of endothelial cells (of the lymphatic capillary) overlap and produce a "flap." Large particles in the interstitial fluid are able to push open these flaps and flow directly into the lymphatic capillary.

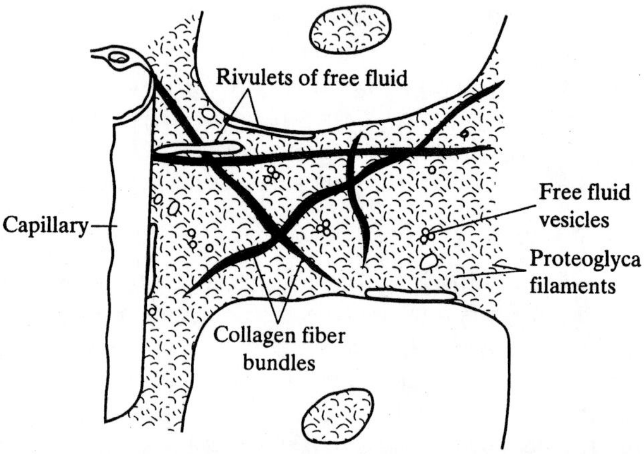

Figure 2.8 Structure of the interstitium, tissue between cells (adapted from Guyton, 1986).

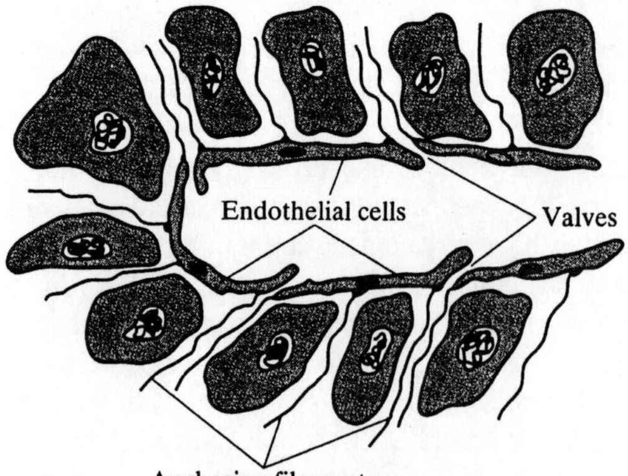

Figure 2.9 Lymphatic capillaries showing structures that enable material of large molecular weight to enter circulation (adapted from Guyton, 1986).

2.2 Respiratory Fluid Mechanics

During normal rest conditions for an average healthy person, 2 to 3 percent of the total energy expended by the body is required for pulmonary ventilation, the inflow and outflow of air between the atmosphere and the lungs. During heavy exercise, breathing demands a somewhat greater portion of the body's energy expenditure; still, only 3 to 5 percent of the total body energy expended is required for pulmonary ventilation during heavy exercise (Guyton, 1986). According to Ganong (1997), the majority (approx. 65%) of this energy is used to expand the lung against elastic resistance forces, while the remainder is used to overcome airway resistance (approx. 28%) and tissue resistance (approx. 7%) The flexible properties of the lung are due to elastic fibers within the lung tissue and the surface tension of the surfactant liquid lining the alveoli surface. The change in volume of the lung with respect to pressure (dV/dP) is called ***compliance***. Compliance is comparable to the elastic coefficient or Young's modulus in solid mechanics or to the isothermal expansion coefficient in thermodynamics. Compliance can be measured for the lungs alone, or for the lungs and thoracic cage together. Figure 2.10 shows some measurements (Guyton, 1986) for the lungs alone; lung volume is plotted as a function of pressure in the space between the lung and rib cage (a space called the ***pleura***). Two things are evident:

- the slope is not constant
- the lung displays ***hysteresis***; expiration and inspiration do not produce coincident curves.

Both effects are believed to be due to the viscoelastic properties of human tissue, the honeycomb structure of alveoli, and the surface tension of the surfactant. A linear relationship connecting the end points in Figure 2.10 yields the compliance, i.e. a slope of approximately 0.22 L/cm water pressure drop for the lungs alone. In other words, when the pleural pressure increases by 1 cm of water, the lungs expand 220 milliliters. The compliance of the lungs and thoracic cage together is much smaller, i.e. 0.13 L/cm of water (Guyton, 1986), the difference being due to the increased resistance of the thoracic cage.

2.2.1 Spirometry

Air is drawn into and out of the lungs by positive or negative pressure in the lung cavity. ***Spirometry*** is the analysis of volumes and volumetric flow rates of air during respiration. Figure 2.11 shows an elementary way to measure the volume of respired air; all values are corrected to STP. ***Total lung capacity*** (V) is the maximum absolute volume of air which the lungs can hold with the greatest possible inspiratory effort; V is about 5800 mL for an average healthy man (about 70 kg). Note that all pulmonary volumes and volume flow rates tend to be greater for athletic persons. In addition, they are

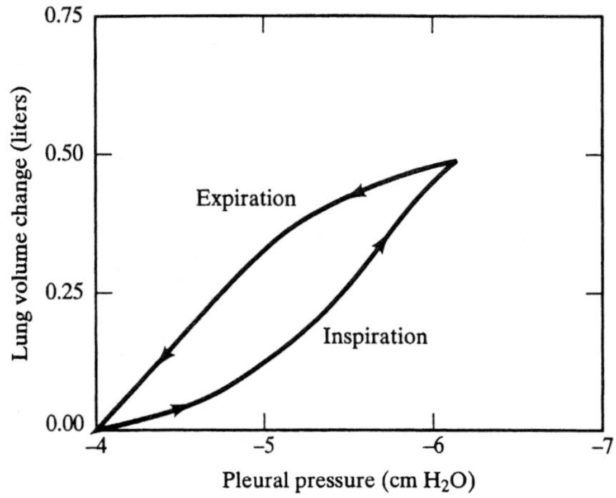

Figure 2.10 Compliance of the lung during inspiration and expiration (adapted from Guyton, 1986).

about 20 to 25 percent smaller in women than in men (Guyton, 1986). There is also great variation in all of the volumes discussed here amongst different individuals; typical values are given, but the reader must understand that a given individual may have much different values. ***Vital capacity*** (V_{vc}), also called ***forced vital capacity*** (FVC), is the maximum volume of air one can expel from the lungs after first filling the lungs to their maximum extent and then expiring to the maximum extent. A typical value of vital capacity is 4600 mL. ***Tidal volume*** (V_t) is the volume of air inhaled per breath during normal breathing; the same volume is subsequently exhaled. At rest, V_t is around 500 mL. ***Functional residual volume***, also called ***functional residual capacity***, (V_{fr}) is the volume of air left in the lungs after normal exhalation, as seen in Figure 2.11. A typical value of V_{fr} is 2300 mL. ***Residual volume*** (V_r) is the volume of air remaining in the lungs following the maximum expiration one can produce, typically around 1200 mL. ***Anatomic dead space*** (V_d) is the volume of the conducting airways from the mouth to the respiratory bronchiole within which there is no gas exchange, in other words the volume of the respiratory system exclusive of the alveoli themselves. Since the bronchiole within the lung expand and contract somewhat during breathing, the anatomic dead space increases with lung expansion, varying from about 140 mL in a collapsed lung to 260 mL after maximum inspiration. Thus, approximately 50% of V_d is located in the upper airways, which do not expand or contract during breathing. For healthy individuals at rest, V_d is approximately 150 mL, and is assumed to be constant in the analyses which follow. ***Total physiologic dead space*** (also called ***wasted ventilation***) is defined as the volume of air that does not exchange gases with blood in the lungs. In healthy individuals, total physiologic dead space is nearly identical to anatomic dead space. However, in diseased individuals, gas exchange between air and blood in the alveoli may be inhibited; thus total physiologic dead space may be much larger than anatomic dead space (Ganong, 1997).

The ***breathing rate*** (f) is defined as the number of breaths per minute. There is some discrepancy among physiologists regarding the typical value of f at rest, but it is usually reported to be around 10 to 15 breaths per minute. The product of the breathing rate (f) and tidal volume (V_t) is a volumetric flow rate called the ***minute respiratory rate*** (Q_t), or simply the ***ventilation rate***:

$$\boxed{Q_t = fV_t} \tag{2-1}$$

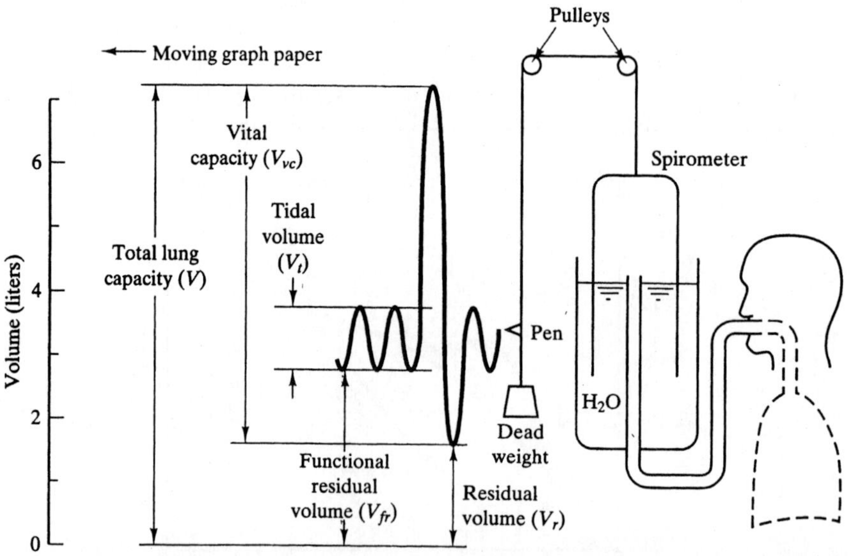

Figure 2.11 Lung volumes and elements of spirometry. A pen records changes in the air volume on graph paper that moves to the left. The residual volume and functional residual volume cannot be measured with the spirometer (from Heinsohn & Kabel, 1999).

The units of Q_t are liters per minute (L/min). It should be pointed out that in many physiology textbooks, volume and volume flow rate are used somewhat interchangeably. Indeed, most physiologists define Q_t as the ***minute volume***, which is actually the minute respiratory rate times one minute, the volume of air breathed in over a period of one minute. In their terminology, Q_t has units of liters, not liters per minute. The ***normal tidal volume*** (V_t) is 500 mL, where the notation "normal" refers to the average man at rest. Using f = 12 breaths/minute, the ***normal minute respiratory rate*** is approximately 6.0 liters per minute,

$$Q_t = V_t f = 500\frac{\text{mL}}{\text{breath}}\left(12.\frac{\text{breath}}{\text{min}}\right) = 6000\frac{\text{ml}}{\text{min}} = 6.0\frac{\text{L}}{\text{min}}$$

Another source of confusion in the literature is a factor of two that comes about when authors define a volume flow rate corresponding to inhalation (or expiration) *only*. Specifically, ***inhalation volume flow rate***, $\dot{V}_{\text{inhalation}}$, is defined as the average volume flow rate during inhalation, and is useful when dealing with velocities, Reynolds numbers, etc. in the lung. Since the entire tidal volume is inhaled during approximately half of the breathing cycle, $\dot{V}_{\text{inhalation}}$ is twice as large as Q_t for the same individual. For example, using the values above, 12 breaths per minute corresponds to 5 seconds per breath cycle, or about half of this (2.5 seconds) for inhalation. At V_t = 500 mL, the inhalation volume flow rate is

$$\dot{V}_{\text{inhalation}} = 500\frac{\text{mL}}{\text{breath}}\left(\frac{1\text{ breath}}{2.5\text{ s}}\right)\left(\frac{60\text{ s}}{\text{min}}\right)\left(\frac{\text{L}}{1000\text{ mL}}\right) = 12.\frac{\text{L}}{\text{min}}$$

which is a factor of two greater than Q_t. ***Exhalation volume flow rate*** ($\dot{V}_{\text{exhalation}}$), is defined in a similar fashion, and also ends up being about a factor of two greater than Q_t.

A person can live for a short period of time at two to four breaths per minute (1 to 2 L/min). The rate at which an individual inhales and exhales depends on the level of physical activity. Under unusual conditions the tidal volume can be as large as the vital capacity, or the breathing rate can rise to as high as 40 to 50 breaths per minute. At these rapid breathing rates, a person usually cannot sustain a tidal volume greater than one half the vital capacity. The ***alveolar ventilation rate*** (Q_a) is the volumetric flow rate of fresh air that actually *reaches* the alveoli, and is thus available for gas exchange. It is equal to the breathing rate times the difference between the tidal volume and the anatomic dead space,

$$\boxed{Q_a = (V_t - V_d)f} \tag{2-2}$$

Using the typical values given above,

$$Q_a = (500-150)\frac{\text{mL}}{\text{breath}}\left(12\frac{\text{breath}}{\text{min}}\right) = 4200\frac{\text{mL}}{\text{min}} = 4.2\frac{\text{L}}{\text{min}}$$

Lung volumes and volumetric flow rates are illustrated schematically in Figure 2.12, in which a higher breathing rate (15 breaths per minute) is assumed. Because of the anatomic dead space, as pointed out by Ganong (1997), rapid shallow breathing results in much less alveolar ventilation than does slow deep breathing, even at the same minute volume. Table 2.3 illustrates this point for a person with V_d = 150 mL.

A great deal of information about lung dysfunction can be gained by measuring the rates of inhalation and exhalation. Of particular importance is the volume of gas exhaled in the first second, which is called ***forced expiratory volume at one second***, and is abbreviated FEV_1 (Menkes et al., 1981). This parameter is particularly reproducible and is a sensitive indication of obstructions in lung airways. Like any other fluid system, volumetric flow rate depends on the pressure difference (δP)

between lung and atmosphere, and frictional forces within the air passageways. The frictional forces can be lumped together and called ***airway resistance*** (AR); the volumetric flow rate (Q) can then be expressed as

$$\boxed{Q = \frac{\delta P}{AR}} \tag{2-3}$$

Airway resistance may vary throughout the respiratory cycle, and the value for inspiration may be different than that for expiration (hysteresis), as in Figure 2.10.

If the instantaneous volume of air is measured versus time by equipment such as that shown in Figure 2.11, graphs similar to Figure 2.13 can be obtained. The normal lung has a vital capacity (V_{vc}) of approximately 4600 mL and a residual volume of 1200 mL. The maximum volumetric flow rate is about 400 L/min and is achieved rather quickly after one begins to exhale. A test to classify respiratory abnormalities is to expel rapidly as much air as one can from maximally filled lungs. The volume so measured is called the forced vital capacity (FVC), as discussed above. Figure 2.13 shows time traces of lung volume for normal lungs, obstructed lungs, and constricted lungs. Individuals with obstructed lungs are not able to inspire quite as much as those with unobstructed lungs, but more importantly it takes a longer time to expel.

Table 2.3 Comparison of tidal volume, ventilation rate, and alveolar ventilation rate for two breathing frequencies (adapted from Ganong, 1997).

respiratory rate (f)	**tidal volume (V_t)**	**ventilation rate (Q_t)**	**alveolar ventilation rate (Q_a)**
30 breaths/minute	200 mL (per breath)	$\left(30\frac{\text{breath}}{\text{min}}\right)\left(200\frac{\text{mL}}{\text{breath}}\right)$ $= 6000\ \frac{\text{mL}}{\text{min}}$	$(200-150)\frac{\text{mL}}{\text{breath}}\left(30\frac{\text{breath}}{\text{min}}\right)$ $= 1500\frac{\text{mL}}{\text{min}}$
10 breaths/minute	600 mL (per breath)	$\left(10\frac{\text{breath}}{\text{min}}\right)\left(600\frac{\text{mL}}{\text{breath}}\right)$ $= 6000\ \frac{\text{mL}}{\text{min}}$	$(600-150)\frac{\text{mL}}{\text{breath}}\left(10\frac{\text{breath}}{\text{min}}\right)$ $= 4500\frac{\text{mL}}{\text{min}}$

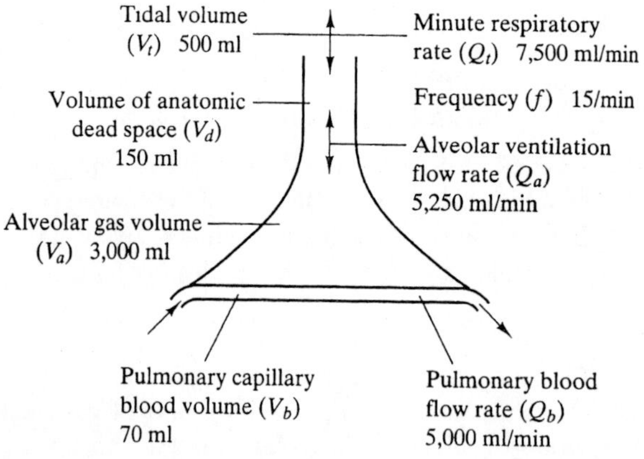

Figure 2.12 Typical lung volumes and volumetric flow rates. There may be considerable individual differences in these values (from Heinsohn & Kabel, 1999).

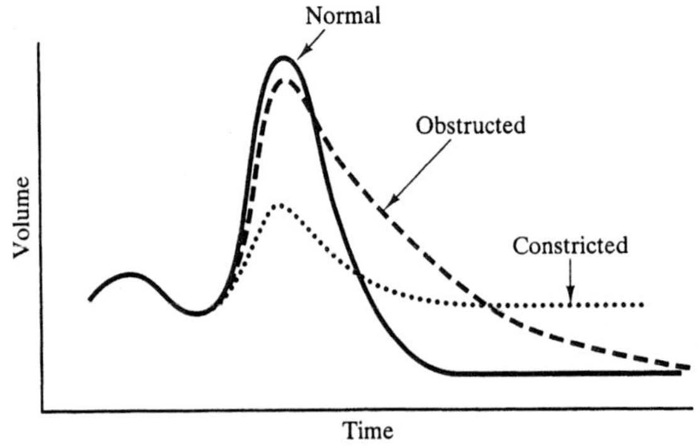

Figure 2.13 Spirometric measurements of normal, obstructed, and constricted lung.

- *Constricted lungs* (sometimes called *restricted lungs*) cannot be fully expanded because they contain lesions or because abnormalities in the chest cage prevent individuals from inflating their lungs fully. A constricted lung is similar to a normal lung except that vital capacity is smaller. Constricted lungs occur in people suffering from fibrosis of the lungs (black lung disease, silicosis, asbestosis), thoracic deformities, large tumors, or other problems that restrict the amount of air that can be put into the lungs.
- Individuals with *obstructed lungs* have a larger airway resistance and have to exert a larger pressure to expel air because the airways are narrower and cannot be emptied as easily as can normal lungs. The residual volume of a restricted lung is about the same as that of a normal lung. Obstructed lungs occur in people suffering from chronic asthma, asthmatic bronchitis, emphysema, mucosal edema, and inflammation of the conducting air passages.

A common way to quantify respiratory disorders is to examine three parameters: the vital capacity (V_{vc}), the forced expiratory volume at one second (FEV_1), and the residual volume (V_r). Some comparisons of these three parameters are given below:

- Normal lung: $$0.7 < \left(\frac{FEV_1}{V_{vc}}\right)_{normal} < 0.75$$
- Constricted lung: $$\left(FEV_1\right)_{constricted} < \left(FEV_1\right)_{normal}, \quad \left(\frac{FEV_1}{V_{vc}}\right)_{constricted} > 0.75,$$ $$\left(V_{vc}\right)_{constricted} < \left(V_{vc}\right)_{normal}, \left(V_r\right)_{constricted} > \left(V_r\right)_{normal}$$
- Obstructed lung: $$\left(FEV_1\right)_{obstructed} < \left(FEV_1\right)_{normal}, \quad \left(\frac{FEV_1}{V_{vc}}\right)_{obstructed} < 0.70,$$ $$\left(V_{vc}\right)_{obstructed} \approx \left(V_{vc}\right)_{normal}, \left(V_r\right)_{obstructed} \approx \left(V_r\right)_{normal}$$

An additional technique to quantify respiratory disorders involves the concept of *moments* (Menkes et al., 1981; Permutt and Menkes, 1979) since it provides insight into the pathophysiology of disease and allows one to have quantitative means to discriminate between curves such as those in Figure 2.13. The *arithmetic mean transient time* (t_m) is defined as the arithmetic average time it takes to expire an amount of air equal to the vital capacity. An equation for conservation of mass of air during the process of expiration is

$$\boxed{Q = -\frac{dV}{dt}} \tag{2-4}$$

where Q is the expiration volumetric flow rate and V is the volume. Eq. (2-4) also assumes that the air density (ρ) is constant. The negative sign is needed because the lung volume decreases during expiration. Both sides of Eq. (2-4) can be multiplied by time (t), and then integrated over the time, t_{vc}, it takes to expire a volume of air equal to the vital capacity (V_{vc}):

$$\int_0^{t_{vc}} tQdt = -\int_{V_{vc}}^{0} tdV$$

The arithmetic mean transient time (t_m) is then defined and rearranged into the following form:

$$t_m \equiv \frac{1}{V_{vc}} \int_0^{V_{vc}} tdV = \frac{1}{V_{vc}} \int_0^{t_{vc}} Qtdt \tag{2-5}$$

The ***n^th^ moment*** (α_n) is defined as

$$\alpha_n = \frac{1}{V_{vc}} \int_0^{t_{vc}} Qt^n dt \tag{2-6}$$

The ***first moment*** (α_1) (n = 1) is the arithmetic mean transient time (t_m) described in Eq. (2-5). The ***second moment*** (α_2) (n = 2) is

$$\alpha_2 = \frac{1}{V_{vc}} \int_0^{t_{vc}} Qt^2 dt \tag{2-7}$$

The square root of the second moment is the ***root mean square*** (RMS) of the transient time. Because of the nature of the tails in Figure 2.13, it can be seen that higher moments reflect the importance of the tails and smaller expired volumes of air. The arithmetic mean transient time (t_m) provides an index of the average time it takes for air to leave a normal lung. It can be shown that for a normal lung,

$$t_m = \alpha_1 = \sqrt{\frac{\alpha_2}{2}} \tag{2-8}$$

For obstructed and constricted lungs, the relationship between α_1 and the square root of α_2 changes. When there is an obstruction in the upper airway, Figure 2.13 shows that the lung volume does not decrease rapidly and $\sqrt{\alpha_2}$ does not fall relative to α_1. In cases of airway obstruction because of emphysema and age, $\sqrt{\alpha_2}$ rises relative to α_1. Permutt and Menkes (1979) report that analysis of 57 nonsmoking males shows that α_1, α_2, and age (in years) are related by

$$\sqrt{\alpha_2} = 0.229 + 1.165\alpha_1 + 0.00347 \cdot (\text{age}) \tag{2-9}$$

with a standard deviation of 0.078. The companion equation for a group of smokers is

$$\sqrt{\alpha_2} = 0.371 + 0.676\alpha_1 + 0.00981 \cdot (\text{age}) \tag{2-10}$$

with a standard deviation of 0.206.

2.2.2 Bohr Model

The Bohr model (Ultman, 1985) shown in Figure 2.14, is a simple but useful analytical model that has been used for many years to illustrate the distribution of inspired air between the rigid conducting airways and the expandable alveolar region. The model assumes that the anatomic dead space (V_d) is constant; it can be thought of as the volume enclosed by a rigid pipe. The model also defines an expandable volume called the ***alveolar region***, $V_a(t)$, which is a function of time. Gas inside the alveolar region is assumed to be well mixed while gas flowing through the dead space is assumed not to mix. During inspiration, a portion of the volume of inspired air, i.e. the tidal volume (V_t), occupies the dead space (V_d) and the remainder (V_t - V_d) mixes with gases in the alveolar region. The

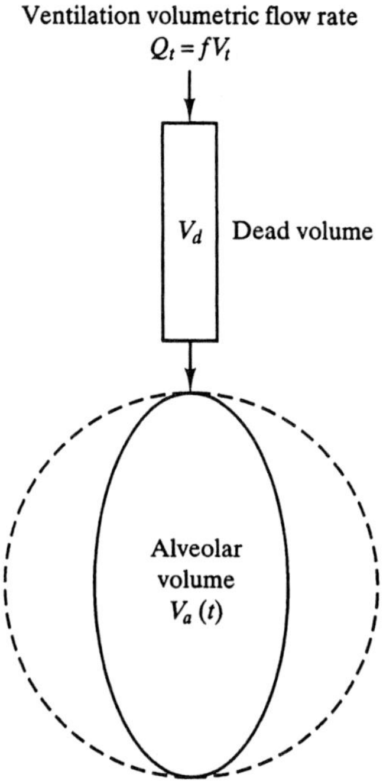

Figure 2.14 The Bohr model, illustrating the rigid anatomic dead volume and the expandable alveolar volume.

maximum alveolar region $(V_a)_{max}$ is thus the volume of air entering the alveolar region during inspiration,

$$\boxed{(V_a)_{max} = (V_t - V_d) = V_t\left(1 - \frac{V_d}{V_t}\right)} \tag{2-11}$$

The alveolar ventilation rate (Q_a), defined by Eq. (2-2), is thus the product of breathing rate (f) and the maximum alveolar region $(V_a)_{max}$ from Eq. (2-11),

$$\boxed{Q_a = (V_a)_{max} f = V_t f\left(1 - \frac{V_d}{V_t}\right)} \tag{2-12}$$

Or, using Eq. (2-1),

$$\boxed{Q_a = Q_t\left(1 - \frac{V_d}{V_t}\right)} \tag{2-13}$$

The Bohr model is useful in understanding the role of pulmonary and cardiac function under varying degrees of exercise. Table 2.4 shows values of ventilation rate (Q_t) and blood flow rate (Q_b) during exercise. Alveolar ventilation can be computed from Eq. (2-12). The ratio of alveolar ventilation rate (Q_a) to blood volumetric flow rate (Q_b) is called the ***ventilation perfusion ratio*** (R_{vp}),

$$\boxed{R_{vp} = \frac{Q_a}{Q_b}} \tag{2-14}$$

It should be noted that the data in Table 2.4 are from Ultman (1998, 1989), based on exercise experiments conducted on nine healthy young men by Gale et al. (1985). The ventilation rate at rest is somewhat higher than that defined previously as "average" for a healthy man, and illustrates the wide

Table 2.4 Ventilation, blood flow, and the ventilation perfusion ratio (R_{vp}) during various activity levels (abstracted from Ultman, 1988 and 1989).

parameter	exercise or activity level			
	rest	light	moderate	heavy
ventilation rate, Q_t (L/min)	11.6	32.2	50.0	80.4
frequency, min^{-1}	13.6	23.3	27.7	41.1
tidal volume, V_t (L)	0.85	1.38	1.81	1.96
V_d/V_t	0.34	0.20	0.16	0.16
blood flow, Q_b (L/min)	6.5	13.8	18.4	21.7
$Q_a = Q_t(1 - V_d/V_t)$ (L/min)	7.66	25.8	42.0	67.5
$R_{vp} = Q_a/Q_b$	1.18	1.87	2.28	3.11

variation in respiratory parameters among individuals. Nevertheless, Ultman's values are useful for comparing rest conditions with those at various levels of exercise. By measuring the ventilation perfusion ratio and monitoring the concentration of specific gases in the inspired air and in the blood, one can obtain quantitative estimates of the effectiveness of the lung in transferring gas to the blood. This transfer is commonly called ***gas uptake***. In the analysis above it has been assumed that the blood absorbs all the material diffusing through the alveolar barrier. If only a fraction of the material is really absorbed, the above can be modified to include the solubility constant of the material in blood (see the extended Bohr model in Section 2.3) The elementary Bohr model shows that exercise increases the ventilation volumetric flow rate by a factor of more than six, while the blood flow rate increases only by a factor of slightly over three. The fraction of inspired air actually reaching the alveolar region increases only slightly (from 0.66 to 0.84) and the ventilation-perfusion ratio (R_{vp}) increases by a factor of 2.6. Thus the uptake of gases by the blood is limited more by the supply of blood than by the amount of air reaching the alveolar region.

The greater the ratio of dead space to tidal volume (V_d/V_t), the smaller the fraction of inspired air reaching the alveoli. Thus, one would expect that effective alveolar ventilation occurs only if the tidal volume (V_t) exceeds the anatomical dead volume (V_d). However, if the respiratory frequency is sufficiently high, there is evidence (Slutsky et al., 1985) that effective gas exchange occurs even when the tidal volume is smaller than the dead space. The phenomenon is called ***high frequency ventilation***; an example of which is panting in dogs.

For years physiologists have contemplated the mechanisms by which gas exchange in the lung is maintained, given that the tidal volume (500 mL) is only a fraction of the vital capacity (4,600 mL). The transport of gases between the mouth (or nose) and the alveoli can be divided into five modes (Haselton and Scherer, 1980; Ultman, 1985; Slutsky et al., 1985; Paiva, 1985; Engle, 1985), each mode pertaining to a different Weibel airway generation:

- direct alveolar ventilation by bulk convection
- convection by high-frequency "pendelluft"
- convective dispersion due to asymmetric inspiratory and expiratory velocity profiles
- Taylor-type dispersion
- molecular diffusion

2.2.3 Bulk Convection

Bulk convection is the conventional flow of air in a passageway. The volumetric flow rate is the cross-sectional area times the average velocity. The flow of air through the trachea is bulk convection. The development of boundary layers and the resulting modification of the velocity profile outside the boundary layer can be analyzed by conventional boundary layer theory. One must be cautious, however; the flow is pulsatile, the airways contain constrictions (such as the vocal cords) and bifurcations, and there is a moving layer of mucus on the walls.

2.2.4 Pendelluft

There is considerable *asymmetry* in the human bronchial tree. Because of the location of the heart and other organs, the right and left lungs are not identical (Figure 2.1). The right lung is cleft by horizontal and oblique fissures into three lobes; the left lung is cleft into two lobes. When a tidal volume of air is inhaled, fresh air may reach upper alveoli but not lower alveoli. The path of the inspired air and the number of alveoli immediately reached is a function of the level of physical activity, number of breaths per minute, and size of the tidal volume. There is asynchronous filling and emptying of the lungs that may lead to an exchange of air between parallel lung units (Figure 2.15).

Inhalation and exhalation can be modeled crudely as two resistance-capacitance circuits in parallel (Otis, 1956) driven by a common square-wave power supply (Figure 2.16). The square-wave power supply corresponds to the positive and negative pressure difference produced by the diaphragm that fills and empties the lungs. The ***pressure drop*** (δP) to force air through a series of branching bronchiole can be thought to be a product of a resistance (R) and volumetric flow rate (Q). Since the lung volume (V) changes, the pressure drop (δP) needed to fill the lung can be thought to be proportional to lung volume (V) and inversely proportional to lung capacitance (C). Thus for a single lung,

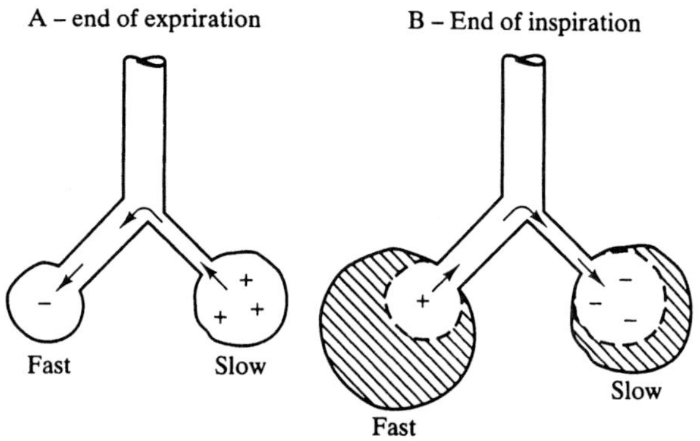

Figure 2.15 Illustration showing that when the breathing frequency is large, lung resistance dominates the rate of filling and emptying for parallel lung units having different time constants (τ = RC). Case A: expired air from the slow unit is transferred to the fast unit. Case B: at the end of inspiration, the fast unit transfers air to the slow unit. The plus (+) and the minus (-) signs indicate the pressure relative to atmospheric pressure (from Heinsohn & Kabel, 1999).

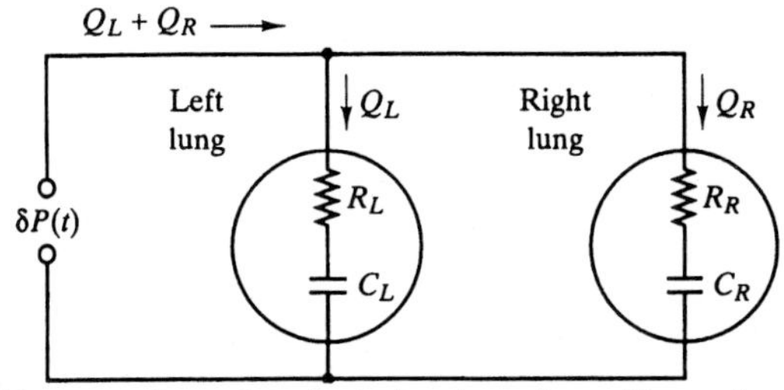

Figure 2.16 Schematic diagram of an electrical circuit analogy illustrating asynchronous filling and emptying of lungs.

$$\boxed{\delta P = RQ + \frac{V}{C}} \tag{2-15}$$

It is recognized that the lumped-parameter circuit in Figure 2.16 is simplistic since lung resistance (R) and capacitance (C) are not truly constant. Nevertheless, assuming R and C constant affords an opportunity to analyze gross behavioral characteristics of pendelluft. Applying the law of conservation of mass for air entering an expandable lung unit, one finds that if the air density is constant, the rate of change of lung volume is equal to the volumetric flow rate of air into the lung,

$$\boxed{\frac{dV}{dt} = Q} \tag{2-16}$$

Combing Eqs. (2-15) and (2-16), one obtains

$$\delta P = R\frac{dV}{dt} + \frac{V}{C}$$

which can be rearranged into standard form for a first-order ordinary differential equation with constant coefficients, as discussed in Chapter 1,

$$\boxed{\frac{dV}{dt} = \frac{\delta P}{R} - \frac{V}{RC}} \tag{2-17}$$

the solution of which is

$$\boxed{V(t) = V_f - [V_f - V(0)]\exp\left(-\frac{t}{RC}\right)} \tag{2-18}$$

where V(0) is the initial lung volume at t = 0, and V_f is the final volume of the inflated lung ($V_f = C\delta P$). The product (RC) is a first-order time constant (τ) equal to the time required for the lung to acquire approximately 63.2% of its final volume increase,

$$\frac{V(\tau) - V(0)}{V_f - V(0)} = 1 - e^{-1} \cong 0.632$$

The volumetric flow rate of air at any instant entering the lung can be obtained by combining Eqs. (2-15) and (2-18),

$$\boxed{Q = \frac{\delta P}{R} - \frac{V}{RC} = \frac{\delta P}{R} - \frac{V_f}{RC} + \left(\frac{V_f - V(0)}{RC}\right)\exp\left(-\frac{t}{\tau}\right)} \tag{2-19}$$

Note that the right and left lung units have different final volumes (V_f) and time constants (τ). If they are linked in parallel as shown in Figure 2.16 and subjected to a common pressure difference (δP), one can compute the volume and volumetric flow rates for each lung. It is clear that the response characteristics of each lung are different. At the end of a rapid expiration of air, the unit with the smaller time constant (the fast unit) on the left (e.g. Figure 2.15) is ready to fill while the slower unit on the right is still emptying. Thus there is a flow of gas from the slower to faster unit where the breathing frequency is large. At the end of a rapid inspiration, air flows from the fast unit to the slow unit which is still filling. This "sloshing" between lung units is known as ***pendelluft*** (Chang, 1984).

2.2.5 Convective Dispersion due to Asymmetric Velocity Profiles

Experiments have shown that the velocity profiles in expiratory flow are flatter than those in inspiratory flow. Figure 2.17 provides a simple explanation (Haselton and Scherer, 1980). Upon inhalation, flow passes into the bronchial tree system as plug flow moving toward the lung (to the right in Figure 2.17a). After a certain downstream distance called the ***entrance length*** (L_e), the flow moves forward as ***fully developed*** or ***fully established*** pipe flow. Figure 2.7 shows that beyond the 7th or 8th

airway generation, $L/L_e > 1$, and the flow is thus fully established. The velocity profile becomes parabolic if the flow is ***laminar***, which occurs at low ***Reynolds number*** (Re), defined by

$$\boxed{Re = \frac{\rho U_{avg} D}{\mu} = \frac{U_{avg} D}{\nu} = \frac{2U_{avg} R}{\nu}} \tag{2-20}$$

where ρ is the air density, U_{avg} is the average speed over a cross-sectional area of the duct ($U_{avg} = Q/A$), $D = 2R$ is the duct diameter, μ is the (dynamic) coefficient of viscosity, and ν is the kinematic coefficient of viscosity ($\nu = \mu/\rho$). For laminar flow in a round duct, Re must be less than about 2000 to ensure laminar pipe flow. Upon exhalation, flow emerges from the bronchial tree system as individual velocity profiles from each of the bronchi. Upon merging, the many velocity profiles flatten into one such that the net profile is nearly ***uniform plug flow***, but with some waviness, moving away from the lung (to the left in Figure 2.17b). Consider a parcel of air at three radial locations in an airway:

- along the centerline, $r = 0$, $U > U_{avg}$
- at the value r_{avg}, where $U(r_{avg}) = U_{avg}$
- in the annular region near the airway wall, $r_{avg} < r < R$, $U < U_{avg}$

Imagine that one could trace the forward and backward displacement of the air parcel during several cycles of inhalation and exhalation. Neglecting any transverse motion of the air parcel, the

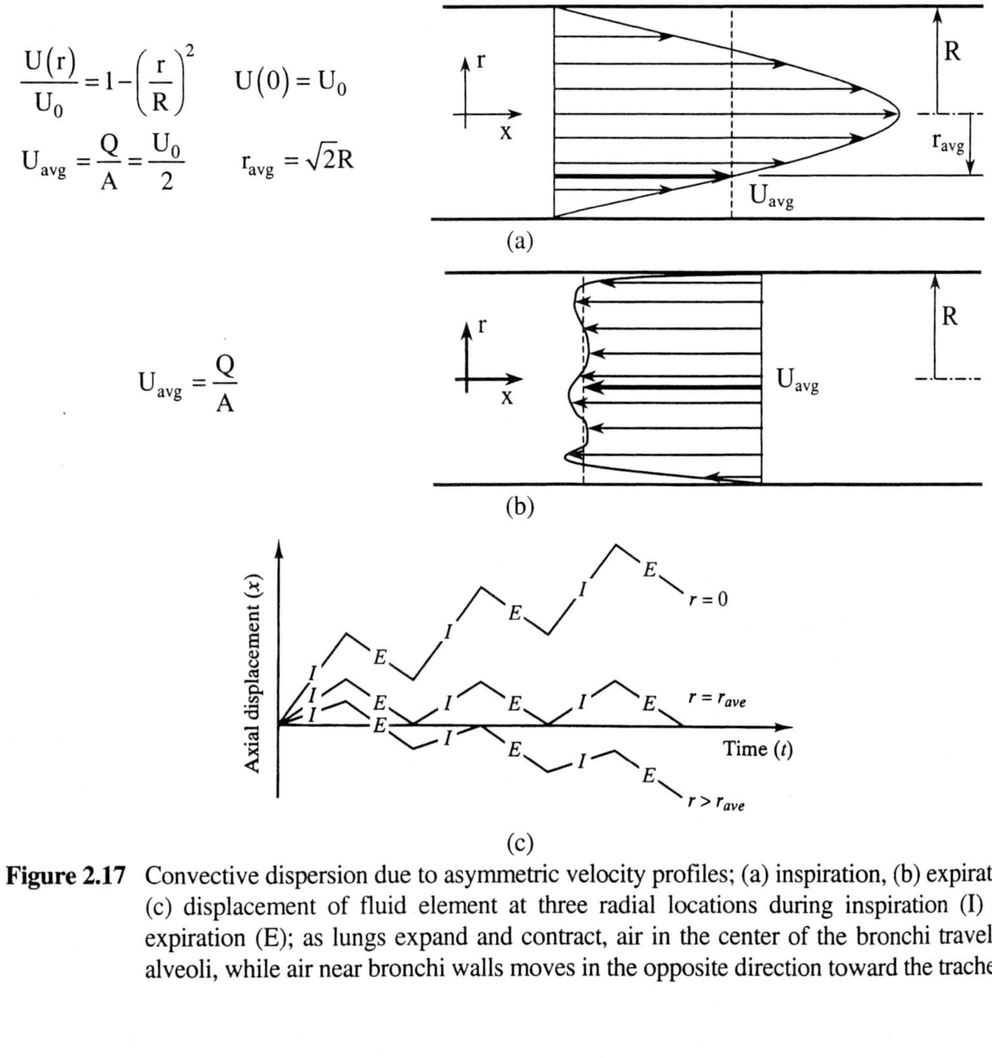

Figure 2.17 Convective dispersion due to asymmetric velocity profiles; (a) inspiration, (b) expiration, (c) displacement of fluid element at three radial locations during inspiration (I) and expiration (E); as lungs expand and contract, air in the center of the bronchi travels to alveoli, while air near bronchi walls moves in the opposite direction toward the trachea.

displacement of the air parcel is equal to the air velocity, U(r), integrated over the elapsed time. Figure 2.17c shows the result of this integration, starting from the beginning of inhalation at t = 0. Initially, air at all three radii move toward the alveoli (upward in the figure) during inhalation. When exhalation starts, the air reverses direction (downward in the figure). As the inhalation-exhalation cycle continues, air in the center of the airway (r = 0) develops net movement toward the alveoli, while air near the airway walls ($r > r_{avg}$) develops net movement *away* from the alveoli, toward the pharynx. At radius $r = r_{avg}$, the air has no net displacement. While the total volumetric flow rate through the airway during inspiration is the same as that during expiration, there is net flow toward the alveoli through the central portion of the airway and net flow away from the alveoli along the outer annular portion of the airway. Recall that cilia and the mucociliary escalation mechanism (Section 2.1.2) cause mucus to move along the bronchi walls toward the epiglottis. Consequently flow in the conducting airways is not only ***pulsatile***, but also ***countercurrent*** during inhalation.

2.2.6 Taylor-Type Dispersion

The transport of gases during breathing is a function not only of the complex oscillating flow described above, but also of the highly complex compliant (expanding/contracting) system of bronchioles. Taylor-type dispersion is the name given to this composite flow. To begin understanding Taylor-type dispersion, consider the original work of Taylor (1953) on fully established laminar flow in a duct of constant radius a. The average velocity is one half the maximum centerline velocity. At time zero imagine that a diffusible material is injected (continuously) into the flow at a velocity equal to the local gas velocity. The material will convect downstream (longitudinally) but will also diffuse radially. Now imagine a reference system moving in the direction of flow at the average velocity U (the subscript avg is dropped for brevity). Taylor showed that the radial transport of mass can be described as ***virtual longitudinal diffusion*** governed by ***Fick's law*** in which the rate of mass transport of material ($\dot{m}$) relative to a moving frame of reference is described by

$$\boxed{\dot{m} = -D' A \frac{dc}{dx'}} \tag{2-21}$$

where

$$\boxed{x' = x - Ut} \tag{2-22}$$

and D' is a ***virtual diffusion coefficient***. For *laminar* flow, the relationship between D' and the molecular diffusion coefficient D is

$$\boxed{D' = \frac{(RU)^2}{D}} \tag{2-23}$$

Taylor (1954) considered such dispersion in *turbulent* flow through a tube. His results show that

$$\boxed{D' = \text{constant R U}} \tag{2-24}$$

In reality, air in bronchioles moves in an oscillatory fashion and one must also cope with inertial and viscous effects. The parameter used to characterize oscillatory flow is the ***Womersley number*** (Wo) (Slutsky et al., 1980), defined as

$$\boxed{\text{Wo} = R\sqrt{\frac{2\pi f}{\nu}}} \tag{2-25}$$

where f is the frequency and ν is the kinematic viscosity. If Wo is less than unity, the flow can be analyzed as quasi-steady viscous flow; if Wo is considerably larger than unity, the flow must be analyzed as unsteady viscous flow. For laminar oscillatory flow (Wo >> 1) the dispersion coefficient, called K by meteorologists for dispersion in the atmosphere, is

$$K \propto \frac{Re^2}{Wo^n} \tag{2-26}$$

where n is a constant (approximately 3) and the Reynolds number is defined here as

$$Re = \frac{2U_{rms}R}{\nu} \tag{2-27}$$

where U_{rms} is the root mean square axial velocity and R is the radius of the bronchiole. Dispersion in the bronchiole is very complex. The flow is often unsteady and it is clear that mass transport based solely on the mechanism of Taylor-type dispersion is inadequate.

2.2.7 Molecular Diffusion

Molecular diffusion is the process by which gas at a region of high concentration moves towards a region of lower concentration due to random molecular motion. This is the mechanism by which oxygen and carbon dioxide are transferred through the alveolar membrane to and from capillaries (Figure 2.18). The overall efficiency of gas exchange is a function of the five modes of gas transport. They are not mutually exclusive and certainly interact; however, for a given set of physical conditions, one mode may be dominant in a certain airway generation. Figure 2.19 shows the dominant modes of transport in the lung. In the trachea and main-stem bronchi, turbulent Taylor-type dispersion should be important. If the tidal volume is large, convective flow can clear this portion of the dead space and ventilate some alveoli directly. In the medium-sized airways, large phase lags and oscillatory convective flow can occur. Mixing in the conducting zone of the lung occurs either by convective dispersion or out-of phase bulk flow. In the small peripheral airways in the respiratory zone, out-of-phase oscillatory motion may be responsible for ventilation of some lung units. Finally, in the alveoli and near the gas exchange surface, molecular diffusion is the dominant mode of gas exchange.

2.3 Analytical Models of Heat and Mass Transfer

Lung ***permeability*** is the rate at which materials penetrate the epithelial lining of the bronchial tree and lung respiratory surfaces. Air contaminants absorbed by the blood migrate to different body organs and affect them differently. Sulfur dioxide is soluble in water and is primarily removed by mucus in the upper respiratory system. Ozone has low solubility in water and penetrates the tracheobronchial region where much of it is absorbed and reacts chemically with mucus; however

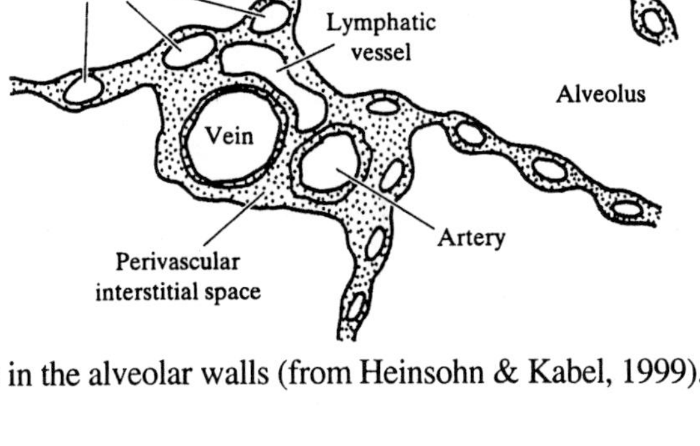

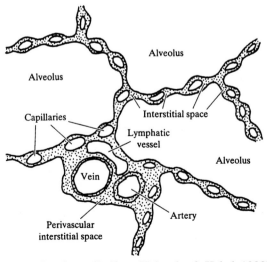

Figure 2.18 Capillaries in the alveolar walls (from Heinsohn & Kabel, 1999).

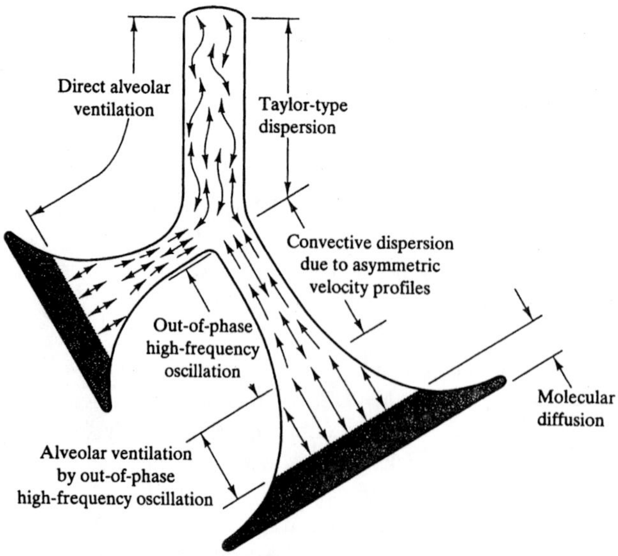

Figure 2.19 Modes of gas transport in the lung (from Heinsohn & Kabel, 1999).

a portion penetrates to attack underlying tissue. Carbon monoxide and volatile hydrocarbons have low solubility in water, and do not react with mucus; thus they penetrate the pulmonary region where they diffuse through the alveolar membrane and are absorbed into the blood. These examples illustrate widely different effects the modeler must address in designing an analytical model:

- Is the contaminant soluble in water?
- Does the contaminant react with mucus or tissue?
- Can the contaminant be absorbed by blood?
- Do the effects of the contaminant vary with airway generation?
- Can well-mixed conditions be assumed?

The functional residual capacity (V_{fr}) of the lungs is approximately 2,300 mL but only 350 mL of inspired air is added to the alveolar volume during inhalation (at normal rest conditions), and 350 mL of alveolar air is expired during exhalation. Table 2.5 shows the typical composition of inspired air, gas in the alveoli, and expired gas. Trace amounts of carbon tetrachloride and benzene produced in metabolism are also transferred to the expired gas. The large volume of alveolar air compared to the volume of inspired gas ensures slowly changing conditions within the alveoli and prevents sudden changes in the gas exchange rates to the blood. An elementary model of alveolar ventilation is useful. Such a model, shown in Figure 2.20, assumes that the alveolar volume is a well-mixed region. At the quasi-steady state, averaged over many inhalation and exhalation cycles, inspired air enters at 4,200 mL/min and mixes with 2,300 mL of alveolar air, resulting in a 4,200 mL/min departing steam of alveolar air. Oxygen is removed from the volume at a rate dictated by the body's metabolic rate (250 mL/min at rest conditions); carbon dioxide enters the volume at a rate also dictated by the metabolic rate (200 mL/min at rest conditions). Note that these oxygen and carbon dioxide volume flow rates do not agree with the difference in oxygen and carbon dioxide concentrations entering and leaving the volume. The disparity arises because materials also enter and leave the body as solids and liquids. Furthermore, steady-state concentrations of oxygen and carbon dioxide vary with ventilation rate as shown in Figures 2.21 and 2.22. These figures show that the normal alveolar partial pressures of O_2 and CO_2 occur at an alveolar ventilation rate of 4,200 mL/min. If individuals exercise, with metabolic processes requiring 1,000 mL of oxygen and producing 800 mL of carbon dioxide per minute, lung ventilation rates given by the dashed lines in Figures 2.21 and 2.22 are needed.

Table 2.5 Mol fraction of respiratory gases that enter and leave the lungs at sea level; inspired air is at 25 °C, 101 kPa, and 50% relative humidity (abstracted from Guyton,1986).

species	inspired air	alveolar air	expired air
N_2	0.741	0.749	0.745
O_2	0.197	0.136	0.157
CO_2	0.0004	0.053	0.036
H_2O	0.062	0.062	0.062

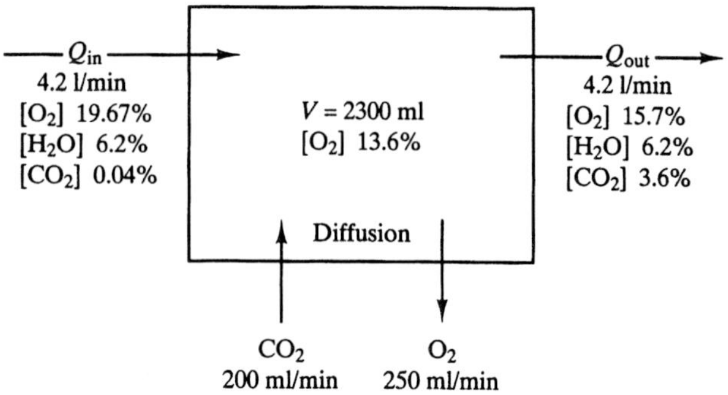

Figure 2.20 Control volume for alveolar gas exchange (adapted from Guyton, 1986).

Figure 2.23 shows components of the alveolar membrane through which gases are exchanged. The labyrinth of capillaries is so dense that the alveolar surface can be thought to be a sheet of blood contained within a thin membrane. The volume of blood in the lung is between 60 and 140 mL and is contained within capillaries about 8 μm in diameter. The movement of blood through the myriad of blood vessels in the alveolar membrane is called ***perfusion***. The rate of gas exchange is expressed in terms of a "diffusing capacity" times the difference in partial pressure of the diffusing species in the blood and gas within the alveoli. Under strenuous exercise (elevated metabolic rate), the pulmonary capillaries dilate, the blood flow rate increases, and the gas exchange rate increases. The diffusing capacity of oxygen is about 21 mL/min per mm Hg under restful conditions, whereas it is about 65 mL/min per mm Hg under strenuous exercise. Thus the diffusing capacity may vary by a factor of more than three.

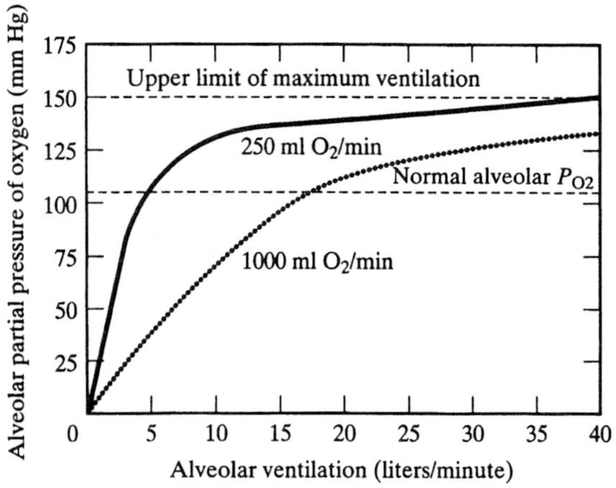

Figure 2.21 Alveolar oxygen partial pressure versus alveolar volume flow rate for two oxygen absorption rates in the blood (adapted from Guyton, 1986).

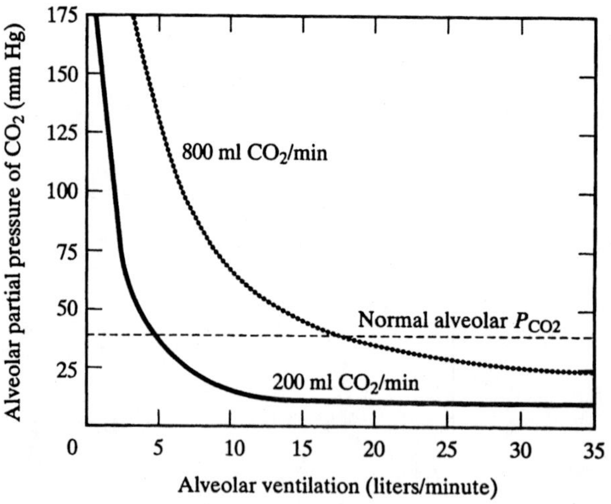

Figure 2.22 Alveolar carbon dioxide partial pressure versus alveolar volumetric flow rate for two carbon dioxide desorption rates in the blood (adapted from Guyton, 1986).

A useful parameter to diagnose dysfunction of gas exchange in the lung is the ventilation-perfusion ratio (R_{vp}), as defined by Eq. (2-14). Experiments can be run to measure oxygen and carbon dioxide partial pressures in the alveolar capillaries for individuals suffering different lung disorders. Table 2.5 shows mol fractions of oxygen and carbon dioxide entering and leaving the normal lung. Ratios of $y_{oxygen}/y_{carbon\ dioxide}$ more than 0.157/0.036 = 4.4 for expired air indicate that the gas exchange is insufficient because O_2 is not reaching the blood.

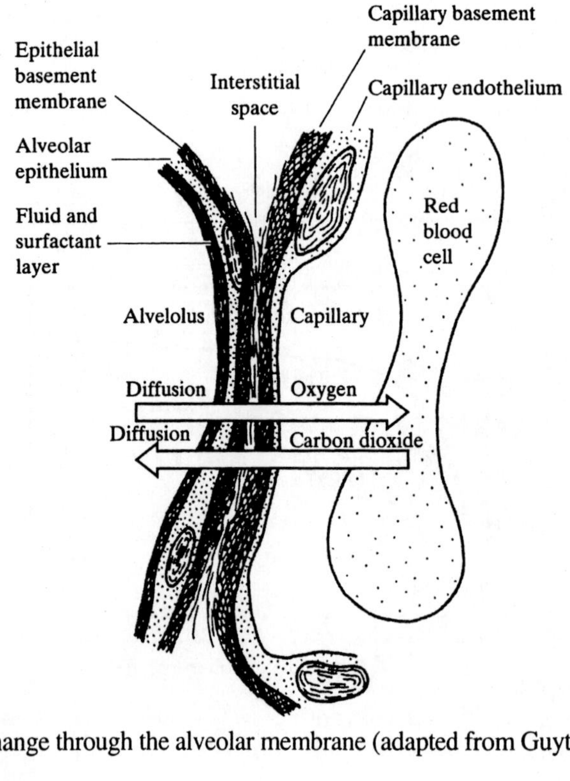

Figure 2.23 Gas exchange through the alveolar membrane (adapted from Guyton, 1986).

Measurements of the ratio of anatomic dead space volume to tidal volume (V_d/V_t) show that it decreases slightly as a result of exercise. Thus, Eq. (2-13) indicates that alveolar ventilation (Q_a) increases with exercise. The volumetric flow rate of blood (Q_b) also increases with exercise, although less rapidly than does Q_a. Consequently the ventilation-perfusion ratio, (Q_a/Q_b) *increases* with exercise and indicates that the transfer of oxygen and carbon dioxide in the lung is more limited by cardiac function than by the availability of fresh air in the alveolar region, as mentioned previously.

The ratio (Q_a/Q_b) is not uniform within the lung. In the upright lung it is larger at the bottom of the lung than at the top. The smaller ventilation-perfusion ratio near the top of the lung enables the concentration of contaminants accumulating in the alveolar tissue to be larger in the top of the lung than in the lower part of the lung. The higher the solubility of the contaminant, the greater the disparity in tissue concentration between the top and bottom of the lung.

Once oxygen enters pulmonary blood it is transferred to the cells. The solubility of oxygen in blood is low, and the principal mechanism of transporting oxygen to the cell is by hemoglobin in red blood cells. From oxygen used in cells, carbon dioxide is formed, transferred to the blood, and carried to the lung. The diffusion coefficient of carbon dioxide is 20 times larger than that of oxygen so that ordinary diffusion is sufficient to exchange it with inspired air in the lungs. Typical partial pressures for oxygen and carbon dioxide in venous and arterial blood are as follows:

- *oxygen*: 40 mm Hg (venous) and 104 mm Hg (arterial)
- *carbon dioxide*: 45 mm Hg (venous) and 40 mm Hg (arterial)

To model the heat and mass transfer processes in the respiratory system, one has to apply the laws of conservation of mass, energy, and momentum. What is less obvious is how to define the region within which the laws apply. A ***control volume*** is the region defined by the user within which the conservation equations are applied. A ***control surface*** is the (closed) surface or boundary surrounding the control volume. Users define a control volume because it encompasses the region of interest and because they are able to define mass and/or energy transferred across the control surface. Thus there are really no right or wrong control volumes, but only useful and not useful ones. Analytical models can be divided into two broad categories: compartmental models and distributed models. ***Compartmental models*** may contain one or several compartments connected in series and/or parallel. Within each compartment, ***well-mixed*** conditions are assumed. Consequently within each compartment, the gas composition is spatially uniform but may vary with time. Along the boundaries of the compartment a variety of processes may occur that add or remove material to or from the compartment. Compartmental models enable engineers to cope with mass and energy transfer processes that may vary with time. Compartmental models can accommodate spatial variations only crudely by judiciously defining control volumes that correspond to Weibel generations of interest. ***Distributed models***, discussed later, deal with spatial variations more rigorously. The Bohr model (Figure 2.14) is a compartmental model that can be extended to describe gas uptake by the blood.

2.3.1 Extended Bohr Model

EPA and OSHA standards prescribe the maximum concentrations of contaminants. Many of these contaminants, for example hydrocarbons, do not react with tissue or the mucous membrane but are soluble in blood. They are not removed in the upper branches of the conducting airways and reach the alveolar region where they diffuse through the alveolar membrane, are absorbed by blood, and are transported to organs. The Bohr model can be extended to model the absorption of hydrocarbons, oxygen, and carbon monoxide by the blood. Figure 2.24 shows the elements of an extended Bohr model. The alveolar region is assumed to be well mixed such that the contaminant molar concentration ($c_{molar,alveolar}$) is uniform. A contaminant with partial pressure $P_{inspired}$ in inspired (inhaled) air is mixed with residual air in the alveolar region, and the partial pressure is reduced to $P_{alveolar}$. (Note that for brevity, and since only one contaminant is considered at a time, no species subscript, such as j or k, is used in the present notation.) The contaminant diffuses across the ***alveolar membrane*** (also called the ***alveolar-capillary barrier***) to a venous bloodstream in which the partial pressure approaching the alveolar region is P_{venous}. Contaminant is absorbed in the blood, and at some point (y) in the alveolar capillaries the partial pressure in the arterial blood is increased to P_y. Blood with volumetric flow rate Q_b and contaminant ***solubility coefficient*** k_b flows through the capillaries. The solubility coefficient is a physical property equal to the ratio of the volume of contaminant absorbed by blood divided by the volume of the blood, typically in units of mL of absorbed gas per mL of blood. The solubility of different gases and vapors in blood varies over several orders of magnitude.

Three transfer processes in series dictate the overall rate at which contaminants enter the bloodstream. Using the ideal gas law, the overall rate of transfer of the contaminant ($\dot{m}$, kg/s) can be written as the product of an overall mass transfer coefficient for the alveolar membrane (K, m/s) times the capillary surface area (S, m^2) times the molecular weight of the contaminant (M, kg/kmol) times the difference between the partial pressure in the alveoli ($P_{alveolar}$, kPa) and the partial pressure in the blood (P_y, kPa) divided by $R_u T_{alveolar}$, where $T_{alveolar}$ is the absolute temperature in the alveolar region in Kelvin:

$$\boxed{\dot{m} = \frac{KSM(P_{alveolar} - P_y)}{R_u T_{alveolar}}} \tag{2-28}$$

The partial pressure of the contaminant in the alveolar air ($P_{alveolar}$) and the contaminant's molar concentration ($c_{molar,alveolar}$, kmol/m^3) are related by another form of the ideal gas law, written in terms of molar concentration:

$$\boxed{c_{molar,alveolar} = \frac{P_{alveolar}}{R_u T_{alveolar}}} \tag{2-29}$$

Derivation of Eq. (2-29) is left as an exercise for the reader. By analogy to Ohm's law, the reciprocal of

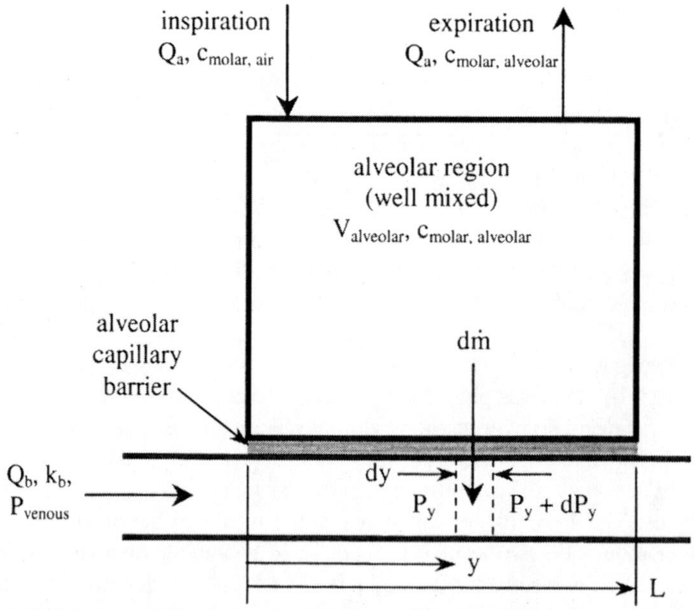

Figure 2.24 Extended Bohr model illustrating mass transfer of a nonreacting gas through the alveolar capillary barrier.

the overall mass transfer coefficient (1/K) can be thought of as a mass transfer resistance equal to the sum of the individual resistances of the components in the alveolar membrane.

A general expression for the resistance through a layer is ($f/k_s k_m$) where k_s is the ***solubility*** of the contaminant in the layer, k_m is the ***mass transfer coefficient*** for the layer, and f is the fraction of the contaminant that remains after depletion because the contaminant may react chemically with materials in the layer. The solubility k_s is a thermodynamic property of the contaminant and materials in the layers, and is independent of the rate of transfer. The mass transfer coefficient k_m is a transport property and depends on the rate of transfer, velocity field of the blood and/or air, geometry of the layer, and diffusivity of the species being transferred. The fraction of the contaminant that diffuses without depletion depends on the chemical kinetics of contaminant and materials in the layer. If the contaminant is transferred without reaction the fraction is unity. For a ***four layer barrier*** as shown in Figure 2.25, the overall resistance to transfer is

$$\frac{1}{K} = \sum_i \left(\frac{f}{k_s k_m} \right)_i \tag{2-30}$$

where subscript i pertains to each of the four layers. For contaminants that do not react with materials in the four layers, fraction f is unity and the overall resistance is governed by the layer with the smallest product ($k_s k_m$). The alveolar membrane consists of four layers, but mucus should be replaced by the pulmonary surfactant.

Consider an element of the capillary and the mass transferred into it. Transfer across the alveolar membrane can be written as

$$d\dot{m} = \frac{KM(P_{alveolar} - P_y)P'dy}{R_u T_{alveolar}} \tag{2-31}$$

where P′ is the perimeter of the capillary into which mass is transferred, and P_y is the contaminant partial pressure at a variable point somewhere in the capillary.

Considering the blood flowing in the capillaries, a mass balance can be written for a differential element of the blood absorbing the contaminant (Figure 2.24):

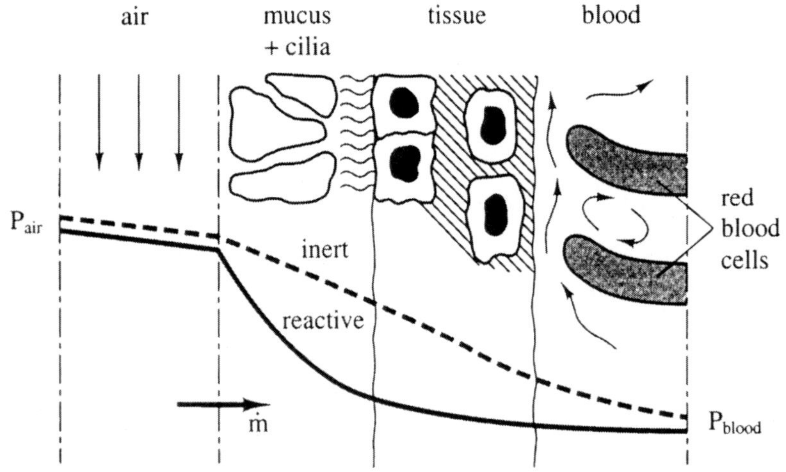

Figure 2.25 Schematic diagram of four-layer diffusion through bronchial walls; in alveoli, the layer of mucus and cilia is replaced by a layer of surfactant (adapted from Ultman, 1988).

$$\frac{d\dot{m}}{M} + \frac{Q_b k_b P_y}{R_u T_{alveolar}} = \frac{Q_b k_b (P_y + dP_y)}{R_u T_{alveolar}}$$

or

$$\boxed{\frac{d\dot{m}}{M} = \frac{Q_b k_b dP_y}{R_u T_{alveolar}}} \tag{2-32}$$

where the contaminant solubility in the blood (k_b) is used since it is assumed that the alveolar membrane does not absorb the contaminant. Eqs. (2-31) and (2-32) can be combined into a first-order ordinary differential equation, which can be expressed in standard form as in Chapter 1 (with time, t, replaced by distance, y). The solution is

$$\boxed{(P_{alveolar} - P_y) = (P_{alveolar} - P_v)\exp\left[-\frac{yKP'}{k_b Q_b}\right]} \tag{2-33}$$

Returning to Eq. (2-31), substituting Eq. (2-33), and integrating over the length (L) of the capillary,

$$\frac{R_u T_{alveolar}}{MKP'\left(P_{alveolar} - P_v\right)} \int_0^m d\dot{m} = \int_0^L \exp\left(\frac{-yKP'}{k_b Q_b}\right) dy$$

or

$$\boxed{\dot{m} = \frac{M(P_{alveolar} - P_v)Q_b k_b}{R_u T_{alveolar}}\left[1 - \exp\left(-\frac{LKP'}{k_b Q_b}\right)\right]} \tag{2-34}$$

Now the alveolar region (Figure 2.24) is considered as the control volume, and a mass balance is written for the contaminant in the entire alveolar region:

$$\boxed{V_{alveolar}\frac{dc_{molar,alveolar}}{dt} = Q_a c_{molar,air} - Q_a c_{molar,alveolar} - \frac{\dot{m}}{M}} \tag{2-35}$$

where $V_{alveolar}$ is the volume of the alveolar region, $c_{molar,air}$ is the contaminant molar concentration in the inspired air, and $c_{molar,alveolar}$ is the contaminant molar concentration in the alveolar region. Assuming quasi-steady state conditions ($dc_{molar,alveolar}/dt = 0$), Eq. (2-35) becomes

$$\boxed{\frac{c_{molar,alveolar}}{c_{molar,air}} = 1 - \frac{\dot{m}}{MQ_a c_{molar,air}}} \tag{2-36}$$

Assuming that $P_{alveolar} >> P_v$, one can divide both sides of Eq. (2-34) by ($Q_a c_{molar,air}$), and simplify. One obtains the ratio of the rate at which the contaminant is absorbed by the blood to the rate at which the contaminant is inhaled; this ratio is called ***uptake absorption efficiency*** (η_u),

$$\boxed{\eta_u = \frac{\dot{m}}{MQ_a c_{molar,air}} = \frac{1}{\left[1 + \frac{R_{vpm}}{1 - \exp\left(-N_D\right)}\right]}} \tag{2-37}$$

where N_D is a dimensionless parameter called the ***diffusion parameter***,

$$\boxed{N_D = \frac{KS}{Q_b k_b}} \tag{2-38}$$

S is the total area over which mass is transferred (useful, working surface area of the alveoli),

$$\boxed{S = P'L} \tag{2-39}$$

and R_{vpm} is a nondimensional parameter called the ***modified ventilation-perfusion ratio*** for the contaminant, which is equal to the previously defined ventilation perfusion ratio (R_{vp}) divided by the solubility coefficient (k_b),

$$\boxed{R_{vpm} = \frac{R_{vp}}{k_b} = \frac{Q_a}{Q_b k_b}} \tag{2-40}$$

Figure 2.26 (Ultman, 1988) is a graph of the uptake absorption efficiency of a highly soluble ***volatile organic compound*** (VOC) in blood for which k_b = 10.7 mL VOC per mL blood and $K = 1.2 \times 10^{-5}$ m/s, versus the modified ventilation-perfusion ratio (R_{vpm}) for several values of the diffusion parameter (N_D) corresponding to four levels of physical activity, each of which is identified as a data point on Figure 2.26.

The following general conclusions can be drawn regarding the influence of exercise on the absorption of contaminants:

- Large values of the diffusion parameter (N_D) imply good diffusion of gas through the alveolar-capillary barrier, and uniformly high uptake absorption efficiency. The location of N_D in the exponential term of Eq. (2-37) ensures that at large values of N_D, uptake absorption efficiency is only a function of R_{vpm}. When R_{vpm} becomes large, the efficiencies decrease, irrespective of N_D, since absorption is now limited by blood flow rate (Q_b).
- Assuming that the overall mass transfer coefficient K and surface area S remain constant, and using the data in Table 2.4 to quantify light, moderate, and heavy exercise, one finds that N_D decreases with exercise since it is inversely proportional to blood flow rate Q_b. On the other hand, R_{vpm} *increases* with exercise; thus the dots move down and to the right with increasing exercise, as clearly seen in Figure 2.26.

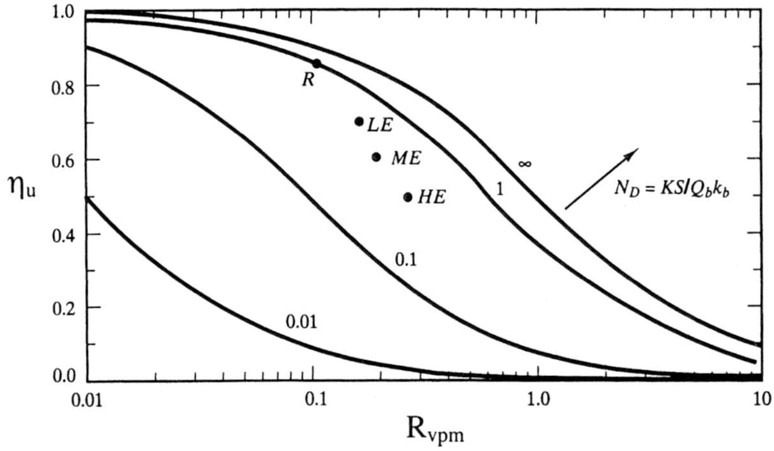

activity level	N_D	R_{vpm}	η_u (%)
rest state (R)	1.0	0.11	85.
light exercise state (LE)	0.49	0.18	69.
moderate exercise state (ME)	0.37	0.21	59.
heavy exercise state (HE)	0.31	0.29	48.

Figure 2.26 Absorption efficiency versus modified ventilation-perfusion ratio for different values of the diffusion parameter corresponding to rest (R), light exercise (LE), moderate exercise (ME) and heavy exercise (HE) (adapted from Ultman, 1988).

- With increasing exercise, Q_a increases, and the total uptake of contaminant (kmol/s) increases. However, the uptake efficiency *decreases*, which implies that the body absorbs a decreasing percentage of the contaminant it inhales (Nadel, 1985). The presumption that bodily dose is proportional to ventilation rate is overly pessimistic regarding injury by pollution. While body dose increases it is not strictly proportional to ventilation rate.

Note that the same conclusions for contaminant absorption by the blood hold for *oxygen* absorption by the blood. The following example examines the effect of a lung disease (emphysema) on the uptake absorption efficiency of oxygen.

Example 2.1 - Effect of Emphysema on Gas Uptake

Given: Individuals suffering from emphysema experience a reduction in the alveoli surface area for gas exchange (S), a reduction in the diffusion parameter for oxygen ($N_{D,oxygen}$), and a reduction in oxygen uptake efficiency ($\eta_{u,oxygen}$), irrespective of the modified ventilation-perfusion ratio for oxygen ($R_{vpm,oxygen}$). Assume the person is an average male, healthy at age 18 (t = 18 years), but emphysema leads to a reduction in alveolar surface area with age as follows:

$$\frac{S(t)}{S(18)} = 1 - \frac{t-18}{120}$$

where t is the person's age in years. Also assume the following:

- $S(18) = 100.\ m^2$
- solubility of oxygen in blood ($k_{b,oxygen}$) is 10.7 mL of oxygen per mL of blood
- Overall mass transfer coefficient (K) is 1.2×10^{-5} m/s

To do: Model the transfer of oxygen through the alveolar membrane, and compute and plot oxygen uptake absorption efficiency ($\eta_{u,oxygen}$) versus age of this individual who suffers from emphysema for 18 yr $\leq t \leq$ 80 yr.

Solution: The volumetric flow rates of air (Q_a) and blood (Q_b) for four levels of activity can be found in Table 2.4. The diffusion parameter ($N_{D,oxygen}$) varies with both age (t) and level of physical activity. For example, from Eq. (2-38), $N_{D,oxygen}$ for the healthy individual at age 18 and at rest is

$$\left[N_{D,oxygen}(18)\right]_{rest} = \left(\frac{K \cdot S(18)}{Q_b k_{b,oxygen}}\right)_{rest} = \frac{1.2\times10^{-5}\ \frac{m}{s}\left(100\ m^2\right)}{6.5\ \frac{L}{min}\left(10.7\frac{L\ O_2}{L\ blood}\right)}\left(\frac{60\ s}{min}\right)\left(\frac{1000\ L}{m^3}\right) = 1.03 \cong 1.0$$

which, using Eq. (2-37), leads to an oxygen uptake absorption efficiency of

$$\left[\eta_{u,oxygen}(18)\right]_{rest} = \frac{1}{1+\dfrac{\left(R_{vpm,oxygen}\right)_{rest}}{1-\exp\left(-\left[N_{D,oxygen}(18)\right]_{rest}\right)}} = \frac{1}{1+\dfrac{\dfrac{\left(Q_a\right)_{rest}}{\left(Q_b\right)_{rest} k_{b,oxygen}}}{1-\exp\left(-\left[N_{D,oxygen}(18)\right]_{rest}\right)}}$$

$$= \frac{1}{1+\dfrac{\dfrac{7.67\ L/min}{(6.5\ L/min)(10.7\ L/L)}}{1-\exp(-1.03)}} = 0.85$$

i.e. about 85%. When the individual is 70 years old, having suffered from emphysema since age 18, the effective alveoli surface area (S) is reduced to

$$S(70) = 100\ m^2\left(1 - \frac{70-18}{120}\right) = 56.67\ m^2 \cong 57.\ m^2$$

which leads to a diffusion parameter ($N_{D,\,oxygen}$) of 0.587, and an oxygen uptake efficiency of only 0.80 (80%). The effect is even more pronounced with increasing activity level.

One can plot oxygen uptake efficiency as a function of age and physical activity. A Mathcad program which does this is available for download from the textbook's website; the results are shown in Figure E2.1.

Discussion: In all cases the reduction in alveolar surface area beyond age 18 reduces the uptake of oxygen. At rest the reduction is only about 5%, but under heavy exercise, the reduction is about 13%. Such a reduction under heavy exercise subjects the body to a serious oxygen deficiency and prevents individuals with emphysema from undertaking heavy exercise.

2.3.2 Distributed Parameter Model

There is considerable experimental evidence that ozone causes short-term biochemical functional changes in the lung. Specifically there is a reduction in the one-second forced expiratory volume (FEV_1) (Folinsbee et al., 1988; Lipmann, 1989) as seen in Figure 2.27, which is discussed in more detail later in this chapter. The distributed parameter model enables one to model these effects. If one is interested in studying the effects of contaminants on portions of the bronchial tree as air pollutants flow in the longitudinal direction, the compartmental model (extended Bohr model) is of little help and one must turn to a distributed parameter model. A single differential equation can be written for the transport of contaminants in the direction of flow. Bifurcations of the bronchial tree can be included by modeling the airway as a single conduit resembling a trumpet, as in Figure 2.28, whose cross-sectional area and perimeter varies with longitudinal distance in accord with the Weibel model.

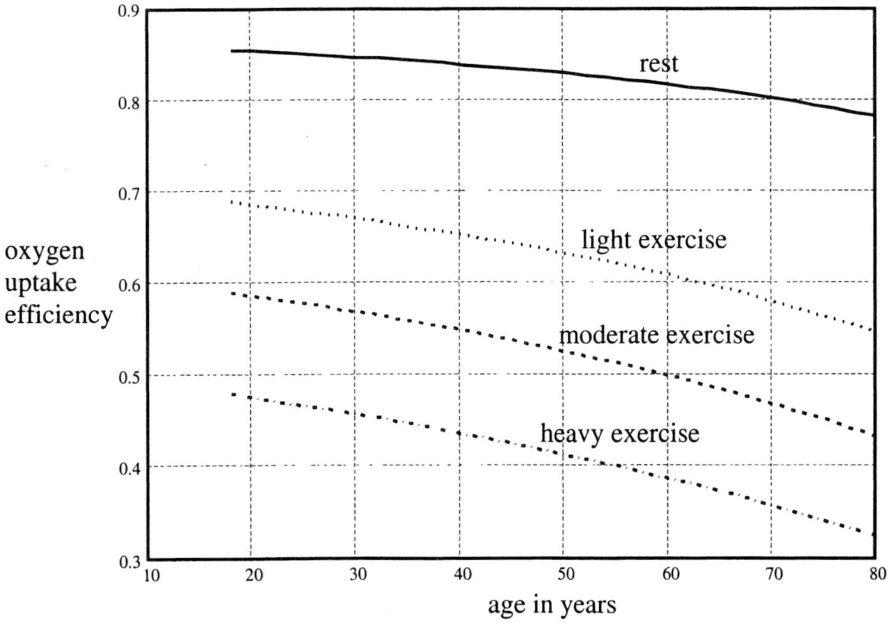

Figure E2.1 Effect of emphysema on oxygen uptake efficiency at four levels of physical activity.

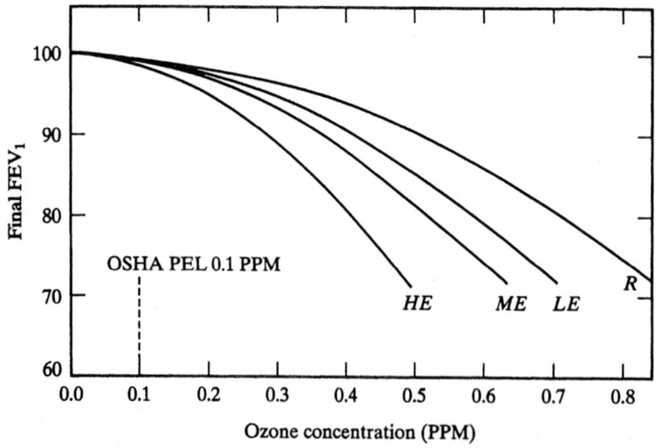

Figure 2.27 Final FEV_1 for healthy young adult males exposed to ozone for two hours, consisting of alternating 15 min periods of exercise and rest at four levels of exercise: rest (R); light exercise (LE): 24 L/min < Q_a < 43 L/min; moderate exercise (ME): 44 L/min < Q_a < 63.1 L/min; heavy exercise (HE): 64.1 L/min > Q_a (redrawn from Tilton, 1989).

Geometric data for the airways are provided in Figure 2.7. Mass transfer between passing air and blood vessels is modeled as mass transfer through several layers in series as shown in Figures 2.23 and 2.25. The parameters describing the transfer across each layer may vary widely depending on the contaminant gas. Some gases such as carbon monoxide diffuse to the blood and are not absorbed or do not react chemically with materials in the layers. Gases such as ozone react chemically with the mucosal layer, while other gases such as sulfur dioxide are mildly soluble in the mucosal layer.

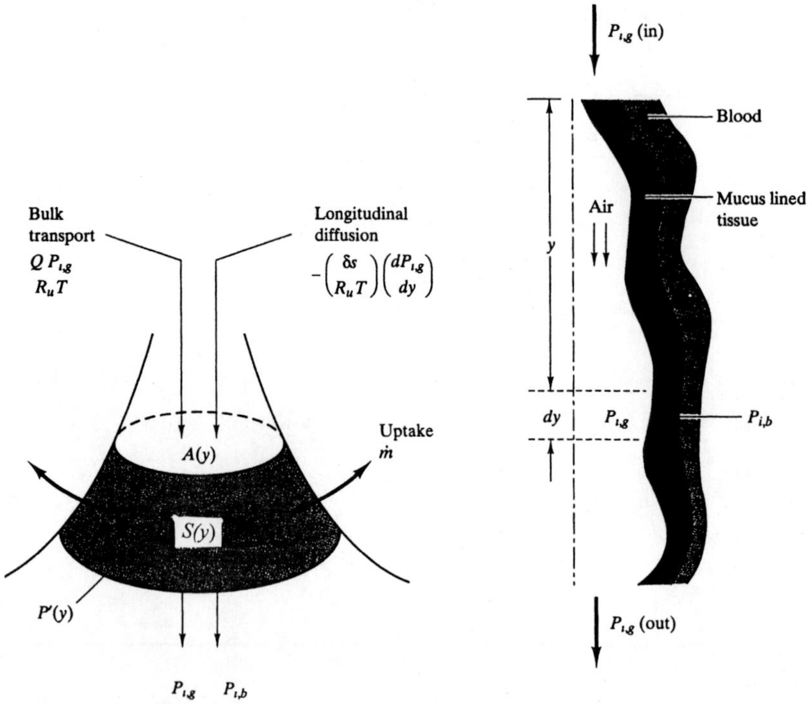

Figure 2.28 Schematic diagram of a distributed parameter model of the conducting airways (redrawn from Ultman, 1988).

The differential equation describing the longitudinal (y-direction) transport of contaminant species j through the airway is

$$\left(\frac{A}{R_u T}\right)\frac{\partial P_j}{\partial t} = -\left(\frac{Q}{R_u T}\right)\frac{\partial P_j}{\partial y} + \left(\frac{1}{R_u T}\right)\frac{\partial\left(DA\frac{\partial P_j}{\partial y} - \frac{P'\dot{m}_j}{S}\right)}{\partial y} \qquad (2\text{-}41)$$

where

- A = cross-sectional area of the airway
- P' = perimeter of the airway; both A and P' vary with y (see Figure 2.7)
- Q = volumetric flow rate
- D = diffusion coefficient
- P_j = partial pressure of contaminant j
- S = total surface area over which mass is transferred to the blood
- $\dot{m}_j$ = rate at which contaminant j is transferred to the blood, i.e. uptake.

The left-hand term in Eq. (2-41) represents the rate of accumulation and can be set to zero for quasi-steady-state solutions. The terms on the right-hand side correspond respectively to bulk transport, longitudinal diffusion, and uptake through the walls of the air passage. A differential equation for uptake is Eq. (2-31), where the overall mass transport coefficient K depends on properties of each of the four layers in Figure 2.25.

Eq. (2-41) can be solved numerically. Miller et al. (1985) and Georgopoulos et al. (1997) have used such a model to describe the effects of ozone inhalation, and predicted the total dose received in branches of the bronchial tree and the dosage penetrating to the tissue. Figure 2.29 shows the relative dose (dose/tracheal ozone concentration) versus location at the inside surface of the airway (mucus-air interface for airway generations up to 16 and surfactant-air interface in the pulmonary region) for several different first-order rate constants (k_r) for the ozone-mucus reaction. Figure 2.30 shows the relative dose to the tissue underneath the mucus layer for four levels of physical activity (rest to heavy exercise) similar but not identical to those in Table 2.4. The difference between the relative dosages in these two figures shows that ozone reacts with the mucus layer or surfactant layers. While the rate constants for the chemical reactions are not known with precision, Figures 2.29 and 2.30 show that the dose at the mucus-air interface is roughly independent of airway generation, but the net dose to tissue increases in the tracheobronchial region (TBL). In both cases the dose decreases in the pulmonary region (P). Thus the tracheobronchial tissue is protected by ***ozone absorption*** and ***chemical reaction with mucus***, and the protection these provide to the tissue is superior to that provided by pulmonary surfactant. The tracheobronchial removal efficiency increases with the rate constant. The highest relative dose experienced by tissue occurs at the 17th airway generation for R, LE and ME physical activity, and at the 20th for HE physical activity, conclusions borne out by experiment. Tissue dose in the tracheobronchial region is affected only slightly by exercise but the point of maximum tissue dosage penetrates into the pulmonary region.

The distributed parameter model has also been used by Hanna and Scherer (1986a, 1986b, 1986c) to model the transfer of energy and water vapor during inhalation and exhalation, as in Figures 2.31 - 2.33. The upper respiratory tract region between the nasal tip to the mid-trachea is very susceptible to inflammation and infection. The upper respiratory tract is also the region within which the majority of energy is transferred to heat the incoming air to body temperature and to add water vapor to the incoming air, i.e. to ***condition*** inhaled air. Some energy and some water are recovered upon expiration. The amount of water and energy transferred to the inspired air depends on the temperature and humidity of the incoming air; the amount transferred during expiration is affected by blood temperature and blood flow rate in the nasal and oral cavity. The bronchi may constrict because

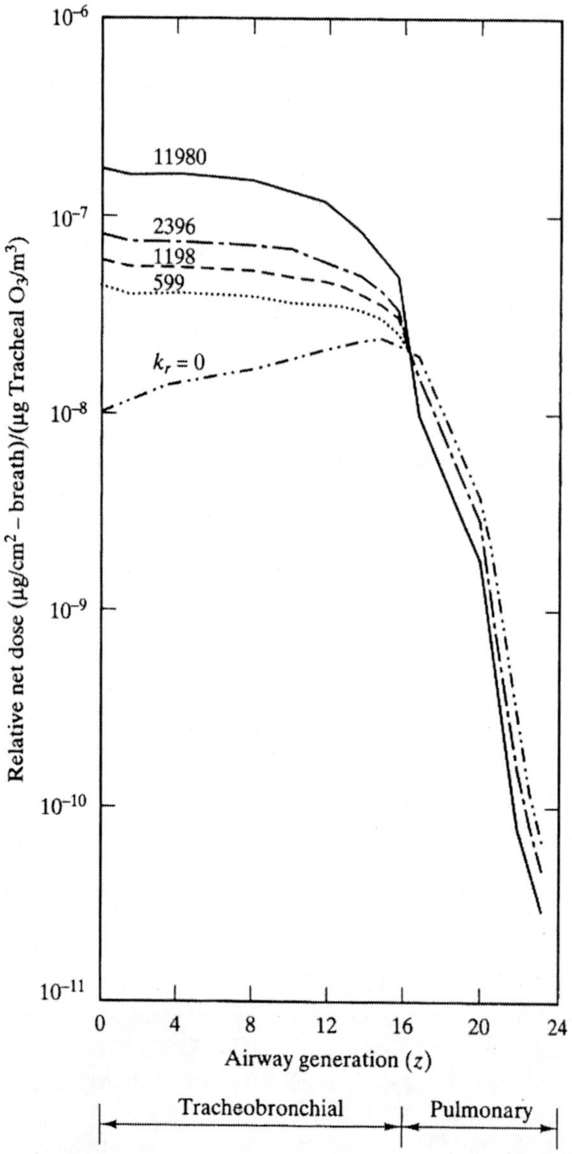

Figure 2.29 Relative ozone dose to the inside airway surface versus airway generation for the ozone-mucus reaction, and for several values of first-order rate constant (k) (redrawn from Miller, et al., 1985).

of the loss of energy and water from the mucosal surface; thus it is important to be able to describe quantitatively the transfer of energy and water from the tracheobronchial passages.

To preserve the functioning of the alveoli, inspired air must be heated to body core temperature and must contain the maximum amount of water vapor. This conditioning is performed upstream in the trachea and bronchi. The mucous membrane is designed to condition incoming air and collect pathogenic organisms and particles. The particles are deposited on the mucus layer and removed by mucociliary escalation. The mucus secretion rate is a function of air temperature and humidity. The rheological properties of the mucus are also functions of net loss of energy and water.

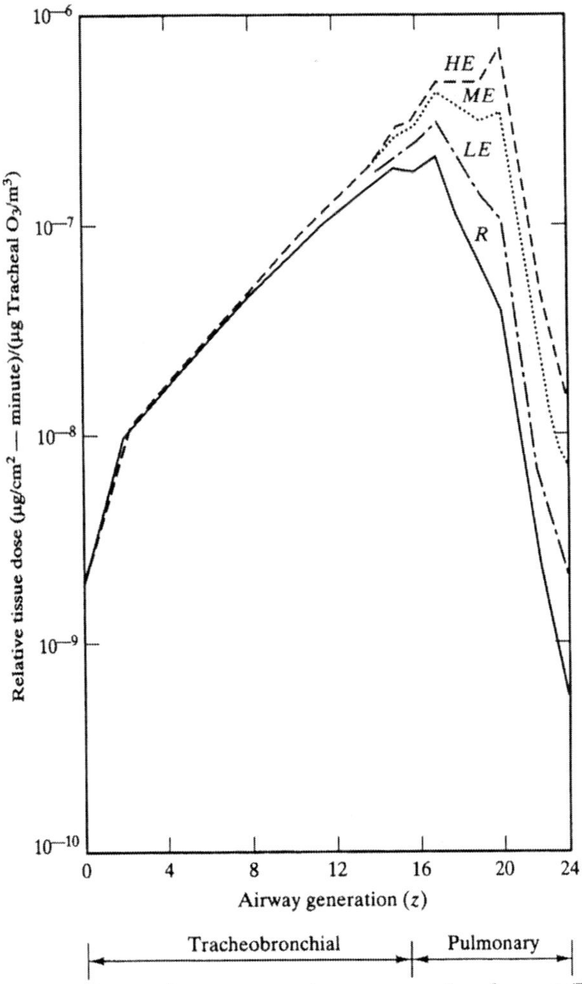

Figure 2.30 Relative ozone dose to tissue versus airway generation for rest (R), light exercise (LE), moderate exercise (ME), and heavy exercise (HE) (redrawn from Miller et al., 1985).

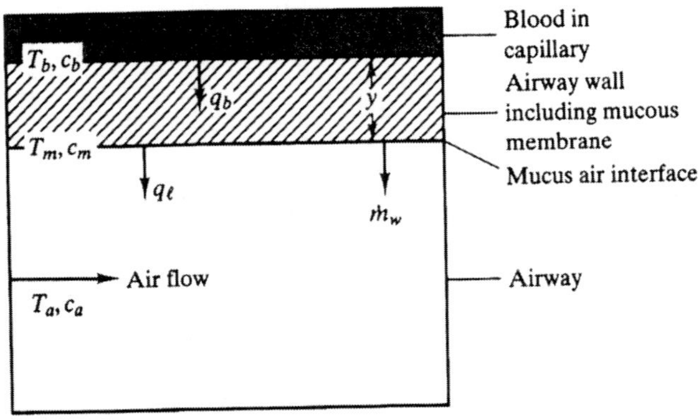

Figure 2.31 Transverse heat and mass transfer across the layers of air passage during inspiration. Quantities are functions of longitudinal distance along the airway (redrawn from Hanna and Sherer, 1986 a, b, c).

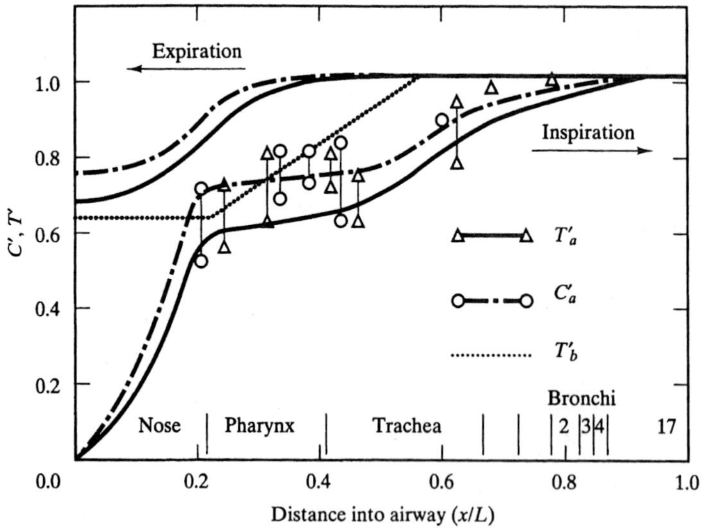

Figure 2.32 Dimensionless air temperature (T_a'), blood temperature (T_b'), and water vapor concentration (c_a') as functions of longitudinal distance in the bronchial system during inspiration and expiration in the rest condition (redrawn from Hanna and Sherer, 1986 a, b, c).

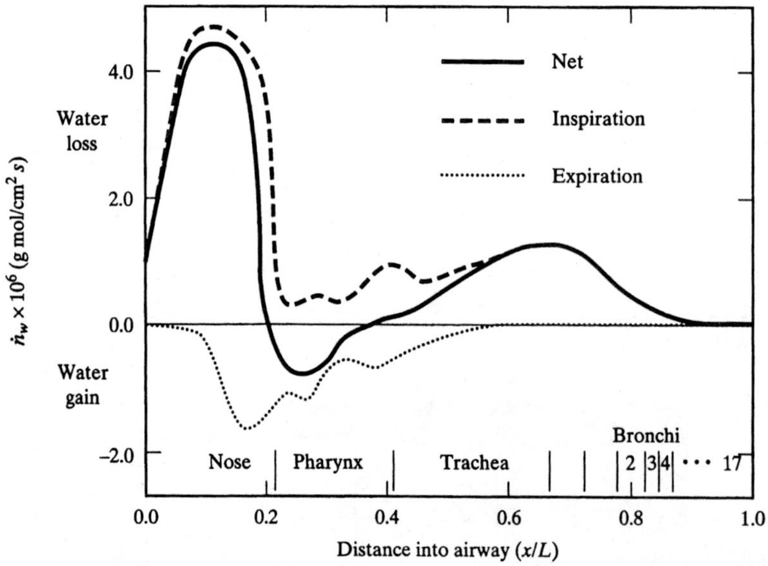

Figure 2.33 Molar transfer rate, $\dot{n}_w$, of water to air as a function of longitudinal distance in the bronchial system during inspiration and expiration in the rest condition. Positive values of $\dot{n}_w$ denote the body's loss of water (redrawn from Hanna and Sherer, 1986 a, b, c).

The majority of conditioning occurs in the nasal cavity where during normal breathing the temperature is raised to 70% of body core temperature (37 °C) and the air becomes saturated with water vapor. The temperature of the mucosal surface is 32-33 °C. After the first third of the bronchial tree, (Figure 2.6) the temperature of the air and mucosal surface are equal to the body core temperature. The energy associated with water evaporation in the respiratory system can account for up to 15% of the entire body's heat loss. The difference in the humidity of the expired and inspired air is the largest source of water loss by the body. During expiration 20 to 25% of the energy lost by evaporation during inhalation is regained by condensation. For quiet breathing, the net heat transfer to the inspired air per day is about 350 kcal, and the net transfer of water is 250-400 mL/day.

Respiratory air flow is oscillatory, but for purposes of analysis the flow can be assumed to be quasi-steady since the Womersley numbers are less than unity. Experimental measurements of air temperature in the nose and trachea show that steady-state temperatures are reached after 20% of the total time for inspiration and expiration. Measurements of relative humidity in air expired through the mouth show steady state values after 20% of the total time. Measurements of the temperature at the interface between the mucus layer and wall of the air passage show that this temperature is nearly constant throughout the respiratory cycle. Beyond the second or third division of the bronchial tree the temperature and humidity of the air do not vary over the respiratory cycle. This point in the bronchial tree is called the ***isothermal saturation boundary*** (ISB). The precise location of ISB varies with ambient temperature and relative humidity, tidal volume, and breathing rate.

Within the mucus layer and tissue, steady-state conditions are achieved within characteristic times equal to the layer thickness divided by the thermal conductivity or mass diffusivity. Assuming the thickness of the layer is 50 μm, these characteristic times are of the order of milliseconds, whereas the duration of inspiration (or expiration) is 2.5 seconds if the breathing rate is 12 breaths/minute. In summary, the assumption of quasi-steady conditions for inspiration and another set of quasi-steady conditions for expiration is reasonable for flow in the trachea and first two generations of the bronchial system.

Figures 2.32 and 2.33 show the results of an analysis based on an elementary quasi-steady model of the transport of mass and heat in the respiratory tract. Air is assumed to pass uniformly through a duct whose walls are a composite of the mucous membrane, bronchial wall, and a layer composed of the tissue containing capillaries. The energy transferred from the blood through the mucus-air interface is assumed to be equal to the latent heat of evaporation of the evaporated water plus energy transferred to the air by convection. The blood is assumed to have constant temperature.

$$\boxed{\frac{k_t}{y_{wall}}(T_b - T_a) = h_c(T_m - T_a) + N_{water} M_{water} \hat{h}_{fg}} \qquad (2\text{-}42)$$

where

- x = distance along the direction of flow
- T_b = blood temperature (constant)
- T_m = mucus-air interface temperature
- T_a = air temperature
- M_{water} = molecular weight of water
- $\hat{h}_{fg}$ = enthalpy of vaporization of water
- N_{water} = evaporation rate of water per unit area, with dimensions of mol/(time area)
- y_{wall} = thickness of airway wall including mucous membrane (constant)
- k_t = thermal conductivity of airway wall at mucus-air interface
- h_c = convection heat transfer coefficient at mucus-air interface

The evaporation rate of water per unit area (N_{water}) a function of x, and is equal to the mass transfer rate of water vapor to air,

$$\boxed{N_{water}(x) = k_c(x)\left[c_{molar,\,m}(x) - c_{molar,\,a}(x)\right]} \qquad (2\text{-}43)$$

The parameter $k_c(x)$ is the convective mass transfer coefficient. The molar concentration of water vapor in air $c_{molar,\,a}(x)$ is a variable but the concentration of water vapor at the mucous-air interface ($c_{molar,\,m}$) is approximately equal to its saturation value at temperature $T_m(x)$, and can be given by the empirical equation (Hanna and Scherer, 1986a)

$$\boxed{c_{molar,\,m}(x) = 22.4\exp\left(-\frac{4{,}900}{T_m}\right)\left(\frac{mol}{cm^3}\right)} \tag{2-44}$$

Alternatively the Clausius-Clapeyron equation can be used. The diffusion of water vapor in the axial direction is small compared to bulk transport and can be neglected (Hanna and Scherer, 1986a). Thus conservation of mass for water vapor is given by

$$\boxed{U\frac{dc_{molar,\,a}}{dx} = P'\frac{k_c\left(c_{molar,\,m} - c_{molar,\,a}\right)}{A}} \tag{2-45}$$

and conservation of energy is given by

$$\boxed{U\frac{dT_a}{dx} = \left(\frac{P'}{A\rho c_{p,air}}\right)\left[h_c\left(T_m - T_a\right) + c_{p,water}M_{water}N_{water}\left(T_m - T_a\right)\right]} \tag{2-46}$$

where

- U = air speed (assumed to be plug flow)
- A, P' = airway cross-sectional area and perimeter respectively
- ρ = air density
- $c_{p,water}$, $c_{p,air}$ = specific heats of water vapor and air respectively

The thermal conductivity of the airway wall at the mucus-air interface, k_t, can be taken as a constant, and the transport parameters $h_c(x)$ and $k_c(x)$ are variables that can be evaluated (Hanna and Scherer, 1986a, b, c) by dimensionless relationships from the literature of heat and mass transfer.

Equations (2-42) through (2-46) constitute a set of coupled ordinary differential equations that can be solved numerically. The values of blood temperature T_b, airway cross-sectional area $A(x)$, and perimeter $P'(x)$ can be taken as constants at the appropriate location (x) in the bronchial tree. The air speed U is equal to the tidal volumetric flow rate divided by $A(x)$.

The computations can be performed for both inspiration and expiration associated with restful room air breathing. Figure 2.32 shows the predicted air temperature, normalized with respect to the temperature of the inspired air [T_a(inspired)],

$$\boxed{T_a'(x) = \frac{\left[T_a(x) - T_a(\text{inspired})\right]}{\left[T_{core} - T_a(\text{inspired})\right]}} \tag{2-47}$$

where T_{core} is the body-core temperature. Water vapor concentration is normalized in similar fashion, and is also shown in Figure 2.32 as a function of normalized airway distance (x/L). Variations in the values of the transport coefficients (by 50%) are shown to have only moderate effect on the results for inspiration and little effect for expiration. Of all the parameters in the analysis, the blood temperature and volume of the nasal cavity were found to have the most importance. The evaporation rate of water is shown in Figure 2.33 for both inspiration and expiration.

Figures 2.32 and 2.33 show vividly that the nasal cavity is the primary organ that ***conditions*** inspired air. Within the nasal cavity the water vapor concentration in air nearly reaches equilibrium during both inspiration and expiration. During expiration, water vapor condenses on the cooler nasal mucosa. The mucosa of the trachea and larynx lose water upon inspiration but regain some of it upon expiration. For the cycle of inspiration and expiration, the nasal cavity, lower trachea, and bronchial tree experience a net loss of water but the pharynx has a net gain of water. Air is predicted to be expired at nearly the nasal blood temperature and nearly saturated. These results are confirmed by experiment. Downstream of the nasal cavity, the air stream reaches 60-70% of body core temperature and is fully saturated during inspiration. Relatively little conditioning of the inspired air occurs within the pharynx and upper trachea. The upper portions of the bronchial tree fully condition inspired air in a relatively short distance. Conditioning in the tracheo-bronchial tree is the result of large surface area due to the numerous bifurcations. As the temperature of the inspired air decreases or the tidal volumetric flow rate increases, less conditioning is accomplished in the nasal cavity and more is accomplished in the tracheobronchial tree.

The health implications of these studies are important. Water must be transported to the pharynx since it lacks cilia and mucus secreting glands and cells. The pharynx is thus particularly vulnerable to drying, disease, organisms, bacterial infiltration, irritation, and assault by air pollutants. This conclusion is also reinforced by exercise experiments involving sulfur dioxide (Kleinman, 1984), in which the dose to the pharynx resulting from mouth breathing is larger than that from nasal breathing owing to the efficient scrubbing that occurs in the well designed nasal cavity.

In the outdoor environment, sulfur dioxide forms sulfuric acid which may be partially neutralized by atmospheric ammonia NH_3 to form ammonium bisulfate NH_4HSO_4 and ammonium sulfate $(NH_4)_2SO_4$. It is reported (Hattis et al., 1987) that sulfuric acid particles are ten times more potent than $(NH_4)_2SO_4$ and 33 times more potent than NH_4HSO_4. When inhaled, small particles impact various surfaces in the tracheobronchial region. Because of the buffering capacity and volume of mucus, it is believed that the pH of the tracheobronchial mucus does not change appreciably (Hattis et al., 1987). However, the acid concentration in very small individual particles may be sufficient to produce localized ***irritant signals*** in the lower airways to increase mucus secretion and contribute to processes involved in chronic bronchitis. Depending on the exact pH depression to produce a signal and the effect of neutralization by NH_3 in the upper respiratory tract, the minimum size acid particle required to produce a signal is believed to lie between 0.4 and 0.7 μm.

In order to predict the deposition of aerosols or the absorption of gases and vapors in the various parts of the respiratory system, it is necessary to develop accurate fluid mechanics mathematical models. These models must account for asymmetry of the respiratory system, expansion and contraction of the air passages, and mucociliary motion. Attempts to secure models that explain observed phenomena have gone on for several decades and have improved steadily. The work of Miller et al. (1985), Ultman (1985, 1988 and 1989), Nixon and Egan (1987), Xu and Yu (1987), Yu and Xu (1987), Eisner and Martomen (1989), Gradon and Yul(1989), Gradon and Orlicki (1990), Muller et al. (1990), Yu and Neretnieks (1990), and Johanson (1991) should be consulted.

Example 2.2 - Loss of Body Water through Respiration

Given: The body loses water through perspiration, elimination, and respiration. Consider a dry environment where the ambient temperature is 25.0 °C and the relative humidity is 5.0%.

To do: Estimate the net loss of water through respiration associated with different physical activities conducted in this dry environment. Assume that expired air has a relative humidity of 83.% based on a temperature of 30.0 °C.

Solution: The saturation (vapor) pressure of water can be found at a given temperature from Appendix A.17. At 25.0 and 30.0 °C, the vapor pressures are $P_v(30.0\ °C) = 4.246$ kPa, and $P_v(25.0\ °C) = 3.169$ kPa. The relative humidity at any temperature, $\Phi(T)$, is defined in Chapter 1,

$$\Phi(T) = \frac{P_{water}}{(P_v)_{water}}$$

Since the relative humidity of the inspired air is 5.0%, the partial pressure of water vapor in the inspired air is

$$P_{water}(\text{inspired}) = \Phi P_v = (0.050)(3.169 \text{ kPa}) = 0.1585 \text{ kPa}$$

Similarly, the partial pressure of water in the expired air is

$$P_{water}(\text{expired}) = (0.83)(4.246 \text{ kPa}) = 3.524 \text{ kPa}$$

The water vapor mass concentrations in the inspired and expired air are, respectively,

$$c_{water}(\text{inspired}) = \frac{P_{water}M_{water}}{R_u T} = \frac{(0.1585 \text{ kPa})18\frac{\text{kg}}{\text{kmol}}}{8.314\frac{\text{kJ}}{\text{kmol K}}(298.15 \text{ K})}\left(\frac{\text{kJ}}{\text{kN m}}\right)\left(\frac{\text{kN}}{\text{m}^2\text{kPa}}\right)\left(\frac{1000\text{g}}{\text{kg}}\right) = 1.15\frac{\text{g}}{\text{m}^3}$$

and

$$c_{water}(\text{expired}) = \frac{(3.524 \text{ kPa})18\frac{\text{kg}}{\text{kmol}}}{8.314\frac{\text{kJ}}{\text{kmol}\cdot\text{K}}(303.15 \text{ K})}\left(\frac{\text{kJ}}{\text{kN}\cdot\text{m}}\right)\left(\frac{\text{kN}}{\text{m}^2\text{kPa}}\right)\left(\frac{1000\text{g}}{\text{kg}}\right) = 25.16\frac{\text{g}}{\text{m}^3}$$

The rate at which the body loses water mass through respiration is

$$\dot{m}_{\text{water loss}} = Q_t\left(c_{water}(\text{expired}) - c_{water}(\text{inspired})\right)$$

where Q_t is the tidal volumetric flow rate of air, given in Table 2.4 for four levels of activity. For the rest condition,

$$\dot{m}_{\text{water loss}} = 11.6\frac{\text{L}}{\text{min}}(25.16-1.15)\frac{\text{g}}{\text{m}^3}\left(\frac{\text{m}^3}{1000 \text{ L}}\right)\left(\frac{60 \text{ min}}{\text{hr}}\right) = 16.7\frac{\text{g}}{\text{hr}}$$

Since the mass of a milliliter of water is 1 gram, this corresponds to a water volume loss of 16.7 mL/hr at rest. Using the respiratory rates of Ultman (1989), a summary is provided in Table 2.6 for an individual at rest (R) and at the three other standard activity levels.

Examples of total water volume loss for some typical activities are shown below:

- Normal day (8 hr R, 16 hr LE): $(8 \text{ hr})\left(\frac{16.7 \text{ mL}}{\text{hr}}\right) + (16 \text{ hr})\left(\frac{46.4 \text{ mL}}{\text{hr}}\right) = 880 \text{ mL}$
- 4-hour round of golf (4 hr LE): $(4 \text{ hr})\left(\frac{46.4 \text{ mL}}{\text{hr}}\right) = 190 \text{ mL}$
- 6-hour hike (6 hr ME): $(6 \text{ hr})\left(\frac{72.0 \text{ mL}}{\text{hr}}\right) = 430 \text{ mL}$
- 3 hours of rock climbing (3 hr HE): $(3 \text{ hr})\left(\frac{116. \text{ mL}}{\text{hr}}\right) = 350 \text{ mL}$

Discussion: These water losses are for respiratory losses only. Additional water is lost through urination and sweating (which also increases with activity level). A rule of thumb is that individuals should drink approximately 1.5 L of water per day. On the basis of the above calculations, physically active individuals should consume more than this, especially if they live in a dry environment.

Table 2.6 Water loss in a typical human being as a function of four standard activity levels.

activity	Q_t (L/min)	water loss (g/hr) or (mL/hr)
rest (R)	11.6	16.7
light exercise (LE)	32.2	46.4
moderate exercise (ME)	50.0	72.0
heavy exercise (HE)	80.4	116.

2.4 Toxicology

2.4.1 Oxygen Deficiency

Respiratory diseases and disorders produce symptoms that are grouped in several categories: cyanosis, hypercapnia, hypoxia, and dyspnea. ***Cyanosis*** refers to the blue hue acquired by skin because of excessive amounts of deoxygenated hemoglobin, and may be a symptom of respiratory insufficiency. ***Hypercapnia*** is a condition of excess carbon dioxide in the blood. If the ventilation rate is abnormally high, both oxygen and carbon dioxide concentrations become excessive. ***Hypoxia*** is an inadequate supply of oxygen to support bodily functions and can be subdivided:

- ***hypoxic hypoxia***: sufficient oxygen does not reach the alveoli; hypoxic hypoxia can be caused by environmental factors
- ***anemic hypoxia***: inadequate hemoglobin prevents sufficient oxygen from reaching the cells
- ***circulatory hypoxia***: blood flow rate carrying oxygen to cells is insufficient
- ***histotoxic hypoxia***: tissues cannot use oxygen properly

Excess carbon dioxide may result in ***dyspnea***, a state of mind, i.e. anxiety, related to the inability to provide the body with sufficient air. It is clear that this symptom may have several causes such as hypoxia, hypercapnia, or purely emotional factors. An ***oxygen deficient atmosphere*** is defined by regulatory agencies as one in which the oxygen concentration is less than 19.5%. Safe practices dictated by OSHA are designed to prevent individuals from accidentally being exposed to an oxygen deficient atmosphere. Table 2.7 shows the effects of insufficient oxygen. The effects of low oxygen concentration are very serious and the cause of many deaths each year.

Unlike the exchange of oxygen and carbon dioxide, the transport of toxic gases occurs throughout all parts of the respiratory tract, as indicated in Table 2.1. Throughout this section, the phrase "gas" includes both gases and vapors. The rates at which gases are taken up and distributed to body organs vary considerably. For example, anesthetics produce their effect rapidly. Some toxic gases such as hydrogen cyanide and hydrogen sulfide are lethal within minutes. On the other hand, carbon monoxide and some hallucinogens take longer to produce physiological effects.

Example 2.3 - Air Containing Large Concentrations of CO_2

Given: You are an officer in the US Navy undergoing training, and have volunteered for submarine duty. The submarine has an air purification system designed to generate oxygen so as to maintain air

Table 2.7 Effects of low oxygen concentration on a typical human being.

O_2 concentration	manifestations
20.9%	normal oxygen concentration in air
17%	hypoxia occurs with deteriorating night vision, increased heart beat, accelerated breathing
14-16%	very poor muscular coordination, rapid fatigue, intermittent respiration
6-10%	nausea, vomiting, inability to perform, unconsciousness
below 6%	spasmodic breathing, convulsive movements, death within minutes

with approximately 80.% N_2 and 20.% O_2, and generates (or removes) water vapor as necessary to keep the relative humidity at 30.%. The air temperature is 25. °C, and the air pressure is 97.6 kPa. You are also told that instruments measure and record both CO and CO_2 on a continuous basis. The CO_2 meter indicates than an alarm will sound if the CO_2 concentration exceeds 45,000 PPM. A fellow officer knows that CO is dangerous and has a PEL of 50 PPM (see Appendix A.1 or A.20), but is puzzled why it is necessary to monitor CO_2 since there is no PEL for CO_2 and even normal atmospheric air contains approximately 350 PPM of CO_2.

To do: Explain why a CO_2 concentration of 45,000 PPM inside the submarine is dangerous.

Solution: Assuming that air inside the submarine contains only O_2, N_2, H_2O, and CO_2, the mol fractions must add to unity,

$$y_{O_2} + y_{N_2} + y_{H_2O} + y_{CO_2} = 1$$

The air purification system maintains a constant temperature, pressure, and relative humidity, from which the mol fraction of water vapor can be calculated,

$$y_{H_2O} = \frac{\Phi P_v(\text{at } T = 25\ ^\circ\text{C})}{P} = \frac{0.30(3.169\ \text{kPa})}{97.6\ \text{kPa}} = 0.00974$$

The submarine air contains approximately 80% N_2 and 20% O_2. This means that the molar ratio of N_2 to O_2 is

$$\frac{y_{N_2}}{y_{O_2}} = \frac{0.80}{0.20} = 4.0, \text{ i.e. } y_{N_2} = 4.0 y_{O_2}$$

Substitution into the above and solving for the mol fraction of oxygen results in

$$y_{O_2} = 1 - y_{N_2} - y_{H_2O} - y_{CO_2} = 1 - 4y_{O_2} - y_{H_2O} - y_{CO_2}$$

or

$$y_{O_2} = \frac{1 - y_{H_2O} - y_{CO_2}}{5}$$

If the CO_2 monitor trips an alarm at 45,000 PPM of CO_2, the oxygen mol fraction at this time is

$$y_{O_2} = \frac{1 - 0.00974 - 0.045}{5} = 0.189 \cong 19.\%$$

The large amounts of CO_2 displace oxygen, and the crew inhales air containing only about 19.% oxygen. Recall that an oxygen deficient atmosphere is defined as one in which the oxygen concentration is less than 19.5%. At a concentration of 19.%, the performance of the crew is impeded. Since a submarine is a military vessel, impediments to the crew are potentially dangerous.

Discussion: The engineers should be looking into the function of the CO_2 removal system even before the alarm sounds. The immediate risk is not from CO_2 "poisoning" but from insufficient O_2.

2.4.2 Ozone Exposure

Ozone is an air pollutant generated by complex atmospheric reactions initiated by the photolysis of nitrogen dioxide (NO_2) in the troposphere. Ozone can be generated indoors by incorrect operation of photocopy machines, the corona from electronic air cleaners, etc. Ozone is a lung irritant producing demonstrable short-term (***acute***) effects and possible long-term (***chronic***) effects (Tilton, 1989; Lipmann, 1989 and 1991). Depending on dose, the acute effects are reduced lung function (FEV_1) for a period up to 42 hours, irritated throat, chest discomfort, cough, and headache. ***Broncho-alveolar lavage*** (BAL) measurements show that ozone produces lung inflammation. There is also a positive association between ambient temperature, ozone concentration, and daily hospital admissions for pneumonia and influenza. For asthmatics, additional effects include increased use of medication and restricted physical activity. Data to support chronic effects are limited and are open to multiple interpretations. Chronic ozone exposure during the summer seems to reduce lung function; this persists for a few months but dissipates by the spring. There is general agreement however, that ozone contributes to premature aging of lungs. Studies with rats and monkeys support this conclusion.

2.4.3 Toxic Gases

Many toxic materials enter the body through the respiratory system but only some affect the lung directly; the others enter the blood system or lymphatic system and harm other body organs. Lung disease is divided into five categories:

- ***Irritation*** - Air passageways become irritated leading to constriction and perhaps edema and secondary infection.
- ***Cell damage*** - Cells lining the air passageways are damaged resulting in necrosis, increased permeability, and edema within the airway.
- ***Allergies*** - Chemicals in pollen, cat hair, etc. excite nerve cells that cause muscles surrounding bronchiole to contract which constricts airways and taxes the cardiovascular system.
- ***Fibrosis*** - Lesions consisting of stiff protein structures appear on lung tissue inhibiting lung function. Fibrosis of the pleura may also occur; this disease restricts movement of the lung and produces pain.
- ***Oncogenesis*** - Tumors are formed in parts of the respiratory system.

Table 2.1 lists the principal occupational diseases of the lung and describes the effects of exposure to several occupational toxic substances.

Toxic gases either react directly on portions of the respiratory tract or are transported to other organs before their effect is registered. Examples of the first type are ozone and sulfur dioxide. Examples of the second type are carbon monoxide and hydrogen cyanide. Exposure to ozone for two hours reduces FVC and FEV_1 in healthy adults by small but statistically significant amounts (4%) and its effects persist for approximately 18 hours (Lipmann, 1987,1989 and 1991), as seen in Figure 2.27. Highly reactive agents soluble in water, e.g. anhydrous acids and strong oxidants, are apt to damage tissue while less reactive gases such as nickel carbonyl diffuse through tissue to react with endothelial cells. Other gases may damage capillaries.

In Section 2.3 it was shown that the nasal cavity is the principal organ that transfers water and energy to and from respired air. Because the surface of the nasal cavity contains a great deal of water, any toxic gas soluble in water is removed. Anhydrous acids and sulfur dioxide are more apt to be removed than ozone owing to the latter's lower solubility in water. The nasal cavity is the body's most efficient wet scrubber. In the tracheobronchial region toxic gases encounter mucus lining the airways. Gases penetrating the mucous lining contact goblet or ciliated cells. Ciliated cells are generally more sensitive to toxins than are goblet cells, and reducing the number of cilia per unit area of passageway impairs the clearance mechanisms.

2.4.4 Aerosols

The body's response to particles contained in inspired air is entirely different than its response to gases. ***Clearance*** is the process by which particles are removed from the lung. ***Mucociliary clearance*** is the process by which the conducting airways of the lung remove deposited particles and carry them to the pharynx on surface mucus propelled by cilia. ***Alveolar clearance*** is the process by which particles are removed by non-ciliated surfaces in the gas exchange region of the lung. Clearance mechanisms include ingestion by macrophage followed by migration from the lung and gradual dissolution of the particle.

Figures 2.34 and 2.35 show that particles deposit themselves throughout all regions of the respiratory system. Deposition occurs through a variety of processes (Hatch and Gross, 1964, Perra and Ahmed, 1979; Rothenberg and Swift, 1984 and Martonen, 1992). In the nasopharyngeal region ***inertial impaction*** is the dominant mechanism and relatively large particles are removed from the air. Particles removed in the trachea and bronchial tree are removed by a combination of inertial impaction and ***gravitational settling*** (also called ***sedimentation***) processes. The larynx affects particle motion because of turbulence created by air passing over the vocal cords. In addition, particles entrained in the air leaving the larynx impact the trachea at localized "hot spots" (Balashazy et al., 1990, Martonen et al., 1992). Elsewhere in the lung, particles are deposited more uniformly over the airway surface. Submicron particles penetrate to the alveolar region and are metabolized by macrophage. Particles that are not metabolized either remain in the alveoli or diffuse to other parts of the body such as the lymphatic system. Particles removed in the nasopharyngeal region enter the digestive tract and pass through the body in a short time. Unless these materials enter the bloodstream through the digestive system, they are of little importance. Particles entering the trachea are removed by cilia and mucus and are transported to the digestive tract. Particles penetrating the bronchial tree are flushed more slowly. Exactly which particles penetrate which region of the respiratory tract depends on the breathing rate and tidal volume. Thus a population of energetic workers in a dusty industrial environment may deposit more particles of a given size to the bronchial tree and alveoli than do individuals in a non-industrial environment under restful breathing. For slender particles of asbestos, cotton, etc., deposition processes must take into account their unusual shape (Kasper et al., 1985; Gallily, 1986).

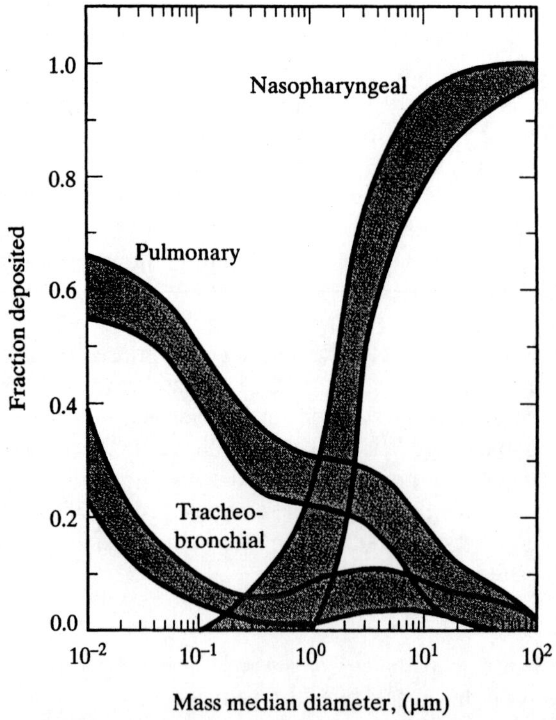

Figure 2.34 Predicted regional deposition of particles in the respiratory system for a tidal volumetric flow rate of 21. L/min. Shaded area indicates the variation resulting from two geometric standard deviations, 1.2 and 4.5 (redrawn from Perra and Ahmed, 1979).

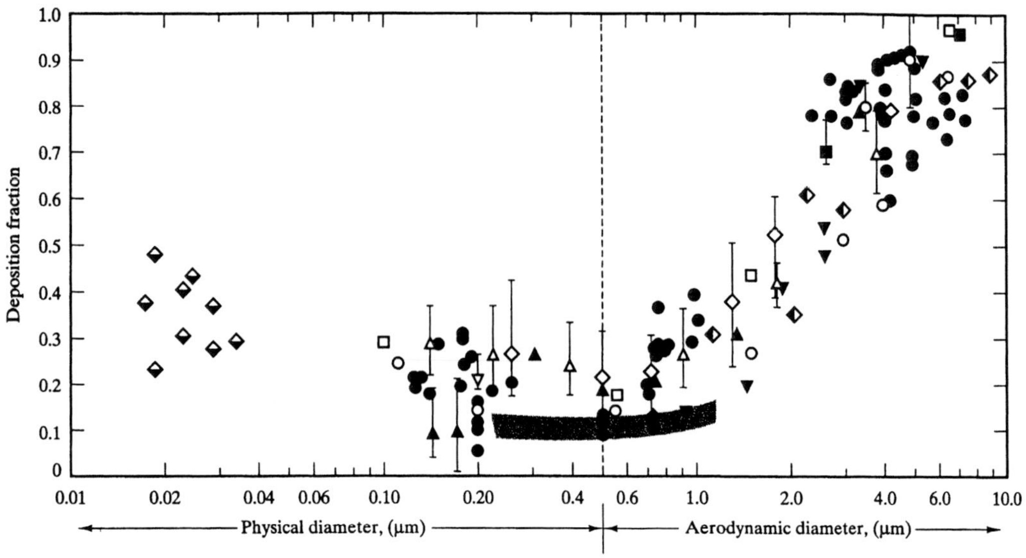

Figure 2.35 Deposition of monodisperse aerosols in the total respiratory tract for mouth breathing as a function of particle diameter. Below 0.5 μm refers to actual physical diameter while above 0.5 μm refers to aerodynamic diameter (adapted from Seinfeld, 1986).

Studies of particle trajectories in the bronchial tree (Balashazy and Hofmann, 1993) reveal that during inspiration, particles deposit themselves preferentially on ***carnial ridges***, the tissue dividing the bronchial airways into two passages. The fraction of particles that are deposited is higher for large particles owing to their larger inertia. Mucociliary clearance is smallest at these airway bifurcations; consequently, the epithelial cells on these carnial ridges suffer a larger dose than do other cells on the airway passages. This helps explain the high occurrence of bronchial carcinomas attributed to cigarette smoke and radon progeny.

The process of ***phagocytosis*** is one of the body's essential protective mechanisms. Organs of the body, including the alveoli, contain special cells called ***macrophage*** that attack foreign matter by a variety of processes. Phagocytosis begins when the macrophage membrane attaches itself to the surface of the particle. Receptors on the surface of macrophage bond with the particle. The edges of the membrane spread outward rapidly and attempt to engulf the particle (mechanism of ***pinocytosis***). The macrophage membrane contracts and pulls the particle into its interior where cell lysosome attach themselves to the particle. ***Lysosome*** are special digestive organelle within cells that contain enzymes called ***hydrolase*** that digest foreign matter. The engulfed particle is called a ***vesicle***. Following digestion, the vesicle containing indigestible material is excreted through the macrophage membrane. Macrophage are generally thought to originate in bone marrow. How they migrate to different parts of the body including alveoli is not fully understood except that they are very flexible and capable of amoebae-like expansion and contraction that enable them to pass through minute openings in tissue.

Several terms are used to categorize particles that pose a hazard to health. The EPA defines ***inhalable*** as particles having an aerodynamic diameter less than 10 μm, while ACGIH defines ***respirable*** as particles having an aerodynamic diameter less than 2.5 μm. Particles with $D_p < 2.5$ μm are called ***fine particles*** and those with $D_p > 2.5$ μm are called ***coarse particles***. Fine particles produced directly by combustion processes are called ***primary particles***, and particles produced by reactions of gaseous products of combustion in the atmosphere are called ***secondary particles***.

Aerodynamic diameter is the diameter of a spherical water drop that has the same settling velocity as the actual particle. To a first approximation, the actual and aerodynamic diameters are related by

$$D_p\left(\text{aerodynamic}\right) = D_p\left(\text{actual}\right)\sqrt{\frac{\rho_p}{\rho_{water}}} \tag{2-48}$$

The definitions of respirable and inhalable may change or new terms may be coined as understanding of the health effects of inhaled particles improves. Particles deposited in the alveoli are acted upon by one or a combination of four processes:

- particles are phagocytized and passed up the tracheobronchial tree by mucociliary escalation
- particles are phagocytized and transferred to the lymphatic drainage system
- particle surface material is dissolved and transferred to blood vessels or lymphatic capillaries
- particles and some dissolved material are retained in the alveoli permanently

There is basic disagreement on interpretation of the epidemiological data on health effects associated with increases in particulate air pollution. Although a strong case has been presented that this association reflects a causal relationship, plausible alternative explanations have also been suggested. (Vedal, 1997). Inhaled fine particles entering the pulmonary region induce ***pulmonary inflammation*** and increase the risk of hospitalization or death of sensitive individuals such as those with chronic lung and heart disease. Inflammatory lung disease is associated with an increase in mortality of people 65 and older. The portion of the increase due to chronic obstructive pulmonary disease is 3.3%, the portion due to ischemic heart disease is 2.1%, and the portion due to pneumonia is 4.0% (Schwartz and Andreae, 1996). Hypotheses describing how inflammation is induced depend on the chemical properties of the particles such as acidity, the presence of transition metals, and the presence of ***ultrafine particles*** ($D_p < 0.02$ μm, i.e. 20 nm). Transition metal ions such as ferric ions catalyze the production of hydroxyl radicals (OH•) via the Fenton reaction. Ultrafine particles are taken up poorly by lung macrophage and are capable of penetrating the pulmonary epithelium and may pass into the interstitium.

2.4.5 Airway Irritation

The cross-sectional area of elements in the bronchial tree is affected by many industrial chemicals. Most chemicals reduce the cross-sectional area of the airway, but a selected few enlarge the airway. Dyspnea, as discussed above, is the anxious feeling that one cannot breath deeply and rapidly enough to satisfy respiratory demand, and may be caused by a narrowing of the airways. Gases such as ammonia, chlorine, and formaldehyde vapor (Alenandesson and Hedenstierna, 1989) are soluble in water and can produce dyspnea. Exposure does not produce chronic respiratory damage, but high concentrations can result in death.

2.4.6 Cellular Damage and Edema

A variety of substances can damage cells of the bronchial tree and alveoli that in turn release fluid into these passageways. The site of the damage depends on the solubility of the material in water; more soluble materials damage the nasal cavity, and less soluble materials damage upper elements of the respiratory tree. The production of fluids (edema) may take considerable time to evidence itself. Phosgene irritates the nasal cavity and upper respiratory passages because it reacts with the abundant supply of water to form carbon dioxide and hydrochloric acid. Ozone and nitrogen dioxide on the other hand are less soluble and penetrate the bronchiole and alveoli. Cadmium oxide fume is a submicron particle that travels to the alveoli and produces edema. Sustained exposure results in an irreversible destruction of alveoli and reduction of oxygen uptake by the blood. Nickel oxide, nickel sulfide fume, and vapors of nickel carbonyl damage the cells of whatever surface on which they reside. Hydrocarbon vapors such as xylene and perchloroethylene have low solubility in water and travel to alveoli where they diffuse to capillaries and are transported to the liver and other organs. Oxygenated intermediaries produce pulmonary edema.

Emphysema is a debilitating disease that reduces vital capacity of the lung and taxes the cardiovascular system. The disease progresses in three stages:

- Chronic infection of alveoli or bronchial tissue produced by tobacco smoke or other irritants increases mucus excretion and paralyzes cilia.
- Infected lung tissue, entrapped air, and obstructing fluids in alveoli and alveolar ducts destroy alveolar walls and their web of capillaries (Figure 2.36).
- Increased airway resistance, decreased diffusing capacity, and decreased ventilation-perfusion ratios occur.

Chronic emphysema progresses slowly and is irreversible. Insufficient oxygen produces hypoxia and takes a toll on the entire cardiopulmonary system.

2.4.7 Pulmonary Fibrosis

Pneumoconiosis is the general class of disease of which pulmonary fibrosis is the central feature. Silica (SiO_2) exists in several crystalline forms, of which ***cristobalite*** and ***tridymite*** induce fibrosis and quartz does not. ***Silicosis*** is formation of silicotic nodules of concentric fibers of collagen that appear in lymphatics around blood vessels beneath the pleura in the lungs and sometimes in mediastinal lymph nodes. The complete explanation of how pulmonary lesions are formed is lacking, even though silicosis has been recognized for centuries. Macrophage attach themselves to silica particles. It is believed that the lysosomal membrane of the macrophage ruptures and releases lysosomal enzymes that "digest" the macrophage. New macrophage go through the same process and the cycle is repeated. The damaged macrophage release some unknown material that stimulates formation of collagen, which is not readily digested by most enzymes. The nodules may fuse and block blood vessels and reduce the flow of blood. Alveolar walls may be destroyed. The size of the alveolar sacs and ducts may enlarge, gas exchange may be reduced, and symptoms of emphysema occur.

Asbestos is the generic name for a group of hydrated silicates existing in the form of clustered fibers. The length, diameter, and number of fibrils (small diameter fibers attached to the main fiber) on each fiber vary. Exposure to asbestos can cause four types of disorders (Mossman et al., 1990):

- asbestosis
- lung cancer
- mesothelioma of the pleura, pericardium, and peritoneum
- benign changes in the pleura

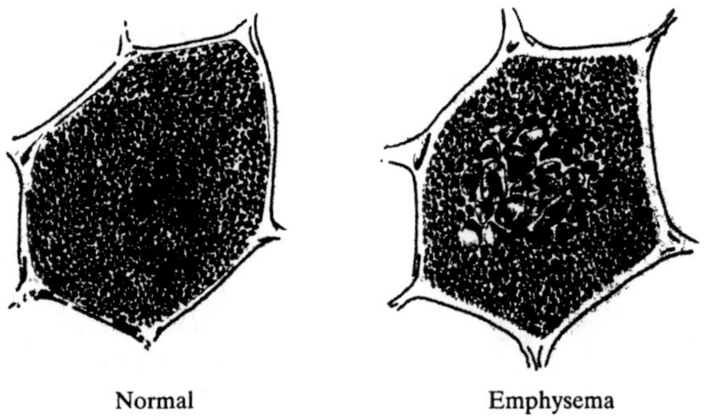

Figure 2.36 Illustration of how emphysema destroys the normal alveolar membrane and leaves large holes in the tissue (adapted from Occupational Health and Safety, 1989).

There are six forms of asbestos but only three forms have been used to any extent in industry, ***chrysotile, crocidolite***, and ***amosite***, and each is capable of causing disease. Chrysotile amounts to 95% of all the asbestos used in the US. Concentrations of chrysotile dust particles in buildings rarely exceed 0.001 fibers/cm^3. There is evidence that chrysotile asbestos does not cause mesothelioma, even after heavy exposure. Other minerals, such as cummingtonite, grunerite, tremolite, anthophyllite, and actinolite are defined as "asbestos" for regulatory purposes, but they have not been shown to cause disease in miners.

Asbestosis is pulmonary interstitial fibrosis associated with an excessive deposition of collagen that stiffens the lung and impairs gas exchange. ***Lung cancers*** are tumors in the tracheobronchial epithelial or alveolar cells. In general, lung cancers have been found in asbestos workers who are smokers, but only rarely in nonsmokers. Diffuse malignant ***mesothelioma*** is a fatal tumor associated with mesothelial cells or underlying mesenchymal cells in the pleura, pericardium, and peritoneum. Mesothelioma is principally caused by crocidolite. The diagnosis of mesothelioma is difficult as the tumor may resemble metastates of other tumor types, and may assume a wide variety of microscopic appearances normally attributed to cancers of the gastrointestinal tract or other organs. A number of benign pleural changes that rarely cause functional impairment may occur in asbestos workers.

As with silicosis, a complete explanation describing how fibrosis occurs does not exist, although a great deal is known about some aspects of the process. The length and diameter of the fiber and the character of the fibrils comprising the total fiber are believed to influence fibrosis. Similar to silicosis, the incidence of disease is strongly enhanced by smoking. Bronchogenic carcinoma can be found in all portions of the bronchial tree. Interstitial fibrosis is most commonly found in the lower lobes of the lung. Asbestos itself is chemically inactive but chemical carcinogens that reside on the fiber surface may also initiate cancer within the lung.

2.4.8 Mercury

Elemental mercury (Hg) and inorganic mercury compounds are used in the manufacture of electronics, lamps for lighting, amalgams with copper, tin, silver, zinc, and/or gold for teeth fillings, and solders with lead and tin. Mercury is used in gold, silver, bronze, and tin plating, tanning, dyeing, felt-making, textile manufacturing, photography, and photoengraving. Mercury is also used to extract gold and silver from ores, as paint pigment, and in the preparation of drugs, disinfectants, fungicides, and bactericides. The permissible exposure limit (PEL) of metallic mercury in air is 0.05 mg/m^3. ***Local toxicity*** of mercury is an irritant to skin and mucous membrane. Alkyl mercury compounds may cause dermatitis. As a ***systemic toxin***, mercury affects the lungs primarily in the form of acute interstitial pneumonitis and bronchitis.

Compounds containing mercury enter the body either through inhalation or ingestion. The adverse effects of compounds of mercury depend on the ***dose***, which is the product of concentration and exposure time. Long-term, low-level exposure produces complex symptoms that vary from individual to individual, and may be confused with symptoms from other illnesses. Symptoms include weakness, fatigue, loss of appetite, loss of weight, insomnia, diarrhea, increased salivation, inflammation of gums, black line on the gums, loosening of teeth, loss of memory, and tremors of fingers, eyelids, lips, and tongue. More extensive doses can produce extreme irritability, anxiety, delirium, hallucinations, melancholia, and manic depressive psychosis. ***Chronic exposure*** produces four classical signs: gingivitis, sialorrhea, increased irritability, and muscular tremors, but rarely are all four symptoms present in an individual case. Either acute or chronic toxicity may produce permanent damage to organs or a system of organs.

Microorganisms in the environment react with metallic mercury and produce organic mercury compounds. The PEL of organic mercury compounds is 0.01 mg/m^3 with a ceiling value of 0.04 mg/m^3. The central nervous system is the principal target for alkyl mercury compounds, and severe poisoning may cause irreversible brain damage. The effects of chronic exposure to alkyl mercury compounds are progressive. In the early stages, there are fine tremors of the hands and perhaps arms and face. For larger exposures the tremors become convulsive, speech is slurred, and there is difficulty pronouncing some words. The worker may develop an unsteady gait that may progress to a spastic gait and to severe ataxia of the arms and legs. Sensory disturbances include tunnel vision, blindness, and deafness. Later symptoms include constriction of visual fields and loss of understanding and ability to reason. Several cerebral effects can be passed to infants born to mothers suffering from alkyl mercury poisoning. To anticipate the effects of mercury exposure on the body, the amount of mercury in the urine should be examined. Opinion varies but levels of 0.1 to 0.5 mg of mercury per liter of urine are considered significant.

2.4.9 Lead

Lead is slightly soluble in water containing nitrates, ammonium salts, and carbon dioxide (all of which are present to a small degree in drinking water). Inorganic lead includes lead oxides, metallic lead, and lead salts. Lead in the form of tetraethyl lead was used for decades as an antiknock agent in gasoline; this practice is now banned. Organic and inorganic lead compounds are used in paint pigments, glass, glazes, and plastics. Metallic lead is used in metallurgy in the manufacture of brass and bronze, and is combined with other metals to form alloys (e.g. steel contains a small amount of lead; solder for electronics and copper pipes is an alloy of lead and tin; pewter, used for mugs, is an alloy of tin, lead, and other metals). There are numerous instances in which organic and inorganic lead compounds are used in the manufacture of rubber, paint glazes, pigments, and chemicals. Exposure to lead dust occurs during mining, smelting, and refining and exposure to lead fumes occurs during high-temperature (above 500 °C) operations such as welding and spray coating of molten metals. Lead-bearing particles may also fall from decaying painted surfaces in older homes. Blast abrasion and welding of old steel structures (bridges, water towers, etc.) generate copious quantities of lead-bearing particles that must be collected (using total or mini enclosures over the structure). Through ventilation and personal protection equipment, workers performing these functions inside the enclosures must also be protected. The airborne PEL of lead and its inorganic compounds is 0.05 mg/m^3.

Like mercury, lead enters the body through inhalation and ingestion. The ***systemic symptoms*** of lead poisoning are nonspecific and difficult to distinguish from symptoms of minor seasonal illnesses such as fatigue, headache, aching bones and muscles, loss of appetite, abdominal pains, and digestive disturbances (particularly constipation). These symptoms of low-level exposure are reversible and complete recovery can be expected. With increased dose, other symptoms include anemia, pallor, "lead line" on gums, and decreased hand-grip strength. Lead colic produces intense periodic abdominal cramping associated with severe constipation, and occasionally nausea and vomiting. Alcohol ingestion and physical exertion may precipitate these symptoms. If the nervous system is affected, it is usually due to ingestion or inhalation of large amounts of lead. The peripheral nerve affected most is the radial nerve. With increased dose, lead exposure manifests itself in "***wrist drop***." Removing the individual from an area of high lead exposure results in a slow recovery that may not always be complete. The results of such a large dose may include severe headache, convulsions, delirium, coma, and possibly death. The kidneys can also be damaged after a long exposure, with the possibility of loss of kidney function.

If the concentration of airborne particles containing lead is not adequately reduced through administrative and engineering controls, personal protection equipment may be required to reduce lead exposure. OSHA requires that dust and fume respirators provided by the employer must be worn if the air concentration at the work site exceeds the PEL. Workers should be supplied with full-body clothing that is collected and cleaned by the company in a prescribed manner. A hat, face shield or goggles, and gloves should also be worn. Lead dust should be removed from an individual's clothing by vacuum (not with an air hose!) before leaving the work site. Showering after each shift prior to changing to street clothes is mandated.

Alkyl lead compounds may penetrate the skin. The most noticeable sign of alkyl lead poisoning is encephalopathy (diseases of the brain) that gives rise to a variety of symptoms beginning with mild anxiety, toxic delirium, hallucinations, delusions, and convulsions. Periodic evaluation of lead levels in the blood is widely used to indicate lead absorption. Diagnosis depends on knowing the dose (concentration multiplied by exposure time), which may not always be possible. The level of lead (in any molecular form) in blood should always be kept below 0.06 mg per 100 g of blood. Unfortunately, blood lead level is usually not elevated in proportion to the degree of intoxification. Analysis of urine shows results even less proportional to the degree of intoxification. When the period of time between termination of exposure and onset of symptoms is long (up to 8 days) recovery is likely. If the period of time is short (of the order of hours) the prognosis is less optimistic. Recovered patients may have permanent damage to the nervous system, and recovery may be prolonged.

2.4.10 Radon

Radon (Rn) is a radioactive gas that exists in several isotopic forms. Only two isotopes occur in significant concentration in the environment – radon-222 and radon-220, radioactive decay products of uranium-238 and thorium-232 respectively. Radon is the only *gaseous* product in the radioactive decay chain of uranium, and it seeps through soil and rock to enter outdoor air. It can also seep from the ground to enter indoor air in buildings (e.g. basements in homes). In regards to toxicity, short-lived decay products of radon are of more importance than the radon gas itself. The radioactive radon daughters attach themselves to dust particles in the air and generate alpha, beta, and gamma particles. When the dust particles are inhaled, they adhere to lung tissue and subject it to alpha, beta, and gamma radiation. The beta and gamma particles possess insignificant energy to damage lung tissue, but the alpha particles are energetic and have the potential to produce cancer.

2.5 Sick Buildings

The average American spends between 87% and 90% of his/her time indoors, only 5% to 7% outdoors, and the remainder of time in transit (Corsi, 2000). People who spend the most time indoors, i.e. children, chronically ill, and elderly, etc., are also those most susceptible to the adverse effects of indoor air pollution. ***Sick building syndrome*** (SBS) is a situation that occurs when the ***indoor air quality*** (IAQ) deteriorates and a statistically significant number of individuals in unique portions of a building complain about the following (Godish, 1995):

- nose and throat: dryness, stuffiness, irritation of the mucous membrane, runny nose, sneezing, odor, taste sensations
- eyes: dryness, itching, irritation, watery eyes
- skin: dryness and irritation, rash, itching, eczema, ***erythema*** (reddening), localized swelling
- general: tight-chest, abnormal fatigue, malaise, headache, feeling heavy-headed, lethargy, sluggishness, dizziness, nausea

SBS symptoms are similar to those of the common flu and asthma, and are thus often misdiagnosed. The symptoms begin within 15 minutes of entering the building and disappear shortly after leaving the building. Symptoms affect women more than men, in part because the work force in the affected portion of buildings is usually predominately women. SBS often accompanies attempts to conserve energy, e.g. reducing the amount of make-up fresh air, reducing infiltration and exfiltration (tighter building with less leaks), and increasing the number of workers in offices.

There are frequent complaints about aliments and diseases attributed to unknown sources inside buildings. The phrase "sick building" has been coined to describe these buildings. There are several reasons for these complaints. A fetish about health and propensity to consider oneself a victim certainly contributes to the problem, as may psychosomatic response to the illness as other workers share their experiences. ***Building related illness*** (BRI) is the term given to an illness brought on by exposure to building air when symptoms of a diagnosable illness are identified and can be attributed directly to environmental agents in the indoor air. ***Chemical hypersensitivity*** is the name given to a chronic multisystem disorder, usually involving symptoms of the central nervous system and at least one other (bodily) system. Affected persons are frequently intolerant to some foods and react adversely to some chemicals or environmental allergens, singly or in combination at levels generally tolerated by the majority of the population. The severity of reactions varies from mild discomfort to total disability. The symptoms are non-specific and multiple, e.g. musculoskeletal or respiratory symptoms, behavioral changes, fatigue, depression, psychiatric disorders, membrane irritations, etc.

One of the agents contributing to SBS is odors generated by the human body. Such odors are called ***bioeffluents*** that contain ***pheromones*** (volatile and semivolatile compounds produced by all primates and humans, designed to elicit specific responses in members of the same species). These scents are present in oils in skin and hair, sweat, semen, and urine.

Whatever its origin (Abt et al, 2000) or however serious the health risks, the issue is of serious economic importance to the owners of buildings who must expend money to alleviate the problem. If not solved, the value of the real estate suffers. Certain gases and vapors above minimal concentrations can reduce the quality of indoor air. See for example the list of ***hazardous air pollutants*** (HAPs) in Table 2.8. However, the culprits are often common molds, yeasts, bacteria, and spores to which individuals are allergic.

There are many sources of allergens that contribute to poor IAQ, but three major changes in housing practice contribute, if not explain sick buildings:

- air conditioners, heat pumps, and evaporative coolers
- wall-to-wall carpets
- inadequate ventilation

Since air conditioning is generally associated with SBS (Godish, 1995), sick building symptoms may be ***biogenic*** and initiate allergic reactions in individuals, or reactions to microbial toxins. Condensate water from air conditioners can become a home for biological organisms. As commercial heat pumps often are used not only to cool, but also to heat and humidify air, the source of water for humidification may also become a breeding place for microorganisms. Eliminating microorganisms in the water of sumps of large, central air conditioning systems is a manageable task, but renovated buildings often have many small room air conditioners mounted in openings in the wall, or in windows. Drip pans in these units often do not receive attention, and allergen colonies develop. In private homes, air conditioners and ***evaporative coolers*** ("swamp coolers") seldom receive necessary attention, and colonies of aquatic organisms breed. Air passing over this water transports organisms into the work place and living quarters.

It is cheaper to install carpets than hardwood floors. Carpets absorb sound, reduce maintenance, and provide a soft visual appearance to interiors. However, it is well known that carpets provide hospitable surroundings for allergens. Because of their size, these small particles are difficult, if not impossible, to remove in spite of meticulous vacuuming. Particles imbedded in carpets are made airborne when people walk on the carpet. Small children ("rug rats") spend considerable time on all fours; it is not surprising that the incidence of respiratory illness is high.

Table 2.8 Hazardous air pollutants (HAPs), from the Clean Air Act Amendments of 1990.

CAS #	chemical name
75070	acetaldehyde
60355	acetamide
75058	acetonitrile
98862	acetophenone
53963	2-acetylamino-fluorene
1007028	acrolein
79061	acrylamide
79107	acrylic acid
107131	acrylontrile
107051	allyl chloride
92671	4-aminodiphenyl
62533	aniline
90040	o-anisidine
1332214	asbestos
71432	benzene
92875	benzidine
98077	benzotrichloride
100447	benzyl chloride
92524	biphenyl
117817	bis(2-ethylhexyl)-phthalate
542881	bis(chloromethyl)-ether
75252	bromoform
106990	1,3 butadiene
156627	calcium cyanamide
105602	caprolactam
133062	captan
63252	carbaryl
75150	carbon disulfide
56235	carbon tetrachloride
463581	carbon sulfide
120809	catechol
133904	chloramben
57749	chlordane
7782505	chlorine
79118	chloroacetic acid
532274	2-chloroacetophenone
108907	chlorobenzene
510156	chlorobenzilate
67663	chloroform
107302	chloromethyl methyl ether
126998	chloroprene
1319773	cresols/cresylic acid (isomers and mixture)
95487	o-cresol
108394	m-cresol
106445	p-cresol
98828	cumene
94757	2,4-d, salts and esters
3547044	DDE
334883	diazomethane
132649	dibenzofurans
96128	1,2-dibromo-3-chloropropane
84742	dibutylphthalate
106467	1,4-dichlorobenzene
91941	3,3-dichlorobenzidene
111444	dichloroethyl ether (bis(2-chloroethyl)ether)
542756	1,3-dichloropropene
62737	dichlorvos
111422	diethanolamine
121697	n,n-diethylaniline
64675	diethyl sulfate
119904	3,3-dimethoxybenzidine
60117	dimethyl aminoazobenzene
119937	3,3′-dimethyl benzidine
79447	dimethyl carbamoyl chloride
68122	dimethyl formamide
57147	1,1-dimethyl hydrazine
131113	dimethyl phthalate
77781	dimethyl sulfate
534521	4,6-dinitro-o-cresol and salts
51285	2,4-dinitrophenol
121142	2,4-dinitrotoluene
123911	1,4-dioxane (1,4-diethyleneoxide)
122667	1,2-diphenylhydrazine
106898	epichlorohydrin (1-chloro-2,3-epoxypropane)
106887	1,2-epoxybutane
140885	ethyl acrylate
100414	ethyl benzene
51796	ethyl carbarmate (urethane)
75003	ethyl chloride (chlorothane)
106934	ethylene dibromide (dibromoethane)
107062	ethylene dichloride (1,2-dichloroethane)
107211	ethylene glycol
151564	ethylene imine (aziridine)
75218	ethylene oxide
96457	ethylene thiourea

CAS #	chemical name
75343	ethylidene chloride (1,1-dichloroethane)
50000	formaldehyde
76448	heptachlor
118741	hexachlorobenzene
87683	hexachlorobutadiene
77474	hexachloro-cyclopentadiene
67721	hexachloroethane
822060	hexamethylene-1,6-diisocyanate
680319	hexamethyl-phosphoramide
100543	hexane
302012	hydrazine
7647010	hydrochloric acid
7664393	hydrogen fluoride (hydrofluoric acid)
123319	hydroquinone
78591	isophorone
58899	lindane (all isomers)
108316	maleic anhydride
67561	methanol
72435	methoxychlor
74893	methyl bromide
74873	methyl chloride (chloromethane)
71556	methyl chloroform (1,1,1-trichloroethane)
78933	methyl ethyl ketone (2-butanone)
60344	methyl hydrazine
74884	methyl iodide (iodomethane)
108101	methyl isobutyl ketone (hexone)
624839	methyl isocyanate
80626	methyl methacrylate
1634044	methyl tert butyl ether
101144	4,4-methylene bis(2-chloroaniline)
75092	methylene chloride (dichloromethane)
101688	methylene diphenyl diisocyante
107779	4,4-methylene-dianiline
91203	naphthalene
98953	nitrobenzene
92933	4-nitrobiphenyl
100027	4-nitrophenol
79469	2-nitropropane
684935	n-nitroso-n-methylurea
62759	n-nitroso-dimethylamine
59892	n-nitroso-morpholine
56382	parathion
82688	pentachloro-nitrobenzene

CAS #	chemical name
87865	pentachlorophenol
108952	phenol
106503	p-phenylenediamine
75445	phosgene
7803515	phosphine
7723140	phosphorus
85449	phthalic anhydride
1336363	polychlorinated biphenyls (aroclors)
1120714	1,3-propane sulfone
57578	beta-propiolactone
123386	propionaldehyde
114261	propoxur (baygon)
78875	propylene dichloride (1,2-dichloropropane)
75569	propylene oxide
75558	1,2-propylenimine (2-methyl aziridine)
91225	quinoline
106514	quinone
100425	styrene
96093	styrene oxide
1746016	2,3,7,8-tetrachlorodibenzo-p-dioxin
79345	1,1,2,2-tetrachloroethane
127184	titanium tetrachloride
108883	toluene
95807	2,4-toluene diamine
584849	2,4-toluene diisocyanate
95534	o-Toluidine
8001352	toxaphene (chlorinated camphene)
120821	1,2,4-trichlorobenzene
79005	1,1,2-trichloroethane
79016	trichloroethylene
95954	2,4,5-trichlorophenol
88062	2,4,6-trichlorophenol
121448	triethylamine
1582098	trifluralin
540841	2,2,4-trimethylpentane
108054	vinyl acetate
593602	vinyl bromide
75014	vinyl chloride
75354	vinylidene chloride (1,1-dichloroethylene)
1330207	xylenes (isomers and mixture)
95476	o-xylene
108383	m-xylene
106423	p-xylene

The economics of construction in the US results in minimizing (to a limit) the number of square feet of floor space and headspace allotted to people in office buildings. Partitions, carpets, and wall hangings are then used to minimize the feeling of being crowded. The economics of design and construction seldom provides space to locate ventilation ducts and outlets to ventilate interior spaces adequately. Even if designed properly, later interior renovations often fail to provide adequate ventilation. While economically beneficial, the reduced amount of fresh air circulated to the workspace increases the chance for concentrations of allergens to accumulate. This is coupled with the fact that the desire to minimize the cost to heat and cool air requires tighter homes, and the natural exchange of air through infiltration and exfiltration is reduced. With the increase in the number and diversity of home entertainment products and an increase in the percentage of air that is conditioned, there has been a decline in the amount of time people spend outdoors. As a result, exposure to indoor allergens has increased.

Environmental tobacco smoke (ETS) consists of exhaled tobacco smoke plus side stream smoke from smoldering tobacco products in ashtrays (see Appendix A.7). ETS is one of the most pervasive contaminants of indoor air pollution, and in years past was one of the principal factors contributing to SBS. ETS is particularly annoying to individuals who recently gave up smoking tobacco products. There are now many industry and government policies that prohibit smoking inside buildings or restrict it to designated ventilated smoking areas. Thus ETS is no longer the principal factor contributing to SBS.

Sources of indoor contamination present in the home or related to office equipment, furnishings, maintenance, and combustion include the following:

Equipment

- carbon black and oxygenated hydrocarbons from toner placed on the rotating drum of laser printers and copy machines
- VOC vapors from duplicating machines (alcohol-based purple dye) still used in some schools
- odors generated by copying machines
- ammonia and acetic acid generated from blueprint machines
- formaldehyde vapor escaping from urea-formaldehyde bonded wood products, and from glues and adhesives
- materials inside HVAC systems decomposing due to elevated temperatures, humidity, and reactions with ozone; acetic acid emitted from silicone calking inside ducts, glass particles escaping from sound absorbing materials, VOCs escaping from resin binders and protective emulsion coatings
- ozone generated (45-158 μg/copy, 0-1350 μg/min) by electrostatic copying machines; the typical ozone concentration in the operator's breathing zone is 0.001-0.15 PPM, and is a problem because copiers are often located in small unventilated rooms
- glass particles eroding from ceiling tiles
- formaldehyde escaping from carbonless copy paper

Furnishings

- fungeal spores generated by moldy plants
- fibers generated by carpet, carpet-backing, and underlying cushion pad
- off-gassing VOCs from dye solvents and chemicals used to withstand stains and resist soiling in carpet; typical values of total VOCs are 0.71-0.315 mg/(m^2-hr), and decrease exponentially (58% disappears within 24 hrs and 91% disappears after one week)

- off-gassing solvents from carpet adhesives (90% disappears within 250 hrs), interior work station partitions, wall coverings, and drapes
- organic matter that can be a sink for microbial contaminants (molds, bacteria, etc.) from carpeting and upholstered furniture
- contaminants introduced by foot/shoe traffic
- VOCs generated by caulks, sealants, insulation, and plasticizers used in vinyl flooring
- pleasant smelling terpene compounds such as pinenes and limonene in wood products

Maintenance

- misapplied office cleaning products, which generate irritating odors
- misapplied cleaning agents in restrooms, which also produce irritating odors
- irritating pesticides on house plants
- irritating eye agents generated by carpet cleaning chemicals (shampoos)
- air "fresheners" and personal care products, which emit VOCs

Combustion systems in private homes

- emissions from cooking food
- indoor combustion systems including gas stoves and ovens, wood burning stoves, and unvented space heaters
- automobile emissions from cars in attached garages

These sources emit varying amounts of carbon monoxide (CO), oxides of nitrogen (NO_x), VOCs, and small particles)

Formaldehyde is a ubiquitous indoor air pollutant characteristic of SBS, and generally present in the range of 0.05-0.5 PPM. It is unlikely that formaldehyde is responsible for SBS, although it may be a causal contributing factor. Shown in Table 2.9 are ***emission factors*** (EFs) for a variety of tested indoor materials used in offices (adapted from Maroni, Seifert, and Lindvall, 1995). Note that there is great variability in data from one product to another. EFs will be discussed in more detail in Chapter 4.

2.5.1 Cross Contamination, Reentry, and Entrainment

Some of the most overlooked sources of SBS are pollutants engineers and facility managers believe are safely discharged to the outdoor environment, but find their way into the building's indoor environment:

- ***Cross contamination***: Make-up fresh air for a building's ventilation system contains pollutants discharged from an adjacent source. Examples include boiler exhaust gases from short stacks, exhausts from local ventilation systems such as laboratory fume hoods and cooking hoods, and specific manufacturing processes such as dip tanks, degreasing tanks, paint booths, etc.
- ***Re-entry***: Pollutants discharged from a building mix inadvertently with fresh air and are drawn into the building's ventilation system instead of being transported to the greater environment. Examples include boiler exhaust gases and auto exhausts from parking facilities beneath the building.
- ***Entrainment***: Pollutants generated inside a building are not discharged as planned but enter the building through open windows and doors, and into elevator shafts; negative pressure inside a building produces reverse flow in process exhaust ducts that then allows exhaust gases to enter the workplace. Other examples include cafeteria food odors, roadway contaminants, sewer gases, and discharges from underground storage tanks and storerooms for hazardous materials.

Table 2.9 Emission factors (EFs) for tested indoor materials; "NA" means not available (abstracted from Maroni, Seifert, and Lindvall, 1995).

source	condition	EF (mg/m^2-hr)
chipboard	NA	0.13
dry-cleaned clothes[4]	0-1 day	1
	1-2 days	0.5
floor adhesive	< 10 hr	220
	10-100 hrs	< 5
floor varnish or lacquer	NA	1
floor wax	< 10 hr	80
	10-100 hrs	< 5
gypsum board	NA	0.026
latex-backed carpet[2]	1 week old	0.15
	2 weeks old	0.08
moth cake[3]	23 °C	14,000
particleboard[1]	2 yrs old	0.2
	new	2
plywood paneling[1]	new	1
polyurethane wood finish	< 10 hrs	9
	10-100 hrs	< 0.1
silicone calk	< 10 hr	13
	10-100 hr	< 2 h
wallpaper	NA	0.1
wood stain	< 10 hr	10
	10-100 hrs	< 0.1

[1]formaldehyde (HCHO), [2]4-phenylcyclohexene, [3]paradichlorobenzene, [4]perchloroethyllene

2.5.2 Avoidance & Mitigation

If engineers could identify the chemical agents responsible for unsatisfactory indoor air quality (IAQ), they could measure their concentrations and correlate them to the symptoms characterizing SBS. Unfortunately the chemical species responsible for SBS are not identified easily; indeed they are not even necessarily hazardous air pollutants (HAPs). Lastly the concentrations of the offending species may be very small and not easily measured. The best way to correct SBS is to avoid using products containing the agents which experience has shown are responsible for inducing symptoms in hypersensitive people. Mitigation involves removing materials identified as sources of contaminants and altering the performance of the ventilation system.

It is obvious that avoidance (also called prevention) is less expensive and a preferred strategy to mitigation. Avoiding SBS begins by reviewing the list of things causing the IAQ deterioration and then taking appropriate countermeasures. These measures can be summarized as follows:

- Use equipment and furnishings constructed of materials which experience has shown generate minimal amounts of vapor and particle contaminants (see Godish, 1995). Similarly, use janitorial materials which experience has shown do not generate copious emissions nor generate them over long periods of time. The truism is obvious, but it is difficult to achieve. Vendors of the equipment, furnishings, and materials are not obliged to describe the emissions with the specificity engineers need, nor are the vendors prevented from changing formulations without notifying buyers.

- Model analytically the air velocity field near fresh air inlets and the concentration field (dispersion models) of plumes from all discharge stacks (large and small) to ensure that there is no cross-contamination or re-entry. Such modeling must also include contaminant emissions from adjacent buildings and or stacks. Secondly, it is necessary to model the distribution of air into and within rooms to ensure that it is appropriate for the number of occupants and activities within the rooms. In many cases new buildings are not designed for specific arrangements of occupants, and the modeling must be conducted after the decisions on occupancy have been made. For rooms within remodeled buildings, it is necessary to analyze the velocity field to ensure that contaminant concentrations are within acceptable levels. The modeling of air distributed within enclosed spaces is one of the principal objectives of this book. The modeling of air transport outside the building will not be covered in this book; the reader should consult Hosker (1982), Turner (1994), Heinsohn and Kabel (1999), or de Nevers (2000).
- Review (and change if necessary) maintenance practices concerning cleaning furniture, floors, washrooms, etc. to ensure that materials used do not generate unacceptable amounts of contaminants.

While avoidance is the desirable way to correct SBS, there are occasions when it is necessary to renovate portions of a building or an entire building. When renovation is called for, the above steps are still necessary, albeit on a smaller scale. Suggestions regarding the renovation of buildings can be summarized as follows:

- ***Product labeling*** - Suggestions that governmental agencies document, rank, or even establish a database of materials, furnishings, etc. that contribute to SBS have not been proposed formally by the EPA. It has been suggested that new products be allowed to "air out" for specified lengths of time before being used, but even this effort has not received acceptance nationally. It would be imprudent to attempt to identify "bad products" having unacceptably large emission rates. It would be wiser to provide incentive for manufacturers to label products as "green", or "healthy building products" or "clean products". Such labeling is controversial and has not gained popular support.
- ***Design factors*** - Buildings should be optimally located, distant from sources of pollution. Exhaust ducts from industrial processes should obviously be located downwind of HVAC fresh air inlets. Less obvious design factors are locating HVAC fresh-air inlets distant from:

    - parking lots and entrances and exits to motor vehicles garages
    - loading docks
    - temporary parking locations for delivery vehicles
    - pedestrian drop-off and pickup points
    - bus stops

  Additional design factors include locating activities such as printing, food preparation, and designated smoking areas distant from HVAC fresh-air inlets.
- ***Barriers*** - Formaldehyde (HCHO) is the single most common pollutant associated with SBS, yet the concentration of HCHO present in sick buildings is often well below values designated by OSHA or EPA as hazardous. One technique to minimize the emission of formaldehyde (and other volatiles) from wood products (pressed board, chip board, plywood, etc.) is to create a ***barrier*** to impede emission by coating the surface with polyurethane. While accepted by some individuals and trade associations, this has not become a state or national standard. Voluntary attempts to reduce formaldehyde emissions, or the emissions of any volatile compound, benefit the manufacturer of the product. Thus it is in the self-interest of vendors of materials and equipment to publicize steps they have taken to reduce emissions. Since ultimately the final cost of the product has an enormous impact on the decision to select

materials and equipment, manufacturers have a marketing task to convince customers to purchase the low emitting products.

- ***Building bake-out*** - One mitigation technique that has received considerable attention is ***bake-out***. Bake-out is based on the knowledge that the vapor pressure of volatile materials increases sharply with temperature. Thus the evaporation of volatile chemicals in materials inside a building can be accelerated by holding the temperature of the building at 30-35 °C for several days. It is not an easy task to raise the temperature and to maintain a high uniform temperature throughout an entire building, or even a portion of a building, for 2 to 3 days, all the while maintaining a specified ventilation rate of fresh make-up air. In addition to the difficult technical task, there is an economic factor for delaying the rental income of a new building. If an existing building is renovated, heating only a portion of the building while the remaining parts of a building remain occupied is very difficult.
- ***Biocides*** - Controlling the source of biological contaminants is the principal method of reducing exposure to bioaerosols that lead to hypersensitive pneumonitis, allergic symptoms, dust mites, mold antigens, fungal volatiles, mycotoxins, and endotoxins. A variety of biocides are effective in treating water in cooling towers, cooling coil condensate water traps, cooling coil drain pans, etc., to control the growth of microorganisms producing legionnaire's disease. Microorganisms can thrive in the sludge of sumps of vessels, even though their concentration in the liquid phase is small. Disinfecting cooling tower water by hyperchlorination is another approach, but it often generates chlorine odors. Also, some microorganisms are immune to chlorination. Heating cooling water sumps with steam or irradiating the water with immersed UV lamps have received some attention.
- ***Source modification*** - Hypersensitive pneumonitis, asthma, and allergic rhinitis are enhanced by exposure to microbial organisms, molds, dust mite antigens, macromolecular organic dust, etc. Dust mites and surface dust accumulating on furnishings and carpeting require more intensive house cleaning practices, e.g. dusting, vacuuming, steam cleaning, and surface wiping; in extreme cases they need to be replaced. A common host to these contaminants may be the porous sound insulation (fiberglass) used to line HVAC ducts and the filters used in the make-up air inlet of HVAC systems. These porous media amplify the microbial population on accumulated dust. While it may seem obvious that the interior surface of HVAC ducts may be a source of SBS, there is no experimental evidence at the time of this writing to support this belief.

2.6 Bioaerosols

Bioaerosols are small particles of biological origin that are toxic, cause infection, or induce an allergic or pharmacological response (Cox and Wathes, 1995). The aerodynamic diameter of bioaerosols is between 0.5 and 100 μm. Bioeaerosls are also called ***biogenic aerosols***. Some of the respiratory problems caused by bioaerosols are discussed below.

2.6.1 Respiratory Allergies and Asthma

The function of the immune system is to recognize and provide protection against agents that are harmful to the body. When the immune systems functions properly, foreign agents are eliminated quickly and efficiently. Cells of the immune system that provide this function are called ***leukocytes***, which are created within bone marrow. An ***allergy*** is a hypersensitivity to a specific substance, called an ***allergen***, which may be harmless to others. Approximately 2% of the population suffers from food allergies associated with cow's milk, eggs, peanuts, wheat, fish, crustaceans, soy, and tree nuts. Many respiratory allergens are finely divided organic dust; others are proteins. An ***antigen*** is an enzyme, toxin, or other substance in the allergen to which the body reacts by producing ***antibodies***. Certain types of white blood cells in allergic persons produce antibodies that attach themselves to mast cells. ***Mast cells*** are heavily granulated wandering cells that are found in areas rich in connective tissue

(Ganong, 1997), usually in the respiratory system, gastrointestinal tract, and the skin. When antibodies detect an allergen, a large number of chemicals is produced, the best known of which is ***histamine,*** which produces watery eyes, runny nose, itching, and sneezing. The type of allergic response, e.g. asthma, hay fever, or hives, etc. depends on which part of the body interacts with the activated mast cells. More than 150 types of airborne allergens are currently recognized (Hamilton 1992). Tree, grass, and weed pollen, as well as molds, are generated in the outdoor environment but may also enter the indoor environment via infiltration or ventilation.

Over 35 million Americans suffer from some kind of allergy. Fifteen million suffer from hay fever and nine million have asthma. ***Hay fever*** is a disorder induced by hypersensitivity that affects the upper respiratory tract. ***Asthma*** is a chronic and debilitating disease causing swollen and inflamed airways that are prone to sudden and violent constrictions, and can increase lung resistance 20-fold. Asthmatic attacks are characterized by shortness of breath and wheezing, and can be life-threatening. Asthmatic airway inflammation is initiated by airborne proteins (allergens). Asthma is associated with the death of over 4,000 people annually (Cookson and Moffatt, 1997). Asthma attacks may also be induced by viral infections or allergic reactions. Mast cells that induce the reaction are located in the bronchial tubes and lungs. Symptoms include swelling of the bronchial tubes and spastic contractions of the muscles surrounding the bronchial tubes. Both actions reduce the internal cross-sectional area of bronchial passages. Simultaneously, sensory nerve endings in the lung release chemicals that stimulate the secretion of excessive amounts of mucus.

Allergens from house pets, insects, microorganisms, and aerosols from tobacco smoke, cooking, and other indoor activities are generated in the indoor environment. With the increase in indoor entertainment products and the increase of US homes that are air-conditioned and weather tight, there has been a decline in outdoor activity as individuals spend more time indoors. As a result, exposure to indoor allergens has increased. Allergens produce rhinitis, asthma, hay fever, and various reactive airway and alveolar diseases. The typical size of allergen particles is shown in Table 2.10.

House pets - Cat allergens are produced by sub-lingual salivary glands and hair root sebaceous glands of the domestic cat. Cat allergens bind strongly to clothing, carpet, upholstered furniture, or inert airborne dust and can be resuspended. Cat antigen is highly persistent and not affected by high temperatures or steam cleaning. Regular vacuuming does little to reduce indoor antigen levels. Bird feathers and droppings shed albumins, proteins, and fungi that are allergens. The domestic dog produces complex protein allergens associated with hair, dander, saliva, feces, and urine. Dog allergens are highly persistent and remain stable in homes for extended periods.

Table 2.10 Particle size of several common indoor allergens.

contaminant	**D_p (μm)**
bacteria	0.3 - 50
dust mites (microscopic anthropods)	> 10
fungal spores	1 - 200
guinea pig pelt	0.8 - 4.9
legionella pneumophila	0.4 - 0.7 (length, 30 - 50 μm)
mycobacteria tuberculosis	0.3 - 0.6 (length, 1 - 4μm)
pollen	1 - 200
ragweed (oblate or prolate spheroids)	20 - 30
rat urine	0.8
tobacco smoke	0.01 - 1
viruses	0.004 - 0.05

Household pests - The most important source of allergens is the feces of dust mites which thrive in warm moist conditions and are ubiquitous in human bedding. Dust mite antigens are excreted in their fecal pellets; antigens are also associated with mite body debris. Dust mites have a life span of 3 to 3.5 months and thrive in a warm humid environment. Dust mites consume human epithelial cells and animal dander in mattresses, pillows, carpets, and furniture. ***Skin flakes*** are shed by the abrasion of clothing and body parts rubbing against each other. An average skin flake has an equivalent diameter of 14 μm. It is estimated that the entire outer layer of skin is shed every day or two (Rothman, 1954) at a rate of 7 million skin flakes per minute (Clark and Cox, 1973). Tests of indoor environmental dust in homes and offices have shown it to be primarily (70-90%) composed of skin flakes (Clark and Cox, 1973; Clark, 1974). Assuming a density of 1 gm/cm^3, 7 million skin flakes per minute corresponds to a mass emission rate of about 20 mg/minute. Rat and mice allergens are generated from rodent hair, skin, feces, and urine. Allergens from cockroaches are produced from their exoskeleton and feces. Air pollution may also aggravate existing asthma, but air pollution is not believed to be solely responsible for the doubling of the disease in the last 20 years (Cookson and Moffatt, 1997).

Bacteria are microscopic organisms of various shapes (spheres, rods, spirals, etc., with characteristic dimension 0.5 to 5 μm), which are generally associated with fermentation, putrefaction, and disease. Bacteria are generated as aerosols from humans, plants, and animals, or from water sources such as showers, toilets, urinals, and HVAC systems. Bacteria can accumulate quickly in humidifier reservoirs. Soil bacteria are tracked into homes on the soles of shoes or attach themselves to airborne inert dust. Each of the approximately 7 million skin flakes which humans shed per minute contains on average 4 viable bacteria. Sneezing, coughing, and singing generate airborne bacteria. ***Thermophilic actinomycetes*** are organisms that grow at elevated temperatures (104 - 158 °F) and high humidity. Farm, sugar cane, tobacco, and mushroom workers are occupational groups at risk. Chronic hypersensitivity may result in inflammation of interstitial tissue leading to lung scarring and fibrosis. In severe cases, the alveolar interstitium becomes fibrotic, bronchiole walls thicken, alveoli widen, and edema and emphysema may occur.

Plant allergens - During the peak pollen season, all kinds of pollen are inhaled each day. Many pollens release antigens within 30 seconds of hydration on mucous membranes or the respiratory tract. Allergic pollens are produced by trees, shrubs, grasses, weeds, mosses, ferns, herbs, and many popular indoor plants. Ragweed allergens consist of pollen and fragments of leaves and stems. Ragweed pollen can transfer antigens to airborne inert dust which can be resuspended. A variety of fungi and molds exists in the outdoor and indoor environment throughout the year.

Farmers lung results from the inhalation of certain spores that inflame alveoli and produce fever and dyspnea. ***Bagassosis*** is a disease having similar symptoms inflicting workers handling dried and partially fermented sugar cane. ***Byssinosis*** is constriction of the bronchial tree producing dyspnea in workers handling cotton, flax, and hemp. ***Black lung*** is a similar disease associated with coal mining. Numerous industrial chemicals, such as toluene diisocyanate (TDI) used in the manufacture of polyurethane plastic and methylisocyanate (MCI) used in the manufacture of insecticide, initiate allergic-like symptoms and more serious symptoms if the exposure and concentration are sufficiently high.

2.6.2 Tuberculosis

Tuberculosis (TB) is the number one killer among infectious diseases and a leading cause of death among women of reproductive age. About 2 million Americans are infected with ***mycobacterium*** (M. tuberculosis). Among these people 10% will develop the disease at some point in their lives; the remaining 90% will never display the symptoms and never realize they have the disease. This 90% is said to have ***latent tuberculosis***. Of the 10% that develop the disease only 5% will have ***primary active tuberculosis***, and they will manifest the disease within a few years of infection. The remaining 5% of infected people will be fine for decades, and become sick long after they were infected. This condition is called ***reactivated tuberculosis***. Unlike other infectious agents, the bacillus does not produce a toxin and does not need oxygen to survive, probably living on lipids in its host. The bacillus has a slow growth cycle and complex cell envelope, colonizes macrophage, and may remain quiescent and then react decades after infection.

When most microorganisms enter the lung they are engulfed by macrophage, and enter a compartment of the macrophage where chemicals and enzymes degrade them. With M tuberculosis this reaction does not occur, and the bacilli survive within a cocoon inside the macrophage. As the bacilli multiply, their number eventually becomes overwhelmingly large, and the macrophage dies, releasing bacilli to be taken up by other macrophage. In 90% of the cases, the bacilli are inactive and remain in the lung for the rest of the person's life. In about 5% of the cases people don't have protective immune response and the bacilli multiply and spread rapidly to other parts of the lung, producing primary active TB within months of the initial infection. In another 5% of the cases the protective immune response produces a latent state that disappears only decades later; at this point, the bacilli proliferate and cause reactivated TB.

The symptoms of pulmonary tuberculosis are persistent cough, fever, chills, night sweats, tiredness, appetite loss, weight loss, and spitting of blood. TB can also cause infections of bone, kidney, brain, and the lymphatic system. TB is highly contagious, and spreads when the host exhales the bacilli, which are then inhaled by others. TB is spread by droplets ($1\ \mu m < D_p < 5\ \mu m$) containing the bacilli. A single laryngeal tuberculosis lesion may produce thousands to millions of organisms whenever an infected person sneezes, coughs, speaks, sings, or laughs. In a typical sneeze, approximately one million TB bacilli are contained in droplets that evaporate before the droplets reach the floor. Usually about 50% of family members of the infected acquire the disease. Persons suffering from HIV have weak immune systems and are particularly susceptible to TB. Tuberculosis is a treatable disease and several drugs are available to treat susceptible strains of M. tuberculosis, but a strict, costly, long-term (6 months) treatment period using several drugs is required. Poor and homeless people often cannot afford the time or money to travel many times to health-care centers for treatment. At the present time, ***multi-drug-resistant*** (MDR) strains of TB are developing, which is an alarming aspect of the disease. These strains of TB occur when patients do not follow the prescribed long-term treatment with drugs, or take the drugs in improper amounts.

2.6.3 Pneumonia and Bronchitis

Pneumonia is an inflammatory condition of the lung in which the alveoli are usually filled with fluid. A common pneumonia is caused by bacterial infection of the alveoli. Lung fluid, largely water but also containing red and white blood cells, enters alveoli and may, in time, fill the entire lung. Reduction of the total alveolar membrane and decrease of the ventilation-perfusion ratio is called ***hypoxemia***. ***Bronchitis*** is an inflammation of the mucous lining of the bronchial tree (see Figure 2.37), and results in an increased production of mucus.

2.6.4 Legionella

Legionella pneumophila are water-borne bacteria that produce a virulent pneumonia. The first outbreak of the disease resulted in the death of several dozen adults attending a convention of the American Legion in 1976 (hence the name ***legionnaire's disease***). The sump water in a wet-cooling tower used in an air conditioning system provided a hospitable environment for the bacteria to breed. The bacteria became air borne when the cooling water passed through the packing in the tower that was discharged to the atmosphere as a plume of warm air and water droplets. After being discharged to

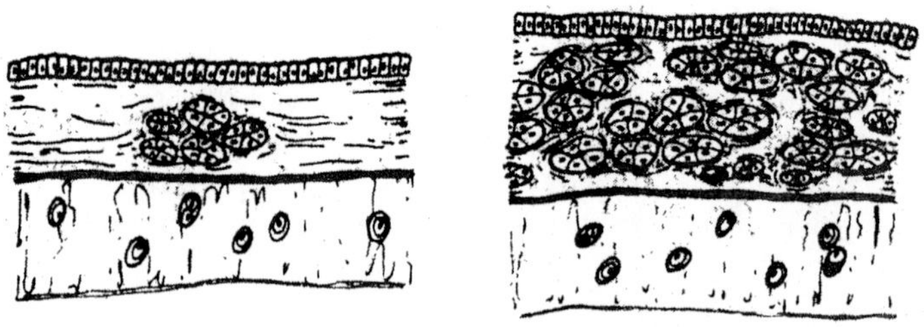

Normal Chronic bronchitis

Figure 2.37 Mucous gland layer in a normal human airway (left) contrasted with one from a patient with chronic bronchitis (right) (adapted from Occupational Health and Safety, 1989).

the atmosphere, the bacteria mixed with the make-up air drawn into the air handling system and was distributed throughout the building. Once the sump water was identified as the host for the bacteria, a concerted effort began to monitor the sumps of all air conditioning systems in public buildings. It was also recognized that drip-pans in window-mounted air conditioners can breed bacteria that can be drawn into the room with the make-up air. ASHRAE has taken the lead in developing guidelines (ASHRAE Guideline 12-2000; Geary, 2000) to control and monitor the bacteria in cooling water.

Cooling towers are the principal source of legionella infections (Meyer, 2000). Legionella are hardy microorganisms that thrive in the slime that coats the interior surfaces in cooling towers. Conventional open cooling towers are evaporative heat transfer devices in which atmospheric air mixes with and cools warm water by evaporating a portion of the water. A fan forces air through the tower containing a packed bed ("fill") that provides a large surface over which flowing water evaporates and is thereby cooled. Cooling towers are used to reject waste heat from a chiller in an air conditioning system. Water that is evaporated is replaced by make-up water generally from a municipal water supply. Since the make-up water likely contains legionella in low concentrations, a small concentration of legionella is always present in the cooling tower. For this reason cooling water needs to be treated, to some degree, all the time. The temperature of the water within a cooling tower is likely to be between 25 °C and 42 °C, which is favorable to causing legionella to multiply. In addition, cooling towers and evaporative condensers move large amounts of unfiltered air and organic matter, and other debris is likely to accumulate inside the tower. Birds that nest near cooling towers provide a varied supply of bacteria; these materials provide nutrients for legionella. Legionella are ***protozoonotic***, in that they live and reproduce within the bodies of other microbes, especially amoebae and paramecia (Meyer, 2000). Legionella growing within protozoa are the primary means by which legionella proliferate. Because protozoa are relatively resistant to both oxidizing and nonoxidizing biocides, it is important that protozoa be controlled by limiting microbial biofilms that provide them with nutrients. Protozoa can, within weeks of cleaning a cooling tower, reestablish a legionella population. The cooling tower needs to be inspected on a regular basis and should be cleaned when there is evidence of a buildup of dirt, organic matter, or other debris.

Following an attack of legionellosis, hot water systems can be decontaminated either by ***thermal shock treatment*** or by ***shock chlorination***. Thermal shock requires raising the water temperature to 71 to 77 °C for at least five minutes and then flushing each outlet in the system. Alternatively, chlorine can be added to the water in the cooling tower sump to achieve a free chlorine residual concentration of at least 2 mg/L for at least 2 hours. In the water storage tank, the chlorine concentration should be between of 20 to 50 mg/L. In either case, the pH of the water should be maintained between 7.0 and 8.0. As a general principle in potable water supplies, cold water should be stored and/or delivered at temperatures below 20 °C (68 °F). In health care facilities, the hot water should be stored at or above 60°C (140 °F).

Conventional water treatment to control bacterial growth includes the addition of chemicals such as chlorine, iodine, ozone, or specialty biocides, or the inclusion of a UV and hydrogen peroxide catalyst system. A water treatment program is needed to:

- minimize scale and corrosion
- control microbial growth
- minimize the deposition of solids (organic and inorganic)

Scaling is minimized by inhibitors such as phosphonates, phosphates, and polymers that keep minerals such as calcium in solution. ***Corrosion*** is controlled by inhibitors such as phosphates, azoles, molybdenum, and zinc. Microbial fouling can influence scaling and corrosion and affect the performance of inhibitors. Microbial biofilms can consume certain inhibitors, prevent inhibitors from migrating to solid surfaces, create localized oxygen-depleted zones, change the pH near surfaces, and accumulate or trap deposits onto surfaces.

Surfactants are used to minimize the deposition of a ***biofilm*** on heat transfer surfaces. The surfactant must be tailored to the dirt, oil, and other materials present, as well as be compatible with the chemicals used to inhibit scale and corrosion. Microbial fouling is controlled by biocides that interfere with microbe cellular functions. Some biocides react with components in some scale and corrosion inhibitors and reduce their effectiveness. Both ***oxidizing biocides*** and ***nonoxidizing biocides*** are used. Oxidizing biocides function in one of two primary ways. Halogen biocides (chlorine, bromine, or iodine) react with protein in the cell membrane to cause the protein to become dysfunctional, thus killing and controlling the organism. Ozone and chloride dioxides are believed to oxidize the components in the microbial cell. On the other hand, non-oxidizing biocides are either enzymes or materials that solubilize cell membranes and precipitate essential proteins in microbial cell walls. Using a variety of biocides over a period of time has proven to be most effective. Individuals responsible for controlling microbial fouling need expertise in water chemistry and water microbiology.

There is currently interest in the US and Europe to curb the wide scale use of biocides to control microbial growth that can be effectively controlled by other techniques. The fear is that some microorganisms may become resistant, rendering the biocide ineffective. It has been demonstrated that in a population of microorganisms, some of the individual microbes are more tolerant of a given biocide than are others in the population, either due to genetic factors or sometimes due to mutations in their genes. Some of these more tolerant individuals may survive the biocide attack, and then breed a new generation of microorganisms which resist the particular biocide (Crabb, 2000). However, the new, more resistant population may be quite susceptible to some other biocide. At the very least, a *variety* of biocides should be used to minimize breeding legionella that are resistant to treatment.

2.6.5 Bioterrorism and Inhalation Anthrax

Since the terrorist attacks on the United States on September 11, 2001, ***bioterrorism*** and ***biological warfare*** have become major concerns throughout the world. Of prominent concern is ***inhalation anthrax***, which is a disease caused by the spore-forming bacterium, ***Bacillus anthracis***. Anthrax spores are cocoon-like structures that the bacteria build around themselves when conditions are unfavorable for their survival. Their tough protective coat allows the spores to survive for decades, and they can be easily distributed in powder form through the mail, as was done in the terrorist attacks in the Fall of 2001. Anthrax spores range in size from about 1 to 5 μm (microns) and are thus easily airborne; some of the smaller spores can pass through the nasal passages and enter the lungs of exposed individuals. Once in the alveoli, the spores are engulfed by macrophage. Some of the spores are destroyed, but others survive and are transported to the lymph system. After an incubation period of 2 to 60 days, the bacteria release several toxins, causing edema, hemorrhage, and local tissue necrosis in the lungs. The victim first experiences fever, malaise, muscular aches, sneezing, and coughing – symptoms similar to the common cold. After several hours to several days of these symptoms, the infected person develops severe breathing problems, followed by shock; the fatality rate at this point is more than 90%.

Antibiotics such as penicillin, tetracycline, and ciprofloxacin hydrochloride can be effective, but only if inhalation anthrax is detected early, and if the antibiotics are taken for up to 60 days (because of the incubation period). There are vaccines such as MDPH-PA and AVA which can prevent inhalation anthrax, but at the time of this writing are available only for those at high risk of contraction, such as veterinarians, lab workers, and military personnel.

2.7 Dose-Response Characteristics

A statement written by Paracelsus (Swiss physician, 1493-1541) is a basic tenet of modern toxicology: "*What is it that is not a poison? All things are poison and nothing is without poison. It is the dose only that makes a thing a poison.*" This axiom has stood the test of time and should be recalled today as the cardinal rule to address the public's anxiety about substances they believe are in the air, water, and food.

Safe exposure limits are derived from experience and experimental studies on humans and animals. Individuals in the workforce comprise a population that is narrow compared to the general population. Individuals in the workforce are, on the whole, healthy adults in the prime of life who are exposed to contaminants for only several hours of the day and can return home to rest and breathe cleaner air for the remainder of the day. Thus standards of ***occupational exposure*** should not be confused or equated to general ***environmental standards*** applicable to the general population. In setting standards for the outside environment, infants, the aged, the ill, and the infirm have to be included in the population. There is no escape from the outside environment and standards are more conservative.

Each individual responds to the exposure of toxins in a unique way. Luckily, groups of individuals respond in ways that can be analyzed statistically, and parameters can be defined that have statistical significance. Compilation of such data constitutes the basis for health standards and public policies prescribing the maximum exposure to which individuals may be subjected in the workplace or the outside environment. Underlying these standards are the "dose-response" characteristics of individuals for specific toxins. The definitions of dose and response depend on the specific toxin but have the following general properties:

- ***Dose*** refers to the total mass of toxin to which the body is subjected. Dose is a function of concentration of the toxin, duration of exposure, and rate at which it is introduced to the body.
- ***Response*** refers to measurable physiological changes produced by the toxin.

Toxicity may be acute or chronic. ***Acute toxicity*** is sudden damage resulting from a single exposure to a large concentration of a toxin. ***Chronic toxicity*** is accumulated damage from repeated exposure to small concentrations of a toxin over long periods of time. Reactions to these two different types of exposure have little resemblance to one another. Chronic toxicity is not predictable from knowledge of acute toxicity for the same chemical. For example, chemicals such as vitamin D, sodium fluoride, and chloride have a large acute toxicity but no known chronic toxicity. On the other hand, a single ingestion of metallic mercury passes through the body without causing significant damage, whereas metallic mercury ingested in small amounts over a long period of time accumulates in the body and is very harmful.

Toxicity may also be local or systemic. ***Systemic toxicity*** is distributed throughout the body, while ***local toxicity*** pertains to particular organs. Entry to the body can be through the skin, through oral intake, or by inhalation. Figure 2.38 is a schematic diagram showing the different pathways a toxin may take. Toxins can be categorized by the response they produce, as shown below, along with an example of each category:

- ***asphyxiant*** (causes insufficient oxygen in blood) - e.g. carbon monoxide
- ***irritant*** (inflames tissue upon contact) - e.g. ammonia
- ***corrosive toxin*** (damages tissue upon contact) - e.g. chromic acid
- ***allergen*** (affects the immune system) - e.g. ragweed pollen
- ***carcinogen*** (produces cancer) - e.g. nickel carbonyl

Shown below are categories of ***site-specific toxins*** (local toxicity), along with the site they attack and some examples:

- ***pulmonary toxin*** (lung) - e.g. asbestos, beryllium, chromium
- ***hepatoxin*** (liver) - e.g. carbon tetrachloride, nitriles
- ***nephrotoxin*** (kidney) - e.g. kepone, lead, allyl chloride
- ***brain toxin*** (brain) - e.g. narcotics, ketones
- ***skin toxin*** (skin) - e.g. benzyl chloride
- ***cardiotoxin*** (heart) - e.g. chloroethane
- ***neurotoxin*** (nervous system) - e.g. mercury, malathion
- ***ocular toxin*** (eyes) - e.g. methyl chloride, methanol, phenol
- ***hematopoietic toxin*** (bloodstream) - e.g. lead, benzene
- ***bone toxin*** (bone) - e.g. inorganic fluorides

Reproductive toxins are materials that produce low sperm production, genetic damage to egg or sperm cells, or menstrual disorders. An example of a reproductive toxin is polychlorinated biphenyl. Some sub-categories are listed below, with an example for each:

- ***mutagen*** - (alters the character of genetic material in cells) - e.g. ethyleneimine
- ***gametoxin*** - (damages sperm or ova) - e.g. benzo[a]pyrene
- ***teratogens*** - (interferes with normal development of the fetus after conception and may result in miscarriage, visible birth defects, or defects not noticeable at birth) - e.g. thalidomide

Finally, ***transplacental carcinogens*** are cancer-producing substances which cross the placenta and reach the fetus, e.g. ethyl nitrosourea.

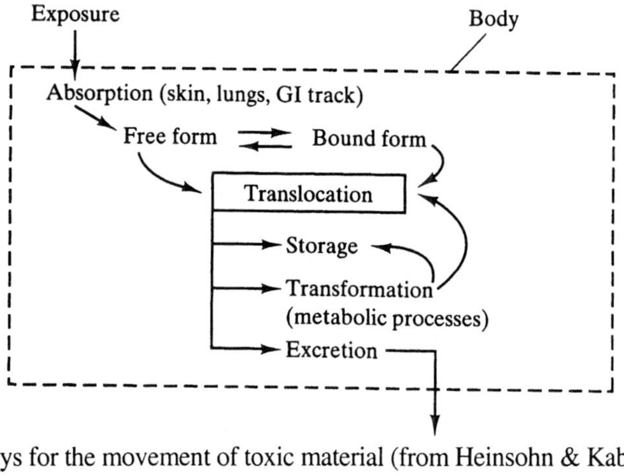

Figure 2.38 Pathways for the movement of toxic material (from Heinsohn & Kabel, 1999).

2.7.1 Biological Exposure Index

Occupational exposure is traditionally assessed by comparing the concentration of the contaminant in inhaled air with concentration standards promulgated by professional health organizations. One could logically ask: why measure the concentration in air presuming that it produces certain concentrations in the blood, urine, soft tissue, or bone; why not measure the concentration of the material or its byproducts in these parts of the body? Certainly measuring the concentration in blood and urine poses no technical difficulty, but the testing is intrusive and requires the cooperation of the individual who has been exposed. Such ***biological monitoring*** may replace air monitoring in the future, but until it does Leung and Paustenbach (1988) recommend that a series of ***biological exposure indices*** (BEIs) should be developed. Biological monitoring measures the actual concentration; the BEI *infers* the value based on measurement of external parameters.

The BEI is a measure of the amount of contaminant within the body resulting from airborne exposure. Specifically, the BEI is defined as the mass of contaminant stored in the body divided by the volume of the component(s) of the body in which the contaminant is stored, e.g. blood, soft tissue, urine, etc. Thus the BEI is a fundamental parameter reflecting the harm the body might endure. Unfortunately, while the concentration of the contaminant in the blood, urine, hair, etc. can be measured, there is no obvious way to extrapolate these measurements to the BEI itself.

Pharmacokinetics is the science that relates the rate processes of absorption, distribution, metabolism, and excretion of chemical substances in a biological system (Kreibel and Smith, 1990). To establish a BEI, one begins by establishing a relationship between the mass of some contaminant in the body, i.e. the ***body burden*** (m_{bb}), and the mass concentration of the contaminant in inspired air ($c_{inspired}$). Note that since only one contaminant is considered at a time, no subscript (such as j) is used in the notation here. Secondly, one establishes a relationship between the body burden and the mass concentration in the blood (c_{blood}). For a first approximation, the relationship can be expressed as first-order kinetics. The mass rate of accumulation of a contaminant in the body is equal to its rate of absorption from inspired air minus its rate of removal (by all forms),

$$\boxed{\frac{dm_{bb}}{dt} = FQc_{inspired} - k_r m_{bb}} \tag{2-49}$$

where

- m_{bb} = body burden, i.e. mass of contaminant in the body at any instant (mass)
- Q = volumetric flow rate of inspired, i.e. inhaled, air (volume/time)
- $c_{inspired}$ = contaminant mass concentration in inspired air (mass/volume)
- k_r = ***first-order rate constant***, which describes the overall process that transforms the contaminant and/or removes the contaminant from the body (1/time)
- F = ***bioavailability***, i.e. the ratio of the mass of contaminant absorbed by the body to the mass of contaminant in inspired air (dimensionless)

Eq. (2-49) is a standard, first-order, ordinary differential equation (ODE) with constant coefficients, as discussed in Chapter 1, which can easily be integrated by separation of variables. If the initial body burden at the start of the test is $m_{bb}(0)$, one obtains

$$\boxed{m_{bb}(t) = m_{bb,ss} - \left(m_{bb,ss} - m_{bb}(0)\right)e^{-k_r t}} \tag{2-50}$$

where the steady-state value of the body burden ($m_{bb,ss}$) is found by setting the left hand side of Eq. (2-49) to zero,

$$\boxed{m_{bb,ss} = \frac{FQc_{inspired}}{k_r}} \tag{2-51}$$

By definition, the BEI is the concentration of the material in the body at any instant. The material may be stored in the blood or urine or in organs of the body, e.g. liver, kidneys, etc. Thus,

$$\boxed{\mathrm{BEI} = \frac{m_{bb}}{V_{store}}} \tag{2-52}$$

where V_{store} is the ***storage volume***, defined here as the sum of the volumes of the fluids or organs of the body in which the material is stored. While blood may be the major component storing the material, it is not the only, or perhaps even the major component storing the material. Nonetheless the concentration of the material in the blood is a convenient parameter to measure. Unfortunately, V_{store} is unknown and must be determined before the BEI can be related to the concentration of the contaminant in blood (c_{blood}). As a first approximation, it is postulated that blood concentration is linearly proportional to BEI,

$$\boxed{c_{blood} = K(\mathrm{BEI}) = K\frac{m_{bb}}{V_{store}}} \tag{2-53}$$

Unfortunately, both the constant of proportionality (K) and the storage volume (V_{store}) are unknown. However, as will be illustrated in the example problem to follow, the ratio of these unknown parameters (K/V_{store}) can be computed from simple experiments at low dosages. Solving Eq. (2-53) for this ratio, one obtains

$$\boxed{\frac{K}{V_{store}} = \frac{c_{blood}}{m_{bb}(t)} = \frac{c_{blood}}{m_{bb,ss} - \left(m_{bb,ss} - m_{bb}(0)\right)e^{-k_r t}}} \tag{2-54}$$

where Eq. (2-51) has been used for $m_{bb,ss}$.

Independent experimental measurements can be conducted to reveal values of the bioavailability (F) and the first-order rate constant (k_r). F is determined by comparing the area under the concentration-time curves following equal doses of a chemical by intravenous injection and by inhalation. From Eq. (2-50), it can be shown that the first-order rate constant is inversely proportional to the ***half-life*** ($t_{1/2}$) of the contaminant in the body,

$$\boxed{k_r = -\frac{\ln(0.5)}{t_{1/2}} = \frac{0.693}{t_{1/2}}} \tag{2-55}$$

where $t_{1/2}$ is the time it takes the body burden to decrease by a factor of two after exposure to the contaminant in inhaled air has ceased. The following example illustrates how the body burden can be determined.

Example 2.4 – Body Burden of Acetone Associated with its PEL

Given: Acetone is used daily at a plant. Some workers are exposed to acetone vapors all day long, and are concerned about the amount of acetone in their blood. To ease their level of anxiety (and to protect the company from potential lawsuits), the supervisor wants to determine the maximum permissible acetone blood concentration. Medical staff at the plant will then monitor the workers' blood, and compare measured blood concentrations with these results.

To do: Estimate the concentration of acetone in the blood of an individual exposed to an 8-hour average acetone concentration equal to the PEL. It will be assumed that this blood concentration is the maximum permissible under OSHA standards.

Solution: The 8-hr PEL for acetone in air is 750 PPM (1,780mg/m^3). Not wanting to expose anyone to the PEL, a series of experiments is conducted in which an individual inhales air containing 545 PPM of acetone (1,293 mg/m^3) at a ventilation volumetric flow rate (Q_t) of 1.25 m^3/hr for 2 hours. The experiments reveal that after 2 hours, the concentration of acetone in the individual's blood (c_{blood}) is 10 mg/L. From toxicology literature, the bioavailability (F) of acetone is 45.%, and its half-life ($t_{1/2}$) is 4.0 hours. The first-order rate constant (k_r) for acetone can be found from Eq. (2-55),

$$k_r = -\frac{\ln(0.5)}{t_{1/2}} = \frac{0.693}{4.0 \text{ hr}} = 0.173 \text{ hr}^{-1}$$

<u>2-hour, 545 PPM Exposure</u>: The steady-state body burden for an ambient concentration of 545 PPM, $(m_{bb,ss})_{545}$, is found from Eq. (2-51),

$$\left(m_{bb,ss}\right)_{545} = \frac{FQc_{inspired}}{k_r} = \frac{0.45\left(1.25\frac{m^3}{hr}\right)\left(1293.\frac{mg}{m^3}\right)}{0.173\frac{1}{hr}} = 4204. \text{ mg}$$

After 2 hours, the body burden for this ambient concentration, $m_{bb}(2 \text{ hr})_{545}$, can be found from Eq. (2-50) with $m_{bb}(0) = 0$,

$$m_{bb}(2 \text{ hr})_{545} = \left(m_{bb,ss}\right)_{545}\left(1-e^{-k_r t}\right) = (4204. \text{ mg})\left[1-\exp\left(-0.173\frac{1}{hr}(2 \text{ hr})\right)\right] = 1230. \text{ mg}$$

Since this exposure for 2 hours produces a blood-level concentration, c_{blood}, of 10 mg/L, the value of (K/V_{store}) can be computed from Eq. (2-54),

$$\frac{K}{V_{store}} = \frac{c_{blood}}{m_{bb,ss}\left(1-e^{-k_r t}\right)} = \frac{10\frac{mg}{L}}{(4204 \text{ mg})\left[1-\exp\left(-0.173\frac{1}{hr}(2 \text{ hr})\right)\right]} = 0.00813\frac{1}{L}$$

<u>8-hour, 750 PPM Exposure</u> *<u>Extrapolation</u>*: The steady state body burden corresponding to an 8-hour exposure from an ambient acetone concentration equal to its PEL (750 PPM, 1,780 mg/m^3) can now be estimated from Eq. (2-51), assuming that the bioavailability (F) and the first-order rate constant (k_r) remain constants regardless of exposure level:

$$\left(m_{bb,ss}\right)_{750} = \frac{FQc_{inspred}}{k_r} = \frac{0.45\left(1.25\frac{m^3}{hr}\right)\left(1780.\frac{mg}{m^3}\right)}{0.173\frac{1}{hr}} = 5788. \text{ mg}$$

After 8 hours, the body burden at this concentration can be predicted from Eq. (2-50),

$$m_{bb}(8 \text{ hr})_{750} = \left(m_{bb,ss}\right)_{750}\left(1-e^{-k_r t}\right) = (5788. \text{ mg})\left[1-\exp\left(-0.173\frac{1}{hr}(8 \text{ hr})\right)\right] = 4337. \text{ mg}$$

Assuming in addition that the ratio (K/V_{store}) remains constant regardless of exposure level, the blood concentration corresponding to this body burden can be estimated from Eq. (2-53),

$$c_{blood}(8 \text{ hr})_{750} = \left(\frac{K}{V_{store}}\right)\left(m_{bb,ss\ 750}\right)\left(1-e^{-k_r t}\right)$$

$$= \left(0.00813\frac{1}{L}\right)(5788. \text{ mg})\left[1-\exp\left(-0.173\frac{1}{hr}(8 \text{ hr})\right)\right] = 35.3\frac{mg}{L}$$

The BEI itself cannot be determined unless the storage volume (V_{store}) is known. For example, if acetone were stored *only* in the blood, V_{store} would be the volume of blood in the worker's body (approximately 5.0 liters), the BEI can then be obtained from Eq. (2-52),

$$BEI = \frac{m_{bb}}{V_{store}} = \frac{4337.\ mg}{5.0\ L} = 870\frac{mg}{L}$$

Since this value is much higher than c_{blood}, it is apparent that V_{store} is much larger than 5.0 L. In other words, the body burden of acetone is stored in other organs or parts of the body in addition to the blood.

Discussion: If a biological testing program reveals blood samples containing an acetone concentration of about 35 mg/liter or higher (to two significant digits), it suggests that an individual was exposed to acetone concentrations equal to or greater than the PEL, and corrective actions should be taken. Note also that the body burden at the beginning of the day may not be zero, particularly for a chemical with a relatively long half-life (the half-life of acetone in this example is 4.0 hours). For example, if the worker goes home on Monday night, his/her body burden will slowly decay with time, but will not reach zero level by Tuesday morning when the workday begins again. Thus, the body burden will continue to increase throughout the workweek. Such an extension of this example problem is left to the reader as an exercise.

Similar correlations can be obtained for other air contaminants that are absorbed primarily in urine. For contaminants that are absorbed in soft tissue or bone, similar correlations can be obtained but the task becomes progressively more difficult. The body's response to a contaminant depends on how the contaminant affects various organs. In the simplest case, the contaminant is a ***mild irritant*** that is metabolized rapidly, and the response is directly proportional to the ambient mass concentration ($c_{ambient}$),

$$\boxed{R_p = kc_{ambient}} \tag{2-56}$$

where R_p is some measurable ***response*** and k is a ***response constant***. At higher concentrations, the initial response may be superseded by a secondary acute or chronic response that may be cumulative. If the material accumulates in the body, the response can be expressed in a form called ***Haber's law***. For an ***acute response***,

$$\boxed{R_p = kc_{ambient}t^n} \tag{2-57}$$

The coefficient (n) is not necessarily unity nor is it necessarily constant and may have higher values for progressive stages of the disease. The body burden (m_{bb}) is large for fat-soluble organics, organic bases, and weak acids. Asbestos, chlorocarbons, and radioactive fluorides may be stored for life, while lead and DDT are retained for years. In contrast, the body burden for water-soluble compounds is small.

At the cellular level, the response is proportional to the rate of change of the effective local concentration (c) with time, i.e. the ***cellar response*** is

$$\boxed{R_p = k\frac{dc}{dt}} \tag{2-58}$$

The ***effective local concentration*** is defined as the concentration within tissue, in contrast to $c_{ambient}$ which is the concentration in the ambient air. The effective local concentration at the site depends on $c_{ambient}$, the rate of metabolism, and the rate of absorption. The ***local response*** is

$$\boxed{R_p = k_{ab}c_{ambient} - k_r c} \tag{2-59}$$

where k_{ab} is the rate of absorption and k_r is the metabolic removal rate. For drugs a moderate value of k_r is sought so that all of the therapeutic agent reaches its target and the dose can be minimized. The respiratory system is an efficient transfer system and many contaminants can be readily absorbed and transmitted to certain organs. Carbon monoxide is a classic example.

Biotransformation can render a contaminant harmless or toxic. For example, formaldehyde is converted by the liver to formic acid which is a normal metabolic by-product and harmless at modest concentrations. On the other hand, aromatic compounds such as benzene and benzo-pyrene are metabolized and form highly reactive products called ***epoxides*** that initiate tumors.

The response of an individual to a given dose cannot be predicted accurately owing to the unique characteristics of individuals. Groups of individuals on the other hand display dose-response characteristics that can be described statistically. Of great use are cumulative distribution functions, median value, and standard deviations. Three commonly used ranges are easily determined by experiment.

- a dose from zero to a ***threshold mass concentration*** value ($c_{threshold}$) at which a response is observed
- a dose at which 100% of the subjects manifest a certain response
- a selected mid-range between (a) and (b), for example a median value in which 50% of the subjects manifest a certain response (e.g. the HERP index discussed in Chapter 1)

It is common to express dose-response data as a cumulative distribution graph and to identify the median 50% response dosage and the standard deviation.

2.7.2 Ehrlich Index

Arsenic and selenium are important trace metals needed by the body. If the daily dose is below some threshold, a deficiency occurs which may cause disease. As the dose increases above the necessary level, the body will eliminate the material. If the dose exceeds the body's ability to metabolize or eliminate the material, it will accumulate and cause another type of disease. In pharmacology, an Ehrlich index, shown in Figure 2.39, is used to elucidate the safety of substances. The ***Ehrlich index*** is the difference in concentration between the 95% point of the curative dose-response curve and the 5% point of the toxic dose-response curve. The larger the Ehrlich index, the safer the material. In the case of mercury that was once used to treat syphilis, the index is small. For anesthetics the index is larger but not large enough to be indifferent to their effects.

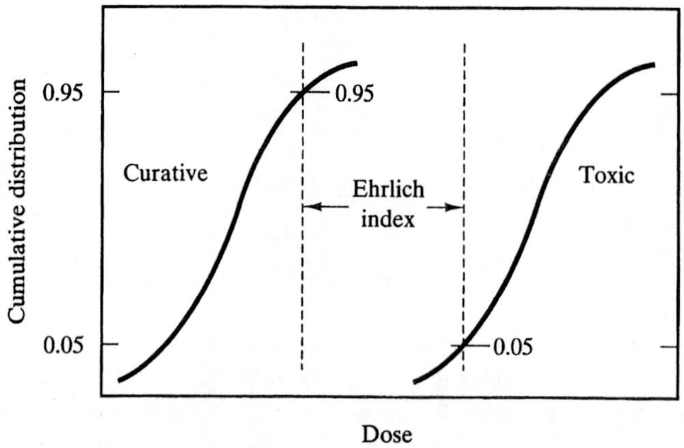

Figure 2.39 The Ehrlich index, defined by the curative and toxic dose-response curves.

2.7.3 Extrapolated Dose-Response Characteristics

Experimental data are often acquired at large doses well in excess of normal exposures, followed by extrapolation of the data to estimate the response at lower dosage. The way extrapolations are made is critical and the cause of much controversy. Figure 2.40 shows three ways extrapolations can be made. Certain highly toxic bacteria such as the tubercle bacillus and certain carcinogens are toxic at very small concentrations and low dosages. Curve I represents such cases. Other air contaminants are quite harmless at low concentrations and require a certain threshold dose before any response is observed. In this case a straight line extrapolation through the threshold (curve II) is reasonable. Naturally occurring materials such as CO, CO_2, NH_3, HCHO, etc. represent intermediate cases (see curve IIII) because there is always a certain amount of such materials in the lungs and airways in excess of what is present in the atmosphere.

Since toxins produce different effects, it is difficult to generalize dose-response characteristics for all toxins. For example, sulfur dioxide constricts the tracheobronchial system. Thus airway resistance is higher than normal (Colucci and Strieter, 1983). Secondly, sulfur dioxide irritates and inflames bronchial tissue. Two parameters are used to quantify exposure: dose rate and total dose. ***Dose rate*** is defined as

$$\boxed{\text{dose rate} = D_{min} = (Qc)_{min}} \tag{2-60}$$

where D_{min} is defined per minute of exposure (the subscript means *minute*, not *minimum*). ***Total dose*** is defined as

$$\boxed{\text{total dose} = D_t = \int_0^t Qc\,dt} \tag{2-61}$$

where t is again measured in minutes, and D_t is the dose in t minutes. The total dose (D_t) is the cumulative (or time-averaged) amount of the chemical introduced to the body; but as an integrated quantity it obscures any possible distinction between breathing frequency, length of exposure period, and concentration that may exist. Previous studies failed to establish a close correlation between response and total dose. The dose rate (D_{min}), on the other hand, represents a more instantaneous assault of the chemical on the body. To study the effect of a chemical (e.g. sulfur dioxide) during exercise when inhaled volumetric flow rates vary considerably, the dose rate is the more attractive

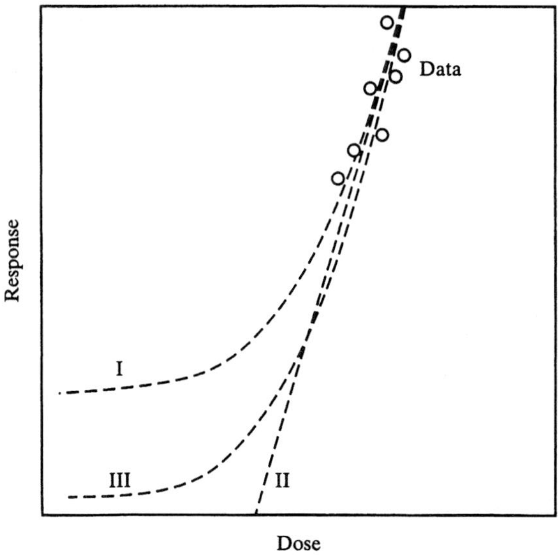

Figure 2.40 Methods to extrapolate dose-response data for three classes of toxins (from Heinsohn & Kabel, 1999).

dose parameter. If the toxins were cumulative and the time for them to be transported to body organs were significant, the total dose might be more attractive. The use of dose rate is logical for toxins like sulfur dioxide, ozone, PAN, and other irritants that produce an immediate response.

Carbon monoxide (CO) is a well-studied toxin (McCartney, 1990). It is toxic because ***hemoglobin*** absorbs carbon monoxide more readily than it absorbs oxygen. The brain fails to receive sufficient oxygen and produces effects such as reduced visual acuity, psychomotor skill, and pulmonary function, and eventually death. For sedentary individuals breathing at approximately the same volumetric flow rate, the total dose is an attractive dose parameter. The response parameter could be altered by behavioral responses such as loss of psychomotor skills, etc., but direct measurement of ***carboxyhemoglobin*** in the blood is a more accurate measure of response. Figure 2.41 is a dose-response curve for carbon monoxide. It should be noted that CO toxicity is serious, resulting in over a thousand deaths (accidental plus suicide) per year in the US (see Table 1.1).

Ozone (O_3) is a highly reactive gas and strong oxidant. It reacts with body fluids and tissue (Figures 2.29 and 2.30) to impair lung function in the short term. In simple terms, ozone limits the ability to take a deep breath. A demonstrative effect of ozone is a reduction in the forced expiratory volume at 1 second (FEV_1). Figure 2.27 is a compilation (Tilton, 1989) of several studies and shows that FEV_1 is reduced for healthy males. Following exposure, lung function is partially restored in 7 days; complete restoration occurs in 2 weeks. In the long term, the effects on health are two-fold. First, there is a transient reduction in resistance to infection, and second, it is believed that there may be chronic damage to the gas-exchange region in the lung.

Dose-response relationships for carcinogens are quite different than those for ozone, carbon monoxide, and sulfur dioxide. Cancer can be thought to progress through three stages: ***initiation, promotion***, and ***progression*** (Ricci and Molton, 1985). Initiation is an irreversible lesion in the DNA that leads to cancer if further attack occurs. The attack can occur through exposure to chemicals or other agents such as viruses. Promotion is a biochemical process that accelerates progression of the uninitiated cell to cancer. If a promoter attacks an uninitiated cell, the damage is thought to be reversible. Progression is the growth and spread of the disease. Cancer can be caused by a ***single hit*** of

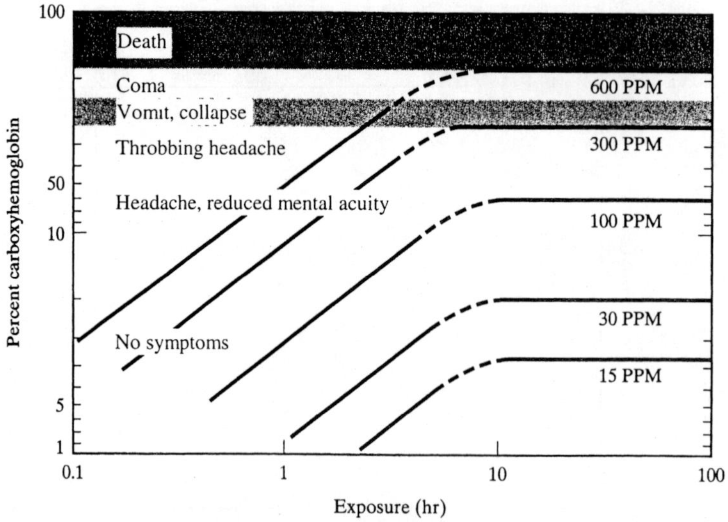

Figure 2.41 Response to carbon monoxide as a function of concentration (PPM) and exposure time (hr). The OSHA 8-hr PEL is 35 PPM and the EPA Primary Air Quality Standard is 9 PPM (redrawn from Seinfeld, 1986).

a toxicant on a DNA molecule in the target organ. The resulting ***point-mutation*** may or may not be reversible. The carcinogen-DNA complex is called an ***adduct***; it may be removed from the cell as new unaffected DNA is produced. If the DNA adduct is retained, however, it is likely to alter cellular control and initiate the cancer process. Adduct formation is an index of exposure but not necessarily an index of initiation since continued attack on the DNA does not mean that cancer necessarily develops.

Some chemicals are transformed into carcinogens through human metabolism. For example, benzo[a]pyrene must be metabolized to an epoxide which reacts with DNA in the cell to induce cancer. Other chemicals such as bis(chloromethyl) ether, do not appear to need metabolic conversion to be reactive. The existence of a threshold for a type of cancer is a point of contention among professionals who assess risk. The argument *against a threshold* assumes that a single hit will lead to uncontrolled growth of a somatic cell and will eventually produce cancer. Arguments *for a threshold* are based on the existence of gene repair mechanisms and immune defenses within the body. The ***one-hit model*** assumes that a single hit causes irreversible damage to DNA and leads to cancer. In the ***multi-stage model,*** a cell line must pass through N stages before a tumor is initiated irreversibly. The rate at which cell lines pass through one or more stages is a function of dose rate. In the ***multi-hit model,*** N dose-related hits to sensitive tissue are required to initiate a cancer. The Weibel model assumes that these hits occur in a single cell line and that different cell lines compete independently to produce tumors. In all these models, it is assumed that the rate at which dose-related hits occur is a linear function of dose rate. The most important difference between the multi-stage model and either the Weibel or multi-hit models, is that in the Weibel and multi-hit models all hits result from the dose, whereas in the multi-stage model, passage through some of the stages can occur spontaneously.

Considerable evidence exists linking exposure to benzene vapor with leukemia. Benzene is metabolized to phenols, quinones, and unstable oxygenated benzene derivatives. These oxygenated metabolic derivatives are reactive and covalently bind to tissue constituents, including genetic elements in bone marrow. In 1969 the government established a ***maximum, average (8-hr) threshold limit value*** (TWA-TLV) of 10 PPM for benzene. The standard produced considerable debate and legal challenge because more conservative interests wanted a TWA-TLV of 1 PPM. White et al. (1982) described how the one-hit model can be used to quantify risk for different benzene exposures and different dose rates, thus providing regulators with a quantitative way to select exposure limits.

The one-hit model presumes no threshold value and is based on the assumption that a single encounter with a carcinogen triggers a series of biological events leading to cancer. The basis of White's analysis consists of 314 white male workers who were exposed to benzene in the production of a rubber-film for a period exceeding 5 years during the years 1940-49. Between 1950-75, five (5) of 314 workers died of leukemia (probability, $P_t = 1.592 \times 10^{-2}$) compared to the national average of 758 per 100,000 (probability, $P_0 = 7.58 \times 10^{-3}$) of comparable workers in other industries who also worked during 1940-49. Death rates from all types of leukemia were taken from the 1975 mortality tables for US white males for 1937-77, and a background probability was found to be 707 per 100,000 (probability, $P_d = 7.07 \times 10^{-3}$). The ***standard mortality ratio*** (SMR) is defined as

$$\boxed{\text{SMR} = \frac{\text{observed deaths}}{\text{expected deaths}} = \frac{P_t}{P_0}} \tag{2-62}$$

For the period 1940-49,

$$\text{SMR}_{1940-49} = \frac{P_t}{P_0} = \frac{0.01592}{0.00758} = 2.10$$

Thus, the rubber-film workers experienced more than twice as many leukemia cancers. Over the 1937-75 period, the TWA-TLV was steadily lowered.

date	TWA-TLV (PPM)
1937-1940	150
1941-1946	100
1947	35
1948-1956	25
1957-1962	25
1963-1968	25
1969-1975	10

Over the period of 1940-49, the average value of the TWA-TLV was 83 PPM. Over the period 1937-75, measurement of actual benzene concentrations in the workplace was on the average 50 PPM. The mathematical expression for the one-hit model is

$$\boxed{P_d = \frac{P_t - P_0}{1-P_0} = \frac{P_t}{1-P_0} - \frac{P_0}{1-P_0}} \qquad (2\text{-}63)$$

which may alternatively be expressed as

$$\boxed{P_d = 1 - e^{-B \cdot D}} \qquad (2\text{-}64)$$

where

- B = constant that needs to be determined from independent data
- D = (dose) = (concentration) times (exposure time) in (PPM-yr)
- P_d = excess cancer probability due to benzene exposure
- P_t = total risk probability of developing leukemia (1.59×10^{-2})
- P_0 = background risk probability in the absence of exposure (7.58×10^{-3})

Eq. (2-63) can be rearranged to obtain an expression for P_t,

$$P_t = P_0 + P_d\left(1 - P_0\right)$$

or, dividing by P_0,

$$\boxed{SMR = \frac{P_t}{P_0} = 1 + P_d \frac{1-P_0}{P_0}} \qquad (2\text{-}65)$$

where Eq. (2-62) for SMR was also used. For the period 1940-49, in which $P_0 = 7.58 \times 10^{-3}$, the SMR can be evaluated using Eq. (2-65):

$$\boxed{SMR_{1940-49} = 1 + \frac{1-P_0}{P_0} P_d = 1 + \frac{1-7.58\times10^{-3}}{7.58\times10^{-3}} P_d = 1 + 130.9 \cdot P_d} \qquad (2\text{-}66)$$

Alternatively, replacing P_d in Eq. (2-65) by Eq. (2-64) yields SMR in exponential form:

$$\boxed{SMR = \frac{P_t}{P_0} = 1 + \frac{1-P_0}{P_0}\left(1 - e^{-B \cdot D}\right)} \qquad (2\text{-}67)$$

The one-hit model presumes that risk depends only on the benzene dose (PPM-yr) to which workers are exposed, and that a certain level of risk is associated with the dose, irrespective of whether exposure occurs over a long period of time at low concentration, or over a short period of time at high concentration. For purposes of calculation, estimating the constant B should be based on two types of exposure:

- actual 5-yr exposure of 314 workers at 83 PPM
- 30-yr exposure of workers at a TWA-TLV of 50 PPM.

Rearranging the exponential expression for P_d yields

$$B = -\frac{\ln(1-P_d)}{D} \tag{2-68}$$

and rearranging Eq. (2-65) to obtain P_d yields

$$P_d = P_0 \frac{SMR - 1}{1 - P_0} \tag{2-69}$$

An expression for B can be obtained by combination of Eqs. (2-68) and (2-69):

$$B = -\frac{1}{D}\ln\left[\frac{1-(P_0)SMR}{1-P_0}\right] \tag{2-70}$$

Using Eq. (2-70), two values of B can be found corresponding to D_{5yr} and D_{30yr}:

(a) 5-yr exposure, high concentration (83 PPM for 5 years, D_{5yr} = 415 PPM-yr):

$$B = -\frac{1}{(415 \text{ PPM-yr})}\ln\left[\frac{1-(7.58\times10^{-3})(2.10)}{1-(7.58\times10^{-3})}\right] = 2.03\times10^{-5} \text{ (PPM-yr)}^{-1}$$

(b) 30-yr exposure, low concentration (50 PPM for 30 years, D_{30yr} = 1500 PPM-yr):

$$B = -\frac{1}{(1500 \text{ PPM-yr})}\ln\left[\frac{1-(7.58\times10^{-3})(2.10)}{1-(7.58\times10^{-3})}\right] = 5.63\times10^{-6} \text{ (PPM-yr)}^{-1}$$

Example 2.5 - Benzene and Risk of Leukemia
Given: The above data concerning benzene exposure.

To do: For the values of B obtained in cases (a) and (b) above, estimate the excess leukemia risks (P_d) associated with the two possible TWA-TLV values considered by NIOSH, i.e. 10 and 1 PPM for individuals working over lifetimes of 45 and 25 years.

Solution: Eq. (2-64) can be used to estimate P_d for the two cases:

Case I: 45-year working lifetime

For TWA-TLV = 10 PPM and a working lifetime of 45 years, the background probability (P_b) (5 year dose) is

$$P_{d,5yr}(45 \text{ yr}) = 1-\exp(-B_{5yr}D) = 1-\exp\left[-\left(2.03\times10^{-5}\frac{1}{\text{PPM-yr}}\right)(10 \text{ PPM})(45 \text{ yr})\right] = 9.11\times10^{-3}$$

Similarly, for the 30 year dose,

$$P_{d,30yr}(45 \text{ yr}) = 1 - \exp[-(5.63 \times 10^{-6})(10)(45)] = 2.53 \times 10^{-3}$$

The standard mortality ratio (SMR) values are:

$$SMR_{5yr}(45 \text{ yr}) = 1 + (130.9)(9.11 \times 10^{-3}) = 2.19$$

$$SMR_{30yr}(45 \text{ yr}) = 1 + (130.9)(2.53 \times 10^{-3}) = 1.33$$

Thus, the *average SMR = 1.76*

For TWA-TLV = 1 PPM and a working lifetime of 45 years, the P_d and SMR values are:

$$P_{d,5yr}(45 \text{ yr}) = 1 - \exp[-B_{5yr}D] = 1 - \exp[-(2.03 \times 10^{-5})(1)(45)] = 9.15 \times 10^{-4}$$

$$P_{d,30yr}(45 \text{ yr}) = 1 - \exp[-(5.63 \times 10^{-6})(1)(45)] = 2.53 \times 10^{-4}$$

$$SMR_{5yr}(45 \text{ yr}) = 1 + (130.9)(9.15 \times 10^{-4}) = 1.12$$

$$SMR_{30yr}\,(45\text{ yr}) = 1 + (130.9)(2.53 \times 10^{-4}) = 1.03$$

Thus, the *average SMR = 1.08*

Case II: 25-yr working lifetime

Similarly, for TWA-TLV = 10 PPM and a working lifetime of 25 years, the values are:

$$P_{d,5yr}\,(25\text{ yr}) = 5.07 \times 10^{-3}$$
$$P_{d,30yr}\,(25\text{ yr}) = 1.45 \times 10^{-3}$$
$$SMR_{5yr}\,(25\text{ yr}) = 1.66$$
$$SMR_{30yr}(25\text{ yr}) = 1.19$$

Thus, the *average SMR = 1.42*

Finally, for TWA-TLV = 1 PPM and a working lifetime of 25 years, the values are:

$$P_{d,5yr}\,(25\text{yr}) = 5.08 \times 10^{-4}$$
$$P_{d,30yr}\,(25\text{ yr}) = 1.41 \times 10^{-4}$$
$$SMR_{5yr}\,(25\text{ yr}) = 1.07$$
$$SMR_{30yr}\,(25\text{ yr}) = 1.02$$

Thus, the *average SMR = 1.04*

Conclusions: For a 45-year working lifetime, reducing the TWA-TLV to 1 PPM results in risks of leukemia that are little different than for the population generally, whereas a 10 PPM standard produces a risk to leukemia comparable to what workers experienced in 1940-49. Reducing the workplace benzene concentration by a factor of ten reduces the risk to leukemia by 39% for a worker exposed for 45 years. If a more traditional working lifetime of 25 years is used as the basis of comparison, the ten-fold reduction of the TWA-TLV reduces the risk to leukemia by 14%.

Discussion: This example demonstrates how risk reduction strategies can be quantified. The question that remains is whether the benefits gained are commensurate with the cost to achieve them. While engineers are comfortable with quantitative discussions, the decision to set a new TWA-TLV is a political decision, and engineers must learn how to cope with the emotional factors entering all political discussions. At times, individuals advocating a particular course of action deliberately mislead by using numerical data irresponsibly. For example someone might suggest that reducing the TWA-TLV by a factor of ten reduces the risk to leukemia by a factor of ten. This is clearly not the case and it is hoped that engineers will rise to the occasion and correct such misleading claims.

2.8 Risk Analysis

Products of incomplete combustion are mixtures of organic and inorganic compounds. Most studies of mutagenicity (DeMarini et al., 1995 and 1996) involve extracting organic material from captured particles with solvents, i.e. ***extractable (particulate) organic material*** (EOM). Mutagenic activity is measured by the number of mutants produced in salmonella per unit mass of EOM. Because salmonella mutagenic assays use mutant cells that revert to phenotypically wild-type cells, the resulting mutants are called ***revertants*** (rev). Mutagenic potency (rev/μg of EOM) can also be expressed as a ***mutagenic emission factor*** in which mutagenic activity is expressed per mass of fuel or the amount of energy released in combustion. Table 2.11 shows mutagenic emission factors for various combustion processes. Expressing the data in these terms shows that open, uncontrolled burning has greater mutagenic activity than controlled burning in combustors designed for this purpose. Even with controlled combustors, mutagenic emission factors enable engineers to rate the carcinogenic risk from different fuels and combustors. Lastly, engineers need to know how these materials are diluted in the atmosphere and are transported downwind to reach an individual's breathing zone.

Table 2.11 Mutagenic emission factors of combustion products (from DeMarini and Lewtas, 1995).

source	rev/μg EOM	rev/(kg fuel) x 10^5	rev/MJ
cigarette smoke condensate	0.5-2.4		
coal-fired power plant	0.5-9.4	0.06	230
coke oven gas	3.4-3.7		
incineration			
municipal waste	1-27	1-12	300-38,000
medical/pathological waste	1-2	0.7	
hazardous waste (pesticide)	0.1-18	1-2	
open burning			
agricultural plastic	0.2-1.8	100	250,000
scrap rubber tires	2-12	800	
residential heating			
wood	1	50	250,000
oil	2-5	1	2,500
roofing coal tar	0.5-78.1		
rotary kiln			
natural gas	0.8	0.01	20
toluene	20	8.97	13,730
polyethylene	400	6.43	19,050
urban air	0.1-8.2		
vehicles			
diesel	1.4-15.1	40	
gasoline with catalyst	8.6	1	

Detecting a contaminant's ability to cause acute effects is relatively straightforward since acute effects appear immediately. Assessing chronic effects is much more difficult to analyze since chronic effects manifest themselves over a long period of time, as much as decades after exposure. Table 1.4 shows that no toxicity information is available for over 46,000 of the 65,000 chemicals used in industry. A complete assessment of health hazards (including chronic effects) has been made for 6,400 (only 9.8%) of the total chemicals used in commerce. Furthermore, it has been estimated that the world's laboratories are capable of studying the toxicity of only 500 compounds a year, woefully less than the 700 to 1,000 new chemicals that are introduced each year. Since a test for carcinogenicity for a single compound may take as long as three years and cost $250,000 or more, it is naïve to believe that society will allocate money to establish the chronic risk of all the chemicals with indisputable certainty. Alternatively, tests to detect a chemical's genotoxicity can be conducted in a short period of time and at considerably lower cost. There is a considerable body of knowledge to suggest that latent diseases such as cancer, birth defects, and genetic disease begin by altering DNA.

For potential carcinogens, the current regulatory attitude (Rich, 1990) is extremely conservative and assumes that even small concentrations can cause cancer. Such an attitude is contrary to the traditional regulatory approach for non-carcinogens in which there is assumed to exist a threshold concentration below which there are no observable effects. For potential carcinogens a variety of mathematical models is used to extrapolate from high to low dose. The EPA currently favors a linearized multistage model that provides a 95 percent upper bound estimate of cancer incidence at a given dose. The slope of the dose-response curve, such as Case I in Figure 2.40, is called the ***carcinogen potency factor***, also called the ***unit risk factor*** ($r_{u,j}$), and is used to estimate the probability of cancer associated with a particular exposure. The unit risk factor is a quantitative estimate of carcinogenic potency and expresses the chance of contracting cancer, though not necessarily life-threatening, from continuous exposure to a contaminant with a mass concentration of 1 μg/m^3 in air for a period of 70 years. For contaminants in water or food, other units may be used. The unit risk factor is a plausible upper estimate of the risk referred to as the 95 percent confidence level. With such an estimate, the true risk is not likely to be higher but it could be lower. Table 2.12 lists the unit risk factors (in units of m^3/μg) for 17 common chemicals of commerce.

The probability of developing cancer following exposure is called the ***excess lifetime cancer risk*** (r_e), which is the estimate of increased risk of cancer above that occurring in the general population. The value of r_e is a measure of the probability that an individual might develop cancer resulting from exposure to a certain contaminant, at a certain concentration, over 70-year exposure. It is calculated by multiplying the unit risk factor by the exposure, i.e. the long-term average mass concentration c_j (in units of μg/m^3):

$$\boxed{r_e = r_{u,j} c_j} \tag{2-71}$$

where subscript j refers to the chemical in question. ***Maximum lifetime individual risk*** is calculated by multiplying the unit risk factor by the highest concentration to which an individual is exposed:

$$\boxed{r_e(\max) = r_{u,j} c_j(\max)} \tag{2-72}$$

Table 2.12 EPA unit risk factors; note that these factors are subject to change and only up-to-date values should be used (abstracted from Rich, 1990).

chemical (CAS number)	**unit risk factor (m^3/μg)**
acetaldehyde (75-07-0)	3.96×10^{-6}
acrylonitrile (107-13-1)	0.48×10^{-4}
arsenic (7740-38-2)	27×10^{-2}
benzene (71-43-2)	2.65×10^{-5}
benzo(a) pyrene	1.75×10^{-2}
beryllium (7440-41-7)	8.85×10^{-4}
1,3-butadiene (106-99-0)	0.19×10^{-4}
cadmium (7440-43-9)	8.28×10^{-3}
carbon tetrachloride (56-23-5)	9.44×10^{-5}
chloroform (67-66-3)	1.12×10^{-4}
chromium VI (7440-47-3)	2.55×10^{-2}
1,2-dichloroethane (75-34-3)	1.05×10^{-4}
1,1-dichloroethylene (540-59-0)	1.98×10^{-4}
epichlorohydrin (106-89-8)	4.54×10^{-6}
ethylene dibromide (106-93-4)	1.69×10^{-3}
ethylene oxide (75-21-8)	1.80×10^{-4}
formaldehyde (50-00-0)	1.60×10^{-5}
gasoline (marketing)	1.69×10^{-6}
hexachlorobenzene	5.71×10^{-3}
methylene chloride (75-09-2)	1.63×10^{-6}
nickel-refinery dust (7440-02-0)	2.36×10^{-3}
nickel-subsulfide	1.18×10^{-3}
perchloroethylene (127-18-4)	3.93×10^{-6}
propylene oxide (75-56-9)	8.79×10^{-5}
styrene (100-42-5)	2.43×10^{-6}
trichloroethylene (79-01-6)	9.28×10^{-5}
vinyl chloride (75-01-4)	1.05×10^{-5}

For example, if $r_{u,j}$ is 4.0 x 10^{-5} ($m^3/\mu g$) and the highest concentration is 3 $\mu g/m^3$, then the maximum individual lifetime risk would be 1.2 x 10^{-4}. This means that there is little more than one chance in 10^4 that a person will contract cancer as a result of exposure to this pollutant for a period of 70 years. Note that nationwide, the risk for all cancers is 2.8 x 10^{-3} and cancer resulting from smoking cigarettes is 3.6 x 10^{-3} (Table 1.5).

For several constituents and exposures, the excess lifetime cancer risk is the sum of the products of the unit risk factor and exposure for each constituent,

$$\boxed{r_e = \sum_{j=1}^{n} \left(r_{u,j} c_j\right)} \tag{2-73}$$

where n represents the total number of contaminants to which one is exposed.

An excess lifetime cancer risk (r_e) of less than 10^{-7} corresponds to the likelihood that one person in 10 million is apt to contract cancer after a period of 70 years continuous exposure. The involuntary risk of 10^{-6} is the risk of death by disease for the entire US population. Figure 1.1 shows that a risk of 10^{-7} is comparable to the risk of fatality riding the railroad. An excess lifetime cancer risk of 10^{-7} is generally considered acceptable. Involuntary excess lifetime cancer risks between 10^{-7} and 10^{-4} resulting from identifiable causes are common targets for remediation.

Cancer risks may be reported in terms of risk estimates for an individual, as already discussed (excess lifetime cancer risk, r_e) or risk estimates for an exposed population (***aggregate risk***, A_j). Assessing a group's risk to cancer from exposure to a contaminant (j) requires three pieces of data:

- unit risk factor ($r_{u,j}$)
- concentration of contaminant j in each community ($c_{k,j}$), where index k refers to a specific community
- number of people exposed to contaminant j in each community ($P_{k,j}$)

Aggregate risk (A_j) pertains to all people within the exposed group. Aggregate risk is equal to the unit risk factor ($r_{u,j}$) times the sum of the products of the number of people ($P_{k,j}$) multiplied by the estimated concentration ($c_{k,j}$) to which they were exposed in their respective communities,

$$\boxed{A_j = r_{u,j} \sum_{k=1}^{N} \left(P_{k,j} c_{k,j}\right)} \tag{2-74}$$

where j is the contaminant in question and N is the number of communities. If the aggregate risk is normalized for the entire population, it can be expressed as the ***normalized aggregate risk*** (a_j):

$$\boxed{a_j = \frac{A_j}{\sum_{k=1}^{N} P_{k,j}} = \frac{r_{u,j} \sum_{k=1}^{N} \left(P_{k,j} c_{k,j}\right)}{\sum_{k=1}^{N} P_{k,j}}} \tag{2-75}$$

Example 2.6 - Cancer Risk to a Population Living in three Different Communities

Given: An industrial park contains several industries that use 1,2-dichloroethane, a colorless flammable liquid with a pleasant sweet odor. It is used in the manufacture of plastics, as a solvent in resins, as a degreaser in textiles, and as an extracting agent for soybean oil and caffeine. For decades, individual companies in the park discharged small amounts of the vapor to the atmosphere unaware that other industries were doing the same thing. Communities A, B, and C downwind of the industrial park experienced the following estimated average annual mass concentrations of 1,2-dichloroethane:

community	population, $P_{k,j}$ (people)	annual avg. concentration, $c_{k,j}$ ($\mu g/m^3$)
A	10,000	4.0
B	30,000	1.0
C	100,000	0.2

For years the PEL of 1,2-dichloroethane was 50 PPM, 200 mg/m^3. The compound irritates the skin and eyes and prolonged exposure is associated with liver and kidney damage. A time arises when community leaders in these communities begin to wonder if the industrial park has subjected their population to life-threatening risks, and they commission a study to consider the issue.

To do: Analyze the risk to see whether the communities should be concerned.

Solution: From Table 2.12, the unit risk factor ($r_{u,J}$) for 1,2 dicholoroethane is found to be 1.05 x 10^{-4} $(\mu g/m^3)^{-1}$. The excess lifetime cancer risks (r_e) for individuals in each community are

$$r_{e,A} = (4\ \mu g/m^3)(1.05 \text{ x } 10^{-4})(\mu g/m^3)^{-1} = 4.2 \text{ x } 10^{-4}$$
$$r_{e,B} = (1\ \mu g/m^3)(1.05 \text{ x } 10^{-4})(\mu g/m^3)^{-1} = 1.0 \text{ x } 10^{-4}$$
$$r_{e,C} = (0.2\ \mu g/m^3)(1.05 \text{ x } 10^{-4})(\mu g/m^3)^{-1} = 0.21 \text{ x } 10^{-4}$$

From Eq. (2-74), the aggregate risk incurred by the three communities is

$$A_j = r_{u,j}\sum_{k=1}^{N}\left(P_{k,j}c_{k,J}\right) = r_{u,j}\left[\left(P_{A,J}c_{A,j}\right)+\left(P_{B,J}c_{B,j}\right)+\left(P_{B,J}c_{B,j}\right)\right]$$

$$= 1.05\times10^{-4}\,\frac{m^3}{\mu g}\left[(10{,}000)(4.0)+(30{,}000)(1.0)+(100{,}000)(0.20)\right]\left(\frac{\mu g}{m^3}\right) = 9.45 \cong 9 \text{ or } 10 \text{ cases}$$

Thus a total of about 10 cases of cancer may occur if the 140,000 individuals were exposed to 1,2 dichloroethane for 70 years. The anticipated number of cases per year would be 9.45/70 or 0.135 cases per year (between one and two cases every 10 years). Finally, using Eq. (2-75), the normalized aggregate risk (a_j) for the entire population of these three communities is

$$a_j = \frac{A_j}{\sum_{k=1}^{N} P_{k,j}} = \frac{9.45 \text{ cases}}{(10{,}000+30{,}000+100{,}000) \text{ people}} = 6.75\times10^{-5}\,\frac{\text{cases}}{\text{person}} \cong 7.\times10^{-5}\,\frac{\text{cases}}{\text{person}}$$

On an annual basis (dividing the above by 70 years),

$$a_j \text{ per year} = \frac{6.75\times10^{-5}\,\frac{\text{cases}}{\text{person}}}{70 \text{ yr}} = 9.64\times10^{-7}\,\frac{\text{cases}}{\text{person-yr}} \cong 1.\times10^{-6}\,\frac{\text{cases}}{\text{person-yr}}$$

where the final result is given to one significant digit, given the overall uncertainty in assessing risk. To put this number into perspective, one can compare to the risk of death in the USA due to any kind of cancer. From Table 1.5 (to one significant digit),

$$a_j \text{ per year (all cancers, US population)} = 3\times10^{-3}\,\frac{\text{cases}}{\text{person-yr}}$$

Thus, the normalized aggregate risk per year for these three communities is more than *three orders of magnitude* smaller than that of any type of cancer for the entire US population!

Discussion: Since many people move into and out of the communities, only a few individuals are exposed for 70 years, and the exposure of people moving into a community is often unknown. The three communities should conclude that their populations were not subjected to life-threatening risks.

2.9 Closure

Engineers are sometimes asked to explain how pollutants affect health. Although physiology is outside the field of engineering, ducking the issue is not necessary since the rudiments of physiology are not difficult to learn; indeed they are satisfying to learn. It is important that readers understand the components of the respiratory system, how they function, and how they may be affected by air pollutants. The literature of physiology is material with which engineers should feel comfortable and have no hesitancy to consult. The manner in which pollutants are transported in the respiratory system and absorbed by blood is inherently interesting to individuals who appreciate thermal science, and lends itself to applying the principles of mass transport.

Chapter 2 - Problems

1. Unlike engineering where knowledge is based on writing and using quantitative relationships, knowledge in physiology and toxicology is based on naming things and explaining their function. The following are *diseases, parts of the body*, or *physiological parameters* that you should be able to recall from memory. If the word is a *disease,* state its symptoms and explain the physiological impairment. If the word is *part of the body*, define it succinctly and explain its physiological function. If the word is a *physiological parameter* describe what it represents.

- aerodynamic particle diameter
- alveoli
- allergic responses
- asthma, bronchitis, and emphysema
- bronchi, bronchiole, and alveolar ducts
- capillary permeability
- cilia
- collagen
- convective flow, pendelluft, Taylor-type diffusion, and molecular diffusion
- dose and dose rate
- dyspnea
- edema
- epithelium and endothelium cells
- hypoxia
- interstitium
- lipids
- lung compliance
- lymphatics
- lysosome and hydrolase
- mucus and surfactant
- phagocytosis and macrophage
- pharynx, trachea, esophagus, and epiglottis
- pneumonia and tuberculosis
- respirable and inhalable particles
- silicosis and asbestosis
- spirometer, tidal volume, vital capacity, residual volume, anatomical dead space, functional residual capacity, and forced expiratory volume in one second
- veins, arteries, arterioles, venules, and capillaries
- ventilation perfusion ratio

2. Write and solve the differential equations that predict the volumetric flow rate in the right and left lung (Q_R and Q_L) arranged in parallel as shown in Figures 2.18 and 2.19. Assume that during exhalation (when the pressure is positive), the pressure (in units of cm of water head) varies as follows:

$$P = 6t/(0.1t_c) \qquad 0 < t < 0.1t_c$$
$$P = 6 \qquad 0.1t_c < t < 0.4t_c$$
$$P = 6\,[1 - (t - 0.4t_c)/(0.1t_c)] \qquad 0.4t_c < t < 0.5t_c$$

where t_c is the length of the cycle (inhalation followed by exhalation) in seconds. During inhalation, the cycle is repeated except the pressure pulse is negative. The frequency of oscillation is to be varied over a range of 5 to 20 times the normal breathing frequency. The frequency for normal breathing is 13.6 cycles per minute (t_c = 4.41 seconds). [For illustrative purposes, you may wish to allow the upper limit to be 300 cycles/min to simulate canine "panting"] For computation purposes assume that the capacitance (C_L) and resistance (R_L) of the left lung are 0.083 L/cm water and 0.8 cm of water per LPM respectively (where LPM is the abbreviation for liters per minute). Assume that the right lung has a capacitance and resistance 30% larger that the left lung. Vary the breathing frequency from 13.6 to 50 cycles per minute and show when pendelluft occurs by plotting the volumetric flow rates in the right and left lungs.

3. If the tail of the forced expiratory capacity curve (Figure 2.13) is an exponential function, show that $\sqrt{\alpha_2} = \sqrt{2\alpha_1}$.

4. Show that in general, the square root of α_2 is the root mean square of α_1.

5. Describe the acute and chronic effects of the following toxicants on the pulmonary system. List the OSHA PEL for each of the substances.

- ammonia
- chlorine
- hydrofluoric acid vapor
- perchloroethylene vapor
- silica
- sulfur dioxide
- toluene vapor

6. The following are characteristic sizes of different airborne particles:

- tobacco smoke	0.01 to 0.1 micrometers
- pollen	10 to 100 micrometers
- sea salt	0.1 to 1 micrometers
- bacteria	0.5 to 50 micrometers
- insecticide dusts	1 to 10 micrometers

On the basis of information in Figures 2.34 and 2.35, estimate the amount (in percentage of inlet particles) you would expect to be deposited in different parts of the respiratory system. Also, in which parts are the particles primarily deposited? If the ventilation rate (volumetric flow rate of inhaled air, m^3/min) tripled, would you expect these values to change? Why or why not?

7. Derive fully the equations for the extended Bohr model to establish the validity of Eq. (2-37).

8. Assume people with impaired cardiovascular function suffer a 10% reduction in blood flow rate (Q_b) (multiply Q_b in the Table 2.4 by 0.9). Using the extended Bohr model, compute the percentage reduction in gas uptake efficiency (η_u) between individuals with healthy and impaired hearts for four levels of activity assuming all other parameters remain the same.

9. Many riding lawn tractors are equipped with exhaust pipes that discharge engine exhaust gases a few feet upwind of the operator's face. Operators inhale carbon monoxide at a concentration of 200 PPM (c_a = 228.6 mg/m^3) over a period of time and may display the effects of carbon monoxide poisoning. Using the extended Bohr model estimate the dose (kmols) of carbon monoxide an operator receives in two hours.

$$\text{Dose} = \int_0^2 \eta_u Q_a c_a dt$$

where t is in hours; uptake efficiency (η_u) and Q_a are the appropriate values for light exercise (LE).

10. Explain briefly the difference between the following pairs of terms or phrases that are sometimes confused with one another.

(a) chronic and acute response
(b) mutagens and teratogens
(c) asthma and bronchitis
(d) tuberculosis and pneumoconiosis
(e) pinocytosis and phagocytosis
(f) obstructed and constricted lungs
(g) respiratory and pulmonary system
(h) residual volume and functional residual volume
(i) vital capacity and tidal volume
(j) epiglottis and esophagus
(k) pharynx and trachea
(l) fibrosis and pneumoconiosis
(m) emphysema and edema and dyspnea
(n) mucociliary clearance and alveolar clearance
(o) biological exposure index and permissible exposure limit
(p) dose and dose rate

11. The annual average ambient concentrations in PPB of several pollutants are measured in New York City (population = 8 x 10^6), Newark, NJ (400,000), and Jersey City, NJ (300,000).

(a) Which city has the highest the excess lifetime cancer risk (r_e) due to chloroform?
(b) Which city has the highest excess lifetime cancer risk (r_e) for all the pollutants?
(c) Which pollutant poses the highest aggregate risk (A_j) to the total population of the three cities?

pollutant (M)	NY City	Newark	Jersey City	PEL (PPM)
benzene (78.1)	18	10	5	10
carbon tetrachloride (153.8)	3	12	10	2
chloroform (119.4)	10	15	8	2
formaldehyde (30)	20	12	15	3
perchloroethylene (166)	5	8	10	25
styrene (104.2)	5	20	18	50

12. Repeat Problem 11 for St. Louis, MO (population = 950,000) and East St. Louis, Ill (65,000). Mass concentration units are $\mu g/m^3$.

pollutant	M	St. Louis	E. St. Louis
benzene	78.1	4.6	10.6
styrene	104.2	2.9	3.3
chloroform	119.4	0.3	0.5
trichloroethylene	131.4	3.5	6.7
carbon tetrachloride	153.8	0.8	1.4
formaldehyde	30.0	5.2	6.8

13. Some scientists believe that production of "greenhouse gases" such as carbon dioxide (CO_2) is too high, and is causing a very slow warming of the world (global warming). In this exercise, the amount of CO_2 produced by people breathing is compared to that produced by combustion and other activities in the United States. From the data in this chapter, one may assume that an average person with an average activity level (some rest, some light exercise, and some moderate exercise) breathes in (inhales) approximately 20. L/min of air, and breathes out (exhales) the same amount of air per minute.

(a) From the data in Table 2.5, estimate the mass concentration (in units of mg/m^3) of CO_2 in exhaled and inhaled air.
(b) Estimate the net gain in mass of CO_2 in the atmosphere caused by one person breathing for one year. Give your answer in metric tons (a metric ton is the weight of 1000 kg).
(c) For comparison, and to give some perspective on the amount of CO_2 produced by breathing, estimate the total mass of CO_2 produced by the world's population (approx. 6 billion people) in one year. Compare to the total amount of CO_2 produced by combustion and other activities in the United States in one year, which is approximately 6.5×10^9 metric tons.

14. A worker arrives for work on Monday morning with no body burden for a certain chemical used at the plant. While on the job, he is exposed to the chemical in the air that he breathes. His average ventilation rate is $Q = 1.3\ m^3/hr$. Use the following data for the chemical: $t_{1/2} = 14.\ hr$, $F = 0.67$. The mass concentration of this chemical in the inspired air at the plant is $500.\ mg/m^3$. The concentration of the chemical outside of the plant (at home, in his car, at a restaurant, etc.) is negligible.

(a) Calculate the first-order rate constant, k_r, for this chemical in hr^{-1}.
(b) If the worker stayed at work for days at a time, what would be the steady-state value of the body burden of the chemical in his body?
(c) Now consider a more typical week: Monday through Friday: Work for 4 hours, lunch outside of the plant for 1 hour (no exposure to the chemical during lunch), work for 4 more hours, and then leave work (go home, shopping, etc.) for the remaining 15 hours of the day. Saturday and Sunday: Not at work all weekend. Estimate and compare the worker's body burden at the following times: Monday at quitting time, Friday at quitting time, Monday of the following week at starting time. Hint: Excel is recommended since there are repetitive calculations ideally suited for a spreadsheet.
(d) Compare the worker's body burden at quitting time on Friday for the case in which he eats his lunch at his desk instead of going out, and is thus exposed to the chemical for 9 hours every day, Monday through Friday. What percentage increase in body burden occurs due to staying at the plant during lunch hour?

15. Some tests are done on a volunteer to determine how much and how rapidly a certain chemical is absorbed by the body. The individual's breathing rate is $1.3\ m^3/hr$, and the mass concentration of the chemical in the air is set to $240\ mg/m^3$ (below the PEL for safety reasons). During the test, a sample of the worker's blood is taken every hour, and the amount of the chemical in his blood sample is measured per unit volume of blood. The body burden is then estimated by multiplying this amount by the total volume of blood in his body. At the start of the test, it is discovered that the worker already has a body burden of 120 mg. The following data (accurate to three significant digits) are obtained during the 6-hour test:

t (hrs)	**m_{bb} (mg)**
0	120
1	167
2	200
3	223

t (hrs)	**m_{bb} (mg)**
4	240
5	252
6	260

(a) Generate a procedure to solve for the ***steady-state body burden***, $m_{bb,ss}$ using any two of the above data points, and calculate $m_{bb,ss}$. Suggestion: use the data points at t = 2 and 4 hours.
(b) Find the ***half-life***, $t_{1/2}$ in units of hours.

(c) Estimate the ***absorption rate constant*** (k_r) of the chemical in the worker's blood, in units of 1/hr.

(d) Estimate the ***bioavailability ratio*** (F) of the chemical.

3

Design Criteria

In this chapter you will learn:

- about established criteria for safe contaminant concentrations
- how contaminant measuring and monitoring instruments work
- how to anticipate (and prevent) fire and explosion
- about hearing and how to calculate safe levels of noise
- about the body's cooling system and how to predict heat stress
- how to design ventilation systems to reduce odors
- about ionizing and nonionizing radiation
- about general safety in the indoor air environment
- how to estimate the cost of controlling contaminants (engineering economics)

Preventing contaminants from entering the work place should be a criterion in the design of all new industrial processes. In many cases it is, and control is adequate. Unfortunately the history of industrial development is replete with examples where health and safety were either ignored or underestimated, and engineering controls were designed after the process equipment was in place. The design of an indoor ***air pollution control system*** (APCS) is the creation of a geometrical configuration that satisfies certain performance specifications, and must be accomplished within a prescribed amount of time and money. The performance specifications, time, and budget are not chosen by engineers; rather they are imposed on engineers by the agency who retains their service. Time, cost, and performance specifications are the criteria by which one judges the success or failure of the venture.

Engineering design is the method by which devices or systems that perform a desired function are created within the constraints of budget and time. Design is the central activity in engineering. It is often assumed that science and engineering are either ends of a continuum. Alas such is not the case. To paraphrase the famous aerodynamicist Theodore von Karman, "Science discovers what is - engineering invents what never was." The goals of science are ***discovery*** and ***explanation***, while the goals of engineering are ***invention*** and ***design***. While science is essential to engineering, so also is finance and marketing. In the final analysis, engineering is as much business as it is science. Another difference between science and engineering is that they have ends and means that are different. The end in science is explaining natural phenomena; the end in engineering is creating a device or system believed to be useful by the agency financing the endeavor. The means of science is the ***scientific method***; the means of engineering is the ***design method***. The scientific method establishes truth and certainty, while the design method establishes usefulness and verifies that the device performs as planned. The scientific method consists of observation, hypothesis, prediction, and verification. In the context of indoor air pollution control, the essential steps in the design method are:

- selecting a geometrical configuration
- predicting its performance based on contaminant generation rate and mass transport
- optimizing the configuration to achieve performance at minimum total cost

- predicting its performance based on contaminant generation rate and mass transport
- optimizing the configuration to achieve performance at minimum total cost
- constructing the system
- verifying its performance through experiment

For the most part, an indoor APCS is designed to specification requirements while ***compliance*** is determined on the basis of achieving performance standards, sometimes called ***performance criteria***. It is very important that engineers understand the difference between specification requirements and performance standards, and are able to explain this difference to others. ***Specification requirements*** prescribe certain critical dimensions and volumetric flow rates, while ***performance standards*** prescribe that a worker's exposure should not exceed certain government standards (or more stringent standards an industry may choose to adopt). It is recommended that in the future, indoor air pollution control systems should be designed to satisfy performance standards rather than specification requirements. The two types of standards are not inimical; indeed they should complement each other. The point to be emphasized is that performance standards make the individual undertaking the design primarily responsible for its satisfactory performance, rather than side-stepping the responsibility and hoping performance will be satisfactory or relying on modifying the system after it is installed. Embracing performance standards is also the tradition in engineering, and essential in order for the US to remain competitive in the world economy.

Design criteria are factors upon which decisions are made in the design of an indoor APCS. Sometimes decisions are made in a deliberate and formal manner involving numerous people having different interests in the outcome of the design. At other times, a designer advances quickly through the steps of the design process with less formality. The formality of the design process depends on administrative practices within the company and the cost of the project. There is a great deal of professional literature on decision making in design (Dixon, 1966; Woodson, 1966). In Chapter 1 much was written about reducing exposures by modifying the process and substituting materials. It is presumed that these steps have been taken and that the decision has been made to design an APCS. The objective of this chapter is to discuss several important design criteria that engineers must address:

- contaminant exposure levels
- fire and explosion
- noise
- heat stress
- odors
- radiation
- safety
- engineering economics

An apocryphal practice used by an enterprising engineering consultant provides insight to designers who are asked to correct problems in safety and health. The consultant first speaks with the highest ranking official in the company to ascertain what management believes the problem to be, and how receptive management will be to accepting changes. Second, the consultant speaks with the workers directly associated with the process to find out what they perceive the problem to be, and what remedies they believe are appropriate. Armed with such information, the consultant is aware of remedies acceptable to the workers whose health and safety the consultant is retained to protect, and how serious management is about implementing advice. If the workers' ideas are sound and can be incorporated in the design of the control system, the consultant has facilitated an exchange of information that should have already taken place. If the workers' ideas cannot be incorporated in the design, the conversations will help designers understand the institutional environment in which they are asked to make recommendations.

An indoor APCS should not impede the worker's movement, visibility, or mental concentration while performing the process for which he/she is responsible. If the design is for a process that has been in operation for quite some time, the designer should learn where workers stand to observe the process, and how they move to add or remove material from the process. It is not that such movement is necessarily the best, but workers operate in their best interest, and if the new design interferes with these activities, the designer can anticipate difficulty. If the worker's movement must change substantially, management must be prepared for a major re-education program. If the process is new, the designer should become familiar with the expected movement of the worker; indeed it is advisable to involve workers in the design process.

3.1 Contaminant Exposure Levels

Contaminant exposure levels pertain to contaminants in air that individuals are apt to inhale. The contaminant itself may consist of particles suspended in the air, or may be a liquid or a gas. For contaminants which normally exist as liquids, the exposure levels discussed here pertain to the contaminant in the vapor phase in air, i.e. ***evaporated*** material. Throughout the world, governmental agencies, professional organizations, and individual corporations establish industrial exposure limits (Paustenbach and Langner, 1986). Only those in the US are discussed, but individuals are urged to be cautious and to seek guidance when working in foreign nations that may have different standards. In the US, four agencies promulgate standards for occupational exposure that the reader is likely to encounter:

- The ***American Conference of Governmental and Industrial Hygienists*** (ACGIH) publishes a group of standards called ***threshold limit values*** (TLVs) which are used widely in the US. These standards are reviewed yearly and revised when necessary. The TLVs published by the ACGIH are recommendations, which should be used as guidelines, and the ACGIH disclaims liability with respect to the use of TLVs.
- The ***Occupational Safety and Health Administration*** (OSHA) in the Department of Labor publishes standards known as ***permissible exposure limits*** (PELs). The standards are not as detailed as the TLVs and in some instances are not as stringent. Nevertheless PELs are backed by the power of the law, i.e. they are enforced by fines, etc.
- The ***American National Standards Institute*** (ANSI) publishes exposure standards called ***maximum acceptable concentrations*** (MACs) which are in general compatible with TLVs and PELs. In many foreign nations MACs are used.
- The ***National Institute for Occupational Safety and Health*** (NIOSH) publishes yet another standard called the ***recommended exposure level*** (REL). The REL values are usually the same as the PEL values, but since NIOSH uses a 10-hour work day of exposure, while OSHA uses an 8-hour work day, and also because two different agencies determine these exposure levels, the values sometimes differ.

Generally both the PEL and the REL are provided on a material safety data sheet (MSDS), which is most easily obtained from the Internet. TLVs are more difficult to obtain, since they must be purchased from the ACGIH.

The Occupational Safety and Health Act of 1970 charged the Department of Labor with the responsibility of enforcing standards that maintain safe and healthy conditions in the workplace. The act also established OSHA to enforce these standards. In addition, the ***Mine Safety and Health Administration*** (MSHA) enforces current ACGIH standards on US coal mining operations. Appendix A.20 contains an abbreviated listing of OSHA PELs, up to date at the time of this writing. (The superscript "d" in this appendix indicates that exposure also involves transfer through the skin, mucous membrane, or the eye.) Agencies that promulgate standards review them regularly and sometimes revise them, as seen in Appendix A.1, where 1989 and 1997 PELs are compared. Some of the values listed in this text may already be out of date. Readers should always use currently approved values contained in up-to-date versions of publications containing these values. The Internet is an excellent resource for obtaining up-to-date PEL values. The US NIOSH web site at **www.cdc.gov/niosh/** is particularly handy (click on Databases and then Pocket Guide to Chemical Hazards).

3.1.1 OSHA and ACGIH Levels

According to the Committee on Industrial Ventilation, ACGIH (1986), the ***threshold limit value-time-weighted average*** (***TLV-TWA***) concentration is the concentration to which nearly all workers can be repeatedly exposed during a normal 8-hour day, 40-hour workweek, week after week, without adverse effect:

$$\boxed{\mathrm{TWA}_{8\text{-hr}} = \frac{1}{8\ \mathrm{hr}} \int_0^{8\ \mathrm{hr}} c(t)dt} \tag{3-1}$$

Less common, but sometimes used is the 40-hour time-weighted average concentration,

$$\boxed{\mathrm{TWA}_{40\ \mathrm{hr}} = \frac{1}{40\ \mathrm{hr}} \int_0^{40\ \mathrm{hr}} c(t)dt} \tag{3-2}$$

Unless noted otherwise, the notation "TLV" by itself implies the TLV-TWA 8-hr average. Similarly "PEL" implies the PEL-TWA 8-hr average. OSHA enforces a daily PEL-TWA 8-hr average, i.e. the average concentration to which the worker is exposed must not exceed the PEL-TWA 8 hr value. Note that it is okay for the worker to be exposed occasionally to concentrations above the TWA, provided that the *average* exposure over 8 hours is below the TWA. However, there are other (higher) limits as discussed below.

The ACGIH defines the ***threshold limit value-short term exposure limit*** (TLV-STEL) as the concentration to which workers can be exposed for a short period of time without suffering:

- irritation
- chronic or irreversible tissue change
- ***narcosis*** (drowsiness) of sufficient degree to increase the likelihood of accidental injury, impair self-rescue, or materially reduce work efficiency

The STEL is defined as a 15-minute time-weighted average exposure which should not be exceeded at any time during a work day, even if the eight-hour time-weighted average is within the TLV.

$$\boxed{\mathrm{STEL} = \frac{1}{15\ \mathrm{min}} \int_0^{15\ \mathrm{min}} c(t)dt} \tag{3-3}$$

In addition, there can be no more than four excursions above the TLV-STEL per 8-hour day, and there must be at least 60 minutes between such excursions. The STEL should not be considered a maximum allowable concentration or absolute ceiling during a 15-minute excursion, but rather the maximum *average* concentration over the 15-minute period.

The ***threshold limit value-ceiling*** (TLV-C) is defined by ACGIH as the concentration that should *never* be exceeded, even instantaneously, during any part of the working exposure. For some substances, e.g. irritant gases, only one category, the TLV-C, may be relevant. For other substances, depending on their physiologic action, two or all three of the categories may be relevant. The ACGIH emphasizes that if any *one* of the three TLVs (TWA, STEL, or C) is exceeded, a potential hazard is presumed to exist. Finally, although Eqs. (3-1), (3-2), and (3-3) are defined in terms of the mass concentration (c), many tables, such as the appendices in this book, list the TLV or PEL values as mol fractions instead, in units of parts per million (PPM). This is more general since mol fraction is independent of temperature and pressure conditions in the ambient air. When these values are listed as mass concentrations (mg/m^3), readers must be aware that they are calculated at standard temperature and pressure (STP) conditions. If either the temperature or air pressure differ significantly from STP values, the mass concentration must be adjusted according to the techniques discussed in Chapter 1.

Shown in Figure 3.1 is a sample record of concentration (actually mol fraction in PPM) versus time for exposure to a pollutant during an 8-hour workday. For this particular pollutant, the TLV-TWA, TLV-STEL, and TLV-C are 60, 100, and 120 PPM respectively, as indicated on the plot. Concentration readings were obtained every 5 minutes, and are marked "instantaneous" on the plot. A running 15-minute average (integrated average over the previous 15 minutes) was calculated from these data, and is also shown on the plot. (The authors employed the trapezoidal rule to perform the integration numerically.) The 15-minute running average "lags behind" the instantaneous data, and by nature of the averaging process has smaller deviations from the mean. From these two curves, one can determine if each of the three ACGIH TLV standards are met:

- TLV-TWA: The average concentration over the 8-hour period (expressed as a mol fraction) is calculated to be 41.5 PPM, which is below the TLV-TWA. This standard is satisfied.
- TLV-STEL: There are three excursions above the TLV-STEL during the 8-hour period. This is acceptable since it is not more than four, and since the excursions are more than one hour apart. In addition, the running 15-minute average never exceeds the TLV-STEL, although it comes close (97.5 PPM at 13:30). Thus, these excursions above the TLV-STEL are acceptable according to ACGIH standards.
- TLV-C: Finally, it is clear that at no point was the concentration of the pollutant ever above the TLV-C. This standard is satisfied.

Since all three standards are satisfied, the conditions in this workplace are not considered to be hazardous.

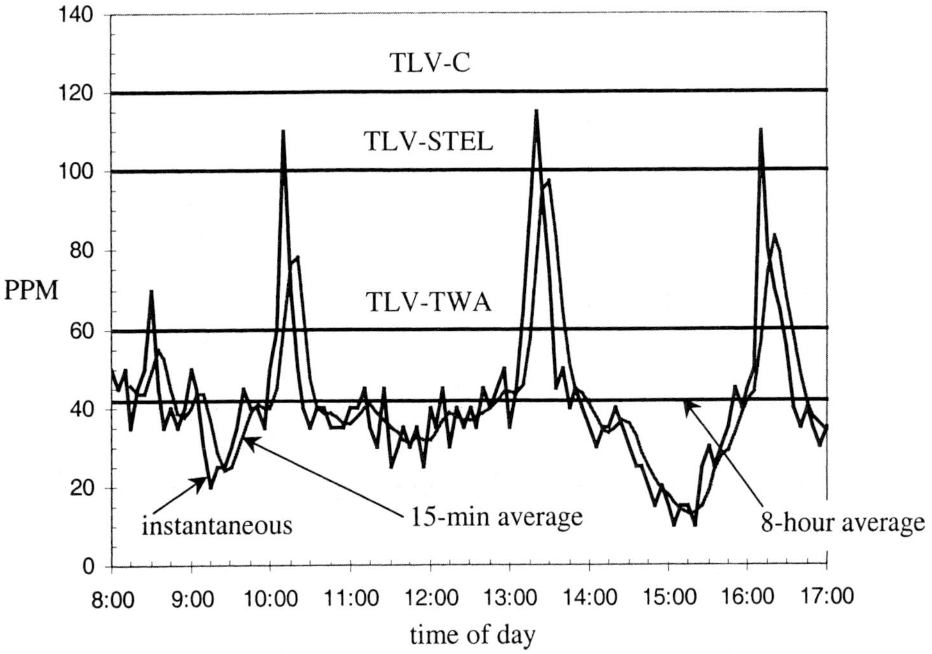

Figure 3.1 Instantaneous and running 15-minute average mol fraction of a contaminant compared to ACGIH standards for an 8-hour work day.

3.1.2 TLV For Mixtures of Liquids and Gases

If a liquid is a mixture of liquids, and the effects of the components are believed to be additive, and it is assumed that the vapor is composed of components in similar proportion to the original liquid, it is useful to define a TLV for the liquid mixture. When a ***liquid mixture*** contains N toxic substances and there is no synergistic relationship between the substances, the ACGIH recommends that an equivalent TLV (TLV-mixture) be computed,

$$\boxed{\left(\text{TLV-mixture}\right) = \frac{1}{\sum_j \frac{f_J}{\text{TLV}_J}}} \tag{3-4}$$

where f_J is the mass fraction of species j in the liquid phase in the mixture, as defined previously in Chapter 1,

$$\boxed{f_J = \frac{m_J}{m_t} = x_j \frac{M_j}{M_{avg}}} \tag{3-5}$$

In Eq. (3-5), x_J is the mol fraction of species j in the liquid phase and M_{avg} is the average molecular weight,

$$\boxed{M_{avg} = \sum_j x_J M_j} \tag{3-6}$$

Typical units of TLV-mixture are the same as those of mass concentration, i.e. mg/m^3.

Alternatively, if a variety of toxic substances is present in the *gas* phase, OSHA assesses hazards by an ***exposure parameter*** (En) defined as

$$\boxed{\text{En} = \sum_J \frac{y_J}{\text{PEL}_J}} \tag{3-7}$$

where both the mol fraction (y_J) and the permissible exposure limit (PEL_J) must be expressed in identical units (PPM, PPB, etc., or simply as fractions). If En is greater than unity, exposure is considered to be beyond acceptable limits; if En is less than unity, exposure is considered to be within acceptable limits. The OSHA standard for a mixture of gaseous contaminants can also be written according to Eq. (3-7), but in terms of *mass concentration* instead of PEL,

$$\boxed{\text{En} = \sum_J \frac{c_J}{L_J}} \tag{3-8}$$

where c_J is the measured mass concentration and L_J is the PEL in comparable units. Finally, Eq. (3-8) can be written in terms of molar concentration as well, provided that the numerator and denominator have the same units.

Example 3.1 - Exposure Limits of Mixtures

Given: At a chemical plant, a mixture of three liquid components was prepared. The amount of each component (% by weight), along with its TLV at the time, is shown in the table below:

material	**% by weight**	**TWA-TLV in PPM (mg/m^3)**
trichloroethylene (TCE)	35.	50 (270)
benzene	25.	10 (30)
methylene chloride	40.	50 (175)

material	concentration (PPM)
methylene chloride	40.
trichloroethylene	30.
benzene	8.0

In another part of the plant, waste liquid hydrocarbons were stored in a vessel. Vapors from the mixture escaped into the workplace atmosphere, and concentration measurements showed that the air contained the chemicals listed above.

To do: Determine the TWA-TLV for the mixture and whether the air in the plant violated OSHA standards.

Solution: Since the mixture composition is known, the TWA-TLV for the mixture can be computed from Eq. (3-4),

$$(\text{TLV-mixture}) = \frac{1}{\sum_j \frac{f_j}{\text{TLV}_j}} = \frac{1}{\left(\frac{0.40}{175} + \frac{0.35}{270} + \frac{0.25}{30}\right)\left(\frac{m^3}{mg}\right)} = 83.93\frac{mg}{m^3} \cong 84.\frac{mg}{m^3}$$

Note that the mass fraction (f_j) of species j is the same as the percent by weight (expressed as a fraction) since mass is simply equal to weight times a constant (the constant of gravitational acceleration). To determine whether the measured workplace atmosphere was in violation of OSHA standards, the parameter En is computed using Eq. (3-7),

$$\text{En} = \sum_j \frac{y_j}{\text{PEL}_j} = \frac{40.}{50} + \frac{30.}{50} + \frac{8.0}{10} = 2.2$$

Thus the workplace atmosphere was in violation of OSHA standards since En is greater than unity.

Discussion: To illustrate how the TWA-TLV for the mixture can be used, suppose 0.50 grams of the liquid mixture were spilled and all of it evaporated in an unventilated storage room of volume 2.5 m^3. Assuming well-mixed conditions in the room, the mass concentration in the air would have been

$$c_{\text{mixture}} = \frac{0.50 \text{ g}}{2.5 \text{ m}^3}\left(\frac{1000 \text{ mg}}{\text{g}}\right) = 200\frac{mg}{m^3}$$

which exceeds the TWA-TLV-mixture value. However, it is unlikely that any workers would be in an unventilated storage room for any appreciable length of time, especially not for 8 hours, the time on which the TWA-TLVs are based. Furthermore, there would be some infiltration of fresh air into the room and exfiltration of contaminated air out of the room, even for unventilated storage rooms, as discussed in more detail in Chapter 5.

The EPA uses criteria called ***ambient air quality standards*** (AAQS) to establish safe environmental conditions in the outdoor (air) environment. A comparison of these standards and PELs shows that for the most part PEL values are higher. The reason for the disparity is that the PEL pertains to the industrial environment (primarily *indoors*), and applies to workers who for the most part are healthy adults in the prime of life who are exposed to contaminants for only a short period (typically 8 hours) of the day. The AAQS pertains to the *outdoor* environment - exposures that may be present for 24 hours per day, and therefore must be more stringent. Secondly AAQS must safeguard the health of infants, the elderly, and the infirm.

Engineering control systems are designed to satisfy performance standards established at the beginning of the project. Since compliance with OSHA regulations requires that the PEL shall not be exceeded, it is logical to use the PEL, or even a more stringent reduced value of it as the performance standard such as PEL/K, where K is a safety factor ($K > 1$). Alternatively the TLV could be used instead of the PEL. The designer's task is then to devise an indoor air pollution control system or to modify a process such that the predicted contaminant concentrations in the worker's breathing zone do not exceed the PEL. It must be emphasized that PELs and TLVs are not precise, indisputable criteria distinguishing "safe" from "unsafe" working conditions; consequently, dividing the PELs and TLVs by a safety factor K is wise. The value of K is an issue to be decided by the designer in consultation with the company at the beginning of the project. In the absence of other agreed upon standards to characterize safe and unsafe conditions, PELs and/or TLVs divided by K are useful parameters to judge the effectiveness of an air pollution control system. Management, insurance carriers, and state or local governmental agencies may devise more stringent standards; it is vitally important that engineers understand what standards and which agencies will judge the performance of the systems they design. Concentrations used to estimate a worker's exposure must be measured close to the face. Such a region is called the ***breathing zone***. Compliance with health standards must be based on concentrations that exist near equipment or processes where workers are expected to be stationed, not in obscure locations where they do not perform their duties.

Because of variations in individual susceptibility, there are often isolated individuals in a population of workers who may experience discomfort at or below TLV or PEL exposures, e.g. alcoholics, heavy smokers, the chronically ill, hypersensitive or allergic individuals, the elderly, pregnant women, diabetics, and individuals on medication. A higher value of the safety factor K would obviously benefit these individuals.

There are materials that are known or suspected to be carcinogens. OSHA is obligated to identify these materials and recommend how these materials are manufactured, processed, repackaged, released, handled, and stored, and what protective equipment workers using them must wear. Shown in Table 3.1 are materials identified by OSHA as cancer-suspect agents.

Table 3.1 OSHA cancer-suspect agents, along with some of their trade names; Chemical Abstract Service (CAS) registry numbers are given in parentheses where appropriate.

2-acetylaminofluorene, 2-fluorenylacetamide, AAF, 2-AAF (53-96-3)
4-aminodiphenyl, 4-phenylaniline (92-67-1)
asbestos fibers longer than 5 micrometers (chrysotile, amosite, crocidolite, tremolite, anthophyllite, actinolite)
benzidine, 4,4'-biphenyldiamine, 4,4'-diaminobiphenyl (92-87-5)
bis-chloromethyl ether, dichloromethyl ether, BCME (542-88-1)
chloromethyl methyl ether, methylchloromethyl ether, CMME (107-30-2)
3,3′ dichlorobenzidine (and its salts), dichlorobenzidine base (91-94-1)
4-dimethylaminoazobenzene, butter yellow; DAB (60-11-7)
ethylenimine, ethyleneimine, dimethyleneimine, dimethylenimine (151-56-4)
4,4′-methylenebis(2-chloroaniline), MBOCA (101-14-4)
1-naphthylamine, alpha-naphthylamine, 1-aminonaphthalene (134-32-7)
2-naphthylamine, beta-naphthylamine, 2-aminonaphthalene (91-59-8)
4-nitrobiphenyl, p-nitrodiphenyl, p-phenylnitrobenzene, PNB (92-93-3)
n-nitrosodimethylamine, dimethylnitrosamine, DMNA (62-75-9)
beta-propiolactone, hydroacrylic acid, beta-lactone, BPL (57-57-8)
vinyl chloride, monochloroethylene, vinyl chloride monomer (VCM) (75-01-4)

3.1.3 Exposure Control Limits

In the pharmaceutical industry, workers may be exposed to pharmacological compounds at concentrations and dosages that may be dangerous. Workers are protected by an exposure limit called the ***exposure control limit*** (ECL). ECLs are mass concentrations which are defined in the same way as ACGIH defines the TLV-TWA 8-hr average, i.e. a "time-weighted average (air) concentration for a normal 8-hr workday and 40-hr workweek to which nearly all workers may be repeatedly exposed day after day without adverse effect" (Sargent and Kirk, 1988). In developing these limits, the pharmaceutical industry follows procedures similar to those used by the ACGIH to develop TLVs. Data needed to compute ECLs are generated as part of the normal investigations required by the ***Food and Drug Administration*** (FDA) in applications for new drugs. To compute an ECL from pharmacological oral dosage data the following is used:

$$\boxed{ECL = NOEL \frac{\alpha m_b F_s}{Q t_s}} \tag{3-9}$$

where the variables (and their common units) are as follows:

- ECL (mg/m^3) is the exposure control limit
- NOEL (mg/kg) is the ***no observable effect level***; defined as the dose in mg per kg of body mass (for one day of exposure) at which there is no observable effect on the person
- m_b (kg) is body mass; 70 kg for males, 50 kg for females
- F_S (dimensionless) is a factor selected by the industry; values range from 10 to 1000 depending on the response produced by the drug
- α (dimensionless) is the fraction of the compound absorbed by the body; determined by laboratory experiment, generally unity
- Q (m^3/day) is the inhalation volumetric flow rate; approx. 10 m^3/day
- t_S (days) is the time to achieve a blood plasma steady-state concentration

Every drug produces a unique response depending on its pharmacokinetics. For chronic, intermittent exposures such as occur in the workplace, the steady-state level is directly proportional to the amount of exposure and the biological half-life of the drug as long as the elimination processes are first order. For most drugs the rate of elimination is rapid, the half-life is of the order of several hours, and there is minimal accumulation in the body. In these cases a value of 1 day would be used for t_S. If the half-life is large, the time to achieve a plasma steady-state (t_S) is determined experimentally by administering an oral dose of the drug once a day and observing the blood plasma concentration as a function of time. Upon administration, the plasma concentration increases to a detectable level and then falls. After the next administration the concentration rises to a higher level before falling. The pattern is repeated over a 5-day period. In many cases the peak plasma concentrations are reached in 1 to 2 days, and hence a value of 2 days is used for t_S. Literally "steady-state" plasma concentrations do not occur. Rather, t_S represents the elapsed time required by the body to exhibit a repeatable maximum concentration produced by a daily dose of the drug.

3.1.4 Biological Exposure Index

Air monitoring assesses a worker's exposure to chemicals by measuring the concentration of the chemicals in the workplace atmosphere. TLVs, PELs, RELs, and ECLs serve as reference values for air monitoring. ***Biological monitoring*** assesses a worker's exposure to chemicals by measuring the concentration of certain materials in the body. The biological exposure index (BEI), defined in Chapter 2, serves as a reference value in biological monitoring to indicate potential health hazards. The BEI for a chemical is the concentration of the chemical (or its metabolite) in the blood, urine, exhaled air, etc. that is likely to be observed in a healthy worker who has been exposed to the chemical at an inhalation exposure equal to the TWA-TLV. In many industries biological monitoring is required by the company and the union.

BEIs do not define a sharp distinction between hazardous and nonhazardous exposures. Due to biological variability, an individual's measured concentrations may be in excess of BEI without incurring increased health risk. The correlation between exposure to air contaminants and their (or their metabolites') appearance in the body is influenced by:

- Physiological and health status: age, body build, diet, enzymatic activity, sex, pregnancy, medication, disease states
- Exposure factors: intensity of physical work, exposure intensity and variation with time, temperature, humidity, skin exposure, exposure to other chemicals
- Environmental factors: community and home pollutants in air, water, and food
- Individual habits: personal hygiene, social activities, working and eating habits, alcohol and drug use, smoking
- Methodological factors: specimen collection, storage, and analysis

Action on unexpected concentrations should not be taken based on a single isolated measurement, but on multiple samplings. If however, numerous measurements for the individual over a period of time are above the BEI, or the majority of measurements from a group of individuals at a specific time are above the BEI, the cause of the excessive values should be investigated and corrective actions should be taken.

Shown below are the BEI, indices, timing, and PEL-TWA for three workplace chemicals (the CAS number is also shown in parentheses):

- carbon disulfide (75-15-0)
  - PEL-TWA: 4 PPM
  - indices: 2-thiothiazolidine-4-carboxylic acid in urine
  - BEI: 5 mg/g creatinine at end of shift
- methyl ethyl ketone (MEK) (78-93-3)
  - PEL-TWA: 10 PPM
  - indices: MEK in urine
  - BEI: 2 mg/L at end of shift
- o-xylene (1330-20-7)
  - PEL-TWA: 100 PPM
  - indices: methylhippuric acids in urine
  - BEI: urination rate of 2 mg/min during the last 4 hours of shift

Individuals who wish to design a biological monitoring program or interpret BEI data should consult the most recent version of ACGIH literature.

3.1.5 Reduced Respiratory Rate

A common physiological response to many chemicals used in the workplace is ***sensory irritations*** consisting of burning sensations of the eyes, nose, or throat. These responses are initiated by nerve endings in the nasal cavity and larynx. Sensory irritants are also called ***upper-respiratory tract irritants*** that emphasize the affected portions of the respiratory system. These irritations may induce a cough. Schaper (1993) reports that male Swiss-Webster (SW) mice exhibit a unique and measurable lengthening of expiration time for air containing certain concentrations of the irritant. The decreased respiratory rate, i.e. percent decrease from the pre-exposure level, is proportional to the concentration of the chemical to which the mice were exposed. Depending on the chemical, the decreased respiratory rate may persist or it may fade after a period of time. In any event, the reduced respiratory rate can be quantified and correlated with the irritant concentration in the inspired air.

The irritant concentration that reduces the respiratory rate by a factor of two, i.e. produces a 50% decrease, is called the ***reduced respiratory rate*** (RD_{50}). Schaper (1993) reports that humans experience intolerable irritation at concentrations corresponding to RD_{50} and minimal or no effect at 1% of the RD_{50}, i.e. 0.01 x RD_{50}. However, a reasonable correlation ($R^2 = 0.90$) was obtained between TLVs and 0.03 x RD_{50}. Investigators in the US, Europe, and Asia have used such sensory-irritation responses as the basis for workplace exposure standards for quite some time. Currently, the method is the basis for Standard E-981-84 of the ***American Society for Testing and Materials*** (ASTM). Schaper (1993) suggests that 0.03 x RD_{50} can serve as an inexpensive screening test of new materials that can be followed by validation tests on humans if it is necessary to do so.

3.2 Instruments to Measure Pollutant Concentration

Instruments that measure the concentration of indoor air pollutants can be divided into three categories:

- laboratory instruments
- continuous monitors and alarms
- portable instruments.

Laboratory instruments measure concentrations with high accuracy and are used in conjunction with laboratory experiments. Laboratory instruments are delicate and often require regulated electrical power, canisters of calibration gases, data-loggers for the storage and processing of data, etc. They may also require a liquid ***reagent***, which is a chemical that, because of the reaction(s) it causes, is used in analysis and synthesis. Laboratory instruments are designed to be used in a clean laboratory environment and are rarely used to obtain measurements at industrial sites. ***Continuous monitors*** are highly reliable and accurate instruments installed permanently at an industrial site to measure and record pollutant concentrations on a continuous basis. Continuous monitors are rugged instruments that are not suited to be carried about for field measurements. ***Alarms*** are accurate rugged instruments designed to measure the concentration of specific pollutants and to register an alarm if concentrations exceed prescribed values. Alarms are designed for permanent installation and must be maintained on a scheduled basis to ensure reliability. The virtue of ***portable instruments*** is that they can be carried to actual sites and can record data with prescribed accuracy. Portable instruments generally provide an instantaneous visual display of the measured concentrations so that engineers can determine at a glance if measurements are needed at additional locations in the interior environment or if longer periods of sampling are needed. With advances in instrumentation some portable instruments possess the capability of storing data, and/or transferring data electronically for later processing. In this chapter no attempt is made to describe laboratory instruments, continuous monitors, or alarms. Readers are encouraged to consult source books on the subject such as ACGIH (1972) and Lodge (1988). Portable instruments for both particle and gas pollutant measurements are discussed below.

There is a variety of portable instruments that sample the air on a continuous basis that can be used to compute the concentration of particles or gases and vapors one may choose to study. The average concentration in the room or workplace can be determined in a number of ways. For example, air can be sampled at several locations in the room or workplace; averaging techniques can then be used to establish an average concentration. While such an average is interesting, it does not establish compliance with OSHA standards. OSHA standards pertain to the exposure that individual workers *experience*, which is related in some (perhaps unpredictable) way to average workplace concentration. To measure an individual's exposure, a worker must wear a portable ***dosimeter*** that samples the air in the vicinity of the worker's face. Figure 3.2 shows an example of such a dosimeter. Air is withdrawn through a tube from a point near the worker's face to the dosimeter by a small battery-powered pump strapped to the worker's waist. The volumetric flow rate of air through the dosimeter (Q) is carefully controlled within prescribed limits. The 8-hr average mass concentration the worker has experienced is

Figure 3.2 Portable dosimeter to measure contaminant concentrations to which workers are exposed (from Sensidyne, Inc.).

equal to the mass of contaminant that has been collected divided by the product of the volumetric flow rate (Q) and the elapsed time.

3.2.1 - Particle Pollutants

The objective of portable particle pollutant measurements is to obtain a value of the mass concentration (mg/m^3) or number concentration ($particles/m^3$) at location(s) of concern. In addition, it is often desirable that instruments segregate the collected particles into categories from which the size distribution can be determined. The most common instrument for this purpose is the ***cascade impactor*** described in more detail in Chapter 9. Since OSHA standards pertain to the maximum concentrations to which a worker can be exposed, the preferred form of a cascade impactor is a portable device, clipped to the worker's lapel, that measures the average concentration and particle size distribution to which the worker is actually exposed. Air is drawn through the impactor by a battery-powered fan strapped to the worker's waist. The measured values can then be compared to OSHA PEL values. An example of a personal cascade impactor is shown in Figure 3.3. Figure 3.4 shows the mass median cut-point, or cut-diameter ($D_{p,50}$ as defined in Chapter 9) of unit density spheres for each stage of the this impactor as a function of the volumetric flow rate of air. Figure 3.4 enables a size distribution to be obtained at whatever volumetric flow rate the user wishes to operate the impactor.

Indoor air standards are written for either ***inhalable particles*** (D_p < 10 μm) or ***respirable particles*** (D_p < 2.5 μm). Consequently, personal samplers often have a cyclone, elutriator, or separator that removes all particles above these sizes while allowing the remaining particles to be drawn into the impactor. For the impactor to work properly the sampled air volumetric flow rate must be controlled within close limits. For this reason, impactors need to be calibrated before samples are taken, and steps must be taken to prevent workers from tampering with the volumetric flow rate during the time air is being sampled. Lastly, administrative steps must be taken to ensure that workers do not alter the aerosol being withdrawn into the impactor. Someone wishing to make the workplace air appear clean might try to filter the air entering the impactor inlet. Conversely someone wishing to make the workplace air appear dirtier than usual might deliberately introduce dust into the impactor inlet.

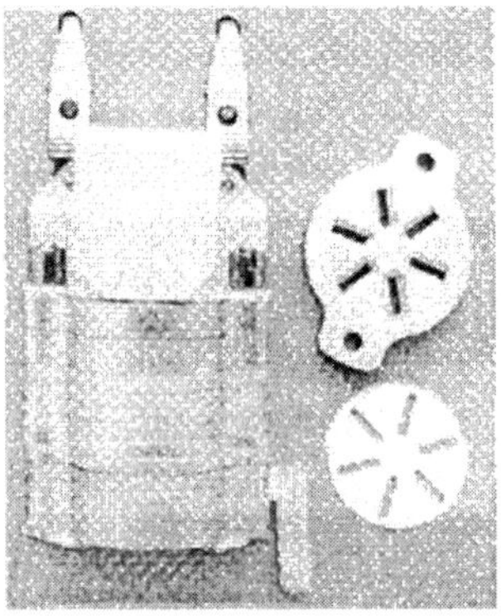

Figure 3.3 Andersen Marple 290 Personal Cascade Impactor worn on worker's lapel, using a battery-powered fan worn on worker's waist (courtesy of Andersen Instruments, Inc.).

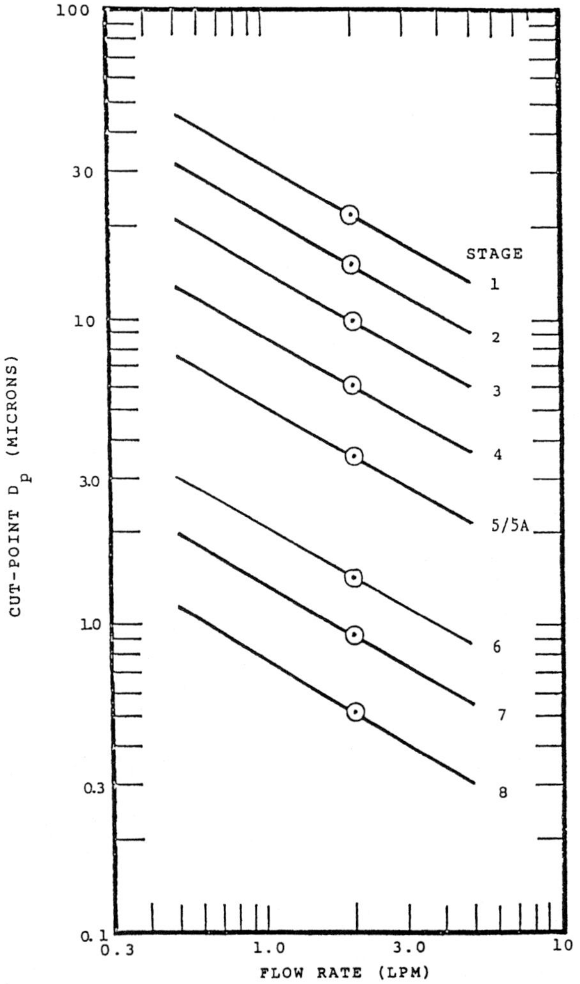

Figure 3.4 Cut-diameters ($D_{p,50}$) for unit density spheres versus volumetric flow rate at STP for Andersen Marple Series 290 Personal Cascade Impactor (courtesy of Andersen Instruments, Inc.).

If it is necessary to measure the *instantaneous* concentration of particles, the light scattering instrument shown in Figure 3.5 can be used. Infrared light (880 nm) is projected through the sensing volume where contact with particles in the moving sampling stream causes the light to scatter. The amount of scatter is proportional to mass concentration, and is measured by the photo-detector.

3.2.2 - Gaseous Pollutants

To assess indoor air quality, the instantaneous concentration of gaseous pollutants can be obtained with several types of portable instruments. Data can be recorded visually and/or electronically - over an 8-hour period, or over some other desired period of time. Alternatively, workers can be asked to wear individual monitors of one sort or another if engineers wish to assess their individual exposure over an 8-hour day.

Passive Badges

Figure 3.6 shows examples of ***passive badges***, also called ***lapel dosimeters***, used to record the total amount of a specific pollutant to which the worker has been exposed over a period of time. Passive badges are direct-reading colorimetric indicators composed of unique chemically treated paper that changes color when exposed to specific pollutants. Estimation of pollutant dose is made by comparing the stain on the treated paper with stains on a calibration chart that accompanies the badge.

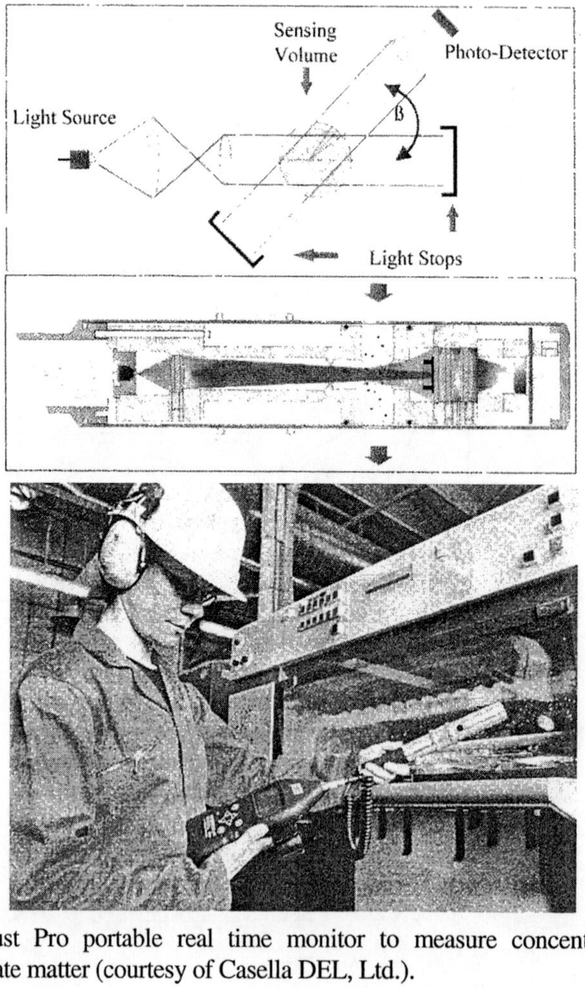

Figure 3.5 Microdust Pro portable real time monitor to measure concentration of suspended particulate matter (courtesy of Casella DEL, Ltd.).

The badge can be freshly prepared and used wet, or stored and used in a dry state. The accuracy of badges is affected by temperature, relative humidity, and the velocity of the air that comes into contact with the badge. Consequently, the accuracy of badges to detect unhealthy exposure is limited. Users have to be aware that pollutants other than the one under study may also discolor the badge. The attraction of passive badges is their versatility, low cost, and ease of use. Administrative controls need to be established so that workers do not tamper with the passive badges to yield false readings.

Detector Tubes

Detector tubes are also called ***indicator tubes*** or ***conduction tubes***. They are sometimes called ***Drager tubes*** or ***Kitagawa tubes***, named after early manufacturers of detector tubes. Detector tubes are sealed glass ampules (50 to 100 mL) containing unique granular solid chemical reagents that discolor when air containing specific pollutants is drawn through the tube. Detector tubes are inexpensive and can be used to estimate the concentration of a large number of indoor air pollutants. To use a detector tube, the ends of the sealed glass tube are broken off and the tube is placed at the entrance of a squeeze-bulb or piston pump. Each squeeze of the bulb or stroke of the piston pump draws a fixed amount of air through the reagent in the tube. Instructions accompanying the tube specify the number of strokes of the pump that should be used for specific pollutants. The manufacturer's instructions must be followed carefully. Within seconds, the reagent in the inlet end of the tube discolors. A well-defined separation appears between the reacted and virgin material. The length of the stained material is measured using a numerical scale printed on the outside of the detector tube. A calibration curve accompanying each tube relates pollutant concentrations to the length of the stained material. Figure 3.7 illustrates a typical detector tube and its displacement pump.

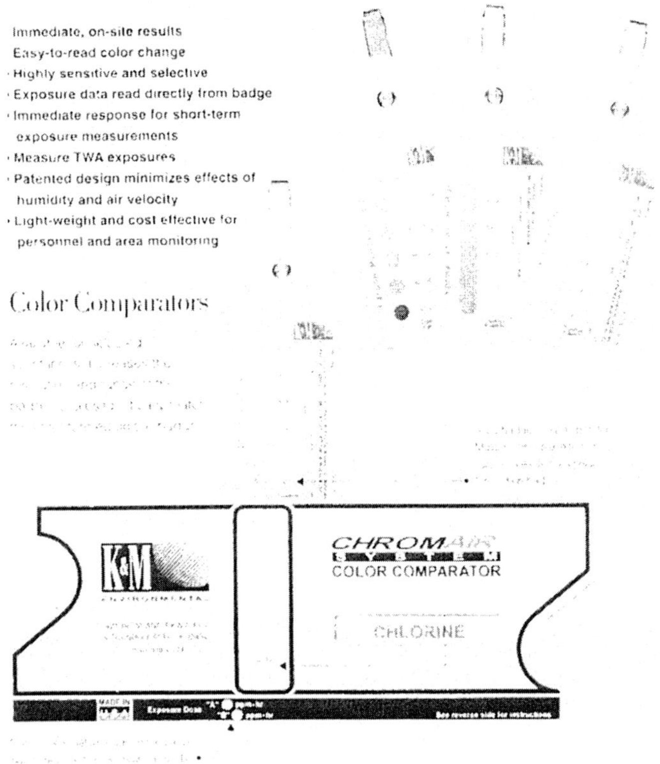

Figure 3.6 Chromair direct read passive monitoring system for gaseous pollutants (courtesy of K&M Environmental Air Sampling Systems).

(a)
(b)
(c)

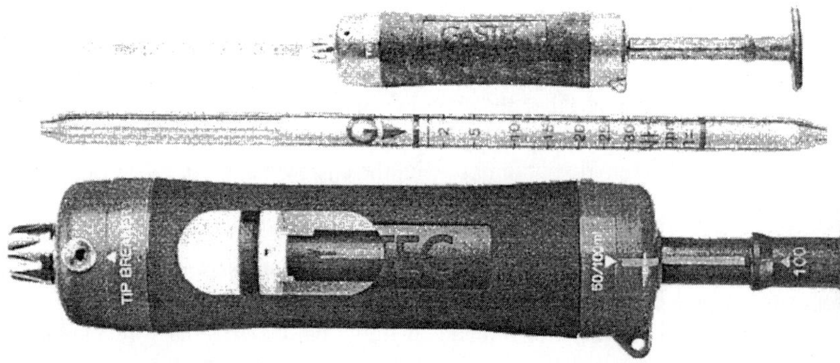

Figure 3.7 Gastec detector tube and hand pump: (a) assembled, (b) detector tube, (c) hand pump (courtesy of Gastec Corp.).

Precautions must be observed when using detector tubes since the amount of water vapor in the sampled air, as well as air temperature, may affect the measurement of some, but not all, pollutants. Users should know which pollutants they are seeking to measure and use specific, appropriate detector tubes for those pollutants. The literature accompanying each detector tube identifies interfering pollutants that can produce false readings. Consequently, an *array* of detector tubes may be needed to characterize pollutants in the indoor environment. In most instances engineers are testing for certain known pollutants since they know the source of these pollutants. While the accuracy of detector tubes is modest, they are a convenient way to test for the presence of certain pollutants in indoor air to ascertain whether concentrations are above or below indoor air quality standards. Detector tubes may be used for purposes of enforcing OSHA regulations.

Hand-Held Meters

Sophisticated ***hand-held meters*** provide instantaneous measurements of gaseous pollutant concentrations. Some meters are restricted to a single pollutant, but shown in Figure 3.8 is an example of a single meter that can be used to measure the concentration of any one of three gases (H_2S, CO, and O_2). Hand-held meters should be rugged and portable, and should provide sufficient accuracy to enable engineers to detect hazardous concentrations and violations of OSHA standards. Obviously there are differences in accuracy between meters; these differences are usually reflected in their prices. Hand-held meters produce a visual display, and some meters may be equipped with connections to other instruments to store data electronically for later processing. Hand-held meters are powered by batteries that require recharging in a docking station on a prescribed schedule.

Principles of Detection

Because gaseous pollutants are of the same phase as air, their concentrations cannot be measured by the same gravimetric principles used to measure particle pollutants. However, pollutant gases have unique spectroscopic and thermo-physical properties whose change can be sensed, and that are proportional to pollutant concentrations. The sensors in some hand-held meters are based on a variety of technologies, e.g. catalytic, electrochemical, and galvanic. Others utilize the properties of nondispersive absorption, chemiluminescence, or photoionization. The challenge is to design detection equipment that produces a unique, reliable, quantitative electrical response that is free of interference from other pollutant gases. Hand-held instruments should not require external electrical power, specialty gases, or recording instruments. Sampled air should be drawn into the meter by a hand pump activated by the user or by a small battery-powered fan/pump. The following summarizes unique physical properties of important gases that can be exploited to measure concentration.

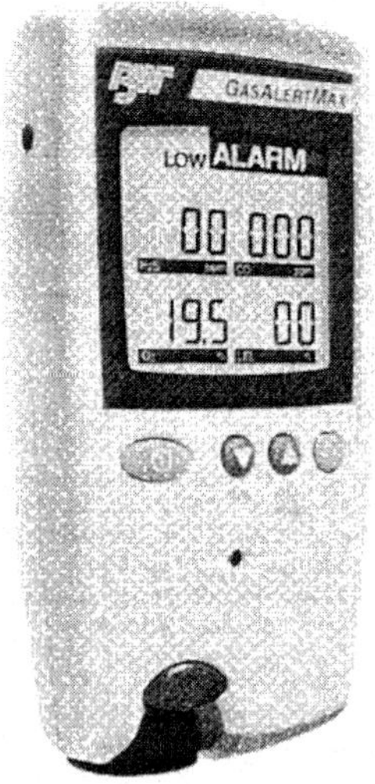

Figure 3.8 GasAlertMax portable three-gas detector for measurement of H_2S, CO, and O_2 (courtesy of BW Technologies, Ltd.).

Oxygen - While not a pollutant, knowing the concentration of oxygen in an indoor environment is very important. Portable instruments need to measure oxygen concentrations accurately only between 10 and 20% (by volume). Individuals entering rooms in which the oxygen concentration is less than 10% would become unconscious in less than a minute and would die within several minutes. Oxygen's magnetic property is commonly the basis of a sensor in an oxygen meter. Oxygen molecules are ***paramagnetic***, i.e. they are attracted by a magnetic field. Oxygen's paramagnetism is caused by its atomic and molecular structure, but is inversely proportional to its absolute temperature. When heated, oxygen becomes ***diamagnetic***, i.e. it is repelled by a magnetic field. Figure 3.9 is a schematic diagram of an analyzing cell for an oxygen meter.

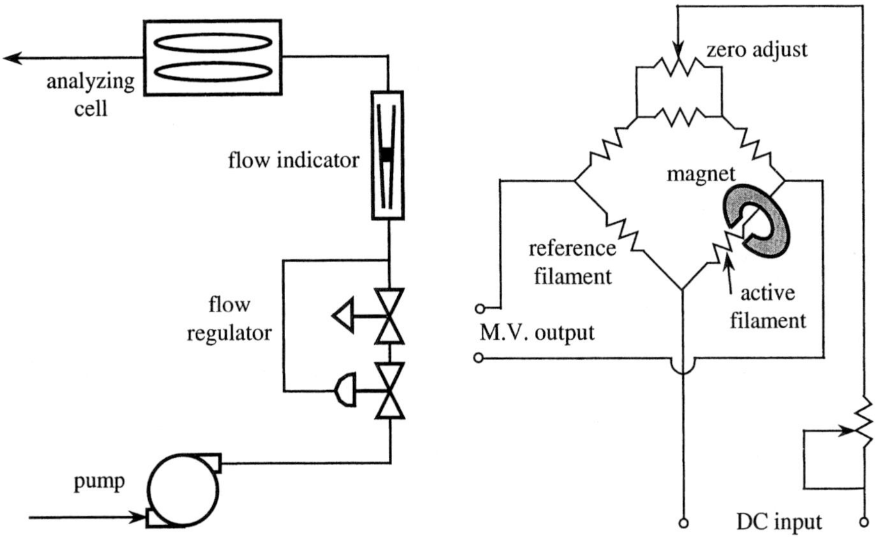

Figure 3.9 Schematic diagram of the sensor of an oxygen meter using the paramagnetic property of oxygen to measure oxygen concentration: (a) flow diagram, (b) bridge circuit (adapted from ACGIH, 1972).

The analyzing cell contains two electrically heated, glass covered, precision resistors. The ***active filament*** and ***reference filament*** form two legs of a Wheatstone bridge circuit. A permanent magnet surrounds the active filament but not the reference filament. A gas sample at room temperature surrounds both filaments, and oxygen molecules are attracted by the magnetic field in the active filament. The gas sample is now heated, oxygen loses its paramagnetism, and is forced out of the magnetic field by cooler, more magnetic oxygen bearing gas. A circulation of sample gas cools the active filament. The cooling effect is proportional to the oxygen content of the gas sample. The reference filament provides compensation for variations in cell temperature, and for other temperature changes that might otherwise introduce measurement error; the resistances of both the active and reference filaments change due to changes in their respective temperatures. The voltage output from the Wheatstone bridge is thus proportional to the amount of oxygen in the sampled gas.

Mercury vapor - Mercury vapor is a very toxic substance, but unfortunately has no odor to characterize it. Several portable mercury vapor meters are commercially available. Figure 3.10 illustrates a common method to measure the mercury concentration in air. The meter is basically a portable ultraviolet (UV) photometer tuned to 253.7 nm, which is the wavelength at which mercury vapor absorbs light. A low-pressure mercury vapor lamp located at one end of the instrument serves as a pulsed UV light source. Mercury vapor in an air sample reduces the amount of radiation received by the sample phototube in proportion to the mercury vapor concentration in the air sample. A sample phototube in the instrument measures the amount of UV radiation passing through the sample space. A second phototube measures the amount of UV radiation from the same source, but passing through air that contains no mercury; this phototube serves as a reference. The meter also contains a series of calibration filters of known absorbance that can be switched into the optical path to provide standards for calibration whenever users need to calibrate the instrument.

NO and NO_2 - While NO_x is not considered to be an indoor air pollutant, there are occasions when the indoor concentration needs to be measured. The oxides of nitrogen are unique in that they absorb visible light. Consequently the reduction in intensity of a visible light beam passed through a stream of air containing NO_x can be measured and compared to a reference, and the NO_x concentration can be calculated. The principle of operation is basically the same as that for mercury vapor, as discussed above, except visible light is used in place of UV light.

Hydrocarbons - Hydrocarbon pollutants are combustible, and ions from the ionization of carbon are some of the combustion products. The ion flow, i.e. current, is proportional to both the number of carbon atoms in the pollutant molecule and the pollutant concentration. The sampled gas is oxidized by a (flameless) catalyst, and the generated ion flow is measured with an ionization detector. The hydrocarbon concentration in the sample is proportional to the measured current. Alternatively, the heated gas can change the resistance of an active filament that is one leg of a Wheatstone bridge

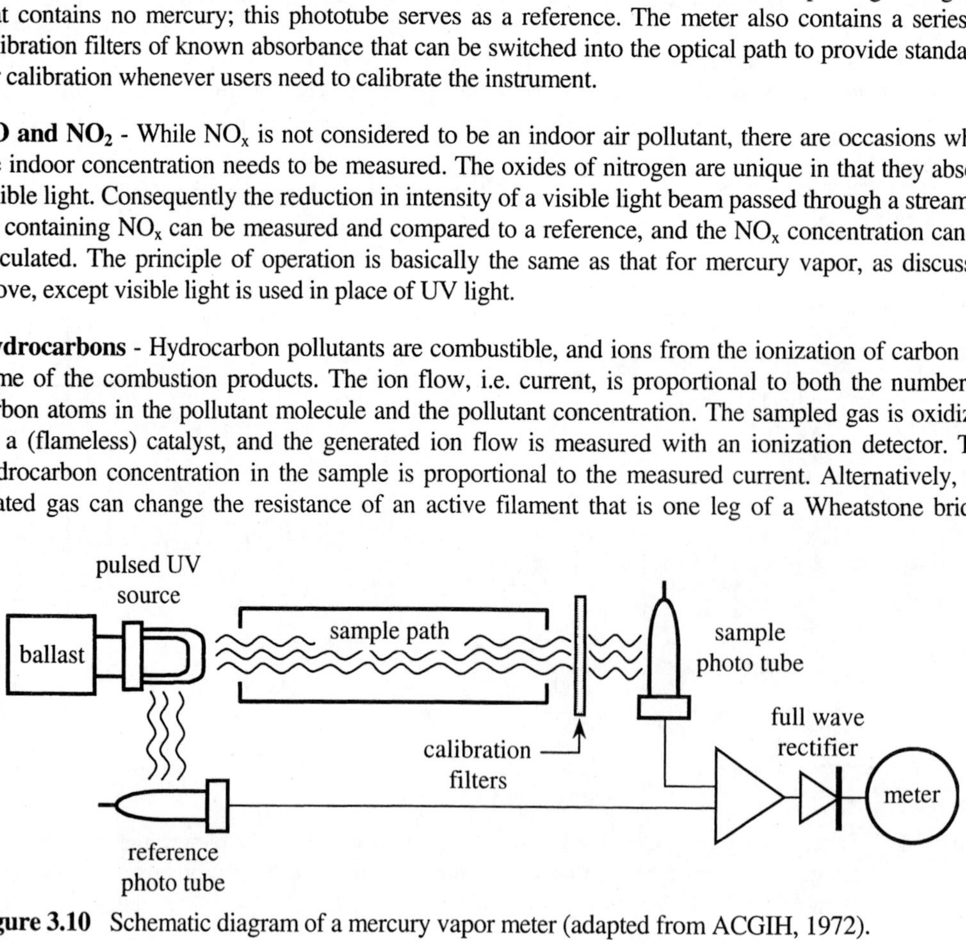

Figure 3.10 Schematic diagram of a mercury vapor meter (adapted from ACGIH, 1972).

(Figure 3.11), and the hydrocarbon content can be deduced from the output voltage. Hand-held pollution meters using ionization report the hydrocarbon concentration as if it were methane because methane is used to calibrate the meter.

Carbon monoxide - A common technique to measure carbon monoxide (CO) concentration is to pass the sampled air through a (flameless) catalyst that oxidizes CO to carbon dioxide (CO_2). The resulting rise in temperature of the sampled air changes the electrical resistance of an active filament that is one leg of a Wheatstone bridge (similar to Figure 3.11). Alternatively, CO fluoresces by infrared radiation, and the fluorescence can be measured with a photodetector sensor whose output is proportional to the CO concentration in the sampled air.

Ozone - Commercially available ozone meters operate on several different principles that exploit the thermophysical properties of ozone. One class of ozone meter is based on the oxidation-reduction reaction of ozone with potassium iodide. Another ozone meter relies on the chemiluminescence generated from a flameless reaction of ethylene gas and ozone. Both these types of meters require a supply of a reagent. A class of ozone meter with a sensitivity of 1 to 10 PPB of ozone that does not require an external supply of reagent exploits the emission of radiation stimulated by a gas-solid chemiluminescent reaction. The meter described in Figure 3.12 is based on the chemiluminescent reaction of ozone with rhodamine B absorbed on silica gel that emits 585 nm radiation, and that is proportional to ozone concentration in an air sample.

3.3 Fire and Explosion

3.3.1 Flammability Limits

A fuel-oxidant mixture that generates sufficient energy upon ignition to cause a flame to propagate throughout the unburned mixture is defined as a ***flammable mixture***. A ***non-flammable mixture*** is one that cannot (of itself) sustain combustion. While the word ***inflammable*** is a synonym of flammable, it is often interpreted to be just the opposite and is not used in this text. (The authors regret that the word *inflammable* is part of the English language.) There is no official word *imflammable*, although it has at times been substituted for *inflammable* to remove the ambiguity. Occasionally a nongaseous explosive material contains both a fuel and oxygen; there are also gaseous materials that

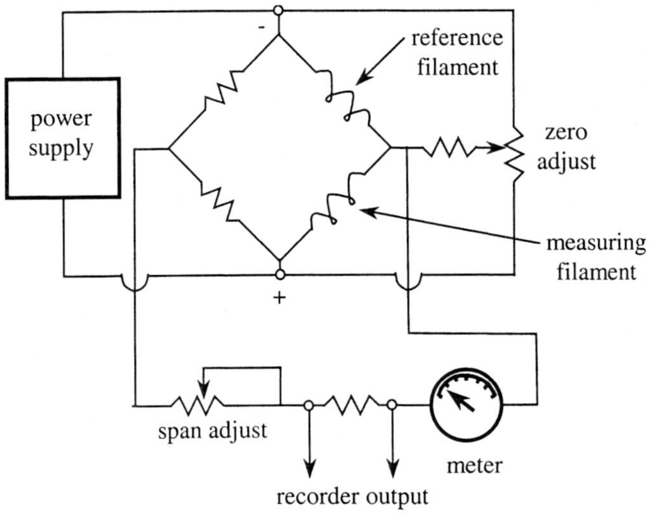

Figure 3.11 Wheatstone bridge used as the sensor unit of a hydrocarbon analyzer (adapted from ACGIH, 1972).

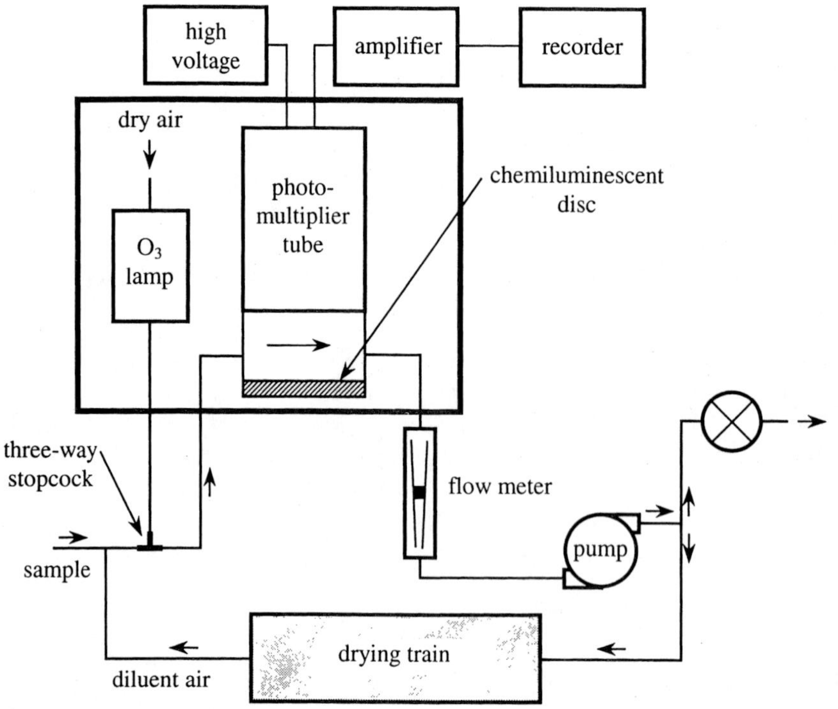

Figure 3.12 Measurement of ozone based on the chemiluminescence reaction of ozone and Rhodamine B absorbed on silica gel (adapted from ACGIH, 1972).

are explosive by themselves, such as azomethane and hydrazine. These types of material will not be discussed. ***Ignition*** is a source of energy that is transferred to the fuel and oxidant at such a rate within a minimal volume to initiate reaction. A ***combustion zone*** is a thin region within which a series of reactions occurs involving the transfer of heat from the burned to unburned region, and simultaneously the diffusion of unique chemical species (chain carriers, free radicals, etc.) from the burned to the unburned region. What one sees and calls a ***flame*** is radiation in the visible spectrum that accompanies the chemical reactions. However, not all reactants, e.g. hydrogen, produce visible radiation. Indianapolis-class racing cars burn a type of alcohol which burns without a visible flame; pit crew personnel can sometimes be seen spraying a fire extinguisher at a car (or driver) which does not look at all like it is on fire.

Flames are categorized as premixed or diffusion flames. In ***diffusion flames***, fuel and oxygen diffuse toward the combustion region from opposite directions, mix, and burn. A match flame and the flame surrounding a burning fuel drop are diffusion flames. In ***premixed flames***, fuel and oxygen are initially mixed and the flame propagates into the unburned mixture. A flame on a Bunsen burner and the flames in a common propane gas cooking grill are examples of premixed flames. The flame speed for premixed flames is a function of fuel-oxidant composition, temperature, and pressure. There are certain fuel-oxidant compositions at which flames cannot be initiated.

Engineers must ensure that the concentrations of flammable materials in air never achieve levels that enable the materials to burn. The ***limit of flammability*** is defined as the lowest percentage of one of the reactive components in a gaseous mixture that can sustain combustion. There are both ***fuel-lean*** and ***fuel-rich*** limits of flammability for both flames and detonation waves. The fuel-lean limit is called the ***lower explosion limit*** (LEL) and the fuel-rich limit is called the ***upper explosion limit*** (UEL). The LEL is also called the ***lean flammable limit***, typically reported as a percent by volume or as a mol fraction. Figure 3.13 shows the flammability limits of mixtures of methane and oxygen at different temperatures diluted with various amounts of nitrogen or carbon dioxide. These curves show that adding a diluent narrows the limits of flammability; increasing the temperature widens the limits, but changing the diluent from nitrogen to carbon dioxide has only a minimal effect on the limits. Figure 3.14 shows that the lean flammability limit is virtually unaffected by pressure whereas the rich flammability limit increases rapidly with pressure. The upper explosion limit is seldom encountered in indoor industrial contaminant control; indeed even the lower explosion limit is generally considerably larger than the TLV or PEL. Unusual cases are materials that are not toxic, but where care must be taken to ensure that LEL levels are never exceeded. As a rule of thumb insurance carriers require that the concentration of a flammable material should never exceed 10% of the LEL. Table 3.2 is an abbreviated table of LEL and UEL for some common industrial materials. Figure 3.15 shows that LEL is inversely proportional to the heat of combustion for many hydrocarbons in air.

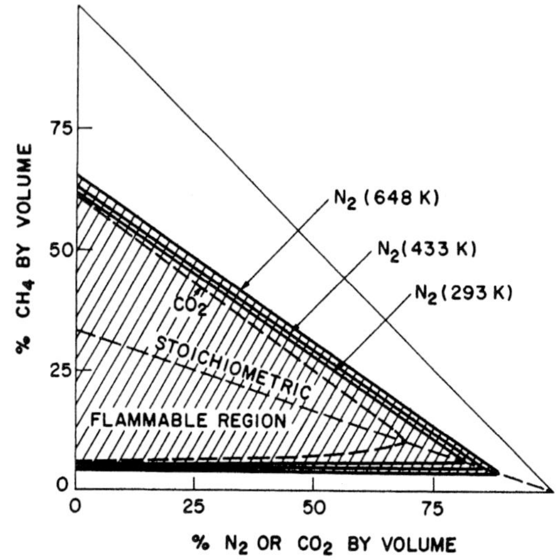

Figure 3.13 Flammability limits for mixtures of methane and oxygen diluted with nitrogen or carbon dioxide and different temperatures (redrawn from Chigier, 1981).

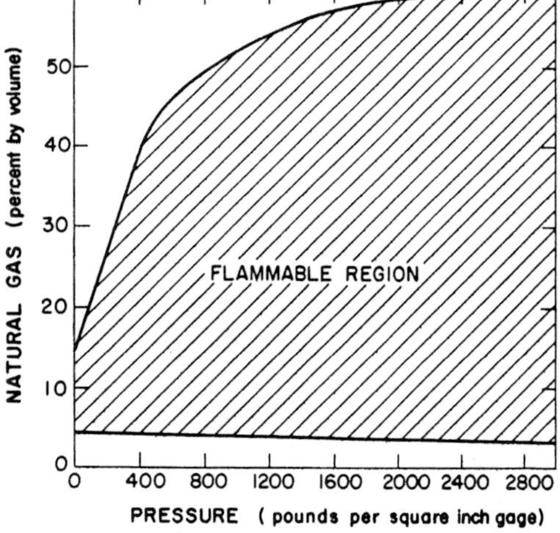

Figure 3.14 Effect of pressure on flammability limits of natural gas and air (redrawn from Lewis and vonElbe, 1951).

Table 3.2 Flammability limits and flash temperatures of common industrial materials (abstracted from Lewis and von Elbe, 1951).

material	M	flash point (°F)+	LEL*	UEL*
acetic acid	60.05	104	5.4	-
acetic anhydride	102.09	121	2.67	10.13
acetone	58.08	0	2.55	12.8
benzene	78.11	12	1.4	7.1
n-butanol	74.12	84	1.45	11.25
butyl acetate	116.16	72	1.39	7.55
carbon tetrachloride	153.84	non-flammable	-	-
chloroform	119.39	non-flammable	-	-
cyclohexane	84.16	1	1.26	7.75
ethane	30.0	gas	3.0	12.5
ethylene	28.0	?	2.75	28.6
ethylene oxide	44.05	-	3.0	80.0
ethyl acetate	88.10	24	2.18	11.4
ethyl alcohol	46.07	55	3.28	18.95
ethyl chloride	64.52	- 58	3.6	14.80
gasoline	86	- 50	1.3	6.0
heptane	100.2	25	1.1	6.7
hexane	86.17	- 7	1.18	7.4
hydrogen sulfide	34.08	gas	4.3	45.5
methane	16.0	gas	5.0	15.0
methanol	32.04	54	6.72	36.5
methyl acetate	74.08	15	3.15	15.6
methyl formate	60.05	- 2	4.5	20
octane	114.22	56	0.95	3.2
iso-propanol	60.09	53	2.02	11.8
propyl acetate	102.13	43	1.77	8.0
propane	44.09	gas	2.12	9.35
toluene	92.13	40	1.27	6.75
trichloroethylene	131.4	non-flammable	-	-
o-xylene	106.16	63	1.0	6.0

*Flammability limits in % by volume
+Flash point measured by closed cup experiments

In order to anticipate (and thus hopefully avoid) fire and explosion, one is required to recognize:

- sources of ignition
- fuel-oxidant concentrations sufficient to support combustion
- unusual combinations of oxygen enrichment and high temperatures and pressures

In order to initiate combustion, a high temperature must be produced in a small volume whose dimensions are comparable to the thickness of a flame. ***Minimum ignition energy*** is inversely proportional to pressure and temperature; thus at elevated pressure and temperature, sparks and short circuits will initiate combustion where they do not at STP.

It is often thought that ignition temperature is the temperature to which a combustible mixture has to be raised before combustion can occur. While raising the temperature of a combustible mixture is not a safe practice, temperature alone is not the determining factor producing ignition. Figure 3.16 is a schematic diagram of a typical temperature profile in a premixed flame. ***Ignition temperature*** for a mixture of gases is defined as the temperature at which there is an inflection point in the flame temperature profile, $\partial^2T/\partial x^2 = 0$. The temperature gradients are very large considering that at STP a premixed flame is approximately 1 mm thick and the burnt temperature is over a thousand degrees centigrade above thc unburnt temperature. On the unburned side of the flame, energy is absorbed by the gas mixture; on the burned side of the flame, energy is transferred to the unburned gas mixture. Thus if one examines the temperature profile across a flame, there is a point where the second derivative is zero corresponding to the point separating the region that absorbs energy from the region that liberates energy. Historically (Lewis and vonElbe, 1951), such a point was called the ignition temperature, and represents the minimum temperature below which the chemical reaction does not occur.

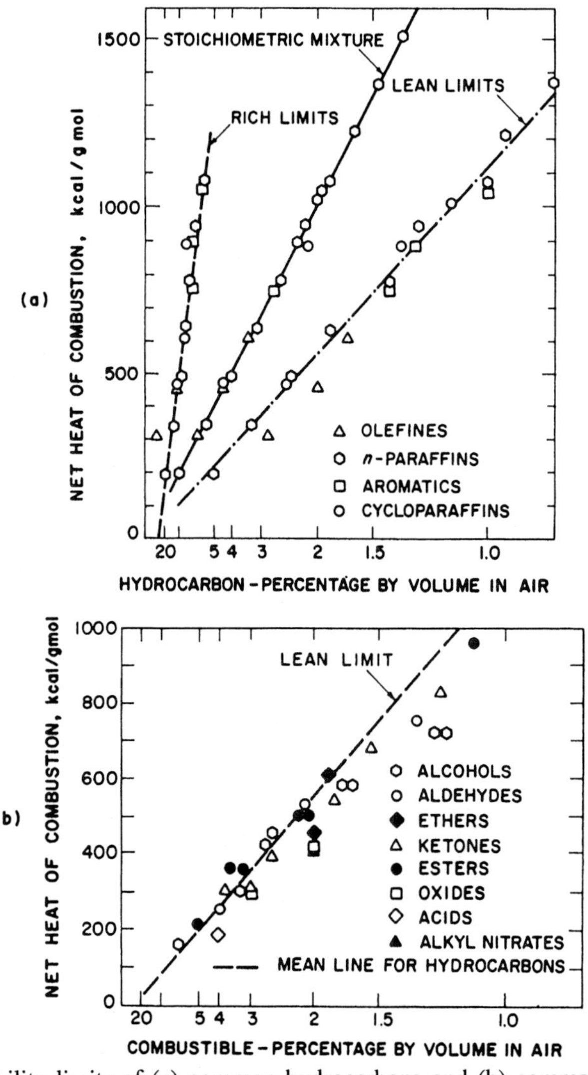

Figure 3.15 Flammability limits of (a) common hydrocarbons and (b) common organic substances (adapted from Penn and Mullins, 1951).

Ignition can also occur due to spontaneous combustion or from heated surfaces. Often cited in the professional literature is the flash-point temperature, which is also listed in Table 3.2. The ***flash-point temperature*** is defined as the minimum temperature at which a flammable liquid generates enough vapor to produce a flame. The flash-point temperature is not an intrinsic property of the fuel and is a function of the experimental apparatus used to measure it. Nonetheless, it provides engineers a way to anticipate when a hot surface can initiate ignition.

Example 3.2 - Estimating the LEL from the Heat of Combustion

Given: A container with a label that has been partially destroyed indicates that it contains the hydrocarbon (HC) cyclohexane (M = 84.16) and that it has a heat of combustion (h_{RP}) of 873.76 kcal/mol.

To do: Estimate the hydrocarbon's lower explosion limit in air expressed as a fuel-air ratio in terms of mass.

Solution: Since cyclohexane is a hydrocarbon, Figure 3.15a is appropriate. The figure indicates that the lower combustion limit at h_{RP} = 873.76 kcal/mol is about 1.4%. This value corresponds to an LEL expressed as a mol fraction of 0.014. Alternatively (and more accurately) from Table 3.2 or from Perry et al. (1984), the LEL of cyclohexane is found to be 1.26%. The agreement is reasonably good. Expressed as a fuel-air ratio based on mass, the estimated lower combustion limit is

$$\frac{F}{A} = LEL\frac{M_{HC}}{M_{air}} = \left(0.0126\frac{\text{mol HC}}{\text{mol air}}\right)\left(84.16\frac{\text{kg HC}}{\text{mol HC}}\right)\left(\frac{\text{mol air}}{28.97\text{ kg air}}\right) = 0.037\frac{\text{kg HC}}{\text{kg air}}$$

Discussion: The mol fraction is approximated above as the ratio of mols of HC to mols of air. More precisely, it is the ratio of mols of HC to *total* mols of the air-fuel mixture. Since the mol fraction corresponding to the LEL is typically very small (less than two percent here), this approximation is reasonable within the overall accuracy of the calculations. The final result is reported to two significant digits.

Combustible dusty air and aerosols can support flames and detonation waves, and the same precautions must be taken as those taken with flammable gases and air. History records fires and explosions of coal dust, sawdust, grain dust, and aerosols of metals such as Al, Zn, Ti, Mn, Sn, and pure Fe. Fires and explosions are a particular hazard in equipment used to collect particles from the indoor work environment, such as filters, cyclones, etc., or in powder handling equipment, such as transfer points of conveyors, pulverizers, etc. Sparks formed by fan motors, fan blades that scrape on fan housings, or electrical discharges produced by static electrification are often sufficient to ignite these materials.

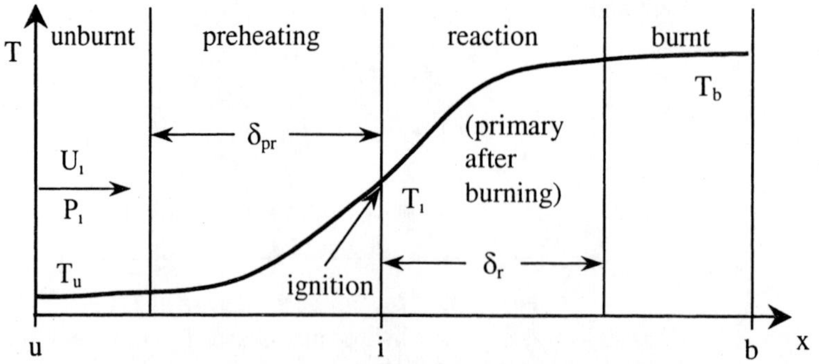

Figure 3.16 Temperature versus distance diagram showing the structure of a premixed flame front.

To prevent the buildup of static electricity, it is common practice to provide a direct electrically conductive path to enable accumulated static charges to pass to ground. It is unwise to assume that bolted parts, with metal-to-metal contact, can provide a sufficient conductive path. When fabric filters (bag house) are used to collect combustible particles, it is common practice to weave a wire mesh into the bag and to ground each bag individually. Another common practice is to maintain a relative humidity no less than 65% so that moisture can be absorbed on the particle surface and provide a conductive path for the discharge of static electricity.

It is more difficult to describe the combustion of ***dust-air mixtures*** than gas-air mixtures because aerosols consist of two phases while gas-air mixtures are of a single phase. The explosion of dust-air mixtures depends on the existence of a flammable mixture and the presence of an ignition source (electrical discharge, chemical exothermic reaction, heated surface, etc.) of sufficient magnitude to initiate combustion. Flammability describes a state of sufficiently high-temperature at which particles will burn in air. Explosion describes the state that in addition to flammability, a source of energy exists to ignite the particles. Thus there are aerosol compositions that can burn but cannot be ignited and explode.

Virtually every industry that processes chemicals handles powders at some stages of production. Among the industries with the most flammable powders are those processing foods, metals, pharmaceuticals, plastics, dyes, and pigments. Examples of such flammable materials include starch, polyethylene, carbon, coal, sulfur, iron, copper, etc. Hertzberg et al. (1991, 1992) claim that the following three sequential processes govern ignition of dust particles:

- heating and pyrolysis
- mixing of the emitted volatiles with air between the particles
- ignition and combustion of the resultant vapor and air mixture

Ignition of dust-air mixtures occurs when a concentrated source of energy exceeding a ***minimum ignition energy*** triggers combustion, or when the air temperature (distributed source of energy) exceeds an ***autoignition temperature***. Ignition followed by combustion also depends on a minimum dust concentration and the dust particle size. There is no fuel-rich explosion limit. Hertzberg contends (1991) that the first step above, pyrolysis, is the rate-controlling step in autoignition. ***Pyrolysis*** is the process in which heat transferred to the particle causes compounds to volatize and escape from the particle. ***Char*** is the name given to the remaining nonvolatile fuel residue. There are four reasons why it is more difficult to describe the combustion limits of heterogeneous dust-air mixtures:

- Dust-air concentration - For purposes of combustion, it is inadequate to describe dust clouds solely in terms of dust mass concentration since the mixture is heterogeneous - it is a two-phase mixture in which compositional properties are discontinuous. In addition, an aerosol may be combustible at one instant of time, but owing to gravimetric settling, the particle mass concentration can decrease or the mixture can segregate.

- Minimum ignition energy - The ignition energy must be sufficiently large to produce a flame front that propagates freely once ignition occurs. Figure 3.17 shows that the minimal ignition energy increases as the particle mass concentration decreases; below some critical value of concentration, the dust cloud does not ignite. For example, consider Pittsburgh bituminous coal. It is seen in Figure 3.17 that at mass concentrations above about 250 g/m^3, the ignition energy is effectively zero; a spark of infinitesimal energy could potentially ignite the coal dust. At a mass concentration around 100 g/m^3, on the other hand, an ignition energy of 1 kJ or more is required to ignite the dust cloud. At concentrations below about 90 g/m^3, the ignition energy curve is nearly vertical, implying that the dust cloud does not ignite, even if a very strong energy source is available.

- Dust volatility - Fuels with high volatility have lower autoignition temperatures than do fuels with low volatility. Fuels with relatively high hydrogen/carbon ratios pyrolyze rapidly at relatively low heat fluxes. With such fuels, the volatiles reach the surface rapidly and are transferred to air, minimizing char formation. On the other hand, fuels with low hydrogen/carbon ratios require higher temperatures to volatize. This can be seen in Figure 3.18. At high concentrations, the autoignition temperature of anthracite coal is around 800 °C, while that of sulfur is only around 300 °C. Also note that below some critical mass concentration, autoignition temperature rises rapidly, due to the sparse concentration of dust particles. Using Pittsburgh bituminous coal again as an example, the autoignition temperature is fairly constant (around 600 °C) for mass concentrations above about 100 g/m^3, but rises rapidly below that concentration.

- Particle size - Figures 3.19 and 3.20 show that the lean flammable limit and autoignitior temperature are constants for a range of small particles, but beyond a certain critical particl size, these properties increase with particle size. This illustrates that it is harder to ignite large dust particles compared to small dust particles. (A 1-kg lump of coal suspended in air inside a container does not explode, but a monodisperse suspension of the same amount of coal with an average particle size of 20 μm diameter is potentially explosive.) As can be seen in Figure 3.19, the lean flammable limit decreases as oxygen concentration increases, as would be expected. The critical particle size, beyond which the lean flammable limit begins to rise, increases as oxygen concentration increases, also as expected. Figure 3.20 shows that higher volatility dusts tend to have not only smaller autoignition temperatures, but also larger critical particle size. It is believed (Hertzberg, 1991) that both phenomena illustrate that volatilization is the rate-controlling step.

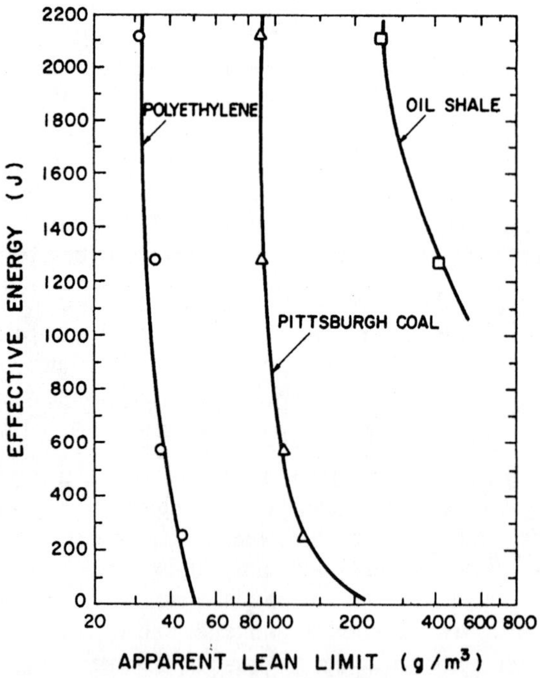

Figure 3.17 Effect of ignition energy on lean flammability limits for polyethylene, coal, and oil shale dusts. Mass mean diameters 37, 50 and 60 μm respectively (redrawn from Hertzberg et al., 1986).

Explosion limits reach their asymptotic limits at energies in the kJ level. Hertzberg et al. (1982) conclude that the product of the lean mass dust concentration limit and the combustible volatile percentage is constant. They conclude that below the critical particle size the rate of devolatization is so rapid that combustion is controlled by the diffusion and combustion of volatiles. For the Pittsburgh coal seam at normal (21%) oxygen levels, the critical particle diameter is approximately 50 μm, as seen in both Figures 3.19 and 3.20. For particles of polyethylene (which are composed entirely of volatile materials), Figure 3.20 shows that the critical particle diameter is somewhat larger, i.e. approximately 80 μm. The critical particle diameter of Pocahontas coal which has a very low volatility, on the other hand, is so small (less than 2 μm) that it is off the chart.

Comparison reveals a considerable difference between the lower explosion limit (LEL) and the TLV (or PEL). For coal dust containing a respirable fraction less than 5% quartz the ACGIH TLV is 2 mg/m^3. The lower explosion limit is approximately 100 g/m^3, a value 50,000 times higher than the TLV. Thus keeping combustible dusts concentrations in the workplace below TLVs and PELs is generally sufficient to prevent fire and explosion. (Note however that inside dust collection equipment, where air is not breathed, the concentrations can be much higher, and may well be combustible.)

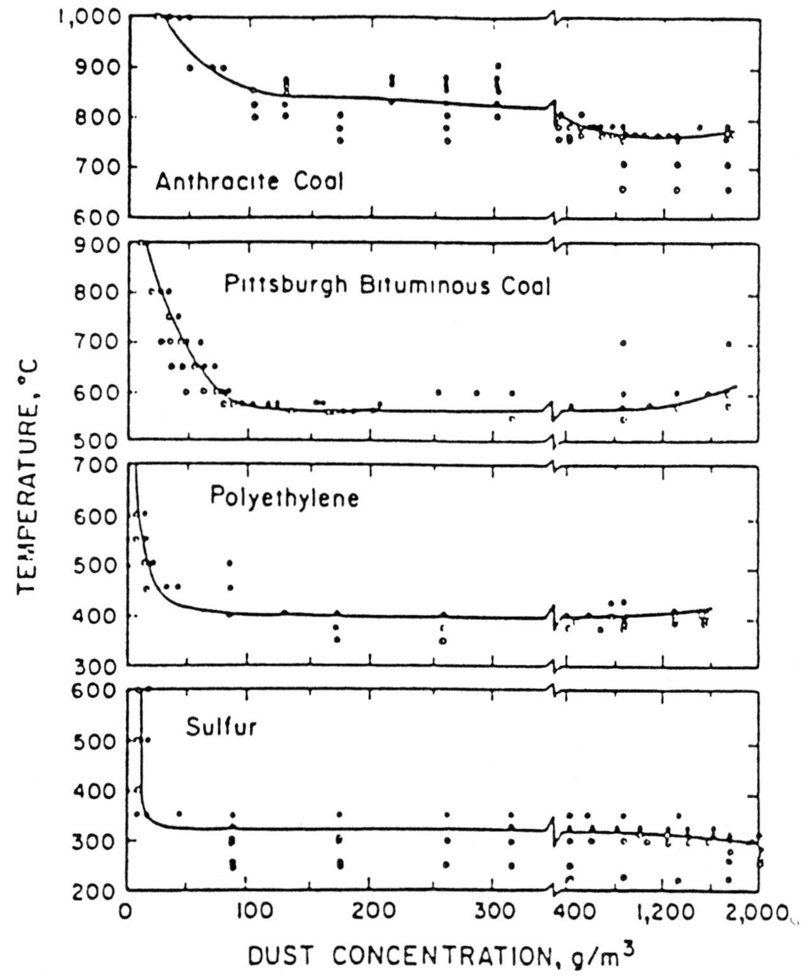

Figure 3.18 Autoignition temperature versus dust concentration of anthracite coal, Pittsburgh bituminous coal, polyethylene, and sulfur (redrawn from Hertzberg et al., 1982).

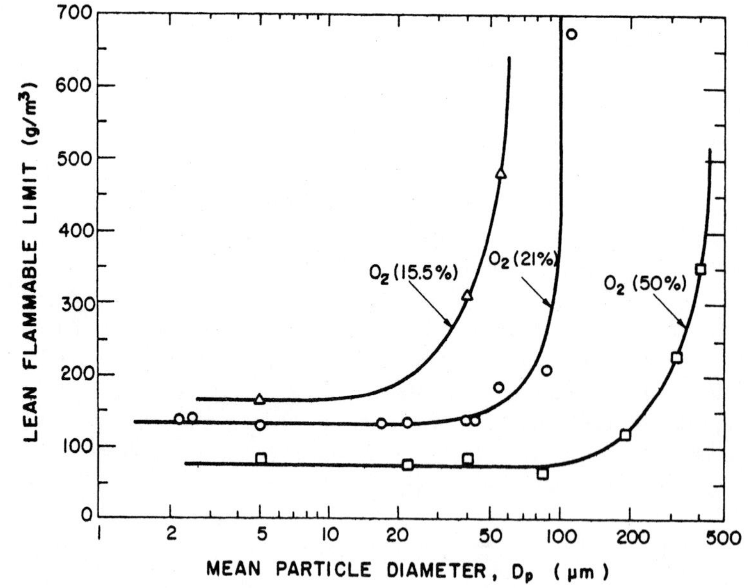

Figure 3.19 Lean flammability limits for Pittsburgh bituminous coal as a function of particle size and three oxygen concentrations (from Zimmerman, 1984).

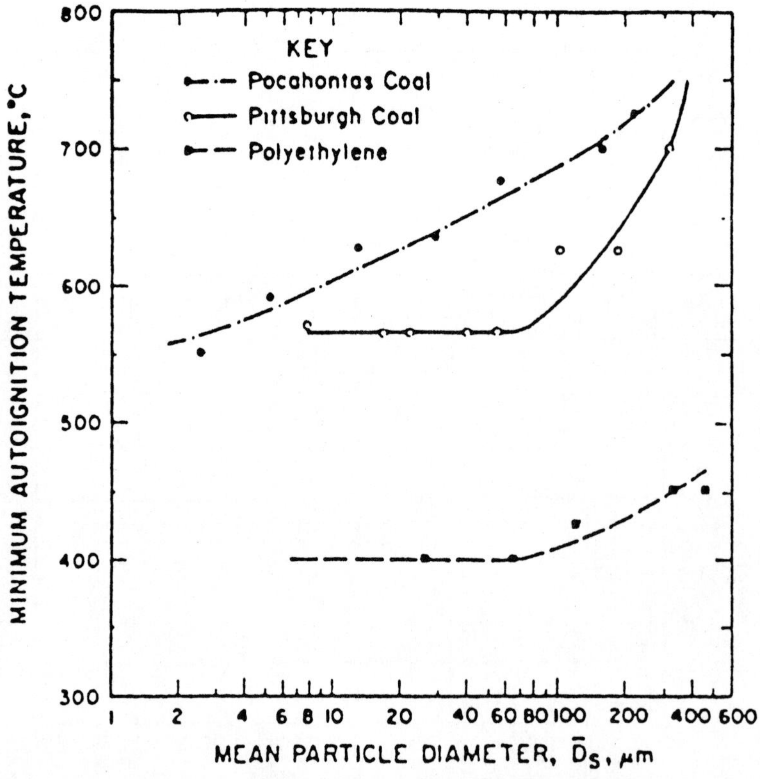

Figure 3.20 Minimum autoignition temperature versus surface area mean particle size of Pocahontas coal, Pittsburgh coal, and polyethylene (from US NIOSH, 1977).

Example 3.3 - Flammability of Coal Dust

Given: An electric utility boiler burns Pittsburgh bituminous coal (ρ_p = 1200 kg/m^3). Coal from the stockpile enters the building and is pulverized into particles 50 μm in diameter (D_p). The pulverized coal is transferred to a small hopper prior to being injected into the boiler. The plant manager is concerned that coal dust suspended in the air space in the hopper might explode if there is a source of ignition. The hopper is equipped with a light-sensing particle counter that monitors the particle number concentration, c_{number} (in particles/m^3), and trips an alarm when combustion might occur. For safety, the alarm is set to trip when the particle concentration is 10% of the minimum flammable concentration.

To do: Using Figure 3.19, estimate the particle number concentration (c_{number}) to trip the alarm.

Solution: From Figure 3.19, it is seen that the minimum flammable particle concentration for D_p = 50 μm is approximately 150 g/m^3. The mass of an individual 50 μm coal particle is

$$m_{particle} = \rho_p \frac{\pi D_p^{\ 3}}{6} = \left(1200 \frac{kg}{m^3}\right) \frac{\pi\left(50\times10^{-6}\ m\right)^3}{6} = 7.854\times10^{-11}\ kg$$

The corresponding flammability particle concentration is

$$c_{number} = \frac{\left(150\frac{g}{m^3}\right)\left(\frac{1\ kg}{1000\ g}\right)}{7.854\times10^{-11}\frac{kg}{particle}} = 1.91\times10^{9}\ \frac{particles}{m^3}$$

If the alarm is set to trip at 10% of this value, the flammability limit based on particle number concentration should be less than 1.91 x 10^8 particles/m^3,

$$c_{number}\ \text{(to trip the alarm)} \geq 1.9\times10^{8}\ \frac{particles}{m^3}$$

Discussion: The answer is given to two significant digits, since that is all that can be expected from Figure 3.19.

To prevent fires and explosions, engineers must eliminate one or more of three essential ingredients:

- combustible materials
- oxygen
- source of ignition

To minimizing damage, engineers must provide ways to:

- isolate fire to prevent its spread (fire walls, flame arresters, barriers)
- relieve high pressure (rupture diaphragms)
- extinguish fires (sprinkler systems)

Engineers should look to the chemical engineering profession for guidance since they have studied loss prevention and industrial health and safety with more rigor than have any of the other engineering professions (AIChE, 1990; Crowl and Louvar, 1990). Engineers should also abide by codes established by the ***National Fire Protection Association*** (NFPA) that are required by many states and insurance carriers. Many times a material being manufactured is inherently combustible and the only option available is to eliminate oxygen and sources of ignition. Ignition can occur from heated surfaces such as overheated bearings. Sparks can come from several sources, including welding and metal cutting, the impact of metal-on-metal collisions, electrical discharges, and metal surfaces that have acquired charge from static electrification; such sparks must be eliminated. Explosion-proof equipment and instruments, along with design procedures to prevent the accumulation of static charge, are discussed in this text and elsewhere, e.g. Crowl and Louvarn (1990).

Inerting is the name given for replacing air with a blanket of inert gas above combustible material contained in a vessel, reactor, etc. Shown in Table 3.3 are maximum oxygen concentrations that can be tolerated safely in nitrogen used to blanket flammable substances. Inerting purges the vessel, and air is replaced with an inert gas before the combustible material is added to the vessel. Continuous monitors are used to control equipment to maintain sufficient amounts of inert gas. To purge a vessel of oxygen, four techniques are used:

- vacuum purging
- pressure purging
- sweep-through purging
- siphon purging

Vacuum purging is the process in which air in a vessel is removed by vacuum until the pressure reaches some predetermined value, whereupon an inert gas (such as nitrogen or carbon dioxide) is added to raise the pressure to a predetermined level. Finally the combustible material is added. Achieving a vacuum is a slow process and not all vessels (even high-pressure vessels) can withstand negative pressures. ***Pressure purging*** is the process of adding an inert gas to a vessel containing air until a homogeneous mixture at a prescribed positive pressure is achieved whereupon the gaseous mixture is vented to the atmosphere. The inert gas is again added to the remaining mixture that is again vented to the atmosphere. The process is repeated several times until the initial amount of oxygen has been diluted to some prescribed value. The process can be automated and accomplished more rapidly than vacuum purging, but requires more inert gas. ***Sweep purging*** is a similar dilution technique, but venting and charging with the inert gas occur simultaneously at essentially atmospheric pressure. Sweep purging requires constant temperature and pressure and well-mixed conditions to occur within the vessel. Sweep purging requires large amounts of inert gas. ***Siphon purging*** is designed to use the minimum amount of inert gas. The vessel is filled with water (or other compatible liquid), and as it is drained, inert gas is added to maintain a constant pressure. The rate of purging is dictated by the liquid discharge rate.

Table 3.3 Maximum oxygen concentrations that can be tolerated safely in nitrogen used to blanket flammable substances (abstracted from the National Fire Protection Association).

material	% by volume
acetone	11.5
benzene	11.4
butadiene	10.4
n-butane	12.0
carbon disulfide	5.0
carbon monoxide	5.5
divinylbenzene	8.5
ethanol	10.5
ethyl ether	10.5
ethane	11.0
gasoline	12.0
n-hexane	12.0
hydrogen	5.0

material	% by volume
hydrogen sulfide	7.5
kerosene	10.0
methane	12.0
methylene chloride	19.0
methanol	10.0
methyl ethyl ketone	11.0
propane	11.5
propylene	11.5
propylene oxide	7.8
styrene	9.0
toluene	9.5
trichloroethylene	9.0
vinyl chloride	13.4

Vessels in which materials are stored or in which they react must be designed so that their contents can be dumped in the event of a runaway reaction that can rupture the vessel or produce excessive temperatures. ***Blowout discs***, ***catch-tanks***, and ***knockout drums*** are vessels designed to temporarily hold material discharged from the vessel after the internal pressure or temperature trips a relief valve, rupture disk, explosion diaphragm, etc. Once contained in one of these temporary vessels, provision must be made to remove the material and treat it by scrubbing, incineration (flares), or condensation.

Sprinkler systems are fire suppression systems that disburse fire extinguishing chemicals automatically when an individual sprinkler head (Figure 3.21) is heated to a specified temperature. Automatic sprinkler heads incorporate a fusible link (or comparable element) that breaks when it is heated to a prescribed temperature. When the link breaks, it opens a port from which the fire extinguishing material (water or chemical) is discharged under pressure. Sprinkler heads are located in places to protect people and equipment, or are placed in ducts and reactors to extinguish fires that may occur. Fire suppressants are chosen to match the type of fire than can be expected, and include water and certain wet or dry chemicals.

Flame arresters are devices that remove energy from a propagating flame or detonation wave to prevent their further propagation. One must be careful because arresting detonation waves requires removing much more energy than arresting flames. Figure 3.22 illustrates the essential features of flame arresters. To appreciate why flame arresters are necessary in engineering design, recall that fuel and air are combustible over a very wide range of fuel-air ratios, e.g. $0.01 < \text{fuel/air} < 0.80$. Thus small amounts of air entering a nearly full tank of fuel, or a small amount of fuel remaining in a tank filled with air are both capable of explosion. Air can surreptitiously enter tanks of combustible material in a number of ways:

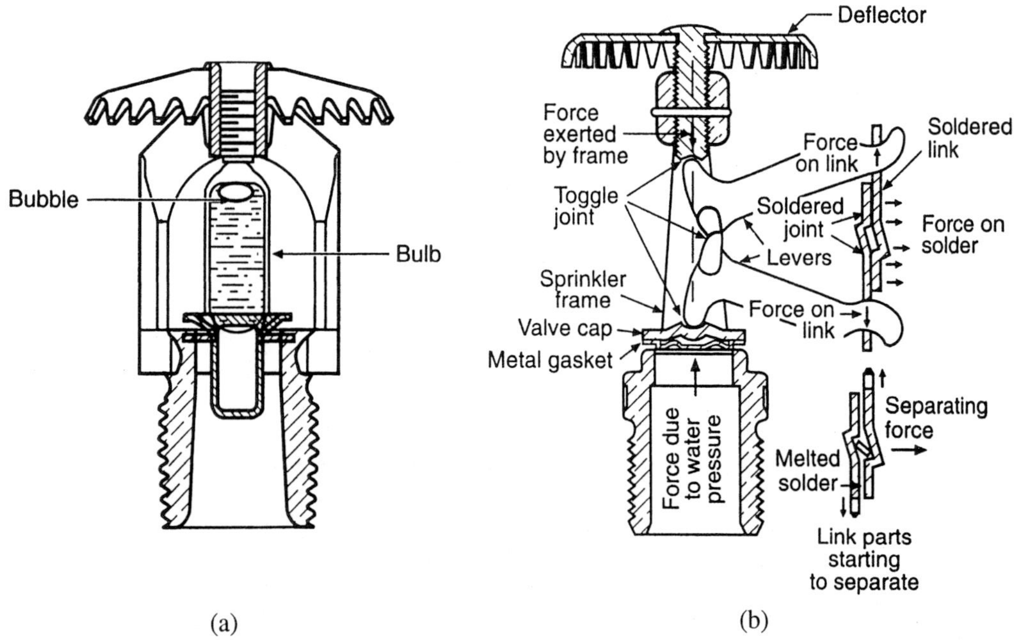

Figure 3.21 Automatic sprinkler heads: (a) bubble-in-bulb automatic sprinkler and (b) solder-type link-and-lever automatic sprinkler.

- Prior to filling new or unused tanks with combustible liquids, the tanks may not have been thoroughly purged of all remnants of the air they formerly contained.
- Tanks that are alternately heated and cooled, for example out of doors tanks exposed to the elements and diurnal temperature variations, expand and contract. Such daily cycles cause the tanks to ***breathe***, that is to say induce air to flow into and out of pressure or vacuum relief valves, seals on pumps or compressors, entrance hatches, etc. Over a period of time, sufficient air can be drawn into the tank to create an explosive mixture.
- Emptying a tank while not adding inert make-up gas at the same or larger volumetric flow rate creates a region of reduced pressure inside the tank which can induce air to enter through vacuum relief valves, faulty seals on relief valves, hatches, pumps, compressors, etc.
- There are occasions when the wind dislodges a flare flame and the flame travels upstream (called ***flashback***) into the oncoming premixed combustible mixture of gases. A flame arrester extinguishes the flame before the moving flame evolves into a detonation wave that can result in an explosion.

Flame arresters (Figure 3.22) and ***bursting discs*** (Figure 3.23) are warranted whenever a combustible mixture of air and hydrocarbons exists within a vessel, and is capable of burning or exploding in the presence of a source of ignition. Two air pollution control systems in which bursting discs and flame arresters are logical additions are the following:

- ***vapor recovery systems*** in which hydrocarbon vapors in air are removed by condensation
- ***activated carbon adsorbers*** in which hydrocarbons are removed from air by adsorbing the vapors on activated charcoal

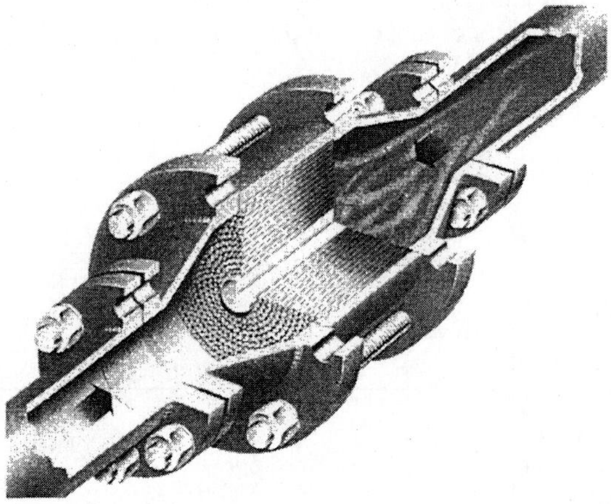

Figure 3.22 Flame arrester quenches flames, preventing propagation (courtesy of IMI Amal, Ltd.).

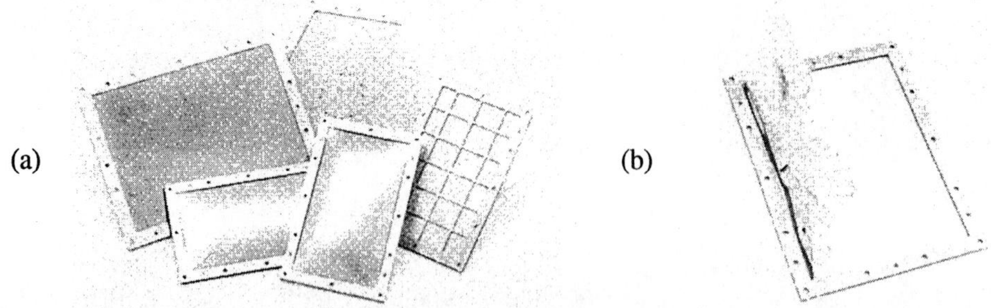

Figure 3.23 Explosion vent panels (bursting discs) provide an economical method of minimizing the effects of an explosion; (a) intact discs, (b) burst disc (courtesy of IMI Marston, Ltd.).

For these two systems it is wise to mount a bursting disc on the vessel. The disc does not prevent an explosion, but if one occurs the bursting disc relieves the internal pressure and prevents the entire vessel from bursting. Similarly, flame arresters do not prevent fire or explosion, but if one occurs, an arrester can quench the flame and prevent it from traveling into other vessels.

3.3.2 Storage of Flammable Materials

The handling and storage of flammable liquids is described in NFPA's Flammable and Combustible Liquid Code (NFPA 30-1989) and OSHA standard 29 CFR 1910.106. The volatility of a liquid is characterized by its vapor pressure, while its flammability is characterized by its flash-point and ignition temperatures. Recall that flash-point temperature (T_f) is the lowest temperature at which a flammable liquid produces enough vapor to cause a momentary fire in the presence of a source of ignition. Ignition temperature is the temperature at which the vapor burns in the presence of air. The ***normal boiling temperature*** (T_b) is the temperature at which vaporization occurs at atmospheric pressure. For example, the flash-point temperature of methyl alcohol is 11 °C, whereas its boiling temperature is 65 °C, and its ignition temperature is 385 °C.

Liquids whose flash-point temperatures are less than 38 °C (100 °F) are defined as ***flammable liquids***, whereas those whose flash-point temperatures are above 38 °C are defined as ***combustible liquids***. Flammable and combustible liquids are divided into three classes (I, II, and III) and three sub-classes (A, B, and C) as shown in Table 3.4.

Transferring the above liquids from one vessel to another is apt to produce spills that may lead to fire. Both OSHA and NFPA require that transfer be performed by:

- closed piping systems
- safety cans
- devices that draw liquid through the top of the container
- gravity through a self-closing valve

Tank-vehicle loading and unloading facilities must be separated from adjoining property lines, above-ground tanks, warehouses, and other plant buildings. In general a 25-foot separation is required for facilities handling Class I liquids and a 15-foot separation for those handling Class II and III liquids.

Flammable liquids can be stored in their original containers or in approved safety cans. Storage in glass or plastic bottles is prohibited. ***Safety cans*** (Figure 3.24) are designed for rugged use, are manufactured of metal or high-density plastic, and have specially designed dual-purpose, spring-loaded spout closures that prevent liquids from spilling if the can is upset. Lids on safety cans also prevent the container from exploding by relieving the internal pressure if the can is exposed to fire.

Table 3.4 Classes of flammable and combustible liquids based on flash point temperature (T_f) and boiling temperature (T_b).

class	**temperature (°C)**	**examples**
flammable liquids		
I-A	$T_f < 23$, $T_b < 38$	ethyl ether, pentane
I-B	$T_f < 23$, $T_b > 38$	acetone, cyclohexane, gasoline
I-C	$23 < T_f < 38$	butyl alcohol, turpentine, o-xylene
combustible liquids		
II	$38 < T_f < 60$	kerosene, fuel oil
III-A	$60 < T_f < 93$	diethyl benzene, isophorone
III-B	$T_f > 93$	castor oil, peanut oil, olive oil

Figure 3.24 Safety can for flammable liquids (courtesy of Eagle Manufacturing Company).

When transferring liquids from one vessel to another, a short jumper ground wire should be used to prevent sparks from occurring due to static electrification. (Readers should observe the procedures followed when gasoline is transferred from a delivery truck to the underground tanks at automobile filling stations.)

Bulk storage of flammable and combustible liquids occurs in specially designed vessels in designated locations. NFPA 30 allows one day's supply to be located within the general plant area, , or the combined sum of the following:

- 25 gal of I-A liquids in containers
- 120 gal of I-B, I-C, II or III-A liquid in containers
- two portable 660-gal tanks of I-B, I-C, II or III-A liquids
- 20 portable 660-gal tanks of III-B liquids

Flammable-liquid cabinets (Figure 3.25) are specially designed cabinets that protect their contents from fire for up to 10 minutes. Ordinary office-supply cabinets do not offer this protection and are not approved. No more than 60 gallons of a Class I or II liquid or 120 gallons of a Class III-A liquid may be stored in a single cabinet. In addition no more than three cabinets are permitted in the same fire area unless they are more than 100 feet apart. A ***fire area*** is a designated portion of a building separated from adjoining areas by construction materials having a one-hour fire-resistance rating. Openings such as doorways, duct openings, and pass-through counters must also be protected by assemblies having at least a one-hour fire-resistance rating.

Since only three cabinets can be kept in a fire area, large quantities of flammable liquids must be stored in a specially designed ***flammable-liquid storage room***. The requirements of such a room are:

- fire-rated construction
- liquid-tight, wall-floor joints
- normally closed "listed" fire door
- non-combustible, four-inch raised door sill to prevent liquids from moving to adjacent areas
- ventilation to remove vapors
- class I, division 2 electrical wiring

Figure 3.25 Storage cabinet for flammable liquids (courtesy of Eagle Manufacturing Company).

Natural ventilation may be used if the room is used solely for storage. If flammable liquids are dispensed within the room, the room must be provided with mechanical ventilation. The NFPA code requires at least one CFM per square foot of floor area, but not less than 150 CFM. The ventilation system must also be provided with an air-flow interlock that activates an audible alarm in the event the ventilation system fails.

Storage rooms are classified as:

- inside storage rooms
- cut-off rooms
- attached buildings

Inside storage rooms are fully enclosed within the building, but share no walls in common with the exterior structural walls of the building. ***Cut-off rooms*** have one or two walls in common with the building's exterior walls. ***Attached buildings*** have three unique exterior walls and share one wall with the main structure. The amount of liquid stored in a storage room is limited to four gallons per square foot. Installation of an automatic sprinkler system enables one to increase the quantity of liquid that can be stored by factors of 2.5, 5.0 and 7.0 for inside storage rooms, cut-off rooms, and attached buildings respectively.

Fixed storage tanks may be located above ground, below ground, or inside buildings. They are classified according to their operating pressure:

- ***Atmospheric tanks*** are designed to operate at internal gauge pressures up to 0.5 psig.
- ***Low-pressure tanks*** are designed to operate at internal pressures between 0.5 and 15 psig.
- ***High-pressure tanks*** are rated at internal pressures in excess of 15 psig.

Horizontal and vertical tanks are designed for materials that are liquids at STP, while tanks with rounded ends are used for materials that are liquid under pressure. Above ground tanks must be equipped with dikes and remote impounding basins to contain leaks if they occur.

3.4 Hearing and Noise

Longitudinal pressure waves from 20 to 20,000 Hz are called ***sound waves***. Sound is often discussed in terms of octaves; an ***octave*** is a doubling of sound frequency; e.g. a tone at 8000 Hz is one octave higher than a tone at 4000 Hz. A vast number of superimposed sound waves in which there is no predetermined relationship between the frequencies and amplitudes of the waves is called ***noise***. Noise is unwanted sound, and is divided into the following categories:

- ***Broad band noise*** - Acoustical energy distributed over a broad range of wavelengths (typically 8 or more octaves) having no distinguishing tone, is called broad band noise. Examples include the noise produced by a large crowd of people, railroad trains, etc.
- ***Narrow band noise*** - Acoustical energy distributed over a narrow range of wavelengths and possessing a dominant tone is called narrow band noise. Examples include jet engines, circular saws, etc.
- ***Impulse noise*** - Acoustical energy distributed over a short period of time (less than 1 second in duration) and generated at a repetition rate less than 200 pulses/min is called impulse noise. Examples include jackhammers (Figure 3.26), punch presses, rappers to loosen dust in clogged hoppers, etc.
- ***White noise*** - White noise is broad band noise with a flat frequency spectrum. The power density (acoustical power at a particular frequency) of white noise can vary with time. An example of time-varying white noise is highway traffic noise.
- ***Pink noise*** – Since each octave represents a frequency doubling, pink noise is weighted at -3 dB/octave such that each octave band has approximately the same power spectrum.

Figure 3.26 Impulse noise and fugitive dust generated by a hydraulic jackhammer (from Valenti, 1999).

3.4.1 Physiology of Hearing

The three components of the ear are shown in Figure 3.27:

- external ear (ornamental outer ear), also called the ***pinna***, ***auricle***, or ***auricular appendage***
- "air-filled" middle ear
- "fluid-filled" inner ear

The ***tympanic membrane*** (eardrum) separates the external and middle ear. The middle ear contains the ***ossicular chain*** (the three tiniest bones in the body), formally called the ***malleus***, ***incus***, and ***stapes***, but more commonly called the hammer, anvil, and stirrup due to their shapes. The malleus and incus vibrate as a unit and move the ***stapedial footplate*** in and out of the ***oval window***. Associated with these three bones are specialized muscles. The ***Eustachian tube*** connects the middle ear to the nasopharynx to equalize air pressure on both sides of the eardrum. To hear properly the air pressure must be the same. The middle ear converts sound waves in air into sound waves in the fluid of the inner ear without appreciable loss of energy or distortion of the frequency spectrum.

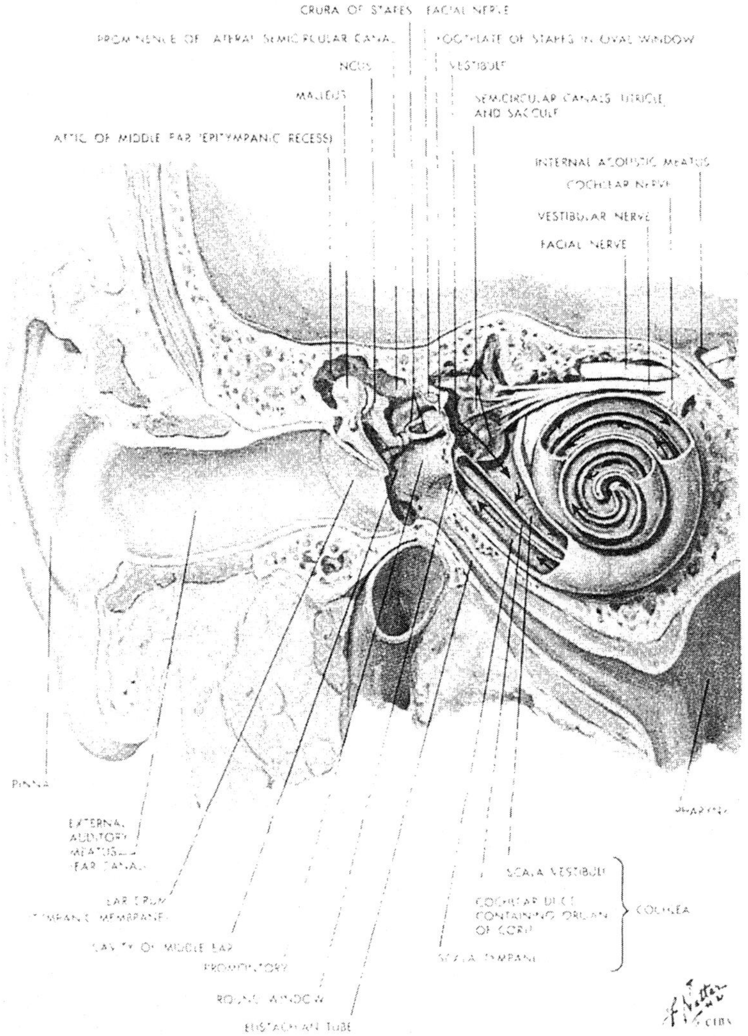

Figure 3.27 Artist's sketch illustrating the anatomy of a human ear (from US NIOSH, 1973).

The inner ear is a complex system of ducts and sacs that house the end organs for hearing and balance. The center of the inner ear, called the ***vestibule***, connects the three semicircular canals and the ***cochlea***. The cochlea resembles a snail shell that spirals about two and three quarter turns. A close-up drawing of a cross-section of the cochlea is shown in Figure 3.28. There are three canals within the cochlea: ***scala vestibuli***, ***scala tympani***, and ***scala media***. The ***basilar membrane*** and ***spiral ligament*** separate the *upper* scala vestibuli from the *lower* scala vestibuli. The scala media is a triangular-shaped duct within which is found the organ of hearing, called the ***organ of Corti***. The basilar membrane is narrowest and stiffest near the oval window and widest at the apex of the cochlea. On the surface of the basilar membrane are phalangeal cells that support the ***critical cilia***, or ***hair cells***. These hair cells are arranged with the inner row containing about 3,500 hair cells and three to five rows of outer hair cells numbering about 12,000. The hair cells extend along the entire length of the cochlear duct and are embedded in the under surface of the gelatinous overhanging ***tectorial membrane***. The vestibular and tympanic canals contain ***perilymph*** and communicate with each other through the tiny opening at the upper-most part of the cochlea. The fluids of the cochlear duct supply nourishment to Corti's organ, a medium for removing waste products, an appropriate medium for the transmission of neural impulses, and a means of eliminating noise that its own flowing blood produces.

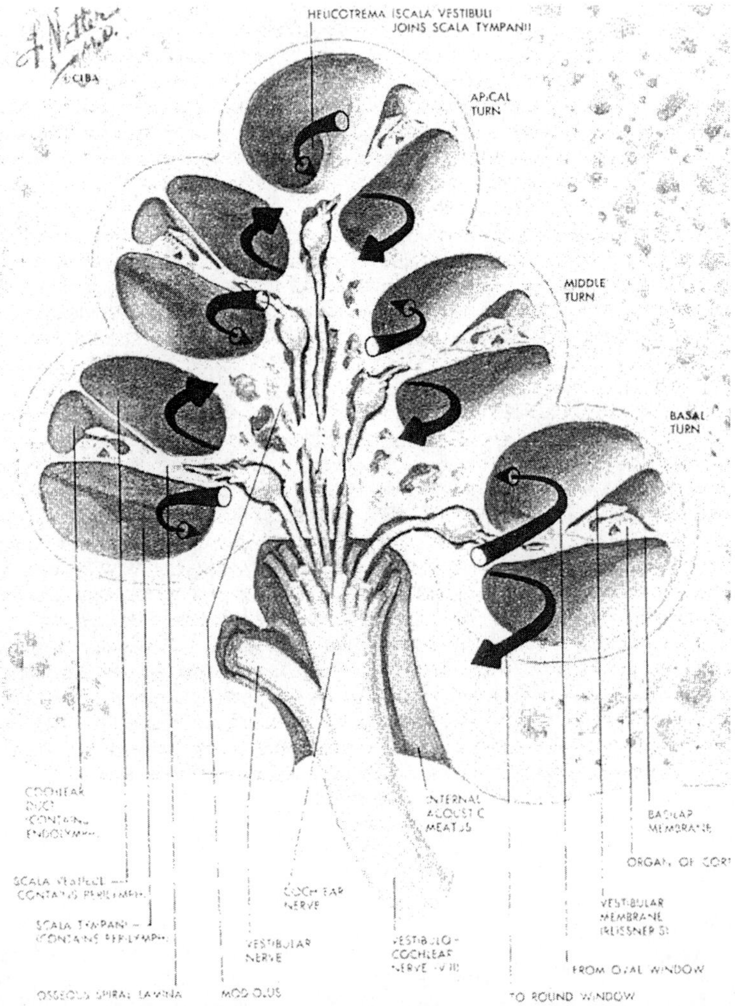

Figure 3.28 Close-up drawing of a cross-section of the cochlea (from US NIOSH, 1973).

The conversion of the mechanical energy of sound into electrochemical activity is called ***transduction***. The two openings afforded by the oval and round windows are essential for sound pressure waves to pass through to the cochlean fluids. The movement of the stapedial footplate in and out of the oval window moves the perilymph of the scala vestibuli. This results in a traveling wave in which there is a specific point of maximum displacement of the basilar membrane. High-pitched sounds travel a short distance along the basilar membrane before they die out, and the opposite is true for low-pitched sounds. The traveling wave moves through the scala vestibuli and causes a downward shift of the cochlear duct. The activity is transmitted through the basilar membrane to the scala tympani. When the oval window is pushed inward, the round window is designed to act as a relief point and bulges outward. The vibration of the basilar membrane causes a shearing force on the hair cells against the tectorial membrane. This to and fro bending of the hair cells activates the neural endings so that sound is transformed into an electrochemical response. In general, the hair cells at the base of the cochlea transmit high frequency sound while those at the apex respond to low frequencies. Each nerve fiber is connected to several hair cells. The hair cells stimulate auditory neural endings and nerve fibers that stream out through small openings in the ***spiral lamina*** into the hollow ***modioulus***. The movement of the hair cells sets up action electrochemical potentials, and coded information from both ears is sent to the ***cochlear nuclei*** and thereafter to the temporal lobe of the brain where cognition and association takes place.

3.4.2 Hearing Loss

Hearing loss or impairment can be classified as follows:

- ***Conductive hearing loss*** (conductive impairment) refers to anything that interferes with the transmission of sound energy to the cochlea. Pure conductive loss damages neither the organ of Corti nor neural pathways. Conductive losses include perforation of the eardrum, blockage of the Eustachian tube, and fluid in the middle ear. Many of these hearing losses are amenable to medical or surgical treatment.
- ***Sensorineural hearing loss*** (sensorineural impairment) pertains to degeneration of the neural elements of the auditory nerve, and is almost always irreversible. Exposure to noise of sufficient intensity and duration causes irreversible sensorineural hearing loss. Damage of the hair cells is of critical importance. Invariably, degeneration of the spiral ganglion cells and peripheral nerve fibers accompany severe injury to the hair cells.
- ***Central hearing loss*** (central impairment) involves difficulty in a person's ability to interpret what he or she hears. The abnormality is localized in the brain between the auditory nuclei and the cortex.
- ***Psychogenic hearing loss*** (psychogenic impairment) is a non-organic basis for an individual's inability to detect high frequencies. For example, ***Tinnitus*** is a hearing impairment that manifests itself as a "ringing or buzzing" in the ears, or as high-pitched whistles or odd sounds resembling bacon frying, horns, or crickets.
- Lastly there are mixed hearing losses that contain components and characteristics of both conductive and sensorineural hearing loss in the same ear.

Hearing loss occurs when a range of frequencies can no longer be detected, referred to as a ***threshold shift***. Threshold shift may be temporary, i.e. ***temporary threshold shift*** (TTS), in which case the body's recuperative powers restore hair cells, and hearing is fully restored; or the threshold shift may be permanent, and hearing is not restored. Unfortunately, permanent hearing loss may progress over an extended period in which it is believed that the noise is within a safe range and individuals believe their hearing loss is only temporary. For example, truck drivers and individuals working near jet aircraft may suffer from only TTS for certain periods of time, but over a period of many years they experience permanent hearing loss. The reader is left to imagine what the future holds for rock music performers, and for fans who attend rock concerts on a regular basis.

A person's hearing declines naturally with age. ***Presbycusis*** is the medical term given for the gradual hearing loss associated with aging. Presbycusis affects more than one third of people over age 75, and is probably due to gradual cumulative loss of cilia and neurons (Ganong, 1997). Higher frequencies are generally affected more than lower frequencies; e.g. some senior citizens cannot hear wristwatch alarms and electronic bells. It is not uncommon for elderly people to be nearly deaf to frequencies above 10,000 Hz, whereas healthy young people can hear frequencies up to 18,000 Hz.

3.4.3 Sound Level

It is important that designers of indoor air pollution control systems be able to predict the noise associated with their designs and take steps to reduce the generation of noise, add material to absorb it, or know when to recommend personal protective devices. All the equations derived in this section pertain to a point source of noise. A companion set of equations can be derived for a line source of noise.

The ***speed of sound*** (a) associated with a sound wave is defined as the square root of the change of pressure with respect to density at constant entropy,

$$a = \sqrt{\left(\frac{\partial P}{\partial \rho}\right)_s} \tag{3-10}$$

For sound traveling in a gaseous medium that can be approximated as a perfect gas, the speed of sound can be expressed as

$$a = \sqrt{\frac{kR_uT}{M}} \tag{3-11}$$

where k is the ratio of specific heats (c_p/c_v), R_u is the universal gas constant, M is the molecular weight of the gaseous medium, and T is the absolute temperature. The speed of sound in air at STP is 344.5 m/s. Sound travels much faster in solids and liquids than it does in gases. The speed of sound in solids and liquids can be derived from Eq. (3-10), and can be shown to be equal to

$$a = \sqrt{\frac{1}{\beta\rho}} \tag{3-12}$$

where β is the ***isothermal expansion coefficient***

$$\beta = \frac{1}{\rho}\left(\frac{\partial \rho}{\partial P}\right)_T \tag{3-13}$$

Consider a sound source located in free space. At a distance (r) from the source, the sound can be characterized by three parameters: ***sound pressure*** (P, force/area), ***sound intensity*** (I, power/area) and ***sound power*** or ***acoustic power*** (W, energy/time). Assuming that pressure waves travel through air without loss (which is consistent with the assumption of constant entropy), the power distributed over a spherical surface is equal to the acoustical power of the source. The power associated with a the sound wave per unit area is called the sound intensity, and is related to the pressure by

$$I = \frac{P^2}{\rho a} \tag{3-14}$$

The sound power (W) is thus equal to sound intensity multiplied by total area (A),

$$W = AI = 4\pi r^2 \frac{P^2}{\rho a} \tag{3-15}$$

where r is the distance from the sound source. If the source of sound is not located in free space, and reflecting surfaces direct the sound in preferential directions, it is convenient to modify Eq. (3-15) as follows:

$$W = \frac{4\pi r^2}{Q} I = \frac{4\pi r^2}{Q} \frac{P^2}{\rho a} \tag{3-16}$$

where the ***directivity factor*** (Q) has been introduced. Q is defined as the ratio of the sound power of a small omni-directional hypothetical source to the sound power of an actual source that produces the same sound pressure level at a point of measurement. For a sound source bounded by acoustically reflecting surfaces, where $(4\pi r^2/Q)$ in Eq. (3-16) represents the actual area over which acoustical power is distributed, the following values of Q can be used:

- Q = 1: source in free space
- Q = 2: source located in an infinite reflecting plane
- Q = 4: source located at the intersection of two mutually perpendicular reflecting planes
- Q = 8: source located at the intersection of three mutually perpendicular reflecting planes

It is convenient to define the ***sound pressure level*** (L_P) related to the sound pressure (P) as follows:

$$L_P = 20\log_{10}\left(\frac{P}{P_0}\right) \tag{3-17}$$

Where the term P_0 is a reference value,

$$P_0 = 2\times10^{-5}\,\frac{N}{m^2} \tag{3-18}$$

which at one time was thought to be an approximate normal threshold value at 1,000 Hz for a young, healthy person. In similar fashion it is useful to define a ***sound intensity level*** (L_I),

$$L_I = 10\log_{10}\left(\frac{I}{I_0}\right) \tag{3-19}$$

where the reference value I_0 corresponds roughly to the reference pressure level (P_0). Specifically, at STP,

$$I_0 = \frac{P_0^2}{\rho a} \approx 1\times10^{-12}\,\frac{\text{watt}}{m^2} \tag{3-20}$$

Readers should note that the unit watt (normally abbreviated as W) is spelled out in these equations to avoid confusion with the symbol for acoustic power or sound power, which is also W. Sound power (W) can also be expressed as a ***sound power level*** or ***acoustic power level*** (L_W),

$$L_W = 10\log_{10}\left(\frac{W}{W_0}\right) \tag{3-21}$$

where

$$W_0 = 1\times10^{-12}\text{ watt} \tag{3-22}$$

The unit used to express the sound pressure level, sound intensity level, and sound-power level is called the ***decibel*** (dB).

Consider an observer at some distance r from the source of sound. Sound intensity (I) and sound pressure (P) are functions of r. Eqs. (3-14) and (3-20) can be combined as follows:

$$\frac{I(r)}{I_0} = \frac{1}{I_0}\frac{P(r)^2}{\rho a} = \frac{P(r)^2}{{P_0}^2} = \left(\frac{P(r)}{P_0}\right)^2$$

Taking the logarithm of both sides yields

$$\log_{10}\left(\frac{I(r)}{I_0}\right) = 2\log_{10}\left(\frac{P(r)}{P_0}\right)$$

Multiplying each side by 10 and recognizing the definitions of Eqs. (3-17) and (3-19), one obtains

$$\boxed{L_I(r) = L_P(r)} \tag{3-23}$$

The sound intensity level $L_I(r)$ and the equivalent sound pressure level $L_P(r)$ define the sound experienced by an observer, and thus depend on the distance (r) between the source and observer. The above derivation shows that *at any distance (r) between source and observer, the sound pressure level (L_P) and the sound intensity level (L_I) have the same value*. On the other hand, sound power level (L_W) defines the acoustical power of the source and is *independent* of the distance (r). It is useful at this point to define a ***reference distance*** (r_0) equal to 1 meter,

$$\boxed{r_0 = 1\text{ m}} \tag{3-24}$$

Then a relationship between L_W and $L_I(r)$ or $L_P(r)$ can be found from Eq. (3-16), also using the definitions of I_0, W_0, and r_0 in Eqs. (3-20), (3-22), and (3-24) respectively:

$$\frac{I(r)}{I_0} = \frac{W}{I_0}\frac{Q}{4\pi r^2}\left(\frac{r_0}{1\text{ m}}\right)^2\left(\frac{I_0}{1\times10^{-12}\,\frac{\text{watt}}{\text{m}^2}}\right)\left(\frac{1\times10^{-12}\text{ watt}}{W_0}\right)$$

Several units in the above equation cancel, and a nondimensional expression results:

$$\boxed{\frac{I(r)}{I_0} = \frac{W}{W_0}Q(4\pi)^{-1}\left(\frac{r}{r_0}\right)^{-2}} \tag{3-25}$$

Taking the logarithm of both sides and then multiplying each term by 10 yields

$$10\log_{10}\left(\frac{I(r)}{I_0}\right) = 10\log_{10}\left(\frac{W}{W_0}\right) + 10\log_{10}Q - 10\log_{10}(4\pi) - 20\log_{10}\left(\frac{r}{r_0}\right)$$

Using Eqs. (3-19) and (3-21), the above reduces to

$$\boxed{L_I(r) = L_W + 10\log_{10}Q - 11.0 - 20\log_{10}\left(\frac{r}{r_0}\right)} \tag{3-26}$$

Because of Eq. (3-23), the left hand side Eq. (3-26) is also equal to $L_P(r)$,

$$\boxed{L_P(r) = L_W + 10\log_{10}Q - 11.0 - 20\log_{10}\left(\frac{r}{r_0}\right)} \tag{3-27}$$

The units of each term in both Eq. (3-26) and Eq. (3-27) are decibels (dB). The healthy human ear is able to hear a limited range of frequencies (approximately 20 to 18,000 Hz). In addition, human hearing does not have a flat frequency response. That is to say, we hear best at midrange frequencies (around a thousand to several thousand Hz), but hear less well at lower frequencies and at higher frequencies. Because of this, readers may also encounter units of dBA in professional literature. The difference is that dB is calculated by giving each frequency equal weighting (no filtering), whereas dBA is calculated by first filtering the sound signal so that the measuring instrument mimics human hearing (it filters low and high frequencies such that the frequency response of the instrument is the same as that of a human ear). Less common are two other frequency ranges, B and C with corresponding units dBB and dBC, to be discussed shortly. All of the equations in this text apply to any of the scales; dBA is the most useful scale since human hearing is of concern here. Figure 3.29 shows the relationship between sound pressure level (in dB), sound power (in watts), and sound pressure (in N/m^2) for several commonplace sources of noise.

It is advantageous to manufacturers to reduce the acoustic power (W) of equipment they manufacture. Consequently the acoustic power of equipment is information that is generally available to the purchaser. The acoustic power can be used to predict the sound pressure level (L_P) at arbitrary locations from the equipment. Using Eq. (3-27), it is a simple matter to show that the sound pressure level changes by approximately 6 dB each time the distance between a listener and a point source doubles or is halved.

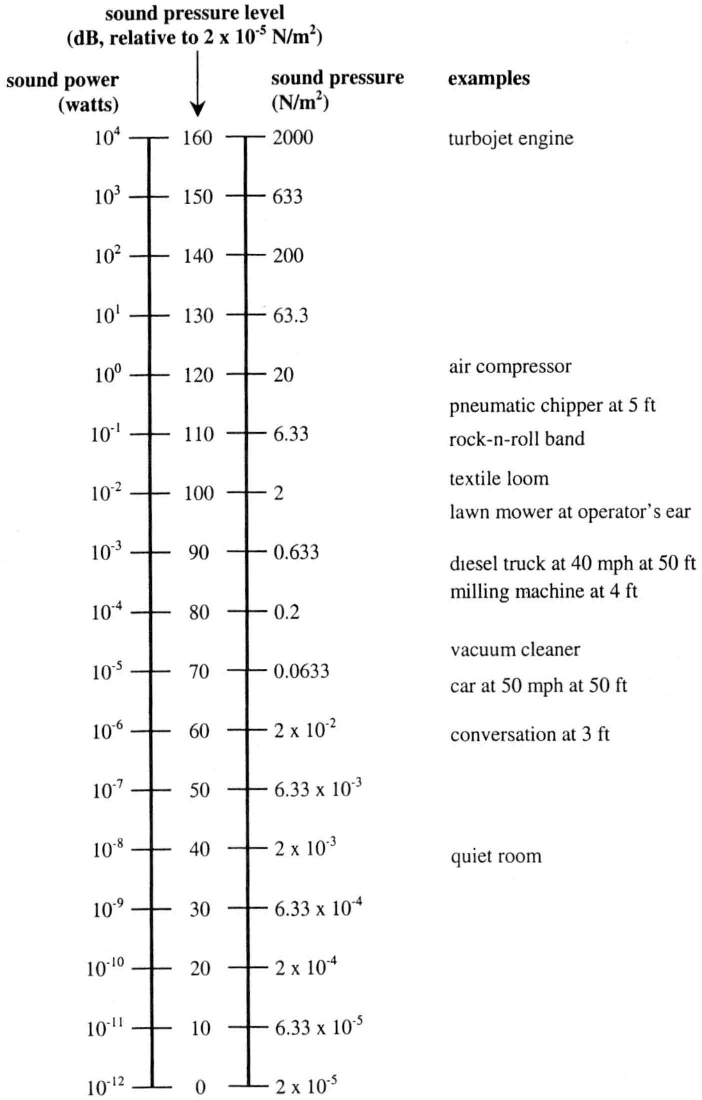

Figure 3.29 Relationship between sound pressure, sound pressure level, and sound power, and some common sources of noise (adapted from US NIOSH, 1973).

Example 3.4 - Sound Pressure Level of a Punch Press
Given: A punch press, located in the corner of a building with highly reflective (hard) walls, is rated as having a free-field sound pressure level of 85. dB at a distance of 10. m.

To do: Calculate the acoustic power of the equipment (in watts) and the sound pressure level (in dB) at workstations 1.5 ft and 6.5 ft from the press.

Solution: The manufacturer's rating is for a free field, i.e. for the source in free space; thus Q = 1 for the quoted sound pressure level of 85. dB. The acoustic power level of the source (L_W) is found by substituting L_P = 85. dB at r = 10. m and Q = 1 in Eq. (3-27),

$$L_W = L_P(r) - 10\log_{10} Q + 11.0 + 20\log_{10}\left(\frac{r}{r_0}\right) = 85. - 10\log_{10}(1) + 11.0 + 20\log_{10}\left(\frac{10.\text{ m}}{1\text{ m}}\right) = 116.\text{ dB}$$

The acoustic power itself is found from Eq. (3-21),

$$W = W_0 \cdot 10^{\frac{L_W}{10}} = \left(1\times10^{-12}\text{ watt}\right)\cdot 10^{\frac{116.\text{ dB}}{10.\text{ dB}}} = 0.398\text{ watt}$$

Recall that acoustic power is independent of observer location. The sound pressure level (L_P) at the two workstation locations is found from Eq. (3-27) again, but this time using L_W = 116. dB and Q = 8 since the punch press is located in the corner of the building (2 walls and a floor, assuming the ceiling is high and irrelevant):

$$L_P(r = 1.5\text{ ft}) = 116. + 10\log_{10}(8) - 11.0 - 20\log_{10}\left(\frac{1.5\text{ ft}}{1\text{ m}}\left(\frac{0.3048\text{ m}}{\text{ft}}\right)\right) = 121.\text{ dB}$$

and

$$L_P(r = 6.5\text{ ft}) = 116. + 10\log_{10}(8) - 11.0 - 20\log_{10}\left(\frac{6.5\text{ ft}}{1\text{ m}}\left(\frac{0.3048\text{ m}}{\text{ft}}\right)\right) = 108.\text{ dB}$$

Discussion: It is typical to give final results to one decibel, which in this case is three significant digits. Because of the logarithms in the equations, this number of significant digits is valid even though some of the parameters are known only to two significant digits.

3.4.4 Multiple Sources of Sound

Since noise consists of several simultaneous sound waves of different frequencies and intensities, the ear is subjected to a composite wave in which the total power is the sum of the power of each individual sound wave. Thus the sound intensity at a point in space subjected to *several* sources of noise is

$$\boxed{\frac{I}{I_0} = \sum_j \frac{I_j}{I_0}} \tag{3-28}$$

where subscript j refers to the jth source of noise, and summation from j = 1 to N is implied (N is the total number of noise sources). The right hand side of Eq. (3-28) can be evaluated by taking the antilogarithm of Eq. (3-19),

$$\boxed{\frac{I_j}{I_0} = 10^{\left(\frac{L_{I,j}}{10}\right)}} \tag{3-29}$$

The total sound intensity level (L_I in dB) of several sources of noise can then be expressed as

$$L_I = 10\log_{10}\left[\sum_j 10^{\left(\frac{L_{I,j}}{10}\right)}\right] \tag{3-30}$$

Example 3.5 – Sound Intensity Level in a Grinding Room
Given: A company wishes to place six pedestal grinders in a room. At a certain point (A) each grinder separately produces the following sound intensity levels:

grinder	L_I (dB) at point A
1	85.
2	92.
3	90.
4	84.
5	93.
6	87.

To do: Estimate the total sound intensity level produced by all six grinders at that same point (A).

Solution: The total sound intensity level (in dB) is found by substitution into Eq. (3-30):

$$L_I = 10\log_{10}\left[10^{\left(\frac{85.}{10}\right)} + 10^{\left(\frac{92.}{10}\right)} + 10^{\left(\frac{90.}{10}\right)} + 10^{\left(\frac{84.}{10}\right)} + 10^{\left(\frac{93}{10}\right)} + 10^{\left(\frac{87}{10}\right)}\right] = 97.52 \text{ dB}$$

Keeping with the convention of answers to one dB of accuracy, L_I = 98. dB.

Discussion: Because of the nature of logarithms and powers of 10 in these acoustics equations, the total sound intensity level (98. dB) is only 5 dB higher than that of the noisiest machine (93. dB) by itself. Sound intensity level certainly does not add linearly!

3.4.5 Noise Standards

Listed in Table 3.5 are both OSHA and ACGIH noise limit standards for the workplace. There are some major differences between these two noise limit standards. OSHA enforces an 8-hr time-weighted average (TWA) limit of 90 dBA while ACGIH recommends an 8-hr TWA of 85 dBA. For several decades OSHA has embraced the belief that an increase of sound intensity of 5 dBA exposes individuals to a risk that is equivalent to the risk if the time of exposure were doubled. This belief is clearly seen in OSHA's TWA limits. However, by applying Eq. (3-27), it can be shown that doubling the sound power level (L_w) increases the sound intensity level (L_p) by 3 dB (all else remaining the same), rather than by 5 dB. This relationship is referred to as the ***equal energy concept***. For example, an individual exposed to 85 dBA for 8 hours is exposed to the same acoustic power as an individual exposed to 88 dBA for 4 hours. From Table 3.5, the ACGIH values follow the equal energy concept, while the OSHA values do not.

When the 5 dB rule was adopted several decades ago, hearing specialists believed that temporary threshold shift (TTS, defined previously) was a reasonable predictor of permanent hearing loss, and that the 5 dB rule would protect against TTS. More importantly, hearing specialists believed that the ability to understand speech required detection of frequencies only in the range of 500 to 2000 Hz. Both of these beliefs are now also known to be false. Detection of frequencies up to at least 4000 Hz is now recognized as important to clear understanding of speech. Because the hearing organ is more sensitive to damage from noise exposure at these higher frequencies, a more protective noise

Table 3.5 ACGIH and OSHA noise limit standards for the workplace (from Internet websites and US Office of the Federal Register, 1988).

sound intensity (dBA)	ACGIH exposure time (hr)	OSHA exposure time (hr)
80	24	32
82	16	24.3
85	8	16
88	4	10.6
90	-	8
91	2	7
92	-	6
94	1	4.6
95	-	4
97	0.5	3
100	0.25	2
102	-	1.5
105	-	1
110	-	0.5
115	-	0.25 or less

exposure limit is needed. Furthermore, scientists now believe that TTS is not a good predictor of permanent hearing loss. It is currently the belief of hearing specialists that noise energy leads to hearing loss. Thus for intermittent noise sources, one must integrate the acoustic power over time to obtain a total exposure to noise energy.

The human function of hearing depends on both frequency and sound level as shown in Figure 3.30. It is convenient to define three sound scales (A, B and C scales, corresponding to dBA, dBB, and dBC respectively) in which there are different weighing factors for selected bands of frequencies as shown in Figure 3.31. The A scale is designed with weighing factors to simulate the human ear while the C scale has nearly uniform weighting factors for all frequencies. The OSHA ***permissible noise exposures*** shown in Table 3.5 are defined in terms of the A scale. While the 8-hour standard is 90 dBA, higher sound levels are allowed for shorter periods of time. When the daily

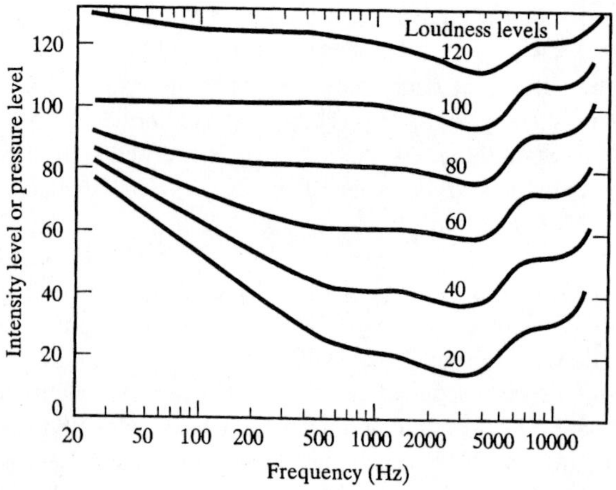

Figure 3.30 Free-field equal-loudness contours of pure tones (redrawn from Zimmerman, 1984).

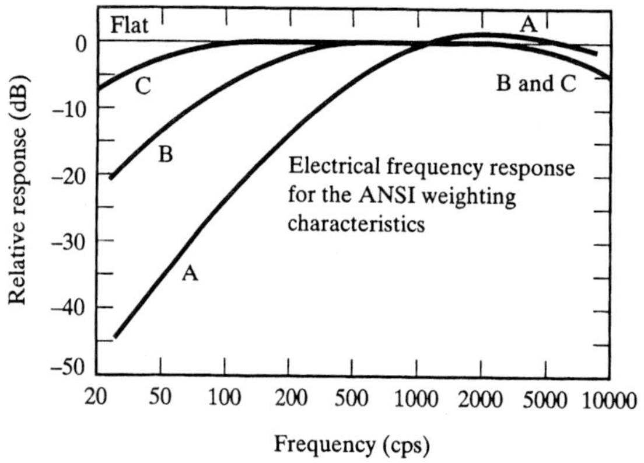

Figure 3.31 Frequency-response characteristics for sound-level meters, A, B and C (from US NIOSH, 1977).

exposure is composed of two or more *different* periods of exposure of *different* sound levels, the collective hazard is evaluated by the following:

$$\text{En} = \sum_j \frac{t_j}{t_{j,\text{permitted}}} \tag{3-31}$$

where the subscript j refers to the jth sound level, t_j is the *actual* duration of exposure at one sound level, $t_{j,\text{permitted}}$ is the *permitted* duration of exposure at that level, and the summation is over the number of periods of exposure (N). If En exceeds unity, the mixed exposure is considered to have exceeded the OSHA limit for safe conditions.

Impact noise is noise produced by a time-varying source in which the maxima occur at intervals greater than one second. If the maxima occur at intervals less than one second, the noise is defined as continuous. Noise produced by impact should not exceed 140 dB.

The punch press in Example 3.4 is so noisy that it should not be used in its present condition because while workers using it may wear personal protective devices, other individuals in its proximity may be affected. The sound level in the grinding room clearly exceeds the 8-hr OSHA standard, and an individual located at the point in question should not remain there for more than 2 hours unless a personal hearing protective device is worn. Hearing impairment develops long after exposure and is irreversible. The susceptibility of individuals to hearing impairment is illustrated in Figure 3.32, which shows that even at 90 dB, a significant portion of a working population will suffer impairment. Individuals who design indoor air pollution control systems must be sensitive to the noise produced by fans, air jets, or indeed any vortices, recirculation, or flow separation that is apt to occur. Not only must OSHA standards be met, but one must minimize noise in general. Worker productivity and morale decrease if annoying noise is present. The noise characteristics of fans can be provided by manufacturers, and information is available from the ***Air Moving and Conditioning Association*** (AMCA) to compute the sound level for a particular installation.

There are indoor air pollution control devices that both remove contaminated air (exhausters) and return make-up fresh air (blowers). Exhausters may have large openings producing low face velocities, i.e. ***low velocity - high volume*** (LVHV) devices, such as canopy hoods (Figure 1.15) and side-draft hoods (Figure 1.16). Other exhausters may have small openings producing large face velocities, i.e. ***high velocity - low volume*** (HVLV) local exhausters (Figure 1.13). In general, higher velocity devices produce more noise. Blowers such as make-up air units typically produce low face

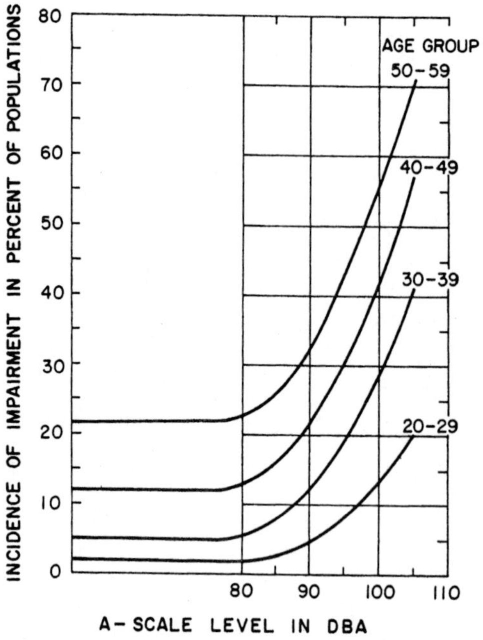

Figure 3.32 Prevalence of impaired hearing and sound-levels for different age groups (from US NIOSH, 1977).

velocities. On the other hand, the face velocities for ***air curtains*** and ***fresh-air islands*** (regions of the workplace bathed by a stream of fresh air) are considerably larger. Noise is generated by all these devices. The noise may be merely distracting, fatiguing, or a nuisance. If the face velocities are large, however, the noise might even exceed OSHA standards. Figure 3.33 shows experimental data (Garrison and Byers, 1980a, 1980b) illustrating relationships between sound level, frequency, face velocity, and distance from the inlet plane for exhausters of various geometries. The two most important parameters affecting noise are face velocity and distance from the inlet plane of the fan. Flanged surfaces act as reflecting surfaces and intensify the noise potential. The aspect ratio and shape of the opening (round, circular, or rectangular) are of minor significance.

A systematic exposition of how to estimate noise generated by the components of an industrial ventilation system is beyond the scope of this text. The chapter, "Sound and Vibration" in the ASHRAE Fundamentals Handbook (1997) describes how to estimate noise generated by fans, duct fittings, etc., and how to attenuate the noise. Much of the discussion concerns hot air heating and air-conditioning systems for public buildings, homes, etc. where noise standards are more stringent than in the workplace. Nonetheless the procedures are applicable to industrial ventilation systems.

3.4.6 Prevention of Hearing Loss

It goes without saying that the safest way to prevent loss of hearing is to prevent noise from being generated, and following this, to suppress noise at its source. Suppression of noise is a large field of technology and is not included in this text. Once one has exhausted ways to minimize noise generation, there are three additional issues to be kept in mind for occupational safety and health:

- Administrative controls or personal protection - Administrative controls such as employee rotation should be avoided when the sound intensity is 103 dBA or more. When noise levels are this high, personal hearing protectors are preferred over administrative controls such as the rotation of workers. It seems incongruous that while the cardinal rule in occupational safety and health is that administrative controls are preferable over personal protective devices (see discussion in Chapter 1), the reverse is true for noise control. Hearing loss is a cumulative response to exposure, and without carefully documenting acoustic energy, duration of exposure, and the length of time workers are rotated, rotation as a general rule is an unreliable way to protect against hearing loss.

- Ototoxic chemicals - While the data are scarce, it is believed that exposure to certain chemicals such as toluene, lead, manganese, and butyl alcohol can cause hearing loss.
- Fetal exposure to noise - There is potential hazard to the fetus if the abdomen of pregnant workers is exposed to noise of sufficient intensity and duration. The suggested limit for an 8-hr average is 115 dBC and the ceiling limit is 155 dBC.

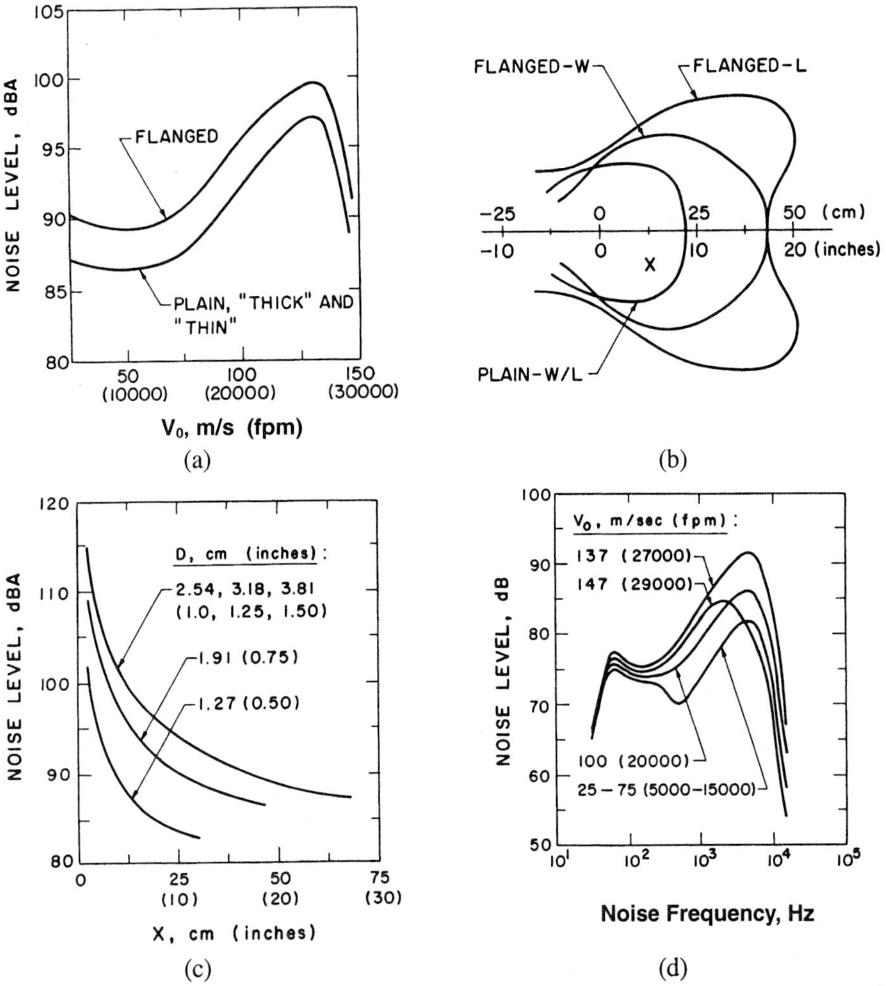

Figure 3.33 Experimental noise data: (a) centerline noise level of flanged and unflanged (plain) circular inlets as a function of face velocity; (b) 90-dBA contours for flanged and unflanged rectangular inlets (width/length = 0.1); (c) centerline noise level as a function of upstream distance for unflanged inlets of different diameters; and (d) centerline noise level as a function of frequency for unflanged circular inlets at different face velocities (redrawn from Garrison & Byers, 1980).

3.5 Thermal Comfort and Heat Stress

Thermal comfort is a frame of mind in which an individual expresses satisfaction with the thermal environment. Fanger (1988) reports that under sedentary conditions thermal comfort can be characterized by the following:

- air temperature between 23 and 26 °C
- difference in vertical air temperature less than 3 °C between 0.1 and 1.1 m
- mean air velocity less than 0.25 m/s

The exposure of humans to a cold environment can cause subnormal body temperatures, a condition called ***hypothermia***. Stages of ***cold stress*** depend on the relative humidity, temperature, and exposure duration. Hypothermia is defined as beginning once the body's core temperature falls below 96 °F (35 °C). Early signs of hypothermia are ***goose bumps*** and skin cooling produced when blood is diverted from the skin to the body's core. Reduction of the flow of blood to the skin is called ***anoxia***. At a core temperature of 96 °F the body begins to shiver in order to increase energy release by contracting muscles. If this condition persists for a long period of time, a low body temperature can bring about an electrolyte imbalance, increased blood acid level, muscle tightening, and dehydration. At 95 °F there are signs of severe numbness and loss of physical coordination. At 94 °F, the individual experiences difficulty in speaking, rigid muscles, disorientation, and reduced visual acuity. Between 90 and 88 °F the individual becomes semi-comatose. Prolonged exposure to a cold environment can produce several chronic injuries:

- ***Skin lesions*** can occur from anoxia.
- ***Trench foot*** can occur when lower extremities are exposed to a cold, wet environment.
- ***Frostbite*** is damage to intercellular organs and body tissue produced by crystal growth and protein denaturation.

At the other extreme, ***heat stress*** can cause abnormally high body temperatures (fevers), producing a condition called ***hyperthermia***, also called ***hyperpyrexia***. When working in a hot environment, the heart beats faster to increase blood circulation in an attempt to cool the body. In addition to general discomfort, workers may experience dizziness, cramps, and nausea that are key indicators that they should cease working. At other times the symptoms may be less obvious, and workers may be unaware of the heat stress to which they are subjecting themselves. In severe cases, workers may lose consciousness and lapse into a coma. Heat stress progresses in stages of increasing seriousness displaying the following symptoms:

- ***Heat rash*** is an uncomfortable skin inflammation resulting from prolonged exposure to high temperature and humidity. Many people experience this in workplaces without air conditioning.
- ***Heat cramps*** are muscle spasms in the lower extremities and abdomen following extended periods of sweating that cause dehydration and an imbalance of the body's electrolytes. Professional tennis players sometimes withdraw from competition in the middle of matches at the US Open Tennis Tournament after several hours in the hot sun if they have not consumed fluids containing sufficient electrolytes.
- ***Heat exhaustion*** is a temporary failure of the body's nervous system, circulatory system, etc. caused by prolonged and unattended dehydration and electrolyte imbalance.
- ***Heatstroke*** occurs when the body can no longer control its temperature. Heatstroke can be fatal. Symptoms of heatstroke begin with the cessation of sweating, followed by fever (T_{body} > 105 °F), rapid pulse, hot and dry skin, headache, confusion, unconsciousness, and convulsions. Individuals suffering heatstroke require immediate medical attention. Illegal immigrants attempting to enter the US across the US-Mexican border often suffer heatstroke crossing the desert in the summer.

Human beings, as well as other mammals who maintain a constant internal temperature, are called ***homeotherms***. Internal body temperatures are regulated within narrow limits by the flow of blood from sites of energy production in muscles and deep tissues to the skin where energy is transferred to the air by radiation, conduction, convection, respiration, and evaporation. When the production of energy is equal to the transfer of energy, internal temperatures are controlled within narrow limits and a state of ***thermal homeostasis*** is said to exist. When heat transfer from the skin is not equal to energy production in the body, internal temperatures do not remain within narrow limits and the heart is stressed; this condition is referred to as heat stress. Heat stress taxes the heart, and beyond certain limits is unsafe and unhealthy. It is important for designers to quantify heat stress so that deleterious effects can be anticipated. By virtue of age, physical fitness, health status, living habits, and ability to acclimatize, an individual's ability to tolerate heat stress varies.

3.5.1 Metabolic Rates Associated with Human Activities

Stored energy (from food and/or body fat) is broken down by chemical reactions that occur in the cells, releasing energy which maintains the human body at a nearly constant temperature of 37. ^{o}C. In more technical terms, energy needed by the body is derived from the enzymatically controlled exothermic oxidation of fats, proteins, and carbohydrates to produce carbon dioxide, water, and nitrogen wastes. This process of "burning" food in our bodies is called ***metabolism***, usually expressed in terms of a rate called the ***metabolic rate*** ($\dot{M}$). The dimensions of metabolic rate are energy per unit time, or power. Physiologists define the ***basal metabolic rate*** ($\dot{M}_b$) as that required for a person reclining at rest, but not asleep (and having been in that state for a minimum of 30 minutes), to maintain normal body functions (breathing, circulating blood, thinking, seeing, hearing, etc.). Accurate measurement of $\dot{M}_b$ also requires that the person is relaxed, has had no food for 12 hours, is in a comfortable temperature environment, and has had a night of restful sleep (Guyton, 1986). The typical average man used in examples throughout this book (30 years old, 1.8 m^2 skin surface area, and 70 kg mass) has a basal metabolic rate of around 84. W.

$$\boxed{\dot{M}_b = 84.\ \text{W}} \tag{3-32}$$

Two organs, the brain and liver, are responsible for nearly half of the basal metabolic rate, even though they represent less than five percent of the body's mass (Cengel, 1997)! Readers should not confuse the basal metabolic rate with that of sleep or with that of sitting at a desk or watching TV. Sleeping, for example, reduces the metabolic rate approximately 14% below $\dot{M}_b$. While there is good agreement in the literature about the value of $\dot{M}_b$ in Eq. (3-32), there is some variation in the literature regarding the numerical values of $\dot{M}$ for various levels of physical activity. Consider for example an average person sitting still, relaxed, quiet, and awake, but doing no work (e.g. reading or watching TV). The metabolic rate under these conditions is somewhat higher than $\dot{M}_b$. The ASHRAE Fundamentals Handbook (1997) reports a metabolic rate of 108. W, while Guyton (1986) reports a value of 116. W for the same average man sitting at rest. For purposes of discussion in this chapter, the metabolic rate of an average man sitting at rest is taken as 108. W, which is about 29% larger than $\dot{M}_b$. To maintain constant body temperature, all of this power must be removed from the body, either by heat loss, by excretion, or by doing work. Note that work here is defined in the thermodynamic sense, i.e. moving against a resisting force such as gravity. At rest, no thermodynamic work is being done; hence all power produced by the metabolic rate must be dissipated to the ambient environment. Our typical man thus heats up the environment at a rate of 108. W, slightly more than the heat generated by a 100-W incandescent light bulb.

The metabolic rate increases dramatically with exercise level, and can approach 1000 W under strenuous exercise! Table 3.6 (from the ASHRAE Fundamentals Handbook, 1997) shows a list of physical activities, each with its corresponding metabolic heat generation rate, for a 70. kg (154. lbm) healthy adult male. Note that although two or three significant digits are shown in the table, there is a wide range of values in the literature, as mentioned previously. Approximately 40% of the energy required for heavy exercise is dissipated as ***sensible heat*** – heat that is felt as an increase in temperature of the surrounding air. The remaining 60% is dissipated as ***latent heat***, which is the energy required to evaporate water by the body's sweating mechanism. It is no surprise, therefore, that gymnasiums and work-out rooms are often unbearably hot and humid.

Thermodynamicists prefer joule (J), kilojoule (kJ), or calorie (cal) as their unit of energy, and watt (W) as their unit of power. On the other hand, physiologists and nutritionists prefer ***kilocalorie*** (kcal, i.e. 1000 cal) as their energy unit, and kcal/hr as their power unit. Furthermore, kilocalorie is often called ***Calorie*** (Cal, using upper case). Unfortunately, packaged food manufacturers often ignore this convention and use a lower case c – their "calorie" is actually a kilocalorie or Calorie, which is 1000 calories. This practice continues even though it is a source of much confusion. The unit kilocalorie is used throughout this discussion to avoid further confusion. One can convert the metabolic rate for an average, seated, relaxed man from watts to kilocalories as follows:

$$\dot{M}_{\text{seated, at rest}} = 108.\ \text{W}\left(\frac{\text{J}}{\text{W s}}\right)\left(\frac{1\ \text{kcal}}{4186.\ \text{J}}\right)\left(\frac{3600\ \text{s}}{\text{hr}}\right) = 93.\ \frac{\text{kcal}}{\text{hr}}$$

Multiplying this by 24 hours, a sedentary person (couch potato) would therefore require approximately 2200 kcal per day to maintain his body weight. In reality, no conscious person is sedentary 24 hours per day – even the act of eating raises the metabolic rate by approximately 6 or 7% compared to sitting still! Most nutritionists estimate that the average healthy adult man needs to ingest 2400 to 2700 kcal/day, while an average healthy adult woman requires 1800 to 2200 kcal/day. In order to better appreciate the metabolic rate, consider a "fun size" chocolate candy bar, which provides the body with 210 kcal of energy. An average person must sit for nearly 2.5 hours to atone for eating that candy bar. If he does some heavy exercise, such as walking up stairs, which requires nearly 800 kcal/hr, he can burn off that candy bar in about 15 minutes.

Table 3.6 Metabolic rate as a function of physical activity for a 70 kg adult man (abstracted from ASHRAE, 1997).

activity	metabolic rate (W)	metabolic rate (kcal/hr)
sleeping	72	62
seated, quiet	108	93
standing, relaxed	126	108
walking about the office	180	155
seated, heavy limb movement	234	201
flying a combat aircraft	252	217
walking on level surface at 1.2 m/s	270	232
housecleaning	284	244
driving a heavy vehicle	333	286
calisthenics/exercise	369	317
heavy machine work	423	364
handling 50-kg bags	423	364
playing tennis	432	372
playing basketball	657	565
heavy exercise	900	774

As a first approximation, smaller or larger individuals can estimate their energy consumption rate by simple proportionality based on their actual body mass. For example, a petite adult woman weighing 110 lbf has a body mass of 50. kg, almost 30% smaller than the average adult male. During very heavy exercise, she burns approximately

$$\frac{774\frac{\text{kcal}}{\text{hr}}}{70.\ \text{kg}}\left(50.\ \text{kg}\right) = 550\frac{\text{kcal}}{\text{hr}}$$

since the metabolic rate for heavy exercise in Table 3.6 is that of a 70 kg man.

Metabolic rate ($\dot{M}$) can be written as an empirical function of oxygen consumption rate (Q_{oxy}) and a parameter called the ***respiration quotient*** (RQ), defined as the ratio of exhaled carbon dioxide to inhaled oxygen. One such empirical formula for metabolic rate, as given in the "Thermal Comfort" chapter of the ASHRAE Fundamentals Handbook (1997), is

$$\boxed{\dot{M} = 352.\left[0.23\left(\text{RQ}\right) + 0.77\right]Q_{oxy}} \tag{3-33}$$

where RQ is dimensionless, Q_{oxy} must be in liters per minute at STP, and $\dot{M}$ is in watts. At rest, Q_{oxy} is around 0.25 L/min. Note that this is the rate at which oxygen is actually *consumed* by the body at rest, and is much smaller than the tidal volume of around 6 L/min, as discussed in Chapter 2, since exhaled air still contains a significant amount of oxygen. Q_{oxy} increases to approximately 2.5 L/min under heavy exercise. Typical values of RQ are in the range 0.83 (resting) < RQ < 1.0 (heavy exercise). Using these values, one can estimate the basal metabolic rate from Eq. (3-33),

$$\dot{M}_b = 352.\left[0.23\left(0.83\right) + 0.77\right]0.25\ \text{W} = 84.6\ \text{W}$$

which agrees well with the standard value of around 84. W for $\dot{M}_b$. Under heavy exercise conditions,

$$\dot{M}_{\text{heavy exercise}} = 352.\left[0.23\left(1.0\right) + 0.77\right]2.5\ \text{W} = 880\ \text{W}$$

which also agrees well with the value of 900 W for heavy exercise in Table 3.6.

The basal metabolic rate for a large number of warm-blooded animals (mice, birds, elephants, etc.) can be approximated by the following empirical equation:

$$\boxed{\dot{M}_b\ (\text{in watts}) = 3.39\left(m_b^{\ 0.75}\right)} \tag{3-34}$$

where the animal's body mass (m_b) must be in kg. One can easily test the applicability of Eq. (3-34) to human beings. For a 70. kg man, Eq. (3-34) yields 82. W, which again agrees well with the value discussed above. It is interesting to rearrange Eq. (3-34) to obtain the basal metabolic rate per mass and surface area of the person or animal, in units of watts/(kg m^2). For convenience the volume and surface area are assumed to be proportional to L^3 and L^2, respectively, where L is a characteristic length of the body. For body mass (m_b) and surface area (A_s),

$$\boxed{\frac{\dot{M}_b}{A_s m_b} = 3.39\frac{1}{A_s m_b^{\ 0.25}} = 3.39\frac{1}{A_s\left(V_b \rho_b\right)^{0.25}} \approx \frac{3.39}{\rho_b^{\ 0.25} L^{2.75}}} \tag{3-35}$$

The value of A_s for the typical man is 1.8 m^2. Eq. (3-35) shows that the basal metabolic rate per unit mass and area is inversely proportional to $L^{2.75}$. Eq. (3-35) suggests that the smaller the person or animal, the higher the relative metabolic rate; thus children have higher relative metabolic rates than adults, and hummingbirds higher than elephants. Eq. (3-35) is consistent with the concepts of heat transfer. The ratio of the rate of heat transfer from a body per mass of the body is proportional to the ratio of the body's surface area (A_s) to its volume (V_b). In terms of a sphere or a cube, this ratio is inversely proportional to the body's characteristic length (L), i.e. $A_s/V_b = 1/L$.

3.5.2 Thermodynamic Analysis of the Human Body

While engineers usually think of engines, heat pumps, and other thermo-mechanical devices when discussing the first and second laws of thermodynamics, these laws must apply to *all* systems and control volumes, even to those associated with living things. The chapter on "Thermal Comfort" in the ASHRAE Fundamentals Handbook (1997) provides an excellent summary of the various types of heat transfer interactions between the body and its surroundings. A rigorous thermal analysis of the human body can be found in the book by Cena and Clark (1981). A simplified discussion of thermodynamic aspects of biological systems can be found in Cengel (1997). Thermodynamic analysis of the human body is difficult because the human body is not a closed system, and is not in a state of thermodynamic equilibrium. There is unsteady mass transfer into and out of the body (respiration, perspiration, food intake, waste disposal, skin flaking, etc.). In addition, the body is continually metabolizing energy in food and stored fat in order to maintain bodily functions, as discussed above. Nevertheless, for the simplest possible analysis, let us consider the human body as a thermodynamic system. The person is at rest; thus there are no significant changes in potential or kinetic energy within the system, i.e. the system is ***stationary***. The first law of thermodynamics for a stationary system, in rate form, is

$$\boxed{\frac{dU}{dt} = \dot{Q} - \dot{W}} \tag{3-36}$$

where U is the internal energy of the system, $\dot{Q}$ is the rate of heat transfer *into* the system, and $\dot{W}$ is the rate of work being done *by* the system, following the standard convention in thermodynamics. For the case at hand, the person is at rest, so no thermodynamic work is being done, eliminating the last term in Eq. (3-36). However, the person's body is continually burning food and fat at metabolic rate, $\dot{M}$. Thermodynamically, the chemical processes which make up metabolism are exothermic, releasing heat inside the body. Since the body's temperature does not change, the body's internal energy decreases at a rate equal to $\dot{M}$ ($dU/dt = -\dot{M}$), and this same rate of change of energy must be transferred from the body in the form of heat transfer to the environment. Thus, the first law of thermodynamics for this system simplifies to

$$\boxed{\dot{M} + \dot{Q} = 0} \tag{3-37}$$

which states that all of the energy associated with metabolism is rejected ultimately into the ambient environment as heat. Note that since $\dot{M}$ is positive, $\dot{Q}$ must be negative in order to satisfy Eq. (3-37). Note also that under conditions in which the person is doing thermodynamic work, $\dot{M}$ in Eq. (3-37) is replaced by $\dot{M} - \dot{W}$.

Five types of heat transfer into the body are considered: conduction, convection, respiration, radiation, and evaporation, which are denoted as $\dot{Q}_{cond}$, $\dot{Q}_{conv}$, $\dot{Q}_{res}$, $\dot{Q}_{rad}$, and $\dot{Q}_{evap}$ respectively. Eq. (3-37) thus becomes

$$\boxed{\dot{M} + \dot{Q}_{cond} + \dot{Q}_{conv} + \dot{Q}_{res} + \dot{Q}_{rad} + \dot{Q}_{evap} = 0} \tag{3-38}$$

All terms in Eq. (3-38) have dimensions of energy per unit time, i.e. power. Each of these heat transfer terms is considered separately below.

Conduction: Conduction heat transfer is by direct contact with solid surfaces, such as chairs, the floor, etc. Since the surface area in such contact is small, and since the materials are generally good insulators, with clothing between the body and the material providing further insulation, conduction heat transfer is negligible in the present analysis.

$$\boxed{\dot{Q}_{cond} \approx 0} \tag{3-39}$$

Convection: Energy transferred into the body by convection depends on the velocity of air passing over the body and the temperature difference between skin and air. Wadden and Scheff (1987) suggest an expression of the following form, for $\dot{Q}_{conv}$ expressed now in units of kcal/min:

$$\dot{Q}_{conv} = KA_s\left(0.0325 + 0.1066U_a^{0.67}\right)\left(T_a - T_s\right) \tag{3-40}$$

where

- K = fraction of skin exposed to atmosphere
- U_a = ambient air velocity (m/s)
- T_a, T_s = ambient and skin temperatures; for a first approximation, T_s = 35.0 °C (308.15 K)

Respiration: The process of breathing in and out involves mass transfer as well as heat transfer, and technically should not be considered as part of thermodynamic closed system analysis. Nevertheless, one can approximate the net rate of heat transfer due to respiration as a steady-state heat transfer term. The chapter on "Thermal Comfort" in the ASHRAE Fundamentals Handbook (1997) provides a simple expression for the heat transfer rate, which takes into account both sensible and latent heat transfer due to respiration,

$$\dot{Q}_{res} = -\dot{M}\left[0.0014\left(307.15 - T_a\right) + 0.0173\left(5.87 - \Phi_a P_v(\text{at } T_a)\right)\right] \tag{3-41}$$

where

- $\dot{M}$ = metabolic rate in watts or kcal/min ($\dot{Q}_{res}$ has the same units as $\dot{M}$)
- T_a = ambient air temperature in Kelvin
- Φ_a = relative humidity of ambient air expressed as a fraction, not a percentage
- $P_v(\text{at } T_a)$ = vapor (saturation) pressure of water at the ambient air temperature in kPa

Radiation: Energy transferred into the body by radiation depends on the skin and radiating surface temperatures, the ***emissivity*** of the radiating surfaces, and various shape factors (Incropera and DeWitt, 1990). In most industrial situations, the identity, location, and temperature of the radiating surfaces are difficult to ascertain. A useful parameter to account for these factors is ***globe temperature*** (T_G), defined as the air temperature inside an enclosure subjected to both convection with ambient air, and radiation from nearby hot or cold surfaces. Figure 3.34 shows a commercial instrument used to measure globe temperature. In the presence of hot radiating surfaces, globe temperature is larger than ambient temperature. Globe temperature depends on the location of the radiating surfaces; thus globe temperature is a physical measurement that must be made where the workers perform their duties. Globe temperature integrates surrounding radiating wall temperatures into a mean effective radiation temperature. To simplify the computation, the conventional difference between temperatures raised to the fourth power is replaced by a first-order difference between a "wall temperature" and skin temperature. Wadden and Scheff (1987) report that the following expression can be used to relate ***wall temperature*** (T_W) and the measured globe and ambient temperatures:

$$T_W = \left[T_G^4 + \left(0.248\times10^9 U_a^{0.5}\right)\left(T_G - T_a\right)\right]^{0.25} \tag{3-42}$$

where the temperatures must be in Kelvin, and U_a is the characteristic air speed which must be in m/s. In lieu of experimental measurement, the observations in Table 3.7 can be used to estimate the speed of moving indoor air. Wadden and Scheff (1987) report that the rate of energy transferred to the body by radiation ($\dot{Q}_{rad}$) in units of kcal/min can be expressed in the following form:

$$\dot{Q}_{rad} = 0.0728A_sK\left(T_W - T_s\right) \tag{3-43}$$

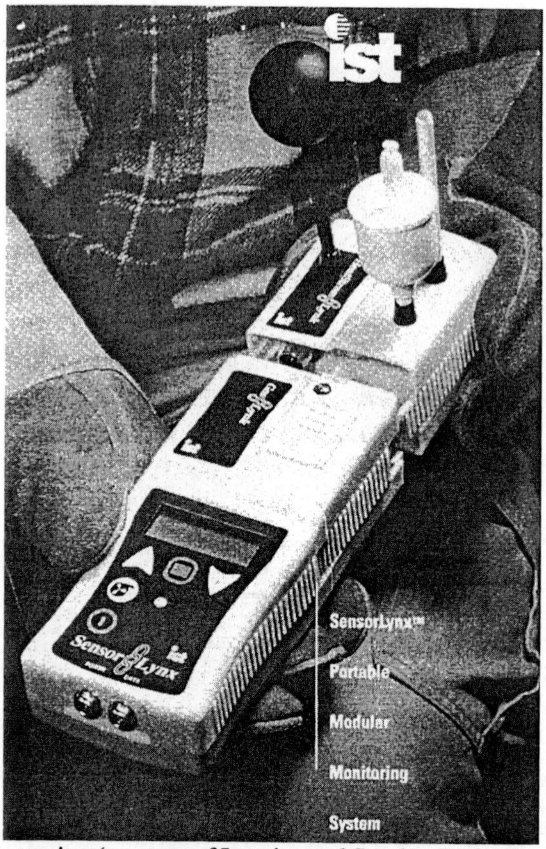

Figure 3.34 Heat stress monitor (courtesy of Imaging and Sensing Technology Corp.).

Evaporation: It is much more difficult to estimate the rate of heat transfer to the body by evaporation. The *required* value of heat transfer by evaporation ($\dot{Q}_{evap,req}$) can instead be found from Eq. (3-38), since all the other terms are now known.

$$\boxed{\dot{Q}_{evap,req} = -\dot{M} - \dot{Q}_{conv} - \dot{Q}_{res} - \dot{Q}_{rad}} \tag{3-44}$$

Table 3.7 Estimates of the characteristic speed of moving indoor air (to two significant digits).

physiological response or observation	**air speed ft/min (m/s)**
observable settling velocity of skin particles	3.0 (0.015)
random indoor air movement, air at lower speed is considered "stale"	20. (0.10)
exposed neck and ankles begin to sense air movement, upper limit of "comfort" at acceptable room temperatures	50. (0.25)
skin on hands begins to sense air movement, moisture on skin increases sensitivity	100 (0.51)
typical walking speed, eddy velocity of a person walking at a brisk speed	250 (1.3)
8-hr, upper comfort level of blown air for cooling, average outdoor air speed	700 (3.6)
30-min, upper comfort level of blown air for cooling, typical fan exit speed for free-standing fans	1,800 (9.1)
10-min, upper comfort level of blown air for cooling	3,500 (18.)

Note that $\dot{Q}_{evap,req}$ is negative implying that heat transfer related to sweat evaporation is from the body to the ambient air. The equations for $\dot{Q}_{conv}$, $\dot{Q}_{res}$, $\dot{Q}_{rad}$ and i.e. Eqs. (3-40), (3-41), and (3-43), presume input heat transfer, and as such the signs in Eq. (3-44) will take care of themselves; the reader should not impose another sign convention.

3.5.3 Heat Stress Index

There are several ways to quantify heat stress so that designers can anticipate hazardous conditions and take steps to reduce risk. The chapter on "Thermal Comfort" in the ASHRAE Fundamentals Handbook (1997) gives the historical development of this technology and presents several methods to quantify heat stress of varying sophistication. What follows is a description of one of these methods, the ***heat stress index*** (HSI), developed by Belding and Hatch (1955). While an old concept and perhaps lacking the detail found in more current technologies, HSI describes the basic physical elements that constitute heat stress, and does so with sufficient accuracy to be a useful design tool for practicing engineers. Belding and Hatch (1955) define HSI as

$$\boxed{\mathrm{HSI} = \frac{\dot{Q}_{evap,req}}{\dot{Q}_{evap,max}} \times 100\%} \tag{3-45}$$

where $\dot{Q}_{evap,max}$ is the maximum amount of evaporation the body can achieve assuming the entire surface of exposed skin is covered with sweat. Wadden and Scheff (1987) report that $\dot{Q}_{evap,max}$ (in units of kcal/min) can be computed from

$$\boxed{\dot{Q}_{evap,max} = 0.198 K A_s U_a^{\,0.63} \left[\Phi_a P_v(\text{at } T_a) - P_v(\text{at } T_s)\right]} \tag{3-46}$$

where Φ_a is the relative humidity of the ambient air expressed as a number rather than as a percentage ($0 < \Phi_a < 1.0$), P_v(at T_a) is the vapor pressure (in mm Hg) of water based on ambient air temperature T_a, and P_v(at T_s) is the vapor pressure (in mm Hg) of water based on skin temperature T_s. Since skin temperature is generally higher than ambient air temperature, and the relative humidity of the ambient air is generally less than unity, Eq. (3-46) yields a negative value of $\dot{Q}_{evap,max}$ which, coupled with the negative value of $\dot{Q}_{evap,req}$, results in a positive value of HSI. It is mathematically possible for $\dot{Q}_{evap,max}$ to be positive under extremely hot and humid conditions, but Eq. (3-46) cannot be used in such cases since the body cannot *absorb* heat due to evaporation. (Perspiration can only cool the body, not heat the body.) In such cases, Eq. (3-45) cannot be used. The ratio of required to maximum evaporative cooling reflects whether the body is capable of maintaining equilibrium by evaporation, and thus is capable of preventing energy from accumulating within the body. Shown in Table 3.8 are the physiological implications of different values of the heat stress index that enable engineers to anticipate hazardous conditions for 8-hour exposures. Values of HSI above 100% are possible, and indicate that the body is unable to cool itself adequately; hyperthermia results. Note that negative values of HSI are also possible; this indicates that the body cannot maintain its constant temperature in cold ambient air, and hypothermia results.

Heat stress can be controlled in a number of ways. Appropriate clothing should be worn that does not impede cooling by evaporation. The length of time an individual works under stressful conditions should be regulated. Lastly, workers may need a specially designed stream of moving air to cool them. The air stream can be permanent and incorporated into the building's return air system (fresh-air islands) or, as is often the case, it can be provided by portable fans called ***spot coolers*** or ***man coolers***. Engineers must be careful because spot coolers direct a stream of fast-moving air over

Table 3.8 Heat stress index (adapted from ASHRAE, 1997).

HSI (%)	consequence of 8-hr exposure
< 0	Indicates varying degrees of stress due to hypothermia.
0	No thermal strain.
10-30	Mild to moderate heat stress. Manual dexterity and mental alertness may suffer but there is little impairment to perform heavy work.
40-60	Severe heat stress. Health may be threatened unless physically fit. This condition should be avoided by people with cardiovascular or respiratory impairment or chronic dermatitis.
70-90	Very severe heat stress. Only specially selected people are capable of sustaining these conditions for 8 hrs. Special care must be taken to replace water and salt.
100	Maximum heat stress. Only acclimated, physically fit young people can withstand this for 8-hrs.
>100	Indicates varying degrees of stress due to hyperthermia.

the worker, and may interfere with the operation of a nearby local ventilation system that captures contaminants, unless spot cooling was incorporated in the design of the ventilation system in the first place.

In many instances, workers are required to wear special apparel which may contribute to heat stress. Examples include workers engaged in rescue operations (e.g. firefighters), workers engaged in hazardous chemical clean-up, individuals exposed to hot environments, maintenance workers performing tasks in confined spaces, workers who must weld or spray paint in confined spaces, etc.

Example 3.6 - Heat Stress Index in a Restaurant Kitchen

Given: The air in the workspace of a restaurant kitchen has a characteristic air speed (U_a) of 0.75 or 1.5 m/s, depending on the fan speed setting. The wet bulb temperature (T_{wb}) is 25. °C and the dry bulb temperature (T_{db}) is 30. °C, which corresponds to a relative humidity of 68.%. Note that the dry bulb temperature is the same as the ambient temperature, i.e. $T_{db} = T_a$. The globe temperature (T_G) is measured and found to be 35. °C. It is estimated that the tasks the workers perform require a net metabolic rate that may be as low as 1.5 kcal/min or as high as 5.0 kcal/min. Assume the worker is clothed, with 60.% of his skin area available for heat transfer.

To do: Estimate the heat stress index (HSI) for workers in the restaurant kitchen under various possible conditions.

Solution: The least dangerous situation is examined first, i.e. the case with the lowest metabolic rate and the highest air speed in the kitchen. The rate of convection heat transfer can be found from Eq. (3-40),

$$\dot{Q}_{conv} = 0.6(1.8)\left(0.0325 + 0.1066(1.5)^{0.67}\right)(303.15 - 308.15) = -0.931\frac{\text{kcal}}{\text{min}}$$

$\dot{Q}_{conv}$ is negative since the ambient air, although warm, is cooler than the worker's skin temperature. The rate of heat transfer due to respiration is found from Eq. (3-41),

$$\dot{Q}_{res} = -1.5\frac{\text{kcal}}{\text{min}}\left[0.0014(307.15 - 303.15) + 0.0173(5.87 - 0.68(4.246))\right] = -0.0858\frac{\text{kcal}}{\text{min}}$$

The wall temperature can be found from Eq. (3-42). Being careful to use absolute temperatures in the calculations,

$$T_w = \left[308.15^4 + \left(0.248\times10^9\left(1.5^{0.5}\right)\right)\left(308.15.-303.15.\right)\right]^{0.25} = 320.4\ \text{K} \cong 320.\ \text{K}$$

Note that this is very high (47. °C) but not unusual in a small kitchen containing grills, ovens, and gas ranges. The radiation heat transfer rate can be found from Eq. (3-43),

$$\dot{Q}_{rad} = 0.0728(1.8)(0.6)(320.4-308.15) = 0.961\frac{\text{kcal}}{\text{min}}$$

$\dot{Q}_{rad}$ is positive, indicating radiation heat transfer *into* the worker's body, as expected. The required evaporative heat transfer can be found from Eq. (3-44). At the lower extreme of the metabolic rate ($\dot{M}$ = 1.5 kcal/min),

$$\dot{Q}_{evap,req} = -\dot{M} - \dot{Q}_{conv} - \dot{Q}_{res} - \dot{Q}_{rad} = -1.5-(-0.931)-(-0.0858)-0.961 = -1.44\frac{\text{kcal}}{\text{min}}$$

In this particular case, the effects of radiation and convection heat transfer nearly cancel each other out, the respiratory heat loss is small compared to the other terms, and the required rate of evaporative cooling is nearly equal to the metabolic rate. The maximum amount of evaporative cooling can be found from Eq. (3-46),

$$\dot{Q}_{evap,max} = 0.198(0.60)(1.8)(1.5)^{0.63}\left[(0.68)31.84-42.21\right] = -5.67\frac{\text{kcal}}{\text{min}}$$

where the water vapor pressures are obtained from thermodynamic tables or Appendix A.17. The heat stress index is thus found from Eq. (3-45),

$$\text{HSI} = \frac{\dot{Q}_{evap,req}}{\dot{Q}_{evap,max}}\times100\% = \frac{-1.44}{-5.67}\times100\% = 25.\%$$

The heat stress values that can be expected for three metabolic rates and two air speeds are shown below (to two significant digits). It is assumed that K = 0.60 for all cases.

metabolic rate	**HSI (%) for U_a = 0.75 m/s**	**HSI (%) for U_a = 1.5 m/s**
1.5 kcal/min	40.	25.
3.0 kcal/min	78.	50.
5.0 kcal/min	130	84.

Discussion: It is clear that if U_a = 0.75 m/s, workers should not work strenuously for 8 hours. If the air speed can be increased to 1.5 m/s (difficult to achieve in a kitchen), stressful conditions can be reduced substantially. Alternatively, if the work is strenuous, the work schedule should be divided by periods of rest, worker assignments should be rotated, or steps should be taken to reduce the room temperature (e.g. air conditioning) to reduce the risks of heat stress. The workers should also be advised to drink plenty of water. An Excel spreadsheet which was used to generate the above table is available on the book's web site.

3.6 Odors

Most complaints about indoor air pollution and inadequate ventilation concern odors (Shusterman, 1992). An odor may be harmless, indeed even pleasant (such as food), but if the odor persists, is intense, or inundates an individual, it constitutes an involuntary intrusion and is a legitimate basis for complaint. An ***odor*** is the physiological response from a particular airborne molecular species and is detected by ***olfactory nerve cells*** (olfactory neurons) located at the top of the nose, just above the bridge of the nose. The concentration of a species that causes a person to detect an odor varies considerably between individuals. An individual's response is affected by temperature, humidity, and exposure to simultaneous odors. In addition, a person's age has significant influence on his/her ability to detect an odor; those under age 30 are up to three times more sensitive to odors than are older individuals (Dravnieks et al., 1986). While an average person might perceive an odor at low concentrations, only a few can *identify* the odor, or even compare it to some other odor. Individuals tend to become accustomed to odors, even those they initially find to be unpleasant. An excellent review of the technical literature about odors presented in ways congenial to engineers can be found in the chapter on odors in the ASHRAE Fundamentals Handbook (2001).

Olfaction is a sensitive physiological response mechanism that is initiated by contaminants at low concentrations, sometimes measured in parts per billion. The olfactory system is capable of discriminating one odor from a background of different odors. Unfortunately the olfaction system exhibits fatigue, and individuals may become insensitive to an odor until the concentration changes. Odors are hedonistic in that they elicit an involuntary response that pleases or offends individuals. While no two individuals are alike, groups of individuals exhibit responses that can be quantified with statistical significance. There is a strong correlation between the senses of smell and taste since the neurological response involves the same organs. Food flavors are generally associated with aromas that are airborne mixtures of molecular species that stimulate the sensory organs. Practices followed in the foods and flavors industry in which the sensory response is described are also used to describe environmental odors. To *quantify* odors, at least three independent properties are needed:

- ***Quality*** (or ***character***): a description of the odor using familiar functional groupings, e.g. fragrant, musky, garlic-like, acrid, oniony, fruity, burnt, mustard-like, sweet, sour, fishy; or reference to common household products, e.g. bleach, mothballs, gasoline, rubber, plastic.
- ***Acceptability*** (or ***hedonic tone***): the sensual pleasure, annoyance, or offense that the odor evokes, e.g. sharp, pungent, irritating, nauseating, repulsive, pleasant.
- ***Intensity***: a quantitative response related to the concentration of odor-causing material.

The intensity or strength of a response can be described in terms of the ***Weber-Fechner law***:

$$\boxed{I = K\log(c)} \tag{3-47}$$

where

- I is the intensity of the odor sensation or the magnitude of the response to an odor
- c is the mass concentration of the material that causes the odor
- K is a constant that varies among odor-causing materials

Although mass concentration is used in Eq. (3-47), mol fraction (y) can be used in place of c, provided that K is adjusted accordingly. (Because it is dimensionless, mol fraction is preferred inside the logarithm in Eq. (3-47) from a mathematical point of view.) An alternative expression for the intensity of an odor sensation is ***Steven's law*** (US EPA, 1992):

$$\boxed{I = Kc^n} \tag{3-48}$$

where n is an exponent whose value depends on the material causing the odor, and constant K in Eq. (3-48) is not necessarily the same as that in Eq. (3-47).

Although there is no satisfactory comprehensive theory of odors, chemicals that elicit an olfactory response tend to have low vapor pressures. Alcohols and acetates have odor thresholds that decrease with increasing carbon chain length, and pungency thresholds decrease exponentially with chain length (Shusterman, 1992; Cometto-Muniz and Cain, 1993). Certain generalities can be stated from reviewing odor threshold values in the literature:

- *mercaptans*: unpleasant, repulsive; detectable at concentrations well below PEL
- *sulfides*: burnt, decayed matter, putrid
- *acetates*: pleasant, fragrant, sweet, fruity

- *aldehydes*: vary from sweet and fruity to pungent
- *amines*: fishy, sharp, pungent

Experiments to establish threshold values involve trained observers (***sniffers***), and ascribe a numerical value, sometimes spanning 6 to 8 orders of magnitude, to diluted samples of the odor. Minimal detectable levels are established by diluting samples until the odor is no longer detected. The classical definition of ***odor detection threshold*** is the minimum concentration which produces a detectable difference in odor from the background. Unfortunately there are different detection thresholds depending on the background. Detection threshold is not reliable and is difficult to reproduce since it relies on the poorly defined judgment of the observer (Leonardos et al., 1969). Even when detected, it may take a significant increase in concentration before individuals can *identify* the odor. A more stringent and more reliable concept of odor threshold is the ***odor recognition threshold***, defined as the minimum concentration at which the odor can not only be detected, but *described*; another name for odor recognition threshold is ***minimum identifiable odor*** (MIA). Because of the large variation in odor detection among individuals, ***odor panels*** (consisting of up to a few dozen trained sniffers) are used to conduct odor threshold experiments, and statistical methods are used to establish the thresholds. Odor researchers define *two* odor recognition thresholds: the ***50% odor recognition threshold*** is the concentration at which 50% of the odor panel recognize the odor; the ***100% odor recognition threshold*** is the lowest concentration at which 100% of trained observers can positively identify an odor which is consistent with the response at all higher concentrations (Hellman and Small, 1974).

Table 3.9 is a comparison of odor detection threshold, 50% odor recognition threshold, and 100% odor recognition threshold for some common petrochemicals. For some chemicals (e.g. cyclohexanone), there is no difference between the detection and 50% recognition thresholds, indicating that more than half of the odor panel recognized the odor as soon as they detected a change from the background. For a number of other chemicals (e.g. amyl alcohol), there is no difference between the 50% and 100% odor recognition thresholds. Repeatability studies were also performed by Hellman and Small (1974) with the same odor panel on different days; results reveal that all three odor thresholds can vary by factors of up to four. Thus, although some of the numbers in Table 3.9 are presented with three significant digits, the data are reliable to perhaps only one digit.

Odor threshold values reported in the literature (Leonardos, et al., 1969; Staub-Reinhault, 1972; Hellman and Small, 1974; Stahl, 1978; Amoore and Hautala, 1983; Verschueren, 1983; Nagy 1995) vary considerably from one research lab to another, sometimes by more than an order of magnitude! A comprehensive review was conducted by Ruth (1986), resulting in a list of the lowest and highest odor recognition threshold values reported in the literature for 450 chemicals. A subset of this list is presented in Appendix A.20, along with each chemical's OSHA PEL for comparison.

If an odor is associated with toxic materials, the mere existence of the odor should be taken seriously, since more than half (nearly 55%) of materials have odor threshold values similar to their PEL or TLV values. Of the remaining materials, approximately 25% have odor threshold values larger than their PEL, and nearly 20% have odor thresholds lower than their PEL. Schaper (1993) found a strong correlation between TLV and materials whose odor produces a 50% reduction in the respiratory rate in mice. Shown in Table 3.10 are examples of contaminants where the odor recognition threshold and the OSHA PEL are vastly different, and one (HCN) where they are comparable.

An additional way to categorize odors is the ***odor index***, defined as the ratio of the chemical's vapor pressure to its odor threshold level,

Table 3.9 Comparison of the three odor thresholds for some common petrochemicals (abstracted from Hellman and Small, 1974).

chemical	odor quality	detection (PPM)	50% recognition (PPM)	100% recognition (PPM)
acetone	sweet, fruity	20.0	32.5	140
acrylic acid	rancid, sweet	0.094	1.04	1.04
amyl alcohol	sweet	0.12	1.0	1.0
n-butanol	rancid, sweet	0.30	1.0	2.0
2-butanol	sweet	0.12	0.41	0.56
butyl acetate	sweet, ester	0.006	0.037	0.037
n-butyl chloride	pungent	8.82	13.3	16.7
n-butyl ether	fruity, sweet	0.07	0.24	0.47
carbitol acetate	sweet	0.026	0.157	0.263
cyclohexanone	sweet, sharp	0.12	0.12	0.24
diacetone alcohol	sweet	0.28	1.1	1.7
diethylamine	musty, fishy, amine	0.02	0.06	0.06
ethyl acetate	sweet, ester	6.3	13.2	13.2
ethyl acrylate	sour, pungent	0.0002	0.00030	0.00036
ethylene	olefinic	260	400	700
ethylene oxide	sweet, olefinic	260	500	500
1-hexanol	sweet, alcohol	0.01	0.09	0.09
isobutanol	sweet, musty	0.68	1.80	2.05
isobutyl acetate	sweet, ester	0.35	0.50	0.50
methanol	sour, sharp	4.26	53.3	53.3
methyl amyl acetate	sweet, ester	< 0.07	0.23	0.40
methyl amyl alcohol	sweet, alcohol	0.33	0.52	0.52
2-methyl butanol	sour, sharp	0.04	0.23	0.23
methyl ethyl ketone	sweet, sharp	2.0	5.5	6.0
n-propanol	sweet, alcohol	< 0.03	0.08	0.13
propylene	aromatic	22.5	67.6	67.6
styrene	sharp, sweet	0.05	0.15	0.15
toluene	sour, burnt	0.17	1.74	1.74
vinyl acetate	sour, sharp	0.12	0.40	0.55
xylene (o-xylene)	sweet	0.08	0.27	0.27

$$\boxed{\text{odor index} = \frac{\text{vapor pressure}}{\text{threshold level}}} \tag{3-49}$$

where the threshold level (expressed as a partial pressure) must be in the same units as the vapor pressure. Alternatively, mass concentration, molar concentration, or mol fraction may be used to define the odor index, as long as the same units are used in the numerator and denominator. Another way to express threshold values is called the ***dilution factor*** (Dravnieks et al., 1986) which is defined as the total number of volumes to which one volume of air saturated with the vapor or odor must be diluted to reach the odor threshold. The odor index and dilution factor are the same, and because the values have a large range, they are often reported in terms of their logarithms. Both terms are useful because they take into account the escaping tendency of the material (its vapor pressure) and the ability of individuals to recognize its odor. For situations in which more than one odor is present at the same time, Striebig (2002) has suggested a multiple odor index that is simply the sum of the odor index of Eq. (3-49) for each odor.

Table 3.10 Comparison of odor recognition threshold and OSHA PEL for various chemicals (maximum value of odor threshold to two significant digits, abstracted from Appendix A.20).

material	odor recognition threshold (PPM)	PEL (PPM)
n-butyl acetate	20.	150
n-butyl mercaptan	9.0×10^{-4}	10
carbon monoxide	no odor	50
ethylene oxide	780	1
hydrogen cyanide	4.5	10
methyl alcohol	20,000	200
methyl bromide	1,000	5
methyl formate	2,800	100
methyl methacrylate	0.34	100
methylene chloride	620	25
nickel carbonyl	30.	0.001
triethylamine	0.27	25
xylidene	0.0049	5

A convention to quantify air quality in spaces occupied by humans was developed in Europe (Commission of European Communities, 1992) and is receiving growing acceptance in the US. Air quality is defined in terms of the percentage of trained individuals who perceive the air quality to be unacceptable. Perception refers to a judgment an individual makes immediately upon entering a space containing ***bioeffluents*** (odors, vapors, gases, etc. emitted by people living or working in the space). After approximately 15 minutes the olfactory senses and mucous membranes of the nose and eyes adapt, and the response to many materials is no longer unacceptable. The convention is based on the fact that sedentary adults in a thermally neutral indoor environment generate bioeffluents that other individuals find objectionable in sufficient concentration. While a generation rate of bioeffluents has dimensions of mass/time, units as mg/s are not commonly used by researchers dealing with odors from humans. Instead, the rate at which such bioeffluents are generated by a single sedentary thermally neutral person is defined by a unit called an ***olf*** (Fanger, 1988). Furthermore, when a sedentary, thermally neutral person is in a room which receives fresh ventilation air at a volumetric flow rate of 10 L/s (2.1 SCFM), the resulting mass concentration in the room is defined as a ***decipol.*** Decipol has the dimensions of mass per unit volume of air, since it is a mass concentration. From the definitions of olf and decipol, the well-mixed steady-state mass concentration (c_{ss}) in a room containing a source of contaminant (S) equal to one olf, and receiving fresh make-up air through the ventilation system at a volumetric flow rate (Q) equal to 10 L/s, is given by

$$c_{ss} = \frac{S}{Q} = \frac{1 \text{ olf}}{10 \frac{L}{s}} = 0.1 \frac{\text{olf} \cdot s}{L} = 1 \text{ decipol} \tag{3-50}$$

Derivation of the above equation for c_{ss} is found in Chapter 5, and c_{ss} is independent of room volume. Three indoor air quality levels are used for design purposes, as seen in Table 3.11. A steady-state mass concentration of 1 decipol lies between categories A and B. As expected, more people are dissatisfied with indoor air quality as ventilation flow rate decreases, corresponding to an increase in mass concentration of bioeffluents. Furthermore, at a given ventilation rate, the concentration is expected to increase linearly with number of persons in the room, since the bioeffluent generation rate is one olf per person. This is one of the reasons why ASHRAE recommends a ventilation rate of at least 6 SCFM per person for offices and general classrooms, as can be seen in Appendix A.16.

Table 3.11 Air quality categories based on dissatisfaction with room odor.

air quality category	% dissatisfied	c_{ss} (decipols)	Q (L/s)	Q (SCFM)
A	10	0.6	16.	3.4
B	20	1.4	7.0	1.5
C	30	2.5	4.0	0.85

3.7 Radiation

While ventilation systems do not capture radiation per se, they may capture radioactive gases (radon) and particles, or use ultraviolet or infrared radiation to initiate chemical reactions to oxidize organic compounds or annihilate airborne pathogens. Humans subjected to electromagnetic radiation may suffer adverse effects depending on the frequency of the radiation. ***Electromagnetic radiation***, henceforth merely called ***radiation***, consist of electric and magnetic waves propagating in mutually perpendicular planes. Radiation may also be thought of as a stream of energetic photons. The radiation frequency (ν) and wavelength (λ) are related by

$$\boxed{c = \nu\lambda} \tag{3-51}$$

where c is the speed of light. The energy of a photon can be expressed as

$$\boxed{E = \frac{\nu}{\lambda} = \frac{c}{\lambda^2}} \tag{3-52}$$

The spectrum of radiation wavelengths varies over a wide range as seen in Figure 3.35. Short wavelengths correspond to gamma rays and X-rays, and long wavelengths correspond to radar, microwaves, and radio waves (off the scale in Figure 3.35). The visible spectrum is a narrow band of wavelengths between 0.3 μm and 0.7 μm. Radiation's effect on matter depends on whether the energy of the photon is capable of affecting molecular structures. The chemical bonds of most molecules are in the range 1 to 15 electron volts (eV), while nuclear binding energies are of 10^6 eV or greater. X-rays are energetic photons of the order 10^8 eV whereas photons associated with microwaves have energies of 10^{-6} eV. It is convenient to divide electromagnetic radiation into two categories:

- ***Ionizing radiation*** is radiation of sufficient energy to produce ions. Ionizing radiation is subdivided into directly or indirectly ionizing radiation. ***Directly ionizing radiation*** consists of charged particles (electrons, protons, alpha particles, etc.) that have sufficient energy to produce ions by collision. ***Indirectly ionizing radiation*** consists of uncharged particles (neutrons, photons, etc.) that generate ionizing particles directly or initiate nuclear transformations.
- ***Nonionizing radiation*** is a stream of photons that do not cause ionization in biological material. A useful rule of thumb is that nonionizing radiation consists of energetic particles less than 10-12 eV since this typifies the energy to ionize oxygen and hydrogen.

3.7.1 Nonionizing Radiation

While nonionizing photons do not produce ions, they have sufficient energy to alter chemical bonds, affect the vibrational and rotational energies of molecules, or dissipate energy in fluorescence or heating. Nonionizing radiation (less than 10-12 eV) includes radio frequency (RF), microwave, infrared, and most ultraviolet radiation. The major source of natural ultraviolet radiation is the sun, although absorption by atmospheric ozone permits only wavelengths greater than 0.29 μm ($\lambda > 290$ nm) to reach the surface of the earth. Ultraviolet radiation is also produced by arc welding, plasma torches, lasers, and mercury, quartz, and xenon discharge lamps. Table 3.12 shows how the transmission and absorption of UV radiation depends on wavelength. The eye is the primary organ at risk to all types of nonionizing radiation. Ultraviolet radiation between 0.26 and 0.32 μm is absorbed in the ***stratum corneum*** of the epidermis and produces ***erythema*** (reddening) of the skin.

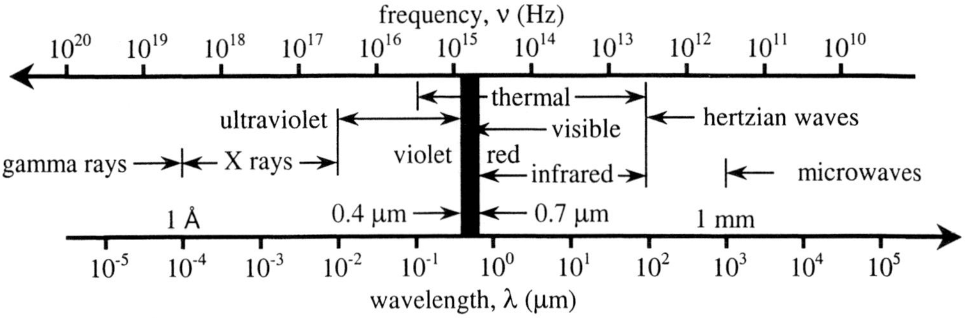

Figure 3.35 Wavelength and frequency of electromagnetic radiation (adapted from Bejam, 1993).

Chemical reactions initiated by ultraviolet light are called ***photolytic*** or ***photochemical*** reactions. A common photolytic reaction in the field of air pollution is the absorption of light by nitrogen dioxide,

$$\boxed{NO_2 + h\nu = NO + O\bullet} \tag{3-53}$$

where h is Planck's constant and ν is the frequency. The symbol O• represents an excited form of the oxygen atom called a ***free radical*** that is capable of reacting with diatomic oxygen to form ozone,

$$\boxed{O\bullet + O_2 = O_3} \tag{3-54}$$

The brown haze of smog is due in part to the absorption of light by NO_2. Another common atmospheric photolytic reaction involves the ***photolysis*** of formaldehyde by ultraviolet light,

$$\boxed{HCHO + h\nu = H\bullet + HCO\bullet} \tag{3-55}$$

which is the reaction that begins the conversion of formaldehyde to CO_2 and H_2O.

Microwave equipment is commonly used in industry for heating and drying. Microwave wavelengths range from 10^{-3} m to 10 m. The photon energy is insufficient to initiate photochemical reactions in biological matter since the energy is dissipated in thermal heating. Thermal heating may nevertheless be a health hazard. Microwave wavelengths less than 0.03 m are absorbed in the outer skin, wavelengths between 0.03-0.10 m penetrate the skin to a depth of 1-10 mm. At wavelengths between 0.25-2.0 m, penetration can cause damage to internal organs. The human body is thought to be transparent to microwave radiation greater than about two meters.

Table 3.12 Transmission and absorption of ultraviolet radiation as a function of wavelength.

wavelength, λ (μm)	transmission, absorption, and/or reflectance
0.3-0.4	transmitted through air, partially transmitted through water and glass
0.2-0.32	transmitted through air and quartz; absorbed by ordinary glass and water
0.16-0.2	partially absorbed by air, quartz, and glass
< 0.16	completely absorbed by air and quartz, can only exist in a vacuum

UV region	wavelength, λ (μm)	photon energy (eV)
vacuum	< 0.16	> 7.7
far or deep	0.16 - 0.26	7.7 - 4.4
middle	0.28 - 0.32	4.4 - 3.9
near	0.32 - 0.4	3.9 - 3.1

3.7.2 Ionizing Radiation

Ionizing radiation includes (a) atomic particles having a variety of physical and electrical properties that may travel at very low velocities or velocities approaching the speed of light, and (b) electromagnetic radiation (photons) having a wide range of energies that travel at the speed of light. In the industrial environment, the types of radiation of primary importance are X-rays, gamma rays, alpha and beta particles, and neutrons.

3.8 General Safety

In addition to the safety issues raised in previous sections there are other safety issues engineers must address. These include:

- guards for rotating machinery and power transmission
- construction details for flooring, scaffolding, guard rails, stairs, and ladders
- tethers and harnesses to prevent falls
- illumination, electrical power, and electrical lockouts
- respirator fit-testing and maintenance
- dip tanks containing flammable liquids
- material handling

Many of these topics involve construction, electrical circuits, manufacturing processes, etc., and it is rare that individual engineers are experts in all of these areas. Nevertheless, engineers should acquaint themselves with OSHA requirements contained in CFR 1910 and 1926.

A particularly hazardous situation related to indoor air pollution is entry into confined spaces. A ***confined space*** is an enclosure not designed for humans to enter or work routinely. Confined spaces are often characterized by low headroom, preventing workers from standing erect and moving about easily. Confined spaces also lack ventilation and adequate illumination. Examples of confined spaces include mixing tanks, reactors, silos, hoppers, large-diameter pipelines, storage vessels, underground utility vaults for electrical conduits and steam pipes, access chambers for wastewaters, etc. Figure 3.36 shows a worker entering an underground conduit that transports wastewaters. Entering confined spaces is not a regular occurrence but is necessary when emergencies arise and for cleaning. When access is necessary, OSHA prescribes a series of steps that must be followed. The most common accident associated with confined spaces is asphyxiation from insufficient oxygen. Toxic gases and vapors may be present also, but insufficient oxygen is the principal cause of death. The first step taken before entering a confined space is to test the composition of the air. Figure 3.8 shows a typical portable meter used to measure air quality. Meters measure the concentration of O_2, CO, H_2S, and possibly other gases, the concentration of which should be compared to the LEL and the PEL. As a regular practice, atmospheric air is blown into the confined space before and during entry, as shown in Figure 3.37. It should be noted that if the blower is powered by a combustion engine, steps must be taken to ensure that the engine's exhaust is not inadvertently drawn into the fresh air being blown into the confined space. If the air in the confined space contains insufficient oxygen, workers must wear respirators equipped with either a portable air supply or an external air supply.

The operational procedure required by OSHA is to equip workers entering a confined space with a harness and tether attached to a tripod straddling the opening of the confined space, as in Figure 3.38. In the event the worker is injured or loses consciousness, he or she can be evacuated without sending another worker into the space. Since workers lose consciousness within minutes of entering a region of deficient oxygen, a second worker entering the confined space to assist an unconscious worker also loses consciousness with the result that both may perish. Workers in major industries are trained in confined space entry, but unfortunately lapses in discipline occur. The situation is far different for small volunteer fire departments, family farms, and small industries where the shortage of

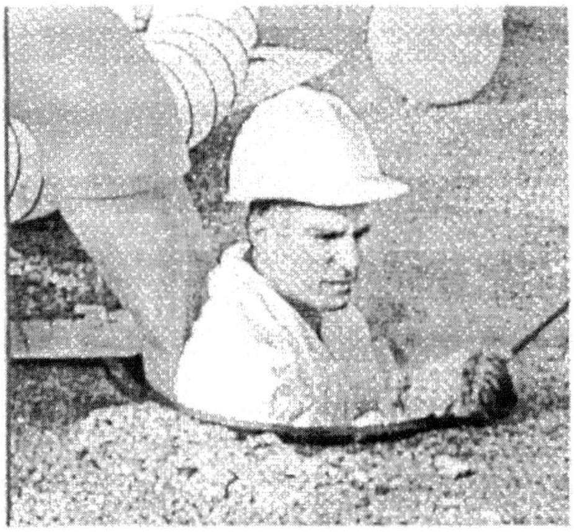

Figure 3.36 Worker entering a confined space (courtesy of Air Systems International).

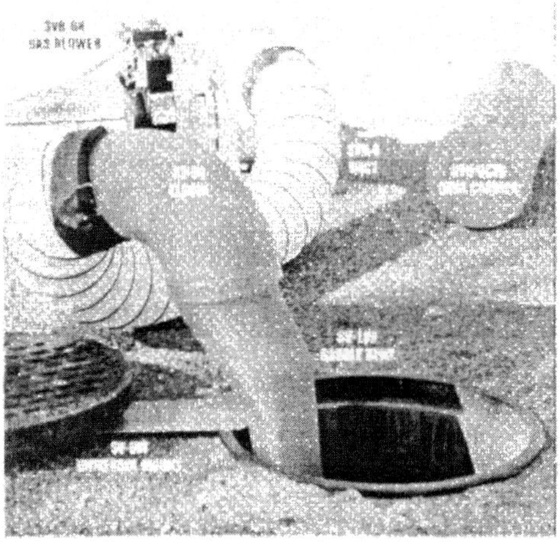

Figure 3.37 Portable ventilation blowers that supply fresh air to a confined space (courtesy of Allegro Industries).

Figure 3.38 Retrieval mechanism and harness used to rescue an unconscious worker from a confined space (courtesy of Gemtor, Inc.).

money causes many to underestimate risk. The events surrounding the need to enter a confined space are often non-threatening and workers do not anticipate insufficient oxygen, e.g. cleaning flooded basements or sump pump pits, unclogging hoppers, cleaning the interior of reactors, etc. On farms, silo unloaders become clogged, manure pumps become clogged, etc.; such incidents have unfortunately caused many unnecessary deaths.

3.9 Engineering Economics

Engineers must be cognizant of the economic ramifications of their design recommendations. There are two classes of costs or expenses that must be clearly distinguished:

- ***Total capital cost*** (TCC) (also called ***total initial cost***) is the initial or ***first cost***. TCC consists of all money spent to design, build, and install an ***air pollution control system*** (APCS). The costs are incurred at the beginning of the project and money must be taken from savings or borrowed to put a system in place. TCC represents all initial expenditures needed to put the system in operation.
- ***Total revenue requirement*** (TRR) is money that must be built into the price of the product to recoup the money spent on TCC, and to operate the APCS. There are both fixed and variable costs associated with TRR, as discussed in more detail below.

TCC is an up-front initial cost borne by the company making the product, while TRR is a cost passed on to the customer. TRR is an additional expense manufacturers must recover and is reflected in the purchase price of the products for which the indoor APCS is required. The objective of this section is to show how these costs can be computed.

The material discussed here follows procedures described by Neveril et al. (1978) and Molburg and Rubin (1983). The reader should also consult established texts on engineering economics, such as Grant and Ireson (1970), for details on capitalization cost, depreciation, capital recovery cost, etc., and Vatavuk (1990) for methods to estimate the costs of air pollution control devices, and utility and basic equipment costs such as:

- equipment (electric motors, fans, pumps)
- energy (electrical power, fresh water, steam, compressed air, natural gas, refrigeration)
- waste disposal (wastewater, sludge, dusts)

3.9.1 Total Capital Cost

TCC consists of ***total direct cost*** (TDC) plus ***total indirect cost*** (TIC).

$$\boxed{\text{TCC} = \text{TDC} + \text{TIC}} \tag{3-56}$$

Table 3.13 shows items to be included in these categories and the amounts (in % of TDC) to be expected. TDC consists of predictable cost to purchase and transport equipment, cost of labor to install the equipment, cost of major materials, engineering consulting fees, and cost of new permanent structures. TIC consists of construction expenses, contractors fees, engineering cost, start-up cost, and interest during construction. TDC is arrived at by negotiations with vendors supplying major pieces of equipment. The cost to install a system depends on the size of the system, whether the system is being installed in a facility under construction or an established facility, and whether pollutants need to be removed before the air is discharged to the outside environment (Neveril, et. al., 1978, Heinsohn and Kabel, 1999). TIC may include internal transfers of funds between divisions of the user's company if assembly and installation of the ventilation system is to be done by the user's personnel. TIC needs to be established by the user. If much of the assembly and installation is done by outside personnel, the associated costs should be included under total direct cost (TDC). Often TIC is represented as a fraction of TDC. The fraction is called the ***indirect cost factor*** (ICF):

Table 3.13 Components of total capital cost (TCC) and total revenue requirements (TRR) (abstracted from Nerveril et al., 1978 and Molburg and Rubin, 1983).

total capital cost, TCC = TDC + TIC = TDC(1 + ICF):	
total direct cost (TDC)	
equipment	
labor	
materials	
structures	
engineering consulting fees	
total indirect cost (TIC)	**indirect cost factor** (ICF = TIC as % of TDC)
construction expense	(10-15%)
contingencies	(5-30%)
contractors fees	(4-5%)
engineering	(4-6%)
interest during construction	(10-25%)
start up costs	(10-15%)
working capital	(2-4%)
total ICF	**(45-100%)**
total revenue requirements, TRR = TVC + TFC = TVC + TCC x FCF:	
total variable cost (TVC)	
administration	
electric, gas, water (utilities)	
maintenance labor	
maintenance material	
operating labor	
raw materials	
supervision	
total fixed cost (TFC)	**fixed cost factor** (FCF = TFC as % of TCC)
capital recovery cost (CRC)	(11-23%) = capital recovery factor (CRF)
taxes	(3-7%)
insurance	(1-3%)
interim replacement	(1-7%)
tax credits	(0-5%)
total FCF	**(16-45%)**

$$\boxed{\text{ICF} = \frac{\text{TIC}}{\text{TDC}}} \qquad (3\text{-}57)$$

which is often multiplied by 100 and listed as a percentage. Using Eq. (3-57), Eq. (3-56) can be rewritten as

$$\boxed{\text{TCC} = \text{TDC}(1 + \text{ICF})} \qquad (3\text{-}58)$$

3.9.2 Total Revenue Requirements

TRR is composed of ***total variable cost*** (TVC), sometimes called ***annual operating cost***, and ***total fixed cost*** (TFC).

$$\boxed{\text{TRR} = \text{TVC} + \text{TFC}} \qquad (3\text{-}59)$$

Examples of fixed and variable costs in TRR are shown in Table 3.13. TVC includes expense to operate the system, cost to provide and heat make-up air, money spent on material and labor to maintain the system, plus associated ***overhead expenses***. Overhead cost depends on how the company is organized and includes the cost to administer and supervise personnel operating the system. In small companies these costs may be small, but in large companies where permanent staff are hired to maintain systems these costs may be substantial. TVC depends on the utilization of the system and the type of contaminant being removed (gas, vapor, or particles), and includes the cost of electricity, water, labor, and raw materials used to capture contaminants. In general, the greater the automation the lower the TVC of labor, supervision, and administration.

TFC includes federal, state, and local taxes, and insurance. Unfortunately, tax laws change rapidly and taxes are difficult to estimate. Installation of an indoor APCS may increase the company's taxable base and hence increase its taxes. On the other hand, purchase of new equipment and installation of air pollution control devices often results in tax credits that reduce the company's taxes. The tax status of the company may change, and the design engineer needs to seek advice from the company tax accountant or tax attorney. Installation of an indoor APCS may also change the cost of insurance. Once again, the design engineer needs to seek advice.

Table 3.13 shows that the major item in TFC is ***capital recovery cost*** (CRC). It is obvious that interest charged by a bank to a company that borrows money to cover TCC is a cost for which revenue must be generated. Suppose however, that TCC was paid entirely from the company's savings. One might assume that an interest charge is now inappropriate. Wrong! By spending its savings, the company has lost the interest that money would have provided had it remained an investment. Thus for purposes of taxes, the ***lost interest*** is a proper annual cost. Lastly the system wears out, and an annual ***depreciation cost*** should be included in TFC. There are many accounting procedures used to compute the cost of interest and depreciation. CRC (Grant and Ireson, 1970; Neveril et al., 1978) can be determined as a fraction of TCC using the ***capital recovery factor*** (CRF),

$$\boxed{\mathrm{CRC} = \mathrm{TCC} \times \mathrm{CRF}} \tag{3-60}$$

where

$$\boxed{\mathrm{CRF} = \frac{\mathrm{i}(1+\mathrm{i})^{\mathrm{t}}}{(1+\mathrm{i})^{\mathrm{t}} - 1}} \tag{3-61}$$

In Eq. (3-61),

- i is the ***annual interest rate*** expressed as a fraction (e.g. i = 0.085 for a percentage interest rate of 8.5%)
- t is the ***capital recovery period*** (system lifetime in years).

For average interest rates of 10% over a recovery period of 10 years, CRF is equal to 0.16275. For a 20-year period, CRF reduces to 0.11746.

A similar but more inclusive (and larger) factor, called the ***fixed cost factor*** (FCF), can be defined in order to calculate TFC as a fraction of TCC,

$$\boxed{\mathrm{FCF} = \frac{\mathrm{TFC}}{\mathrm{TCC}}} \tag{3-62}$$

As can be seen in Table 3.13, since CRC represents the majority of TFC, CRF represents the majority of FCF. In some cases, CRC is used in place of TFC. Using FCF, Eq. (3-59) can be re-written as

$$\boxed{\mathrm{TRR} = \mathrm{TVC} + \mathrm{TCC} \times \mathrm{FCF}} \tag{3-63}$$

Example 3.7 – Total Revenue Requirements of an Indoor Air Pollution Control System

Given: A company manufactures 25,000 electronic circuit boards per year. The circuit boards are prepared by chemical etching processes involving open vessels containing volatile toxic liquids. The open vessels are presently not equipped with any kind of local control system, the general ventilation of the work area is woefully ineffective, and OSHA standards are not being met. An engineering firm has recommended installation of several lateral exhaust systems operated by independent fans. Three lateral exhausts can be purchased ready-made; others will be constructed from PVC sheet by company personnel. Shown below is a list of the materials and equipment to be bought, and the prices negotiated with the vendor. Shown also is total indirect cost (TIC) estimated by the manufacturer for construction and assemblage of the system by company personnel. Total variable cost (TVC) to operate the system has also been estimated. The lifetime of the system is to be 10 years during which time the average interest rate is expected to be 10%.

total capital cost (TCC):	
total direct cost (TDC)	
equipment	
three ready-built PVC lateral exhausters	3000
two 4 HP, 3000 SCFM PVC fans	4000
two 5 HP, 4500 SCFM PVC fans	4500
materials	
160 ft2, 3/8-inch PVC sheet	1000
40 ft x 24-inch diameter, PVC duct	1500
45 ft x 12-inch diameter, PVC duct	500
PVC cement	50
elbows, tees, other fittings	1500
electrical controls	600
TDC total	**$16,500**
total indirect cost (TIC)	
construction expense	2500
contingencies	3500
engineering	1000
TIC total	**$ 7,000**
total revenue requirements (TRR):	
total variable cost (TVC)	
electricity	6000
maintenance labor	400
maintenance material	500
make-up air heating	5000
TVC total	**$11,900**
total fixed cost (TFC)	
capital recovery cost (for 10 years at 10%)	3848
insurance	100
taxes	300
interim replacement	100
TFC total	**$ 4,348**

To do: Estimate the total revenue requirements (TRR) of the new contaminant capture system. What additional revenue per board will be needed to cover TRR?

Solution: Total capital cost (TCC) can be computed from Eq. (3-56),

$$TCC = TDC + TIC = \$16,500 + \$7,000 = \$23,650$$

Total indirect cost (TIC) is only 42% of TDC, somewhat below what is suggested in Table 3.13. Because the system is small and company personnel will undertake much of the construction and installation, it is realistic. TRR can be calculated from Eq. (3-59),

$$TRR = TVC + TFC = \$11,900 + \$4,348 = \$16,248$$

TFC is 18.4% of TCC and within the range suggested by Table 3.13. If 25,000 circuit boards are manufactured each year, \$0.65 should be added to the price of each board to provide the \$16,248 of needed revenue.

Discussion: Eqs. (3-60) and (3-61) were used to find CRC in the above list.

3.9.3 Time Value of Money

Engineers recommend actions that involve the ***time value of money***. The ultimate cost of an undertaking must obviously take into account the initial cost, lifetime (t years) of the device, taxes, and yearly operating and maintenance cost. It must also include the cost of borrowing money at interest rate (i) or the return that would be gained if available money were invested, as discussed previously. To compute the ultimate cost and to compare the ultimate costs of several alternative proposals on an equivalent basis, the following parameters can be used:

- P = ***present worth***: cost, value, or payment at the present time
- A = ***annual*** or ***annualized cost***: cost or payment distributed equally over a number of years
- F = ***future worth***: cost, value, or payment realized at a prescribed time in the future

If a technical recommendation involves considering alternatives in which the lifetime (t) is the same, comparisons can be made using either A, P, or F. If however, the time periods (t) are different, an easily understood basis to establish the equivalency of several alternatives is the annualized cost (A). The relationships between A, P and F can be found in handbooks (e.g. Perry et al., 1984) and texts on engineering economics (e.g. Grant and Ireson, 1970). For interest rate (i) and lifetime (t), the results are summarized below:

- Compound interest:

$$\boxed{\frac{F}{P} = (1+i)^t} \tag{3-64}$$

- Present value of an annuity:

$$\boxed{\frac{P}{A} = \frac{(1+i)^t - 1}{i(1+i)^t}} \tag{3-65}$$

- Future worth of an annuity:

$$\boxed{\frac{F}{A} = \frac{(1+i)^t - 1}{i}} \tag{3-66}$$

Note that Eqs. (3-64) - (3-66) are related to each other as well. For example,

$$\frac{F}{A} = \left(\frac{F}{P}\right)\left(\frac{P}{A}\right)$$

The use of these equations to perform engineering economic analyses can be seen in Examples 3.8 and 3.9.

Example 3.8 - Buy or Lease?

Given: An electronics company needs to install an exotic electronic air cleaner to purify air for a clean room. The air cleaner can either be bought or leased from the vendor. If it is leased, the company pays the vendor an annual leasing fee for 5 years and the vendor provides yearly maintenance and attends to unexpected malfunctions in the equipment for this period. At the end of 5 years, the vendor removes the equipment. If purchased, the company provides it's own maintenance and service but owns the equipment that has a salvage or trade-in value at the end of 5 years. A summary of the terms for these arrangements and the variables they represent are given below:

Lease Agreement:

yearly rental fee = \$3,500 (A)
yearly operating expenses = \$1,000 (A)
period of use = 5 years (t)

Purchase Agreement:

initial purchase price = \$10,000 (P)
yearly operating expense + maintenance = \$2,000 (A)
salvage or trade-in = \$2,000 (F)
period of use = 5 years (t)

To do: If the interest rate (i) during the period is 10%, determine which agreement is the least expensive.

Solution: One basis of equivalence is to compare the future cost (F) at the end of 5 years:

$$\text{F (lease)} = (\$3{,}500)\frac{F}{A} + (\$1{,}000)\frac{F}{A} = (\$4{,}500)\frac{F}{A}$$

and

$$\text{F (purchase)} = (\$10{,}000)\frac{F}{P} + (\$2{,}000)\frac{F}{A} - \$2{,}000$$

For 5 years at 10% interest, Eq. (3-64) yields

$$\frac{F}{P} = (1+i)^t = (1+0.10)^5 = 1.6105$$

and Eq. (3-66) yields

$$\frac{F}{A} = \frac{(1+i)^t - 1}{i} = \frac{(1+0.10)^5 - 1}{0.10} = 6.105$$

Thus,

$$\text{F (lease)} = (\$4{,}500)(6.105) = \$27{,}472 \cong \$27{,}500$$

compared to

$$\text{F (purchase)} = (\$10{,}000)(1.6105) + (\$2{,}000)(6.105) - \$2{,}000 = \$26{,}315 \cong \$26{,}300$$

(The numbers have been rounded off to the nearest \$100 since speculation into the future is certainly no more precise than this.) Thus purchasing is about 4% less expensive than leasing. Alternatively, a comparison can be made on the basis of annualized costs (A). For 5 years at 10% interest, Eq. (3-65) yields

$$\frac{A}{P} = \frac{i(1+i)^t}{(1+i)^t - 1} = \frac{0.10(1+0.10)^5}{(1+0.10)^5 - 1} = 0.2638$$

The annualized cost for the lease option is

$$A\text{ (lease)} = \$3,500 + \$1,000 = \$4,500$$

compared to that for the purchase option,

$$A\text{ (purchase)} = (\$10,000 - \$2,000)\frac{A}{P} + (\$2,000)(0.10) + \$2,000$$
$$= (\$8,000)(0.2638) + \$200 + \$2,000 = \$4,310 \cong \$4,300$$

In the above calculation one annualizes the net price (purchase minus salvage) and then adds the salvage value times the interest rate as an additional annual cost. The additional \$2,000/year represents the yearly operating and maintenance cost. Again, purchasing is the less expensive option by the same 4%. Alternatively, the annual cost can be calculated by annualizing the full purchase price and then subtracting the annualized value of the future salvage payment,

$$A\text{ (purchase)} = (\$10,000)\frac{A}{P} - (\$2,000)\frac{A}{F} + \$2,000$$
$$= (\$10,000)(0.2638) - \frac{(\$2,000)}{6.105} + \$2,000 = \$4,310 \cong \$4,300$$

which yields the same results, i.e. A (lease) = \$4,500 compared to A (purchase) = \$4,300, and purchasing is less expensive.

Discussion: The two analyses illustrate how to compare two different offers on an equivalent basis. Limitations lie with omissions and inadequacies in the assumptions, not in the analyses themselves. For example, the above analysis shows that the future cost of purchasing, F(purchasing), is less expensive than the future cost of leasing, F(leasing), by about \$1200, providing maintenance and repairs cost no more than \$2000 per year. If for any reason serious repairs have to be made, or there is a lack of trained personnel within the company to provide them, the savings could disappear. Leasing removes many of these difficulties since the terms of the contract require the vendor to attend the company's needs upon request. In many small firms where there are no skilled personnel to perform maintenance, the leasing arrangement is clearly more attractive since the plant manager knows that skilled help is always available. Lastly the issue of taxes has been ignored. Leasing provides a yearly business expense that reduces the company's taxable gross income, while owning the equipment increases its capital investment and property taxes but also affords a deduction for depreciation. Since it is uncertain whether the company's tax status may change in 5 years, taxes may have a bearing on the ultimate cost of operation not revealed by the above analysis. Thus while it is relatively easy to compute which alternative is least expensive, many more factors (often non-quantifiable ones) have to be considered to determine which alternative is better.

Example 3.9 - Value of Quality Equipment

Given: A company discharges noxious odors to the environment. The state environmental pollution control agency requires the company to stop. Incineration appears to be the only feasible solution, and the company has applied for a state permit. After consulting vendors, the company learns that incineration produces exhaust products that corrode the incinerator. One vendor proposes an incinerator made of ordinary steel costing \$10,000 that lasts 25 years and has yearly maintenance and operating costs of around \$1,000, while another vendor recommends a stainless steel and ceramic incinerator that costs \$15,000, lasts 50 years, and has yearly maintenance and operating costs \$200 less than those of the incinerator made of ordinary steel. The company president likes first-rate equipment and prefers the stainless steel incinerator.

To do: Compare the two options. Which incinerator should be purchased?

Solution: Since the time periods are different, a useful way to determine which alternative is less expensive is to compare annualized costs (A). Assume also that the interest rate is 10% for both proposals. From Eq. (3-65),

$$\frac{A}{P}\left(25 \text{ years at } 10\%\right) = 0.11017$$

and

$$\frac{A}{P}\left(50 \text{ years at } 10\%\right) = 0.10086$$

Comparing the annual cost for the ordinary steel incinerator and the stainless steel incinerator,

- A (ordinary steel) = (present worth) (A/P, 25 yrs at 10%) + yearly maintenance and operating cost
- A (stainless steel) = (present worth) (A/P, 50 yrs at 10%) + yearly maintenance and operating cost - $200

These equations result in the following:

$$\text{A (ordinary steel)} = (\$10,000)(0.11017) + \$1,000 = \$2,102 \cong \$2,100$$

compared to

$$\text{A (stainless steel)} = (\$15,000)(0.10086) + \$1,000 - \$200 = \$2,313 \cong \$2,300$$

Thus the ordinary steel incinerator is less expensive.

Discussion: The difference in annual cost between stainless steel and ordinary steel is only about $200. From the point of view of expenditure only, the president's preference for high quality costs $200 more a year. The amount is not large, and as in the previous example, the best incinerator to buy should be determined by considering factors not included in the analysis, such as taxes or whether an incinerator will even be needed beyond 25 years. In a situation like this, the president himself/herself should make the final decision, after being shown the cost analysis.

3.10 Closure

There are many design criteria engineers have to face when designing systems to control indoor contaminants. Some criteria are obvious, others are less obvious (if not obscure), but all may be important to the individuals for which the system is designed. Engineers must discuss issues such as those in this chapter with the agency sponsoring the design to clearly establish what operating criteria are to be used to evaluate performance of the system. No one wants additional criteria imposed after the design is finished. It is wise practice for engineers to look up the PEL (Appendix A.20 or the Internet), odor threshold (Appendix A.20), vapor pressure (Appendix A.8), and flammability limits (Table 3.2) *every time* they analyze substances involved in indoor air pollution. In this way they will develop a working knowledge about pollutants and improve their ability to anticipate hazards based on a substance's volatility, odor, toxicity, and flammability.

Chapter 3 - Problems

1. Individuals engaged in occupational safety and health engineering must be familiar with the OSHA General Industry Standard (33). There is no better way of doing this than by locating specific passages in the document to determine whether certain industrial situations are or are not in compliance with the OSHA standard. Study the standard, state the page and identify the part, subpart, section, paragraph, subparagraph, division, and subdivision, etc. of the OSHA standard that applies to each situation described below. Determine whether a violation exists.

(a) Approximately one gallon of a flammable liquid used by a company is kept in an uncovered container on a shelf in the equipment repair room, on the first floor of the building. The flash point of the fluid is 20 °F.
(b) A motor/generator is mounted on an open-sided platform 10 feet above the main floor of a factory. There is no guard railing (or its equivalent) nor is there a toe board to protect workers passing beneath the platform.
(c) Respirators worn by workers in a plant that inadvertently discharges ethylene oxide into the workplace are found to contain cartridges that only capture particles.
(d) A metal stamping firm has not maintained noise exposure records for a period of 2 years.
(e) Full term sampling is conducted of workers operating a dip tank containing soluble salts of chromium. It is found that the 8-hour TWA is 0.9 mg/m^3 at STP.
(f) A worker removing asbestos from a public building is exposed to asbestos particles. A personal monitor shows that the 8-hour TWA for particles greater than 5 micrometers is 16 fibers/cm^3.
(g) A table top circular saw in the university shop does not have a guard that completely encloses the blade when it is operating.

2. Workers use isopropyl alcohol (normal household rubbing alcohol) to clean parts of electric motors prior to their assembly. What is the OSHA PEL for isopropyl alcohol (in PPM)? What is the OSHA PEL in units of mg/m^3 on a day in which the workplace temperature is 5 °C?

3. Workers in a leather finishing firm spray solvents on leather to produce various colors, textures, water proofing, etc. What is the 8-hr TWA-TLV (mg/m^3) of the solvent if it is composed of the following materials (if TLV data are not available, use PEL data from the MSDS for each chemical, available on the Internet).

material	mass fraction (%)
ethyl acetate	15
isopropyl alcohol	25
toluene	25
isobutyl acetate	15
isobutyl alcohol	10
nontoxic materials	10

4. Cotton swabs are used to apply carbon tetrachloride to clean small metal parts prior to applying a finish. The process is such that the concentration of carbon tetrachloride in the workers breathing zone varies with time as a ramp function during an hour-long event,

- $c(PPM) = 5$ for $0 < t < 0.5$ hr
- $c(PPM) = 5 + 95(t - 0.5)/0.5$ for 0.5 hr $< t < 1.0$ hr
- $c(PPM) = 5$ for $t > 1.0$ hr

These hour-long events are repeated throughout the day. Compute a 8-hr TWA. Are OSHA standards violated?

5. [This problem requires use of the CFR] What is the maximum allowable dust concentration (in units of mg/m^3) in a limestone quarry? The particle size distribution is log-normal. Eighty percent (80%) of the particles (by mass) are inhalable (diameter less than or equal to 10 micrometers, and thus able to enter the bronchi). The composition of the inhalable portion of the particle sample and the composition of the entire particle sample are as follows:

inhalable particles	
quartz	8 %
other silicates	5 %
inert material	80 %
total particle sample	
quartz	4 %
other silicates	6 %
inert material	80 %

6. Individuals engaged in occupational safety and health engineering encounter numerous acronyms and medical terms. It is necessary to understand the phrases shown below. Many of these terms can be found in the Merck Manual (334).

(6.1) Consider the acronyms PEL, TLV, 8-hr TWA-TLV, and STEL-TLV. What do the acronyms mean? Identify the organization using the acronym and discuss the similarities and differences in the concepts represented by the acronym.

(6.2) What is homeostasis?

(6.3) What are alveoli?

(6.4) With regard to pulmonary ventilation, what is the tidal volume?

(6.5) Define metabolic rate. Estimate its value for walking.

(6.6) Estimate the volume of blood is the body of an adult.

(6.7) Define the following:

- teratogens
- carcinogens
- mutagens
- allergens

(6.8) Contrast acute and chronic toxicity. Contrast local and systemic toxicity.

(6.9) Contrast hepatoxins and nephrotoxins.

(6.10) Identify the following acronyms for organizations and contrast the functions the organizations perform:

- OSHA
- NIOSH
- EPA
- MSHA

(6.11) Contrast sound level scales A, B, and C.

(6.12) What is the difference between the sound intensity level of a free-field source and the sound pressure level at a distance r meters away?

(6.13) Discuss the major components in the ear canal, middle ear, and inner ear.

(6.14) Discuss ways in which the respiratory system reacts to airborne particles and noxious gases.

(6.15) What is the OSHA 8-hr PEL for trichloroethylene?
(6.16) Discuss the differences between ionizing and nonionizing radiation.
(6.17) What do the acronyms RAD and REM denote? Discuss similarities and differences.
(6.18) Contrast asthma and bronchitis.
(6.19) What is heat stress and why is it a hazard to health?
(6.20) What is isometric work? Contrast it with conventional work associated with raising a weight.
(6.21) Which of the following occupational disorders is most common?

- lower back pain
- bursitis
- tendonitis
- Raynaud syndrome
- carpal tunnel syndrome

(6.22) What is the difference between a syndrome and a disease?
(6.23) Discuss the role of mucous, cilia, macrophage, and lymphatic fluid in the respiratory system.
(6.24) Contrast emphysema and pulmonary fibrosis.
(6.25) Contrast carcinomas, sarcomas, lymphomas and mesotheliomas.
(6.26) What is meant by a dose-response curve?
(6.27) Which of the following occupational diseases are the top three? List in order with the most serious disease first:

- traumatic death, amputations, eye loss
- musculoskeletal injury (back, arthritis, etc.)
- lung disease (cancer, pneumoconiosis, asthma)
- noise-induced hearing impairment
- occupational cancers (other than lung)
- cardiovascular disease

(6.28) Contrast viruses and bacteria.
(6.29) Contrast the scientific fields toxicology, physiology, and epidemiology.
(6.30) What is the physiological response to heat stress?
(6.31) Contrast hypothalamus with hypothermia.
(6.32) Discuss the pathways by which toxic material travels through the body.
(6.33) Discuss individuals who may not be protected by TLVs and PELs.

7. In a factory manufacturing wallpaper, a sample of the air shows the presence of the following hydrocarbons:

material	**concentration (PPM)**
ethyl acetate	80
isopropyl alcohol	100
toluene	10
isobutyl acetate	75
isobutyl alcohol	20

Is this environment in compliance with OSHA standards?

8. An underground tank of methyl isocyanate (molecular weight = 61.25) begins to leak and discharge vapor into the interior of an unheated storage building in which the temperature is –5.0 °C. The initial concentration inside the building is 0.001 PPM and the mass concentration increases linearly with time according to the following equation: $c(t) = c_0 + At$, where A = 0.0033 PPM/hr and c_0 = 0.001 PPM. If a worker is inside the building for 8 hours each day, are OSHA standards met? What is the mass concentration (in units of mg/m^3) at the end of 8 hours?

9. A worker in a refrigerated enclosure (-20 °C, 95 kPa) is repairing a pump used to transport ethylene oxide (M = 44.1). The initial concentration of ethylene oxide in the enclosure is negligible but a leak in the pump seal allows the gas to enter the enclosure and to cause the concentration (measured at -20 °C, 95 kPa) to increase linearly with time at a rate of 5.0 mg/m^3 per hour. What is the concentration at the end of an 8-hour day? Does the worker's exposure over an 8-hour day constitute a violation of OSHA standards?

10. Ethylene oxide (molecular weight 44.1) stored in a holding vessel begins to leak (as a vapor) into an unheated storage building in which the temperature is -5 °C. The initial concentration inside the building is 0.1 PPM and the concentration increases linearly,

$$c(t)\ (\text{PPM}) = 0.1\ (\text{PPM}) + 0.6\ (\text{PPM/hr})\ t(\text{hr})$$

(a) If workers are present in this building for 8 hours a day, what is the 8-hr time-weighted average? Are they within OSHA standards?
(b) What is the concentration in mg/m^3 inside the building at the end of 8 hours?
(c) Can the odor of ethylene oxide be detected after 8 hours?

11. Explain the difference between the lower explosion limit (LEL) and the upper explosion limit (UEL).

12. To illustrate the difference in value between TLV and LEL, compute the ratio of TLV/LEL for the materials listed in Table 3.2.

13. The maximum allowable concentration for inert dust (OSHA standards) is 5 mg/m^3 for respirable dust and 15 mg/m^3 for total dust. To illustrate the difference between TLV and LEL for inert dusts, compute the ratio of TLV/LEL for coal dust, polyethylene dust and oil shale dust shown in Figures 3.17 and 3.18. If OSHA standards are satisfied what are the chances that fire or explosion will occur? Assume that the three types of dust are inert.

14. The dust concentration in many dusty industrial environments is often measured as the number of particles per unit volume of air (c_{number}). Review Figure 3.19. Assuming that the oxygen concentration is 21% by volume and that the density of coal particles is 1200 kg/m^3, re-plot the LEL of coal dust in units of number of particles per cm^3 versus particle size D_p in micrometers for particles between 2 and 100 micrometers.

15. A stamping machine is rated as having a free-field sound pressure level of 75 dB at 10 m. (a) What is the output acoustic power (watts)? (b) What is the sound pressure level (L_p) at a distance 0.5 meters from the machine when the machine is located along an outside wall of the building, roughly in the middle of the wall? (Assume the ceiling is high.) (c) Repeat for a worker 1.5 meters from the machine.

16. Six pedestal grinders are located in a foundry. The noise (sound pressure level) at point P in the room for each machine is found to be 85 dBA, 92 dBA, 90 dBA, 84 dBA, 93 dBA, and 87 dBA. What is the sound pressure level at point P when all six machines are working at the same time?

17. A plant manager buys 5 swing grinders each of which is rated as having an acoustic power output of 0.01 watts. The five machines will be located 5 meters apart midway along the long wall of a building. The building dimensions are 30 m by 100 m by 20 m high. Estimate the sound pressure level of the operator at the enter machine working 0.2 m from his machine.

18. Three identical stamping machines are located in the center of a large building. The layout of the machines is in the form of an equilateral triangle, 2.0 meters on a side. The free-field sound pressure level of each machine is 85. dBA at a distance of 0.10 m from the machine. When all three machines are in use, what sound power level (in dBA) can be anticipated for an operator?

19. Estimate the sound pressure level experienced by an aircraft ground crew chief who guides a 4-engine jet aircraft into it's berth at an airport. Each engine produces 10,000 watts of acoustical power. At the closest point to the aircraft, the crew chief stands directly in front of the aircraft's nose, 10 m from the inboard engines and 20 m from the outboard engines.

20. Repeat problem 19, but with the worker standing 10 m directly in front of one of the inboard engines.

21. Show that Eq. (3-34) can be manipulated to yield Eq. (3-35).

22. Workers are asked to conduct an EPA Method 5 particle sampling test on a tall stack from a power plant. The work is not strenuous but it is tedious and one must be attentive. The workers stand on a narrow platform surrounding the stack over a hundred feet above the ground. They place a long sampling probe at various positions inside the stack and operate equipment that withdraws a gas sample from the stack. The wind speed is 60. FPM and the air temperature is 96. °F. The wet bulb temperature (T_{wb}) is 80. °F and the dry bulb temperature (T_{db}) is 96. °F. The globe temperature is 100 °F. Stack sampling requires a metabolic rate of 500 BTU/hr. It is estimated that the test requires the workers to be on the platform continuously for 4.0 hours. Can this be done without unhealthy heat stress? Is the work any safer if the air speed is 10. ft/sec, air temperature is 90. °F, and the wet bulb temperature is 65. °F?

23. A ventilation system is run by a fan that exposes a worker to 80 dBA when standing 5 meters from the fan. The plant manager wishes to install an identical fan 3 meters on the other side of the worker. To what sound level (dBA) is the worker exposed?

24. It is necessary to replace the ties on a railroad line. Conventional ties, costing $5.00 in-place, have a life of six years. What expenditure per tie is warranted for using a tie containing an improved preservative, if the life of the tie is extended to nine years? Assume interest rates are 10%.

25. Consider Example 3.8. What is the yearly maintenance cost at which leasing and purchasing are equivalent?

26. A floor in a food plant is designed for ease of cleaning and non-skid characteristics. The present floor surface costs $1500 per hundred square feet and has to be replaced every 5 years. A new floor material has been proposed that costs twice as much. In terms of annual total revenue required, how long must the new material last to warrant the larger expenditure if the interest is (a) 10%, (b) 7.5%?

27. Two 25 HP electrical motors are being considered for purchase. The first costs $200 and has an efficiency of 85%. The second costs $150 and has an efficiency of 82%. All annual expenses, such as depreciation, insurance, maintenance, etc., amount to 15% of the original cost annually. How many hours of full-load operation per year are necessary to justify purchasing the more expensive motor if electricity costs (a) $0.051 per kW-hr, (b) $0.065 per kW-hr, (c) $0.084 per kW-hr?

28. An incinerator is to be installed in a plant that "coats" (applies color, texture, etc.) leather used in wearing apparel, book binding, upholstery, etc. The process emits hydrocarbons and gummy particles that can be removed only by thermal (direct flame) incineration. The following costs are contained in a quotation from the incinerator manufacturer:

base unit price:	$1,055,000
(incinerator and heat exchanger, fans, ducts, fans, motors and drives, flow control valves, burner controls, hydraulics, service platforms, control cabinet, painting, drawings, operating instructions)	
operating costs:	
operation	2500 hr/year
exhaust volume	26,500 SCFM
inlet exhaust temperature	70 °F
contaminant rate	120 lbm/hr
incineration temperature	1500 °F
incineration residence time	0.5 seconds
percent thermal energy recovery	95%
natural gas consumption rate	1.383 million BTU/hr
energy costs:	
fuel cost (estimated)	$7.00/million BTU
fan electric motor (80% efficiency)	300 HP
other electrical drives	10 HP
yearly maintenance	$2000
electricity	$0.051/kW-hr
other initial costs borne by company:	
labor to erect	$5,000
concrete foundation	$35,000
consulting engineering fees	$10,000
legal fees	$10,000
permits	$5,000
connections for utilities	$10,000
temporary shelter for materials	$8,000
yearly taxes and fees (state and local):	$7,500

Assume that the incinerator will operate for 25 years. Estimate the additional cost per square foot of leather that will be passed on the customers if the plant's production rate is one million square feet of leather per year. Assume an interest rate of 10%.

29. An electrical motor is needed to operate a ventilation system. The features of two available motors are listed below. The total variable cost for maintenance is 15% of the total capital cost and the cost of electricity is $0.050 per kW-hr. If the indirect cost factor is 45% and interest is 10%, what is the total revenue requirement to operate the motors for 2500 hrs?

motor number 1	motor number 2
efficiency = 85%	efficiency = 80%
lifetime = 10 yrs	lifetime = 5 yrs
initial cost = $300	initial cost = $150
salvage = 0	salvage = 0

30. A worker inadvertently leaves uncovered a can containing a mixture of waste solvents stored in an unvented (OSHA violation) paint storage room. The waste solvent consists of equal mol fractions of butyl acetate, toluene and acetone. A large amount of the solvent evaporates filling the air in the storage room with solvent vapors. A worker enters the room to use a hand drill that is not explosion-proof. The air contains 80 PPM of each of the solvents.

(a) How many kilograms of each liquid solvent are there if the mixture fills a 5-gallon can? The specific gravity of acetone, butyl acetate, and toluene are 0.792, 0.882, and 0.866 respectively.
(b) What is the average TLV (in units of mg/m^3) that can be assigned to the liquid mixture?
(c) Is the vapor-air mixture "safe" by OSHA standards?
(d) Can you expect the worker to smell any of the solvents? How could the worker characterize the odor?
(e) Is there apt to be an explosion from sparks produced by the motor in the hand drill?

31. To control dust generated in a furniture factory, all machines are equipped with local ventilation systems that direct dust and air to a baghouse (fabric filter air cleaner). The air also contains a variety of hydrocarbon vapors generated throughout the plant. What is the likelihood that this aerosol (dust particles suspended in a vapor-laden air stream) will burn or explode, and if so, is it the dust or hydrocarbon vapor that is the cause? The following conditions exist inside the baghouse:

- hydrocarbons:

  enthalpy of combustion = 400 kcal/gmol
  total hydrocarbon concentration (at STP) = 1,000 mg/m^3
  typical molecular weight = 100

- dust:

  assume combustion properties of Pittsburgh bituminous coal
  concentration = 150 g/m^3
  mean particle diameter (D_p) = 20 μm

32. Dollops of molten glass fall into molds on a rotating table and are pressed into television picture tubes. There are many radiating heat transfer surfaces, and workers (press molders) must be fully clothed even though they are bathed by a stream of clean air. Press molding is stressful. The environment is hot and noisy and the workers must be alert to what is going on. To improve working conditions the union suggests that the company adopt an administrative control strategy to rotate job assignments so that workers spend no longer than 20 minutes per hour as press molders. Your supervisor asks you to estimate the heat stress index (HSI) of press molding and respond to the suggestion. The following conditions exist for press molders:

- air speed: U_a = 1.5 m/s
- temperatures: $T_a = T_{db}$ = 30.0 °C, T_{wb} = 27.0 °C, T_G = 60.0 °C
- net metabolic rate: M-W = 500. kcal/hr

33. Figure 1.24 shows five open vessels, each 6 feet wide, 4 feet high, and 30 feet long. The distance between each vessel is 8 feet and each vessel is equipped with a lateral suction exhauster. Each exhauster has a 15,000 ACFM fan at the center of the near end of the vessel. The sound power level (L_w) of each fan is 80 dBA. Where would a 6-foot worker stand to be subjected to the largest sound pressure level (L_p)?

34. Restaurant chefs and food preparers work at a rapid pace 8 hours a day. There are many hot, radiating surfaces. The stoves are equipped with hoods similar to the canopy hoods shown in Figure 1.27. City health regulations require the workers to be fully clothed. The environment is hot, humid, and noisy. Measurements of the air in the kitchen are as follows:

- air speed: 0.30 m/s,
- relative humidity: 50%,
- air temperature: $T_a = T_{db} = 30.0$ °C,
- globe temperature: $T_G = 38.0$ °C,
- net metabolic rate (M-W): 200. kcal/hr

What is the heat stress index (HSI)? Do these conditions produce unhealthy heat stress?

35. A military maintenance facility repaints vehicles and discharges the solvent vapors directly to the atmosphere. To comply with state air pollution regulations, a consulting firm recommends purchase of a new paint booth (see Figure 1.21) equipped with a solvent capture device. The cost of two systems that can be permitted are as follows:

system A: booth and a direct flame thermal oxidizer
first cost: $50,000
life: 20 years
salvage value: $10,000
annual O&M disbursements: $9,000

system B: booth and an activated charcoal adsorber
first cost: $120,000
life: 40 years
salvage value: $20,000
annual O&M disbursements: $6,000

The annual O&M (operating and maintenance) disbursements include operation, maintenance, property taxes, and insurance. There is an extra annual disbursement for income taxes for system B (but not system A) of $1,250/year. Assume the interest rate is 8%.

(a) Compare the equivalent uniform annual disbursement for systems A and B.
(b) What is the present worth of net disbursements for 40 years for systems A and B?

36. Estimate the daily requirement of food, in kilocalories, for a 130-pound woman who does the following in a typical 24-hour period:

- 7 hours — sleep
- 9.5 hours — seated at rest (reading, attending meetings, watching TV, etc.)
- 2.5 hours — walking about the office
- 0.5 hours — housecleaning
- 1 hour — medium level of exercise such as tennis
- 0.5 hours — walking
- 3 hours — miscellaneous activities (cooking, eating, showering, etc.) which average to a metabolic rate of 200 W

37. It is mentioned in the text that the lower explosion limit (LEL) is generally much higher than the permissible exposure limit (PEL). Verify that this is true by comparing the LEL to the PEL for the following chemicals. Convert all values to PPM for consistency. Note: Use MSDS data on the Internet to obtain the most recent values of PEL.

(a) carbon disulfide

(b) formaldehyde
(c) ethyl mercaptan
(d) propane
(e) sewer gas

38. Fred likes to relax in the sauna at the local health club. The air temperature in the sauna is 60. °C, and the relative humidity is 20.%. The globe temperature is estimated to be 55. °C. There is not much air movement in the sauna, but natural convection generates an average air speed of about 0.30 m/s in the room. Consider Fred to be an average healthy male, seated and at rest in the sauna, with only a swimming suit on (K = 0.9).

(a) Estimate Fred's heat stress index (HSI) in percent.
(b) Should Fred stay in the sauna very long? Explain.

39. A leaf-blower engine is tested in an anechoic chamber to simulate a noise source in free space. The free field sound level is 85. dB at a distance of 1.5 m from the engine. Professor Cimbala uses the leaf blower to clean the leaves off of his driveway. Estimate the sound pressure level in decibels, assuming his ears are located 0.60 m from the engine.

40. On the MSDS sheets on the Pocket Guide on the Internet, there is a listing of something called "Fl.P".

(a) Look up what the abbreviation "Fl.P" means, and provide a definition.
(b) Consider benzene. What is the "Fl.P" of benzene? What precautions should be taken around tanks of liquid benzene to prevent explosion?

41. An ammonia detector is adjusted to set off an alarm at a mol fraction of 30. PPM. Ammonia leaks into an air stream in a pipe. The air is at 80. °C and 200. kPa.

(a) Calculate the mass concentration of ammonia (in mg/m^3) at which the alarm will sound.
(b) Does this concentration exceed the PEL?
(c) For comparison, what is the lower explosion limit (LEL) for ammonia in PPM? (Hint: Obtain this value from the MSDS for ammonia.)

4

Estimation of Pollutant Emission Rates

In this chapter you will learn how pollutant emission rates are estimated:

- physical measurements and empirical equations
- mass balances
- emission factors

The ability to prevent and/or capture pollutants is no better than the degree to which engineers can describe and estimate the generation and emission of pollutants. A distinction must be made between pollutant generation rate and pollutant emission rate. ***Generation rate*** refers to the rate at which pollutants are *created* by some process (combustion, evaporation, grinding, etc.). ***Emission rate***, on the other hand, refers to the rate at which pollutants actually *enter* (are emitted into) room air. Generation rate and emission rate are not necessarily equivalent since not all of the generated contaminants enter the room air. Some contaminants may be absorbed into liquids, adsorbed onto surfaces, removed by filters, etc. If there is no pollution removal system whatsoever, the source of pollution is said to be ***uncontrolled***, and the generation rate and emission rate are the same. In this book, the phrase ***source strength*** (S) is used synonymously with emission rate, and has dimensions of mass per unit time. Typical units are mg/min, kg/s, etc.

Indoor air quality engineers need to: (a) know the actual emission rate and how it varies over the surface of a polluting source, and (b) understand the physical relationships governing the rate of emission. Some factors of importance for the two main categories of air pollution are listed below:

Gases and Vapors

- how the source strength varies with location and time
- chemical composition of the pollutant (molecular species that comprise the contaminant mixture and the mol fraction of each species)
- local contaminant concentration at an interface between the source and the air

Particles

- how the source strength varies with location and time
- particle shape, density, and size distribution
- initial particle velocities, and how they vary with particle size, location, and time

It is important to choose consistent symbols and notation to represent physical properties. In this chapter (and throughout the book) the following conventions are adopted:

- molecular species: In the development of general relationships between species in a gas mixture, molecular species are designated with a subscript such as "j" or "k" (e.g. c_j), when necessary to distinguish from other species. If only one species is under consideration, the subscript is dropped for brevity.

- *Thermodynamic properties*: For a mixture of gases and liquids, phase diagrams (P-v, P-T, T-s) for a component (j) are valid if P is the ***partial pressure*** (P_j) of molecular species j. To avoid confusion between a functional form and multiplication, the partial pressure at a particular temperature, say T_1, is denoted as P_j(at T_1), as shown in Figure 4.1.
- *vapor pressure* or *saturation pressure*: The maximum partial pressure of species j is the pressure of the saturated vapor ($P_{sat,j}$), also known as vapor pressure ($P_{v,j}$), which is a function of T, and is shown as $P_{v,j}$(at T_1), in Figure 4.1. Throughout this chapter only the phrase *vapor pressure* is used. At a given temperature (T_1), the partial pressure of species j is always less than or equal to the vapor pressure at that temperature, i.e. $P_j(\text{at } T_1) \leq P_{v,j}(\text{at } T_1)$. The vapor pressure is tabulated for many pure chemical species (Perry et al., 1984 and thermodynamics textbooks); P_v is listed in Appendix A.8 for several species.
- *interface*: the interface between liquid and gas phases is designated by the subscript "i".
- *far-field properties*: physical properties in the far field are designated by the subscript "0", e.g. c_0, or "∞", e.g. c_∞.
- *initial values*: the initial value of a variable which is a function of time, such as the mass concentration, c(t), is designated as c(0).
- *mass concentration*: The mass concentration of species j is designated as c_j. When it is clear that only one species of contaminant is under consideration, the subscript "j" is dropped, and mass concentration is denoted simply as c.
- *molar concentration*: The molar concentration of species j is designated as $c_{molar,j}$, or simply as c_{molar}.
- *molar density*: The molar density of a liquid is designated as $c_{molar,L}$. The total molar density of a gas mixture is designated as $c_{molar,t}$. In the liquid phase, ***molar density*** is defined as $c_{molar,L} = (\Sigma n_j)_{liquid}/V$, while in the gas phase, $c_{molar,t} = (\Sigma n_j)_{gas}/V$, where n_j represents the number of mols of species j in the appropriate phase.

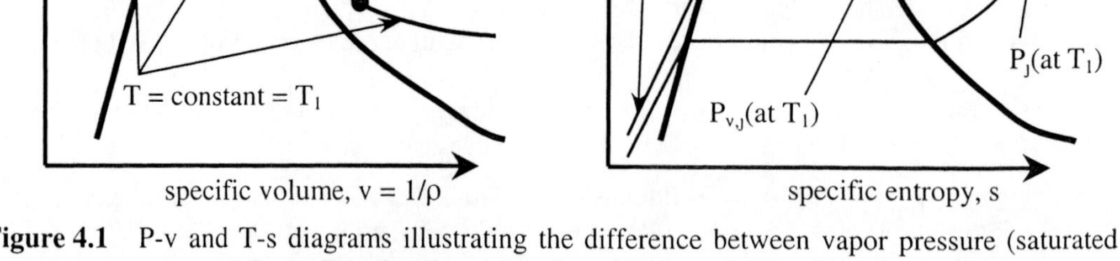

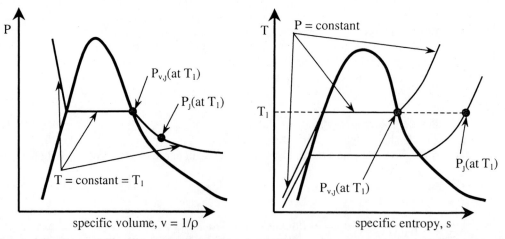

Figure 4.1 P-v and T-s diagrams illustrating the difference between vapor pressure (saturated pressure) $P_{v,j}$(at T_1) of pollutant j and partial pressure P_j(at T_1) of pollutant j.

4.1 Experimental Measurements

The most useful information about pollution emission is what can be measured at the source. Unfortunately, these measurements may be difficult, may be expensive, and may require equipment not readily available. One should keep in mind that the purpose of these measurements is to obtain input data for design computations and thus one should use techniques that provide accuracy commensurate with this objective. Determining particle size, gas composition, interface mol fraction, etc., involves sophisticated equipment and sampling techniques that are difficult to master and that have to be adapted to each source.

Measurement of the actual emission rate is not difficult in principle. For example, one can directly measure the mass lost by evaporation or the mass of material removed by grinding over a period of time,

$$\boxed{\dot{m} = \frac{m(t_2) - m(t_1)}{t_2 - t_1}} \tag{4-1}$$

However, the accuracy obtained with Eq. (4-1) is apt to be poor since one may be subtracting two quantities of nearly equal value in the numerator. However, Eq. (4-1) has been used with success to estimate the emission rate of methylene chloride (CH_2Cl_2), of hydrocarbons from aerosol spray cans for paint stripping and removing, and for applications of metallic aerosol finishes and clear polyurethane finishes (Girman and Hodgson, 1986).

In the case of evaporation of volatile liquids, one must also realize that the evaporation rate is dependent on the speed of air passing over the tank or pool. In particular, the evaporation rate of a volatile liquid is smaller if the air is quiescent than if the air is moving. Lastly Eq. (4-1) yields the average mass emission rate during the period (t_2 - t_1). The actual mass emission rate at any instant might vary over the period t_1 to t_2.

There are many instances in which pollutants are generated by a process and discharged as a well-defined exhaust stream. Such emissions are called ***point source emissions***. Emissions that are generated by a continuous set of points distributed along a line are called ***line source emissions***. Emissions that are distributed widely over an area and originate from numerous ill-defined locations are called ***fugitive emissions*** or ***area source emissions***.

Consider for the moment point source emissions. The ***mass emission rate*** ($\dot{m}_j$) of molecular species j across some area (A) can be found by integration of mass concentration (c_j) and velocity ($\vec{U}$) over the cross-sectional area:

$$\boxed{\dot{m}_j = \int_A c_j \vec{U} \cdot d\vec{A} = \int_A c_j U dA} \tag{4-2}$$

where U is the magnitude of velocity normal to the differential area. If the concentration is uniform across the cross-sectional area of the process gas stream, the mass emission rate is

$$\boxed{\dot{m}_j = Q c_j} \tag{4-3}$$

where Q is the actual volumetric flow rate of the process gas stream (in units such as ACFM), computed from the actual temperature (T) and actual pressure (P) of the gas stream. Alternatively, in terms of molar concentration,

$$\boxed{\dot{m}_j = Q M_j c_{molar,j}} \tag{4-4}$$

where M_j is the molecular weight of species j and $c_{molar,j}$ is the measured molar concentration, which is the number of mols of molecular species j per unit volume of exhaust gas at temperature

(T) and pressure (P). It is not necessary to adjust the values of c_j (or $c_{molar,j}$) and Q to STP. If standard volumetric flow rates are to be used, however, one must use values of c_j (or $c_{molar,j}$) in Eqs. (4-3) and (4-4) that have also been adjusted to standard temperature and pressure..

Example 4.1 - CO Emission Rate from an Internal Combustion Engine
Given: An internal combustion engine powers a fork-lift and discharges CO inside a warehouse. The CO concentration is measured to be 300. PPM in an exhaust stream leaving the engine at 80.0 °C and 1.0 atm. The exhaust duct diameter is 3.0 cm, and the average gas velocity is 0.355 m/s.

To do: Compute the actual volumetric flow rate of exhaust gas and the mass flow rate of CO.

Solution: The actual gas volumetric flow rate is

$$Q = UA = U\frac{\pi}{4}D^2 = 0.355\frac{\text{m}}{\text{s}}\frac{\pi}{4}(0.030\text{ m})^2 = 2.509\times10^{-4}\frac{\text{m}^3}{\text{s}} \cong 0.00025\frac{\text{m}^3}{\text{s}}$$

The mass concentration of CO can be calculated from Eq. (1-31),

$$c_{CO} = \frac{(PPM)_{CO}M_{CO}}{24.5}\left(\frac{P_{actual}}{P_{STP}}\right)\left(\frac{T_{STP}}{T_{actual}}\right) = \frac{300.(28)}{24.5}\left(\frac{1\text{ atm}}{1\text{ atm}}\right)\frac{298.15\text{ K}}{(80.0+273.15)\text{K}} = 289.5\frac{\text{mg}}{\text{m}^3} \cong 290\frac{\text{mg}}{\text{m}^3}$$

The emission rate of CO is then

$$\dot{m}_{CO} = c_{CO}Q = 289.5\frac{\text{mg}}{\text{m}^3}\left(2.509\times10^{-4}\frac{\text{m}^3}{\text{s}}\right) = 0.07264\frac{\text{mg}}{\text{s}} \cong 0.073\frac{\text{mg}}{\text{s}}$$

Discussion: If desired, one could also calculate the *molar* concentration ($c_{molar,CO}$) by dividing the mass concentration by the molecular weight of CO,

$$c_{molar,CO} = \frac{c_{CO}}{M_{CO}} = \frac{289.5\frac{\text{mg}}{\text{m}^3}}{28.0\frac{\text{g}}{\text{mol}}}\left(\frac{\text{g}}{1000\text{ mg}}\right) = 1.034\times10^{-2}\frac{\text{mol}}{\text{m}^3} \cong 0.010\frac{\text{mol}}{\text{m}^3}$$

The final answers are given to two significant digits since some of the variables are known to an accuracy of only two digits.

Often the velocity and concentration in a process gas stream are not uniformly distributed across the duct and it is necessary to integrate Eq. (4-2) by numerical methods. For example, fans, elbows, and tees located upstream of a sampling port create distorted velocity profiles. Also, the process that generates pollutants may not operate steadily. Even after measurements have been made, it may be necessary to report data on the basis of standard temperature and pressure, or to correct the data for standard amounts of water vapor or carbon dioxide, etc. Under these conditions care must be taken to acquire and report data in ways acceptable (if not dictated) by the agencies requesting the information.

The EPA has promulgated ***EPA Reference Methods*** for analyzing process gas streams and reporting data. The methods are described in 40 CFR Part 60 and the AMTIC and EMTIC bulletin boards of the TTNBBS. Readers should learn which method is used to measure certain emissions. The abridged titles of these methods are listed below.

Method 1 - Velocity traverses (profiles) for stationary sources
Method 2 - Stack gas velocities and volumetric flow rates
Method 3 - Gas analysis for CO_2, O_2, excess air, dry molecular weight

Method 4 - Moisture content in stack gases
Method 5 - Particulate mass emissions
Method 6 - SO_2 emissions
Method 7 - Nitrogen oxide emissions
Method 8 - H_2SO_4 mist and SO_2 from stationary sources
Method 9 - Opacity of visible emissions
Method 10 - CO emissions
Method 11 - H_2S emissions
Method 12 - Inorganic lead emissions
Method 13&14 - Total fluoride emissions
Method 15 – H_2S, carbonyl sulfide (COS), and carbon disulfide (CS_2) emissions
Method 16 - Total reduced sulfur emissions from stationary sources
Method 17 - Particulate emissions using in-stack filtration methods
Method 18 - Measuring gaseous organic compounds using gas chromatography
Method 19 - SO_2 removal efficiency
Method 20 - Nitrogen oxides, SO_2, and dilute emissions from gas turbines
Method 21 - VOC leaks
Method 22 - Opacity of fugitive emissions from material sources and flares
Method 24 - VOC, H_2O, density, solids fraction of surface coatings
Method 25 - Total gaseous nonmethane organic emissions
Method 27 - Vapor tightness of gasoline tanks
Method 28 - Certification and auditing of wood heaters

Many of the methods above are really a family of methods (each of which is distinguished from other methods in the same family by addition of a suffix to the method number) that pertains to specific devices or processes. Engineers conducting field measurements must be proficient in these methods. Professional organizations such as ASTM, ACGIH, ASHRAE, and professional organizations in other nations have established similar procedures that engineers may be asked to follow.

4.1.1 Mass Balance

There are occasions when pollutants are formed in a predictable fashion. In these situations the pollutant mass flow rate can be estimated from a simple mass balance. Some examples follow.

VOC from surface coating: When metal, wood, paper, textiles, etc. are coated, the coating contains solvents used to convey the pigment, solids, etc. to the surface. The solvent may be water (water-based coatings) or it may be a ***volatile organic compound*** (VOC) ("oil-based coatings). Virtually all of the solvent leaves the coating between the time it is applied (spraying, dipping, etc.) and when it has dried. In terms of the instantaneous emission rate, it is important to know the rate at which solvent evaporates, but in terms of the total amount of solvent discharged, engineers need to know only the amount of solvent in the coating as it was applied and the rate (kg/s) at which the coating was used. The key element that makes this elementary analysis valid is the knowledge that the coating retains a negligible amount of solvent; thus all the solvent in the coating ultimately enters the air.

Halogenated hydrocarbons: As a first approximation, halogenated hydrocarbons do not react or become absorbed or adsorbed on particles in the troposphere. Thus there is no depletion or production of them in the troposphere. The yearly emission rate can be estimated by the masses of each that are produced in a year minus the corresponding masses that industry recovers per year.

Carbon dioxide from stationary and mobile sources: The carbon in hydrocarbon fuels consumed by stationary sources and mobile sources such as automobiles, buses, trucks, railroads, aircraft, etc. is eventually converted to CO_2. Even unburned hydrocarbons and CO eventually react in the atmosphere to form CO_2. To a first approximation, then, the average yearly rate at which CO_2 is generated can be estimated from the mass of carbon in the fuels times the ratio of molecular weights (approximately 44/12) representing its conversion to CO_2,

$$\boxed{C + O_2 = CO_2} \tag{4-5}$$

If other air pollutants enter the atmosphere directly without reacting, their emission can be estimated in ways similar to these. Unfortunately there are countless pollutants whose emission rate depends on a complex chemical reaction mechanism containing many intermediate chemical species that cannot be analyzed as easily. Thus the emission rates of particles, oxides of nitrogen, and many ***hazardous air pollutants*** (HAPs) have to be estimated by other means.

As an interesting side note related to automobile emissions, Holmen et al. (2001) report that particles attributable to roadways are more likely to be resuspended roadway dust, e.g. particles of tires, brakes, asphalt, and aged tail pipe emissions, rather than combustion products from tailpipes. Furthermore, trucks generate three times more roadway dust than do cars, even though there are approximately 20 cars for every truck on the road.

Example 4.2 - VOC Emission from a Can of Paint

Given: A can contains 4.6 kg of paint, half of which is a VOC, and the remaining half of which is composed of pigments, binders, and other nonvolatile materials. The entire can is used to paint a bedroom.

To do: Compute the mass of the VOC emitted to the room, and ultimately to the atmosphere.

Solution: When the paint has dried, a negligible amount of the VOC solvent remains on the wall. In other words, virtually *all* of the VOC eventually evaporates into the air in the room, and ultimately into the atmosphere. Since the VOC accounts for half of the 4.6 kg of paint, the amount of VOC emitted to the atmosphere is

$$m_{VOC} = \frac{m_{paint}}{2} = \frac{4.6 \text{ kg}}{2} = 2.3 \text{ kg}$$

Discussion: Much of the VOC evaporates during the painting process, which is why one should always have adequate ventilation while painting; the rest evaporates much more slowly as the paint dries, which is why a freshly painted room still smells of paint after a couple days. Ultimately however, *all* of the VOC is emitted to the atmosphere.

4.1.2 Flux Chambers

Leakage of gases and vapors from valve packing, pump seals, flange gaskets, etc., escape of vapors or gases from contaminated soil, or evaporation of vapor from the free surface of a liquid are difficult quantities to measure directly (Shen, 1982). A simple experimental method (Figure 4.2) for estimating these rates is to enclose the source in an air-tight container of volume V, pass a clean stream of air through the container at a measured volumetric flow rate Q_a, mix the contents of the chamber thoroughly, and measure the concentration of the gas or vapor either inside the container or leaving the container. Enclosures used to make these measurements are called ***flux chambers*** (Eklund, 1992).

If the emission source strength (S, typically in mg/s or kg/s) is constant (e.g. leaks in packing or flanges), the user should maintain a constant volumetric flow rate of ambient air (Q_a) and either monitor the mass concentration, c(t) or record its steady-state value (c_{ss}) with an appropriate gas analyzer. If S decreases with time (e.g. solvent evaporation from painted surfaces), the user may want to isolate the chamber ($Q_a = 0$) and merely record c(t) inside the chamber. In either event, the source strength can be determined by writing a mass balance for the escaping fluid: the rate of change of mass inside the chamber must equal the source strength (rate of mass emission), plus the rate at which mass enters the chamber, minus the rate at which mass leaves the chamber. Mathematically,

$$\boxed{V\frac{dc}{dt} = S + Q_a c_a - Q_a c} \tag{4-6}$$

where V is the volume of the flux chamber, M is the molecular weight of the contaminant, and c_a is the mass concentration of the contaminant in the outside ambient air. For brevity, no subscript (such as "j") is used here following the convention established previously, since only one contaminant is under consideration at any given time.

Measuring the steady-state mass concentration: If steady-state conditions ($c = c_{ss}$ = constant) are reached within a reasonable time, the source strength (S) can be found by setting the left hand side of Eq. (4-6) to zero. If the volumetric flow rate (Q_a) is controlled and the values of c_{ss} and c_a are measured, a constant source strength can be computed from

$$\boxed{S = Q_a\left(c_{ss} - c_a\right)} \quad . \tag{4-7}$$

If the escaping gas or vapor is adsorbed by the walls of the container, Eq. (4-6) must be modified to incorporate wall losses. The experiment should be repeated for several values of Q_a to ensure that Eq. (4-7) is valid. No inaccuracy will occur if small values of Q_a are used (so long as Q_a can be measured accurately) but there is a danger that if Q_a is too large, air velocities sweeping past the source may influence the rate of emission, S, of the escaping gas or vapor. Lastly, large values of Q_a produce small values of c_{ss} that may be below the low-limit accuracy of the gas measuring equipment.

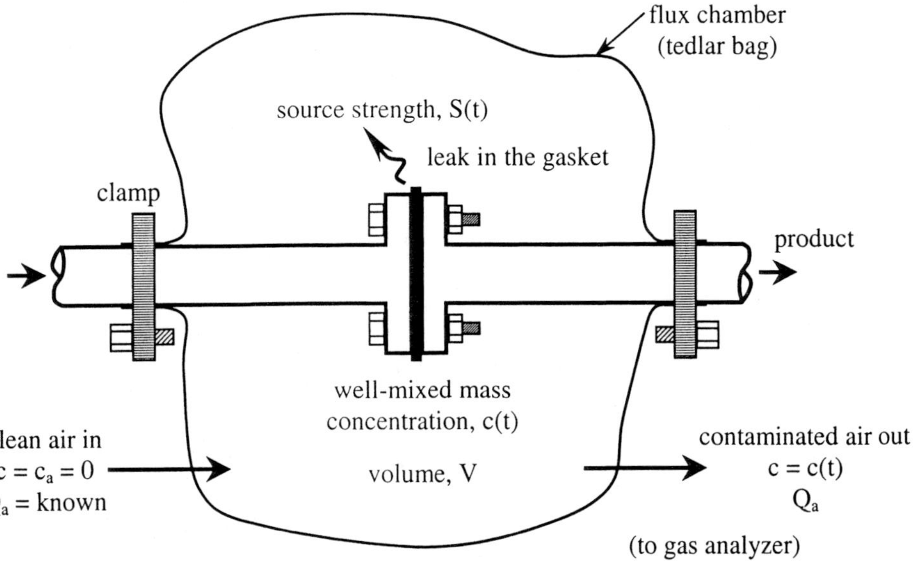

Figure 4.2 Simple tedlar bag flux chamber used to measure leak rate in a flange and gasket.

Measuring the instantaneous concentration: Consider a flux chamber of volume V_{iso} which is isolated from the outside air ($Q_a = 0$), but the source strength S(t) varies with time. Equation (4-6) reduces to

$$\boxed{V_{ISO} \frac{dc}{dt} = S(t)} \tag{4-8}$$

At time t = 0, just before the source is turned on, the mass concentration in the flux chamber is the same as that of the outside air, $c(t=0) = c_a$. Thus, the variables c and t in Eq. (4-8) can be separated and then integrated from t = 0 to t, which corresponds to c varying from c_a to c(t):

$$\boxed{\int_{c_a}^{c(t)} dc = c(t) - c_a = \int_0^t \left(\frac{S}{V_{ISO}} \right) dt} \tag{4-9}$$

From either Eq. (4-8) or (4-9), the user can determine the source strength (S).

Example 4.3 - Emission Rate of Legionella Bacteria from a Humidifier Reservoir

Given: A reservoir for a building's HVAC humidifier is located on the roof of the building. The water is agitated slightly, and atmospheric air passes over the top of the reservoir. Unfortunately the aeration brought on by agitation coupled with solar radiation enables Legionella pneumophila bacteria to breed in the water, and some of the bacteria enter the building air. (This microorganism and the disease it causes are discussed in Chapter 2.) Experiments are conducted to measure the number of bacteria per unit time emitted from the surface of the water. A flux chamber that is 1.0 m high in the center of the reservoir, and has a cross-sectional area equal to 4.0 m^2 is constructed. The chamber is covered with a transparent material that allows the water to receive the same amount of sunlight as the rest of the reservoir. The water temperature and air humidity inside the flux chamber are nearly the same as those outside the flux chamber. Ambient air is drawn into the chamber at volumetric flow rate $Q_a = 1.50$ m^3/min. The incoming air contains assorted ambient particles (e.g. dust, bacteria, etc.). The inlet mass concentration (c_a) of Legionella pneumophila bacteria is 0.20 mg/m^3. The mass concentration is measured in the air exiting the chamber. A fan produces well-mixed conditions inside the flux chamber. The following are measurements of the exiting particle mass concentration (c) versus time:

t (minutes)	c (mg/m^3)
0	0.20
1.0	5.44
2.0	9.88
3.0	13.54
4.0	16.52
5.0	18.96

t (minutes)	c (mg/m^3)
6.0	20.96
7.0	22.60
8.0	23.94
9.0	25.04
10.0	15.94
15.0	28.50

t (minutes)	c (mg/m^3)
20.0	29.44
30.0	29.92
60.0	29.98
90.0	29.99
120.	30.00
150.	30.00

To do: Estimate the ***source strength per unit area*** (s) from the reservoir.

Solution: From the data above, it can be seen that the steady-state mass concentration is 30.0 mg/m^3. From Eq. (4-7), the particle emission rate inside the flux chamber is

$$S = Q_a \left(c_{ss} - c_a \right) = 1.50 \frac{m^3}{min} (30.0 - 0.20) \frac{mg}{m^3} = 44.7 \frac{mg}{min}$$

The source strength per unit area (s) is simply the value of S divided by the surface area of the water surface covered by the flux chamber,

$$s = \frac{S}{A} = \frac{44.7 \frac{mg}{min}}{4.0 \text{ m}^2} = 11.2 \frac{mg}{m^2 \text{ min}} \cong 11. \frac{mg}{m^2 \text{ min}}$$

Discussion: The result is reported as source strength per unit area (s). To compute the actual source strength (S), s must be multiplied by the total surface area of the water in the humidifier.

4.2 Empirical Equations

While not possessing the sophistication researchers may desire or the detail designers may desire, empirical expressions enjoy widespread industrial use and offer an expedient way to characterize the source. Empirical equations express contaminant emission in terms of easily measurable quantities. The expressions often contain constants that require the use of certain units, and designers must be careful. The acceptance of a design is enhanced if it is based on empirical expressions describing contaminant emission data that are used by industry.

4.2.1- Vapors from Filling and Opening Vessels

The EPA has the task of assessing the hazard from a countless number of new chemical substances developed in the US each year. They use several elementary expressions to estimate the rate at which vapors enter the air due to the following (EPA, 1986):

- filling containers
- obtaining samples from vessels containing liquids
- cleaning and maintaining containers
- open vessels

Vapor emitted by a liquid filling an initially empty vessel: When filling a vessel, the liquid being moved evaporates and vapor is emitted to the surrounding air. The rate at which this vapor is emitted to the air can be expressed as

$$\boxed{S_{f,J} = f \frac{V P_{v,J} M_J L_r}{R_u T}} \qquad (4\text{-}10)$$

where each variable is listed below, along with its dimensions:

- $S_{f,J}$ = emission rate of molecular species j into the air (mass/time); (f denotes "filling" and j is the species that is entering the vessel)
- f = ***filling factor***, defined by $f = P_j / P_{v,J}$ (dimensionless, $0 < f < 1$)
- V = volume of the container (volume)
- $P_{v,j}$ = vapor pressure of species j (see Appendix A.8) (pressure)
- M_j = molecular weight of species j (mass/mol)
- L_r = ***loading rate*** = filling volumetric flow rate (Q) divided by container volume (V); L_r can be thought of as the number of containers loaded per unit time (1/time)
- R_u = universal gas constant [energy/(mol-temperature)]
- T = liquid temperature (absolute temperature)

Figure 4.3 illustrates three common methods of adding a liquid to a vessel. If the liquid is poured into the vessel (***splash filling***, Figure 4.3a), there are many small drops of liquid j splashed into the vapor portion of the tank. These droplets quickly evaporate, and the partial pressure P_j inside the tank is nearly equal to the vapor pressure $P_{v,J}$; the filling factor is approximately 1.0. If the liquid enters the vessel through a pipe beneath the liquid surface (***submerged filling***, Figure 4.3b or ***bottom filling***, Figure 4.3c), the evaporation rate is lower since the liquid surface is more smooth. Experiments show that f is approximately 0.5 in this case.

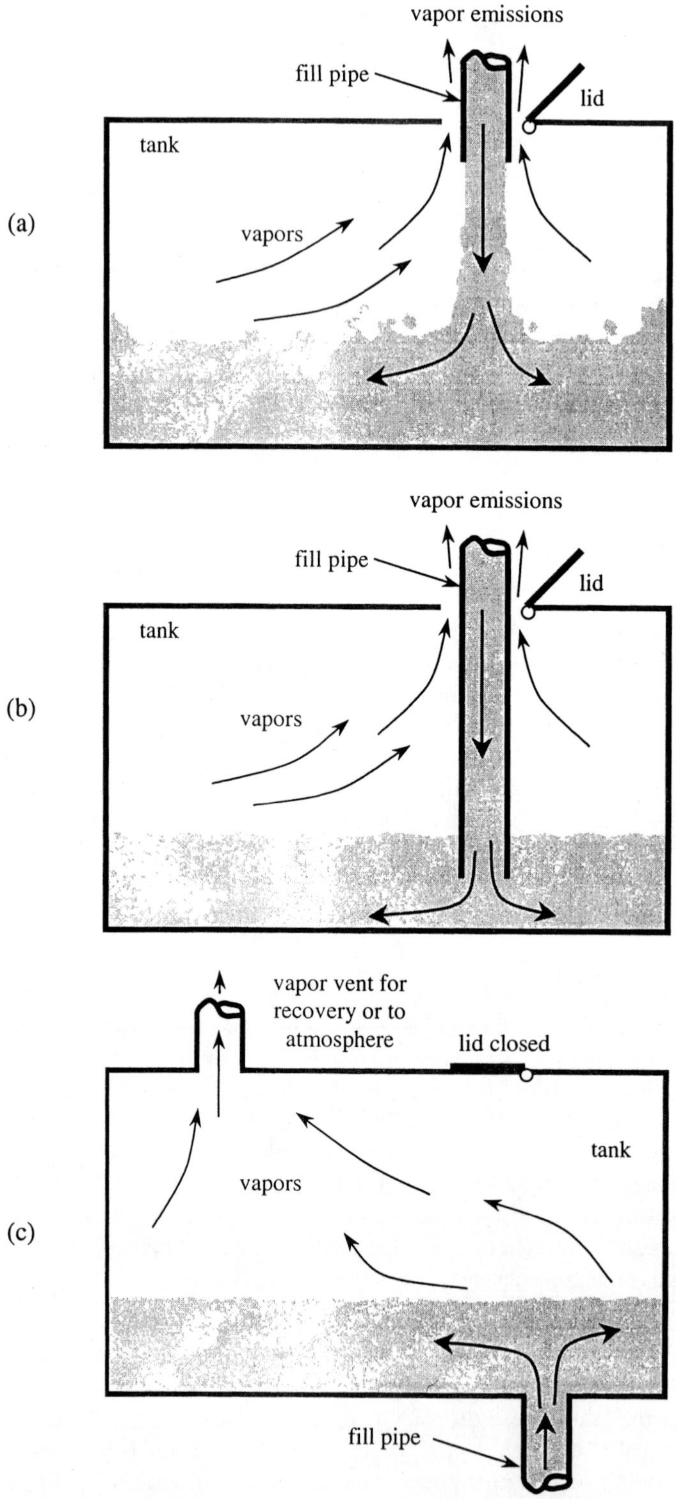

Figure 4.3 Methods to fill vessels with liquids; (a) splash filling, (b) submerged filling, and (c) bottom filling (redrawn from AWMA Handbook on Air Pollution Control, 2000).

Vapor emitted by refilling a vessel: When a closed container is nearly empty, vapor fills the air space above the liquid (species k) in the container. If sufficient time is available, the partial pressure of the contaminant vapor is equal to the contaminant's vapor pressure, i.e. the air is ***saturated*** with the contaminant vapor; $P_k = P_{v,k}$. When such a vessel is being filled, air and the vapor that formerly occupied the vessel are ***displaced*** by the liquid entering the vessel, and are emitted into the air.

The rate at which vapor of the former liquid is displaced can be expressed as

$$S_{d,k} = \frac{V P_{v,k} M_k L_r}{R_u T} \tag{4-11}$$

where $S_{d,k}$ is the emission rate of vapor (k) into the air. In this case M_k and $P_{v,k}$ are the molecular weight and vapor pressure of the *former* liquid (species k) which may or may not be different from the *present* liquid (species j) being added to the tank. If j and k are the same, the total rate of emission is equal to Eq. (4-11), which is identical to Eq. (4-10) when f = 1. If j and k are different, the total rate at which *both* vapors (species j and k) enter the air is ($S_{f,j} + S_{d,k}$).

Vapor from opening a vessel: When workers open a lid, hatch, port, etc. of a vessel, vapor of the liquid (species j) within the vessel escapes, even when no new liquid is being added. The EPA estimates the emission rate by the following:

$$S_{e,j} = K \frac{P_{v,j} M_j A}{R_u T} \tag{4-12}$$

where

- K = mass transfer coefficient, typically in units of velocity, such as m/s, cm/s, etc.
- A = area over which evaporation occurs

and the remaining terms are similar to those of the expressions above. The EPA suggests that the mass transfer coefficient (K) in Eq. (4-12) is proportional to the diffusion coefficient to the power 2/3, and the diffusion coefficient is assumed to be inversely proportional to the square root of the molecular weight M_j. Using water as a reference value, K can be expressed as

$$K \left(\text{in units of } \frac{\text{cm}}{\text{s}} \right) = 0.83 \left(\frac{18}{M_j} \right)^{\frac{1}{3}} \tag{4-13}$$

4.2.2 Spills

Engineers may have to analyze evaporation of a volatile liquid from an ***industrial spill***, defined as a shallow pool in which the surface to volume ratio is large and heat transfer is sufficient to compensate for the cooling effect of evaporation. Kunkel (1983) provides an expression for cases where the wind speed is low and the Reynolds number based on length is less than 20,000,

$$S_s = 0.3 \left(\frac{U_a}{T_a} \right)^{0.8} A^{0.9} M_j P_v \left[\frac{\left(3.1 + c_{\text{molar,L}}^{\ -0.33}\right)^2}{T_a^{\ 0.5} \left(\frac{1}{29} + \frac{1}{M_j} \right)^{0.5}} \right]^{-0.67} \tag{4-14}$$

where the variables and their units are

- S_s = evaporation rate from the spill (kg/hr)
- U_a = air speed (m/s)
- A = area of spill (m^2)
- M_j = molecular weight of volatile liquid (species j)
- T_a = air temperature (K)
- $c_{molar,L}$ = average molar density of the volatile liquid, ($gmol/cm^3$)
- P_v = vapor pressure of the volatile liquid at the liquid-air interface (mm Hg)

Kunkle (1983) also discusses several analytical models that predict the evaporation rate for spills in which heat transfer is of varying importance.

Example 4.4 – Estimation of the Evaporation Rate of a Spill of Trichloroethylene (TCE)
Given: Trichloroethylene (TCE, CAS 79-01-6) is a commonplace solvent used to clean materials in the manufacture of electronic equipment. The liquid has a sweet odor, a PEL of 100 PPM, and an odor threshold of about 400 PPM. The air temperature is 20.0 °C and the wind speed is 2.00 m/s. Some other properties of TCE at this temperature are:

- $P_v = 60.0$ mm Hg
- $M = 131.4$ g/mol
- $\rho = 1460$ g/cm^3
- $c_{molar,L} = \rho/M = 1460/131.4 = 11.14$ $gmol/cm^3$

A 55-gallon drum of TCE is handled roughly and a leak develops along a seam. A circular pool approximately 10. m in diameter develops. You have been asked to estimate the vapor concentration at various points in the workplace, and as a first step you need to estimate the evaporation rate from the spill.

To do: Estimate the evaporation rate (S_S in kg/hr) from the spill.

Solution: Substitution into Eq. (4-14) yields

$$S_S = 0.3\left(\frac{2.00}{293.15}\right)^{0.8}\left(\pi\frac{10^2}{4}\right)^{0.9}(131.4)(60.0)\left[\frac{\left(3.1+11.14^{-0.33}\right)^2}{(293.15)^{0.5}\left(\frac{1}{29}+\frac{1}{131.4}\right)^{0.5}}\right]^{-0.67} = 943.3\frac{kg}{hr}$$

Discussion: No more than two significant digits can be expected from these empirical equations, so the final result should be reported as $S_S = 940$ kg/hr.

Often, unconventional applications and increasingly rigorous requirements necessitate mathematical modeling based on the principles of fluid flow, heat transfer, and diffusive and/or convective mass transport. Several practical examples of such modeling follow.

4.2.3 Grinding

Hahn and Lindsay (1971) suggest that the total rate at which metal is removed by grinding (V_g, volume/min) can be expressed as

$$\boxed{V_g = \Lambda F_n} \tag{4-15}$$

where F_n is the normal force between the work piece and grinding wheel and Λ is the ***metal removal parameter***, which is a function of the wheel speed and the physical properties of the work piece. Properties of the grinding wheel have only a slight affect on the metal removal rate.

Grinding particles are primarily metal particles; only a small amount of the wheel becomes airborne. Values of Λ between 0.0001 and 0.025 $in^3/(min\text{-}lbf)$ can be expected for the majority of industrial grinding operations (Hahn and Lindsay, 1971). The metal removal parameter (Λ) can be measured experimentally or estimated analytically. Figure 4.4 shows values of Λ for several alloys, wheel speeds, and normal forces. According to Heinsohn and Johnson (1982), the dimensionless particle generation rate at points around the wheel's periphery can be expressed as

$$\boxed{g(\alpha) = 0.17 - 3.53\times10^{-3}\alpha + 2.3\times10^{-5}\alpha^2 - 4.0\times10^{-8}\alpha^3} \quad (4\text{-}16)$$

where α is the angular displacement ($0^\circ < \alpha < 90^\circ$) measured from the point of contact with the work piece. The function $g(\alpha)$ is a dimensionless generation rate equal to the ratio of the local particle mass generation rate at angle α to the total ***metal removal rate*** ($\dot{m}_t$). Thus the particle mass emission rate at the point on the grinding wheel, $S_g(\alpha)$ is

$$\boxed{S_g(\alpha) = \rho_m V_g g(\alpha)} \quad (4\text{-}17)$$

The particles can be assumed to leave the wheel with a velocity equal to the tangential velocity of the wheel (Bastress et al., 1974). Figure 4.5 shows the ***particle size distribution*** for different metal removal rates. See Chapter 8 for details about the cumulative number distribution function (N) and the aerodynamic particle diameter.

4.2.4 Pouring Powders

There are countless industrial operations in which workers pour small quantities of powdered materials into vessels. Such operations require skill to minimize dust generation, but nonetheless the operations are notoriously dusty. Several laboratory techniques are available to estimate the amount of dust that is generated (Cowherd et al., 1989). For representative drop heights of 25 cm, the authors suggest the following empirical equation to predict the mass of airborne dust (in mg) per kg of material poured:

$$\boxed{L = 16.6 f_w^{-0.75} \sigma_g^{3.9} \rho_b^{-1.2} \left(D_{p,50}(\text{mass})\right)^{-0.45}} \quad (4\text{-}18)$$

where the parameters and the specific units that must be used are the following:

- L = mass of airborne dust per kg poured powder (mg/kg)
- f_w = powder moisture content (%)
- σ_g = geometric standard deviation of particle size distribution (unitless)
- ρ_b = powder bulk density (g/cm^3)
- $D_{p,50}(\text{mass})$ = mass median particle diameter of powder (μm)

Mass median particle diameter and geometric standard deviation are discussed in Chapter 8.

4.2.5 Spray Finishing and Coating

Surface coatings are applied to countless products by spray nozzles (O'Brien and Hurley, 1981). Workplace contaminants are generated by the evaporating solvents and from ***overspray*** - atomized finish particles that do not land on the intended surfaces. The total mass of a volatile organic compound (VOC) entering the workplace environment per unit area is equal to the VOC content in the finish (kg/gallon) (a value listed on the MSDS) times the application rate (gallons/m^2 of surface coated). The VOC emission rate is divided between the VOC generated during spraying and that generated during drying. The former depends on the type of sprayer and spraying rate, while the later depends on the drying time, a value that can be provided by the coating supplier.

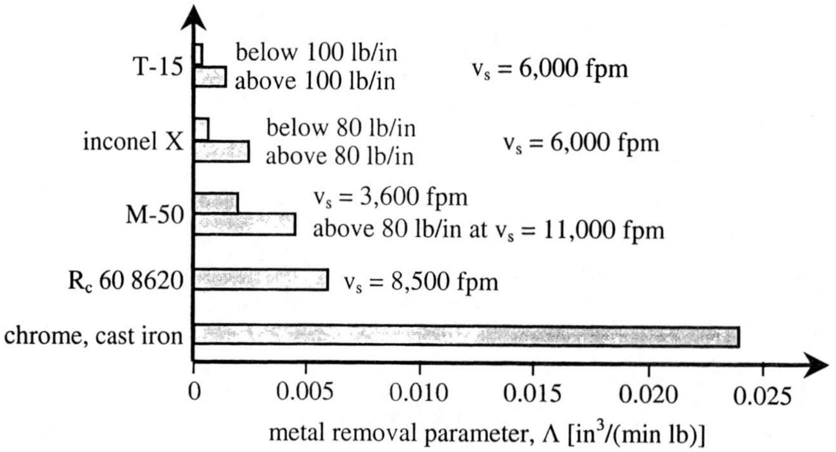

Figure 4.4 Metal removal parameter of five metals for different normal forces and grinding wheel speeds (v_s); T-15 and inconel X: typically difficult to grind steel alloys; chrome and cast iron: easy to grind material (adapted from Heinsohn, 1991).

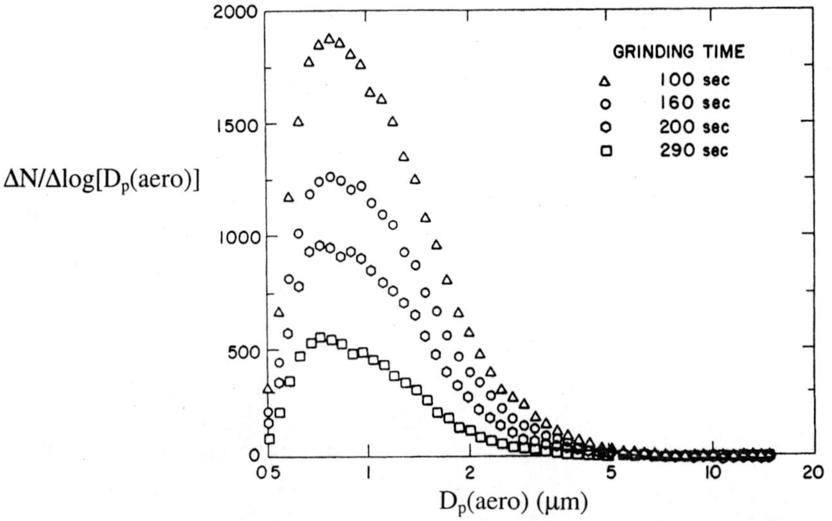

Figure 4.5 Particle size distribution in the boundary layer of a grinding wheel 270 degrees from the point of contact after grinding has begun (adapted from Heinsohn, 1991).

Examples of spray finishing include automobile refinishing and coatings for furniture, machinery, and appliances. The major constituents in finishes are binders, pigments, solvents, and additives. Liquid coatings are classified broadly as those that use VOCs as solvents for the binders, pigments, and additives, and those that use water. Coatings can also be categorized as either thermoplastic or convertible. ***Thermoplastic coatings*** are soft when heated and harden upon cooling. ***Convertible coatings*** undergo irreversible chemical changes as the film forms. ***Binders*** are non-volatile materials that cement the pigments in the film to the material to be finished. Synthetic resins, natural resins, and drying oils are used as binders. ***Pigments*** are fine insoluble powders dispersed in a liquid medium. Pigments are used to impart color, to inhibit corrosion, and to produce a certain opacity or absorptivity of light. ***Additives*** are used in relatively small amounts to increase the finish shelf-life and to impart certain features of the finish not obtained by pigments (e.g. gloss or texture).

Industrial finishes are applied by sprayers that use compressed air or high pressure liquid. Compressed air is the most widely used method to atomize the finish and transport it to the intended surface. Unfortunately a great many of the particles fail to land on the intended surface, and become overspray. Finer control of the particle size and a higher percentage of particles can be applied by ***airless spraying*** in which atomization is achieved as the liquid is forced through the nozzle at high liquid pressure. ***Electrostatic spraying*** consists of applying an electrical charge to the particle and an opposite charge to the surface to be painted.

The mass emission rate associated with overspray cannot be calculated accurately, but can only be estimated from manufacturers' literature. The amount of solvent that evaporates can be estimated on the basis of finish used and the percent solvent, but the rate at which the solvent evaporates is not easily calculated. The particle size distribution generated by the spray nozzles has to be provided by the vendors of the spray equipment.

4.3 Emissions Factors

In the absence of actual measurements of the rate at which pollutants are generated or emitted, engineers often use ***AP-42 emission factors*** (EFs). Since the original publication of EPA emission factors was in a document noted as AP-42, The notation "AP-42" has become a well-used phrase in the vocabulary of the air pollution control community; readers should not forget it. AP-42 emission factors express the typical amount of pollutant generated in an uncontrolled process per unit quantity of material being processed. If the process has no air pollution control system (APCS) to remove pollutants, the emission rate (the rate at which pollutant is discharged into the air), also called the ***pollutant discharge rate*** ($\dot{m}_d$), is equal to the pollutant generation rate ($\dot{m}_g$). Rarely however are processes uncontrolled, and when an air pollution control device with a ***removal efficiency*** (η), also called ***control efficiency*** or ***capture efficiency***, cleans the exhaust process gas stream, the rate at which the pollutant is emitted to the air is

$$\boxed{\dot{m}_d = (1-\eta)\,\dot{m}_g} \tag{4-19}$$

Emission factors are typically listed in terms of generation rate rather than emission rate, and are thus called "emission factors from uncontrolled sources". To compute the pollutant generation rate ($\dot{m}_g$), engineers need to know the rate at which raw materials are processed. The rate at which a product is processed (***production rate***) is a quantity of which the source operator is keenly aware (e.g. fuel firing rate, mass of product being stored, etc.) since it is a key item to the company's profitability. It must be stressed that emission factors typify only the rate or amount of pollutant generated by a class of industrial operations and may not necessarily quantify any specific source in the category. Within a category, individual sources may generate quite different amounts of contaminant. Nevertheless, in lieu of actual experimental data, empirical expressions for a particular process, or analytical expressions that can be derived from first principles, emission factors may be the only data available to the designer.

AP-42 emission factors have been compiled by the EPA for a large variety of industries, and the EPA updates and expands emission factors on a regular basis. A full selection of AP-42 emission factors can be accessed through the Internet TTNBBS bulletin board called CHIEF (see **www.epa.gov/ttn/chief**). In addition, AP-42 emission factors are available in commercial computer databases that are reviewed in professional journals (Bare, 1988). AP-42 emission factors are also available on a CD-ROM (AIR CHIEF, EFIG/EMAD/OAQPS/EPA) that can be purchased from the Government Printing Office at a modest cost. Appendices A.2 to A.7 provide an abridged list of AP-42 emission factors relevant to the indoor environment. Contained also in AP-42 are emission factors for industrial processes equipped with a variety of air pollution control devices. These data are of great value in preparing state permit applications and environmental impact statements. Other emission factors are published from time to time in professional journals (e.g. Miller et al., 1996) and handbooks (e.g. Bond and Staub,1972). While not strictly AP-42 emission factors, they fulfill the same function. For example see Jenkins et al. (1996) for emission factors for polycyclic aromatic hydrocarbons from biomass burning. If the data are accurate and fulfill the same purpose, readers should not hesitate to use them.

4.3.1 Household Products

Since the mid 1970s, the public has become increasingly concerned about contaminants generated by household products and appliances. Several books have been written on indoor air pollution and indoor air quality (Formica, 1978; Committee on Indoor Air Pollution, 1981; Meyer, 1983; Wadden and Scheff, 1983 and 1987; Burgess et al., 1989; and Heinsohn, 1991). In addition, countless articles have been published in the professional literature. These books and articles describe the rate at which contaminants are emitted, and the environmental factors affecting the emission rate. These data are useful to engineers for estimating indoor contaminant concentrations and for designing ventilation systems. The articles may be exhaustive survey papers, research monographs, or merely timely reports of important discoveries. The following is a sample of such literature for several common household activities:

- Organic vapors emitted by consumer products (Pickrell et al., 1984; Meyer and Hermans, 1985; Hawthorne, 1987; Matthews et al., 1987; Stock, 1987; Tichenor and Mason, 1988; Tichenor et al., 1990; Weschler et al, 1990).
- Kerosene space heaters (Leaderer B P, 1982; Ryan et al., 1983; Traynor et al., 1983; Traynor et al., 1988).
- Wood and coal stoves (Hall and DeAngelis, 1980; Butcher and Ellenbecker, 1982; Traynor, 1987).
- Gas kitchen ranges (Traynor et al., 1982; Traynor et al., 1985; Leaderer et al., 1986).
- Tobacco (Repace and Lowery, 1982; Sterling et al., 1982; Cain et al., 1983; Leaderer et al., 1984; Ryan et al., 1986; Guerin et al., 1987; Oladakar et al., 1987; Crawford, 1988; Davies et al., 1988; Ingebrethsen et al., 1988; Benner et al., 1989; Eatonough et al., 1989; Lofroth et al., 1989).
- Volatile organic compounds in tap water (McKone, 1987; McKone and Knezovich, 1991; Giardion et al., 1992, 1996; Moya et al., 1999).

Shown below is an example of the range of indoor average source strengths of carbon monoxide (CO), total suspended particulate (TSP), and benzo[a]pyrene (BaP) for several "airtight" and "non-airtight" wood-burning stoves operating steadily in a home (Traynor et al., 1987):

stove	**CO (cm^3/hr)**	**TSP (mg/hr)**	**BaP (μg/hr)**
airtight	10-140	2.5-8.7	0.02-0.76
non-airtight	220-1800	16-320	2.2-57

It is impractical to provide a complete and current summary of such literature for every household or industrial activity, and designers must keep abreast of the field themselves. Computer-aided literature search software can simplify the search considerably. Appendix A.7 summarizes emissions from several household devices.

4.3.2 Manufacturing Processes

Large manufacturing companies assemble data about emissions associated with specific manufacturing processes. Such data are used to design industrial ventilation systems to maintain health and safety standards. In some cases companies belonging to trade associations pool these data for their members and sell the data to the public. An example of this information is the publications of the American Welding Society on the characteristics of fume and gases produced in welding (American Welding Society, 1983).

4.3.3 Combustion Processes

Emission factors are often the only way to estimate the generation rates of pollutants from combustion processes, since material balances are of no value owing to the complicated ways in which the pollutants are formed during combustion. In fact, sulfur dioxide is the only pollutant that can be estimated by a material balance if the sulfur content of the fuel and lubricating oil and their respective consumption rates are known.

Example 4.5 - Pollutants Generated by a Gasoline Powered Internal Combustion Engine
Given: Consider the 30 HP gasoline powered internal combustion engine of Example 4.1.

To do: Using AP-42 emission factors, estimate the rate at which the following pollutants are generated: total particulate matter ($\dot{m}_p$), sulfur dioxide ($\dot{m}_{SO_2}$), oxides of nitrogen ($\dot{m}_{NO_x}$), carbon monoxide ($\dot{m}_{CO}$), aldehydes ($\dot{m}_{aldehydes}$), and hydrocarbons ($\dot{m}_{HC}$).

Solution: AP-42 emission factors for small gasoline engines can be obtained from the Internet:

- total particulate matter:	0.327 g/HP-hr
- SO_2:	0.268 g/HP-hr
- NO_x:	5.16 g/HP-hr
- CO:	199 g/HP-hr
- aldehydes:	0.22 g/HP-hr
- crankcase hydrocarbons:	38.3 g/hr
- evaporative hydrocarbon losses:	62.0 g/hr
- exhaust hydrocarbon losses:	6.68 g/HP-hr

Thus, for a 30 HP engine,

- $\dot{m}_p$ = 30 HP(0.327 g/HP-hr) =	9.8 g/hr (2.7 mg/s)
- $\dot{m}_{SO_2}$ = 30 HP(0.268 g/HP-hr) =	8.0 g/hr (2.3 mg/s)
- $\dot{m}_{NO_x}$ = 30 HP(5.16 g/HP-hr) =	150 g/hr (43. mg/s)
- $\dot{m}_{CO}$ = 30 HP(199 g/HP-hr) =	6,000 g/hr (1.7 g/s)
- $\dot{m}_{aldehydes}$ = 30 HP(0.22 g/HP-hr) =	6.6 g/hr (1.8 mg/s)
- $\dot{m}$ (crankcase hydrocarbons) =	38. g/hr (11. mg/s)
- $\dot{m}$ (evaporative hydrocarbon losses) =	62. g/hr (17. mg/s)
- $\dot{m}$ (exhaust hydrocarbon losses) =	(30 HP)(6.68 g/HP-hr) = 200 g/hr (5.6 mg/s)

Discussion: It must be emphasized that these values represent the pollutant generation rate, not the rate at which pollutants are emitted to the air. If no steps are taken to capture the pollutants, the above values are the rates at which the pollutants enter the workplace environment. Emission factors, although sometimes listed to three significant digits, are typically only accurate to one or two significant digits for any particular process; the final emission rates above are given to two significant digits.

4.3.4 Limitations

Whenever empirical equations or emission factors are used, there is concern that they may be used under conditions that do not warrant their use. Unfortunately, these conditions may not be provided by the individuals who generated the equations or emission factors. If equations can be derived from fundamental engineering principles, engineers are better served, but such derivations are not always possible. The following example problem illustrates the dilemma by contrasting emissions predicted from fundamental principles with predictions obtained by empirical equations and emission factors.

Example 4.6 - Emissions From Filling a 5-Gallon Can with Gasoline

Given: A new (empty) 5-gallon can (0.019 m^3) is filled with gasoline on a day when the temperature of the air and gasoline are 31.5 °C (304.65 K). Gasoline enters the can at a volumetric flow rate of 2.5 GPM. Assume that the properties of gasoline can be approximated by the properties of octane, M = 114, P_v (at 31.5 °C) = 20 mm Hg (2.657 kPa). Your supervisor asks you to estimate the rate at which gasoline vapor is emitted to the air. Your inclination is to use emission factors (Appendix A.3) and to check the results with Eq. (4-10), but your supervisor distrusts these methods. Since you both graduated from the same university and are proud of its high academic standards, she asks you to predict the emissions from an equation derived from fundamental engineering principles, and then to compare to the prediction from emission factors.

To do: Use two methods to estimate the rate at which gasoline vapor is emitted to the air: (a) from fundamental engineering principles, and (b) from emission factors.

Solution:

(a) As an upper limit, assume that as soon as filling begins, the new empty tank contains air with gasoline vapor at a partial pressure equal to the gasoline vapor pressure ($P_{v,j}$). Gasoline (denoted as species j) entering the can displaces air and vapor from within the tank. Envision the rising level of gasoline as an upward moving piston. The mass flow rate of vapor leaving the can ($\dot{m}_j$) is equal to the velocity of the rising liquid level (v) times the cross-sectional area of the can (A) times the mass concentration of vapor (c_j):

$$\dot{m}_j = vAc_j$$

The velocity of the rising liquid (v) is equal to the volumetric flow rate of gasoline entering the can divided by the cross-sectional area (A). Thus

$$\dot{m}_j = \frac{Q}{A} A c_j = Q c_j$$

The mass concentration of gasoline vapor can be expressed as

$$c_j = \frac{P_j M_j}{R_u T} = \frac{P_{v,j} M_j}{R_u T}$$

since it has been assumed that the vapor partial pressure P_j is equal to the gasoline vapor pressure $P_{v,j}$. Thus,

$$\dot{m}_j = Q M_j \frac{P_{v,j}}{R_u T}$$

Multiplying and dividing by volume V and rearranging,

$$\boxed{\dot{m}_j = \left[\frac{Q P_{v,j} M_j}{R_u T}\right]\left(\frac{V}{V}\right) = \left[\frac{V P_{v,j} M_j \left(\frac{Q}{V}\right)}{R_u T}\right] = \frac{V P_{v,j} M_j L_r}{R_u T}} \qquad (4\text{-}20)$$

where L_r is the loading rate,

$$\boxed{L_r = \frac{Q}{V}} \qquad (4\text{-}21)$$

which can be thought of as the number of containers loaded per unit time. Here, L_r is

$$L_r = \frac{2.5\dfrac{\text{gal}}{\text{min}}}{5\text{ gal}} = 0.50\frac{1}{\text{min}}$$

If Eq. (4-20) is multiplied by an empirical (unitless) loading factor f, reflecting the unique physical conditions pertaining to the filling operation, an equation identical to Eq. (4-10) results,

$$\dot{m}_J = S_{f,J} = f\frac{VP_{v,J}M_JL_r}{R_uT}$$

For example, it can be assumed that f = 1 when the can is filled from the top (splash loading, as in Figure 4.3a), and the falling liquid mixes the air and liquid inside the tank, which suggests that the partial pressure (P_j) is equal to the gasoline vapor pressure, $P_{v,J}$. Alternatively, f = ½ represents "bottom filling" when gasoline enters the can gently from the bottom and causes the air above the liquid level to be only partially saturated with vapor.

Thus Eq. (4-10), which was introduced as an empirical equation, is in fact borne out by fundamental principles. If gasoline is assumed to be octane, $P_{v,J}$ (at 31.5 °C) = 20.0 mm Hg (2.657 kPa), the emission rate is calculated from the above equation,

$$\dot{m}_J = 1.0\left[\frac{(2.657\text{ kPa})(0.019\text{ m}^3)\left(114.\dfrac{\text{kg}}{\text{kmol}}\right)}{8.314\dfrac{\text{kJ}}{\text{kmol}\cdot\text{K}}(304.65\text{ K})}\right]0.50\frac{1}{\text{min}}\left(\frac{\text{kN}}{\text{m}^2\text{kPa}}\right)\left(\frac{\text{kJ}}{\text{kN}\cdot\text{m}}\right)\left(\frac{1000\text{ g}}{\text{kg}}\right) = 1.1\frac{\text{g}}{\text{min}}$$

(b) Appendix A.3 indicates that the emission factor for normal splash loading of gasoline is equal to 1.4 kg gasoline vapor per m^3 of liquid. Thus the vapor emission rate is

$$\dot{m} = \left(1.4\frac{\text{kg}}{\text{m}^3}\right)(0.019\text{ m}^3)\left(0.50\frac{1}{\text{min}}\right)\left(\frac{1000\text{ g}}{\text{kg}}\right) = 13.\frac{\text{g}}{\text{min}}$$

This emission rate predicted by emission factors is more than an order of magnitude larger than that predicted by Eq. (4-10)!

Discussion: The disparity in the two methods is discomforting, but clearly demonstrates the inherent inaccuracy and uncertainty of emission factors. The principal factor affecting the emission rate is the vapor pressure of the gasoline. If gasoline were more volatile than octane, or if the temperature were higher, such that its vapor pressure was larger than what was assumed, the disparity in predictions would be smaller. Secondly it is not clear that the emission factor for "transfer of hydrocarbons by tank cars and trucks - splash loading, normal service" accurately describes filling a 5-gallon gasoline can. Certainly if the stream of gasoline entering the can possessed a large surface from which evaporation could occur before the liquid even enters the can, the additional evaporation could perhaps account for the disparity. Lastly, nothing in the emission factor accounts for temperature, and since vapor pressure is very sensitive to temperature, the emission factor should not be constant over a wide range of temperatures. Nevertheless, emission factors are useful for obtaining "ballpark estimates."

4.4 Puff Diffusion

Contaminants may enter the air in the form of bursts. ***Bursts***, or ***instantaneous sources*** as they are called formally, occur when a vessel containing gas or vapor ruptures, a safety valve pops open, a mass of powder falls on the floor, etc. The movement of particles or gas through the surrounding air is called ***dispersion***. The overall phenomenon of the burst followed by its mixing with air and traveling downstream is called ***puff diffusion***. During its early stages, puff diffusion is distinctly different from the dispersion of contaminants from continuous sources. ***Atmospheric dispersion*** from continuous point sources concerns the plumes from smoke stacks, volcanic plumes, forest fires, and plumes from other large-scale sources that transport contaminants hundreds of kilometers for periods measured in hours or even days. Nevertheless, a great deal of the theory of atmospheric dispersion can be applied to puff diffusion in the workplace.

4.4.1 Puff Diffusion of Gases

Once in the air, contaminants are transported by bulk motion of the carrier gas (***advection***, often called ***convection***) and ***diffusion***. The mass concentration downwind of the source is given by a transport equation for conservation of mass of the contaminant,

$$\frac{\partial c}{\partial t} + \vec{\nabla} \cdot \left(\vec{U} c \right) = \nabla^2 \left(D_{12}\, c \right) \tag{4-22}$$

In quiescent air (no bulk motion of air, U = 0), the coefficient D_{12} is the ***bimolecular diffusion coefficient*** (species 1 diffusing through species 2). The dimensions of D_{12} are length^2/time. If there is convection or turbulent mixing in addition to molecular diffusion, coefficient D_{12} must reflect turbulence in the carrier gas. In quiescent air, the solution of Eq. (4-22) for a point source in free space is the Gaussian formula (Hanna et al., 1982),

$$c(r,t) = \frac{m_{puff}}{(2\pi)^{3/2}\sigma^3} \exp\left[-\frac{1}{2}\left(\frac{r}{\sigma}\right)^2 \right] \tag{4-23}$$

where m_{puff} is the mass released instantaneously for each puff, r is the distance from the center of the puff to a point in space and σ is the standard deviation. From the field of atmospheric dispersion, sigma is also called the ***dispersion coefficient***, which depends on time. Hanna et al. (1982) report that when the elapsed time is short and the puff dimensions are small with respect to the size of eddies in the air, σ is linearly proportional to time. When the puff has grown and is larger than the eddies, σ is proportional to the square root of time. Cooper and Horowitz (1986) assumed that the dispersion coefficient can be given by

$$\sigma = \sqrt{2 D_{12}\, t} \tag{4-24}$$

Example 4.7 - Instantaneous Release of Gas in a Laboratory

Given: A pressure vessel of methyl chloride (CH_3Cl, CAS 74-87-3) located in the center of a small laboratory develops a leak, and 400 g of methyl chloride vapor suddenly enters the air. The vessel is 2.0 meters from a laboratory technician. Methyl chloride is a toxic gas with a PEL of 100 PPM (207 mg/m^3); it has a sweet odor and its molecular diffusion coefficient is 1.3 x 10^{-5} m^2/s. There is no perceptible motion to the air in the laboratory.

To do: Calculate the concentration at the location of the laboratory technician and assess the hazard to the technician.

Solution: The concentration at 2.0 meters can be found by direct substitution into Eqs. (4-23) and (4-24). The following results are obtained:

time (hr)	concentration at 2.0 m (mg/m^3)
3.7	80
9.3	200
18.6	360
37	200
186	40
370	8

Discussion: These predictions show that the technician is exposed to concentrations well above the PEL for tens of hours, but that the predicted times for the puff to pass seem way too high. Intuition or experience suggest that the times should be two or more orders of magnitude smaller! The anomaly lies with the assumption that the room air is quiescent, i.e. that there is no bulk motion. Such an environment does not exist in the workplace; air currents are always present and are more important than molecular diffusion in the transport of contaminants. In addition, even low velocity air currents can lead to turbulence, which can increase the diffusion coefficient by orders of magnitude.

Equations (4-23) and (4-24) suffer from two serious deficiencies:

- A momentary release of material produces a jet that disturbs the surrounding air. Assuming quiescent conditions during the first few moments following discharge is unrealistic.
- The workplace environment is not strictly quiescent. There is always movement of air whether one is aware of it or not. Thus, molecular diffusion is small in comparison with bulk motion and turbulence.

To account for these two realities, the dispersion coefficient must be related to the turbulent properties of the carrier gas in addition to time. It is believed that the general form of Eq. (4-23) remains the same, but the dispersion coefficients have to be computed in another fashion. Hanna et al. (1982) report that in experiments involving a wide variety of instantaneous point sources in the air, σ correlates well with time (even for a short duration of several seconds) by

$$\frac{\sigma^2}{2Kt} = \frac{t}{t_{Lv}} - \left[1 - \exp\left(\frac{-t}{t_{Lv}}\right)\right] - 0.5\left[1 - \frac{v_o^{\,2} t_{Lv}}{K}\right]\left[1 - \exp\left(\frac{-t}{t_{Lv}}\right)\right]^2 \tag{4-25}$$

where K is the ***eddy diffusivity***, v_0 is an initial velocity of the puff, and t_{Lv} is the Lagrangian turbulence time scale. Hanna et al. (1982) summarize numerous experiments concerning large-scale instantaneous releases in the atmosphere in which good agreement with experiment was obtained when

- K = 50,000 m^2/s
- v_0 = 0.15 m/s
- t_{Lv} = 10,000 s

4.4.2 Puff Diffusion of Particles

For the case of particles diffusing in air, the ***particle diffusion coefficient*** (*D*) is used instead of the bimolecular diffusion coefficient. For particles of diameter D_p diffusing into air, *D* can be estimated from the following expression from Fuchs (1964):

$$D = \frac{kTC}{3\pi\mu D_p} \tag{4-26}$$

where k is the Boltzmann constant, μ is the viscosity of the carrier gas, and C is the unitless ***Cunningham correction factor*** (see Chapter 8). For particles larger than 2 μm, C is essentially unity. Values of the diffusion coefficient for particles in air at STP are given in Table 4.1.

The concentrations at various radii from the origin of the puff were studied by Cooper and Horowitz (1986) assuming σ is given by Eq. (4-24); this assumption ignores gravimetric settling of particles, which is discussed in Chapter 8. Values predicted from Eq. (4-23) are shown in Figure 4.6 as functions of *D*t. The graph predicts two interesting features about puff diffusion:

- As the puff moves radially outward, the concentration (at any radius) increases with time, reaches a maximum, and then decreases. For larger radii the pattern remains the same except the peak concentration occurs later and is smaller than what it was closer to the source. Both results are consistent with conservation of mass for a spherically propagating wave. For large values of *D*t, the exponential function approaches unity and the concentration varies as ($4\pi D$t) to the -3/2 power at any radius from the source.
- The second interesting feature predicted by Figure 4.6 concerns the maximum values of the concentration. Drawing a line through the maximum values, one finds that they occur for values of radius and time such that (r^2/Dt) = 6.

The contaminant ***exposure*** (E) of individuals affected by a puff can be computed from

$$\boxed{E(t) = \int_0^t c(t)dt} \tag{4-27}$$

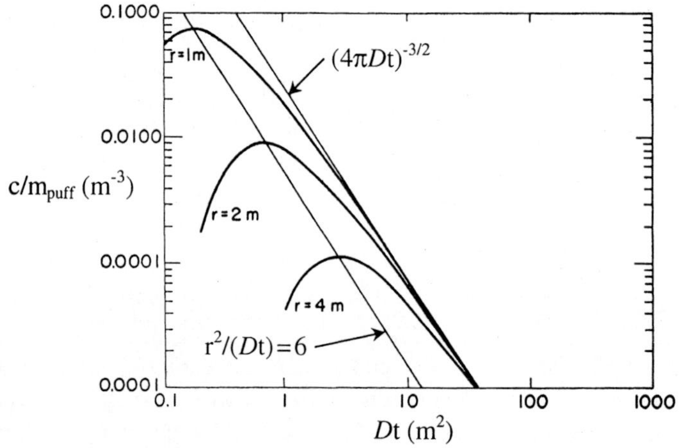

Figure 4.6 Ratio of concentration to released mass (c/m_{puff}, m^{-3}) versus diffusion coefficient times time (*D*t, m^2) for puff diffusion (adapted from Heinsohn, 1991).

Table 4.1 Particle diffusion coefficients in air at STP (abstracted from Fuchs, 1964).

D_p (μm)	D_{ja} (cm^2/s)
0.01	1.35×10^{-4}
0.05	6.82×10^{-6}
0.10	2.21×10^{-6}
0.50	2.74×10^{-7}
1.00	1.27×10^{-7}
5.00	3.38×10^{-8}
10.0	1.28×10^{-8}

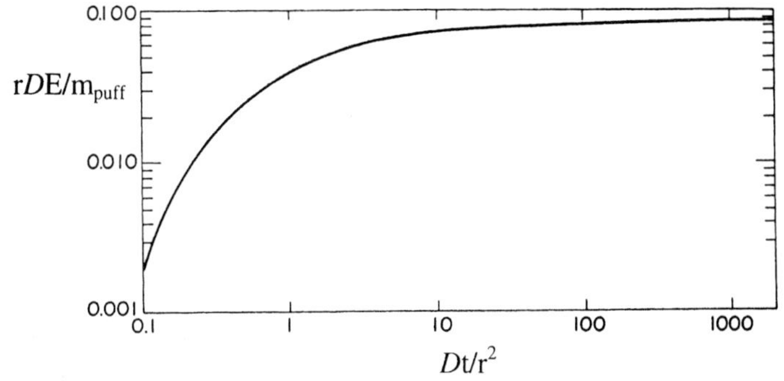

Figure 4.7 Normalized exposure parameter, rDE/m_{puff}, versus dimensionless parameter, Dt/r^2, for puff diffusion (adapted from Heinsohn, 1991).

From the above, the dimensions of E are (mass time)/volume. The integration has been evaluated numerically (Cooper and Horowitz, 1986) and a normalized exposure parameter, rDE/m_{puff}, is shown in Figure 4.7 as a function of a normalized time parameter, Dt/r^2. At large values of Dt/r^2, the exposure approaches approximately 0.085 $m_{puff}/(rD)$.

4.5 Evaporation and Diffusion

A variety of volatile liquids is used in industrial operations. Unless precautions are taken, vapors from these liquids may enter the indoor or outdoor environment. To predict pollutant concentrations in air, it is necessary to know the rate at which the vapors evaporate. In certain cases one can estimate these rates accurately from the principles of mass transfer and obtain more accurate rates than those predicted by AP-42 emission factors. Texts on the theory of mass transfer (Hirschfelder et al., 1954; Bird et al., 1960; Treybal, 1968; Sherwood et al., 1975; and McCabe et al.,1993) should be consulted to acquire a thorough understanding of the physical concepts and principles of mass transfer. Mackay and Paterson (1986) developed a comprehensive model describing the rates of mass transfer at the air-liquid interfaces for organic chemical volatilization, absorption in water, dissolution in rainfall, and wet and dry particle deposition.

In order to model the generation of contaminants, one must know the thermodynamic and physical properties of air and the contaminant. The properties of air are readily available, but the properties of contaminants are less abundant. Appendix A.8 lists the vapor pressures of many organic liquids for a variety of temperatures. Primary references and handbooks (Vargaftik 1975; Mackay et al., 1982; Perry et al., 1984) should be consulted for data pertaining to specific contaminants or for more accurate data at different temperatures.

4.5.1 Vapor Pressure

The vapor pressure (P_{v2}) of a pure species at an untabulated temperature (T_2) can be estimated from its value (P_{v1}) at a tabulated temperature (T_1) by using the ***Clausius-Clapeyron equation***:

$$\boxed{\ln\left(\frac{P_{v2}}{P_{v1}}\right) = \left(\frac{M\hat{h}_{fg}}{R_u}\right)\left(\frac{(T_2 - T_1)}{T_2 T_1}\right)} \tag{4-28}$$

where R_u is the universal gas constant, M is the molecular weight, and $\hat{h}_{fg}$ is the ***enthalpy of vaporization*** (typically in units of kJ/kg), also called the ***heat of evaporation***, averaged over the two temperatures, T_1 and T_2. The reference values P_{v1} and T_1 are often one atmosphere and the saturation temperature at atmospheric pressure respectively. The saturation temperature at atmospheric pressure is called the ***normal boiling temperature***. The accuracy of this calculation is high if T_1 and T_2 are not vastly different, but can be poor if T_1 differs significantly from T_2.

Example 4.8 - Estimation of the Vapor Pressure at an Arbitrary Temperature
Given: You are employed in a firm that manufactures pharmaceuticals. For some products the firm uses pyridine as a solvent. Pyridine (C_5H_5N, CAS 110-86-1) is a colorless, toxic (PEL = 5 PPM), flammable liquid with an unpleasant odor. For purposes of design you need to know the vapor pressure at 38. °C and 75. °C. The data can be obtained from tables of thermophysical data available on the internet (see Chapter 1) or perhaps Perry et al. (1984), but unfortunately all you have at your disposal is an old data sheet which states that the vapor pressure at 77. °F is 20.0 mm Hg, and the BP (boiling point at atmospheric pressure) is 240. °F.

To do: Estimate the vapor pressure of pyridine at 38. and 75. °C using the Clausius-Clapeyron equation, and compare to exact values.

Solution: For clarification, the four temperatures used in this problem are numbered as follows:

- $T_1 = 77.$ °F (= 25. °C = 298.15 K)
- $T_2 = 38.$ °C (= 311.15 K)
- $T_3 = 75.$ °C (= 348.15 K)
- $T_4 = 240.$ °F (= 115.56 °C = 388.71 K)

In thermodynamic terms the data correspond to P_v(at 25. °C) = 20.0 mm Hg and P_v(at 115.56 °C) = 760. mm Hg. To use Eq. (4-28), it is necessary to know the average value of the enthalpy of vaporization per mol between temperatures T_1 and T_2. If this cannot be found in the available literature, an approximate average value can be computed by applying Eq. (4-28) between the temperatures 298.15 K and 388.71 K.

$$\frac{M\hat{h}_{fg}}{R_u} = \ln\left(\frac{P_{v,4}}{P_{v,1}}\right)\left[\frac{T_4 T_1}{T_4 - T_1}\right] = \ln\left(\frac{760.}{20.0}\right)\left[\frac{(388.71\text{ K})(298.15\text{ K})}{(388.71\text{ K} - 298.15\text{ K})}\right] = 4655.\text{ K}$$

Note that $\hat{h}_{fg}$ can be calculated by multiplying the above value by R_u/M, but it is not necessary to do so. The vapor pressure P_v (at 311.15 K) can now be found by applying Eq. (4-28) between 298.15 K and 311.15 K, using the above ratio:

$$\ln\left(\frac{P_{v,2}}{P_{v,1}}\right) = \frac{M\hat{h}_{fg}}{R_u}\left[\frac{T_2 - T_1}{(T_2)(T_1)}\right] = (4655.\text{ K})\left[\frac{311.15\text{ K} - 298.15\text{ K}}{(311.15\text{ K})(298.15\text{ K})}\right] = 0.6523$$

from which

$$P_{v,2}(\text{at } 38.\text{ °C}) = P_{v,1} \cdot e^{0.6523} = (20.0\text{ mm Hg})(1.920) = 38.4\text{ mm Hg}$$

To find the vapor pressure at 75. °C, the calculation can be repeated using temperatures T_3 and T_4. The computation yields

$$P_{v,3}\ (\text{at } 348.15\text{ K}) = 188.\text{ mm Hg}$$

For comparison, exact values of pyridine vapor pressure are found from Appendix A.8:

method	**38. °C**	**75. °C**
estimation from Eq. (4-28)	38.4 mm Hg	188. mm Hg
exact value from Appendix A.8	40.0 mm Hg	200. mm Hg
percentage error	4.0%	5.9%

Discussion: The errors arise because the estimations use two temperatures that are quite different from one another (25. °C and 115.56 °C). The value of $M\hat{h}_{fg}$ for pyridine at 114.1 °C is 8,492.2 cal/gmol (Perry et al., 1984). If this value were used in the above calculation, the computed values of P_v (at 38. °C) and P_v (at 75. °C) would be even less than those shown in the table. These disparities illustrate the sensitivity of predictions obtained from the Clausius-Clapeyron equation.

4.5.2 Diffusion Coefficient

Appendix A.9 contains a listing of the ***molecular diffusion coefficients*** or ***diffusivities*** for several hydrocarbons in air and water. Handbooks or source books should be consulted to find diffusion coefficients for other materials. ***Gas phase binary diffusion coefficients*** (D_{12} in cm^2/s) can also be estimated with an equation developed by Chen and Othmer (Vargaftik, 1975):

$$D_{12} = \frac{0.43\left(\frac{T}{100}\right)^{1.81}\sqrt{\frac{1}{M_1}+\frac{1}{M_2}}}{P\left(\frac{T_{c1}T_{c2}}{10000}\right)^{0.1405}\left[\left(\frac{v_{c1}}{100}\right)^{0.4}+\left(\frac{v_{c2}}{100}\right)^{0.4}\right]^2}\ \frac{cm^2}{s} \tag{4-29}$$

where v_c is the ***critical specific volume*** (which must be in units of cm^3/mol), T_c is the ***critical temperature*** (which must be in K), M is the molecular weight, and P is the total pressure (which must be in atmospheres). The convention for D_{12} is that species 1 is the contaminant and species 2 is the carrier gas, but it is clear from the symmetry of Eq. (4-29) that $D_{12} = D_{21}$. Since the carrier gas is air in indoor air pollution problems, the diffusion coefficient of contaminant j into air (a) is denoted by D_{ja}. Other equations to predict the diffusion coefficient can be found in Bird et al. (1960), Treybal (1968), Sherwood et al. (1975), and Perry et al. (1984). Evaluations of the best methods are provided in Reid et al. (1977). Appendix A.10 lists values of critical temperature and pressure. Values of critical specific volume can be computed from

$$v_c = Z_c \frac{R_u T_c}{P_c} \tag{4-30}$$

where Z_c is the ***compressibility factor*** at the critical point (T_c, P_c). Recall from Chapter 1 that Z is nearly unity for gases near STP. However, at the critical point, Z is *not* close to unity. In fact, values of Z_c vary from about 0.23 to 0.33 for different substances, and a value of 0.27 is a good average for many substances encountered in indoor air pollution.

The diffusion coefficient (D_{12}) depends on temperature and pressure. At temperatures and pressures other than those for which D_{12} is known, one can estimate the diffusion coefficient with the following:

$$D_{12}(P_2,T_2) = D_{12}(P_1,T_1)\left(\frac{P_2}{P_1}\right)\left(\frac{T_2}{T_1}\right)^{1.81} \tag{4-31}$$

where the pressures and temperatures must be in consistent *absolute* units, such as atmospheres (atm) and Kelvin (K). In the absence of the necessary thermodynamic data, a rough estimate of the diffusion coefficient of a molecular species j in air (subscript a) is

$$D_{ja} = D_{water,a}\sqrt{\frac{M_{water}}{M_j}} \tag{4-32}$$

where M_{water} is the molecular weight of water (18.0) and $D_{water,a}$ is the diffusion coefficient of water vapor in air (2.2 x 10^{-5} m^2/s at zero °C and one atm).

A widely recognized correlation used to estimate the ***liquid phase binary diffusion coefficient***, D_{12}, of contaminant (1) at very low concentration in liquid (2) (in units of cm^2/s) is the ***Wilke-Chang equation*** (Reid et al., 1977),

$$D_{12} = \frac{7.4\text{x}10^{-8} T\sqrt{\phi M_2}}{\mu_2 v_1^{0.6}} \frac{cm^2}{s} \tag{4-33}$$

where each variable is defined below, along with its required units:

- M_2 = molecular weight of the liquid (g/mol)
- T = absolute temperature (K)
- μ_2 = viscosity of the liquid (centipoise)
- v_1 = molar specific volume of the contaminant (1) at its normal boiling temperature (cm^3/mol)
- ϕ = ***association factor***, a dimensionless empirical constant (unitless); e.g. ϕ(water) = 2.6, ϕ(methanol) = 1.9, and ϕ(ethanol) = 1.5

Figure 4.8 shows the comparison between Eq. (4-33) and measured data for several hydrocarbons in water. The agreement is quite good.

A review of tabulated diffusion coefficients for a variety of materials reveals that the range of values is small. For design purposes, inaccuracies in the diffusivity of 10% or so are not significant because actual flow fields are nearly always turbulent and the ***turbulent diffusivity*** is vastly larger than the molecular diffusivity. Uncertainty in computed turbulent diffusivities is substantial, so one should be content using molecular diffusivities of modest accuracy.

Prediction of diffusion coefficients for the liquid phase is more difficult than for the gas phase. Diffusivities in concentrated solutions are different than diffusivities in dilute solutions and they display more irregularity. Diffusion in electrolytes is particularly complex. Primary reference sources should be consulted for liquid phase diffusivities.

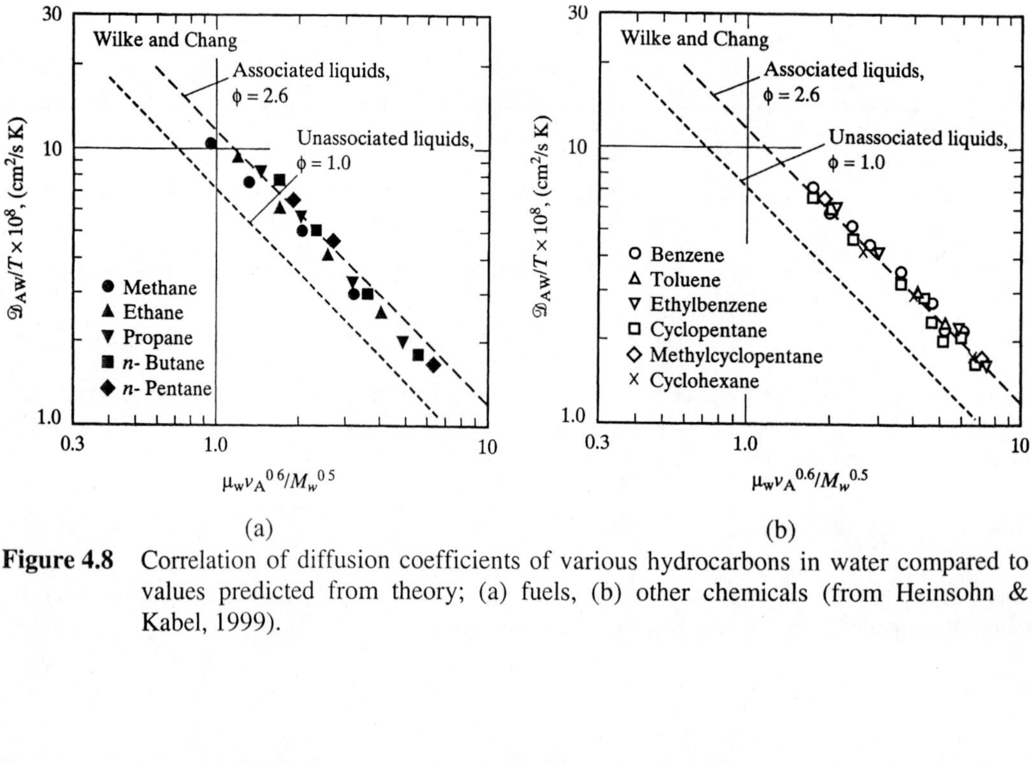

Figure 4.8 Correlation of diffusion coefficients of various hydrocarbons in water compared to values predicted from theory; (a) fuels, (b) other chemicals (from Heinsohn & Kabel, 1999).

Example 4.9 – Estimation of the Diffusion Coefficient of Toluene in Air

Given: Toluene (C_6H_5OH, CAS 108-88-3) is a common solvent with a sweet, pungent odor and a PEL equal to 200 PPM. From handbooks and Appendix A.10 the following data can be found:

air (use nitrogen)	**toluene**
$T_c = 126.2$ K	$T_c = 593.75$ K
$v_c = 89.9$ cm^3/mol	$v_c = 315.5$ cm^3/mol
M = 28.02 g/mol	M = 92.13 g/mol

To do: Compute the diffusion coefficient of toluene in air at STP using Eq. (4-29).

Solution: Substitution of values from the above table into Eq. (4-29) results in

$$D_{12} = 0.0766\frac{\text{cm}^2}{\text{s}} \cong 7.7\times10^{-6}\frac{\text{m}^2}{\text{s}}$$

Discussion: The value of diffusivity tabulated in Appendix A.9 is 7.1 x 10^{-6} m^2/s. The computed value can also be compared to values obtained from Figure 4.8.

4.5.3 Evaporation

Figure 4.9 depicts a cylindrical vessel open to the air at the top. The vessel contains a small amount of liquid contaminant (j) at the bottom, which evaporates and diffuses through a column of air, molecular species (a). The motion of the contaminant is expressed in terms of mols of j crossing a unit area (normal to the direction of diffusion) per unit time. It is assumed that there are no transverse gradients but only gradients in the vertical direction. The motion of the contaminant vapor can be described in terms of two fluxes (along with their typical units):

- $\vec{N}_j$ = molar flux of j relative to a fixed observer, kmol/(m^{2}s)
- $\vec{J}_j$ = molar flux of j relative to the molar average velocity of all the constituents (to be defined shortly), kmol/(m^{2}s)

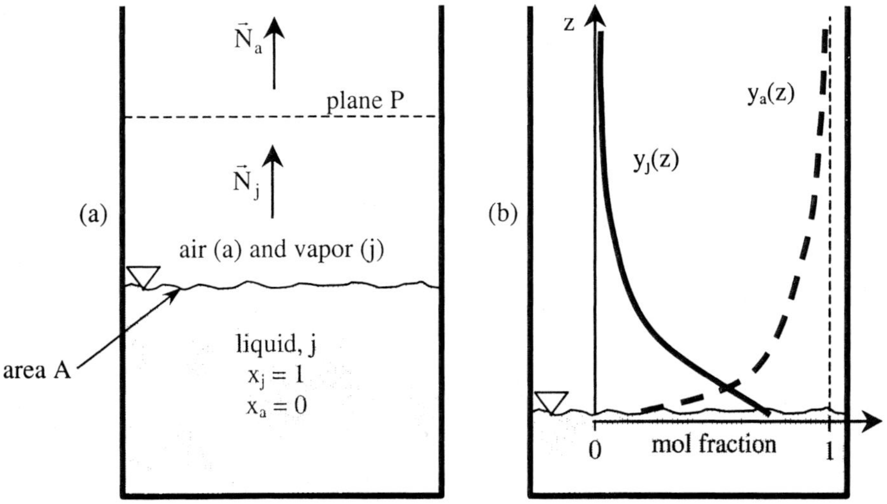

Figure 4.9 Evaporation of a volatile liquid through quiescent air, with $\vec{N} = \vec{N}_a + \vec{N}_j$ across any plane P; (a) some arbitrary time, (b) some later time.

The first molar flux $\vec{N}_j$ is important for the purpose of design and performance of equipment. The second flux $\vec{J}_j$ reflects the nature of the diffusing species and the medium through which it moves; $\vec{J}_j$ is expressed by ***Ficks law***:

$$\boxed{\vec{J}_j = -D_{ja}\,\overrightarrow{\text{Grad}}\left(\frac{c_j}{M_j}\right) = -D_{ja}\,\vec{\nabla}\left(\frac{c_j}{M_j}\right)} \tag{4-34}$$

where c_j is the mass concentration and M_j is the molecular weight of species j. $\overrightarrow{\text{Grad}}$ is the gradient vector operator, $\vec{\nabla}$. Consider the column of gas in Figure 4.9, of unit cross-sectional area, in which high concentrations of contaminant (j) and air (a) exist at opposite ends of the element. At plane P somewhere between the ends of the element, air and contaminant diffuse in opposite directions, each moving from a region of high concentration to low concentration. The rate at which mols of air (species a) pass the stationary observer per unit area is $\vec{N}_a$,

$$\boxed{\vec{N}_a = \vec{U}_a \frac{c_a}{M_a} = \vec{U}_a c_{molar,a}} \tag{4-35}$$

where c_a is the mass concentration of air, $c_{molar,a}$ is the molar concentration of air, and M_a is the molecular weight of air. Similarly, the rate at which mols of the contaminant (species j) pass the same observer per unit area is $\vec{N}_j$,

$$\boxed{\vec{N}_j = \vec{U}_j \frac{c_j}{M_j} = \vec{U}_j c_{molar,j}} \tag{4-36}$$

The volatile liquid evaporates and the vapor moves upward with speed $|\vec{U}_j|$. Meanwhile, air moves downward with speed $|\vec{U}_a|$ to fill the void from the evaporated liquid. These velocities are barely observable but nonetheless exist relative to a stationary observer. The net molar transfer rate per unit area past the observer is

$$\vec{U}_m c_{molar,t} = \vec{N}_a + \vec{N}_j = \vec{U}_a c_{molar,a} + \vec{U}_j c_{molar,j}$$

where the ***molar average velocity*** ($\vec{U}_m$) is defined as

$$\boxed{\vec{U}_m = \frac{\vec{N}_a + \vec{N}_j}{c_{molar,t}} = \frac{\vec{U}_a c_{molar,a} + \vec{U}_j c_{molar,j}}{c_{molar,t}}} \tag{4-37}$$

where $c_{molar,t}$ is the ***total molar concentration***,

$$\boxed{c_{molar,t} = c_{molar,a} + c_{molar,j}} \tag{4-38}$$

The molar flux of species j relative to the bulk gas can be written as

$$\boxed{\vec{J}_j = c_{molar,j}\left(\vec{U}_j - \vec{U}_m\right)} \tag{4-39}$$

and the molar flux of air relative to the bulk gas can be written as

$$\boxed{\vec{J}_a = c_{molar,a}\left(\vec{U}_a - \vec{U}_m\right)} \tag{4-40}$$

Replacing $\vec{J}_j$ by Eq. (4-34) and $\vec{U}_m$ by Eq. (4-37) and rearranging, one obtains

$$\boxed{\vec{N}_J = \vec{U}_J c_{molar,J} = \frac{c_{molar,J}}{c_{molar,t}}\left(\vec{N}_a + \vec{N}_J\right) - D_{ja}\,\vec{\nabla} c_{molar,j}} \tag{4-41}$$

where Eq. (4-36) has also been used. Similarly,

$$\boxed{\vec{N}_a = \vec{U}_a c_{molar,a} = \frac{c_{molar,a}}{c_{molar,t}}\left(\vec{N}_a + \vec{N}_j\right) - D_{ja}\,\vec{\nabla} c_{molar,a}} \tag{4-42}$$

Equations (4-41) and (4-42) relate molar concentration to molar flow rate. To obtain an equation involving the concentration only, consider a volume element in space through which a fluid composed of several molecular species flows. A mass balance results in the following expression, a more general form of Eq. (4-22), hereafter called the ***species continuity equation***:

$$\boxed{\frac{\partial c_J}{\partial t} + \overrightarrow{\mathrm{Div}}\left(\vec{U} c_J\right) = \frac{\partial c_J}{\partial t} + \vec{\nabla}\cdot\left(\vec{U} c_J\right) = D_{ja}\,\nabla^2 c_j + S_{j,vol}} \tag{4-43}$$

where $\overrightarrow{\mathrm{Div}}$ is the divergence operator and $S_{J,vol}$ represents the ***rate of production*** (per unit volume) of molecular species j by a chemical reaction, i.e. a chemical source of species j. The dimensions of $S_{J,vol}$ here are mass/(volume time). In a nonreacting gas mixture, $S_{j,vol}$ is zero. Equation (4-43) can be written in terms of molar concentration instead of mass concentration by dividing each term by M_J.

Consider diffusion in the z-direction in which $\vec{N}_a$, $\vec{N}_J$, and D_{ja} are all constant. Since there is only one spatial coordinate (z, with unit vector $\vec{k}$), components of vectors in the z-direction such as $N_{J,z}$ and $J_{J,z}$, etc. are written simply as N_j and J_j etc. for the sake of brevity. Additionally, the gradient operator has a non-zero component only in the z-direction,

$$\vec{\nabla} c_{molar,j} = \frac{dc_{molar,j}}{dz}\vec{k}$$

Eq. (4-41) can then be simplified and, after some rearrangement, integrated between any two z locations (z_1 and z_2) to yield

$$\boxed{\int_{c_{molar,J,1}}^{c_{molar,J,2}} \frac{dc_{molar,J}}{c_{molar,t} N_J - c_{molar,J}\left(N_a + N_J\right)} = -\int_{z_1}^{z_2} \frac{dz}{c_{molar,t}\, D_{ja}}} \tag{4-44}$$

which when integrated and rearranged becomes

$$\boxed{N_j = \frac{N_J}{N_a + N_j}\,\frac{c_{molar,t}\, D_{ja}}{z_2 - z_1}\ln\left[\frac{\dfrac{N_J}{N_j + N_a} - \dfrac{c_{molar,J,2}}{c_{molar,t}}}{\dfrac{N_j}{N_J + N_a} - \dfrac{c_{molar,j,1}}{c_{molar,t}}}\right]} \tag{4-45}$$

Finally, since N_J has dimensions of mols per unit area per time, one must multiply N_j by both the molecular weight (M_J) and the cross-sectional area (A) of the evaporating surface to obtain an expression for the evaporation mass flow rate (mass per time),

$$\boxed{\dot{m}_{evap,j} = N_J M_J A} \tag{4-46}$$

The above general expressions can be used to predict the rate of evaporation for several types of processes of interest, such as evaporation of a pure liquid through stagnant air and evaporation of a volatile species from a homogeneous liquid mixture through stagnant air.

4.5.4 Evaporation and Diffusion through Quiescent Air

Consider diffusion of the vapor of a volatile liquid (denoted by subscript j) through a stagnant layer of air (denoted by subscript a). Such mass transfer is also called "pure diffusion" because no convective effects exist in the medium through which diffusion occurs. An example of such diffusion is the evaporation of liquid from a partially filled tank or container that is open at the top, as in Figure 4.10. Assume that room air currents over the top of the container sweep the vapors away from the opening of the container and sweep air into the container, but do not induce any motion of the air and vapor inside the container. Assume also that air is not absorbed by the volatile liquid, and that air currents carrying away vapors establish a vapor concentration at the mouth of the container that is negligibly small. The temperature and pressure are assumed to be constant. Assume that radial gradients are zero and that there is only one spatial coordinate (z). Thus, vector quantities are understood to be in the z-direction. Since air is not absorbed by the liquid, $N_a << N_j$, N_J is essentially constant, and $N_j/(N_J + N_a) \cong 1$. Equation (4-45) simplifies to

$$N_J = \frac{c_{molar,t}\, D_{Ja}}{z_2 - z_1} \ln\left[\frac{1-\dfrac{c_{molar,j,2}}{c_{molar,t}}}{1-\dfrac{c_{molar,j,1}}{c_{molar,t}}}\right] \tag{4-47}$$

Assuming that both the air and the vapor can be described by the ideal gas equation, the mol fraction of the vapor (y_J) is related to molar concentrations, total pressure, and partial pressure by the following set of equations:

$$P = c_{molar,t} R_u T \tag{4-48}$$

$$P_J = c_{molar,j} R_u T \tag{4-49}$$

$$y_J = \frac{n_J}{n_t} = \frac{c_{molar,J}}{c_{molar,t}} = \frac{P_J}{P} \tag{4-50}$$

and Eq. (4-38) can be rearranged as

$$1 = \frac{c_{molar,a}}{c_{molar,t}} + \frac{c_{molar,j}}{c_{molar,t}} = y_a + y_b \tag{4-51}$$

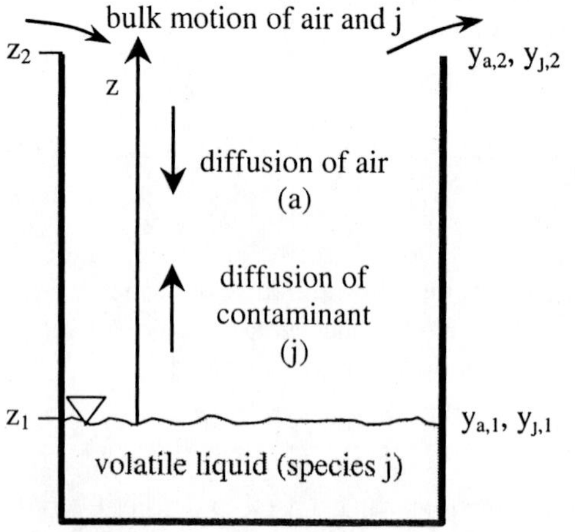

Figure 4.10 Evaporation of a volatile liquid (species j) from a container open at the top, partially filled with the volatile liquid.

Equation (4-47) can then be rewritten as

$$N_j = \left[\frac{P\,D_{Ja}}{R_uT(z_2 - z_1)}\right]\ln\left(\frac{y_{a,2}}{y_{a,1}}\right) \tag{4-52}$$

A ***log mean mol fraction*** (y_{am}) and a ***log mean partial pressure ratio*** (P_{am}) are defined as

$$y_{am} = \frac{y_{a,2} - y_{a,1}}{\ln\left(\frac{y_{a,2}}{y_{a,1}}\right)} \tag{4-53}$$

and

$$P_{am} = \frac{(P_{a,2} - P_{a,1})}{\ln\left(\frac{P_{a,2}}{P_{a,1}}\right)} \tag{4-54}$$

respectively. y_{am} and P_{am} are related as follows:

$$y_{am} = \frac{y_{a,2} - y_{a,1}}{\ln\left(\frac{y_{a,2}}{y_{a,1}}\right)} = \frac{\dfrac{P_{a,2} - P_{a,1}}{\ln\left(\frac{P_{a,2}}{P_{a,1}}\right)}}{P} = \frac{P_{am}}{P} \tag{4-55}$$

furthermore, since

$$y_{a,2} - y_{a,1} = (1 - y_{a,1}) - (1 - y_{a,2}) = y_{J,1} - y_{J,2}$$

y_{am} can be written as

$$y_{am} = \frac{y_{j,1} - y_{j,2}}{\ln\left(\frac{y_{a,2}}{y_{a,1}}\right)} \tag{4-56}$$

The natural log term in Eq. (4-52) can be replaced by use of Eq. (4-56), resulting in the molar flux N_j being expressed as

$$N_J = P\left[\frac{D_{Ja}}{R_uT(z_2 - z_1)\,y_{am}}\right](y_{J,1} - y_{J,2}) \tag{4-57}$$

or, using Eq. (4-55),

$$N_J = P\left[\frac{P\,D_{Ja}}{R_uT(z_2 - z_1)\,P_{am}}\right](y_{J,1} - y_{J,2}) \tag{4-58}$$

Because of concepts that are developed in subsequent sections, it is instructive to define the terms within the square brackets as a ***mass transfer coefficient*** (k_G). For pure diffusion, k_G is defined as

$$k_G = \frac{D_{Ja}}{R_uT(z_2 - z_1)\,y_{am}} \tag{4-59}$$

from Eq. (4-57). Or, from Eq. (4-58),

$$k_G = \frac{P D_{ja}}{R_u T (z_2 - z_1) P_{am}} \tag{4-60}$$

Convenient units of k_G are kmol/(m^2 s kPa). Using the mass transfer coefficient (k_G), the molar flux (N_J) in Eq. (4-57) or (4-58) can be expressed as

$$N_j = k_G P (y_{j,1} - y_{j,2}) \tag{4-61}$$

Alternatively, the driving potential ($y_{J,1}$ - $y_{J,2}$) can be expressed in terms of molar concentration difference ($c_{molar,J,1} - c_{molar,J,2}$) or as the difference in partial pressures ($P_{j,1}$ - $P_{j,2}$). Using Eqs. (4-48) and (4-50), the molar flux (N_J) can be written instead as

$$N_J = k_G R_u T (c_{molar,J,1} - c_{molar,J,2}) = k_G (P_{j,1} - P_{j,2}) \tag{4-62}$$

Points 1 and 2 are chosen by the engineer to identify two different locations in the particular application. Often, points 1 and 2 correspond to the liquid-gas interface (subscript i) and the top of the container or tank (subscript 2 in Figure 4.10) respectively.

When the interfacial contaminant vapor pressure is considerably less than the total pressure ($P_{J,I}$ << P) and the far-field contaminant vapor pressure is negligible, ($P_{j,\infty} \cong 0$ and $P_{a,\infty} \cong$ P), P_{am} may be difficult to evaluate. Help can be gained by expressing the logarithm of ($P_{a,i}/P_{a,\infty}$) as a truncated Taylor series,

$$\ln\left(\frac{P_{a,I}}{P_{a,\infty}}\right) = \ln\left(\frac{P - P_{J,I}}{P}\right) = \ln\left(1 - \frac{P_{j,I}}{P}\right) \approx -\frac{P_{J,i}}{P}$$

Thus

$$P_{am} = \frac{(P_{a,i} - P_{a,\infty})}{\ln\left(\frac{P_{a,i}}{P_{a,\infty}}\right)} \approx \frac{(P - P_{j,I}) - P}{-\frac{P_{j,I}}{P}} \approx P \tag{4-63}$$

Similarly when $1 >> y_{J,I} > y_{j,\infty} \cong 0$,

$$y_{am} = \frac{(y_{a,I} - y_{a,\infty})}{\ln\left(\frac{y_{a,i}}{y_{a,\infty}}\right)} \approx \frac{(y_{a,i} - 1) - P}{\ln(y_{a,I})} \approx \frac{(y_{a,i} - 1)}{(y_{a,i} - 1)} = 1 \tag{4-64}$$

In summary, when the contaminant vapor pressure at the liquid-air interface is small and the far-field contaminant partial pressure is negligible, it can be assumed that P_{am} = P.

The concentration profile that accompanies diffusion through a stagnant air layer can be found by substituting $N_a \cong 0$ into Eq. (4-41),

$$N_J = -\left[\frac{c_t D_{ja}}{(1 - y_j)}\right] \frac{dy_j}{dz}$$

Since N_J, D_{ja}, and c_t are constants, these can be conveniently combined into one new constant, C_1,

$$C_1 = -\frac{N_J}{c_t D_{ja}} \tag{4-65}$$

Integration yields

$$-\ln\left(1-y_j\right)=C_1 z+C_2$$

The two constants may be eliminated by knowledge of the concentration at locations z_1 and z_2, i.e. $y_j(z_1) = y_{j,1}$ and $y_j(z_2) = y_{j,2}$. After some rearranging, the concentration profile becomes

$$\boxed{\frac{1-y_j}{1-y_{j,1}}=\left[\frac{1-y_{j,2}}{1-y_{j,1}}\right]^{\left(\frac{z-z_1}{z_2-z_1}\right)}} \tag{4-66}$$

Example 4.10 - Evaporation of PERC from a Nearly Empty Drum

Given: A dry cleaning establishment uses perchloroethylene, commonly called PERC or "perk" ($Cl_2C{=}CCl_2$, CAS 127-18-4), as the fluid to clean clothes. PERC is a colorless liquid with a chloroform-like odor and a PEL of 100 PPM. The liquid arrives in a 55-gallon drum and is transferred to the cleaning equipment. A nearly empty drum with an open top is improperly stored inside the facility and a small pool of PERC in the bottom of the drum evaporates into the workplace. The drum has a cross-sectional area of 0.25 m^2 and a height, $(z_2 - z_1)$, of 0.813 m. From the MSDS and Appendices A.8, A.9, and A.20, PERC has the following properties:

- $M = 165.8$, $\rho = 1.62$ g/cm^3
- odor recognition threshold (high value) = 69. PPM
- boiling point (BP) = 250. °F
- vapor pressure, P_v(at 20 °C) = 14.0 mm Hg
- $D_{ja} = 0.74 \times 10^{-5}$ m^2/s

To do: Estimate the rate at which PERC evaporates.

Solution: In the far-field, $P_{a,2} = P$, since $P_{j,2} \approx 0$. At the liquid interface, the partial pressure of the PERC is equal to the vapor pressure at standard atmospheric conditions, $P_{j,1}$ = 14.0 mm Hg. Thus,

$$P_{a,1}=P-P_{j,1}=\left(760.-14.0\right)\text{ mm Hg}=746.\text{ mm Hg}$$

From Eq. (4-54),

$$P_{am}=\frac{\left(P_{a,2}-P_{a,1}\right)}{\ln\left(\frac{P_{a,2}}{P_{a,1}}\right)}=\frac{\left(760.-746.\right)\text{ mm Hg}}{\ln\left(\frac{760.}{746.}\right)}=753.0\text{ mm Hg}$$

The mass transfer coefficient (k_G) can be obtained from Eq. (4-60),

$$k_G=\frac{P D_{ja}}{R_u T\left(z_2-z_1\right)P_{am}}=\frac{\left(760.\text{ mm Hg}\right)\left(0.74\times10^{-5}\,\frac{m^2}{s}\right)}{8.314\frac{kJ}{kmol\cdot K}\left(293.15.\text{ K}\right)0.813\text{ m}\left(753.0\text{ mm Hg}\right)}\left(\frac{kJ}{kN\cdot m}\right)\left(\frac{kN}{kPa\cdot m^2}\right)$$

$$k_G=3.769\times10^{-9}\,\frac{kmol}{kPa\cdot m^2\cdot s}$$

The molar flux is then found from Eq. (4-62),

$$N_j=k_G\left(P_{j,1}-P_{j,2}\right)=$$

$$=3.769\times10^{-9}\,\frac{kmol}{kPa\cdot m^2\cdot s}\left[\left(14.0-0\right)\text{ mm Hg}\right]\left(\frac{101.3\text{ kPa}}{760.0\text{ mm Hg}}\right)=7.034\times10^{-9}\,\frac{kmol}{m^2 s}$$

Finally, the mass flow rate of the PERC is obtained from Eq. (4-46),

$$\dot{m}_{evap,j} = N_j M_j A = \left(7.034\times10^{-9}\ \frac{kmol}{m^2 s}\right)\left(166.0\frac{kg}{kmol}\right)\left(0.25\ m^2\right) = 2.92\times10^{-7}\ \frac{kg}{s}$$

which is approximately 1.03 g/hr.

If the drum contains a different volatile liquid, the evaporation rate can be found in similar fashion as that obtained for perchloroethylene in Example 4.10. A compilation of vapor pressures, diffusion coefficients, and evaporation rates for other volatile liquids is shown in Table 4.3. Also listed in Table 4.3 are the evaporation rates relative to carbon tetrachloride (CCl_4), since AP-42 emission factors for evaporation of volatile hydrocarbons are listed as a fraction of the evaporation rate of carbon tetrachloride, as shown in the last portion of Appendix A.3. The rates of evaporation in Table 4.3 relative to carbon tetrachloride are in fair agreement with the emission factor data of Appendix A.3. From Eqs. (4-60) and (4-62), evaporation rate is directly proportional to both the molecular diffusion coefficient and the vapor pressure of the contaminant. However, while P_v varies by more than an order of magnitude for the chemicals listed in Table 4.3, the diffusion coefficient varies by less than a factor of 2. Hence, vapor pressure has the larger influence on evaporation rate.

Example 4.11 - Emissions of HCl from Aqueous HCl Solutions

Given: Consider the emission of hydrogen chloride (HCl) from drums of aqueous HCl solutions inadvertently left open. Assume the drums are half full and that the distance (z_2 - z_1) is 50. cm.

To do: Estimate the evaporation rate for several concentrations and temperatures.

Solution: Assuming room air sweeps the acid fumes away from the opening such that $y_{j,2}$ is very small with respect to unity, the emission of HCl can be estimated from Eqs. (4-60) and (4-62). In attempting to perform the calculations two questions present themselves:

- what is the diffusion coefficient of HCl in air?
- can the aqueous solution of HCl be considered to be homogeneous, i.e. is the acid concentration at the air-liquid interface equal to the overall concentration?

The diffusion coefficient of polar compounds in air is difficult to predict, but for a first approximation Eq. (4-29) is used. The computation results in

$$D_{HCl,a} = 1.864\times10^{-5}\ \frac{m^2}{s}$$

Table 4.3 Vapor pressure, diffusion coefficient, and evaporation rate for several liquids; evaporation rate is also compared to that of carbon tetrachloride.

hydrocarbon	**P_v(at T_{room}) (mm Hg)**	**D_{ja} (10^{-5} m^2/s)**	**evap. rate (g/hr)**	**evap. rate relative to CCl_4**
acetone	180	0.83	7.95	1.11
carbon tetrachloride	91	0.62	7.16	1.00
benzene	75	0.77	3.90	0.54
toluene	20	0.71	1.20	0.17
acetic acid	11	1.06	0.75	0.10
chlorobenzene	12	0.62	0.23	0.03

Uncertainty about the value of the acid concentration at the liquid-air interface is more difficult to resolve. The rigorous approach would be to do a simultaneous diffusional analysis on the transport in the liquid phase. In the case of strong electrolytes dissolved in water, however, the diffusion rates are those of the individual ions that move rapidly (McCabe et al., 1993). Thus a good estimate of the maximum emission rate is to neglect diffusional resistance in the liquid film and assume that the liquid HCl concentration at the liquid-air interface is equal to the overall liquid concentration. The equilibrium partial pressure of HCl over an aqueous HCl solution can be obtained from Perry et al. (1984). The table below shows the estimated HCl emission rate at 25 °C for a variety of HCl concentrations:

% HCl (by mass)	$P_{j,1}$ (mm Hg)	N_j [kmol HCl/(m^2 s)]
20	0.32	6.30×10^{-10}
24	1.49	2.95×10^{-9}
28	7.05	1.40×10^{-8}
32	32.5	6.57×10^{-8}
36	142	3.11×10^{-7}
40	515	1.70×10^{-6}

To illustrate the sensitivity of HCl emissions to temperature, the above computations can be repeated for a 30% acid concentration but varying temperature:

T (°C)	$P_{j,1}$ (mm Hg)	N_j [kmol HCl/(m^2 s)]
15	7.6	1.51×10^{-8}
20	10.6	2.11×10^{-8}
25	15.1	3.00×10^{-8}
30	21.0	4.21×10^{-8}
35	28.6	5.77×10^{-8}

Discussion: From these calculations it should be apparent that the vapor pressure of the volatile component is the dominant factor controlling the emission rate of the contaminant, as mentioned previously. Thus whether due to the concentration or the temperature, *an increase of vapor pressure results in an increase of the rate of evaporation.*

4.5.5 Evaporation of Single Component Liquids

This section develops the fundamental equations describing evaporation of a volatile liquid consisting of a single molecular species, i.e. a pure species or ***single component liquid***. Sections 4.5.6 and 4.5.7 are devoted to ***multi-component liquids***, evaporation of one or more volatile liquid(s) dissolved in other liquid(s). Evaporation of multi-component liquids is also called ***gas scrubbing***, ***gas stripping*** (see Figure 4.11), ***desorption***, or more generally ***distillation***. For the most part the liquid from which the volatile material evaporates is water although the analysis is not necessarily restricted to water. The opposite effect, pollutants absorbed by water, is discussed in Chapter 9 when describing air pollution control systems called ***scrubbers***.

Consider evaporation of a single molecular species into air passing over it at velocity U_∞. When air passes over a volatile liquid, the volatile material evaporates if its partial pressure in the gas phase is sufficiently low. Figure 4.12 is a diagram of the velocity and concentration profiles above the air-liquid interface. Evaporation is an endothermic process, but it is assumed that there is sufficient heat transfer from the surroundings to maintain a constant liquid temperature. When evaporation occurs in conjunction with the motion of air over the liquid surface, the molar flux of molecular species j normal to a unit area is expressed as

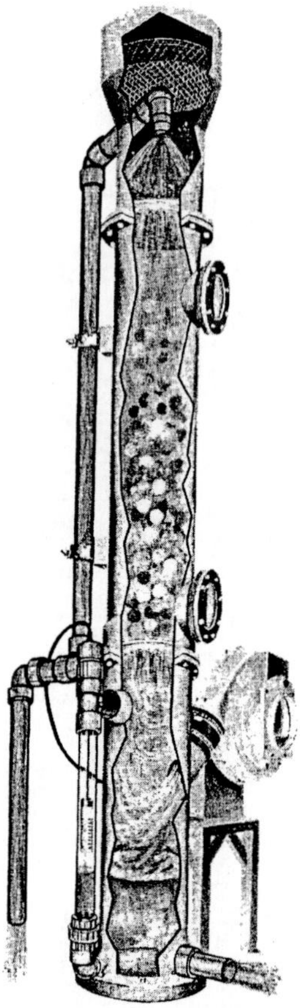

Figure 4.11 Gas scrubbing using non-clogging fluidized non-spherical packing (from Diversified Remediation Controls, Inc.).

$$\boxed{N_J = k_G\left(P_{J,I} - P_{J,\infty}\right)} \quad (4\text{-}67)$$

where the variables are defined (along with typical units) as

- N_J = molar flux [kmol/(m^{2}s)]
- k_G = ***gas phase mass transfer coefficient*** [kmol/(m^{2}s kPa)]
- $P_{J,I}$ = partial pressure of species j at the liquid-gas interface (mm Hg or kPa)
- $P_{J,\infty}$ = partial pressure of species j in the far-field (mm Hg or kPa)

For a pure liquid of species j, $P_{J,I}$ is equal to the vapor pressure $P_{v,J}$.

For mass transfer by bulk flow it is useful to express the gas-phase mass transfer coefficient in dimensionless form by introducing the ***Sherwood number***,

$$\boxed{Sh = C_1\left(Re\right)^{a_1}\left(Sc\right)^{b_1}} \quad (4\text{-}68)$$

where C_1, a_1, and b_1 are dimensionless quantities that depend on the geometry of the evaporating surface and the range of Reynolds numbers under consideration. The three dimensionless parameters, Sherwood, Reynolds, and ***Schmidt number*** are defined by Eqs. (4-69) through (4-71).

$$\boxed{\text{Sh (Sherwood number)} = \frac{k_G R_u T P_{am} L}{P\, D_{Ja}}} \quad (4\text{-}69)$$

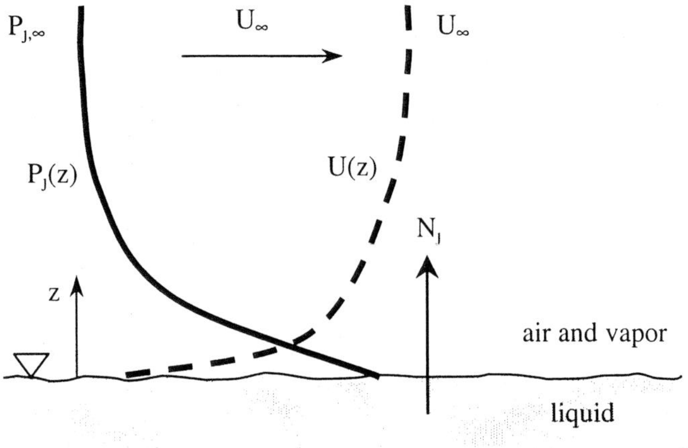

Figure 4.12 Mass transfer through a single-film; concentration and velocity profiles in air passing over a volatile single molecular species liquid.

$$\text{Re (Reynolds number)} = \frac{LU_\infty \rho}{\mu} = \frac{LU_\infty}{\nu} \tag{4-70}$$

$$\text{Sc (Schmidt number)} = \frac{\mu}{D_{Ja}\,\rho} = \frac{\nu}{D_{Ja}} \tag{4-71}$$

Characteristic length (L) is chosen by the user or defined in a particular empirical equation. It is the practice in heat transfer to also express the film coefficient in dimensionless form as

$$\text{Nu} = C_2 \left(\text{Re}\right)^{a_2} \left(\text{Pr}\right)^{b_2} \tag{4-72}$$

where C_2, a_2, and b_2 are dimensionless constants that depend on the geometry of the surface and the range of Reynolds numbers under consideration. The two new parameters, ***Nusselt number*** and ***Prandtl number***, are also dimensionless and are defined as

$$\text{Nu (Nusselt number)} = \frac{Lh}{k} \tag{4-73}$$

$$\text{Pr (Prandtl number)} = \frac{\mu c_P}{k} \tag{4-74}$$

where h is the ***heat transfer coefficient***. In all of the above equations, the dynamic viscosity (μ), kinematic viscosity (ν), density (ρ), and thermal conductivity (k) pertain to the *air*, not the liquid. Since the contaminant concentration is usually very small, the properties of pure air can be used.

From the theory of mass and heat transfer (see Treybal, 1980 and Incropera and DeWitt, 1990 for details) it can be shown that the local ***velocity boundary layer thickness*** (δ_u) is proportional to the local ***thermal boundary layer thickness*** (δ_T), with the proportionality constant determined by the Prandtl number (Pr),

$$\delta_u = \text{Pr}^n \, \delta_T \tag{4-75}$$

where n is a positive number, typically 3. Similarly the velocity boundary layer thickness (δ_u) is proportional to the concentration boundary layer thickness (δ_m), with the proportionality constant determined by the Schmidt number (Sc),

$$\delta_u = \text{Sc}^n \delta_m \tag{4-76}$$

The ratio of the Schmidt and Prandtl numbers is called the ***Lewis number*** (Le),

$$\text{Le} = \frac{\text{Sc}}{\text{Pr}} = \frac{k}{\rho c_P D_{ja}} \tag{4-77}$$

From the above equations,

$$\frac{\delta_T}{\delta_m} = \text{Le}^n \tag{4-78}$$

Heat and mass transfer processes can be described by fundamental equations containing convection and diffusion terms of similar form. When these equations are made dimensionless (Incropera and Dewitt, 1990) one finds that each equation is related to the velocity through the Reynolds number (Re), while the Prandtl and Schmidt numbers depend on the temperature and composition of the phase of interest, albeit differently in detail. Because the dimensionless equations are similar, it is qualitatively useful to think of heat and mass transfer as analogous processes, an analogy called ***Reynolds analogy***. Indeed if the respective dimensionless parameters are equal, the solution of one equation is interchangeable with the solution of the other providing the boundary conditions are analogous. The Reynolds analogy is useful as a starting point, in which case Eqs. (4-68) and (4-72) are comparable to each other, i.e. $C_1 = C_2$, $a_1 = a_2$, and $b_1 = b_2$, which implies

$$\text{Sh} = \text{Nu}\left(\frac{\text{Sc}}{\text{Pr}}\right)^{b_1} \tag{4-79}$$

and the gas-phase mass transfer coefficient k_G can be written as

$$k_G = \text{Nu}\frac{D_{ja}}{L}\left(\frac{\text{Sc}}{\text{Pr}}\right)^{b_1}\frac{P}{P_{am}}\frac{1}{R_u T} \tag{4-80}$$

In this way one could estimate mass transfer behavior from heat transfer data.

Analogies between heat and mass transfer have been emphasized here because heat transfer correlations are often more readily available. This is because research to generate them has been well-supported and sustained, and experiments in heat transfer that generate correlations are easier to conduct. Table 4.3 represents a variety of heat transfer relationships, and more relationships can be obtained from heat transfer texts. Bird et al. (1960) analyze the equivalence of heat and mass transfer theory and point out that true equivalence requires:

- constant physical properties
- small rate of mass transfer
- no chemical reactions in the fluid
- negligible viscous dissipation
- negligible emission or absorption of radiant energy
- no pressure diffusion, thermal diffusion, or forced diffusion

These restriction are not constraining in most air pollution calculations, but users of analogies should be aware of their limitations.

Example 4.12 - Estimating Evaporation Rate from Fundamental Principles

Given: Ethyl mercaptan (CH_3CH_2SH, CAS 75-08-1) is a liquid with an unpleasant skunk-like odor. The material is a strong oxidizer with a PEL of 0.5 PPM. A 55-gallon drum of the material is handled roughly and a small leak develops in a seam. A circular pool, approximately 10. m in diameter develops. The wind speed over the pool is estimated to be around 3.0 m/s.

Table 4.3 Convection heat transfer and mass transfer relationships (abstracted from Bird et al., 1960 and Incropera and DeWitt, 1990).

1. *Air flowing over a flat surface* $\quad Nu = \frac{Lh}{k} \quad Pr = \frac{c_P \mu}{k} \quad Re = \frac{\rho U_\infty L}{\mu}$

Laminar flow: $Pr > 0.6 \quad Re < 5 \times 10^5 \quad Nu = 0.664(Re)^{0.5}(Pr)^{0.33}$

Turbulent flow: $0.6 < Pr < 60 \quad 5 \times 10^5 < Re < 10^8 \quad Nu = \left(0.037(Re)^{0.80} - 871\right)(Pr)^{0.33}$

2. *Air flowing through a cylindrical duct* $\quad Sh = k_G \frac{R_u T L P_{am}}{D_{ja} P}$

Turbulent flow: $0.6 < Sc < 300 \quad 4{,}000 < Re < 60{,}000 \quad Sh = 0.023(Re)^{0.83}(Sc)^{0.33}$

3. *Air flowing over a stationary sphere* (L = D)

Laminar or turbulent flow: $0.71 < Pr < 380 \quad 3.5 < Re < 7.6 \times 10^4 \quad 1.0 < \mu_\infty/\mu_S < 3.2$

$$Nu = 2 + \left[0.4\sqrt{Re} + 0.06(Re)^{2/3}\right](Pr)^{0.4}\left(\frac{\mu_\infty}{\mu_s}\right)^{1/4}$$

where viscosity μ_∞ is evaluated at the far-field temperature and viscosity μ_s is evaluated at the average surface temperature.

4. *External flowing over plates and cylinders* (Pr > 0.7) $\quad Nu = C\,(Re)^{a_2}(Pr)^{0.33}$

	Re	C	a_2
Horizontal cylinder (D = L):	0.4-4	0.989	0.330
	4-40	0.911	0.385
	40-4000	0.683	0.466
	4000-40,000	0.193	0.618
	40,000-400,000	0.027	0.805
Square (L by L), flow 90° to flat surface:	5000-100,000	0.246	0.588
Square (L by L), flow 45° to a flat surface:	5000-100,000	0.102	0.675
Semi-infinite vertical plate (height L) flow 90° to flat surface:	4000-15,000	0.288	0.731

5. *Air flowing through a fixed bed of pellets* $\quad Sc = 0.6 \quad \varepsilon = 1 - \frac{D_p a_s}{6}$

Low Re flow: $90 < Re < 4{,}000 \quad Nu = \frac{2.06}{\varepsilon}(Re)^{0.422} Pr (Sc)^{-0.67}$

High Re flow: $5{,}000 < Re < 10{,}300 \quad Nu = \frac{20.4}{\varepsilon}(Re)^{0.185}(Sc)^{-0.67} Pr$

where a_s = total pellet surface area per volume of fixed bed

To do: Estimate the evaporation rate from the spill (S_s, in kg/hr), using first principles rather than empirical equations such as Eq. (4-14).

Solution: It is assumed that the partial pressure of the contaminant at the liquid-gas interface is equal to the vapor pressure of the contaminant, $P_{v,J}$ the value of which can be found in Appendix A.8. It is also assumed that the characteristic length is the diameter of the pool, i.e. L = D. The temperature is 17.7 °C (290.85 K) and the total pressure is 1.00 atm (760. mm Hg).

ethyl mercaptan	**air**
$P_{J,I} = P_{v,J} = 400.$ mm Hg	$\nu = 15.89 \times 10^{-6}$ m²/s
$D_{Ja} = 0.9 \times 10^{-5}$ m²/s	Pr = 0.707
$M_J = 62.1$	$U_\infty = 3.0$ m/s
	$P_{a,I} = 760. - P_{J,I} = 360.$ mm Hg

The presence of ethyl mercaptan in air just above the interface influences the Prandtl (Pr) and Schmidt (Sc) numbers. For a first approximation, this influence is neglected because the concentrations are not large. The following parameters are computed:

$$P_{am} = \frac{(760. - 360.)\ \text{mm Hg}}{\ln\left(\frac{760.}{360.}\right)} = 535.3\ \text{mm Hg} \qquad \text{Re} = \frac{U_\infty D}{\nu} = \frac{3.0\frac{\text{m}}{\text{s}}(10.0\ \text{m})}{1.589\times10^{-5}\ \frac{\text{m}^2}{\text{s}}} = 1.888\times10^6$$

The criteria in Table 4.3 show that the flow is turbulent. Table 4.3 indicates that the Nusselt number can be calculated as

$$\text{Nu} = \left[0.037(\text{Re})^{0.8} - 871\right](\text{Pr})^{0.33} = \left[0.037\left(1.888\times10^6\right)^{0.8} - 871\right](0.707)^{0.33} = 2685.$$

The Schmidt number is

$$\text{Sc} = \frac{\mu}{D_{Ja}\rho} = \frac{\nu}{D_{Ja}} = \frac{1.589\times10^{-5}\ \frac{\text{m}^2}{\text{s}}}{9.0\times10^{-6}\ \frac{\text{m}^2}{\text{s}}} = 1.766$$

The mass transfer coefficient (k_G) can be computed from Eq. (4-80), where $b_l = 0.33$,

$$k_G = \text{Nu}\frac{D_{Ja}}{D}\left(\frac{\text{Sc}}{\text{Pr}}\right)^{0.33}\frac{P}{P_{am}}\frac{1}{R_u T}$$

$$= 2685\frac{9.0\times10^{-6}\ \frac{\text{m}^2}{\text{s}}}{10.0\ \text{m}}\left(\frac{1.766}{0.707}\right)^{0.33}\frac{760.\ \text{mm Hg}}{535.3\ \text{mm Hg}}\frac{1}{8.314\frac{\text{kJ}}{\text{kmol}\cdot\text{K}}(290.85\ \text{K})}\left(\frac{\text{kJ}}{\text{m}^3\text{kPa}}\right)$$

which yields

$$k_G = 1.92\times10^{-6}\ \frac{\text{kmol}}{\text{m}^2\text{s}\cdot\text{kPa}}$$

The molar flux (N_J) can be computed from Eq. (4-67), assuming that the ethyl mercaptan partial pressure in the far field is zero ($P_{J,\infty} = 0$),

$$N_J = k_G\left(P_{J,I} - P_{J,\infty}\right) = 1.92\times10^{-6}\ \frac{\text{kmol}}{\text{m}^2\text{s}\cdot\text{kPa}}(400.\ \text{mm Hg} - 0)\left(\frac{101.3\ \text{kPa}}{760.\ \text{mm Hg}}\right) = 1.02\times10^{-4}\ \frac{\text{kmol}}{\text{m}^2\text{s}}$$

Finally, the total evaporation rate from the pool is obtained from Eq. (4-46),

$$\dot{m}_{evap,J} = N_J M_J A = 1.02\times10^{-4}\frac{kmol}{m^2 s}62.1\frac{kg}{kmol}\frac{\pi}{4}\left(10.0\ m\right)^2\left(\frac{3600\ s}{hr}\right) = 1.79\times10^{3}\frac{kg}{hr}$$

or approximately 1800 kg/hr.

Discussion: If the empirical equation, Eq. (4-14), were used instead to estimate the evaporation rate, a value of 1670 kg/hr would have been obtained. The evaporation rate computed from fundamental principles is about 5% larger than the value computed from the empirical equation. This agreement is excellent, considering the numerous approximations and assumptions! An explanation for the discrepancy may lie in the fact that the equation used to compute the Nusselt number assumes that the velocity and concentration boundary layers begin at the leading edge of the pool. In the actual case, air passing over the ground has a fully established boundary layer at the leading edge of the pool, and the equation for the Nusselt number may not apply to the actual case.

4.5.6 Single Film Theory for Multi-Component Liquids

Consider air passing over a liquid that contains several molecular components of varying volatility. Let the characteristic, far-field air velocity and temperature be denoted by U_∞ and T_∞. The most volatile components evaporate fastest. For the ***single film model*** it is assumed that there is a mixing mechanism inside the liquid such that the liquid-phase concentration is uniform with regard to space but variable with respect to time,

$$\boxed{c_J(x, y, z, t) = c_J(t)} \qquad (4\text{-}81)$$

The mass transfer is similar to that shown in Figure 4.12 and discussed in the previous section, except that now there are several molecular species (j) in the liquid phase. The theory is called "single film" because the primary resistance to mass transfer is in the concentration boundary layer in the air phase. The task for the engineer is to estimate the mass transfer rate of any one of the molecular species and the total evaporation rate of all species as a function of time.

For the analysis that follows (Drivas, 1982 and Stiver et al., 1984), it is assumed that the liquid is an ***ideal solution*** (see Chapter 1) in the form of a pool of diameter D_0 and depth d_p. The depth decreases very slowly with time. ***Raoult's law*** is invoked to describe the equilibrium that exists at the interface between a multi-component liquid and a multi-component vapor. The partial pressures of molecular species j in the gas phase (P_J) and at the interface ($P_{J,i}$) are

$$\boxed{P_J = y_J P \qquad P_{J,i} = y_{J,i} P = x_{J,i} P_{v,j}} \qquad (4\text{-}82)$$

where $P_{v,J}$ is the vapor pressure (see Appendix A.8) of pure species j at the temperature of the liquid, and $x_{J,i}$ is the mol fraction at the interface in the liquid phase. Note that $x_{J,i} = x_J$ because of the liquid phase uniformity assumption made earlier. It is also assumed that the total molar density ($c_{molar,L}$), i.e. the total number of mols per volume of liquid, is constant. Lastly it is assumed that the system is isothermal ($T_{liquid} = T_{air} = T_\infty$ = constant), and that the far-field partial pressure of each volatile molecular species is zero ($P_{J,\infty} = 0$). The rate of evaporation of molecular species j, in units of kmol/(m^2 s), can be found from Eq. (4-67),

$$\boxed{N_J = k_G\left(P_{J,i} - P_{J,\infty}\right) \approx k_G P_{J,i} = k_G x_J P_{v,J} = k_G\left(\frac{c_{molar,J}}{c_{molar,L}}\right)P_{v,J}} \qquad (4\text{-}83)$$

The reader must be mindful that subscript i refers to conditions at the interface; a particular molecular species is denoted by subscript j.

If the mol fraction of species j in the liquid were unchanging, the evaporation rate of species j would be constant. However, in a multi-component liquid pool, the most volatile liquids evaporate faster than the less volatile. If molecular species j is the most volatile, its (spatially uniform) concentration decreases with time and the liquid becomes increasingly rich in the less volatile species. If the pool is of finite mass, the ratio of the mol fraction of the most volatile species to the mol fraction of the least volatile species decreases with time. Consequently, while the total number of mols of liquid decreases slowly with time, the mol fraction of the least volatile species increases slowly with time simultaneously with a slow decrease in the mol fraction of the most volatile species. The following analysis shows how to estimate the instantaneous evaporation rate of each molecular species (N_J).

A control volume is defined in a layer of liquid of cross-sectional area A and depth d_p. From conservation of mass,

$$\frac{dc_{molar,J}}{dt} = -\frac{k_G P_{v,J}}{d_p c_{molar,L}} c_{molar,j} \tag{4-84}$$

where the variables (and their typical units) are:

- $P_{v,J}$ = vapor pressure (atm) of pure species j evaluated at temperature T_∞
- $c_{molar,J}$ = molar concentration of species j, i.e. number of mols of species j per volume of liquid (mol/m^3)
- k_G = mass transfer coefficient [mol/(m^2 atm s)], assumed to be the same for every species
- $c_{molar,L}$ = molar concentration of the liquid, i.e. total number of mols per volume of liquid (mol/m^3), assumed to be constant
- d_p = depth of the liquid pool (m), assumed to decrease very slowly with time

Eq. (4-84) is a first-order ordinary differential equation for $c_{molar,J}$ as a function of time, with constant coefficients. It is written in the standard form discussed and solved in Chapter 1. The solution (integration from time t = 0 to some arbitrary time t) yields an expression for molar concentration of species j at any instant of time,

$$c_{molar,J}(t) = c_{molar,J}(0)\exp\left[\frac{-k_G P_{v,J}}{d_p c_{molar,L}} t\right] \tag{4-85}$$

where $c_{molar,J}(0)$ is the initial molar concentration of species j in the liquid. Note also that the steady-state molar concentration (as t approaches ∞) is zero since there is no constant term (called B in Chapter 1) in Eq. (4-85).

For a mixture of liquid species, j, it is useful to define the total mass of liquid per unit area (m_t/A in units of kg/m^2):

$$\frac{m_t}{A} = \sum_J \left(d_p c_{molar,J} M_J\right) = \sum_J \left(d_p c_J\right) \tag{4-86}$$

where the summation is taken over the total number of molecular species, c_J is the mass concentration of contaminant species j, and M_J is the molecular weight of species j. The initial mass per unit area of liquid, $m_t(0)/A$, is

$$\frac{m_t(0)}{A} = d_p c_{molar,L} M_{avg}(0) \tag{4-87}$$

where

$$M_{avg}(0) = \sum_{J} x_J(0) M_J \qquad (4\text{-}88)$$

The total rate of evaporation ($\dot{m}_{evap,t}$) per unit area, in which several species evaporate at individual rates, is obtained by differentiating Eq. (4-86) with time,

$$\frac{\dot{m}_{evap,t}}{A} = -\frac{\dot{m}_t}{A} = -\frac{d\left(\frac{m_t}{A}\right)}{dt} = -\sum_{J}\left(d_p M_J \frac{dc_{molar,J}}{dt}\right)$$

The negative sign is introduced since $\dot{m}_{evap,t}$ is a positive quantity (evaporation rate), while $\dot{m}_t$ is negative since the mass of the liquid is decreasing with time. Substituting Eq. (4-84),

$$\frac{\dot{m}_{evap,t}}{A} = \sum_{J} k_G P_{v,J} M_J \frac{c_{molar,J}}{c_{molar,L}} \qquad (4\text{-}89)$$

Replacing the instantaneous molar concentration of species j by Eq. (4-85),

$$\frac{\dot{m}_{evap,t}}{A} = \sum_{J} k_G P_{v,J} M_J \left(\frac{c_{molar,J}(0)}{c_{molar,L}}\right) \exp\left[\frac{-k_G P_{v,J}}{d_p c_{molar,L}} t\right] \qquad (4\text{-}90)$$

Defining $x_J(0)$ as the initial mol fraction of species j,

$$x_J(0) = \frac{c_{molar,J}(0)}{c_{molar,L}} \qquad (4\text{-}91)$$

and recognizing that k_G can be moved outside the summation since it is constant, Eq. (4-90) reduces to

$$\frac{\dot{m}_{evap,t}}{A} = k_G \sum_{J} P_{v,J} M_J x_J(0) \exp\left[\frac{-k_G P_{v,J}}{d_p c_{molar,L}} t\right] \qquad (4\text{-}92)$$

The evaporation rate of a single species (j) per unit area is the same as Eq. (4-92), but without the summation sign,

$$\frac{\dot{m}_{evap,J}}{A} = k_G P_{v,J} M_J x_J(0) \exp\left[\frac{-k_G P_{v,J}}{d_p c_{molar,L}} t\right] \qquad (4\text{-}93)$$

For ***nonideal solutions***, Raoult's law can be modified by including liquid ***activity coefficients*** (Γ_J), so that Eq. (4-84) can be rewritten as

$$\frac{dc_{molar,J}}{dt} = -\Gamma_J \frac{k_G P_{v,J}}{d_p c_{molar,L}} c_{molar,J} \qquad (4\text{-}94)$$

The remainder of the derivation proceeds in the same fashion, with the activity coefficient carried through as a known parameter.

The mass transfer coefficient (k_G) can be obtained from Eq. (4-80). For large spills of petroleum on bodies of water, spills of the order of hundreds of meters in diameter, Drivas (1982) recommends the expression

$$k_G = \frac{0.0292 U_\infty^{0.78}}{R_u T_\infty D_0^{0.11} Sc^{0.67}} \qquad (4\text{-}95)$$

for all the evaporating species. The parameters in Eq. (4-95) are listed below (with their required units):

- k_G = mass transfer coefficient [mol/(m^2 atm hr)]
- U_∞ = air velocity (m/hr)
- D_0 = diameter of the spill (m)
- Sc = gas-phase Schmidt number, using a mass weighted average for the liquid mixture (unitless)
- R_u = universal gas constant [8.206 x 10^{-5} (atm m^3)/(mol K)]
- T_∞ = air temperature (K)

Example 4.13 - Evaporation of Volatile Compounds from a Waste Water Lagoon

Given: A lagoon is 10.0 m in diameter (D_0), and 3.0 m deep (d_p). The land is flat surrounding the lagoon. Dry air (zero water vapor in the far-field, $P_{water,\infty} = 0$) at 25.0 °C passes over the surface of the lagoon at a velocity of 3.0 m/s. An agitator in the lagoon ensures that the composition in the liquid phase is uniform (well mixed). Initially the mol fractions of benzene (C_6H_6), methyl alcohol (CH_3OH), carbon tetrachloride (CCl_4), and toluene ($C_6H_5CH_3$) are each 0.050. The remainder of the liquid is water.

To do: Estimate the evaporation rate of each species as a function of time.

Solution: Let subscript j denote the particular species. Some useful approximate properties of the air are Pr = 0.708, ν = 1.5 x 10^{-5} m^2/s, and ρ = 1.2 kg/m^3. The required properties of the other species are listed below:

species (j)	**M**	**$x_j(0)$**	**D_{ja} (m^2/s)**	**Sc_j**	**$P_{v,j}$ (at 25 °C) (kPa)**
C_6H_6	78.	0.050	0.77 x 10^{-5}	1.948	12.77
CH_3OH	32.	0.050	1.33 x 10^{-5}	1.128	15.72
CCl_4	154.	0.050	0.62 x 10^{-5}	2.419	14.40
$C_6H_5CH_3$	92.	0.050	0.71 x 10^{-5}	2.113	4.19
H_2O	18.	0.80	2.64 x 10^{-5}	0.568	3.17

The average molecular weight of the liquid is

$$M_{avg} = 0.050(78.+32.+154.+92.)+0.80(18.) = 32.2\frac{kg}{kmol}$$

The mass fraction of each species (f_j) in the liquid phase is given by

$$\boxed{f_j = x_j \frac{M_j}{M_{avg}}} \tag{4-96}$$

To compute the total molar density ($c_{molar,L}$), it is assumed that the mixture is an ideal liquid mixture. Assume 1,000 kg of the liquid mixture. The mass fraction f_j, density of the pure species ρ_j, mass m_j, number of mols n_j, and volume V_j of each species are:

species (j)	**M_j**	**f_j**	**ρ_j (kg/m^3)**	**m_j (kg)**	**n_j (n_j/M_j)**	**V_j (m^3) = m_j/ρ_j**
C_6H_6	78	0.1210	879	121	1.55	0.1376
CH_3OH	32	0.0496	790	49.6	1.55	0.0628
CCl_4	154	0.2391	1595	239.1	1.55	0.1499
$C_6H_5CH_3$	92	0.1428	866	142.8	1.55	0.1649
H_2O	18	0.4472	1000	447.2	24.84	0.4472

Summing the last two columns,

$$n_t = \sum_J n_J = 31.04$$

and

$$V_t = \sum_J V_J = 0.9624$$

The total molar density $c_{molar,L}$ is

$$c_{molar,L} = \frac{n_t}{V_t} = \frac{31.04\ \text{kmol}}{0.9624\ \text{m}^3} = 32.25\frac{\text{kmol}}{\text{m}^3}$$

The initial mol-weighted average Schmidt number for the pollutants in air (including water vapor) is

$$Sc_{avg} = 0.050(1.948) + 0.050(1.128) + 0.050(2.419) + 0.050(2.113) + 0.80(0.568) = 0.8349$$

An initial mol-weighted average diffusion coefficient for the pollutants in air (including water vapor) is

$$D_{Ja,avg} = 10^{-5}\left[0.050(0.77) + 0.050(1.33) + 0.050(0.62) + 0.050(0.71) + 0.80(2.64)\right]$$
$$= 2.284\times10^{-5}\ \frac{\text{m}^2}{\text{s}}$$

These initial values of Schmidt number and average diffusion coefficient are assumed to remain as constants. The Reynolds number (characteristic length, D_0 = 10. m) of the air passing over the lagoon is

$$\text{Re} = \frac{U_\infty D_0}{\nu} = \frac{3.0\frac{\text{m}}{\text{s}}(10.0\ \text{m})}{1.5\times10^{-5}\ \frac{\text{m}^2}{\text{s}}} = 2.0\times10^6$$

From Table 4.3, such a Reynolds number indicates that the flow is turbulent and that the following Nusselt number correlation can be used for the air:

$$\text{Nu} = \left(0.037(\text{Re})^{0.8} - 871.\right)(\text{Pr})^{0.33} = \left(0.037\left(2.0\times10^6\right)^{0.8} - 871.\right)(0.708)^{0.33} = 2850$$

The agitator in the lagoon ensures uniform concentration within the aqueous phase. Thus mass transfer from the liquid surface is governed by "single film" theory. The mass transfer coefficient can be found from Eq. (4-80), where P_{am} is computed from Eq. (4-54). In the far-field it has been assumed that there is no water vapor (Φ = 0%), thus $P_{a,\infty}$ = 101. kPa. At the air-liquid interface, $P_{a,I}$ is equal to

$$P_{a,I} = P - \left[0.05P_v(C_6H_6) + 0.05P_v(CH_3OH) + 0.05P_v(CCl_4) + 0.05P_v(C_6H_5CH_3) + 0.80P_v(H_2O)\right]$$
$$= 101.\ \text{kPa} - 4.89\ \text{kPa} = 96.1\ \text{kPa}$$

Thus,

$$P_{am} = \frac{P_{a,\infty} - P_{a,I}}{\ln\left(\frac{P_{a,\infty}}{P_{a,I}}\right)} = \frac{101. - 96.11}{\ln\left(\frac{101.}{96.11}\right)} = 98.5\ \text{kPa}$$

The mass transfer coefficient is calculated from Eq. (4-80), using the mol-weighted average diffusion coefficient and the mol-weighted average Schmidt number:

$$k_G = Nu\frac{D_{Ja,avg}}{D}\left(\frac{Sc_{avg}}{Pr}\right)^{0.33}\frac{P}{P_{am}}\frac{1}{R_uT}$$

$$= 2850\left(\frac{2.284\times10^{-5}\ \frac{m^2}{s}}{10.0\ m}\right)\left(\frac{0.8349}{0.708}\right)^{0.33}\frac{101.\ kPa}{98.5\ kPa}\frac{1}{8.314\frac{kJ}{kmol\cdot K}(298.\ K)}\left(\frac{kJ}{m^3kPa}\right)$$

which yields

$$k_G = 2.844\times10^{-6}\ \frac{kmol}{m^2s\cdot kPa}$$

At t = 0, when evaporation is assumed to begin, the total mass per unit area of the lagoon is $m_t(0)/A$, which is calculated from Eq. (4-87),

$$\frac{m_t(0)}{A} = d_p c_{molar,L} M_{avg}(0) = (3.0\ m)\left(32.25\ \frac{kmol}{m^3}\right)\left(32.2\ \frac{kg}{kmol}\right) = 3115.\frac{kg}{m^2}$$

The evaporation rate per unit area for each species (j), as a function of time, is found from Eq. (4-93). The evaporation rate of each species can be determined by repetitive computations using commercially available mathematical computer programs. The results are seen in Figure E4.13. The Mathcad program producing Figure E4.13 can be found on the text's web site. The evaporation rates appear linear on the log scale in Figure E4.13, due to the exponential decay rate of Eq. (4-93); each species has a different slope.

Discussion: The above equations assume that d_p and $c_{molar,L}$ remain constant throughout the period of evaporation. Such an assumption is clearly untrue since d_p has to decrease as materials evaporate from the pool, and $c_{molar,L}$ will also vary slightly owing to the fact that the most volatile species evaporate fastest and their mol fraction and partial molar concentrations decrease to zero while the partial molar concentrations of the less volatile species are non-zero. To estimate the total evaporation rate for such a case, the user should use Eq. (4-92) to estimate the evaporation rate during a time step $t = \delta t$, and then update the values of d_p and $c_{molar,L}$ in Eq. (4-92) to compute the evaporation rate for the next time step. By marching forward through a series of many time steps, the user can estimate the evaporation rate over a long period of time. The most straightforward marching scheme is to set the values of d_p and $c_{molar,L}$ at the beginning and of each time step equal to the values at the end of the previous time step,

$$c_{molar,J}(t) = c_{molar,J}(t-\delta t) \qquad d_p(t) = d_p(t-\delta t) - \frac{\dot{m}_{evap,t}\delta t}{\rho(t-\delta t)}$$

and

$$c_{molar,L}(t) = \sum_J c_{molar,J}(t-\delta t)$$

Other more sophisticated marching schemes, such as the Runge-Kutta technique described in Chapter 1, can also be applied.

4.5.7 Two-Film Evaporation of Multi-Component Liquids

Section 4.5.5 was concerned with evaporation of a single component liquid through a column of stagnant air. Such mass transfer is called ***pure (molecular) diffusion***. If the air is moving, evaporation is more rapid because the moving air enhances mass transfer. Such mass transfer is called ***convective diffusion***. The present section concerns convective diffusion of multi-component liquids.

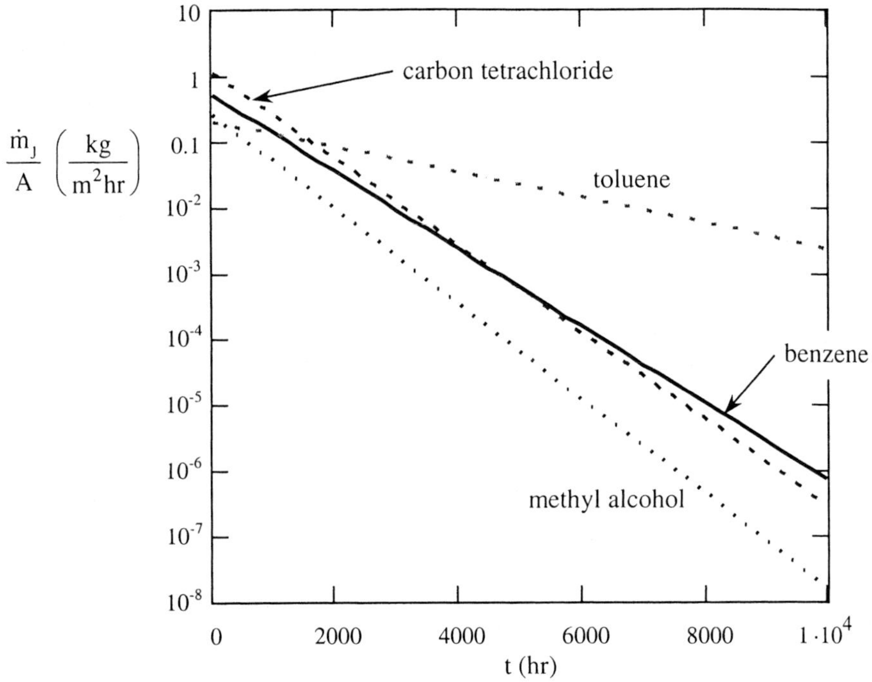

Figure E4.13 Rate of evaporation of volatile compounds from a well-mixed waste-water lagoon as a function of time.

Desorption is the name of the physical process in which a dissolved gas (or vapor) is transferred from the liquid phase to the gas phase. Carbon dioxide bubbles in carbonated soft drinks and beer are familiar examples of desorption. Stripping is the name of the process in which air, steam, or some other stripping medium is used to enhance desorption. An example (Figure 4.11) is the removal of volatile organic compounds (VOCs) from contaminated groundwater. Less familiar examples of desorption follow. Trace amounts of trichloroethylene (TCE) in tap water desorb from hot water passing through a bathroom shower fixture. Little (1992) found that the daily indoor inhalation exposure to materials in tap water from a 10-minute shower in a home is equivalent to 1.5 times that incurred by ingesting 2 liters (typical daily consumption) of the same water. Giardino et al. (1992) found that approximately 60% of TCE in water is volatized from the spray drops and water deposited on the surface of the shower enclosure. Shepherd et al. (1996) found that chloroform is generated in washing machines that use bleach containing sodium hypochlorite. Once generated, chloroform enters the indoor environment. Chloroform in the discharge water could constitute a significant fraction of chloroform mass loading in a municipal wastewater plant. On the larger scale, Jones et al. (1996) found that volatiles in municipal waste waters escaping through curb drains and holes in manholes in municipal sewer systems are a potentially significant source of hazardous air pollutants (HAPs) in the urban environment.

In the Sections 4.5.5 and 4.5.6 it was assumed that the concentration in the liquid phase was spatially uniform. In general this is not the case; it certainly should not be assumed without evidence to support it. The evaporation of pollutants from a multi-component liquid mixture is complicated by the fact that each species must be transported from the liquid interior through a concentration boundary layer (film) in the liquid before it reaches the air-liquid interface to evaporate. Once leaving the surface as a vapor, it must be transported through another boundary layer (film) on the air side. Thus mass is transferred through ***two films*** or ***two resistances*** as shown in Figure 4.13. If one film offers much more resistance than the other, the analysis reduces to a single-film analysis, but until this is known for a fact, both films have to be dealt with as if they were equally important. As a general rule, the resistance of the liquid film controls mass transfer for very volatile contaminants (contaminants with large values of Henry's law constant, H, to be discussed shortly), while the resistance of the gas-phase film becomes increasingly important as the volatility decreases.

If the masses of liquid and air are large, such as evaporation of a VOC from an industrial spill or from a wastewater lagoon, a steady-state mass transfer rate can be achieved. If, on the other hand, the mass of the liquid is small, the analysis is complicated by depletion of material in the liquid phase and the subsequent decrease in evaporation rate with time as shown in Figure 4.14. Only steady-state mass transfer through two films is considered in the analysis that follows. As in single film theory, isothermal conditions are presumed. Physically this means that the liquid phase is in contact with an infinite heat source that transfers sufficient energy to the liquid to replace the energy consumed in evaporation.

Before postulating equations to describe the mass transfer, it is essential to understand equilibrium of a liquid/air-pollutant system. Consider a cylindrical vessel equipped with a frictionless piston as shown in Figure 4.15. Water containing a dissolved contaminant (species j) is placed in the cylindrical vessel along with (pure) air. The vessel is immersed in a constant temperature bath (T), as in Figure 4.15. Some of the contaminant evaporates from the water and enters the air. Constant overall air pressure is maintained by the frictionless piston. Eventually the system comes to equilibrium, at which the mol fraction of species j in water (x_j) and the partial pressure of species j in air (P_j) are measured. The experiment is repeated for several values of x_j at temperature T, and partial pressure is plotted as a function of liquid mol fraction at equilibrium. The resulting curve is called an ***equilibrium isotherm***, the shape of which depends on the chemistry of the particular contaminant. If the temperature is varied and the experiment is repeated, a series of equilibrium isotherms can be plotted. Two typical isotherms for the same contaminant at two different temperatures are sketched in Figure 4.16.

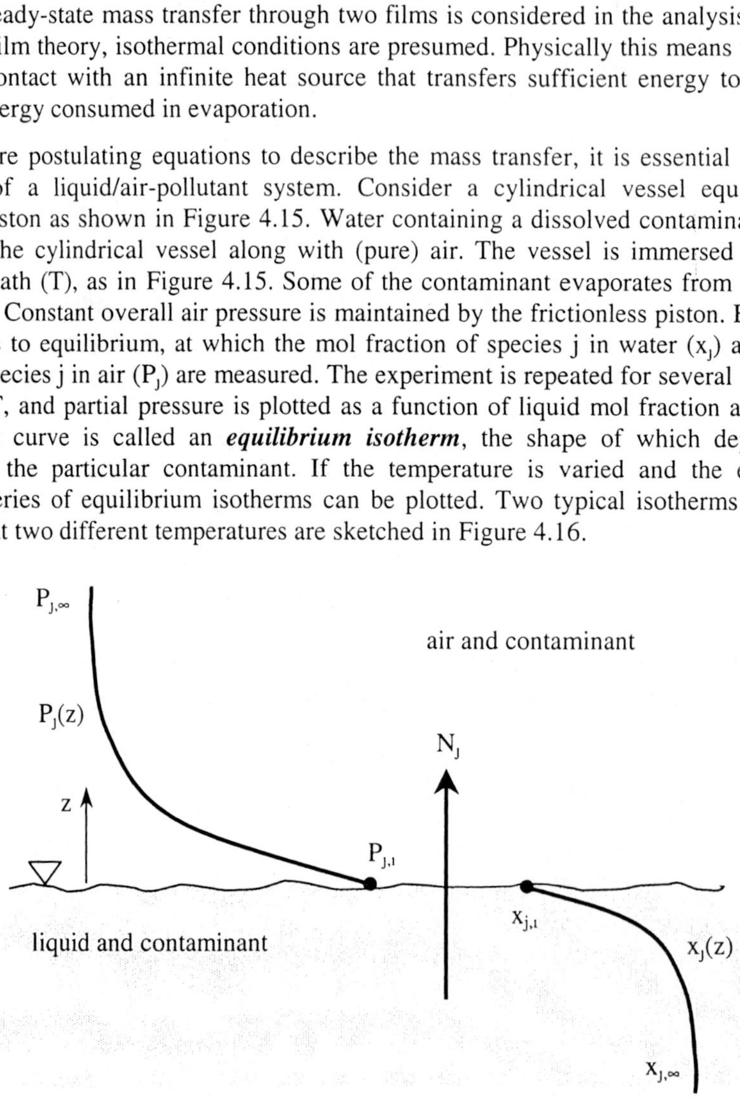

Figure 4.13 Concentration profiles for two-film mass transfer; x_j is the mol fraction in the liquid phase, and P_j is the partial pressure in the vapor phase.

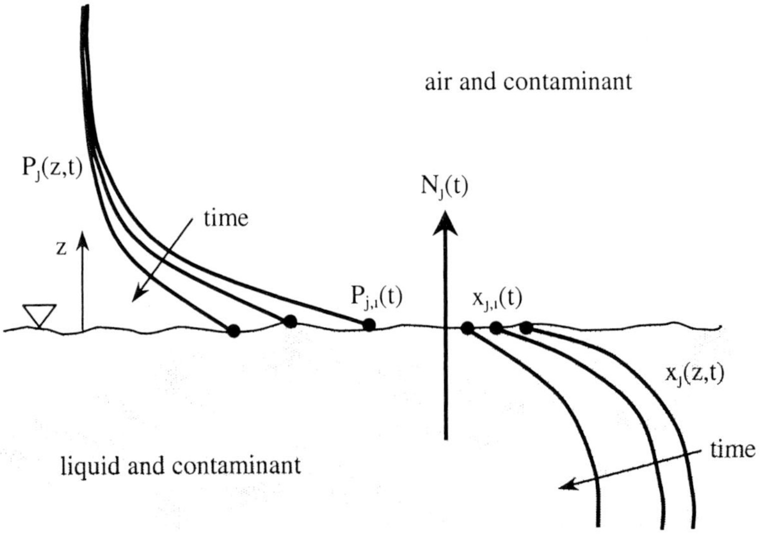

Figure 4.14 Instantaneous concentration profiles for two-film mass transfer showing depletion in the liquid phase with time, and the resulting decrease in evaporation rate with time.

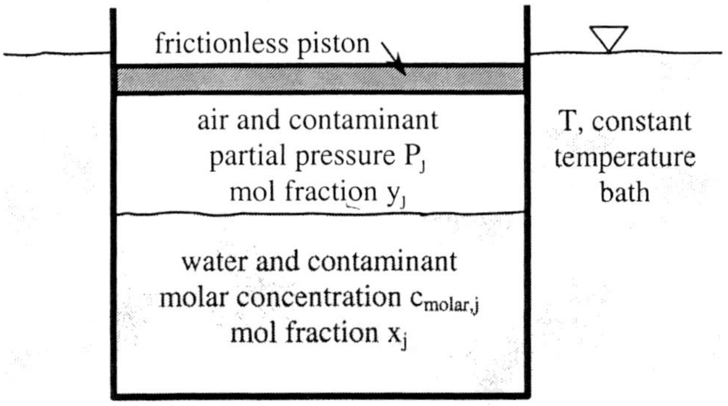

Figure 4.15 Multi-component equilibrium for contaminate in air and water; experimental technique to obtain equilibrium isotherms.

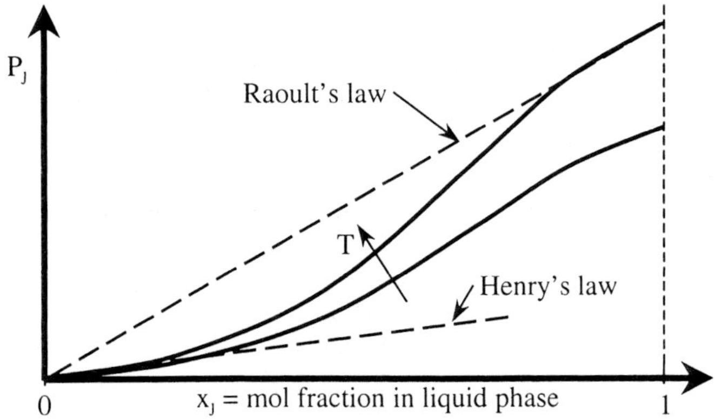

Figure 4.16 Typical absorption equilibrium isotherms for contaminant in air and water; Henry's law is valid for small x_j, while Raoult's law is valid for large x_j.

As sketched in Figure 4.16, at any given liquid mol fraction, partial pressure increases with temperature. Likewise, at any given partial pressure, equilibrium liquid mol fraction decreases with increasing temperature. Raoult's law, Eq. (4-82), is an excellent approximation as x_J approaches unity (high concentrations), as shown in Figure 4.16.

For purposes of understanding indoor air pollution, in which solutes often exist at very *low* concentrations, the portion of the equilibrium curve in the vicinity of the origin must be considered. In this range the equilibrium isotherms do not follow Raoult's law, but instead satisfy ***Henry's law***, which can be expressed in several forms depending on the units chosen to represent the solute concentration.

$$\boxed{P_J = Hc_{molar,J} = Hx_J c_{molar,L} = H'x_J} \tag{4-97}$$

where

$$\boxed{H' = c_{molar,L} H} \tag{4-98}$$

and $c_{molar,J}$ and $c_{molar,L}$ are the solute and total liquid molar concentrations, as defined previously. The parameters H and H′ are the slopes of the equilibrium curves near $x_J = 0$, and are called ***Henry's law constants*** (Mackay and Shiu, 1981). (Readers should note that H and H′ are not truly constants, since they depend on temperature.) The units of H are typically (atm m^3/kmol) and the units of H′ are the same as that of pressure (typically atm, kPa, mm Hg, or N/m^2). The right-most form of Eq. (4-97) is sketched in Figure 4.16; the agreement is excellent, but only for very small values of x_J.

Some authors (e.g. Little, 1992) normalize Henry's law constant by writing Eq. (4-97) in terms of mol fraction (y_J),

$$\boxed{y_J = \frac{P_J}{P} = \frac{H'}{P}x_J = H''x_J} \tag{4-99}$$

where P is the ambient pressure and H'' is a nondimensional Henry's law constant

$$\boxed{H'' = \frac{H'}{P}} \tag{4-100}$$

Note that some authors use the symbol M or m for the nondimensional Henry's law constant, but to avoid confusion with molecular weight or mass, H'' is used in this text. Henry's law constant is a physical property that reflects how a molecular species *partitions* itself between air and water. Chemicals with low values of Henry's law constant are those that partition more of the solute in the liquid phase than in the air phase, while higher values of the Henry's law constant imply the opposite. Henry's law constant can be predicted from thermodynamic data (Nirmalakhandan and Speece, 1988) or computed from experimental measurements of vapor pressure in air and solubility in water (Yaws et al., 1999). Henry's law constants for many organic compounds can be found in books such as Yaws et al. (1991) and on the Internet (search for "Henry's law constant"). Appendix A.9 is a tabulation of H′ and diffusion coefficients (D_{Ja} in air and D_{Jw} in water) for many important pollutants at STP.

Example 4.14 - Computation of Henry's Law Constant from Solubility Data

Given: Toluene (M = 92.0) is mixed in water until saturated. From Perry et al. (1984), the solubility of toluene is 0.050 g of toluene per 100. g of water at 16.0 °C and 1.0 atm.

To do: Estimate Henry's law constant for toluene in water.

Solution: The mol fraction of toluene in water is

$$x_{toluene} = \frac{c_{molar,toluene}}{c_{molar,L}} = \frac{\dfrac{m_{toluene}}{M_{toluene}V}}{\dfrac{1}{V}\sum_J\left(\dfrac{m_j}{M_J}\right)} = \frac{\dfrac{m_{toluene}}{M_{toluene}}}{\dfrac{m_{toluene}}{M_{toluene}} + \dfrac{m_{water}}{M_{water}}} = \frac{\dfrac{0.050\ g}{92.0\ \dfrac{g}{mol}}}{\dfrac{0.050\ g}{92.0\ \dfrac{g}{mol}} + \dfrac{100.\ g}{18.0\ \dfrac{g}{mol}}} = 9.78\times10^{-5}$$

From Appendix A.8, the vapor pressure of toluene is found to be 40.0 mm Hg at 31.8 °C and 20.0 mm Hg at 18.4 °C. The vapor pressure at 16.0 °C is not given. Since H′ depends on vapor pressure ($P_{v,J}$) which in turn is keenly dependent on temperature, the vapor pressure at 16.0 °C is estimated using the Clausius-Claperon equation. The following notation is used:

- T_1 = 16.0 °C = 289.15 K $P_{v,1}$(at T_1) = unknown (to be found)
- T_2 = 18.4 °C = 291.55 K $P_{v,2}$(at T_2) = 20.0 mm Hg
- T_3 = 31.8 °C = 304.95 K $P_{v,3}$(at T_3) = 40.0 mm Hg

From Eq. (4-28),

$$\frac{\hat{h}_{fg}}{R_u} = \ln\left(\frac{P_{v,3}}{P_{v,2}}\right)\frac{T_3T_2}{T_3-T_2} = \ln\left(\frac{40.0\ mm\ Hg}{20.0\ mm\ Hg}\right)\frac{(304.95\ K)(291.55\ K)}{304.95\ K-291.55\ K} = 4599.\ K$$

and

$$\ln\left(\frac{P_{v,2}}{P_{v,1}}\right) = \frac{\hat{h}_{fg}}{R_u}\frac{T_2-T_1}{T_2T_1} = \ln\left(\frac{20.0\ mm\ Hg}{P_{v,1}}\right) = (4599.\ K)\frac{291.55\ K-289.15\ K}{(291.55\ K)(289.15\ K)}$$

which can be solved for $P_{v,1}$,

$$P_{v,1} = 17.55\ mm\ Hg$$

Assuming that P_J is equal to the vapor pressure $P_{v,1}$ at the solubility condition of 0.050 g of toluene per 100. g of water, Henry's law constant at 16.0 °C can then be estimated from Eq. (4-97),

$$H' = \frac{P_{toluene}}{x_{toluene}} = \frac{17.55\ mm\ Hg}{9.78\times10^{-5}}\left(\frac{1.0\ atm}{760.\ mm\ Hg}\right)\left(\frac{1.013\times10^5\ \dfrac{N}{m^2}}{1.0\ atm}\right) = 2.39\times10^7\ \frac{N}{m^2}$$

Discussion: The value listed in Appendix A.9 is 3.72 x 10^7 N/m^2. The agreement is not so good, but of the correct order of magnitude, illustrating that the (linear) Henry's law approximation is not very accurate, at least for this example. The main reason for the discrepancy is the *nonlinear* curve of P_J versus x_J, as illustrated in Figure 4.16. Namely, even though the concentration of toluene in water is small (of order 10^{-4}), the slope of the P_J versus x_J curve is not constant as x_J increases, and hence the estimated value of H′ is too low.

Vapor-liquid equilibrium embodies a wide spectrum of chemical behavior. For components present in very small amounts, Henry's law is often a satisfactory approximation. For components in very large amounts, e.g. solvents, Raoult's law may suffice. In general an entire range of non-ideal behavior would need to be quantified by methods available in the chemical engineering literature, but beyond the scope of this book.

Example 4.15 - Comparison and Warnings about Henry's and Raoult's Laws
Given: An aqueous solution of toluene, with a liquid mol fraction, $x_{tol} = 9.78 \times 10^{-5}$ (same mol fraction as in the previous example, but at a different temperature). The temperature is 25.0 °C.

To do: Calculate the partial pressure of toluene over the aqueous solution at 25.0 °C using Henry's Law and Raoult's law.

Solution: From the tables,

- toluene vapor pressure at 25.0 °C: $P_{v,toluene} = 29.8$ mm Hg
- Henry's law constant: $H' = 3.72 \times 10^7$ N/m^2

Henry's law, Eq. (4-97): The partial pressure at the interface is predicted by Henry's law to be

$$P_{toluene} = H'x_{toluene} = 3.72\times10^7 \frac{N}{m^2}\left(9.78\times10^{-5}\right) = 3640\frac{N}{m^2}$$

Raoult's law, Eq. (4-82): The partial pressure at interface is predicted by Raoult's law to be

$$P_{toluene} = x_{toluene}P_{v,toluene} = 9.78\times10^{-5}\left(29.8 \text{ mm Hg}\right)\left(\frac{1.013\times10^5 \frac{N}{m^2}}{760. \text{ mm Hg}}\right) = 0.388\frac{N}{m^2}$$

Discussion: This nearly 10,000-fold discrepancy illustrates the dramatic errors that are possible when Raoult's Law is applied to minor components in non-ideal liquid solutions. At a mol fraction of 10^{-4}, toluene certainly meets the criterion of a minor component. However, as an aromatic hydrocarbon (C_7H_8) in water, the solution is non-ideal. Consequently the reader should use Henry's law to compute the partial pressure of a contaminant at low concentration in solution with water. One characterization of non-ideal solutions is activity coefficients (Γ) that differ from unity. Chemically one might expect a toluene molecule, surrounded by water, to try to leave the solution, i.e. to exert a much greater partial pressure than that implied by its very low mol fraction. Equally gross errors may be expected when Henry's law is applied to components present in large proportion.

If a liquid containing a pollutant is brought into contact with air containing the pollutant, the pollutant will desorb and transfer to the air if the coordinates (P_J, x_J) lie below the equilibrium line on Figure 4.16. If the coordinates lie above the equilibrium line, the pollutant will leave the air phase and be absorbed in the liquid. In terms of the variables shown in Figure 4.17, the molar transfer rate can be expressed as

$$\boxed{N_J = k_L\left(x_{J,\infty} - x_{J,I}\right)c_{molar,L} = k_G\left(P_{J,I} - P_{J,\infty}\right)} \qquad (4\text{-}101)$$

Unfortunately both the properties at the interface ($x_{J,I}$ and $P_{J,I}$) and the mass transfer coefficients (k_L and k_G) are unknown. An alternative expression for the mass transfer can be postulated that is often more useful,

$$\boxed{N_J = K_L\left(x_{J,\infty} - x_J^*\right)c_{molar,L} = K_G\left(P_J^* - P_{J,\infty}\right)} \qquad (4\text{-}102)$$

where K_L and K_G are called ***overall mass transfer coefficients*** in the liquid and gas phases, and the ***star states*** (x_J^* and P_J^*) are hypothetical mol fraction and partial pressure values defined by the equilibrium diagram, as indicated in Figure 4.17:

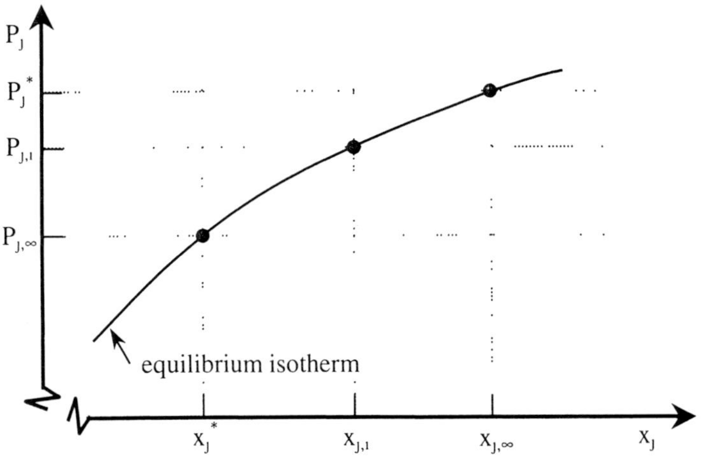

Figure 4.17 Relationship between a star state and actual state along an equilibrium isotherm.

- x_J^* = ***hypothetical mol fraction in the liquid phase*** (liquid-phase mol fraction which would be in equilibrium with the actual far-field gas-phase partial pressure $P_{J,\infty}$),
- P_J^* = ***hypothetical partial pressure in the gas-phase*** (gas-phase partial pressure which would be in equilibrium with the actual liquid phase mol fraction $x_{J,\infty}$).

It is important to recognize that the star states are hypothetical states defined by the graphical relationship shown in Figure 4.17. The overall mass transfer coefficient in the liquid phase (K_L) can be expressed using the following construction:

$$\boxed{\left(x_{J,\infty}-x_J^*\right)=\left(x_{J,\infty}-x_{J,I}\right)+\left(x_{J,I}-x_J^*\right)} \tag{4-103}$$

From Eq. (4-97) it can be seen that Henry's law allows one to relate $x_{J,I}$ to $P_{J,I}$, and x_J^* to $P_{J,\infty}$,

$$\boxed{P_{J,I}=H'x_{J,I} \quad \text{and} \quad P_{J,\infty}=H'x_J^*} \tag{4-104}$$

Thus

$$\boxed{\left(x_{J,\infty}-x_J^*\right)=\left(x_{J,\infty}-x_{J,I}\right)+\frac{P_{J,I}-P_{J,\infty}}{H'}} \tag{4-105}$$

The differences contained in the three terms within parentheses in Eq. (4-105) can be replaced by expressions involving the molar flux N_J using Eq. (4-101). In so doing one obtains

$$\frac{N_J}{c_{molar,L}K_L}=\frac{N_J}{c_{molar,L}k_L}+\frac{P_{J,I}-P_{J,\infty}}{Hc_{molar,L}}$$

which, after simplification reduces to

$$\boxed{\frac{1}{K_L}=\frac{1}{k_L}+\frac{1}{Hk_G}} \tag{4-106}$$

One can make an analogy here to an electric circuit, in which current is impeded by resistors. In particular, it is useful to think of each one of the terms in Eq. (4-106) as a resistance to mass transfer. Specifically, one can define

- R_L = ***liquid film mass transfer resistance*** = $1/k_L$
- R_G = ***gas film mass transfer resistance*** = $1/(Hk_G)$

The overall liquid resistance to mass transfer ($1/K_L$) is thus equal to the resistance of the liquid film (R_L) plus the resistance of the gas film (R_G). Returning to Eq. (4-102), and introducing Eqs. (4-98) and (4-104), the mol transfer rate can be written as

$$N_J = K_L c_{molar,L}\left(x_{J,\infty} - x_J^*\right) = K_L\left(c_{molar,L}x_{J,\infty} - c_{molar,L}x_J^*\right)$$

$$= K_L\left(c_{molar,J,\infty} - \frac{c_{molar,L}P_{J,\infty}}{H'}\right) = K_L\left(c_{molar,J,\infty} - \frac{c_{molar,L}P_{J,\infty}}{Hc_{molar,L}}\right)$$

or

$$\boxed{N_J = K_L\left(c_{molar,J,\infty} - \frac{P_{j,\infty}}{H}\right)} \tag{4-107}$$

Expanding the analogy between mass transfer and electric circuits, one can think of Eq. (4-107) as analogous to Ohm's law. In an electric circuit, current equals the driving potential (voltage) divided by electric resistance. Here, the number of mols being transferred (N_J) equals the driving potential ($c_{molar,J,\infty}$ - $P_{J,\infty}/H$) divided by the resistance to mass transfer ($1/K_L$),

$$\boxed{N_J = \frac{c_{molar,J,\infty} - \frac{P_{J,\infty}}{H}}{\frac{1}{K_L}} = \frac{c_{molar,J,\infty} - \frac{P_{J,\infty}}{H}}{\frac{1}{k_L} + \frac{1}{Hk_G}} = \frac{c_{molar,J,\infty} - \frac{P_{J,\infty}}{H}}{R_L + R_G}} \tag{4-108}$$

where Eq. (4-106) has been used. In a circuit consisting of two resistances in series in which one resistance is much larger than the other, the larger resistance dictates the current. Similarly if mass transfer is governed by boundary layers (films) in the liquid and gas phases, the film having the larger resistance dictates the rate of mass transfer. Quite often the liquid film has the largest resistance since air passing over the liquid surface transfers vapor from the liquid in a vigorous manner.

Equation (4-107) has been used to estimate the rate of evaporation of volatile organic hydrocarbons from ocean spills, industrial waste lagoons, and for stripping operations to remove dissolved pollutants from underground water. The mass transfer coefficient (k_G) has been extracted from empirical expressions given by Mackay and Yeun (1983), which also include the mass transfer coefficient (k_L):

$$\boxed{k_G\left(\text{in units of } \frac{\text{mol}}{\text{m}^2\text{s}\cdot\text{atm}}\right) = 4.1\times10^{-2} + 1.9U^*\left(Sc_G\right)^{-0\,67}} \tag{4-109}$$

$$\boxed{k_L\left(\text{in units of } \frac{\text{m}}{\text{s}}\right) = 1.0\times10^{-6} + \left(34.1\times10^{-4}\right)U^*\left(Sc_L\right)^{-0\,5} \quad \text{if } U^* > 0.3} \tag{4-110}$$

$$\boxed{k_L\left(\text{in units of } \frac{\text{m}}{\text{s}}\right) = 1.0\times10^{-6} + \left(144.\times10^{-4}\right)\left(U^*\right)^{2.2}\left(Sc_L\right)^{-0.5} \quad \text{if } U^* \le 0.3} \tag{4-111}$$

where subscript G refers to the gas phase and L to the liquid phase. The parameter U* (m/s) is called the ***friction velocity*** which is defined as

$$\boxed{U^* = \sqrt{\frac{\tau_a}{\rho_a}}} \tag{4-112}$$

where τ_a is the shear stress of air passing over the liquid surface. In the outdoor environment where the liquid surface is large and evaporation is driven by the wind, Mackay and Yeun (1983) recommend that the friction velocity (U*) be expressed as

$$\boxed{U^* \left(\text{in units of } \frac{m}{s}\right) = U_{10}\sqrt{6.1+0.63U_{10}}} \tag{4-113}$$

where U_{10} is the wind speed (units must be in m/s) a distance 10 m above the liquid surface, a common practice in meteorology.

Example 4.16 - Evaporation of Volatile Compounds from a Stagnant Waste Lagoon
Given: A waste lagoon is 25. m x 40. m x 3.5 m deep. It contains 100. mg/L of benzene (M_b = 78.0) and 100. mg/L of chloroform (M_c = 119.0) in water. The air and liquid temperatures are 25.0 °C and the wind speed is 1.70 m/s at z = 10. m.

To do: Estimate the evaporation rate (kg/hr) of benzene and chloroform.

Solution: First, the Schmidt numbers are listed:

Sc (benzene-water) = 1,000 Sc (benzene-air) = 1.76

Sc (chloroform-water) = 1,100 Sc (chloroform-air) = 2.14

The total molar density ($c_{molar,L}$) is

$$c_{molar,L} = \frac{\rho_{total}}{M_{total}} \approx \frac{\rho_{water}}{M_{water}} = \frac{1000.\frac{kg}{m^3}}{18.0\frac{kg}{kmol}}\left(\frac{1000.\text{ mol}}{kmol}\right) = 5.556\times10^4\ \frac{mol}{m^3}$$

Henry's law constants for the two chemicals in the water can be looked up:

- benzene: $H'_b = 3.05 \times 10^7$ N/m² $H_b = H'_b/c_{molar,L} = 5.5 \times 10^{-3}$ (atm m³)/gmol
- chloroform: $H'_c = 2.66 \times 10^7$ N/m² $H_c = H'_c/c_{molar,L} = 3.39 \times 10^{-3}$ (atm m³)/gmol

The overall mass transfer coefficient $K_{L,J}$ is given by Eq. (4-106), and the mass transfer coefficients for the gas and liquid phases are calculated from Eqs. (4-109) through (4-111). The friction velocity U* can be found from Eq. (4-113),

$$U^* = U_{10}\sqrt{6.1+0.63U_{10}} = \left(1.7\ \frac{m}{s}\right)\sqrt{6.1+0.63(1.7)} = 4.55\ \frac{m}{s}$$

Since U^* is greater than 0.3 m/s, Eq. (4-110) is relevant for this example problem. Summarizing the mass transfer coefficients for the two chemicals,

<u>Benzene (subscript b)</u>:

$k_{G,b} = 4.1 \times 10^{-2} + (1.9)(4.55)/1.76^{0.67} = 5.959 \cong 5.96$ mol/(m² s atm)
$k_{L,b} = 10^{-6} + 34.1 \times 10^{-4}(4.55)/1{,}000^{0.5} = 10^{-6} + 490.6 \times 10^{-6} = 4.92 \times 10^{-4}$ m/s
$K_{L,b} = 4.85 \times 10^{-4}$ m/s

<u>Chloroform (subscript c)</u>:

$k_{G,c} = 4.1 \times 10^{-2} + (1.9)(4.55)/2.14^{0.67} = 5.233 \cong 5.23$ mol/(m² s atm)
$k_{L,c} = 10^{-6} + 34.1 \times 10^{-4} (4.55)/1{,}100^{0.5} = 10^{-6} + 467.8 \times 10^{-6} = 4.69 \times 10^{-4}$ m/s
$K_{L,c} = 4.57 \times 10^{-4}$ m/s

The molar flux (N_J) for species j is given by Eq. (4-107) where $c_{molar,J,\infty}$ is the molar concentration of the species in the liquid far-field and $P_{J,\infty}$ is the partial pressure of species j in the gas far-field. In the liquid far-field (near the bottom of the lagoon),

$$c_{molar,b,\infty} = 100.\frac{mg}{L}\frac{1}{M_b} \qquad c_{molar,c,\infty} = 100.\frac{mg}{L}\frac{1}{M_c}$$

but in the gas far-field (distances far above the liquid-air interface) there is only pure air, so

$$P_{b,\infty} = P_{c,\infty} = 0$$

Eq. (4-107) then simplifies to

$$N_J = K_{L,J}\left(c_{molar,J,\infty} - \frac{P_{J,\infty}}{H}\right) \approx K_{L,J}c_{molar,J,\infty}$$

The mass transfer rate of benzene (j = b) and that of chloroform (j = c) from the lagoon (in kg/hr) for surface area A = 1,000 m^2 are thus

$$\dot{m}_b = N_b M_b A = K_{L,b}c_{molar,b,\infty}M_b A$$

$$= 4.85\times10^{-4}\frac{m}{s}\left(100.\frac{mg}{L}\frac{1}{M_b}\right)M_b\left(1000\ m^2\right)\left(\frac{kg}{10^6\ mg}\right)\left(\frac{3600\ s}{hr}\right)\left(\frac{1000\ L}{m^3}\right) = 175.\frac{kg}{hr}$$

and

$$\dot{m}_c = N_c M_c A = K_{L,c}c_{molar,c,\infty}M_c A$$

$$= 4.57\times10^{-4}\frac{m}{s}\left(100.\frac{mg}{L}\frac{1}{M_c}\right)M_c\left(1000\ m^2\right)\left(\frac{kg}{10^6\ mg}\right)\left(\frac{3600\ s}{hr}\right)\left(\frac{1000\ L}{m^3}\right) = 164.\frac{kg}{hr}$$

Discussion: Comparing the resistance of the liquid and gas films to mass transfer, it is clear that the liquid film resistance is nearly one hundred times larger than the gas film resistance (for either chemical).

4.5.8 Evaporation in Confined Spaces

An interesting use for Eq. (4-107) is to predict the rate of evaporation of the constituents from a multi-component liquid when the liquid is contained in a confined space and the vapors accumulate in the air above the air-liquid interface. On rare occasions individuals enter confined spaces; e.g. utility workers enter underground vaults to repair power or telephone lines, farmers enter manure pits under animal confinement buildings to remove manure, chemical workers enter reactors to make repairs of mechanical components, etc. Entering confined spaces is very a hazardous activity, and workers often underestimate the danger since everything looks peaceful and they are unaware of the presence of toxic gas or inadequate oxygen. Inadequate oxygen can cause one to pass out and die in a matter of minutes. Workers entering the space to assist a comrade may face the same fate. Operating electrical equipment in the presence of combustible gases can cause explosion even when there is no detectable odor. For these reasons, OSHA prescribes detailed procedures for entering a confined space, as discussed in Chapter 3. Individuals are risking their lives if these procedures are ignored. Equation (4-107) can be used to predict the equilibrium vapor concentration of volatile liquids in confined spaces and to predict the rate at which these equilibrium values are achieved.

Example 4.17 - Combustible Organic Compounds in a Confined Space

Given: A reactor is filled with a solution of water and ethyl alcohol (ethanol, abbreviated here as EA) with an initial mol fraction, $x_{EA}(0)$, equal to 0.0010. The temperature is 26.0 °C. Initially

75.% of the reactor is filled with the liquid solution; the remaining space is air. There is concern that the lower explosion limit (LEL) may eventually be reached in the air space and there is interest in how fast the alcohol evaporates.

To do: Determine if there is an equilibrium partial pressure for the alcohol (and if so, find its value). Second, find an expression for the rate of evaporation at any instant of time.

Solution: Equation (4-107) specifies the evaporation rate. Evaporation occurs at a decreasing rate until equilibrium is achieved, whereupon $N_{EA} = 0$. Thus at steady state,

$$c_{molar,EA,ss}(liq) = \left[\frac{P_{EA,\infty}}{H_{EA}}\right]_{ss}$$

For clarity, the parenthetical abbreviation "liq" is used to denote the liquid phase, and subscript "EA" is used for ethyl alcohol. The partial pressure of the ethyl alcohol is uniform when steady-state conditions are reached, hence $P_{EA} = P_{EA,\infty}$ everywhere in the gas phase in the tank. Since 75.% of the volume contains liquid, the mol fraction of alcohol in water, $x_{EA}(t)$, remains essentially constant with time, and equal to the initial value, $x_{EA}(0) = 0.0010$. The partial pressure of ethyl alcohol in the air is thus

$$\left(P_{EA,\infty}\right)_{ss} = H_{EA} c_{molar,EA,ss}(liq) = H_{EA} x_{EA}(0) c_{molar,L} = H'_{EA} x_{EA}(0)$$

where $c_{molar,L}$ is the total molar concentration of the liquid phase, which is essentially water. From Appendix A.9, $H'_{EA} = 4.56 \times 10^4$ N/m^2. Thus

$$\left(P_{EA,\infty}\right)_{ss} = 4.56\times10^4 \frac{N}{m^2}(0.0010)\left(\frac{1 \text{ kPa}}{1000\frac{N}{m^2}}\right) = 0.0456 \text{ kPa}$$

Assuming an ambient pressure of 100. kPa, this ethyl alcohol partial pressure corresponds to a gas phase mol fraction of 0.000456 (456 PPM). Since the LEL of ethyl alcohol is 3.28% (a mol fraction of 0.0328 or 32,800 PPM), and the OSHA PEL is 1000 PPM, there is no chance of fire, explosion, or asphyxiation in the air space above the liquid. Nevertheless other confined entry procedures still have to be followed.

At any instant the evaporation rate of alcohol is given by Eq. (4-107), where $P_{EA,\infty}$ refers to the partial pressure of alcohol in the far-field gas phase. It is obvious that while P_{EA} is initially zero, it increases with time until the above equilibrium value is achieved. To predict the partial pressure at any instant, the air space above the liquid is defined as a control volume (V), and conservation of mass is employed to predict the change in the molar gas phase alcohol concentration, $c_{molar,EA,\infty}(gas)$, with time,

$$\frac{dm_{EA}}{dt} = M_{EA} V \frac{dc_{molar,EA,\infty}(gas)}{dt} = M_{EA} N_{EA} = M_{EA} K_L \left(c_{molar,EA,ss}(liq) - \frac{P_{EA,\infty}}{H}\right)$$

Using the ideal gas law,

$$c_{molar,EA,\infty}(gas) = \frac{P_{EA,\infty}}{R_u T}$$

and the differential equation becomes

$$\frac{V}{R_u T}\frac{dP_{EA,\infty}}{dt} = K_L\left[c_{molar,EA,ss}(liq) - \frac{P_{EA,\infty}}{H}\right]$$

The above first-order ODE for $P_{EA,\infty}$(gas) as a function of time can be rearranged into the standard form of Chapter 1, for which the solution is known analytically since the coefficients are constant. The result is an expression for the instantaneous alcohol partial pressure in the air space:

$$P_{EA,\infty}(t) = Hc_{molar,EA,ss}(liq)\left[1-\exp\left(\frac{-K_L R_u T}{HV}t\right)\right] = P_{EA,ss}\left[1-\exp\left(\frac{-K_L R_u T}{HV}t\right)\right]$$

Finally, an expression for the rate of evaporation (molar flux) as a function of time is obtained,

$$N_{EA}(t) = K_L\left[c_{molar,EA,ss}(liq) - \frac{P_{EA,\infty}(t)}{H}\right]$$

Discussion: If the original volume of liquid were small compared to the tank volume, the approximation that $c_{molar,EA}$(liq) remains constant with time would not be valid. In such a case, the ODE would not have constant coefficients, and an analytical solution would not be possible; the problem would need to be solved numerically (e.g. using Runge-Kutta).

4.6 Drop Evaporation

Atomizers, spray nozzles, etc. create drops of liquid and are used in countless industrial processes such as

- spray drying applications or coatings
- transferring liquids
- scrubbing, washing, and cleaning
- aeration
- fuel atomization

Liquid drops are also generated naturally as sea spray and when gas bubbles traveling upward through sea water collapse and break through the liquid-air interface. Every drop formed by this process has a liquid-air interface through which the liquid and species dissolved in the liquid evaporate.

Consider the evaporation of drops of a liquid consisting of a single molecular species, at uniform drop particle temperature (T_p) moving or suspended in air at a constant temperature (T_∞). In the analysis that follows it is assumed that the number of drops per unit volume of air is low such that one drop does not influence another. Thus the total evaporation rate from a cloud of drops is equal to the evaporation from one drop times the total number of drops. More complete analyses can be found in meteorology texts on the microphysics of clouds (e.g. Pruppacher and Klett, 1978). While the assumption of independent drops is simplistic, valuable intuition can be gained that will aid readers studying more complex issues such as clouds (see Chapter 8).

Before modeling the process, the physical processes that occur must be considered. Vapor escapes from the surface of the drop as shown in Figure 4.18 because the vapor pressure of the saturated liquid (based on drop surface temperature) exceeds the partial pressure of the vapor in the far field. As the liquid evaporates, the drop diameter decreases, which in turn influences the rate of evaporation. Simultaneously, the evaporating liquid extracts energy from the drop (a process called ***evaporative cooling***) that lowers the temperature of the drop, which in turn lowers the saturation vapor pressure at the drop-air interface. Because evaporation tends to lower the

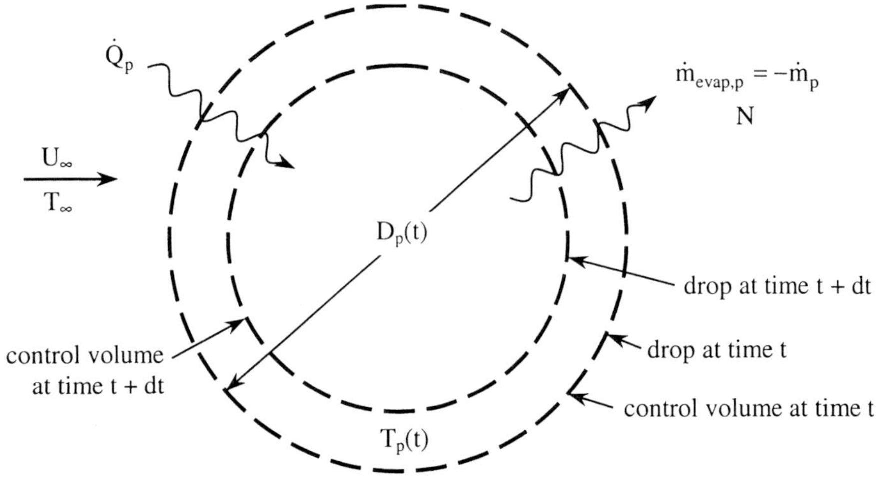

Figure 4.18 Control volume for evaporating drop at two times, t and t + dt.

drop temperature, energy is transferred to the drop from the air by convection. Thus both mass and heat transfer are coupled, and together they control the rate at which the drop evaporates. Since the total surface area of a sphere of diameter D_p is πD_p^2, the evaporation rate from a drop of diameter D_p is

$$\boxed{\dot{m}_{evap,p} = M_p \pi D_p^2 N} \qquad (4\text{-}114)$$

where the subscript p refers to properties of the drop (p refers to "particle"), and the molar flux N, in units of kmol/(m^{2}s), can be expressed using Eq. (4-62),

$$\boxed{N = k_G R_u T_\infty \left(c_{molar,i} - c_{molar,\infty}\right)} \qquad (4\text{-}115)$$

Note that since the drop contains only one species of molecular weight M_p, the subscript for species (j) has been dropped. The nomenclature has been simplified so that the concentration subscripts pertain only to the liquid-air interface (subscript i) and to the far-field concentration (subscript ∞). The molar concentrations can be converted to partial pressures using the ideal gas law. Thus, Eqs. (4-114) and (4-115) become

$$\boxed{\dot{m}_{evap,p} = M_p \pi D_p^2 k_G T_\infty \left[\frac{P_v}{T_{p,i}} - \frac{P_\infty}{T_\infty}\right]} \qquad (4\text{-}116)$$

where P_∞ is the pollutant partial pressure in the far-field where the temperature is T_∞, and P_v is the liquid saturation pressure based on drop interface temperature $T_{p,i}$. The convection mass transfer coefficient (k_G) can be found from heat transfer relationships using the Reynolds analogy described in Section 4.5.5. The binary diffusion coefficient of the contaminant into air is D_{ja}. Combining Eqs. (4-116) and (4-80) in which the characteristic length L is the drop diameter (D_p), one obtains

$$\boxed{\dot{m}_{evap,p} = \pi M_p D_p \frac{D_{ja}}{R_u}\left(\frac{Sc}{Pr}\right)^{b_1} \frac{P}{P_{am}} Nu \left[\frac{P_v}{T_{p,i}} - \frac{P_\infty}{T_\infty}\right]} \qquad (4\text{-}117)$$

The Nusselt number (Nu) for a sphere can be expressed as a function of the Reynolds and Prandtl numbers, as in Eq. (4-72), where the Reynolds number for the drop is based on drop diameter (D_p) and the velocity of the drop (U_∞) relative to the surrounding air,

$$\boxed{Re = \frac{\rho D_p U_\infty}{\mu}} \tag{4-118}$$

Note also that the density and viscosity in Eq. (4-118) are those of the *air*, not of the drop.

As the drop evaporates, its diameter decreases. To compute the instantaneous drop diameter, $D_p(t)$, an equation for conservation of mass (m_p) can be written for the drop. First, the mass, m_p, of the spherical drop at any instant is

$$\boxed{m_p = \rho_p \frac{\pi}{6} D_p^{\,3}} \tag{4-119}$$

Then, the rate of change of mass of the drop can be written by differentiation with respect to time,

$$\boxed{\dot{m}_p = \frac{dm_p}{dt} = \frac{d\left(\rho_p \dfrac{\pi D_p^{\,3}}{6}\right)}{dt} = \rho_p \frac{\pi D_p^{\,2}}{2}\frac{dD_p}{dt}} \tag{4-120}$$

Since the drop is decreasing in size, $\dot{m}_p$ is negative. If the drop temperature at the liquid-air interface is known, the diameter can be found as a function of time by equating Eq. (4-116) with the negative of Eq. (4-120), since the mass rate of evaporation from the drop must equal the negative of the rate of change of drop mass:

$$\boxed{\dot{m}_{evap,p} = -\frac{dm_p}{dt} = -\dot{m}_p = -\rho_p \frac{\pi D_p^{\,2}}{2}\frac{dD_p}{dt}} \tag{4-121}$$

Unfortunately, one does not know the liquid drop interface temperature ($T_{p,\imath}$), nor can one assume that the drop surface temperature is equal to the air temperature. A separate calculation is needed to compute the drop temperature. From the field of heat transfer (Kreith, 1973; Incropera and DeWitt, 1990) the ***Biot number*** is defined as

$$\boxed{Bi = \frac{hD_p}{2k_p}} \tag{4-122}$$

where h is the convection heat transfer film coefficient and k_p is the liquid particle's thermal conductivity. Note that while the Biot and Nusselt numbers are composed of similar variables, the thermal conductivity (k_p) in the Biot number refers to the liquid, while the value of (k) in the Nusselt number refers to the surrounding air. If Bi is less than approximately 0.1, drop temperature does not vary appreciably with radius. In other words, if $Bi < 0.1$, drop temperature is essentially uniform throughout the drop, $T_p(r,t) = T_{p,\imath}(t) = T_p(t)$, and one can assume that the drop has ***lumped heat capacity***. Physically this means that resistance to conduction heat transfer within the drop is negligible; whenever heat is transferred to the drop, it is distributed throughout instantly so that the temperature of the drop does not vary with radius, but only with time. More generally, the ***lumped parameter approximation*** applies when *all* properties (temperature, density, enthalpy, etc.) of the drop are uniform throughout the drop. For small drops that constitute an evaporating cloud, the lumped parameter approximation is generally valid and is adopted in this analysis.

The drop temperature, $T_p(t)$, can be found from conservation of energy (first law of thermodynamics) using the drop itself as the control volume and the drop surface as the control surface. In its simplest integral form, making the lumped parameter approximation and neglecting changes in kinetic energy, potential energy, and viscous shear work, the first law of thermodynamics for a control volume is

$$\frac{d}{dt}(m\hat{u}) + \sum_{out} \dot{m}\hat{h} - \sum_{in} \dot{m}\hat{h} = \dot{Q} - \dot{W}_{shaft} \tag{4-123}$$

where $\hat{h}$ is the specific enthalpy (enthalpy per unit mass), $\hat{u}$ is the specific internal energy, and $\dot{Q}$ is the rate of heat transfer, positive by convention for heat transferred into the control volume. Pressure work has been included in the enthalpy flux terms, as is the convention in thermodynamics. The last term represents shaft work done by the control volume, which is zero for an evaporating drop. The unsteady term is important because both the mass and specific internal energy within the control volume change with time. Defining $\dot{Q}_p$ as the rate of heat transfer into the drop, $\dot{Q} = \dot{Q}_p$, and recognizing that there is no mass transfer *into* the control volume, but there is mass transfer (in the form of vapor) *out of* the control volume, Eq. (4-123) simplifies to

$$\dot{Q}_p = \frac{d\left(m_p \hat{u}_f\right)}{dt} + \dot{m}_{evap,p}\hat{h}_g \tag{4-124}$$

where $\hat{u}_f$ is the specific internal energy of the saturated liquid, and $\hat{h}_g$ is the specific enthalpy of the saturated vapor; both $\hat{u}_f$ and $\hat{h}_g$ are evaluated at drop temperature T_p. The change in specific internal energy of the saturated liquid can be expressed as the liquid specific heat at constant volume (c_v) times the change in temperature,

$$d\hat{u}_f = c_v dT_p \tag{4-125}$$

Using Eq. (4-125) and the product rule, Eq. (4-124) can be re-written as

$$\dot{Q}_p = m_p c_v \frac{dT_p}{dt} + \hat{u}_f \frac{dm_p}{dt} + \dot{m}_{evap,p}\hat{h}_g \tag{4-126}$$

The time rate of change of drop mass in the middle term on the right hand side of Eq. (4-126) can be replaced by the negative of the evaporation rate using Eq. (4-121). Heat transfer to the drop is assumed to be via convection heat transfer only; the rate of heat transfer $\dot{Q}_p$ is thus

$$\dot{Q}_p = h\pi D_p^2 \left(T_\infty - T_p\right) \tag{4-127}$$

where h is the convection heat transfer coefficient, and πD_p^2 is the surface area of the spherical drop. Upon substitution, Eq. (4-126) becomes

$$h\pi D_p^2 \left(T_\infty - T_p\right) = \rho_p \frac{\pi}{6} D_p^3 c_v \frac{dT_p}{dt} + \dot{m}_{evap,p}\left(\hat{h}_g - \hat{u}_f\right) \tag{4-128}$$

where again, $\hat{h}_g$ and $\hat{u}_f$ are evaluated at the drop temperature (T_p). The quantity $\left(\hat{h}_g - \hat{u}_f\right)$ is approximately equal to $\hat{h}_{fg}$, the enthalpy of vaporization. $\hat{h}_{fg}$ is also evaluated at T_p. Since T_∞ is a constant,

$$\frac{dT_p}{dt} = -\frac{d\left(T_\infty - T_p\right)}{dt}$$

and Eq. (4-128) can be rewritten as

$$\frac{d\left(T_\infty - T_p\right)}{dt} = \frac{6\dot{m}_{evap,p}\hat{h}_{fg}}{\pi\rho_p D_p{}^3 c_v} - \frac{6h}{\rho_p D_p c_v}\left(T_\infty - T_p\right) \tag{4-129}$$

Equations (4-117), (4-121), and (4-129) constitute a set of simultaneous equations whose solution yields the evaporation rate, drop temperature, and drop diameter as functions of time. Since the equations are coupled, numerical techniques (such as Runge-Kutta) should be used.

A significant simplification can be employed if the ***thermal time constant*** (τ) is small compared to the characteristic time of the process. The thermal time constant is analogous to the time constant for an electrical R-C circuit. One can think of a *thermal* R-C circuit in which the reciprocal of the time constant is the product of a thermal resistance and a thermal capacitance,

$$\tau = \left(\frac{k_p}{h\pi D_p{}^2}\right)\left(\frac{\rho_p \pi D_p{}^3 c_v}{6k_p}\right) = \frac{\rho_p D_p c_v}{6h} \tag{4-130}$$

If the thermal time constant (τ) is small, the system responds quickly to changes in the thermal environment and the drop achieves a quasi-steady-state temperature. If such equilibrium is achieved quickly, the left-hand side of Eq. (4-129) can be assumed to be zero, and at any time the quasi-steady-state temperature can be found by equating the two remaining terms:

$$T_p\left(\text{quasi-steady-state}\right) = T_\infty - \frac{\dot{m}_{evap,p}\hat{h}_{fg}}{\pi h D_p{}^2} \tag{4-131}$$

The temperature difference, (T_p - T_∞), is called the ***temperature depression***. Finally, the evaporation rate from the drop can be found by solving Eqs. (4-117) and (4-121), with drop temperature given by Eq. (4-131). The approximation of small τ has replaced one of the original differential equations, Eq. (4-129), with a simpler algebraic expression, Eq. (4-131). The equations are still coupled, however, and must be solved simultaneously.

Example 4.18 - Drop Evaporation

Given: Consider a cloud of water droplets in quiescent humid air at STP. The relative humidity (Φ) is 80.%. Initially the drop diameter (D_p) is 500. μm. Assume that the temperature depression (T_p - T_∞) is initially zero. Thus $T_p = T_{STP} = 25.0$ °C.

To do: Analyze the evaporation of water droplets smaller than 500 μm (0.5 mm). Specifically, compute and plot the evaporation rate ($\dot{m}_{evap,p}$ in mg/s) of a drop of water. Also compute and plot the instantaneous drop diameter, $D_p(t)$ (in μm), versus time.

Solution: The relevant properties of air and water (to at least three digits of accuracy) are:

- Air: T_∞ = 25.0 °C, P_∞ = 101. kPa, $\nu = 1.55 \times 10^{-5}$ m²/s, k = 0.0258 W/(m K), Pr = 0.707, $D_{wa} = 0.260 \times 10^{-4}$ m²/s, Sc = 0.5769, and $\mu = 1.83 \times 10^{-5}$ kg/(m s).
- Water: k_p = 0.613 W/(m K), ρ_p = 1,000 kg/m³, c_v = 4.18 kJ/(kg K), P_v (at 25.0 °C) = 3.169 kPa, $P_{p,\infty}$ = partial pressure of water vapor in far field = ΦP_v = (0.80)(3.169 kPa) = 2.535 kPa, and $\hat{h}_{fg}$ (at 25.0 °C) = 2,442.3 kJ/kg.

The settling velocity of a drop of water falling freely in quiescent air is given by the following (see Chapter 8 for details):

$$v_t = \frac{g\rho_p D_p^{\ 2}}{18\mu}$$

The Reynolds number is calculated on the basis of the settling velocity and is thus very small. Using the appropriate equation from heat transfer, Table 4.3, the Nusselt number can be computed for flow over a sphere, from which the convection heat transfer coefficient (h) can be computed. The Biot numbers are computed and are seen to be less than 0.1 for particles 100 μm and smaller; the lumped heat capacity assumption is reasonable. On the other hand, the computed values are suspect for D_p greater than 100 μm. The time constant (τ) can be computed directly from Eq. (4-130). The time constant is less than 0.1 s for particles smaller than 100 μm and one would expect that the drop temperature is equal to its equilibrium value predicted by Eq. (4-131). To further simplify the analysis, drop temperature could be assumed to remain constant during the evaporation process.

D_p (μm)	**v_t (m/s)**	**Re**	**Nu**	**h (W/m^{2}K)**	**Bi**	**τ (s)**
5	0.0008	0.001	2	10,320	0.042	0.0003
10	0.003	0.002	2	5,160	0.042	0.001
50	0.078	0.258	2.18	1,125	0.046	0.031
100	0.255	1.7	2.54	655	0.054	0.106
500	2.0	66.7	5.81	300	0.122	1.162

For purposes of reference one can crudely categorize the above particles as follows:

- fog: 1(μm) < D_p < 10 (μm)
- mist: 10 (μm) < D_p < 100 (μm)
- drizzle: 100 (μm) < D_p < 1,000 (μm)
- rain: D_p > 1,000 (μm) = 1 mm

The calculations that follow are tedious and the reader must be very careful to check the units used in the equations. Throughout the calculations, drop diameter D_p is expressed in units of μm. The log mean pressure difference (P_{am}) is

$$P_{am} = \frac{P_{a,l} - P_{a,\infty}}{\ln\left(\frac{P_{a,l}}{P_{a,\infty}}\right)} = \frac{(101.-3.169)\text{ kPa} - (101.-2.535)\text{ kPa}}{\ln\left(\frac{97.831}{98.469}\right)} = 98.16\text{ kPa}$$

The Reynolds number (Re) is

$$\text{Re} = \frac{D_p v_t}{\nu} = \frac{g\rho_p D_p^{\ 3}}{18\mu\nu} = \frac{9.81\frac{\text{m}}{\text{s}^2}\left(1000\frac{\text{kg}}{\text{m}^3}\right)D_p^{\ 3}\left(\frac{\text{m}}{10^6\ \mu\text{m}}\right)^3}{18\left(1.83\times10^{-5}\frac{\text{kg}}{\text{m}\cdot\text{s}}\right)\left(1.55\times10^{-5}\frac{\text{m}^2}{\text{s}}\right)} = \left(1.92\times10^{-6}\right)D_p^{\ 3}$$

where D_p must be in μm. Plugging all the values into Eq. (4-117), and using a value of 0.4 for the exponent b_l in that equation, the evaporation rate is equal to

$$\dot{m}_{evap,p} = \left(3.56\times10^{-7}\right)D_p \text{Nu}\ \frac{\text{mg}}{\text{s}}$$

The Nusselt number in the above equation is obtained from Table 4.3, and is a function of drop diameter. Neglecting the difference in air viscosity between the far field and the drop surface, i.e. assuming constant viscosity,

$$Nu = 2 + \left[0.4\sqrt{Re} + 0.06(Re)^{2/3}\right](Pr)^{0.4}$$

The differential equation expressing the rate of change of drop diameter in terms of evaporation rate is given by Eq. (4-121). After some rearranging,

$$\frac{dD_p}{dt} = -\frac{2\dot{m}}{\rho_p \pi D_p^{\ 2}} = -227.\frac{Nu}{D_p}\ \frac{\mu m}{s}$$

where again, D_p must be in microns. Since Nusselt number is a function of Reynolds number [Eq. (4-72)], and Reynolds number is in turn a function of drop diameter [Eq. (4-118)], Nusselt number varies with drop diameter. Hence, the differential equation above cannot be integrated in closed form, but is instead solved numerically using the 4th-order Runge-Kutta technique. Figure E4.18 shows the results of these computations, namely the initial evaporation rate as a function of drop diameter, and the instantaneous drop diameter as a function of time. Note that for simplicity, the temperature depression is assumed to be zero at all times, not just initially. The Mathcad program to solve the above is provided on the textbook's web site.

Discussion: The above analysis can be applied to steam plumes produced by industry, comparing the lifetimes above to the lifetime of water droplets that comprise steam plumes. Water vapor (invisible) from a process condenses as it leaves an exhaust stack and is cooled with ambient air. Thus a "steam" plume is really a plume of small liquid water drops. After a short period the drops evaporate and the plume disappears. A distinctive feature of steam plumes, in contrast to particle plumes, e.g. fly ash, is that steam plumes have distinct boundaries and end abruptly. Particle plumes on the other hand have diffuse boundaries and persist for a considerable distance downwind of the exhaust stack. The results shown here indicate that even relatively large (500 μm) water droplets shrink to extinction in less than two minutes.

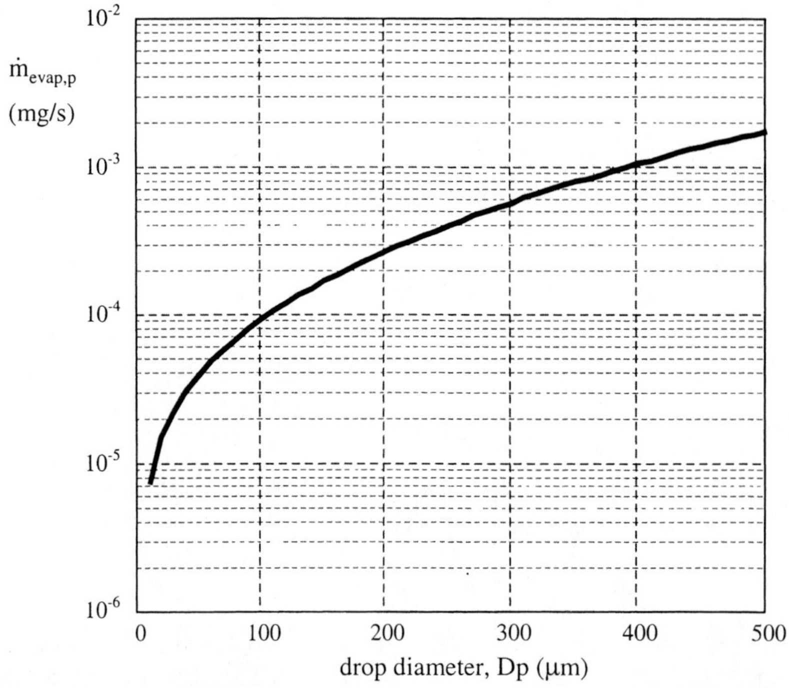

Figure E4.18a Evaporation rate as a function of drop diameter for evaporating water droplets in quiescent humid air at STP.

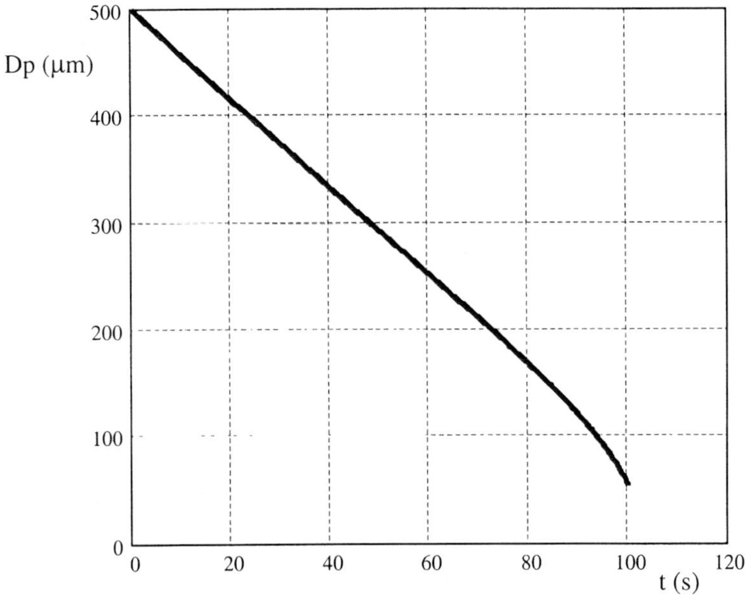

Figure E4.18b Decay of drop diameter with time; evaporating water droplets in quiescent humid air at STP.

4.7 Leaks

Hazardous materials may escape as leaks of gas, vapor, or liquid from vessels, piping, pumps, flanges, valves, etc. Figure 4.19 illustrates leaks that one may expect. Leaks may occur through holes in the vessels or piping, or through cracks in valve packing, pump seals, or gaskets; such locations are called ***leakers***. Leakers are generally the result of broken seal joints and have a characteristic opening larger than 100 μm and emission concentrations larger than 10,000 PPM. Emissions may also occur through small pores 1 μm or smaller, by the process of capillary flow. For the same pressure drop across the leak site, the average emission rate is higher if the emitting fluid is a volatile liquid than if it is a permanent gas (Choi et al., 1992). If the concentration of the escaping vapor can be measured with a portable organic analyzer, the total emission rate (kg/hr) of an organic compound can be estimated by EPA emission factors shown in Appendix A.3. Leak rates can also be measured experimentally using flux chambers (Section 4.1.2).

The term ***aperture*** is used in this section to denote leakers and openings larger than 100 μm. Liquid leaks may produce conspicuous puddles while escaping gases and vapors may produce a conspicuous noise. Often however, no such symptoms are evident and the leak may be undetected for a considerable time. While leaks occur through openings of very small cross-sectional area, the length (in the direction of flow) of the hole or crack is often several times larger than the characteristic dimension perpendicular to the direction of flow. In a sense, leaks occur through a "conduit" which has both entrance and exit pressure losses as well as frictional losses through the conduit. The analysis in this section treats an aperture as a sharp edged hole with loss coefficients associated with rapid contractions. A more sophisticated analysis should treat the aperture as a conduit. The upstream temperature and pressure are noted by T_1 and P_1 with the final atmospheric temperature and pressure noted by T_a and P_a. The temperature and pressure of the leaking fluid at the exit plane of the aperture are noted by T_2 and P_2. The pressure at the exit plane of the aperture is equal to P_a if the exit velocity is less than the speed of sound, or possibly larger than P_a if the exit velocity exceeds the speed of sound and the crack opening can be considered a converging-diverging (supersonic) nozzle.

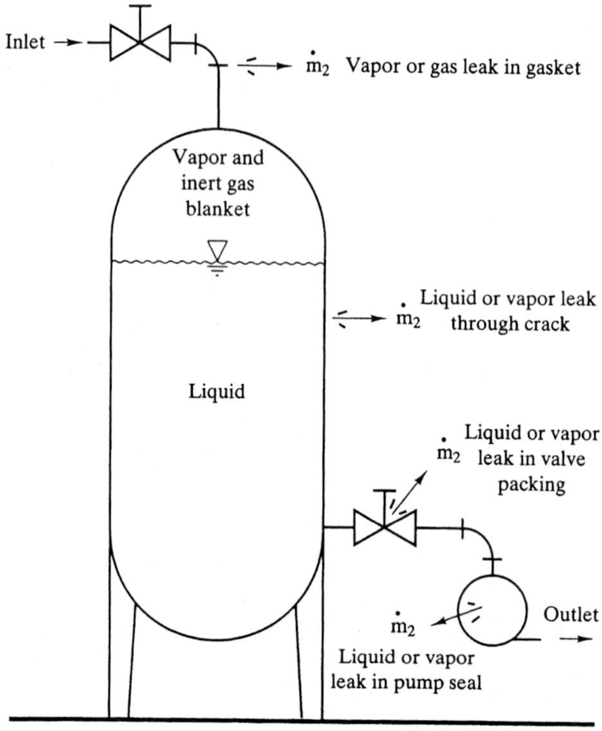

Figure 4.19 Locations of potential leaks in a pressure vessel and its inlet and outlet piping.

- subsonic flow: $P_2 = P_a$
- sonic or supersonic flow: $P_2 > P_a$

Shown in Table 4.4 is a convenient way to categorize leaks in terms of the hardware in which they occur and the mass flow rate of the leak. These criteria define the thermodynamic and heat transfer analyses.

A ***puncture*** of a storage vessel or ***rupture*** of a transmission pipeline may produce a discharge with large mass flow rate that is more properly called a process ***dump***, and will not be analyzed in this section. Leaks, consisting of small mass flow rates, are addressed. Material leaks through an aperture when upstream pressure P_1 exceeds atmospheric pressure P_a. Thermodynamics enters the analysis by relating aperture temperature and pressure (T_2, P_2) to upstream temperature and pressure (T_1, P_1) and to heat transferred into or out of the fluid.

High-pressure gases and vapors produce a jet as they escape, and the escaping material can often be assumed to be an ideal gas as it enters the air. Liquids stored under pressure produce different types of leaks depending on the relationship between the aperture temperature (T_2) and the ***normal boiling temperature*** (T_b), which is the boiling temperature at atmospheric pressure.

Table 4.4 Examples of the two categories of leaks: isothermal slow leak and adiabatic rapid discharge leak.

example	**isothermal slow leak**	**adiabatic rapid discharge**
storage tank	T_1 = constant	T_1 & P_1 decrease
transmission pipeline	P_1 & T_1 = constant, steady flow	rupture

A leak from the space above the liquid level can produce a stream of vapor or a two-phase stream composed of liquid and vapor. If the space above the liquid contains an inert blanket of gas, the inert gas will also be present in the leak. The rapid vaporization (boiling) of the liquid due to a sudden decrease in pressure is called ***flash boiling*** or ***flashing*** for short. A leak below the liquid level produces a liquid stream that flashes (partially or totally) into vapor as it exits. If the exit temperature (T_2) is less than T_b, flashing does not occur and the escaping liquid merely drips downward.

The three types of leaks engineers may encounter can be shown on a temperature-entropy diagram, as in Figure 4.20.

I. Liquid leak: P_1, T_1 correspond to a compressed liquid and $T_2 < T_b$.
II. Gas or vapor leak: P_1, T_1 correspond to a superheated vapor and $T_2 > T_b$.
III. Flash boiling leak: P_1, T_1 correspond to a compressed liquid and $T_2 \geq T_b$.

A ***compressed liquid*** is a liquid at a temperature (T_1) less than the saturation temperature (T_{sat}) corresponding to the pressure (P_1). As the leak progresses, if there is heat transfer to the upstream fluid, then point 1 remains constant. If however the flow is adiabatic, then point 1 in Figure 4.20 moves downward with time as the temperature T_1 decreases.

Leaks associated with the steady flow of a fluid in a pipeline are analyzed as either isothermal or adiabatic steady flow through an aperture. Leaks associated with storage vessels may be isentropic, isothermal, or nearly adiabatic. On Figure 4.20 isothermal leaks map as horizontal lines and isentropic flow maps as vertical lines. The paths shown in Figure 4.20 correspond to leaks that are adiabatic but not isentropic. Owing to the small cross-sectional area of the aperture, a leak by definition has low mass flow rate. If the leak is a liquid, the density and temperature of the fluid are essentially constant and no heat transfer issues arise. If the fluid is a gas or vapor or if flashing occurs, the temperature of the fluid passing through the aperture decreases since there is inadequate time for heat to be transferred to the escaping fluid. If the mass flow rate is low there may be adequate time for heat to be transferred from the vessel, pipe, or from outside agents to the upstream fluid. If this occurs, the upstream fluid may be assumed to be isothermal. If however a discharge occurs because of a dramatic event, e.g. a worker drives a fork-lift vehicle into a high-pressure vessel, a pipe fractures due to thermal fatigue, etc., the mass flow rate will be large, and it is unwise to assume that there is adequate time for heat transfer. Such

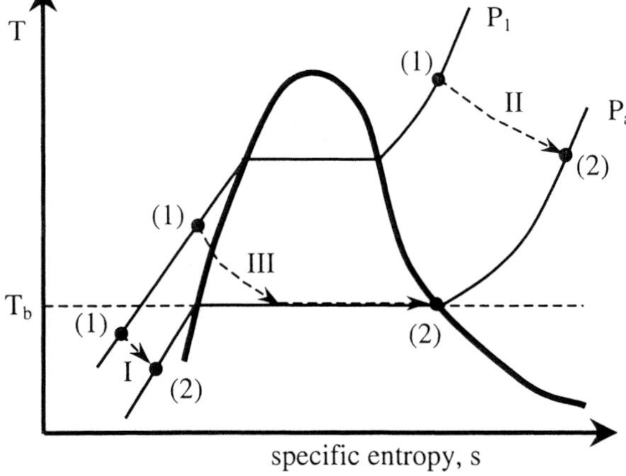

Figure 4.20 Three types of adiabatic but irreversible leaks: I - liquid leaks, II - gas or vapor leaks, III - adiabatic flashing leaks.

dramatic events are not discussed in this section; the reader is encouraged to consult Crowl and Louvar (1990) for a more thorough discussion of the subject.

4.7.1 Type I - Liquid Leaks

Liquid leaks refer to the discharge of a fluid that remains a liquid as it passes through an aperture. Liquid leaks pertain to incompressible flow (or more specifically constant density flow). Figure 4.21 shows the essential features of liquid leaks. The velocity of an incompressible liquid through an aperture is governed by the incompressible steady-state, steady-flow energy equation for a control volume from location 1 to location 2,

$$\frac{P_1}{\rho g}+\frac{U_1^2}{2g}+z_1=\frac{P_2}{\rho g}+\frac{U_2^2}{2g}+z_2+h_{LT} \tag{4-132}$$

where subscript 1 designates upstream flow conditions and subscript 2 designates conditions at the exit plane of the aperture. Each term in Eq. (4-132) has dimensions of length, which engineers call ***head***, defined in terms of the equivalent column height of fluid. For liquid leaks, $P_2 = P_a = 1$ atm. The term h_{LT} is the ***total head loss***, which for the leak is equal to the ***minor head loss***, h_{LM} (head loss comprising entry and exit losses), since frictional loss, h_f, on the inside wall of the aperture (***major head loss***) is negligible for the geometries of interest here. Using standard fluid mechanics notation,

$$h_{LM}=C_0\frac{U_0^2}{2g} \tag{4-133}$$

The term C_0 is a ***loss coefficient*** for the aperture, and U_0 is a characteristic velocity defined in the table specifying C_0. Note that some fluid mechanics textbooks use K instead of C_0. In the absence of other data, one can assume that the aperture is a rapid contraction. The ASHRAE Fundamentals Handbook shows that for a rapid contraction (θ = 180°) and a circular opening considerably smaller than the upstream area, $C_0 = 0.43$ where U_0 is the velocity through the exit plane of the opening (U_2). The mass flow rate of the leak ($\dot{m}_2$) is given by

$$\dot{m}_2=\rho_2 A_2 U_2 \tag{4-134}$$

where

$$U_2=\left[\frac{2(P_1-P_2)}{\rho_2(1+C_0)}\right]^{1/2} \tag{4-135}$$

Crowl and Louvar (1990) suggest using a discharge coefficient to characterize flow through a sharp edge orifice and another discharge coefficient to account for the ***vena contracta*** of the flow through the aperture. It is not obvious that this is more accurate owing to the fact that the flow

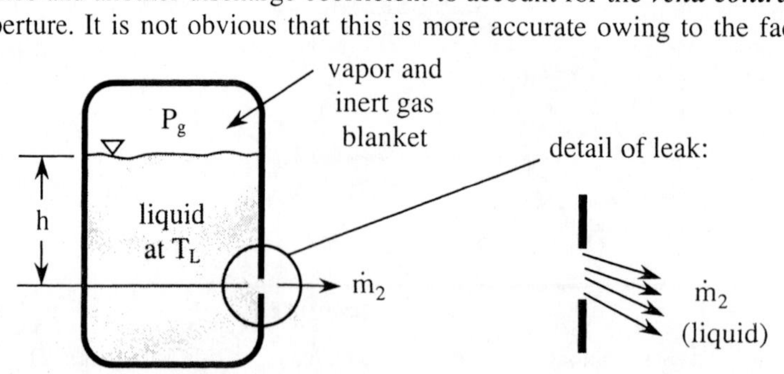

Figure 4.21 Type I liquid leak for the case in which $T_1 = T_L < T_b$.

through the aperture is really flow through a narrow conduit of length considerably larger than the dimension perpendicular to the flow. In this section only the single coefficient C_0 is used.

Example 4.19 - Leak through Valve Packing

Given: The packing around the stem of a shut-off valve mounted in a 6.0-inch pipe carrying 500. GPM of trichloroethylene, TCE, ($\rho_2 = 1{,}460\ kg/m^3$) at 25.0 °C and at 2.0 atm pressure leaks at the rate of 1.0 kg/day. The ambient air pressure is 1.0 atm.

To do: Estimate the area of the aperture.

Solution: The elevation change (z_1 - z_2) is negligible. While the velocity inside the pipe is appreciable, the fluid located in the valve housing upstream of the leak is relatively stagnant, but its pressure is 2.0 atm. Equation (4-135) is used to estimate U_2,

$$U_2 = \left[\frac{2(P_1 - P_2)}{\rho_2(1+C_0)}\right]^{1/2} = \left[\frac{2(2.0-1.0)\ \text{atm}}{1460\dfrac{\text{kg}}{\text{m}^3}(1+0.43)}\left(\frac{101.3\ \text{kPa}}{\text{atm}}\right)\left(\frac{1000\ \text{kg}}{\text{s}^2\text{m kPa}}\right)\right]^{1/2} = 9.85\frac{\text{m}}{\text{s}}$$

Thus, from Eq. (4-134),

$$A_2 = \frac{\dot{m}_2}{\rho_2 U_2} = \frac{\dfrac{1.0\ \text{kg}}{\text{day}}\left(\dfrac{\text{day}}{24\ \text{hr}}\right)\left(\dfrac{\text{hr}}{3600\ \text{s}}\right)}{1460\dfrac{\text{kg}}{\text{m}^3}\left(9.85\dfrac{\text{m}}{\text{s}}\right)} = 8.05\times10^{-10}\ \text{m}^2$$

Discussion: The area is quite small, i.e. about 0.0008 mm^2. If the aperture were a circular hole, its diameter would be 0.032 mm (32. μm). Note that neither the pipe diameter nor the volume flow rate through the pipe were required for the solution.

Suppose a leak develops in a pressurized storage vessel containing an incompressible liquid protected by a blanket of inert gas at pressure P_g as shown in Figure 4.21. If the leaking liquid does not flash, Eq. (4-135) can be used to compute its velocity. Flow inside the tank is stagnant. However, because there is a hydrostatic head, i.e. elevation (h) between the surface of the liquid and the aperture in the vessel wall, the pressure upstream of the leak (P_1) must be found by application of Eq. (4-132) for this simplified stagnant case (all velocities are zero),

$$P_1 = P_g + \rho_2 g h$$

and thus

$$\boxed{P_1 - P_2 = P_g - P_a + \rho_2 g h} \tag{4-136}$$

Substitution of Eq. (4-136) into Eq. (4-135) yields a modified expression for the velocity of the leaking fluid,

$$\boxed{U_2 = \sqrt{\frac{2\left[\left(P_g - P_a\right) + \rho_2 g h\right]}{\rho_2\left(1+C_0\right)}}} \tag{4-137}$$

Unique to the configuration in Figure 4.21 is the fact that as the leak progresses, both the fluid level (h) and the pressure of the inert gas blanket (P_g) decrease, which in turn reduces the leak mass flow rate. Leaks in storage vessels occur slowly and there is sufficient time for heat transfer to establish isothermal conditions; thus gas temperature T_g is constant. Applying conservation of mass for the leaking liquid shows that at any instant, the liquid level h(t) is

$$\boxed{h(t) = h(0) - \frac{1}{\rho_2 A_2}\int_0^t \dot{m}_2 dt} \tag{4-138}$$

where h(0) is the liquid head at time t = 0. At any instant, the pressure of the gas blanket, $P_g(t)$ is

$$\boxed{P_g = \frac{R_u T_g}{M_g}\frac{m_g}{V_t - Ah(t)}} \tag{4-139}$$

where T_g and m_g are the inert gas blanket temperature and mass, and V_t is the total volume of the vessel. A is the cross-sectional area of the vessel, which is assumed in Eq. (4-139) to be constant over the height of the vessel. Over time, T_g and P_g change as the volume of the compressed liquid decreases, and the volume of the inert gas increases.

4.7.2 Type II - Vapor and Gas Leaks

For a fluid stored at a pressure P_1 several times larger than atmospheric pressure (P_a), if the temperature T_1 is greater than T_{sat} at P_1, the fluid will be a superheated vapor. If leaks occur in a pipe, pipe fittings, or a storage vessel containing gases or vapors under high pressure as shown in Figure 4.22, the density of the material does not remain constant as it escapes, and the equations in the previous section cannot be used. For analysis, it is assumed that the escaping material can be approximated as an ideal gas and that the flow through the aperture, up to the aperture exit plane, is isentropic. Such a leak can be depicted on the temperature entropy diagram in Figure 4.23. Note that beyond the exit plane, the flow is *not* isentropic.

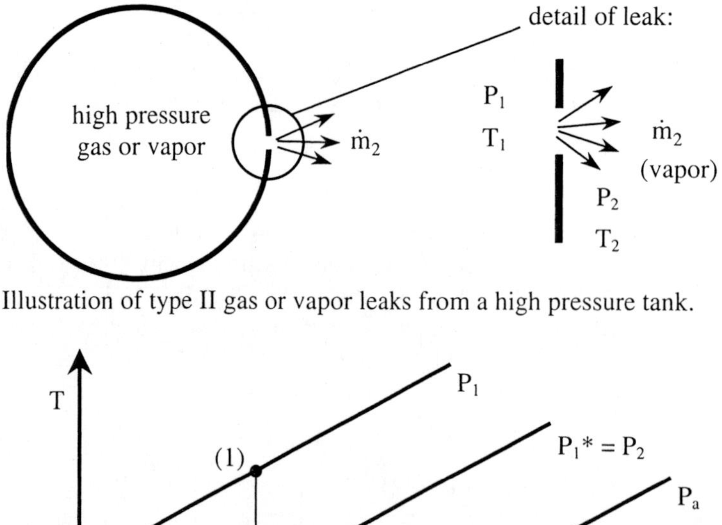

Figure 4.22 Illustration of type II gas or vapor leaks from a high pressure tank.

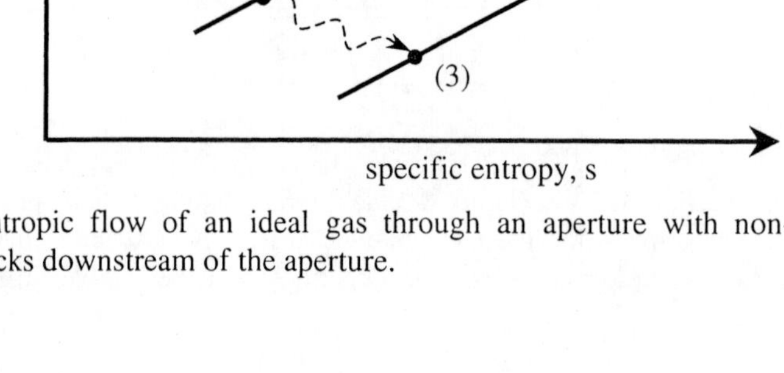

Figure 4.23 Isentropic flow of an ideal gas through an aperture with non-isentropic oblique shocks downstream of the aperture.

If the leak is in the wall of a high-pressure vessel, the upstream velocity U_1 is zero. Even if the leak is in the wall of a pipe containing a flowing gas or vapor, the velocity U_1 is small compared to the sonic velocity in the aperture and can be neglected. For those familiar with compressible flow, it is to be noted that the upstream temperature and pressure T_1 and P_1 are the stagnation temperature and pressure, often denoted as T_0 and P_0. If

$$\frac{P_1}{P_a} > \left(\frac{\gamma+1}{2}\right)^{\frac{\gamma}{\gamma-1}} \tag{4-140}$$

the flow is ***choked*** and the aperture velocity (U_2) equals the speed of sound, $U_2 = a$. Here, γ is the ***ratio of specific heats***, often written in thermodynamics books as k instead of γ,

$$\gamma = \frac{c_p}{c_v} \tag{4-141}$$

where c_p is the specific heat at constant pressure, and c_v is the specific heat at constant volume. γ, which is also sometimes called the ***isentropic coefficient***, is approximately 1.4 for air. The speed of sound is defined by

$$a = \sqrt{\left(\frac{\partial P}{\partial \rho}\right)_s} \tag{4-142}$$

For an ideal gas, it can be shown that the speed of sound (a) is equal to

$$a = \sqrt{\gamma R T_2} \tag{4-143}$$

where R is the ***specific gas constant*** = R_u/M (Shapiro, 1953). T_2 is the aperture temperature, which for isentropic flow is related to upstream temperature T_1 by

$$T_2 = \frac{2T_1}{\gamma+1} \tag{4-144}$$

The pressure corresponding to choked conditions is noted as P_1^* in Figure 4.23. For isentropic flow, the relationship between P_1 and P_1^* is

$$P_1^* = P_1\left(\frac{\gamma+1}{2}\right)^{\frac{-\gamma}{\gamma-1}} \tag{4-145}$$

If atmospheric pressure is less than P_1^*, the flow through the exit of the aperture is sonic (choked), and there is an irreversible pressure adjustment (oblique shocks) external to the aperture that has no bearing on the mass flow rate of the leak. If the atmospheric pressure is larger than P_1^*, the flow through the aperture is compressible but *subsonic* and governed by another set of equations. Since gas and vapor leaks are generally associated with high-pressure sources ($P_1 >> P_a$), choked flow is common for such leaks. Leak mass flow rate ($\dot{m}_2$) is given by Eq. (4-134), where the density is the actual density of the leak at the exit of the aperture ($\rho_2 = \rho_2^*$). From the theory of isentropic flow (Shapiro, 1953),

$$\rho_2 = \rho_2^* = \rho_1\left(\frac{\gamma+1}{2}\right)^{\frac{-1}{\gamma-1}} \tag{4-146}$$

Example 4.20 - Natural Gas Leak through a Cracked Gasket
Given: A gasket in a flange in a high-pressure natural gas (assume methane) transmission pipeline develops a crack 10. μm in diameter (area = 7.85×10^{-11} m^2). The pressure (P_1) and temperature

(T_1) in the pipeline are 20.0 atm and 25.0 °C. From Perry et al. (1984) the molecular weight (M) and isentropic coefficient (γ) for CH_4 are 16.04 and 1.299 respectively.

To do: Estimate the leak mass flow rate.

Solution: From Eq. (4-145),

$$P_1^* = P_1\left(\frac{\gamma+1}{2}\right)^{-\frac{\gamma}{\gamma-1}} = (20.0\text{ atm})\left(\frac{1.299+1}{2}\right)^{-\frac{1.299}{1.299-1}} = 10.92\text{ atm}$$

Since $P_1^* > P_a$, the gas velocity at the aperture exit is equal to the speed of sound. Assuming the ambient air pressure to be at STP (P_a = 101.3 kPa), the gas pressure at the aperture exit plane is

$$P_2 = P_1^* = 10.92\text{ atm}\left(\frac{101.3\text{ kPa}}{\text{atm}}\right) = 1106.\text{ kPa}$$

Thus there is considerable pressure adjustment to be achieved, and a complex shock pattern occurs downstream of the crack. A detectable noise can be expected. The gas density at the aperture exit plane can be found from the ideal gas law. The gas density at upstream point 1 is

$$\rho_1 = \frac{P_1}{RT_1} = \frac{P_1 M}{R_u T_1} = \frac{(20.0\text{ atm})16.04\dfrac{\text{kg}}{\text{kmol}}}{8.3143\dfrac{\text{kJ}}{\text{kmol}\cdot\text{K}}(298.15\text{ K})}\left(\frac{101.3\text{ kPa}}{\text{atm}}\right)\left(\frac{\text{kJ}}{\text{m}^3\text{kPa}}\right) = 13.11\frac{\text{kg}}{\text{m}^3}$$

From Eq. (4-146), the gas density at the exit plane of the aperture is

$$\rho_2 = \rho_1\left(\frac{\gamma+1}{2}\right)^{-\frac{1}{\gamma-1}} = 13.11\frac{\text{kg}}{\text{m}^3}\left(\frac{1.299+1}{2}\right)^{-\frac{1}{1.299-1}} = 8.23\frac{\text{kg}}{\text{m}^3}$$

The temperature at the exit plane can be obtained from Eq. (4-144),

$$T_2 = \frac{2T_1}{\gamma+1} = \frac{2(298.15\text{ K})}{1.299+1} = 259.4\text{ K}$$

The exit gas velocity is equal to the speed of sound; thus, from Eq. (4-143),

$$U_2 = a = \sqrt{\gamma R T_2} = \sqrt{\gamma\frac{R_u}{M}T_2} = \sqrt{1.229\frac{8.3143\dfrac{\text{kJ}}{\text{kmol}\cdot\text{K}}}{16.04\dfrac{\text{kg}}{\text{kmol}}}(259.4\text{ K})\left(\frac{1000\text{ m}^2\text{kg}}{\text{s}^2\text{kJ}}\right)} = 406.5\frac{\text{m}}{\text{s}}$$

Finally, the leak rate ($\dot{m}_2$) is obtained from Eq. (4-134),

$$\dot{m}_2 = \rho_2 A_2 U_2 = 7.85\times10^{-11}\text{ m}^2\left(8.23\frac{\text{kg}}{\text{m}^3}\right)406.5\frac{\text{m}}{\text{s}} = 2.63\times10^{-7}\frac{\text{kg}}{\text{s}}\left(\frac{1000\text{ g}}{\text{kg}}\right)\left(\frac{3600\text{ s}}{\text{hr}}\right) \cong 0.95\frac{\text{g}}{\text{hr}}$$

Discussion: The final result is given to two significant digits; the accuracy of such calculations, given all the assumptions, is certainly no better than this.

4.7.3 Type III – Flash Boiling Leaks

Consider a high-pressure storage vessel as in Figure 4.24, containing a compressed liquid at pressure P_1 and temperature T_1. If a hole enables the compressed liquid to escape, the liquid undergoes rapid reduction in pressure as it passes through the aperture. If the aperture temperature (T_2) or the atmospheric temperature (T_a) is larger than the normal boiling temperature (T_b), the

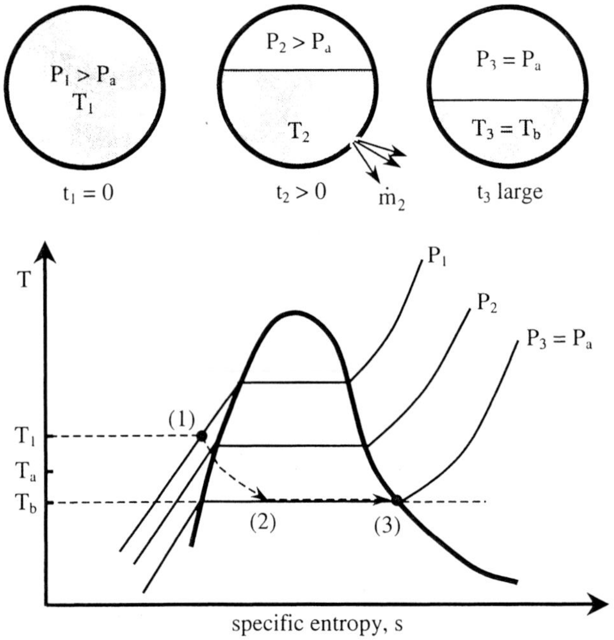

Figure 4.24 Adiabatic flash boiling of a compressed liquid leaking from a storage vessel (type III leak).

liquid "flashes" into the vapor phase. Flashing may not be instantaneous and one can expect some liquid droplets to exist for a short period of time before they evaporate. To model the process, it is assumed that vaporization is complete and that the escaping vapor exists at the boiling temperature (T_b). Eventually the vapor mixes with atmospheric air, and the temperature rises to T_a. One can also assume that the velocity of the vapor leaving the aperture is equal to the speed of sound (choked flow), $U_2 = a$.

To model flash boiling, the process is assumed to be adiabatic; the energy to vaporize the leaking fluid is transferred from the compressed liquid upstream of the aperture. During the increment of time (dt) when the mass flow rate of vapor escaping the vessel is $\dot{m}_2$, the mass of liquid inside the vessel changes by dm. Application of conservation of mass of the compressed liquid in the vessel yields

$$\boxed{dm = -\dot{m}_2 dt} \qquad (4\text{-}147)$$

The energy per unit mass required to vaporize this mass is equal to the specific enthalpy of vaporization ($\hat{h}_{fg}$) based on the instantaneous temperature of the compressed liquid. The energy is obtained from the compressed liquid causing its temperature to change. Thus,

$$\boxed{\hat{h}_{fg} dm = mc_p dT} \qquad (4\text{-}148)$$

where c_p is the specific heat of the compressed liquid. For liquids in general, $c_p \cong c_v$ and is often written simply as c. c_p is used here to avoid confusion with mass concentration (c). Integrating both equations from the time the leak begins ($t = 0$, $T = T_1$, and $P = P_1$) to the time when the leak ceases ($t = t_3$, $T = T_3 = T_b$, and $P = P_3 = P_a$),

$$m_1 - m_3 = -\int_0^{t_3} \dot{m}_2 dt$$

and

$$\int_{m_l}^{m_3} \frac{dm}{m} = \int_{T_l}^{T_b} \left(\frac{c_p}{\hat{h}_{fg}} \right) dT$$

or

$$\boxed{\frac{m_3}{m_l} = \exp\left[-\frac{c_p (T_l - T_b)}{\hat{h}_{fg}} \right]} \tag{4-149}$$

where $m_2 = m_l - m_3$. Equation (4-149) can be rearranged as follows:

$$\boxed{\frac{m_l - m_3}{m_l} = \frac{m_2}{m_l} = f_{LK} = 1 - \exp\left[-\frac{c_p (T_l - T_b)}{\hat{h}_{fg}} \right]} \tag{4-150}$$

where f_{LK} is the fraction of the original liquid that ultimately leaks.

The leak mass flow rate ($\dot{m}_2$) can be found from conservation of mass. So long as atmospheric pressure (P_a) satisfies

$$\boxed{\frac{P_l}{P_a} > \left(\frac{\gamma + 1}{2} \right)^{\frac{\gamma}{\gamma - 1}}} \tag{4-151}$$

one can expect that the flow through the aperture is choked, with U_2 equal to the speed of sound. Thus

$$\boxed{\dot{m}_2 = \rho_2 A_2 a} \tag{4-152}$$

where A_2 is the aperture area, a is the speed of sound, and ρ_2 is the density of the escaping vapor which is assumed to be equal to the reciprocal of the specific volume of the saturated vapor based on the boiling temperature T_b. For a two-phase, liquid-vapor mixture, Crowl and Louvar (1990) have shown that the speed of sound is

$$\boxed{a = \frac{\hat{h}_{fg}}{\sqrt{c_p T_2}}} \tag{4-153}$$

where c_p is the specific heat of the saturated liquid.

The temperature, pressure, and mass of the compressed liquid remaining in the vessel decrease with time as the leak continues. Thus the leaking mass flow rate ($\dot{m}_2$) is not constant even though the flow may be choked. Furthermore, as soon as the tank pressure no longer satisfies the choked criterion, the velocity through the aperture falls below the speed of sound (the flow is no longer choked), and another set of equations has to be used.

Example 4.21 - Chlorine Leak through a Cracked Vessel

Given: Liquid chlorine is stored in a 1,000-gallon vessel at 80.0 °F and 500. psia. A crack 1.0 mm in diameter ($A_2 = 7.85 \times 10^{-7}$ m^2 = 8.45×10^{-6} ft^2) develops in a gasket and chlorine escapes to the air.

To do: Determine the mass flow rate during the initial portion of the leak, and estimate the fraction of the original chlorine that ultimately escapes to the air after a long period of time.

Solution: From Perry et al. (1984) the following data on chlorine are obtained:

- At T_1 = 80.0 °F = 26.7 °C and P_1 = 500. psia, chlorine is a compressed liquid: P_{sat} = 116.46 psia, the specific volume of saturated liquid (v_f) = 0.01154 ft^3/lbm, and $\hat{h}_{fg}$ = 107.32 Btu/lbm. The liquid chlorine specific heat is c_p = 0.229 cal/(g K) = 0.7426 Btu/(lbm R).
- At atmospheric pressure, P_a = 14.7 psia, chlorine is an ideal gas: T_{sat} = -33.8 °C = -28.84 °F = 430.85. R, the specific volume of saturated vapor, v_g = 4.335 ft^3/lbm, $\hat{h}_{fg}$ = 123.67 Btu/lbm, and $\gamma = c_p/c_v$ = 1.355.

The mass of chlorine originally in the vessel (m_1) is

$$m_1 = \frac{V}{v_f} = \frac{1000 \text{ gal}}{0.01154 \frac{\text{ft}^3}{\text{lbm}}}\left(\frac{0.1336 \text{ ft}^3}{\text{gal}}\right) = 11{,}600 \text{ lbm}$$

The fraction of chlorine that leaks is obtained from Eq. (4-150),

$$f_{LK} = 1 - \exp\left[-\frac{c_p\left(T_1 - T_b\right)}{\hat{h}_{fg}}\right] = 1 - \exp\left[-\frac{0.7426 \frac{\text{Btu}}{\text{lbm R}}\left(80.0 - (-28.84)\right)\text{R}}{115.5 \frac{\text{Btu}}{\text{lbm}}}\right] = 0.5033$$

A bit more than half of the initial chlorine leaks out. The value of $\hat{h}_{fg}$ used above (115.5 Btu/lbm) is the average value between T_1 and T_b. The mass of chlorine leaked into the air outside the tank is

$$m_2 = f_{LK} m_1 = 0.5033(11{,}600 \text{ lbm}) = 5838 \text{ lbm}$$

During the initial period of escape, when the liquid temperature is T_1, the acoustic velocity (a) is found from Eq. (4-153),

$$a = \frac{\hat{h}_{fg}}{\sqrt{c_p T_2}} = \frac{115.5 \frac{\text{Btu}}{\text{lbm}}}{\sqrt{0.7426 \frac{\text{Btu}}{\text{lbm R}}(430.85 \text{ R})\left(\frac{\text{Btu}}{778 \text{ ft lbf}}\right)\left(\frac{\text{s}^2\text{lbf}}{32.174 \text{ ft lbm}}\right)}} = 1022. \frac{\text{ft}}{\text{s}}$$

The mass flow rate is a maximum during this initial period,

$$\dot{m}_2 = \frac{A_2 a}{v_g} = \frac{\left(8.45\times10^{-6} \text{ ft}^2\right)1022. \frac{\text{ft}}{\text{s}}}{4.336 \frac{\text{ft}^3}{\text{lbm}}} = 0.00199 \frac{\text{lbm}}{\text{s}} \cong 0.0020 \frac{\text{lbm}}{\text{s}}$$

Discussion: The final result is given to two significant digits. If the flash boiling mass flow rate is very small, the process may be isothermal; leak mass flow rate and flash fraction f_{LK} need to be calculated by another set of equations. Even at this maximum flow rate, the leak is slow; it would take over 800 hours for half of the chlorine to escape!

4.7.4 Leaks in Process and Storage Vessels

The objective of this section is to analyze leaking of a vented incompressible liquid from chemical storage and batch process vessels. ***Vented*** means that the upper surface of the liquid is open to the air, and is hence at atmospheric pressure throughout the leak. Large vessels of varied geometry are used, such as

- spheres

- upright vessels of constant circular cross-sectional area
- horizontal vessels of constant circular cross-sectional area

To compute the leak rate, time to drain a vessel, etc., it is necessary to compute the volume of the stored liquid as a function of the location of the leak in the vessel and the top surface of the liquid. The computation is complicated for several reasons:

- Cylindrical vessels of circular cross-sectional area have dome-shaped ends for strength.
- Vessels often contain heating or cooling coils, blenders, agitators, etc., which reduce the usable cross-sectional area.
- The vessel may initially be less than totally filled with liquid, in which case there would be a gas mixture of air (or an inert gas) and contaminant vapor above the liquid.

Leaks occur because seams may rupture, the vessel may be accidentally punctured, or gaps may be present in valves, pipe fittings, etc. If a leak occurs in an upright vessel of constant circular cross-sectional area containing no internal members, the time to drain the vessel from liquid level z_1 above the leak to z_2 can be computed to be

$$t = \frac{\pi D^2}{A_2 C_0 \sqrt{8g}}\left(\sqrt{z_1} - \sqrt{z_2}\right) \tag{4-154}$$

If, on the other hand, a leak occurs in a spherical or horizontal vessel of circular cross-sectional area, the task is more difficult (Crowl, 1991).

Consider the vented spherical vessel shown in Figure 4.25. Initially, the height of the liquid is z_1. After the leak occurs, the liquid level is z_2. The leak occurs through an opening of area A_2 located a vertical distance z^* above the bottom of the vessel. At any time t, let h be the distance between the upper surface of the liquid and the leak opening,

$$h = z - z^* \tag{4-155}$$

As the leak progresses, h decreases with time. Engineers need to know how to compute the leak mass flow rate at any instant, and the amount of time it takes to either drain the vessel or for the liquid level to drop from z_1 to z_2. Assuming that the tank is vented to the air, the pressure above the liquid surface is always equal to atmospheric pressure. Equation (4-137) is valid for this kind of leak, but with $P_g = P_a$; the speed at which the liquid leaks from the aperture is thus

$$U_2 = \sqrt{\frac{2gh}{1 + C_0}} \tag{4-156}$$

The above can be combined with an expression of conservation of mass to obtain

$$\dot{m} = -\rho \frac{dV}{dt} = \rho A_2 U_2 = \rho A_2 \sqrt{\frac{2gh}{1 + C_0}} \tag{4-157}$$

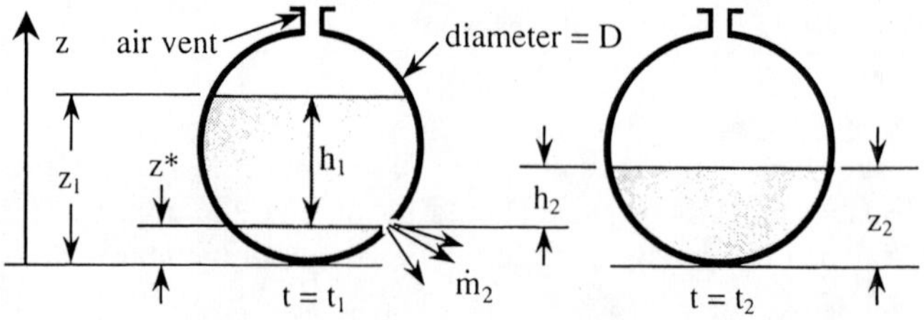

Figure 4.25 Liquid leaking from a spherical tank vented to the room air at times t_1 and t_2.

where V is the volume of the liquid. The volume of a spherical segment defined by the upper free surface of liquid and a horizontal plane passing through the leak is

$$V = \pi\left[h\left(Dz^* - \left(z^*\right)^2\right) + h^2\left(\frac{D}{2} - z^*\right) - \frac{h^3}{3}\right] \tag{4-158}$$

When a leak occurs, the volume of liquid inside the tank decreases with time. One can calculate the time derivative of Eq. (4-158) using the chain rule,

$$\frac{dV}{dt} = \frac{dV}{dh}\frac{dh}{dt} = \pi\left[Dz^* - \left(z^*\right)^2 - 2hz^* + hD - h^2\right]\frac{dh}{dt}$$

To estimate the time (t) required for the liquid surface to fall from z_1 to z_2, the above equation can be combined with Eq. (4-157), the variables (h and t) can be separated, and the equation can be integrated:

$$\int_0^t dt = -\int_{z_1}^{z_2} \frac{\pi\sqrt{1+C_0}\left[Dz^* - \left(z^*\right)^2 - 2hz^* + hD - h^2\right]dh}{A_2\sqrt{2gh}} \tag{4-159}$$

Example 4.22 - Liquid Leak through a Punctured Vessel

Given: A vented spherical storage vessel 12. m in diameter is filled to capacity with a toxic hydrocarbon (density 800. kg/m^3). A forklift operator inadvertently strikes the vessel but does not inform his supervisor. Sometime later, another worker notices hydrocarbon leaking through a hole 1.8 m above the bottom of the vessel and reports it to you. You need to submit a "spill" report to the state Department of Environmental Resources. When reported to you, the upper level of liquid is 4.8 m below the top of the vessel and measurements reveal that the cross-sectional area of the puncture is 0.80 cm^2, and the current leak rate is 0.40 kg/s.

To do:

- calculate the volume and mass of liquid that has been spilled
- calculate the discharge coefficient (C_0)
- calculate the leak discharge rate (kg/s) at the time of the puncture
- estimate the time that has elapsed between the accident and when it was reported to you

Solution: In terms of the parameters shown in Figure 4.25 and described above, D = 12. m and z* = 1.8 m.

$t = t_1 = 0$:	$z_1(0) = D = 12.$ m,	$h_1 = z_1 - z^* = (12 - 1.8) = 10.2$ m
$t = t_2$.	$z_2(t) = (12 - 4.8)$ m $= 7.2$ m,	$h_2 = z_2 - z^* = (7.2 - 1.8) = 5.4$ m

The initial volume of liquid is

$$V_1 = V(0) = \pi\frac{D^3}{6} = \pi\frac{(12\text{ m})^3}{6} = 904.8\text{ m}^3$$

The volume of liquid in the vessel when the spill was reported is found from Eq. (4-158),

$$V_2 = \pi\left[5.4\text{ m}\left(12.\text{ m}(1.8\text{ m}) - (1.8\text{ m})^2\right) + (5.4\text{ m})^2\left(\frac{12.\text{ m}}{2} - 1.8\text{ m}\right) - \frac{(5.4\text{ m})^3}{3}\right] = 531.3\text{ m}^3$$

The volume of fluid spilled is

$$V_{spill} = V_1 - V_2 = 904.8\text{ m}^3 - 531.3\text{ m}^3 = 373.4\text{ m}^3 \cong 370\text{ m}^3$$

and its mass is

$$m_{spill} = \rho V_{spill} = 800.\frac{kg}{m^3}\left(373.4\ m^3\right) = 298,760\ kg \cong 300,000\ kg$$

The discharge coefficient C_0 can be estimated from Eq. (4-157) using the measured mass flow rate at the time of the report and $h_2 = 5.4$ m,

$$C_0 = \frac{2gh_2\rho^2 A_2^{\ 2}}{\dot{m}_2^{\ 2}} - 1 = \frac{2\left(9.81\frac{m}{s^2}\right)5.4\ m\left(800.\frac{kg}{m^3}\right)^2\left(0.80\times10^{-4}\ m^2\right)^2}{\left(0.40\frac{kg}{s}\right)^2} - 1 = 1.71$$

At the time of the puncture, the mass flow rate can be estimated using Eq. (4-157) with $h_1 = 10.2$ m:

$$\dot{m}_1 = \rho A_2\sqrt{\frac{2gh}{1+C_0}} = 800.\frac{kg}{m^3}\left(0.80\times10^{-4}\ m^2\right)\sqrt{\frac{2\left(9.81\frac{m}{s^2}\right)(10.2\ m)}{1+1.71}} = 0.55\frac{kg}{s}$$

To estimate the amount of time that elapsed between when the puncture occurred and when it was observed, Eq. (4-159) can be used. The equation can be integrated using a mathematical computer program. This is left as an exercise for the reader.

Discussion: Final answers are given to only two significant digits, which is the most one can expect, given the scope of the approximations.

4.8 Closure

It is essential that engineers be able to estimate the rates of emission of pollutants from a variety of sources. Securing operating permits from environmental regulatory agencies requires these estimates. In some cases, it may be possible to measure the emission rates. In other cases, empirical equations can be used if the physical limitations on their use are satisfied. Emission factors developed by the EPA lack sophistication, but are available for hundreds of different sources and processes, and have been used by the EPA and industry for decades. Since state and federal environmental operating permits often require the use of emission factors, engineers must know where these values can be obtained and how to use them. Use of the EPA electronic bulletin board grows steadily and it is essential that engineers know how to use it. There are times when the methods mentioned above are unavailable or when greater accuracy is needed. Accordingly engineers must be able to model emission processes and estimate rates from the principles of mass transfer found in numerous reference works and textbooks. The professional journals regularly add to this literature and should be consulted.

Chapter 4 - Problems

1. Using emission factors, estimate the contaminant generation rate in units of kg/s for each of the following:

(a) Your firm makes office furniture and paints the metal surfaces with an enamel paint (8.0 lbm/gal). The consumption of paint is 5 gal per minute. What is the hydrocarbon emission rate?
(b) You work for a slaughterhouse and local hunters bring venison for smoking. Estimate the emission of particles (smoke) if meat is processed at the rate of 50 lbm/hr.
(c) You use an unvented kerosene space heater in your hunting cabin. Estimate the rate at which carbon monoxide is emitted while you sleep. Kerosene (density 7.5 lbm/gal, heating value 40,000 kJ/kg) is burned at the rate of 0.1 gal/hr.
(d) Six people smoke simultaneously during a meeting in a small conference room. Estimate the emission rate of carbon monoxide assuming only sidestream smoke.
(e) You are in charge of loading concrete mix into trucks. The space where this is done is small. Estimate the particle emission rate if trucks are loaded at a rate of one every 10 minutes. Each truck carries 10 tons of concrete mix.
(f) Estimate the emission rate of gasoline fumes from a filling station (having no fume controls) if on the average 10 gallons of gasoline are added to each auto and if tanks are filled ("splash filling") at the rate of one every 5 minutes.

2. Use emission factors for each of the following. Report your results in units of kg/s unless specified otherwise.

(a) Estimate the generation rate (kg/hr) of particles, SO_2, NO_x, CO, VOCs, and CH_4 from a hand-fired stoker that burns anthracite coal at a rate of 5 tons/hr. The coal assay is 12,550 HHV (higher heating value), 3% sulfur, and 8% ash by mass. Compare the SO_2 generation rate with the value estimated by a mass balance.
(b) Estimate the hourly generation of particles, SO_2, CO, NO_2, and organics from a municipal multiple chamber refuse incinerator processing 10 tons refuse/hr.
(c) Estimate the number of miles a new automobile must be driven to emit an amount of NO_x equal to what is emitted by a Boeing 747 in one landing and take-off cycle. Repeat for unburned hydrocarbons.
(d) Compare the fugitive emission rate (lbm/hr) from a flange in a 6-inch high-pressure pipeline carrying C_2H_6 at 10 atm. Use data from Appendix A.3.
(e) Estimate the NO_x generation rate from a rotary lime kiln and an asphalt mix plant, each handling 10 tons/material per hour.
(f) What is the minimum particle collection efficiency needed to bring a plant that manufactures flat glass into compliance with the New Source Performance Standards?
(g) Estimate the generation rate (kg/hr-mile) of fugitive dust from a 10-lane paved highway carrying 100 passenger vehicles/minute traveling at an average speed of 50 MPH.

3. An electric generating station burns coal at a rate of 200 tons per hour. The plant operates continuously (8,760 hours per year). The coal has a higher heating value of 12,500 BTU/lbm, 3% sulfur, 8% ash, and 0.01% mercury (by weight). An electrostatic precipitator (ESP) removes sufficient fly ash to satisfy the New Source Performance Standards. If all the mercury condenses on minuscule fly ash particles that are not captured by the ESP, estimate the mercury emissions (in units of tons/yr.)

4. "Sour natural gas" is methane containing odorous gases such as H_2S. If sour natural gas (assume M = 18) containing 5% (mol fraction) of H_2S, is burned at a rate of 10 kg/min, what is the SO_2 emission rate (kg/min) assuming all the sulfur is converted to SO_2?

5. In an automotive body shop, autos are "refinished" in a paint booth at a net rate that corresponds to 5 square meters per hour. Refinishing consists of applying all the following materials:

- primer: 100 g/m^2 per coat, one coat applied
- enamel: 200 g/m^2 per coat, two coats applied
- lacquer: 150 g/m^2, one coat applied

Exhaust from the paint booth is vented to the atmosphere. Using emission factors, estimate the rate (kg/hr) at which hydrocarbon vapor enters the atmosphere.

6. Using the empirical equations provided in this chapter, estimate the rate at which trichloroethylene vapor is emitted (kg/hr) when a conventional closed 55-gal drum is filled at a rate of 2 GPM under conditions called "splash loading".

7. Using the empirical equations provided in this chapter, estimate the rate at which gasoline (assume octane) vapor is emitted to the atmosphere (tons/yr) when US automobile gasoline tanks (assume 20 gal/car) are filled at a rate of 5 gal/min. Assume that 80 x 10^6 US autos are filled with 5,000 gallons of gasoline each year.

8. Crushed limestone is transferred from one conveyor to another. To minimize the generation of fugitive dust, an enclosure is built to surround the operation. Using emission factors for screening, conveying, and handling in stone quarries, estimate the mass of inhalable dust generated per ton of stone. For calculation purposes define inhalable dust as particles 10 μm or less. Assume that the density of limestone is 105 lbm/ft^3 and that the inhalable mass fraction is 15%.

9. Trichloroethylene is a common nonflammable hydrocarbon used to clean metal surfaces. The chemical formula is C_2HCl_3 and the molecular weight is 131. The enthalpy of vaporization at 85.7 °C is 57.24 cal/g. The saturation temperatures and pressures can be obtained from Appendix A.8. Using the Clausius-Clapeyron equation, estimate the vapor pressure at 25 °C.

10. An open 55-gal drum contains a thin layer of trichloroethylene at the bottom. The drum is left uncovered in a storage room.

(a) Using first principles, estimate the rate (kg/hr) at which vapor enters the room air.
(b) Using emission factors, estimate the rate (kg/hr) at which vapor enters the room air.

11. Benzene (MW = 78) at 28 °C (83 °F) lays in a shallow pool in a flat bottomed cylindrical reactor 3 meters in diameter and 5 meters high. The normal boiling temperature is 80 °C, the heat of vaporization is 433.5 kJ/kg at 25 °C and the diffusion coefficient of benzene in air is 0.77 x 10^{-5} m^2/s. The reactor has a circular opening (diameter 0.5 meter) at the top through which materials are added. The lid on the opening has been removed and benzene vapor escapes to the workplace air (25 °C). Estimate the rate (kg/hr) at which benzene vapor escapes through the opening.

12. An artist is casual about handling chemicals in his small, hot, unvented studio. To save money the artist cleans his brushes with used allyl alcohol brought to him by a friend who obtains it from his place of work and is unaware of its toxicity. (Unfortunately neither the friend nor the artist are aware of MSDS literature!) Uncapped one-gallon cans (4.0 by 6.5-inch base, 9.5-inch high, 1.0-inch diameter cylindrical opening, 0.50 inches high) with pools of alcohol at the bottom are stashed in corners of the room. Near his easel he keeps a 6.0-inch cup filled to the rim with alcohol for cleaning his brushes. The artist uses a room fan to cool himself while he works. Air passes over the open cup at a speed of 3.0 m/s. Assume the diffusion coefficient for allyl alcohol in air is 1×10^{-5} m^2/s. See Appendix A.20 and the Internet for other properties of allyl alcohol.

(a) Estimate the vapor pressure (kPa) and enthalpy of vaporization (kJ/kg) at 25 °C.
(b) Using first principles, estimate the emission rate of fumes (kg/hr) from an open gallon can assuming:

- Evaporation is governed by diffusion through a stagnant air column 1-inch in diameter, 9.5 inches high.
- The concentration of alcohol in the air inside the can is dictated by its vapor pressure and diffusion through the opening can be approximated as diffusion through a stagnant air column ½ inches high.

(c) Using first principles, estimate the emission rate of fumes (kg/hr) from the cup.

13. A pie plate is filled with water. The upper diameter of the pie plate is 10 inches. Water evaporates at a rate of 20 mg/s when the air and water are at 25 °C and the air is absolutely dry (relative humidity is zero).

(a) What is the mass transfer coefficient (k_G)?
(b) Estimate the evaporation rate if the relative humidity is 70%.

14. Cylindrical brass stock, one-fourth inches in diameter, leaves a machine with a thin (0.10 mm) layer of water coating its surface. The speed with which the brass stock moves is 0.10 m/s. Room air at 25. °C and 20.% relative humidity passes over the brass perpendicular to the axis at a velocity of 10. m/s, causing the water to evaporate. The temperature of the water film is constant (30. °C) due to the large heat capacity and thermal conductivity of brass. Estimate the following:

(a) overall convection mass transfer coefficient (k_G),
(b) local evaporation rate (kg/s m^2) at a point 5 seconds after it has left the machine,
(c) water film thickness at the point in (b) above.

15. Moth repellent (paradichlorobenzene) in the form of a long cylinder of circular cross-sectional area is hung in an air stream (at STP) moving with uniform velocity U. The moth repellent has uniform chemical composition, density (ρ), and sublimates.

(a) Show that the sublimation rate ($\dot{m}$, kg/s) varies as $\dot{m}$ = constant U^m, where "m" is a constant given in a table of heat transfer correlations.
(b) Show that the rate of change in diameter (dD/dt) is dD/dt = - constant D^m.

16. Ethyl alcohol (C_2H_5OH) is accidentally spilled on the floor of a store room in which the temperature and pressure are 23. °C, 1.0 atm. The average air velocity is 2.0 m/s and the spill is 2.0 m in diameter. Estimate the rate of evaporation in kg/m^2 from first principles and compare the results with those obtained from empirical equations.

17. A large open vessel contains an aqueous mixture of volatile waste liquids. The vessel is 10. m in diameter, 3.0 m high, and filled to the rim. The vessel is stored outdoors in a remote part of an industrial plant at 25. °C. Air passes over it at a velocity of 3.0 m/s parallel to the surface. Initially the waste liquid consists of the following:

hydrocarbon	**mol fraction**
benzene	0.050
methyl alcohol	0.050
carbon tetrachloride	0.050
toluene	0.050

The remainder of the liquid is water. Estimate the evaporation rate (kg/hr) of volatile materials after 1 hr, 10 hr, 100 hr, and 1000 hr assuming that the wastes are agitated to make the concentration within the vessel uniform. Does the evaporation rate change with time? Does the composition of the remaining liquid remain constant?

18. Questions are sometimes raised (McKone, 1987) about the exposure humans experience due to volatile organic compounds (VOCs) that evaporate from household tap water. If the concentration of trichloroethylene (TCE) in household tap water is 1.0 mg/liter of water, estimate the rate of evaporation of TCE from a bathtub of water at 25. °C using two-film theory. For a first approximation, neglect any TCE in the air.

19. Using two-film theory, estimate the evaporation rates (kg/hr) of benzene and chloroform from a waste lagoon containing 100 mg/liter of benzene (M = 78.) and 100 mg/liter of chloroform (M = 119). The lagoon is 25. m by 40. m and 3.5 m deep. The air and liquid temperatures are 25. °C and the wind speed is 3.0 m/s. The Henry's law constants are:

- benzene: 5.5×10^{-3} atm-m^3/gmol
- chloroform: 3.39×10^{-3} atm-m^3/gmol

and the Schmidt numbers are:

- benzene and water: 1000 — benzene and air: 1.76
- chloroform and water: 1100 — chloroform and air: 2.14

20. Grain stored in silos, transported in ships' holds, etc. is sprayed with a fumigant to prevent deterioration. A spray containing ethyl formate and other dissolved fumigant materials is used. Your supervisor wishes to manually spray this material on grain as it is transferred to a vessel. You suspect that spraying will produce an unhealthy exposure to ethyl formate vapor. To prove your point you wish to estimate the concentration to which workers might be exposed. As a first step you need to estimate the evaporation occurring during the time a drop leaves a spray nozzle until it lands on the grain. Using Runge-Kutta techniques, solve the simultaneous differential equations and predict the diameter (D_p), temperature (T_p) and evaporation rate (g/s) of a drop as function of time. As a first step, consider only a single drop falling through motionless air at its terminal settling velocity. Assume that the ethyl formate vapor concentration in air (far removed from the drop) is zero at all times.

- <u>air</u>: 25. °C, 101. kPa, 20.% relative humidity
- <u>drop</u>: initial drop diameter is 500 μm, initial drop temperature is 25. °C, see Appendix A.8 for ethyl formate vapor pressures, estimate the enthalpy of vaporization ($\hat{h}_{fg}$) with the Clausius-Clapeyron equation and, if other property data about ethyl formate cannot be found, assume its properties are those of water

21. Waste isopropyl alcohol, also called rubbing alcohol or isopropanol, with a molecular weight of M = 60.1, leaks into a pond that is 10. m in diameter. Because of its infinite solubility, the mol fraction of the alcohol in the aqueous phase is uniform and equal to 0.0010. The air and pond are at a temperature of 23.8 °C. Atmospheric air passes over the pond at a (far-field) velocity of 3.0 m/s. Estimate the alcohol evaporation rate in units of mg/(m^{2}s). The overall pressure is 101. kPa

(760. mm Hg). The properties of the air at this temperature and pressure are: $\nu = 1.589 \times 10^{-5}$ m^2/s, $\rho = 1.186$ kg/m^3, $\mu = 1.88 \times 10^{-5}$ kg/(m s), and Pr = 0.707.

22. Imagine that pure isopropyl alcohol is contained in a constant pressure (P_0), constant temperature (T_0) vessel. Initially 75% of the vessel is filled with alcohol and the space above (V_0) is filled with air. The surface area of the air-liquid interface is A_i. Initially, the evaporation rate is equal to the value developed in Example E.4.17, but as the space above fills with alcohol vapor the alcohol evaporation rate changes.

(a) What is the alcohol mol fraction in air at equilibrium?
(b) Write and integrate a differential equation describing how the alcohol evaporation rate varies with time, assuming the mass transfer coefficient(s) are constant.

23. Xylene (o-xylene) is to be transferred from one vessel to an overhead vessel. The volumetric flow rate is 100 GPM, the temperature is 25 °C and the pressure inside the pump is 1.5 atm. The pump is inexpensive and the seal surrounding the pump shaft is poor and can be characterized as having a clearance (open) area of 0.0010 mm^2. Will xylene leak through the pump seal? If so estimate the leak rate (lbm/hr) and compare the value to the value estimated by emission factors. Does a leak constitute a potential fire hazard?

24. Ammonia (NH_3) is used in a selective noncatalytic reduction process (SNCR) to remove NO_x from the exhaust gas of an electric utility boiler. The ammonia gas is stored in a shed in large, high-pressure vessels at 2,500 kPa. The air in the shed is at 25. °C. There is evidence that small holes (of area 10^{-6} mm^2) may occur due to corrosion, allowing the ammonia to escape to the air in the shed.

(a) What is the leak rate (g/hr) at a time when the upstream temperature and pressure of the ammonia are 25. °C and 2,500 kPa?
(b) As the ammonia escapes to the air, the upstream pressure decreases. Since the leak is slow, there is adequate heat transfer to maintain the upstream temperature at 25. °C. Write an expression to predict the leak rate as a function of time. Compute and plot the leak rate and upstream pressure as a function of time.

25. Hydrogen cyanide (HCN) is stored in a laboratory in high-pressure vessels at 25. °C and 1,000 kPa. Initially the HCN in each vessel is 80.% liquid and 20.% vapor (by mass) at this temperature and pressure. The vessels stand upright, each with a shut-off valve at the top. Unfortunately a small leak develops in the valve seat of the shut-off valve (leak area = 10^{-6} mm^2) and HCN escapes to the air. HCN is a lethal gas.

(a) By means of a T-s phase diagram, describe and show the thermodynamic state of the hydrogen cyanide in the storage vessel.
(b) Estimate the leak rate (g/s) under these conditions.

26. [*Design Problem]* Automobile parts are forged from heated steel. After forging, the parts have a surface temperature of 150 °C and are placed on an overhead conveyor where they are transported to several machining operations. While on the conveyor they cool by forced convection (air velocity 10. m/s) with room air at 25. °C. After being cooled to 25. °C and after several machining operations the parts are dipped in trichloroethylene (TCE) to remove oils, cutting fluids, etc. After being dipped in TCE the parts have a 0.010-mm film of TCE over the entire surface. The parts are placed on an overhead conveyor. While being conveyed, TCE evaporates. You have been asked to estimate the initial heat transfer rate (kJ/s) after forging and the evaporation rate (kg/s) immediately after being removed from the TCE cleaning tank. The part can be assumed to resemble a sphere of diameter 0.50 m. Your supervisor claims that within 5 seconds 80% of the requisite cooling has occurred and 80% of the TCE film has evaporated. Is you supervisor correct?

27. [*Design Problem]* You work in a research facility that models rivers and flood control projects, tests the design of ship hulls, etc. Topographically scaled models are laid out on the floor of the building to simulate a river, bay, lake, impounded water behind a dam, etc. Thus a large water surface is exposed to room air and evaporation occurs. At other times an open tank of water is used and scale models of ship hulls are towed through the water so that drag and wave patterns can be studied. In all these experiments certain values of the Reynolds, Prandtl, Froude, etc. numbers are needed so that researchers can achieve proper similitude. To control the water's viscosity, special hydrocarbons are added to the water. Unfortunately researchers do not read the MSDS literature and select volatile toxic materials. Toxic vapors escape into the room. Using two-film theory, estimate the evaporation rate of toxic vapor from the water in units of kmol/(m^2s) under the following conditions:

- the characteristic length of water surface is 100 m
- room air has an average velocity of 2.0 m/s
- there is negligible water movement
- Henry's law constant, diffusivities and other physical properties of the additives are similar to perchloroethylene
- the initial mol fraction ($x_{c,0}$) in water is 0.010

What precautions should be taken to keep the vapors from entering the work place?

28. [*Design problem]* Fiberglass is sprayed into molds to manufacture caps used to enclose the beds of pick-up trucks. Periodically, workers flush their spray nozzles with xylene (o-xylene). To minimize vapor entering the workplace air, workers direct the spray downward into a container equipped with a lateral exhauster that prevents vapors from leaving the container. In spite of this practice, you believe that too much xylene evaporates and enters the air. Your supervisor contends that only a small amount of vapor enters the air so long as the particles are captured by the exhauster within 0.5 seconds. Estimate the following:

(a) the initial evaporation rate per drop (g/s)
(b) the initial temperature depression in the drop
(c) the percent of the original drop that enters the workplace before the drop enters the container (neglecting the temperature depression)
(d) the time required for 99% of the drop to evaporate (neglecting the temperature depression); use the following:

- initial drop diameter, $D_p(0)$ = 100 μm
- M(xylene) = 106
- $\hat{h}_{fg}$ (xylene) = 409 kJ/kg
- P_v (saturation) xylene - see Appendix A.8
- diffusion coefficient of xylene in air is 0.70 x 10^{-5} m^2/s
- drop velocity = 0.261 m/s
- use Appendix A.11 for the properties of air

(e) Is this practice a hazard to the workers' health? If so what precautions should be taken to eliminate, or at least minimize, the hazard?

29. Use emission factors to solve each of the parts below.

(a) A plastics manufacturing plant has a machine which produces polyvinyl chloride (PVC). The machine uses 160 kg of raw material per hour. Estimate the emission rate (in units of kg/hr) of vinyl chloride gas from the machine.
(b) A distilling machine produces whiskey at a rate of 250. gallons per hour. The density of the whiskey is 3.0 kg/gallon. Estimate the emission rate of hydrocarbon vapors (in units of kg/hr) during the fermentation process.
(c) Professor Smith burns about 3 tons of wood in his woodburning stove each heating season. Estimate the emission rate of carbon monoxide from his wood-burning stove (in units of kg/season).

30. You will need to do some Internet surfing for this problem. It is designed to familiarize you with the EPA web site, particularly CHIEF and AP-42 emission factors.

(a) Search and explore the EPA web site at www.epa.gov. On the CHIEF portion of the web site, go to AP-42 Emission Factors. Which chapter and section deals with transportation and marketing of petroleum liquids?
(b) Read that section of AP-42 (you may wish to print it out, but you are not required to do so). Sketch and briefly explain how a "***vapor balance***" form of emission control works for the case of a tank truck unloading gasoline from the tank to a service station's underground fuel tank. In particular explain how gasoline vapor is recovered by this system.
(c) When people fill up their automobile gasoline tanks, gasoline sometimes spills over. According to the EPA, what is the average spilling loss of dispensed gasoline at filling stations (in mg/L)?
(d) According to the EPA, what is the average amount of uncontrolled emissions from gasoline vapors displaced during vehicle refueling (in mg/L)? If there is a vapor recovery system installed, these emissions are reduced dramatically. What is the average amount of controlled emissions from gasoline vapors displaced during vehicle refueling (in mg/L)? ***Recovery factor***, or ***recovery efficiency*** is defined as

$$\eta_{\text{recovery}} = \frac{\text{uncontrolled emissions} - \text{controlled emissions}}{\text{uncontrolled emissions}}$$

Calculate the recovery factor for emissions from gasoline vapors displaced during vehicle refueling.
(e) Explain briefly how a balance vapor control system works for the case of a person filling his/her gas tank. Draw a sketch and explain the natural pressure differentials between the car's fuel tank and the gasoline station's underground fuel tank.
(f) For a typical gasoline station equipped with an average vapor recovery system, how many grams of gasoline vapor are emitted to the atmosphere for a 15-gallon automobile tank fill-up? (Include both spillage and displacement losses.)
(g) Finally, if gasoline costs $1.30 per gallon, how much money is wasted, on average, with each 15-gallon fill-up? Hint: The density of gasoline is 680 kg/m^3. Is this a significant amount compared to the cost of the fill-up?

5

General Ventilation and the Well-Mixed Model

In this chapter you will learn:

- about dilution and displacement ventilation
- to estimate maximum contaminant concentrations in unventilated spaces
- to model well-mixed spaces containing sources, wall losses, recirculation, and air cleaners
- about wall losses, plate-out adsorption, infiltration and exfiltration, and desorption
- to use the well-mixed model as an experimental tool
- about partially mixed spaces
- about clean rooms, split-flow booths, floor sweeps, and concentrators
- to quantify and rank the effectiveness of a ventilation system
- to model ventilation in tunnels

5.1 Definitions

The phrase ***general ventilation*** suffers from ambiguity. It has one meaning as an engineering control strategy, and another meaning as an analytical concept. As a control strategy, it denotes the practice of removing air from the entire workplace and replacing it with cleaned air or outside air. General ventilation can be divided into ***dilution ventilation*** and ***displacement ventilation***. The essential features of these two strategies are shown in Figure 5.1. In dilution ventilation, fresh air enters the room from overhead diffusers and mixes with room air; the mixture is withdrawn through exhaust registers in the ceiling. In displacement ventilation, fresh air enters the base of the room, moves upward without much mixing with room air, and "pushes" room air out through registers in the ceiling. In its simplest form, dilution ventilation is a strategy to dilute polluted air with fresh air so that the contaminant concentration is kept to an acceptable level, while displacement ventilation is a strategy in which the fresh air *displaces* the polluted air. In the final analysis, it is the location and design of the air registers and diffusers which control mixing, determine the local contaminant mass concentrations c(x,y,z,t), and establish whether healthy conditions exist.

Construction costs associated with ventilation are a prime reason that both inlet and outlet ducts are located in the ceiling; dilution ventilation is therefore the most common strategy used in the US. In Scandinavian countries, displacement ventilation is more widely used than it is in the US. Displacement ventilation requires installation of exhaust ducts in the ceiling space and inlet ducts at floor level, which necessitates additional vertical ductwork. Ducts mounted on the outside of walls produces an unsightly appearance; thus, displacement ventilation requires vertical ducts to be located inside walls, which reduces useable floor space. With displacement ventilation, room furnishings cannot be allowed to block floor-level inlets or inhibit movement of the fresh air inlet stream. Considering the cramped quarters in many offices, it is obvious that displacement ventilation is more costly to install than dilution ventilation, although under some conditions displacement ventilation can improve indoor air quality, and may be more energy efficient.

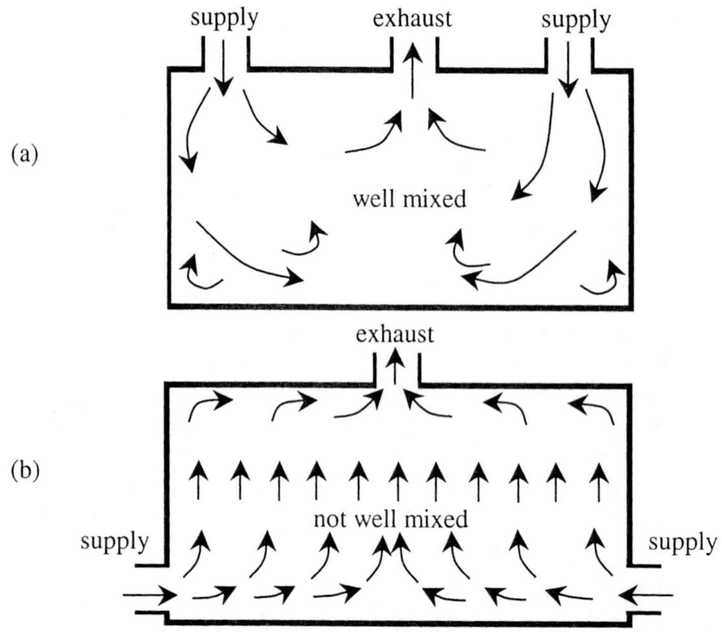

Figure 5.1 Two types of room ventilation: (a) dilution ventilation, (b) displacement ventilation.

As an analytical concept, general ventilation denotes the assumption that the concentration is uniformly distributed throughout the enclosure. Note that the concentration may vary with time (t) but it does not vary with location (x,y,z), i.e. it achieves ***spatial uniformity***; for general ventilation, it is assumed that

$$\boxed{c(x, y, z, t) = c(t)} \tag{5-1}$$

where c designates a mass concentration, in units such as mg/m^3. When the phrase "general ventilation" is used, individuals must be careful to indicate whether they are referring to the strategy or the analysis, because adoption of the strategy does not ensure that the analysis is valid. One can always add to and remove air from a room, but only experiment can verify that spatial uniformity, as in Eq. (5-1), has been achieved. It is suggested that the phrase "general ventilation" be used only to refer to the strategy of adding and removing air from the entire workplace, in contrast to ***local ventilation***, in which air is removed close to a contaminant source (using hoods, as discussed in Chapter 6). Equation (5-1) is not valid physically unless air is added and removed in a unique fashion such that mixing removes any spatial variations of the contaminant concentration. In other fields of engineering, enclosures in which the concentration is spatially uniform are called ***well mixed***, ***perfectly stirred***, or ***well stirred***. It is suggested that indoor air pollution control adopt the same language. The phrase "well mixed" is used in this text to describe enclosures in which Eq. (5-1) is valid. In outdoor air pollution control the well-mixed model is called the ***box model*** or ***top hat model***.

5.1.1 ASHRAE Recommendations

ASHRAE does not use general and dilution (ventilation) as synonyms. For heating and ventilation purposes, the objective of general ventilation is to satisfy personal comfort, and the objective of dilution ventilation is to reduce contaminant concentrations to acceptable levels. For gaseous contaminants, the ASHRAE HVAC Applications Handbook (1999) posits a relationship between dilution volumetric flow rate (Q_d) and mass concentration in the zone occupied by workers (c_{oz}). After modification to the present nomenclature,

$$Q_d = Q_E + \frac{S - Q_E\left(c_{oz} - c_a\right)}{\left[\frac{c_E - c_a}{c_{oz} - c_a}\right]} \tag{5-2}$$

where Q_E is the total volumetric flow rate of air exhausted from volume V, S is the source strength defined as the mass rate of contaminant emission (assumed constant), c_E and c_a are contaminant mass concentrations in the exhausted air and ambient make-up air respectively, and the subscript "oz" refers to the occupied zone.

For there to be analytical rigor to Eq. (5-2), conservation of mass for air and contaminant within the room must be satisfied. Accordingly it is necessary that Q_d and Q_E are the same if conditions within the volume (V) are not to vary with time. Similarly, owing to the assumption of well-mixed conditions that is usually made when analyzing dilution ventilation, the contaminant mass concentration in the exhaust air (c_E) and in the occupied zone (c_{oz}) must be the same, ($c_E = c_{oz}$). Under these conditions Eq. (5-2) reduces to the familiar expression

$$Q_d = \frac{S}{c_{oz} - c_a} \tag{5-3}$$

which is derived and discussed in the material that follows.

Throughout the analyses of indoor ventilation systems, the phrases control volume, control surface, open system, and closed system are used.

- A ***control volume*** is a volume in space defined by users, through which they know or can calculate the mass flow rates and the transfer of energy and work. Control volumes are also called ***open systems*** by some authors. The mass of material within the control volume may change, but the boundaries of the control volume must be definable. The surface of a control volume is called a ***control surface*** or ***open system boundary***.
- A ***closed system*** is a *fixed* mass of material defined by users. Heat may be transferred into or out of a closed system, and work may be performed on the mass within a closed system, but no mass may enter or leave the system. The surface of a closed system is also called the ***system boundary***. ***Control mass*** is another name for the mass inside a closed system.

In this chapter the contaminant concentration within a room, workplace, etc., is analyzed. The actual volume of such spaces may be small or it may be large. So long as the volume in question can be defined as a control volume, the phrase ***enclosed space*** or ***enclosed volume*** (V), is used irrespective of the magnitude of the actual volume.

Another way to describe general ventilation is to adopt a ***global*** or ***lumped parameter*** perspective. Such an analytical approach ignores details of the velocity field within the control volume, but concentrates on mass entering and leaving the control volume. Only global information about the flow field is required and the finite-sized control volume becomes a lumped mass of air containing uniformly distributed contaminant. Lumped parameter models involve only ordinary differential equations for the conservation of mass. Such mathematical models are called ***macro-models*** by some because details of the flow field are obscured, and because the size of the control volume is considerably larger than what would be called a ***differential volume***. Macro-models are easy to analyze mathematically to predict time-varying and steady-state concentrations. Macro-models can be applied to a single system or to one composed of several components that exchange air with each other. Examples include modeling ventilation in automobile compartments (Heinsohn et al., 1989, 1991, 1993; Engelmann et al., 1992), oil storage vessels (Haberlin and Heinsohn, 1993), and the cabins in commercial airliners (Ryan et al., 1986; National Research Council, 1986), modeling coating practices inside railroad tank cars (Bruno and Heinsohn, 1992), and modeling airflow in buildings (Dols, 2001).

In contrast to the macro-model is the ***micro-model*** that is the subject of study in subsequent chapters. Micro-models are mathematical models based on a very large number of differential volumes (i.e. ***distributed parameters***) and partial differential equations describing the conservation of mass, momentum, and energy. Details of the flow field are not assumed as they are in macro-models but are unknowns to be computed along with temperatures and concentrations throughout the space under study. Micro-models may be either two- or three-dimensional, depending on the physical phenomena being modeled. Micro-models are conceptually easy to analyze, but involve large sophisticated computer programs and inventive numerical techniques to solve (Kurabuchi and Kusuda, 1987; Lemaire and Luscuere, 1991; Busnaina, 1988). Micro-models employ ***computational fluid dynamics*** (CFD) to study a vast array of issues in fluid mechanics. Examples of applications of CFD to controlling contaminants in the indoor environment are found in Gran et al. (1993), Anastas (1993), Flynn and Taeheung (1993), Fontaine et al. (1993), and in Chapter 10 of the present book.

A common ventilation practice is to select the volumetric flow rate (Q) of make-up air in terms of the ***number of room air changes per unit time*** (N). The contaminant concentration is directly related to the number of room air changes only if well-mixed conditions exist. If there are spatial variations in concentration, the well-mixed model should not be used since it does not ensure that PELs (or RELs or TLVs) are satisfied. If the time scale is large, the assumption of well mixed may be valid because diffusion and slow mixing within the enclosure distribute the contaminant (albeit slowly) to ensure a uniform concentration. Unfortunately one does not always know the mixing time scale to know whether several hours or several days is the appropriate time scale. The number of room air changes is an anachronism and it is unfortunate that it is used in building codes with such aplomb. Its general use should be abandoned.

ASHRAE Standard 62 provides guidance on ventilation system requirements for building design, construction, and operation. First issued in 1973 as Standard 62-1973: *Standards for Natural and Mechanical Ventilation*, it was renamed *Ventilation for Acceptable Indoor Air Quality* in the 1981 revision and has retained that title through subsequent revisions. Standard 62-1989, superseding 62-1981, was the last revision under periodic maintenance, in which the entire standard is reviewed and approved. In 1997, Standard 62 was converted to continuous maintenance, a process in which revisions are made through stand-alone changes and addenda that are reviewed and approved separately. While the requirements of Standard 62 have changed somewhat over the years, its purpose, "...to specify minimum ventilation rates and indoor air quality that will be acceptable to human occupants," and intent, "...to minimize the potential for adverse health effects," (ASHRAE, 1999) have not.

ASHRAE Standard 62 provides two compliance procedures for ventilation system design: the ventilation rate procedure and the indoor air quality procedure. The ventilation rate procedure is based on the assumption that sufficient dilution of indoor air with acceptable outdoor air (which is defined in the standard) will produce conditions that are acceptable to most occupants of a space. Research cited in the standard showed that 80% of *visitors* to an occupied space express satisfaction when the air ventilation rate is 15 ACFM/person, and 80% of the *occupants* express satisfaction at 5 ACFM/person. Since people adapt quickly to many odors, *visitors* judge odor acceptability within 15 seconds after entering a space while *occupants* judge acceptability 15 minutes or more after entering the space. Ongoing research confirms the long-standing people-odor dilution requirements of 5 ACFM/occupant, 15 ACFM/visitor, and 20 ACFM/per person for offices. In support of the ventilation rate procedure, ventilation requirements are listed for 83 types of commercial and institutional spaces; an abstract of this table may be found in Appendix A.16. The requirements vary from 15 ACFM per person for spaces such as classrooms in which there is no smoking, to as much as 60 ACFM per person for smoking lounges.

The indoor air quality procedure is based on the control of specific contaminants to levels considered acceptable, either through ventilation or the use of air cleaning devices, or a combination of the two. The ventilation rate procedure, because of its prescriptive nature and because it requires no contaminant monitoring, is almost always the choice of designers. Standard 62 does not address residential or industrial ventilation; a separate standard on residential ventilation has been proposed, and the ACGIH handbook is referenced for industrial ventilation.

It is implicit in basing ventilation rates on occupancy that occupants are the source of contaminants. In some cases, however, building furnishings, finishes, and contents are significant sources. In recognition of this fact, some of the Table 2 requirements in Standard 62 are based on floor area rather than number of occupants. Examples are included in the extract in Appendix A.16, A proposed addendum to the standard, Addendum n, goes beyond this to propose ventilation rates that are a combination of "people" and "building" components. The procedure described in Addendum n would replace the existing ventilation rate procedure published in Standard 62-2001. An abstract of the second public review draft of this addendum is also included in Appendix A.16.

An obvious situation in which well-mixed conditions occur is a room in which the inlet and outlet ventilation ducts are located uniformly around the room, the room is not subdivided by partitions, and there is a great deal of activity within the room that mixes the air thoroughly. The contaminant emission rate and ventilation flow rate can be constants or they may vary with time. If there is a circulation fan or if individuals are moving about, such activity can mix the contaminants rapidly so that at any instant, Eq. (5-1) is valid. Under these conditions, the well-mixed assumption is physically reasonable. The well-mixed model may also be appropriate in enclosures where mixing occurs more rapidly than the time scale of the problem. For example, if one were concerned about the hydrocarbon concentration at points within a new mobile home whose furnishings liberate hydrocarbon vapors at a decreasing rate over a period of months, the well-mixed model would be appropriate if measurements were concerned with the concentration at weekly intervals for a period of many months.

The well-mixed model represents the upper limit of a real process because it predicts what would happen if mixing were instantaneous, and helps engineers place the results from more accurate models in perspective. If the designer is expressly concerned with spatial variations, the well-mixed model should never be used since the model assumes explicitly that there are no spatial variations. For example, consider contaminant concentrations related to an open vessel containing a volatile liquid located in a poorly ventilated enclosed space. Design engineers and OSHA are particularly interested in the concentration at the edge of the tank where workers stand rather than directly above the tank where the concentration is expected to be larger. For such applications the well-mixed model is inappropriate, and the designer needs to use some other analytical model to predict the concentration at the edge of the tank. If however one wishes to predict the concentration of tobacco smoke in a well ventilated cocktail lounge in which people are moving about, the well-mixed model would be a reasonable physical model to choose. In general, when the source is concentrated or small with respect to the size of the enclosure, when there is not vigorous mixing, or when the time scale of the problem is short, the well-mixed model should not be used.

5.1.2 Gravitational Settling and Floor Sweeps

Particulate contaminants (dust, smoke, aerosols) in air settle to the floor via a process called ***gravimetric settling***. Dust in air settles because (a) the dust density is typically of order a thousand times larger than the density of air, and (b) dust particles are of order a million times larger than molecules of oxygen and nitrogen. Dust suspended in air is a mixture of two phases, dust (solid) and air (gas). The motion of dust is governed by the principles of convection (bulk motion) of air, diffusion, inertial forces on the particles, and gravimetric forces that cause the dust to settle. As a second example, smoke particles are typically 0.01 to 1.0 μm in diameter while oxygen or nitrogen molecules are approximately 0.0003 μm. Thus smoke particles are 30 to 3000 times larger than air molecules. In addition, the density of smoke particles is nearly 1000 times larger than that of air. Thus if smoke particles enter a quiescent enclosure, gravity causes them to settle, producing regions of high concentration near the floor and low concentration near the ceiling. The alert reader may point out that smoke from a candle, cigarette, etc. is observed to *rise*, not settle, in apparent contradiction to the present discussion. However, this observed behavior is temporary, due to the high temperature of the smoke, and the resulting buoyant plume of air which carries smoke particles upward. In a quiescent enclosure, after sufficient time for the temperatures to equalize, there will be a region near the ceiling virtually devoid of particles. If the enclosure is well mixed, however, the particle concentration will be uniform but will decrease with time as particles stick to walls and floor.

It is often alleged that contaminant *vapors* heavier than air also settle so that their concentration near the floor is larger than near the ceiling. This is an "old wives' tale" that cannot be supported by principles in the thermal sciences. In contrast to solid or liquid aerosol particles in air, contaminant gases and vapors have (a) molecular weights only 4 to 5 times larger than those of oxygen and nitrogen, and (b) molecules which are only somewhat larger than those of oxygen and nitrogen. Vapor contaminants in air are a single phase. The motion of contaminant molecules relative to air is governed only by molecular diffusion, which is independent of gravity. Molecular diffusion dictates that contaminant molecules move in all directions to places where their concentration is less. To illustrate that heavy molecules do not settle, consider two commonplace examples:

- If gravity causes heavy gas molecules to settle, why is carbon dioxide (M = 44.0) uniformly mixed in the air (M = 28.97) within a room and in the outdoors?
- Why is it that Freon R-11 (CCl_3F, M = 136.0) rises from the earth's surface to the stratosphere (altitude = 15 km) where it reacts with ozone?

From these examples (and there are countless others), and from the principles of thermal sciences, there is no basis to believe that heavy molecules fall when mixed with air.

The reason why odors are more intense near the floor when liquids spill on the floor is because the spilled liquid evaporates from the floor, and diffuses upward. Diffusion is always from a region of high concentration to a region of lower concentration. For example, consider an undetected spill of TCE (M = 131.4) in a closed storage cabinet. The concentration is always larger near the floor because this is where the puddle lies. Recall from Chapter 4 that the partial pressure of contaminant vapor in the air just above the liquid-air interface is equal to the saturation pressure (vapor pressure) of the liquid, and decreases upward. If all the liquid evaporates, and adequate time is allowed to achieve steady state, the concentration of TCE in the cabinet air will be the same throughout the entire cabinet, e.g. the concentration at the top and bottom of the cabinet will be the same.

Under certain physical conditions, the observation that heavier vapors concentrate near the floor may be correct. For example, if a jet of gas or vapor heavier than air enters a room, it is deflected downward by gravity. Thus for a short period of time the lower portion of the room contains a higher concentration. But soon thereafter, diffusion produces a homogeneous mixture throughout the room. If the jet of gas or vapor is lighter than air, the deflection is upwards. However, this gas or vapor also eventually mixes uniformly with room air. While the spontaneous separation of fluids based on molecular weight may occur due to gravimetric settling, documented examples in nature are rare. Two examples that are sometime cited are:

- segregation of liquids in deep, quiescent wells, e.g. abandoned oil wells
- preferential escape to space of atmospheric hydrogen that is said to occur in the upper reaches of the stratosphere

These examples are slow processes that occur in unusual places and are accompanied by very small gradients and very long times. Thus, it is safe to conclude that spontaneous separation based on molecular weight is insignificant in indoor air problems where the time scale is short, distances are small, and mass transport by thermal gradients and bulk motion is the dominant factor.

To illustrate the issue further, consider two cases: contaminant added to an enclosure containing only quiescent uncontaminated air, and the equilibrium state of an initially uniform mixture of air and contaminant:

(a) <u>*Contaminant added to quiescent air*</u> - If a volatile liquid leaks onto the floor of a room, vapors rise as the liquid evaporates. The process is dynamic, so the rate of evaporation and the concentrations vary with time (see Chapter 4 for details). Mass is transported upward; the concentration is highest near the spill and decreases with height. This situation is true even when the molecular weight of the contaminant is greater than that of air, although the molecular weight of the contaminant affects the rate of evaporation and upward diffusion. Once evaporation ceases, the contaminant distributes itself until the concentration gradients become zero and the concentration is uniform throughout the room. The mixing of gas molecules can be described in a simple fashion by kinetic theory, or if one wishes more detail, by more sophisticated analyses. In any case, the molecular weight of each molecular species influences the rate of mixing, but not the equilibrium state of uniform concentration. If the contaminant consists of liquid or solid particles the situation is entirely different because two phases are involved, as discussed above.

(b) <u>*Initially uniform conditions*</u> - Now consider the opposite condition to (a) above in which the initial concentration inside the enclosure is uniform. What is the final state of equilibrium? Is it logical to expect gaseous constituents to segregate themselves spontaneously into layers within which the molecular weight is the same? Wouldn't the chemical industry love to use inexpensive gravity to separate gaseous mixtures into their constituent parts? In reality, spontaneous separation does not occur in an industrial or domestic environment. Gases and vapors remain uniformly distributed while only particles (solid or liquid) settle.

A ***floor sweep*** is an exhaust register, placed at floor-level in storage rooms, that is designed to capture the vapor of flammable contaminants. Many people erroneously believe that floor sweeps are near the floor because heavier-than-air hydrocarbon vapors settle to the floor, and can thus be removed more effectively. However, as discussed above, contaminant vapors do *not* settle appreciably in air. The reason why storage rooms have exhaust registers near the floor (floor sweeps) is because a spill of flammable liquid initially produces high concentrations near the floor. It is the momentary condition after a spill that is of concern; an exhauster near the floor removes vapor and lessens the chance of fire. As a general proposition, hazardous conditions occur because air and contaminants are not uniformly mixed in the first place, not because they separate after mixing.

5.1.3 Applicability of Dilution and Displacement Ventilation

As is shown in subsequent sections in this text, it is unwise and uneconomical to attempt to control contaminant concentration in a room or workplace by either type of general ventilation. The amount of air that has to be exhausted and the amount of make-up air that has to be added to the volume to keep contaminant concentrations below their PEL values are huge. Furthermore the cost to control the temperature and humidity of the make-up air is enormous. To control contaminants, it is wiser to use local ventilation, i.e. to install hoods which capture contaminants at locations close to where they are emitted but before they enter the workspace. General ventilation is attractive when it is necessary to improve the overall quality of the air in the workplace, to remove odors, to control the relative humidity, and to heat (or cool) the air. Local ventilation is necessary when contaminants are emitted at known locations in the workplace.

A low-cost way to create well-mixed conditions is to install large-diameter, low-speed circulating fans above the work floor, as shown for example in Figure 5.2 (DeGasperi, 1999). Such fans are ideal for diary barns, hog enclosures, manufacturing production lines, and possibly offices and public buildings. In winter the fans circulate warm air that would otherwise accumulate below the ceiling. In summer, if the building is air-conditioned, the circulating fans produce a uniformly cool environment. Even if the interior air is not air conditioned, the fans produce a quiet, gentle breeze (see Table 3.7) and more uniform temperatures, which workers find beneficial. The fans can be mounted in the vertical plane to circulate air to a mezzanine. High-speed fans are in general noisy and move only small columns of air that dissipate quickly due to viscous effects, whereas large-diameter, low-speed fans are quiet and move a large amount of air. A sufficient number of fans can eliminate the need for ductwork to distribute air throughout the workplace. In the case of processes that generate heated plumes, odors, or nontoxic gases and vapors, the fans can be used to draw air upward to a space below the ceiling where exhaust fans can discharge the air from the building.

Figure 5.2 Low-speed large-volume circulating fans in an industrial environment (courtesy of HVLS Fan Company).

Ceiling fans such as the one shown in Figure 5.2 vary in diameter from 8 to 20 feet and produce a downward shaft of slow moving air that radiates outward when it strikes the work floor, covering 10,000 to 15,000 ft^2. Such fans are powered by ¾ to 1 HP electrical motors. The blades are made of extruded aluminum with an airfoil cross-sectional area at an 8° angle of attack. The fans rotate at 25 to 125 RPM; larger diameter fans use lower rotational speed.

5.2 Thermodynamics of Unventilated Enclosures

Unventilated enclosures are unlikely to be encountered in industry because it is virtually impossible to prevent air from entering or leaving an enclosure, and secondly OSHA and fire codes prohibit them. Nonetheless it is instructive to consider unventilated enclosures in which volatile liquids are stored, and to determine whether there is an upper limit to the amount of a volatile liquid that can evaporate, since such a value represents the upper limit to the concentration in a ventilated enclosure. The material here also applies directly to unvented storage tanks themselves. The rate at which the maximum concentration is approached depends on the particularities of the problem, but the maximum concentration is a simple matter of thermodynamics. In an unventilated enclosure one might assume that all volatile liquid completely evaporates when there is sufficient time to do so. This assumption is not necessarily correct, because the initial mass of liquid and principles from thermodynamics impose limiting conditions. Thermodynamics dictates that the maximum partial pressure a vapor may have when the vapor is mixed with other gases is equal to its vapor pressure (also called its saturation pressure), as discussed in previous chapters. *A single component-liquid evaporates until the partial pressure of its vapor (in air) is equal to its vapor pressure.*

To understand the issue, one must compare room temperature and pressure (T_{room}, P_{room}) to the ***critical temperature*** and ***critical pressure*** (T_c, P_c) of various contaminants. Appendix A.10 is a table of common workplace contaminants and shows that all the materials have critical pressures in excess of atmospheric pressure and most have critical temperatures greater than room temperature. Thus contaminants fall into two categories:

- Category 1: $P_c > P_{room}$ and $T_c > T_{room}$
- Category 2: $P_c > P_{room}$ and $T_c < T_{room}$

Figure 5.3 shows these two categories on P-v and T-s phase diagrams. Note that in the vapor phase, the pressure on such diagrams is the *partial* pressure of the contaminant.

Category 1: $P_c > P_{room}$ and $T_c > T_{room}$: Figure 5.3a shows the possible states of such an air-vapor mixture. The maximum amount of contaminant that can exist as a vapor has a partial pressure equal to its vapor pressure based on room temperature. Thus, providing there is sufficient liquid to evaporate, the maximum vapor mol fraction ($y_{maximum}$) is the ratio of the contaminant vapor pressure, P_v(at T_{room}) divided by the total pressure, P_{room},

$$\boxed{y_{maximum} = \frac{P_v\left(\text{at } T_{room}\right)}{P_{room}}} \qquad (5\text{-}4)$$

Whether this value is ever achieved depends on the initial amount of volatile liquid and air. If the number of mols of volatile liquid contaminant (n_{liquid}) divided by the number of mols of air (n_{air}) is less than the value predicted by Eq. (5-4), then the actual maximum concentration is determined by the amount of contaminant available for evaporation,

$$\boxed{y_{maximum} = \frac{n_{liquid}}{n_{liquid} + n_{air}}} \qquad (5\text{-}5)$$

In summary, for Category 1, the actual maximum concentration is the smaller of Eqs. (5-4) and (5-5).

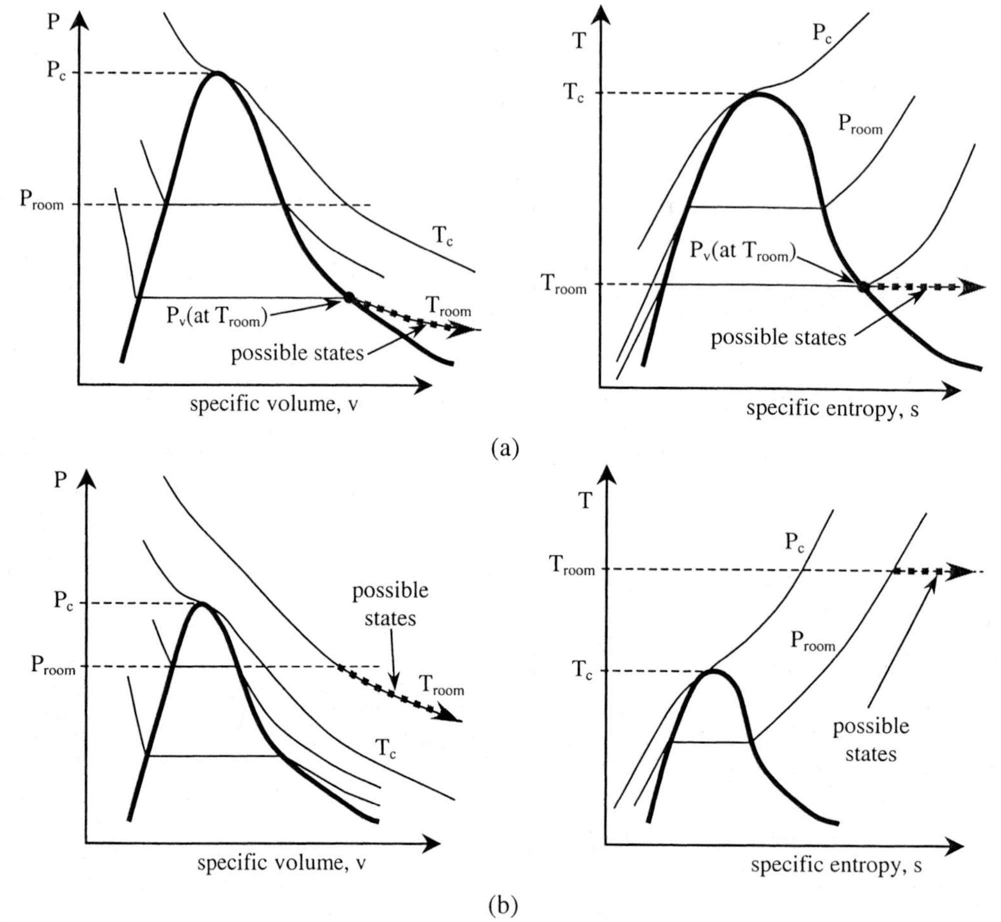

Figure 5.3 Possible thermodynamic states for contaminant vapor at temperature T_{room} and pressure P_{room}: (a) Category 1: $P_c > P_{room}$, $T_c > T_{room}$; (b) Category 2: $P_c > P_{room}$, $T_c < T_{room}$.

Category 2: $P_c > P_{room}$ and $T_c < T_{room}$: Figure 5.3b shows the possible states of such an air-vapor mixture. In this case thermodynamics does not establish an upper limit on contaminant mol fraction; the maximum mol fraction is dictated only by Eq. (5-5).

Example 5.1 illustrates the use of the above equations for Category 1 and Category 2.

Example 5.1 - Maximum Concentration in an Unventilated Enclosure
Given: Chemicals are stored in an unventilated outdoor shed. The shed volume is 5.0 m^3 and ambient temperature and pressure are 18.4 °C (291.55 K) and 1.0 atm.

To do: Find the maximum contaminant mol fraction if the following accident scenarios occur (separately) to materials left in the shed:

(a) an open container containing 1.2 kg of toluene is left for several days
(b) same as (a), but only 0.12 kg of toluene is originally in the container
(c) a 1-liter (0.0010 m^3) cylinder containing nitric oxide with an initial pressure of 2.0 atm develops a leak

Solution:

(a) 1.2 kg of Toluene:

From Appendix A.10, toluene is found to have a critical temperature and pressure of 320.8 °C and 41.6 atm. Thus, $P_c > P_{room}$ and $T_c > T_{room}$ (Category 1). From Figure 5.3a, it can be seen that while there is a wide range of possible toluene vapor concentrations in air, the maximum value is dictated by the thermodynamic consideration that toluene vapor cannot exist with a partial pressure greater than its vapor pressure based on room temperature. From Appendix A.8 the vapor pressure of toluene at 18.4 °C is 20.0 mm Hg. Thus if the total pressure of toluene and air is 1.0 atm (760. mm Hg), the maximum mol fraction of toluene allowed by thermodynamic considerations is found from Eq. (5-4),

$$y_{maximum} = \frac{P_v\left(\text{at } T_{room}\right)}{P} = \frac{20.0 \text{ mm Hg}}{760 \text{ mm Hg}} = 0.0263$$

The mass of toluene (m_{tol}) in the air corresponding to the above mol fraction can be found from the Dalton's law version of the ideal gas law (written for partial pressures):

$$m_{tol} = \frac{M_{tol} P_{v,tol} V}{R_u T} = \frac{92.0 \frac{\text{kg}}{\text{kmol}} (20.0 \text{ mm Hg})(5.0 \text{ m}^3)}{8.314 \frac{\text{kJ}}{\text{kmol} \cdot \text{K}} (291.55) \text{K}} \left(\frac{\text{kJ}}{\text{m}^3 \text{kPa}}\right)\left(\frac{101.3 \text{ kPa}}{760. \text{ mm Hg}}\right) = 0.506 \text{ kg}$$

or approximately 0.51 kg (to two significant digits). Since 1.2 kg of toluene is available initially, it is clear that not all of it evaporates because thermodynamics limits the amount that can exist in the vapor phase. At equilibrium, 0.51 kg of toluene exists as a vapor at a partial pressure equal to the vapor pressure (20.0 mm Hg) and the remainder (0.69 kg) remains in the liquid phase. Thermodynamics prevents the complete evaporation of all the liquid toluene, and the final mol fraction is 0.0263 (approx. 26,000 PPM). Such a value is cause for alarm because not only does it exceed the PEL (200 PPM) by a considerable amount, but it also exceeds the lower explosion limit (LEL). Table 3.2 shows that the mol fraction at the lower explosion limit is 0.0127 (1.27% or 12,700 PPM). Thus there is a strong possibility that an explosion could occur caused by a spark in a light switch or any other source of ignition.

(b) 0.12 kg of Toluene:

If the original amount of toluene is only one-tenth of the amount analyzed in part (a), the above calculation indicates that *all* the toluene would evaporate. The toluene partial pressure and mol fraction can then be found from Dalton's law:

$$P_{tol} = \frac{m_{tol} R_u T}{M_{tol} V} = \frac{0.12 \text{ kg}\left(8.314 \frac{\text{kJ}}{\text{kmol} \cdot \text{K}}\right)(18.4 + 273.15) \text{K}}{92.0 \frac{\text{kg}}{\text{kmol}} (5.0 \text{ m}^3)} \left(\frac{\text{m}^3 \text{kPa}}{\text{kJ}}\right) = 0.632 \text{ kPa}$$

and

$$y_{tol} = \frac{P_{tol}}{P} = \frac{0.632 \text{ kPa}}{101.3 \text{ kPa}} = 0.00624 \cong 6,200 \text{ PPM}$$

While this mol fraction is below the LEL, it is still well above the PEL, and presents a hazardous situation.

(c) Leaking Cylinder of Nitric Oxide:

The amount of air (n_{air}) in the shed can be found from the ideal gas law:

$$n_{air} = \frac{PV}{R_u T} = \frac{101.3\ \text{kPa}\left(5.0\ \text{m}^3\right)}{8.314\frac{\text{kJ}}{\text{kmol}\cdot\text{K}}\left(291.55\ \text{K}\right)}\left(\frac{\text{kJ}}{\text{m}^3\text{kPa}}\right) = 0.209\ \text{kmol}$$

Appendix A.10 shows that the critical temperature and pressure of nitric oxide are -93. °C and 64. atm; thus, $P_c > P_{room}$ and $T_c < T_{room}$ (Category 2). Under these conditions Figure 5.3b shows that thermodynamics does not limit the amount of nitric oxide that can be mixed with air and that the final mol fraction depends only on the amount of nitric oxide leaking into the air. The nitric oxide stops leaking when the pressure inside the cylinder is 1.0 atm. Since the leak is slow, it is reasonable to assume that heat transfer between the cylinder and air maintains the final temperature equal to the room temperature. Thus,

$$y_{maximum} = \frac{n_{NO}}{n_{NO} + n_{air}}$$

Since the critical pressure is so much larger than the cylinder pressure, one may assume that the nitric oxide is an ideal gas. The amount of NO escaping into the room is

$$n_{NO,\ escape} = n_{NO,\ initial} - n_{NO,\ final} = \frac{\left(P_{NO,\ initial} - P_{NO,\ final}\right)V}{R_u T}$$

$$= \frac{\left(202.6 - 101.3\right)\text{kPa}\left(0.0010\ \text{m}^3\right)}{8.314\frac{\text{kJ}}{\text{kmol}\cdot\text{K}}\left(291.55\ \text{K}\right)}\left(\frac{\text{kJ}}{\text{m}^3\text{kPa}}\right) = 4.18\times10^{-5}\ \text{kmol}$$

The final nitric oxide mol fraction is thus

$$y_{final} = \frac{n_{NO}}{n_{NO} + n_{air}} = \frac{4.18\times10^{-5}\ \text{kmol}}{4.18\times10^{-5}\ \text{kmol} + 0.209\ \text{kmol}} = 1.9996\times10^{-4} \cong 200\ \text{PPM}$$

This mol fraction is in excess of the PEL (25 PPM). At room temperature nitric oxide and oxygen react to form nitrogen dioxide and hence one is apt to find 200 PPM of nitrogen dioxide.

Discussion: Final results are given to two significant digits. The values calculated here are *upper limits* since some fresh air inevitably leaks into the shed, even though it is unventilated.

5.3 Dilution Ventilation with 100% Make-up Air

Consider an enclosure of volume V in which the mass concentration of some contaminant is spatially uniform, as in Figure 5.4. (The ceiling fan in Figure 5.4 indicates well-mixed conditions in the room.) Initially (t = 0) the mass concentration in the enclosed space is c(0), and a source begins to emit contaminant at a constant emission rate S (mass/time). Outside ambient air containing contaminant at a mass concentration of c_a is added to the enclosed space at constant volumetric flow rate Q (volume/time), and contaminated air is removed from the enclosed space at the same rate. The density of the air entering and leaving the enclosure is assumed to be constant and equal to the density in the enclosed space. Using the enclosed space as a control volume and writing an equation for conservation of mass of the contaminant,

$$\boxed{\frac{d(Vc)}{dt} = V\frac{dc}{dt} = Qc_a + S - Qc} \tag{5-6}$$

In words, the time rate of change of mass of contaminant equals the rate at which contaminant enters the volume, plus the rate at which contaminant is emitted into the volume, minus the rate at which contaminant leaves the volume. Since only one contaminant is being considered at a time, no subscript

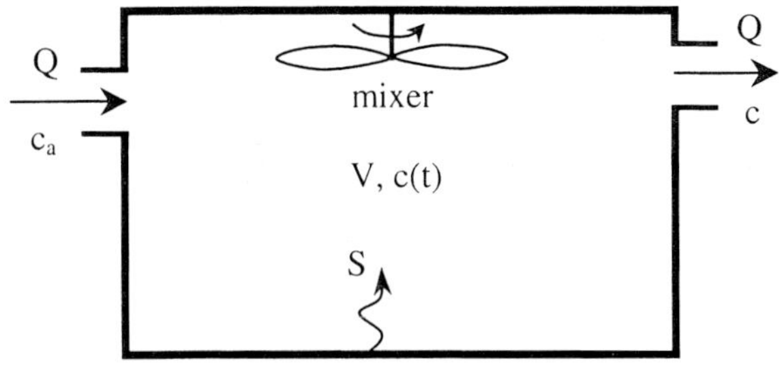

Figure 5.4 Schematic diagram of a typical general ventilation system, with 100% make-up air.

(such as j) is necessary here, as per the convention established in earlier chapters. The volume (V) is constant, which is why it could be taken outside the derivative. Because of the well-mixed assumption, the mass concentration leaving the control volume in the last term in Eq. (5-6) is the same as the mass concentration in the volume itself (c). Rearranging Eq. (5-6),

$$\boxed{\frac{dc}{dt} = B - Ac} \tag{5-7}$$

where

$$\boxed{B = \frac{Qc_a + S}{V}} \tag{5-8}$$

and

$$\boxed{A = \frac{Q}{V}} \tag{5-9}$$

The form of Eq. (5-7) shall hereafter be referred to as the ***standard form***. Parameters A and B are constants in this particular problem, but can be functions of time in general. There is no need to write an equation expressing conservation of mass for the air since the inlet flow rate is equal to the outlet flow rate, and the density is constant. Since Eq. (5-7) is of the same form as the standard first-order ODE of Eq. (1-59), the solution is given by Eqs. (1-61) and (1-63),

$$\boxed{\frac{c_{ss} - c(t)}{c_{ss} - c(0)} = \exp(-At)} \tag{5-10}$$

and

$$\boxed{c_{ss} = \frac{B}{A}} \tag{5-11}$$

where c_{ss} is the ***steady-state mass concentration***, i.e. the value of c after a very long time has elapsed. Note that c_{ss} can also be found by setting the left hand side of Eq. (5-7) to zero. Using Eq. (1-67), the time (t) can be found for some desired value of c(t),

$$\boxed{t = -\frac{1}{A}\ln\left[\frac{c_{ss} - c(t)}{c_{ss} - c(0)}\right]} \tag{5-12}$$

If the initial concentration in the room, c(0) is zero, Eq. (5-10) reduces to

$$\boxed{\frac{c(t)}{c_{ss}} = 1 - \exp(-At)} \tag{5-13}$$

A great deal of indoor air pollution literature expresses the ventilation rate (Q) in terms of the number of room air changes per unit time (N), defined as

$$\boxed{N = \frac{Q}{V}} \tag{5-14}$$

The dimensions of N are 1/time. For the case of zero initial concentration, Eq. (5-13) can be written in terms of N as

$$\boxed{\frac{c(t)}{c_{ss}} = 1 - \exp(-Nt)} \tag{5-15}$$

For example, if the number of room air changes per unit time is ten room air changes per hour (N = 10./hr), this represents one room change every 6.0 minutes. Equation (5-15) predicts that in 6.0 minutes, the mass concentration ratio is

$$\frac{c(t = 6 \text{ min})}{c_{ss}} = 1 - \exp\left[-\left(10.\frac{1}{\text{hr}}\right)(6.0 \text{ min})\left(\frac{\text{hr}}{60 \text{ min}}\right)\right] = 0.632$$

In other words, the mass concentration in the room grows to about 63.2% of its steady-state value in 6.0 minutes. From the discussion about first-order ODEs in Section 1.5, it is seen that the first-order time constant (τ) is

$$\boxed{\tau = \frac{1}{A} = \frac{1}{N} = \frac{V}{Q}} \tag{5-16}$$

for this example, τ is

$$\tau = \frac{1}{N} = \frac{1}{10.\frac{1}{\text{hr}}} = 0.10 \text{ hr} = 6.0 \text{ min}$$

The quantity 1/N = V/Q (the reciprocal of the number of room air changes per unit time) has units of time and is visualized by some as the time it takes a mass of make-up air equal to the volume of the room to traverse the room (as a piston pushes out exhaust gases in an automotive cylinder), and displace all the contaminated air. Of course, with dilution ventilation, make-up air does not traverse a room as a continuous mass like a piston, but the visualization is often used in indoor air ventilation analyses. The visualization of make-up air displacing contaminated air is actually more in line with the other strategy of general ventilation, displacement ventilation. It is easy to see how the phrase *number of room air changes per unit time* came into being, but the practice of assessing the worth of general ventilation or of setting building standards in terms of room air changes per unit time should perhaps be abandoned. It is also unwise to assume that enclosed spaces are well mixed until some independent means shows that they are. Rarely is the activity within an enclosed space so robust and the inlet and outlet ducts so judiciously located to enable one to assume that the air is well mixed.

Example 5.2 - When is it Safe to Enter a Room Containing HCN?

Given: A fire has occurred in a motel room (V = 85. m^3) in which an upholstered sofa containing polyurethane foam has burned and filled the room with smoke containing HCN. Immediately after the fire has been extinguished, the smoke concentration is 100. g/m^3, of which 10.% by mass is HCN. Firemen evacuate the smoke with an exhaust fan and blow outside air into the room at the same rate, 1000 ACFM (28.3 m^3/min). Because of the fire, the outside air contains HCN with a concentration (c_a) of 1.0 mg/m^3.

To do: On the basis of the well-mixed model, estimate the length of time that must elapse before individuals can enter the room after the fire has been extinguished. The fire company uses a short time exposure limit (STEL) of 5 mg/m^3 as the HCN concentration that must be achieved before individuals can enter the room.

Solution: The initial HCN mass concentration is

$$c(0) = 0.10\left(100.\frac{g}{m^3}\right)\left(\frac{1000\ mg}{g}\right) = 10{,}000\frac{mg}{m^3}$$

Figure 5.4 depicts the flow of HCN in and out of the room. A mass balance for HCN is

$$\frac{d(Vc)}{dt} = V\frac{dc}{dt} = Qc_a - Qc$$

which can be put into standard form for a first-order ODE, i.e. the form of Eq. (5-7), with

$$A = \frac{Q}{V} \qquad B = \frac{Qc_a}{V}$$

from Eqs. (5-8) and (5-9) with S = 0 (there is no source term since the fire is out). The steady-state mass concentration is eventually that of the ambient air, $c_{ss} = c_a$, which can be obtained from both common sense and application of Eq. (5-11). The solution of the ODE is given by Eq. (5-10), which was solved for time (t) in Eq. (5-12),

$$t = -\frac{1}{A}\ln\left[\frac{c_{ss} - c(t)}{c_{ss} - c(0)}\right] = -\frac{V}{Q}\ln\left[\frac{c_a - c(t)}{c_a - c(0)}\right] = -\frac{85.\ m^3}{28.3\frac{m^3}{min}}\ln\left[\frac{(1.0-5.0)\frac{mg}{m^3}}{(1.0-10{,}000)\frac{mg}{m^3}}\right] = 24.\ min$$

Thus, the firemen should wait about a half hour before entering the room.

Discussion: Considering that HCN is highly toxic, the room may not be well mixed, and there may be regions where the HCN concentration is large, it is clear that respirators should be worn by the firemen, even if they wait a half hour before entering the room.

Example 5.3 combines elements from this chapter plus all previous chapters, namely risk assessment (Chapter 1), interaction of contaminants with the body (Chapter 2), permissible exposure levels (Chapter 3), and evaporation rate (Chapter 4).

Example 5.3 - TCE Volatilization in a Bathroom Shower Stall

Given: Trichloroethylene (CAS 79-01-6), commonly called TCE, is a commonplace toxic solvent (airborne PEL = 100 PPM) with a sweet odor (odor threshold = 400 PPM). Trace amounts of TCE are present in ground water and tap water at concentrations well below EPA allowable values. The EPA is concerned that TCE might volatilize in bathroom showers since hot water increases the volatility of dissolved TCE. Your morning newspaper reports that some researchers (Little, 1992; Giardino et al., 1992) found that individuals taking a 10-minute shower have an inhalation exposure to TCE 1.5 times greater than the exposure from ingesting two liters of tap water. You are enrolled in an in indoor air quality course and the instructor asks the class to examine the validity of the claim. The instructor points out that the report is misleading because it compares risks associated with inhalation with risks associated with ingestion. The two routes of entry into the body have different ramifications. For example, the dose is small and the body experiences no permanent damage from the publicity stunt of swallowing a gram of liquid mercury, because the mercury passes through the GI tract in a few hours. (Do not try this at home.) On the other hand, inhaling one gram of mercury vapor over several years constitutes a huge dose. The vapor enters the respiratory system, is transferred to the body's organs, and remains in the body. Continuous exposure to mercury vapor for months and years can damage the central nervous system.

Fresh air enters the stall at volumetric flow rate (Q_{air}), and the same amount of air containing TCE vapor leaves the stall. Other parameters of the problem are as follows:

- M_{TCE} (molecular weight of TCE) = 131.4 kg/kmol = 131.4 g/mol
- V (volume of air in the shower stall, which is the control volume) = 0.20 m^3
- T_{air} (air temperature in the shower stall) = 25.0 °C (298.15 K)
- T_{water} (water temperature) = 40.0 °C (313.15 K)
- $c_{molar,air}$ (molar concentration of air),

$$c_{molar,air} = \frac{P}{R_u T} = \frac{101.3\ \text{kPa}}{8.314\dfrac{\text{kJ}}{\text{kmol}\cdot\text{K}}(298.15\ \text{K})}\left(\frac{\text{kPa}\cdot\text{m}^3}{\text{kJ}}\right) = 0.0409\frac{\text{kmol}}{\text{m}^3} = 0.0409\frac{\text{mol}}{\text{L}}$$

- $c_{molar,water}$ (molar concentration of water), $c_{molar,water} = \dfrac{1000\dfrac{\text{kg}}{\text{m}^3}}{18.0\dfrac{\text{kg}}{\text{kmol}}} = 55.6\dfrac{\text{kmol}}{\text{m}^3} = 55.6\dfrac{\text{mol}}{\text{L}}$
- Q_{air} (volumetric flow rate of air through shower stall) = 50. L/min = 0.050 m^3/min
- Q_{water} (volumetric flow rate of water) = 10. L/min = 0.010 m^3/min
- H'' (nondimensional Henry's law constant for TCE) = 0.32 (Little, 1992)
- $c_{TCE,water}$ (TCE mass concentration in water) = 3.0 μg/L (0.0030 g/m^3)
- PEL_{TCE} (PEL of airborne TCE) = 100 PPM (540 mg/m^3)
- $K_L A_s$ (overall mass transfer coefficient times water surface area) Little (1992) states that $K_L A_s \approx$ 10. L/min
- $y_{TCE,air}$ (mol fraction of TCE in the air phase)

To do: Compare the TCE exposure from a 10-minute shower to that of ingesting two liters of tap water.

Solution: In analyzing such a problem, it is useful to divide the task into several parts:

(a) Calculate the TCE consumed by drinking two liters of tap water: It is important to comprehend a TCE mass concentration of 3.0 μg/L in tap water. This concentration is comparable to adding 350. milligrams (mass of an aspirin tablet) to an Olympic-size swimming pool! The mol fraction (x_{TCE}) of TCE in the water (aqueous phase) can be found as follows:

$$x_{TCE} = \frac{c_{TCE,water}}{M_{TCE}c_{molar,water}} = \frac{0.0030\dfrac{\text{g}}{\text{m}^3}}{131.4\dfrac{\text{g}}{\text{mol}}55{,}600\dfrac{\text{mol}}{\text{m}^3}} = 4.11\times10^{-10}$$

which is equal to 0.411 PPB. The mass of TCE consumed by drinking two liters of water is thus

$$m_{TCE} = 3.0\frac{\mu\text{g}}{\text{L}}(2.0\ \text{L}) = 6.0\ \mu\text{g} = 0.0060\ \text{mg}$$

(b) Estimate the TCE concentration in the air, assuming that *all* the TCE in the shower water is transferred to the air in the shower stall: This is a worst-case scenario. The source (S) is the mass flow rate of TCE into the air in the shower stall, and can be computed by assuming that *all* of it evaporates into the air:

$$S = \dot{m}_{TCE} = c_{TCE,water}Q_{water} = 0.0030\frac{\text{g}}{\text{m}^3}0.010\frac{\text{m}^3}{\text{min}} = 3.0\times10^{-5}\ \frac{\text{g}}{\text{min}} = 0.030\ \frac{\text{mg}}{\text{min}}$$

A mass balance equation for the *airborne* TCE mass concentration in the shower can be written as follows, assuming that the shower stall is a well-mixed chamber:

$$V\frac{dc}{dt} = c_a Q_{air} + S - Q_{air}c$$

which is the same as Eq. (5-6). The ambient air flowing into the shower stall is assumed to have no TCE, i.e. $c_a = 0$. Dividing by V, the above equation can be written in the standard form of Eq. (5-7), with coefficients

$$A = \frac{Q_{air}}{V} \qquad B = \frac{S}{V}$$

The time constant (τ) for this first-order system is given by Eq. (5-16),

$$\tau = \frac{1}{A} = \frac{V}{Q_{air}} = \frac{0.20\ m^3}{0.050\frac{m^3}{min}} = 4.0\ min$$

The steady-state TCE concentration can be found from Eq. (5-11):

$$c_{ss} = \frac{B}{A} = \frac{S}{Q_{air}} = \frac{0.030\ \frac{mg}{min}}{0.050\frac{m^3}{min}} = 0.60\frac{mg}{m^3}$$

At this point, the analysis can stop, because even if one stayed in the shower all day, the steady-state TCE mass concentration in the air is nearly three orders of magnitude smaller than the PEL; there is no need to be concerned! Nevertheless, further calculations can be performed for illustration. After 10 minutes in the shower, the TCE mass concentration rises to only about 92% of the steady-state value, since 10 minutes is equal to 2.5τ. Using Eq. (5-13), and noting that $\tau = 1/A$,

$$c(10.\ min) = 0.60\frac{mg}{m^3}\left[1 - \exp\left(-\frac{10.\ min}{4.0\ min}\right)\right] = 0.55\frac{mg}{m^3}$$

(c) Calculate the mass of TCE entering the lungs, assuming that *all* the TCE in the shower water is transferred to the air in the shower stall: While in the shower for 10 minutes, an individual inhales air at a rate of 32.2 L/min (light exercise; see Table 2.4). The mass of TCE entering the lungs can be estimated by integration as follows:

$$m_{TCE,inhaled} = \int_{t=0}^{t=10\ min} cQ_{inhale}dt = c_{ss}Q_{inhale}\int_{t=0}^{t=10\ min}\left[1 - \exp\left(-\frac{t}{\tau}\right)\right]dt$$

The integration is straightforward, and yields

$$m_{TCE,inhaled} = c_{ss}Q_{inhale}\left[t + \tau\exp\left(-\frac{t}{\tau}\right)\right]_{t=0}^{t=10\ min} = 0.60\frac{mg}{m^3}\left(32.2\frac{L}{min}\right)6.33\ min\left(\frac{m^3}{1000\ L}\right) = 0.12\ mg$$

which is about 20 times the amount ingested by drinking two liters of tap water. Note, however, that this is a worst-case scenario. In addition, not all of the inhaled TCE actually diffuses into the bloodstream (refer to the discussion in Chapter 2 about uptake efficiencies, etc.)

(d) Calculate the steady-state airborne TCE concentration using Henry's law: This represents a more realistic scenario. The actual airborne TCE concentration should be much less than in analysis (b) above, because all the TCE in the water cannot be transferred to the air. From Eq. (4-

99), the airborne mol fraction of TCE in equilibrium with 3.0 μg/L of TCE in water can be expressed through Henry's law:

$$y_{TCE} = H''x_{TCE} = H''\frac{n_{TCE}}{n_{water}} = H''\frac{\frac{n_{TCE}}{V}}{\frac{n_{water}}{V}} = H''\frac{c_{molar,TCE}}{c_{molar,water}} = H''\frac{c_{TCE}}{c_{molar,water}M_{TCE}}$$

where n_{TCE} is the number of mols of TCE in water and H'' is the nondimensional Henry's law constant ($H'' = H'/P_{atm}$). In the above equation, the numerator and denominator were divided by the same volume in order to convert to concentrations. The equilibrium mol fraction of TCE in air is thus

$$y_{TCE} = 0.32\frac{3.0\frac{\mu g}{L}}{55.6\frac{mol}{L}131.4\frac{g}{mol}}\left(\frac{g}{10^6\ \mu g}\right) = 1.31\times10^{-10} \cong 1.3\times10^{-4}\ \text{PPM}$$

The equilibrium airborne concentration of TCE (in mg/m^3) is given by Eq. (1-30),

$$c_{TCE,\,air} = \frac{[PPM_{TCE}]M_{TCE}}{24.5}\frac{mg}{m^3} = \frac{(1.31\times10^{-4})(131.4)}{24.5}\frac{mg}{m^3} = 7.05\times10^{-4}\frac{mg}{m^3} = 0.705\frac{\mu g}{m^3}$$

This is almost 800 times smaller than the worst-case TCE concentration calculated in part (b) above, and it is more than five orders of magnitude smaller than the PEL!

(e) Calculate the time required to reach steady-state concentration: Initially, air in the shower stall contains no TCE, i.e. y(0) = 0, and it is not known how long it takes for the TCE concentration to achieve its steady-state value. If it takes longer than 10 minutes, the dose would be considerably less than if it achieves steady state quickly; accordingly the health risk would be considerably less. To compute how the airborne TCE concentration varies with time, the shower stall can be modeled as a well-mixed chamber. Volatilization involves a mass transfer resistance in both the liquid and air phase. For volatile compounds, the liquid-phase resistance is much larger than the air-phase resistance. Consequently the molar transfer rate of TCE per unit area (N_{TCE}) can be obtained from evaporation rate equations derived in Chapter 4,

$$N_{TCE} = K_L\left(c_{molar,TCE} - \frac{P_{TCE}}{H}\right) = K_L\left(x_{TCE}c_{molar,water} - \frac{P_{TCE}}{P}\frac{P}{H'}\frac{H'}{H}\right) = K_L\left(x_{TCE} - \frac{y_{TCE}}{H''}\right)c_{molar,water}$$

which can be multiplied by A_s to yield the molar transfer rate (mols/min), where A_s is the total surface area of water, i.e. shower stream, drops, and pool at the bottom of the shower stall. The TCE concentration in the water decreases only slightly and it is assumed that $x_{TCE} \cong$ constant. A mass balance can be written for TCE in the shower stall,

$$c_{molar,air}V\frac{dy_{TCE}}{dt} = c_{molar,water}K_LA_s\left(x_{TCE} - \frac{y_{TCE}}{H''}\right) - Q_{air}c_{molar,air}y_{TCE}$$

or

$$\frac{dy_{TCE}}{dt} = B - Ay_{TCE}$$

where

$$A = \frac{Q_{air}}{V} + \frac{c_{molar,water}K_LA_s}{H''c_{molar,air}V} \qquad B = \frac{c_{molar,water}K_LA_sx_{TCE}}{c_{molar,air}V}$$

The above ODE is of the same form as that of Eq. (5-7), but in terms of $y_{TCE}(t)$ rather than c(t). Substitution of data into the above yields

$$A = \frac{0.050 \frac{m^3}{min}}{0.20\ m^3} + \frac{55.6 \frac{mol}{L}\left(10. \frac{L}{min}\right)}{0.32\left(0.0409 \frac{mol}{m^3}\right)0.20\ m^3} = 2.12 \times 10^5 \frac{1}{min}$$

and

$$B = \frac{55.6 \frac{mol}{L}\left(10. \frac{L}{min}\right)\left(4.11 \times 10^{-10}\right)}{\left(0.0409 \frac{mol}{m^3}\right)0.20\ m^3} = 2.79 \times 10^{-5} \frac{1}{min}$$

From Eq. (5-11), the steady-state mol fraction is

$$y_{TCE,ss} = \frac{B}{A} = 1.32 \times 10^{-10} \cong 1.3 \times 10^{-4}\ \text{PPM}$$

which agrees nicely with the equilibrium mol fraction of TCE in air calculated in Part (d) above, and corresponds to a TCE mass concentration of 0.705 $\mu g/m^3$, which is almost 800,000 times smaller than the PEL! From Eq. (5-16), the time constant is

$$\tau = \frac{1}{A} = 4.71 \times 10^{-6}\ \text{min} \cong 2.8 \times 10^{-4}\ \text{s}$$

The solution is thus given by Eq. (5-15), since $y_{TCE}(0)$ is assumed to be zero (no TCE at the start of the person's shower),

$$y_{TCE}(t) = y_{TCE,ss}\left[1 - \exp(-At)\right]$$

but since the time constant is so small, it is clear that the steady-state TCE mol fraction occurs nearly instantly. Consequently the individual in the shower stall experiences the steady-state TCE concentration for the entire duration of the shower. The mass of TCE inhaled during the 10-minute shower is

$$m_{TCE,inhaled} = c_{TCE,air} Q_{inhale} t = \left(0.705 \frac{\mu g}{m^3}\right)\left(32.2 \frac{L}{min}\right)(10.\ min)\left(\frac{m^3}{1000\ L}\right) = 0.23\ \mu g$$

Using this more realistic assumption, the amount of TCE inhaled is about 26 times *less* than the amount ingested by drinking two liters of tap water.

Discussion: It is clear that the amount of TCE inhaled by individuals taking a 10-minute shower (0.23 μg) is considerably less than the amount of TCE ingested by drinking two liters of tap water (6.0 μg). Furthermore, the concentration of TCE in the air is orders of magnitude smaller than the PEL. Finally, as discussed in Chapter 3, PEL is prescribed for an 8-hour work day; the shower is almost 50 times shorter than this! In conclusion, there is no need for anxiety about TCE inhalation during a shower. Enjoy your shower; you might even wish to sing while doing so.

5.4 Time-Varying Source, Ventilation Flow Rate, or Make-up Air Concentration

In many practical applications, source strength (S), volumetric flow rate (Q), and/or make-up air concentration (c_a) are not constant, but are functions of time. Thus, even though the equation for conservation of mass can be written in the form of Eq. (5-7), it may not be possible to obtain an analytical solution to the ODE. Under these conditions Eq. (5-7) must be solved numerically. Some examples include:

- variable number of smokers in a room, S(t)
- leaky valve, faulty kerosene space heater with an unsteady source strength, S(t)
- erratic ventilation controller, Q(t)
- poorly designed duct system to supply make-up air, Q(t)
- emission of volatile organic compounds (VOC) within the home from waxes and cleaners for floors and furniture, moth balls and moth flakes, dry cleaned clothes, drapes, new carpets, furniture, wall paper, etc., S(t) (Tichenor and Mason, 1988)
- variable pollutant concentration in the make-up air, $c_a(t)$
- indoor radon, S(t) (Nazaroff and Teichman, 1990; Fleischer, 1988)

The next three examples show how to solve Eq. (5-7) for three situations engineers are apt to encounter in which the terms A and/or B are functions of time and take on certain form:

(a) Source (S) and/or volumetric flow rate (Q) are discontinuous ***step functions*** (changing with time, but remaining constant over specified time intervals); Eq. (5-7) can be solved analytically for *each* time interval. [Example 5.4]
(b) Source (S) and/or volumetric flow rate (Q) are analytical functions of time; Eq. (5-7) is nonintegrable, and must be solved numerically, e.g. via the 4th-order Runge-Kutta marching scheme. [Example 5.5]
(c) Source (S) and/or volumetric flow rate (Q) are obtained from experiment; a curve-fitting scheme (e.g. cubic spline) is employed, followed by the 4th-order Runge-Kutta scheme. [Example 5.6]

5.4.1 A and B are Discontinuous Step Functions

Suppose the source of contamination is individuals who smoke at different times in a ventilated conference room. Thus there is no simple analytical function for the source term S(t), but it remains constant during each time interval in which the number of smokers is constant. By integrating Eq. (5-7) over each such time interval, the concentration can be predicted over a long period of time. The solution begins at t = 0 when the initial concentration is c(0). Coefficients A and B on the right hand side of Eq. (5-7) are calculated from Eqs. (5-8) and (5-9), and steady-state concentration, c_{ss}, is calculated from Eq. (5-11) based on the value of S during the first time interval. Since A and B are constants during this time interval, the analytical solution, Eq. (5-10), is valid. The concentration is thus known at any time during this first time interval; steady-state is not reached because the time interval is finite. At the end of the first time interval, the concentration at that time is used as a new initial value, A, B, and c_{ss} are recalculated for the new time interval, and the concentration is calculated for the second time interval. The calculations are repeated until the elapsed time has been spanned. A simple computer program to perform these repetitive calculations can be written in any of the common computer programming languages, such as BASIC, C, or FORTRAN. Alternatively, Excel, Mathcad or some other spreadsheet or mathematical program can be used. The same procedure can be applied if any of the other variables, e.g. Q, c_a, or V, are discontinuous step functions in time.

Example 5.4 - Conference Room with 100% Make-up Air

Given: A conference room of volume 33.3 m^3 contains 6 people who smoke at irregular times as follows, over a 90 minute period:

t (minutes)	number of smokers (n)
0 < t < 10	2
10 < t < 20	4
20 < t < 30	0
30 < t < 40	6
40 < t < 50	4
50 < t < 90	0

Assume that the smoke particles are emitted at a rate of 1100 μg/min while each cigarette is smoked (Repace and Lowery, 1980). A ventilation system removes air from the room at a rate of 6.3 m^3/min and replaces it with ambient air at the same rate. Such a condition corresponds to 100% make-up air and 0% recirculated air. The smoke concentration in the ambient air and the initial smoke concentration in the room are both 20. μg/m^3.

To do: Assuming well-mixed conditions, compute and plot smoke concentration as a function of time for the 90-minute period. Also compute the maximum concentration and the average concentration. Compare the average concentration to the ACGIH standard of 3 mg/m^3 (3000 μg/m^3) for "***not otherwise classified***" (***NOC***) respirable particles (particles with diameter D_p less than about 2 or 3 μm).

Solution: The source S(t) is a step function of time, but volume flow rate (Q) and ambient concentration (c_a) are constant. The equation for conservation of mass of the smoke is given by Eq. (5-7), and the problem is solved analytically during each time interval as discussed above. The authors solved this problem with Excel, the results of which are shown in Figure E5.4. The Excel file for this problem is also available on the book's Internet site. The maximum smoke concentration is around 920 μg/m^3, and the average smoke concentration over the period is around 330 μg/m^3, both of which are within the ACGIH standard.

Discussion: A ventilation rate of 6.3 m^3/min corresponds to 37. ACFM/person, similar to the old standards of 35 ACFM/person for public facilities containing smokers. The ASHRAE 62-1989 standard calls for 60 ACFM per person for an office smoking lounge. Cain et al. (1983) report that 75% of a sample (mixed smokers and non smokers) found that 35 ACFM per person reduced tobacco odor to an acceptable level. Similar results were obtained by Leaderer et al. (1984), who also found that this ventilation rate was adequate to maintain safe levels of CO. Figure E5.4 shows that a ventilation rate of 37. ACFM/person results in particulate concentrations that exceed the EPA 24-hr and Annual Primary Air Quality Standard (260 μg/m^3). The reader must keep in mind, however, that the EPA standard is an *outdoor* standard that pertains to continuous exposure of long duration. The

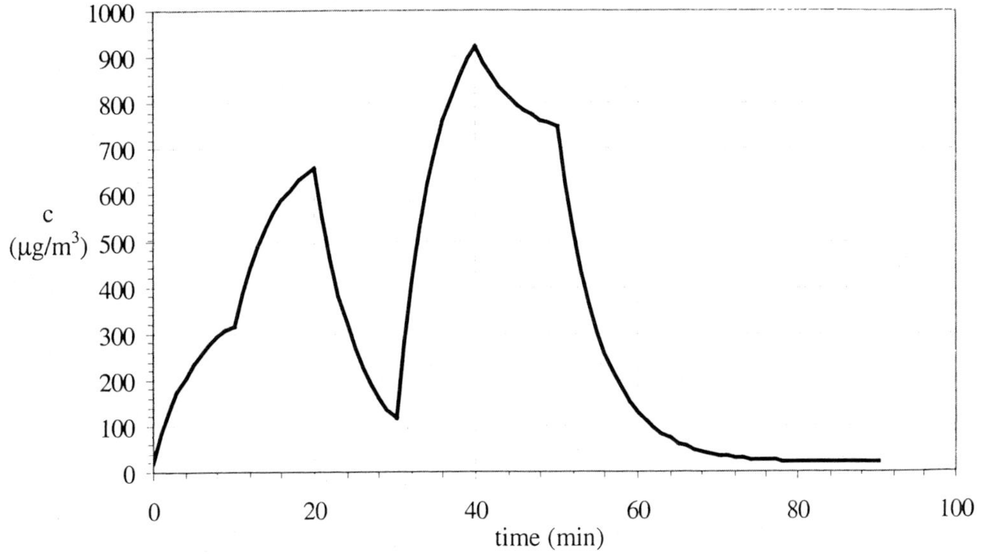

Figure E5.4 Particulate mass concentration (smoke particles) due to smokers in a room as a function of time.

EPA standard is designed to protect people "at risk," whereas OSHA and ACGIH standards pertain to short duration exposure designed to protect workers enjoying good health. An alternative method to the above is to solve Eq. (5-7) numerically, using a marching scheme such as the modified Euler method (Heinsohn and Kabel, 1999).

5.4.2 A and B are Nonintegrable Functions

Consider the situation in which any combination of the source (S), volumetric flow rate (Q), or inlet ambient concentration (c_a) is a function of time such that a closed form solution is not feasible. Equation (5-7) is still valid, but contains non-constant coefficients A = A(t) and/or B = B(t). The differential equation can be solved numerically using the 4th-order Runge-Kutta approximation. (see Appendix A.12 for details). Mathcad contains a function that executes the 4th-order Runge-Kutta scheme with a minimal amount of input data. The following example illustrates the technique.

Example 5.5 - Clever Outdoorsman

Given: An outdoorsman with a bent toward engineering lives in a small cabin 12. ft x 12. ft x 8.0 ft (V = 32.7 m^3). The cabin is heated by a single kerosene space heater that is not vented to the outside. Outside air leaks into the cabin (infiltration) at the rate of 0.30 room air changes per hour (N). Air inside the cabin escapes (leaks) to the atmosphere at the same rate. The concentration of CO in the outside air is 10. PPM (11.4 mg/m^3). The outdoorsman considers himself to be clever, and designs a thermostat for the kerosene heater. Without knowing it, he inadvertently alters the combustion process so that CO is emitted in the following fashion:

$$S(t) = 1500\left[1 + \sin\left(0.80\pi t\right)\right]$$

where the units of S are mg of CO per hour, and t is measured in hours.

Hemoglobin in the blood has a stronger affinity for CO than for oxygen. Carbon monoxide absorbed by hemoglobin produces carboxyhemoglobin (COHb), while oxygen absorbed by hemoglobin produces oxyhemoglobin (O_2Hb). Both quantities can be measured. When the ratio of (COHb) to (O_2Hb) in the blood exceeds 10%, a person is apt to suffer a headache and experience reduced mental acuity (see Chapter 2). If exposure continues, a person is apt to die of asphyxiation. For purposes of analysis, it is assumed that at any instant the ratio of carboxyhemoglobin to oxyhemoglobin in the blood is linearly related to the concentration of CO in the air,

$$R = \frac{\text{COHb}}{\text{O}_2\text{Hb}} \approx \frac{\text{CO (in PPM)}}{1000}$$

To do: Compute the concentration of CO in the air as a function of time. Also compute the maximum value of R. Will the clever outdoorsman suffer from CO poisoning? Will he survive the night?

Solution: The equation for conservation of mass of CO in the cabin is given by Eq. (5-7). Since the source strength (S) is a function of time, the problem is solved numerically. The authors solved this problem using the fourth-order Runge-Kutta function in Mathcad, the results of which are shown in Figure E5.5. The maximum concentration of CO is around 160 mg/m^3, which corresponds to

$$\text{PPM}_{\text{CO}} = 24.5\frac{c_{\text{CO}}\left(\frac{\text{mg}}{\text{m}^3}\right)}{M_{\text{CO}}} = 24.5\frac{160}{28.01} = 140 \text{ PPM}$$

which yields a ratio of carboxyhemoglobin to oxyhemoglobin in the blood of

$$R = \frac{\text{CO (in PPM)}}{1000} = \frac{140}{1000} = 0.14 = 14.\%$$

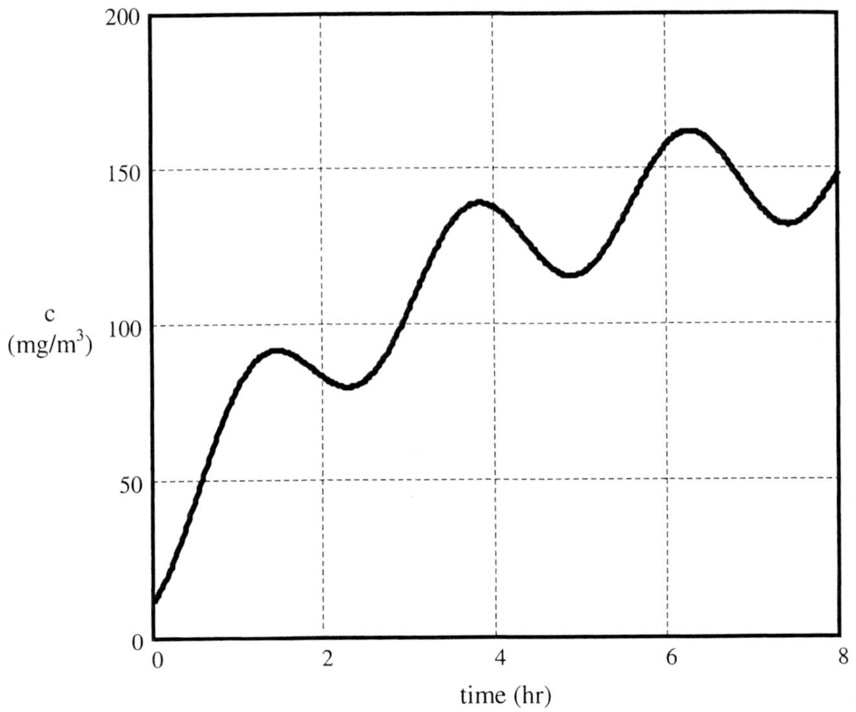

Figure E5.5 CO mass concentration as a function of time in a cabin with a faulty heater.

According to CO data in Chapter 2, the outdoorsman will experience a headache and some reduced mental acuity at these CO concentrations, but he is not in danger of collapsing until the concentration of CO reaches about 300 PPM. He will survive the night.

Discussion: If R exceeds 5%, individuals experience serious changes in cardiac and pulmonary function. Concentrations in excess of 80% may result in death. This outdoorsman will display symptoms of CO poisoning within a few hours, but will survive; the thermostat will fortunately not be the last device he ever designs! The authors used Mathcad, and the Mathcad file for this problem is available on the book's web site; it is also possible to solve this problem with a spreadsheet program such as Excel.

5.4.3 A and B are Obtained from Experimental Data

Consider a situation in which the source (S), volumetric flow rate (Q), and/or inlet ambient concentration (c_a) are experimental tabulated data (functions of time). Suppose also that it is necessary to use these data to solve Eq. (5-7) in order to predict the contaminant concentration as a function of time. Mathcad contains several functions which generate curve fits of tabulated data. One such function is the cubic spline fit, which connects a smooth curve through each data point. Once created, the curve fit function can be used within the 4th-order Runge-Kutta scheme to solve the equation.

Example 5.6 - Renovated Conference Room with 100% Make-up Air

Given: The room in Example 5.4 is renovated by adding a carpet and replacing the ventilation fan with one whose output volumetric flow rate unfortunately varies with time. The concentration of the hydrocarbon in the make-up air entering the room is 10. mg/m^3. The new carpet is secured to the floor by an adhesive that emits an objectionable hydrocarbon (HC) odor whose mass emission rate varies with time. Both S and Q are measured over a period of nine hours. The results are tabulated below:

t (hr)	S (mg/hr)	Q (SCFM)
0	80	120
0.5	112	140
1.0	220	150
1.5	240	140
2.0	240	120
2.5	160	60
3.0	140	42
3.5	120	40
4.0	100	45
4.5	80	60
5.0	80	80
5.5	77	85
6.0	75	90
6.5	73	95
7.0	71	97
7.5	70	98
8.0	70	99
8.5	70	100
9.0	70	100

To do: Predict and plot the hydrocarbon mass concentration in the conference room as a function of time. At what time does the mass concentration reach its peak, and what is the peak value?

Solution: The appropriate equation for conservation of mass of the hydrocarbon is Eq. (5-7). Since the source strength (S) and volume flow rate (Q) are functions of time, the problem is solved numerically. The authors used the cubic spline function in Mathcad to generate curve fits to replicate the tabulated empirical values of source strength (S) and volumetric flow rate (Q). Then, the Runge-Kutta function was employed, the results of which are shown in Figure E5.6. The peak hydrocarbon concentration is about 6.8 mg/m^3, which occurs at about 3.5 hours.

Discussion: From the plot of concentration versus time, it is seen that although the source strength reaches its maximum value in less than 2 hours, the concentration does not reach its peak until around 3.5 hours. This fact could not have been estimated by any means other than numerical integration.

5.5 Removal by Solid Surfaces

In the preceding sections it was presumed that the exhaust stream was the only way contaminants were removed from the enclosed spaces, i.e. that there were no ***wall losses***. In the case of tobacco smoke, anyone who has cleaned house knows that various surfaces in a home adsorb tobacco smoke and odors. Furthermore, when smoking ceases, desorption occurs so that the room reacquires an odor of tobacco. Adsorption such as this occurs for many combinations of contaminant and surface material. The adsorption of contaminants on walls is also called ***plate-out***. As a crude first approximation, one may assume that the mass removal rate (mass/time) is equal to the product of an effective surface area (A_s), the mass concentration (c) and a ***wall loss coefficient*** (k_w), also called a ***plate-out coefficient***. Since this coefficient has units of length per time, some authors call k_w a ***deposition velocity***. The statement may also be phrased that plate-out is *first order* with respect to the concentration. If the contaminant is a gas or vapor, plate-out occurs by the physical process called ***adsorption*** and k_w is truly an ***adsorption coefficient***, sometimes called an ***adsorption rate constant***. If however the contaminant is in the form of aerosol particles, removal occurs because particles travel to

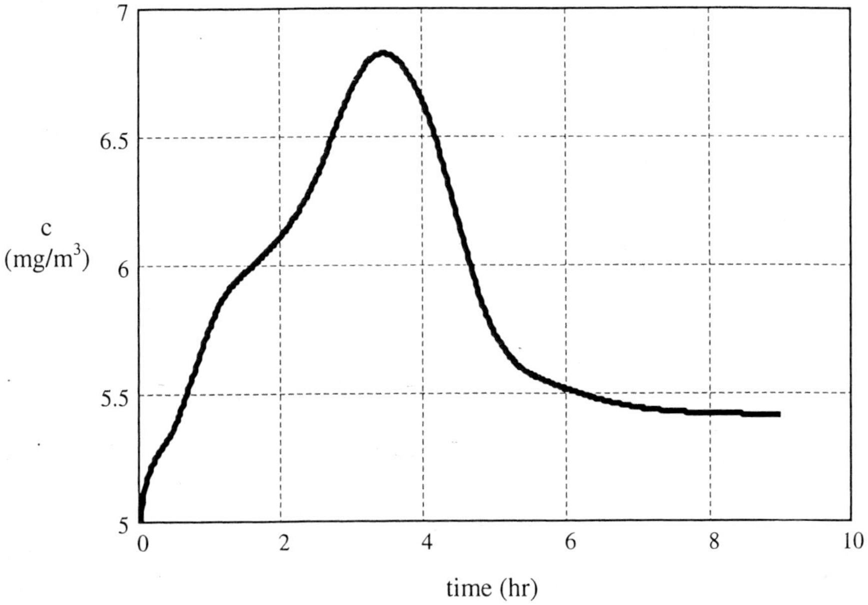

Figure E5.6 Hydrocarbon mass concentration as a function of time in a renovated conference room.

a solid surface with a unique deposition velocity, strike it and adhere to it. k_w is a parameter that can be expressed as a function of the gas transport properties, or it may be determined experimentally. The magnitude of k_w depends on:

- orientation of the surface, e.g. vertical, upward-facing horizontal, downward-facing horizontal
- particle size
- natural convection for enclosure surfaces that have a different temperature than the gas
- bulk motion and homogeneous turbulence in the gas

For the deposition of gases within an enclosure, Cano-Ruiz et al. (1993) suggest values for k_w of 0.20 cm/s for well-mixed conditions and 0.070 cm/s when the air motion is governed by natural convection. For ozone that plates-out on indoor residential surfaces, Reiss et al. (1994) report deposition velocities from 0.023 to 0.041 cm/s.

Crump and Seinfeld (1981), Nazaroff and Cass (1989), and Xu et al (1994) recommend the following expressions to estimate plate-out coefficients (k_w) for small particles:

- walls:

$$\boxed{k_w = \frac{2}{\pi}\sqrt{D K_e}} \tag{5-17}$$

- ceiling:

$$\boxed{k_w = \frac{v_t}{\exp\left[\dfrac{\pi}{2}\dfrac{v_t}{\sqrt{D K_e}}\right] - 1}} \tag{5-18}$$

- floor:

$$k_w = \frac{v_t}{1-\exp\left[-\frac{\pi}{2}\frac{v_t}{\sqrt{D K_e}}\right]} \tag{5-19}$$

where v_t is the settling velocity, D is the diffusion coefficient from Eq. (4-26), and K_e is the turbulent intensity. The turbulent intensity depends on the air velocity field. Xu et al., (1994) suggest that K_e can be set equal to 0.26 s^{-1} if room air movement is governed by natural convection and upwards to 0.45 s^{-1} if air movement is governed by a vigorous circulating fan. For particles above the respirable range (D_p greater than 2 or 3 μm), gravimetric settling (sedimentation) on upward-facing horizontal surfaces is the dominant mechanism. In this case, k_w can be set equal to the gravimetric settling velocity (Chapter 8). Particles smaller than this may be removed nearly equally by surfaces of any orientation since the processes of thermophoresis, Brownian motion, eddy diffusion, and bulk motion may be as important, or more important than gravity.

Consider the enclosed volume (V) shown in Figure 5.5 containing a contaminant source and inside surfaces that remove contaminants by adsorption. Conservation of mass for the contaminant can be written (Nazaroff and Crass, 1989; Dunn and Tichenor, 1988) as

$$V\frac{dc}{dt} = Qc_a + S - cQ - cA_s k_w \tag{5-20}$$

where A_s is the surface area of the respective surface (wall, ceiling, floor), and V is the volume of the enclosure. Equation (5-20) can be re-written in standard form, i.e. in the form of Eq. (5-7), with

$$A = \frac{Q + A_s k_w}{V} \qquad B = \frac{Qc_a + S}{V} \tag{5-21}$$

Note that Eq. (5-21) for B is identical to that defined previously, in Eq. (5-8), but A in Eq. (5-21) has an extra term in the numerator compared to the previous definition, Eq. (5-9), which accounts for wall losses. For constant values of Q, S, and k_w, and for initial condition c(0), the solution of the ODE is the same as previously, Eqs. (5-10) and (5-11). The steady-state concentration is found by substitution of Eqs. (5-21) into Eq. (5-11),

$$c_{ss} = \frac{B}{A} = \frac{Qc_a + S}{Q + A_s k_w} \tag{5-22}$$

Wall losses are seen to be a removal mechanism working in parallel but independently of the exhaust ventilation rate. Both removal mechanisms reduce the steady-state concentration, and both enable steady state to be reached more rapidly.

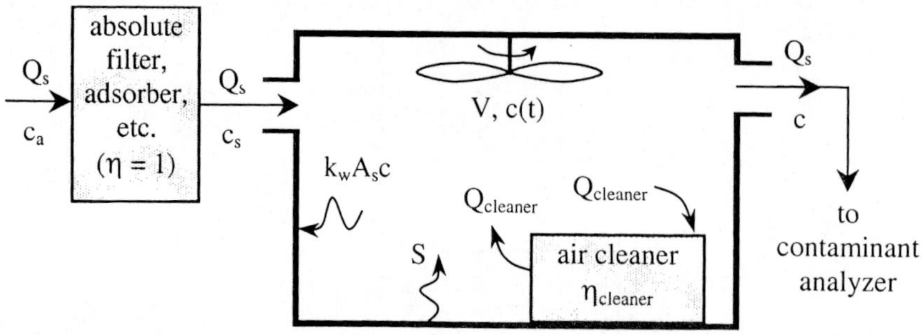

Figure 5.5 Test apparatus to measure source strength, chamber wall loss (plate-out), and performance of air-cleaning device.

To illustrate the importance of adsorption of tobacco smoke, consider the results of experiments of Repace and Lowery (1980), shown in Figure 5.6. A single cigarette was burned steadily and then extinguished in an isolated, small room (V = 22. m^3). The total mass of suspended particulate matter was measured during the entire period. In one experiment, well-mixed conditions were produced by fans within the room. In another experiment, the only mixing was by natural air currents within the isolated room. The emission of smoke is expressed by

$$\boxed{S = n\frac{m_p}{t_b}} \tag{5-23}$$

where n is the number of cigarettes, t_b is the length of time burning occurred, and m_p is the total mass of particle matter produced per cigarette determined by tests conducted by the Federal Trade Commission (FTC).

The effect of mixing is clearly seen in Figure 5.6. Following extinction of the cigarette, the concentration in the well-mixed experiment falls rapidly. The slope of the curve enables one to estimate the removal of contaminants due to adsorption by solid surfaces. For the case in which Q and S are zero, c_{ss} is also zero by Eq. (5-22). Letting the initial concentration c(0) be denoted as c_{max}, setting $c_{ss} = 0$, and substituting Q = 0 in Eq. (5-21), Eq. (5-10) reduces to

$$\boxed{\frac{c(t)}{c_{max}} = \exp(-At) = \exp\left(-\frac{A_s k_w}{V} t\right)} \tag{5-24}$$

Repace and Lowery (1980) found that the quantity ($V/A_s k_w$) has a value of 10. min which corresponds to an adsorption rate constant (k_w) of 0.078 cm/s, which is in general agreement with Cano-Ruiz et al. (1993). If the surface area of the room (A_s) is 30. m^2, the rate of adsorption is equivalent to an exhaust ventilation flow rate of 1.4 m^3/min (50. ACFM). Thus adsorption in this small room lowered the concentration in a fashion equivalent to a ventilation flow rate of 50. ACFM or 4.2 room air changes per hour.

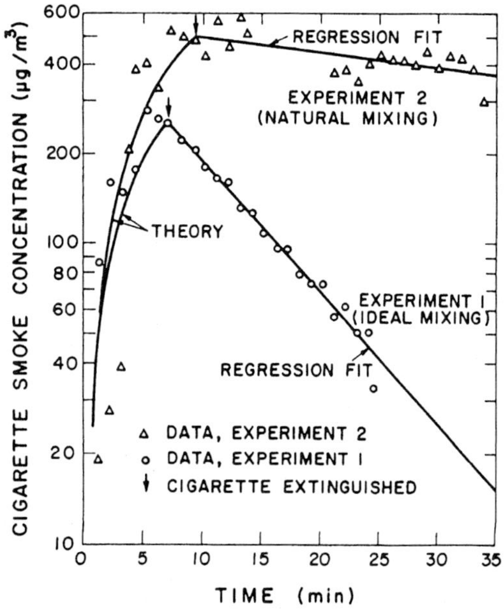

Figure 5.6 Smoke concentration from one smoldering cigarette in an isolated room (22 m^3) with natural mixing and ideal internal mixing (redrawn from Repace and Lowery, 1980).

Contaminants may also *desorb* from a surface. If this occurs, ***desorption*** becomes a source term in Eq. (5-20). The rate at which contaminants desorb [mass/(time area)] from a surface is expected to be proportional to the concentration of contaminant on the surface of the material (mass/area). In such a case, the ***desorption rate constant*** (k_{desorb}) has units of length/time and needs to be determined from experiment.

Up to this point, it has been presumed that the contaminant was removed in a first-order process described by a single differential equation, Eq. (5-20). It may be argued that the way in which wall losses and the source strength are modeled is too simplistic and that more refined expressions are needed. If refined expressions exist, they can be included in the differential equations for the well-mixed model without fundamentally changing the methods used to solve them. For example, if the contaminant consists of a range of submicron particles, separate equations have to be written for each range of particles since deposition velocities depend on particle diameter. In addition, particles may coagulate to form a single particle of a new size. Thus the physical process of ***coagulation*** becomes a significant removal process. The coagulation process involves a removal term proportional to the product of particle concentrations, i.e. $c(D_{p,1})c(D_{p,2})$. Thus instead of a single equation of the form of Eq. (5-20), one now needs to write a set of simultaneous equations, one equation for each particle size, and solve a coupled set of nonlinear ordinary differential equations. This can be done easily with the Runge-Kutta technique.

5.6 Recirculation

Supply air flow rates in HVAC systems are usually dictated by heating, cooling, and dehumidification requirements rather than by ventilation requirements. Make-up air requirements are established separately by code (e.g. by reference to ASHRAE Standard 62). Because the cost of heating or cooling make-up air is high, only rarely is there 100% make-up air in a ventilation system unless separate systems are used to provide ventilation air and space conditioning. Instead, only a portion of the supply air in most applications is fresh make-up air; the remainder is ***recirculated air*** that has been cleaned. Consider Figure 5.7 which shows an enclosed space of volume V in which a fraction (f) of the return air flow is fresh make-up air (Q_m), and the remainder (Q_r) is recirculated air. The contaminant concentration in the supply air is reduced by an air cleaning device.

The ***cleaning efficiency*** (η) of the air-cleaning device is defined as the ratio of the rate of removed mass to the rate of incoming mass,

$$\eta = \frac{\text{rate of mass of material removed by the device}}{\text{rate of mass of material entering the device}}$$

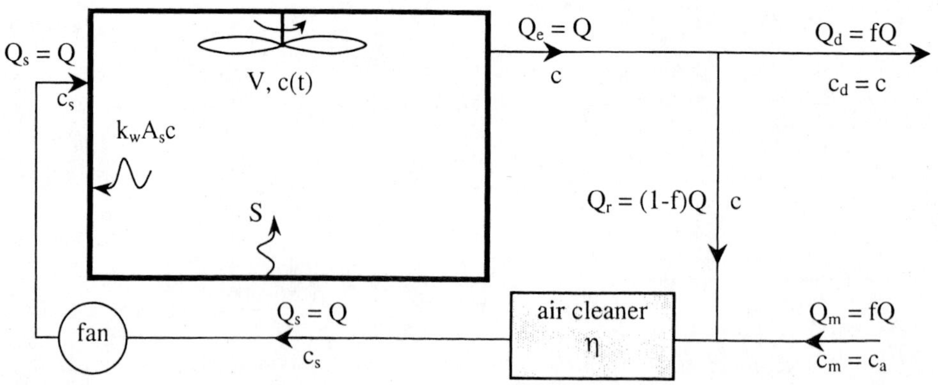

Figure 5.7 Schematic diagram of a ventilation system with recirculation and make-up air.

or

$$\eta = \frac{Qc_{in} - Qc_{out}}{Qc_{in}} = \frac{c_{in} - c_{out}}{c_{in}} = 1 - \frac{c_{out}}{c_{in}} \tag{5-25}$$

In engineering practice, the cleaning efficiency is obtained from the air-cleaning equipment manufacturer, and one needs to solve for c_{out}. From Eq. (5-25),

$$c_{out} = (1-\eta)c_{in} \tag{5-26}$$

For the case of *particulate* contaminants, the air cleaner is referred to as a ***filter***, and η depends also on particle size. Particle filtration is discussed in more detail in Chapters 8 and 9.

Since the absolute pressure and temperature are virtually constant at all points in Figure 5.7, it is reasonable to assume that the air density is also constant at all points, i.e. $\rho_m = \rho_r = \rho_s = \rho$(enclosed space) = constant. In steady-state conditions, the mass flow rate of air into and out of the enclosure is the same, and since air density is constant, $Q_s = Q_e = Q$. Application of conservation of mass of air at the junction of the make-up and return air upstream of the air cleaner yields

$$Q_r = Q_e - Q_m = Q_e(1-f) = Q(1-f) \tag{5-27}$$

The factor (f) is the ***make-up air fraction***, defined as the ratio of make-up air to supply air:

$$f = \frac{Q_m}{Q_e} = \frac{Q_m}{Q} \tag{5-28}$$

Note also that to conserve mass, the volume flow rate of air discharged into the outdoors (Q_d) must equal that of the make-up air (Q_m). Assuming no wall losses, and applying conservation of mass to the contaminant in the enclosed space of volume V,

$$V\frac{dc}{dt} = Q_s c_s + S - Q_e c = Qc_s + S - Qc \tag{5-29}$$

where c_s is the mass concentration in the ***supply air*** entering the enclosure. The value of c_s can be found by using the definition for air cleaner efficiency, Eq. (5-25),

$$\eta = 1 - \frac{c_{out}}{c_{in}} = 1 - \frac{Qc_s}{cQ_r + fQc_a}$$

which when solved for c_s yields

$$c_s = (1-\eta)\left[fc_a + c(1-f)\right] \tag{5-30}$$

Substitution of c_s into Eq. (5-29) yields the following first-order differential equation:

$$V\frac{dc}{dt} = S + Qfc_a(1-\eta) - Q\left[1-(1-f)(1-\eta)\right]c \tag{5-31}$$

which once again can be written in the standard form of Eq. (5-7), with A and B defined by

$$A = \frac{Q}{V}\left[1-(1-f)(1-\eta)\right] \qquad B = \frac{S + Qfc_a(1-\eta)}{V} \tag{5-32}$$

As with all of the previous ventilation problems, if A and B are constant, an analytical solution is obtainable from Eq. (5-10), with a steady-state concentration (c_{ss}) inside the enclosed volume obtained from Eq. (5-11),

$$\frac{c_{ss} - c(t)}{c_{ss} - c(0)} = \exp[-At] = \exp\left[-\frac{Q}{V}\left[1-(1-f)(1-\eta)\right]t\right] \tag{5-33}$$

which yields

$$\boxed{c_{ss} = \frac{B}{A} = \frac{S + Qf(1-\eta)c_a}{Q[1-(1-f)(1-\eta)]}} \tag{5-34}$$

Note that c_{ss} can also be found by setting dc/dt = 0 in the differential equation, Eq. (5-31).

If any of the parameters in Eqs. (5-32) is not constant, i.e. if Q, S, f, η, and/or c_a are functions of time, causing A and/or B to vary with time, Eq. (5-33) is no longer valid. In such a case, Eq. (5-7) must be integrated numerically as in previous examples.

Figure 5.7 is a schematic diagram of one particular ventilation system that utilizes recirculation; the location of the air cleaning devices and the location of fresh air make-up ducts can be arranged differently. For example, since the make-up air often contains no contaminant ($c_m = c_a \cong 0$), it is economical to place the air-cleaning device in the return leg carrying return volumetric flow rate Q_r, since Q_r is smaller than the supply volumetric flow rate Q_s. If the arrangement of inlet, recirculation ducts, and air cleaners is not the same as shown in Figure 5.7, Eqs. (5-29) through (5-34) are not necessarily valid. However, the concepts leading to their derivation are valid; the user needs to apply conservation of mass for the control volume encompassing the enclosure to derive a new differential equation. So long as the terms in the differential equation can be organized in the form of Eq. (5-7) in which parameters A and B can be defined, the solution can be obtained.

Example 5.7 - Hospital Operating Room

Given: The ventilation system for a hospital operating room is shown in Figure E5.7. The maintenance department inadvertently installs the exhaust from a chemical laboratory hood within a few feet of the make-up air inlet. A laboratory technician evaporates ethyl alcohol in the laboratory hood on a day in which an operation is in progress. The following parameters characterize the situation:

- A_s = total area of adsorbing surfaces in operating room = 85. m^2
- c(0) = initial alcohol concentration inside operating room = 3.0 mg/m^3
- c_a = constant concentration of ethyl alcohol entering the make-up air duct = 100. mg/m^3
- f = Q_m / Q_s = make-up air fraction = 0.90
- k_w = wall loss rate coefficient = 0.10 cm/s = 0.060 m/min
- Q = Q_e = Q_s = overall volumetric flow rate = volume flow rate of exhausted air = volume flow rate of supply air = 20. m^3/min
- Q_m = volumetric flow rate of make-up air = fQ
- Q_r = volumetric flow rate of recirculated air = (1-f)Q
- S = source strength = rate at which ethyl alcohol is vaporized inside operating room = 1000 mg/min
- V = volume of operating room = 50. m^3
- η_1, η_2 = efficiencies of the activated charcoal filters = 0.95

To do: Compute the steady-state concentration of ethyl alcohol that will eventually exist in the operating room and the length of time before the operating personnel begin to smell the alcohol. The odor threshold for ethyl alcohol is 10 PPM (19 mg/m^3) (Leonardos et al., 1969).

Solution: Assuming the operating room to be well mixed, a mass balance for the alcohol in the operating room results in the following first-order ODE:

$$V\frac{dc}{dt} = Q_s c_s + S - Q_e c - A_s k_w c$$

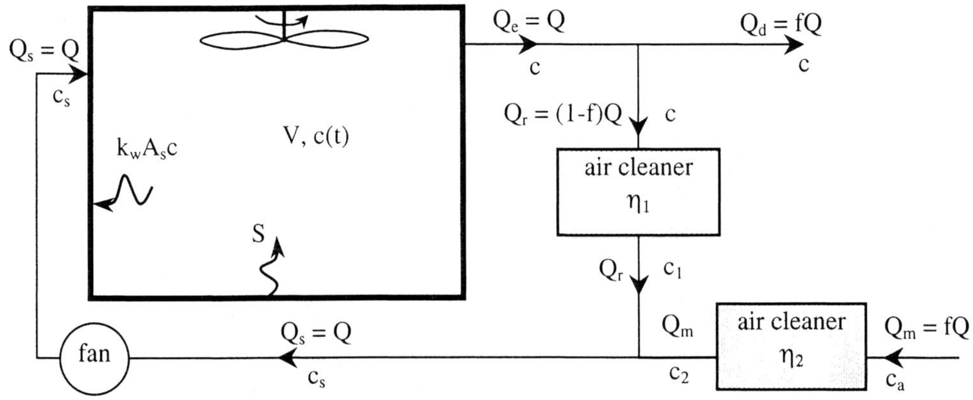

Figure E5.7 Schematic diagram of a ventilation system for a hospital operating room.

A mass balance for the air yields

$$Q_s = Q_m + Q_r \qquad Q_e = Q_s$$

At the inlet to the fan, a mass balance can be written for the alcohol,

$$c_a Q_m (1-\eta_2) + c Q_r (1-\eta_1) = c_s Q_s$$

From the definition of (f), the concentration (c_s) can be written as

$$c_s = c_a f(1-\eta_2) + c(1-\eta_1)(1-f)$$

Upon substitution, the differential equation can be arranged in standard form, i.e. in the form of Eq. (5-7), with

$$A = \frac{Q_s + A_s k_w - Q_s(1-\eta_1)(1-f)}{V}$$

$$= \frac{20.\frac{m^3}{min} + (85.\ m^2)0.060\frac{m}{min} - 20\frac{m^3}{min}(1-0.95)(1-0.9)}{50.\ m^3} = 0.50\frac{1}{min}$$

and

$$B = \frac{S + fQ_s c_a (1-\eta_2)}{V} = \frac{1000\frac{mg}{min} + 0.9\left(20.\frac{m^3}{min}\right)100\frac{mg}{m^3}(1-0.95)}{50.\ m^3} = 21.8\frac{mg}{min \cdot m^3}$$

The steady-state concentration is thus

$$c_{ss} = \frac{B}{A} = 43.6\frac{mg}{m^3}$$

Finally, to find the time until the alcohol is smelled, Eq. (5-12) is used with $c(t) = 19.\ mg/m^3$ (the odor threshold) and with the given value of $c(0) = 3.0\ mg/m^3$,

$$t = -\frac{1}{A}\ln\left[\frac{c_{ss} - c(t)}{c_{ss} - c(0)}\right] = -\frac{1}{0.50\frac{1}{min}}\ln\left[\frac{43.6 - 19.}{43.6 - 3.0}\right] = 1.0\ min$$

Discussion: The answer is given to two significant digits. The hospital workers in the operating room should smell the alcohol in about a minute, assuming well-mixed conditions in the room.

5.6.1 - Contaminant Concentrators

To convey powders and dry products in the bulk powder industry, it is useful to use pneumatic means. Consider for the moment foodstuffs, dry chemicals, pharmaceuticals, etc., indeed any industry in which it is necessary to move large quantities of powder material cheaply over considerable distances without contaminating the powder or allowing powder to enter the workplace. In order to process the powder, it is necessary to separate it from the air stream. Devices that separate powder from air are called ***concentrators***. Concentrators bleed a portion of the inlet flow and increase the concentration in the bled air; a straight-through cyclone (see Chapter 9) can be used for this purpose. The ***bleed ratio*** (b) is the ratio of the bled air volumetric flow rate to the volumetric flow rate input to the concentrator,

$$\boxed{b = \text{bleed ratio} = \frac{\text{bleed air flow rate}}{\text{input air flow rate to the concentrator}}} \tag{5-35}$$

Concentrators are often used in combinations to achieve a high degree of separation.

Figure 5.8 represents an unusual application of recirculation used in industries that manufacture powder material. In such industries, the objective is to *concentrate* rather than *remove* the powder. Powders are transferred pneumatically from the workstations to the first concentrator (with collector efficiency η_1) whereupon most of the powder is separated from the air stream for further processing. The discharge from the first separator is added to the inlet of a second concentrator (with collector efficiency η_2), with the objective of further concentrating the powder. To analyze the process of separation systems, engineers employ the same concepts of conservation of mass as used in the previous examples of pollution control systems. The difference in analysis is a matter of emphasis. Pollution control emphasizes minimizing the concentration in the discharge air stream with little concern about how the collected particles are handled. Concentrator technology seeks to maximize collection of the powder, expressed as a mass flow rate ($\dot{m}_c$), while using as little air as possible.

Example 5.8 - Gold Dust Recovery System

Given: A system is to be designed to recover gold dust generated by a buffing wheel used in making gold jewelry. To maximize recovery, a feedback system is proposed which consists of a particle concentrator of efficiency (η_1) and a second particle concentrator (particle collector) of efficiency (η_2), as shown in Figure 5.8.

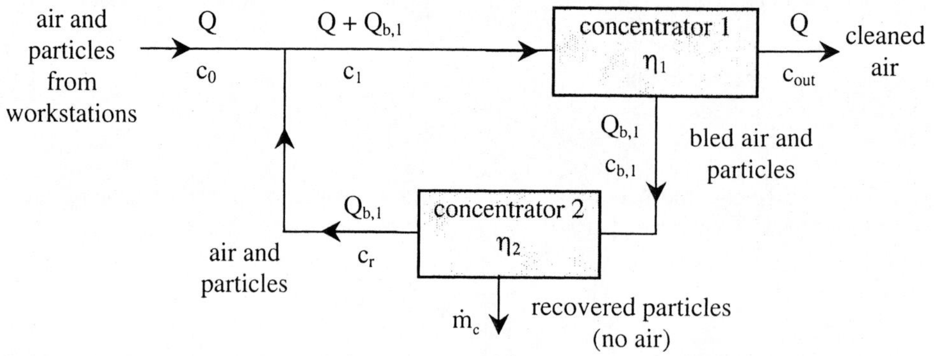

Figure 5.8 Schematic diagram of a particle recovery system for the powder manufacturing industry.

To do: Derive an expression for the mass recovery rate ($\dot{m}_c$) in terms of the inlet conditions, bleed ratio, and efficiencies, and show that the overall particle collection efficiency for the system (η_0) is equal to

$$\eta_0 = \frac{\eta_1 \eta_2}{1 - \eta_1 + \eta_1 \eta_2}$$

Solution: From an overall point of view, the two collectors can be combined into one, with an overall efficiency equal to η_0. From the definition of collection efficiency, Eq. (5-25),

$$\eta_0 = \frac{\text{rate of mass removed}}{\text{rate of mass entering}} = \frac{\dot{m}_c}{Qc_0}$$

The mass recovery rate can thus be expressed as

$$\dot{m}_c = \eta_0 c_0 Q$$

where Q is the volumetric flow rate entering the system. From Eq. (5-35), the concentrator bleed ratio is

$$b_1 = \frac{Q_{b,1}}{Q_{b,1} + Q}$$

or, solving for $Q_{b,1}$,

$$Q_{b,1} = \frac{b_1 Q}{1 - b_1}$$

The efficiencies of collector 1 (also called a concentrator) and 2 (a conventional particle collector such as a filter or high-efficiency cyclone) are defined by Eq. (5-25). Substituting data from Figure 5.8, the efficiencies can be expressed as

$$\eta_1 = \frac{Q_{b,1} c_{b,1}}{Qc_0 + Q_{b,1} c_r} \qquad \eta_2 = \frac{\dot{m}_c}{Q_{b,1} c_{b,1}} = \frac{Q_{b,1} c_{b,1} - Q_{b,1} c_r}{Q_{b,1} c_{b,1}} = 1 - \frac{c_r}{c_{b,1}}$$

where $c_{b,1}$ is the particle concentration in the bleed flow leaving concentrator 1 and c_r is the particle concentration leaving concentrator 2. The volumetric flow rate of the bleed flow leaving concentrator 1 is $Q_{b,1}$. Since there is no bleed flow from collector 2, the volumetric flow rate entering and the volumetric flow rate leaving concentrator 2 are the same, and equal to $Q_{b,1}$. Solving for $c_{b,1}$ from the equation for η_1 above,

$$c_{b,1} = \frac{\eta_1 \left(Qc_0 + Q_{b,1} c_r \right)}{Qb_1}$$

Similarly, solving for $c_{b,1}$ from the equation for η_2 above,

$$c_{b,1} = \frac{c_r}{1 - \eta_2}$$

c_r can be found by equating the two equations above. After some algebra,

$$c_r = \frac{\eta_1 c_0 Q (1 - \eta_2)}{Q_{b,1} \left[1 - \eta_1 (1 - \eta_2) \right]}$$

From the equation for η_2 above,

$$\dot{m}_c = \eta_2 Q_{b,1} c_{b,1}$$

The above equations can be combined and simplified to yield

$$\dot{m}_c = \eta_2\eta_1\left(Qc_0 + Q_{b,1}c_r\right)$$

Further simplification yields

$$\dot{m}_c = \frac{\eta_1\eta_2 Qc_0}{1-\eta_1\left(1-\eta_2\right)}$$

Finally, the overall efficiency η_0 can be expressed as

$$\eta_0 = \frac{\dot{m}_c}{Qc_0} = \frac{\eta_1\eta_2}{1-\eta_1+\eta_1\eta_2}$$

Discussion: The equations in this example problem apply only to this particular configuration. However, the methods used here are applicable to a wide variety of recirculating air configurations.

5.7 Partially Mixed Conditions

Sections 5.3 to 5.6 are analyses in which the concentration is uniform throughout the enclosed space, although it may vary in time, i.e. *spatial* uniformity but not *temporal* uniformity. If the ventilation volumetric flow rate (Q), source strength (S), and adsorption rate (k_w) are constant, the mass conservation equations can be integrated in closed form. If these parameters vary with time, the equations can be integrated numerically. It must be emphasized that the notion of spatial uniformity is critical to the validity of the well-mixed model and the solutions that follow from it. Unfortunately in many situations, both spatial and temporal variations in concentration occur simultaneously, i.e. the enclosed space is not well mixed. Analysis of these situations is difficult since the equations of both mass and momentum transfer have to be solved simultaneously. Numerical computational procedures are available for this and are discussed in Chapter 10.

Over the years an alternative computational technique has arisen that many workers in indoor air pollution find useful. The technique employs using a scalar constant called a ***mixing factor*** (m) to modify the equations of the well-mixed model to account for non-uniform concentrations brought on by poor mixing. Consider the ventilated enclosed space with 100% recirculation shown in Figure 5.9. Other geometric configurations can be modeled in comparable fashion. Assuming well-mixed conditions and neglecting adsorption on the walls, the following expression for the contaminant can be written:

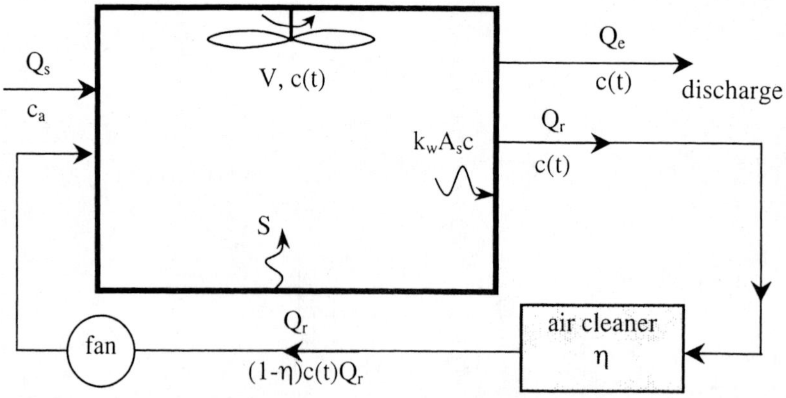

Figure 5.9 Schematic diagram of a typical ventilation system with 100% recirculation and separate make-up air.

$$V\frac{dc}{dt} = S + Qc_a - Qc + (1-\eta)Q_r c - Q_r c$$

which reduces to

$$\boxed{V\frac{dc}{dt} = S + Qc_a - Qc - \eta Q_r c} \tag{5-36}$$

To account for non-uniform mixing, mixing factor (m) is adopted, and Eq. (5-36) can be rewritten as

$$\boxed{V\frac{dc}{dt} = S + mQc_a - mQc - m\eta Q_r c} \tag{5-37}$$

Eq. (5-37) can be written in the standard form of Eq. (5-7) as usual, with

$$A = \frac{m(Q+\eta Q_r)}{V} \qquad B = \frac{S + mQc_a}{V}$$

If m, S, c_a, Q, Q_r, and η are constants, the ODE can be solved in closed analytical form using Eqs. (5-10) and (5-11),

$$\boxed{\frac{c_{ss} - c(t)}{c_{ss} - c(0)} = \exp[-At] = \exp\left[-\frac{m(Q+\eta Q_r)}{V}t\right]} \tag{5-38}$$

where

$$\boxed{c_{ss} = \frac{B}{A} = \frac{S + mQc_a}{m(Q+\eta Q_r)}} \tag{5-39}$$

Esmen (1978) states that values of m are normally 1/3 to 1/10 for small rooms and possibly less for large spaces. Table 5.1 contains values of m referenced by Repace and Lowery (1980). If m is less than unity, the concept of mixing factor suggests that a fraction of each flow, mQ and mQ_r, is well mixed while another fraction, (1 - m)Q and (1 - m)Q_r, bypasses the enclosure. Consequently

- $\underline{m = 1}$ implies well-mixed model and concentration that is spatially uniform
- $\underline{m < 1}$ implies nonuniform mixing and spatial variations in concentration

The parameter "m" is a discount rate or handicap factor. It implies that the enclosed space is a well-mixed region in which the effective ventilation rate is a fraction m times the actual volumetric flow rate. Conversely, the reciprocal of m could be called a "safety factor," i.e. the actual flow rate is equal to the well-mixed value times the safety factor.

The difficulty in selecting the proper value of "m" can be seen in Figure 5.10 taken from Ishizu (1980). Six cigarettes were allowed to smolder in the center of a room of volume 70. m^3. Ventilation consisted of 32. m^3/min of ambient air and 8.0 m^3/min of cleaned recirculated air. No information was given on the location of the inlet and outlet ducts. The concentration of smoke was measured in the center of the room before, during, and after the cigarettes were burned. During the

Table 5.1 Mixing factors (m) for various enclosed spaces.

enclosed space	**m**
perforated ceiling	1/2
trunk system with anemostats	1/3
trunk system with diffusers	1/4
natural draft and ceiling exhaust fans	1/6
infiltration and natural draft	1/10

burning phase, a steady-state concentration was not reached even though the well-mixed model predicted adequate time for one to occur. Figure 5.10 shows the sensitivity of the calculations of concentration on the choice of m. The maximum concentration exceeded the steady-state, well-mixed value (m = 1.0 in Figure 5.10) by a factor of about two, clearly indicating non-uniform conditions within the enclosed space. During the smoldering period, m ≈ 0.4, but after extinction, a single value of m could not explain the data; m appears to decrease with time from around 0.4 to less than 0.3.

Uniform mixing is synonymous with the well-mixed model. It is not possible to insert a constant, scalar multiplier into the equations for the well-mixed model and expect to acquire equations appropriate for non-uniform concentrations. There are several fundamental flaws in the concept of mixing factor:

- The principles of science governing the motion of air and contaminants do not justify the use of a scalar multiplier m.
- Experimental values of m are unique to the volumetric flow rates, geometry of the enclosed space, location of inlet and outlet duct openings, and location of the point where the contaminant is measured.
- The value of m cannot be predicted with any precision. Once it is found experimentally for a particular enclosure, it can't be generalized for other enclosed spaces.
- The range of values used for m is so large as to make it an ineffective parameter for design and economic analysis.

In the final analysis, modifying an equation based on the well-mixed model to account for non-uniform concentrations is a contradiction in terms. Either the concentration is uniform in space or it is not; and if it is not, no amount of fudging can yield meaningful answers. Nevertheless, arcane practices that have been used for a considerable time have a sizable following and are not going to be changed simply because they are illogical. The well-mixed model implies something concrete, i.e. c(x,y,z,t) = c(t). Non-uniform mixing and the concept of mixing factor mean that the equality does not

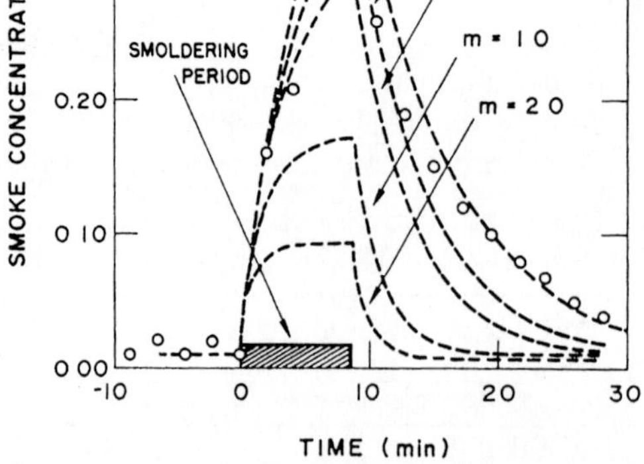

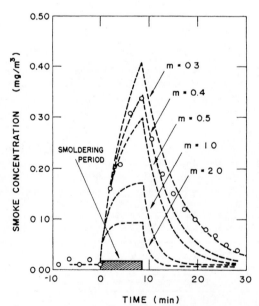

Figure 5.10 Comparison of measured smoke concentrations (circles) in a ventilated room (V = 71. m^3) with analytical predictions (dashed lines) for different mixing factors (m); Q = 32. m^3/min (redrawn from Ishizu, 1980).

hold, but the concept does not predict how, where, or in what way the concentration varies. The present authors recommend that the use of mixing factors be abandoned.

5.8 Well-Mixed Model as an Experimental Tool

While general ventilation is of narrow and limited use to control contaminants in the workplace, the well-mixed model is ideally suited as a laboratory technique to measure the following (Whitby et al., 1983; Donovan et al., 1987):

- wall-loss coefficient (k_w)
- contaminant emission rate, i.e. source strength (S)
- efficiency of room air cleaners (η_{room}), defined as the contaminant mass removal rate divided by the contaminant mass flow rate entering the cleaner

Consider the test apparatus shown schematically in Figure 5.5. Air inside the chamber is sampled at rate Q_s, and the concentration (c) of the contaminant (particle or gas) is measured by a suitable analyzer. Fresh air at the same volumetric flow rate (Q_s) is added to the chamber after passing through an "absolute filter" (η_m = 100%) or adsorber, etc. that removes all but a negligible amount of contaminant. Inside the chamber there is an internal room air-cleaning device with its own value of efficiency ($\eta_{cleaner}$). It is assumed that no contaminant enters the chamber except via the source (S) in the room. It is also assumed that there is no extraneous air entering or leaving the chamber, i.e. air infiltration or exfiltration. Inside the chamber a mixing fan ensures well-mixed conditions. The mass concentration inside the chamber (c) satisfies the following conservation of mass equation:

$$\boxed{V\frac{dc}{dt} = S - \left(A_s k_w + Q_s + \eta_{cleaner} Q_{cleaner}\right) c} \tag{5-40}$$

The known, easily measurable quantities are chamber volume (V), elapsed time (t), internal surface area (A_s), volumetric flow rate through the sampling instrument (Q_s), and volumetric flow rate through the room air cleaner ($Q_{cleaner}$). The remaining quantities in Eq. (5-40), i.e. wall-loss coefficient (k_w), source strength (S), and efficiency of the room air-cleaning device ($\eta_{cleaner}$) are to be obtained through the analysis discussed here. It is assumed that only the concentration varies with time, c = c(t); all the other parameters above have constant (but perhaps unknown) values.

5.8.1 Wall-Loss Coefficient

To use the test chamber to measure source characteristics or room air cleaner performance, the wall-loss coefficient (k_w) must be known for each contaminant to be studied. Alternatively the researcher's explicit goal may be to study the adsorption characteristics of wall hangings, furniture, etc. Clean air is allowed to enter the chamber at a volumetric flow rate equal to the sampling rate Q_s. To measure k_w, the air-cleaning device is shut off ($Q_{cleaner}$ = 0) and the source is allowed to fill the chamber with contaminant. Once a satisfactory concentration is achieved (which need not be the steady-state value), the source is shut off (S = 0), and the decreasing contaminant concentration is measured over a period of time. Under these conditions Eq. (5-40) becomes

$$\boxed{\frac{1}{c}\frac{dc}{dt} = \frac{d(\ln c)}{dt} = -\left[\frac{A_s k_w + Q_s}{V}\right]} \tag{5-41}$$

Assuming the sampling rate (Q_s) is known and is constant, the slope of (ln c) versus time enables one to determine the wall-loss coefficient (k_w). Consider formaldehyde emissions from building materials and home products such as:

- urea-formaldehyde foam insulation
- fiberglass, sealants, and adhesives
- gypsum wallboard and pressed wood products, such as wood paneling and particle board
- carpeting, wall coverings, and upholstery

The net emission of formaldehyde depends on formaldehyde emitted by the material minus wall losses due to adsorption. Unfortunately adsorption depends on the temperature and the concentration of formaldehyde in the material (called ***bulk concentration***, c_{bulk}), the concentration of formaldehyde in air, temperature (T), and relative humidity (Φ) of the air.

Rather than deal with a separate source strength and wall loss, some researchers (Hawthorne and Matthews, 1987; Matthews et al., 1987; Tichenor and Mason, 1988) suggest using a ***net source strength*** (S'), defined as

$$\boxed{S' = k_0(T,\Phi)\left[r_1(t)\cdot r_2(T)\cdot r_3(\Phi)\right]\left(c_{bulk} - c\right)} \tag{5-42}$$

where $k_0(T,\Phi)$ is the air transport property that reflects dependence on temperature and humidity. Hawthorne and Matthews (1987) suggest that the term can be expressed as

$$\boxed{k_0(T,\Phi) = k_0(T_0,\Phi_0)\left[1 - a_1\left(T - T_0\right)\right]\left[1 - a_2\left(\Phi - \Phi_0\right)\right]} \tag{5-43}$$

where a_1 and a_2 are model constants unique to the application. The functions r_1, r_2, and r_3 are functional relationships that account for the fact that the rate of formaldehyde emission depends on the age of the material, the temperature, and the relative humidity respectively:

- age dependence:

$$\boxed{r_1(t) = \exp\left[-\frac{t - t_0}{t'}\right]} \tag{5-44}$$

where $(t - t_0)$ is the age of the material since measurement of the emission rate, and t' is a characteristic time of the order of 1 to 5 years.

- temperature dependence:

$$\boxed{r_2(T) = \exp\left[-B\left(\frac{1}{T} - \frac{1}{T_0}\right)\right]} \tag{5-45}$$

where B is a coefficient to be determined and T_0 is a reference temperature.

- relative humidity dependence:

$$\boxed{r_3(\Phi) = \left(\frac{\Phi}{\Phi_0}\right)^{a_3}} \tag{5-46}$$

where the exponent a_3 in is a coefficient to be determined, and Φ_0 is a reference value.

Before one can use the above expressions, the model constants (a_1, a_2, a_3, and B) and reference values (k_0, T_0, Φ_0, and t_0) have to be determined from data obtained either from the literature or from experiment. Silberstein et al. (1988) suggest alternative equations that include relative humidity, age, and temperature. But like the above equations, they also include a number of parameters and reference states that have to be determined experimentally. Kelly et al. (1999) describe the initial emission rate of formaldehyde (HCOH) from 55 diverse common-place materials and consumer products. The authors report experimental data for products in which HCOH is contained in a dry product, and products in which HCOH is applied to a surface as a wet coating.

5.8.2 Source Strength

The well-mixed model in Figure 5.5 can also be used to determine the ***source emission rate***, i.e. the source strength (S) (Matthews, Hawthorne, and Thompson,1987). The experiment is begun by running the air-cleaning device over a long period of time without the source (S = 0). When a steady minimum concentration is obtained, the air cleaner is turned off ($Q_{cleaner} = 0$), the source is turned on, and the rising concentration is measured and recorded. The mass conservation equation, Eq. (5-40) becomes

$$V\frac{dc}{dt} = S - (A_s k_w + Q_s)c \tag{5-47}$$

Immediately after the source is activated, and while the concentration (c) is still small, the second term on the right-hand side is small with respect to S. Thus, initially

$$S \approx V\frac{dc}{dt} \tag{5-48}$$

and the initial source strength can be found from the slope of concentration versus time. Obtaining S by Eq. (5-48) is inherently inaccurate owing to the difficulty of computing a derivative from a few concentration values obtained over a short period of time. If the concentration rises slowly, the accuracy improves. There are two other ways to measure a constant value of S: (a) S can be calculated from measured values of concentration obtained over a period of time. Specifically, S can be found by integrating Eq. (5-47) between elapsed times t_1 and t_2, which yields

$$S = \frac{(A_s k_w + Q_s)\left\{c_2 - c_1 \exp\left[-\frac{(A_s k_w + Q_s)(t_2 - t_1)}{V}\right]\right\}}{1 - \exp\left[-\frac{(A_s k_w + Q_s)(t_2 - t_1)}{V}\right]} \tag{5-49}$$

where c_1 and c_2 are the concentrations at times t_1 and t_2, respectively. (b) One can wait until equilibrium (steady-state) conditions occur, so that the left-hand side of Eq. (5-47) is zero and $c = c_{ss}$. Under these conditions the constant source strength and steady-state concentration c_{ss} are related by

$$S = (A_s k_w + Q_s)c_{ss} \tag{5-50}$$

If steady state is achieved in a reasonable time, Eq. (5-50) should be used, since it represents the most accurate (and simplest) solution. If achievement of steady-state conditions requires a large amount of time, Eq. (5-49) can be used instead. The reader can verify that as t_2 - t_1 gets very large, and c approaches c_{ss}, the exponential terms in Eq. (5-49) become negligible, and Eq. (5-49) reduces to Eq. (5-50).

If the source strength S is not constant, Eq. (5-47) cannot be integrated in closed form. The instantaneous value of S(t) at some instant (t) can instead be found from a graph of mass concentration (c) versus time (t). From Eq. (5-47),

$$S(t) = V\frac{dc}{dt} + c(A_s k_w + Q_s) \tag{5-51}$$

where dc/dt is the slope of c(t) at the instant of time t. Since it is inherently difficult to measure slopes from experimental data, Eq. (5-51) may not yield highly accurate values of S(t).

Example 5.9 - "New Car Smell": Emission Rate of a Hydrocarbon in a New Automobile

Given: Most people enjoy the "new car smell" produced by hydrocarbon emissions from the interior coverings inside a new car. An automobile manufacturer is concerned about how long the odor lasts, and needs to measure the decaying source strength, S(t), of a particular hydrocarbon inside the car. The interior volume of the car is $V = 4.0\ m^3$. An experiment is run in which the car is sealed tight, but a small amount of fresh air is added to its interior ($Q_s = 200\ cm^3/hr$, $c_a = 0$), and the same volumetric flow rate of air from inside the automobile is withdrawn. A small circulating fan is placed inside the automobile interior to ensure well-mixed conditions. The air in the car is purged just prior to the experiment so that the initial hydrocarbon concentration is small. The concentration is then measured and recorded twice per month for nine months. Shown below is the instantaneous hydrocarbon mass concentration (in units of mg/m^3) as a function of elapsed time (in months).

time (mo)	c(mg/m^3)	time (mo)	c(mg/m^3)	time (mo)	c(mg/m^3)
0	20.	3.5	96.	6.5	15.
0.5	238	4.0	80.	7.0	11.
1.0	235	4.5	61.	7.5	8.
1.5	215	5.0	42.	8.0	6.
2.0	162	5.5	31.	8.5	4.
2.5	135	6.0	21.	9.0	3.
3.0	108				

Adsorption of hydrocarbon vapors on interior coverings in the automobile is negligible, i.e. $k_w = 0$.

To do: From these data compute and plot the instantaneous source strength, S(t), over the elapsed time.

Solution: The authors used Mathcad (the file is available on the book's web site) to generate a cubic spline fit of c(t) from the empirical data, and to differentiate c(t) to obtain dc/dt. The source strength S(t) can then be found from Eq. (5-51), which simplifies to

$$S(t) = V\frac{dc}{dt} + Q_s c$$

Figures E5.9a and b show hydrocarbon mass concentration and source strength as functions of time.

Discussion: The computed source strength decreases as expected when hydrocarbons desorb from a surface; it decays to zero, but the curve is not smooth. The reason for this is inaccuracies in the measurement of concentration that become magnified when taking derivatives of experimental data. Smoother data can be generated by using a least-squares polynomial fit rather than a cubic spline fit.

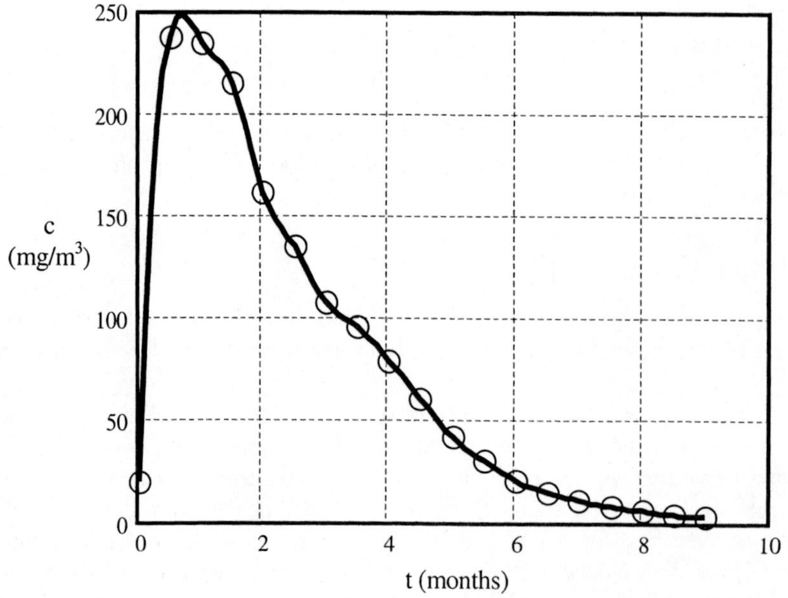

Figure E5.9a Mass concentration of hydrocarbon vapor in a car as a function of time.

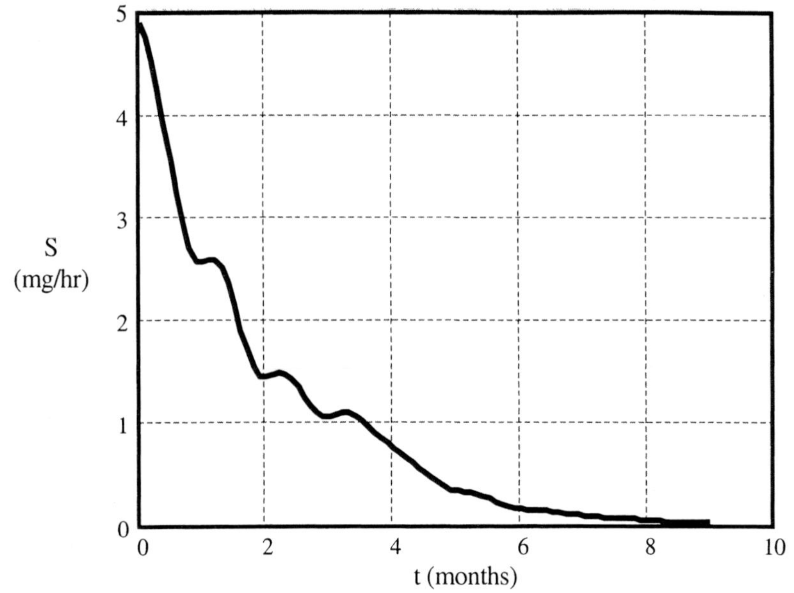

Figure E5.9b Source strength of hydrocarbon vapor in a car as a function of time.

5.8.3 Efficiency of an Air Cleaning Device

To find the efficiency ($\eta_{cleaner}$) of a room air-cleaning device, the source and cleaning device are run at steady rates for a long period of time until a steady-state concentration (c_{ss}) is obtained. Under these conditions Eq. (5-40) reduces to

$$\eta_{cleaner} = \frac{\frac{S}{c_{ss}} - Q_s - A_s k_w}{Q_{cleaner}} \tag{5-52}$$

Alternatively, the source is allowed to produce a significant concentration (although not necessarily its steady-state value), the source is then removed or turned off (S = 0), and the air-cleaning device is turned on. The concentration begins to fall and the efficiency can be obtained from

$$\eta_{cleaner} = \frac{-V\frac{d(\ln c)}{dt} - Q_s - A_s k_w}{Q_{cleaner}} \tag{5-53}$$

As discussed previously, if enough time is available for steady-state conditions to be reached, the method leading to Eq. (5-52) is recommended because it yields more accurate results. The time derivative term in Eq. (5-53) is inherently inaccurate.

5.9 Clean Rooms

Clean rooms (see Figure 1.17 and Figure 5.11) are enclosed spaces in which individuals work and in which the following atmospheric properties are controlled within stringent limits: temperature, humidity, concentration of particles, and concentration of contaminant gases and vapors. Fredrickson (1993) discusses the design criteria for clean rooms. The geometry and operation of clean rooms vary, but all are designed to provide an environment that protects a manufactured product from contamination. Unfortunately many materials used in clean rooms are toxic. In the manufacture of semiconductors, the principal concern is to remove small airborne particles that can short-circuit the minute integrated circuits on silicon wafers. Often overlooked however, are emissions of vapors of corrosive, reactive, and toxic materials used to fabricate the wafers. The amount of these materials is

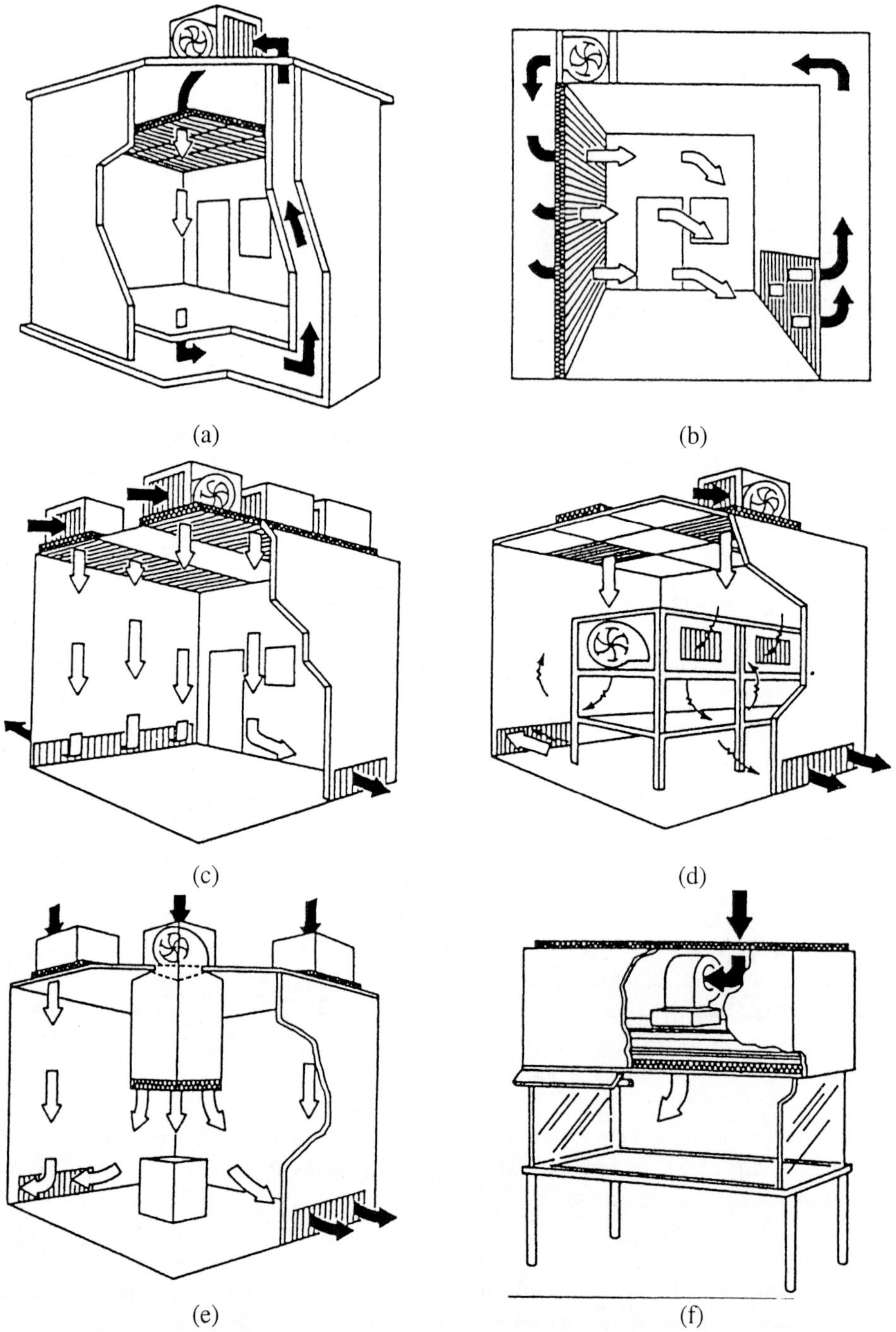

(a) (b) (c) (d) (e) (f)

Figure 5.11 Clean rooms: (a) vertical laminar-flow, (b) horizontal laminar-flow, (c) tunnel laminar-flow, (d) tabletop tunnel laminar-flow, (e) island laminar-flow, and (f) unitary work station (miniature) (from Canon Communications, 1987).

small, but handling them can produce spills, splash, and airborne emissions. The fabrication process begins by applying photoresist to the wafer. ***Photoresist*** is an ultraviolet-sensitive, polymeric material mixed in a solvent carrier. Following curing in ovens, the wafer is exposed to ultraviolet light and placed in an alkaline developer. The exposed resist dissolves in the developer leaving the open surface for subsequent processing. A ***wet etching*** process using bases or corrosive acids such as hydrofluoric acid may be employed to remove unwanted material. Alternatively a ***dry etching*** process involving an RF plasma can remove unwanted material, but in so doing a variety of gaseous compounds may be formed that must be controlled. Thin films of material such as silicon nitride, silicon dioxide, etc. are then deposited on the wafer by liquid and gaseous processes involving silane, tetraethylorthosilicate, phosphine, diborane, ammonia, etc. Next, highly toxic or reactive materials called ***dopants*** (arsenic, phosphorous, arsine, phosphine, or boron trifluoride), which have unique electrical properties, are imbedded into the surface of the silicon wafer. Liquid solutions of dopants pass into high temperature furnaces by bubblers using inert gases whereupon the dopant atoms diffuse to the silicon surface. Because of the acute toxicity of dopant materials, safety procedures must be strictly adhered to, and sophisticated controls must be used. Layers of noble or common metals such as gold, aluminum, titanium, tantalum, or tungsten are next deposited as thin films by evaporative or sputtering processes. In between all steps in the process, wafers are cleaned by a variety of solvents such as carbon tetrachloride, methylene chloride, and trichloroethylene, which have long-term toxicity.

Clean rooms should not be confused with ***laboratory fume hoods*** (Figure 1.19), ***biological cabinets***, or ***glove-boxes***. The objective of clean rooms is to protect a product that is being manufactured, as distinct from protecting the worker. Standards for the purity of air in clean rooms are considerably more stringent than those to ensure the health and safety of workers. Air entering clean rooms is cleaned and conditioned continuously. Well-mixed conditions are achieved because of the unique ways air enters and leaves the clean room rather than because there is a vigorous mixing mechanism within the room.

The cleanliness of a clean room is classified by Federal Standard 209E according to its ***class***, which is based on particle number concentration (c_{number}). Specifically, ***class limits*** are based on the total number of particles 0.5 μm and larger permitted per cubic foot of air. For example, c_{number} for a class 10 clean room cannot exceed 10 particles/ft^3. Other particle diameters can alternatively be used to determine the class of a clean room, as listed in Table 5.2. SI (metric) classes have also been defined based instead on the *exponent* of the total number of 0.5 μm or larger particles permitted per cubic *meter* of air. For example, c_{number} for a class M2 clean room cannot exceed $10^2 = 100$ particles/m^3. Intermediate SI classes have also been defined to correspond to the older English classifications. For example, SI class M2.5 has been designated as the equivalent to class 10, even though more precise unit conversion would yield class M2.548. Similarly, class M3.5 is the same as class 100, etc. The data of Table 5.2 are plotted in Figure 5.12. Since the slope for each class is the same on a log-log plot of c_{number} versus D_p, extrapolation to other particle sizes is also possible.

Workers in clean rooms are clothed in garments designed to prevent particles from being emitted into the room from clothing and the body. The humidity is set to values appropriate for the product being manufactured and equipment being used. The temperature is normally set at 68 °F. Floor, ceiling, and wall surfaces are designed so as not to generate particles. In addition, floor coverings and garments are designed so as not to generate static electricity.

As requirements for high performance filters have become more demanding, new international classifications have been developed. Two main classifications are the following (ASHRAE HVAC Applications Handbook, 1999):

(a) A ***high efficiency particulate air (HEPA) filter*** is defined as a filter with an efficiency in excess of 99.97% for 0.3 μm particles
(b) An ***ultra low penetration air (ULPA) filter*** is defined as a filter with a minimum efficiency of 99.999% for 0.12 μm particles

Table 5.2 Clean room class limits; maximum permissible c_{number} in English and SI units; **bold** c_{number} indicates number concentration on which the corresponding **bold** class name is based (abstracted from ASHRAE HVAC Applications Handbook, 1999.)

class name		$D_p \geq 0.1$ μm		$D_p \geq 0.2$ μm		$D_p \geq 0.5$ μm		$D_p \geq 5$ μm	
SI	English	#/m³	#/ft³	#/m³	#/ft³	#/m³	#/ft³	#/m³	#/ft³
M1		350	9.9	75.0	2.14	**10^1**	0.283	-	-
M1.5	**1**	1240	35	265	7.5	35.3	**1**	-	-
M2		3500	99.1	757	21.4	**10^2**	2.83	-	-
M2.5	**10**	12400	350	2650	75.0	353	**10**	-	-
M3		35000	991	7570	214	**10^3**	28.3	-	-
M3.5	**100**	-	-	26500	750	3530	**100**	-	-
M4		-	-	75700	2140	**10^4**	283	-	-
M4.5	**1000**	-	-	-	-	35300	**1000**	247	7.00
M5		-	-	-	-	**10^5**	2830	618	17.5
M5.5	**10000**	-	-	-	-	353000	**10000**	2470	70.0
M6		-	-	-	-	**10^6**	28300	6180	175
M6.5	**100000**	-	-	-	-	3530000	**100000**	24700	700
M7		-	-	-	-	**10^7**	283000	61800	1750

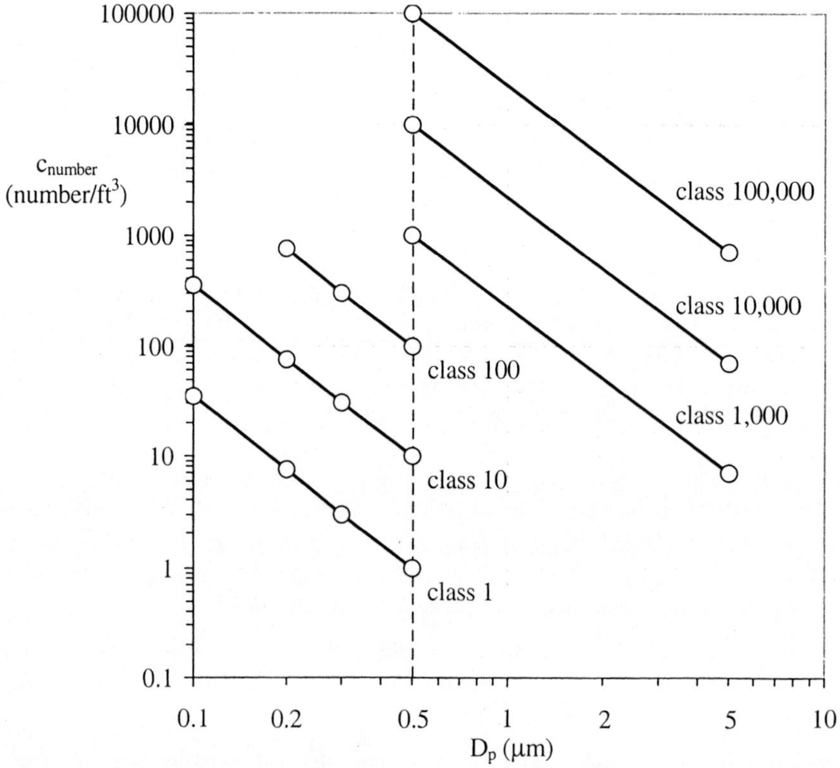

Figure 5.12 Class definitions for clean rooms in the US; class based on cubic feet – conversion: 1.00 particles/ft³ = 35.3 particles/m³.

The efficiencies of HEPA and ULPA filters are based on 0.3 and 0.12 μm particles, respectively, because the ***most penetrating particle size*** (MPPS) of fibrous filters is typically between these two values. MPPS is discussed in more detail in Chapter 9. In vertical laminar flow clean rooms (Figure 5.11a), the entire ceiling is a high efficiency (HEPA or ULPA) filter and the floor is the receiving plenum. Typical air velocities entering a vertical laminar flow clean room are 60-100 FPM (ft/min). Temperature and humidity control are achieved by a separate air handling system. Class 100 conditions can be achieved by such designs. The performance of laminar-air flow rooms is hampered by wake regions downstream of equipment and personnel. Such wakes are recirculation regions that tend to accumulate airborne particles and prevent their removal.

It must be emphasized that while air may enter a laminar-flow room in a laminar fashion, the existence of wakes and recirculation regions produces limited degrees of turbulence that are unavoidable. In addition, Reynolds numbers for the rooms themselves or the obstacles around which the air passes can be considerably large (several thousand), and thus the assumption of laminar flow may be incorrect.

Example 5.10 - Time to Achieve Clean Room Conditions

Given: Consider a vertical laminar-flow clean room similar to Figure 5.11, and assume that its schematic diagram is given by Figure E5.7. The clean room will be operated using existing equipment with the following specifications:

- η_1 = 98.%, η_2 = 98.% (air cleaner efficiencies)
- f = 0.050 (fresh make-up air fraction)
- c_a = 10^3 particles/m^3 (particle concentration in the ambient make-up air)
- D_p = 1.0 μm (particle size of concern)
- Q_s = 20. m^3/min (supply ventilation rate into the clean room)
- S = 300 particles/min (particle emission rate within the clean room)
- V = 300 m^3 (volume of the clean room)
- A_s = 320 m^2 (total surface area of the clean room)
- k_w = 0.030 m/min (wall loss coefficient)
- c(0) = 10^5 particles/m^3 (initial particle concentration within the clean room)

The goals of management are to achieve a class 1 clean room.

To do: Advise the company if a class 1 clean room can be achieved. Estimate how long it will take to achieve class 10,000, class 1,000, class 100, class 10, and class 1 conditions.

Solution: Using Figure 5.12, maximum permissible particle concentration at D_p = 1.0 μm can be tabulated as a function of class. Note that for classes 1, 10, and 100, extrapolation is required:

class	c_{max} (particles/ft^3) for D_p = 1.0 μm	c_{max} (particles/m^3) for D_p = 1.0 μm
10,000	2,100	74,000
1,000	210	7,400
100	21.	740
10	2.1	74.
1	0.21	7.4

Conservation of mass for the particles in Figure E5.7 can be written in standard form, as in Eq. (5-7):

$$\frac{dc}{dt} = B - Ac$$

where

$$A = \frac{Q_s + k_w A_s + Q_s(1-f)(1-\eta_1)}{V} \qquad B = \frac{S + Q_s c_a f(1-\eta_2)}{V}$$

The solution of the differential equation is given by Eq. (5-10), with steady-state concentration (c_{ss}) given by Eq. (5-11). For the conditions given above,

$$A = 0.0999\frac{1}{\text{min}} \qquad B = 1.07\frac{\text{particles}}{\text{min}\cdot\text{m}^3}$$

The steady-state particle concentration within the room will thus be

$$c_{ss} = \frac{B}{A} = \frac{1.07\dfrac{\text{particles}}{\text{min}\cdot\text{m}^3}}{0.0999\dfrac{1}{\text{min}}} = 10.7\frac{\text{particles}}{\text{m}^3}$$

Thus, the minimum possible particle concentration is about 11. particles/m^3, where two significant digits of precision is the most that can be expected from these calculations. Therefore, the company can achieve a class 10 clean room, which allows a maximum of 74. particles/m^3, but cannot achieve a class 1 clean room with the existing equipment, although it can come close. Any increase in the particle emission rate (S) or reduction in the volumetric flow rate (Q_s) will worsen the situation.

If the initial particle concentration, c(0), is 10^5 particles/m^3, Eq. (5-12) can be used to calculate the time to achieve the various classes of clean rooms:

class	**time (min)**
10,000	3.0
1000	26.
100	49.
10	74.
1	∞

Discussion: The only way to achieve a class 1 clean room with the existing equipment is to operate the manufacturing process in ***batch process*** mode, i.e. intermittently (operate for some time period, during which the particle concentration in the room rises to nearly the class 1 limit of 7.4 particles/m^3, and then shut down the process (S = 0) for a while to allow the concentration to drop.) Alternatively, minor improvements to one or more of the components may be just enough to achieve a class 1 clean room, e.g. if η_2 can be increased from 98.% to 99.%.

5.10 Infiltration and Exfiltration

The transfer of air into and out of an enclosed space is equal to deliberate input and removal of air (forced ventilation) plus uncontrolled air leakage through cracks, holes, etc. Uncontrolled flow of air into a building is called ***infiltration***, and uncontrolled removal of air is called ***exfiltration*** (Perera et al., 1986). Infiltration and exfiltration are produced primarily by pressure differences between the building interior and the atmosphere resulting from the aerodynamic flow of air around and over the building. To a lesser extent they are also due to temperature differences between the building interior and the atmosphere, and to diffusion processes. To a first approximation one may assume that the volumetric flow rates of infiltration and exfiltration are equal. The relative air leakage of a typical building is distributed as in Table 5.3. Infiltration can be expressed in three ways:

- empirical estimates of air changes per hour
- equations based on construction details
- empirical equations

Table 5.3 Sources of air leakage in a typical building (from ASHRAE, 1997).

source of leakage	relative leakage (%)
walls (top and bottom joints, plumbing and electrical penetrations)	18 to 50, avg. 35
ceiling	3 to 30, avg. 18
heating system	3 to 28, avg. 18
windows and doors	6 to 22, avg. 15
fireplaces	0 to 30, avg. 12
vents in conditioned spaces	2 to 12, avg. 5
diffusion (conduction) through walls	<1

Table 5.4 Infiltration and exfiltration; air changes per hour occurring under average conditions in residences exclusive of air provided for ventilation (abstracted from ASHRAE, 1981).

room description	single glass, no weather-stripping	storm sash or weather stripping
no windows or exterior doors	0.5	0.3
windows or exterior doors on one side	1.0	0.7
windows or exterior doors on two sides	1.5	1.0
windows or exterior doors on three sides	2.0	1.3
entrance halls	2.0	1.3

Table 5.4 is a condensation of ASHRAE's 1981 estimates of infiltration of air into a room in terms of number of air changes per hour for average buildings under average conditions. In the newer editions of the ASHRAE Fundamentals Handbook (e.g. ASHRAE, 2001), infiltration is estimated as a function of building construction details such as wall, ceiling, and floor construction, window and door specifications, etc., along with even finer details such as number of recessed ceiling lights, electrical outlets, etc. Each component source of infiltration is assigned an ***effective leakage area***, A_L (typically in units of cm^2); these components can be summed to obtain the total effective air leakage area of the building. Infiltration volumetric flow rate ($Q_{infiltration}$) is then calculated according to

$$\boxed{Q_{infiltration} = A_L\sqrt{C_s \Delta T + C_w V^2}} \tag{5-54}$$

where the symbols and their typical units are:

- $Q_{infiltration}$ = infiltration volumetric flow rate (L/s)
- A_L = total building effective leakage area (cm^2)
- C_s = ***stack coefficient*** ($L^2cm^{-4}s^{-2}K^{-1}$); varies with number of stories
- C_w = ***wind coefficient*** ($L^2cm^{-4}m^{-2}$); varies with number of stories and amount of shielding (trees, shrubbery, sheds, other buildings, etc.)
- ΔT = average absolute value of indoor-outdoor temperature difference (K or °C)
- V = average wind speed (m/s)

Tables for calculating these parameters, along with examples, are provided in the ASHRAE Fundamentals Handbook (ASHRAE, 2001), and are too lengthy to duplicate here. A typical modern two-story single-family home, for example, has a total volume of 340. m^3, with a total effective air leakage area of about 500 cm^2. Consider the following winter design conditions for Lincoln, Nebraska: wind speed = 6.7 m/s, ΔT = 39. °C, C_s = 0.000290 $L^2cm^{-4}s^{-2}K^{-1}$, and C_w = 0.000231$L^2cm^{-4}m^{-2}$. Equation (5-54) yields

$$Q_{infiltration} = \left(500 \text{ cm}^2\right)\sqrt{0.000290\frac{L^2}{s^2cm^4K}(39.\ K) + 0.000231\frac{L^2}{cm^4m^4}\left(6.7\frac{m}{s}\right)^2} = 73.6\frac{L}{s}$$

Eq. (5-14) can be used to convert the infiltration rate to number of air changes per hour (N),

$$N = \frac{Q}{V} = \frac{73.6\frac{L}{s}}{340.\ m^3}\left(\frac{m^3}{1000\ L}\right)\left(\frac{3600\ s^2}{hr}\right) = 0.78\frac{1}{hr}$$

Building infiltration rates have improved (decreased) significantly over the past several decades, prompted largely by the increasing cost of energy. The infiltration rate calculated above for a typical modern home is less than one air change per hour, even in severe winter design conditions. It would be even lower in less severe weather. While this is good for the family budget, it is not so good in terms of indoor air quality, the spread of airborne contaminants and diseases, etc., as discussed in Chapter 2.

For quick estimates, Wadden and Scheff (1983) report the following empirical equation for the number of air changes per hour (N):

$$\boxed{N = \frac{Q}{V} = 0.315 + 0.0273U + 0.0105\left|T_{outside} - T_{inside}\right|} \tag{5-55}$$

where the units of N are hr^{-1}, U is the wind speed in miles per hour, and $T_{outside}$ and T_{inside} are the outside and inside temperatures in degrees Fahrenheit. The absolute value signs in Eq. (5-55) ensure a component of N due to any temperature difference between $T_{outside}$ and T_{inside}, regardless of which temperature is greater.

Example 5.11 – Did the Professor Suffer Mercury Poisoning?

Given: The sons and daughters of a deceased faculty member have sued his university because they believe their father died from complications related to failure of his central nervous system caused by hazardous airborne concentrations of mercury vapor in his university office. Unknown to everyone at the time, liquid mercury lay under the floor boards of his office. In 1900 the university's chemistry laboratory was built containing a small storeroom for chemical supplies. The room was supported by 8-inch floor joists separating the storeroom from the ceiling of the room one floor below. The floor of the room was constructed of un-joined boards. Over time, narrow spaces developed between the boards. Stored in the room were 5-pound bottles of mercury, mercury thermometers, glass barometers; and U-tube manometers containing mercury used by students in their experiments. From time to time the barometers and manometers broke and mercury was spilled on the floor. Mercury was also spilled by students trying to fill glass manometers. Some of the liquid mercury fell through the spaces between the floor boards, and remained there. No record was ever kept of the mercury that was spilled or swept up afterwards. In 1940, all the mercury was removed from the storeroom and the room was used to store laboratory glassware. In 1945 the storeroom was remodeled into an office for a new faculty member, and he used the room for the next 35 years until he retired in 1980. When he retired he displayed symptoms indicating failure of his central nervous system. The symptoms became progressively worse and contributed to his death in 1985. A ventilation system, including air conditioning, was installed in 1982, but there are no records about how the office had been ventilated before 1982. Discussions with some of the older employees revealed that there was no forced air ventilation system at all prior to 1982. An exterior window was added to the office sometime in the 1950s after the professor received tenure, but there is some dispute as to the date. In 1991 the building was again remodeled. The old floor was removed, and approximately 40. kg of liquid mercury was found lying at the bottom of the dead-air space beneath the floor. The clean-up crew reported that the mercury was dispersed in puddles of various sizes, but most of it was in the form of small nearly spherical balls; they estimated the average size of the mercury balls to be around a half centimeter.

Upon hearing of the discovery of mercury, the professor's family filed suit against the university, claiming that his death was caused by exposure to mercury vapor during the 35 years he occupied his office. The university claimed that the failure of his central nervous system was a genetic predisposition, unrelated to mercury. Toxicologists were called to testify about the health issues (see Chapter 2 for a discussion of mercury poisoning), but their testimony depends on information about the concentration of mercury vapor in the office during the period between 1945 and 1980. Since no mercury vapor concentration measurements were ever made, you have been called as an expert witness on indoor air quality.

To do: Prepare three analyses:

1. Analysis 1 – Compute the maximum possible mercury vapor concentration: Estimate the maximum airborne concentration of mercury vapor that could possibly occur in the office, and determine if the concentration exceeds safe levels.
2. Analysis 2 – Estimate the amount of mercury in the room between 1945 and 1991: Since liquid mercury was found beneath the floor in 1991, even *more* liquid mercury would have been there in earlier years, since some of it would have evaporated during the period. The evaporation rate must be determined, and the mass of liquid mercury must be extrapolated back in time. As a worst-case scenario, assume the maximum possible evaporation rate.
3. Analysis 3 – Estimate realistic mercury vapor concentrations for different ventilation rates for the period 1945-1980: The vapor concentration depends on the evaporation rate, how the room was ventilated, and how much mercury remained below the floor. Estimate the mercury vapor concentration for three types of ventilation:

   (a) infiltration if the room had no exterior windows (conditions prior to sometime in the 1950s), which gives an upper bound for the vapor concentration
   (b) infiltration if the room had one exterior window containing single-plane glass without weather stripping (conditions since sometime in the 1950s), which gives a lower bound for the vapor concentration
   (c) forced ventilation assuming 62-1989 standards of 20 SCFM/person in addition to the infiltration rate of case (a); this condition did not exist until 1982, but is calculated for educational purposes

Solution: The room dimensions are measured: floor area (A_{floor}) = 20 m^2 (4m by 5m), height (h) = 3.5m, volume (V) = 50 m^3, and height of the dead-air space under the floor ($z_2 - z_1$) = 8 inches (0.203 m). Note that V is less than the total room volume because the room was partially filled with furniture, books, etc. The appropriate properties of mercury (Hg) are tabulated:

PEL = 0.1 mg/m^3 (1.2 x 10^{-2} PPM) $\quad$ $P_{v,Hg}$ = vapor pressure at 300 K = 0.0012 mm Hg
ρ_{Hg} = liquid density = 13,530 kg/m^3 $\quad$ M_{Hg} = molecular weight = 200.6 kg/kmol

Analysis 1 – As an *upper limit* of mercury concentration, consider the room to be totally isolated, receiving *no* fresh air ventilation whatsoever. The evaporation of mercury is a slow process, so one can assume there is sufficient time to achieve well-mixed conditions in the room. Mercury evaporates until its partial pressure is equal to its vapor pressure, whereupon evaporation ceases. Under these conditions, the steady-state mercury vapor mol fraction is given by Eq. (5-4),

$$y_{ss} = \frac{P_{v,Hg}}{P} = \frac{0.0012 \text{ mm Hg}}{760. \text{ mm Hg}} = 1.579 \times 10^{-6} \cong 1.6 \text{ PPM}$$

The concentration of mercury vapor corresponding to this mol fraction can be obtained from Eq. (1-30),

$$c = \frac{[PPM]M_{Hg}}{24.5}\left(\frac{mg}{m^3}\right) = \frac{(1.6)(200.6)}{24.5}\frac{mg}{m^3} = 13.\frac{mg}{m^3}$$

which is well in excess of the PEL. Since the office was *not* totally sealed off, the actual concentrations would have been much lower than this; therefore, further analysis is warranted. (Note that if this upper limit turned out to be less than the PEL, no further analysis would be necessary – it is unlikely that the university would be liable.)

Analysis 2 – The air in the space under the floor boards is stagnant; the discussion in Chapter 4 about evaporation in stagnant air is therefore relevant here. It is assumed that evaporation progresses at its maximum possible rate. This occurs when the far-field mercury vapor mol fraction (just above the floor boards) is zero; i.e. following the notation in Chapter 4, $y_{Hg,2} = 0$. This is a reasonable approximation if the room was adequately ventilated with fresh air, which is highly unlikely for a storeroom. Nonetheless assuming $y_{Hg,2} = 0$ yields the maximum evaporation. It is also assumed that the spilled liquid mercury is in the form of spheres approximately 5.0 mm in diameter, uniformly dispersed in the dead space. A differential equation of mass balance for the liquid mercury beneath the floor between 1940 and 1991 can be written:

$$\frac{dm_{liquid\ Hg}}{dt} = S_{liquid\ Hg} - \dot{m}_{evap\ Hg} = S_{liquid\ Hg} - A_{drop} n_{drops} M_{Hg} N_{Hg}$$

where

- t = elapsed time (yr)
- $m_{liquid\ Hg}$ = mass of accumulated liquid mercury (kg)
- $S_{liquid\ Hg}$ = source of liquid mercury into the room due to breakage and spillage = 0 beyond 1940
- A_{drop} = surface area of a 5.0-mm spherical drop of liquid mercury = 7.854×10^{-5} m^2
- $\dot{m}_{evap\ Hg}$ = rate of evaporation of liquid mercury (kg/yr)
- M_{Hg} = molecular weight of mercury = 200.6 kg/kmol
- n_{drops} = number of spherical drops of mercury in the room
- N_{Hg} = molar evaporation rate of liquid mercury into mercury vapor [kmol/(m^2 yr)]

Since the number of drops of liquid mercury is

$$n_{drops} = \frac{m_{liquid\ Hg}}{m_{drop}}$$

the mass balance can be written as

$$\frac{dm_{liquid\ Hg}}{dt} = -\frac{A_{drop} M_{Hg} N_{Hg}}{m_{drop}} m_{liquid\ Hg}$$

where the mass of a 5.0-mm spherical drop of mercury is

$$m_{drop} = \rho_{Hg}\frac{\pi D_{drop}^{\ 3}}{6} = 13530\frac{kg}{m^3}\frac{\pi(0.005\ m)^3}{6} = 8.855\times10^{-4}\ kg$$

Since the space below the floor boards is quiescent, the molar evaporation rate can be estimated from Eq. (4-57),

$$N_{Hg} = P\frac{D_{Hg,a}}{R_u T(z_2 - z_1) y_{am}}\left(y_{Hg,1} - y_{Hg,2}\right)$$

where $(z_2 - z_1) = 0.203$ m. The diffusion coefficient of mercury in air ($D_{Hg,a}$) can be estimated from Eq. (4-32),

$$D_{Hg,a} = D_{water,a}\sqrt{\frac{M_{water}}{M_{Hg}}}$$

where M_{water} is the molecular weight of water (18.0 kg/kmol), and $D_{water,a}$ is the diffusion coefficient of water in air (2.2 x 10^{-5} m^2/s).

$$D_{Hg,a} = 2.2\times10^{-5}\,\frac{m^2}{s}\sqrt{\frac{18.0}{200.6}}\left(\frac{3600\ s}{hr}\right) = 2.372\times10^{-2}\,\frac{m^2}{hr}$$

The maximum evaporation rate occurs when the far-field mercury vapor mol fraction ($y_{Hg,2}$) is zero, i.e. $y_{a,2} = 1$. From Eq. (4-53),

$$y_{am} = \frac{y_{a,2} - y_{a,1}}{\ln\left(\frac{y_{a,2}}{y_{a,1}}\right)} = \frac{1-\left(1-1.579\times10^{-6}\right)}{\ln\left[\frac{1}{\left(1-1.579\times10^{-6}\right)}\right]} = 0.999999 \cong 1$$

The partial pressure of mercury vapor at the interface between liquid mercury and air is equal to the vapor pressure of mercury. Thus, the mol fraction of mercury vapor at the surface of each drop is

$$y_{Hg,1} = y_{Hg,i} = \frac{P_{v,Hg}}{P} = \frac{0.0012\ mm\ Hg}{760.\ mm\ Hg} = 1.579\times10^{-6}$$

Thus, the molar evaporation rate is

$$N_{Hg} = (101.3\ kPa)\frac{2.372\times10^{-2}\,\frac{m^2}{hr}}{8.3143\frac{kJ}{kmol\ K}(300\ K)(0.203\ m)(1)}\left(1.579\times10^{-6} - 0\right)\left(\frac{kJ}{m^3kPa}\right)$$

$$= 7.49\times10^{-9}\,\frac{kmol}{m^2hr}\left(\frac{8766\ hr}{yr}\right) = 6.57\times10^{-5}\,\frac{kmol}{m^2yr}$$

After substitution of N_{Hg} and the other parameters into the mass balance,

$$\frac{dm_{liquid\ Hg}}{dt} = -\frac{\left(7.854\times10^{-5}\ m^2\right)\left(200.6\frac{kg}{kmol}\right)6.57\times10^{-5}\,\frac{kmol}{m^2yr}}{8.855\times10^{-4}\ kg}m_{liquid\ Hg}$$

which reduces to

$$\frac{dm_{liquid\ Hg}}{dt} = -1.169\times10^{-3}\,\frac{1}{yr}m_{liquid\ Hg}$$

The above ODE is of the same form as Eq. (5-7), but with mass concentration (c) replaced by mass ($m_{liquid\ Hg}$),

$$\frac{dm_{liquid\ Hg}}{dt} = B - Am_{liquid\ Hg}$$

with coefficients

$$A = -1.169\times10^{-3}\,\frac{1}{yr} \qquad B = 0$$

Thus, the solution is given by an equation similar to Eq. (5-10) with the steady-state mass equal to zero, but with the "initial" mass of liquid mercury set to the mass discovered in 1991, considering time relative to the year 1991. The mass of mercury under the floorboards during the period from 1940 to 1991 is thus

$$m_{\text{liquid Hg}}(t_{\text{years}}) = m_{\text{liquid Hg}}(1991)\exp\left(-A\left(t_{\text{years}} - 1991\right)\right)$$

where t_{years} is the year number (1945, 1946, etc.). Thus in 1945, when the professor moved into the office,

$$m_{\text{liquid Hg}}(1945) = (40.\ \text{kg})\exp\left(-1.169\times10^{-3}\frac{1}{\text{yr}}(1945-1991)\right) = 42.2\ \text{kg}$$

When he retired in 1980, the mass of liquid mercury remaining below the floorboards was

$$m_{\text{liquid Hg}}(1980) = (40.\ \text{kg})\exp\left(-1.169\times10^{-3}\frac{1}{\text{yr}}(1980-1991)\right) = 40.5\ \text{kg}$$

Mercury does not evaporate rapidly; even if one assumes the maximum possible evaporation rate, the amount of liquid mercury under the floor was fairly constant during the entire period (35 years) in which the professor occupied his office. It must be kept in mind that several assumptions were made in the above analysis. For example, as the spheres of liquid mercury evaporate, their diameter decreases; this was not taken into account. In addition, the evaporation rate would be somewhat less than its maximum value and less mercury would have therefore existed beneath the floor in 1945. However, since the evaporation rate is so small, these assumptions are reasonable.

Analysis 3 – An accurate estimate of the mercury vapor concentration in the professor's office since the year 1945 can be made only if:

(a) the ventilation rate is known
(b) account is taken of the fact that the amount of liquid mercury decreases as it evaporates
(c) the far-field vapor mol fraction ($y_{Hg,2}$) is not zero, but varies with time; thus the driving potential for evaporation ($y_{Hg,1} - y_{Hg,2}$) is not constant

Each of these points is examined: (a) Unfortunately, the ventilation rate between 1945 and 1980 is *not* known; two possible values, with and without an exterior window, are used in the calculations to determine the upper and lower bounds of vapor concentration respectively. A third ventilation rate is also used to see the effect of forced ventilation, even though it did not exist until 1982. (b) The amount of liquid mercury during the period has already been calculated in Analysis 2 above. (c) An equation needs to be solved describing how the vapor mol fraction varies with time. This is accomplished by writing a mass balance for mercury vapor in the ventilated room,

$$V\frac{dc}{dt} = Qc_a + S - Qc$$

where Q is the ventilation rate, assumed to be constant, c_a is the mass concentration of mercury vapor in the ambient air, assumed to be zero, and S is the source of mercury vapor, which is equal to the evaporation rate previously calculated,

$$S = \dot{m}_{\text{evap Hg}} = \frac{A_{\text{drop}} M_{Hg} N_{Hg}}{m_{\text{drop}}} m_{\text{liquid Hg}}$$

Since the far-field concentration varies slowly (months and years), it is realistic to assume that at any instant the mass concentration, c (mg/m^3), can be expressed as the quasi-steady-state value given by Eq. (5-11), with coefficients B = S/V and A = Q/V. Thus at any time between 1945 and 1980 (t_{years}),

$$c_{ss}(t_{years}) = \frac{\dot{m}_{evap\,Hg}}{Q} = \frac{A_{drop} M_{Hg} N_{Hg}}{m_{drop} Q} m_{liquid\,Hg}(t_{years})$$

Converting steady-state mass concentration (c_{ss}) to steady-state mol fraction (y_{ss}),

$$y_{ss}(t_{years}) = \frac{R_u T c_{ss}(t_{years})}{M_{Hg} P} = \frac{A_{drop} N_{Hg} R_u T}{m_{drop} QP} m_{liquid\,Hg}(t_{years})$$

where $y_{ss} = y_{Hg,2}$ Substitution of the equation for N_{Hg} from Analysis 2 above yields

$$y_{ss}(t_{years}) = \frac{A_{drop}}{m_{drop} Q} \frac{D_{Hg,a}}{(z_2 - z_1) y_{am}} (y_{Hg,1} - y_{ss}) m_{liquid\,Hg}(t_{years})$$

For simplicity, the mass of liquid mercury under the floorboards ($m_{liquid\,Hg}$) at any year (t_{years}) between 1945 and 1980 is assumed to decrease according to the rate calculated in Analysis 2 above. Simplifying and rearranging the above, and solving for y_{ss} gives

$$y_{ss}(t_{years}) = \frac{C_1(t_{years}) y_{Hg,1}}{1 + C_1(t_{years})}$$

where $C_1(t_{years})$ is a collection of parameters from the above equation,

$$C_1(t_{years}) = \frac{A_{drop}}{m_{drop} Q} \frac{D_{Hg,a}}{(z_2 - z_1) y_{am}} m_{liquid\,Hg}(t_{years})$$

The above equation can be solved for the three different ventilation conditions given in the problem statement. From Table 5.4 and the ASHRAE handbook, the rates are as follows:

(a) infiltration when there are no exterior windows or doors; number of room air changes N $= Q/V = 0.50\ hr^{-1}$, $Q = NV = (0.50\ hr^{-1})(50.\ m^3) = 25.\ m^3/hr$
(b) infiltration when there is one exterior window containing a single-plane glass and no weather stripping; number or room air changes $N = Q/V = 1.0\ hr^{-1}$, $Q = 50.\ m^3/hr$
(c) ASHRAE Standard 62-1989 ventilation rate, i.e. Q = 20. SCFM (34. m^3/hr) in addition to the infiltration rate of case (a) above; total $Q = 34. + 25. = 59.\ m^3/hr$

As a sample calculation, consider the mercury vapor mol fraction in the year 1945, when the professor first moved into his office. For ventilation case (a),

$$C_1(1945) = \frac{7.854\times10^{-5}\ m^2}{8.855\times10^{-4}\ kg\left(25.\frac{m^3}{hr}\right)} \frac{2.372\times10^{-2}\frac{m^2}{hr}}{(0.203\ m)0.999999} 42.2\ kg = 1.75\times10^{-2}$$

and

$$y_{ss}(1945) = \frac{1.75\times10^{-2}(1.579\times10^{-6})}{1 + 1.75\times10^{-2}} = 2.72\times10^{-8} \cong 0.027\ PPM$$

This value is well above the PEL (0.012 PPM). Similar calculations must be performed for each year and for each of the three ventilation rates. A plot of y_{ss} (in PPM) versus time (in years) is shown in Figure E5.11. The authors used Excel to generate this plot; a copy of the Excel spreadsheet is available on the book's web site. The reader is encouraged to experiment with different values of flow rates to see the effect on mercury vapor concentration in the room. Recall that case (a) is an upper limit and case (b) is a lower limit, reflecting conditions of the office without and with an exterior window, respectively. The actual concentration should lie somewhere between these two limits, depending on when the window was added. From the figure,

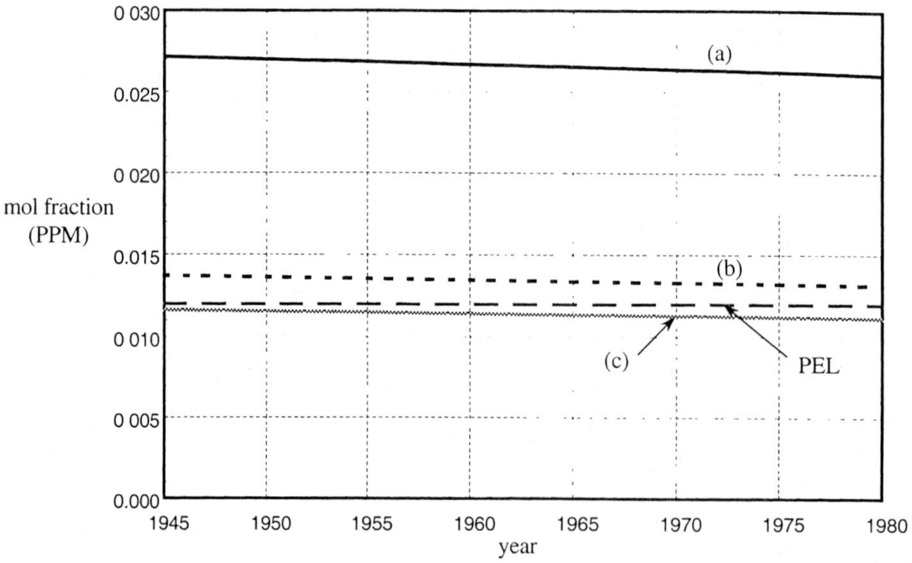

Figure E5.11 Mercury vapor mol fraction in the professor's office versus year since 1945, for three values of ventilation flow rate: (a) 25. m^3/hr, (b) 50. m^3/hr, and (c) 59. m^3/hr.

it is seen that the mercury vapor concentration decreased very slowly between 1945 and 1980, but the concentration was always above the PEL for either ventilation rate (a) or (b). The case with forced room ventilation, case (c), would have reduced the mercury vapor concentration below its PEL, but unfortunately forced ventilation was not added until after the professor retired. In conclusion, the professor was exposed to mercury vapor of hazardous concentration for 35 years. Since mercury is a cumulative toxin that accumulates in the body, it can be concluded that the dose associated with this exposure constitutes a hazardous condition.

Discussion: This example illustrates why stringent precautions are taken concerning liquid mercury. Certainly the air in rooms or buildings that formerly contained liquid mercury should be sampled at a variety of points to determine if hazardous mercury vapor concentrations are below the PEL (or whatever other standard is used) before the space is used for human occupancy. The example also illustrates why one's intuition about mercury can be misleading. Since evaporation is very slow, hazardous mercury vapor concentrations persist for a much longer time than most people would suspect. Finally, ventilation conditions (b) existed for the majority of time, and the predicted concentrations are only about 20% higher than the PEL. One may argue that since PELs are rather conservative, the professor may not have been in hazardous conditions after all. Is the university to blame for the professor's illness and death? The final answer to this question is left to the attorneys.

5.11 - Split-Flow Ventilation Booths

If an industrial operation generates and emits toxic material, it is logical to determine if the operation can be performed by robots and conducted inside an enclosure in which minimal amounts of air containing high concentrations of contaminants can be tolerated. For example, applying a surface coating (paint) to automobiles and furniture can occur inside specially designed booths by robots. Unfortunately, there are instances in which workers must apply the coating, e.g. automobile repair or laying up fiberglass on molds to manufacture small boats, bathtubs, and truck caps. There are agricultural endeavors such as poultry and hog finishing (final stages of feeding before slaughter) in which the health of the animals and workers requires that the work be conducted inside specially designed enclosures. For these applications engineers can expect that tens of thousands of SCFM of air need to be drawn into the enclosure, and equal amounts of polluted air need to be removed and cleaned before being discharged to the atmosphere. A ***ventilated booth*** is the generic name given to these enclosures. Figure 1.21 shows an enclosed booth for painting autos, and Figure 1.23 is a water-wall paint spray booth.

There are instances when contaminants are emitted near the floor, near the ceiling, or in one corner, etc. of these enclosures, and contaminant concentrations are stratified, i.e. are not well mixed. If circumstances prevent engineers from capturing the contaminants as they are emitted (using local ventilation techniques), ***split-flow ventilation*** is a design practice that presents an opportunity to minimize costs by taking advantage of natural stratification. This enables engineers to reduce the amount of air that needs to be treated, and as a result reduces the cost of emission control. Instead of cleaning all the air passing through the booth, air is removed from two portions of the enclosure as two distinct exhaust streams, and each is treated in a fashion appropriate to the contaminant concentration in the stream.

As a first step, engineers must know if stratification occurs. Split-flow ventilation does not create stratification, it only capitalizes on the stratification that occurs because of how contaminants are emitted. Stratification can be determined by experimental measurements or it can be predicted by numerical techniques. However it is done, engineers must be certain that they know the region where the contaminant concentration is high and where it is low. Furthermore, they must know the concentrations in these two regions. In the discussion that follows, an enclosed paint booth used in applying coatings to vehicles is analyzed since split-flow ventilation techniques have been successfully applied to an enclosed, side-draft, dry-filter paint booth (Ayer and Hyde, 1990).

Large paint booths are used when workers (or robots) apply coatings to automotive vehicles, aircraft assemblies, or other large pieces of equipment. Let Q_t represent the total volumetric flow rate of air entering and leaving the booth, and let A_{booth} represent the cross-sectional area of the booth. The phrase "process air stream" is used since air is the carrier gas. If no contaminants are present, the phrase "fresh air", or "atmospheric air" is used. Because spraying occurs in the lower 6 feet of the booth, contaminant concentrations in the lower portion of the booth are often several orders of magnitude greater than they are in the upper portion of the booth. OSHA regulations prescribe that the average velocity through the booth (Q_t/A_{booth}) should be between 75-150 FPM. For years conventional ventilation practice required that this velocity be achieved by removing air (Q_t) from the entire booth and replacing it fully (100%) with fresh air.

For paint booths the primary contaminants are solvents, i.e. ***volatile organic compounds*** (VOCs) in the coating, and to a lesser extent small particles that are not removed by the filter material inside the booth. The yearly amount of solvent exhausted to the atmosphere is large, but because the amount of air exhausted from the paint booth is huge, the VOC concentration in the exhaust stream is generally small, of order hundreds of PPM. In virtually every locality in the US, the paint booth exhaust must be treated to remove VOCs before it can be discharged to the atmosphere. The cost of VOC removal is proportional to the volumetric flow rate that is treated and is only slightly affected by the inlet VOC concentration. Because the volumetric flow rates are of the order of many thousands of ACFM, the cost is large.

Schematic diagrams of two configurations of split-flow ventilation booths are shown in Figure 5.13.

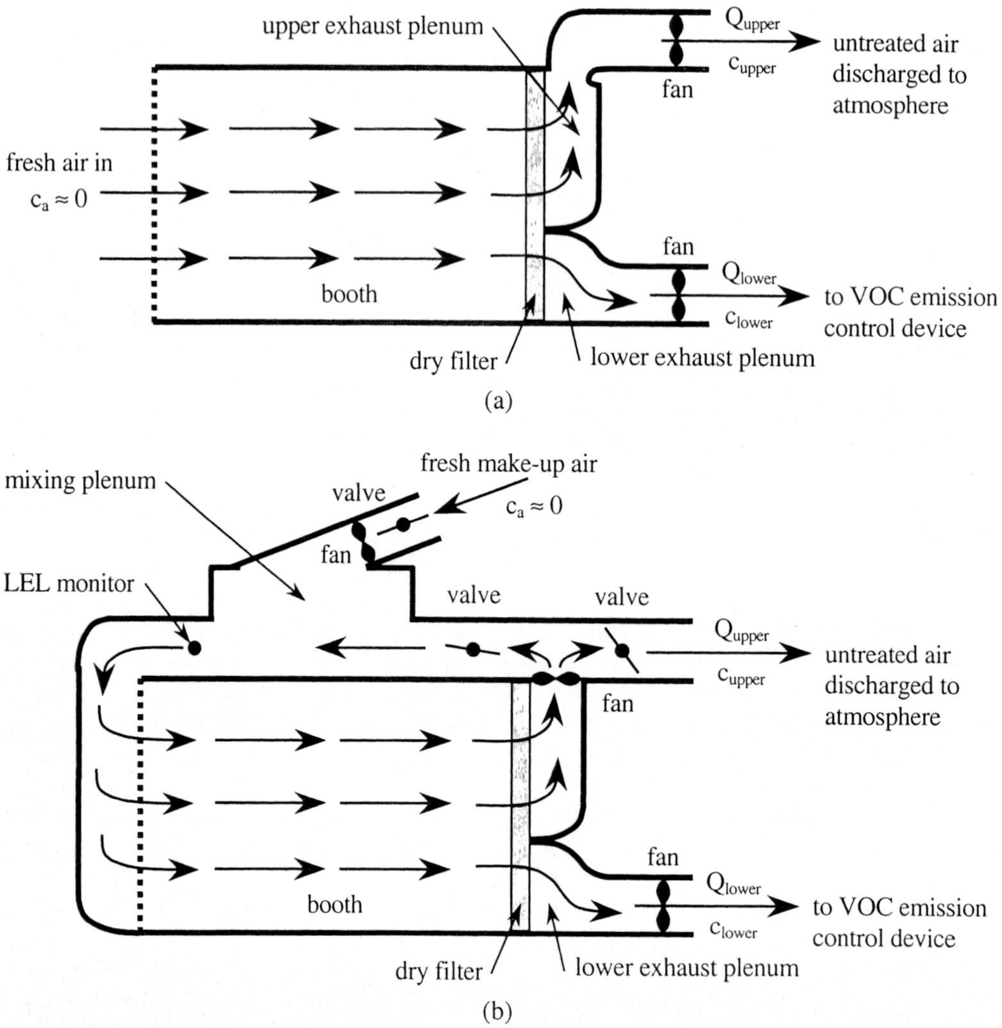

Figure 5.13 Split-flow ventilation booths: (a) untreated air discharged to the atmosphere, and (b) untreated air recirculated to the booth.

(a) Figure 5.13a shows that exhaust from the upper portion of the booth where the VOC concentration is low is untreated, and is discharged to the atmosphere. Air from the lower portion of the booth is treated to remove VOCs and then discharged to the atmosphere.

(b) Figure 5.13b shows that exhaust from the lower portion of the booth is treated in the same way as (a), but exhaust from the upper portion of the booth is mixed with fresh make-up air and recirculated to the booth. In both cases the analysis that follows assumes that the concentrations in the upper and lower portions of the booth are different but uniform within the gas stream. In the model to be described, only steady-state concentrations are analyzed. A transient analysis can be formulated, although it is more complicated.

Split-flow paint booths can be designed using intuition to select the volumetric flow rate at which air is withdrawn from the upper and lower sections of the booth, and then experiments can be undertaken to verify that performance is satisfactory. Such seat-of-the-pants engineering has always been practiced and sometimes works. Unfortunately it is unreliable, is an imprudent expenditure of resources, and cannot be recommended. A better strategy is to employ the design method in which several proposed configurations are analyzed and winnowed, and the most promising is chosen for construction and testing.

To more accurately model split-flow paint booths it is necessary to use CFD techniques because the essential feature of the booths is that the concentration varies with regard to height as well as laterally. To divide the booth into two well-mixed sections and to guess how contaminant is transported from one section to the other is an imprudent leap of faith. Assigning values to the transfer of contaminant inside the booth is tantamount to assuming the solution of the problem. A half-way measure is to divide the booth into a number of internal boxes and assign parametric values describing this transfer. This analytical technique, called the ***sequential box model*** (Ryan et al., 1986, 1988; Heinsohn et al., 1989, 1991, 1993) will not be discussed here because CFD analysis is far superior (see Chapter 10).

The principal chore in designing split-flow paint booths is selecting the volumetric flow rates at which air should be withdrawn from the upper and lower sections of the booth. The design requires selection of judicious openings inside the booth, ductwork, and fans, and should not be approached casually. The total volumetric flow rate of air passing through the booth (Q_t) is dictated by the cross-sectional area of the booth that accommodates the equipment to be painted; the velocity of air passing through the booth is dictated by OSHA standards, which prescribe a velocity of 75-150 FPM. The rate at which VOC is emitted to the air in the booth (S) is dictated by the painting schedule and is a given condition. Inside an enclosed paint booth, workers wear respirators equipped with an external supply of fresh air; consequently the VOC concentration does not have to satisfy PEL requirements. Nonetheless the VOC concentration must be considerably below the lower explosion limit (LEL), and care must be taken that pockets of high concentration do not inadvertently occur. Lastly, all electric equipment used inside the booth must be explosion-proof. The minimum overall VOC removal efficiency ($\eta_{overall}$) is dictated by state regulation. While analysis of booth performance should be performed with CFD, it is instructive to analyze a surrogate problem that provides insight into the larger problem. The surrogate problem reflects the booths seen in Figures 5.13a and 5.13b.

Example 5.12 – Split-Flow Ventilation Paint Booth

Given: You are an engineer employed at military base that performs routine maintenance on military trucks and tanks. The final maintenance operation is repainting. The vehicles are painted inside three identical enclosed paint booths. Workers wearing respirators equipped with external fresh air for breathing (see Chapter 1) apply unique coatings of camouflage paint. Airless sprayers apply coatings that use VOCs to convey unique pigments to the metal because it must be able to withstand attack by biological and chemical agents. Painting is currently accomplished in enclosed, straight-through booths in which fresh air enters the booth at a volumetric flow rate of 312. m^3/min, and an equal amount of air containing VOCs is exhausted from the booth and sent directly to a direct-flame thermal incinerator that removes 85% of the VOC before discharging the products to the atmosphere. Painting emits VOCs into the air in the booth at a rate (S) of 0.112 kg/min. The state regulatory agency now requires 95% of the VOC to be removed from the paint booths.

The cost of a new direct-flame thermal oxidizer is high, and is proportional to the volumetric flow rate of treated air rather than removal efficiency. The Director of Maintenance recommends that instead of purchasing three new thermal oxidizers, two of the booths be modified for split-flow. One new thermal oxidizer is purchased for Booth A; Figure E5.12a is a

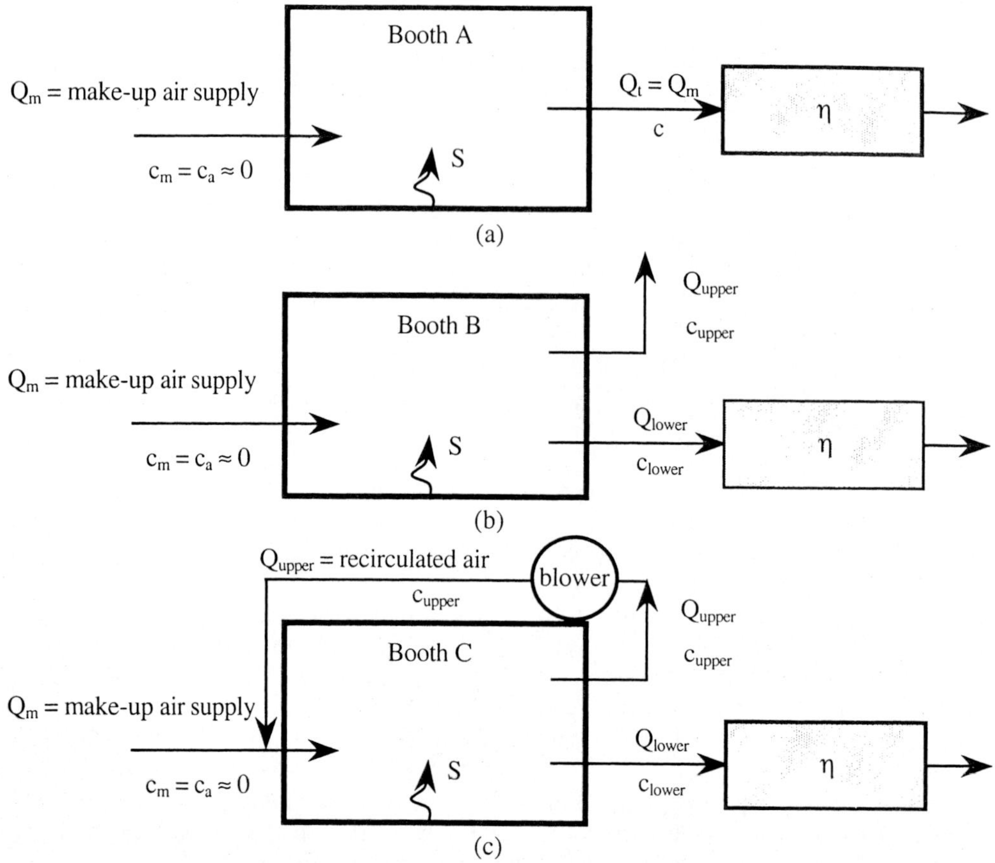

Figure E5.12 Three ventilation schemes: (a) Booth A: original configuration, (b) Booth B: no recirculation, no treatment of upper flow, and (c) Booth C: recirculation of untreated upper flow.

schematic diagram of Booth A, which is the original configuration. Figure E5.12b is a diagram of the modified Booth B and Figure E5.12c is a diagram of the modified Booth C. The Director of Maintenance also recommends that the current (85%) thermal oxidizers be retained and used on the modified booths (B and C), with the naïve assumption that at reduced volumetric flow rate, the efficiency might improve.

The base commander orders that the work be done. A 95% efficient oxidizer with a capacity of 400 m^3/min is installed for Booth A. New ducts are constructed and new fans are purchased and installed in Booths B and C. Unfortunately, no one analyzes the booths to determine how much air should be withdrawn from the upper and lower portions of each booth and the required VOC removal efficiency for each booth. The Director assumes (incorrectly) that the existing thermal oxidizers will be adequate. After the work is completed, Booth A is in compliance, but neither Booth B nor Booth C is in compliance with the new state regulations. The Director of Maintenance calls you for help. Specifically you are asked to recommend one of the options below:

(a) the new ductwork should be removed, and two new direct-flame thermal oxidizers with a minimum removal efficiency of 95.% should be purchased for Booths B and C
(b) the new ductwork should be retained, but the volumetric flow rate and VOC removal efficiency for two new direct-flame thermal oxidizers must be prescribed

The spraying practices and characteristics of the original booths and contaminants before modification are as follows.

- volume = V = 254. m^3
- cross-sectional area = A_{booth} = 23.8 m^2 (16.0 ft by 16.0 ft)
- total exhaust volumetric flow rate = Q_t = 11,000 ACFM (312. m^3/min)
- average velocity through current booth = Q_t/A_{booth} = 43.0 FPM (13.1 m/min)
- VOC source strength = S = 6.72 kg/hr (0.112 kg/min)
- OSHA PEL for the VOC = 540 kg/m^3
- the VOC concentration in make-up air is negligible

The first thing you do is ask that the volumetric flow rate and VOC concentration in the air withdrawn from the upper and lower portions of each modified booth be measured. The following data are reported to you:

Booth B (Figure E5.12b)

Q_{upper} = 50. m^3/min, c_{upper} = 40. mg/m^3
Q_{lower} = 262. m^3/min, c_{lower} = 419. mg/m^3

Booth C (Figure E5.12c)

Q_{upper} = 188. m^3/min, c_{upper} = 100. mg/m^3
Q_{lower} = 248. m^3/min, c_{lower} = 480. mg/m^3

To do:

(a) Consider the original paint booth with its new thermal oxidizer (Booth A, Figure E5.12a), which treats the entire volumetric flow rate of air, and verify that the new VOC removal system achieves compliance.
(b) Analyze Booth B (Figure E5.12b), which discharges untreated air withdrawn from the top of the booth, but treats air withdrawn from the bottom of the booth; recommend a new direct-flame thermal oxidizer.
(c) Analyze Booth C (E5.12c), which discharges treated air withdrawn from the bottom of the booth, but recirculates air from the top of the booth; recommend a new direct-flame thermal oxidizer.

Solution:

(a) Treat the entire volumetric flow through the booth (Booth A, Figure E5.12a): If 95% of the VOC is to be removed at 312. m^3/min, the maximum allowable VOC emission rate to the atmosphere is

$$\dot{m}_{VOC} = S(1-\eta) = 0.112\frac{kg}{min}(1-0.95) = 0.0056\frac{kg}{min} = 0.34\frac{kg}{hr}$$

If gases in the booth are well mixed, the average VOC concentration in Booth A is

$$c_{booth} = \frac{S}{Q_t} = \frac{0.112\frac{kg}{min}}{312.\frac{m^3}{min}}\left(\frac{10^6\ mg}{kg}\right) = 359.\frac{mg}{m^3}$$

Obviously, a 95.% efficient direct-flame thermal oxidizer with a volume flow rate rating of at least 312 m^3/min is required for Booth A. The newly installed thermal oxidizer is adequate.

(b) Discharge untreated air from the upper part of the booth and discharge treated air from the lower part of the booth (Booth B, Figure E5.12b): The total volumetric flow rate through Booth B is

$$Q_t = Q_{lower} + Q_{upper} = 262.\frac{m^3}{min} + 50.\frac{m^3}{min} = 312.\frac{m^3}{min}$$

The overall average velocity of VOC-laden air inside the booth remains at 13.1 m/min; however, unless measurements were made, it would not be possible to assign an average velocity in the upper and lower portions of the booth. The efficiency of a VOC removal system to achieve compliance for Booth B can be found from an equation for mass balance,

$$\dot{m}_{VOC} = (1-\eta)c_{lower}Q_{lower} + c_{upper}Q_{upper}$$

Using the same maximum allowable VOC emission rate to the atmosphere as in Part (a) above,

$$\eta = 1 - \frac{\dot{m}_{VOC} - c_{upper}Q_{upper}}{c_{lower}Q_{lower}} = 1 - \frac{0.0056\frac{kg}{min}\left(\frac{10^6\ mg}{kg}\right) - 40.\frac{mg}{m^3}50.\frac{m^3}{min}}{419.\frac{mg}{m^3}262.\frac{m^3}{min}} \cong 96.7\%$$

The direct-flame thermal oxidizer for the lower stream should be sized to remove VOC from air at 262. m^3/min with removal efficiency of no less than 96.7%. The average VOC concentration inside Booth B is

$$c_{booth} = \frac{c_{lower}Q_{lower} + c_{upper}Q_{upper}}{Q_t} = \frac{419.\frac{mg}{m^3}262.\frac{m^3}{min} + 40.\frac{mg}{m^3}50.\frac{m^3}{min}}{312.\frac{m^3}{min}} = 358.\frac{mg}{m^3}$$

As expected, this is the same as that in Booth A (within round-off error), since conditions inside the booth itself have not been modified by the ventilation scheme of Figure E5.12b.

(c) Discharge treated air withdrawn from the lower portion of the booth, but recirculate untreated air withdrawn from the upper portion of the booth (Booth C, Figure E5.12c): The total volumetric flow of VOC-laden air through Booth C is

$$Q_t = Q_{lower} + Q_{upper} = 248.\frac{m^3}{min} + 188.\frac{m^3}{min} = 436.\frac{m^3}{min}$$

An overall average velocity through the booth is

$$U_{booth} = \frac{Q_t}{A_{booth}} = \frac{436.\frac{m^3}{min}}{23.8\ m^2} = 18.3\frac{m}{min}$$

The minimum required VOC removal efficiency can be found from a mass balance,

$$\dot{m}_{VOC} = (1-\eta)c_{lower}Q_{lower}$$

from which

$$\eta = 1 - \frac{\dot{m}_{VOC}}{c_{lower}Q_{lower}} = 1 - \frac{0.0056\frac{kg}{min}\left(\frac{10^6\ mg}{kg}\right)}{480.\frac{mg}{m^3}248.\frac{m^3}{min}} = 0.9530 \cong 95.3\%$$

The direct-flame thermal oxidizer for the lower stream should be sized to remove VOC from air at 248. m^3/min with a removal efficiency no less than 95.3%. Finally, the average VOC concentration inside Booth C is

$$c_{booth} = \frac{c_{lower}Q_{lower} + c_{upper}Q_{upper}}{Q_t} = \frac{480.\frac{mg}{m^3}248.\frac{m^3}{min} + 100.\frac{mg}{m^3}188.\frac{m^3}{min}}{436.\frac{m^3}{min}} = 316.\frac{mg}{m^3}$$

Discussion: It is clear that it would be unwise to remove all the ductwork from the modified booths to return them to their original state and purchase two new thermal oxidizers to remove VOC from the total volume flow rate of 312. m^3/min with a minimum removal efficiency of 95%. Not only would there be an additional expense to remove the new ductwork, but the capital cost and operational cost of the new thermal oxidizers would be greater than those needed for the modified booths.

Booth B would achieve compliance without changing the overall average air velocity or VOC concentration inside the booth that workers may have become accustomed to. In addition, the size of the new thermal oxidizer for Booth B is considerably (16%) smaller than what would be needed to bring the original into compliance (262. m^3/hr rather than 312. m^3/hr). However, the required efficiency of the oxidizer is higher than that required for (unmodified) Booth A. But since the cost of the oxidizer is more dependent on volume flow rate than on efficiency, the savings in capital and operating cost should be significant.

Booth C would achieve compliance with a thermal oxidizer smaller in size (248. m^3/hr rather than 262. m^3/hr) and at a lower required efficiency (95.3% rather than 96.7%) compared to the thermal oxidizer for Booth B. The average concentration inside Booth C is smaller than in either Booth A or Booth B, so this option seems at first to be the best choice. However, the average air velocity is considerably (40%) higher than that to which workers are accustomed. Increased air velocity is significant because it can affect the quality of the finished painted surface; this option is therefore not recommended. Booth C should be converted to the ventilation scheme of booth B by bypassing (rather than removing) the recirculation duct.

Even greater financial savings might be possible if details of the split-flow booths were based on design recommendations obtained from CFD calculations. Certainly there is nothing to suggest that the modifications that were made were the optimum ones. Nonetheless, as luck would have it in this example, design "by the seat of the pants" produced changes that can not only achieve compliance, but can lead to substantial financial savings.

5.12 Mean Age of Air and Ventilation Effectiveness

Only rarely do workplaces and offices serviced by dilution or displacement ventilation systems fully achieve their desired performance. To assess the effectiveness of a ventilation system, engineers need a method to evaluate ventilation systems. Simply measuring contaminant concentrations at selected points inside the workplace does not fully assess the effectiveness of the overall system. At first reading, one suspects that displacement ventilation systems have the edge over dilution systems because individuals receive a constant stream of supply air, and contaminants are carried upward to the exhaust registers. With dilution ventilation systems, the best an individual can expect is to have the pollutant concentration diluted to an acceptable level. The ideal displacement system consists of ***plug flow*** in which supply air moves upward (floor to ceiling) at a velocity Q/A_c as a plug of fresh air sweeping out pollutants in front of it, where A_c is the cross-sectional area of the room. In practice, however, only rarely does the velocity field achieve plug flow. Similarly, dilution ventilation systems generate internal recirculation that mixes polluted air with fresh supply air until minimally acceptable concentrations are achieved. In practice, only rarely are uniform concentrations achieved, as specified by Eq. (5-1). Displacement ventilation is considered to be the most efficient ventilation system (Holmberg et al., 1993; Breum and Orhede, 1994) because if the air inlets and outlets are placed properly, the entire volume of air in an enclosure is replaced by fresh make-up air after an elapsed time called the ***characteristic time*** (t_N). The characteristic time is defined as the reciprocal of the number of room air changes per unit time (N),

$$t_N = \frac{1}{N} = \frac{V}{Q} \tag{5-56}$$

If the circulation of air in an enclosure is known to produce *nonuniform* concentrations, i.e. ***partially mixed conditions***, it would be useful to know which parts of the enclosure receive too little of the incoming air and which parts receive too much. For example, point P in Figure 5.14 may be located in a low-velocity eddy region in which the concentration changes slowly, whereas point Q may be located in a high-velocity region in which the concentration changes rapidly. However, just because the concentration changes slowly does not mean the concentration is large or small, but only that it changes slowly. The ***local mean age***, also called the ***age of the make-up air*** at point P is defined as the elapsed time ($t_{age,P}$) between entering the room and reaching point P. After passing point P, the air parcel travels to the exhaust. The ***residence time*** of make-up air at point P ($t_{residence,P}$) is defined as the time to reach the exhaust after passing through point P. The ***average age of air in an enclosure*** is defined by considering the age of air parcels passing through every point in the enclosure (room).

Sandberg (1983) has developed a comprehensive theory to analyze nonuniform conditions within enclosures based on measuring the concentration over a period of time (t′) and computing a ***dose*** (or ***zero moment*** of the concentration) at some location in the room under consideration

$$\text{dose} = \int_0^{t'} c(t)\,dt \tag{5-57}$$

and a ***first moment*** of the concentration at the same location,

$$\text{first moment} = \int_0^{t'} t c(t)\,dt \tag{5-58}$$

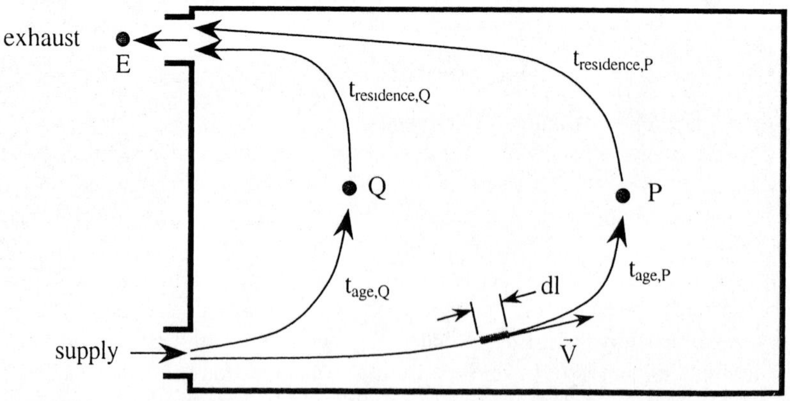

Figure 5.14 Local mean age and residence time in a room for two points, P and Q, in the room.

where t' is the time period of interest. Partially mixed conditions can be characterized by an ***effectiveness coefficient*** (e) (Skaret and Mathisen 1983; Niemela et al., 1986; Perera et al., 1986; Fontaine et al., 1989; Breum and Orhede, 1994; ASHRAE Fundamentals Handbook, 2001), defined as

$$\boxed{e = \frac{t_N}{t_{age}}} \tag{5-59}$$

where the local mean age (t_{age}) at the point is approximated by the ratio of the first moment at the point to the dose at the point as t' goes to infinity in Eqs. (5-57) and (5-58),

$$\boxed{t_{age} = \frac{\int_0^\infty tc(t)dt}{\int_0^\infty c(t)dt}} \tag{5-60}$$

A tracer gas decay ("step-down") experiment is assumed such that c approaches zero as t becomes large; otherwise Eq. (5-60) would yield an infinite value for t_{age}. For well-mixed conditions, one can show mathematically that the local mean age (t_{age}) is equal to t_N at every point in the room. If the effectiveness coefficient is unity (e = 1) at a point in a room, mixing at that point is similar to what would be predicted if the entire space were well mixed. If $t_{age} > t_N$ at some other point in the room, the effectiveness coefficient (e) is less than unity, and it means that the point is located in a somewhat stagnant region where mixing is poor. Finally, if the effectiveness coefficient (e) is greater than unity, then mixing at that point in the room is very vigorous and one may visualize that make-up air passes quickly through the point and "short-circuits" other points in the room. In summary,

- $e < 1$ requires $t_{age} > t_N$ which implies poor mixing at that point
- $e = 1$ requires $t_{age} = t_N$ which implies the same level of mixing at that point as if the entire room was well mixed
- $e > 1$ requires $t_{age} < t_N$ which implies good mixing at that point

The reader must be careful to understand the difference between two rather overlapping terms:

- m = mixing factor, defined by Eq. (5-37)
- e = effectiveness coefficient, defined by Eq. (5-59)

The mixing factor (m) is an empirical constant included in the differential equation for a well-mixed model to try to account for the fact that only a portion of the inlet air mixes with air inside the enclosure. Even when a mixing factor is included in the differential equation, the concentration remains *spatially uniform*, i.e. c(x,y,z,t) = c(t). On the other hand, the effectiveness coefficient (e) is a scalar quantity evaluated at a specific point in space; the effectiveness coefficient reflects the level of mixing occurring at that point, and the concentration is *not* spatially uniform, i.e. c = c(x,y,z,t).

Suppose the flow field in an enclosure is as shown in Figure 5.14. If the velocity field is known, the age of the make-up air ($t_{age,P}$) reaching some point (P) in the room could be computed from the integral

$$\boxed{t_{age,P} = \int_{\text{pathline from inlet to P}} \frac{dl}{\left|\vec{V}\right|}} \tag{5-61}$$

where l (lower case L) is the length of the fluid pathline from the inlet to point P, and $\left|\vec{V}\right|$ is the speed of the air along the pathline, as illustrated in Figure 5.14. In a similar fashion, the residence time ($t_{residence,P}$) passing through point P can be computed from

$$t_{residence,P} = \int_{\text{pathline from P to outlet}} \frac{dl}{|\vec{V}|} \tag{5-62}$$

If some other point (Q) receives make-up air faster than does point P, the age of the make-up air at point P is larger than at point Q and the effectiveness coefficient at point P is less than it is at point Q. The greater the disparity of effectiveness coefficients (e) throughout the room, the more uneven the distribution of make-up air and the less valid an assumption of well mixed becomes. By determining the effectiveness coefficients (e) at various points in the room, one can evaluate partially mixed conditions on a quantitative basis and characterize the enclosure in a more detailed fashion than by merely using a mixing factor (m).

A similar integral taken over the entire pathline from inlet to outlet is called the ***total elapsed time*** ($t_{elapsed}$), which is defined as the time it takes for the make-up air to pass through the room, and can be found by integrating from the inlet to the outlet,

$$t_{elapsed} = \int_{\text{pathline from inlet to outlet}} \frac{dl}{|\vec{V}|} = t_{age} + t_{residence} \tag{5-63}$$

If the flow through a room is in one direction and spatially uniform (*plug flow*), the pathline in Eq. (5-63) is simply a straight line from the floor of the room to the ceiling (assuming the supply and exhaust are on the bottom and top of the room respectively). Also, the speed of the air along the pathline is a constant equal to the plug flow velocity (Q/A_c) through the room. Thus, for plug flow,

$$t_{elapsed} = \int_{\text{pathline from inlet to outlet}} \frac{dl}{|\vec{V}|} = \int_{z=0}^{H} \frac{dz}{(Q/A_c)} = \frac{HA_c}{Q} = \frac{V}{Q} = t_N$$

In other words, for plug flow, the total elapsed time is the same as the characteristic time (t_N). If plug flow conditions exist (idealized displacement ventilation), the conventional time constant for well-mixed conditions is equal to the physical time in which air resides in an enclosure. The reader should be careful not to conclude that well-mixed means plug flow because it certainly does not.

The local mean age of air at point P, ($t_{age,P}$) can be determined experimentally by timing the movement of neutral density bubbles (small helium filled spheres of overall density equal to that of air) traveling between the inlet and point P. This experimental procedure is tedious; a more efficient experimental procedure can be made using a ***tracer gas*** injected into the supply air. A common tracer gas is SF_6 because it is nontoxic, noncombustible, and has a distinctive molecular weight that can be detected easily. The mean age of air at some point P or Q can also be determined experimentally by ***step-up***, ***step-down***, or ***pulse injection*** tracer experiments. In step-up experiments, a tracer gas is added to the make-up air entering an enclosure at a constant rate and the (rising) concentration at a point P or Q in the room is measured over a period of time; Figure 5.15 shows the concentration ratio as a function of time in a step-up experiment. In step-down experiments, the tracer gas is added to the room and uniformly distributed. At time zero, the introduction of tracer gas is stopped and only fresh air is added to the room. The (decreasing) concentration is measured at a point P or Q in the room over a period of time, and t_{age} at point P or Q is computed.

From plots like the one shown in Figure 5.15, the local mean age at point P can be determined mathematically by evaluating the manner in which the concentration $c_P(t)$ approaches its equilibrium value, $c_P(\infty)$ (which can also be called c_{Pss}). If the local mean age is small, the concentration approaches its equilibrium value quickly and both the dose, Eq. (5-57), and first moment, Eq. (5-58), are small. The local mean age is essentially the distance on the horizontal

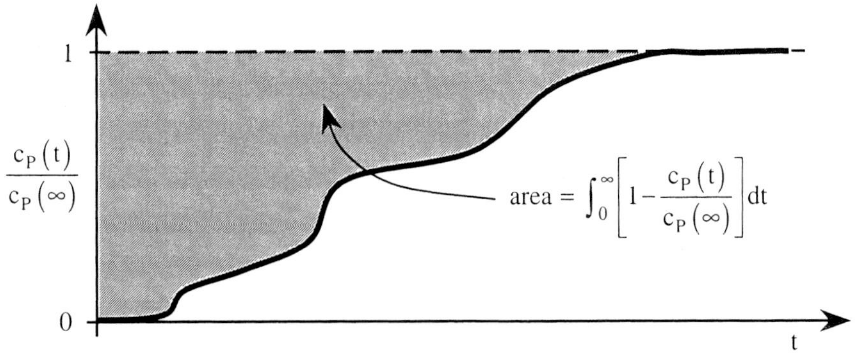

Figure 5.15 Normalized mass concentration at point P as a function of time for a step-up experiment. The shaded area represents the denominator of Eq. (5-64).

axis (time) to the centroid of the shaded area in Figure 5.15. Specifically, for a step-up tracer experiment, at some point P in the room, Eq. (5-60) is modified as follows:

$$t_{age,P} = \frac{\int_0^\infty t\left[1-\frac{c_P(t)}{c_P(\infty)}\right]dt}{\int_0^\infty \left[1-\frac{c_P(t)}{c_P(\infty)}\right]dt} \tag{5-64}$$

where $c_P(\infty)$ is the tracer concentration at point P at time equal to infinity. For a step-down tracer experiment,

$$t_{age,P} = \frac{\int_0^\infty t\frac{c_P(t)}{c_P(0)}dt}{\int_0^\infty \frac{c_P(t)}{c_P(0)}dt} \tag{5-65}$$

where $c_P(0)$ is the initial tracer concentration at point P. Since cP(0) is a constant, Eq. (5-65) is identical to Eq. (5-60). The experiments described above can be conducted without much difficulty because for most rooms and buildings, the concentration changes slowly over a period of many minutes, and rapid-response instruments are not needed. In addition, measurements can be made at several points and effectiveness coefficients can be computed for several points in a single experiment. The technique has been found to be very useful in characterizing ventilation in buildings and rooms within buildings (Breum 1988). Constant-injection and pulse-injection tracer techniques can also be used to analyze air flow in ducts and in heating, ventilating, and air conditioning (HVAC) systems (Cheong and Riffat, 1993).

Example 5.13 - Local Mean Age of Air at Points in an Enclosure

Given: Points P, Q, and R correspond to three locations in an enclosure of volume V = 6000 ft^3. The enclosure receives fresh air at volumetric flow rate Q = 600. ACFM, and discharges room air at the same rate. A step-down tracer experiment has been run and the mass concentrations at points P, Q, and R are found to vary in the following way:

- at point P, $\quad \frac{c_P(t)}{c(0)} = 1.0 \text{ for } t \le t_1 \qquad \frac{c_P(t)}{c(0)} = 0 \text{ for } t > t_1$

- at point Q, $\frac{c_Q(t)}{c(0)} = 1.0 - \frac{t}{t_2}$ for $t \le t_2$ $\frac{c_Q(t)}{c(0)} = 0$ for $t > t_2$

- at point R, $\frac{c_R(t)}{c(0)} = \exp\left(-\frac{Q}{V}t\right)$ for all t

where c(0) is the initial (uniform) concentration in the room, t_1 = (0.20)V/Q = 2.0 min, and t_2 = (0.50)V/Q = 5.0 min.

To do: Calculate and compare the effectiveness coefficient ($e = t_N/t_{age}$) at each of the three points.

Solution: From Eq. (5-56), the characteristic time is

$$t_N = \frac{V}{Q} = \frac{6000\ \text{ft}^3}{600\ \frac{\text{ft}^3}{\text{min}}} = 10.\ \text{min}$$

Using Eq. (5-65) at point P,

$$t_{age,P} = \frac{\int_0^\infty t \frac{c_P(t)}{c_P(0)} dt}{\int_0^\infty \frac{c_P(t)}{c_P(0)} dt} = \frac{\int_0^{t_1} (1.0)t dt}{\int_0^{t_1} dt} = \frac{t_1}{2} = \frac{2.0\ \text{min}}{2} = 1.0\ \text{min} \qquad \underline{e_P = 10.}$$

at point Q,

$$t_{age,Q} = \frac{\int_0^\infty t \frac{c_Q(t)}{c_Q(0)} dt}{\int_0^\infty \frac{c_Q(t)}{c_Q(0)} dt} = \frac{\int_0^{t_2} \left[1 - \frac{t}{t_2}\right] t dt}{\int_0^{t_2} \left[1 - \frac{t}{t_2}\right] dt} = \frac{\frac{t_2^{\,2}}{6}}{\frac{t_2}{2}} = \frac{t_2}{3} = \frac{5.0\ \text{min}}{3} = 17.\ \text{min} \qquad \underline{e_Q = 0.59}$$

and at point R,

$$t_{age,R} = \frac{\int_0^\infty t \frac{c_R(t)}{c_R(0)} dt}{\int_0^\infty \frac{c_R(t)}{c_R(0)} dt} = \frac{\int_0^\infty t \exp\left(-\frac{tQ}{V}\right) dt}{\int_0^\infty \exp\left(-\frac{tQ}{V}\right) dt} = \frac{\left(\frac{V}{Q}\right)^2}{\left(\frac{V}{Q}\right)} = \left(\frac{V}{Q}\right) = t_N = 10.\ \text{min} \qquad \underline{e_R = 1.0}$$

Discussion: All results are reported to two significant digits. Point P is a location where the mean age ($t_{age,P}$) is considerably smaller than the characteristic time (t_N); such a location has very good mixing. Point Q, on the other hand, is a location where the average age ($t_{age,Q}$) is somewhat larger than t_N; mixing is poor at point Q. Point R is a location where $t_{age,R} = t_N$; mixing is what one would predict if the entire volume were well mixed. The well-mixed assumption is certainly *not* valid for this room.

While the ventilation effectiveness at points in a room is useful to distinguish one point from another, a better parameter for engineering design is to use the *exhaust opening* (point E in Figure 5.14) for the measurements. Since all tracer gas entering the room ultimately has to pass through the exhaust opening, the mean age computed at the exhaust is essentially integrated over the entire room and can be called the ***room mean age*** ($t_{room,avg}$). For a step-up experiment, Eq. (5-64) becomes

$$t_{room,avg} = t_{age,E} = \frac{\int_0^\infty t\left[1-\frac{c_E(t)}{c_E(\infty)}\right]dt}{\int_0^\infty \left[1-\frac{c_E(t)}{c_E(\infty)}\right]dt} \quad (5\text{-}66)$$

where c_E is the concentration at the exhaust opening (E). Similarly, for a step-down experiment, $t_{room,avg}$ is defined by replacing subscript P in Eq. (5-65) by E. To assign an effectiveness for the whole room, room mean age is compared to the characteristic time (t_N); the ***room ventilation effectiveness coefficient*** (e_{room}) is defined similar to Eq. (5-59),

$$e_{room} = \frac{t_N}{t_{room,avg}} \quad (5\text{-}67)$$

To place a numerical value on this parameter, the engineer needs to conduct step-up or step-down tracer experiments and measure the concentration in the exhaust air leaving the room (point E in Figure 5.14). To compute the ventilation effectiveness of a room, Eq. (5-64) or (5-65) is used, with the understanding that the point in question is now the exhaust opening. Another virtue of computing the room ventilation effectiveness is that ideal displacement ventilation produces an effectiveness of 200%, and well-mixed dilution ventilation produces an effectiveness of 100%:

- $e_{room} < 1.0$ — not well-mixed, response slower than well-mixed dilution ventilation system
- $e_{room} = 1.0$ — well-mixed, dilution ventilation system
- $1.0 < e_{room} < 2.0$ — ventilation system performance is somewhere between displacement and dilution system
- $e_{room} = 2.0$ — displacement ventilation system (most efficient)

Mathematical purists would divide Eq. (5-67) by 2 and rename the quantity "ventilation efficiency", because this would ensure that the most efficient (displacement) ventilation system has an efficiency of 100%. Unfortunately, this is not done in the ventilation literature.

Example 5.14 - Ventilation Effectiveness of three Beauty Shops

Given: Three beauty shops have been built. Each has a ventilation system which satisfies ASHRAE ventilation standards (see Appendix A.16), but the location of the inlet and outlet registers in each shop are vastly different. The owner is concerned that vapors and odors from the beauty products may accumulate at certain points in the shop. You have been asked by the owner of the three shops to compare their ventilation systems. In each shop the volume is 16,000 ft^3 and the ventilation volumetric flow rate is 1,250 ACFM. You sample air in the exhaust duct, and conduct a step-up tracer experiment in each shop. The following results are obtained:

- Shop 1: $\frac{c_E(t)}{c_E(\infty)} = 0 \text{ for } t \le 12.8 \text{ min} \qquad \frac{c_E(t)}{c_E(\infty)} = 1.0 \text{ for } t > 12.8 \text{ min}$
- Shop 2: $\frac{c_E(t)}{c_E(\infty)} = 1.0 - \exp\left(-\frac{t}{12.8 \text{ min}}\right) \text{ for all t}$
- Shop 3: $\frac{c_E(t)}{c_E(\infty)} = \frac{t}{30.0 \text{ min}} \text{ for } t \le 30.0 \text{ min} \qquad \frac{c_E(t)}{c_E(\infty)} = 1.0 \text{ for } t > 30.0 \text{ min}$

To do: Compare the performance of the ventilation systems of the three shops.

Solution: Since the shops have the same volume and ventilation flow rate, they have the same characteristic time constant:

$$t_N = \frac{V}{Q} = \frac{16,000\ \text{ft}^3}{1250\ \frac{\text{ft}^3}{\text{min}}} = 12.8\ \text{min}$$

Using the above equations for the three shops, the results are tabulated below:

beauty shop	$t_{room,avg}$ (min)	e_{room}
1	6.4	2.0
2	12.8	1.0
3	10.0	1.3

Thus Shop 1 has a ventilation system that performs as a displacement ventilation system; contaminants can be expected to be removed from every point in the room within about 13. minutes after being emitted *anywhere* inside the shop. Shop 2 has a ventilation system that performs as a dilution ventilation system (well-mixed model); contaminants emitted anywhere inside the shop are removed very slowly. Shop 3 has a ventilation system that does not perform as well as a displacement system (Shop 1) nor as slowly as a dilution system (Shop 2).

Discussion: In reality, e_{room} is never exactly 1.0 nor exactly 2.0 because no real-life ventilation system can be completely well-mixed (there are always some stagnant areas), nor behave completely as a displacement system (there are always room currents from heat sources, people walking, etc.). Real-life data would not be so simple; numerical integration would most likely be necessary. Nevertheless, the above example is useful as a learning tool since the integrations are straightforward.

5.13 Make-up Air Operating Costs

Make-up air must be heated (or cooled) before it enters the workplace. In some cases it may also have to be cleaned and/or have water vapor added or removed. The cost associated with conditioning make-up air is significant, and is of key consideration in the design of HVAC (general ventilation) systems. Unfortunately, make-up air operating costs are sometimes ignored in the design of contaminant removal (local ventilation) systems. Such an omission is serious, and care must be taken to avoid it. For purposes of illustration, only the yearly cost to *heat* make-up air is considered here. The rate at which energy is required to heat outside air from the local ambient outdoor air temperature ($T_{outdoor}$) to the building temperature ($T_{building}$), typically 70 °F, is equal to the make-up air mass flow rate (ρQ) times the enthalpy change $c_p(T_{building} - T_{outdoor})$,

$$\boxed{\dot{Q}_{\text{make-up air}} = \rho Q c_p \left(T_{building} - T_{outdoor}\right)} \tag{5-68}$$

where c_p is the specific heat at constant pressure (of the make-up air). The energy added throughout the year is found by integrating Eq. (5-68) over the time in which energy is added. Since the outdoor temperature ($T_{outdoor}$) varies throughout the year, such an integral would be difficult to compute, and once done would have to be repeated for other years in which $T_{outdoor}$ differs. Instead, HVAC engineers use the concept of ***heating degree days*** (DD_h), defined as

$$\boxed{DD_h = (1\ \text{day}) \sum_{\text{days}} \left(T_{bal} - \overline{T_{outdoor}}\right)^+} \tag{5-69}$$

where T_{bal} is the balance point temperature, defined as the outdoor ambient temperature ($T_{outdoor}$) at which the total heat loss from the building is equal to the heat gain from the sun, occupants, lights, etc., and $\overline{T_{outdoor}}$ is the daily average outdoor temperature. The plus sign in Eq. (5-69) indicates that only positive values are to be counted. The summation is over the total number of days in the heating season. The units of DD_h are (°F day) or (Kelvin day). Values of DD_h for specific communities have been measured and tabulated; e.g. tables of DD_h for communities throughout the US can be found in the ASHRAE Fundamentals Handbook (2001). Shown in Table 5.5 is a sample summary of degree days (DD_h) (in units of °F days and Kelvin days) for selected cities, based on a balance point temperature (T_{bal}) of 18.3 °C.

If energy is acquired from a heater, the yearly fuel cost (annual cost) is equal to the energy gained times the number of hours in which heating is required times the cost per unit of fuel, called the ***unit fuel cost*** (C_{fu}), divided by the ***unit fuel energy***, also called ***available energy per unit of fuel*** (q_{fu}), which is related to the efficiency of conversion. The 23rd ACGIH manual (1998) recommends the following equation:

$$\boxed{\text{annual cost (in dollars)} = 0.154\frac{(DD_h)\, t_{operating} C_{fu} Q}{q_{fu}}} \tag{5-70}$$

where

- Q = volumetric flow rate (ACFM)
- DD_h = annual heating degree days (°F day) for building air at 70 °F
- $t_{operating}$ = operating time (hours/week)
- C_{fu} = unit fuel cost ($/unit of fuel)
- q_{fu} = unit fuel energy = available energy per unit of fuel (BTU per unit of fuel)

The constant (0.154) in Eq. (5-70) incorporates all the necessary unit conversions; the variables *must* be in the units specified above. Typical values of available energy per unit of fuel are listed in Table 5.6.

Table 5.5 Heating degree days for several cities in the United States (abstracted from ASHRAE Fundamentals Handbook, 1997).

city	DD_h (°F days)	DD_h (K days)
Los Angeles, CA	1245	692
Denver, CO	6016	3342
Miami, FL	206	114
Chicago, IL	6127	3404
Albuquerque, NM	4292	2384
New York City, NY	4909	2727
Bismarck, ND	9044	5024
Nashville, TN	3696	2053
Dallas/Ft. Worth, TX	2290	1272
Seattle, WA	4727	2626

Table 5.6 Available energy per unit of fuel for several heating options.

fuel	efficiency	available energy per unit of fuel (q_{fu})
coal	50%	6,000 BTU/lbm
oil	75%	106,500 BTU/gal
natural gas (heat exchanger)	80%	800 BTU/ft^3
natural gas (direct-fired)	90%	900 BTU/ft^3
electricity (resistance heating)	100%	3,415 BTU/kWh

A similar concept is used for cooling (air conditioning) cost estimates. HVAC engineers define ***cooling degree days*** (DD_c) in a similar way as Eq. (5-69), but with cooling rather than heating requirements.

Example 5.15 - Cost to Heat Make-up Air
Given: A firm that manufactures circuit boards is located in Nashville, TN. The following data are available:

- $Q = 34{,}500$ ACFM
- $t_{operating} = 60$ hours/week
- $C_{fu} = \$0.007/ft^3$

To do: Estimate the annual cost to heat make-up air to 70. °F, assuming that a direct-fired modular natural gas heater is used.

Solution: From Table 5.5 (for Nashville, TN) and Table 5.6 (for a direct-fired natural gas heater),

- $DD_h = 3696$ °F days
- $q = 900$ BTU/ft³

The annual cost is found by application of Eq. (5-70):

$$\text{annual cost } (\$) = 0.154 \frac{(DD_h) t_{operating} C_{fu} Q}{q_{fu}}$$

$$= 0.154 \frac{(\text{min})(\text{week})(\text{BTU})}{(\text{ft}^3)(\text{hr})(^\circ\text{F})(\text{day})} \frac{(3696\ ^\circ\text{F day})\left(60 \frac{\text{hr}}{\text{week}}\right)\frac{\$0.007}{\text{ft}^3} 34{,}500 \frac{\text{ft}^3}{\text{min}}}{900 \frac{\text{BTU}}{\text{ft}^3}} = \$9{,}160$$

Discussion: The units of the constant (0.154) are shown above; the user can verify that all the units cancel except ($), as required. The annual cost is reported to the nearest $10; there is insufficient precision for predictions of any better accuracy.

The HVAC systems in many industrial and office buildings do not supply constant make-up air flow, and have both heating and cooling loads in the winter. In such cases the degree day method described above is inadequate. More sophisticated methods to estimate the cost to heat (or cool) and (de)humidify replacement air are described in the ASHRAE Fundamentals Handbook (2001).

Finally, the advent of inexpensive micro-computer-based controls has opened the door to many exciting new options in HVAC engineering. For example, controllers can now be designed to adjust the make-up air flow rate as a function of CO_2 concentration in a room (Emmerich and Persily, 1997). The advantage of such a system is obvious – more make-up air is required when there are more people in the room. Another area of current HVAC research involves design of a ***dedicated outdoor air system*** (DOAS) to deliver the proper make-up air and (de)humidification to each individual room (Mumma, 2001); heating and cooling are accomplished with a parallel but independent system, with fan coils, heat pumps, radiant panels, or other terminal devices. While the concept of DOAS is not new, recent advances in sensors and controls have made DOAS more attractive. These and other advanced technologies (such as computational fluid dynamics, discussed in Chapter 8) will enable HVAC engineers to design systems that ensure adequate make-up air ventilation, while keeping the cost of conditioning that make-up air to a minimum.

5.14 Tunnel Ventilation

5.14.1 Automotive Tunnels

Tunnels for automotive and railway traffic, or for underground mining and underground passageways for communication lines or piping, or for wastewater transport, are places where unhealthy air is apt to occur, because ventilation is inadequate and/or emissions are greater than expected. Tunnels for automotive vehicles are places where passengers in cars are exposed to tunnel air for only short periods of time, but public employees (police, highway workers, etc.) work in tunnels for longer periods of time. The hazards of greatest concern are carbon monoxide, fire, and explosion. Steps must be taken to provide fresh air and remove automotive exhaust emissions. In the event of emergencies, fans are needed to remove smoke and provide fresh air for rescue work. For this reason the ventilating fans are often reversible so as to provide fresh air or remove smoke depending on where the accident occurs. In 1988 the EPA revised its previous ventilation guidelines for tunnels located at or below an altitude of 1500 m (ASHRAE HVAC Applications Handbook, 1999):

- maximum 120 PPM for CO exposures of 15 minutes
- maximum 65 PPM for CO exposures of 30 minutes
- maximum 45 PPM for CO exposures of 45 minutes
- maximum 35 PPM for CO exposures of 60 minutes

Similar guidelines exist for enclosed parking garages which require ventilation (Krarti and Ayari, 2001). Above an elevation of 1500 m, automotive emission of CO increases, human tolerance decreases, and special precautions must be undertaken in the design of tunnel ventilation systems.

The ventilation system for tunnels must be designed for a mixture of private and commercial vehicles traveling at constant speed, accelerating, decelerating, and traveling on roads with up and down grades. Typical spacing between vehicles is listed below:

speed (MPH)	distance between vehicles (ft)
25	72.8
20	54.4
15	36.0

Provision must also be made in the event that traffic is halted because of stalled cars or accidents. Lastly the specific design of the ventilation system must account for the number of lanes in the tunnel and whether the tunnel contains one-way or two-way traffic.

By assuming that steady-state conditions occur and that the only spatial coordinate in which quantities change is the longitudinal direction (x), tunnels can be modeled by one-dimensional ordinary differential equations, and the concentration of gaseous contaminants can be predicted as a function of location in the tunnel (Chang and Rudy, 1990). Figure 5.16 depicts four classes of ventilation, applicable to one-way automotive tunnels which engineers are likely to encounter. The four classes of tunnel ventilation are described below, where Q_m is the volume flow rate of make-up air forced into the tunnel, and Q_e is the volume flow rate of exhaust air withdrawn from the tunnel:

(a) ***Natural ventilation*** occurs by moving autos that draw air into the tunnel by viscous forces. No fans introduce or withdraw air from the tunnel ($Q_m = Q_e = 0$). Automotive and railway vehicles traveling through the tunnel in a single direction draw air through the tunnel, as if such vehicles were loose-fitting "pistons." Natural ventilation also occurs due to wind and turbulence. If the tunnel is constructed on a slant, thermal buoyancy causes air movement as well. Natural ventilation is typically used for tunnels less than 300 m long.

(b) ***Local make-up air ventilation*** refers to fresh air introduced into the tunnel at a single point (usually at the inlet) at volumetric flow rate Q_m. No fans withdraw air from the tunnel (Q_e = 0). Local make-up air ventilation is typically used for tunnels less than 600 m long.

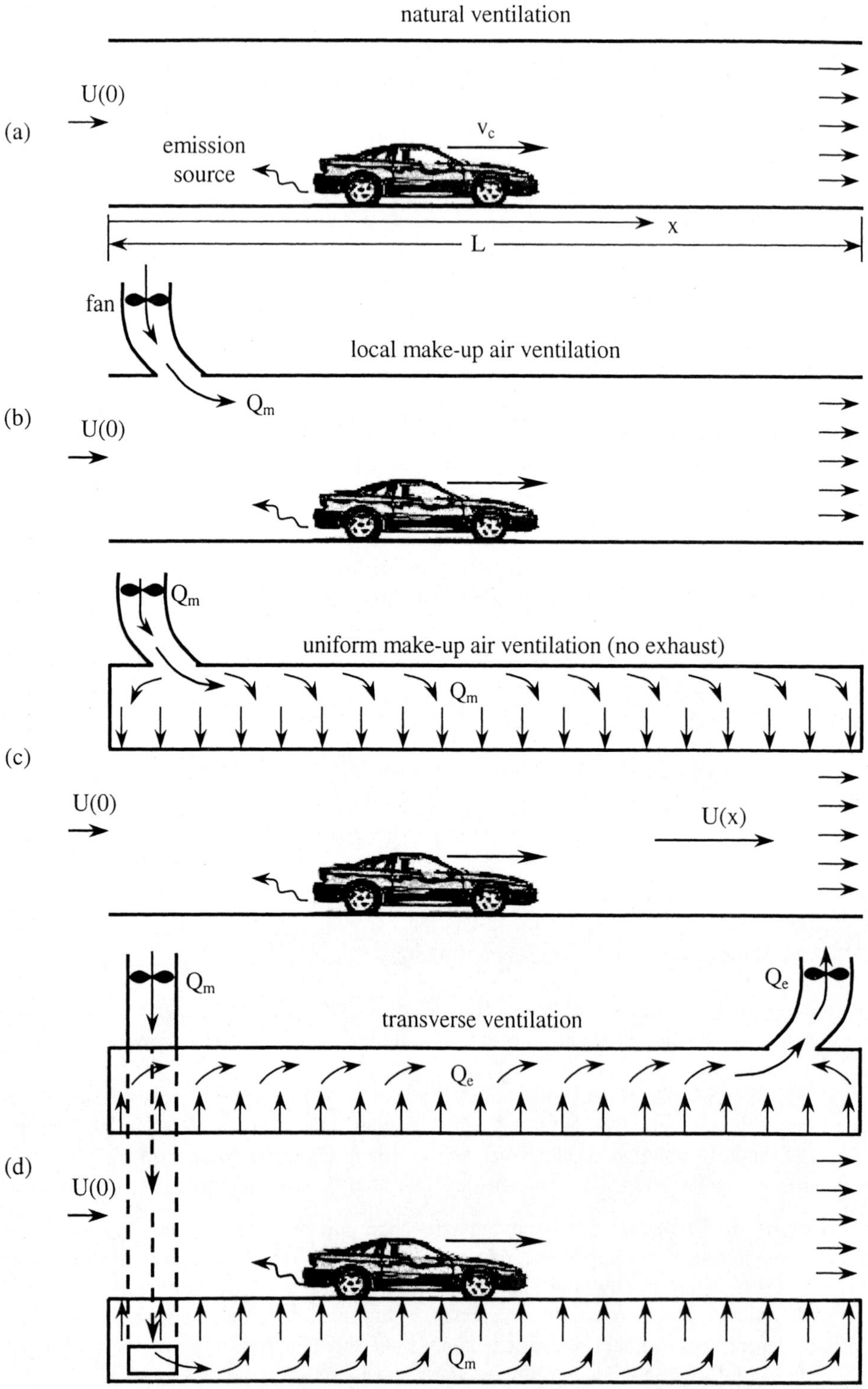

Figure 5.16 Classes of ventilation in an automotive tunnel: (a) natural ventilation, (b) local make-up ventilation, (c) uniform make-up, no exhaust ventilation, and (d) transverse ventilation.

(c) ***Uniform make-up air ventilation***, also called ***no exhaust ventilation***, refers to plenums and fans that introduce outside air ($Q_m > 0$) but withdraw no contaminated air ($Q_e = 0$). Uniform make-up air with no exhaust ventilation is typically used for tunnels less than 1500 m, because all the supplied air has nowhere to go but out the tunnel exit; the air speed through the tunnel can get too large with this ventilation scheme if the tunnel is very long.

(d) ***Transverse ventilation*** involves plenums and fans to introduce outside air (Q_m) uniformly along the length of the tunnel and plenums and fans to withdraw contaminated air (Q_e) uniformly along the tunnel length.

- If $Q_e = Q_m$, the ventilation is called ***balanced transverse ventilation***.
- If $Q_e \neq Q_m$, the ventilation is called ***unbalanced transverse ventilation***.

It is common for $Q_m > Q_e$, but it is rare that the reverse is true. Transverse ventilation is typically used for long tunnels in excess of 1500 m. For example, the Holland Tunnel under the Hudson River connects New York City and New Jersey, and is 9250 ft long. It was completed in 1927; at the time it was the longest tunnel built for automobiles (McKay, 1988). The tunnel consists of two separated parallel tubes 29.5 ft in diameter, each tube carrying two lanes of traffic in a single direction. Traffic in the tunnel averages 80,000 vehicles per day. Transverse ventilation is used to supply almost 3,800,000 ACFM of fresh air, and to withdraw the same amount of exhaust air. Fresh air enters the tunnel through narrow slots above the curb, ten to fifteen feet apart along the length of the tunnel. Exhaust air is withdrawn through openings in the tunnel ceiling located ten to twenty feet apart along the length of the tunnel.

If automobiles travel through a tunnel, the total rate at which pollutants are emitted (S) is

$$\boxed{S = (EF)_c\, n_c v_c L} \tag{5-71}$$

where typical units of S are mg/hr and the variables in Eq. (5-71) are described below:

- $(EF)_c$ is an emission factor (typical units are mg/vehicle-km).
- n_c is called the ***traffic density***, which is the number of moving vehicles per distance traveled (typical units are vehicles/km)
- L is the tunnel length (typically in units of m)
- v_c is the automobile speed, considered to be constant (in units of km/hr)

For consistency with the steady-state analysis described here, the unsteady emissions from each automobile are averaged uniformly throughout the tunnel length. In other words, total source (S) in Eq. (5-71) is distributed uniformly from x = 0 to L, so that over some segment of tunnel length (dx), the source is equal to $S\frac{dx}{L}$. Make-up and exhaust air are handled in similar fashion.

Figure 5.17 is a schematic diagram representing air and contaminant entering and leaving an elemental control volume within a tunnel in which there is uniform make-up air (Q_m) and uniform exhaust ventilation (Q_e). The volume of the control volume in Figure 5.17 is equal to $A_c dx$, where A_c is the tunnel cross-sectional area (typical units are m^2). The model applies to *all* of the above classes of tunnel. Within the elemental volume, conservation of mass for air results in the following:

$$\boxed{\frac{dU}{dx} = q_m - q_e} \tag{5-72}$$

where x is the distance along the tunnel axis, and q_m and q_e are defined as

$$q_m = \frac{Q_m}{A_c L} \qquad q_e = \frac{Q_e}{A_c L} \tag{5-73}$$

If q_m and q_e are constant, or vary in a known fashion with x, Eq. (5-72) can be integrated. When q_m and q_e are constant, the air velocity (U) in the tunnel at any location x in the tunnel can be expressed as

$$\frac{U(x)}{U(0)} = 1 + \frac{(q_m - q_e)x}{U(0)} \tag{5-74}$$

where U(0) is the air velocity entering the tunnel. So long as $q_m > q_e$, U(x) increases linearly in the direction of travel. If a tunnel has only an exhaust system or if $q_m < q_e$, U(x) *decreases* in the direction of travel. If the tunnel is long enough, a stagnation point occurs at some location x', where $U(x') = 0$. At locations between x' and the end of the tunnel ($x' < x < L$), the air reverses direction.

It is rare for tunnels to be designed with only an exhaust system, but some tunnels have only make-up air systems. Within the control volume, conservation of mass for the contaminant is

$$c\frac{dU}{dx} + U\frac{dc}{dx} = s + q_m c_m - c q_e - kc \tag{5-75}$$

where c_m is the contaminant mass concentration in the make-up air, s is the source strength per tunnel volume,

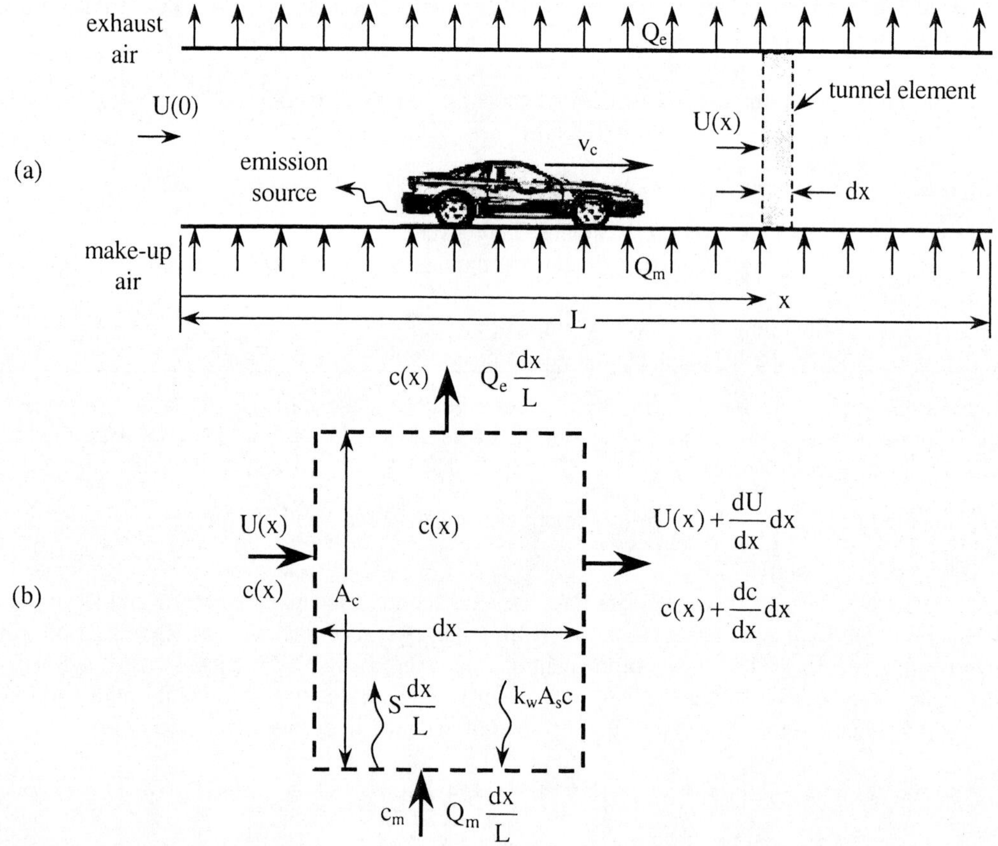

Figure 5.17 Air and contaminant transport in a tunnel: (a) schematic diagram of tunnel, with slice of length dx at location x taken as a control volume, and (b) details of the control volume.

$$s = \frac{S}{A_c L} \tag{5-76}$$

and k is a ***tunnel wall loss parameter***,

$$k = \frac{A_s}{A_c dx} k_w \tag{5-77}$$

where A_s is the surface area in the interior of the tunnel onto which contaminant is adsorbed or settles, and k_w is the wall loss coefficient (typically in m/s) representing the rate at which the contaminant is adsorbed or settles. For a round tunnel, assuming the entire inner tunnel wall adsorbs contaminant,

$$k = 4\frac{k_w}{D} \tag{5-78}$$

where D is the tunnel diameter (typically in units of m). Expressions of k for tunnels with rectangular or other cross-sectional shapes can be derived from Eq. (5-77).

5.14.2 Tunnels with Unequal Exhaust and Make-up Air ($q_e \neq q_m$)

Consider tunnels where exhaust and make-up air are provided but are unequal to each other, i.e. ***unbalanced ventilation***. A combination of Eqs. (5-72) and (5-75) yields the following equation describing the variation of contaminant concentration in the tunnel:

$$U\frac{dc}{dx} = s + q_m c_m - (k + q_m)c \tag{5-79}$$

The maximum contaminant concentration occurs when dc/dx = 0. Thus,

$$c_{max} = \frac{s + q_m c_m}{k + q_m} \tag{5-80}$$

Combining Eqs. (5-74) and (5-79),

$$\frac{dc}{(q_m c_m + s) - (k + q_m)c} = \frac{dx}{U(0) + x(q_m - q_e)} \tag{5-81}$$

If q_m and q_e are constant, unequal, and nonzero, Eq. (5-81) can be integrated to yield the following expression for the concentration at any location, $0 < x < L$, in the tunnel:

$$c(x) = c_{max} - [c_{max} - c(0)]\left[\frac{U(x)}{U(0)}\right]^{-b} \tag{5-82}$$

where

$$b = \frac{k + q_m}{q_m - q_e} \tag{5-83}$$

and U(x)/U(0) can be replaced by Eq. (5-74). If q_m and q_e are equal or identically zero, Eq. (5-82) cannot be used, and one must return to Eq. (5-79) to obtain a solution. If q_m and q_e and other parameters in the above equations vary with x, numerical techniques such as Runge-Kutta are needed to integrate the expressions.

5.14.3 Tunnels with Equal Exhaust and Make-up Air ($q_e = q_m$)

Consider tunnels with ***balanced ventilation*** such that the exhaust and make-up air volumetric flow rates are equal, in which case Eq. (5-82) is invalid. When the volumetric flow rates are equal, Eq. (5-72) requires dU/dx = 0; thus U = constant. Equation (5-74) shows that the constant is simply U(0).

$$U = U(0) = \text{constant}$$

Eq. (5-79) then becomes a simple first-order ODE with constant coefficients, which can be written in the standard form of Eq. (5-7) (with x instead of t as the independent variable). The constants A and B in Eq. (5-7) are

$$A = \frac{k + q_m}{U(0)} \qquad B = \frac{s + q_m c_m}{U(0)}$$

Thus, for initial (inlet) concentration c(0), the solution at any location in the tunnel ($0 \le x \le L$) is given analytically by Eqs. (5-10) and (5-11) (again using x instead of t, and using c_{max} instead of c_{ss}),

$$\frac{c_{max} - c(x)}{c_{max} - c(0)} = \exp(-Ax)$$

or, solving for c(x),

$$\boxed{c(x) = c_{max} - [c_{max} - c(0)]\exp\left(-\frac{k + q_m}{U(0)}x\right)} \tag{5-84}$$

The maximum concentration (c_{max}) is found from Eq. (5-11):

$$\boxed{c_{max} = \frac{B}{A} = \frac{s + q_m c_m}{k + q_m}} \tag{5-85}$$

which also agrees with Eq. (5-80). For the special case of *natural ventilation* ($q_m = q_e = 0$), Eqs. (5-84) and (5-85) simplify to

$$\boxed{c(x) = c_{max} - [c_{max} - c(0)]\exp\left(-\frac{k}{U(0)}x\right)} \tag{5-86}$$

and

$$\boxed{c_{max} = \frac{s}{k}} \tag{5-87}$$

which implies that if the tunnel is long enough, wall adsorption eventually balances the source such that a maximum mass concentration is reached.

Example 5.16 - Formaldehyde from Methanol-Fueled Vehicles in Tunnels

Given: Methanol-fueled autos receive increasing attention because of their potential to reduce ozone levels in urban areas. While methanol combustion produces fewer unburned hydrocarbons that ultimately produce ozone, combustion of methanol produces more *formaldehyde* than does combustion of gasoline. The PEL for formaldehyde is currently 0.75 PPM (920 $\mu g/m^3$), but the EPA is concerned that outdoor concentrations as low as 150 $\mu g/m^3$ may cause irritation for some individuals. One may assume that the conditions given by Chang and Rudy (1990) apply to roadway tunnels under severe conditions of traffic congestion, poor tunnel ventilation, and engines with high rates of formaldehyde emission. Four tunnels are analyzed, as shown in Table E5.16. The diameter and amount of traffic are the same in each tunnel, but the tunnels are of various lengths; each tunnel also employs a different type of ventilation:

(a) a short tunnel with natural ventilation
(b) a moderate length tunnel with uniform make-up air ventilation
(c) a long tunnel with balanced transverse ventilation
(d) the same long tunnel, but with unbalanced transverse ventilation

To do: Estimate the formaldehyde concentration in the four tunnels listed above.

Table E5.16 Parameters for the four tunnels of Example 5.16.

parameter (units)	**(a) natural** ($q_m = q_e = 0$)	**(b) uniform make-up air** ($q_e = 0$, q_m = constant)	**(c) balanced transverse** ($q_m = q_e$)	**(d) unbalanced transverse** ($q_m \neq q_e$)
L (m)	300	1000	2000	2000
D (m)	7.57	7.57	7.57	7.57
U(0) (m/min)	60.	60.	60.	60.
q_m (min^{-1})	0	0.20	0.20	0.20
q_e (min^{-1})	0	0	0.20	0.18
c_m ($\mu g/m^3$)	5.0	5.0	5.0	5.0
c(0) ($\mu g/m^3$)	7.4	7.4	7.4	7.4
n_c (autos/km)	100	100	100	100
v_c (km/hr)	8.0	8.0	8.0	8.0
$(EF)_c$ [mg / (auto km)]	100	100	100	100
k (min^{-1})	0.020	0.020	0.020	0.020

Solution: The source term (s) is common to several of the equations above, and can be calculated from Eq. (5-76), using Eq. (5-71),

$$s = \frac{S}{A_c L} = \frac{(EF)_c n_c v_c L}{\frac{\pi D^2}{4} L} = \frac{4(EF)_c n_c v_c}{\pi D^2}$$

which upon substitution of the values provided in the table yields

$$s = \frac{4\left(100.\frac{mg}{auto \cdot km}\right)100.\frac{auto}{km}8.0\frac{km}{hr}}{\pi(7.57\ m)^2}\left(\frac{hr}{60\ min}\right)\left(\frac{1000\ \mu g}{mg}\right)\left(\frac{km}{1000\ m}\right) = 29.6\frac{\mu g}{m^3\ min}$$

Now the various ventilation cases can be calculated:

(a) Natural ventilation ($q_m = q_e = 0$):
For natural ventilation, Eqs. (5-86) and (5-87) apply. The air velocity through the tunnel is constant (U = U(0)) and the maximum mass concentration of formaldehyde is thus

$$c_{max} = \frac{s}{k} = \frac{29.6\frac{\mu g}{m^3\ min}}{0.020\frac{1}{min}} = 1480\frac{\mu g}{m^3}$$

The reader should note that c increases with x, and that this maximum concentration is predicted for a very *long* tunnel. As will be seen, tunnel (a) is so short that the actual concentration never goes above about 10% of this value. The mass concentration of formaldehyde at any x location along the tunnel is

$$c(x) = c_{max} - [c_{max} - c(0)]\exp\left(-\frac{k}{U_0}x\right) = 1480\frac{\mu g}{m^3} - [1480 - 7.4]\frac{\mu g}{m^3}\exp\left(-\frac{0.020\frac{1}{min}}{60.\frac{m}{min}}x\right)$$

where x must be in meters for unit consistency. The concentration at the tunnel exit (x = L = 300 m for this tunnel) is

$$c(L) = 150\frac{\mu g}{m^3}$$

(b) Uniform make-up ventilation, no exhaust (q_m = constant, $q_e = 0$):
When $q_m = 0.20\ \text{min}^{-1}$ and $q_e = 0$, Eq. (5-82) can be used, where the value of c_{max} is obtained from Eq. (5-80),

$$c_{max} = \frac{s + q_m c_m}{k + q_m} = \frac{29.6\frac{\mu g}{m^3\ \text{min}} + 0.20\frac{1}{\text{min}}5.0\frac{\mu g}{m^3}}{0.020\frac{1}{\text{min}} + 0.20\frac{1}{\text{min}}} = 139.\frac{\mu g}{m^3}$$

and the exponent b is obtained from Eq. (5-83),

$$b = \frac{k + q_m}{q_m - q_e} = \frac{(0.020 + 0.20)\frac{1}{\text{min}}}{(0.20 - 0)\frac{1}{\text{min}}} = 1.1$$

The concentration at any location x inside the tunnel, $0 \le x \le L$, is obtained by substituting these values into Eq. (5-82):

$$c(x) = 139.\frac{\mu g}{m^3} - [139. - 7.4]\frac{\mu g}{m^3}\left[1 + \frac{0.20\frac{1}{\text{min}}}{60.\frac{m}{\text{min}}}x\right]^{-1.1}$$

where Eq. (5-74) has been used for U(x)/U(0), and x must be in meters in order for the units to be consistent. At the tunnel exit (x = L = 1000 m for this tunnel), the above yields

$$c(L) = 110\frac{\mu g}{m^3}$$

Again, as in tunnel (a), this tunnel is too short for the concentration to reach the predicted maximum value.

(c) Balanced transverse ventilation ($q_m = q_e$ = const):
When the system is *balanced*, $q_m = q_e = 0.20\ \text{min}^{-1}$; Eqs. (5-84) and (5-85) apply, and can be used to calculate the concentration at any x location. The maximum concentration (c_{max}) is the same as that calculated in Part (b) above, i.e. $c_{max} = 139.\ \mu g/m^3$, and

$$c(x) = 139.\frac{\mu g}{m^3} - [139. - 7.4]\frac{\mu g}{m^3}\exp\left(-\frac{(0.020 + 0.20)\frac{1}{\text{min}}}{60.\frac{m}{\text{min}}}x\right)$$

where again x must be in meters in order for the units to be consistent. At the tunnel exit (x = L = 2000 m for this tunnel), the above yields

$$c(L) = 138.9\frac{\mu g}{m^3} \cong 139.\frac{\mu g}{m^3}$$

In this case, the tunnel is long enough that the mass concentration of formaldehyde at the tunnel exit has nearly reached its maximum possible value (the exponential term in the above equation is negligibly small).

(d) Unbalanced transverse ventilation ($q_m \neq q_e$) (q_m = const, q_e = const):
If $q_e = 0.18 \text{ min}^{-1}$ and $q_m = 0.20 \text{ min}^{-1}$, the system is *unbalanced* and Eq. (5-82) can be used, with exponent b determined from Eq. (5-83),

$$b = \frac{k + q_m}{q_m - q_e} = \frac{(0.020 + 0.20)\frac{1}{\text{min}}}{(0.20 - 0.18)\frac{1}{\text{min}}} = 11.0$$

The maximum concentration (c_{max}) is obtained from Eq. (5-80), and is the same as that calculated in Part (b) above, i.e. $c_{max} = 139.\ \mu g/m^3$. Thus, at any x location in the tunnel,

$$c(x) = 139.\frac{\mu g}{m^3} - [139. - 7.4]\frac{\mu g}{m^3}\left[1 + \frac{(0.20 - 0.18)\frac{1}{\text{min}}}{60.\frac{m}{\text{min}}} x\right]^{-11.0}$$

At the tunnel exit (x = L = 2000 m for this tunnel), the above yields

$$c(L) = 138.5\frac{\mu g}{m^3} \cong 139.\frac{\mu g}{m^3}$$

Again, as in Part (c) above, the tunnel is long enough that the mass concentration of formaldehyde at the tunnel exit has nearly reached its maximum possible value.

Discussion: Comparing the four tunnels, the maximum concentration (at the end of the tunnel in each case) lies between 110 and 150 $\mu g/m^3$. This is below the value of 150 $\mu g/m^3$, the concentration at which the EPA expressed concern. Thus, one can conclude that workers in the tunnel are not in any great danger from formaldehyde vapors. Drivers moving through the tunnel are only inside the tunnel for a short period, and should have even less concern.

It is straightforward to generate plots of formaldehyde concentration as a function of tunnel length (x or L), using the above values and equations for each of the four ventilation cases. A plot generated by Mathcad is shown in Figure E5.18a. The Mathcad file can be downloaded from the book's web site. It is clear from the plot that as the tunnel length increases, natural ventilation (curve a) becomes inadequate. The plot clearly shows why it is necessary to use some type of tunnel ventilation scheme to supply fresh make-up air for long tunnels. There is not much difference between balanced and unbalanced transverse ventilation (curves c and d respectively), and both yield somewhat higher concentrations than does uniform make-up air ventilation with no exhaust. For the values in the above example problem, ventilation scheme (b), i.e. uniform make-up air ventilation with no exhaust appears to be the best scheme for any tunnel length. However, as mentioned above, there is a physical limit to the length of a tunnel with this ventilation scheme. Namely, without forced exhaust, the only place for all of the make-up air to go is out the end of the tunnel; thus U(x) grows linearly with x as given by Eq. (5-74).

It is interesting to also plot U(x) for the four cases, as shown in Figure E5.18b. From this plot, it is clear that the two balanced ventilation schemes (cases a and c - natural ventilation and balanced transverse ventilation) maintain a constant U(x) regardless of tunnel length, but U(x) grows with x for the other two (unbalanced) schemes. Ventilation scheme (b) - uniform make-up air ventilation with no exhaust, has the higher slope, and leads to very large air velocities for long tunnels. (At x = 1500 m for the values used in the above example problem, U = 360 m/min, or 6.0 m/s!) Thus, even though the lowest contaminant concentration inside the tunnel is predicted for the uniform make-up air ventilation scheme with no exhaust, *balanced transverse ventilation* (case c) is the best choice for very long automobile tunnels.

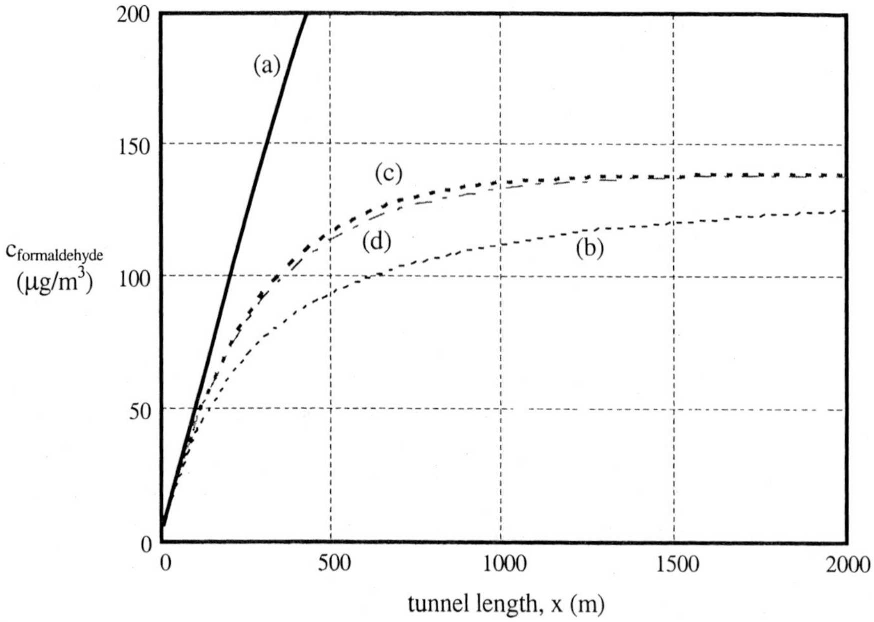

Figure E5.18a Formaldehyde mass concentration versus tunnel length in four types of automobile tunnels: (a) natural, (b) uniform make-up air, (c) balanced transverse, and (d) unbalanced transverse ventilation.

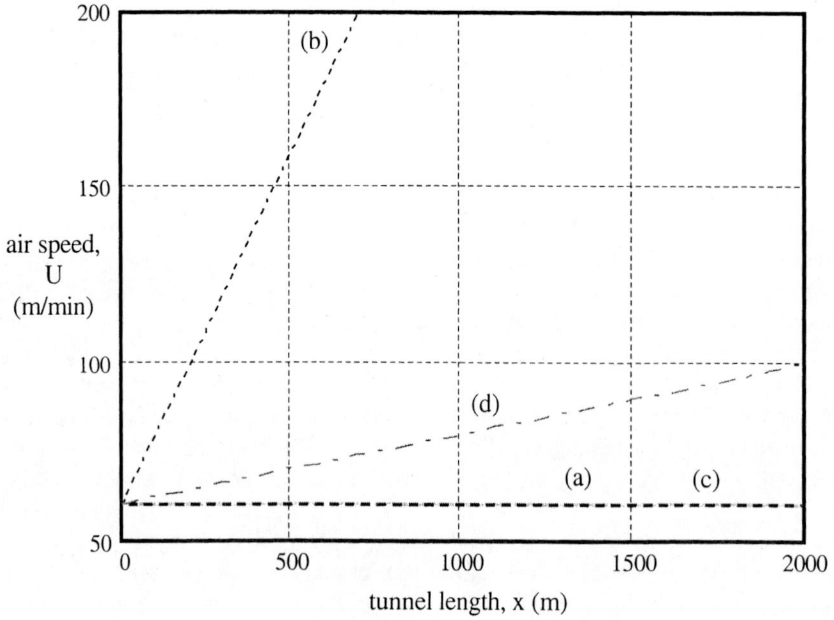

Figure E5.18b Air speed versus tunnel length in four types of automobile tunnels: (a) natural, (b) uniform make-up air, (c) balanced transverse, and (d) unbalanced transverse ventilation.

Details of ventilation systems designed for large tunnels throughout the world can be found in the literature. For example, a discussion of the ventilation system for the Sydney (Australia) Harbor

Tunnel can be found in Bendelius and Hettinger (1993). Another example is the English Channel Tunnel ("Chunnel") connecting England and France. Figure 5.18 is a diagram of the 50-km Chunnel. The structure consists of two parallel 7.6 m diameter tunnels carrying rail trains, called "shuttles." Located between the large rail tunnels is a 4.8 m diameter service tunnel that delivers make-up air and also provides an emergency tunnel in the event of fire, explosion, etc. Because it is a tunnel for rail transportation, there is not a large emission of pollutants (principally CO) as would occur if the tunnel carried automobiles. Make-up air (187,000 ACFM) enters each end of the service tunnel (between the rail tunnels) by 525-HP fans and is distributed uniformly to each rail tunnel through cross-passage ducts 3.3 m in diameter occurring every 375 m. In addition, relief ducts 2.0 m in diameter are located every 250 m directly connecting each rail tunnel. As the shuttle passes the opening of each relief duct, air is "pumped back and forth" between the rail tunnels by the moving shuttle, which acts as a piston (Henson and Fox, 1974; Henson and Lowndes, 1978). A single 635,000 ACFM (1200 HP) fan is available to provide exhaust or fresh air to the rail tunnels in the event of an emergency (Dodge, 1993).

5.14.4 - Rapid Transit Systems

Tunnels for rapid transit systems are different than automotive tunnels for several reasons. The contaminants that are encountered are dust, graphite, and metal particles. Because the trains occupy a majority of the tunnel cross section, the moving train acts a piston in a tube. The pumping action of the moving train is more significant than in an automotive tunnel where the vehicle cross-sectional area is much smaller than the tunnel cross-sectional area. The air pressure in front of a moving train is higher than atmospheric pressure, while behind in the train's wake the pressure is less than atmospheric pressure. The higher pressure at the front end of the train expels air from the tunnel through portals called ***blast shafts*** connecting the tunnel to the atmosphere. Blast shafts are placed near the entrance of stations. In the wake of the train the lower pressure draws atmospheric air into the tunnel through portals called ***relief shafts***. Relief shafts are placed near the station exit. In addition, a considerable amount of tunnel air is shuffled back and forth as trains pass one another. To assist ventilating rapid transit systems, blast and relief shafts may also be placed at strategic points between stations. Rapid transit tunnels also require ventilation to remove heat generated by the train's electric motors and air conditioning units on the trains.

The action of the shuttle to "pump" air is a practice used in many urban rail mass transportation (subway) systems. In many cities the subway tunnels are equipped with ubiquitous grated ducts that open onto city streets and sidewalks. As the train approaches the duct opening, tunnel air is pumped upwards and discharged. After the train passes, the train's wake region sucks fresh air into the tunnel. Such subway grates are often places where a city's homeless people congregate to keep warm.

There are several design concepts employed in rapid transit tunnels. Make-up air is introduced at stations by fans that introduce the air through ducts under the platform. These fans can be reversed to function as exhaust fans in the event of emergency. Typical filtered make-up air requirements for rapid transit systems are approximately 7.5 ACFM per person. Reversible emergency fans that can provide fresh air or draw off smoke in the event of an emergency are located at selected points in the tunnel.

5.15 - Closure

General ventilation is the practice of removing room air and replacing it with fresh make-up air with the goal of controlling the indoor air quality inside an enclosed space. As an air pollution strategy to satisfy OSHA standards, it is expensive and ineffective. The practice is useful for controlling humidity, temperature, and bothersome things such as odors. The assumption of well-

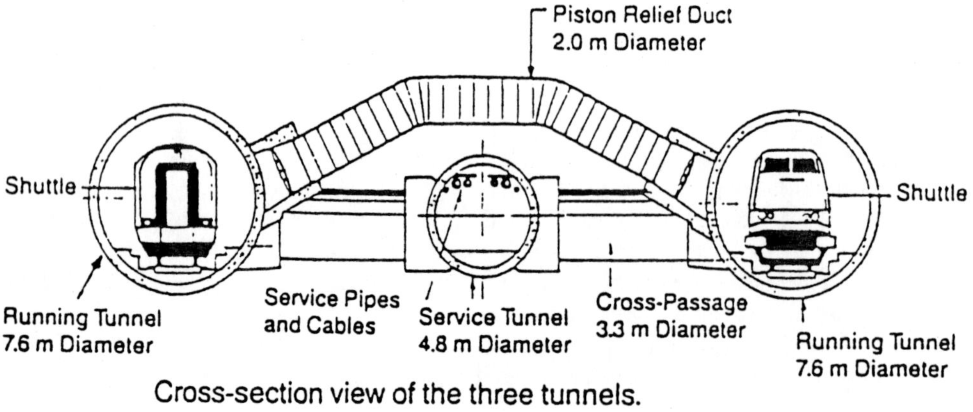

(a)

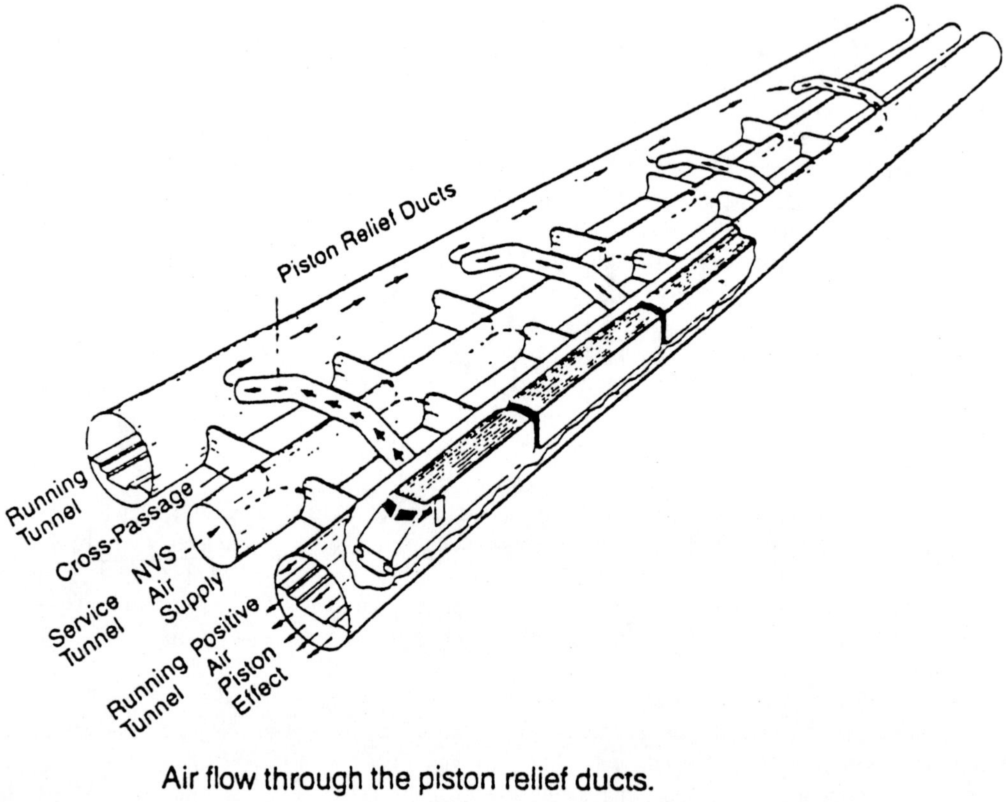

(b)

Figure 5.18 Ventilation of the English Channel tunnel: (a) cross-sectional view, and (b) perspective view; from Dodge (1993).

mixed conditions is a powerful analytical tool to predict how pollutant concentrations inside an enclosed space vary with time. However, the model is poorly suited to predict how the concentration varies with both location *and* time inside the enclosed space, although steps can be taken to achieve some degree of spatial resolution.

Chapter 5 - Problems

1. A closed storage room (V = 5.0 m^3) is used to store hydrocarbons. A leak occurs in one of the vessels containing 1.0 kg of liquid acetone, and there is no transfer of air into or out of the room.

(a) Compute the maximum mol fraction of acetone. How does this concentration compare with the PEL and LEL for acetone?
(b) Repeat for benzene, carbon tetrachloride, and methyl alcohol.
(c) repeat (a) if the initial amount of liquid acetone is 5.0 kg.

2. General ventilation may be a viable control strategy when the contaminant is not toxic, the emission rate (S) is low, and room volume (V) and ventilation rate (Q) are large. If a room is ventilated as per Figure 5.10, show that if V and Q are constant, and the outside concentration and wall adsorption are negligible, the concentration at any instant is linearly proportional to source strength (S) during the first few moments of the process.

3. One kg of paint remover, composed of 60.% methyl alcohol and 40.% methylene chloride by mass, is stored in an unventilated closed closet (4.0 m^3) in a home. If the can is left open for a long period of time, find the maximum concentration of alcohol and methylene chloride in the air in the closet.

4. An underground conduit carries waste liquids to a central disposal process. The conduit contains a trap to remove solids that can accumulate and stop the flow of liquid. Every few months workers clean the trap by entering an underground chamber (8.0 by 8.0 ft and 10. deep) normally covered by a manhole cover. Unfortunately, gases and vapors dissolved in the waste water escape to the air in the chamber. For a period of time the conduit has carried an aqueous mixture containing dissolved carbon monoxide, hydrogen cyanide, and benzene. The concentrations of these materials in the waste water, expressed as mass fractions, are as follows:

- CO: 4.0 μg per 100 g of water
- HCN: 12.0 mg per 100 g of water
- C_6H_6: 0.18 mg per 100 g of water

Estimate the equilibrium partial pressure of these materials in the air using Henry's law. If the there is no exchange of air between air in the underground chamber and the atmosphere, estimate the maximum concentration of carbon monoxide, hydrogen cyanide, and benzene in the air in the underground chamber. Discuss these values in terms of OSHA and ACGIH standards.

5. An underground access pit (3.0 by 3.0 by 4.0 m) in a petrochemical plant contains five valves (with flanges on either end) for pipes carrying phosgene gas (carbonyl chloride, Cl_2CO, molecular weight 98.92) under high pressure. The access pit has a manhole at the top, and small amounts of air pass between the pit and the outside at a rate of 0.10 m^3/hr. The average temperature of air inside the pit is 12. °C. Using emission factors for fugitive emissions for pipeline valves and flanges for gas streams in petroleum refineries, estimate the steady-state phosgene concentration using the well-mixed model. Is the concentration below the PEL for phosgene? Is it safe to send a workman into the pit?

6. A degreaser with an open top (1.0 by 2.0 m) containing trichloroethylene (see Chapter 4, Problems 17 and 18 for properties) is located inside a special room (2.0 by 3.0 by 2.5 m) in an automotive repair company. Outside air is drawn into the room and air from within the room is withdrawn by a small fan. Your supervisor wishes to know if a 300 ACFM fan is adequate. When the facility is built the initial concentration of trichloroethylene is 1.0% of the PEL. Using emission factors and the well-mixed model determine whether the steady-state concentration is below the PEL. The vessel is closed every evening (5:00 PM); estimate the concentration at 12:00 noon, 5:00 PM, and 8:00 AM.

7. A collection system is used to recover particles generated by a buffing wheel used in making gold jewelry. The feed-back system shown in Figure 5.8 is installed, but your supervisor is not satisfied with the mass recovery rate, and asks you to improve the performance. You have only enough money to improve one of the concentrators. Which one should you improve? Explain why.

8. Consider a paint spray booth. Paint with a density of 12. lbm/gal containing 90.% hydrocarbons (assume toluene) by weight is used at a rate of 0.514 gal/min. Unpainted objects enter the booth on a conveyor, receive a coating of paint and leave in a painted (but wet) condition. The rate at which paint is applied to the objects inside the both is 0.514 gal/min. The rate at which the paint (solids plus toluene) leaves the booth on the surface of the object objects is 0.36 gal/min (transfer efficiency = 0.36/0.514 = 70.%). A water spray in the rear of the booth removes the bulk of the unused paint particles but the hydrocarbons are discharged outside the building. Hydrocarbon laden air is withdrawn from the booth at a volumetric flow rate of 7,000 SCFM and an equal amount of air containing 0.020 PPM of toluene is supposed to enter the booth from the room. Your supervisor suggests that 1,650 ACFM, T = 300. °C, from a nearby dryer containing 5,000 PPM of toluene (molecular weight = 92.1) can be discharged into the booth to reduce the amount of room air. Such a step reduces the amount make-up air for the building and lowers heating costs. The fire insurance carrier requires that the hydrocarbon concentration in the booth never exceed 1% of the lower explosion limit (LEL = 1.27 % by volume, i.e. LEL mol fraction = 0.0127). What is the toluene concentration in the booth under normal circumstances? What is the toluene concentration if your supervisor's suggestion is adopted?

9. Derive an equation that expresses the time-varying particle concentration for the clean room shown in Figure P5.9. Write an expression that gives the class of the clean room as a function of the operating parameters of the problem (Q_1, Q_2, V, S, f_1, f_2, η_1, and η_2). Assume that $c_a = 0$, and that the particles are 1.0 μm in diameter. Neglect gravimetric settling and neglect wall adsorption. If initially there are 10^8 particles per m^3, draw graphs of mass concentration c versus time and class versus time.

- Q_1/V = 0.50 room air changes per hour
- Q_2/V = 2.0 room air changes per hour
- V = 600. m^3
- S = 2.0 gm/s
- $f_1 = f_2$ = 0.10
- $\eta_1 = \eta_2$ = 0.95

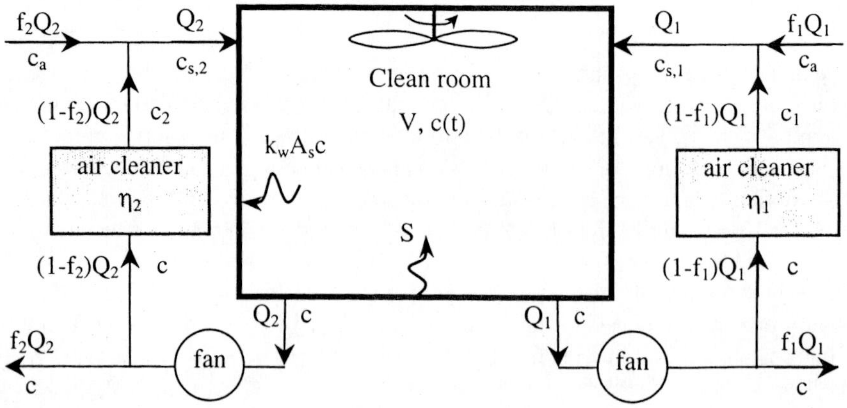

Figure P5.9 Schematic diagram of the ventilation system for the clean room of Problem 5.9.

10. New homes and new automobiles contain many materials that emit hydrocarbons for a period of time i.e. the "new car smell". Solvents are contained in adhesives, dyes, and finishes in carpets, furniture, wall paper, paint, drapes, laminated wood paneling, thermal insulation, etc. People normally cope with these hydrocarbons but some people are sensitive to very low concentrations and suffer serious discomfort. Assume the hydrocarbon is formaldehyde (HCOH) which has a 8-hr PEL of 3 PPM and an odor threshold of approximately 1 PPM. Compute and plot the formaldehyde concentration (PPM) as a function of time in a home based on the well-mixed model assuming the following conditions:

- After construction, the home is sealed shut for 10 days whereupon it is opened and inside air is exchanged for outside air at the rate of 60. m^3 /hr. The concentration of formaldehyde in the ambient air is 10 PPB.
- Area of formaldehyde emitting surfaces (A_g) = 464. m^2, home volume (V) = 284. m^3 and area of formaldehyde adsorbing (and desorbing surfaces) (A_s) = 93. m^2.
- Formaldehyde emission rate S (mg/day) = $S = K_1 A_g \exp(-t/K_2)$, where K_1 = 75. mg/(day m^2) and K_2 = 500. hr.
- For simplicity, assume that formaldehyde is adsorbed by surfaces (or desorbed by the same surfaces) at a rate S_a (mg/s) given by $S_a = K_3 A_s(c - c_0)$, where K_3 = 0.00050 m/s, c = formaldehyde concentration at any instant, and c_0 = 10. PPB. Thus adsorption occurs whenever $c > c_0$ and desorption occurs whenever $c < c_0$.

11. Toluene 2,4-diocyanate (M= 171.5) called TDI is a highly toxic material. A sample of a process gas stream containing TDI has been taken and shipped to a laboratory for analysis. The sample was placed in a rigid, spherical, stainless steel container 0.50 meters in diameter. Adsorption of TDI on the walls of the container occurs at a rate given by kA_sc (in units of mg/hr), where k = 0.00277 m/hr, A_s = container internal surface area (m^2), and c = TDI concentration (mg/m^3). After a period of 30 hours, the laboratory measures a TDI concentration of 2.0 mg/m^3 at STP. If the pressure and temperature of the original sample were 1.5 atm, 80. °C, what was the original TDI concentration? Is this value below the PEL for TDI?

12. Oil mist is used to lubricate the dies in a plant that produces metal stampings. Unfortunately some of the mist enters the workplace atmosphere. Activity in the workplace produces a uniform (but not constant) mist concentration. Airborne oil mist is emitted at a rate of 0.010 kg/s. The workplace volume is 1.0 x 10^5 m^3. An oil mist eliminator withdraws 1.0 m^3/s of room and air, removes 95.% of the oil mist from the withdrawn air, and recirculates the cleaned air back into the room. A fan in the roof withdraws 1.0 m^3/s of room and discharges it outdoors. Another fan adds 1.0 m^3/s of tempered outside make-up air (containing no oil mist). Adsorption and settling processes within the workplace remove mist at a rate equal to the concentration times the product of the surface area and adsorption rate constant; the quantity ($A_s k_{ad}$) is equal to 5.0 m^3/s. If the initial oil mist concentration is 0.010 g/m^3, what is the concentration at the end of an 8-hour day?

13. An enclosure of volume (V) contains processes that emit toxic vapors at rate S (mass/time). The enclosure is equipped with a ventilation system similar to Figure 5.7 except that fresh make-up air containing no contaminant is added directly to the enclosure using outside air at a rate of one room air change per hour. Air is removed from the room at a rate of 10. room air changes per hour and cleaned. A portion of the cleaned air is returned to the room and a portion (equal to the volumetric flow rate of make-up air) is discharged. Neglecting wall losses and settling, write an expression that predicts the concentration as a function of time. What is the steady-state concentration?

14. Consider the general ventilation configuration of Figure 5.7. Suppose there has been a great deal of smoking in the enclosure such that initially, the smoke concentration is a high value c(0). Everyone leaves the room but the ventilation system continues to run. In terms of the parameters in the figure, write an expression for the time it takes to reduce the concentration by a factor of two. Neglect adsorption and desorption of smoke by the walls of the room.

15. The ventilation system for a conference room is shown in Figure 5.7. Initially the room is empty and the concentration of total suspended particulate matter (TSP) is 10. μg/m^3. A meeting is called and a number of people begin to smoke such that throughout the meeting, smoke is emitted at a nearly constant rate of 0.50 g/hr. Physical properties of the problem are:

$V = 100.\ m^3$	$Q_e = 25.\ m^3/min$
$S = 0.50$ g/hr	$c_a = 0$
f (make-up air fraction) = 10.%	efficiency of each collector 80.%

How long will it take before the TSP concentration reaches the Primary National Ambient Air Quality Standard of 260 μg/m^3? Is there a steady-state concentration, and if so, what is it?

16. Welding fume is produced in a metal fabrication facility. Activity in the workplace produces uniform but not constant fume concentration throughout the room. Welding fume is emitted at a rate of 1.0 g/sec. The room volume is 1000 m^3. A fume eliminator is placed in the room that withdraws 1.0 m^3/s, removes 85.% of the fume, and recirculates the cleaned air back into the room (see Figure 5.5). If the initial fume concentration is negligible, what is the concentration after one hour?

17. The rooms of a "smoke-free" building are served by a single air conditioning system. Examine the smoke concentration that will occur if one person smokes in an office. For simplicity, assume the building is divided into two rooms of volumes V_1 and V_2. Air is withdrawn from each room at volumetric flow rates Q_1 and Q_2. The withdrawn air enters a common header that leads to a recirculating fan. Upstream of the fan a volumetric flow rate of air [$f(Q_1 + Q_2)$, where $0 < f < 1$] is discharged to the atmosphere and the same amount of outside "make-up air" is added to the remaining air. The fan returns air at volumetric flow rates Q_1 and Q_2 to rooms V_1 and V_2. Initially, the smoke concentration in rooms V_1 and V_2 is the same as the outside air (c_a). At $t = 0$, someone begins to smoke at a steady rate S (g/s) in room V_1. Because of the design of the air conditioning system it is believed that some of the smoke will travel to room V_2. Write an expression that shows how the concentration in room V_2 varies with time and the other parameters of the design. Will the concentration in V_2 reach a steady-state value? If so, what is it?

18. Assume that a two-person office has the general ventilation configuration of Figure 5.5. The total volume of the office is 40. m^3, and the general ventilation system supplies 0.60 room air changes per hour; the supply air is clean. One of the individuals is a chain smoker who smokes a filtered cigarette every 6.5 minutes. He installs a small portable air cleaner in the room so as not to annoy the other person in the office. The air cleaner discharges the cleaned air back into the room as in Figure 5.5. It is claimed that the air cleaner has an efficiency of 80.% and cleans tar particles from the air at a volumetric flow rate of 6.0 CFM. Write an expression for the instantaneous mass concentration of tar particles inside the office as a function of time, assuming that the air in the room is well mixed. What is the steady-state mass concentration of tar particles in the room in units of mg/m^3? Hint: Use the sidestream (smoldering) emission factor for cigarettes, as found in Appendix A.7.

19. A volatile liquid is stored in an open vessel inside a small, unventilated, well-mixed room. Liquid evaporates and enters the air where it remains since there is no mechanism to remove it. Show that the vapor concentration (in air) increases in an exponential fashion, $c = a\,[1 - \exp(-bt)]$ where a and b are constants that do not depend on the concentration.

20. You have been asked to evaluate the ventilation system for a new clean room (Figure 5.11c) used to manufacture integrated circuits. One of the operations is the removal of unwanted material from the circuits by a dry etching process using RF plasma. Unfortunately gaseous products are emitted. The ventilation system is designed to achieve well-mixed conditions but diffusion and adsorption of these materials to the wall (wall losses), the rate at which these materials are emitted, and the air cleaner efficiency are unknown. A series of experiments is run to evaluate these quantities during which time the discharge and make-up air flow rates are set to zero and the concentration is measured. The volumetric flow rate through the sampling device (Q_s) is negligible. Properties of the ventilation system are as follows.

- volume = 60. m^3
- surface area over which adsorption occurs = 150. m^2
- discharge and make-up air volumetric flow rates = 5.0 m^3/min
- volumetric flow rate through air cleaner = 10. m^3/min
- efficiency of air cleaner = unknown
- wall loss coefficient = unknown
- contaminant emission rate = unknown

(a) The air cleaner, discharge and make-up air flow are shut off and a large amount of contaminant is emitted until the concentration is 0.30 mg/m^3. Emission is halted, and the concentration is measured experimentally. What is the wall-loss coefficient?

concentration (mg/m^3)	time (min)
0.300	0
0.148	5
0.060	15
0.030	21
0.020	26
0.007	35

(b) The make-up air and discharge are shut off and the air cleaner is run to reduce the contaminant concentration to a relatively low value. The air cleaner is then shut off. The etching process is begun at the desired production level and maintained at a constant value. Calculate the contaminant emission rate if experimental measurements show that the concentration varies as follows:

concentration (mg/m^3)	time (min)
0.010	0
0.020	1
0.063	5
0.082	10
0.089	15

(c) The make-up air and discharge are shut off, contaminant emission is maintained at the constant rate in (b) above, and the air cleaner is turned on. If the concentration decreases and achieves a steady-state value of 0.053 mg/m^3, what is the efficiency of the air cleaner?

21. Contaminants are emitted at rate S inside a well-mixed enclosure equipped with a ventilation system similar to Figure 5.7. There is no adsorption on the walls of the enclosure. Air from the enclosure is withdrawn at rate Q and a fraction is discharged to the atmosphere. Make-up fresh air (Q_m) from the outside (contaminant concentration c_a) is added to the recirculated air and the mixture is cleaned by an air cleaner with removal efficiency η_3. The mixture is returned to the enclosure. A second supply of make-up air (Q_m) from the outside is sent through an air cleaner of removal efficiency η_1 and discharged directly into the enclosure. Lastly an air cleaner is placed inside the enclosure to withdraw air at rate $Q_{cleaner}$ from inside the enclosure, clean it with an efficiency η_2, and return all of it to the enclosure. The mass concentration inside the enclosure varies with time given by the differential equation dc/dt = B - Ac, where A and B are constants. Evaluate the constants A and B in terms of the parameters in the problem.

22. A house painter uses an indoor air cleaner to remove paint odors in homes in which he works. The efficiency of the indoor air cleaner is 95% and the volumetric flow rate is $Q_{cleaner}$. A schematic diagram of the ventilation system for the room without the indoor air cleaner is similar to Figure 5.5. Assuming well-mixed conditions, calculate the volumetric flow rate $Q_{cleaner}$ that produces a steady-state concentration one-half the value that would occur if the indoor air cleaner were not used.

$V = 50.\ m^3$	$S = 100.\ g/min$
$c(0) = 10.\ mg/m^3$	$c_a = 10.\ mg/m^3$
$k_{ad} = 0.0010\ m/s$	$A_s = 85.\ m^2$
$f = 0.10$	$\eta_1 = \eta_2 = 75.\%$
$Q = 20.\ m^3/min$	Q_a (make-up air) = $2.0\ m^3/min$

23. A smoldering cigarette emits particles in an unventilated well-mixed chamber of volume 22. m^3 and surface area 47. m^2. At different times particle concentrations are measured. Figure 5.6 shows that the concentration rises during the first 8 minutes in which the smoldering cigarette emits particles: after 8 minutes the cigarette ceases to emit particles and the concentration decreases as particles plate-out on the chamber walls.

(a) Estimate the source strength (mg/min) of the smoldering cigarette.
(b) Estimate the wall loss coefficient in the chamber (m/hr).
(c) In another experiment an air cleaning device is placed in the chamber and tested. The volumetric flow rate of the device is 300. ACFM. After the cigarette has ceased emitting particles, the device is turned on and the concentration is measured.

time (min)	concentration (mg/m^3)
0	0.240
2	0.122
4	0.067
6	0.020
8	0.018
10	0.010

Estimate the removal efficiency of the room air cleaner.

24. An enclosure is not well mixed. Step-down tracer experiments are conducted. Initially, the concentration at a point P is c(0). Fresh air is added at volumetric flow rate Q and air containing the tracer is removed at the same rate. The concentration at point P decreases linearly with time and becomes negligible at time t_1. Show that the exchange effectiveness is equal to $Q(t_1)/3V$.

25. An enclosure is not well mixed. Initially, the concentration of a tracer is a low value, c(0). Tracer is added to the incoming air at a constant rate and the concentration is measured at point P. The concentration is found to rise linearly until time t_1 when it is equal to the value in the incoming air. Compute the ventilation effectiveness coefficient.

26. A new office complex (V = 600. m^3) has been designed with uniquely located inlet and outlet ventilation registers; the designer claims that cigarette smoke from one work station will not travel to another work station. The total volumetric flow rate for the entire volume is 10. m^3/min. Step-up tracer experiments are conducted to verify the designer's claim. In these experiments the entire work space (without workers) is ventilated with fresh air and suddenly a tracer gas is added to the inlet air supply for every register and a constant tracer gas concentration is maintained. The concentration is measured experimentally at three work stations.

Station A: $c_A = 0$ for $t < t_A$, $c(\infty)$ for $t > t_A$

Station B: $c_B = c(\infty)\dfrac{t}{t_b}$ for $t < t_B$; $c(\infty)$ for $t > t_B$

Station C: $\dfrac{c_c}{c_\infty} = 1 - \exp\left(-\dfrac{t}{t_C}\right)$ for all t

From these measurements discuss the validity of the designers claim if values for t_A, t_B and t_C are as follows:

(a)	$t_A = 0.20$ hr	$t_B = 5.0$ hr	$t_C = 1.0$ hr
(b)	$t_A = 0.50$ hr	$t_B = 0.50$ hr	$t_C = 2.0$ hr
(c)	$t_A = 2.0$ hr	$t_B = 1.0$ hr	$t_C = 0.50$ hr
(d)	$t_A = 2.0$ hr	$t_B = 0.20$ hr	$t_C = 0.20$ hr

27. Your automobile is stuck in a snow drift. You start the car but open the windows a crack to allow fresh air into the car and air from inside the car interior to escape to the atmosphere. You then turn on the hot air heater and sit inside the car to warm up before attempting to dig yourself out of the snow drift. Unknown to you, the snow surrounds the bottom of the car such that the auto exhaust is drawn directly into the air inlet for the heating system. The following properties are known:

- V (car interior) = 2.0 m^3
- Q_H (hot air heater) = 0.50 m^3/min
- Q_I (infiltration) = 0.50 m^3/min
- Q_E (exfiltration) = 1.0 m^3/min
- c_H (CO in heater air inlet) = 1000 mg/m^3
- c_A (CO in ambient air) = 10. mg/m^3

Assume that the above conditions are constant and that well-mixed conditions exist inside the car.

(a) What is the steady-state CO concentration inside the car?
(b) How long will it take for the CO concentration to reach 400. mg/m^3
(c) How long could you remain in the car before you'd experience serious cardiac and pulmonary functional changes?

28. Ammonia is generated by a chemical etching process at a rate of 10. mg/min and escapes into a room (10. by 10. by 4.0 m). The temperature and pressure are 300. K and 1.0 atm. Fresh air leaks into the room (infiltration) and room air leaks out (exfiltration); a steady-state ammonia concentration is established. The room is equipped with a large circulating fan that distributes the ammonia fumes uniformly throughout the room and a steady-state ammonia concentration of 100. PPM is established. Workers complain about the odor. Your supervisor suggests that fresh air be ducted into the room and a similar amount of room air be discharged to the outside. What volumetric flow rate (ACFM) of new fresh air is needed (infiltration and exfiltration remaining the same) to reduce the concentration of to one-half the odor threshold? (Use the odor threshold listed in Appendix A.20.)

29. A room of volume (V) is ventilated by adding air at volumetric flow rate (Q) at one location and withdrawing an equal amount of air at another location. Initially the concentration of a tracer gas in the room is zero. For t > 0, a tracer gas is added to the incoming air. The concentration in the inlet air supply is c_a. Such an experimental technique is called a "step-up tracer input". If the instantaneous tracer concentration is found to be given by

$$\frac{c(t)}{c_a} = 1 - \exp(-at)$$

where a is a constant, show that the mixing factor (m) is equal to (aV/Q).

30. Air is added to a room of volume (V) at one location at volumetric flow rate (Q) and an equal amount air is withdrawn at another location. Initially the concentration of a tracer gas in the incoming air and in the room is equal to c(0). The tracer gas in the inlet air is suddenly eliminated. Such an experimental technique is called a "step-down tracer input". If the instantaneous tracer concentration is found to be given by

$$\frac{c(t)}{c(0)} = \exp(-at)$$

where a is a constant, show that the mixing factor (m) is equal to (aV/Q)

31. Consider fully developed flow in a tunnel of radius R. The velocity u(r) at any radius is $u(r)/U_0 = 1 - (r/R)^2$ where U_0 is the centerline velocity that is twice the average velocity (Q/A). Show that in a step-up tracer experiment in which a tracer gas is injected at distance L upstream of the sampling station, the effectiveness coefficient (e) one would measure at any radius satisfies

$$e = 4\left[1 - \left(\frac{r}{R}\right)^2\right]$$

32. Redo Example 5.2 using the 4th-order Runge-Kutta method (see Appendix A.12).

33. Redo end-of-chapter problem 5.10 using the 4th-order Runge-Kutta method (see Appendix A.12).

34. Describe an experiment to measure the source strength related to the emission of volatile hydrocarbons that occurs in spray painting. Address both vapor emitted during spraying and during the time the paint dries.

35. Radon gas seeps into your basement. The daughters of radon gas generate radioactivity of 50. pCi per hour per liter of air in your 300. m^3 basement. The radioactivity is directly proportional to the concentration of radon gas in your basement. You wish to remove some basement air and dilute the remainder with outside air (containing background radioactivity of 1.0 pCi per liter) such that the average steady-state radioactivity in your basement is 4.0 pCi per liter of air. Using the well-mixed model estimate the volumetric flow rate of outside air necessary to accomplish this goal. Using the degree days for Chicago, estimate the cost to heat the make-up air, assuming direct-fire natural gas heat with the cost per unit of fuel at $\$0.011/ft^3$ (1.1 cents per cubic foot).

36. You are employed by a manufacturer of a home air purifier who asks you to prepare technical material to support an advertising campaign that claims their product reduces household radon. Since radon is chemically inert and its concentration (c) is very low, it is easier to measure its radioactive decay (I = - dc/dt) rather than its concentration. The decay is linearly proportional to the concentration, i.e. $I = -dc/dt = \beta c$, where $\beta = 2.1 \times 10^{-6}\ s^{-1}$. Radioactivity is measured in picocuries per liter (pCi/L) where 1 pCi/L = 2.2 radioactive decays per minute. As a first approximation you plan to model the home as a single, well-mixed air space of volume (V) in which radon enters in the following ways:

- through the floor from the soil (S_s)
- from drinking water (S_w)
- from home building materials (S_b)
- from infiltrated air at volumetric flow rate $Q_{infiltration}$, ambient radioactivity (I_0), where I_0 = 0.25 pCi/L

Radon leaves the air space by the following processes:

- air discharged from the home at a volumetric flow rate (Q)
- internal room air purifier, that cleans room air at a volumetric flow rate (Q_r), with an efficiency (E_r)
- by radioactive decay, decay constant (β)

(See Nazaroff and Teichman, 1990, for an interesting discussion about the health risk associated with radon, its relation to smoking and the costs associated with radon control.)

(a) From EPA data, it is known that for typical homes in the US,

  - ($\beta S_s/Q$) = 1.4805 pCi/L
  - ($\beta S_w/Q$) = 0.0540 pCi/L
  - ($\beta S_b/Q$) = 0.0108 pCi/L

  Show that the steady-state radioactivity can be expressed as I_{ss} (pCi/L) = $82.5/[1 + \beta(V/Q) + E_r(Q_r/Q)]$.

(b) Show that the mol fraction of radon in air (at STP) associated with the EPA recommended maximum radon radioactivity of 4 pCi/L is equal to 2.8 x 10^{-18} ! [Note: there is one decay for each radon molecule; at STP there are 6.02 x 10^{23} molecules of air in one mol and one mol of occupies 24.5 cm^3.]

(c) Using typical values of the number of room air changes (Appendix A.16), estimate the room air volumetric flow rate (Q_r) needed to reduce the steady-state radon radioactivity by 50%. How practical is the company's air purifier that requires such a volumetric flow rate even if the air purifier has an efficiency approaching 100%?

37. Workers apply an epoxy coating to the inside surface of a closed chemical reactor, Figure 5.4. The volume of the reactor is 300. ft^3. A light coating is applied for 20 minutes and then it is allowed to dry for 20 minutes. A heavy coat is then applied for 20 minutes and then it is allowed to dry for 20 minutes. The hydrocarbon evaporation rate, S(t) (g/min), during the process is as follows.

- S(t) = 150. g/min for $0 \le t \le 20$ min.
- S(t) decreases linearly from 100. g/min to zero $20 \le t \le 40$ min.
- S(t) = 250. g/min for $40 \le t \le 60$ min.
- S(t) decreases linearly from 150. g/min to zero $60 \le t \le 80$ min.
- S(t) = zero for $t \ge 80$ min.

Throughout the period, a fan removes air and vapor from within the reactor and another fan introduces an equal amount of fresh air into the reactor. Because of a malfunction in the electrical circuit, the volumetric flow rate of air is not constant and fluctuates as follows:

- Q = 130. ACFM; $0 \le t \le 20$, $40 \le t \le 60$, $80 \le t \le 100$ min.
- Q = 65. ACFM; $20 \le t \le 40$, $60 \le t \le 80$, $100 \le t \le 120$ min.

Using the well-mixed model, predict the vapor concentration (mg/m^3) inside the reactor and plot the values for the period of zero to 120 minutes. Use Runge-Kutta methods to integrate the differential equation. Compute the average concentration to which the workers are exposed during the time they are inside the reactor, i.e. $0 \le t \le 80$ minutes.

38. The operating room described in Example 5.7 is used on an occasion when the source strength S(t) and supply volumetric flow rate, $Q_s(t)$, vary as shown below. Using the initial conditions shown in Example 5.7, plot the contaminant concentration versus time for 36 minutes.

time (min)	**S(g/min)**	**$Q_s(t)$ (m^3/min)**
0	2.0	30.0
2	2.8	29.0
4	5.5	17.5
6	6.0	10.0
8	6.0	10.0
10	4.0	20.0
12	3.5	30.0
14	3.0	34.0
16	2.5	34.5
18	2.2	35.0
20	2.0	35.0
36	2.0	35.0

39. Two identical office rooms (V = 40. m^3) have been designed for displacement ventilation with volumetric flow rate (Q) equal to 4.0 m^3/min. Interior decorators select different office furnishings and place them in different locations in the rooms, perhaps even blocking floor registers. Tenants complain that rooms are stuffy and unventilated. As a consulting engineer, you have been asked to advise the building owner on the validity of the complaints. You ask that step-up tracer experiments be conducted. The ratio of exhaust opening tracer mass concentration to the steady state value [$c_e/c_e(\infty)$] is shown as a function of time for both rooms below. Compute the ventilation effectiveness for each room and comment on the claim that the rooms have poor ventilation.

time (min)	**$c_e/c_e(\infty)$, room 1**	**$c_e/c_e(\infty)$, room 2**
0	0	0
2	0.03	0.01
4	0.10	0.02
6	0.20	0.06
8	0.40	0.08
9	0.55	0.10
10	0.60	0.12
12	0.72	0.25
13	0.75	0.50
14	0.80	0.80
15	0.82	0.94
16	0.86	0.95
18	0.92	0.98
20	0.96	0.99
22	0.99	1.00
30	1.00	1.00

40. A truck paint booth (Figure 1.21) currently discharges paint fumes directly to the atmosphere. The booth volume (V) is 15,000 ft^3 and the volumetric flow rate (Q) is 50,000 ACFM. Coatings are applied at a rate of 0.514 gal/min and have a density of 12. lbm/gal of which 90.% is the solvent n-propyl alcohol (M = 200.). The LEL of n-propyl alcohol is 22,000 PPM. The state air pollution agency requires the installation of a thermal incinerator to oxidize the alcohol. The capital cost of a thermal incinerator is $35.00/ACFM. To reduce the capital cost, it is proposed to recirculate a portion of the untreated exhaust back into the booth (see Figure 5.7 but eliminate the air cleaner shown in the figure). Workers inside the booth are fully clothed and use an external air supply for breathing.

(a) Write a general expression for the transient alcohol mass concentration, c(t), assuming that the source strength (S), volumetric flow rate (Q) and inlet air mass concentration, c_a, are constant.
(b) Plot the cost and steady-state alcohol mass concentration as functions of the make-up air fraction (f) for $0<f<1$, and find the value of f when the steady-state mass concentration is 10.% of LEL. Assume $c_a = 0$.

41. Consider an automobile tunnel of rectangular cross section. W is the tunnel width and H is the tunnel height. The tunnel ceiling and floor are sealed concrete, which have negligible contaminant adsorption. However, the side walls of the tunnel, which are designed to absorb sound, also adsorb contaminant with wall adsorption coefficient k_w.

(a) Generate an expression for k, the tunnel wall loss parameter, in terms of W and k_w. (H should drop out.)
(b) Consider a rectangular tunnel, 10. m wide and 5.0 m high, with adsorption only on the side walls, not the ceiling or floor, and with k_w = 0.070 cm/s. Calculate tunnel wall loss parameter k in units of 1/min. Your answer should be 0.0084 1/min. Verify that your answer is correct before proceeding to the next problem.

42. An engineering firm is designing an automobile tunnel. The following design information is provided:

$(EF)_c$ = emission factor for carbon monoxide = 8000 (mg)/(auto km)	PEL of carbon monoxide = 55 mg/m^3
U(0) = inlet air speed into tunnel = 40. m/min	n_c = traffic density = 60. auto/km
c_m = mass concentration of make-up air = 1.5 mg/m^3	v_c = average automobile speed = 90. km/hr
k_w = wall adsorption coefficient = 0.18 cm/s	c(0) = mass concentration at inlet = 5.3 mg/m^3
wall adsorption on side walls (not on ceiling or floor)	W = tunnel width = 12. m
Q_m = make-up air volumetric flow rate = 25,000 m^3/min	H = tunnel height = 5.0 m
Q_e = exhaust air volumetric flow rate = 25,000 m^3/min	L = tunnel length = 2500 m

A ***balanced transverse ventilation*** system is chosen since the tunnel is long.

(a) Calculate k, the tunnel wall loss parameter, in units of 1/min.
(b) Plot carbon monoxide mass concentration, c (in mg/m^3), as a function of distance through the tunnel, x (in m). (Excel or Mathcad is recommended.) Also indicate the PEL and c_{max} on the same plot for comparison.
(c) Is the CO level below the PEL? If not, recommend a remedy.

43. [*Design Problem*] Consider the painting booths of Example 5.12. Explore the possibility of using the existing 85% efficient thermal oxidizers to save money. The problem is open-ended –

you may analyze combinations of the existing units in series or in parallel with each other and/or with new units, and with or without split-flow ventilation. The goal is to decrease as much as possible the required volume flow rate (and therefore the cost) of new thermal oxidizers. Hint: Look ahead to Chapter 8 for expressions for the overall efficiency of air cleaners in series and in parallel.

6

Present Local Ventilation Practice

In this chapter you will learn:

- contemporary hood design standards and practices
- how to control emissions from handling of bulk materials
- how to design air curtains for buoyant plumes
- about the importance of proper placement of building air inlets and exhausts
- how to design exhaust duct systems
- how to select and fans for a system of interconnecting ducts
- how to use fan scaling laws

Local ventilation systems are designed to remove contaminants at the point(s) of emission. The most common type of local ventilation system is a ***hood*** – a shaped inlet designed to capture contaminated air and conduct it into the exhaust duct system. The options available to the designer are the geometry of the shaped inlet, its location with respect to the source, and the exhaust volumetric flow rate. Local ventilation systems are classified in the following broad categories:

- annular exhaust hoods and lateral exhausters (Figure 6.1)
- canopy hoods (Figure 6.2)
- close-fitting canopies and receiving hoods (Figure 6.3)
- downdraft hoods (Figure 1.28)
- flanged openings and side-draft hoods (Figure 1.16)
- full enclosures (Figures 1.19 and 1.21)
- high velocity - low volume (HVLV) hoods (Figure 1.13)
- partial enclosures and booths (Figures 1.22 and 6.4)
- plain openings, also called unflanged openings (Figure 1.26)
- push-pull systems (Figure 1.27)
- receiving hoods and partial enclosures (Figure 6.5)

These photographs and sketches illustrate features of each category. Features of one category can often be combined with those of another category. The reader is urged to examine the figures to become familiar with the state-of-the-art technologies.

At present, the design of local ventilation systems follows practices recommended by the ACGIH and published in "Industrial Ventilation, A Manual of Recommended Practice" (current edition at the time of this writing, 2001). The manual is updated every few years, and readers should always use the current edition. Hereafter this publication is referred to as the ***ventilation manual.*** The ventilation manual does not develop design configurations from fundamental concepts in the thermal sciences nor recommend design configurations using these principles. Rather the ventilation manual posits design configurations for a variety of industrial applications, and is widely used throughout industry. While the practices do not enable designers to predict contaminant concentrations to ensure

(a)

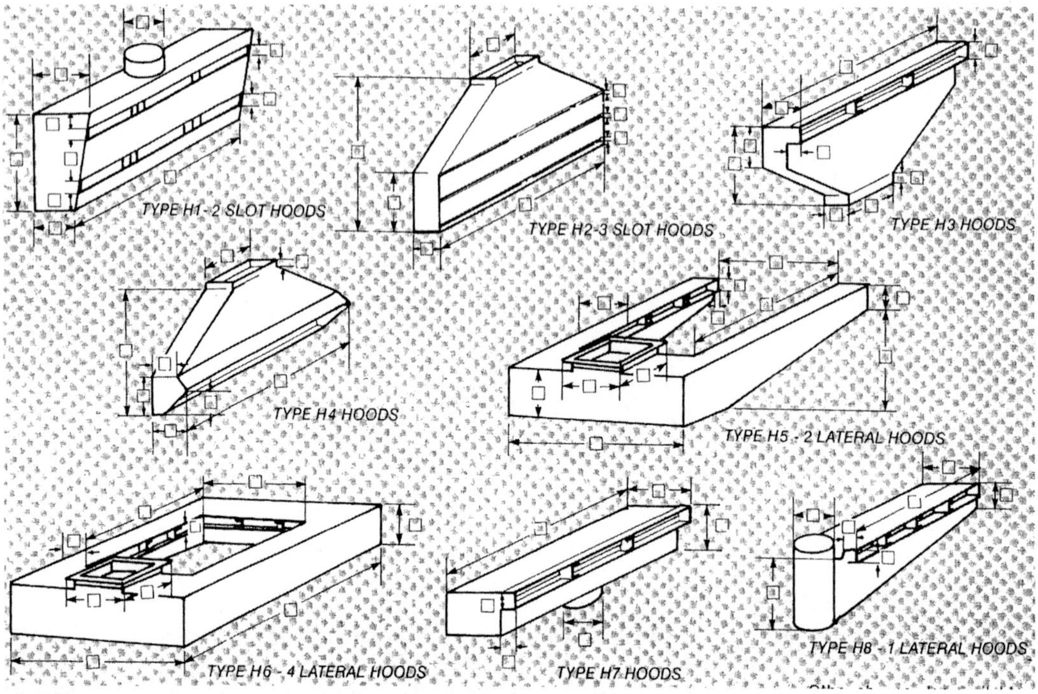

(b)

Figure 6.1 Lateral exhausters (fume hoods): (a) designs for various types of vessels and processes. Corrosion resistant materials may be PVC, polypropylene, or fiberglass. (from Vanaire Ltd.); (b) lateral rim exhauster for a crucible (from Heinsohn, 1991).

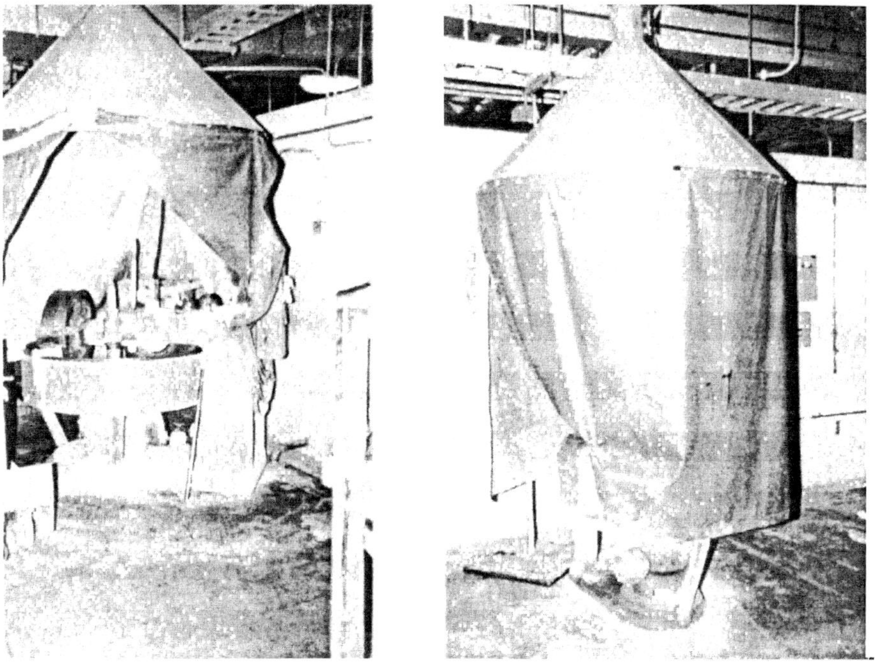

(a) (b)

Figure 6.2 Canopy hood: (a) curtains open, (b) curtains closed (from Heinsohn, 1991).

Figure 6.3 Close-fitting canopy and receiving hood for a reheat furnace (from Heinsohn, 1991).

that mandatory health standards are met, the ventilation manual provides reasonable assurance that the ventilation system will control the transfer of contaminants to the workplace environment. Ultimately, compliance must be confirmed by experimental measurements. The ventilation manual is written in uncomplicated language and contains principles that do not require the reader to possess a baccalaureate degree in engineering; there is a deliberate effort to write for a diverse class of users rather than for specialists. The ACGIH conducts training seminars several times each year, in which ACGIH methods are taught. A useful self-instruction manual (Burton, 1984) can also be used.

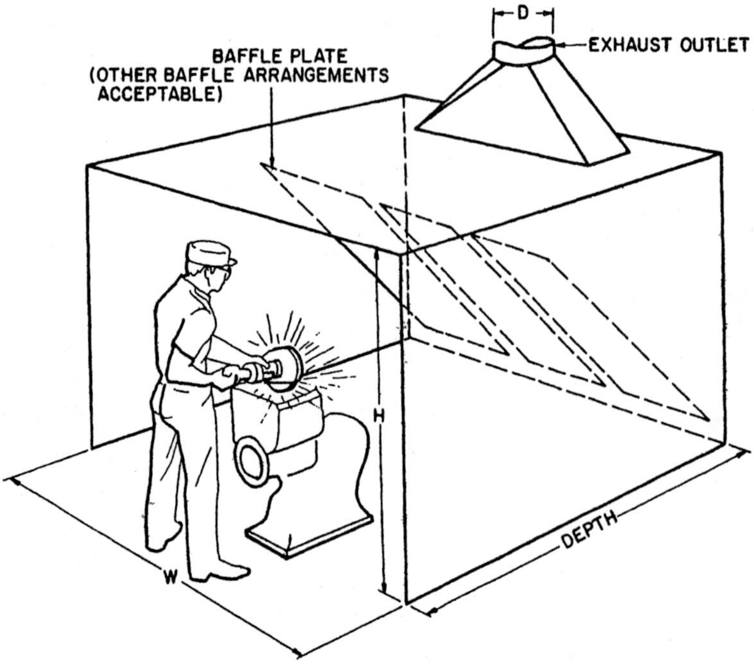

Figure 6.4 Booth for portable grinding, polishing, and buffing: Q = volume flow rate = HWV; W = equipment width + 6 ft; H = equipment height + 3 ft, with H(min) = 7 ft; D = exhaust outlet diameter; Depth = equipment depth + 6 ft; A_{baffle} = baffle area = 0.4WH; V = face velocity = 100 FPM for class 1 contaminants, 200 FPM for class 2 contaminants, and 400 FPM for class 3 contaminants. Entry loss for tapered outlet plus baffles = (1.78 slot VP) + (entry loss factor for tapered outlet) times duct VP. Entry loss for plain outlet plus baffles = (1.78 slot VP) + (0.5 duct VP). (Adapted from Hagapain et al., 1976.)

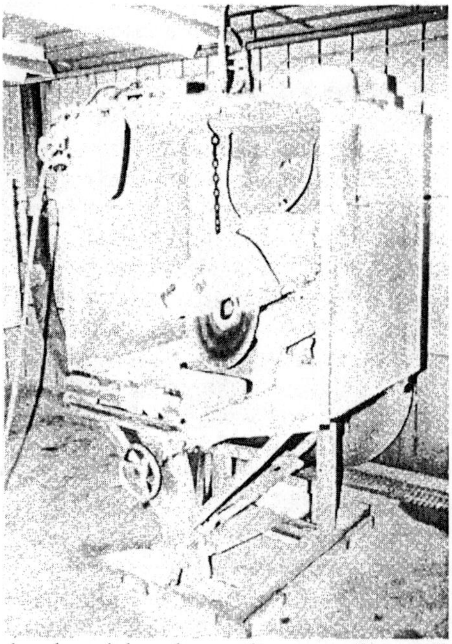

Figure 6.5 Receiving hood and partial enclosure for a cut-off saw (from Heinsohn, 1991).

The ASHRAE HVAC Applications Handbook (1999) is updated every four years and describes procedures to design local ventilation systems. Design practices recommended by ASHRAE are written in language used in the engineering profession and in four-year baccalaureate engineering education programs. The practices recommended by ASHRAE are in basic agreement with those recommended by ACGIH; what follows is a summary of these practices. If individuals wish to use them, they should consult the original sources for a full discussion of details they must address, and for references to the original material upon which the design practices and recommendations are based.

6.1 Control of Particles

The central concept in contemporary ventilation practice for control of particles is the ***capture velocity*** (v_c), defined as the air velocity at any point in front of the hood or at the hood opening necessary to overcome air currents and to capture the contaminated air at that point by causing it to flow into the hood (ACGIH, 2001). The velocity of air at the plane of the hood inlet is called the ***face velocity*** (U_{face}). The area of the opening is called the ***face area*** (A_{face}). [Note that technically, velocity is a *vector* quantity, but it is common practice in the literature to use the term "velocity" when discussing the *magnitude* of velocity, which should more properly be called ***speed***. Hence, the velocities referred to here are not vectors, but simply speeds.]

The numerical value of the capture velocity is a function of the particles to be removed. Table 6.1 shows the range of values recommended by ACGIH. Once the capture velocity is selected, the designer chooses an inlet shape and a suction flow rate such that the air velocity is equal to the capture velocity at the point where the contaminant is to be removed. Such a point is called the ***null point***. Whether the particles are removed near the point of emission or at another location is a decision to be made by the designer. It is desirable to capture the particle as close to the point of emission as possible.

Table 6.1 Capture velocities (abstracted from ACGIH, 2001).

characteristics of contaminant emission	examples	capture velocity (FPM)
1. contaminant enters quiescent air with negligible velocity	degreasing tank, evaporation	50-100
2. contaminant enters slightly moving air with a low velocity	welding, vessel filling	100-200
3. contaminant actively generated and enters rapidly moving air	spray painting, stone crushers	200-500
4. contaminant air enters rapidly at high velocity	grinding, abrasive blasting	500-2000

Lower values of capture velocity:

- room air movement minimal or conducive to capture
- contaminants of low toxicity
- intermittent use or low production rates
- large hood and large mass of air moved

Upper values of capture velocity:

- adverse room air movement
- contaminants of high toxicity
- heavy use and high production rates
- small hood and small mass of air moved

In the absence of wakes or jets, it is accurate to model an inlet as a sink in a potential flow field. For example, consider a slot of width 2w and length L along the edge of perpendicular flanges, above which is a quiescent air mass, as in Figure 6.6. If air is withdrawn uniformly along the length of the slot at volumetric flow rate Q, the radial air velocity U(r) entering normal to the surface of a quarter cylinder of radius (r) above the slot can be estimated by dividing Q by the surface area of the quarter cylinder,

$$\boxed{U(r) = \frac{Q}{\frac{1}{4}(2\pi r L)} = \frac{2Q}{\pi r L}} \tag{6-1}$$

where r is the distance from the center of the slot to the point on the quarter-cylinder surface. Equation (6-1) is only valid for $r > w$, since $r < w$ is within the interior of the slot. Slot width (w) is unimportant at distances (r) much greater than w since the slot appears to the flow as a line sink; hence, w does not appear as a parameter in Eq. (6-1).

Figure 6.7 shows a sketch of a circular opening of diameter D at the junction of two flat perpendicular flanges or walls. Imagine a quarter sphere of radius r between the two walls, as sketched. The radial velocity, U(r), can be calculated in similar fashion as above, where the surface area is now one-fourth of the area of a sphere rather than a cylinder,

$$\boxed{U(r) = \frac{Q}{\frac{1}{4}(4\pi r^2)} = \frac{Q}{\pi r^2}} \tag{6-2}$$

Similar analyses can be used to estimate the air velocity into openings of various other shapes. For example, one can easily calculate the radial velocity, U(r), upstream of the center of a circular opening of diameter D, located in a flat surface through which air is withdrawn at volumetric flow rate Q. When the point in space under consideration gets close to the inlet, the velocities can no longer be predicted by simple equations like Eq. (6-1) or (6-2), but the flow field can still be modeled as potential flow. Potential flow is discussed in more detail in Chapter 7.

An ***unflanged inlet*** is also called a ***plain inlet***. Velocity gradients are very large near the lip of an unflanged inlet, and room air is drawn into the opening from along the outside surface of the inlet duct, which contains little contaminant and therefore reduces the effectiveness of the suction. A ***flange***

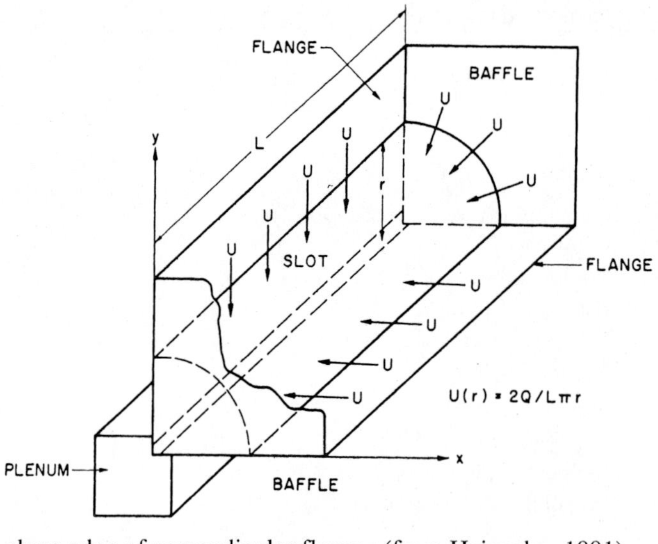

Figure 6.6 Slot along edge of perpendicular flanges (from Heinsohn, 1991).

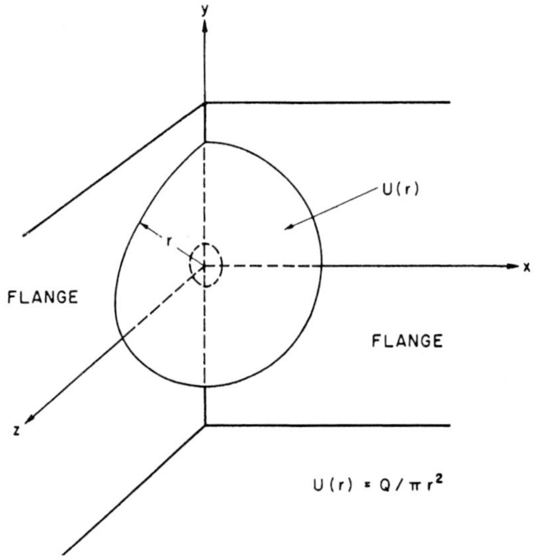

Figure 6.7 Circular opening at edge of perpendicular flanges (from Heinsohn, 1991).

is a flat (usually metal) plate around the rim of the inlet in the form of a collar. The flange normally lies in the plane of the inlet area. The width of a flange varies; it may be large and give the inlet the appearance of a hole in a plate, but more often the flange is narrow with a width equal to approximately half of the diameter (or width) of the inlet. A flange serves two purposes: it provides structural strength to the inlet, and it increases the size of the region in front of the inlet from which air is withdrawn. Figure 1.26 is an example of an unflanged inlet (***snorkel***), and Figure 1.16 is an example of a ***flanged inlet*** (side-draft hood used in foundries).

For several decades, designers have used empirical equations to predict the velocity along the centerline in front of plain and flanged circular and rectangular inlets that withdraw air from a quiescent environment. During that time nearly two dozen empirical equations were proposed for each case (Yousefi and Annegarn, 1993). Braconnier (1988) reviewed these equations, compared their results with experimental measurements, and suggested that the following empirical expressions are most accurate for speed U(x,0) along the centerline (y = 0) at distance x from the inlet plane for inlets with no transition regions (straight-sided inlets). The average inlet velocity is designated as U_{face}, the face velocity, and is equal to volumetric flow rate (Q) divided by the face area of the opening (A_{face}),

$$U_{face} = \frac{Q}{A_{face}} \tag{6-3}$$

- Rectangular, unflanged (plain), free-standing inlet (a by b, where a > b) (Figure 6.8):

$$\frac{U(x,0)}{U_{face}} = \frac{1}{0.93 + 8.58p^2}$$
$$p = \frac{x}{\sqrt{A_{face}}}\left(\frac{a}{b}\right)^q \quad q = 0.2\left(\frac{x}{\sqrt{A_{face}}}\right)^{-0.33} \quad \text{for } 1 < \frac{a}{b} < 16 \text{ and } 0.05 < \frac{x}{\sqrt{A_{face}}} < 3 \tag{6-4}$$

- Rectangular, flanged, free-standing inlet (a by b, where a > b):

$$\frac{U(x,0)}{U_{face}} = 1 - \frac{2}{\pi}\arctan\left\{\frac{2x}{a}\left[\left(\frac{2x}{b}\right)^2 + \left(\frac{a}{b}\right)^2 + 1\right]^{1/2}\right\} \tag{6-5}$$

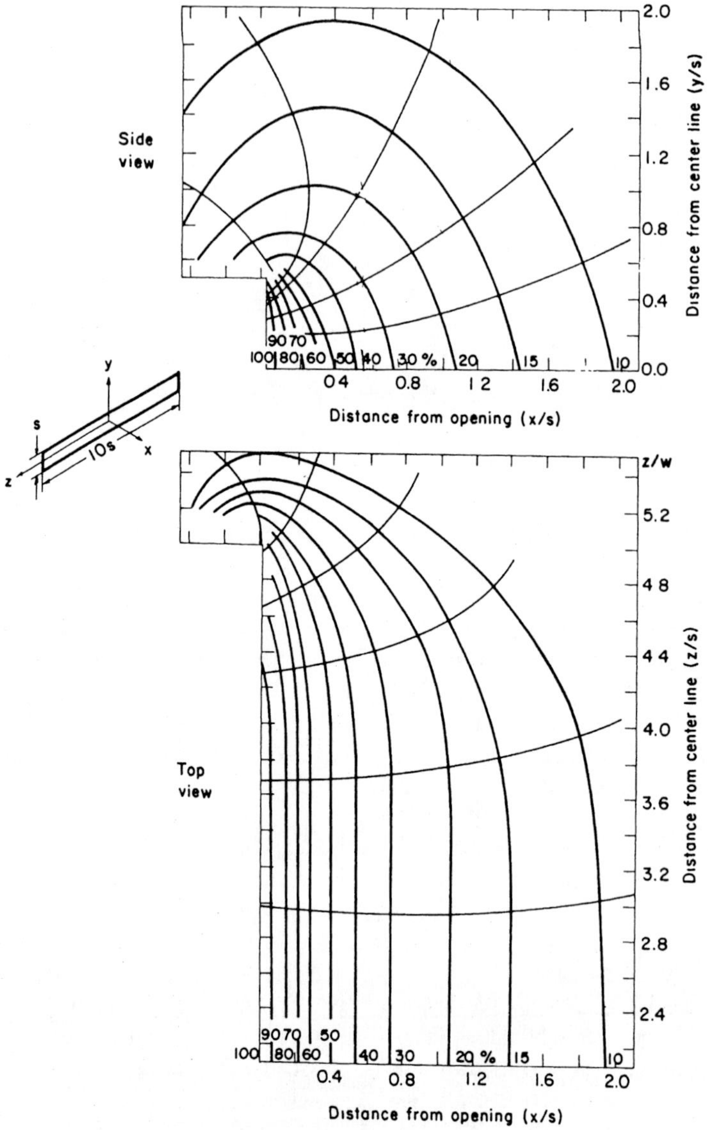

Figure 6.8 Velocity isopleths (curves of constant U/U_{face}, %) for an unflanged rectangular opening, aspect ratio 1:10 (adapted from Baturin, 1972).

- Circular, unflanged (plain), free-standing inlet of diameter D (Figure 6.9):

$$\frac{U(x,0)}{U_{face}} = \frac{1}{1+10\frac{x^2}{A_{face}}} = \frac{1}{1+\frac{40}{\pi}\left(\frac{x}{D}\right)^2} \tag{6-6}$$

- Circular, flanged, free-standing inlet of diameter D (flange width approximately equal to inlet radius) (Figure 6.10):

$$\frac{U(x,0)}{U_{face}} = 1.1(0.070)^{x/D} \text{ for } 0 < \frac{x}{D} < 0.5 \qquad \frac{U(x,0)}{U_{face}} = 0.1\left(\frac{x}{D}\right)^{-1.6} \text{ for } 0.5 < \frac{x}{D} < 1.5 \tag{6-7}$$

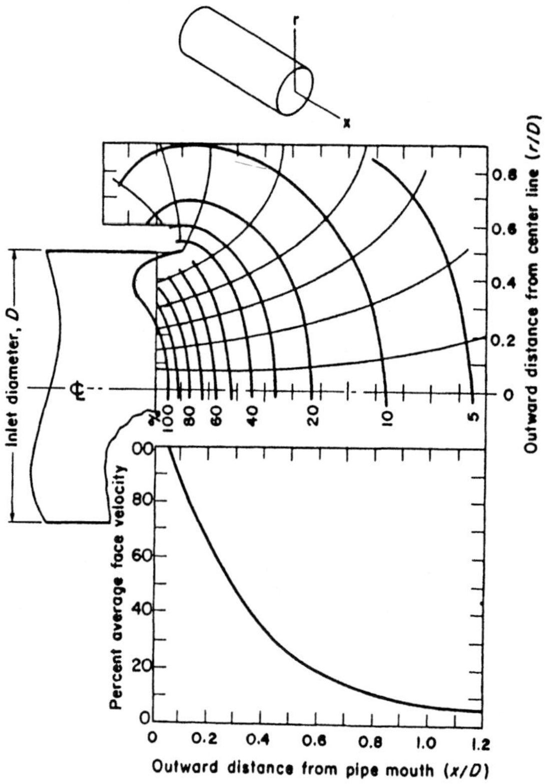

Figure 6.9 Velocity isopleths (curves of constant U/U_{face}, %) and decay of $U(x,0)/U_{face}$ (along the centerline, %) for a plain circular opening (adapted from ASHRAE HVAC Applications Handbook, 1995).

Unfortunately the empirical equations above are of limited value since seldom is an industrial environment quiescent, and for the most part designers want to know the velocity at arbitrary points in front of an inlet, not just at points along the centerline. The professional literature contains numerous graphs such as Figures 6.8 through 6.10 which show empirically generated ***isopleths***, defined as contours of constant relative air speed (U/U_{face}) at arbitrary points in front of flanged and unflanged inlets. For design purposes, analytical expressions (curve fits) can be generated numerically to express these contours. The data in Figure 6.10 are valid for a circular opening with a flange width approximately equal to half of the diameter (D) of the opening. It is safe to assume that the data are still valid for flanges larger than this, but the isopleths for openings with smaller flanges would lie somewhere in between those of Figure 6.9 (unflanged) and 6.10 (flanged).

Comparing the velocity contours for flanged and unflanged circular inlets (Figures 6.9 and 6.10), it is clear that at any point in front of the opening, a flanged inlet produces higher air speeds than does an unflanged inlet of equal diameter. The term ***reach*** is used to characterize the size of the region in front of the inlet within which the incoming air is capable of drawing contaminants into the inlet. Flanged inlets therefore have greater reach than do unflanged inlets. Unflanged inlets draw air into the inlet from near the inlet rim, including regions behind the inlet plane that are of little importance if one wishes to capture contaminants emitted in front of the inlet plane. These conclusions are supported by experiments conducted by Garrison and Erig (1988) who measured velocity gradients along an inlet's centerline for flanged and unflanged openings that faced an infinite solid surface one to four inlet diameters (or slot widths) away. The centerline velocity gradients were only slightly affected by the

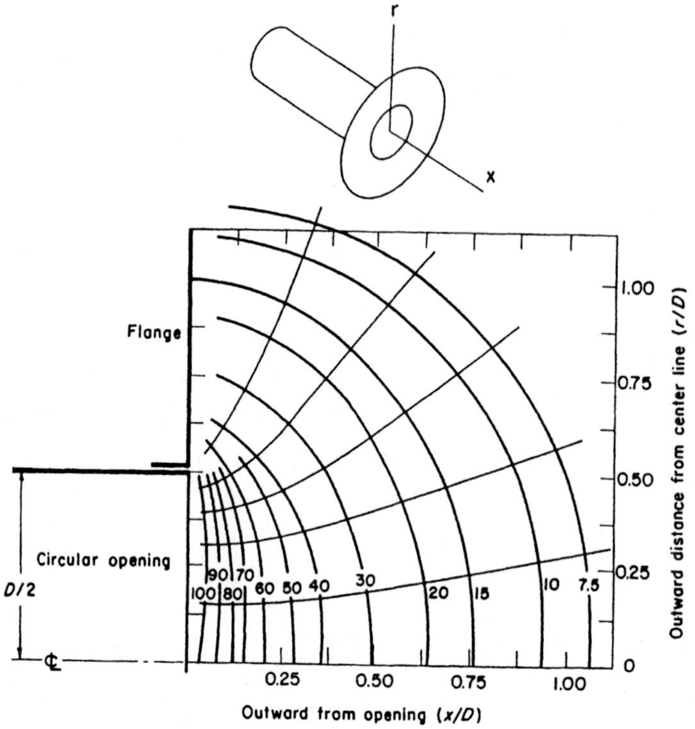

Figure 6.10 Velocity isopleths (curves of constant U/U_{face}, %) for a flanged circular opening (adapted from ASHRAE HVAC Applications Handbook, 1995).

solid surface for unflanged inlets that drew much of their air from regions close to and behind the plane of the inlet rim. The centerline velocity gradients were affected significantly for flanged inlets that drew more of their air from regions in front of the inlet.

To use Figures 6.8 through 6.10 (or other isopleth plots found in the literature) to calculate the required volume flow rate (Q) through the inlet, the following procedure is recommended:

1. Use Table 6.1 (or some other appropriate information) to determine the required capture velocity, v_c.
2. Determine the location of the source of contaminant particles (x, y, and z for rectangular inlets; x and r for circular inlets) relative to the center of the inlet face. (For most engineering analyses, this location is known, e.g. the location of the emissions from a grinding wheel relative to a snorkel inlet.)
3. For a rectangular opening, calculate the normalized axial distance (x/s) and the normalized off-axis distance (y/s or z/s) at the point of interest. For a circular opening, calculate the normalized axial distance (x/D) and the normalized radial distance (r/D) at the point of interest.
4. Using the appropriate isopleths, estimate the normalized velocity (U/U_{face}), noting that the values in Figures 6.8 through 6.10 are *percentages*.
5. Calculate U_{face} by setting U equal to v_c,

$$\boxed{U_{face} = \frac{U}{U/U_{face}} = \frac{v_c}{U/U_{face}}} \tag{6-8}$$

6. Calculate Q for the known values of U_{face} and A_{face} using Eq. (6-3).

Example 6.1 - Control of Fume from a Robotic Welder
Given: An industrial robot performs arc welding on an assembly line for automotive chasses.

To do: Recommend a local ventilation system of circular cross section to control welding fume. There are several design constraints:

- inlet can be no closer than 6.0 inches (0.50 ft) from the welding zone
- cross-sectional area of the inlet must be less than 0.79 ft^2
- overall maximum diameter (D_{max}) of the inlet (including flange) must be less than 12. inches (1.0 ft) so as not to interfere with the robot's movement

Solution: OSHA stipulates a workplace exposure limit for welding fume of 5 mg/m^3, although it does not stipulate an exhaust volumetric flow rate. Table 6.1 recommends a capture velocity of 100 - 200 FPM (approx. 0.5 - 1. m/s). The upper value is chosen because of the toxicity of welding fume and because of air currents that are known to exist in the workplace.

A local ventilation system consisting of a flanged or unflanged opening is considered, with welding performed along the centerline at x = 6.0 inches. The air velocity U at this location is set equal to the capture velocity at that point, v_c = 200 FPM. Under production line conditions, there is always an obstruction at the point of welding; it is unavoidable. The way in which the obstruction affects the velocity field is hard to predict - it can lower the velocities or raise them depending on the geometry of the surfaces to be welded. The required volumetric flow rate for an unflanged opening of different diameters (D) is computed according to the following procedure:

1. Choose several values of D.
2. Calculate the face area (A_{face}) for each value of D.
3. Compute x/D for x = 6.0 inches for each value of D.
4. For each flanged case, calculate flange size = smaller of (D/2) or (D_{max} – D/2) to keep total diameter less than D_{max}.
5. Use Eqs. (6-6) (unflanged) and (6-7) (flanged) to calculate U/U_{face} for each diameter (alternatively, Figures 6.9 and 6.10 can be used, but with more difficulty and less accuracy).
6. For U = 200 FPM, compute U_{face} for each diameter and for both unflanged and flanged inlets using Eq. (6-8).
7. Compute the required volumetric flow rate (Q), using Eq. (6-3), for all cases.

Shown below are the required volumetric flow rates for several unflanged and flanged entry diameters predicted by the above procedure. Because of the repetitive nature of the calculations, the authors used an Excel spreadsheet to perform the calculations; the file is available for downloading from the book's web site. All values are given to two significant digits.

		unflanged openings		**flanged openings**		
D (in)	**A (ft^2)**	**Q (ACFM)**	**U_{face} (FPM)**	**flange (in)**	**Q (ACFM)**	**U_{face} (FPM)**
4.0	0.087	520	5,900	2.0	330	3800
6.0	0.20	540	2,700	3.0	390	2000
8.0	0.35	570	1,600	2.0	440*	1300*
10.	0.55	610	1,100	1.0	480*	880*
12.	0.79	660	840	0.	-	-

*For these cases, the flange is smaller than D/2; the data in Figure 6.10 are therefore not reliable.

From the results above, it can be seen that for either the flanged or unflanged case, the required suction flow rate increases (albeit slowly) with diameter; the smaller diameter inlets are thus more attractive.

The reader must keep in mind, however, that if the welding location is off centerline, this conclusion may no longer be true. In addition, while smaller entries require less air, they have higher face velocities and produce more noise. It can also be seen that for a given diameter, a flanged entry affords the best control since it requires approximately 30 to 40% less air for the same capture velocity. As seen in the footnote below the table of results, because of the restriction on overall diameter the flanges on the 8.0 and 10. inch inlets are smaller than those used to generate data for Eq. (6-7). There is still some benefit of a small flange, but the required Q lies somewhere between the unflanged and flanged values above. On the basis of these results, a 6.0-inch inlet with a 3-inch flange is a sound recommendation for this application.

Discussion: The ACGIH ventilation manual (2001) recommends exhaust volumetric flow rates for flanged and unflanged openings for welding. For a 6-inch inlet, the manual recommends 250 ACFM and a flange of 3 inches; for an unflanged (plain) opening, it recommends 335 CFM. In both cases the recommended face velocity is 1500 FPM. The agreement between this and the recommendations above is reasonably good considering that a capture velocity of 200 FPM, rather than 100 FPM, was chosen for the computation. Since there is such a large range in the capture velocity (a factor of 2!), it is presumptuous and misleading to tabulate values to more than two significant digits.

While the design constraints require the entry to be no closer than 6 inches from the point of welding, the designer should consider a high velocity - low volume (HVLV) local ventilation device, such as that in Figure 1.13. The ACGIH ventilation manual (2001) provides some details on the design of such systems, which consist of a small entry located very close to the point of welding that produces a very large face velocity but requires considerably less exhaust air. The systems may produce considerable noise because of the high face velocity.

Human workers need an unobstructed view of the welding region, but robots do not. Furthermore, the robot performs prescribed welding operations that may tolerate a small entry located close to the point of welding. The designer should examine the constraint of 6.0 inches and investigate the possibility of using such a system. If a clear advantage can be realized, the designer should recommend this alternative design approach.

When a plain or flanged inlet is used to capture contaminants produced by a source directly in front of the inlet, air from all around the inlet region is drawn into the inlet, thereby reducing the reach of the inlet and its ability to capture contaminants. Several years ago a Danish engineer named J. Aaberg proposed a remedy, as sketched in Figure 6.11. In Aaberg's design, a high-velocity jet blows air at velocity v_B perpendicular to the inlet, which sucks in air at velocity v_S. The blowing jet is used to prevent air from approaching the exhaust slot from above the inlet, thereby channeling the suction flow (Q_S) into the inlet. In so doing, the reach of the inlet increases without increasing the suction volumetric flow rate (Q_S). When the blowing (v_B) and suction (v_S) velocities are adjusted properly, the reach can be nearly doubled for the same amount of withdrawn air (Q_S). The flow field has been modeled using computational fluid dynamics (CFD) by Kulmula (2000), and recast empirically by Burton (2000). The practical and theoretical results achieved by an Aaberg inlet can be found in Pedersen and Nielsen (1993).

Burton (2000) suggests the following relationships for Aaberg inlets for bench tops and open surface tanks. Readers must be careful to use dimensions and velocities within the applicable range. The design procedure is as follows:

1. Determine the height (H) and width (W) of the volume that is to be controlled (2.0 ft $< W <$ 4.0 ft, and $2.0 \geq W/H \geq 0.5$).

2. Select the dimensions of the openings, i.e. L, S_S, and S_B, where L is the dimension perpendicular to the plane of Figure 6.11 (L > 4 ft), S_S is the width of the suction slot (S_S = 2.0 in), where the slot is in the center of the rear wall, and S_B is the width of the blowing slot (0.25 in < S_B < 0.5 in).
3. Select a value of the capture velocity (v_c) at the edge of the bench, as illustrated in Figure 6.11. v_c should be appropriate for the contaminant to be captured (Aaberg inlets are recommended for 75 ft/min < v_c < 150 ft/min).
4. Estimate the ratio of suction velocity to capture velocity ($R = v_S/v_c$) from the dimensions of the control volume (W/H), using the following table:

W/H	0.5	0.75	1.0	1.51	2.0
$R = v_S/v_c$	8.0	10.0	11.1	14.3	16.7

5. Calculate the suction velocity (v_S) from

$$v_S = v_c\left(\frac{v_S}{v_c}\right) = v_c R \tag{6-9}$$

where R is taken from the above table.

6. Calculate the suction volumetric flow rate (Q_S) from

$$Q_S = S_S v_S L \tag{6-10}$$

7. Calculate the volumetric flow rate of the high-velocity blowing jet (Q_B, in units of ACFM) by the following, where H is expressed in feet:

$$Q_B = Q_S \frac{0.479\sqrt{\text{ft}}}{\sqrt{H}} \tag{6-11}$$

While algebraically correct, Eq. (6-11) is disturbing because it suggests that the ratio of blowing to suction volume flow rates (Q_B/Q_S) is dependent only on the height of the region (H); it is independent of the width (W) of the region, and independent of the ratio of suction to capture velocities (R). Unfortunately, empiricism often produces such vagueness. The above equations are valid only for rectangular Aaberg inlets; similar concepts can be employed to design Aaberg inlets of *circular* cross section.

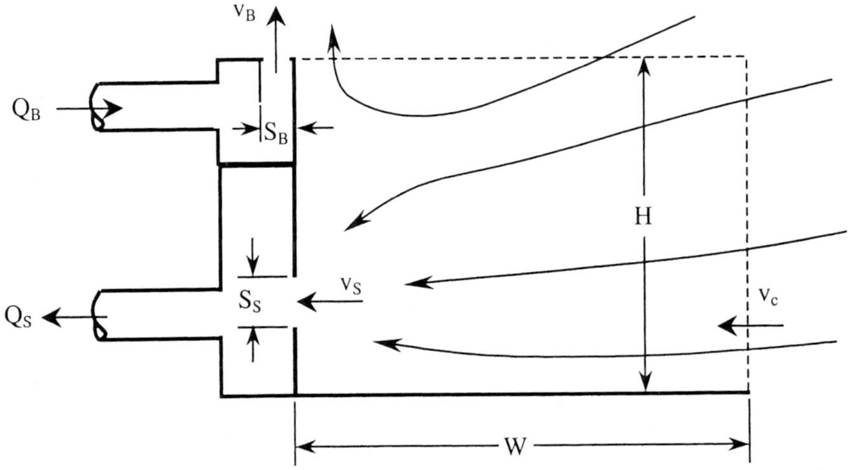

Figure 6.11 Schematic diagram of an Aaberg inlet for welding benches or open surface vessels. The suction inlet is placed in the center of the rear wall, and side baffles are used at the ends of the bench or tank (adapted from Burton, 2000).

Figure 6.12, from Hunt and Ingham (1992), shows a schematic diagram of a rectangular Aaberg inlet, along with isopleths and streamlines for the case in which $a/S_S = 2$, where a is the distance between the upper and lower blowing jets, as defined in Figure 6.12a. The cross-hatched region in Figure 6.12c defines the region of flow entering the inlet. Compared to a plain rectangular opening (Figure 6.8), it is clear that the axial reach of the Aaberg inlet is greatly enhanced.

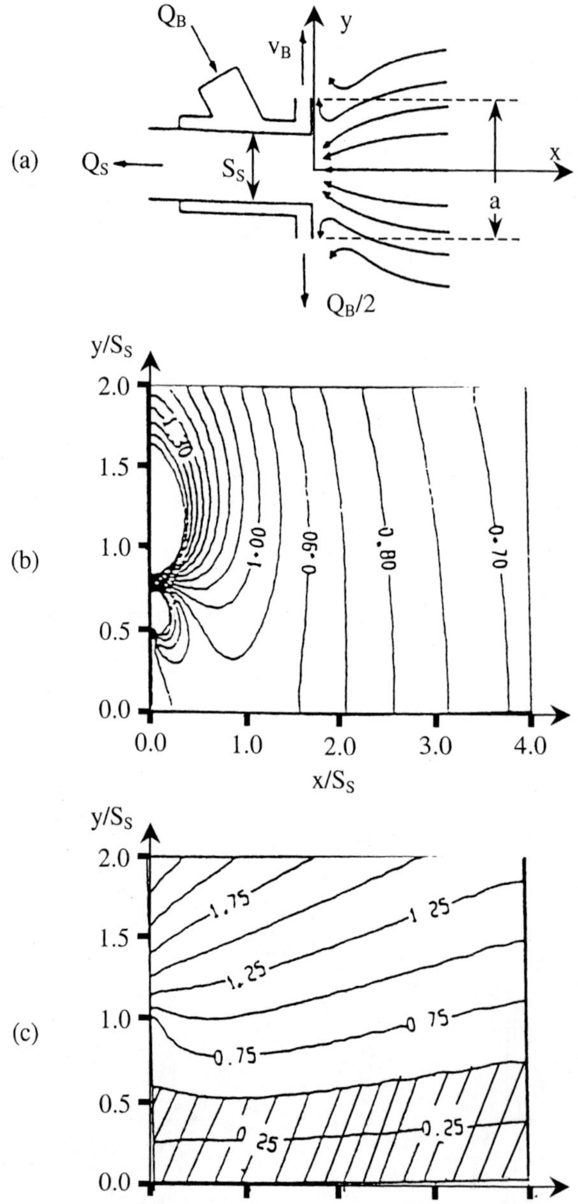

Figure 6.12 Details of Aaberg inlet for $S_S/a = 1/2$: (a) schematic diagram; (b) predicted isopleths, U/U_{face}; (c) predicted streamlines, where streamlines $0 < \psi < 0.5$ define the region of flow entering the inlet; volume flow rate of the shaded region shown is $Q_S/2$ by symmetry (adapted from Hunt and Ingham, 1992).

A physical explanation for this improvement is as follows: air flowing outward from the blowing jets entrains air from upstream of the inlet. This entrained air is coupled with the air that is withdrawn, thereby significantly increasing the axial reach of the inlet. This attractive result must be conditioned by three facts. The Aaberg inlet requires fresh air to be discharged outwardly; the total required volumetric flow rate is thus larger than the withdrawn flow rate alone, and requires an additional blower. Secondly, depending on the orientation of the device, the high speed blowing jets may irritate workers in their path, or my even interfere with the very process being controlled. Finally, it can be seen from the streamlines in Figure 6.12c, that there exist regions outside the cross-hatched region, in which no capture is possible. Thus to use an Aaberg inlet, the source must be close to the centerline; in other words, although axial reach is enhanced, lateral reach is reduced.

Example 6.2 - Welding Snorkel

Given: A snorkel (such as the one shown in Figure 1.26) is to be installed to control welding fumes, and you have been asked to recommend the volumetric flow rate necessary to capture fumes for the case in which the snorkel is located almost directly above the welding zone, with the face of the snorkel nearly horizontal. To encourage workers to use the device, a bright light will be placed in the snorkel, and will turn on only when the suction is turned on. It has been observed that welding is performed at points anywhere within a circular region 1.2 ft in diameter in a plane 1.0 ft from the mouth of the snorkel. For analysis, assume that the snorkel is a circular opening 1.0 ft in diameter.

To do: Estimate the volumetric flow rate of air that needs to be withdrawn through the snorkel assuming three possible designs:

(a) plain circular inlet
(b) flanged circular inlet
(c) Aaberg inlet

Solution:

(a) Plain circular inlet: Table 6.1 recommends a capture velocity of 100-200 ft/min (FPM) for welding. The upper value is used because drafts are expected. Velocity isopleths for a plain circular inlet are shown in Figure 6.9. It should be noticed that at any upwind distance (x) the highest air speed lies along the centerline (r = 0). From Figure 6.9 at x/D = 1 and r = 0, $U/U_{face} = 0.080$ (8.0%). For U = 200 FPM, U_{face} = (200 FPM)/.080 = 2,500 FPM, and the required volumetric flow rate (Q) is

$$Q = U_{face}\frac{\pi D^2}{4} = \left(2500\frac{\text{ft}}{\text{min}}\right)\frac{\pi(1.0\ \text{ft})^2}{4} = 1960\ \text{ACFM} \cong 2000\ \text{ACFM}$$

The designer cannot stop at this point because the above only establishes the *lower limit* on Q. To establish the upper limit, suppose the worker wishes to weld at the furthest observed point, i.e. at x/D = (1.0 ft)/(1.0 ft) = 1.0 and r/D = (0.60 ft)/(1.0 ft) = 0.60. Using the velocity isopleths, it is seen that at this point, $U/U_{face} = 0.050$. Thus if U = 200 FPM, U_{face} = 200/0.050 = 4,000 FPM and volumetric flow rate is

$$Q = U_{face}\frac{\pi D^2}{4} = \left(4000\frac{\text{ft}}{\text{min}}\right)\frac{\pi(1.0\ \text{ft})^2}{4} = 3140\ \text{ACFM} \cong 3100\ \text{ACFM}$$

Thus the designer should recommend the higher value since this will produce an air speed no lower than 200 FPM over the entire welding region, namely a circular region 1.2 ft in diameter on a plane 1.0 ft upwind of the snorkel face.

(b) Flanged circular inlet: The procedure in Part (a) above is repeated using Figure 6.10 instead of Figure 6.9. To achieve a velocity of 200 FPM along the centerline 1.0 ft in front of the inlet, the required volumetric flow rate is

$$Q = U_{face}\frac{\pi D^2}{4} = \left(\frac{200\frac{ft}{min}}{0.090}\right)\frac{\pi(1.0\ ft)^2}{4} = 1745\ ACFM \cong 1700\ ACFM$$

To achieve a velocity of 200 FPM at a location x/D = 1.0 and r/D = 0.60,

$$Q = U_{face}\frac{\pi D^2}{4} = \left(\frac{200\frac{ft}{min}}{0.075}\right)\frac{\pi(1.0\ ft)^2}{4} = 2094\ ACFM \cong 2100\ ACFM$$

The required suction volumetric flow rate is thus approximately 30% smaller for the flanged inlet compared to the plain inlet. In addition, the bright light can be incorporated into the flange. The disadvantage is that the flange sticks out an extra six inches from the centerline, and may interfere with the workers' movements and/or line of sight to the welding area.

(c) Rectangular Aaberg inlet: The Aaberg design shown in Figure 6.12 pertains to a rectangular opening. Comparing a rectangular opening with a circular one is poor practice, yet in the absence of similar geometry some insight can be gained on the benefit of combining blowing and suction flows to increase the reach of an inlet. A rectangular inlet, 1.0 ft by 4.0 ft, is used for comparison. The procedure in Part (a) above can be repeated using Figure 6.12b. To achieve a velocity of 200 FPM at a location of $x/S_S = 1.0$ and $y/S_S = 0.60$, U/U_{face} is about 0.97. The required suction volume flow rate is

$$Q_S = U_{face}A_{face} = \frac{200\frac{ft}{min}}{0.97}\left(4.0\ ft^2\right) = 825.\frac{ft^3}{min} \cong 820\ ACFM$$

The required blowing volume flow rate is obtained from Eq. (6-11); by comparing Figures 6.11 and 6.12, H is approximately $a/2 = S_S = 1.0$ ft.

$$Q_B = Q_S\frac{0.479}{\sqrt{H}} = 825.\frac{ft^3}{min}\frac{0.479\sqrt{ft}}{\sqrt{1.0\ ft}} \cong 400\ ACFM$$

The required suction volumetric flow rate for the Aaberg inlet is approximately a factor of 2.5 smaller than that of the flanged round inlet, which indicates that the combination of a blowing jet and sucking inlet is a very effective way to capture contaminants. However, upon closer examination of Figure 6.12c, the location $x/S_S = 1.0$ and $y/S_S = 0.60$ is outside of the cross hatched area, indicating that *particles in this region are not captured*! Thus, this Aaberg inlet is not recommended unless the workers restrict their welding to an off-axis distance of about 1.0 ft rather than 1.2 ft. In other words, the workers must perform their work nearer the inlet centerline. If the welding is performed *on the centerline* (at $x/S_S = 1.0$ and $y/S_S = 0.0$), U/U_{face} is smaller than the value used above (about 0.92 compared to 0.97). Thus, for this case, the centerline represents the upper limit of suction. Re-calculating, the required $Q_S = 870$ ACFM. Furthermore, because of the large flange, the Aaberg inlet suffers from the same disadvantages discussed above for the flanged inlet.

Discussion: All results have been rounded to two significant digits, due to the inherent inaccuracy of reading the isopleth plots, and also due to the large range of capture velocity (100 to 200 FPM) in Table 6.1. It is seen that a flanged inlet achieves control with a smaller volumetric flow rate than the unflanged inlet. If an Aaberg inlet is used, the capture of particles can be achieved at a considerably smaller volumetric flow rate, but the reach in the radial direction is greatly reduced, and the workers would need to modify their welding habits. Both the flanged inlet and Aaberg inlet have large flanges which may interfere with the workers' movements and their view. The final recommendation is thus the plain circular inlet, even though it requires the largest exhaust fan. A small flange can perhaps be added to support the light, and to improve the performance somewhat.

6.2 Control of Vapors from Open Surface Vessels

Vapors and gases emitted by industrial processes are more difficult to control than particles, which owing to their large density have inertia that affects their motion, and allows them to settle by gravity. If the process can be enclosed and the evolving gases and vapors can be removed before they enter the workplace environment, it is obviously wise to do so. The enclosure must truly isolate the source of contamination. Provision must be made to introduce make-up air into the workplace at a rate equal to the rate at which contaminated air is removed from the enclosure. Designers must be absolutely sure that the velocity at the inlet of the enclosure is directed into the enclosure at every point across the inlet face of the enclosure. There are industrial operations in which an open vessel contains a volatile liquid with which the worker must maintain visual contact, and therefore cannot be enclosed.

The ACGIH uses the concept of ***control velocity*** to design control systems to prevent vapors and gases from entering the workplace. Control velocity (not to be confused with capture velocity!) is not the physical velocity of air at a specific point in space; it is a design parameter. The emission of vapors by a process is rated in terms of a control velocity that depends on the ***hazard potential*** of the vapor and the ***rate of contaminant evolution***. A local ventilation system is rated in terms of the control velocity it can produce. The designer chooses a ventilation system that is capable of producing the control velocity appropriate to the escaping vapor.

Shown in Table 6.2 is a letter (A through D) that corresponds to the hazard potential of a contaminant and a number (1 through 4) that indicates the rate of evolution of the contaminant, i.e. the source strength (mass per unit time). The combination of a letter and a number defines the ***class*** of an industrial process. Table 6.3 shows the control velocity needed to cope with different classes of contaminant for enclosing, lateral, and canopy ventilation systems. Table 6.4 shows the exhaust volumetric flow rate needed to produce the control velocities for the lateral hoods listed in Table 6.3. Control velocities lie between 50 and 175 FPM, an order of magnitude smaller than the actual air velocities through the slot, which are typically between 1,000 and 2,000 FPM. Table 6.5 shows the classifications for a variety of metal surface treatment processes obtained by this method.

Table 6.2 Hazard potential and rate of contaminant evolution (abstracted from ACGIH, 2001).

hazard potential	health standard for gas or vapor (PPM)	health standard for mist (mg/m^3)	flash point (°F)
A	0 to 10	0 to 0.1	-
B	11 to 100	0.11 to 1.0	under 100
C	101 to 500	1.1 to 10	100 to 200
D	over 500	over 10	over 200

rate	liquid temperature (°F)	degrees below boiling (°F)	evaporation time[1] (hr)	gassing[2]
1	over 200	0 to 20	0 to 3 (fast)	high
2	150 to 200	21 to 50	3 to 12 (medium)	medium
3	94 to 149	51 to 100	12 to 50 (slow)	low
4	under 94	over 100	over 50 (nil)	nil

[1] time for 100% evaporation

[2] extent to which gas or vapor are generated: rate depends on the physical process and the solution concentration and temperature

Table 6.3 Minimum control velocities (FPM) for undisturbed locations (abstracted from ACGIH, 2001).

class	enclosing hood		lateral hood[1]	canopy hood[4]	
	1 side open	2 sides open		3 sides open	4 sides open
A1[2], A2[2]	100	150	150	do not use	do not use
A3[2], B1, B2, C1	75	100	100	125	175
B3[3], C2[3], D1[3]	65	90	75	100	150
A4[2], C3[3], D2[3]	50	75	50	75	125
B4, C4, D3[3], D4	adequate general room ventilation required				

[1] use Table 6.4 to compute the volumetric flow rate
[2] do not use a canopy hood for hazard potential A processes
[3] where complete control of hot water is desired, design as next highest class
[4] use Q = 1.4(PD) control velocity, where P is hood perimeter and D is distance between vessel and hood face (27)

Table 6.4 Minimum volumetric flow rates per unit surface area (CFM/ft^2) for lateral exhaust systems (abstracted from ACGIH, 2001).

control velocity (FPM)	[2]aspect ratio = tank width/tank length (W/L)				
	0 - 0.09	0.1 - 0.24	0.25 - 0.49	0.5 - 0.99	1.0 - 2.0
tank against wall or baffled[1]					
50	50	60	75	90	100
75	75	90	110	130	150
100	100	125	150	175	200
150	150	190	225	250[3]	250[3]
free-standing tank[1]					
50	75	90	100	110	125
75	110	130	150	170	190
100	150	175	200	225	250
150	225	250[3]	250[3]	250[3]	250[3]

[1] use half width to compute W/L for inlet along tank centerline or two parallel sides of tank
[2] inlet slot along the long side (L); if $6 < L < 10$ ft, multiple takeoffs are desirable; if $L > 10$ ft, multiple takeoffs in plenum are necessary if:

- W = 20 inches: slot on one side is suitable
- $20 < W < 36$ inches: slots on both sides are desirable
- $36 < W < 48$ inches: slots on both sides are necessary unless all other conditions are optimum
- $W > 48$ inches: lateral exhausts are not usually practical, use push-pull or enclosures
- it is undesirable to use lateral exhaust when $W/L > 1$ and not practical when $W/L > 2$

[3] while control velocities of 150 FPM may not be achieved, 250 CFM/ft^2 is considered adequate for control

Table 6.5 Metal surface treatment processes (adapted from ACGIH, 2001).

process	bath	emission[1]	T (°F)	class
anodizing:				
aluminum	H_2SO_4	H_2, acid mist	60-80	B1
etching:				
aluminum	NaOH, NA_2CO_3, Na_3PO_4	alkaline mist	160-180	C1
copper	HCl	HCl	70-90	A2
pickling:				
aluminum	HNO_3	oxides of nitrogen	70-90	A2
copper	H_2SO_4	acid mist, steam	125-175	B3, B2
monel & nickel	HCl	acid mist, steam	180	A2
stainless	H_2SO_4	acid mist, steam	180	B1
cleaning:				
alkaline	sodium salts	alkaline mist	160-210	C2, C1
degreasing	trichloroethylene	vapor	188-250	B
degreasing	perchloroethylene	vapor	188-250	B
electroplating:				
platinum	NH_4PO_4 & NH_3(g)	NH_3(g)	158-203	B2
copper	NaOH, cyanide salts	cyanide, alkaline	110-160	C2
chromium	chromic acid	chromic acid mist	90-140	A1
nickel	HF, NH_4F	acid mist	102	A3
stripping:				
gold	H_2SO_4	acid mist	70-100	B3, B2
nickel	HCl	acid mist	70-90	A3
silver	HNO_3	oxides of nitrogen	70-90	A1

[1] at high temperature, an alkaline or acid bath produces a mist of similar composition

Example 6.3 - Pickling Copper in Sulfuric Acid

Given: Copper plate is to be immersed in a bath of water and sulfuric acid prior to coating the metal. The bath is 10. ft long and 3.0 ft wide (W/L = 0.30), and the bath is to either be placed against a wall with a lateral exhaust along one side, or be a free-standing installation in which lateral exhausts will be placed along both long sides of the tank. The bath temperature is 175. °F and acid fume is apt to exist in the form of a mist. The acid concentration is such that the liquid mixture boils at 225. °F.

To do:

(a) Estimate the ventilation requirements, i.e. estimate the required volumetric flow rate, Q.
(b) Recommend volumetric flow rates if an Aaberg inlet (Figure 6.11) is used instead to control the acid mist.

Solution:

(a) Sulfuric acid has a PEL of 1 mg/m^3. Using the upper part of Table 6.2, the mist is borderline between hazard potentials B and C. To be conservative, hazard potential B is picked. From the lower part of Table 6.2, for a bath temperature of 175. °F and a temperature difference between the boiling temperature and the bath temperature of 50. °F, the rate of contaminant evolution is 2. The bath is thus classified as class B2 (which also agrees with Table 6.5 for pickling of copper plate). Table 6.3 indicates that the control velocity for a lateral hood and a class B2 contaminant is 100 FPM.

Tank against wall (suction on one side):

At a control velocity of 100 FPM, Table 6.4 indicates that the ventilation volumetric flow rate per area of liquid surface (Q/A) is 150 ACFM/ft^2 for the case of the tank against the wall, in which W/L = 0.30. The ventilation volumetric flow rate is found by multiplying Q/A by the surface area of the liquid, A = WL,

$$Q = \left(\frac{Q}{A}\right)A = \left(\frac{Q}{A}\right)WL = \left(150\frac{\text{ACFM}}{\text{ft}^2}\right)(3.0\ \text{ft})(10.0\ \text{ft}) = 4500\ \text{ACFM}$$

Free-standing tank (suction on both sides):

For the case of suction along both sides of the free-standing tank, the value of W/L must be divided by two according to the footnote in Table 6.4. For this case, Q/A = 175 ACFM/ft^2, and

$$Q = \left(\frac{Q}{A}\right)WL = \left(175\frac{\text{ACFM}}{\text{ft}^2}\right)(3.0\ \text{ft})(10.0\ \text{ft}) = 5300\ \text{ACFM}$$

(b) For the Aaberg inlet, the following geometry dimensions are assumed:

L = 10. ft W = 3.0 ft S_S = 2.0 inches S_B = 5/8 inch H = 2.0 ft

Since the acid is a mist (small liquid particles), capture velocity can be used in place of control velocity. From Table 6.1 for evaporation processes, capture velocity (v_c) lies between 50 and 100 FPM. The computation is performed using the higher value to be conservative. For a value of W/H = 3.0/2.0 = 1.5, the Aaberg table indicates that the value of R = v_S/v_c is 14.8. The suction volumetric flow rate is obtained by combination of Eqs. (6-9) and (6-10):

$$Q_S = S_S v_S L = S_S v_c R L = \left(\frac{2.0}{12}\text{ft}\right)\left(100\frac{\text{ft}}{\text{min}}\right)(14.8)(10.\ \text{ft}) = 2467.\frac{\text{ft}^3}{\text{min}} \cong 2500\ \text{ACFM}$$

The blowing volumetric flow rate (Q_B) can then be found from Eq.(6-11). Substitution of the above data yields a blowing volumetric flow rate of

$$Q_B = Q_S\frac{0.479\sqrt{\text{ft}}}{\sqrt{H}} = \left(2467.\frac{\text{ft}^3}{\text{min}}\right)\frac{0.479\sqrt{\text{ft}}}{\sqrt{2.0\ \text{ft}}} = 840\ \text{ACFM}$$

Discussion: Comparing the values of the suction volumetric flow rates (Q_s) it is seen that having suction on both sides of a free-standing tank requires about 17% more volumetric flow rate than does the case of the tank against the wall. The Aaberg inlet enables users to control the acid fume using only about 56% of the suction volumetric flow rate required for a conventional open tank suction slot with the tank against the wall. The Aaberg inlet is therefore recommended unless the high-speed blowing jet would interfere with the pickling operation.

A control strategy called ***push pull*** consists of placing an exhaust slot on one side of the vessel and a smaller slot on the other side that blows air across the surface of the liquid in the open vessel. Figure 6.13 shows the elements of a push pull system and provides guidance on selecting design features for the system. The combination of blowing and sucking simulates a jet pump in which the bowing jet induces ambient air into itself as it passes over the surface of the liquid, which aids the suction slot on the opposite side to withdraw the vapors. The total amount of air withdrawn is reduced considerably by using the push pull concept compared to a purely suction system. Numerical analyses of the performance of a push pull system can be found in Heinsohn et al. (1986) and Braconnier et al. (1993). The application of a push pull system to control contaminants from plating tanks is described by Sciloa (1993).

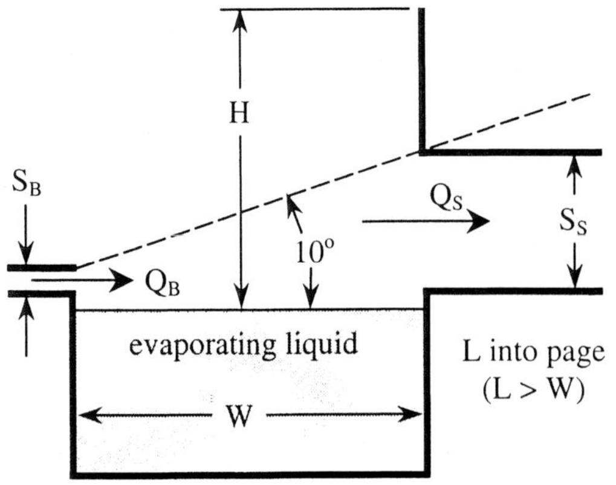

blowing jet area (A_B, ft^2)
blowing plenum cross sectional area > $3A_B$
blowing slot width (S_B, in): $1/8" \le S_B \le 1/4"$
or, 1/4" diameter holes, spaced holes 3/4" to 2" apart
blowing volumetric flow rate: $Q_B/L = 243(A_B/L)^{0.5}$ ACFM/ft, where A_B/L is in units of (ft^2/ft)
suction opening: $A_S = LS_S$ ft^2
suction slot width: $S_S = 0.14W$
suction volumetric flow rate: depends on liquid temperature, i.e.
$T \le 150$ °F, $Q_S/LW = 75.$ ACFM/ft^2
$T > 150$ °F, $Q_S/LW = [0.40T(°F) + 15.]$ ACFM/ft^2

Figure 6.13 Push pull ventilation system for an open surface vessel for widths up to 10 ft (abstracted from ACGIH, 1988 and 1998).

The selection of the dimensions of the blowing and suction slots and the volumetric flow rate of air through each slot is of immense importance. Inadequate volumetric flow rates are incapable of capturing vapor, and excessive blowing has a detrimental effect on capture. Figure 6.13 provides guidance on selecting these quantities.

Example 6.4 - Control of Vapor from an Open Vessel
Given: You have been asked to study lateral ventilation systems to control the vapor entering the workplace from a rectangular, free standing open vessel (1.4 m wide by 2.5 m long) containing trichloroethylene (TCE) at an elevated temperature (150 °F). The width of the vessel is the smaller of the two dimensions and normal to the lateral exhaust. The open vessel is used for degreasing operations in which workers must have access to the liquid; thus it is not possible to enclose the vessel. Two lateral ventilation systems are compared:

(a) suction system (Figure 6.1b is a circular version) - lateral exhaust slots along both (long) sides of the vessel following ACGIH procedures, Tables 6.1 to 6.4
(b) push pull system - an exhaust slot along one side of the vessel and a blowing slot along the opposite side following procedures shown in Figure 6.13

To do: Analyze both systems and compare.

Solution:

(a) Suction System:
The PEL for trichloroethylene is 100 PPM and the normal boiling temperature is 189. °F (87.2 °C). Table 6.2 shows that for this PEL, the hazard potential of TCE is between category B and C; for this boiling temperature, the rate of contaminant evolution is between levels 2 and 3. Current environmental concerns for volatile organic compounds (VOCs), particularly halogenated hydrocarbons such as TCE, suggest that engineers should be cautious. For this reason, the lower (most hazardous) hazard potential and rate are chosen for the design, i.e. class B2. For class B2, Table 6.3 recommends a minimum control velocity of 100 FPM (0.51 m/s) for a lateral hood. Table 6.4 must now be used to find the minimum volumetric flow rate per unit area. The ratio of the tank width to length is W/L = 0.56, but the footnote in Table 6.4 indicates that if inlets are on two parallel sides of a free-standing tank, as in this design, half of this W/L (i.e. 0.28) should be used. Thus, the minimum

exhaust volumetric flow rate per unit area is 200 ACFM/ft^2 (61. m^3/min per m^2) of liquid surface. For a vessel 1.4 by 2.5 m, the total exhaust volumetric flow rate is about 7,500 ACFM (about 210 m^3/min).

(b) ACGIH Push-Pull System:

Select a continuous 1/4-inch (S_B) slot that runs the length (L = 2.5 m) of the vessel. ACGIH recommendations in Figure 6.13 assume that both S_B and L are in units of feet; the equations suggest a blowing volumetric flow rate (Q_B) of

$$\frac{Q_B}{L} = 243\sqrt{\frac{A_B}{L}} = 243\sqrt{\frac{S_B L}{L}} = 243\sqrt{S_B} = 243\sqrt{(0.25 \text{ in})\frac{\text{ft}}{12 \text{ in}}} = 35.1\frac{\text{ACFM}}{\text{ft}}$$

From the given information, L = 2.5 m = 8.2 ft. Thus,

$$Q_B = 35.1\frac{\text{ACFM}}{\text{ft}}(8.2 \text{ ft}) = 288. \text{ ACFM} \cong 290 \text{ ACFM} = 8.2\frac{\text{m}^3}{\text{min}}$$

The figure also suggests a suction volumetric flow rate [Q_S/(LW)] of 75 ACFM/ft^2. Thus,

$$Q_S = 75 \cdot L \cdot W = 75(8.2 \text{ ft})(4.59 \text{ ft}) = 2800 \text{ ACFM} = 80.\frac{\text{m}^3}{\text{min}}$$

where L and W must be in units of feet for the equation to be valid. The suction system requires 2.7 times more air than does the ACGIH push-pull system. Thus the push-pull system is the preferred system. Finally, the "suction slot width", S_S, which is actually the suction slot *height*, is found from Figure 6.13:

$$S_s = 0.14W = 0.14(4.59 \text{ ft}) = 0.64 \text{ ft} \cong 8. \text{ in}$$

The *width* of the slot is the same as that of the vessel, i.e. W, which is about 4.6 ft.

Discussion: Figure 6.13 shows empirical equations from the ACGIH ventilation manual (2001). Readers are cautioned that empirical equations requiring users to select operating parameters between prescribed upper and lower limits often yield initial results that seem impractical. In such cases, users must repeat the process several times, increasing or decreasing these operating parameters each time until a solution is obtained that agrees with their intuition and experience or that of colleagues.

6.3 Control Systems for Specific Applications

The ACGIH ventilation manual contains dozens of ***design plates*** of ventilation systems for specific industrial applications that have been used to control emissions. An example is shown in Figure 6.14, which is the ACGIH design plate for a welding bench with a lateral exhauster. If the process for which control is sought can be found in the ventilation manual, the designer should use the recommended control system. At the least, the recommendations should be used as a preliminary design that can later be refined as conditions warrant. The design plates are intended as guidelines and do not take into account special conditions such as cross drafts, motion of workers or machinery, strong buoyant plumes, etc. Even though the ventilation manual does not contain a design plate for every industrial process, the design plates stimulate the reader's imagination and enable him or her to choose design concepts that have a high probability of being successful.

Engineers may find sources of inspiration from the existing design plates to design their own improved local ventilation systems. Each design plate contains the following:

- drawing of the system with instructions for the construction of key elements
- recommended volumetric flow rates
- recommended minimum duct transport velocities
- entry pressure losses

- other information such as slot velocities, slot dimensions, face velocities, baffles, and clearance dimensions between source and inlet

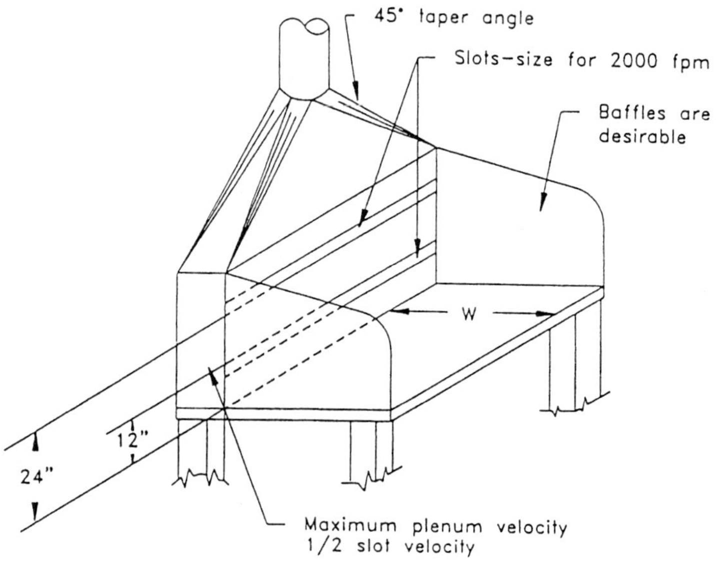

Q = 350 cfm/ft of hood length
Hood length = required working space
W = 24" maximum, if W>24" see chapter 3
Minimum duct velocity = 2000 fpm
$h_e = 1.78\ VP_s + 0.25\ VP_d$

General ventilation, where local exhaust can not be used:

Rod, diam.	Cfm/welder
5/32	1000
3/16	1500
1/4	3500
3/8	4500

or

A. For open areas, where welding fume can rise away from the breathing zone: cfm required = 800 x lb/hour rod used

B. For enclosed areas or positions where fume does not readily escape breathing zone: cfm required = 1600 x lb/hour rod used

For toxic materials higher airflows are necessary and operator may require respiratory protection equipment.

Other types of hoods
Local exhaust: See VS-90-02
Booth: For design see VS-90-30
Q = 100 cfm/ft² of face opening
MIG welding may require precise air flow control

AMERICAN CONFERENCE OF GOVERNMENTAL INDUSTRIAL HYGIENISTS	*WELDING VENTILATION BENCH HOOD*	
	DATE *1-91*	FIGURE *VS-90-01*

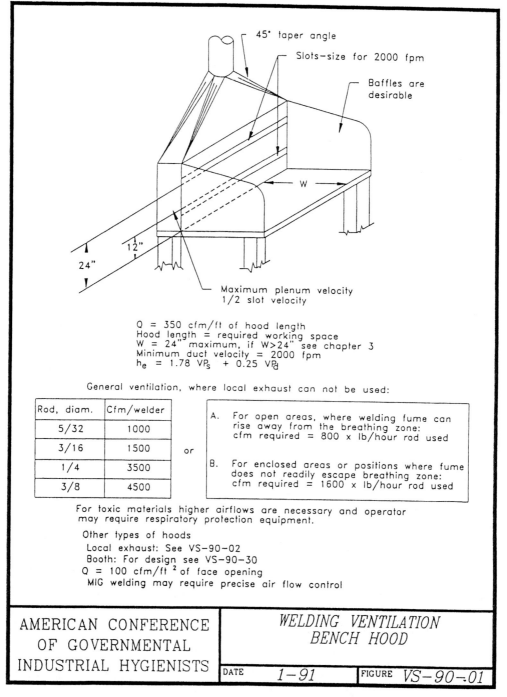

Figure 6.14 Sample design plate: lateral exhauster for a welding bench. From American Conference of Governmental Industrial Hygienists (ACGIH®), *Industrial Ventilation: A Manual of Recommended Practice*; copyright 2001; reprinted with permission.

Unless stated explicitly, the design data should not be applied to materials of high toxicity, radioactive materials, or any unusual materials in which higher volumetric flow rates are required. Details about specific control systems are also published in many professional journals, government reports, and trade publications, and provide designers with advice and inspiration. For example, Figure 6.4 is similar to the ACGIH design plates in the ventilation manual; it describes a booth for portable grinding, polishing, or buffing operations. The ***OSHA General Industry Standards*** do not as a rule provide design information and operating parameters about specific control systems. An exception is information in the CFR on pedestal grinders and swing grinders.

Control systems that have received a great deal of attention recently are laboratory fume hoods, biological safety cabinets, and biohazard cabinets.

- ***Laboratory fume hoods*** are five-sided partial enclosures used to handle toxic, corrosive, flammable, or odorous materials in the laboratory. The devices are sometimes called ***chemical fume hoods*** or ***fume cupboards***. A typical example is shown in Figure 1.19. The purpose of the device is to prevent materials used or produced in experiments from entering the workplace. The performance of these devices is affected by the face velocity, drafts in the workplace, and obstructions (chemical apparatus, etc.) inside the hood. The face velocity should not be skewed beyond 20% of the rated values. If the face velocity is too high, regions of high velocity or recirculation regions may be produced inside the hood that may affect the chemical experiments, processes, etc. being conducted inside the hood. OSHA requires a face velocity of 150 FPM for handling 13 carcinogens. Standard laboratory fume hoods use 100% air drawn from the laboratory. ***Auxiliary-air laboratory fume hoods*** (also called ***supply-air hoods***) recirculate up to 70% of the total air flow, using as low as 30% air from the laboratory. Doors on laboratory fume hoods may open horizontally or vertically and produce ***recirculating eddies*** inside the fume hood that may affect experiments. For particularly corrosive materials, one should use a laboratory fume hood composed of stainless steel with smooth (untaped) seams. Such hoods may also contain a water-wash system.
- ***Biological safety cabinets*** are similar to laboratory fume hoods but are used to provide sterile air for experiments with cells and tissue cultures. At the same time, the cabinet protects the worker and prevents any material from being transferred to the workplace. Two currents of air are used. There is a low-velocity, steady downward current of air directed over the material to be protected. Particles are removed by HEPA filters. When the moveable doors are open, a second current of room air flows inward at such a velocity that no biologic material enters the workplace.
- ***Bio-hazard cabinets*** are enclosures for handling very hazardous biological materials. Class II cabinets are total enclosures under negative pressure and have at least one double-door lock.

An ***enclosure*** is a volume in which there is no opening through which air can pass into or out of the volume other than openings explicitly constructed for this purpose. People enter or leave enclosures through doors that are closed after them. Examples of enclosures are industrial clean rooms (Figure 1.17), motel rooms, elevators, rest rooms, etc. A ***confined space*** is defined by OSHA as a "relatively small or restricted space such as a tank, boiler, pressure vessel, or small compartment of a ship." Thus confined spaces may resemble enclosed spaces but they contain unobstructed openings through which an unknown amount of air can pass in addition to the openings and air supplies explicitly constructed for this propose. The OSHA definition also suggests that confined spaces are not intended for normal human use and that generally fresh air must be supplied by auxiliary (or external) means to people working in confined spaces.

A hazardous atmosphere in a confined space is categorized as flammable, toxic, asphyxiating, irritating, and/or corrosive:

- A ***flammable*** atmosphere exists when the concentration of a combustible material exceeds 10% of the lower explosion limit (LEL).
- A ***toxic*** atmosphere exists when the concentration of a toxic material exceeds OSHA's permissible exposure limit (PEL).
- An ***asphyxiating*** atmosphere exists when the oxygen mol fraction is less than 19.5%.
- An ***irritating*** or ***corrosive*** environment exists when the air contains material that aggravates the face, upper respiratory system, eyes, or mucous membranes of the nose or mouth.

The velocity and concentration fields inside enclosed and confined spaces are not readily apparent but can be predicted by the sequential box model (Heinsohn, 1989) and CFD techniques. With such predictions engineers can anticipate whether hazardous conditions will or will not occur for personnel performing certain jobs at different locations in these spaces. ***Confined space asphyxiations*** are tragic since they often involve multiple deaths as workers attempt to provide assistance to each other. Workers entering confined spaces must follow an OSHA protocol, as discussed in Chapter 3. Equipment associated with entering confined spaces is also discussed in Chapter 3. Workers may have the presence of mind to anticipate or test for toxic gases, but often fail to anticipate an environment where the oxygen concentration is below 19.5%, since this is only slightly below the usual content of air. If the "air" contains insufficient oxygen, workers will pass out in just a few minutes and unless removed immediately, they will die in just a few more minutes. The following is a list of common enclosures in which workers may be asked to inspect, weld, paint, clean, or perform other ordinary activities in which insufficient oxygen is apt to occur:

- ships' holds
- farm silos and manure pits
- underground utility vaults for electricity, telephone, steam, pumps, and/or waste water
- storage vessels for fuel, liquids, or granular material
- tanker trucks and railroad cars
- chemical reactors
- closed storage vessels whose free space is ***inerted*** (filled with a non-reacting, non-combustible gas like nitrogen or carbon dioxide)
- containers used to transport fruits and vegetables which may be inerted with nitrogen or carbon dioxide to maintain freshness
- large-diameter pipelines
- elevator shafts
- unventilated tunnels
- fuel storage vessels being removed in accordance with TSCA regulations

In these cases contaminants are emitted and more stringent ventilation standards need to be met. The OSHA standards for welding in confined spaces provides guidance:

> *"In spaces less than 10,000 ft^3/welder or rooms with ceilings less than 16 ft, general ventilation must be provided at a rate of 2,000 CFM/welder except if local hoods are provided that produce velocities of 100 FPM in the welding zone or approved airline respirators are worn."*

General ventilation is questionable since 2,000 ACFM of fresh air may never reach a welder located in a remote part of an unusually shaped confined space. The "well-mixed" presumption underlying the statement may not be valid and the air duct should be located in close proximity of the welder's breathing zone. If possible, welding snorkels (Figure 1.26) or a welding gun incorporating a local ventilation system (Figure 1.13) should be used.

6.4 Bulk Materials Handling

The transfer of bulk solids (powders, granular material, etc.) generates particles which, if not controlled, are emitted to the workplace atmosphere. Particles can be emitted from the following:

- crushing, grinding, milling, screening, classifying, size reduction, and cleaning
- conveyor belts, chutes, vibratory feeders, elevators, and augers

If processes can be enclosed, it is wise to do so. Figure 6.15 shows covers that are placed over drums when they are filled. The large flexible duct directs material from a screener to the drum, and another flexible hose (not shown) withdraws air through a short exit pipe to a vacuum system shown to the left. Figure 6.2 shows a canopy hood with curtains placed around a rolling mill to provide a partial enclosure. Figure 6.16 shows an enclosure for a transfer point on a conveyor.

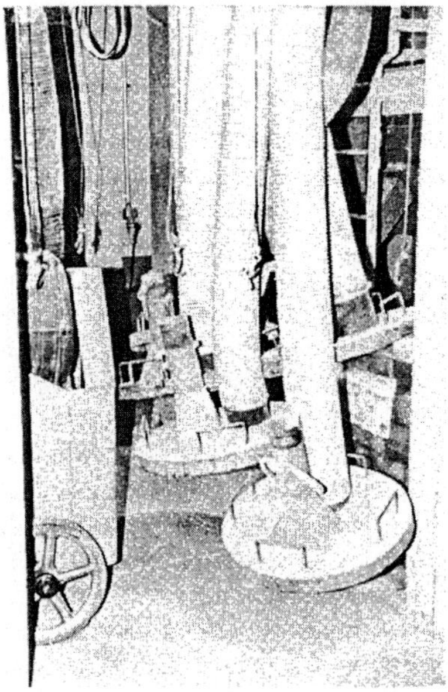

Figure 6.15 Lids to draw-off dust generated in filling barrels; barrels not shown (from Heinsohn, 1991).

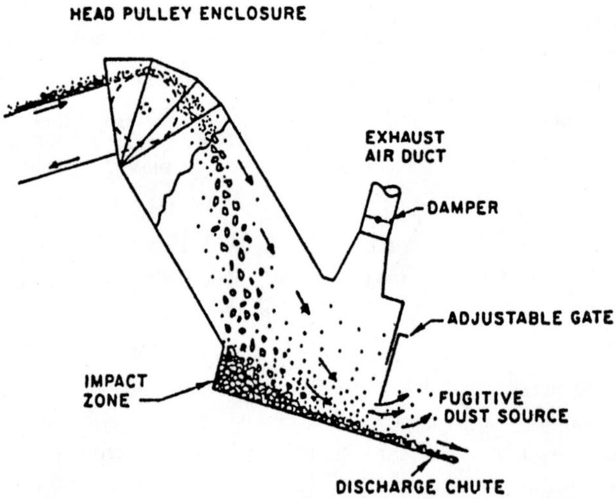

Figure 6.16 Enclosure at a transfer point on a conveyor (adapted from Dennis et al., 1983).

Emissions from vaguely defined transfer points are called ***fugitive emissions*** because they enter the workplace atmosphere at vaguely defined points. Fugitive emissions that occur over an area are called ***area sources***. If emissions are contained within a duct they are called ***point source emissions*** because their origin is well defined. Because the momentum of the stream of moving bulk solids is large, these sources of particulate emissions are often referred to as ***inertial sources***.

When bulk materials are transferred from a conveyor to a pile in the outdoors, the generated fugitive dust is proportional to the wind speed and inversely proportional to its moisture content. Thus it is a common practice to spray water to suppress the generation of dust. For continuous dumping of powdered coal through a distance of 1.3 to 1.55 m and particle size less than 500 μm, Visser (1992) found that a fugitive dust emission factor could be expressed as

$$\boxed{EF = 10^{6\ 2} M^{-9\ 2} + 10^{3\ 2} U^{4\ 0} M^{-4\ 5}} \tag{6-12}$$

where

- EF = emission factor (mg dust/kg coal)
- M = moisture content of the bulk material (%)
- U = wind speed (m/s)

The only viable method known to control inertial sources is to control the face velocities across the inlets and outlets within prescribed values. There has been scant attention paid to inertial sources. Analytical models proposed by Hemeon (1963) have been modified and improved upon (Dennis and Bubenick, 1983; Kashdan et al., 1986a&b) and are presently used. The transfer of bulk materials generates particles by four mechanisms:

- induced air
- splash
- displaced air
- secondary air

Induced air is the name of the process whereby moving bulk material transfers momentum to surrounding air, causing the air to move in the same direction as the bulk materials. As an illustration, consider the amount of air set into motion by the stream of water from a bathroom shower (Kashdan et al., 1986a&b). When bulk materials enter a hopper, mixer, or merely an enclosure surrounding a transfer point, etc., this air is ***entrained*** and drawn into the vessel with the bulk materials. Unless precautions are taken, this incoming air can stir up the particles and generate a substantial amount of particulate matter. A commonly used equation to estimate the volumetric flow rate of induced air surrounding a falling stream of bulk materials is reported by Kashdan et al. (1986a&b) to be

$$\boxed{Q_1 = 0.631 \left[\frac{\dot{m}_b H^2 A_b^{\ 2}}{\rho_b D_{p,m}} \right]^{1/3}} \tag{6-13}$$

where

- Q_1 = volumetric flow rate of induced air (m^3/s)
- $\dot{m}_b$ = bulk mass flow rate (kg/s)
- H = drop height (m)
- A_b = cross-sectional area of falling stream (m^2)
- ρ_b = bulk solids density (kg/m^3)
- $D_{p,m}$ = particle mass median diameter (m)

Splash is the term used to describe the generation of particles as the stream of bulk materials strikes a solid surface. The initial velocity of the particles may be modest and their range may be small, but if the velocity of the surrounding air is large, the particles may travel a considerable distance. At this time the only way to estimate the particle generation rate by splash is by means of emission factors.

Displaced air is the air displaced within the vessel by the incoming bulk material. The volumetric flow rate of the displaced air (Q_d) can be found by defining the enclosure as a control volume and applying conservation of mass for the enclosure in which the bulk solids are being stored, as sketched in Figure 6.17.

The displaced air volumetric flow rate is

$$Q_d = -\frac{dV_a}{dt} = \frac{dV_b}{dt} \tag{6-14}$$

where V_a is the volume of air in the enclosure and V_b is the volume of bulk material in the enclosure; both of these are functions of time in general, as the enclosure is filled. As bulk material enters the enclosure, V_b increases with time. The top surface of the bulk material rises with time, and can be thought of as a "piston" pushing air out. (Readers may recall a similar concept for filling of tanks with liquids, as discussed in Chapter 4.) The mass of bulk material in the container at any instant of time is denoted by $m_b(t)$, and the mass of air in the container at any instant of time is denoted by $m_a(t)$, as in Figure 6.17. Equation (6-14) can be re-written in terms of either of these masses,

$$Q_d = -\frac{1}{\rho_a}\frac{dm_a}{dt} = -\frac{\dot{m}_a}{\rho_a} = \frac{1}{\rho_b}\frac{dm_b}{dt} = \frac{\dot{m}_b}{\rho_b} \tag{6-15}$$

If bulk material is simultaneously being removed from the bottom of the container at the same mass flow rate as it is being added to the top of the container ($\dot{m}_b$), such as at a transfer point, steady-state conditions are achieved and Q_d is zero. In other words, when the mass (or volume) of bulk material inside the container does not change with time, no air is displaced by bulk material, and hence $Q_d = 0$.

Secondary air (Q_s) is moving air, produced by independent means, which enters the enclosure and causes particles made airborne by splash to be carried throughout the workplace. Sources of secondary air include spurious air currents (cross drafts) present in the workplace, air discharged from pneumatic equipment, etc. Secondary air is sometimes called ***entrained air*** (Kashdan et al., 1986a&b), although such a phrase tends to be confused with induced air. A distinction can be made that induced air (Q_i) refers to air entrained by the stream of bulk solids before splash while secondary air refers to air entrained after splash.

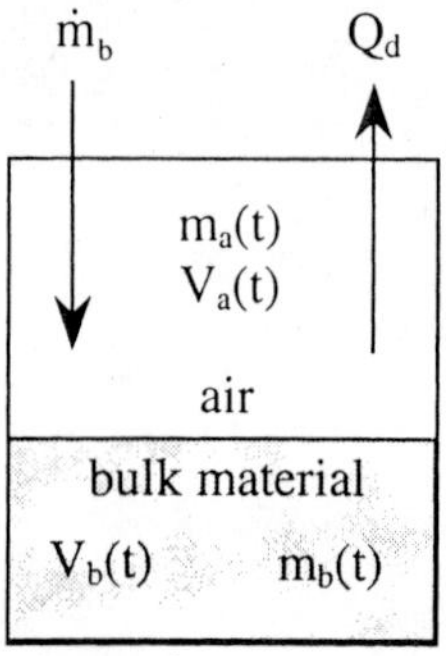

Figure 6.17 Displaced air due to filling a vessel.

6.4.1 Filling a tank

Figure 6.18 shows a bin that is being filled, i.e. bulk material is added at some mass flow rate ($\dot{m}_b$), induced air enters at an induced air volumetric flow rate (Q_I), and secondary air enters with some flow rate (Q_s). Suppose air is withdrawn from the enclosure by a vacuum at some withdrawal flow rate (Q_w) through a port in the upper part of the enclosure. Finally, let A_{face} be the area of the face, defined as the opening to the room air through which the bulk material is introduced into the container. The volume flow rate of air entering the face is equal to

$$\boxed{Q_f = U_{face} A_{face}} \tag{6-16}$$

Applying conservation of mass for air,

$$\boxed{\frac{dm_a}{dt} = \rho_a \frac{dV_a}{dt} = \rho_a \left(Q_I + Q_s + Q_f - Q_w \right)} \tag{6-17}$$

But the left hand side of Eq. (6-17) is equal to $-\rho_a Q_d$ by Eq. (6-14); thus, Eq. (6-17) can be solved for the required withdrawal flow rate,

$$\boxed{Q_w = Q_I + Q_s + Q_f + Q_d} \tag{6-18}$$

Finally, using the right hand side of Eq. (6-15) for Q_d, Eq. (6-18) can be re-written in terms of the (usually known) mass flow rate of the bulk material,

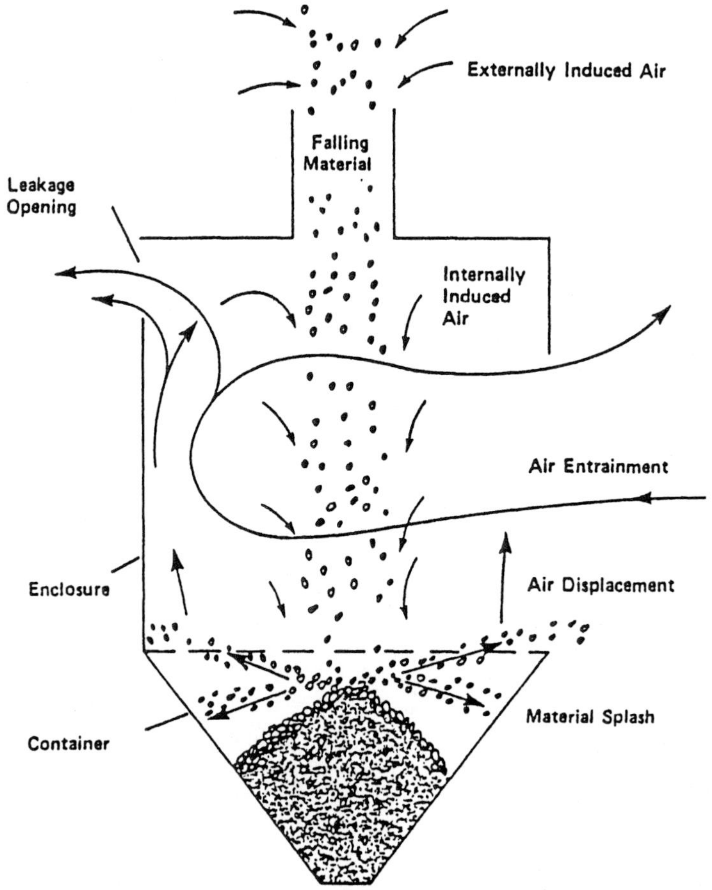

Figure 6.18 Dust generation from filling a vessel (adapted from Kashdan et al., 1986).

$$\boxed{Q_w = Q_I + Q_s + Q_f + \frac{\dot{m}_b}{\rho_b}} \tag{6-19}$$

6.4.2 Transfer point

Figure 6.16 shows an enclosure around a transfer point between two conveyors. For steady-state conditions, there is no accumulation of bulk material inside the enclosure, and no change in the volume of air (V_a) in the enclosure. Hence the displaced air volumetric flow rate is zero, and the required withdrawal flow rate is

$$\boxed{Q_w = Q_I + Q_s + Q_f} \tag{6-20}$$

Eq. (6-20) is a simplification of Eq. (6-18) since the fan does not have to accommodate the additional volumetric flow rate of the displaced air. The induced air flow (Q_I) contains a large amount of particles; if the fan is not sized to accommodate this additional flow of air and particles, they will spill over at the inlet, allowing particles to escape to the workplace atmosphere. To prevent particles from escaping the enclosure, it is commonly believed that the atmospheric air entering the enclosure should have a face velocity (U_{face}) of 0.5 to 1. m/s (approximately 100 to 200 FPM).

6.5 Canopy Hoods for Buoyant Sources

Plumes from hot sources rise because of buoyancy. Exothermic industrial processes that produce buoyant plumes are called ***buoyant sources***. In most industrial hot processes, the initial momentum is negligible in contrast to process gas streams that are discharged to the atmosphere through tall stacks external to the building. The velocity field of the buoyant plume (with or without a control device) can now be predicted by computational fluid dynamics (CFD). In lieu of such solutions, empirical expressions can be used to guide engineers in the design of devices to control buoyant plumes. Figure 6.19 contains an empirical equation from ACGIH (2001) to predict the volumetric flow rate of exhausted air. There are also methods to experimentally model buoyant plumes, and there are analytical techniques to study full-scale plumes within plants (Hayashi et al., 1985 and Goodfellow, 1986).

Figure 6.20 shows an example of a buoyant source, namely buoyant emissions associated with the ***basic oxygen process***, used to make steel. In the metallurgical industry, the volumetric flow rates of buoyant plumes are huge and vary with time owing to the nature of intermittent batch processes. When metal scrap is added to a furnace (Figures 1.15 and 1.18), or the furnace is skimmed or tapped, large amounts of fume (vapor and particles) are emitted at volumetric flow rates that are

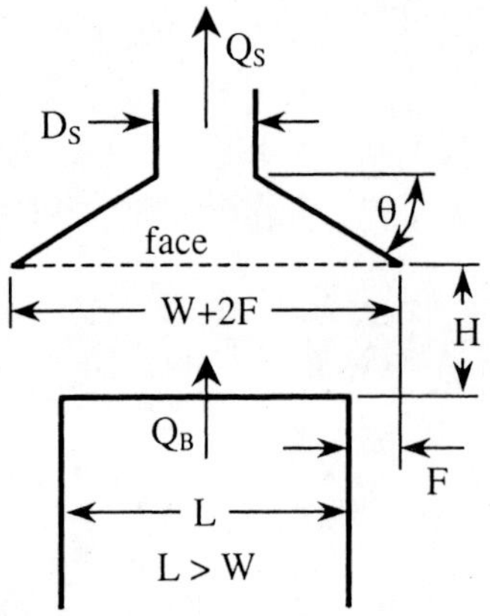

$\theta = 45°$

$F = 0.4H$

$Q_s = 1.4[2v_cH(W + L)]$

50 FPM < v_c < 500 FPM
where v_c = capture velocity

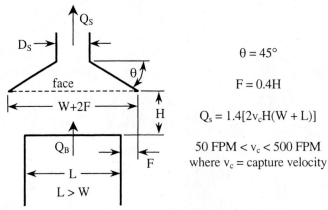

Figure 6.19 Canopy hood open on all four sides, with dimensions specified (adapted from ACGIH, 1998).

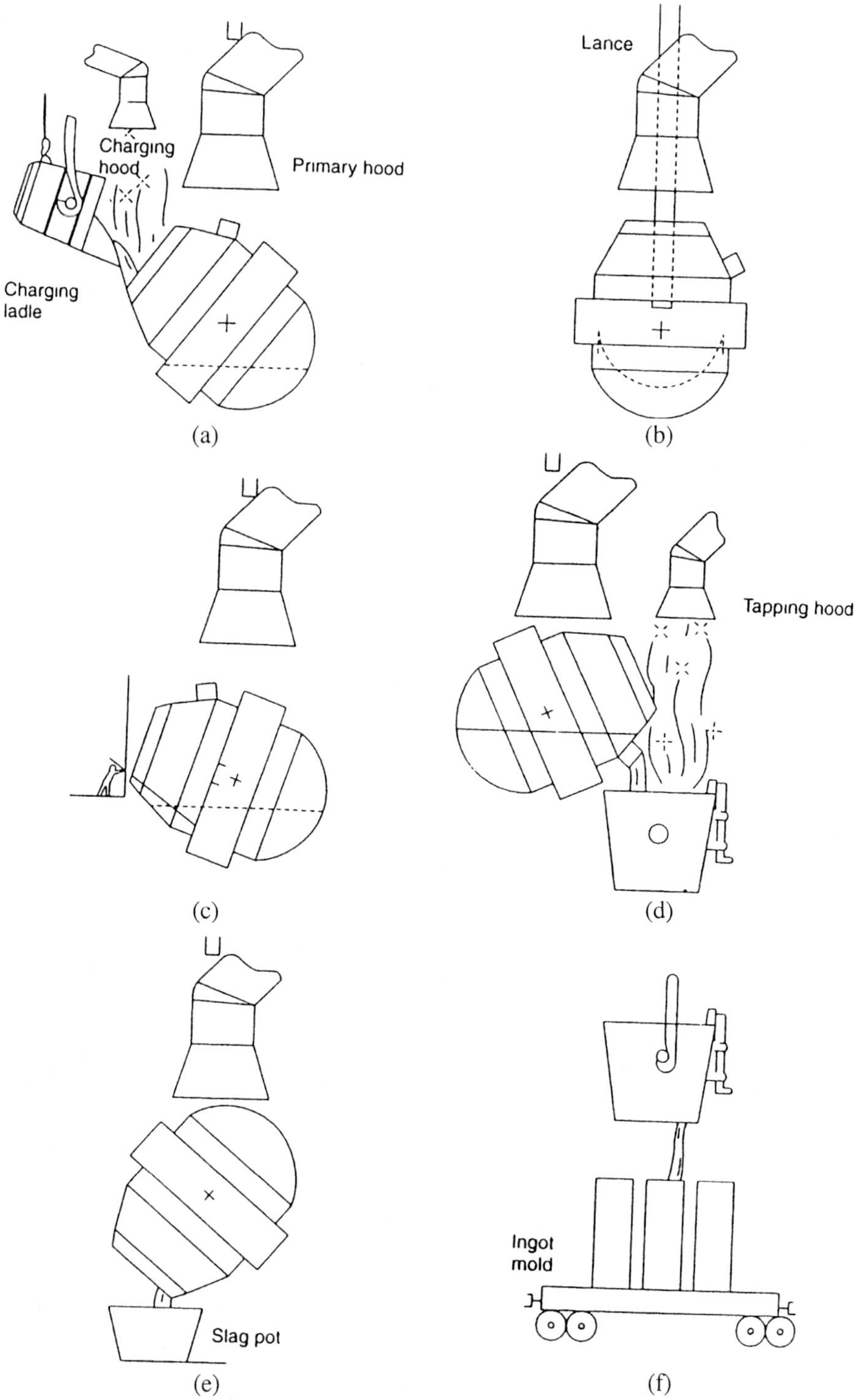

Figure 6.20 Emissions associated with steps to make steel by the basic oxygen process: (a) charging, (b) O_2 blowing, (c) turndown, (d) tapping, (e) deslagging, and (f) teeming (adapted from AWMA Handbook on Air Pollution Control, 2000).

several times the values that exist for the rest of the cycle. The issue is doubly difficult because state and federal agencies do not allow these fumes to be discharged to the atmosphere, and OSHA requires that concentrations within the workplace remain within safe values. While a steady process such as the reheat furnace shown in Figure 6.3 can be equipped with a close-fitting canopy hood that functions as a receiving hood to capture the hot plume, batch processes are more difficult to control. When oven doors are opened, vapors escape but may be captured by a canopy hood. Unfortunately, workers stand under these hoods when they open the oven doors and are engulfed by these vapors.

Based on work done by Morton et al. (1956), Morton (1959), Turner (1973), and Kashdan et al. (1986a&b), the following expression is suggested to predict the volumetric flow rate (Q_u) induced by a steady buoyant plume a distance (z) above the virtual origin of the plume. Note that a steady plume is interpreted in the literature as a plume lasting more than 30 seconds.

$$\boxed{Q_u = 0.166 z^{5/3} F^{1/3}} \tag{6-21}$$

where

- Q_u = volumetric flow rate (m^3/s) induced by the buoyant plume
- z = virtual source height (m), which is equal to the distance between the hood and source plus three halves of the source diameter,

$$\boxed{z = h + \frac{3}{2} D} \tag{6-22}$$

- h = distance (m) between the actual source and the hood inlet plane
- D = diameter (m) of the actual source
- F = buoyant flux parameter (m^4/s^3),

$$\boxed{F = \frac{g\dot{Q}_c}{c_p T_a \rho_a}} \tag{6-23}$$

- g = acceleration of gravity (9.81 m/s^2)
- c_p = specific heat [cal/(g K)] of air
- ρ_a, T_a = density (kg/m^3) and absolute temperature (K) of the ambient air
- $\dot{Q}_c$ = rate of convective heat transfer (kcal/s) from the source

$$\boxed{\dot{Q}_c = h_c A_s (T_s - T_a)} \tag{6-24}$$

- h_c = convection film coefficient [kcal/(s m^2 K)]
- A_s, T_s = source surface area (m^2), and temperature (K)

It is appropriate to neglect heat transfer by radiation unless it can be shown that radiation contributes significantly to the buoyant movement of air (Hemeon, 1963).

To capture buoyant plumes, analyses and experiments by Kashdan et al. (1986a&b) have shown that the volumetric flow rate with which to withdraw gas from such hoods should be about 20% larger than Q_H. Specifically,

$$\boxed{Q_w = 1.2 Q_H} \tag{6-25}$$

where Q_H is the volumetric flow rate measured at the inlet face of the hood in the absence of the buoyant source. ***Spillage*** is the phrase used to describe portions of the buoyant plume that are not captured by the hood and spill outward to other parts of the workplace. Spillage is due to cross drafts, obstructions from machinery (overhead cranes, ducts, building structural elements, etc.) that block the path of the hot plume to the canopy hood, and turbulent recirculating eddies within the hood causing gas to leave the hood before being withdrawn.

The dimensions of the hood are critically important, particularly in the case of canopy hoods for large-scale metallurgical processes that are intermittent in nature. For steady flow, the face velocity at the hood inlet should be at least 1.5 m/s (about 300 FPM) or else the plume may overturn and spill from the hood. Canopy hoods for *steady* processes should have a diameter equal to one-half the effective height (z). Alternatively if there are no impediments or cross drafts, the diameter can be computed assuming that the included angle of the hot plume is 18 degrees (Kashdan et al., 1986a&b and Bender, 1979).

Canopy hoods for intermittent processes pose unique design problems. As a first step, the designer should estimate the plume volumetric flow rate (Q_u) throughout the cycle of the process. If the maximum and minimum values are close to each other and the economics allow it, a single canopy hood run at a constant withdrawal volumetric flow rate (Q_w) may suffice. If however, the variations, i.e. ***surges***, in Q_u are large, such as shown in Figure 6.21, a single hood designed for the maximum value may be very costly. A common practice (Kashdan et al., 1986a&b and Bender 1979) in these cases is to design a canopy hood that captures and retains the gas for a short period of time until the surge has ended. Such canopy hoods are called ***pool hoods*** and ***hopper hoods***, as sketched in Figure 6.22. Internal baffles (Bender, 1979) are recommended to ensure that only room air is displaced as the plume fills the hood and that the plume does not overturn and spill from the hood.

The volume of the pool hood or hopper hood (V_h) should be greater than the integral of a surge in Figure 6.21,

$$\boxed{V_h \geq \int_{t_{su}} Q_u dt} \tag{6-26}$$

To be conservative, V_h can be approximated by

$$\boxed{V_h = t_{su}\left(Q_{su} - Q_w\right)} \tag{6-27}$$

where t_{su} is the duration of the surge and Q_{su} is the maximum volumetric flow rate associated with the surge. The steady volumetric flow rate at which gas is withdrawn from the hood (Q_w) must be sufficient so that the entire volume of surge gas stored in the hood can be *completely* withdrawn before the next surge occurs. Thus

$$\boxed{\frac{V_h}{Q_w} < \frac{1}{f_{su}}} \tag{6-28}$$

where f_{su} is the frequency of the surges. If the hood is not completely emptied, the new surge will displace gas in the hood containing remnants of the previous surge that will then enter the workplace.

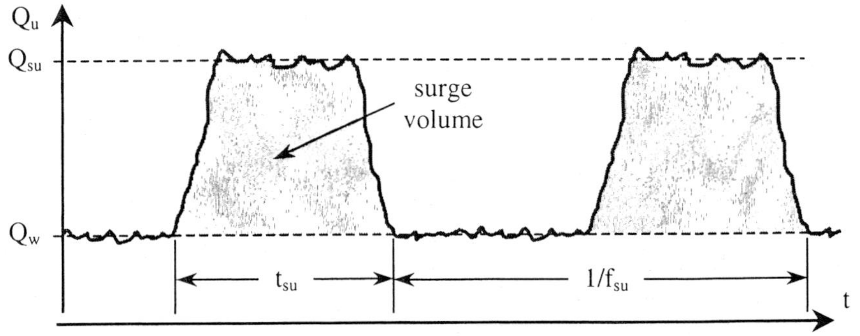

Figure 6.21 Plume volumetric flow rate for an intermittent buoyant source; two surges of duration t_{su} are shown, with surge frequency f_{su}.

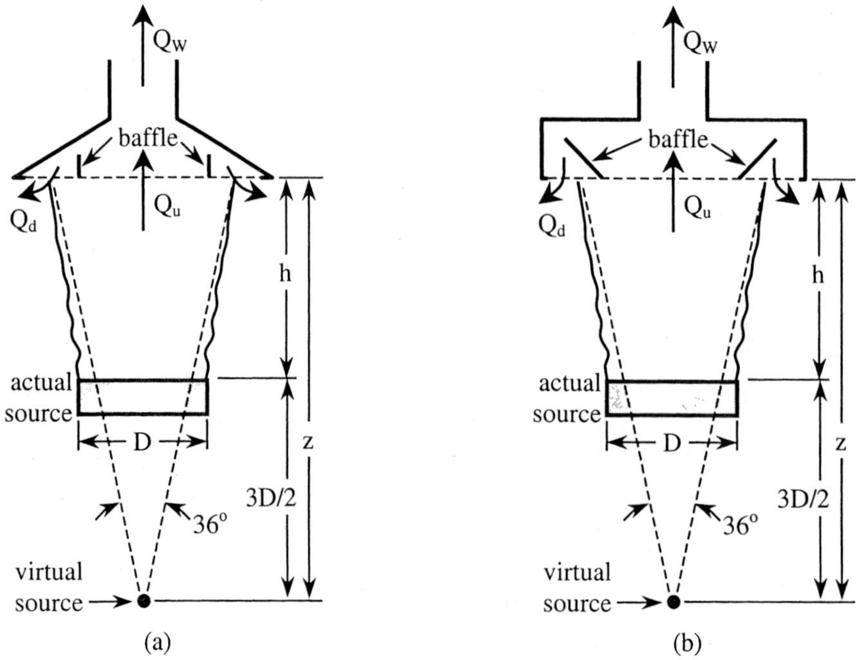

Figure 6.22 Sketch of (a) pool hood and (b) hopper hood showing displaced air (Q_d) during initial portion of buoyant plume surge (Q_u) (adapted from Kashdan et al., 1986).

Often a pool and hopper hood is formed by partitioning off the space between building roof trusses. With deep pool hoods, a face velocity of 0.5 m/s (approximately 100 FPM) is adequate (Kashdan et el., 1986a&b).

If cross drafts are present within the building, the volumetric flow rate of the withdrawn gas must be larger than predicted by Eq. (6-25). The following is suggested by Bender (1979):

$$\boxed{Q_w = Q_H\left(1 + 4.7\frac{V_{cross}}{U_{max}}\right)} \tag{6-29}$$

where V_{cross} is the cross draft velocity and U_{max} is the plume centerline velocity. For small values of V_{cross}, Q_w should be taken as the larger value of Eq. (6-25) and Eq. (6-29). A cross draft also deflects the plume, so the hood has to be either off-center or made large enough to accommodate the deflection.

6.6 Air Curtains for Buoyant Sources

Canopy hoods are located some distance above a source, and the amount of air to be moved would be less if the contaminants were captured closer to the source. ***Air curtains*** (Van and Howell, 1976; Hayes and Stoecker, 1969a&b) offer this opportunity, and there has been significant success in using them to control fumes in the metallurgical (Kashdan et al., 1986a&b) and rubber milling (Hampl et al., 1988) industries. An air curtain (jet) in combination with an oppositely located exhaust is called a ***push-pull control system***. The phrases *push pull* and *air curtain* are used interchangeably.

Figure 6.23 illustrates the concept. In lieu of applying the full set of conservation equations to the interaction of the plume and air curtain, an elementary model proposed by Bender (1979) has been used by Kashdan et al. (1986a&b) and Cesta (1989). The model assumes that the upward momentum of the buoyant plume (M_u) interacts with the push jet momentum (M_j) inclined downward at an angle

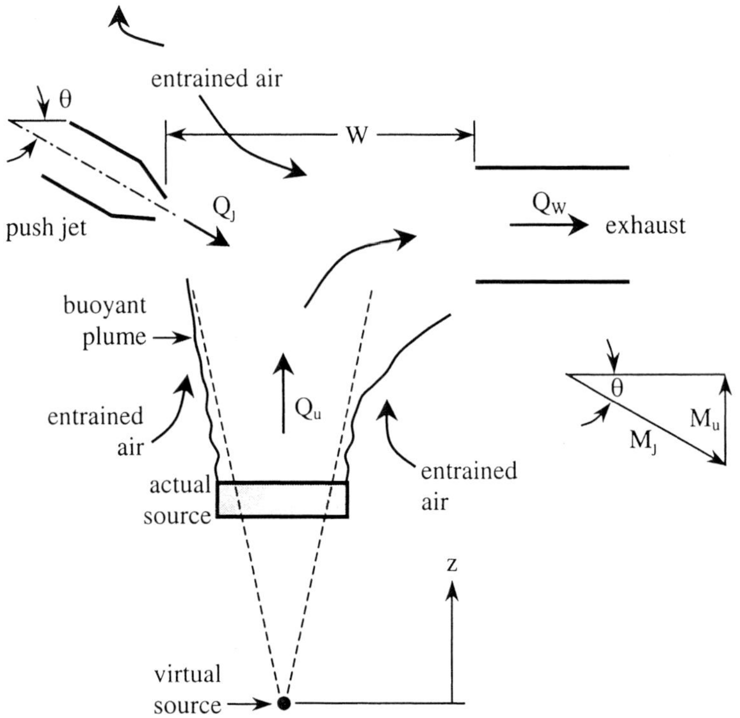

Figure 6.23 Schematic diagram of an air curtain control system to capture a buoyant plume.

(θ), such that the resultant momentum is horizontal and the combined flow enters the exhaust opening.

Elementary control volume analysis of a jet shows that while viscosity entrains surrounding air, and there is significant viscous dissipation, the momentum of the push jet is conserved downstream of the jet exit. The magnitude of the momentum of the push jet in the direction θ below the horizon is

$$\boxed{M_J = \rho_J U_J^2 A_J = \text{ constant}} \tag{6-30}$$

where A_J and U_J are the area and velocity at the exit of the push jet, and ρ_J is the density of the push jet air. By multiplying Eq. (6-30) by sinθ, one can obtain the component of momentum of this jet in the vertically *downward* direction.

Meanwhile, the upward moving buoyant stream entrains air as it rises; the model assumes that at the plane where the plume and push jet interact, the plume has an *upward* momentum, with magnitude given by

$$\boxed{M_u = \rho_u U_u^2 A_u} \tag{6-31}$$

where A_u and U_u are the plume cross-sectional area and the (mass) average velocity at the plane where the plume and push jet interact, and ρ_u is the density of the rising plume gas (a mixture of air and contaminant). To achieve a horizontal resultant momentum from the interaction of these two jets, the downward momentum of the push jet must exactly cancel the upward momentum of the buoyant plume. Thus,

$$\rho_u U_u^2 A_u = \rho_J U_J^2 A_J \sin\theta$$

which can be rearranged to express the required push jet velocity (U_J),

$$U_J = U_u \sqrt{\frac{\rho_u}{\rho_J} \frac{A_u}{A_J} \frac{1}{\sin\theta}} \tag{6-32}$$

The model also assumes that the volumetric flow rate of the rising plume (Q_u) is given by Eq. (6-21), and that the average velocity is the plug flow velocity given by dividing Q_u by the cross-sectional area of the plume at the plane between the push jet and exhaust opening. Kashdan et al. (1986a&b) report that for rectangular push-pull systems, the required volumetric flow rate of the withdrawn gas (Q_w) per unit length of the exhaust opening can be expressed by

$$\frac{Q_w}{L} = 0.88\sqrt{\frac{Q_J}{L} U_J W} \tag{6-33}$$

where

- Q_w = withdrawn (exhaust) volumetric flow rate (ACFM)
- Q_J = push jet volumetric flow rate (ACFM)
- U_J = jet velocity (ft/min, i.e. FPM)
- W = distance separating push jet and inlet opening (ft)
- L = length of exhaust opening (ft) (perpendicular to the page in Figure 6.23)

Example 6.5 - Capture of Secondary Emissions from a Copper Converter

Given: Copper converting is the process of transforming copper matte (produced in a smelter) into blister copper. A copper converter is a refractory-lined cylindrical vessel with an opening in the center. The vessel is rotated to various positions during blowing, charging, and skimming as shown in Figure E6.5a, with more details shown in Figure E6.5b (Kashdan et al., 1986a&b). For most of the cycle (12 hours) the vessel is in the blowing phase in which oxygen-enriched air is blown through the melt and the emissions are drawn off through a close-fitting primary hood and sent to a sulfur recovery plant. During the charging and skimming phases (approximately 30 minutes total, in periods of 4 minutes each) the converter is rotated away from the primary hood, and controls are needed to prevent fume from entering the workplace.

To do: Design an appropriate push-pull control system

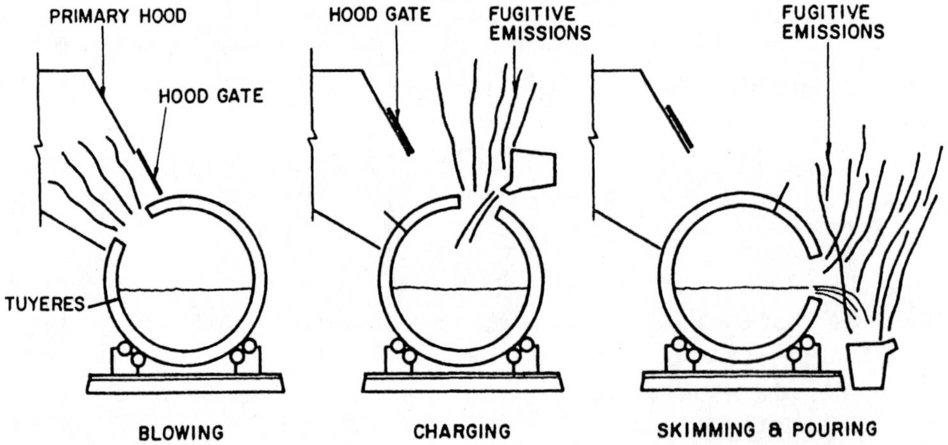

Figure E6.5a Fume from a copper converter, showing blowing, charging, and skimming processes (adapted from Kashdan et al., 1986).

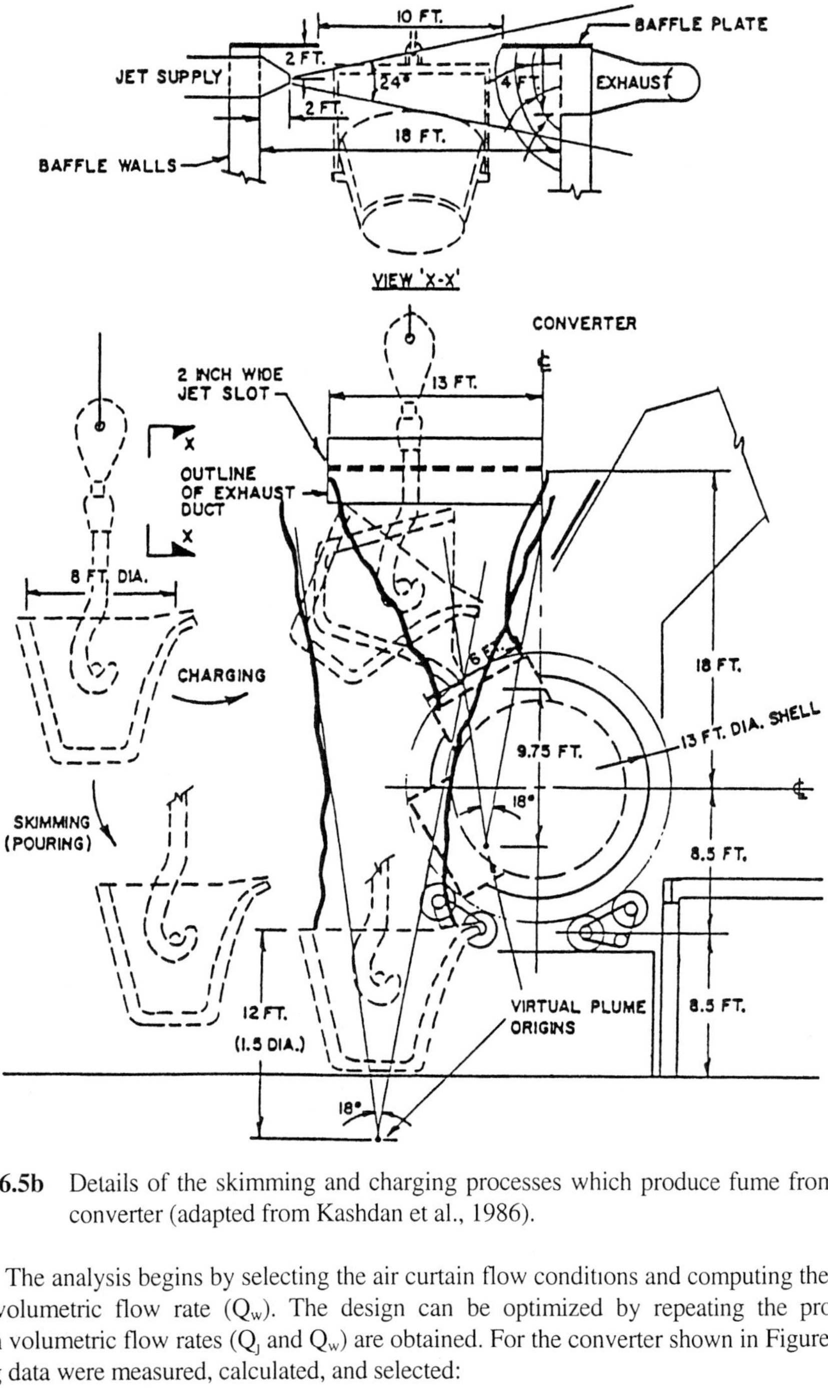

Figure E6.5b Details of the skimming and charging processes which produce fume from a copper converter (adapted from Kashdan et al., 1986).

Solution: The analysis begins by selecting the air curtain flow conditions and computing the necessary exhaust volumetric flow rate (Q_w). The design can be optimized by repeating the process until minimum volumetric flow rates (Q_J and Q_w) are obtained. For the converter shown in Figure E6.5b the following data were measured, calculated, and selected:

- D_{ladle} = diameter of ladle = 8.0 ft
- S_B = jet slot width = 0.050 m (2.0 inches)
- L = jet slot length = 3.96 m (13.0 ft)
- W = distance separating push jet and exhaust = 3.66 m (12.0 ft)
- T_J = jet air temperature = 21.0 °C (294.15 K)

- U_J = jet velocity = 41.7 m/s (8210 FPM)
- Q_J = jet volumetric flow rate = 8.40 m³/s (17,800 ACFM)
- θ = jet tilt angle from the horizon = 15 degrees
- T_a = ambient temperature = 21.0 °C (294.15 K or 529.47 R)
- ρ_a = ambient air density = 1.2 kg/m³ (0.075 lbm/ft³)
- T_u = plume temperature during skimming = 85.0 °C (358.15 K or 185. °F)
- T_c = plume temperature during charging = 249. °C (522.15 K or 939.87 R or 480.2 °F)
- $\dot{Q}_c$ = rate of convective heat transfer from the source = 150,000 BTU/min

From Eq. (6-23), the buoyancy flux parameter (F) is

$$F = \frac{g\dot{Q}_c}{c_p T_a \rho_a} = \frac{32.2\frac{\text{ft}}{\text{s}^2}150{,}000\frac{\text{BTU}}{\text{min}}}{0.24\frac{\text{BTU}}{\text{lbm}\cdot\text{R}}(529.47\ \text{R})0.075\frac{\text{lbm}}{\text{ft}^3}}\left(\frac{\text{min}}{60\ \text{s}}\right)\left(\frac{0.3048\ \text{m}}{\text{ft}}\right)^4 = 73.0\frac{\text{m}^4}{\text{s}^3}$$

<u>Charging</u>: During charging, the plume spreads around the charging ladle, and the cross-sectional area of the plume cannot be based on the idealized 18-degree plume spreading angle. Instead the diameter of the ladle (8.0. ft) is used to estimate the diameter (D_u) of the plume, which is used to calculate the area. The cross-sectional area (A_u) of the plume is thus

$$A_u = \frac{\pi D_u^{\,2}}{4} = \frac{\pi D_{\text{ladle}}^{\,2}}{4} = \frac{\pi(8.0\ \text{ft})^2}{4} = 50.3\ \text{ft}^2 = 4.67\ \text{m}^2$$

The height of the virtual source (z) is assumed to be 21. ft (6.4 m) during charging. Consequently, from Eq. (6-21),

$$Q_u = 0.166 z^{5/3} F^{1/3} = 0.166(6.4\ \text{m})^{5/3}\left(73.0\ \frac{\text{m}^4}{\text{s}^3}\right)^{1/3} = 15.3\frac{\text{m}^3}{\text{s}}$$

at temperature T_c = 249. °C. In English units, this is equivalent to 32,400 ACFM at 480.2 °F. The mass-averaged plume velocity (U_u) at this cross-sectional area is thus

$$U_u = \frac{Q_u}{A_u} = \frac{32{,}400\frac{\text{ft}^3}{\text{min}}}{50.3\ \text{ft}^2} = 644.\ \text{FPM} = 3.27\frac{\text{m}}{\text{s}}$$

<u>Skimming</u>: During skimming, the plume rises from the skimming ladle (which has an overall diameter of 15. ft) and spreads under the influence of buoyancy. From Figure E6.5b the height of the virtual source (z) is found to be

$$z = 12. + 8.5 + 18.\ \text{ft} = 38.5\ \text{ft} = 11.7\ \text{m}$$

Consequently, using the same equations as above, but with this new value of z,

$$Q_u = 0.166(11.7\ \text{m})^{5/3}\left(73.0\ \frac{\text{m}^4}{\text{s}^3}\right)^{1/3} = 41.8\frac{\text{m}^3}{\text{s}}$$

at temperature T_u = 85.0 °C, which is equivalent to 88,600 ACFM at 185. °F in English units. The plume area is

$$A_u = \frac{\pi(15.\ \text{ft})^2}{4} = 176.7\ \text{ft}^2 = 16.4\ \text{m}^2$$

and

$$U_u = \frac{88,600 \frac{ft^3}{min}}{176.7\ ft^2} = 501.\ FPM = 2.55 \frac{m}{s}$$

Since skimming produces the largest plume, the air curtain is designed (conservatively) for 90,000 ACFM at 185. °F. The flow rate of gas entering the exhaust opening is computed from Eq. (6-33):

$$Q_w == 0.88(13.\ ft)\sqrt{\frac{17,800 \frac{ft^3}{min}}{13.\ ft} 8210 \frac{ft}{min}}(12.0\ ft) = 132,900 \frac{ft^3}{min} \cong 130,000\ ACFM = 63. \frac{m^3}{s}$$

Note that Q_w is the actual volumetric flow rate, rounded to two significant digits.

The process can be repeated for other values of push jet volume flow rate (Q_J), slot dimensions (S_B and L), and angle (θ) until a minimum Q_J and Q_w are obtained.

Discussion: The analysis employs several untested assumptions of which users must be cognizant:

- The model uses a push-pull equation to describe the interaction of a jet directed into a hot, low-speed plume.
- The engineer cannot select an exhaust's dimensions (here 4 ft wide) arbitrarily, and be assured that the desired volumetric flow rate (Q_w = 130,000 CFM) can be withdrawn. The engineer has an obligation to prove or measure that the desired volumetric flow rate is truly withdrawn. It is unwise to presume that spillage at the exhaust opening does not exist.
- The model assumes incorrectly that the mass-average velocity is the same as the momentum-average velocity. This may in fact not be the case.

Nevertheless, measurements made in the plant confirm (Kashdan et al., 1986a) that the above values are reasonable.

6.7 Surface Treatment

The surface of most manufactured products is treated at various times during manufacturing. Examples of surface treatment are:

- grinding, blasting, de-burring, polishing, buffing, and tumbling
- cleaning and stripping
- printing
- drying
- metal plating, anodizing, and phosphating
- applying a coating or bonding agent
- electroplating
- dipping to apply a coating or to enable a surface chemical reaction to occur

Applying a surface finish or cleaning a surface involves equipment engineers are apt to think too mundane to warrant attention. Wrong. It may be mundane, but significant emissions are apt to occur unless proper care is taken. The control method of choice is to enclose the process, and exhaust only enough contaminated air from the enclosure to keep the region under negative pressure. If the amount and nature of the emission are within categories for which OSHA or air pollution regulations apply, an air pollution control device will be necessary before the gas stream can be exhausted to the atmosphere or to the workplace air.

Drying, printing, and chemical etching are performed in uniquely designed equipment that essentially encloses the emission source. Grinding, polishing, and spray painting are performed in booths that are discussed in Chapter 1. Mechanical operations such as blasting, tumbling, de-burring, etc. are performed in special equipment from which no emission should occur. Metal plating, anodizing, electroplating, and dipping are accomplished in open vessels that can be fitted with lateral exhausters or push-pull ventilation systems as discussed earlier in this chapter.

Paint stripping is a common industrial operation requiring the removal of a coating of lacquer, epoxy, urethane, etc. The task is complicated by the fact that the surface beneath the coating, called the ***substrate***, may be fragile. In the past, numerous solvents have been used at room or elevated temperature, methylene chloride at room temperature being the most common agent. The Clean Air Act Amendments (CAAA) of 1990 required the gradual phase-out of this and other chlorinated hydrocarbons within a decade, and alternative techniques had to be devised. Some of these blasting techniques are:

- novel media can be used; listed below are some examples:
  - wheat starch can be used for delicate substrates such as fiberglass, graphite composites, and aluminum
  - crystalline ice for aluminum, plastic, graphite composites, concrete, and titanium
  - bicarbonate (baking soda) for restoration of copper structures, such as the Statue of Liberty
  - dry ice (CO_2) or plastic media for removing paint from aircraft
- cryogenic removal uses liquid nitrogen below -50 °F to make the coating brittle; the coating is then removed by blasting with a plastic media
- high-pressure water blasting uses water at 35,000 to 50,000 psi to produce a high-velocity jet (> 2000 ft/s) to remove hard, tough coatings from ship hulls, deposits on heat exchangers, and rubber liners on metal

Of all the surface treatment processes, cleaning surfaces with solvents is the most common. ***Solvent cleaning***, as it is called, is common in the metalworking industries (automotive, electronics, plumbing, aircraft, refrigeration, and business machines) and the printing, chemicals, plastics, textiles, glass, and paper industries. Solvent cleaning is always present in maintenance and repair activities associated with transportation, electrical components, tools, etc.

Surface cleaning poses unique problems warranting attention. Many metal, glass, and plastic products must be washed and degreased before electroplating, painting, or receiving a bonding agent. The generic term ***degreasing*** is used for these cleaning operations. Most degreasing is conducted in specially designed packaged units available commercially. For many decades chlorinated hydrocarbon solvents were used in either the liquid or vapor state since unwanted dirt, cutting oil, tar, grease, wax, fat, and oil are removed easily at temperatures from 75 °F to 200 °F. These units emit hydrocarbon vapors, which while not producing concentrations above TLV, produce a yearly emission of hundreds of kilograms per year. If the cleaning agent is a chlorinated hydrocarbon, the EPA is concerned because it depletes stratospheric ozone and contributes to photochemical reactions in the troposphere resulting in smog.

In the near future, cleaning agents, coatings, and finishes involving solvents will be subjected to increasingly strict controls as the EPA and state agencies strive to reduce the amount of volatile organic compounds (VOCs) emitted to the atmosphere. The VOCs are not particularly hazardous or reactive, but once in the atmosphere they participate in photochemical reactions that produce ozone and objectionable oxygenated hydrocarbons (smog) that exceed EPA ambient air quality standards. The CAAA of 1990 prescribed a schedule to phase out the production of hydrochloroflouro-carbon solvents used in cleaning metal surfaces including the most popular cleaning agent 1,1,1-

trichloroethane (methyl chloroform). Alternative cleaning methods that have gained favor include those using ultrasonics, liquid nitrogen, and aqueous alkaline solutions followed by iron phosphate treatment of metal surfaces.

In the years to come the development of aqueous cleaning solutions, aqueous coatings, and powder coatings will receive considerable attention as substitutes for solvent based materials. Water-based coatings and cleaning agents illustrate the control strategy called "substitution" discussed in Chapter 1, since eliminating the solvent eliminates VOCs without the need to install an air pollution control device. Powder coatings also illustrate the substitution control strategy since they emit no VOCs. Powders can be applied to a surface by electrostatic techniques. Once applied, the powder is heated (direct or infrared heating) to cause a surface reaction that forms a contiguous coating. No VOCs are emitted; assuming that there are no toxic materials in the small emission of vapor from the bonding agents, there are no emissions to control.

For the ubiquitous chore of removing grease, dirt, and oil from surfaces, the term ***solvent degreasing*** (or merely *degreasing*) is commonly used. Equipment to accomplish these tasks is divided into three categories:

- cold cleaners
- open top vapor degreasers
- degreasers equipped with conveyors to introduce and remove parts

Cold cleaners are vessels of nonboiling solvents that accommodate immersed parts and baskets of parts. Cold cleaning operations include soaking with or without agitation, spraying, brushing, flushing, and immersion. Used primarily in maintenance operations, the solvents are usually petroleum solvents such as mineral spirits. After they are cleaned, the parts are suspended over the tank or in racks to allow solvent draining from parts to return to the tank. The vessel is generally equipped with a lid to reduce emissions to the workplace.

Open top solvent degreasers vary in size from simple heated washbasins to large heated conveyorized vapor units. In either case, parts are cleaned by hot solvent vapor. Figure 6.24 is a sketch of a typical vapor degreaser. The parts to be cleaned are placed in a basket and suspended above the pool of heated solvent. Solvent is vaporized in the bottom of the vessel by an external heater, and its vapor fills the lower portion of the vessel containing the parts to be cleaned. On the walls of the tank, but at a distance (freeboard) below the top of the tank, are located cooling coils (using water or refrigerant) which condense VOCs and return them to the sump below. The cooling coils also produce a region of low temperature extending across the vessel to condense upward rising solvent vapor. Water vapor and VOC vapors pass over finned condenser tubes that condense the vapors that then pass to a water separator. The condensed VOC is returned to the reservoir. Objects to be cleaned are lowered into the tank in metal baskets or by a conveyor system. Many units are equipped with a hose and nozzle so that parts can be sprayed manually with hot solvent. If this is done it is imperative that spraying be done below the cooling coils; otherwise the hot solvent would flash and emit large amounts of vapor to the workplace.

While the cooled region prevents the bulk of hot vapor from entering the workplace, significant amounts of vapor nevertheless escape to the workplace. Vapor escaping from the tank is called ***boil over***. To increase the capture of hot vapor, degreasing tanks are often equipped with metal covers, slot exhausters, or push-pull ventilation systems. Vapor emitted from the surface of hot parts removed from the degreaser is called ***carryout***. It is prudent to enclose the degreaser in such a way that the removed parts can be cooled and dried in an enclosure preventing carryout from entering the workplace.

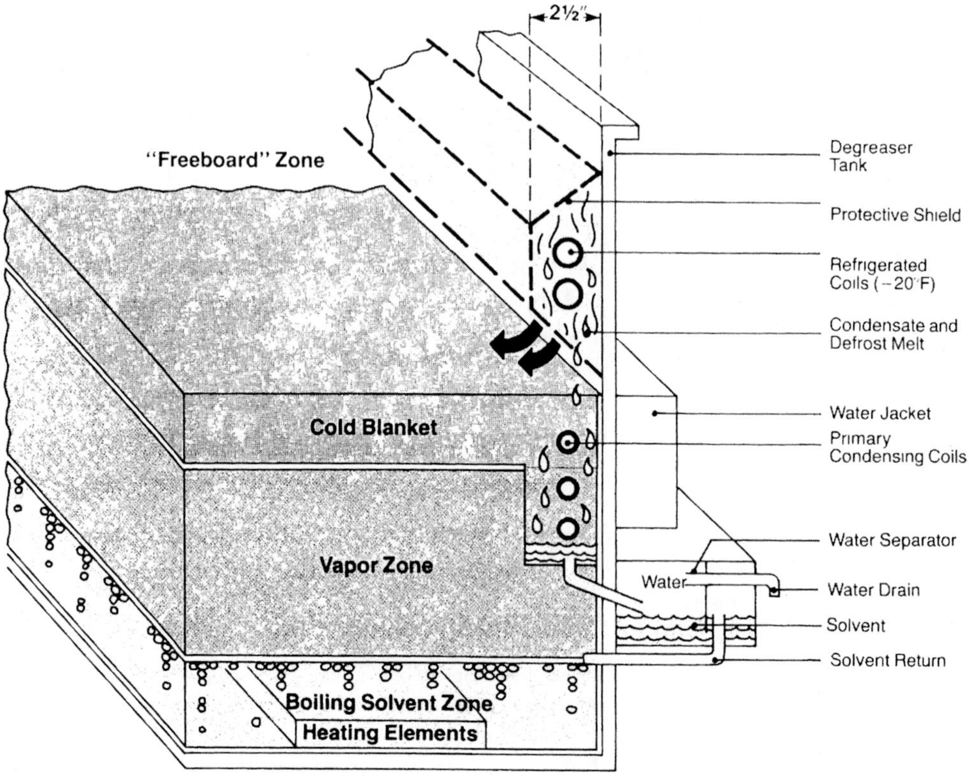

Figure 6.24 Typical freeboard chiller for vapor degreasing.

Nonaqueous solvents such as petroleum distillates, chlorinated hydrocarbons, ketones, and alcohols are used in degreasing operations. The selection of a specific solvent depends on the solubility of the substance to be removed and the toxicity, flammability, flash point, evaporation rate, boiling temperature, and cost of the solvent. The preferred and principal solvents used in degreasers are the nonflammable chlorinated hydrocarbons PERC, TCE, and methyl chloroform,

- perchloroethylene ($Cl_2C{=}CCl_2$), also called tetrachloroethylene, perchloroethylene, PERK, PERC, or tetrachloroethylene
- trichloroethylene ($ClHC{=}CCl_2$), also called ethylene trichloride, TCE, trichloroethene, or trilene
- 1,1,1-trichloroethane (CH_3CCl_3), also called chlorothene or methyl chloroform

The selection of the solvent is generally dictated by the nature of the dirty coating that needs to be removed. Most greases and tars dissolve readily at the 189 °F boiling point of TCE. PERC, which boils at 249 °F, is used for more stubborn material requiring higher temperatures. Shown below are EPA emission factors for solvent loss for different types of degreasers:

type of degreaser	**uncontrolled emissions/unit**
cold cleaner	
boil off	0.165 metric ton/yr
carryout	0.075 metric ton/yr
bath and spray	0.060 metric ton/yr
open top	9.5 metric ton/yr or 0.7 kg/(hr m^2)

Some degreasers are equipped with lids or covers that prevent boil-off, but unless precautions are taken, opening the lids to add and remove parts is apt to produce a significant emission to the workplace. If covers are used they should be opened by sliding, rolling, or guillotine action rather than by hinged solid lids. Even when slid open, lateral exhausts should be used to control the momentary hot plume that is apt to escape.

Efforts are in progress throughout the US to eliminate altogether the use of hydrocarbon solvents and to achieve a desired surface treatment by aqueous solutions or by utterly different methods such as fluidized beds, ceramic tumbling, ultrasonics, etc. Note that aqueous solutions can be used; both the application of the cleaning agent and its removal are accomplished inside negative pressure enclosures that ensure that there are no emissions to the workplace.

An emerging technology that is receiving increasing attention to replace solvents used in cleaning and in the manufacture of pharmaceuticals is the use of supercritical carbon dioxide. The critical temperature and pressure of carbon dioxide are 304.2 K (31.05 °C) and 7,390 kPa (73.17 atm). At temperatures and pressures above the critical point, carbon dioxide exists in a single phase as a very dense fluid. Under these conditions carbon dioxide displays unique chemical and physical properties. Its surface tension and viscosity are considerably smaller than values possessed by liquids at STP, while its diffusivity is considerably higher than that of liquids at STP. The absorption of solutes and oxidation processes are considerably higher than at STP. For this reason supercritical carbon dioxide possesses attractive properties to act as a cleaning agent to replace conventional halogenated hydrocarbons. Supercritical carbon dioxide also has attractive properties as a solvent for the extraction of organics from soil (Erkey et al., 1993) or in the manufacture of chemicals that presently use hydrocarbon solvents. Since elevated temperatures are not required, parts can be cleaned at temperatures comparable to what are currently used in solvent degreasers. The high pressures that are needed may restrict some parts, but more troublesome are the vessels and materials handling procedures that are required to clean parts in batches. Specially designed high-pressure vessels are needed to contain the cleaning process, and unique means will need to be devised to add and remove the parts from the high-pressure vessel. Residues absorbed by the supercritical carbon dioxide need to be removed (just as they need to be removed from conventional solvent cleaning agents), but since carbon dioxide becomes a gas as the pressure is reduced, it is quite possible that wastes can be handled in a simpler fashion than is presently the case with hydrocarbon solvents that need to be distilled and condensed to purify them.

6.8 Building Air Inlets and Exhaust Stacks

One of the most common errors made in industrial ventilation is to place a make-up air inlet too close to an exhaust stack. The error usually arises because the locations of the inlet and exhaust are decided at different times by different people unaware of each other's actions. Errors are also made because individuals are unaware of the size and consequences of the ***aerodynamics*** associated with buildings. In the vicinity of a building, air velocities are *not* equal to freestream values. Figures 6.25 and 6.26 illustrate the ***vortices*** (turbulent recirculating eddies) one can expect from a block-like building on level terrain immersed in a deep terrestrial boundary layer. If the building lies in the wakes of other buildings or if the building lies on terrain that is not level, the vortices are somewhat different in shape and location but exist nonetheless.

Two types of flow regions can be defined:

- A ***recirculating eddy region*** is a region in which a relatively fixed amount of air moves in circular fashion and there is little air transported across the eddy boundaries.
- A ***turbulent shear region*** is one in which there is a net convective flow, but the turbulent shear stresses are much larger than freestream values.

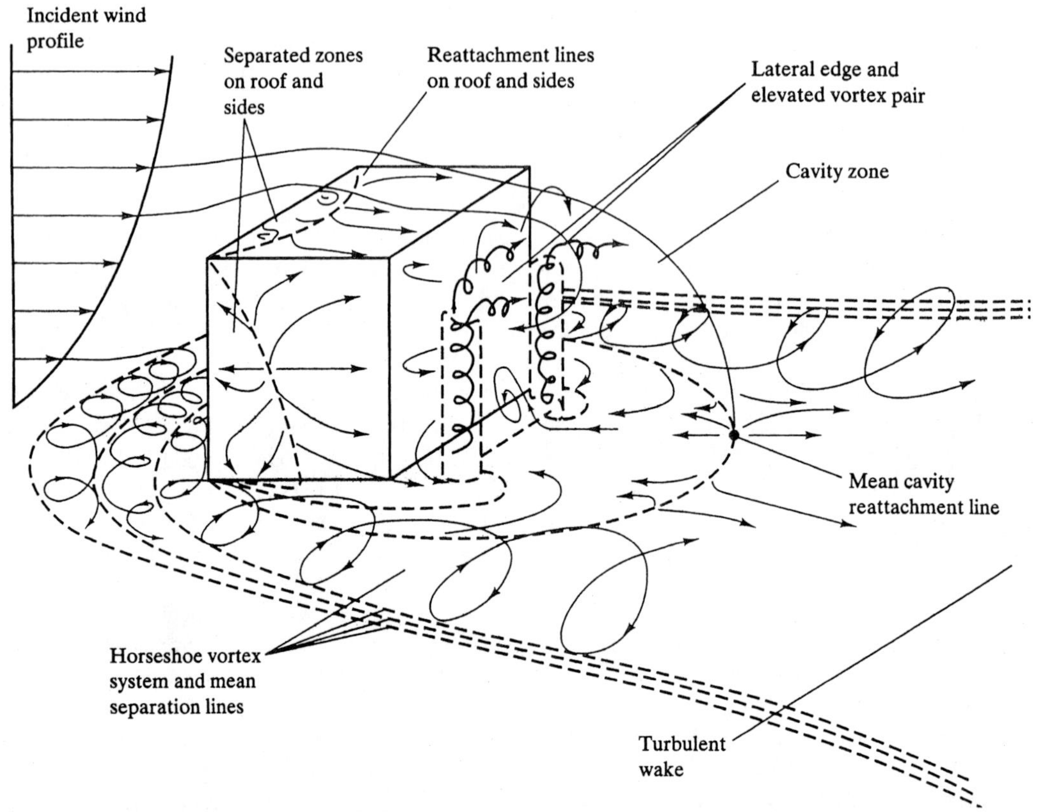

Figure 6.25 Wakes and vortices of a block-like building in a terrestrial boundary layer (adapted from Hosker, 1982).

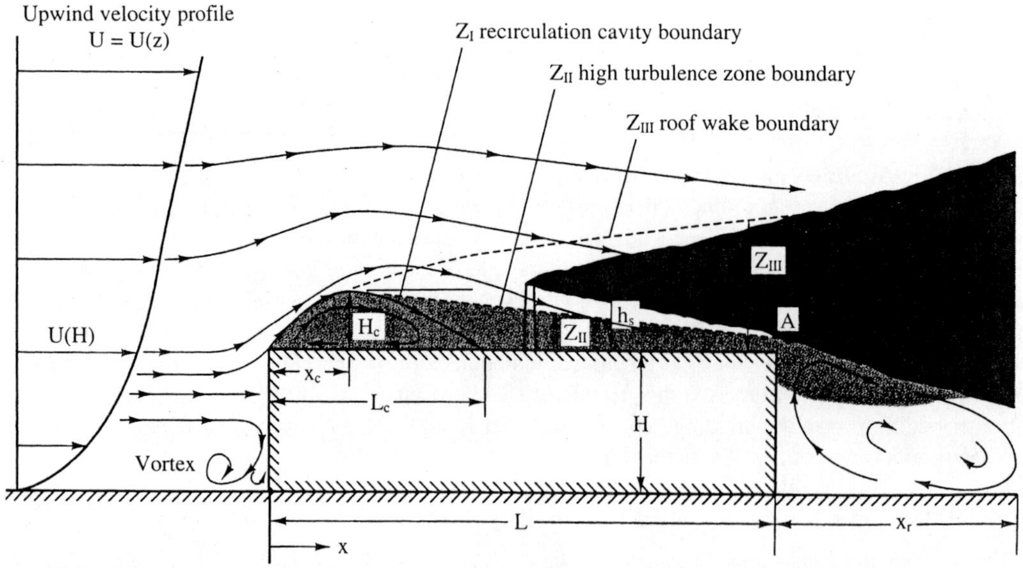

Figure 6.26 Wakes and recirculation eddies of a two dimensional flat building in a terrestrial boundary layer (adapted from Hosker, 1982).

Relationships predicting the location and velocity field of recirculating eddy regions, turbulent shear regions, and wakes are not readily available, but a series of empirical equations defining the boundaries of these regions is widely used (Hosker, 1982; Petersen, 1993). In the discussion that follows it is assumed that the building is a block-like structure of height (H), cross-wind width (W), and length (L) in the direction of the wind. In selecting the location and dimensions of an air inlet or exhaust stack, two general principals should be followed:

- air inlets should not be located in recirculating eddy regions or at other locations susceptible to contamination from exhaust gases
- the effective stack height above the building roof should be above the roof wake boundary

As good engineering practice, the EPA recommends the following ***stack height*** (h_s), defined as the height above ground level:

$$\boxed{h_s = H + 1.5L_S} \tag{6-34}$$

where L_S is the smaller of the width (W) or height (H). EPA approved models for stack height can be found in the "Industrial Source Complex" (ISC) dispersion models. The stack height (h_s) is a geometric dimension. The ***effective stack height*** is equal to the geometric stack height (h_s) plus the plume rise due to the buoyancy and the momentum of the exhaust gas (Hanna et al., 1982). If zoning laws or other considerations prevent Eq. (6-34) from being followed, it may be necessary to estimate the location of the wakes. There are two regions to be concerned about:

- the recirculation eddy, i.e. the ***roof eddy***, and wake produced by the leading edge of the building roof
- the large recirculation eddy, i.e. the ***building eddy***, directly downwind of the building

Key dimensions of the roof eddy are shown in Figure 6.26, and can be expressed in terms of length parameters L_L, L_S, and R', where L_L is defined as the larger of H or W, L_S is defined as the smaller of H or W, and

$$\boxed{R' = L_S^{2/3} L_L^{1/3}} \tag{6-35}$$

The height (H_c), length (L_c) and center (x_c) of the roof eddy can be estimated from the following:

$$\boxed{H_c = 0.22R' \qquad L_c = 0.90R' \qquad x_c = 0.50R'} \tag{6-36}$$

The roof eddy is bounded above by a turbulent shear region (zone II) and above that by a wake region (zone III). The vertical distances Z_I, Z_{II}, and Z_{III} are measured from the roof. The height of the turbulent shear zone (Z_{II}) can be estimated from

$$\boxed{\frac{Z_{II}}{R'} = 0.27 - 0.10\left(\frac{x}{R'}\right)} \tag{6-37}$$

where x is measured in the downwind direction from the roof lip. The height of the roof wake (Z_{III}) can be estimated from

$$\boxed{\frac{Z_{II}}{R'} = 0.28\left(\frac{x}{R'}\right)^{1/3}} \tag{6-38}$$

The safest design is one in which the expanding plume remains above the roof wake boundary (Z_{III}). The manner in which the plume expands depends on atmospheric stability conditions, which in turn depend on wind speed and solar radiation. If the plume cannot be kept above Z_{III}, a plume that is kept in the turbulent shear region (zone II) is the next best choice.

Directly downwind of a structure, a very large building eddy forms. Contaminants trapped in this region are apt to enter building windows or affect people and vehicles on the ground. The size of the building eddy is shown in Figures 6.25 and 6.26. Schulman and Scire (1993) suggest that the height (H_r) and width (W_r) of the building eddy are related to the building dimensions by the following:

$$\begin{array}{lll} H_r = H, & W_r = W & \text{if } L_c < L \\ H_r = H + H_c, & W_r = 0.60H + 1.1W & \text{if } L_c > L \end{array} \qquad (6\text{-}39)$$

The length of the building eddy (x_r) can be estimated from

$$x_r = \frac{1.8W}{\left(\frac{L}{H}\right)^{0.3}\left(1 + 0.24\left(\frac{W}{H}\right)\right)} \qquad (6\text{-}40)$$

provided that $0.3 \le L/H \le 3.0$. If the value of (L/H) is smaller than 0.3, the authors suggest that one should set (L/H) equal to 0.3. If the value of (L/H) exceeds 3.0, one should set (L/H) equal to 3.0.

Schulman and Scire (1993) provide an equation for the maximum mass concentration ($c_{r,max}$) in the building eddy, assuming that the entire plume is captured by the eddy:

$$c_{r,max} = \frac{B_0 \dot{m}_j}{B_0 Q + s^2 U_H} \qquad (6\text{-}41)$$

where

- $\dot{m}_j$ = mass emission rate of species j through the stack
- s = "*stretch string distance*" = distance between base of stack and the point (x,y)
- Q = actual volumetric flow rate of gas exiting the stack
- U_H = wind speed at height of building
- B_0 = a nondimensional constant; a value of 16 is recommended

It is assumed in Eq. (6-41) that plume dispersion is Gaussian, and that at any downstream distance its radius is given by

$$r_{plume} = 0.21(R')^{1/4} x^{3/4} \qquad (6\text{-}42)$$

where x is the distance downwind of the stack. The actual pollutant mass concentration (c_r) at points in the building eddy will be smaller than the maximum value given by Eq. (6-41), since not all of the plume is captured by the building eddy,

$$c_r = f_r c_c \qquad (6\text{-}43)$$

where f_r is the fraction of the plume captured by the building eddy.

The previous material describes building wake regions and provides guidance about where to place exhaust stacks or fresh air inlets. If for some reason a vertical stack is placed on a large horizontal surface and the wind speed is constant, it is possible to be more definitive and actually predict the plume gas concentration at arbitrary downwind points, including points on the ground. ***Dispersion modeling*** is the name given to this body of knowledge, and excellent, concise presentations suitable for engineers can be found in the literature (Hanna et al., 1982; Heinsohn and Kabel, 1998).

Gases leave the stack and rise due to momentum and buoyancy. Once the exit momentum has been dissipated and the gases have cooled to the local atmospheric temperature, the gases become

neutrally-buoyant; the plume is carried downwind by the prevailing wind. In the lateral direction, mixing is governed by molecular and turbulent diffusion. In the vertical direction, mixing is governed in addition by the ***lapse rate*** (the negative value of the atmospheric temperature gradient), which in turn is governed by the solar intensity and the wind speed. Studies of how plumes are transported in the atmosphere have been conducted for many years. Since the solution of the governing transport equation is the Gaussian distribution function, the plumes are called ***Gaussian plumes***. There are several expressions (Hanna et al., 1982; Seinfeld and Pandis, 1997; Heinsohn and Kabel, 1998) to predict the height to which a plume rises above the physical stack, i.e. ***plume rise***, before being transported downwind.

Predictions of ground-level concentrations from Gaussian plumes are valid for regions no closer than approximately 100 m from the stack. There is a paucity of information for regions closer to the stack where the plume is rising due to buoyancy and momentum. For neutrally-buoyant plumes, Halitsky (1989) provides expressions to predict mass concentrations in regions downwind of the stack where the jet velocities have decayed and are essentially equal to the freestream values.

In the wake of a stack, the pressure is lower than the freestream value, and under some conditions may cause the exiting stack gases to fall rather than rise. Such a fall is called ***downwash***. Downwash may be a serious consideration in the design of short stacks. Downwash can be prevented by keeping the ratio of the duct exit velocity to the wind speed greater than 1.5, i.e. $U_j/U_0 > 1.5$ (Goodfellow, 2000). If the ratio is less than 1.5 it is suggested (Hanna et al., 1982) that the downwash distance (H_{dw}) can be estimated from

$$\boxed{\frac{H_{dw}}{D} = 2\left(\frac{U_J}{U_0} - 1.5\right)} \tag{6-44}$$

where D is the stack exit diameter. The negative value computed by Eq. (6-44) implies that downwash causes the plume centerline to fall below rather than rise above the stack exit plane.

6.9 Unsatisfactory Performance

Troublesome to readers may be the fact that design data in previous sections recommend a range of volumetric flow rates or a range of capture velocities rather than specific values. Furthermore, the range is often very large. A range acknowledges variables in the workplace (drafty or still air, dirty or clean environment, toxicity of the contaminant, etc.) and deliberately forces engineers to decide whether the upper or lower end of the range is to be selected. One of the principal variables is drafts, or more formally, spurious air currents. Seldom do engineers know the ambient air velocity at points throughout the workplace unless measurements are made to acquire them. Individuals are generally unaware of drafts, e.g. a velocity of 100 FPM (0.51 m/s) is similar to blowing on the hand, and most people would be unaware of such a draft unless their attention was called to it. While small, such a velocity influences the capture of contaminants. Later in the book procedures are described that incorporate the velocity of the ambient air in the design process in a quantitative way. The following physical conditions produce spurious air currents:

- air issuing from internal room air heaters, conditioned make-up air entering through registers
- portable fans used by workers for comfort
- windows or doors workers choose to leave open for comfort
- wakes produced by workers and machinery
- drafts produced by moving workers or vehicles

A well-designed industrial ventilation system may perform badly because the designer is unaware of actions taken after the system is installed. The following is a partial list of actions that adversely affect performance:

- the volumetric flow rate may decrease because someone has blocked off part of the inlet area of the capture device or the inlet to the exhaust fan
- someone may tap into the exhaust duct for another control device
- large amounts of material may collect in the exhaust duct and reduce the volumetric flow rate
- holes may be cut into a hood or enclosure to improve a worker's visibility
- workers may stand between the source and the inlet
- a more volatile or toxic material may be substituted

An exhaustive list would be long and perhaps embarrassing. Such blunders can be avoided if designers observe the movement of workers and talk to them about their work before the design is begun. After the system is installed it behooves the designer to inspect the site and observe if any of the above conditions occur or if any unanticipated changes have occurred that adversely affect performance.

Inadequate make-up air is another common reason why ventilation systems fail to perform adequately. The following are symptoms of inadequate make-up air:

- exhaust gases from ovens, furnaces, water heaters, cookers, dryers, etc. (any device having a natural draft flue) have insufficient draft, and the exhaust gases enter the workplace; this event is often preceded by complaints of odors
- pilot flames on ovens fail to remain lit
- face velocities on ventilation systems and volumetric flow rates are well below their design values even though fan horsepower and RPM may be within specifications
- doors opening into the workplace do not stay shut
- large overhead doors try to collapse inward when closed

The above symptoms often occur because equipment such as ovens, exhaust fans, ventilation systems, etc. are installed in the workplace in a piecemeal, uncoordinated fashion, generally over a period of time, without regard for make-up air. As a rule, each time a device that exhausts air from a building is installed, a comparable amount of make-up air must be provided. The symptoms are most severe in the wintertime when doors and windows are closed.

Present design practices are primitive. They lack the precision brought to engineering by advances in analytical modeling and numerical computation. These advances have greatly benefited other fields of engineering such as heat transfer, aerodynamics, mass transfer, etc. Nevertheless, while sophistication is lacking, present design practices are not wrong nor do they lead to unsafe designs. Present practices should continue to be followed, but used as initial estimates. Analytical techniques and modeling concepts can then be used to optimize the design for the particularities of an industrial site, to determine whether mandatory health standards will be met, and to estimate the performance for a variety of operating conditions. Irrespective of how a ventilation system is conceived, its performance must be verified by measurements at the site. The objectives of improving analysis are to:

- make design decisions authoritatively in a minimum amount of time
- withdraw a minimum amount of air and use a minimum amount of electrical power
- increase the probability that the design will require few changes in the field

6.10 Exhaust Duct System Design

Collection hoods are connected to an exhaust fan (or fans) by a network of ducts whose dimensions must be selected by the designer. Some hoods are to be connected to a single exhaust fan, while other hoods are to be connected in parallel. In the latter case, workers should be able to isolate each hood when it is not in use. Discussions about such important matters must be made at the beginning of the design process in close consultation with the users of the system. Factors entering the decision are:

- space available to locate ducts, fans, and air cleaning equipment (if needed)
- cost (capital and operating costs)
- ease of operation, e.g. how opening and closing of blast gates affect other hoods
- maintenance, repair, and duct cleaning

Once the decision has been made about which hoods are going to be connected to a particular exhaust fan, the designer should prepare a line sketch that identifies each hood, length of each duct segment, contaminant removal device, blast gate, fan, and number and type of duct fittings (elbow, tee, expansion, contraction, fitting, etc.). The volumetric flow rates for each collection hood are determined by the designer based on the nature of the process to be controlled and type of collection hood selected. Fans or blowers are selected based on the required volumetric flow rate and the overall pressure drop, which can be predicted. The pressure drop in ducts and fittings varies as a function of duct diameter, transport velocity, particle size distribution, density, and concentration. What is generally unknown is the diameter of each duct segment. (For air cleaning purposes, it is wise to use ducts of circular cross section.)

If particles are to be collected, it is important to achieve duct velocities above certain minimum values to minimize gravimetric settling, which is discussed in more detail in Chapter 8, and illustrated in Figure 6.27. Operating at a low transport velocity may result in serious particle deposition that reduces the duct cross-sectional area, slows down the flow because of the increased blockage, and thus contributes to further deposition. Within a very short time the system can become plugged and immobilize all collection hoods in the line. Plugged ducts sag under the weight of the deposited particles and have to be removed. ACGIH design plates specify minimum duct velocities which should be followed. Lacking any specific instructions, the ***duct transport velocities*** shown in Table 6.6 should be used. Duct systems should contain numerous ***clean-out ports*** easily accessible by workers for routine cleaning. In addition, drop out boxes (Figure 6.27a) should be placed at duct junctions, hoods, etc. so as to collect large particles before they enter the ducts. A ***drop out box*** is a device to remove very large particles as a consequence of their mass. There are no moving parts, the overall pressure drop is small, and an access port is available to remove collected particles.

If vapors are to be collected, provision should be made for clean-out ports. Vapors condense on cool duct surfaces. Flammable condensate may catch fire whenever there is a source of ignition, e.g.

- discharge from statically electrified surfaces
- sparks from welding
- burning particles from dryers
- sparks from sanding machines that encounter wood containing metal (e.g. nails)

Since ducts are located near roofs and ceilings, initially minor duct fires often lead to serious building fires.

Unless a hood is to operate continuously, it should be equipped with a blast gate. ***Blast gates*** are dampers or slide valves that workers use to allow air to be withdrawn into the hood only when they wish to control some industrial activity. If there is no industrial activity, the blast gate should be closed. If left open, air will be exhausted unnecessarily, fan operating costs will increase, and the exhaust volumetric flow rate may be reduced at hoods where there *is* industrial activity. Selection of an appropriate exhaust fan requires knowledge of the volumetric flow rate and overall pressure drop. The volumetric flow rate is the sum of the volumetric flow rates (corrected for density) of the separate collection hoods, taking into account which hoods will operate full-time and which will be activated by blast gates.

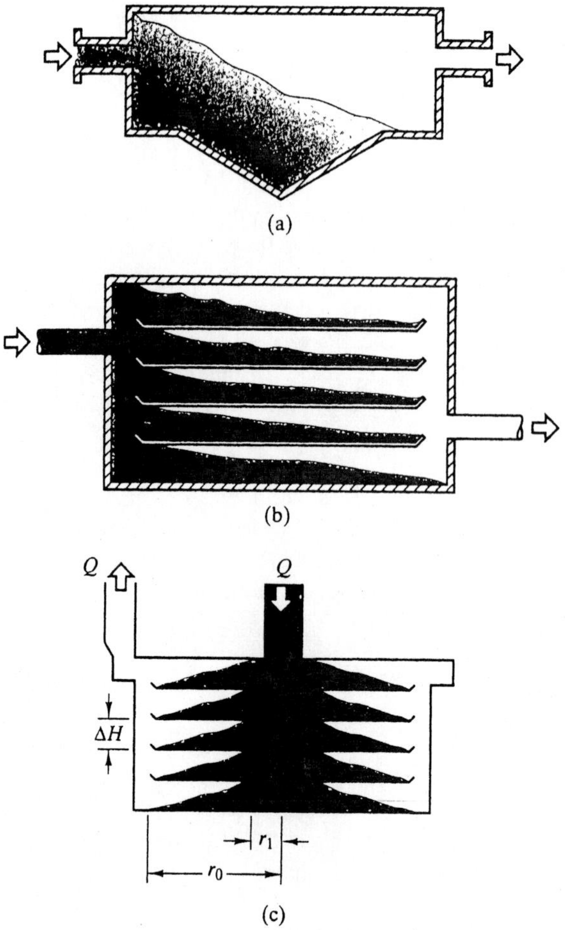

Figure 6.27 Gravimetric settling devices: (a) drop out box, (b) horizontal settling chamber with five trays, and (c) radial flow settling chamber with five trays (from Crawford, 1976).

There are several detailed procedures for designing duct systems. The most authoritative discussion is contained in the ASHRAE Fundamentals Handbook (2001). These procedures are general in nature and can be used for heating and air conditioning as well. Procedures recommended by the ACGIH ventilation manual (2001) pertain to contaminant control, are widely used in the field of industrial ventilation, and are compatible with ASHRAE procedures.

Table 6.6 Duct transport velocities (abstracted from ACGIH, 2001).

contaminant	**velocity (FPM)**
gases, vapors, smoke	1,000 - 2,000
fumes	2,000 - 2,500
very fine light duct (cotton lint, wood flour)	3,000 - 4,000
dry dust and powder (soap, rubber, Bakelite molding powder)	2,500 - 3,500
average industrial dust (general materials handling)	3,500 - 4,000
heavy dust (foundry shakeout, metal working)	4,000 - 4,500
very heavy and sticky dust (quick-lime, moist cement, metal chips)	4,500 and above

Computing the pressure drop in a duct is the basis of all duct design procedures. Since the velocity and pressure drop in a duct are sufficiently low, one may assume that the air is a constant density viscous fluid. Thus the relationship between the pressures (P_1 and P_2) between two arbitrary points (upstream point 1 to downstream point 2) in a duct can be expressed as

$$\boxed{P_1 + \frac{1}{2}\rho U_1^2 + \rho g z_1 = P_2 + \frac{1}{2}\rho U_2^2 + \rho g z_2 - \delta P_{fan} + \rho g h_{LT}} \tag{6-45}$$

where U_2 and U_1 are the average velocities at points 1 and 2, δP_{fan} is the pressure rise across a fan or blower in line in the ductwork between points 1 and 2, and h_{LT} is the ***total head loss*** between points 1 and 2. This last term is called a head loss because it affects the flow as if the flow involved an additional elevation change. In the present book, head has dimensions of length, which can also be thought of as energy per unit weight. This is the definition of head used in turbomachinery, and by civil engineers. Note that some textbooks on fluid mechanics define head differently, including the gravitational acceleration (g); their "head" has dimensions of energy per unit mass or length^2/time^2 (typical units being kJ/kg or m^2/s^2). Some authors define head a third way as pressure, including both ρ and g into their definition of head. These various definitions of head have been, and continue to be, the source of much confusion. The definition of head used in this book is the one adopted by ASHRAE, and is the one most commonly used in duct design.

Eq. (6-45) is sometimes referred to as a generalized form of the ***Bernoulli equation***, but in actuality, it is a simplification of the more general conservation of energy equation for a control volume, the derivation of which can be found in any elementary fluid mechanics textbook (e.g. White, 1999). The grouping of three terms at location 1 or at location 2 is called the ***total pressure***,

$$\boxed{P_t = P + \frac{1}{2}\rho U^2 + \rho g z} \tag{6-46}$$

In terms of the total pressure, Eq. (6-45) becomes

$$\boxed{P_{t,1} = P_{t,2} - \delta P_{fan} + \rho g h_{LT}} \tag{6-47}$$

which shows that in the absence of a fan in the line, total pressure always *decreases* along the axis of a duct due to irreversible head losses (h_{LT} is always positive definite). In fact, the only way to *increase* total pressure is by addition of external energy, e.g. by a fan or blower.

The total pressure (P_t) in Eq. (6-46) is often split into two components, namely the ***stagnation pressure***,

$$\boxed{P_{stagnation} = P + \frac{1}{2}\rho U^2} \tag{6-48}$$

and the ***hydrostatic pressure***,

$$\boxed{P_{hydrostatic} = \rho g z} \tag{6-49}$$

The stagnation pressure itself is further split into two components: the ***static pressure*** (SP),

$$\boxed{SP = P} \tag{6-50}$$

and the ***velocity pressure*** (VP),

$$\boxed{VP = \frac{1}{2}\rho U^2} \tag{6-51}$$

which is also called the ***dynamic pressure***. SP and VP are the symbols adopted for static pressure and velocity pressure by ACGIH, whereas ASHRAE uses the notation p_s for static pressure and p_v for velocity pressure. For air at STP (25.0 °C, 101.3 kPa) the velocity pressure (VP) in Pascals (Pa) is equal to

$$\boxed{VP = \left(\frac{U}{1.29}\right)^2} \tag{6-52}$$

where the average duct velocity (U) must be in meters per second (m/s). Use of a "unit-dependent" equation such as Eq. (6-52) is discouraged by the present authors, who prefer the corresponding "unit-independent" equation, Eq. (6-51), which is valid for any system of units.

The above equations are valid for flows of both gases and liquids. When designing ducts for an indoor air pollution control system (APCS) in which air is the primary carrier gas, the hydrostatic pressure terms dealing with elevation changes can be neglected unless there is a huge change from elevation z_1 to elevation z_2. For example, an elevation change of 10 m corresponds to a pressure difference of only 0.12 kPa (0.4 inches of water) for air at STP, which is negligible compared to the pressure drop due to friction in the ducts, or pressure losses due to fittings, expansions, etc. If elevation changes are insignificant, Eq. (6-45) reduces to

$$\boxed{P_1 + (VP)_1 = P_2 + (VP)_2 - \delta P_{fan} + \rho g h_{LT}} \tag{6-53}$$

The total head loss (h_{LT}) is the sum of the ***friction head losses*** (h_{Lf}) in constant area ducts (often referred to as ***major losses***) plus the ***minor head losses*** (h_{LM}) due to other irreversible head losses such as turbulent swirling eddies and flow separations occurring in

- duct fittings (elbows, tees, dampers, etc.)
- duct area changes (expansions and contractions)
- inlet losses at entrances to equipment.

Thus,

$$\boxed{h_{LT} = \sum_{major} h_{Lf} + \sum_{minor} h_{LM}} \tag{6-54}$$

6.10.1 Duct Friction - Major Head Loss

It is customary to express the drop in total pressure due to major losses in terms of a nondimensional ***friction factor*** (f), defined as

$$\boxed{f = \frac{h_{Lf}}{\frac{L}{D}\frac{U^2}{2g}}} \tag{6-55}$$

where L is the pipe length and D is the inner pipe diameter. The major pressure drop through a section of fully developed straight duct is thus

$$\boxed{\Delta P_{t,\ major\ loss} = \rho g h_{Lf} = \rho g\left(f\frac{L}{D}\frac{U^2}{2g}\right) = f\frac{L}{D}\left(\frac{1}{2}\rho U^2\right) = f\frac{L}{D}(VP)} \tag{6-56}$$

Friction factor, f, is a function of the equivalent sand-grain roughness height of wall deposits or material roughness (***roughness height***), ε, and the Reynolds number, Re, as seen in the ***Moody chart*** (Perry et al., 1984; White, 1999; ACGIH manual, 2001). Readers are encouraged to use the following equations in place of the Moody chart (White, 1999):

$$\boxed{\begin{array}{lll} \text{laminar:} & f = \dfrac{64}{Re} & (Re < \text{ approx. } 2300) \\ \text{turbulent:} & \dfrac{1}{\sqrt{f}} = -2.0\log_{10}\left(\dfrac{\varepsilon/D}{3.7} + \dfrac{2.51}{Re\sqrt{f}}\right) & (Re > \text{ approx. } 2300) \end{array}} \tag{6-57}$$

where

$$\boxed{Re = \frac{\rho U D}{\mu} = \frac{UD}{\nu}} \quad (6\text{-}58)$$

The turbulent part of Eq. (6-57) is implicit, but easily solved by Newton's method or by mathematical or spreadsheet programs. The authors used Eqs. (6-57) and (6-58) to generate an abbreviated Moody chart, as shown in Figure 6.28 for four different roughness values. The Excel spreadsheet used to generate this chart is available on the book's web site, and can be used in place of the Moody chart to calculate f for known values of Re and ε/D.

Equations (6-57) are valid for any newtonian fluid (air, water, oil, etc.) and for ducts of any size. Alternatively, for applications specific to HVAC engineering, ASHRAE publishes friction charts that can be used for air at 20 °C and 101 kPa flowing through specific types of ducts, e.g. round galvanized ducts of various diameters, with roughness height $\varepsilon = 0.09$ mm, having approximately 40 slip joints per 30 m of length (ASHRAE Fundamentals Handbook, 1997). Duct friction loss on these charts is provided in units of Pascals per meter of duct (Pa/m), and the charts can also be used for other duct materials of comparable roughness.

If the duct cross section is not round, the diameter should be replaced by its circular equivalent called the ***hydraulic diameter*** (D_y, also denoted as D_h in many reference books), which is equal to 4 times the duct's cross-sectional area divided by its perimeter,

$$\boxed{D_y = \frac{4A_c}{\text{Perimeter}}} \quad (6\text{-}59)$$

6.10.2 Fittings - Minor Head Loss

It is customary to express the drop in total pressure due to a minor head loss in terms of a dimensionless ***local loss coefficient*** (C_0), which depends on the geometry of the fitting. C_0, also

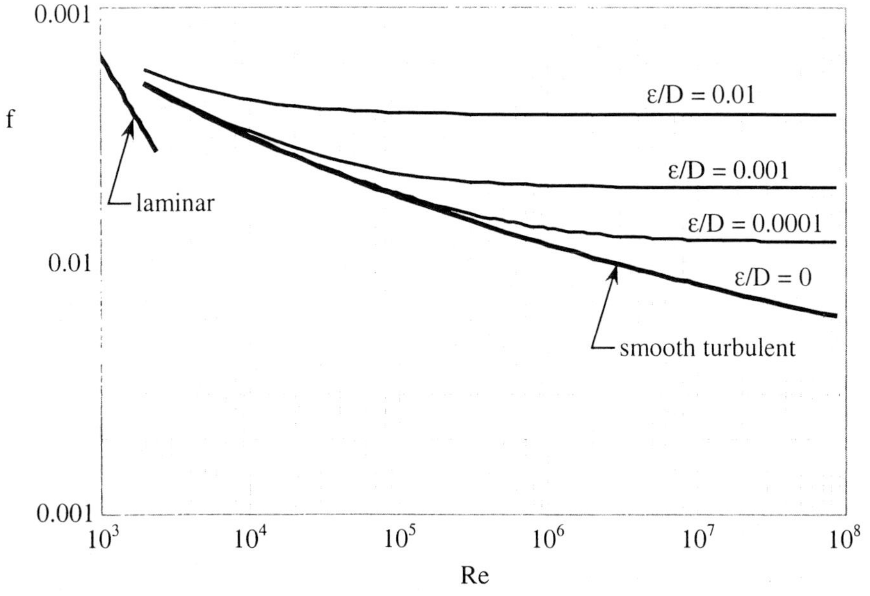

Figure 6.28 Moody chart for laminar flow (at any roughness) and for turbulent flow with $\varepsilon/D = 0$ (smooth), 0.0001, 0.001, and 0.01.

denoted as C, K, or F in some texts, is defined as the ratio of total pressure loss to the velocity pressure at some reference cross section in the duct,

$$C_0 = \frac{\Delta P_{t,\ \text{minor loss}}}{(VP)_0} \tag{6-60}$$

where The subscript "0" refers to the point in the fitting at which the velocity pressure is to be calculated, generally taken as the location either upstream or downstream of the fitting at which the average velocity is *highest* (cross-sectional area is *smallest*). The point is usually specified on the table of loss coefficients for the fitting. In most practical applications, there are several minor losses through the duct; the total minor loss is then the sum of each individual minor loss. In terms of the minor head loss (h_{LM}), Eq. (6-60) can be rearranged as

$$\Delta P_{t,\ \text{minor loss}} = C_0 (VP)_0 = \rho g h_{LM} \tag{6-61}$$

Local loss coefficients represent losses due to flow separation, swirling eddies, and other irreversible losses within and downstream of the fitting. The ASHRAE Fundamentals Handbook should be consulted for a detailed listing of local loss coefficients for a large variety of fittings. The local loss coefficients for several common fittings are shown in Figures 6.29 and 6.30.

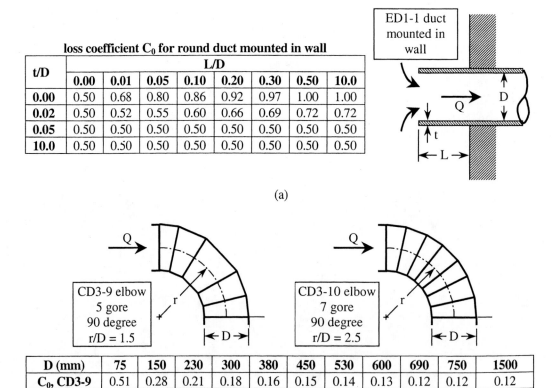

loss coefficient C_0 for round duct mounted in wall

t/D	L/D							
	0.00	**0.01**	**0.05**	**0.10**	**0.20**	**0.30**	**0.50**	**10.0**
0.00	0.50	0.68	0.80	0.86	0.92	0.97	1.00	1.00
0.02	0.50	0.52	0.55	0.60	0.66	0.69	0.72	0.72
0.05	0.50	0.50	0.50	0.50	0.50	0.50	0.50	0.50
10.0	0.50	0.50	0.50	0.50	0.50	0.50	0.50	0.50

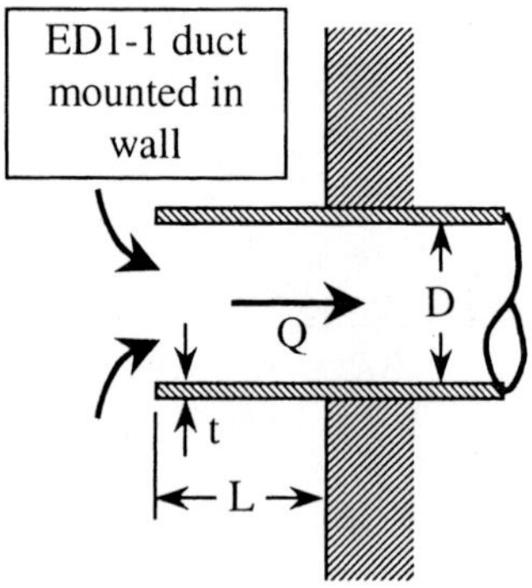

(a)

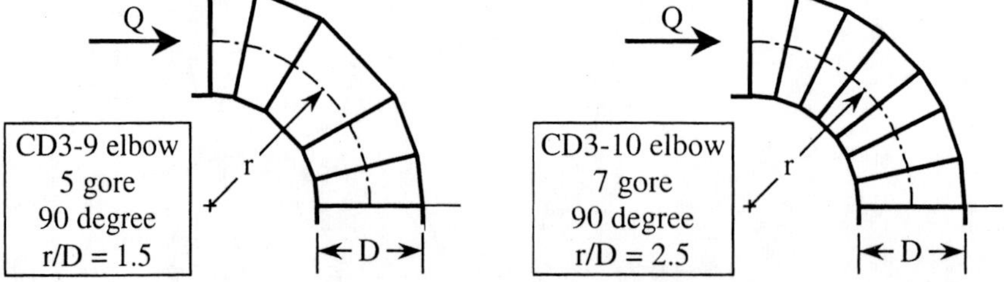

D (mm)	75	150	230	300	380	450	530	600	690	750	1500
C_0, CD3-9	0.51	0.28	0.21	0.18	0.16	0.15	0.14	0.13	0.12	0.12	0.12
C_0, CD3-10	0.16	0.12	0.10	0.08	0.07	0.06	-	-	0.05	-	0.03

(b)

Figure 6.29 Fitting loss coefficients: (a) duct mounted in a wall (also called a ***re-entrant inlet***); (b) round 90-degree elbows: 5 gore, r/D = 1.5 and 7 gore, r/D = 2.5 (abstracted from ASHRAE Fundamentals Handbook, 1997).

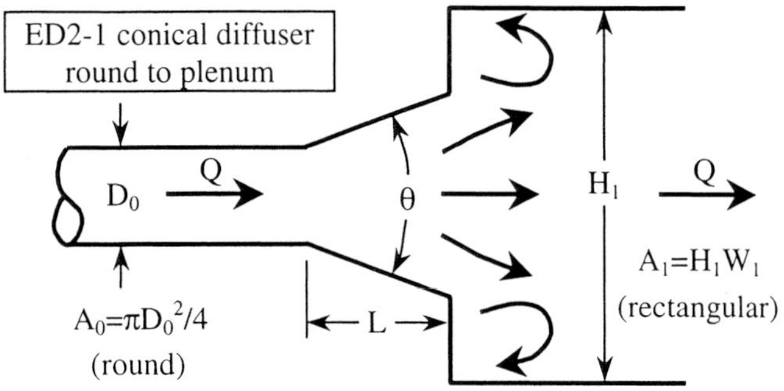

loss coefficient C_0 at listed values of A_1/A_0 and L/D_0

A_1/A_0	L/D_0								
	2.0	**3.0**	**4.0**	**5.0**	**6.0**	**8.0**	**10.0**	**12.0**	**14.0**
1.5	0.03	0.03	0.04	0.05	0.06	0.08	0.10	0.11	0.13
2.0	0.04	0.04	0.04	0.05	0.05	0.06	0.08	0.09	0.10
2.5	0.06	0.06	0.06	0.06	0.06	0.06	0.07	0.08	0.09
3.0	0.09	0.07	0.07	0.06	0.06	0.07	0.07	0.08	0.08
4.0	0.12	0.10	0.09	0.08	0.08	0.08	0.08	0.08	0.08
6.0	0.16	0.13	0.12	0.10	0.10	0.09	0.09	0.09	0.08
8.0	0.18	0.15	0.13	0.12	0.11	0.10	0.09	0.09	0.09
10.0	0.20	0.16	0.14	0.13	0.12	0.11	0.10	0.09	0.09
20.0	0.24	0.20	0.17	0.15	0.14	0.12	0.11	0.11	0.10

optimum value of angle θ (degrees) at listed values of A_1/A_0 and L/D_0

A_1/A_0	L/D_0								
	2.0	**3.0**	**4.0**	**5.0**	**6.0**	**8.0**	**10.0**	**12.0**	**14.0**
1.5	13	9	7	6	4	3	2	2	2
2.0	17	12	10	9	8	6	5	4	3
2.5	20	15	12	11	10	8	7	6	5
3.0	22	17	14	12	11	10	8	8	6
4.0	26	20	16	14	13	12	10	10	9
6.0	28	22	19	16	15	12	11	10	9
8.0	30	24	20	18	16	13	12	11	10
10.0	30	24	22	19	17	14	12	11	10
20.0	32	26	22	20	18	15	13	12	11

(c)

Figure 6.30 Fitting loss coefficients for an ED2-1 conical diffuser (abstracted from ASHRAE Fundamentals Handbook, 1997).

The total pressure loss through a converging or diverging ***branch section*** (a ***wye*** composed of a branch joining a straight main section) is more complicated owing to momentum exchange between the air streams. *The pressure loss through the main section is not necessarily equal to that through the branch section.* The total pressure loss through the main section is equal to

$$\boxed{\Delta P_{t,\ \text{main section loss}} = C_s \left(VP\right)_c = \rho g h_{LM,\ \text{main section loss}}} \tag{6-62}$$

and the total pressure loss through the branch section is

$$\boxed{\Delta P_{t,\ \text{branch section loss}} = C_b \left(VP\right)_c = \rho g h_{LM,\ \text{branch section loss}}} \tag{6-63}$$

where $(VP)_c$ is the velocity pressure at the common section, and C_s and C_b are the loss coefficients for the straight (main) and branch flow paths respectively; an example is provided in Figure 6.31. Each loss coefficient is in reference to the velocity pressure at the common section shown in Figure 6.31. For brevity, values are shown here only for $A_s/A_c = 0.2$; values for other area ratios can be obtained from the ASHRAE Fundamentals Handbook. Converging flow junctions may have either positive or negative local loss coefficients because when two streams of different velocity converge, the momentum exchange may increase or decrease the total pressure.

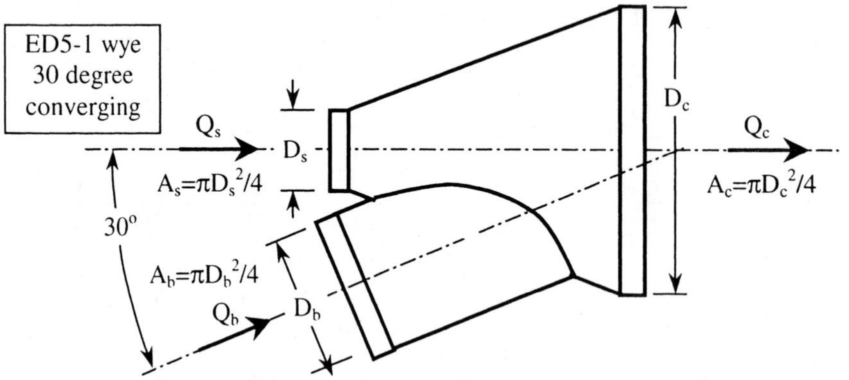

straight section loss coefficient C_s at listed values of A_b/A_c and Q_s/Q_c; $A_s/A_c = 0.2$

A_b/A_c	Q_s/Q_c								
	0.1	**0.2**	**0.3**	**0.4**	**0.5**	**0.6**	**0.7**	**0.8**	**0.9**
0.2	-16.02	-3.15	-0.80	0.04	0.45	0.69	0.86	0.99	1.10
0.3	-11.65	-1.94	-0.26	0.32	0.60	0.77	0.90	1.01	1.10
0.4	-8.56	-1.20	0.05	0.47	0.68	0.82	0.92	1.02	1.11
0.5	-6.41	-0.71	0.25	0.57	0.73	0.84	0.93	1.02	1.11
0.6	-4.85	-0.36	0.38	0.63	0.76	0.86	0.94	1.02	1.11
0.7	-3.68	-0.10	0.48	0.68	0.79	0.87	0.95	1.03	1.11
0.8	-2.77	0.10	0.56	0.71	0.81	0.88	0.95	1.03	1.11
0.9	-2.04	0.26	0.62	0.74	0.82	0.89	0.95	1.03	1.11
1.00	-1.45	0.38	0.66	0.76	0.83	0.89	0.96	1.03	1.11

branch section loss coefficient C_b at listed values of A_b/A_c and Q_b/Q_c; $A_s/A_c = 0.2$

A_b/A_c	Q_b/Q_c								
	0.1	**0.2**	**0.3**	**0.4**	**0.5**	**0.6**	**0.7**	**0.8**	**0.9**
0.2	-24.17	-3.78	-0.60	0.30	0.64	0.77	0.83	0.88	0.98
0.3	-55.88	-9.77	-2.57	-0.50	0.25	0.55	0.67	0.70	0.71
0.4	-99.93	-17.94	-5.13	-1.45	-0.11	0.42	0.62	0.68	0.68
0.5	-156.5	-28.40	-8.37	-2.62	-0.52	0.30	0.62	0.71	0.69
0.6	-225.6	-41.13	-12.30	-4.01	-0.99	0.20	0.66	0.78	0.75
0.7	-307.3	-56.14	-16.90	-5.61	-1.51	0.11	0.73	0.90	0.86
0.8	-401.4	-73.44	-22.18	-7.44	-2.08	0.04	0.84	1.06	1.01
0.9	-508.1	-93.02	-28.15	-9.49	-2.71	-0.03	0.99	1.27	1.20
1.00	-627.4	-114.9	-34.80	-11.77	-3.39	-0.08	1.18	1.52	1.43

Figure 6.31 Supply and branch loss coefficients for 30-degree converging wye with $A_s/A_c = 0.2$ (abstracted from ASHRAE Fundamentals Handbook, 1997).

6.10.3 Hood Entry Losses

In industrial ventilation, engineers use a variety of devices such as plain inlets, flanged inlets, slotted inlets, and canopy hoods, as well as partial enclosures. Head losses occur as air passes through these devices and enters the duct. The drop in total pressure caused by the minor loss associated with a ***hood entry loss*** can be expressed as

$$\boxed{\Delta P_{t,\text{ hood entry loss}} = C_o (VP)_{ref} = \rho g h_{LM,\text{ hood entry loss}}} \tag{6-64}$$

where C_0 is the local entry loss coefficient, and the subscript "ref" refers to the fact that some loss coefficients are based on the slot velocity pressure as the reference, while others are based on the duct velocity pressure. ACGIH design plates and comparable design data published in the professional literature generally state equations similar to Eq. (6-64). In the absence of such data, loss coefficients from the ASHRAE Fundamentals Handbook, such as those in Figures 6.29 through 6.31 can be used where appropriate. For example, the loss coefficient for a *flanged* inlet can be estimated from Figure 6.29a, with t/D = 0 and L/D = 0. For this case, C_0 = 0.50. The loss coefficient for a *plain* inlet can be approximated from the same figure, but with t/D = 0 and L/D = large (the maximum L/D shown is 10, but C_0 does not change beyond L/D = 5). For this case, C_0 = 1.0. Readers should note the factor of two difference between these two cases – the plain inlet wastes about twice as much flow energy due to flow separation around the sharp edges. Plain and flanged inlets are modeled and compared in Chapter 7.

The static pressure difference between the ambient air ($P = P_a$) and a reference location downstream of an entry where the velocity (U_{ref}) is known can be found from Eq. (6-53), using the fact that at a stagnant point in the room, far from the hood's influence, the velocity is negligible and thus $(VP)_a$ is approximately zero.

$$P_a + \cancel{(VP)_a} = P_{ref} + (VP)_{ref} - \cancel{\delta P_{fan}} + \rho g h_{LM,\text{ hood entry loss}}$$

where the fan pressure rise term has also been crossed out since there is no fan in the control volume between points a and ref. Using Eq. (6-64), the above simplifies to

$$\boxed{P_a - P_{ref} = (VP)_{ref} + C_o (VP)_{ref} = (1 + C_0)(VP)_{ref}} \tag{6-65}$$

The addition of unity to the local entry loss coefficient (C_0) is sometimes called the ***acceleration factor*** since it accounts for the pressure drop associated with accelerating still air into the hood.

6.10.4 Fan Inlet and Outlet Losses

If the velocity profile entering or leaving a fan is skewed because of elbows or some other fittings, ***fan inlet loss coefficients*** may be needed. No such loss coefficients are listed in this book; the reader should consult the ASHRAE Fundamentals Handbook (2001) for details if such losses must be accounted for. The drop in total pressure due to such minor head losses is sometimes expressed in a manner similar to that for major losses, Eq. (6-56),

$$\boxed{\Delta P_{t,\text{ fan inlet loss}} = \rho g h_{LM,\text{ fan inlet loss}} = f \frac{L_e}{D} (VP)} \tag{6-66}$$

where L_e is an ***equivalent length*** of straight duct which would yield the same head loss as that of the actual inlet. The literature on duct design is immense and several conventions are used to compute the pressure drop in duct components and in duct systems. Rather than seeking to standardize conventions, users should be flexible and accommodate themselves to these different conventions (and notations).

6.10.5 Design Procedure

The following is a general design procedure that should be followed when designing an exhaust duct system:

1. Select the collection hood appropriate for the industrial process to be controlled. Select the exhaust volumetric flow rate for each hood.
2. Prepare a line sketch of the duct system showing the location of each component: hood, type of fitting in each duct segment, drop out box, contaminant removal system (if needed), and exhaust fan.
3. Determine the length of each duct segment between fittings. Identify each point where a component is to be placed and designate by number or letter each point in the duct system where junctions, fittings, etc. are to be placed.
4. Select the minimum duct velocity for the contaminant to be collected.
5. Determine the loss factor for each of the components in Step 2 above.
6. Choose a diameter for each segment of duct. Compute the static pressure drop for each duct segment. Ensure that whenever two or more ducts meet at a common point (P), the pressure drop between atmospheric pressure and the static pressure at point P is the same. If the pressure drop for each duct segment is not the same, select another duct diameter for one or each of the segments and recompute the pressure drop. Repeat the process until the pressure drop in each of the duct segments is the same.
7. Select an exhaust fan of a size, speed, and horsepower (HP) such that the overall pressure drop and volumetric flow rate occur as near as possible to the optimum point on the fan performance curve. Use the ***fan laws*** (discussed below) to estimate the horsepower, pressure drop, and volumetric flow rates at off-design conditions.

The design of a system of ducts is a tedious job requiring repetitive computations and is ideally suited for computers. The requirement that the designer make certain critical decisions about duct length, transport velocity, etc., must be preserved, but computer programs can facilitate the acquisition of loss coefficients and the processing of numerical data. In 1990 there were nearly two dozen computer-aided duct design programs available commercially (Howell and Sauer, 1980 and ASHRAE, 1990) and many proprietary programs have also been developed within companies for their own use. Several duct design computer programs are available for purchase from the ACGIH.

A simple, but instructive computer program, "Duct Design Program", as well as detailed instructions for its use, can be downloaded from the web page for this book, for use on a personal computer. An example problem is also included on the web site, but is not included here.

Example 6.6 – Duct Head Loss

Given: A local ventilation system is used to remove air and contaminants produced by a dry-cleaning operation. The duct is round, and is constructed of galvanized steel with longitudinal seams and with joints every 30. inches (0.76 m). The inner diameter (ID) of the duct is D = 9.06 inches (0.230 m), and its total length is L = 44.0 ft (13.4 m). There are five CD3-9 elbows along the duct. The volume flow rate through the duct is Q = 750. SCFM (0.354 m^3/s). Literature from the hood manufacturer lists the hood entry loss coefficient as 1.3 based on duct velocity pressure.

To do: Estimate the pressure drop through the duct, in units of kilopascals (kPa) and inches of water.

Solution: From the ASHRAE Fundamentals Handbook (2001), the equivalent roughness height of this duct is 0.15 mm. The dimensionless roughness factor is thus

$$\frac{\varepsilon}{D} = \frac{0.15 \text{ mm}}{230. \text{ mm}} = 6.52\times10^{-4}$$

For air at STP, $\nu = 1.546 \times 10^{-5}$ m^2/s; the Reynolds number of flow through the duct is

$$\text{Re} = \frac{DU}{\nu} = \frac{D}{\nu}\frac{Q}{A} = \frac{D}{\nu}\frac{4Q}{\pi D^2} = \frac{4Q}{\nu\pi D} = \frac{4\left(0.354\frac{m^3}{s}\right)}{1.546\times10^{-5}\frac{m^2}{s}\pi(0.230\ m)} = 1.27\times10^5$$

From Eq. (6-57) at this Re and roughness factor, the friction factor is f = 0.0204 (note that some iteration is required to solve for f). From Figure 6.29, each elbow has a local (minor) loss coefficient of $C_0 = 0.21$; the reference velocity pressure $(VP)_0$ for the elbows is everywhere equal to the duct velocity pressure VP since the duct diameter is constant. For air at STP, $\rho = 1.184\ kg/m^3$; the duct velocity pressure can be calculated from Eq. (6-51),

$$VP = \frac{1}{2}\rho U^2 = \frac{1}{2}\rho\left(\frac{4Q}{\pi D^2}\right)^2 = \frac{1}{2}\left(1.184\frac{kg}{m^3}\right)\left(\frac{4\left(0.354\frac{m^3}{s}\right)}{\pi(0.230\ m)^2}\right)^2\left(\frac{N\ s^2}{kg\ m}\right) = 42.9\frac{N}{m^2}$$

Finally, the pressure drop through the duct is the sum of major and minor losses, from Eqs. (6-56) and (6-61) respectively,

$$\Delta P_t = \Delta P_{t,\ major\ loss} + \Delta P_{t\ minor\ loss} = f\frac{L}{D}(VP) + \sum_{minor\ losses} C_0(VP)_0 = (VP)\left(f\frac{L}{D} + \sum C_0\right)$$

$$= 42.9\frac{N}{m^2}\left(0.0204\frac{13.4\ m}{0.230\ m} + 1.3 + 5(0.21)\right)\left(\frac{kPa\ m^2}{1000\ N}\right) = 0.152\ kPa$$

The pressure drop through the duct can be converted to inches of water column height as follows:

$$h_{water} = \frac{\Delta P_t}{\rho_{water}g} = \frac{0.152\ kPa}{998.\frac{kg}{m^3}\left(9.81\frac{m}{s^2}\right)}\left(\frac{1000\ N}{kPa\ m^2}\right)\left(\frac{kg\ m}{N\ s^2}\right)\left(\frac{in}{0.0254\ m}\right) = 0.61\ in$$

Discussion: The final answer is given to two significant digits; more than this cannot be expected for these kinds of problems. If there are expansions, wyes, etc. in the duct, such that duct diameter is not constant, the major and minor losses must be computed separately for each section of constant diameter, being careful to use the appropriate reference velocity pressure for each section of duct. Duct design computer programs are ideally suited for such laborious calculations.

6.11 Fan Performance and Selection

Once the volumetric flow rate (Q) and required fan pressure rise (δP_{fan}) in the duct system have been calculated, it is necessary to select the fan ("air mover") to operate the system. If the ventilation system is being changed and requires a larger volumetric flow rate and/or pressure rise, the engineer will need to estimate the new fan speed (ω) or input power (HP) of the existing fan. If the existing fan cannot accommodate these new conditions, a new and larger fan will need to be installed.

6.11.1 Fan Characteristics

Fan characteristics are presented in either tabular form or graphical form (hereafter called "***fan performance curves***" or simply "***fan curves***") showing the pressure rise across the fan (δP_{fan}), the fan efficiency (η), and the required input power to the fan (HP) as functions of volumetric flow rate (Q, usually corrected to STP using units such as SCFM) and rotational velocity (ω). Figure 6.32 shows examples of fan curves for three classes of centrifugal fans and for one type of axial fan.

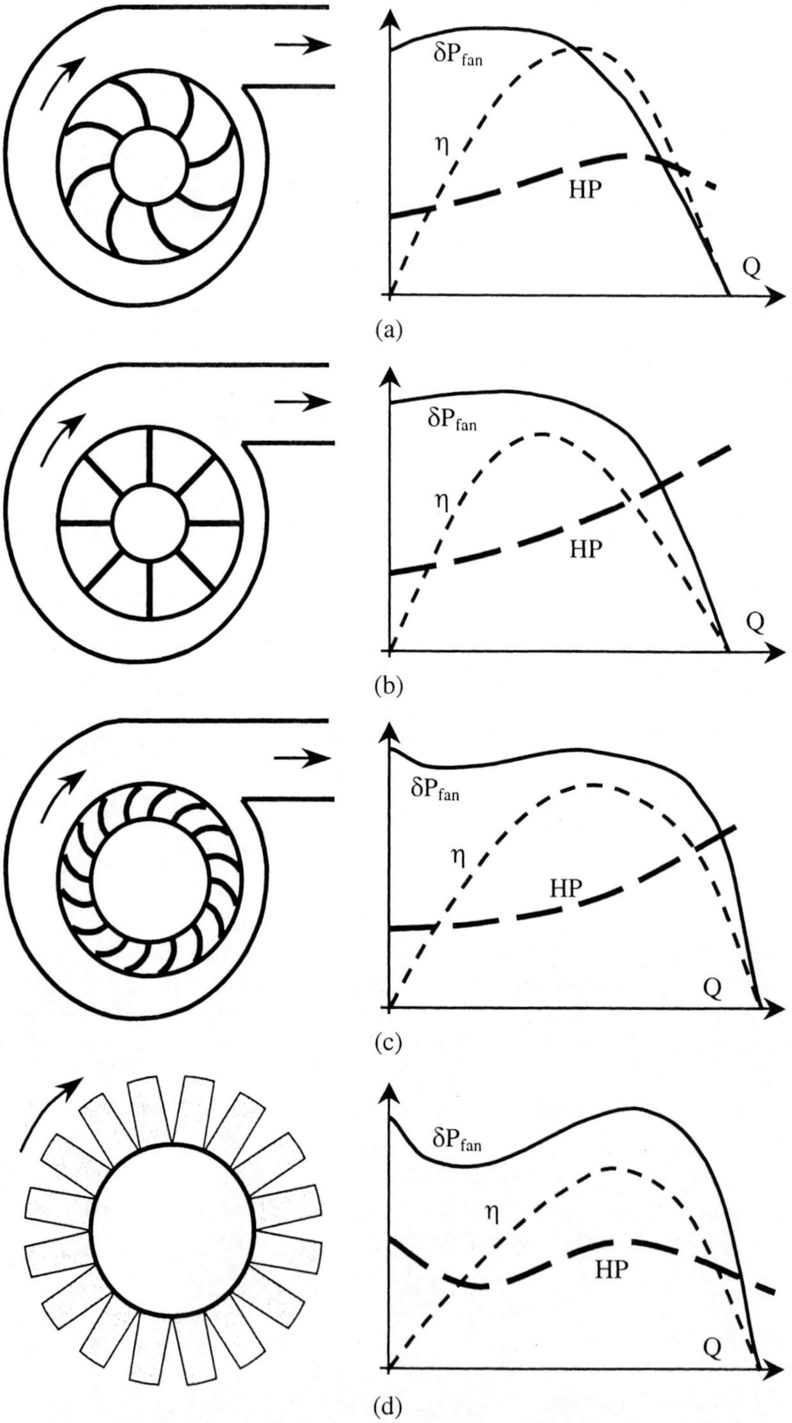

Figure 6.32 Types of fans and typical fan performance curves: (a) backward-inclined centrifugal fan, (b) straight-bladed centrifugal fan, (c) forward-inclined centrifugal fan, and (d) vaneaxial fan (rotors shown; stators not shown) – flow into page.

Fan characteristics are typically provided by the fan manufacturer. Some vendors may use the actual volumetric flow rate (ACFM) and other vendors leave users in the dark and simply state volumetric flow rate in CFM; fortunately the difference between actual and corrected volumetric flow rates is usually small for APCS applications. The pressure rise across the fan (δP_{fan}) is often given in units of ***equivalent water head*** (h_{water}), even though the fluid through the fan is air, not water,

$$\boxed{\delta P_{fan} = \rho_{water} g h_{water}} \tag{6-67}$$

where h_{water} is the height of a stagnant water column which yields a hydrostatic pressure equal to δP_{fan} as in Eq. (6-67).

The maximum volumetric flow rate of a fan or blower occurs when the pressure rise is zero, $\delta P_{fan} = 0$, and is called the fan's ***free delivery***. The free delivery condition is achieved when there is no flow restriction at the fan inlet or outlet. At this operating point, the fan's efficiency is zero because the fan is doing no useful work. At the other extreme, the ***shutoff pressure rise*** is the pressure rise established when the volumetric flow rate is zero, Q = 0, and is achieved when the outlet port of the fan is blocked off. Under these conditions, the fan's efficiency is also zero, again because the fan is doing no useful work. Between these two extremes, from shutoff to free delivery, the fan's pressure rise (δP_{fan}) may increase from its shutoff value somewhat as Q increases, but then decreases to zero as the volumetric flow rate increases to its free delivery value. The fan's efficiency reaches its maximum value somewhere between the shutoff condition and the free delivery condition; this operating point is appropriately called the ***best efficiency point*** (BEP).

6.11.2 Types of Fans

There are two basic types of fans: axial and centrifugal, which describe how air passes through the fan. Within each type there are several basic subdivisions and in some cases, esoteric subdivisions. ***Axial fans*** move air and achieve a pressure rise (δP_{fan}) by passing air in the axial direction through a series of rotating (often followed by stationary) blades. ***Centrifugal fans*** move air and achieve a pressure rise by passing air in the *radial* direction through a series of rotating blades. Readers are encouraged to study the performance curves in Figure 6.32 while reading the descriptions below.

Centrifugal fans: Centrifugal fans are generally used when the required pressure rise is large. The pressure rise (δP_{fan}), efficiency (η), and power characteristics of centrifugal fans depend on blade geometry and volumetric flow rate (Q). There are three classes of centrifugal fans that warrant discussion, based on blade geometry: backward-inclined blades, radial blades, and forward-inclined blades. ***Backward-inclined blades*** yield the highest efficiency of the three classes of centrifugal fans because air flows into and out of the blade passages with the least amount of turning. Sometimes the blades are airfoil shaped, yielding similar performance but even higher efficiency. The pressure rise is intermediate between the other two types of centrifugal fans, and backward-inclined centrifugal fans achieve their BEP at approximately 50 to 60% of free delivery. Centrifugal fans with ***radial blades*** (also called ***straight blades***) have the simplest geometry and produce the largest pressure rise of the three for a wide range of volumetric flow rates, but the pressure rise decreases rapidly after the point of maximum efficiency. Straight-bladed centrifugal fans achieve their BEP at approximately 40 to 50% of free delivery. ***Forward-inclined blades*** produce a pressure rise that is nearly constant, albeit lower than that of radial blades, over a wide range of volumetric flow rates. Forward-inclined centrifugal fans have about the same maximum efficiency as do straight-bladed fans, and achieve their BEP at approximately 50% of free delivery. Radial and backward-inclined centrifugal fans are preferred for applications where one needs to provide volumetric flow rate and pressure rise within a narrow range of values. If a wider range of volumetric flow rates and/or pressure rises are desired, the performance of radial fans and backward inclined fans may not be able to satisfy the new requirements; these types of fan are less forgiving (less robust). The performance of forward inclined fans is more forgiving and accommodates a wider variation, at the cost of lower efficiency and less pressure rise per unit of input power. If a fan is needed to produce large pressure rise over a wide range of volumetric flow rates, the forward-inclined fan is attractive.

Axial fans: The function of axial fans is to move a large amount of air with a minimum amount of power. Consequently the pressure rise is small. The pressure rise can be increased by encasing the fan in a cowling and using straightening vanes at the fan inlet and/or outlet; such fans are called ***vaneaxial fans***. Examples of axial fans are room-air exhaust fans, ubiquitous domestic window fans, ceiling circulating fans, "flat" fans, etc. The efficiency of axial fans is fairly low, and the best efficiency point (BEP) occurs at approximately 60% of free delivery.

A summary of fan performance characteristics is provided in Table 6.7 for four types of fans.

Table 6.7 Fan performance characteristics.

centrifugal fans

backward-inclined blades

- highest efficiency of all centrifugal fans
- 9 to 16 blades curved away from the direction of rotation, deep blades produce an efficient expansion of air as it passes radially outward between the blades
- air leaves the impeller with a velocity less than the blade tip speed
- maximum efficiency requires a small clearance between blades and inlet
- efficiency improves if blades have an airfoil contour
- power reaches maximum near peak efficiency and tapers off near free delivery
- typically used for low pressure rise, large volumetric flow rates in HVAC

radial blades

- higher pressure characteristics than backward-inclined blades
- power rises continually to free delivery
- scroll may be somewhat narrower than that of backward-inclined blades
- dimensions are not as critical as backward-inclined blades
- typically used for material handling; not commonly used for HVAC

forward-inclined blades

- flatter pressure curve and lower efficiency
- power increases continually to free-delivery
- scroll is similar to the other centrifugal fans but clearance is not as critical as backward-inclined blades
- usually smaller blades than backward-inclined blades, but more blades
- typically used for low pressure rise, large volumetric flow rates in HVAC

axial fans

- low efficiency, low pressure rise applications in which large volumes of air are to be moved
- 4 or more blades, each having an airfoil cross section
- additional pressure rise and efficiency can be obtained by close clearances, upstream and downstream guide vanes, and a cylindrical extension downwind of the impeller
- maximum efficiency nearer to free delivery
- typically used for make-up air applications, general ventilation, drying ovens, spray booths

6.11.3 Fan Laws

Application of the ***Buckingham Pi Theorem*** of dimensional analysis (Fox and McDonald, 1985) to the relationship between pressure rise (δP_{fan}) and fan properties such as volume flow rate (Q), blade diameter (D), and rotational velocity (ω), along with fluid properties density (ρ) and dynamic viscosity (μ), results in the following relationship involving dimensionless parameters:

$$\frac{\delta P_{fan}}{\rho\omega^2 D^2} = \text{function of}\left(\frac{Q}{\omega D^3}, \frac{\rho\omega D^2}{\mu}\right); \quad \text{i.e. } \Pi_1 = \text{function of}\left(\Pi_3, \Pi_4\right) \tag{6-68}$$

A similar analysis with net input power (HP) results in

$$\frac{HP}{\rho\omega^3 D^5} = \text{function of}\left(\frac{Q}{\omega D^3}, \frac{\rho\omega D^2}{\mu}\right); \quad \text{i.e. } \Pi_2 = \text{function of}\left(\Pi_3, \Pi_4\right) \tag{6-69}$$

where the nondimensional parameters (the Πs) are defined and labeled as follows:

$$\Pi_1 = C_H = \frac{\delta P_{fan}}{\rho\omega^2 D^2} = \text{head coefficient} \tag{6-70}$$

$$\Pi_2 = C_P = \frac{HP}{\rho\omega^3 D^5} = \text{power coefficient} \tag{6-71}$$

$$\Pi_3 = C_Q = \frac{Q}{\omega D^3} = \text{capacity coefficient} \tag{6-72}$$

$$\Pi_4 = Re = \frac{\rho\omega D^2}{\mu} = \text{Reynolds number} \tag{6-73}$$

Other variables, such as roughness height of blades or other surfaces inside the fan, gap thickness between blade tips and fan housing, etc. can be added to the dimensional analysis if necessary. Fortunately, these variables typically are only of minor importance and are not considered here. Relationships derived via dimensional analysis, such as those given by Eq. (6-68) and (6-69), can be interpreted as follows: If two fans, A and B, are geometrically similar (fan A is geometrically proportional to fan B, even if they are of different sizes), and if the *independent* Πs are equal to each other (in this case if $\Pi_{3,A} = \Pi_{3,B}$ and $\Pi_{4,A} = \Pi_{4,B}$), then the *dependent* Πs are guaranteed to also be equal to each other, in this case $\Pi_{1,A} = \Pi_{1,B}$ from Eq. (6-68), and $\Pi_{2,A} = \Pi_{2,B}$ from Eq. (6-69). If such conditions are established, the two fans are said to be ***dynamically similar***. The requirement of equality of the independent dimensionless parameters can be relaxed somewhat for parameter Π_4, since it is a Reynolds number. If the Reynolds number of fan A and that of fan B exceed several thousand, as surely they will for fans of interest to contaminant control, turbulent flow conditions exist. It turns out that for turbulent flow, even if the values of $\Pi_{4,A}$ and $\Pi_{4,B}$ are not equal, dynamic similarity between the two fans is still a good approximation. This fortunate condition is called ***Reynolds number independence***, and is assumed in the material that follows. Under these conditions, Eq. (6-68) establishes that curves of δP_{fan} versus Q for geometrically similar fans can be reduced to a single curve of Π_1 versus Π_3. Equation (6-68) does not predict the *shape* of the curve, but once a curve of δP_{fan} versus Q is obtained for a particular fan, it can be generalized for geometrically similar fans that are of different diameter (D), operate at a different speed (ω) and flow rate (Q), and even operate with a fluid of different density (ρ) and viscosity (μ). A similar conclusion can be made for a curve of input power (HP) versus flow rate (Q) from Eq. (6-69).

A fifth dimensionless parameter, called the ***specific speed,*** can be formed by combination of parameters Π_3 and Π_1:

$$\Pi_5 = N_S = \frac{\Pi_3^{1/2}}{\Pi_1^{3/4}} = \frac{\left(\frac{Q}{\omega D^3}\right)^{1/2}}{\left(\frac{\delta P_{fan}}{\rho \omega^2 D^2}\right)^{3/4}} = \frac{\omega Q^{1/2}}{\left(\frac{\delta P_{fan}}{\rho}\right)^{3/4}} = \text{specific speed} \tag{6-74}$$

Readers should note that some authors define specific speed only at the best efficiency point of the fan; the result is a single dimensionless number which characterizes the fan. This convention is not followed here. Rather, at any operating condition, specific speed is given by Eq. (6-74), and defines a unique physical state of flow through the fan. When the values of the specific speed are equal, the flow through geometrically similar fans is aerodynamically similar and the physical state is dynamically similar regardless of fan diameter, rotational speed, etc. A curve of constant Π_5 appears as an upward sloping curve on a graph of δP_{fan} versus Q, as seen in Figure 6.33. The plot in this figure superimposes values of the constant Π_5 on the fan curves for two geometrically similar fans. For example, consider the points of intersection, A and B. At these two points, $\Pi_{5,A} = \Pi_{5,B}$, and dynamic similarity is achieved between the two fans. In such a case, operating points A and B are said to be ***homologous***. The following relationships can then be established for two homologous states A and B:

$$\frac{Q_A}{Q_B} = \left(\frac{D_A}{D_B}\right)^3 \left(\frac{\omega_A}{\omega_B}\right) \tag{6-75}$$

$$\frac{(\delta P_{fan})_A}{(\delta P_{fan})_B} = \left(\frac{\rho_A}{\rho_B}\right)\left(\frac{\omega_A D_A}{\omega_B D_B}\right)^2 \tag{6-76}$$

$$\frac{(HP)_A}{(HP)_B} = \left(\frac{\rho_A}{\rho_B}\right)\left(\frac{\omega_A}{\omega_B}\right)^3 \left(\frac{D_A}{D_B}\right)^5 \tag{6-77}$$

Similar relationships can be written for any pair of dynamically similar operating points on the curves, such as points C and D in Figure 6.33. If the two operating points are not dynamically similar (e.g. points A and C, in which $\Pi_{5,A} \neq \Pi_{5,C}$), no such relationships can be written. The above relationships are called ***fan laws***, ***scaling laws***, or ***affinity laws***, and relate flow through geometrically similar fans. These fan laws are extremely useful for design. Specifically, knowing δP_{fan}, Q, HP, ρ, and

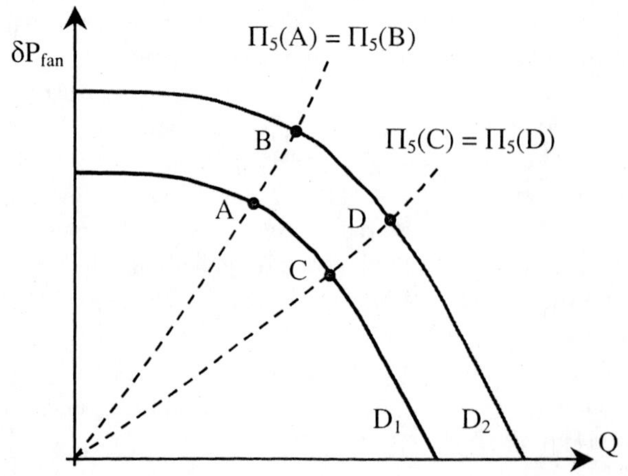

Figure 6.33 Typical fan curves for two geometrically similar fans, showing two pairs of homologous points (A and B) and (C and D).

ω at one point on a fan curve enables one to predict values of these variables at a dynamically similar operating point along a different fan curve, provided that Π_5 is the same for both operating points.

Finally, the fan efficiency (η) is itself a dimensionless parameter. As might be expected, at two homologous operating points, i.e. at two points on the fan curve where dynamic similarity is achieved, the efficiency of fan A equals that of fan B. In general, efficiency is always defined as the ratio of useful output to required input. In the case of fans, fan efficiency is defined as the useful output power (sometimes called ***water horsepower***, even if the fluid is not water) divided by the required input power (sometimes called ***brake horsepower***),

$$\boxed{\eta = \frac{Q\delta P}{HP}} \tag{6-78}$$

It can be shown that η is a combination of three of the other dimensionless parameters, namely

$$\boxed{\eta = \frac{\Pi_1 \Pi_3}{\Pi_2} = \frac{C_H C_Q}{C_P}} \tag{6-79}$$

Example 6.7 - Performance of a Fan under New Operating Conditions

Given: A radial centrifugal fan with an 18-inch diameter blade (D = 18. in) has performance characteristics at 600 RPM, listed in the table below, supplied by the fan manufacturer. The air is at STP. The fan is an element in an industrial ventilation system, and it has been determined that changes need to be made in the design of the duct system. In particular, the redesigned duct system requires a volumetric flow rate of 12,000 SCFM at a pressure rise of 2.0 inches of water. The fan motor can be re-geared to achieve 750 RPM if necessary. A second, geometrically similar centrifugal fan with a 20-inch diameter blade is also available, and can be mated to the existing motor running at either 600 or 750 RPM.

Q (SCFM)	**δP_{fan} (inches of water)**	**η (%)**
0	1.00	0
2000	1.18	30
4000	1.25	47
6000	1.20	65
8000	1.05	73
10,000	0.85	74
12,000	0.60	61
14,000	0.25	35
15,500	0	0

To do: Determine whether the fan speed or input power can be changed to satisfy the new design, or whether a new, larger fan is needed. Specifically, provide the following:

(a) Plot δP_{fan} and η versus Q for the 18-inch fan at 600 RPM, and determine the shutoff pressure (in inches of water) and free delivery (in SCFM). What are the pressure rise (in inches of water) and efficiency at 9,000 SCFM and 600 RPM?

(b) With the 18-inch fan operating at 600 RPM and at its best efficiency point (BEP), what fan horsepower (HP) is needed?

(c) Using the data for the 18-inch fan, generate the normalized performance curve for a class of geometrically similar fans, i.e. express the head coefficient as a function of the capacity coefficient. Ignore the dependence on Reynolds number (Reynolds number independence is assumed).

(d) Using the fan laws, estimate and plot δP_{fan} (in inches of water) versus Q (in SCFM) for the 18-inch fan operating at a rotational speed (ω) equal to 750 RPM. In addition, estimate and plot δP_{fan} versus Q for the geometrically similar 20-inch fan operating at 750 RPM.

(e) Can the 18-inch fan provide the required flow rate and pressure rise at 600 or 750 RPM? Could the fan speed be increased to achieve this performance? Can the 20-inch fan at 750 RPM meet the requirements?

(f) Rather than run the 18-inch fan at a higher speed and power, or replace it with the 20-inch fan, the company is also considering the option of purchasing a new geometrically similar fan which would operate at its maximum efficiency. What diameter and RPM would you recommend?

Solution: Assume that the air density (ρ) is constant. The authors used Excel to create third-order polynomial curve fits and to generate plots. A copy of the spreadsheet file is available on the book's web site.

(a) The fan performance curves, δP_{fan} and η as functions of Q for the 18-inch fan at 600 RPM, are shown in Figures E6.7a and b respectively. The authors used third-order polynomial curve fits; higher-order polynomial fits would pass more closely through the data, but the higher the order, the more wiggly the curve. From the original data, it is seen that at 600 RPM the free delivery volumetric flow rate is 15,500 SCFM and the shutoff pressure is 1.0 inches of water. By interpolation with the polynomial fit, the pressure rise and efficiency at 600 RPM, 9000 SCFM are 0.967 inches of water and 75.5% respectively. These values can also be read (less accurately) from the plot.

(b) From the curve fits (or the fan performance plots), the best efficiency point at 600 RPM is at a flow rate of approximately 9,000 SCFM and a pressure rise of 0.967 inches of water. For convenience with the units later on, this pressure rise is first converted to lbf/ft^2 using the pressure conversion that

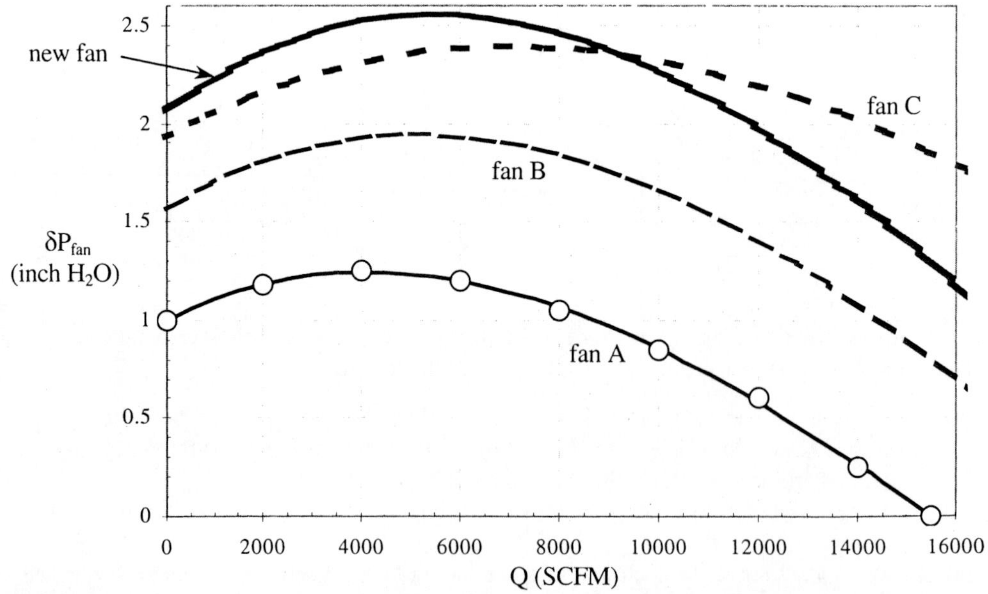

Figure E6.7a Performance curves (pressure rise versus volumetric flow rate) for fan A (original 18-inch fan at 600 RPM), fan B (original 18-inch fan at 750 RPM), fan C (geometrically similar 20-inch fan at 750 RPM), and a proposed new geometrically similar fan..

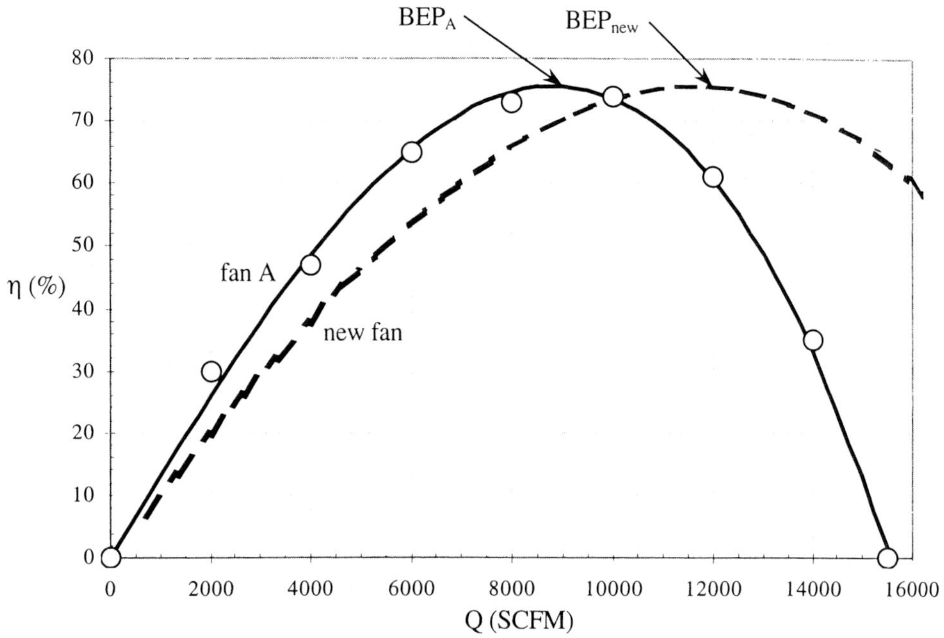

Figure E6.7b Efficiency as a function of volumetric flow rate for fan A (original 18-inch fan at 600 RPM) and the new fan (17.3-inch fan at 896. RPM). The best efficiency point (BEP) is marked for both cases.

one atmosphere of pressure (14.7 lbf/in^2) is equivalent to a 33.91 ft column height of water. At the BEP,

$$\delta P_{fan} = 0.967 \text{ in } H_2O \frac{14.7 \frac{lbf}{in^2}}{(33.91 \text{ ft } H_2O)\left(\frac{12 \text{ in } H_2O}{\text{ft } H_2O}\right)} \left(\frac{12 \text{ in}}{\text{ft}}\right)^2 = 5.03 \frac{lbf}{ft^2}$$

The input power (from an electrical motor) to operate the fan at 9,000 SCFM and 0.967 inches of water can now be found from Eq. (6-78),

$$HP = \frac{Q\delta P_{fan}}{\eta} = \frac{9{,}000 \frac{ft^3}{min} 5.03 \frac{lbf}{ft^2}}{0.755} \left(\frac{\text{horsepower} \cdot s}{550 \text{ ft} \cdot lbf}\right)\left(\frac{min}{60 \text{ s}}\right) = 1.82 \text{ horsepower}$$

(c) From the data given for the 18-inch fan at 600 RPM, the dimensionless groups Π_1 and Π_3 can be computed using Equations (6-70) and (6-72). The calculations were performed in Excel, and the results are shown in Figure E6.7c. An example calculation at 9,000 SCFM is shown below.

$$\Pi_1 = \frac{\delta P_{fan}}{\rho \omega^2 D^2} = \frac{5.03 \frac{lbf}{ft^2}\left(32.174 \frac{lbm \cdot ft}{s^2 lbf}\right)\left(\frac{60 \text{ s}}{min}\right)^2}{0.0739 \frac{lbm}{ft^3}\left(600 \frac{rot}{min}\right)^2 \left(\frac{2\pi \text{ rad}}{rot}\right)^2 \left(\frac{18.}{12} ft\right)^2} = 0.247$$

and

$$\Pi_3 = \frac{Q}{\omega D^3} = \frac{9000\dfrac{\text{ft}^3}{\text{min}}}{\left(600\dfrac{\text{rot}}{\text{min}}\right)\left(\dfrac{2\pi\ \text{rad}}{\text{rot}}\right)\left(\dfrac{18.}{12}\text{ft}\right)^3} = 0.707$$

Readers should note the conversion from rotations to radians; there are 2π radians per rotation. Radian has units of 1 (dimensionless), and it is necessary mathematically to use radians rather than rotations. Efficiency, η, is also plotted as a function of capacity coefficient in Figure E6.7c.

(d) Letting A represent a point on the performance curve of the18-inch fan in which (Q_A, $\delta P_{fan,A}$) have particular values taken from the data given at 600 RPM, the fan laws, Eqs. (6-75) through (6-77), can be used to predict (Q_B, $\delta P_{fan,B}$) at homologous point B on a companion performance curve for the 18-inch fan at 750 RPM. Sample calculations are shown below for the BEP conditions of fan A.

$$\delta P_{fan,B} = \delta P_{fan,A}\left(\frac{\rho_B}{\rho_A}\right)\left(\frac{\omega_B}{\omega_A}\frac{D_B}{D_A}\right)^2 = 0.967\ \text{in}\ H_2O\left(\frac{0.0739\dfrac{\text{lbm}}{\text{ft}^3}}{0.0739\dfrac{\text{lbm}}{\text{ft}^3}}\right)\left(\frac{750\dfrac{\text{rot}}{\text{min}}}{600\dfrac{\text{rot}}{\text{min}}}\frac{18\ \text{in}}{18\ \text{in}}\right)^2 = 1.51\ \text{in}\ H_2O$$

$$Q_B = Q_A\left(\frac{D_B}{D_A}\right)^3\frac{\omega_B}{\omega_A} = 9000\ \text{SCFM}\left(\frac{18\ \text{in}}{18\ \text{in}}\right)^3\frac{750\dfrac{\text{rot}}{\text{min}}}{600\dfrac{\text{rot}}{\text{min}}} = 11250\ \text{SCFM}$$

Note that although the density of air for fans A and B is the same, the above equations would still be valid if the densities were different. The process is repeated for all the curve-fitted data; the results are shown in Figure E6.7a, as calculated in Excel. The same process can be used to estimate the performance of a 20-inch fan run at 750 RPM, which will be called fan C. The results are also shown in Figure E6.7a. All three fan performance curves are similar in shape, but the maximum pressure rise is different for each case, and occurs at different volumetric flow rates.

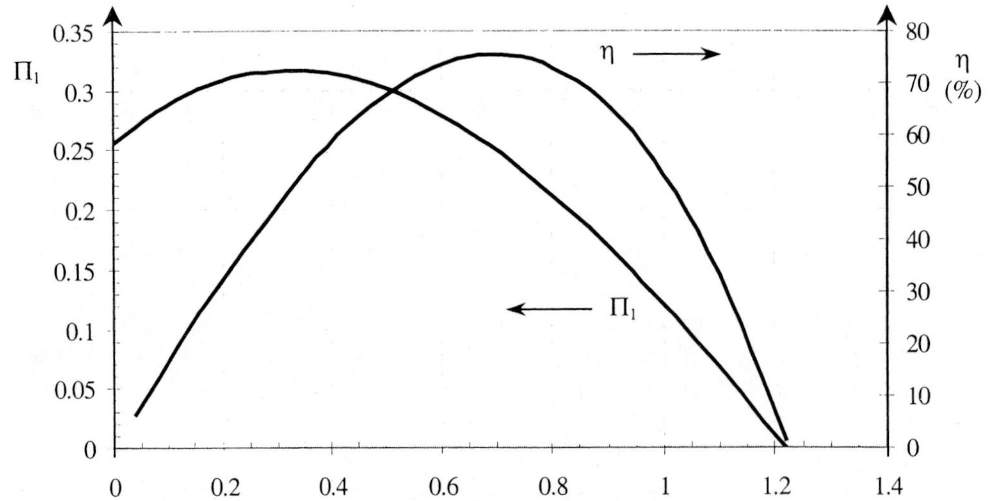

Figure E6.7c Normalized fan performance curves (head coefficient, Π_1, and efficiency, η, versus capacity coefficient, Π_3) as calculated from original 18-inch fan at 600 RPM, but valid for all four dynamically similar fans.

(e) Examination of Figure E6.7a shows that the 18-inch fan cannot achieve 12000 SCFM at 2.0 inches of water at either 600 or 750 RPM. If the bearings can handle it, even higher values of fan speed (ω) can be selected and Part (d) above repeated until the required performance is achieved. This is left as an exercise for the reader. Figure E6.7a shows that the 20-inch fan running at 750 RPM *can* meet the requirements (at 12,000 SCFM, the pressure rise is about 2.20 inches of water). In fact, it exceeds the requirements; the flow can be cut back by installing a damper in the duct. Alternatively, the fan speed can be lowered, if possible. The latter choice would save electricity.

(f) The BEP (75.5%) for the existing 18-inch rotor at ω = 600 RPM occurs at Q = 9,000 SCFM and δP_{fan} = 0.967 inches of water. If the flow through the new fan is to be aerodynamically similar, the maximum efficiency will occur at a homologous point, when the dimensionless parameters match. It is simplest to match specific speed (Π_5), Eq. (6-74), to find the value of ω_{new} needed to achieve this flow at maximum efficiency,

$$\Pi_{5,\,new} = \frac{\omega_{new} Q_{new}^{\ 1/2}}{\left(\dfrac{\delta P_{fan,new}}{\rho_{new}}\right)^{3/4}} = \Pi_{5,\,existing} = \frac{\omega_{existing} Q_{existing}^{\ 1/2}}{\left(\dfrac{\delta P_{fan,existing}}{\rho_{existing}}\right)^{3/4}}$$

Since the air density is the same for the new and existing fan, the above can be solved for the rotational speed of the new fan:

$$\omega_{new} = \omega_{existing}\left(\frac{Q_{existing}}{Q_{new}}\right)^{1/2}\left(\frac{(\delta P)_{fan,new}}{(\delta P)_{fan,existing}}\right)^{3/4} = 600\ \text{RPM}\left(\frac{9{,}000\ \text{ACFM}}{12{,}000\ \text{ACFM}}\right)^{1/2}\left(\frac{2.0\ \text{in}\ H_2O}{0.967\ \text{in}\ H_2O}\right)^{3/4}$$
$$= 896.\ \text{RPM}$$

The fan diameter (D) needed to achieve the required performance can be found in similar fashion using Eq. (6-75):

$$D_{new} = D_{existing}\left(\frac{Q_{new}}{Q_{existing}}\frac{\omega_{existing}}{\omega_{new}}\right)^{1/3} = 18.\ \text{in}\left(\frac{12{,}000}{9{,}000}\frac{600.}{896.}\right)^{1/3} = 17.3\ \text{in}$$

The reader can verify that the same values for ω_{new} and D_{new} can alternatively be obtained by simultaneous solution of Eqs. (6-75) and (6-76) at the two homologous points. Thus, a new 17.3-inch fan running at approximately 900 RPM would meet the requirements, and would operate near its best efficiency point. The fan performance curve is plotted in Figure E6.7a along with the performance curves of the other fans. Indeed, it is seen that at Q = 12,000 SCFM, the pressure rise is 2.0 inches of water. Note that since all four fans are dynamically similar, the four fan performance curves are similar in shape, but stretched in both the horizontal and vertical directions because of the differences in fan rotational speed and blade diameter. The required power to run this new fan can be obtained from Eq. (6-77), where again the density is constant:

$$(HP)_{new} = (HP)_{existing}\left(\frac{\omega_{new}}{\omega_{existing}}\right)^3\left(\frac{D_{new}}{D_{existing}}\right)^5 = 1.82\ \text{horsepower}\left(\frac{896.}{600.}\right)^3\left(\frac{17.3}{18.}\right)^5$$
$$= 4.97\ \text{horsepower}$$

which can be rounded to 5.0 horsepower. Alternatively, the required power can be found from Eq. (6-78),

$$(HP)_{new} = \frac{Q_{new}\left(\delta P_{fan,new}\right)}{\eta_{new}} = \frac{12{,}000\dfrac{\text{ft}^3}{\text{min}}\left(10.4\dfrac{\text{lbf}}{\text{ft}^2}\right)}{0.755}\left(\frac{\text{horsepower}\cdot\text{s}}{550\ \text{ft}\cdot\text{lbf}}\right)\left(\frac{\text{min}}{60\ \text{s}}\right) = 5.01\ \text{horsepower}$$

where the pressure rise has been converted to lbf/ft^2. Again, the final result should be rounded to 5.0 horsepower. The small difference between the two power calculations is due to round-off error, and should not be of concern since one cannot realistically expect more than two significant digits of accuracy from these kinds of calculations. As a final check, the power and pressure rise at various values of Q are computed in the spreadsheet for the new fan. From these, the efficiency of the new fan is calculated and plotted in Figure E6.7b. Indeed, the best efficiency point of the new fan occurs at Q = 12,000 SCFM; the efficiency curve is stretched to the right because of the new diameter and rotational speed, but the maximum efficiency is 75.5%, the same as that of the original fan.

Discussion: The horsepower calculated above represents the *shaft* power driving the fan. The *electrical* power required to drive the motor (which is in turn driving the fan) would be even higher; its value depends on the efficiency of the motor. Assuming a conservative value of 80.%, the electrical motor draws approximately 6.3 HP of electrical power (4.7 kW). As a final comment, the normalized fan performance curve of Figure E6.7c was prepared based on data from the original fan. However, an identical curve would be generated by using data from *any* dynamically similar fan (such as fans B, C, or the new fan discussed above). Verification of this is left as an exercise for the reader.

6.11.4 Fans in Series and Parallel

When faced with the need to increase volumetric flow rate or pressure rise, a designer might be tempted to add an additional fan in series or in parallel with the original fan. While this arrangement is acceptable for certain applications, arranging fans in series or parallel is generally not recommended for indoor air quality control. A better course of action is to increase the fan speed and/or input power (larger electric motor), and if that is not sufficient, to replace the fan with a larger one.

The logic for this decision can be seen from the fan curves, realizing that the pressure rise and volumetric flow rates are related. Arranging fans in series creates problems because the volumetric flow rate through each fan is the same but the overall pressure rise is equal to the pressure rise of one fan plus that of the other. If the fans are not sized properly, one fan could inadvertently reduce the total volumetric flow rate. Arranging fans in parallel creates problems because the overall pressure rise must be the same. If the fans are not sized properly, one fan could inadvertently reduce the overall pressure rise.

6.11.4 Matching Fan to Ventilation System Requirements

One rarely has the good fortune of finding an available fan with *exactly* the required pressure rise at the required volumetric flow rate. Even if such a fan is found, it is unlikely that the operating point would turn out to be the BEP of the fan. The far more common situation is that an engineer selects a fan which is somewhat heftier than actually required. The volume flow rate through the ventilation ductwork would then be larger than needed; in such a case it is wise to install a damper in the line, so that the flow rate can be decreased as necessary. It is straightforward to match a fan's performance curve to the requirements of the ventilation system. Specifically, if the control volume from point 1 to point 2 includes the fan, Eq. (6-53) can be solved for the required pressure rise across the fan,

$$\boxed{\delta P_{fan} = [P_2 - P_1] + [(VP)_2 - (VP)_1] + \rho g h_{LT}} \qquad (6\text{-}80)$$

Eq. (6-80) agrees with intuition, i.e. the pressure drop across the fan does three things:

- it increases the static pressure from point 1 to point 2 (first term in square brackets)
- it increases the velocity pressure (kinetic energy) from point 1 to point 2 (second term in square brackets)
- it overcomes irreversible losses in the duct

The first two terms in the above list can be either positive or negative, while the third term is always positive, as discussed previously. If elevation effects are included, an additional term would appear on the right hand side of Eq. (6-80), namely $\rho g(z_2 - z_1)$, which represents the pressure rise across the fan needed to raise the elevation (potential energy) of the fluid. This term can be positive or negative. In civil engineering problems in which the fluid is water, this elevation term cannot be neglected, but in air pollution control problems, it is almost always negligible. If one plots the required pressure rise (δP_{fan}) in Eq. (6-80) as a function of volumetric flow rate (Q) on the same plot as the fan performance curve for a particular fan, the point of intersection of these two curves will be the operating point of the system. This is illustrated more clearly in Example 6.8 and Figure E6.8.

Example 6.8 – Matching a Fan to a Local Ventilation System

Given: A fan is to be purchased for the local ventilation system of Example 6.6. A centrifugal fan with a 9.0 inch inlet is chosen. Its performance data are listed by the manufacturer as follows:

Q (SCFM)	$(\delta P)_{fan}$ (inches of water)	Q (SCFM)	$(\delta P)_{fan}$ (inches of water)
0	0.9	750	0.75
250	0.95	1000	0.4
500	0.9	1200	0

To do: Predict the actual operating conditions of the local ventilation system, and draw a plot of required and available fan pressure rise as functions of volume flow rate.

Solution: The required fan pressure rise is a function of volume flow rate through the duct. The equations of Example 6.6 are used here, but with variable flow rate, Q. The authors solved these equations for a range of Q from 0 to 1000 SCFM, and plotted the results using Excel. The results are shown in Figure E6.8, and the spreadsheet is available on the book's web site. The actual operating point is approximately 800 SCFM at 0.70 inches of water pressure rise.

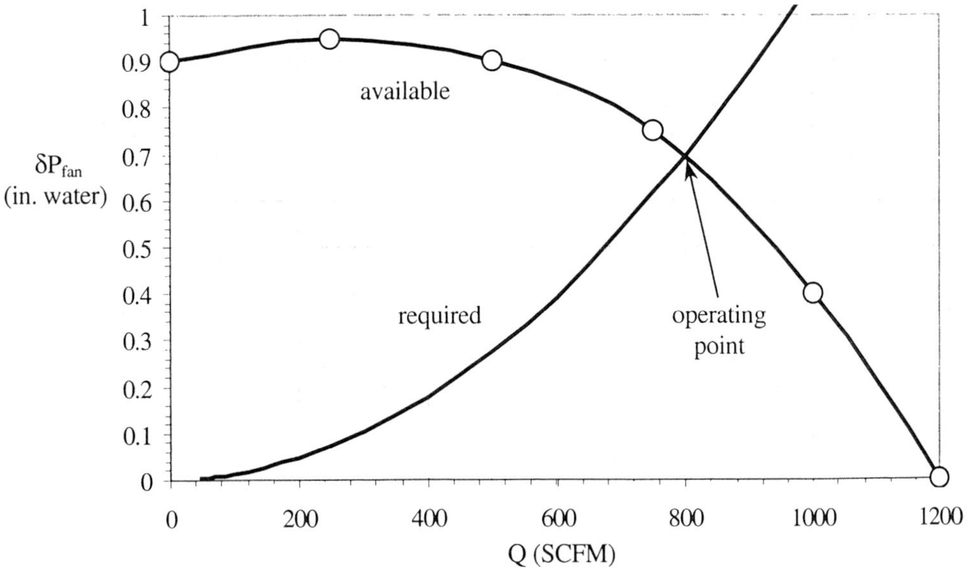

Figure E6.8 Required verses available pressure rise for the local ventilation system and fan of Example 6.8.

Discussion: The required flow rate through the duct was given in Example 6.6 as 750 SCFM. The purchased fan is somewhat more powerful than required, yielding a higher flow rate. The difference is small, and is probably acceptable; a butterfly valve could be installed somewhere in the duct to cut back the flow rate to 750 SCFM if desired. It is clearly better to oversize than undersize a fan when used with an air pollution control system.

6.12 Closure

Local ventilation, consisting of judiciously designed and located inlets, is the most effective way to capture contaminants before they enter the workplace environment. The total amount of exhaust (and make-up) air would be considerably larger if general ventilation were used instead. There is a considerable body of professional literature on the selection and design of inlets for local ventilation. The designs are conservative, and designers can be reasonably assured that they will capture contaminants unless there are strong drafts, etc. Designers should also realize that improvements can be made to reduce the volumetric flow rate of the withdrawn air by modeling the flow field analytically and then verifying the effectiveness by experiment. The design of the system of branching ducts that connect the local inlets to an exhaust fan must be conducted carefully to ensure that the desired amount of air is withdrawn through each inlet. There are many duct design computer programs to assist engineers in accomplishing this task. Lastly the engineer must select an exhaust fan that possesses an appropriate fan performance curve (δP_{fan} versus Q) to accompany the system of ducts. Engineers should proceed with an eye that at some time in the future, situations may arise in which it is necessary to increase the exhaust air volumetric flow rate and/or pressure drop; they should therefore select a fan which is sufficiently robust to accomplish this task.

Chapter 6 - Problems

1. An acid dipping tank contains chromic and hydrochloric acid and is used to clean the surface of steel sheet prior to dipping into a molten zinc alloy. The vessel is 2.0 m wide and 8.0 m long. The tank has no side baffles and is not located near any walls. It is not possible to enclose the process, so lateral exhausts are needed along each long side. Estimate the volumetric flow rate and slot dimensions for

(a) lateral suction system along both of the long sides
(b) push-pull system along the long sides.

2. Describe a way to control dust generated when 80-lbm bags are filled with a powder containing lead. Estimate the required ventilation volumetric flow rate. If the horsepower required to run a fan is approximately 1/4 HP per 1,000 CFM, estimate the cost per bag associated with dust control. Electrical power costs $0.05 per kW-hr.

3. In a small chemical plant several bags of different powders are added to each batch of material in a large reactor. The reactor is a 5,000-gal, glass-lined, 8.0 ft diameter cylindrical vessel with hemispherical ends. In the center of the reactor is a helical mixer. Powders in 80-lbm bags are added by workers through a 2.0 ft diameter access port on the top of the vessel after the reactor is 3/4 full with various liquids. A cloud of dust escapes through the port and into the face of the worker each time the contents of a bag are added to the reactor. Contaminants also enter the workplace air. The company has been cited by OSHA because of unhealthy exposure of the worker and others in the workplace. Recommend engineering controls to prevent escape of dust. For a variety of reasons, it is not possible to eliminate the need for individual workers to add the powder directly through the access port.

4. When the reaction has ended, a slurry is drained through a 6.0 inch valve in the bottom of the reactor. The valve is 7.0 feet above the floor. The slurry travels over a rectangular, slightly inclined screen where liquid is extracted by applying a vacuum to the bottom side of the screen. Highly irritating vapor escapes from the top surface of the slurry burning the eyes and skin of the workers. Vapors also enter the workplace generally which irritate eyes and produce obnoxious odors. Unfortunately the workers need to be able to see the process and scrape the material into a second vessel that is later sent to a standard dryer a few hundred feet away. The screen is wheeled under the reactor when the reactor is unloaded. Recommend an engineering control.

5. Devise a way to control dust generated from a 1/2 HP hand-held, surface sanding machine used to smooth the surface of fiberglass boats. It is occasionally necessary to sand *internal* surfaces. Estimate the ventilation volumetric flow rate and discuss limitations you expect the system may possess.

6. One step in the manufacture of TV picture tubes requires workers to remove the tube from an overhead conveyor, dip it in an aqueous hydrogen fluoride bath for a few moments, remove and inspect it and place it back on the conveyor. Recommend a ventilation system to control hydrogen fluoride vapor and estimate the volumetric flow rate. The production rate for the bath is 20 tubes per minute. The vessel is 1.1 m wide and 8.0 m long and workers stand along both long sides of the vessel to perform their duties. Estimate the cost per tube if fans require 0.75 HP per 1,000 SCFM and electrical power costs $0.05 per kW-hr.

7. A vertical stack is to be located on the upstream side of a building 20. m high by 50. m wide (in the direction of the wind) by 100 m (cross wind length). The wind is perpendicular to the 20. m by 100 m face. What is the minimum height of the stack exit above the roof if there is to be reasonable assurance that the exhaust will not be trapped in the upwind recirculation eddy? What is the EPA recommended stack height?

8. An unflanged circular inlet 6.0 inches in diameter is used to capture grinding particles. What should be the volumetric flow rate (CFM) if the minimum capture velocity of 200 ft/min is to be achieved over a circular region 12. inches in diameter in a plane lying 3.0 inches in front of the inlet?

9. An open vessel containing chromic acid at 470 °F is used for electroplating chromium. The tank is 3.0 ft wide and 10. ft long, and is placed with its long dimension against a wall. If the vessel is equipped with a lateral exhauster, what suction volumetric flow rate (CFM) would you recommend if ACGIH design standards are to be followed? Use the PEL-C value as the health standard.

10. You wish to control particles escaping from an open barrel being filled with powder. The barrel is 2.0 ft in diameter. Define a "controlled region" as one in which the air velocity along a streamline is 100 FPM or more. Management asks you to choose either a flanged round circular opening (Figure 6.10) or a semicircular slot similar to Figure 6.1. Since the slot width is small compared to barrel radius, the curvature can be ignored, and Figure 6.8 can be used as an approximation. On the basis of the data in Figures 6.8 and 6.10, which type of exhaust affords the maximum control per CFM?

11. It is common practice to cut concrete pavements prior to excavation so as to minimize the amount of pavement that has to be replaced. To accomplish this, large and small disk cutters are used. Large machines are equipped with water for lubrication and dust control. Small hand-held machines (resembling chain saws fitted with cutting disks) often make no attempt to control dust (or noise). Recommend engineering controls to suppress the generation of dust.

12. Clay from a conveyor is discharged into a chute that leads to a blender where the clay is mixed with water and other ingredients in the first step in making vitreous china (sinks, toilet bowls, etc.). A partial enclosure is built to surround the transfer point. Air is withdrawn through the exhaust duct to control fugitive emissions. Estimate the volumetric flow rate (CFM) needed to select an exhaust fan.

$\dot{m}_b$ = 40. kg/s	H = 0.50 m
A_b = 0.10 m^2	A_{face} = face area near adjustable gate = 0.50 m^2
D_p = 100 μm	Q_S = secondary air) = 0.050 m^3/s
ρ_b = 2,000 kg/m^3	

13. After graduation, you work for a company that uses acetone for cleaning. You are asked to design the duct system for an exhaust hood used to remove acetone vapors from a laboratory. Analysis of the vapor source and hood reveal that a volume flow rate of 850 ft^3/minute (CFM) of air through the duct is required to remove enough of the vapors to maintain a healthy air environment in the lab. The hood inlet has a face area of 3.3 ft^2. Some sheet metal ducts are available for the project. They are round with an 8.0 inch inner diameter and an average surface roughness of 0.0020 inches. You install an air cleaner above the hood, and route the ducting up through the ceiling tiles and across to the mechanical room, where a fan is to be installed. The total length of all the pipe sections is 50. ft. The outlet of the fan will exhaust into the atmosphere as sketched in Figure P6.13. From manufacturers literature and from the *Industrial Ventilation* manual (ACGIH), the minor loss coefficients for the various components of the duct system are found. The minor loss coefficients are as follows:

- hood loss coefficient = 0.49 (from the room to the outlet of the hood)
- air cleaner loss coefficient = 8.1
- elbow loss coefficient = 0.55 each (there are three identical elbows)

Some properties and constants in the English system are provided here for your convenience:

- kinematic viscosity of air = 1.61 x 10^{-4} ft^2/s
- density of air = 2.34 x 10^{-3} $slug/ft^3$

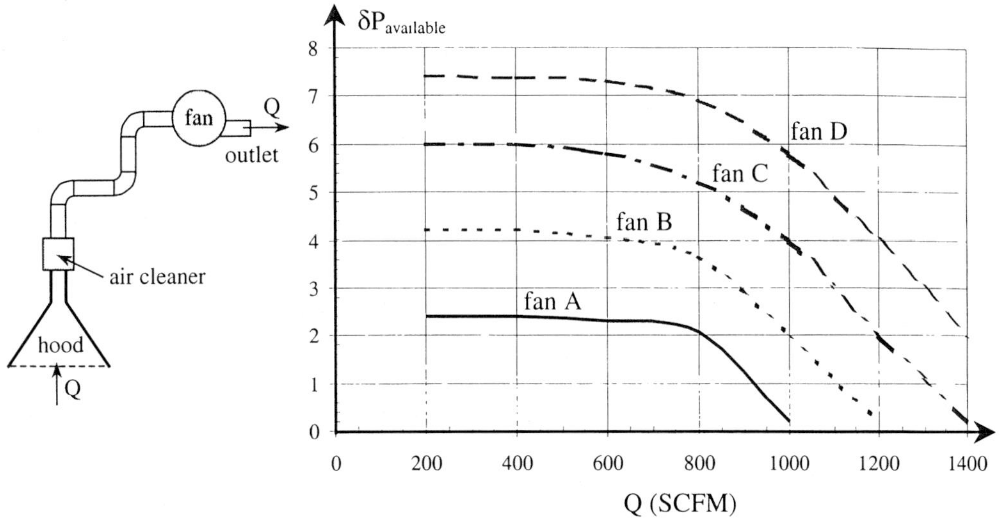

Figure P6.13 Schematic diagram of hood configuration; fan performance curves for Problem 6.13.

- density of water = 1.94 slug/ft^3
- $g = 32.174$ ft/s^2
- conversion factor: 1 slug = 1 lbf s^2/ft

(a) Calculate the major and minor losses in the duct system. Calculate the total head loss through the system. Give your answer both in feet of air and in inches of water. Hint: The pressure in the middle of the room is atmospheric, and the velocity is negligibly small there.

(b) Perform a control volume analysis of the duct system, and determine the required net head of the fan ($\delta P_{\text{fan required}}$) in units of inches of water.

(c) Now you search through a catalog of fans. One reliable company has a family of four fans which seem to fit your budget. The fan performance curves ($\delta P_{\text{available}}$) for these four fans are plotted in Figure P6.13. Fan A is the smallest and least expensive, followed by fan B, and then fan C; fan D is the largest and most expensive. Which fan should you purchase? Hint: Choose the smallest fan that performs as well or better than necessary. If the fan delivers more net head than required at the desired flow rate, a damper can be installed later to increase the head loss, thereby decreasing the volume flow rate.

14. You have been asked to design a lateral exhauster for an open vessel 3.0 ft wide, 20. ft long and 3.5 ft high. The exhausters will run along both (long) sides of the vessel. The exhauster plenum should protrude above the vessel to a minimal amount. It has been decided that the volumetric flow rate per square foot of liquid surface is 175 CFM/ft^2. Use the duct-design computer program to compute the dimensions of the plenum chamber. It has been decided that air will be withdrawn through 3.0 inch holes located along 6-inch centers a minimal distance above the surface of the liquid. Use the program to compute the dimensions of the plenum chamber such that the pressure is constant throughout the chamber and the volumetric flow rate through each 3.0 inch hole is 233 CFM. Because the holes are 3.0 inches in diameter, it has been decided that the width of the plenum chamber should be 6.0 inches. The vertical dimension is of course unknown. You may select a plenum that withdraws air in a single direction for the full length of 20. ft or a symmetrical plenum that withdraws air in opposite directions for 10. ft.

15. An electric arc furnace is opened for a period of three minutes every 30 minutes in order to add scrap metal or minerals or for pouring. Management suggests installing a hopper hood 30 feet above the furnace to collect fume which in turn will be sent to a baghouse. The furnace diameter is 25. feet. Its heated surface is at a temperature of 600 K. When the furnace lid is opened the volumetric flow rate of hot gas is approximately 300,000 CFM. The baghouse is designed to handle 100,000 CFM on a steady basis. You have been asked to recommend the volume and dimensions of a hopper hood. What recommendations would you make to management?

16. A hopper-type canopy hood is to be located above an electric arc furnace used to melt scrap metal. Every 30 minutes, the lid is removed and a 65,000 CFM surge of hot gas is produced that lasts approximately 30 seconds. The lid is removed to recharge the furnace with scrap, to add special minerals, and to tap the molten metal. Recommend the dimensions of the hood and the volumetric flow rate of an exhaust fan if the hopper hood is to accommodate both the steady-state and surge plumes. Shown below are key parameters of the furnace:

- furnace diameter = 5.0 m
- furnace surface temperature = 800 K
- natural convection film coefficient = 10. watts/m^2 K
- distance between furnace lid and canopy face = 10. m
- draft velocity in workplace = 0.50 m/s
- maximum centerline velocity of buoyant plume (exclusive of the surge) = 5.0 m/s

17. A ship's hold can be considered a enclosed container with two openings on the top surface. Through one opening the product enters the hold and through the second air is withdrawn to prevent fugitive dust from being discharged to the atmosphere. Estimate the volumetric flow rate (Q_W) needed to control fugitive dust generated when filling a ships hold with grain (bulk density = 5,000 kg/m^3). The grain is loaded at the rate ($\dot{m}_b$) of 50. kg/s, the distance through which the grain falls before entering the hold (H) is 1.0 m. The cross-sectional area (A_b) of the falling grain is 0.10 m^2, the cross-sectional area surrounding the falling grain through which air is drawn downward (A_f) is 0.20 m^2. The diameter of the fugitive dust is typically 10. μm. No secondary air is drawn into the ships hold, $Q_S = 0$.

18. Consider an unflanged rectangular inlet 3.0 inches by 300 inches and a plane 3.0 inches upstream and parallel to the entrance plane. Air is withdrawn through the inlet at a volumetric flow rate of 100 CFM. Define the dimensions of a surface on this upstream plane on which the velocity is never less than 32. FPM.

19. Consider an unflanged circular inlet 6.0 inches in diameter and a plane 6.0 inches upstream and parallel to the entrance plane. Air is withdrawn through the inlet at a volumetric flow rate of 30. CFM. Define the dimensions of a surface on this upstream plane on which the velocity is never the less than 35. FPM.

20. In laminar flow the friction factor is equal to 64 divided by the Reynolds number based on diameter, whereas in turbulent flow (at high enough Reynolds numbers) the friction factor is essentially constant. Show that for a long duct of constant diameter, the cost of operating a fan to draw air through the duct is proportional to the *cube* of the volumetric flow rate for turbulent flow. Show that for laminar flow the cost is proportional to *square* of the volumetric flow rate.

21. You need to produce a capture velocity no less than 100 FPM along a plane 24. inches high lying 12. inches upwind of the inlet plane of

(a) two unflanged rectangular inlets, each 6.0 inches wide and 180 inches between centers
(b) a single unflanged rectangular inlet 12. inches wide.

Estimate the total volumetric flow rate for configurations (a) and (b). The aspect ratio of the inlets is 1:10.

22. Figure 6.3 shows a close-fitting canopy hood over a reheat furnace. The following are key features of the furnace.

- length (L) = 4.0 m
- width (W) = 3.0 m
- clearance (H) = 0.20 m
- overhang (F) = 0.080 m
- temperature difference (delta T) = 50. °C

Estimate the suction volumetric flow rate Q_S needed to cope with the buoyant plume plus other air needed to control the furnace.

23. An open vessel 3.0 ft wide and 10 ft long contains a hot (180 °F) aqueous H_2SO_4 solution used to pickle stainless steel. The tank will be situated next to a wall.

(a) Your supervisor wishes to use a lateral exhauster such as those in Figure 6.1 to control the vapors. Using Tables 6.2 to 6.5, estimate the total volumetric flow rate for a lateral exhauster. Can one suction slot be used or is it wiser to use two slots? If the suction slot face velocity is 2,000 FPM, estimate the suction slot width (S_S).

(b) You believe that a push pull system would be superior to the lateral exhauster. Using the equations shown in Figure 6.13, estimate the velocity and volumetric flow rates of the suction flow (Q_S) and the push flow (Q_b) through 1/4-inch holes, spaced 3/4 inches between centers.

24. Swine eat granular bulk feed from feeders that are filled from above by an overhead auger 2.0 meters above the top of the feeder. For simplicity, assume that the feeder is a box 2.0 meters high with a square cross section 1.0 m by 1.0 m. The feed falls freely from a point 2.0 meters above the top of the feeder and enters the box with a velocity of 4.4 m/s. Properties of the feed are as follows:

- bulk density of feed in the box, $\rho_b = 700\ kg/m^3$
- bulk density of the falling feed stream, $\rho_f = 500\ kg/m^3$
- particle mass mean diameter, $D_{p,m} = 10.\ \mu m$
- feed mass flow rate, $\dot{m}_b = 10.\ kg/s$

Using a representative emission factor of 0.5 kg/Mg for the falling feed, estimate the particle emission rate and particle concentration (g/m^3) in the air stream leaving the feeder.

25. An unflanged rectangular inlet (2.0 by 20. inches) is used to capture welding fume. Estimate the volumetric flow rate (in units of CFM) required to achieve a minimum capture velocity of 200 FPM over a flat surface of dimensions 2.0 by 20. inches, located 1.5 inches in front of the inlet plane.

26. A vertical exhaust stack is to be placed on the upwind side of a manufacturing building that is 30. ft high, 100 ft wide and 1,500 ft long. The stack will be 750 ft from one end of the building. The prevailing winds are perpendicular to the 30. ft by 1,500 ft face of the building. What is the minimum height of the stack if there is to be reasonable assurance that the exhaust will not be trapped in the recirculation eddy on top of the building nor in the wake region immediately downwind of the building? What is the EPA recommended stack height?

27. Estimate the volumetric flow rate for a canopy hood for an outdoor, 3.0 ft by 6.0 ft barbecue grill, open on all four sides, that is part of food stand located in a public park. The lower edge of the canopy hood is 6.5 ft above the ground and the grill is 3.5 ft above the ground.

28. A hopper hood similar to that shown in Figure 6.22 is used in a foundry. Every 20 minutes the furnace is opened for charging and pouring. During these periods a 30-second surge of hot fume is discharged at a volumetric flow rate of 200,000 ACFM. Air and fume are removed steadily from the hopper hood at a volumetric flow rate of 5,000 ACFM. Estimate the minimum volume of a hopper hood to accomplish these tasks.

29. An adsorption bed of activated charcoal is used as an air pollution control system to remove an assortment of airborne volatile organic compounds (VOCs). The bed has cross-sectional area (A), length in the direction of the flow (L) and thickness (t). The bed dimensions are unique such that upstream and downstream plenum chambers ensure that

- the face velocity of air entering and leaving the bed (Q/A) is uniform,
- the pressure drop (δP) across the bed is uniform.

So long as the Reynolds number (defined in terms of a characteristic bed granule of diameter D′) is less than 10, the pressure drop across the bed can be expressed as

$$\delta P = \frac{200\mu t c^2}{(D')^2 \phi^2 (1-c)^3} \frac{Q}{A}$$

where

- δP = pressure drop (units defined by the variables above)
- Q = gas volumetric flow rate
- A = bed cross-sectional area
- μ = actual gas viscosity
- t = bed thickness
- c = packing density, bed bulk density/granule density
- D′ = diameter of a sphere having a volume of equal to the volume of a granule
- φ = surface area of a sphere of diameter D′ divided by the actual surface area of a solid granule (A_g) neglecting its interior pore structure, $\phi = 4.836\,(V')^{2/3}/A_g$
- A_g = actual surface area of a granule having the gross shape of the porous particle neglecting its interior pore structure
- V′ = volume of the granule neglecting its pore structure

Assume the following flow conditions: carrier gas is air at STP, A = 10. m^2, D′ = 0.10 mm, Q = 1.2 m^3/s, t = 0.50 m, L = 5.0 m, c = 0.50.

(a) Specify the dimensions of the upstream and downstream plenums using analytical methods.
(b) Specify the dimensions of the upstream and downstream plenums using the Duct Design computer program, or some other duct pressure drop computer program.

30. Air flows through a ventilation duct of diameter D and length L at a volumetric flow rate Q and average velocity U. The flow is laminar. Your supervisor wishes to increase the duct diameter D but keep the volumetric flow rate (Q) and duct length (L) constant. A senior engineer informs your supervisor that the major head loss per unit length (h_L/L) is proportional to D^n. Your supervisor wishes to know the value of n. Find the value of n.

31. A hood is needed to remove vapors from a chemical process. The source of contaminant is located 6.0 inches axially from the hood inlet, and 3.0 inches off the axis of symmetry. Two hoods are available, at nearly identical cost. Hood A is a flanged round inlet of 6.0 inches diameter, with a volume flow rate of 500 CFM. Hood B is also a flanged round inlet, but of 12. inches diameter, with a volume flow rate of 300 CFM. Assume the data of Figure 6.10 are valid for both cases. Which hood is the better choice for this application?

32. A hood is needed to remove particles from a grinding operation. It is determined that the required capture velocity is 300 feet per minute for particles located 6.0 inches axially from the hood inlet, and 3.0 inches off the axis of symmetry (i.e. *radially*). Two hoods are available, at nearly identical cost. Hood A is a flanged round inlet of 6.0 inches diameter. Hood B is an unflanged round inlet, 8.0 inches diameter. Assume the air is at STP.

(a) *To one significant digit*, calculate the required volume flow rate for *hood A* in CFM.
(b) *To one significant digit*, calculate the required volume flow rate for *hood B* in CFM.
(c) Which hood is better for this application? Why?

7

Ideal Flow

In this chapter you will learn:

- fundamentals of ideal flow
- how to predict the velocity field upstream of:
    - plain and flanged inlets in quiescent air
    - plain and flanged inlets immersed in moving air
    - multiple inlets

Chapter 6 describes contemporary practices and design standards that provide guidance for designing systems to control indoor air pollution. Unfortunately, in most instances engineers cannot predict the effectiveness of the systems they design, nor can they optimize the dimensions and volumetric flow rates prior to construction and field-testing. Engineering today requires one to be able to optimize designs based on analytical models of performance. There is no reason why attempts to control indoor air pollution should not do the same (Heinsohn, 1991). The essential ingredient is accurate prediction of the velocity field upwind of the inlet. Figures 6.8 through 6.10 provide some details about the velocity field upwind of rectangular and circular inlets; a larger selection of air speed contours can be found in Baturin (1972). Since these figures show contours along which air speed $U = \left|\vec{U}\right|$ is constant (isopleths), engineers can generate analytical expressions for air speed by fitting curves to the empirical data. Curve-fitting procedures can be accurate but are tedious, the tedium increasing with the level of desired accuracy. The objective of this chapter is to describe methods to predict the movement of contaminants upwind of an inlet of arbitrary geometry withdrawing air and pollutants that pass by it, based on fundamental concepts of fluid mechanics.

7.1 Fundamental Concepts

General ventilation describes how air from the entire workspace is withdrawn and replaced, whereas local ventilation describes the capture of contaminants by withdrawing air at a selected rate through a uniquely shaped inlet, opening, entry, port, etc. located close to where the pollutants are generated. ***Close capture hood*** is the generic name given to this opening. Chapter 1 describes how local ventilation hoods are classified. Local ventilation is preferred over general ventilation because the volumetric flow rates of withdrawn air are considerably smaller, equipment and operating costs are smaller, and the cost of heating and air conditioning of make-up air is less.

7.1.1 Comparison of Jets (Blowing) and Inlets (Suction)

The uninitiated might imagine that the velocity field near the suction inlet or intake of a close capture hood is a mirror image of the flow field near a supply outlet or blowing jet. Nothing could be further from the truth; one is not the inverse of the other as can be seen in Figure 7.1. A vivid illustration of the disparity is contained in the book by Hayashi et al. (1985) in which two individuals try to extinguish a candle, one by exhaling and the other by inhaling. Even if both produce the same

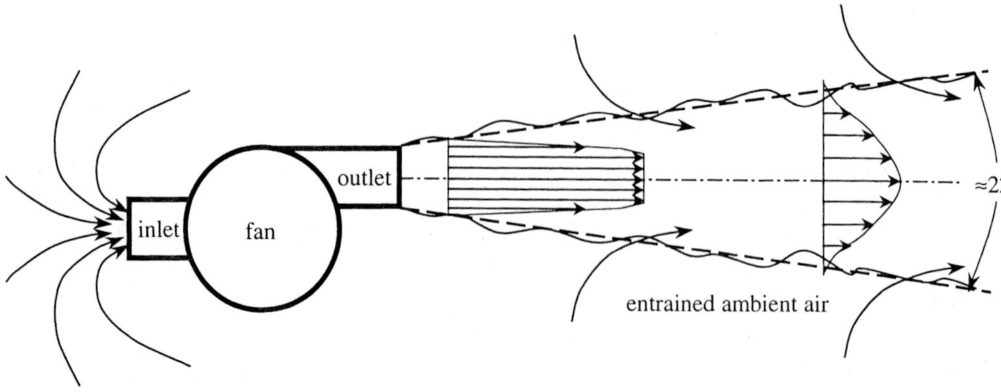

Figure 7.1 Contrast between the reach of a suction inlet and the throw of a blowing jet.

volumetric flow rate and air speed (at their mouths) and both stand the same distance from the flame, exhaling produces a flow field that extinguishes the flame while the flow field from inhalation does not. In the vicinity of a blowing jet, viscosity and turbulence profoundly affect the spread and penetration of the jet, whereas the flow in the vicinity of an inlet is not greatly affected by these parameters. The jet entrains surrounding air due to turbulence and viscosity, and the width of the jet increases with an included angle of approximately 22 degrees. Within the core of the jet the velocity remains equal to the exit velocity, but after approximately seven exit diameters ($x \approx 7D$) the centerline velocity begins to decrease. At approximately thirty exit diameters downstream ($x \approx 30D$), the centerline velocity is about ten percent of the exit velocity. Thus, the influence of a jet (sometimes called ***throw***) extends approximately thirty exit diameters downwind of the exit plane, whereas the influence of an intake (defined as ***reach*** in this book) extends to only approximately *one* inlet diameter upwind of the inlet plane, at which the air speed is approximately ten percent of the inlet velocity.

The reason that the flow fields associated with a blowing jet and a suction inlet are not mirror images of each other is because a jet possesses momentum that induces ambient air into the jet. Velocity gradients produced by the jet are large, which coupled with the air's viscosity, produce viscous shear stresses that diffuse the jet's momentum. An inlet on the other hand has no momentum to diffuse - quiescent air from the surroundings enters the inlet due to suction. Since viscosity does not affect the inlet velocity field appreciably, some authors call this kind of flow ***inviscid***. However, even in this type of flow, there are appreciable viscous stresses which cause fluid elements to distort. However, while there are non-zero viscous stresses, the *net* viscous force on a fluid element entering an inlet is negligible compared to the net inertial force. This can be more clearly understood by considering the concepts of vorticity and rotationality.

7.1.2 Vorticity and Rotationality

Of great importance to the analysis of flows into hoods are the concepts of vorticity and rotationality. ***Vorticity***, a vector, is defined mathematically as the curl of the velocity vector $\vec{U}$,

$$\boxed{\vec{\zeta} = \vec{\nabla} \times \vec{U} = \text{curl}\left(\vec{U}\right)} \tag{7-1}$$

The symbol used above for vorticity (ζ) is the Greek letter *zeta*. Readers should note that this symbol for vorticity is not universal among fluid mechanics textbooks; some authors use $\vec{\omega}$ while still others use $\vec{\Omega}$. In this book, $\vec{\Omega}$ is used to denote the ***angular velocity vector***, sometimes called the ***rotation vector***, of a fluid element. It turns out that the rotation vector is equal to half of the vorticity vector,

$$\vec{\Omega} = \frac{1}{2}\left(\vec{\nabla}\times\vec{U}\right) = \frac{1}{2}\text{curl}\left(\vec{U}\right) = \frac{\vec{\zeta}}{2} \tag{7-2}$$

Thus, regardless of the symbol used, *vorticity is a measure of rotation of a fluid particle*. Specifically, vorticity is equal to twice the angular velocity of a fluid particle. If the vorticity at a point in a flow field is non-zero, the fluid particle which happens to occupy that point in space is rotating; the flow at that location is ***rotational***. Likewise, if the vorticity at a point is zero, the fluid particle at that point is not rotating; the flow is ***irrotational***. Physically, fluid particles in a rotational flow rotate end over end as they move along in the flow. For example, fluid particles within the boundary layer near a solid wall are rotational (and thus have non-zero vorticity), while fluid particles outside the boundary layer are irrotational (and their vorticity is zero). Both of these cases are illustrated in Figure 7.2.

Rotation can be produced by wakes, shock waves, boundary layers, flow through air-moving equipment (fans, turbines, compressors etc.), and thermal gradients. The vorticity of a fluid element can not change except through the action of viscosity, nonuniform heating (temperature gradients), or other nonuniform phenomena. Thus if a flow originates in an irrotational region, it remains irrotational until some dissipative process alters it. Flow entering an inlet from a quiescent environment is irrotational and remains so unless a wake or nonuniform heating transforms it. This is why the irrotational flow approximation is generally applicable to flow into inlets such as hoods.

7.1.3 Equations of Motion

In this text an ***Eulerian frame of reference*** is adopted in which attention is focused on a point in space at a particular instant of time. Similarly, attention may be focused on a particular volume in space (called the ***control volume***) through which a fluid passes. With the Eulerian description, the properties of the flow ***field***, including contaminants it may contain, are described as functions of space and time. For example, the pressure field is given by P(x,y,z,t).

If it cannot be assumed that the contaminant concentration within an enclosed space is nearly spatially uniform, i.e. if $c(x,y,z,t) \neq c(t)$, the well-mixed model described in Chapter 5 should not be used; one must instead solve a series of differential equations to find the contaminant concentration field. A coupled set of differential equations describing contaminant transport can be derived, including the contaminant concentrations and the velocity of the carrier gas throughout the region. This is illustrated in Chapter 10. However, there are conditions wherein the problem can be *uncoupled*, with the air velocities computed first and the pollutant concentration computed later. This is the approach taken in this chapter. The instantaneous, compressible, differential ***conservation equations*** for an ideal gas are:

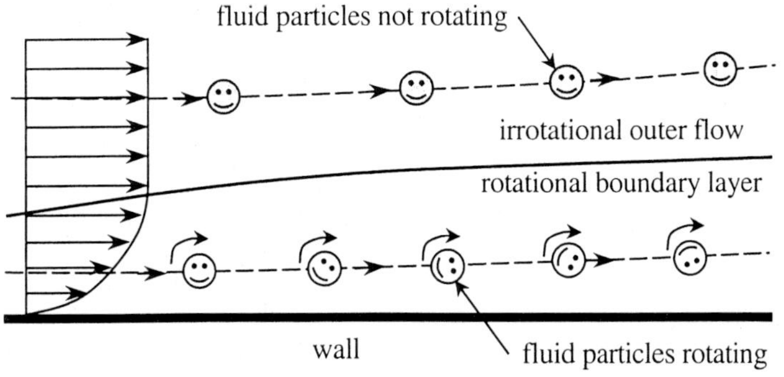

Figure 7.2 Illustration of the difference between rotational and irrotational flow; fluid elements in a rotational region of the flow rotate.

- Equation of state (ideal gas law):

$$\boxed{P = \rho RT} \tag{7-3}$$

- Conservation of mass (compressible form of the continuity equation):

$$\boxed{\frac{\partial \rho}{\partial t} + \vec{\nabla} \cdot \left(\rho \vec{U}\right) = 0} \tag{7-4}$$

- Conservation of linear momentum (compressible form):

$$\boxed{\frac{\partial \vec{U}}{\partial t} + \left(\vec{U} \cdot \vec{\nabla}\right)\vec{U} = -\frac{1}{\rho}\vec{\nabla}P + \vec{g} + \frac{1}{\rho}\vec{\nabla} \cdot \tau_{ij}} \tag{7-5}$$

where ρ is the density of air, P is the pressure, $\vec{g}$ is the gravitational acceleration vector, and τ_{ij} is the ***viscous stress tensor*** acting on the element (White, 1999). Readers are encouraged to locate a handbook containing the above vector expressions written in rectangular, cylindrical, and spherical coordinates. One can also write an equation for conservation of energy, but this is not needed in the material presented here. In fact, the above equations can be further simplified for inlet flows. The following simplifications are made for flows near the inlet of a hood or other local ventilation device:

Newtonian - Air is assumed to be a Newtonian fluid. A ***newtonian fluid*** is one in which the ***shear stress*** on a fluid element is linearly proportional to its ***deformation rate***. The constant of proportionality is the dynamic viscosity (μ). As a result of this assumption, the viscous stress tensor in Eq. (7-5) is greatly simplified.

Incompressible - ***Mach number*** is defined as the speed of a fluid divided by the speed of sound in that fluid. ***Subsonic flows*** are flows in which the Mach number is less than unity. ***Supersonic flows*** are those in which the speed exceeds the speed of sound; the Mach number is greater than unity. The speed of sound in air at room temperatures is approximately 1100 ft/s (67,000 FPM, 760 mph, or 340 m/s). Clearly, flows encountered in indoor air pollution control are subsonic since typical air speeds are much smaller than the speed of sound. While air is by definition compressible, it turns out that at low speeds (more precisely at low Mach numbers, i.e. at Mach numbers less than about 0.3), density changes are insignificant. Thus, at low speeds, air can be thought of as approximately ***incompressible***. Flow into the intake of a hood or other local ventilation device easily satisfies this criterion for incompressibility. Thus for inlet flows, the equation of state, Eq. (7-3), reduces to

$$\boxed{\rho \approx \text{constant}} \tag{7-6}$$

and the equation of conservation of linear momentum, Eq. (7-5), reduces to

$$\boxed{\frac{\partial \vec{U}}{\partial t} + \left(\vec{U} \cdot \vec{\nabla}\right)\vec{U} = -\frac{1}{\rho}\vec{\nabla}P + \vec{g} + \nu\nabla^2\vec{U}} \tag{7-7}$$

where ν is the kinematic viscosity, $\nu = \mu/\rho$, assumed to be constant. Eq. (7-7) is the incompressible form of the ***Navier-Stokes equation***, the most famous (and most useful) equation in fluid mechanics.

Steady - The analysis here assumes that the flow is steady-state, i.e. not a function of time. For incompressible steady flow, the continuity equation, Eq. (7-4), reduces to

$$\boxed{\vec{\nabla} \cdot \vec{U} = 0} \tag{7-8}$$

Irrotational - It turns out that the flow of air into an intake of some kind, such as a hood, can be approximated as irrotational to appreciable accuracy. If a process that generates contaminant is close to the inlet, an aerodynamic wake is apt to be produced as the air passes over and around the source. Conditions in the wake region are rotational, much like in the boundary layer near a wall. If however the work piece is sufficiently far away, its wake may be outside the velocity field of the inlet and the assumption of irrotational flow is still valid. Aerodynamic wakes from any other solid body must also be outside the velocity field of the inlet if the assumption of irrotational flow is to be appropriate. Wakes have a large effect on the motion and concentration of contaminants and cannot be neglected. The size of the wake region is a matter of computation, but as a first approximation a value twice the characteristic dimension of the bluff body perpendicular to the flow field can be used. As discussed above, if a flow is irrotational, its vorticity is zero. Thus, Eq. (7-1) becomes

$$\boxed{\vec{\nabla}\times\vec{U}=0} \tag{7-9}$$

From the study of vector mathematics, it is known that any vector $\vec{U}$ which satisfies Eq. (7-9) can be written as the gradient of a scalar function called a ***potential function***. Here, since the vector is the velocity vector, the potential function is called the ***velocity potential function***, or simply the ***velocity potential***, ϕ. The velocity field itself is called a ***conservative velocity field***. Thus, for an irrotational flow,

$$\boxed{\vec{U}=-\vec{\nabla}\phi} \tag{7-10}$$

Readers should note that *the sign convention for velocity potential is not universal; many authors do not include the negative sign in this definition*. The notation convention of Streeter et al. (1998) has been adopted in the present book. Because of the existence of the velocity potential in irrotational flow, irrotational flow is also called ***potential flow***.

Substitution of Eq. (7-10) into the steady, incompressible continuity equation, Eq. (7-8), yields

$$\boxed{\nabla^2\phi=0} \tag{7-11}$$

which is the well-known ***Laplace equation***, where the ***Laplacian operator*** is defined as

$$\boxed{\nabla^2=\vec{\nabla}\cdot\vec{\nabla}} \tag{7-12}$$

The reader can verify that substitution of Eq. (7-10) into the right-most term of Eq. (7-7) renders that term identically zero for incompressible, irrotational flow. Equation (7-7) without the last term (the viscous term) is known as ***Euler's equation***. Furthermore, the second term on the left-hand side of Eq. (7-7) can be modified via the following mathematical identity:

$$\boxed{\left(\vec{U}\cdot\vec{\nabla}\right)\vec{U}=\vec{\nabla}\left(\frac{U^2}{2}\right)-\vec{U}\times\left(\vec{\nabla}\times\vec{U}\right)} \tag{7-13}$$

where U (without the arrow) is the magnitude of the velocity vector (the ***speed***). The last term on the right-hand side of Eq. (7-13) is zero for irrotational flow; thus, for steady, incompressible, irrotational flow, Eq. (7-7) reduces to

$$\boxed{\vec{\nabla}\left(\frac{U^2}{2}\right)=-\frac{1}{\rho}\vec{\nabla}P+\vec{g}} \tag{7-14}$$

For simplicity, the z-direction is taken as "up", i.e. opposite to the direction in which gravitational acceleration acts. Thus,

$$\boxed{\vec{g}=-g\vec{\nabla}(z)=-\vec{\nabla}(gz)} \tag{7-15}$$

where g is the magnitude of the gravitational acceleration vector. Finally then, after some further rearrangement, the incompressible Navier-Stokes equation, originally Eq. (7-7), and reduced to Eq. (7-14), becomes

$$\boxed{\vec{\nabla}\left(\frac{P}{\rho}+\frac{U^2}{2}+gz\right)=0} \tag{7-16}$$

From fundamental mathematical principles, if the gradient of a quantity is zero everywhere, that quantity must be a constant. Here, the quantity in parentheses in Eq. (7-16) must therefore be a constant everywhere in the flow field,

$$\boxed{\frac{P}{\rho}+\frac{U^2}{2}+gz=C=\text{constant everywhere}} \tag{7-17}$$

The above equation is the well-known ***Bernoulli equation***, and the constant (C) is often called the ***Bernoulli constant***. The dimensions of C are velocity squared. Readers should note that some authors prefer to multiply each term in Eq. (7-17) by the density, such that their Bernoulli constant has dimensions of pressure. Other authors prefer to divide each term in Eq. (7-17) by the gravitational constant (g), such that their Bernoulli constant has dimensions of length or ***head***.

There are some flows (such as solid body rotation) in which the flow field is rotational, but viscous forces are still negligible. In such flows, a modified Bernoulli equation can be derived,

$$\boxed{\frac{P}{\rho}+\frac{U^2}{2}+gz=C=\text{constant along streamlines}} \tag{7-18}$$

This equation is similar to Eq. (7-17), except that the Bernoulli "constant" is not constant everywhere, but only along streamlines of the flow. The Bernoulli constant changes from one streamline to the next in a rotational flow in which viscous forces are negligible.

Example 7.1 - Velocity Field near a Rotating Disk

Given: A hand-held surface grinder (see Figure 1.11) of radius R is rotating at a constant angular velocity Ω in the z-direction. Assume that the density of air is constant and that the flow is steady. Gravitational effects are also neglected. Assume a velocity traverse has been made and indicates that only the tangential velocity component (U_θ) is important; the velocity components in the radial and axial directions are negligible. The velocity field is given by:

- Region I: $U_\theta = \Omega r \quad \text{for} \quad r < R \qquad \left(\frac{U_\theta}{\Omega R}=\frac{r}{R} \quad \text{for} \quad \frac{r}{R}<1\right)$

- Region II: $U_\theta = \frac{\Omega R^2}{r} \quad \text{for} \quad r > R \qquad \left(\frac{U_\theta}{\Omega R}=\frac{R}{r} \quad \text{for} \quad \frac{r}{R}>1\right)$

A sketch of the flow field is shown in Figure E7.1a. The tangential velocity and radius have been normalized (nondimensionalized) for convenience.

To do: To illustrate the previous fundamental concepts, analyze the flow field with respect to the following:

(a) vorticity and rotationality
(b) the pressure field needed to satisfy conservation of momentum
(c) applicability of the Bernoulli equation

Solution:

(a) Vorticity - In cylindrical coordinates, the velocity vector is $\vec{U}=(U_r, U_\theta, U_z)$ for unit vector

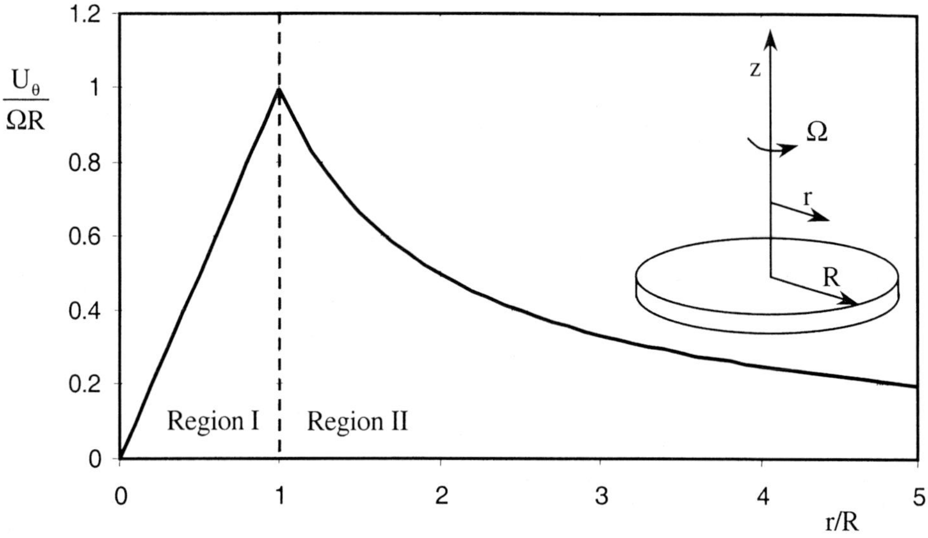

Figure E7.1a Normalized tangential velocity above a rotating disk as a function of normalized radial location.

$(\hat{r}, \hat{\theta}, \hat{z})$. The vorticity, Eq. (7-1), is

$$\vec{\zeta} = \vec{\nabla} \times \vec{U} = \hat{r}\left(\frac{1}{r}\frac{\partial U_z}{\partial \theta} - \frac{\partial U_\theta}{\partial z}\right) + \hat{\theta}\left(\frac{\partial U_r}{\partial z} - \frac{\partial U_z}{\partial r}\right) + \hat{z}\frac{1}{r}\left(\frac{\partial (rU_\theta)}{\partial r} - \frac{\partial U_r}{\partial \theta}\right)$$

Applying this to Region I,

$$\vec{\zeta} = \hat{z}\frac{1}{r}\left(\frac{\partial (r^2\Omega)}{\partial r}\right) = 2\Omega\hat{z}$$

and to Region II,

$$\vec{\zeta} = \hat{z}\frac{1}{r}\left(\frac{\partial (\Omega R^2)}{\partial r}\right) = 0$$

Thus the flow is irrotational in the outer region (Region II) and rotational with constant vorticity in the inner region (Region I). In addition, since the vorticity is constant in Region I, the rotational flow field is said to undergo ***solid body rotation***, i.e. to rotate as a solid body such as a rotating compact disk (CD). The analysis illustrates clearly that just because air revolves about a center of rotation, it is not necessarily rotational; it can be either rotational or irrotational, that is to say it may or may not have vorticity. All curvilinear flows revolve about a center of rotation, but irrotation or rotation in the fluid dynamic sense depends on the vorticity.

(b) Pressure distribution - To find the pressure distribution consistent with the conservation equations, the continuity and momentum equations are applied to the velocity profile. Equation (7-8) becomes

$$\vec{\nabla} \cdot \vec{U} = \frac{1}{r}\frac{\partial (rU_r)}{\partial r} + \frac{1}{r}\frac{\partial U_\theta}{\partial \theta} + \frac{\partial U_z}{\partial z} = 0$$

Substituting the observed velocity distributions, it is seen that the continuity equation is identically satisfied in both Region I and Region II. While the continuity equation is a scalar equation, the momentum equation (Navier-Stokes equation) is a vector equation. The radial (r-direction) and tangential (θ-direction) components of Eq. (7-7) are

$$\left(\vec{U}\cdot\vec{\nabla}\right)U_r - \frac{1}{r}U_\theta^{\,2} = -\frac{1}{\rho}\frac{\partial P}{\partial r}$$

and

$$\left(\vec{U}\cdot\vec{\nabla}\right)U_\theta + \frac{1}{r}U_r U_\theta = -\frac{1}{\rho r}\frac{\partial P}{\partial \theta}$$

Note that the flow is steady, gravity is neglected, and the viscous terms have dropped out. In Region I, although the fluid is viscous, viscosity does not affect solid body rotation. In Region II, the flow is irrotational and thus the viscous terms drop out as discussed previously. Upon substituting the observed velocity distribution, it is found that in both Region I and Region II,

$$\frac{\partial P}{\partial r} = \frac{\rho U_\theta^{\,2}}{r}$$

in the radial direction, and

$$\frac{\partial P}{\partial \theta} = 0$$

in the tangential direction. Thus the pressure is only a function of r. The r-momentum equation becomes an ordinary differential equation in both Region I and Region II, and the partial derivative operator in the above equation can be replaced by the total derivative operator,

$$\frac{dP}{dr} = \frac{\rho U_\theta^{\,2}}{r}$$

First consider Region II (r > R). Substituting $U_\theta = \Omega R^2/r$, and integrating the above equation between r and infinity where U_θ is zero and the pressure is the ambient pressure P_0, one obtains

$$P(r) = P_0 - \frac{\rho\Omega^2 R^4}{2r^2} \quad \text{for} \quad r > R \qquad \left(\frac{P_0 - P(r)}{\rho\Omega^2 R^2} = \frac{1}{2}\frac{R^2}{r^2} \quad \text{for} \quad \frac{r}{R} > 1\right)$$

At the interface between Region I and Region II (r = R) the pressure must be continuous, thus P(R) can be found by evaluating the above equation at r = R,

$$P(R) = P_0 - \frac{\rho\Omega^2 R^2}{2} \quad \text{at} \quad r = R$$

Now consider Region I (r < R). Substituting $U_\theta = \Omega r$, and integrating between r and R,

$$P(r) = P_0 - \rho\Omega^2\left(R^2 - \frac{r^2}{2}\right) \quad \text{for} \quad r < R \qquad \left(\frac{P_0 - P(r)}{\rho\Omega^2 R^2} = 1 - \frac{1}{2}\frac{r^2}{R^2} \quad \text{for} \quad \frac{r}{R} < 1\right)$$

Figure E7.1b is a graph of the pressure distribution, and shows that the pressure at the center of the flow is one half the value at the edge of the rotational region (Region I). Again, for convenience, the pressure has been normalized.

(c) Bernoulli equation - The Bernoulli constant (C) from Eq. (7-17) or (7-18) is equal to

$$C = \frac{P}{\rho} + \frac{U^2}{2}$$

since gravitational effects have been neglected. Since the velocity and pressure distributions have been computed for Region I and Region II, the Bernoulli constant can be found. In Region I,

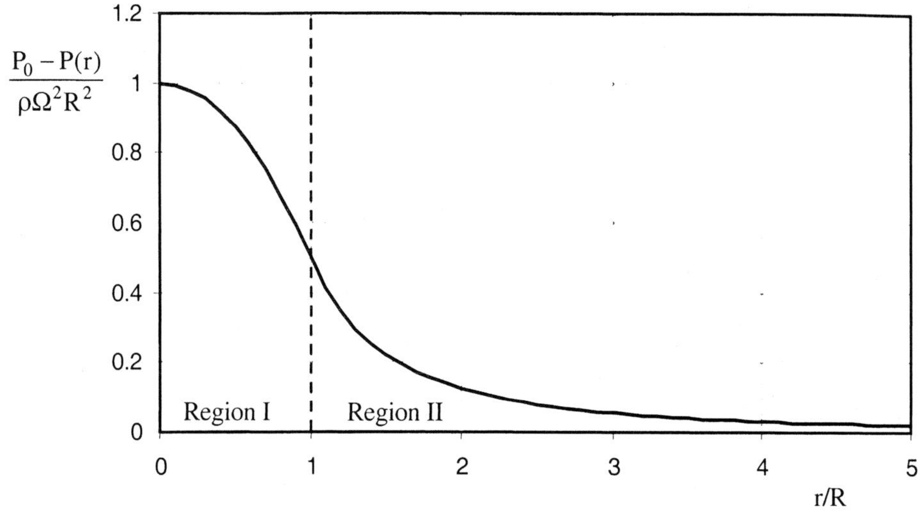

Figure E7.1b Normalized pressure above a rotating disk as a function of normalized radial location.

$$C = \frac{P_0}{\rho} - \Omega^2\left(R^2 - r^2\right) \quad \text{for} \quad r < R$$

which is obviously *not* constant. In Region II,

$$C = \frac{P_0}{\rho} \quad \text{for} \quad r > R$$

which *is* constant. Thus Bernoulli's equation is valid everywhere in the irrotational region (Region II), but is only valid along streamlines in the rotational region (Region I). In Region I the constant has a unique value for each radius.

Discussion: The above results are consistent with the assumptions made at the beginning of the analysis. If a more detailed analysis were desired, engineers would have to consider the three-dimensional features of the flow. Viscous effects are also significant close to the rotating disk where a boundary layer exists. Secondly, the predicted low pressure region in the center of the disk initiates an axial flow of air toward the disk which in turn produces a radially outward flow of air. Under these conditions, U_z and U_r are no longer zero and the full set of Navier-Stokes equations would have to be solved.

7.1.4 Ideal Flow

The above set of equations, (7-10), (7-11), and (7-17), form what many authors call the equations of ideal flow. ***Ideal flow*** is defined as incompressible and irrotational, and usually also steady, although unsteady ideal flow analysis is possible. Ideal flow is *not* inviscid, a subtle point often missed by students of fluid mechanics. As discussed previously, viscous stresses are still present in an irrotational flow, and the viscosity is certainly not zero. However, in irrotational flow, viscous stresses around a fluid element combine in such a way that the net viscous force is negligible compared to the inertial force. The terms involving the viscosity in the Navier-Stokes equation are negligible - not because the viscosity itself is negligible, but because the flow is irrotational. Analysis of ideal flow fields is not overly difficult, and a vast body of literature exists (Streeter, 1948; Milne-Thomson, 1950; Kirchhoff, 1985; Kundu, 1990; Panton, 1996; Streeter et al., 1998) that can be employed. Many complicated analytical expressions from technical literature once considered quaint or anachronistic are important for local ventilation systems. Complicated mathematics once considered troublesome can now be handled easily with computers. Steady ideal flow is assumed in the remainder of this chapter.

7.1.5 Stream Function

Consider two-dimensional incompressible flow in rectangular coordinates (x,y), (U_x,U_y), where U_x and U_y represent components of the air velocity in the x and y directions. (Note that U_x and U_y are *not* derivatives of $\vec{U}$ with respect to x and y, which is an editorial practice used in some texts.) Similar terminology is used for cylindrical and spherical coordinates. A ***streamline*** is defined as a line (actually a *curve*) everywhere tangent to the local velocity vector. In steady flow, streamlines are identical to ***pathlines***, which are the paths followed by individual fluid elements. From this definition, at any point along a streamline,

$$\frac{U_y}{U_x} = \frac{dy}{dx}$$

which can be written as

$$\boxed{0 = U_y dx - U_x dy} \tag{7-19}$$

A continuous scalar function $\psi(x,y)$ called the ***stream function*** can be defined such that for any steady two-dimensional flow field,

$$\boxed{U_x = -\frac{\partial \psi}{\partial y} \qquad U_y = \frac{\partial \psi}{\partial x}} \tag{7-20}$$

Readers must be aware that the sign convention shown in Eq. (7-20) and adopted throughout this text is not universal. Some fluid mechanics texts define the stream function with an opposite sign (the negative sign on U_y instead of U_x). The notation convention of Streeter et al. (1998) has been adopted in the present book. The total derivative of ψ can be expressed by the chain rule:

$$d\psi = \frac{\partial \psi}{\partial x}dx + \frac{\partial \psi}{\partial y}dy$$

The partial derivatives of ψ with respect to x and y in the above equation can be replaced by velocity components from Eq. (7-20). Thus,

$$\boxed{d\psi = U_y dx - U_x dy} \tag{7-21}$$

Comparing Eqs. (7-19) and (7-21), it can be concluded that along a streamline ψ is constant. From the definition of streamline given earlier, it can also be concluded that between any two streamlines the mass flow rate is constant, since fluid cannot pass through a streamline. Thus if the density is also constant, the volumetric flow rate per unit depth between two streamlines is

$$\boxed{Q = \int_1^2 d\psi = (\psi_2 - \psi_1)} \tag{7-22}$$

The region bounded by two streamlines in two-dimensional flow is called a ***stream tube***. The volumetric flow rate in a stream tube (per unit depth) is constant.

Substituting the definition of the stream function, Eq. (7-20), into the continuity equation for steady incompressible flow, Eq. (7-8), results in

$$\frac{\partial U_x}{\partial x} + \frac{\partial U_y}{\partial y} = 0 \qquad \text{or} \qquad -\frac{\partial^2 \psi}{\partial x \partial y} + \frac{\partial^2 \psi}{\partial y \partial x} = 0$$

which is identically satisfied as long as ψ is a smooth, continuous function of x and y (the order of differentiation does not matter). In other words, the very definition of stream function is consistent with the steady incompressible form of the continuity equation, in that the continuity equation is automatically satisfied.

If the steady, incompressible, two-dimensional flow is also *irrotational*, Eq. (7-9) holds. In rectangular coordinates, the z-component of Eq. (7-9), which for 2-D flow in the x-y plane is the only non-zero component, becomes

$$\frac{\partial U_y}{\partial x} - \frac{\partial U_x}{\partial y} = 0$$

Substitution of Eq. (7-20) into the above yields

$$\boxed{\frac{\partial^2 \psi}{\partial x^2} + \frac{\partial^2 \psi}{\partial y^2} = 0} \tag{7-23}$$

which is the two-dimensional form of the Laplace equation for the stream function in rectangular coordinates,

$$\boxed{\nabla^2 \psi = 0} \tag{7-24}$$

The reader may recall that a similar equation, Eq. (7-11), was obtained previously for the velocity potential function for irrotational (potential) flow. It must be pointed out that Eq. (7-11) is valid for any irrotational flow, whether two- or three-dimensional. Equation (7-24), on the other hand, is only valid for two-dimensional, irrotational flows.

7.2 Two-Dimensional Potential Flow Fields

Many practical flow fields in indoor air pollution control can be approximated as ideal (incompressible and irrotational), steady, and two-dimensional. Such flow fields, described by Eqs. (7-11) and (7-24), are called ***potential flow fields***, and the functions ψ and ϕ are called ***potential functions***.

7.2.1 Relationship between ψ and ϕ

Curves of constant values of ψ define fluid streamlines, while curves of constant values of ϕ define ***equipotential lines***. In two-dimensional potential flow fields, streamlines are everywhere normal to equipotential lines, a condition known as ***mutual orthogonality***. In two-dimensional potential flows, the potential functions ψ and ϕ are intimately related to each other - both satisfy the Laplace equation, and from either ψ or ϕ, one can determine the velocity field. Mathematicians call solutions of ψ and ϕ ***harmonic functions***, and ψ and ϕ are ***harmonic conjugates*** of each other. Although ψ and ϕ are related, their origins are somewhat opposite; it is perhaps best to say that ψ and ϕ are *complimentary* to each other:

- *The stream function is defined by continuity; the Laplace equation for ψ results from irrotationality.*
- *The velocity potential is defined by irrotationality; the Laplace equation for ϕ results from continuity.*

Some authors use the phrase "equipotential lines" to refer to both streamlines and lines of constant ϕ.

Solutions of Laplace's equation have been found for a variety of phenomena encountered in engineering (Pipes and Harvill, 1970). Solutions of Laplace's equation can be found elegantly and simultaneously using complex variables (Churchill et al., 1976). In this book, however, only real variables are used. Solutions to Eqs. (7-11) and/or (7-24) must satisfy boundary conditions dictated by the physical constraints of the particular ventilation system. The equation can be solved by analytical or numerical methods. Graphical or experimental techniques, often called ***flux plotting***, can also be used. Since the accuracy of these techniques is poor, they are omitted in the present discussion.

The concept of "two-dimensional" flow is not limited to rectangular coordinates, but also includes cylindrical coordinates and spherical coordinates in which only two directions of motion are important. Equations for these three coordinate systems are summarized below.

7.2.2 Planar Flow with Rectangular Coordinates:

Two-dimensional planar flows are generally flows in the x-y plane, with no variation in the z-direction. The directional components are (x,y) and the velocity components are (U_x,U_y). In terms of velocity potential and stream function,

$$U_x = -\frac{\partial \phi}{\partial x} = -\frac{\partial \psi}{\partial y} \quad \text{and} \quad U_y = -\frac{\partial \phi}{\partial y} = \frac{\partial \psi}{\partial x} \tag{7-25}$$

7.2.3 Planar Flow with Cylindrical Coordinates:

The same two-dimensional planar flows can be written in cylindrical rather than rectangular coordinates. This is often more convenient, especially for flows which have circular streamlines or equipotential lines. The directional components are (r,θ) and the velocity components are (U_r,U_θ). In terms of velocity potential and stream function,

$$U_r = -\frac{\partial \phi}{\partial r} = -\frac{1}{r}\frac{\partial \psi}{\partial \theta} \quad \text{and} \quad U_\theta = -\frac{1}{r}\frac{\partial \phi}{\partial \theta} = \frac{\partial \psi}{\partial r} \tag{7-26}$$

One can convert between rectangular coordinates and cylindrical coordinates using the following conversions:

$$x = r\cos\theta \qquad y = r\sin\theta \qquad r = \sqrt{x^2 + y^2} \tag{7-27}$$

With a bit of trigonometry, a relationship for rectangular velocity components in terms of cylindrical velocity components can be derived,

$$U_x = U_r \cos\theta - U_\theta \sin\theta \qquad U_y = U_r \sin\theta + U_\theta \cos\theta \tag{7-28}$$

7.2.4 Axisymmetric Flow with Polar Coordinates:

Axisymmetric flows, such as flow around a navy torpedo, have an axis of symmetry, taken here as the x-axis. The directional components are again (r,θ) and the velocity components are (U_r,U_θ). In this coordinate system, radius r is the distance from the origin, and polar angle θ is the angle of inclination between the radial vector and the x-axis, as sketched in Figure 7.3.

Axisymmetric flow is a type of two-dimensional flow in that there are only two independent spatial variables, r and θ, since the flow is rotationally symmetric about the x-axis. In other words, regardless of the inclination of the r-θ plane from the vertical x-z plane, the velocity is determined only by r and θ. A plot of streamlines in *any* r-θ plane is therefore sufficient to characterize the axisymmetric flow field. In terms of velocity potential and stream function,

$$U_r = -\frac{\partial \phi}{\partial r} = -\left(\frac{1}{r^2 \sin\theta}\right)\frac{\partial \psi}{\partial \theta} \quad \text{and} \quad U_\theta = -\frac{1}{r}\frac{\partial \phi}{\partial \theta} = \left(\frac{1}{r\sin\theta}\right)\frac{\partial \psi}{\partial r} \tag{7-29}$$

A summary of potential flow functions is provided in Table 7.1 for the above three coordinate systems.

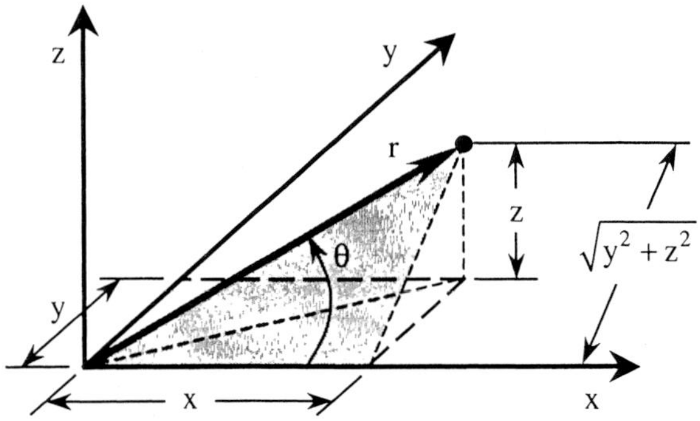

Figure 7.3 Coordinate system for axisymmetric flow with the x-axis as the axis of symmetry.

Table 7.1 Velocity components in terms of potential functions.

coordinate system	velocity component 1	velocity component 2
two-dimensional (rectangular)	$U_x = -\frac{\partial\phi}{\partial x} = -\frac{\partial\psi}{\partial y}$	$U_y = -\frac{\partial\phi}{\partial y} = \frac{\partial\psi}{\partial x}$
two-dimensional (cylindrical)	$U_r = -\frac{\partial\phi}{\partial r} = -\frac{1}{r}\frac{\partial\psi}{\partial\theta}$	$U_\theta = -\frac{1}{r}\frac{\partial\phi}{\partial\theta} = \frac{\partial\psi}{\partial r}$
axisymmetric	$U_r = -\frac{\partial\phi}{\partial r} = -\left(\frac{1}{r^2\sin\theta}\right)\frac{\partial\psi}{\partial\theta}$	$U_\theta = -\frac{1}{r}\frac{\partial\phi}{\partial\theta} = \left(\frac{1}{r\sin\theta}\right)\frac{\partial\psi}{\partial r}$

7.2.5 Physical Significance

In order to map a particle's trajectory or to compute the transport of contaminant vapor in a moving air mass, it is necessary to compute the air velocity, $\vec{U}$, at any location (x,y). If analytical expressions for either ψ or ϕ are known as functions of (x,y), one can differentiate to find the velocity components in the x and y directions, using Eqs. (7-25), (7-26), or (7-29), depending on the coordinate system chosen. Expressions for ψ and ϕ have been derived (Streeter 1948; Milne-Thomson 1950; Kirchhoff, 1985) for several useful ventilation inlets:

- point sinks and line sinks in a plane
- uniform flow over a surface (streaming flow)
- flow into a rectangular opening (flanged slot)
- flow into a plain or flanged rectangular duct
- flow into a flanged circular inlet

Table 7.2 is a summary of the harmonic functions for several of the above flow conditions. If one is faced with a flow field vastly different from these, it is possible to solve Laplace's equation numerically to obtain tabulated values of ψ and ϕ at any location in the flow field.

Many potential functions have ***singular points*** (also called ***singularities***) at which the function and/or its derivatives approach infinity. For example, point sinks and the lip of an inlet, both with and without flanges, are singular points. Infinite velocities do not occur in reality, and hence users should not expect physically meaningful results in regions near singular points. Since the size of these regions is small with respect to the large region for which the solutions are valid, avoiding these regions is a small limitation on the analysis. A summary of the physical significance of the velocity potential and stream function is provided here:

Table 7.2 Potential functions for various inlets.

flow description	sketch	potential functions
oblique streaming flow (x,y) coordinate system	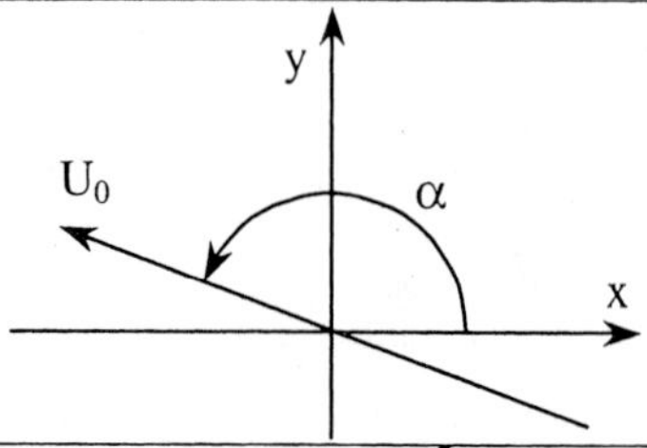	$\phi = -U_0 x\cos\alpha - U_0 y\sin\alpha$ $\psi = -U_0 y\cos\alpha + U_0 x\sin\alpha$
point sink in a plane (r,θ) coordinate system	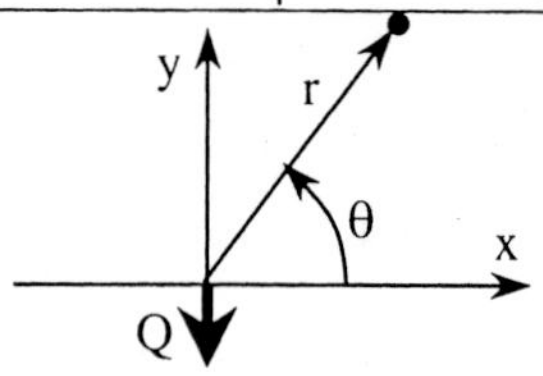	$\phi = -\frac{Q}{2\pi}\frac{1}{r}$ $\psi = -\frac{Q}{2\pi}\cos\theta$
line sink in a plane (r,θ) coordinate system	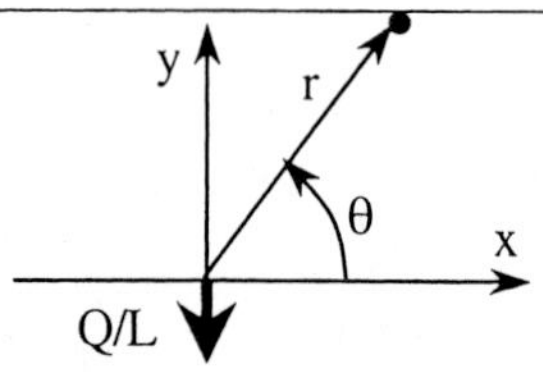	$\phi = \frac{Q}{L\pi}\ln r$ $\psi = \frac{Q}{L\pi}\theta$
flanged rectangular inlet (x,y) coordinate system	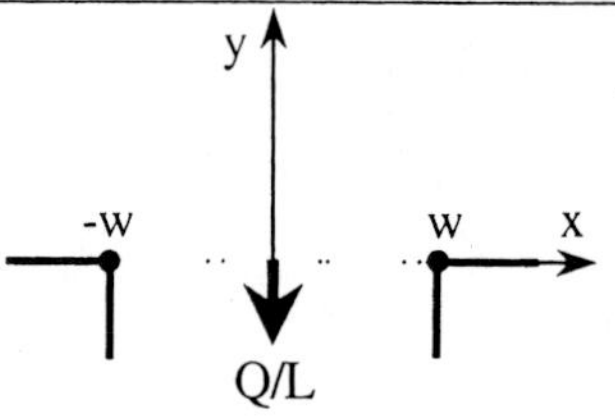	$\frac{x}{w} = \cosh\left(\frac{\phi}{k}\right)\cos\left(\frac{\psi}{k}\right)$ $\frac{y}{w} = \sinh\left(\frac{\phi}{k}\right)\sin\left(\frac{\psi}{k}\right)$ $k = \frac{Q}{\pi L}$
unflanged rectangular inlet (x,y) coordinate system	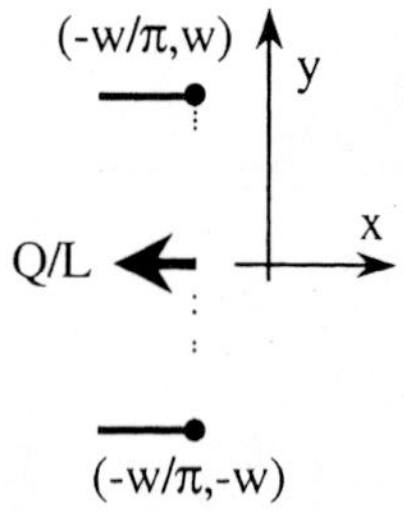	$\frac{x}{w} = \frac{1}{\pi}\left[2\frac{\phi}{k} + \exp\left(2\frac{\phi}{k}\right)\cos\left(2\frac{\psi}{k}\right)\right]$ $\frac{y}{w} = \frac{1}{\pi}\left[2\frac{\psi}{k} + \exp\left(2\frac{\phi}{k}\right)\sin\left(2\frac{\psi}{k}\right)\right]$ $k = \frac{Q}{\pi L}$
Flanged circular inlet (r,z) coordinate system	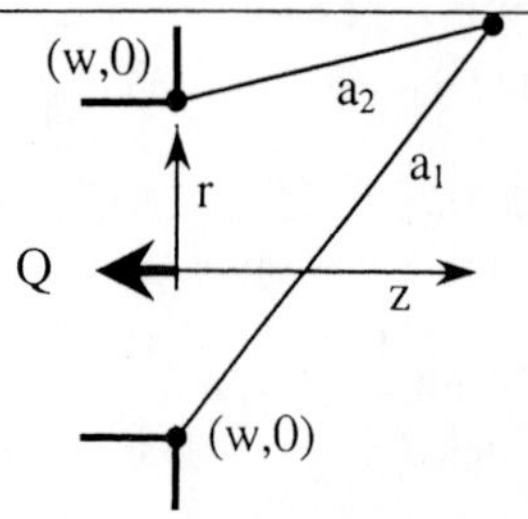	$\phi = \frac{Q}{2\pi w}\arcsin\left(\frac{2w}{a_1 + a_2}\right)$ $\psi = \frac{Q}{4\pi w}\sqrt{4w^2 - (a_1 - a_2)^2}$ $a_1 = \sqrt{z^2 + (w+r)^2}$ $a_2 = \sqrt{z^2 + (w-r)^2}$

- A curve along which the stream function is constant is called a streamline.
- A curve along which the velocity potential is constant is called an equipotential line.
- Fluid streamlines are lines everywhere tangent to the local velocity vector. For steady flow, streamlines are equivalent to pathlines.
- Fluid particles do not cross streamlines.
- In planar flow, the volumetric flow rate (Q) per unit depth between any two streamlines ψ_1 and ψ_2 is equal to $(\psi_2 - \psi_1)$
- Streamlines and equipotential lines are mutually orthogonal, i.e. they cross at 90° angles, except at singularities in the flow.
- The velocity components can be obtained from either ψ or ϕ, using Eqs. (7-25), (7-26), or (7-29), depending on the geometry of the flow and the coordinate system chosen.
- Once the velocity field has been obtained, the *pressure* field can be calculated using the Bernoulli equation, Eq. (7-17). In problems related to indoor air pollution control, the gravity term in the Bernoulli equation is usually ignored without appreciable loss of accuracy.

7.2.6 Superposition

Since Laplace's equation is a *linear* homogeneous differential equation, the combination of two or more solutions to the equation is also a solution. For example, if ϕ_1 and ϕ_2 are each solutions to Laplace's equation, then $A\phi_1$, $(A + \phi_1)$, $(\phi_1 + \phi_2)$, and $(A\phi_1 + \phi_2)$ are also solutions, where A is an arbitrary constant. Thus, one may combine several harmonic functions and solutions to Laplace's equation, and the combination is guaranteed to also be a solution. If a flow field is known to be the sum of two separate flow fields, e.g. a sink located in a streaming flow, one can combine the harmonic functions for each to describe the combined flow field. This process of adding two or more known solutions to create a third, more complicated solution is known as ***superposition***.

Superposition reveals an important feature of ideal flow fields. Suppose a flow field is produced by two independent flow fields, each of which can be treated as ideal. Then the potential function of the composite field is the sum of the potential functions of the component flow fields, and the velocity at any point in the composite field is the vector sum of the velocities of the component flow fields. Consider an ideal planar flow field that is the superposition of two independent flow fields denoted by the subscripts 1 and 2. The composite velocity potential function is given by

$$\boxed{\phi = \phi_1 + \phi_2} \tag{7-30}$$

Using Eq. (7-25), the x-component of the velocity of the composite field is then

$$\boxed{U_x = -\frac{\partial \phi}{\partial x} = -\frac{\partial(\phi_1 + \phi_2)}{\partial x} = -\frac{\partial \phi_1}{\partial x} - \frac{\partial \phi_2}{\partial x}} \tag{7-31}$$

But for each component flow, Eq. (7-25) is also valid,

$$U_{x,1} = -\frac{\partial \phi_1}{\partial x} \qquad U_{x,2} = -\frac{\partial \phi_2}{\partial x}$$

Thus, Eq. (7-31) becomes

$$\boxed{U_x = U_{x,1} + U_{x,2}} \tag{7-32}$$

with a similar expression for the y component of velocity (U_y). Thus, superposition enables one to simply add the velocity fields vectorially,

$$\boxed{\vec{U} = \vec{U}_1 + \vec{U}_2} \tag{7-33}$$

A similar analysis can be performed using the stream function rather than the velocity potential function, with the identical result, Eq. (7-33). The concept of superposition is useful, but is valid only for *irrotational* flow fields. Readers must be careful to ensure that the two flow fields they wish to add vectorially are both irrotational. For example the velocity profile of a jet should never be added to the velocity profile of an inlet or streaming flow because the velocity field associated with a jet is strongly affected by viscosity, is not irrotational, and cannot be described by potential functions.

Superposition is the concept upon which ***capture velocity*** (see Chapter 6) seems to be based. If the contaminants to be removed are vapors (gases), then the velocity of the air stream in which they are contained can be added to the velocity produced by an inlet to predict the velocity of the composite flow field. Hence the underlying notion of capture velocity, i.e. "... velocity ... necessary to overcome opposing air currents ..." has validity. However, the concept of capture velocity is applied in practice to capture of *particles*, not vapors. If the contaminant is a particle whose velocity is not equal to the air velocity or a particle whose Reynolds number is much larger than 1, the concept of superposition is not appropriate because the particle velocity and gas velocity are not the same; the particle's motion is governed by its inertial effects and by viscosity, as discussed in Chapter 8.

7.3 Elementary Planar Ideal Flows

Superposition enables one to add two or more simple ideal flow solutions to create a more complex (and hopefully more physically significant) flow field. It is therefore useful to establish a collection of elementary "building block" ideal flows from which one can construct a variety of useful flows. Elementary two-dimensional planar ideal flows are described in x-y and/or r-θ coordinates, depending on which pair is more useful in a particular problem.

7.3.1 Line Sink

Consider a line segment of length L in space through which an ideal fluid flows inward at uniform volumetric flow rate (Q). As L approaches infinity, the flow becomes two-dimensional in planes perpendicular to the line's axis, and the line into which flow is drawn is called a ***line sink***. Several streamlines and equipotential lines are sketched for a line sink in Figure 7.4. For an infinite line, Q also approaches infinity; thus, it is more convenient to consider the volumetric flow rate per unit depth, Q/L. *In the equations in this book, Q is positive (Q > 0) if there is suction, i.e. if air is withdrawn into a sink.* This is the sign convention adopted by engineers who design close capture hoods. Readers should note that this sign convention is opposite to the traditional convention found in most fluid mechanics textbooks, in which positive Q signifies a source and negative Q signifies a sink. (A source is opposite to a sink.)

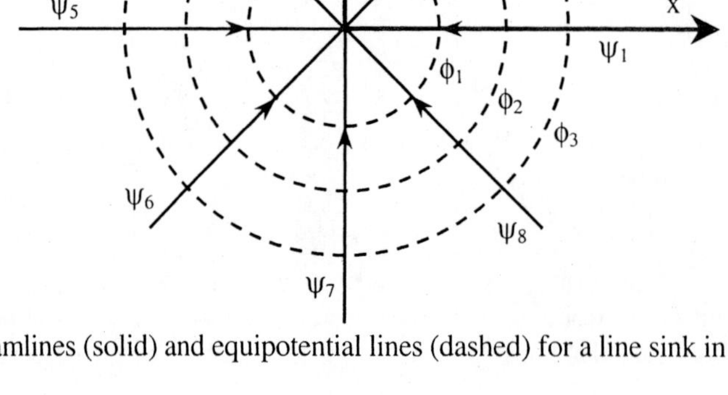

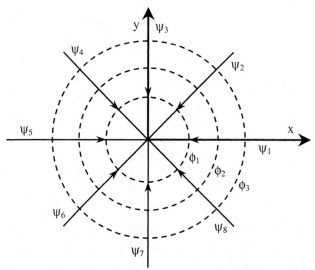

Figure 7.4 Streamlines (solid) and equipotential lines (dashed) for a line sink in a plane.

The volumetric flow rate through a cylindrical surface of radius r and length L surrounding the line sink is

$$Q = -2\pi r L U_r$$

which can be solved for the radial velocity component (U_r),

$$\boxed{U_r = -\frac{Q}{2\pi L}\frac{1}{r}} \tag{7-34}$$

The negative sign for U_r occurs because the flow is inward, i.e. radial velocity is in the direction of decreasing values of r. Due to symmetry, the tangential component of velocity (U_θ) is zero. Equation (7-26) is used to relate U_r to the potential functions ϕ and ψ,

$$\boxed{U_r = -\frac{\partial\phi}{\partial r} = -\frac{1}{r}\frac{\partial\psi}{\partial\theta} = -\frac{Q}{2\pi L}\frac{1}{r}} \tag{7-35}$$

Similarly,

$$\boxed{U_\theta = -\frac{1}{r}\frac{\partial\phi}{\partial\theta} = \frac{\partial\psi}{\partial r} = 0} \tag{7-36}$$

To find the velocity potential, Eq. (7-35) is integrated:

$$\phi = \int \frac{Q}{2\pi L}\frac{1}{r}dr = \frac{Q}{2\pi L}\ln(r) + g(\theta)$$

A function of the other variable (θ), rather than a constant, has been added to the integration since this is a *partial* integration of a function of two variables. Now the θ derivative of the above equation is taken:

$$\frac{\partial\phi}{\partial\theta} = g'(\theta) = \frac{d[g(\theta)]}{d\theta} = 0$$

where the prime denotes the derivative with respect to θ, and the zero on the right hand side comes from Eq. (7-36). Integrating again, the function $g(\theta)$ is an arbitrary constant which can be set equal to zero since constants can be added and deleted from harmonic functions. Finally then,

$$\boxed{\phi = \frac{Q}{2\pi L}\ln r} \tag{7-37}$$

In similar fashion (details left for the reader), Eq. (7-35) can be integrated to find the stream function. The result is

$$\boxed{\psi = \frac{Q}{2\pi L}\theta} \tag{7-38}$$

As a check of the algebra, the reader should verify that the correct velocity components are obtained by differentiation of Eqs. (7-37) and/or (7-38).

7.3.2 Uniform Streaming Flow

Consider an ideal fluid moving uniformly in the negative x-direction with speed U_0. This is called ***streaming flow***. In terms of the velocity potential and stream function, from Eq. (7-25),

$$\boxed{U_x = -\frac{\partial\phi}{\partial x} = -\frac{\partial\psi}{\partial y} = -U_0} \tag{7-39}$$

and

$$\boxed{U_y = -\frac{\partial \phi}{\partial y} = \frac{\partial \psi}{\partial x} = 0} \tag{7-40}$$

Integrating Eq. (7-39) with respect to x,

$$\phi = U_0 x + g(y)$$

and then differentiating with respect to y,

$$\frac{\partial \phi}{\partial y} = \frac{d[g(y)]}{dy} = g'(y) = 0$$

where the zero comes from Eq. (7-40). Therefore g(y) is a constant that may be set equal to zero, and

$$\boxed{\phi = U_0 x} \tag{7-41}$$

Similarly the stream function can be found,

$$\psi = U_0 y + f(x) \qquad \text{and} \qquad \frac{\partial \psi}{\partial x} = \frac{d[f(x)]}{dx} = f'(x) = 0$$

Here, f(x) also must therefore be a constant which can be set to any value. Zero is again chosen for simplicity since any arbitrary constant can be added later if necessary or convenient. Thus,

$$\boxed{\psi = U_0 y} \tag{7-42}$$

Figure 7.5 shows streamlines and equipotential lines for streaming flow in the negative x-direction.

It is often convenient to express the stream function and velocity potential function in cylindrical coordinates rather than rectangular coordinates, particularly when superposing streaming flow with some other ideal flow(s). Using Eq. (7-27), the equations for ϕ and ψ become

$$\boxed{\phi = U_0 x = U_0 r \cos\theta} \tag{7-43}$$

and

$$\boxed{\psi = U_0 y = U_0 r \sin\theta} \tag{7-44}$$

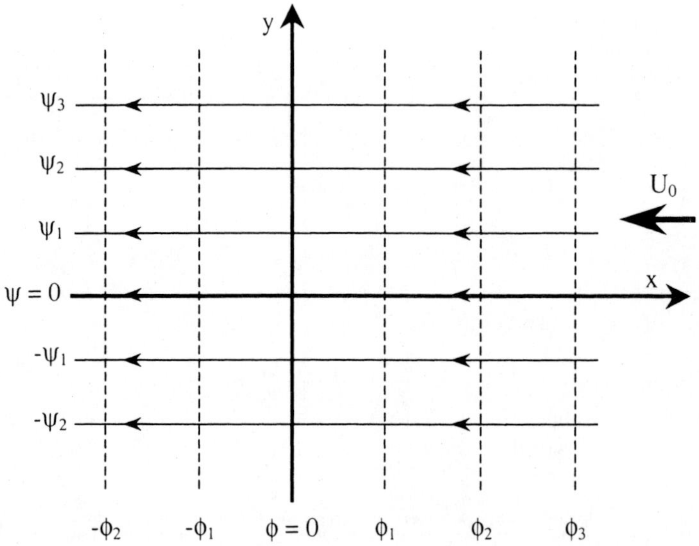

Figure 7.5 Streamlines (solid) and equipotential lines (dashed) for streaming flow in the negative x-direction; streamlines are solid, and lines of constant ϕ are dashed.

Table 7.2 shows equations for ϕ and ψ for the case of oblique streaming flow at an arbitrary angle (α) as measured from the x-axis.

7.3.3 Example of Superposition - Line Sink in a Plane in Streaming Flow

As a simple example of superposition, the above two building block flows (line sink and uniform stream) are combined. Consider an ideal fluid passing over a horizontal surface at uniform velocity U_0 in the negative x-direction (angle $\alpha = \pi$, Table 7.2). The surface contains a line sink through which fluid is withdrawn at volumetric flow rate Q. The potential functions for the composite flow are the sum of the potential functions for a streaming flow and a line sink. To account for the effect of the wall, the volumetric flow rate of the line sink needs to be twice that of a line sink in space (since Q is here defined as the entire flow rate withdrawn into the line sink, but it is coming from only half of space). The potential functions are

$$\boxed{\phi = U_0 r\cos\theta + \left(\frac{Q}{\pi L}\right)\ln r} \tag{7-45}$$

and

$$\boxed{\psi = U_0 r\sin\theta + \left(\frac{Q}{\pi L}\right)\theta} \tag{7-46}$$

The velocity components at any point (r, θ) are obtained from Eq. (7-26). In cylindrical coordinates,

$$\boxed{U_r = -\frac{\partial\phi}{\partial r} = -\left[U_0\cos\theta + \frac{Q}{\pi L r}\right]} \tag{7-47}$$

and

$$\boxed{U_\theta = -\frac{1}{r}\frac{\partial\phi}{\partial\theta} = U_0\sin\theta} \tag{7-48}$$

Using the conversions for velocity components derived earlier, namely Eqs. (7-28), along with the conversions for x and y themselves, Eqs. (7-27), one can also write expressions for the velocity components of this flow field in rectangular coordinates:

$$U_x = -\left[U_0\cos^2\theta + \frac{Q}{\pi L r}\cos\theta\right] - U_0\sin^2\theta = -U_0 - \frac{Q}{\pi L}\frac{x}{\sqrt{x^2+y^2}}\frac{1}{\sqrt{x^2+y^2}}$$

or

$$\boxed{U_x = -U_0 - \frac{Q}{\pi L}\frac{x}{x^2+y^2}} \tag{7-49}$$

Similarly,

$$U_y = -\left[U_0\cos\theta + \frac{Q}{\pi L r}\right]\sin\theta + U_0\sin\theta\cos\theta = -\frac{Q}{\pi L r}\sin\theta = -\frac{Q}{\pi L}\frac{y}{\sqrt{x^2+y^2}\sqrt{x^2+y^2}}$$

or

$$\boxed{U_y = -\frac{Q}{\pi L}\frac{y}{x^2+y^2}} \tag{7-50}$$

Examination of Eqs. (7-49) and (7-50) shows that the velocity of a composite ideal flow is the sum of the velocities of its component parts, as was mentioned previously. This is a direct result of the linear nature of Laplace's equation, which enables superposition to work.

Far above the line sink, the flow is deflected downward by the line sink as it passes from right to left but it is not drawn into the line sink. Closer to the horizontal surface, fluid is drawn into the line sink. Consequently there must a unique streamline that differentiates the flow that enters the line sink from that which does not. The ***dividing streamline*** (ψ_d) is the streamline that divides the two regions. Figure 7.6 shows the location of the dividing streamline. The region between the x-axis and the dividing streamline represents the region in which all the fluid passes through the line sink. It is useful to define the ***reach*** of an inlet as the region in space upwind of the inlet from which all the fluid ultimately enters the inlet. Thus the reach of a line sink in streaming flow is defined by the region between the x-axis and the dividing streamline (ψ_d).

Just as there is a dividing streamline that distinguishes two regimes of flow, so also must there be a ***stagnation point*** at some location on the x-axis, where $U_r = 0$ and $U_\theta = 0$. It turns out that the stagnation point is at the intersection of the dividing streamline and the x-axis ($r = r_0$, $\theta = \pi$), as labeled in Figure 7.6. Setting $U_r = 0$ in Eq. (7-47) at this point,

$$U_r\left(r = r_0,\ \theta = \pi\right) = -\left[U_0 \cos\pi + \frac{Q}{\pi L r_0}\right] = 0$$

from which one can solve for r_0,

$$\boxed{r_0 = \frac{Q}{\pi L U_0}} \tag{7-51}$$

The dividing streamline passing through the stagnation point has the value

$$\boxed{\psi_d = \psi\left(r_0, \pi\right) = U_0 r_0 \sin\pi + \frac{Q}{\pi L}\pi = \frac{Q}{L}} \tag{7-52}$$

The streamline that coincides with the positive x-axis has the value $\psi = 0$. The volumetric flow rate per unit depth between ψ_d and $\psi = 0$ is equal to (Q/L) in agreement with Eq. (7-22). The height of the dividing streamline at $x >> 0$, i.e. h_∞, can be found from Eq. (7-46) by taking the limit of the following expression as x approaches infinity and θ approaches zero:

$$\psi_d = \frac{Q}{L} = \underset{\theta\to 0}{\text{limit}}\left[-U_0 r \sin\theta - \frac{Q}{\pi L}\theta\right] = -U_0 h_\infty$$

from which

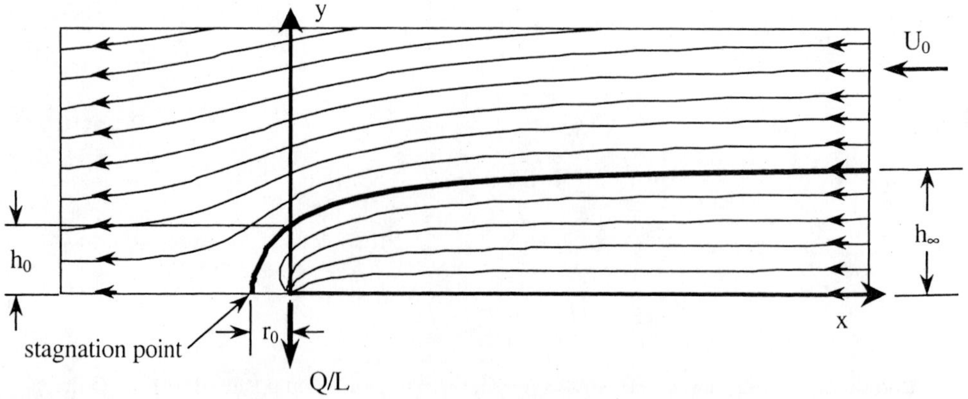

Figure 7.6 Streamlines resulting from superposition of a line sink and a streaming flow; the dividing streamline is shown in bold; flow is from left to right.

$$\boxed{h_\infty = \frac{Q}{LU_0}} \qquad (7\text{-}53)$$

The intersection of the dividing streamline and the y-axis ($r = h_0$, $\theta = \pi/2$) can be found from

$$\psi_d = \frac{Q}{L} = U_0 h_0 \sin\frac{\pi}{2} + \frac{Q}{\pi L}\frac{\pi}{2}$$

from which

$$\boxed{h_0 = \frac{Q}{2LU_0}} \qquad (7\text{-}54)$$

In summary,

$$\boxed{h_\infty = \frac{Q}{LU_0} = \pi r_0 = 2h_0} \qquad (7\text{-}55)$$

h_∞, r_0, and h_0 are labeled in Figure 7.6.

Example 7.2 - Reach of a Flanged Line Sink in a Streaming Flow

Given: A line sink of strength (Q/L) equal to 30.0 m^2/min (129. SCFM/ft) lies in a horizontal plane over which air passes from right to left, as in Figure 7.6. The sink is located at the origin. Far upwind of the line sink the air has a uniform velocity (U_0) of 30.5 m/min (100. FPM or 0.508 m/s).

To do:

(a) Compute and plot the shape and location of the dividing streamline.
(b) Compute and plot the values of U_x and U_y at points along the dividing streamline.
(c) Describe the region within which the air speed is 200 FPM or greater. Repeat for air speeds of 150 and 120 FPM.

Solution:

(a) A plot of the dividing streamline can be generated from Eq. (7-46) by setting ψ = Q/L and computing a series of radii (r) for a range of θ from a small value (such as 1°, 0.0174 radians) to 180° (π radians). The coordinates (x,y) are computed from Eq. (7-27). The authors used Excel to do the calculations; the file is available on the web site. Figure E7.2a shows the dividing streamline ($\psi = \psi_d$). The reach is the region between the dividing streamline and the x-axis. Using Eqs. (7-51), (7-53), and (7-54), the following distances are computed:

$$r_0 = \frac{\frac{Q}{L}}{\pi U_0} = \frac{30.0\frac{m^2}{min}}{\pi\, 30.5\frac{m}{min}} = 0.313\text{ m} \qquad h_\infty = \frac{\frac{Q}{L}}{U_0} = \frac{30.0\frac{m^2}{min}}{30.5\frac{m}{min}} = 0.984\text{ m}$$

and

$$h_0 = \frac{\frac{Q}{L}}{2U_0} = \frac{30.0\frac{m^2}{min}}{2\left(30.5\frac{m}{min}\right)} = 0.492\text{ m}$$

(b) At the coordinates (x,y) along the dividing streamline obtained in Part (a), U_x and U_y are computed using Eqs. (7-49) and (7-50). Figure E7.2b was created with the spreadsheet, and shows

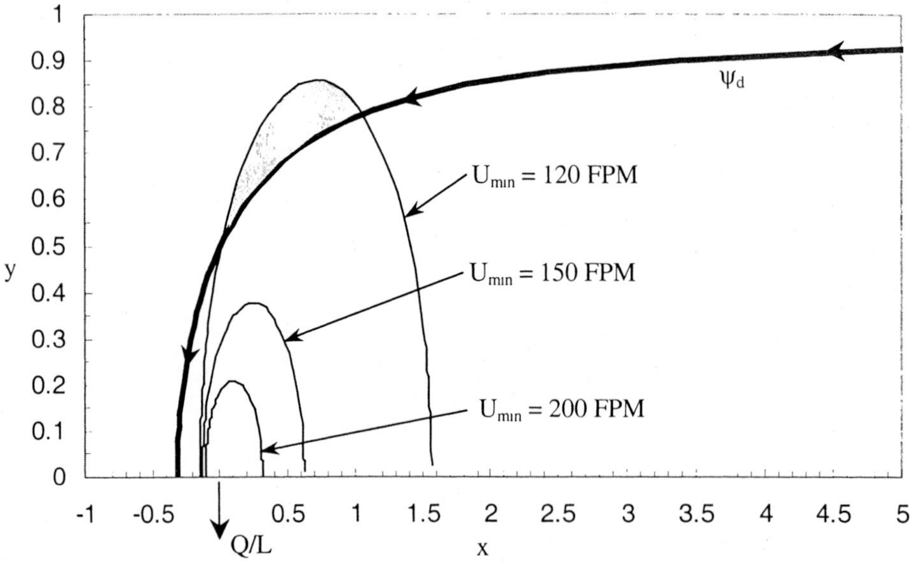

Figure E7.2a Dividing streamline formed by superposition of a line sink and streaming flow; flow is from left to right. Also shown are isopleths at three air speeds; the shaded region indicates locations above the dividing streamline where air speed exceeds 120 FPM.

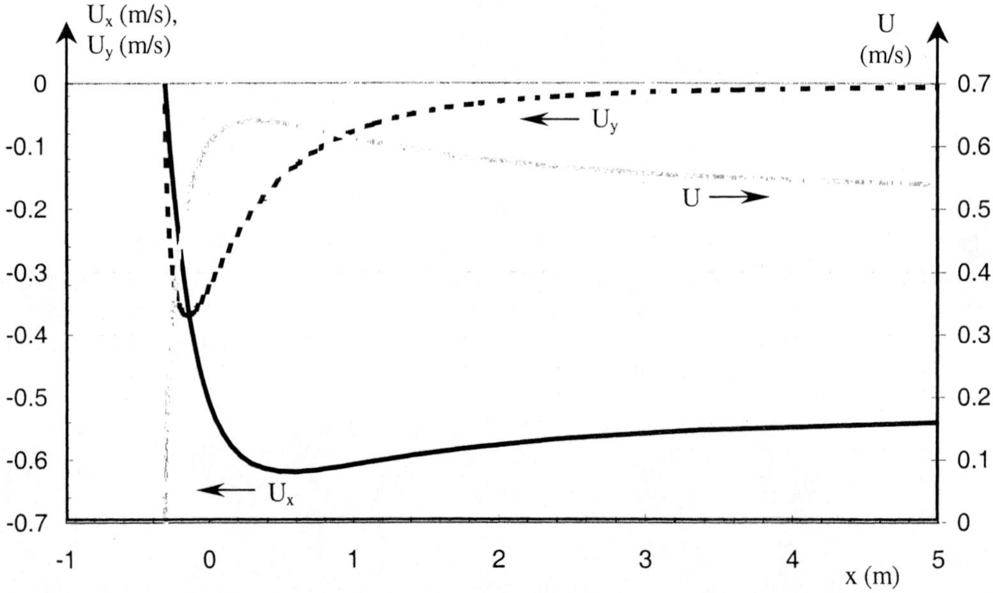

Figure E7.2b Velocity components U_x and U_y and air speed, U, along the dividing streamline for the data of Example 7.2.

these velocity components as functions of x. The speed of the air along the dividing streamline varies from U_0 at x >> 0 to zero at the stagnation point. U_x and U_y are both negative along the dividing streamline, indicating flow downward and to the left. The magnitude of the air velocity, i.e. air speed (U) along the dividing streamline is also plotted in Figure E7.2b. The maximum air speed along the dividing streamline is about 0.64 m/s (126. FPM), which occurs at x ≈ 0.30 m. The region near the maximum air speed along the dividing streamline is sometimes called the "shoulder".

(c) Suppose an industrial operation is conducted for which the capture velocity (see Chapter 6) is 200 FPM (1.02 m/s). The ventilation engineer wants to locate the region within which the air velocity is equal to or larger than this value, which will be called U_{min}. The ***capture region*** is the locus of points (r, θ) within which the air speed exceeds U_{min}. From Eqs. (7-47) and (7-48),

$$\left|\vec{U}\right|^2 = U_{min}^{\ 2} = U_r^{\ 2} + U_\theta^{\ 2} = \left[-\left(U_0 \cos\theta + \frac{Q}{\pi L r}\right)\right]^2 + \left(U_0 \sin\theta\right)^2$$

which can be manipulated into quadratic form for unknown r as follows:

$$\left(U_{min}^{\ 2} - U_0^{\ 2}\right) r^2 - \frac{2U_0 \cos(\theta)}{\pi} \frac{Q}{L} r - \left(\frac{Q}{L}\right)^2 \frac{1}{\pi^2} = 0$$

In the spreadsheet, r is calculated from the above equation using the quadratic formula for various values of θ. Because of the quadratic nature of the equation, r is double-valued. Fortunately, it turns out that one of the r values is negative and unphysical; that value of r is rejected. The values of r versus θ are converted to y versus x, and are plotted as an isopleth of value U_{min} = 200 FPM in Figure E7.2a. The air speed below this isopleth is greater than 200 FPM, and that above the isopleth is smaller than 200 FPM. The capture region is therefore the area enclosed by the isopleth and the x-axis. The calculations are repeated for U_{min} = 150 and 120 FPM; these isopleths are also plotted in Figure E7.2a.

Discussion: The capture region grows as U_{min} decreases, as expected. The isopleths at U_{min} = 200 and 150 FPM lie entirely within the dividing streamline, indicating that the air speed along the dividing streamline is everywhere less than 150 FPM (0.762 m/s), which agrees with the results shown in Figure E7.2b. Particles under these isopleths are predicted to get removed by the suction system, according to the concept of capture velocity. The isopleth at U_{min} = 120 FPM crosses the dividing streamline, indicating that there is a region (shaded in Figure E7.2a) above the dividing streamline in which the air speed exceeds 120 FPM. If the capture velocity of the process is 120 FPM, particles in this region are predicted to be removed by the suction. However, this is unphysical since even the air itself in the shaded region does not get sucked into the slot! This illustrates a fundamental flaw in the concept of capture velocity – particles in the shaded region are unlikely to be removed, even though the air speed in that region is greater than the capture velocity. Furthermore, capture or non-capture of a particle is dependent on the initial *direction* of the particle, as well as its location. The trajectory of a particle in a flow field is calculated in a more sophisticated manner in Chapter 8.

7.4 Elementary Axisymmetric Ideal Flows

Consider three-dimensional flow of an ideal fluid in which there is symmetry about an axis of rotation. The use of symmetry reduces the three-dimensional problem to a two-dimensional problem and enables one to use a stream function in addition to a velocity potential. The relationships between the velocity components and the derivatives of the potential functions in spherical coordinates were given in Eqs. (7-29). Just as was done for planar ideal flows, a collection of elementary axisymmetric ideal flows can be used as "building blocks" to construct more complicated flows via superposition.

7.4.1 Point Sink in Space for Axisymmetric Flow

Consider a point in space through which a constant density fluid is withdrawn at uniform volumetric flow rate (Q is again a positive quantity for suction). Applying conservation of mass to a spherically symmetric system in which the radius is the only spatial coordinate results in the following expression for the radial velocity:

$$U_r = -\frac{Q}{4\pi r^2} \tag{7-56}$$

The tangential velocity U_θ is zero because of symmetry. Since Q and r are positive quantities, the negative sign indicates that the air travels toward the sink. Using Eqs. (7-29) and (7-56),

$$U_r = -\frac{\partial\phi}{\partial r} = -\left(\frac{1}{r^2 \sin\theta}\right)\frac{\partial\psi}{\partial\theta} = -\frac{Q}{4\pi r^2} \tag{7-57}$$

and

$$U_\theta = -\frac{1}{r}\frac{\partial\phi}{\partial\theta} = \left(\frac{1}{r\sin\theta}\right)\frac{\partial\psi}{\partial r} = 0 \tag{7-58}$$

To find the velocity potential, Eq. (7-57) can be integrated:

$$\phi = \int \frac{Q}{4\pi r^2} dr + g(\theta) = -\frac{Q}{4\pi r} + g(\theta)$$

but since $\partial\phi/\partial\theta$ is zero by Eq. (7-58), $g(\theta)$ is equal to an arbitrary constant which may be taken as zero. In similar fashion, Eq. (7-57) can be integrated to obtain the stream function ψ. In conclusion, the stream function and velocity potential for a ***point sink*** are

$$\phi = -\frac{Q}{4\pi r} \qquad \psi = -\frac{Q}{4\pi}\cos\theta \tag{7-59}$$

A scale drawing of lines of constant ψ and ϕ is shown in Figure 7.7. If the point sink lies along a wall, Eq. (7-59) can still be used, but one must replace the integer 4 by 2 since the volumetric flow rate through the point sink in the plane is half the value in free space. Another way to say this is that since the entire volumetric flow rate (Q) is sucked from only half of free space, the potential functions must be increased by a factor of two to account for the "missing" half.

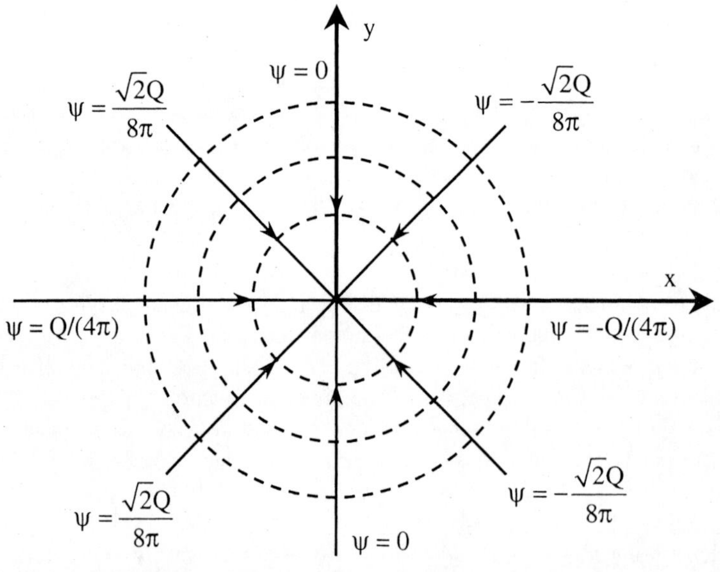

Figure 7.7 Streamlines (solid) and equipotential lines (dashed) for a point sink in a plane; several streamlines are labeled, as calculated from Eq. (7-59).

7.4.2 Uniform Streaming Axisymmetric Flow

Consider flow of an ideal fluid moving uniformly in the negative x-direction with velocity U_0. Using procedures similar to those used above, the velocity potential ϕ is found to be

$$\boxed{\phi = U_0 r \cos\theta} \tag{7-60}$$

To find the stream function, U_r is computed as an intermediate step. Using Eq. (7-29),

$$U_r = -\frac{\partial\phi}{\partial r} = -U_0\cos\theta = -\left(\frac{1}{r^2\sin\theta}\right)\frac{\partial\psi}{\partial\theta}$$

After some algebraic manipulation (left as an exercise for the reader), it can be shown that

$$\boxed{\psi = \frac{U_0\left(r\sin\theta\right)^2}{2}} \tag{7-61}$$

7.4.3 Axisymmetric Streaming Flow over a Point Sink in a Plane

Consider an ideal fluid passing over an x-y plane containing a point sink at the origin through which fluid is withdrawn at volumetric flow rate Q. The z-axis is perpendicular to the x-y plane. Let U_0 be the speed of the streaming flow in the negative x-direction, as in Figure 7.8. The plane containing the point sink (the x-y plane) is a plane of symmetry, and the flow is rotationally symmetric about the x-axis. The potential functions for this flow can be constructed by adding the potential functions of streaming flow and a point sink at the origin. Figure 7.9 shows streamlines in the r-θ plane. As discussed previously, the x-z plane is shown, but the streamline shapes would be identical in any other r-θ plane, such as the x-y plane, since the flow is axisymmetric.

Because fluid withdrawn from above the plane never mixes with fluid withdrawn from below the plane, the sink strength used for a point sink in space is twice the volumetric rate of fluid withdrawn from above the plane, as previously discussed. The potential functions associated with streaming flow over a point sink in a plane are thus written by superposition:

$$\boxed{\phi = U_0 r\cos\theta + \frac{Q}{2\pi r}} \tag{7-62}$$

and

$$\boxed{\psi = \frac{U_0\left(r\sin\theta\right)^2}{2} - \frac{Q}{2\pi}\cos\theta} \tag{7-63}$$

The velocity components U_r and U_θ are

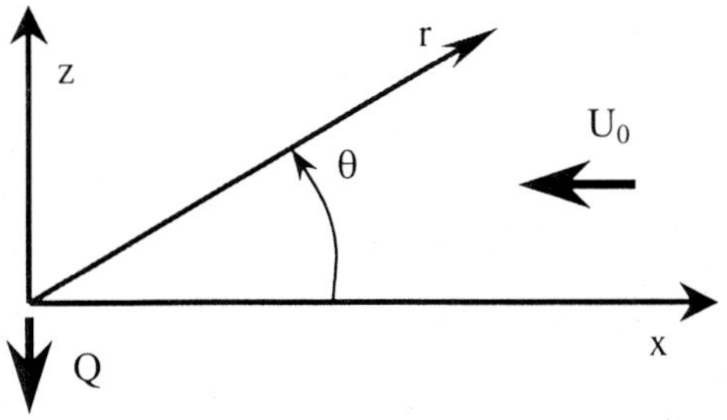

Figure 7.8 Axisymmetric three-dimensional coordinates used to describe air streaming in the negative x-direction over a plane (x-y) containing a point sink.

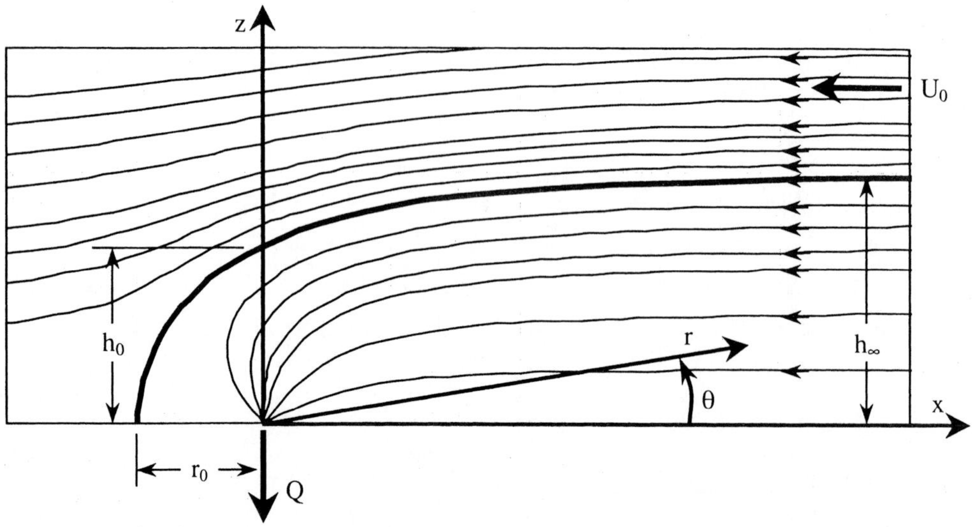

Figure 7.9 Streamlines for air streaming in the negative x-direction over a plane (the x-y plane) containing a point sink at the origin; the dividing streamline is bold.

$$\boxed{U_r = -\frac{\partial \phi}{\partial r} = -U_0 \cos\theta + \frac{Q}{2\pi r^2} \qquad U_\theta = -\frac{1}{r}\frac{\partial \phi}{\partial \theta} = U_0 \sin\theta} \tag{7-64}$$

In two-dimensional planar ideal flow, a particular value of the stream function ψ corresponds to a unique fluid streamline. In axisymmetric flow, a numerical value of the stream function ψ corresponds instead to an axisymmetric *surface* along which the fluid flows. The space between two surfaces ψ_1 and ψ_2 describes a space through which fluid passes, and no fluid can cross a surface of constant ψ. Along a surface of constant ψ there is a myriad of fluid streamlines (which are also pathlines if the flow is steady). The unique correlation between a particular fluid streamline and a particular value of ψ is valid only in two-dimensional planar flow. When viewing an axisymmetric flow in the direction of the y-axis, it is useful to speak of specific streamlines which are the intersections of specific stream surfaces (constant values of ψ) and the x-z plane, as in Figure 7.9.

At a point along the axis of rotation, namely at $r = r_0$, $\theta = \pi$, there exists a stagnation point where $U_r = U_\theta = 0$. The location of the stagnation point can be found by setting the velocity components in Eqs. (7-64) equal to zero. It turns out that the stagnation point is located to the left of the origin at

$$\boxed{r_0 = \sqrt{\frac{Q}{2\pi U_0}}} \tag{7-65}$$

The value of ψ that passes through the stagnation point is also the value of ψ along the dividing streamline, ψ_d, which can be found by substituting the coordinates of the stagnation point ($r = r_0$, $\theta = \pi$) into Eq. (7-63):

$$\boxed{\psi_d = \frac{Q}{2\pi}} \tag{7-66}$$

The height of the dividing streamline directly above the sink (h_0) can be found from Eq. (7-63) by setting ψ equal to $Q/(2\pi)$, $r = h_0$, and $\theta = \pi/2$:

$$\frac{Q}{2\pi} = \frac{1}{2}U_0\left[h_0 \sin\left(\frac{\pi}{2}\right)\right]^2 - \frac{Q}{2\pi}\cos\left(\frac{\pi}{2}\right)$$

from which

$$\boxed{h_0 = \sqrt{\frac{Q}{\pi U_0}}} \qquad (7\text{-}67)$$

Far upwind of the sink (x >> 0), the distance between the dividing streamline and the x-axis (h_∞) can be found by rewriting Eq. (7-63) as

$$\psi_d = \frac{U_0 y^2}{2} - \frac{Q}{2\pi}\cos\theta$$

Using Eq. (7-66) and taking the limit as θ approaches zero,

$$\frac{Q}{2\pi} = \frac{U_0 h_\infty^{\ 2}}{2} - \frac{Q}{2\pi}$$

from which

$$\boxed{h_\infty = \sqrt{\frac{2Q}{\pi U_0}}} \qquad (7\text{-}68)$$

Figure 7.9 shows the distances r_0, h_0, and h_∞. Finally, the stream function along the positive x-axis can be calculated by setting θ = 0 in Eq. (7-63), resulting in

$$\boxed{\psi_{\text{along x-axis}} = -\frac{Q}{2\pi}} \qquad (7\text{-}69)$$

Recall that in two-dimensional flow, the dividing streamline, $\psi = \psi_d = Q/L$, defines the region within which the flow enters the line sink at volumetric flow rate per unit length Q/L. In axisymmetric flow, the dividing streamline, $\psi = \psi_d = Q/(2\pi)$, defines the upper boundary below which the flow enters the point sink at volumetric flow rate Q. The streamline that passes through the stagnation point is the dividing streamline. Flow below the dividing streamline enters the sink, and flow above the dividing streamline passes over it. The dividing streamline therefore defines an axisymmetric region within which gaseous contaminants are removed by the sink. The coordinates (x,y) or (r,θ) for points along the dividing streamline can be found by setting Eq. (7-63) equal to Q/(2π) and computing values of y for selected values of x (or vice versa).

7.5 Flanged and Unflanged Inlets in Quiescent Air

So far, only line sinks (planar flow) and point sinks (axisymmetric flow) have been considered, in which flow is sucked into an infinitesimally small entity. The inlet faces of real-life close capture hoods, however, are not lines or points, but rather rectangular or round openings of finite dimensions, with or without flanges. The ideal flow approximation is still reasonable for flows into such inlet geometries, and there are techniques available to analyze these flows analytically.

7.5.1 Flanged Slot

Consider the flow (Q/L) drawn into a flanged rectangular slot of width 2w and length L, lying in a horizontal plane as in Figure 7.10. It is assumed that the aspect ratio of the slot, L/(2w), is very large, so that the slot can be analyzed as if it were of infinite length. Expressions for the stream function and velocity potential have been derived by Streeter (1948), Milne-Thomson (1950), and Kirchhoff (1985):

$$\frac{x}{w} = \cosh\left(\frac{\phi}{k}\right)\cos\left(\frac{\psi}{k}\right) \qquad \frac{y}{w} = \sinh\left(\frac{\phi}{k}\right)\sin\left(\frac{\psi}{k}\right) \tag{7-70}$$

where

$$k = \frac{Q}{\pi L} \tag{7-71}$$

After some algebraic simplification, namely squaring and adding Eqs. (7-70), the variable ψ can be eliminated, and the following implicit equation for $\phi(x,y)$ can be obtained:

$$1 = \left[\frac{\frac{x}{w}}{\cosh\left(\frac{\phi}{k}\right)}\right]^2 + \left[\frac{\frac{y}{w}}{\sinh\left(\frac{\phi}{k}\right)}\right]^2 \tag{7-72}$$

Similarly, the squares of Eqs. (7-70) can be subtracted from one another to eliminate the variable ϕ in order to produce the following implicit equation for $\psi(x,y)$:

$$1 = \left[\frac{\frac{x}{w}}{\cos\left(\frac{\psi}{k}\right)}\right]^2 - \left[\frac{\frac{y}{w}}{\sin\left(\frac{\psi}{k}\right)}\right]^2 \tag{7-73}$$

If one is fortunate enough to have equations such as the two above, one can find the value of ϕ (or ψ) at particular point (x,y) by numerical methods such as Newton's method or the bisection method described in Appendices A.14 and A.15. Alternatively, Mathcad programs or Excel spreadsheets can be written to compute values of ϕ and ψ at particular points (x,y).

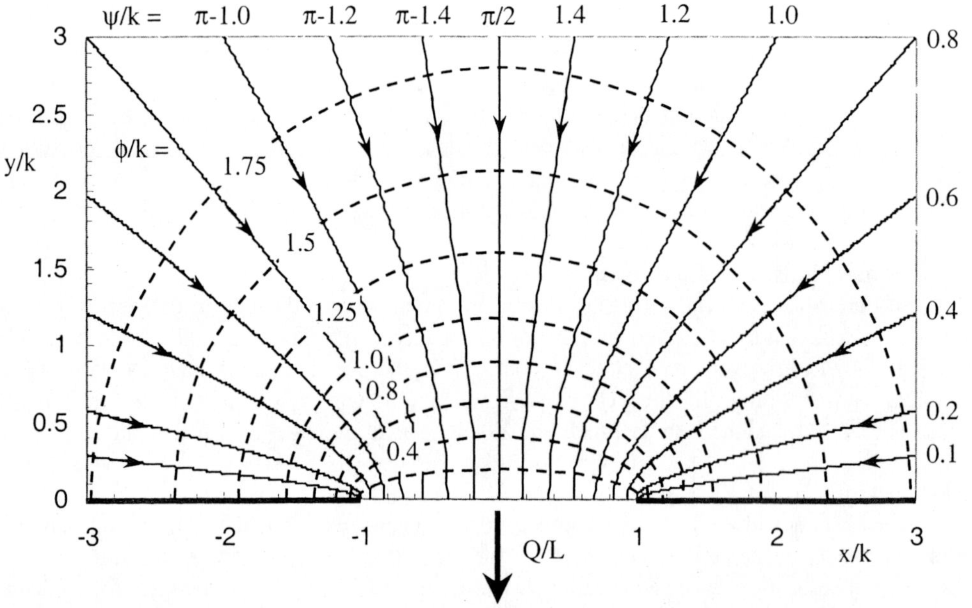

Figure 7.10 Streamlines (solid) and equipotential lines (dashed) for flow through a flanged inlet of width 2w in quiescent air.

For the particular problem under consideration here, i.e. a flanged rectangular slot, equipotential surfaces in which values of ϕ are constant are semi-ellipses with foci at (w,0) and (-w,0). Streamlines in which values of ψ are constant are hyperbolae which have the same foci. Figure 7.10 shows these curves and certain limiting values of ψ and ϕ. The streamlines $\psi = 0$, Q/(2L), and Q/L ($\psi/k = 0$, $\pi/2$, and π) correspond to the positive x-axis, positive y-axis, and negative x-axis respectively. The line $\phi = 0$ corresponds to the x-axis between x = -w and x = w. Curves of $\phi > 0$ correspond to semi-ellipses above this line. Since the curves in Figure 7.10 are symmetric with respect to the y-axis, Eqs. (7-72) and (7-73) can also be used to describe flow through a slot in a surface that meets another surface at 90 degrees. Figure 7.11 illustrates two such configurations which can be analyzed with these same equations; the first configuration can model flow through baffles, while the second can model flow through a slot in the back wall of a workbench.

The velocity components at any point (x,y) upstream of the inlet can be obtained from Eq. (7-25) by computing partial derivatives of ψ and ϕ. Unfortunately, obtaining these derivatives from Eqs. (7-72) and (7-73) is difficult to say the least. If one wants to determine the value of ψ and ϕ at a point (x,y), these derivatives can be found by ***implicit differentiation***. Equations (7-70) can be expressed in the following way:

$$\boxed{x = F(\phi, \psi) \qquad y = G(\phi, \psi)} \tag{7-74}$$

What is desired is the inverse of the above,

$$\boxed{\phi = f(x, y) \qquad \psi = g(x, y)} \tag{7-75}$$

x and y in Eq. (7-74) can be differentiated with respect to x, remembering that x and y are independent variables. Employing the chain rule, one obtains

$$\boxed{\frac{\partial x}{\partial x} = 1 = \frac{\partial F}{\partial \phi}\frac{\partial \phi}{\partial x} + \frac{\partial F}{\partial \psi}\frac{\partial \psi}{\partial x} \qquad \frac{\partial y}{\partial x} = 0 = \frac{\partial G}{\partial \phi}\frac{\partial \phi}{\partial x} + \frac{\partial G}{\partial \psi}\frac{\partial \psi}{\partial x}} \tag{7-76}$$

After evaluating the partial derivatives ($\partial F/\partial \phi$, $\partial F/\partial \psi$, $\partial G/\partial \phi$, and $\partial G/\partial \psi$) from Eqs. (7-70), Eqs. (7-76) yield

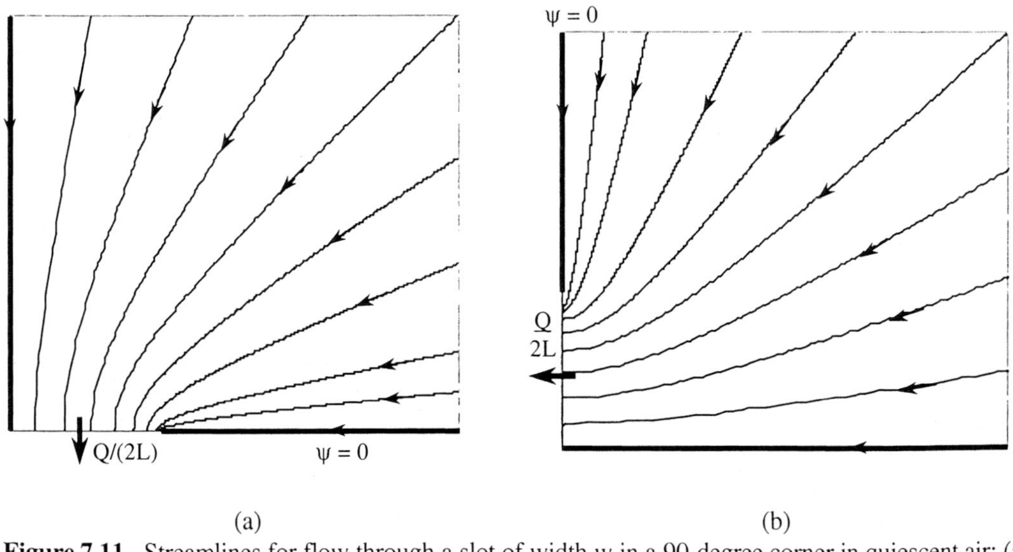

Figure 7.11 Streamlines for flow through a slot of width w in a 90-degree corner in quiescent air; (a) horizontal slot, and (b) vertical slot.

$$\left[\sinh\left(\frac{\phi}{k}\right)\cos\left(\frac{\psi}{k}\right)\right]\frac{\partial\phi}{\partial x} - \left[\cosh\left(\frac{\phi}{k}\right)\sin\left(\frac{\psi}{k}\right)\right]\frac{\partial\psi}{\partial x} = \frac{k}{w} \tag{7-77}$$

and

$$\left[\cosh\left(\frac{\phi}{k}\right)\sin\left(\frac{\psi}{k}\right)\right]\frac{\partial\phi}{\partial x} + \left[\sinh\left(\frac{\phi}{k}\right)\cos\left(\frac{\psi}{k}\right)\right]\frac{\partial\psi}{\partial x} = 0 \tag{7-78}$$

This pair of simultaneous equations can be solved for $\partial\phi/\partial x$ and $\partial\psi/\partial x$, provided that the ***Jacobian***, $J(\phi,\psi)$, is not zero. The Jacobian is defined as

$$J(\phi,\psi) = \begin{vmatrix} \dfrac{\partial F}{\partial \phi} & \dfrac{\partial F}{\partial \psi} \\ \dfrac{\partial G}{\partial \phi} & \dfrac{\partial G}{\partial \psi} \end{vmatrix} \tag{7-79}$$

The solution is

$$U_x = -\frac{\partial\phi}{\partial x} = -\frac{k}{w}\frac{\sinh\dfrac{\phi}{k}\cos\dfrac{\psi}{k}}{\left(\sinh\dfrac{\phi}{k}\cos\dfrac{\psi}{k}\right)^2 + \left(\cosh\dfrac{\phi}{k}\sin\dfrac{\psi}{k}\right)^2} \tag{7-80}$$

and

$$U_y = \frac{\partial\psi}{\partial x} = -\frac{k}{w}\frac{\cosh\dfrac{\phi}{k}\sin\dfrac{\psi}{k}}{\left(\sinh\dfrac{\phi}{k}\cos\dfrac{\psi}{k}\right)^2 + \left(\cosh\dfrac{\phi}{k}\sin\dfrac{\psi}{k}\right)^2} \tag{7-81}$$

A check of the limits of the above equations for U_x and U_y shows that along the right flange surface (where $y = 0$, $x > w$, and $\psi = 0$), and along the left flange surface (where $y = 0$, $x < -w$, and $\psi = Q/L$), $U_y = 0$ and U_x increases as the flow approaches the slot. Along the y-axis (where $x = 0$, $y > 0$, and $\psi = Q/[2L]$), U_x is zero and U_y increases as the flow approaches the slot. Across the inlet ($y = 0$, and $-w < x < w$), U_x is zero and $U_y = -(k/w)/\sin(\psi/k)$. Thus, while one may think in terms of an average face velocity equal to $Q/(2Lw)$, the actual velocity is not constant and approaches infinity at the slot lips ($y = 0$, $x = \pm w$), which are singular points.

Example 7.3 - Flanged Slot of Finite Width in Quiescent Air

Given: Consider a slot in an infinite plane as shown in Figure 7.10. The width of the slot is 2w (w = 0.020 m) and the aspect ratio of the slot is very large, $L/w >> 1$. Air is withdrawn through the slot at a volumetric flow rate (Q/L) of 12.0 m^2/min, which produces an average face velocity, U_{face} = 300. m/min. This corresponds favorably with typical slot velocities recommended by the ACGIH for different types of industrial ventilation hoods.

To do:

(a) Predict and plot several streamlines for this flow.
(b) Predict the normalized velocity components (U_x/U_{face} and U_y/U_{face}) and the normalized air speed,

$$\frac{U}{U_{face}} = \frac{\sqrt{U_x^{\,2} + U_y^{\,2}}}{U_{face}}$$

at points along the streamline $\psi = Q/(2L)$. This is the streamline along the y-axis (x = 0), i.e. along the axis of the slot (axis of symmetry), normal to the slot face. Plot U/U_{face} as a function of the normalized radial distance to the center of the slot (r/w) along this streamline,

$$\frac{r}{w} = \frac{\sqrt{x^2 + y^2}}{w}$$

Compare with the air speed that would exist if the slot were a line sink of strength Q/L.

(c) Predict the velocity components (U_x and U_y) at an arbitrary point (x,y). Find the velocity components at the particular point x = 0.63 m, y = 0.025 m.

Solution:

(a) Eq. (7-73) is solved for x/w,

$$\frac{x}{w} = \cos\left(\frac{\psi}{k}\right)\sqrt{1 + \left[\frac{\left(\frac{y}{w}\right)}{\sin\left(\frac{\psi}{k}\right)}\right]^2}$$

A value of the stream function ψ is selected. An range of y/w values is generated (from 0 to 1.4), corresponding to the flanged surface. A value of x/w is computed for each y/w value using the above equation. The computation is repeated for several values of ψ, and the results (the streamlines) are plotted in Figure E7.3a. The authors used Excel, and the spreadsheet file is available on the web site.

(b) Along the axis of symmetry of the flow, x = 0 and $\psi = Q/(2L)$ since half of the total volumetric flow rate enters from the right half of the slot. An array of y/w values from 0 to 5.0 is generated. Using the second part of Eq. (7-70), the value of ϕ at each y value is computed along this streamline. Using Eqs. (7-81) and (7-80), the velocity components, U_x and U_y, are computed at each value of y. (U_x is verified to be zero along the axis of symmetry.) The normalized air speed (U/U_{face}) is then computed at each y value. The normalized radius (r/w) is computed from the center of the slot upward; in this case, r/w = y/w since x = 0. The result is plotted in Figure E7.3b, which also shows how the speed varies if the slot were to be replaced by a line sink, as given by Eq. (7-34). It is seen that for both cases the air speed decreases away from the inlet. At radial distances less than about three slot half-widths, the air speed for a slot of width 2w is lower than that predicted by a line sink with the same volumetric flow rate. Beyond about three slot half-widths, the air speeds of the two flows are virtually the same. Thus for practical purposes, at radial distances much greater than the slot width (2w), a slot of finite width and a line sink yield similar results. Indeed, if the streamlines in Figure E7.3a were viewed from "far away", they would be indistinguishable from those of a line sink, i.e. they would look like rays to the origin. Finally, the empirical expression for U/U_{face} along the symmetry axis of a rectangular flanged inlet, Eq. (6-5), is also plotted in Figure E7.3b. An aspect ratio of 10:1 is assumed, and y is used in place of x in Eq. (6-5) because of the coordinate system used here. The results are between the line sink and the finite width flanged slot calculations. All three curves merge beyond about three slot half-widths.

(c) To find the velocity components U_x and U_y at an arbitrary point (x,y) it is first necessary to know the values of the potential functions ϕ and ψ at this point. Equations (7-70) must be solved simultaneously for ϕ and ψ at (x,y). There are several ways to do this – trial and error, Newton's

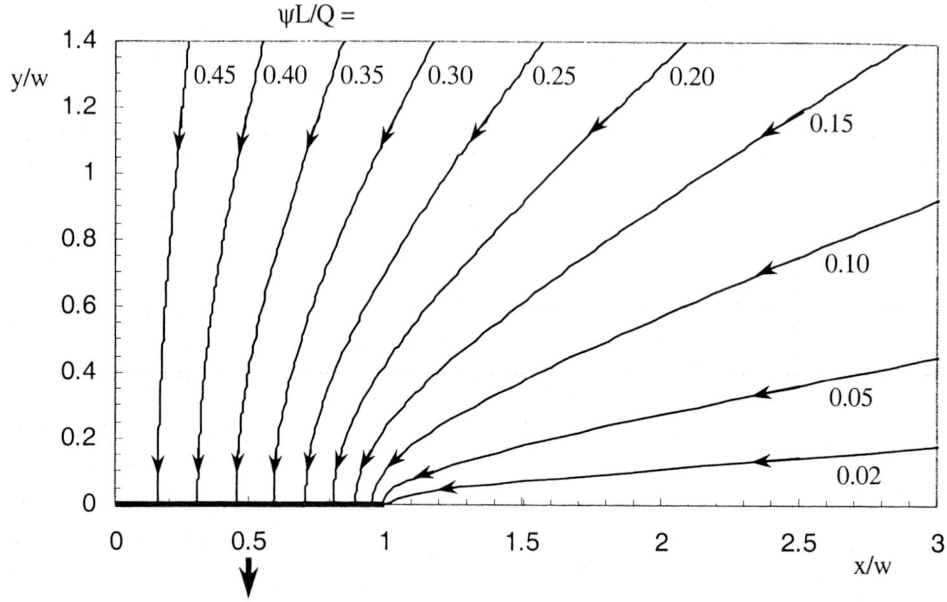

Figure E7.3a Streamlines generated by a two-dimensional flanged slot of half-width w, for the data of Example 7.3. The x and y axes are normalized, and are not to scale. Only half of the flow is shown; streamlines are symmetric about the y-axis.

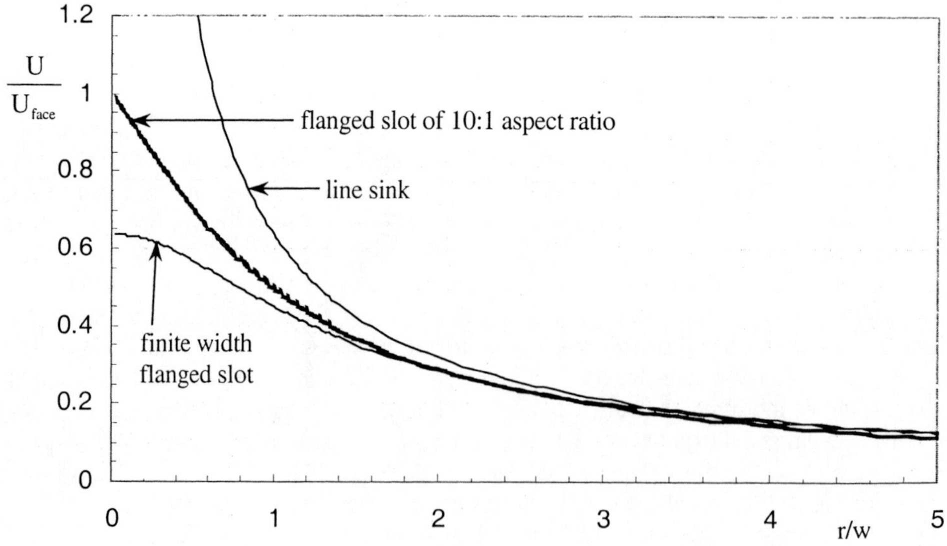

Figure E7.3b Normalized air speed versus normalized radial location along the axis of symmetry for a two-dimensional flanged slot of half-width w compared to that generated by a line sink, and the empirical curve fit for a flanged slot of 10:1 aspect ratio.

method, etc.; the authors chose to use Mathcad's *Solve Block* feature. The following results are obtained at x = 0.063 m and y = 0.025 m:

$$\psi(0.063, 0.025) = 1.503 \frac{m^2}{min} \qquad \text{and} \qquad \phi(0.063, 0.025) = 7.249 \frac{m^2}{min}$$

From these results, Eqs. (7-80) and (7-81) are used to find the velocity components:

$$U_x(0.063, 0.025) = -55.4 \frac{m}{min} \qquad \text{and} \qquad U_y(0.063, 0.025) = -23.2 \frac{m}{min}$$

The Mathcad program can be found on the web site for this book.

Discussion: The potential flow equations used here for a flanged slot do a better job than those for a simple line sink at predicting the velocity field near the inlet face, and the infinite velocity at the origin is removed. However, both potential flow approximations do an excellent job far from the inlet.

7.5.2 Approximating a Flanged Slot as N Parallel Line Sinks

An alternative approach for finding the velocity potential function for a flanged slot is to assume that the flanged slot consists of an infinite number of differential line sinks, each of which has a volumetric flow rate per unit depth of

$$\frac{Q}{L}\frac{dx}{2w}$$

where dx is an infinitesimal distance along the slot face, as in Figure 7.12. For the differential line sink which lies at the origin (center of the slot), its velocity potential is given by Eq. (7-37), but with an additional factor of 2 due to the fact that all of Q comes from only half of space, as discussed previously. The differential of the velocity potential for this differential sink is thus

$$d\phi = \frac{Q}{2\pi L w} \ln(r) dx$$

Now consider a differential line sink in the slot face at some distance x from the origin, where x lies between -w and w. At some arbitrary point P upwind of the inlet whose coordinates are (r_P,θ_P) or (x_P,y_P), the differential of the velocity potential can be expressed as

$$\boxed{d\phi = \frac{Q}{2\pi L w} \ln(r_P) dx} \qquad (7\text{-}82)$$

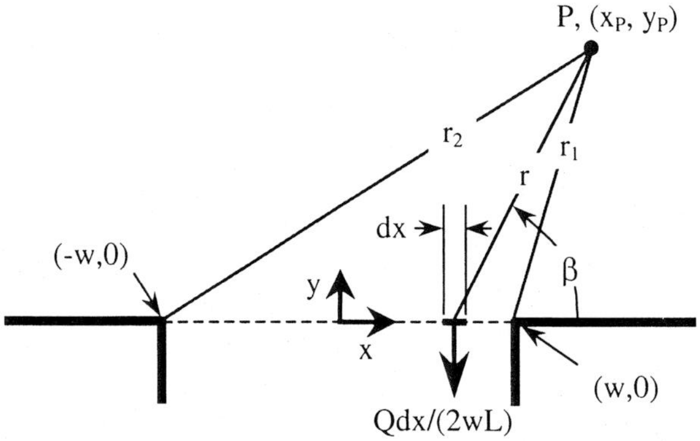

Figure 7.12 Flanged slot in quiescent air, modeled as a large number of differential line sinks along the dashed line (only one line sink is shown).

where r_P is the radial distance from the differential line sink to point P,

$$r_P = \sqrt{(x_P - x)^2 + y_P^2} \tag{7-83}$$

To compute the velocity potential at point P, Eq. (7-82) must be integrated over the width of the slot (from -w to w):

$$\phi = \frac{Q}{2\pi L w} \int_{-w}^{w} \ln\left(\sqrt{(x_P - x)^2 + y_P^2}\right) dx \tag{7-84}$$

Once integrated, the expression can be differentiated with respect to x_P and y_P to find the velocity components at point P.

This approach also lends itself to numerical computation in which the flanged slot is replaced by N line sinks spaced a distance 2w/N apart across the distance (-w to +w). Each line sink has a volumetric flow rate per unit depth of Q/(LN). The velocity potential at point P is then found by summing over the N line sinks,

$$\phi = \sum_{i=1}^{N} \left[\frac{Q}{\pi L N} \ln(r_i) \right] \tag{7-85}$$

where again Eq. (7-37) has been used with the factor of 2 included. In Eq. (7-85), r_i is the radial distance from an individual line sink to point P. To perform the sum, a unique value of r_i has to be calculated for each line sink. With this method, tabulated values of the velocity potential can be computed for an array of upwind points P. The velocity components at any point P can then be obtained by numerically differentiating ϕ with respect to x and y.

The velocity at point (x_P,y_P) can also be found directly by vectorially adding the velocities produced by the N line sinks. The radial velocity produced by one of the individual line sinks can be obtained from Eq. (7-34), again remembering to account for the factor of two. Since each of the N line sinks produces only a radial velocity component relative to the line sink, the velocity vector at point P contributed by the ith line sink is given by

$$\vec{U}_i = -\hat{r}_i \frac{Q}{\pi L N r_i} \tag{7-86}$$

where r_i is the radial distance from the ith line sink to point P, and $\hat{r}_i$ is the unit vector from the line sink to point P. The velocity produced by all N line sinks can be obtained by summing these vectors over the N line sinks,

$$\vec{U} = -\sum_{i=1}^{N} \left[\hat{r}_i \frac{Q}{\pi L N r_i} \right] \tag{7-87}$$

It must be remembered that when performing this summation, the unit vectors ($\hat{r}_i$) change direction, and the magnitude of distance r_i also changes. To compute the velocity components U_x and U_y at point P, the appropriate vector components of Eq. (7-87) must be calculated. After some algebra,

$$U_x(x_P, y_P) = -\sum_{i=1}^{N} \left[\frac{\frac{Q}{\pi L N}(x_P - x_i)}{y_P^2 + (x_P - x_i)^2} \right] \tag{7-88}$$

and

$$U_y\left(x_P, y_P\right) = -\sum_{i=1}^{N}\left[\frac{\dfrac{Q}{\pi L N}\,y_P}{y_P^{\,2} + \left(x_P - x_i\right)^2}\right] \tag{7-89}$$

The summation is performed over the N line sinks lying equidistant between one edge of the slot to the other, i.e. $-w \le x \le w$. In this process all the terms in Eq. (7-89) for U_y are the same sign (negative), leading to flow in the negative y-direction, while some of the terms in Eq. (7-88) for U_x are positive and others are negative, depending on where point P lies with the respect to the origin.

It is expected that at distances far upwind of the inlet ($y \gg w$), simulating a flanged slot by a large number of line sinks will produce velocities reasonably close to those predicted by the analytical expressions for a flanged slot, i.e. Equations (7-81) and (7-80). Along a plane close to the inlet itself, differences are expected. Equation (7-81) predicts infinite negative velocities at the slot lip and a mid-plane velocity less than the slot face velocity, $Q/(2wL)$, whereas the numerical simulation is expected to yield velocities U_y closer to the actual face velocity.

7.5.3 Unflanged Slot

Consider the flow of an ideal fluid into an unflanged rectangular inlet of width 2w as shown in Figure 7.13. The velocity potential and stream function satisfy the following equations:

$$\frac{x}{w} = \frac{1}{\pi}\left[2\frac{\phi}{k} + \exp\left(2\frac{\phi}{k}\right)\cos\left(2\frac{\psi}{k}\right)\right] \tag{7-90}$$

and

$$\frac{y}{w} = \frac{1}{\pi}\left[2\frac{\psi}{k} + \exp\left(2\frac{\phi}{k}\right)\sin\left(2\frac{\psi}{k}\right)\right] \tag{7-91}$$

where

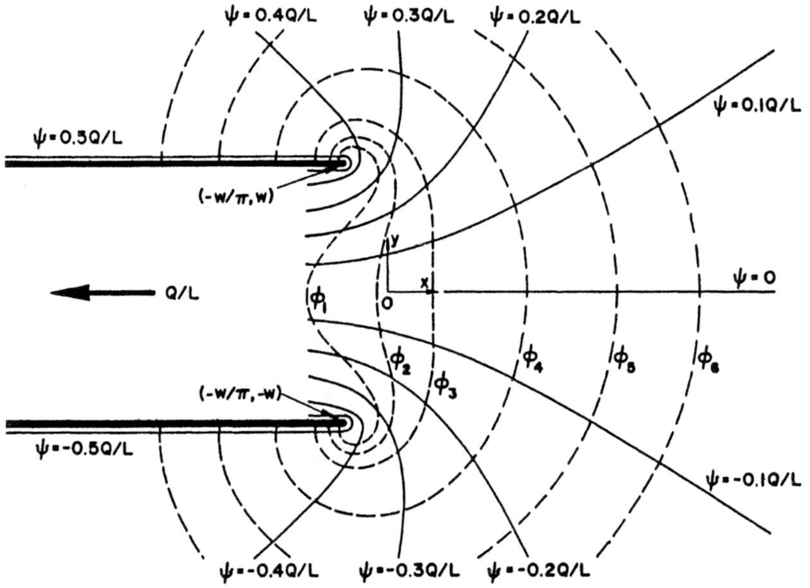

Figure 7.13 Streamlines and equipotential lines for flow through an unflanged inlet of width 2w in quiescent air (from Heinsohn, 1991).

$$k = \frac{Q}{\pi L} \tag{7-92}$$

Equipotential lines and streamlines are shown in Figure 7.13. To use these equations wisely it is important to know the location of certain key streamlines. To begin with, note that the origin is at a distance (w/π) upwind of the inlet, and consider the positive x-axis (y = 0, x > 0). For Eq. (7-91) to be zero for arbitrary values of φ, ψ must equal zero. Thus the positive x-axis corresponds to ψ = 0. Now consider the location where ψ is equal to plus or minus Q/(2L). Substitution of these values into Eqs. (7-90) and (7-91) shows that streamline ψ = Q/(2L) corresponds to the upper surface of the inlet while streamline ψ = - Q/(2L) corresponds to the lower surface. This agrees with a simple analysis using Eq. (7-22); namely, between the upper surface of the inlet and the x-axis (axis of symmetry), half of the total suction volumetric flow rate, i.e. Q/(2L) must pass. Thus the difference between ψ values along these two streamlines must equal Q/(2L). Values of constant φ are orthogonal to values of constant ψ and increase as one moves in the direction of increasing x.

The velocity components U_x and U_y can be found from Eqs. (7-90) and (7-91) by the implicit differentiation process described earlier. The results are

$$U_x = -\frac{k\pi}{2w}\frac{1+\exp(2\phi/k)\cos(2\psi/k)}{\left[\exp(2\phi/k)\sin(2\psi/k)\right]^2+\left[1+\exp(2\phi/k)\cos(2\psi/k)\right]^2} \tag{7-93}$$

and

$$U_y = -\frac{k\pi}{2w}\frac{\exp(2\phi/k)\sin(2\psi/k)}{\left[\exp(2\phi/k)\sin(2\psi/k)\right]^2+\left[1+\exp(2\phi/k)\cos(2\psi/k)\right]^2} \tag{7-94}$$

A check of the limits on the equations for U_x and U_y shows that along the positive x-axis, where x > -(w/π), y = 0, and ψ is equal to zero, U_y is zero, U_x is negative, and the magnitude of U_x increases as the flow approaches the inlet. Along the upper surface of the inlet where x < -(w/π), y = w, and ψ is equal to Q/(2L), U_y is zero, U_x is positive, and U_x increases as the flow approaches the inlet. Since the upper and lower lips on the inlet are singular points, the velocities there approach infinity. Along a streamline where ψ is equal to Q/(4L) (quadrant where both x and y are positive), both U_x and U_y are negative since the flow enters the slot from the right. The magnitudes of U_x and U_y increase as the flow approaches the inlet.

Example 7.4 - Unflanged Slot of Finite Width in Quiescent Air

Given: Air enters an unflanged rectangular inlet that has a large aspect ratio, L/w >> 1 (Figure 7.13). In the far-field the air is motionless. The width of the inlet is 2w (w = 0.020 m) and the volumetric flow rate per unit length entering the inlet is Q/L = 12.0 m^2/min. The face velocity, U_{face} = Q/(2Lw), is 300 m/min which is typical of the slot velocities recommended by the ACGIH for industrial ventilation systems. The numerical values used in this example problem are the same as those used in the previous problem; the only difference here is that the inlet is unflanged.

To do:

(a) Select several values of the stream function (ψ) and compute and plot streamlines near the inlet face.

(b) Compute and plot the normalized air speed, U/U_{face}, along the line of symmetry, i.e. the x-axis. Compare to the air speed predicted for a line sink and from the empirical results of Chapter 6 for an unflanged slot.

(c) Compute and plot several isopleths along which the normalized air speed (U/U_{face}) is constant.

Solution:

(a) The locus of points (x,y) of a streamline can be found as follows: A particular value of ψ is selected, and a range of y/w values is defined (here, $0 < y/w < 2$). The value of ϕ is calculated at this ψ and at each y by solving Eq. (7-91) for $2\phi/k$,

$$\frac{2\phi}{k} = \ln\left(\frac{\frac{\pi y}{w} - \frac{2\psi}{k}}{\sin\left(\frac{2\psi}{k}\right)}\right)$$

Eq. (7-90) is then used to solve for x/w for each value of y/w. The process is repeated for several values of ψ. The authors used Excel to generate the streamlines; the file is available on the book's web site. Figure E7.4a shows ten streamlines in the upper half of the flow field. Unlike the flanged slot example above in which the axis of symmetry was the y-axis, the x-axis is the axis of symmetry here.

(b) Along the axis of symmetry of the flow (x-axis), y = 0 and $\psi = 0$. Using Eq. (7-90), the value of x/w is computed for several values of ϕ along this streamline. Using Eqs. (7-93) and (7-94), the velocity components, U_x and U_y, are computed at each value of x/w. (U_y is verified to be zero along the axis of symmetry.) The normalized air speed (U/U_{face}) is then computed at each x/w value. The normalized radius (r/w) is computed from the center of the slot forward; in this case, $r/w = (x + 1/\pi)/w$ since the face is located to the left of the origin at normalized distance $1/\pi$, and y = 0 along the axis of symmetry. The result is plotted in Figure E7.4b, which also shows how the speed would vary if the slot were to be replaced by a line sink, Eq. (7-34).

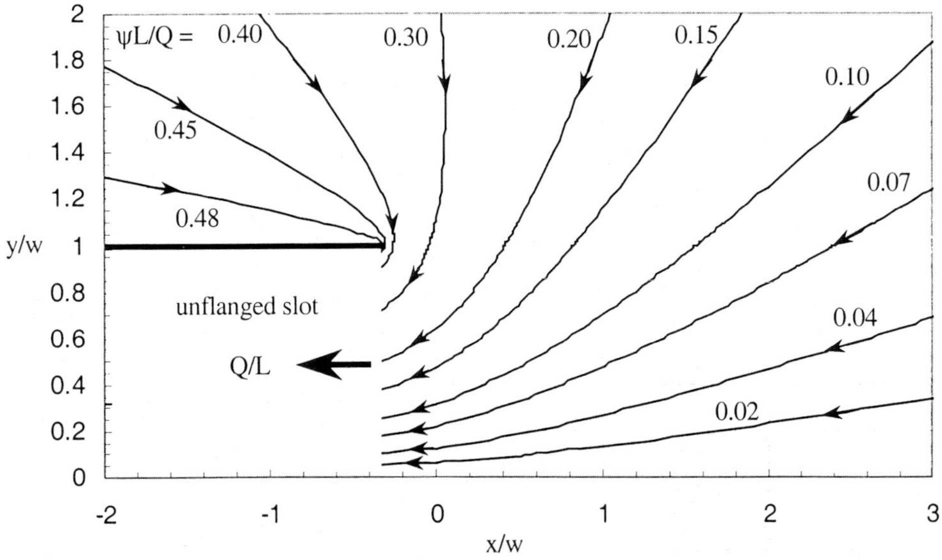

Figure E7.4a Streamlines generated by a two-dimensional unflanged slot of half-width w, for the data of Example 7.4; the x and y axes are normalized, and are not to scale. Only half of the flow is shown; streamlines are symmetric about the x-axis.

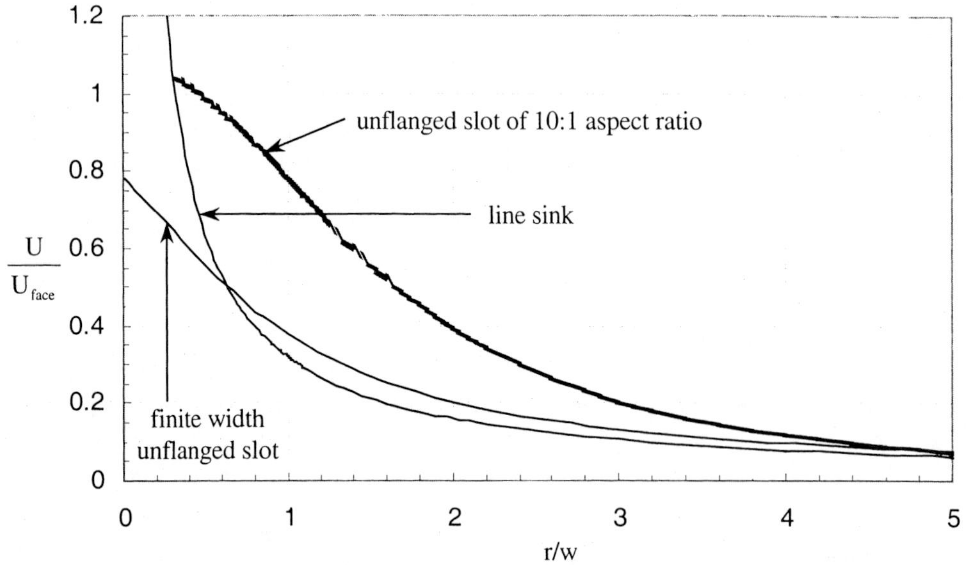

Figure E7.4b Normalized air speed versus normalized radial location along the axis of symmetry for a two-dimensional unflanged slot of half-width w compared to that generated by a line sink, and the empirical curve fit for an unflanged slot of 10:1 aspect ratio.

The empirical expression for U/U_{face} along the symmetry axis of a rectangular unflanged inlet, Eq. (6-4), is also plotted in Figure E7.4b. An aspect ratio of 10:1 is assumed. The results show much disagreement among the three curves for radial distances less than about four slot half-widths. Beyond this, the three curves show reasonably good agreement.

(c) Equations (7-93) and (7-94) can be used to compute normalized velocity components U_x and U_y, from which air speed (U) can be found. Unfortunately the values of ϕ and ψ for coordinates (x,y) are not known explicitly. To compute these values, implicit Eqs. (7-90) and (7-91) must be solved numerically. The authors used Mathcad's *Solve Block* feature. Once computed, the velocity components U_x and U_y are calculated at (x,y) using Eqs. (7-93) and (7-94), and normalized air speed (U/U_{face}) is calculated for an array of points (x,y) upwind of the inlet. Mathcad's contour plotting routine is used to generate the contour plot; Figure E7.4c shows the results of this computation.

Discussion: The shape of the isopleths in Figure E7.4c compare favorably with those of Figure 6.8, although the magnitudes differ. This is not surprising since the centerline air speed also differs significantly, as seen in Figure E7.4b, especially near the inlet face. The reader should note the factor of two, however; x and y are normalized by the slot half-width in Figure E7.4c, while they are normalized by the full slot width (s = 2w) in Figure 6.8.

7.5.4 Flanged Circular Inlets

The velocity field produced by a circular inlet of radius w in a flanged surface in quiescent air has been studied by Flynn, Ellenbecker et al. (1985, 1987) and Flynn and Miller (1988) who report that the potential functions are

$$\phi = \frac{Q}{2\pi w}\arcsin\left(\frac{2w}{a_1 + a_2}\right) \qquad \psi = \frac{Q}{4\pi w}\sqrt{4w^2 - \left(a_1 - a_2\right)^2} \tag{7-95}$$

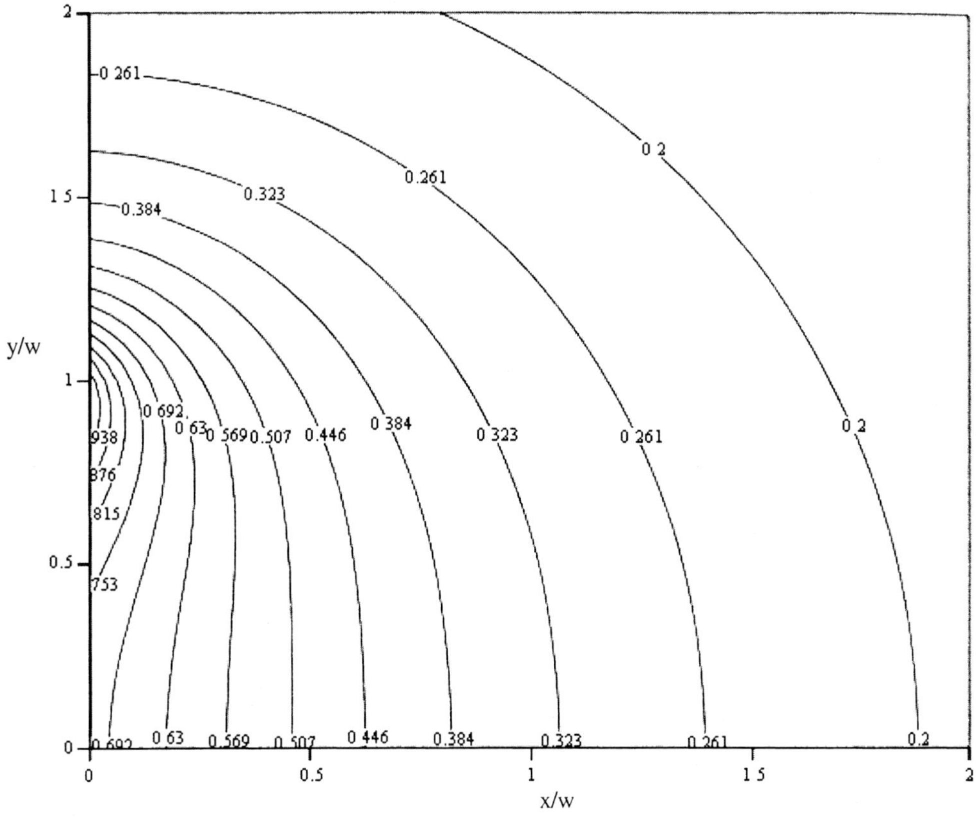

Figure E7.4c Isopleths (contours of constant normalized air speed U/U_{face}) for a two-dimensional unflanged slot of half-width w.

where the lengths a_1 and a_2 lie in the r-z plane and are defined as

$$a_1 = \sqrt{z^2 + (w + r)^2} \qquad a_2 = \sqrt{z^2 + (w - r)^2} \tag{7-96}$$

and r and z are the radial and axial distances measured from the origin which lies in the center of the inlet plane as in Figure 7.14.

The radial and axial velocity components at any point upstream of the inlet are given by

$$U_z = -\frac{\partial \phi}{\partial z} = -\frac{zQ}{\pi a_1 a_2 a_3} \tag{7-97}$$

and

$$U_r = -\frac{\partial \phi}{\partial r} = -\frac{\frac{Q}{\pi}\left[(r + w)a_2 + (r - w)a_1\right]}{(a_1 + a_2)a_1 a_2 a_3} \tag{7-98}$$

where

$$a_3 = \sqrt{(a_1 + a_2)^2 - 4w^2} \tag{7-99}$$

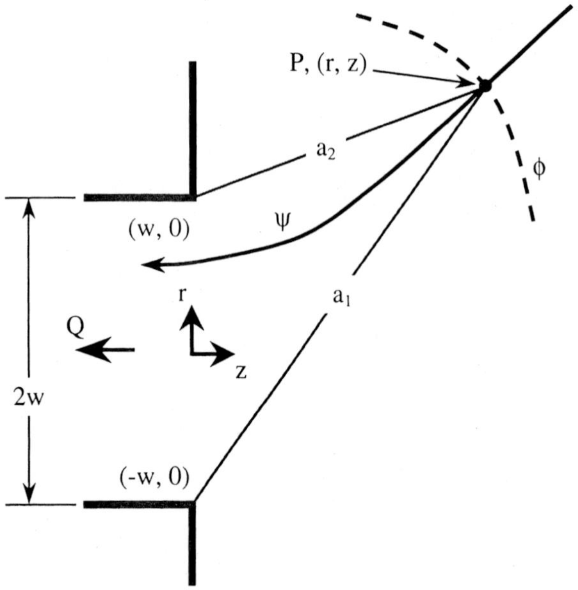

Figure 7.14 Flanged circular inlet of radius w; the streamline (solid) and equipotential line (dashed) passing through point P are shown.

Values of constant ϕ and ψ are similar to their counterparts for the rectangular flanged slot. Singular points again exist at the rim of the inlet where the velocity is infinite. The region influenced by the singularity is small and can be neglected. Thus for the majority of points upstream of the circular inlet, Eqs. (7-97) and (7-98) can be used to predict the radial and axial velocity components.

If the surrounding air passes over the inlet plane at uniform velocity U_0, the velocity potential for streaming flow can be added to Eqs. (7-97) and (7-98) to obtain the composite velocity potential. The velocity at an arbitrary point upwind of the inlet can be found by an analysis similar to that described later in Section 7.6. Alternatively, the velocity can be found by adding the vector U_0 to U_r and U_z above. The collection efficiency of a flanged circular opening to capture gaseous contaminants released from a point source in a streaming flow was studied analytically and experimentally by Flynn and Ellenbecker (1986).

7.5.5 Flanged Elliptical Inlets

The ideal flow field associated with an elliptical opening in a flanged surface was studied by Conroy et al. (1988) who report that the three-dimensional velocity potential is

$$\boxed{\phi = \pm \frac{Q}{4\pi} \int_0^m \frac{dm}{\sqrt{m\left(a^2 + m\right)\left(b^2 + m\right)}}} \tag{7-100}$$

where a and b are one half the major and minor diameters of the ellipse, Q is the volumetric flow rate and m is the positive root of the equation

$$\boxed{\frac{x^2}{a^2 + m} + \frac{y^2}{b^2 + m} + \frac{z^2}{m} = 1} \tag{7-101}$$

Following considerable mathematical manipulation, Conroy et al. (1988) report that the velocity components are

$$U_x = \frac{Qx\left(a^2+m\right)^{0.5}\left(b^2+m\right)^{1.5} m^{1.5}}{2\pi a_4} \qquad U_y = \frac{Qy\left(a^2+m\right)^{1.5}\left(b^2+m\right)^{0.5} m^{1.5}}{2\pi a_4} \tag{7-102}$$

and

$$U_z = \frac{Qz\left(a^2+m\right)^{1.5}\left(b^2+m\right)^{1.5} m^{0.5}}{2\pi a_4} \tag{7-103}$$

where

$$a_4 = (mx)^2\left(b^2+m\right)^2 + (my)^2\left(a^2+m\right)^2 + z^2\left(a^2+m\right)^2\left(b^2+m\right)^2 \tag{7-104}$$

A useful extension of the above model lies in the ability to simulate a flanged rectangular inlet of various aspect ratios (length/width) by approximating the opening by the inscribed ellipse of major and minor diameters 2a and 2b, that is to say rectangular inlets of length 2a and width 2b. Conroy et al. (1988) measured the velocities in front of such a rectangular inlet and found that the above expressions yielded reasonably accurate results so long as one does not get close to regions near the rims of the inlet. Tim Suden et al. (1990) combined the equations for a flanged elliptical inlet with the equations from Chapter 6 for buoyant plumes to simulate the capture of arc welding fume.

There are many industrial applications in which an unflanged pipe or rectangular duct is used to capture contaminants, e.g. welding fume, local sources of obnoxious odors, etc. On a smaller scale, the same physical principles are involved when a sample of a dust laden gas stream is withdrawn through a probe and sampling nozzle. In the first instance the engineer wants to know the reach of the inlet, especially when room air currents are present. In the second instance, one must be sure that the gas sample represents the aerosol in the gas stream from which the sample is taken, i.e. that the dust concentration and size distribution are the same as those of the sampled gas. The velocity field of the gas sample entering the inlet can be modeled by the potential functions discussed in this chapter.

7.6 Flanged and Unflanged Inlets in Streaming Flow

7.6.1 Flanged Slot in Streaming Flow

Consider the withdrawal of air through an infinitely long rectangular slot of width 2w lying in a plane over which air passes from right to left. The volumetric flow rate per length of slot is Q/L. Air approaches the slot in the negative x-direction with uniform speed U_0. The potential functions for the flow field are equal to the sums of the potential functions for streaming flow and flow through a flanged slot,

$$\phi = \phi_{\text{streaming flow}} + \phi_{\text{flanged slot}} = \phi_{sf} + \phi_{slot} \tag{7-105}$$

and

$$\psi = \psi_{\text{streaming flow}} + \psi_{\text{flanged slot}} = \psi_{sf} + \psi_{slot} \tag{7-106}$$

where the abbreviations are used for conciseness. For streaming flow the potential functions are given by Eqs. (7-41) and (7-42), but unfortunately ϕ and ψ for the flanged slot can not be written as explicit functions of x and y. Thus procedures used earlier cannot be used in their entirety. Alternatively the velocities U_x and U_y can be computed directly, using Eqs. (7-39), (7-81), and (7-80):

$$U_x = -\frac{\partial \phi}{\partial x} = -U_0 - \frac{k}{w}\frac{\sinh\left(\frac{\phi_{slot}}{k}\right)\cos\left(\frac{\psi_{slot}}{k}\right)}{\left[\sinh\left(\frac{\phi_{slot}}{k}\right)\cos\left(\frac{\psi_{slot}}{k}\right)\right]^2 + \left[\cosh\left(\frac{\phi_{slot}}{k}\right)\sin\left(\frac{\psi_{slot}}{k}\right)\right]^2} \tag{7-107}$$

and

$$U_y = \frac{\partial\psi}{\partial x} = -\frac{k}{w}\frac{\cosh\left(\frac{\phi_{slot}}{k}\right)\sin\left(\frac{\psi_{slot}}{k}\right)}{\left[\sinh\left(\frac{\phi_{slot}}{k}\right)\cos\left(\frac{\psi_{slot}}{k}\right)\right]^2 + \left[\cosh\left(\frac{\phi_{slot}}{k}\right)\sin\left(\frac{\psi_{slot}}{k}\right)\right]^2} \tag{7-108}$$

where constant k is given by Eq. (7-71).

Mapping the location of the streamlines is not difficult for a flanged inlet in streaming flow. The process begins by selecting the value of the stream function one wishes to map. For example, along the positive x-axis for $x > w$, it can be seen that ($\psi = \psi_{sf} + \psi_{slot}$) is zero. At the left lip of the inlet ($y = 0$, $x = -w$), the dividing streamline intersects the lip and must have a value of Q/L since the flow being sucked into the inlet must be confined between streamlines $\psi = 0$ and $\psi = Q/L$, as seen from Eq. (7-22). Thus at the point $y = 0$, $x = -w$, $\psi = \psi_{sf} + \psi_{slot} = Q/L$. For a streamline which enters the inlet, the procedure begins at a small value of y near the inlet and marches upwind, computing the coordinates (x,y) of the streamline at points progressively upwind of the inlet. The following algorithm can be used:

(1) With ψ known, select a value of y just above the inlet, for example at 0.001w.
(2) Compute $\psi_{sf} = U_0 y$.
(3) Compute $\psi_{slot} = \psi - U_0 y$.
(4) Using Eq. (7-73), solve for x:

$$x = w\cos\left(\frac{\pi L \psi_{slot}}{Q}\right)\sqrt{1 + \left[\frac{\frac{y}{w}}{\sin\left(\frac{\pi L \psi_{slot}}{Q}\right)}\right]^2} \tag{7-109}$$

(5) Repeat steps (1) to (4) for several values of y.
(6) Connect the locus of x and y points – this represents the streamline.

The algorithm is then repeated for several streamlines.

Example 7.5 - Flanged Slot in Streaming Flow

Given: Air flows in the negative direction over a plane containing a slot (width 2w, w = 0.020 m) through which air is withdrawn at a volumetric flow rate (Q/L) equal to 30.0 m^2/min. The far field air velocity (U_0) is uniform and equal to 30.5 m/min (100 FPM). These values of Q/L and U_0 are identical to those of Example 7.2 for comparison.

To do: Compute and plot several streamlines using the algorithm, and compare to the results of Example 7.2.

Solution:
The dividing streamline ($\psi = \psi_d$) is

$$\psi_d = \frac{Q}{L} = 30.0\frac{m^2}{min}$$

The procedure described in Section 7.6.1 was used to generate the streamlines. The authors used Excel, and the file is available on the book's web site. Figure E7.5a shows several streamlines in the vicinity of the slot.

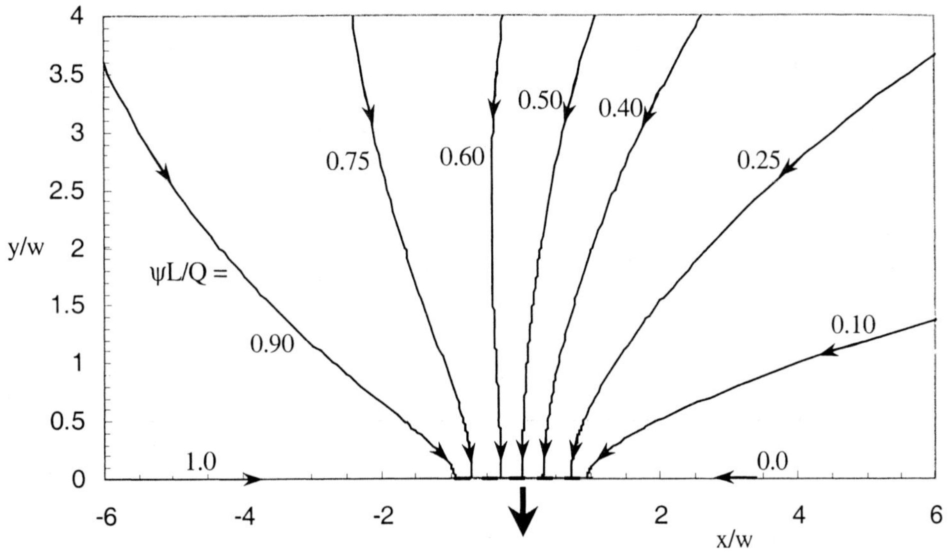

Figure E7.5a Streamlines generated by a two-dimensional flanged slot of half-width w superposed with a streaming flow; the x and y axes are normalized, and are not to scale.

Compared to the flanged slot without a streaming flow (Figure E7.3a), it is clear that the streamlines here are not symmetric about the slot, but are tilted to the right, reflecting the distortion in the slot streamlines due to the streaming flow. A view of the same streamlines from "further away," and in dimensional coordinates (x and y) is shown in Figure E5.7b. These streamlines can be compared to those of Figures 7.6 and E7.2a.

The dividing streamline looks similar to the dividing streamline of Example 7.2 in which the air was withdrawn through a line sink of equal magnitude to that in the present problem, 30.0 m^2/min. In fact all aspects of the dividing streamline correspond favorably, as indicated in the comparison below:

parameter	line sink (Example 7.2)	finite slot (Example 7.5)
h_∞	0.984 m	0.984 m
h_0	0.492 m	0.490 m
r_0	0.313 m	0.314 m

Discussion: The streamlines of a line sink in streaming flow and those of a slot in streaming flow are very similar, especially when viewed from afar. Thus, at distances more than a few slot widths away from the inlet, users may assume that the finite slot can be replicated by a line sink. Since the mathematics for a line sink is far simpler than for a slot, the approximation affords a great deal of simplification.

Using an infinite source of smoke traveling over flanged square inlets, Fletcher and Johnson (1986) found that all ***capture envelopes*** could be reduced to a single curve when the variables are grouped in dimensionless form. The dimensionless variables reduce to $\{y\ [U_0/Q]^{1/2}\}$, and $\{x\ [U_0/Q]^{1/2}\}$. Flynn and Miller (1989) suggest an alternative approach to predict three-dimensional flow fields for plane and flanged rectangular inlets; it is called the ***boundary integral equation method***. They report that the results are in excellent agreement with accepted values.

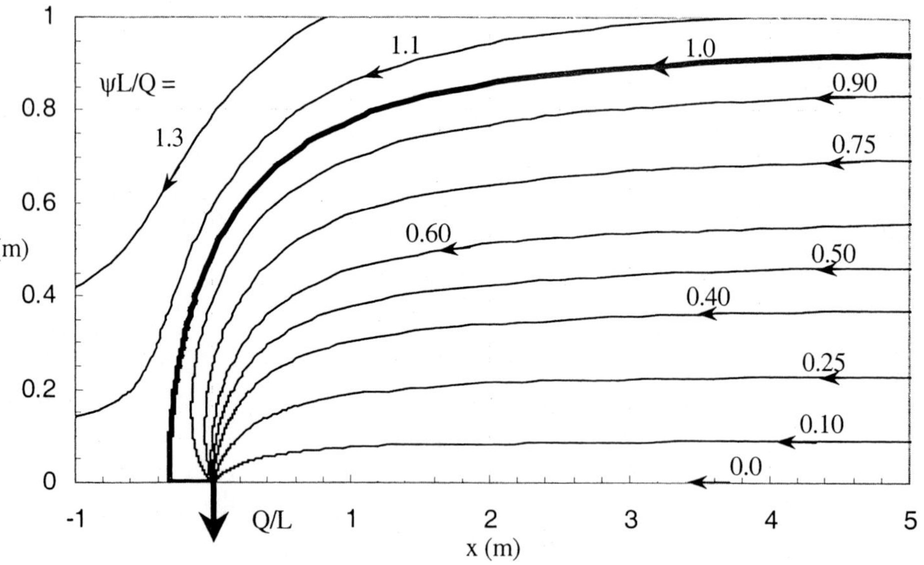

Figure E7.5b Streamlines generated by a two-dimensional flanged slot of half-width w superposed with a streaming flow. The dividing streamline is shown in bold; the x and y axes are not to scale.

7.6.2 Calculating Velocities using Newton-Raphson

The principal value of modeling inlets with potential functions is to enable users to predict velocity components U_x and U_y at arbitrary coordinates (x,y) upwind of the inlet. Equations (7-107) and (7-108) give the velocity components as functions of ϕ_{slot} and ψ_{slot}, not x and y as one would like. ϕ and ψ can be found at arbitrary (x,y) using the ***Newton-Raphson method*** and the following algorithm. Readers should consult Appendix A.13 for details about the Newton-Raphson method.

(1) Define the functions F and G:

$$\boxed{F = \frac{x}{w} - \cosh\frac{\phi_{slot}}{k}\cos\frac{\psi_{slot}}{k} \qquad G = \frac{y}{w} - \sinh\frac{\phi_{slot}}{k}\sin\frac{\psi_{slot}}{k}} \tag{7-110}$$

(2) Partially differentiate Eqs. (7-110) with respect to both ϕ_{slot} and ψ_{slot}, to obtain

$$\frac{\partial F}{\partial \phi_{slot}} = -\frac{1}{k}\sinh\frac{\phi_{slot}}{k}\cos\frac{\psi_{slot}}{k} \qquad \frac{\partial F}{\partial \psi_{slot}} = \frac{1}{k}\cosh\frac{\phi_{slot}}{k}\sin\frac{\psi_{slot}}{k}$$

$$\frac{\partial G}{\partial \phi_{slot}} = -\frac{1}{k}\cosh\frac{\phi_{slot}}{k}\sin\frac{\psi_{slot}}{k} \qquad \frac{\partial G}{\partial \psi_{slot}} = -\frac{1}{k}\sinh\frac{\phi_{slot}}{k}\cos\frac{\psi_{slot}}{k}$$

(3) Form the following two Taylor series expansions (to first order, neglecting higher-order terms):

$$F_{i+1} = F_i + \left(\frac{\partial F}{\partial \phi_{slot}}\right)_i \left(\phi_{slot,\,i+1} - \phi_{slot,\,i}\right) + \left(\frac{\partial F}{\partial \psi_{slot}}\right)_i \left(\psi_{slot,\,i+1} - \psi_{slot,\,i}\right)$$

$$G_{i+1} = G_i + \left(\frac{\partial G}{\partial \phi_{slot}}\right)_i \left(\phi_{slot,\,i+1} - \phi_{slot,\,i}\right) + \left(\frac{\partial G}{\partial \psi_{slot}}\right)_i \left(\psi_{slot,\,i+1} - \psi_{slot,\,i}\right)$$

(4) Assuming F_{i+1} and G_{i+1} are the correct values, solve the above equations simultaneously to obtain the following:

$$\phi_{slot,\,I+1} = \phi_{slot,\,I} + \frac{G_I\left(\dfrac{\partial F}{\partial \psi_{slot}}\right)_I - F_I\left(\dfrac{\partial G}{\partial \psi_{slot}}\right)_I}{J} \qquad (7\text{-}111)$$

and

$$\psi_{slot,\,I+1} = \psi_{slot,\,I} + \frac{F_I\left(\dfrac{\partial G}{\partial \phi_{slot}}\right)_I - G_I\left(\dfrac{\partial F}{\partial \phi_{slot}}\right)_I}{J} \qquad (7\text{-}112)$$

where

$$J = \left(\frac{\partial F}{\partial \phi_{slot}}\right)_I\left(\frac{\partial G}{\partial \psi_{slot}}\right)_I - \left(\frac{\partial F}{\partial \psi_{slot}}\right)_I\left(\frac{\partial G}{\partial \phi_{slot}}\right)_I \qquad (7\text{-}113)$$

(5) Select locations (x,y) at which the velocity is wanted. Note that these values do not change in the iterations that follow.

(6) Guess initial values of $\phi_{slot,\,I}$ and $\psi_{slot,\,I}$, and compute the values of F, G, the partial derivatives, and J from the above equations. Then compute $\phi_{slot,\,I+1}$ and $\psi_{slot,\,I+1}$ from Eqs. (7-111) and (7-112).

(7) Using $\phi_{slot,\,I+1}$ and $\psi_{slot,\,I+1}$ as initial values, repeat step (6) until the difference between successive values of ϕ_{slot} (i.e. $\phi_{slot,\,I+1} - \phi_{slot,\,I}$) and ψ_{slot} (i.e. $\psi_{slot,\,I+1} - \psi_{slot,\,I}$) are as small as desired. The resulting values of ϕ_{slot} and ψ_{slot} should be interpreted as $\phi_{slot}(x,y)$ and $\psi_{slot}(x,y)$. With these values the velocity components U_x and U_y can be computed using Eqs. (7-107) and (7-108).

7.6.3 Unflanged Slot in Streaming Flow

Figure 7.15 is a schematic diagram of a plain (unflanged) rectangular opening of width (2w) immersed in a gas stream of uniform velocity U_0 inclined at angle α with respect to the duct walls. Gas is withdrawn through the duct at uniform volumetric flow rate Q/L per unit length of slot. The average inlet velocity at the inlet plane is U_{face} which is equal to Q/(2Lw). The face velocity (U_{face}) is sometimes called the ***suction velocity***.

The coordinate system and orientation of the streaming flow is consistent with earlier sections. The overall potential functions for the flow field are

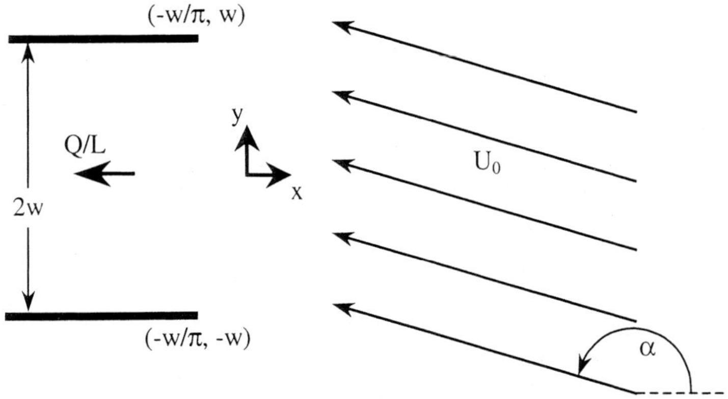

Figure 7.15 Unflanged inlet of width 2w in an oblique streaming flow; the origin is offset to the right of the slot face by w/π.

$$\phi = \phi_{\text{streaming flow}} + \phi_{\text{unflanged slot}} = \phi_{sf} + \phi_{slot} = -U_0 x \cos\alpha - U_0 y \sin\alpha + \phi_{slot} \tag{7-114}$$

and

$$\psi = \psi_{\text{streaming flow}} + \psi_{\text{unflanged slot}} = \psi_{sf} + \psi_{slot} = -U_0 y \cos\alpha + U_0 x \sin\alpha + \psi_{slot} \tag{7-115}$$

where subscripts "sf" and "slot" are used for the streaming flow and the unflanged slot respectively for simplicity. Analytical expressions for these functions were listed previously. Since the potential functions (ϕ_{slot} and ψ_{slot}) for the unflanged rectangular inlet cannot be written as explicit functions of x and y, procedures similar to those used to derive Eqs. (7-111) and (7-112) must be used. The velocity components are

$$U_x = -U_0 \cos\alpha - \frac{\dfrac{k\pi}{2w}\left[1+\exp\left(2\dfrac{\phi_w}{k}\right)\cos\left(2\dfrac{\psi_w}{k}\right)\right]}{\left[\exp\left(2\dfrac{\phi_w}{k}\right)\sin\left(2\dfrac{\psi_w}{k}\right)\right]^2 + \left[1+\exp\left(2\dfrac{\phi_w}{k}\right)\cos\left(2\dfrac{\psi_w}{k}\right)\right]^2} \tag{7-116}$$

and

$$U_y = -U_0 \sin\alpha - \frac{\dfrac{k\pi}{2w}\left[\exp\left(2\dfrac{\phi_w}{k}\right)\sin\left(2\dfrac{\psi_w}{k}\right)\right]}{\left[\exp\left(2\dfrac{\phi_w}{k}\right)\sin\left(2\dfrac{\psi_w}{k}\right)\right]^2 + \left[1+\exp\left(2\dfrac{\phi_w}{k}\right)\cos\left(2\dfrac{\psi_w}{k}\right)\right]^2} \tag{7-117}$$

where $k = Q/(\pi L)$ as previously. Equations (7-116) and (7-117) show that the concept of superposition is indeed valid. The velocity at any point upwind of an unflanged inlet in streaming flow is the sum of the streaming flow velocity (the first term on the right hand side) and the velocity produced by the inlet alone in a quiescent flow (the second term on the right hand side). Note that the walls of the slot are ignored by this model; results are useful only upstream of the inlet.

The streamlines can be mapped by integrating the velocities and computing the displacements by the Newton-Raphson procedure discussed in the previous section (Heinsohn and Choi, 1986). A forward time step can be used and the Newton-Raphson method can be used to compute ϕ_{slot} and ψ_{slot} at arbitrary values of x and y. The dividing streamlines can be computed by the same trial-and-error procedure discussed previously, but since there is no symmetry with respect to the x-axis, both the upper and lower dividing streamlines have to be computed.

Example 7.6 - Unflanged Slot in Streaming Flow at Non-Zero Angle

Given: A plain (unflanged) slot of width 2w (w = 0.020 m) withdraws air that streams past it as seen in Figure 7.15. In the far-field the air velocity (U_0) is uniform with a speed equal to 1.5 m/min. Air approaches the inlet from right to left at an angle (α) of 150° from the positive x-axis. Air enters the slot with a volumetric flow rate (Q/L) of 3.6 m^2/min and face velocity (U_{face}) of 90. m/min.

To do:

(a) Predict and plot several streamlines of the air entering the inlet.
(b) Predict the air speed along these streamlines.
(c) Predict the components of the air velocity, U_x and U_y, at some arbitrary point.

Solution:

(a) The stream function for the composite flow is given by Eq. (7-115). The stream function for a leftward streaming flow is given in Table 7.2, and the stream function for the plain inlet is expressed implicitly by Eqs. (7-90) and (7-91). The dividing streamlines are given by $\psi_{dividing} = \pm 1.8$. The streamlines between the dividing streamlines are given by Eq. (7-115). The algorithm consists of selecting values of the stream function ψ, specifying a value of ϕ_{slot}, and then using a trial and error method to calculate the values of y, x, and ψ_{slot} that satisfy Eqs. (7-90) and (7-91). Once the guessed values have been computed, the process is repeated for the next value of ϕ_{slot}. The process is repeated until a selected region upstream of the inlet has been covered. Five streamlines, $\psi L/Q$ = -0.499, -0.25, 0, 0.25, and 0.40, are shown in Figure E7.6a, as generated by Mathcad; the file is available on the book's web site. The authors were not able to get Mathcad to converge for the values of –0.50 or 0.50, which are the dividing streamlines. The reader should note that Mathcad's *Solve Block* feature is capable of solving three (or more) simultaneous algebraic equations.

(b) The coordinates (x,y) of points along a particular streamline are contained in Mathcad's *Find* matrix obtained in Part (a). A new matrix is created from Part (a) from which values of ϕ_{slot} and ψ_{slot} exist at the coordinates (x,y) along streamline ψ. The velocity components produced by the inlet, $U_{x,\,slot}$ and $U_{y,\,slot}$, are computed along the streamline using Eqs. (7-93) and (7-94). The actual velocity components (U_x and U_y) that exist at point (x,y) are found by superposition,

$$U_x = U_{x,\,slot} + U\cos\alpha \qquad U_y = U_{y,\,slot} + U\sin\alpha$$

Graphs of the velocity components U_x and U_y along the same streamlines as in Figure E7.6a are shown in Figures E7.6b and E7.6c.

(c) The concept of superposition is applied in which the velocity components (U_x and U_y) for the unflanged inlet alone are computed from Eqs. (7-93) and (7-94), and then vectorially added to the velocity components of the streaming flow (U_0). To begin the calculation, one needs to compute the value of the potential functions ϕ_{slot} and ψ_{slot} at an arbitrary point (x,y), and then substitute these values into Eqs. (7-93) and (7-94). The *Solve Block* feature of Mathcad can be used to obtain the solution of Eqs. (7-90) and (7-91). At the point (x = 0.23 m, y = 0.211 m), U_x = -2.7 m/s and U_y = -0.60 m/s.

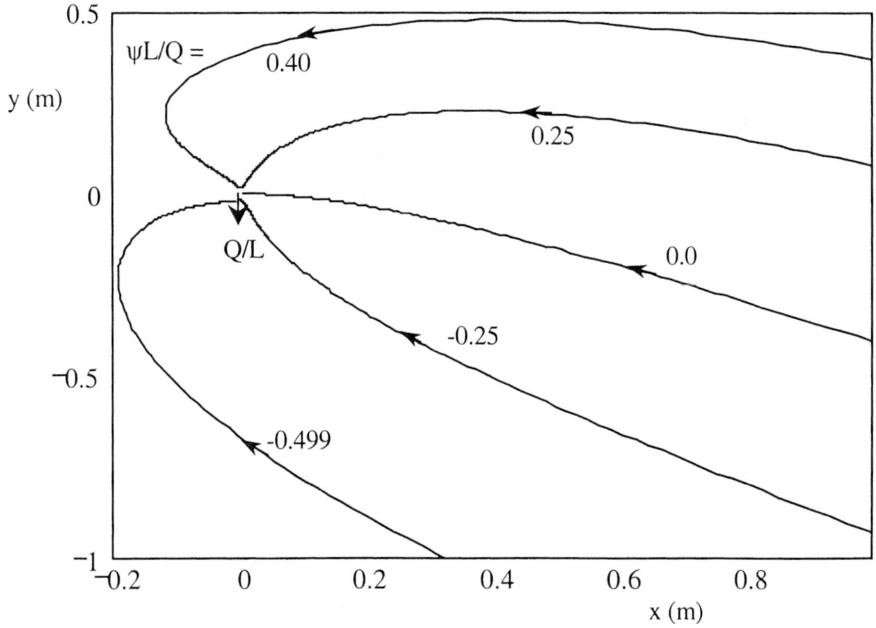

Figure E7.6a Streamlines generated by a two-dimensional unflanged slot of half-width w superposed with a streaming flow at angle α; the x and y axes are not to scale.

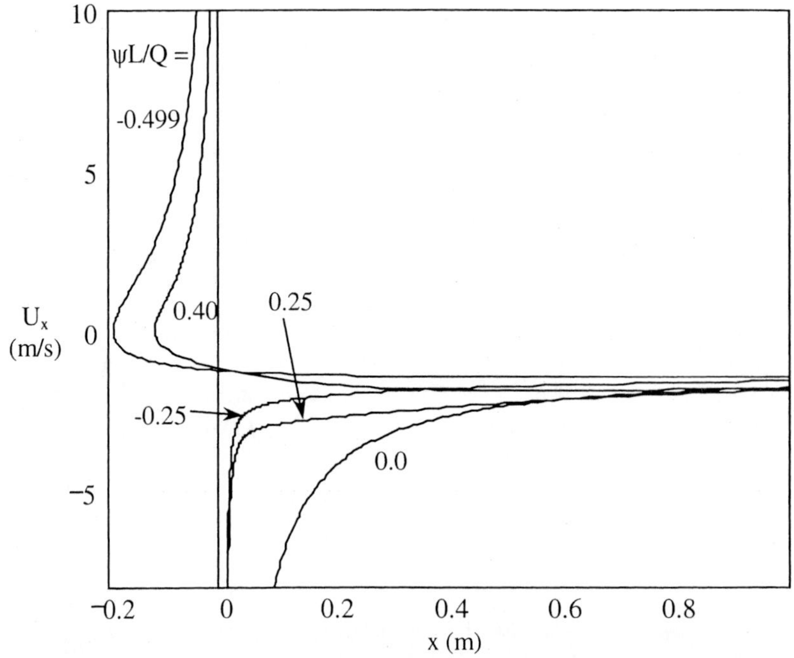

Figure E7.6b Velocity component U_x along streamlines generated by a two-dimensional unflanged slot of half-width w superposed with a streaming flow at angle α.

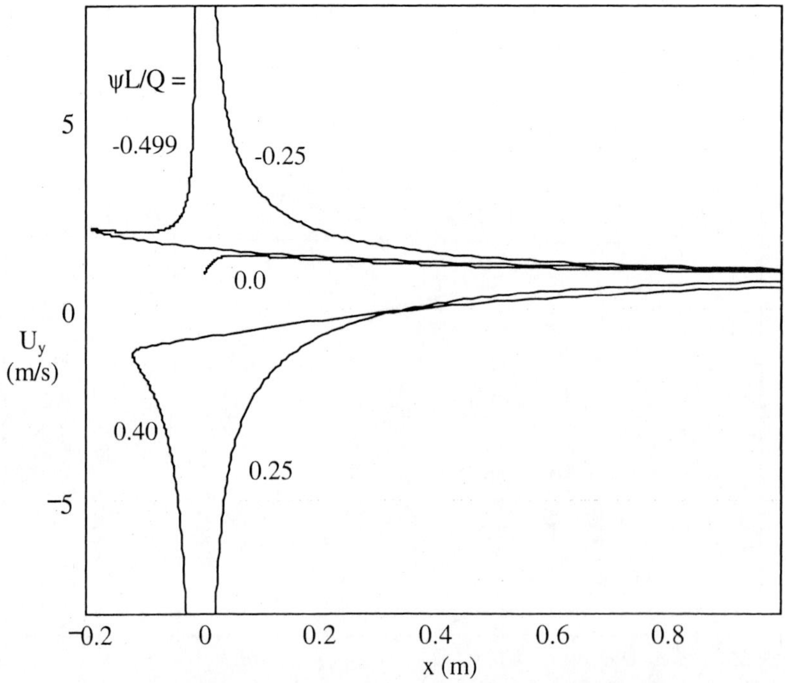

Figure E7.6c Velocity component U_y along streamlines generated by a two-dimensional unflanged slot of half-width w superposed with a streaming flow at angle α.

Discussion: The lowest streamline in Figure E7.6a is very close to the lower dividing streamline. The point x = 0.23 m, y = 0.211 m was chosen so that the predicted velocity components U_x and U_y can be compared to the values shown in Figures E7.6b and E7.6c. At these coordinates, the normalized stream function ($\psi L/Q$) is equal to 0.25; the values of U_x and U_y in Figures E7.6b and E7.6c correspond favorably with the values shown in Part (c).

7.6.4 Mapping Dividing Streamlines by Shooting

Another approach for locating the dividing streamlines associated with inlet(s) in streaming flows is called ***shooting*** by some because it involves selecting an upstream point (x_0,y_0), and following a streamline with the hope that it intersects the lip of the inlet. If it doesn't, another location (x_0,y_0) is selected following a rational method; shooting is then repeated until the selected streamline intersects the lip within tolerable limits. The locations of streamlines can be determined by integrating the velocity and computing a displacement. By repeating the process, one can map a fluid pathline from an arbitrarily selected initial upstream location (x_0,y_0) until the pathline either enters the slot or passes it. The process uses a forward time step, such that the location of the pathline at the end of time step (δt) is based on the air velocity components $U_x(t)$ and $U_y(t)$ at the beginning of the time step,

$$\boxed{x(t+\delta t) = x(t) + U_x(t)\delta t \qquad y(t+\delta t) = y(t) + U_y(t)\delta t} \tag{7-118}$$

The magnitude of the time step (δt) should be small, and progressively smaller values should be used in the close proximity of the inlet. To compute U_x and U_y, the user must be able to compute ϕ_{slot} and ψ_{slot} at arbitrary values of x and y. Since these parameters are not explicit functions of x and y, a Newton-Raphson method for two simultaneous equations can be used. Details of this method can be found in Appendix A.13.

The upper dividing streamline can be determined by a trial-and-error procedure, consisting of the following steps:

(1) An upstream location x_0 is chosen to begin the analysis. This coordinate will not change.
(2) A guess is made for the distance y_1 through which the dividing streamline is believed to pass.
(3) Using the Newton-Raphson method, $\phi_{slot}(x_0,y_1)$ and $\psi_{slot}(x_0,y_1)$ are computed.
(4) Using superposition, the overall stream function for the flow is computed, i.e. $\psi = \psi_{sf}(x_0,y_1) + \psi_{slot}(x_0,y_1)$. This value of ψ is compared to the value of $\psi_{dividing}$, the value of the stream function on the dividing streamline, which must be equal to Q/(2L). If the values do not agree, another value of y_1 is selected, and step (3) is repeated until satisfactory agreement is obtained.
(5) Once the correct value of y_1 is obtained, step (3) is repeated. Once the correct values of ϕ_{slot} and ψ_{slot} have been found, $U_x(x_0,y_1)$ and $U_y(x_0,y_1)$ can be computed from Eqs. (7-116) and (7-117).
(6) The location of the air parcel after an interval of time (δt) is computed using Eqs. (7-118).
(7) Using the new location found in step (6), the process is repeated, beginning with step (4). In doing so repeatedly, the solution marches toward the inlet after each increment of time until the streamline passes through an arbitrarily small region surrounding the lip ($-w/\pi$, w).

To accomplish the above procedure, one has to understand the relative location of streamlines and the general relationship between the initial location and the point where the streamline intersects the x-axis.

If a streamline other than the dividing streamline is studied, and if the numerical value of the stream function is known, the above process may be used intact. If the numerical value of ψ is *not* known, but the location (x_0,0) of the streamline as it passes through the plane of the inlet *is* known, the above process must be preceded by two steps to compute the numerical value of ψ:

(1) Using the Newton-Raphson method, $\phi_{slot}(x_0,0)$ and $\psi_{slot}(x_0,0)$ are computed.
(2) The value of ψ is computed from Eq. (7-115).

Steps (1) to (7) above are then repeated. To compute the velocity at an arbitrary point (x,y) upwind of the inlet, the Newton-Raphson method can be used in a manner similar to that described for the flanged slot to analyze multiple inlets and flanged inlets with adjacent planes (Esmen and Weyel, 1989).

7.7 Multiple Flanged Rectangular Inlets

Inlets are generally designed with flanges because they increase the reach. To a first approximation the reach of an inlet is equal to the smaller of its characteristic dimensions. The isopleths of Figure 6.8 show that at upstream distances equal to the smaller of the inlet characteristic dimensions, the air velocity is between 10% and 20% of the face value. Thus for the same volumetric flow rate, a single large inlet influences contaminants farther from the inlet plane than do a series of small inlets of the same total cross-sectional area. However, the cross-sectional area of the region of influence is larger for the multiple inlets. Esmen et al. (1986) and Esmen and Weyl (1986) have expanded the use of potential functions to analyze multiple inlets. The analysis presumes that each inlet is a quadrilateral opening lying in the y-z plane. Figure 7.16 shows the dimensions and orientation of the surfaces used in their analysis. Coordinates x and z are measured from the center of the inlet; note that their coordinate system is left-handed. The z-y plane containing the inlets is bounded by a single flanking plane parallel to the x-z plane and a single flanking plane parallel to the x-y plane.

Consider an arbitrary point located at coordinates (x,y,z) in front of a single quadrilateral opening of dimensions A by B through which air is withdrawn at a volumetric flow rate Q. The magnitude of the velocity (the speed) is

$$\left|\vec{U}(x,y,z)\right| = \frac{\frac{Q}{AB}}{f(R)} \tag{7-119}$$

where f(R) is the following function:

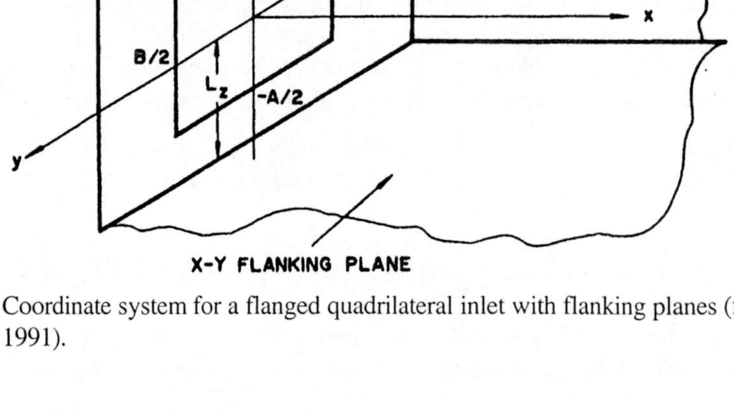

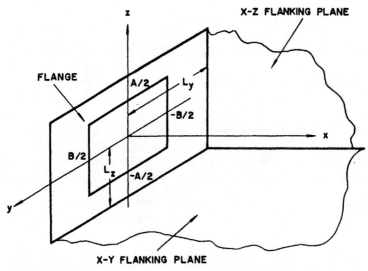

Figure 7.16 Coordinate system for a flanged quadrilateral inlet with flanking planes (from Heinsohn, 1991).

$$f(R)=\frac{1}{2}\left[\frac{R}{B}\left(\pi+\arcsin k_1\right)+\frac{R}{A}\left(\pi+\arcsin k_2\right)\right]+\sqrt{1+4\frac{|xy|+|zx|}{A^2+B^2}+\frac{\pi}{AB}\left(k_1+k_2\right)R^2} \quad (7\text{-}120)$$

The parameter R is called a ***space parameter*** which is defined as

$$R^2=x^2+k_3\left(|y|-\frac{B}{2}\right)^2+k_4\left(|z|-\frac{A}{2}\right)^2 \quad (7\text{-}121)$$

where

$$k_1=1 \text{ if } R<L_y, \quad k_1=\frac{L_y}{R} \text{ if } R\geq L_y \quad (7\text{-}122)$$

$$k_2=1 \text{ if } R<L_z, \quad k_2=\frac{L_z}{R} \text{ if } R\geq L_z \quad (7\text{-}123)$$

$$k_3=0 \text{ if } |y|\leq\frac{B}{2}, \quad k_3=1 \text{ if } |y|>\frac{B}{2} \quad (7\text{-}124)$$

and

$$k_4=0 \text{ if } |z|\leq\frac{A}{2}, \quad k_4=1 \text{ if } |z|>\frac{A}{2} \quad (7\text{-}125)$$

The velocity at point (x,y,z) has a ***directional vector*** defined as

$$\vec{D}=\frac{1}{k_5}\left\{\hat{i}\cdot 4x+\hat{j}\cdot\left[2y\left(1+k_3\right)-Bk_3\right]+\hat{k}\cdot\left[2z\left(1+k_4\right)-Ak_4\right]\right\} \quad (7\text{-}126)$$

where

$$k_5=\sqrt{16x^2+\left[2y\left(1+k_3\right)-Bk_3\right]^2+\left[2z\left(1+k_4\right)-Ak_4\right]^2} \quad (7\text{-}127)$$

and $\hat{i}$, $\hat{j}$, and $\hat{k}$ are unit vectors in the x, y, and z directions respectively.

When several quadrilateral inlets of different dimensions are located in the same plane, the velocity at any point (x,y,z) in front of the plane can be found by adding the vector velocities predicted by Eqs. (7-119) through (7-127). As shown in earlier sections, a useful property of ideal flows is that the net velocity field produced by a composite of sinks is the vector sum of the velocities each sink produces. The computations are tedious because the origin of the coordinate system used in the equations above is at the center of each opening, thus the coordinates (x,y,z) are unique for each opening. One computes the velocity that would exist at the desired point for each inlet, defining (x,y,z) for that inlet. As an example, Figure 7.17 shows surfaces of constant air speed in front of a flanged inlet containing three openings; two values of U/U_{face} are shown.

7.8 Flanged Inlets of Arbitrary Shape

Up to this point, the discussion has involved inlets of conventional geometry, i.e. circular openings or rectangular openings of finite or infinite aspect ratio. Occasions may arise when a flanged opening of face area A_{face} but of unusual shape is to be used. In this case, the following analysis offers the ability to compute the velocity potential $\phi(x,y,z)$ at arbitrary points upwind of the inlet. Figure 7.18 illustrates the geometrical features one must take into account. Let point P, at location (x,y,z) be an arbitrary point upwind of an inlet of area A_{face} lying in an infinite x-y plane. The essential feature of the analysis is to assume that the inlet is composed of elemental point-sinks lying in the x-y plane through which air is withdrawn at a volumetric flow rate equal to $(Q/A_{face})dxdy$, where Q is the overall

(a)

(b)

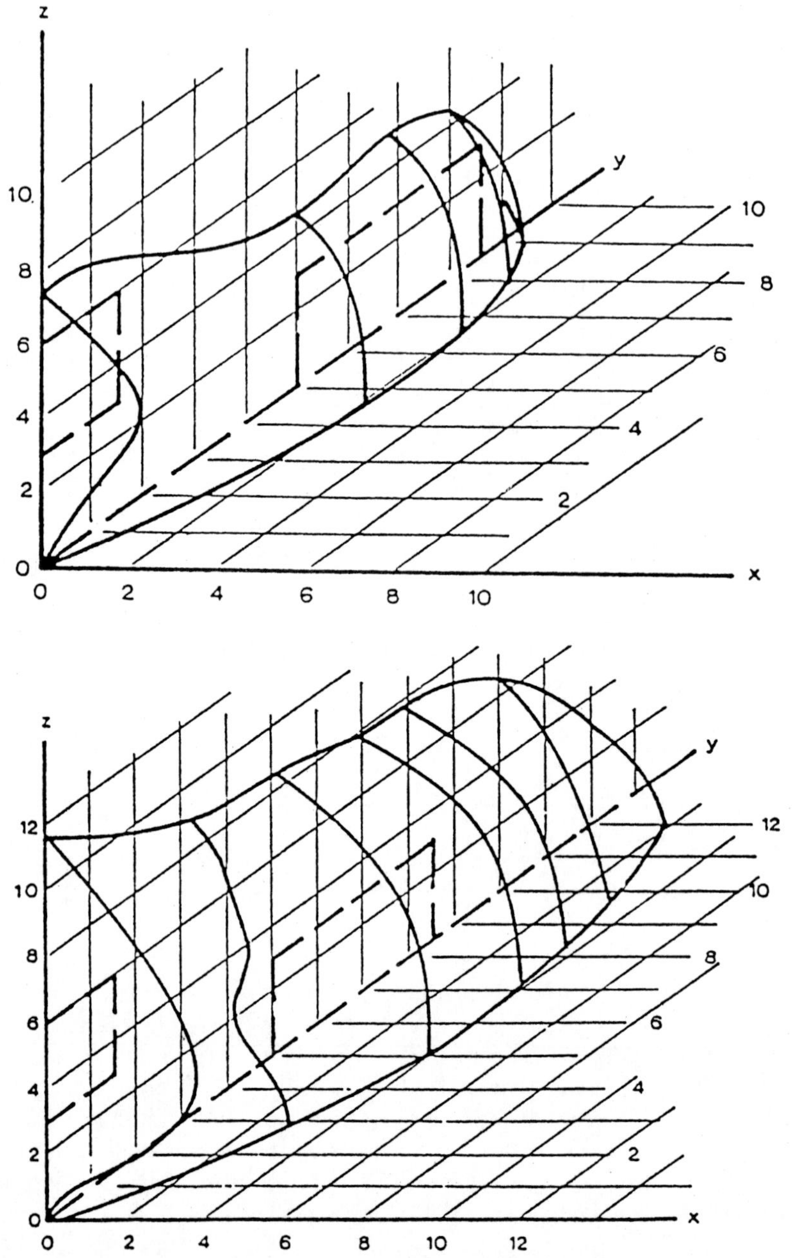

Figure 7.17 Surfaces of constant speed for a flanged inlet containing three openings: (a) U/U_{face} = 0.3, (b) U/U_{face} = 0.1 (redrawn from Esmen et al., 1986).

volumetric flow rate. The contribution to the overall velocity potential due to one elemental point sink is

$$\boxed{d\phi = -\frac{Q}{2\pi A_{face} s} dxdy} \tag{7-128}$$

where s is the distance between point P at (x,y,z) and the elemental area dxdy. The overall velocity potential $\phi(x,y,z)$ is obtained by integrating Eq. (7-128) over the inlet area:

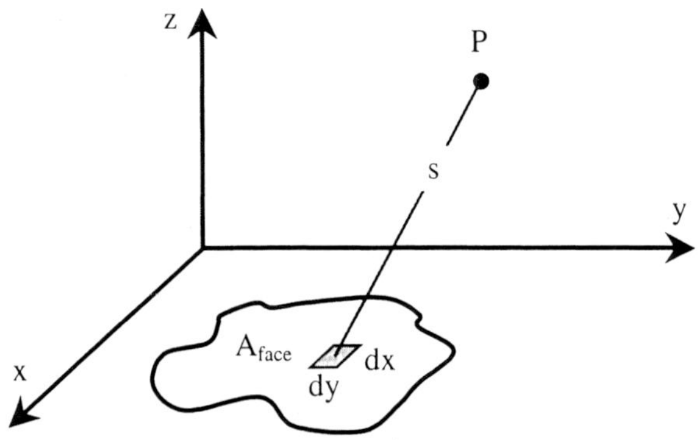

Figure 7.18 Point P at (x,y,z) upwind of a flanged inlet of arbitrary shape and area A_{face} in quiescent air; point P is distance s from elemental area element dxdy.

$$\boxed{\phi = \oiint_{A_{face}} d\phi = -\oiint_{A_{face}} \frac{Q}{2\pi A_{face} s} dxdy} \qquad (7\text{-}129)$$

The crucial term in the integrand is the distance s, which is the straight line distance between the elemental area dxdy and point P. As one integrates over the inlet area the distance s changes. Thus this seemingly benign integral is all but impossible to integrate in closed form except for elementary inlets that have been analyzed by other means in earlier sections. However, the integral can be solved numerically when repeated for an array of points P, providing tabulated data of the velocity potential for inlets that can not be analyzed by other means.

Anastas and Hughes (1988) obtained partial solutions for Eq. (7-129) at points along the centerline for a circular opening and along the mid-plane of slots. With these expressions, they differentiated with respect to z to obtain centerline velocities. The results agreed well with measurements. To illustrate the technique, Eq. (7-129) is applied to find the centerline velocity upwind of the flanged circular inlet of radius R shown in Figure 7.19. Because of symmetry in this example, Eq. (7-129) reduces to the following:

$$\phi(0,0,z) = -\int_0^R \int_0^{2\pi} \frac{Q}{2\pi^2 R^2} \frac{r d\theta dr}{\sqrt{z^2 + r^2}} = -\frac{Q}{\pi R^2} \int_0^R \frac{r dr}{\sqrt{z^2 + r^2}}$$

which after integration yields

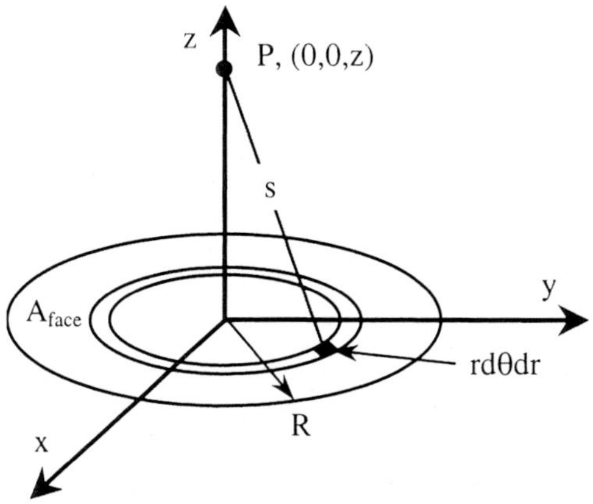

Figure 7.19 Point P at (0,0,z) along the centerline upwind of a flanged circular inlet in quiescent air.

$$\phi(0,0,z) = -\frac{Q}{\pi R^2}\left(\sqrt{z^2 + R^2} - z\right) \qquad (7\text{-}130)$$

The air velocity along the centerline can be found by differentiating the velocity potential with respect to z. Thus

$$U_z = -\frac{\partial \phi}{\partial z} = -\frac{Q}{\pi R^2}\left(1 - \frac{\frac{z}{R}}{\sqrt{1+\left(\frac{z}{R}\right)^2}}\right) \qquad (7\text{-}131)$$

It can be seen that at large z, U_z approaches zero, and as z approaches zero, U_z approaches the face velocity $Q/(\pi R^2)$.

For a slot, one can repeat the above analysis, but now using an infinite number of line sinks lying along the slot. Using this approach, Anastas and Hughes (1988) found that the velocity along the centerplane upwind of a slot of width 2w can be written as

$$U_z = \frac{Q}{\pi w L} \arctan \frac{1}{2z} \qquad (7\text{-}132)$$

For a rectangular slot of width A and height B (aspect ratio $A_r = A/B$), they found that the centerline velocity can be written as

$$U_z = \frac{2Q}{\pi A_{face}} \arctan \frac{1}{C} \qquad (7\text{-}133)$$

where

$$C = \frac{2z}{1 + A_r}\sqrt{1 + A_r^{\,2} + \frac{4z^2 A_r^{\,2}}{(1 + A_r)^2}} \qquad (7\text{-}134)$$

Equations (6-4) to (6-6) are analytical expressions for the centerline velocities for plain and flanged inlets abstracted from an extensive literature review by Braconnier (1988). The use of centerline velocities is limited since contaminants are apt to be at any location upwind of an inlet. To determine the reach of inlets and to assess their effectiveness, users need to be able to compute the velocity at arbitrary points upwind of an inlet.

7.9 Closure

ACGIH and ASHRAE design procedures provide useful guidance to engineers interested in capturing pollutants from sources in an indoor environment. Unfortunately, these procedures do not delineate the exact region in space from which the inlet captures the pollutant gases and particles. What is needed are predictive methods that define this region as functions of the inlet geometry and the volumetric flow rate of withdrawn air. Fortunately, the velocity fields associated with air entering inlets satisfy the physical conditions that allow engineers to approximate such flows as ideal flows, for which analytical solutions are possible. Accordingly the velocity components, U_x and U_y, are analytical functions of potential functions ϕ and/or ψ. The vast literature on potential flow is available to engineers to predict analytical functions for ϕ and ψ. The principles of ideal flow can be used to predict the reach of the inlet and the path that pollutant particles and gases take as they enter (or do not enter) the inlet. As computational fluid dynamics (CFD) becomes more readily available to engineers, ideal flow solutions are becoming less popular (see Chapter 10 for a discussion of CFD). Nevertheless, much can be learned from the ideal flow solutions presented here.

Chapter 7 - Problems

1. Each of the questions below has a single correct answer. Find it.

[**a**] Which of the following defines an ideal fluid?

- constant density
- irrotational
- negligible viscous effects
- all of the above

[**b**] Which of the following defines irrotational flow?

- incompressible
- divergence of the velocity is zero
- vorticity is zero
- frictionless

[**c**] What is the necessary and sufficient condition in steady incompressible flow without work or heat transfer to ensure that the quantity ($P/\rho + U^2/2 + gz$) is constant along a streamline?

- net force due to shear stresses is zero
- irrotational
- vorticity is zero
- the streamlines are not curved

[**d**] What is the necessary and sufficient condition in steady incompressible flow without work or heat transfer to ensure that the quantity ($P/\rho + U^2/2 + gz$) is constant everywhere in the flow field?

- the streamlines are not curved
- frictionless
- irrotational
- all of the above

[**e**] Irrotational flow requires streamlines to be straight lines. True or false?

[**f**] The vorticity is nonzero only if the streamlines are curved. True or false?

[**g**] The vorticity is nonzero only if the curl of the velocity is nonzero. True or false?

[**h**] Euler's equation is a simplified form of the Navier-Stokes equation, in which

- the flow is incompressible
- net viscous forces in the flow are negligible
- the flow is rotational
- the first two choices above

[**i**] The steady flow equation

$$\left(\frac{P}{\rho}\right)_2 - \left(\frac{P}{\rho}\right)_1 + \frac{U_2^{\,2} - U_1^{\,2}}{2} + g(z_2 - z_1) = (\text{head gain})_{\text{shaft work}} - (\text{head loss})_{\text{friction}}$$

comes from which conservation law?

- conservation of mass
- conservation of momentum
- conservation of energy
- conservation of angular momentum

[j] Which of the following conditions is necessary in order to define the stream function for a steady incompressible flow?

- two-dimensional flow
- irrotational flow
- inviscid flow
- all of the above

[k] Which of the following conditions is necessary in order to define the velocity potential, ϕ?

- two-dimensional flow
- irrotational flow
- inviscid flow
- all of the above

[l] Answer true or false for each of the following:

- Vorticity is a measure of the angular deformation of a fluid element.
- Vorticity is always zero in a rotational flow field.
- All curvilinear flow has vorticity.
- Vorticity is a measure of the viscosity of the fluid.
- Vorticity is always zero for incompressible flow.
- Vorticity is a measure of rotation of a fluid element.

[m] For irrotational flow, the velocity is equal to the negative of the gradient of the velocity potential (ϕ), but under what conditions is the Laplacian of ϕ equal to zero?

[n] For two-dimensional flow, $U_x = -\partial\psi/\partial y$ and $U_y = \partial\psi/\partial x$, but under what conditions is the Laplacian of ψ equal to zero?

2. Prove that if a flow is steady and the divergence of the velocity is zero, then the flow is incompressible.

3. Derive analytical expressions for the potential functions (where appropriate) in Example 7.1.

4. Consider a stationary circular cylinder of radius (a) immersed in an ideal liquid that is moving in the negative x-direction with a speed U_0. The stream function for the flow around the cylinder is given by

$$\psi = U_0 y\left[1-\left(\frac{a}{r}\right)^2\right]$$

where r is the radial distance from the center of the cylinder.

(a) Show that the velocity potential is

$$\phi = U_0\left[r+\frac{a^2}{r}\right]\cos\theta$$

(b) Prepare a graph showing lines of constant stream function and velocity potential.

(c) Show that the pressure coefficient C_P is equal to $C_P = 1 - 4\,(\sin\theta)^2$, where $C_P = \dfrac{2(P-P_0)}{\rho U_0^{\,2}}$ and P_0 is the far-field pressure.

5. A circular cylinder of radius (a) moves through a stationary ideal fluid with a velocity U_0 normal to its axis. The stream function is

$$\psi = U_0 \frac{a^2}{r} \sin\theta$$

Derive an expression for the velocity potential (ϕ).

6. The flow of a fluid through a vertical row of identical circular cylinders of radius (a) is modeled as an ideal fluid flow. The cylinders are placed one above the other, separated by distance (h) between their centers. The axes of the cylinders are normal to the x-y plane. The upwind velocity (U_0) is uniform and in the negative x-direction. The stream function for such a flow is

$$\psi = U_0 a \left(\frac{y}{a}\right) - \frac{\pi \frac{a}{h} \sin\left(\frac{y}{m}\right)}{\cosh\left(\frac{x}{m}\right) - \cos\left(\frac{y}{m}\right)}$$

where $m = h/(2\pi)$; $x = 0$, $y = 0$ corresponds to the center of one of the cylinders. If $h/a = \pi$, find the velocity components (U_x and U_y) midway between the cylinders, i.e. find U_x and U_y at ($x = 0$, $y = h/2$).

7. The velocity potential for a sphere of radius (a) traveling with velocity U_0 in a stationary infinite ideal fluid is given by

$$\phi = -a\frac{U_0}{2}\left(\frac{a}{r}\right)^2 \cos\theta$$

(a) Show that the stream function is given by

$$\psi = \frac{U_0}{2r} a^3 (\sin\theta)^2$$

(b) Are the velocity and pressure distributions on the sphere's surface different than the distributions when the roles are reversed and the fluid passes over a stationary sphere? Are the far-field velocity and pressure distributions different?

8. Examine the validity of a statement that appears in Chapter 8 concerning circumstances under which spherical particles A and B (of radii a and b respectively) influence each other. It is asserted that particle B does not significantly affect the flow field around an adjacent particle A when the distance between the particles is greater than approximately than ten times the particle diameter.

(a) Case I - Particle A is to the rear of particle B and both move along their line of centers with a velocity U_0 through quiescent air. If the particle centers are separated by distance c, the velocity potential in the vicinity of particle A is

$$\phi = \frac{aU_0}{2}\left(\frac{a}{r}\right)^2 \cos\theta + \left(\frac{ab}{c}\right)^3 \frac{U_0}{2c^2}\left(1 + \frac{2r}{c}\cos\theta\right) + \frac{U_0}{2}\left(\frac{a}{c}\right)^6 b^3 \frac{\cos\theta}{r^2}$$

Show that the effect of particle B on the velocity at the surface of particle A is negligible if $c/b > 20$.

(b) Case II - The centers of particles A and B are distance c apart and both particles move through quiescent air at a velocity U_0 at right angles to their line of centers. If the velocity potential in the vicinity of A is

$$\phi = \frac{aU_0}{2}\left(\frac{a}{r}\right)^2 \left(1 + \frac{3}{2}\left(\frac{b}{c}\right)^3\right) \cos\theta$$

Show that the effect of particle B on the velocity at the surface of particle A is negligible if c/b > 20.

9. The stream function

$$\psi = -\frac{aU_w}{m}\left(\frac{r}{a}\right)^m \sin(m\theta), \text{ where } m = \frac{\pi}{\alpha}$$

describes the flow of fluid entering a corner (from right to left) of angle α. When α is greater than $\pi/2$ the stream function describes flow over a corner. The velocity (U_w) is the velocity adjacent to either wall at a radius (a) from the origin. Derive an expression for the velocity potential ϕ.

10. Show that the mass conservation equation reduces to Eq. (7-8) for a constant temperature, constant density fluid.

11. In cylindrical coordinates, show that if the vorticity is constant, liquid rotates about its axis at an angular speed N as if it were a solid body (hence the phrase solid body rotation) and as a result, $U_\theta/r = N$. If the fluid motion is irrotational, show that $U_\theta r$ = constant.

12. Prove the vector identity Eq. (7-13).

13. A vessel of circular cross section and outer radius R is filled with liquid to a uniform depth H. The vessel is rotated about its vertical axis at constant angular speed N and the surface of the liquid is no longer horizontal. The pressure on the upper surface of the liquid is constant and equal to atmospheric pressure. Write an equation that describes the shape of the liquid surface, z = f(r, N, R, H, fluid properties) assuming (a) the motion is rotational, and (b) the motion is irrotational.

14. Fully established flow in a circular duct of outer radius R has the following (axial) velocity profile,

$$U_x = \frac{2Q}{\pi R^2}\left(R^2 - r^2\right) \qquad U_r = 0 \qquad U_\theta = 0$$

Is this irrotational flow? Is it rotational flow of constant vorticity?

15. Derive the stream function ψ that corresponds to the following velocity potentials for two-dimensional flows:

(a) $\phi = A\, r \cos\theta$
(b) $\phi = (A \cos\theta)/r$
(c) $\phi = A \ln r$

where A is a constant. Describe these flows in physical terms.

16. For a two-dimensional flow field show that curves of constant ψ and curves of constant ϕ are orthogonal.

17. Show that the velocity potential ϕ satisfies both the continuity and Navier-Stokes equations.

18. If a two-dimensional flow field has velocity components $U_x = Kx$, $U_y = -Ky$ where K is a constant, derive an analytical expression for the stream function and describe the flow field physically.

19. A line sink Q_1 lies in a plane located at (c,0) and a second line sink Q_2 ($Q_1 = 2\ Q_2$) lies in the plane at (-c,0). Air streams above the surface from right to left at uniform velocity U_0. Write expressions for the potential functions ψ and ϕ for the composite flow field. Find the velocity components at (0,c).

20. Consider a novel welding bench consisting of a line source having coordinates (0,c,z) lying in the solid vertical surface (z-y plane) and a line sink, coordinates (0,0,z), lying in at the intersection of the solid vertical (z-y) plane and the solid horizontal (z-x) plane. You have been asked to devise a way to predict the velocity at any point (x,y), where $x > y > 0$. If the input volumetric flow rate of air per unit length of the source is equal to the exhaust volumetric flow rate per unit length of the sink, write the potential functions for the composite flow field and describe the flow field physically. What are the velocity components at (c,c/2)?

21. Describe the physical flow field represented by the following potential functions:

$$\phi = Ar^2\cos(2\theta) \qquad \phi = Ar^2\sin(2\theta)$$

What does A represent in physical terms?

22. Describe the physical flow field represented by the potential functions

$$\psi = A\sinh\left(\pi\frac{x}{a}\right)\sin\left(\pi\frac{y}{a}\right) \qquad \text{and} \qquad \phi = A\cosh\left(\pi\frac{x}{a}\right)\cos\left(\pi\frac{y}{a}\right)$$

Describe the flow field in the region $x > 0$, $0 < y < a$.

23. A process produces an unusual ideal, two-dimensional flow described by the following:

$$\psi_s = \frac{K}{2\pi}\ln(r) \qquad \phi_s = -\frac{K}{2\pi}\theta$$

where $K = 2\pi R_1 U_0$ and R_1 is a constant. The process is placed in a streaming flow in the negative x-direction whose potential functions are

$$\psi_{\text{streaming flow}} = U_0\sin\theta \qquad \phi_{\text{streaming flow}} = U_0\cos\theta$$

(a) Write general equations that express the velocity components U_r and U_θ at arbitrary points in the flow field.
(b) Describe the flow field in physical terms.
(c) Write an equation that will enable one to plot the values of r and θ at which the air speed is constant and equal to the magnitude of U_0
(d) If the far-field pressure is P_0 and the velocity is U_0, find the pressure at the following points: ($r = R_1$, $\theta = 0$), ($r = R_1$, $\theta = \pi/2$).

24. It is claimed that the stream function

$$\psi = U_0\left[y - \frac{\frac{c}{4}\sin\left(2\pi\frac{y}{c}\right)}{\cosh\left(2\pi\frac{x}{c}\right) - \cos\left(2\pi\frac{y}{c}\right)}\right]$$

describes the flow around a circular cylinder (diameter c/2) located at point (0,0) lying midway between parallel plates at distances c/2 above and below the x-axis. The velocity U_0 is the approach velocity. Verify that the statement is true and find the air velocity at (-c,c/2).

25. The following stream functions describe two-dimensional flow fields. Compute the x and y velocity components for each case. Which of these flow fields (if any) are irrotational?

(a) $\psi = A\, r \sin\theta$
(b) $\psi = A\, r\,(1 - B/r^2) \sin\theta$
(c) $\psi = -Ay + B \sin x\, (e^{-y})$

26. Consider the velocity field produced by a uniform flow streaming over an infinite slot inlet (width 2w) in an infinite flanged surface. The uniform velocity of the streaming flow (U_0) is equal to the average velocity through the inlet [$U_0 = Q/(4Lw)$]. The values of w and U_0 are 1.0 cm and 1.0 m/s respectively.

(a) Find the coordinates (x,y) of points along the dividing streamline.
(b) Draw a graph that locates the streamline that passes through the center of the inlet, i.e. that passes through (0,0).
(c) Find the numerical value of U_x, and U_y at $x = y = w/2$.

27. Overburden from a quarry is piled in a large mound that has a semicircular cross section (radius a = 30. m and length L = 200. m). During the winter months, freezing and thawing make the pile's surface susceptible to erosion by the wind. For a first approximation assume that particles in the pile have a uniform diameter of 100 microns ($D_p = 1.0 \times 10^{-4}$ m). Assume that air approaches the pile with a uniform velocity (U_0 = 10. km/hr) perpendicular to the long dimension. Assume that soil is made airborne at a rate (g_p) proportional to the cube of local wind speed in excess of a critical value (U_c),

$$g_p = 12.8\left[U(a,\theta) - U_c\right]^3 \qquad \left(\text{in units of } \frac{\text{mg}}{\text{m}^2\text{s}}\right)$$

where U and U_c must be written in units of m/s and

$$U_c = 116.\sqrt{D_p} \qquad \left(\text{in units of } \frac{\text{m}}{\text{s}}\right)$$

where D_p must be written in units of m. $U(a,\theta)$ is the tangential velocity adjacent to the surface of the pile. Neglect gravimetric settling of the dust after it is made airborne. Estimate the fugitive emission from the overburden, in units of g/s. Assuming dry conditions for 100 days per year during which fugitive emissions occur, compare the loss with emission factors for stockpiles which are typically 5 to 20 kg/MG per year.

28. For irrotational flow of a constant density fluid, explain why the Laplacian of the velocity potential is zero, i.e. $\nabla^2\phi = 0$.

29. For two-dimensional flow of a constant density fluid, explain why the Laplacian of the stream function is equal to zero, i.e. $\nabla^2\psi = 0$.

30. Ideal flow about a stationary circular cylinder of radius (a) is given by the stream function

$$\psi = U_0\left(r - \frac{a^2}{r}\right)\sin\theta - aU_0 \ln\left(\frac{r}{a}\right)$$

where U_0 is the far-field air velocity. Compute the velocity potential ϕ and the radial and tangential velocity components. Describe the physical attributes of the flow including the location of the stagnation points and the location of the streamlines $\psi = 0$ and $\psi = +/- U_0/2$. Derive an expression that predicts the pressure distribution on the surface of the cylinder.

31. Prove that the stream function below describes the flow of an ideal fluid through a semi-infinite aperture of width 2w by defining the location of several key streamlines.

$$\left(\frac{x}{w}\frac{1}{\cos\psi}\right)^2-\left(\frac{y}{w}\frac{1}{\sin\psi}\right)^2=1$$

Derive an expression for the velocity component U_y at an arbitrary point in the aperture inlet plane, i.e. $-w < x < w$ and $y = 0$.

32. A prairie dog is a small animal that lives in the western US. Colonies of prairie dogs live in underground burrows (tunnels). One end of the tunnel has an opening level with the ground several meters from the other opening that is in the center of a hemispherical mound (assume 2.0 m in diameter) above the ground. An interesting question arises as to how the tunnel is ventilated since several prairie dogs may remain in the burrow for many hours. By instinct, prairie dogs use the venturi effect to draw fresh air into the ground level opening and exhaust air through the elevated opening (the pressure at the ground-level opening is atmospheric while the pressure at the top of the hemispherical mound is less than atmospheric because the air velocity reaches a maximum there. Using the potential flow function for flow over a stationary sphere and neglecting boundary layers, estimate the ventilation volumetric flow rate (m^3/sec) through a single tunnel. When applying the energy equation, define beginning and end points carefully and be sure be to include friction and entry and exit losses properly.

(a) each tunnel has a circular cross section (diameter 0.10 m) and two 90-degree elbows, each of which produces a pressure drop equivalent to 3.0 meters of tunnel
(b) total tunnel length (exclusive of elbows) is 20 meters
(c) far-field wind speed is 4.0 m/s
(d) the pressure drop due to the viscous wall shear stress can be expressed by the Moody chart or by Eq. (6-57)

33. Compute the air velocity (speed and direction) along a line perpendicular to the plane containing two rectangular openings and intersecting the plane at a point midway between the two rectangular openings. The openings are located near two flanking planes (see Figure 7.16). If a computer-graphics program is available, plot the surface over which the air speed is 200 FPM (typical capture velocity).

- Opening 1: $B_1 = 0.50$ m ; $A_1 = 0.25$ m, slot velocity = 10. m/s, $L_{Y1} = 0.75$ m ; $L_{Z1} = 0.50$ m
- Opening 2: $B_2 = 1.0$ m ; $A_2 = 0.125$ m, slot velocity = 10. m/s, $L_{Y2} = 0.75$ m ; $L_{Z2} = 0.125$ m

34. Consider a point sink and a flanged circular opening of diameter 0.010 m. The volumetric flow rate in both cases is 0.040 m^3/min. Compare the air speed along two radial lines (from the center of the opening) inclined 30 and 60 degrees to the flange. Also compare the results to the isopleths of Figure 6.10.

35. Compute the location of the dividing streamline for streaming flow parallel to a circular opening (diameter 0.010 m) in a flanged surface. The volumetric flow rate is 0.040 m^3/min and the far-field velocity (U_0) is 5.0 m/s. Compare the dimensions of this dividing streamline with what would be predicted if the opening were a point sink with the same volumetric flow rate.

36. Consider an infinite surface lying in the x-y plane that intersects an infinite surface lying in the y-z plane. Air approaches the y-z plane with far-field velocity U_0 in the –x direction. Assume the air flow is incompressible and irrotational. The air velocity on the ground a distance (a) upwind of the wall is measured and found to be:

$$U_x(a,0) = -3U_0 \qquad U_y(a,0) = 0$$

The wall has a small hole (point sink) in it a distance a/2 above the ground and air is withdrawn through the hole at volumetric flow rate Q. Estimate the air velocity at the point x = a/2, y = a/2.

37. The circular cylinder of Problem 7.5 moves through a stationary incompressible fluid with velocity $\vec{U}_0$ in a direction perpendicular to its axis of symmetry. Assume irrotational flow. Show that the speed of the air is equal to $|\vec{U}_0| = U_0$ at all points on the surface of the cylinder (i.e. for all values of angle θ). Find the pressure on the cylinder surface as a function of angle θ.

38. Compute the pressure distribution on the surface of a stationary circular cylinder over which an irrotational incompressible flow passes in a direction perpendicular to its axis of symmetry with far-field speed U_0. Compare the results with the pressure distribution found in the previous problem and reconcile any differences.

39. Write a computer program (Excel or Mathcad is recommended) that computes the coordinates of an arbitrary streamline for the following inlets in streaming flow:

(a) flanged slot
(b) plain slot
(c) flanged circular inlet

For each case, plot the location of the dividing streamline for the situation in which $U_0 = Q/A$.

40. Write a computer program that allows you to compute the velocity components at arbitrary points upwind of the following inlets in streaming flow:

(a) flanged slot
(b) plain slot
(c) flanged circular inlet

For each case, compute the velocity at a point one opening distance directly in front of the center of the inlet for the situation in which $U_0 = Q/A$.

41. Using the methods discussed in Section 7.8, compute the centerline velocity along the centerplane for an infinite slot of width 2w.

42. To illustrate the usefulness of ideal flow, estimate the ratio of air speed to face velocity $[U(r,x)/U_{face}]$ for flanged and unflanged circular openings at two points (a) x/D = 1, r/D = 0 and (b) x/D = 1, r/D = 0.45 using the isopleths of Figures 6.9 and 6.10 and the potential function for a point sink in a plane.

43. To illustrate the usefulness of ideal flow, estimate the ratio of air speed to face velocity $[U(x,y,z)/U_{face}]$ for a line sink in a plane at (a) x/s = 2.0, y/s = 0, z/s = 0 and (b) x/s = 1.2, y/s = 0.60, z/s = 0. How do these values compare to the values for an unflanged slot shown Figure 6.8? Is it clear that flanged slots have greater ability to capture contaminants than unflanged slots?

44. Using Eqs. (7-80) and (7-81), prove that the slot face velocity approaches infinity at the lip (x = w, or -w) and is equal to (2/π) times the average face velocity, $U_{avg} = Q/(2wL)$, at the center of the slot (x = 0).

45. In this chapter is a description of how to approximate a flanged inlet of arbitrary cross-sectional area as a series of point sinks distributed across the inlet plane. Thus at any point upwind of the inlet, the potential function and velocity are the superpositions of potential functions and velocities of a series of point sinks distributed across the inlet plane. Consider a flanged rectangular inlet 4.0 cm by 40. cm in quiescent air that withdraws air at a rate of 32. m^3/min. Using this method write a computer program that computes the velocity along a line perpendicular to the inlet plane that passes through a point in the inlet plane 2.0 cm from each of the long sides and 10.0 cm from the end of the inlet. Draw a graph of the air speed as a function of distance from the inlet plane and compare the results with

(a) Equation (6-5)
(b) contours in Figure 6.8
(c) results predicted from equations in this chapter
(d) Table 7.2, assuming an infinite flanged rectangular inlet

46. It has been suggested that an irrotational, constant density flow striking a wall can be approximated by the stream function, $\psi = 2xy$.

(a) Compute the velocity components U_x and U_y. Sketch the streamlines $\psi = 0$ and $\psi = 1$ in the region $x > 0$, $y > 0$ and show the direction of the flow.
(b) Does the velocity potential exist? If it does, sketch some equipotential lines on the same plot.

47. Air is withdrawn through a flanged circular inlet 0.10 m in diameter (D = 0.10 m) at a volumetric flow rate of 1.0 m^3/s from quiescent air. Find the velocity components U_z and U_r at point z = 0.10 m, r = 0.050 m.

48. From the results of Problem 7.13 (a) and (b), predict the pressure distribution on the bottom surface of the vessel. If the vessel is placed on a spring scale that records the total downward force, will the reading vary with the angular rotational speed? Prove your answer analytically.

49. A welding and soldering bench consists of a rectangular 20. cm slot in a 90-degree corner (Figure 7.11b). The planned volumetric flow rate is 3.0 m^3/s per meter of slot (Q/L = 3.0 m^2/s). Your supervisor asks you to prepare a computer program that computes U_x and U_y at arbitrary points above the bench top. (Express the velocity in m/s and dimensions x and y in centimeters.) Using this program plot U_x and U_y along the horizontal plane coincident with the upper lip of the slot. Assume the flow field is ideal and characterized by the potential functions ψ and ϕ. Estimate the potential functions in the following two ways and plot U_x and U_y using each method:

(a) analytical functions; note the unusual coordinate system needed to use the analytical equations, i.e. increasing x = vertically downward, and increasing y = bench top. Thus find $U_x(-20,y)$ and $U_x(-20,y)$ for $0 < y < 60$.
(b) an approximate method in which the slot is assumed to consist of a large number of finite line sinks.

50. Air moves uniformly in the negative x-direction over a flat plate containing an infinite slot of width 2w (w = 0.010 m) at a uniform speed equal to 1.0 m/s (U_0 = 1.0 m/s in the –x-direction). Air is withdrawn through the slot at volumetric flow rate 0.10 m^3/s per meter of slot (Q/L = 0.10 m^2/s). Prepare a computer program that accomplishes the following:

(a) Compute and map the coordinates of the dividing streamline beginning at x = 0.60 m.
(b) Compute and map the coordinates of the streamline passing through the center of the slot (0,0).
(c) Beginning at x = 0.60 m, plot U_x and U_y along the streamline that ultimately passes through the center of the slot (0,0).

51. Repeat problem 7.13 for fluid motion that can be described as n-degree rotational flow.

52. Consider a slot (w = 1.0 cm), through which air is withdrawn at volumetric flow rate Q/L = 0.30 m^2/s.

(a) Write a computer program that computes the velocity components U_x and U_y at all points lying in a plane above the inlet at y = 5.0 cm, and –5.0 cm ≤ x ≤ 5.0 cm. Use Eq. (7-80), Eq. (7-81), and the Newton-Raphson method.
(b) Compute the velocity components U_x and U_y along the plane described in Part (a) by approximating the total withdrawal rate by 50 equally spaced line sinks.
(c) Plot, compare, and discuss the results of Parts (a) and (b) above.

53. Approximate a push-pull system as a combination of a line source (Q_B/L) located at (x,y) = (0,c) and a line sink (Q_S/L) located at (x,y) = (c,c). Consider a surface at (x,c/2) directly above the x-axis between the source and sink. Plot the velocity components U_x and U_y along this line for c = 5.0 ft, Q_S/Lc = 75. CFM/ft^2 and Q_B/Q_S = 5.0, 7.5, and 10. Evaluate the results in light of the ACGIH guidelines shown in Figure 6.13.

54. You need to produce a capture velocity no less than 100 FPM along a plane 24.0 inches high lying 12.0 inches upwind of the inlet plane of two configurations:

(a) two unflanged rectangular inlets, 6.0 inches wide and 18. inches between centers
(b) a single unflanged rectangular inlet, 12. inches wide

Which configuration requires the smallest volumetric flow rate?

8

Motion of Particles

In this chapter you will learn how to:

- describe an aerosol in statistical terms
- write and solve equations describing the trajectory of particles moving in a viscous gas stream

8.1 Particle Size

The purpose of this chapter is to provide an analytical basis to describe the motion of particles in a moving air stream so that engineers can design collection systems and predict their performance. The motion of gaseous and vaporous contaminants is affected only by the velocity field of the air through which they move. The motion of *particle* contaminants on the other hand is also affected by the inertia of the particle, and by the particle's aerodynamic drag. The most important parameter in the prediction of particle motion is particle size. Shown in Figure 8.1 are the sizes of particles commonly encountered in the home and in industry.

It is important for the reader to develop intuition about the size of small particles. Water droplets can be used as a familiar basis for comparison of the relative size of aerosol particles. The following are rough diameters (to one significant digit):

- fog and clouds: 2 μm $< D_p <$ 70 μm
- mist: 70 μm $< D_p <$ 200 μm
- drizzle: 200 μm $< D_p <$ 700 μm
- rain: 700 μm $< D_p <$ 10,000 μm

Fine particles are particles with diameter less than approximately 2.5 μm. Particles whose characteristic size is less than 1 μm are called ***submicron particles***, and for the most part owe their origin to combustion processes in which vapor products of combustion condense to form well-defined spheres. For example Kleeman et al. (1999) found that submicron particles generated by the combustion of wood products (wood burning, meat charbroiling, etc.) have a single mode at approximately 0.1 - 0.2 μm. Cigarette smoke has a single mode that peaks at 0.3 - 0.4 μm.

Submicron particles may also grow to become particles greater than 1 μm through a process called ***agglomeration***. For the most part however, particles larger than about 1 μm owe their origin to size reduction processes such as grinding, pulverizing, etc. Examples include roadway dust, pulverized coal, milled flour, etc. Naturally formed particles such as sea salt, pollens, and bacteria are generally larger than 1 μm, whereas viruses are very small (submicron) particles, in the range 0.003 - 0.05 μm.

Some "rules of thumb" are presented in Table 8.1; these are useful as benchmarks for categorizing particle sizes.

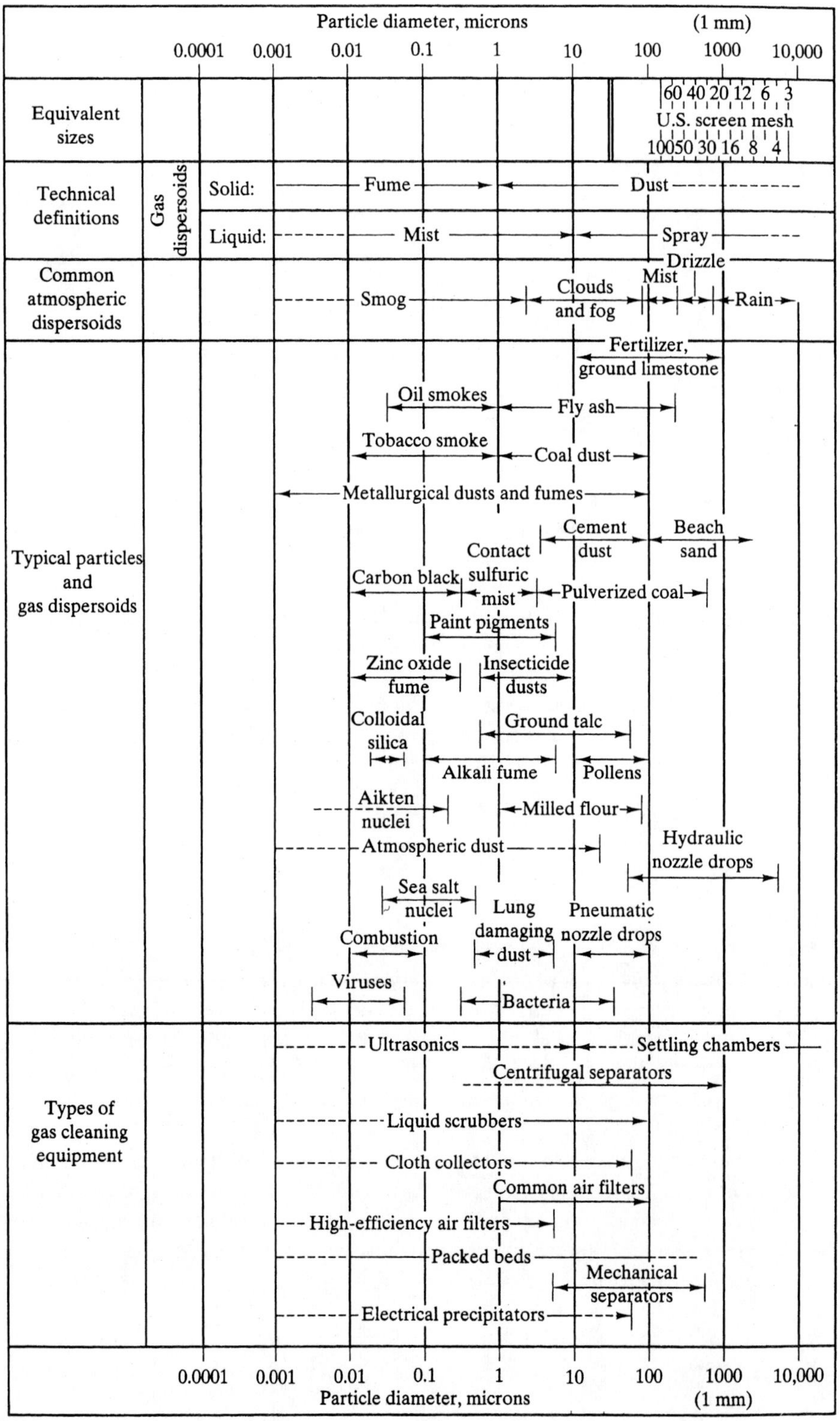

Figure 8.1 Characteristic size of particles and particle dispersoids, and the size range of particles removed by various types of gas cleaning equipment (from Lapple, 1961).

Table 8.1 Useful categorization of particle size; readers are encouraged to memorize these.

description	particle size
particles produced by *combustion* processes	$D_p < 1$ μm
particles produced by *mechanical* processes	$D_p > 1$ μm
respirable particles (able to reach the alveoli)	$D_p < 2$ or 3 μm
inhalable particles (able to enter the bronchi)	$D_p < 10$ μm
particles individually *visible* to the naked eye	$D_p > 10$ μm

Readers must keep in mind that the values in Table 8.1 are orders of magnitude only, to one significant digit, and are not to be taken as exact numerical values; actual values depend on the particular individual and/or process. Nevertheless, readers are urged to memorize these generalizations, so that they can quickly categorize the origin and potential hazard of aerosol particles. For example, if an individual particle is visible to the naked eye, it is not inhalable. Particles produced by combustion processes are not only inhalable, but also respirable, and thus potentially more hazardous.

8.2 Statistical Analysis of Aerosols

A suspension of particles in air is called an ***aerosol***. The particles may be solid or liquid, and the aerosol may be stationary or moving. Properties of the aerosol depend on properties of both the particles and the air. An aerosol in which all particles have the *same* diameter is called a ***monodisperse aerosol***. Monodisperse aerosols are rarely encountered in engineering practice. An aerosol in which particles have *different* diameters is called a ***polydisperse aerosol***.

8.2.1 Size Distribution

In previous chapters, mass concentration was defined as the mass of a contaminant per unit volume of carrier gas (typically in units of mg/m^3). For simplicity, mass concentration is expressed here by the symbol c without any subscript, since only one species of contaminant is considered at a time. The above definition for c is valid for both gases and particles. For gases, mass concentration is *sufficient* to fully describe the contaminant in the air. For aerosols, however, particle mass concentration alone is *not* sufficient to distinguish one aerosol from another; a second physical property is required, namely the ***size distribution*** of the aerosol. In order to select appropriate particle control devices to satisfy EPA performance standards, it is necessary that the size distribution of the source be known. If two aerosols have the same mass concentration (c) but different size distributions, the dynamic behavior of one aerosol may be quite different from the other.

Example 8.1 - Clouds of Water Droplets

Given: Consider four different monodisperse aerosols consisting of water droplets ($\rho = 10^9$ mg/m^3) in air, as in Table E8.1. As can be seen in the table, the number concentration gets higher as the aerosol particles get smaller, such that the mass concentration of each aerosol is the *same*, c = 0.5236 mg/m^3.

To do: Verify that mass concentration is insufficient to distinguish one aerosol from another.

Table E8.1 Four sample monodisperse aerosols of water droplets, each with a different number concentration, but with the same mass concentration.

aerosol	D_p (μm)	c_{number} (particles/m^3)	c (mg/m^3)
A	1,000	1	0.5236
B	100	10^3	0.5236
C	10	10^6	0.5236
D	1	10^9	0.5236

Solution: Aerosol A consists of *one* large water drop (D_p = 1,000 μm, about the size of a rain drop) suspended in a cubic meter of air, yet produces the same mass concentration as aerosol D which consists of 10^9 water droplets (D_p = 1 μm) in a cubic meter of air! Obviously mass concentration alone does not satisfactorily differentiate one aerosol from another.

Discussion: Other factors such as average mass, average diameter, or median diameter, along with other statistical properties are required in order to discriminate one polydisperse aerosol from another.

To illustrate the difference between the ***arithmetic mean*** ($D_{p,am}$) of an ensemble and the ***median diameter*** ($D_{p,median}$) of the ensemble, consider an aerosol containing a single water drop 1,000 μm in diameter and 10^9 water droplets 1 μm in diameter in a total volume (V) of 1 m^3. The total number of particles (n_t) is

$$n_t = 1 + 10^9 = 1{,}000{,}000{,}001$$

For J sets of monodisperse particles, with each set (j) having n_j particles of diameter $D_{p,j}$, the total mass concentration (c) is found by summing the mass concentration of each set. From the definitions of mass concentration and the volume of a sphere, the total mass concentration for a polydisperse aerosol made up of spherical particles is

$$\boxed{c = \frac{\pi \rho_p}{6V} \sum_{j=1}^{J} \left(n_j D_{p,j}^{\ 3} \right)} \tag{8-1}$$

where V is the volume of air and ρ_p is the density of the particles. For the example at hand, there are two sets of particles (J = 2), and

$$c = \frac{\pi \left(10^9 \frac{mg}{m^3} \right)}{6\left(1\ m^3\right)} \left((1)\left(1000 \times 10^{-6}\ m\right)^3 + (10^9)\left(1 \times 10^{-6}\ m\right)^3 \right) = 1.047 \frac{mg}{m^3}$$

The ***arithmetic mass mean diameter***, $D_{p,am}$(mass), is the diameter that each of these 1,000,000,001 particles would need to have in order to produce this same mass concentration (1.047 mg/m^3). From the definition of mass concentration,

$$c = n_t \frac{\pi \rho_p \left[D_{p,am}(\text{mass}) \right]^3}{6V}$$

one can solve for $D_{p,am}$(mass),

$$\boxed{D_{p,am}(\text{mass}) = \left(\frac{6cV}{\pi n_t \rho_p} \right)^{\frac{1}{3}}} \tag{8-2}$$

For the concentration above,

$$D_{p,am}(\text{mass}) = \left(\frac{6\left(1.047 \frac{mg}{m^3}\right)\left(1\ m^3\right)}{\pi(1{,}000{,}000{,}001)\left(10^9 \frac{mg}{m^3}\right)} \right)^{\frac{1}{3}} = 1.26 \times 10^{-6}\ m \cong 1.3\ \mu m$$

Although there is only one large (1,000 μm) particle amongst a billion small (1 μm) particles, the single large particle greatly influences the mass of the aerosol.

The median particle diameter with respect to mass, $D_{p,median}$(mass), is defined as the diameter at which 50% of the particles have a mass less than this diameter, and 50% of the particles have a mass greater than this diameter. With respect to the 1,000,000,001 particles in the example aerosol, $D_{p,median}$(mass) is 1.0 μm, whereas the arithmetic mass mean diameter is 1.3 μm. Clearly, large particles have an enormous influence on such calculations, and tend to skew upwards the size distribution based on mass.

8.2.2 Log-Normal Size Distribution

If the physical properties of the particles in two polydisperse aerosols are the same, one way to distinguish one aerosol from the other is to compare their size distributions. What follows is a synopsis of a commonly encountered statistical description called the ***log-normal distribution***. For a systematic study of particle size distributions, readers should consult the following texts: Cadle (1965), Hinds (1982), Waters et al. (1991), and Willeke and Baron (1993).

For example, suppose an aerosol is analyzed on the basis of size. One thousand particles (n_t = 1000) are captured, and sorted into bins or intervals based on particle diameter; each such interval is formally called a ***class***, where each class (j) consists of a range of particle sizes. The aerosol is found to have the distribution shown in Table 8.2, in which the total number of classes is equal to J = 19. The particle size distribution of Table 8.2 is in the form of ***grouped data***. In other words, instead of a list of individual particle diameters, the distribution is given in terms of n_j particles in each class (j). The ***arithmetic mean diameter based on number***, $D_{p,am}$, can be computed for grouped data from the following equation:

$$D_{p,am} = D_{p,am}(\text{number}) = \frac{1}{n_t}\sum_{j=1}^{J}\left(n_j D_{p,j}\right) \tag{8-3}$$

Table 8.2 Sample particle size distribution (grouped data); 1000 particles divided into 19 groups or classes (j); summations of n_j and n_j/n_t are given in bold at the bottom of the table.

j	$D_{p,min,j}$ (μm)	$D_{p,max,j}$ (μm)	$\Delta D_{p,j}$ (μm)	$D_{p,j}$ (μm)	n_j	n_j/n_t (%)	$N(D_{p,j})/n_t$ (%)
1	0.01	0.5	0.49	0.255	16	1.6	1.6
2	0.5	1.0	0.5	0.75	159	15.9	17.5
3	1.0	1.5	0.5	1.25	235	23.5	41.0
4	1.5	2.0	0.5	1.75	200	20.0	61.0
5	2.0	2.5	0.5	2.25	133	13.3	74.3
6	2.5	3.0	0.5	2.75	97	9.7	84.0
7	3.0	3.5	0.5	3.25	55	5.5	89.5
8	3.5	4.0	0.5	3.75	36	3.6	93.1
9	4.0	4.5	0.5	4.25	24	2.4	95.5
10	4.5	5.0	0.5	4.75	15	1.5	97.0
11	5.0	5.5	0.5	5.25	10	1.0	98.0
12	5.5	6.0	0.5	5.75	6	0.6	98.6
13	6.0	6.5	0.5	6.25	5	0.5	99.1
14	6.5	7.0	0.5	6.75	3	0.3	99.4
15	7.0	7.5	0.5	7.25	2	0.2	99.6
16	7.5	8.0	0.5	7.75	1	0.1	99.7
17	8.0	8.5	0.5	8.25	1	0.1	99.8
18	8.5	9.0	0.5	8.75	1	0.1	99.9
19	9.0	9.5	0.5	9.25	1	0.1	100.0
					1000	**100**	

Standard deviation (σ) is defined as

$$\sigma = \sqrt{\frac{\sum_{j=1}^{J} n_j \left[D_{p,j} - D_{p,am}\right]^2}{n_t - 1}} \tag{8-4}$$

for grouped data. ***Variance*** is defined as σ^2. The notation $D_{p,am}$(number) in Eq. (8-3) is introduced here to distinguish $D_{p,am}$ based on number of particles, $D_{p,am}$(number), and $D_{p,am}$ based on mass, $D_{p,am}$(mass), as discussed later. Alternatively, in terms of a smoothed distribution function,

$$D_{p,am} = \frac{1}{n_t} \int_0^\infty D_p n(D_p) dD_p \tag{8-5}$$

where the variable $n(D_p)$ is called the ***number distribution function*** (sometimes called the ***size distribution function***), with units of number of particles per micron. The quantity $n(D_p)dD_p$ represents the number of particles between sizes D_p and $(D_p + dD_p)$. The variable n_j on the other hand is the number of particles in class "j" where the mid-range particle has diameter, $D_{p,j}$. The symbol n_t is the total number of particles, given by

$$n_t = \sum_{j=1}^{J} n_j = \int_0^\infty n(D_p) dD_p \tag{8-6}$$

The reader must be careful to distinguish between the discrete number of particles, n_j, within a certain range, which has units of number of particles, and the continuous number distribution function $n(D_p)$, which has units of number of particles per micron. For a set of discrete grouped data as in Table 8.2, $n(D_{p,j})$ can be approximated for any class j as

$$n(D_{p,j}) \approx \frac{n_j}{\Delta D_{p,j}} \tag{8-7}$$

If the number of classes (J) and the number of particles (n_t) approach infinity, while class width ($\Delta D_{p,j}$) approaches zero, $n(D_{p,j})$ approaches the continuous function $n(D_p)$. Variables n_j and $n(D_p)$ may instead be defined in terms of number *concentration*, i.e. number of particles/m^3 and number of particles/(m^3 μm) respectively. In most cases, the number distribution function is normalized by dividing by the total number of particles in the sample (n_t); the resulting fractional distribution function is called a ***probability distribution function*** or PDF. Most statistics books use the symbol f for a PDF; here,

$$f(D_p) = \frac{n(D_p)}{n_t} \approx \frac{n_j}{\Delta D_{p,j} n_t} \tag{8-8}$$

Figure 8.2a shows a histogram of the ***discrete number fraction*** (n_j/n_t) versus class number (j); the distribution is not symmetric, and is certainly not a gaussian distribution. Figure 8.2b shows the smoothed fractional number distribution function as defined by Eq. (8-7), and normalized by the total number of particles (n_t). The area under the curve of Figure 8.2b is unity by definition. The shape seen here – a peak skewed towards smaller diameters and a long tail towards larger diameters – is typical of many particle size distributions encountered in air pollution. The normalized number distribution function of Figure 8.2b is duplicated in Figure 8.3, but with a logarithmic abscissa. Although the particle distribution is sparse on the smaller diameter side and there is some experimental scatter, the distribution in Figure 8.3 is much more symmetric, and looks like a standard normal (gaussian) distribution. This kind of distribution is typical of many particle size distributions encountered in air pollution, and is called ***log normal***. A log-normal distribution is defined as one that behaves as a normal (gaussian) distribution when the abscissa is a log scale as in Figure 8.3. Similar results can be obtained if the abscissa is a linear scale of $\log_{10}(D_p)$ or $\ln(D_p)$.

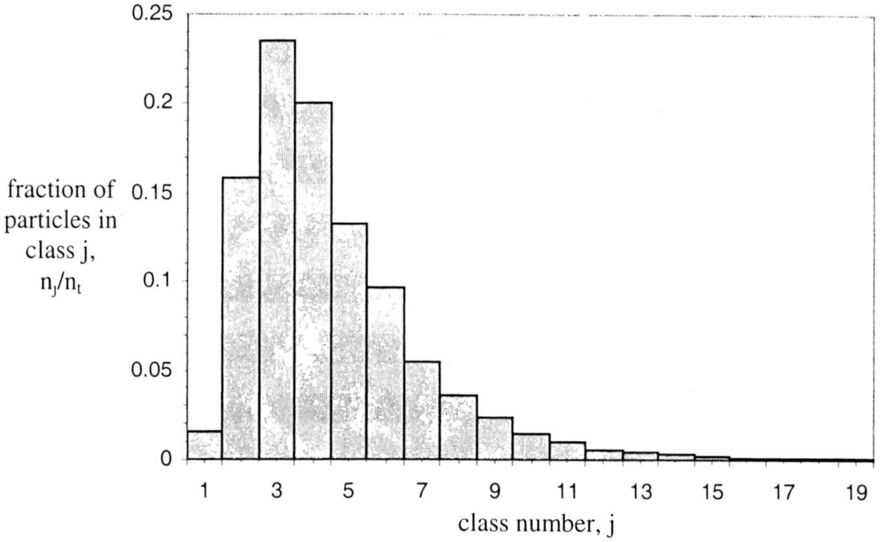

(a)

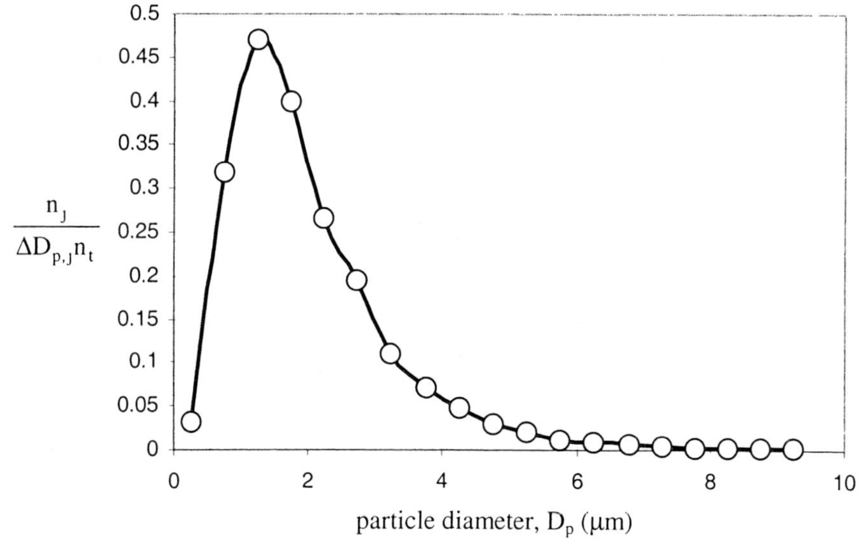

(b)

Figure 8.2 Size distribution of the aerosol particle data of Table 8.2; (a) histogram, and (b) normalized fractional number distribution function.

A statistic which is useful for log-normal distributions is the ***geometric mean diameter***, $D_{p,gm}$,

$$\boxed{D_{p,gm} = \left[D_{p,1}^{\ n_1} D_{p,2}^{\ n_2} \ldots D_{p,J}^{\ n_J} \right]^{1/n_t}} \tag{8-9}$$

Readers can show that Eq. (8-9) can also be expressed (exactly) by

$$\boxed{D_{p,gm} = \exp\left[\frac{1}{n_t} \sum_{j=1}^{J} n_j \ln\left(D_{p,j}\right) \right]} \tag{8-10}$$

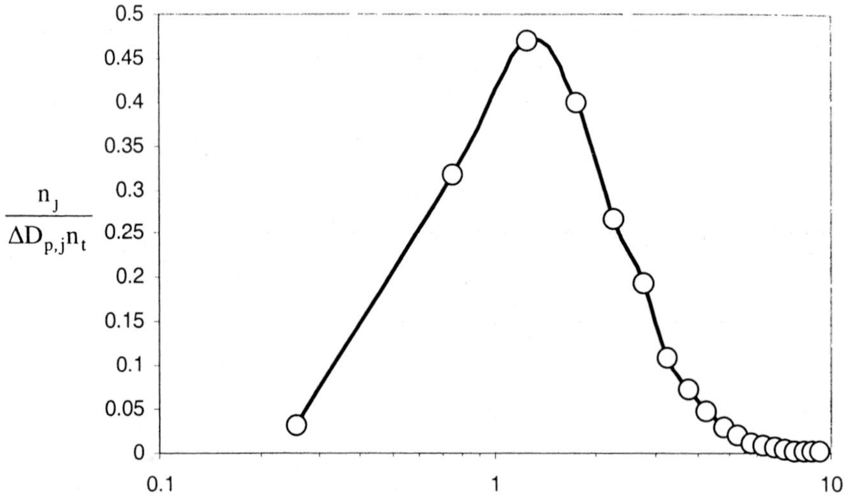

Figure 8.3 Normalized number distribution function versus particle diameter (log scale) for the grouped data of Table 8.2.

An analytical function can be generated to describe a log-normal distribution; the function can then be incorporated into computer programs to calculate parameters needed in air pollution studies. Log-normal distributions are characterized by two properties: geometric mean diameter based on number ($D_{p,gm}$), as defined above, and geometric standard deviation (σ_g).

If an aerosol has a log-normal size distribution, the normalized number distribution function can be expressed as

$$\frac{n(D_p)}{n_t} = \frac{1}{\sqrt{2\pi} D_p \ln \sigma_g} \exp\left\{ -\frac{1}{2} \left(\frac{\ln\left(\frac{D_p}{D_{p,gm}}\right)}{\ln \sigma_g} \right)^2 \right\} \tag{8-11}$$

where σ_g is called the ***geometric standard deviation***. For grouped data, $\ln\sigma_g$ is defined in the same manner as σ, but the natural logarithm of particle diameter is used everywhere in place of D_p itself:

$$\ln\sigma_g = \sqrt{\frac{\sum_{j=1}^{J} n_j \left(\ln(D_{p,j}) - \left[\ln(D_p)\right]_{,am} \right)^2}{n_t - 1}} \tag{8-12}$$

σ_g is then simply $\exp(\ln\sigma_g)$. In Eq. (8-12), the term $[\ln(D_p)]_{,am}$ is defined in the same way as $D_{p,am}$, namely Eq. (8-3), except the algebra is performed on $\ln(D_p)$ instead of D_p:

$$\left[\ln(D_p)\right]_{,am} = \frac{1}{n_t} \sum_{j=1}^{J} n_j \ln(D_{p,j}) \tag{8-13}$$

Eqs. (8-12) and (8-13) are valid for any set of grouped data; for the particular case of log-normal distributions, geometric standard deviation is related to the regular standard deviation by the following expression:

$$\sigma_g = \exp\left\{\sqrt{\ln\left[1+\left(\frac{\sigma}{D_{p,am}}\right)^2\right]}\right\} \tag{8-14}$$

Geometric mean diameter ($D_{p,gm}$) is defined by Eqs. (8-9) and (8-10). It can be shown that $D_{p,gm}$ is related to arithmetic mean diameter and geometric standard deviation as follows:

$$D_{p,gm} = D_{p,am} \exp\left[-\frac{(\ln \sigma_g)^2}{2}\right] \tag{8-15}$$

It can also be shown that

$$D_{p,gm} = \exp\left\{\left[\ln(D_p)\right]_{,am}\right\} \tag{8-16}$$

Readers should note that the above expressions are for distributions based on particle number or count, but are also valid for distributions based on area or mass instead of number, as discussed later. The ***cumulative distribution function*** $N(D_p)$ is the total number of particles *smaller* than D_p. The normalized cumulative distribution function for a log-normal distribution can be expressed in terms of the ***error function*** (erf), by integration of Eq. (8-11); the result is

$$\frac{N(D_p)}{n_t} = \frac{1}{2}\left[1+\text{erf}\left(\frac{\ln\left(\frac{D_p}{D_{p,gm}}\right)}{\sqrt{2}\ln \sigma_g}\right)\right] \tag{8-17}$$

For experimental data such as those of Table 8.2, the normalized continuous function $N(D_p)/n_t$ can be approximated by the discrete normalized cumulative distribution function, $N(D_{p,j})/n_t$, which is the rightmost column in Table 8.2. These data are plotted in Figure 8.4 as a function of D_p, using a log scale for particle size. Readers must note that *the diameter used in generating a plot of cumulative distribution function is $D_{p,max,j}$ (maximum diameter in class j), not $D_{p,j}$ (middle diameter in class j)*, because of the definition of cumulative. The fraction of particles between any two particle sizes, $D_{p,1}$ and $D_{p,2}$, can be found by subtracting the values of the respective normalized cumulative distribution functions,

$$\frac{n_{(D_{p,1}<D_p<D_{p,2})}}{n_t} = \frac{N(D_{p,2})}{n_t} - \frac{N(D_{p,1})}{n_t} \tag{8-18}$$

There exists a unique particle diameter called the ***median particle diameter*** ($D_{p,median} = D_{p,50}$) such that when the value is substituted into Eq. (8-17), the value of the normalized cumulative number distribution function is one half,

$$\frac{N(D_{p,50})}{n_t} = \frac{1}{2} \tag{8-19}$$

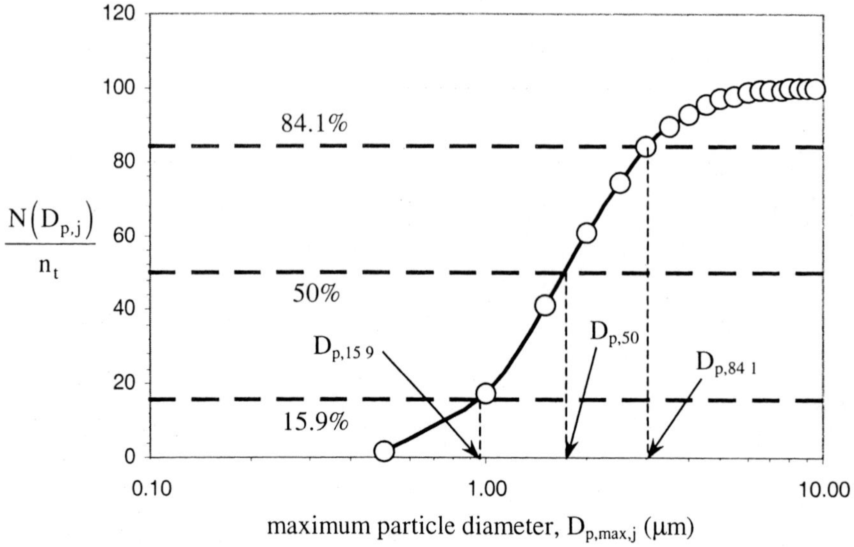

Figure 8.4 Normalized cumulative distribution function versus maximum particle diameter in each class j (log scale) for the grouped data of Table 8.2.

which is to say that half of the particles are smaller than $D_{p,50}$. Equation (8-19) is valid only if

$$\text{erf}\left\{\frac{\ln\left(\frac{D_{p,50}}{D_{p,gm}}\right)}{\sqrt{2}\ln\sigma_g}\right\} = 0 \tag{8-20}$$

Since ln(1) = 0 and erf(0) = 0, Eq. (8-20) shows that

$$D_{p,gm} = D_{p,50} = D_{p,median} \tag{8-21}$$

Thus for a log-normal size distribution, the median particle diameter based on number is equal to the geometric mean particle diameter based on number and *the terms median and geometric mean can be used interchangeably.*

There also exists another unique particle diameter ($D_{p,84.1}$) defined by

$$\ln\left(\frac{D_{p,84.1}}{D_{p,gm}}\right) = \ln\sigma_g$$

from which

$$\sigma_g = \frac{D_{p,84.1}}{D_{p,gm}} = \frac{D_{p,84.1}}{D_{p,50}} \tag{8-22}$$

When $D_{p,84.1}$ is substituted into Eq. (8-17),

$$\frac{N(D_{p,84.1})}{n_t} = \frac{1}{2}\left[1 + \text{erf}\left(\frac{1}{\sqrt{2}}\right)\right] \cong 0.841 \tag{8-23}$$

The term $D_{p,84\,1}$ corresponds to a particle whose normalized cumulative number distribution function is 84.1%, which means that 84.1% of the particles are smaller than $D_{p,84\,1}$. A similar analysis can be performed to show that

$$\sigma_g = \frac{D_{p,50}}{D_{p,15\,9}} = \frac{D_{p,gm}}{D_{p,15\,9}} \tag{8-24}$$

and

$$\sigma_g = \sqrt{\frac{D_{p,84\,1}}{D_{p,15\,9}}} \tag{8-25}$$

On the basis of number, approximately 34.1% of the total number of particles are between $D_{p,gm}$ and $D_{p,84\,1}$. Similarly, approximately 34.1% of the particles are between $D_{p,15\,9}$ and $D_{p,gm}$.

One can determine whether a size distribution is log normal by fitting the data to the equations above, or by plotting the data on special graph paper, called ***log probability paper***, designed for this purpose. The graph paper contains a conventional logarithmic axis for particle diameter and a specially scaled axis for the cumulative distribution. *Particle size distributions which are log normal become straight lines on such graph paper*, and the values of $D_{p,15\,9}$, $D_{p,50}$, and $D_{p,84\,1}$ can be read directly from the plot. Plotted in Figure 8.5 are the data of Table 8.2, specifically $N(D_{p,J})/n_t$ (abscissa) as a function of $D_{p,max,J}$ (ordinate) on log probability paper.

From Figure 8.5, it is reasonable to conclude that the sample number distribution of Table 8.2 is log normal with the following values:

- $D_{p,15\,9}$ = 0.98 μm
- $D_{p,50}$ = 1.70 μm
- $D_{p,84\,1}$ = 3.00 μm

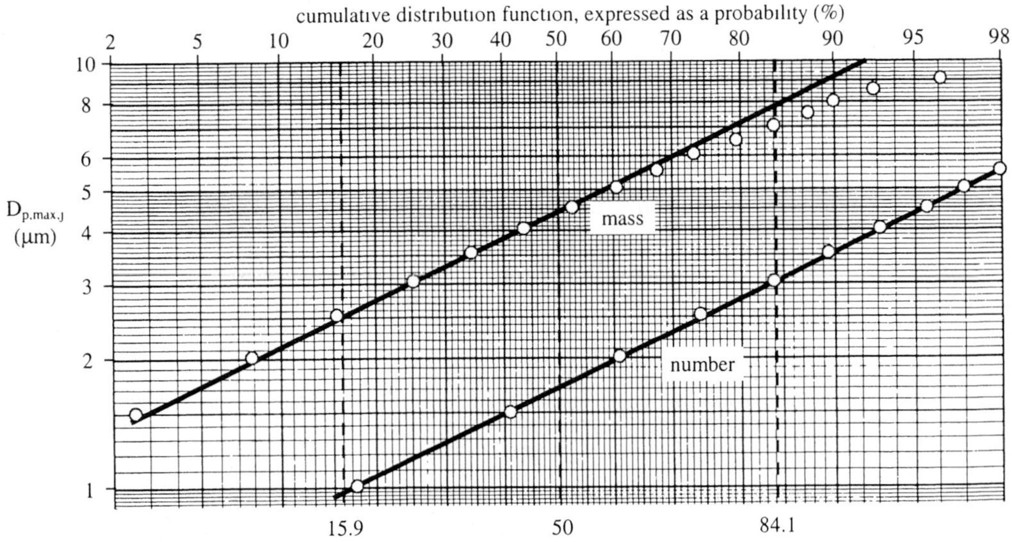

Figure 8.5 Log probability plot: normalized cumulative number distribution and cumulative mass distribution as functions of maximum particle diameter in each class for the grouped data of Table 8.2.

From Eq. (8-3), one can compute the arithmetic mean diameter based on number for the data of Table 8.2:

- $D_{p,am} =$ 2.01 μm

The standard deviation and geometric standard deviation based on number can be calculated from Eqs. (8-4) and (8-14) respectively:

- $\sigma =$ 1.23 μm
- $\sigma_g =$ 1.76

There are six equations available for calculation of geometric mean diameter based on number:

- $D_{p,gm} =$ 1.69 μm Eq. (8-10) (*exact*, definition)
- $D_{p,gm} =$ 1.71 μm Eq. (8-15)
- $D_{p,gm} = D_{p,50} =$ 1.70 μm Eq. (8-21)
- $D_{p,gm} = D_{p,84.1}/\sigma_g = 3.00/1.76 =$ 1.70 μm Eq. (8-22)
- $D_{p,gm} = D_{p,15.9}\sigma_g = 0.98(1.76) =$ 1.72 μm Eq. (8-24)
- $D_{p,gm} = (D_{p,84.1}/D_{p,15.9})^{1/2} = (3.00/0.98)^{1/2} =$ 1.75 μm Eq. (8-25)

These six values show remarkable agreement, implying that these sample data are very close to true log normal; this is not always the case due to experimental uncertainty and scatter in the data.

8.2.3 Mass Distribution

The discussion to this point has concerned only the *number* of particles of particular sizes. Many instruments used to analyze aerosols are based on light scattered by particles. Light scattering depends on particle surface area, which is proportional to the square of particle diameter. Other instruments discriminate among particles on the basis of mass, which depends on the cube of particle diameter. Consequently particle area distributions and particle mass distributions can be defined. Equations (8-3) to (8-25) have counterparts based on mass in which the normalized number distribution, $n(D_p)/n_t$, is replaced by a ***normalized mass distribution***, $m(D_p)/m_t$, where m_t is the total mass of the aerosol, and in which the normalized cumulative number distribution, $N(D_p)/n_t$, is replaced by the ***normalized cumulative mass distribution***, $M(D_p)/m_t$. Analogous to Eq. (8-7), the mass distribution function, $m(D_p)$ can be approximated for discrete data as

$$\boxed{m\left(D_{p,J}\right) \approx \frac{m_J}{\Delta D_{p,J}}} \tag{8-26}$$

The cumulative mass distribution for discrete data is found by adding the probability for class (j) and the probabilities for all classes less than j, similar to what was done previously with number distributions. $M(D_p)/m_t$ is plotted as a function of $D_{p,max,J}$ in Figure 8.5 for the sample data of Table 8.2. Hatch and Choate (1929) derived a conversion equation that relates number geometric mean diameter and mass geometric mean diameter for log-normal distributions:

$$\boxed{\ln\left(\frac{D_{p,gm}(\text{mass})}{D_{p,gm}(\text{number})}\right) = 3\left(\ln \sigma_g\right)^2} \tag{8-27}$$

or, after some rearrangement,

$$\boxed{\ln\left(D_{p,gm}(\text{mass})\right) = \ln\left(D_{p,gm}(\text{number})\right) + 3\left(\ln \sigma_g\right)^2} \tag{8-28}$$

Equation (8-27) or its equivalent (8-28) is used frequently and is called the ***Hatch-Choate equation*** (Hinds, 1982). Since D_p has units of length and one can not obtain the logarithm of a dimensional quantity, the reader should think of $\ln D_p$ as $\ln[D_p/(1\ \mu m)]$, where 1 μm is a "reference" particle size even though it is not stated explicitly. Thus in using Eq. (8-28), D_p should always be in units of micrometers (μm). Equation (8-28) can also be written in terms of $D_{p,50}$ instead, since $D_{p,50} = D_{p,gm}$ for a log-normal distribution:

$$\ln\left(D_{p,50}(\text{mass})\right) = \ln\left(D_{p,50}(\text{number})\right) + 3\left(\ln\sigma_g\right)^2 \tag{8-29}$$

For a log-normal distribution, the mass distribution function $m(D_p)$ is defined as

$$m(D_p) = \left(\frac{\pi}{6}\right)\frac{n_t D_p^{\,3}}{D_p \ln\sigma_g \sqrt{2\pi}\exp\left[-\frac{\left(\ln\left(\frac{D_p}{D_{p,gm}}\right)\right)^2}{2\left(\ln\sigma_g\right)^2}\right]} \tag{8-30}$$

Using the identity $D_p^{\,3} = \exp(3 \ln D_p)$, expanding the exponential portion of Eq. (8-30), and completing the square within the exponent, the above can be written as

$$m(D_p) = \frac{n_t}{\sqrt{2\pi}D_p \ln\sigma_g}\exp\left[3\ln D_{p,gm} + \frac{9}{2}\left(\ln\sigma_g\right)^2\right]\exp\left[-\frac{\left(\ln D_p - \left(\ln D_{p,gm} + 3\left(\ln\sigma_g\right)^2\right)\right)^2}{2\left(\ln\sigma_g\right)^2}\right] \tag{8-31}$$

While the above expression is complicated, it should be noted that ln σ_g appears in ways that indicate that Eq. (8-31) is yet another log-normal expression that has the same geometric standard deviation as Eq. (8-14). Consequently, *if $n(D_p)$ is log normal, then $m(D_p)$ is also log normal with the same geometric standard deviation σ_g*. On log probability paper, the slope of the mass line is the same as the slope of the number line (the two lines are parallel, with the mass line shifted to the left). Median particle size with respect to mass is related to the corresponding median particle size with respect to number by Eq. (8-28). Using Eq. (8-28) and the data in Table 8.2, one can predict the geometric mean diameter based on mass, $D_{p,gm}$(mass) = 4.40 μm. With this value, the predicted mass distribution is shown in Figure 8.5 as the solid line labeled "mass". The actual cumulative mass distribution is also plotted on the graph as data points. It is seen that for diameters below about 6 or 7 μm, the agreement between the predicted line and the data points is excellent. However, at the larger particle diameters the data deviate from the straight line. This is also typical of particle distributions, as discussed by Hinds (1982). The explanation is that there is a particle size above which particles are aerodynamically unable to enter the sampling apparatus; as D_p approaches this upper limit, fewer particles enter the apparatus than should, causing the deviation from the straight line in the log probability plot of cumulative distribution. From the straight line for mass distribution in Figure 8.5, one can estimate the percent of the mass between any two particle diameters $D_{p,1}$ and $D_{p,2}$. For example, the percent of the mass that is inhalable (D_p < 10 μm) is about 92.0%, and the percent of the mass that is respirable (D_p < 2 μm) is about 8.0%. Thus the percent of the mass between 2 μm and 10 μm is around 84.0%.

The relationship between mass concentration (c = the mass of particles per unit volume of air), and ***number concentration*** (c_{number}), which is defined as the number of particles per unit volume, can be found from

$$c = c_{number}\frac{\pi\rho_p\left[D_{p,am}(\text{mass})\right]^3}{6} \tag{8-32}$$

The arithmetic mean diameter based on mass can be found from Table 8.2 by compiling a column based on $D_{p,j}^3$,

$$D_{p,am}(\text{mass}) = \left[\frac{\sum_{j=1}^{J}\left(n_j D_{p,j}^{\ 3}\right)}{n_t}\right]^{\frac{1}{3}} \tag{8-33}$$

Arithmetic mean diameter based on mass is not the same as arithmetic mean diameter based on number, which was defined by Eq. (8-3). The difference can be significant since a few large particles have considerably greater mass than the same number of smaller particles. By similar reasoning, $D_{p,50}$(mass) is not the same as $D_{p,50}$(number).

Computing the cumulative distribution may mask whether the aerosol has one or several modes. For example, if an aerosol consists of two slightly overlapping log-normal distributions, the cumulative distribution of the entire aerosol will not be log normal and one may not be able to discern the separate modes.

The reader should develop an appreciation of particle distributions based on mass and number. A few large particles can have an enormous influence on the mass distribution, since mass varies with the cube of particle diameter. To dramatize this fact, consider the following example.

Example 8.2 - Particle Size Distribution

Given: Air in a foundry is dusty because of the handling of sand used to make molds, shaking castings out of the sand molds, etc. A sample of the workplace air was drawn through a filter at a rate of 0.15 liter/min for a period of 100 seconds. The sampled air contained 240 particles which were counted and sized optically on the basis of diameter. Table E8.2 shows the results.

To do: Answer the following questions:

(a) Can the size distribution be described as log normal?
(b) What is the mass concentration (c, in mg/m^3) of particles in the workplace, assuming the particle density (ρ_p) is 1,800 kg/m^3?

Table E8.2 Foundry dust size distribution (grouped data) for Example 8.2.

class, j	range (μm)	$D_{p,j}$ (μm)	n_j	n_j/n_t	N_j/n_t (%)	$n_jD_{p,j}$	$n_jD_{p,j}^3/n_t$	m_j/m_t	M_j/m_t (%)
1	5-6	5.5	0	0	0	0	0	0	0
2	6-9	7.5	0	0	0	0	0	0	0
3	9-13	11.0	2	0.008	0.80	22	11	0.0002	0.02
4	13-18	15.5	29	0.121	12.9	450	450	0.0083	0/85
5	18-26	22.0	54	0.225	35.4	1,188	2,396	0.0441	5.26
6	26-37	32.0	84	0.350	70.4	2,688	11,469	0.2110	26.36
7	37-52	42.5	54	0.225	92.9	2,295	17,272	0.3178	58.14
8	52-73	62.5	14	0.058	98.7	875	14,241	0.2620	84.34
9	73-103	88.0	3	0.012	99.9	264	8,518	0.1567	100.0
totals:			**240**			**7,782**	**54,357**		

Solution:

Part (a): The cumulative number distribution (N_J/n_t) and the cumulative mass distribution (M_J/m_t) in Table E8.2 are plotted on log probability paper in Figure E8.2 as a function of $D_{p,max,J}$.

There is some experimental scatter, as expected, and only four of the data points lie within the limits of the axes for each case, but these data fit reasonably to parallel straight lines, showing that the assumption of log normality is acceptable. From the plot,

- $D_{p,15.9}$(number) = 19.3 μm
- $D_{p,50}$(number) = 29.0 μm
- $D_{p,84.1}$(number) = 43.5 μm

A further test can be performed when the predicted median particle diameter $D_{p,50}$ and geometric standard deviation σ_g are compared with the plotted data. The arithmetic mean diameter based on number and the standard deviation can be computed from the data in Table E8.2. Using Eq. (8-3),

$$D_{p,am}(\text{number}) = \frac{\sum_{J=1}^{J}\left(n_J D_{p,J}\right)}{n_t} = \frac{7782\ \mu\text{m}}{240} = 32.4\ \mu\text{m}$$

and using Eq. (8-4),

$$\sigma = \sqrt{\frac{\sum_{J=1}^{J} n_J\left[D_{p,am} - D_{p,J}\right]^2}{n_t - 1}} = 13.3\ \mu\text{m}$$

From these values, the geometric standard deviation and geometric mean diameter based on number are found from Eqs. (8-14) and (8-15) respectively,

$$\sigma_g = \exp\left\{\sqrt{\ln\left[1+\left(\frac{13.3}{32.4}\right)^2\right]}\right\} = 1.484$$

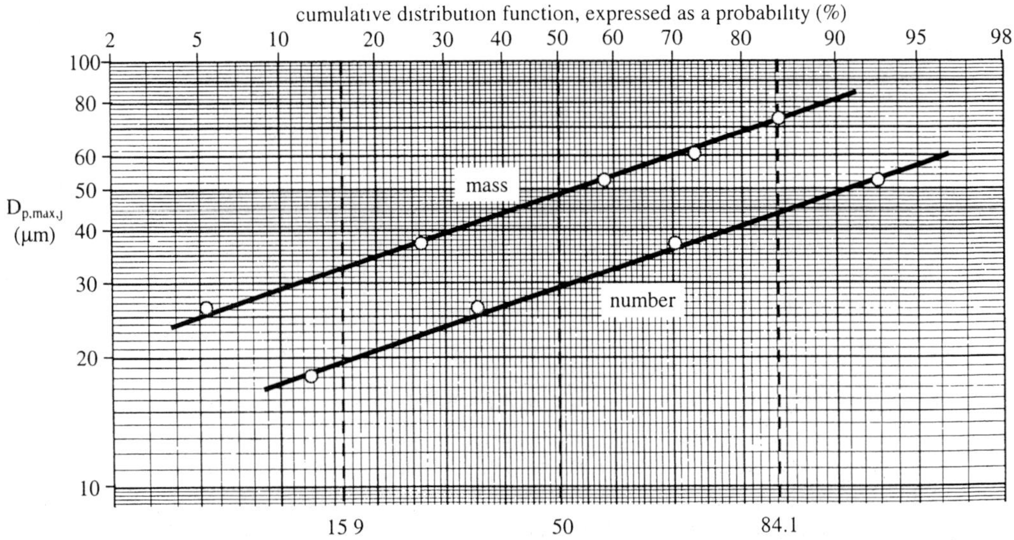

Figure E8.2 Log probability plot of the cumulative number distribution and cumulative mass distribution for the aerosol of Table E8.2.

and

$$D_{p,gm} = D_{p,am} \exp\left[-\frac{(\ln \sigma_g)^2}{2}\right] = (32.4\ \mu m)\exp\left[-\frac{(\ln 1.484)^2}{2}\right] = 30.0\ \mu m$$

For a log-normal distribution, $D_{p,50} = D_{p,gm}$; thus the predicted median particle diameter of 30.0 μm is seen to agree reasonably well with that obtained from the plotted data (29.0 μm). The values of $D_{p,84\,1}$ and $D_{p,15\,9}$ can be predicted from Eqs. (8-22) and (8-24),

$$D_{p,84\,1} = \sigma_g D_{p,50} = 1.484(30.0\ \mu m) = 44.5\ \mu m$$

and

$$D_{p,15\,9} = \frac{\sigma_g}{D_{p,50}} = \frac{30.0\ \mu m}{1.484} = 20.2\ \mu m$$

Compared to the values of 43.5 and 19.3 obtained from Figure E8.2, the agreement is also found to be reasonable. The geometric mean diameter based on mass can be predicted from Eq. (8-28),

$$D_{p,gm}(\text{mass}) = \exp\left[\ln\left(D_{p,gm}(\text{number})\right) + 3(\ln \sigma_g)^2\right] = \exp\left[\ln(30.0) + 3(\ln 1.484)^2\right] = 47.9\ \mu m$$

This can be compared to the value of $D_{p,gm}$(mass) obtained from Figure E8.2,

$$D_{p,gm}(\text{mass}) = D_{p,50}(\text{mass}) = 48.5\ \mu m$$

Again the agreement is good (within one micron). Thus, one can conclude that the particle distribution is log normal.

Part (b): The total mass of particles in the sample is

$$m_t = n_t \frac{\pi \rho_p}{6} \sum_{j=1}^{J} \left(\frac{n_j D_{p,j}^{\ 3}}{n_t}\right) = 240 \frac{\pi\left(1800 \frac{kg}{m^3}\right)}{6} \left(54357\ \mu m^3\right) \left(\frac{m}{10^6\ \mu m}\right)^3 = 1.23 \times 10^{-8}\ kg$$

The volume of the air sample is

$$V = Qt = 0.15 \frac{\text{liter}}{\text{min}} (100\ s) \left(\frac{1\ \text{min}}{60\ s}\right) \left(\frac{1\ m^3}{1000\ \text{liter}}\right) = 2.50 \times 10^{-4}\ m^3$$

The mass concentration of particles in the air sample is

$$c = \frac{m_t}{V} = \frac{1.23 \times 10^{-8}\ kg}{2.50 \times 10^{-4}\ m^3} \left(\frac{10^6\ mg}{kg}\right) = 49.2 \frac{mg}{m^3}$$

Discussion: This particle concentration is higher than the OSHA standard, which specifies that the maximum allowable mass concentration for nonrespirable nuisance dust is 15 mg/m^3.

8.2.4 Physical Explanation of Log-Normal Distributions

Ott (1990) provides a straightforward method called ***successive random dilutions*** to explain why many natural processes produce log-normal distributions. These processes consist of a series of random events that follow one another over a period of time, e.g. operations where the size of particles is reduced by successive grinding operations, or where particles in a monodisperse aerosol grow by the process of agglomeration. If the successive dilutions involve fixed amounts, the final state can be predicted with certainty. Such predictions are called ***deterministic***. If the successive dilutions involve randomness, the final state can be predicted only by using probability theories and methods. Random processes are called ***stochastic***.

To illustrate the concept of successive random processes, consider the concentration of a contaminant (particles, vapor, or gas) in an enclosure of volume V_0, as illustrated in Figure 8.6. At time zero, the contaminant concentration is c_0. A circulating fan distributes the contaminant uniformly so that at any instant the concentration is spatially uniform. The enclosure contains no contaminant sources or sinks and there are no chemical reactions. The enclosure is flexible such that small amounts of air can be removed and added without changing the pressure inside the enclosure. Consider the following sequence of events:

- Remove contaminated air at concentration c_0 from the enclosure at volumetric flow rate Q_1 over elapsed time t_1. Simultaneously add fresh air at the same volumetric flow rate Q_1 over the same elapsed time t_1.
- Wait for well-mixed conditions to occur and measure the new concentration c_1.
- Repeat the process, i.e. remove contaminated air at concentration c_1 from the enclosure at volumetric flow rate Q_2 over elapsed time t_2. Simultaneously add fresh air at volumetric flow rate Q_2 for elapsed time t_2.
- Wait for well-mixed conditions to occur and measure the new concentration c_2.
- Repeat the above steps m times.

Initially (j = 0) the concentration is c_0 and the mass of contaminant in the enclosure is m_0, thus

$$\boxed{c_0 = \frac{m_0}{V_0}} \tag{8-34}$$

Following the first dilution (j = 1), the concentration in the enclosure is

$$\boxed{c_1 = \frac{m_1}{V_0} = \frac{m_0 - Q_1 t_1 c_0}{V_0} = \frac{V_0 c_0 - V_1 c_0}{V_0} = c_0\left(1 - \frac{V_1}{V_0}\right)} \tag{8-35}$$

where V_1 is the volume of air added and removed from the enclosure ($V_1 = Q_1 t_1$). A ***dilution factor*** (D_1) is defined as

$$\boxed{D_1 = 1 - \frac{V_1}{V_0}} \tag{8-36}$$

Thus

$$\boxed{c_1 = c_0 D_1} \tag{8-37}$$

Following the second dilution (i = 2), the concentration in the enclosure is c_2, where after similar algebra,

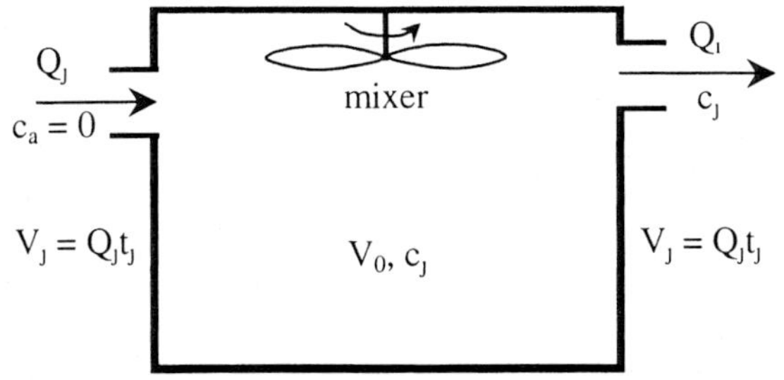

Figure 8.6 Successive random dilutions; j = dilution number, c(0) = initial concentration = $c_0 > 0$.

$$c_2 = \frac{m_2}{V_0} = \frac{m_1 - Q_2 t_2 c_1}{V_0} = \frac{V_0 c_1 - V_2 c_1}{V_0} = c_1\left(1 - \frac{V_2}{V_0}\right)$$

and thus

$$\boxed{c_2 = c_0\left(1 - \frac{V_1}{V_0}\right)\left(1 - \frac{V_2}{V_0}\right) = c_0 D_1 D_2} \tag{8-38}$$

where V_2 is the volume of contaminated air added and removed from the enclosure and D_2 is a second dilution factor,

$$\boxed{D_2 = 1 - \frac{V_2}{V_0}} \tag{8-39}$$

The process is repeated m times and the dilution factors are selected randomly; the concentration after m dilutions is

$$c_m = c_0 D_1 D_2 \ldots D_m$$

Taking the logarithm of the above gives

$$\boxed{\ln c_m = \ln c_0 + \left(\ln D_1 + \ln D_2 + \ldots + \ln D_m\right)} \tag{8-40}$$

Using the additive form of the central limit theorem, Ott showed that since each dilution factor D_i is an independent random variable, the sum is *also* an independent random variable. Random variables have frequency distributions that are gaussian or normal. Thus since the logarithm of the concentration is normally distributed, it can be said that c_m is log-normally distributed.

Graphic proof of why successive dilutions produce log-normal distributions can be obtained by making 1,000 computations (henceforth called a trial) each involving five dilutions (m = 5). In each dilution the dilution factors (D_j) are selected by a random number generator, keeping in mind that $0 < D_j \leq 1.0$. The results of the 1,000 trials are tabulated, and the cumulative distribution is calculated and plotted for the final concentration on log probability paper. The results can be approximated in a reasonable manner by a straight line. The analysis of 1,000 trials of five dilutions is only *approximately* log normal since the number of dilutions and the number of trials are not infinite. If the number of dilutions and trials are increased and the above computation repeated, the log-normal fit improves. In conclusion, the example shows that if a process consists of a succession of random events, the final result has a distribution that can be characterized as log normal.

8.3 Overall Collection Efficiency

Particles are removed from a gas stream by a collection device in which particles either settle or impact and adhere to the surfaces of other bodies. The removal process is traditionally characterized by a ***grade efficiency***, or ***fractional efficiency*** in which the fraction of particles of a particular size that are removed (η) is expressed as function of particle size (D_p),

$$\boxed{\eta(D_p) = 1 - \frac{c(D_p)_{out}}{c(D_p)_{in}}} \tag{8-41}$$

where the subscripts in and out mean ***inlet*** and ***outlet*** states respectively. Fractional efficiency curves are obtained from experiment or predicted from first principles. Deriving these curves is a large subject in itself; what is important now is how such data and knowledge of the aerosol size distribution can be used to predict the overall collection efficiency. Typical grade efficiency curves are shown in Figure 8.7. In general, high-efficiency collectors have higher $\eta(D_p)$ at any given D_p and set of inlet conditions.

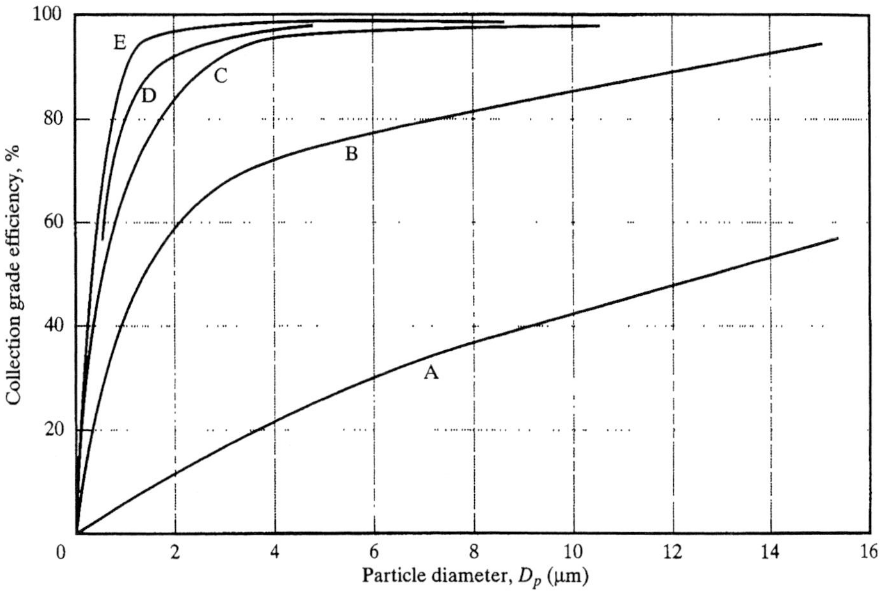

Figure 8.7 Typical collection grade efficiencies of different dust collectors; A: low-efficiency high-flow rate cyclones, B: high-efficiency low-flow rate cyclones, C: spray towers, D: electrostatic precipitators, E: venturi scrubbers.

If an aerosol enters a device or process with a size distribution that can be expressed statistically, the aerosol will leave with another size distribution in which there is a higher percentage of small particles. The overall mass concentration is lower than when it entered but the size distribution is skewed toward the smaller particles. The ***overall removal efficiency***, also called the ***overall collection efficiency***, is defined as

$$\boxed{\eta_{\text{overall}} = 1 - \frac{c_{\text{out}}}{c_{\text{in}}}} \tag{8-42}$$

where c_{out} and c_{in} are the overall mass concentrations at the outlet and inlet of the device respectively. The size distribution can be described by the number fraction or mass fraction of particles of a given size. For this analysis it is preferable to use the *mass fraction*. The variable $g(D_{p,j})_{in}$ is called the ***inlet mass fraction***, defined as the concentration of particles of a particular size range "j" at the inlet $c(D_{p,j})_{in}$ divided by the total mass concentration at the inlet c_{in},

$$\boxed{g(D_{p,j})_{\text{in}} = \frac{(m_j)_{\text{in}}}{(m_t)_{\text{in}}} = \frac{\dfrac{(m_j)_{\text{in}}}{V}}{\dfrac{(m_t)_{\text{in}}}{V}} = \frac{c(D_{p,j})_{\text{in}}}{c_{\text{in}}}} \tag{8-43}$$

The overall collection efficiency, from Eq. (8-42), can be expressed as

$$\boxed{\eta_{\text{overall}} = 1 - \frac{\sum\limits_j c(D_{p,j})_{\text{out}}}{c_{\text{in}}}} \tag{8-44}$$

where $c(D_{p,j})_{out}$ refers to the concentration of particles within class j at the outlet of the device, and j is summed over the total number of classes. The fractional efficiency defined in Eq. (8-41) can be rearranged as follows:

$$c(D_{p,j})_{out} = c(D_{p,j})_{in}\left[1-\eta(D_{p,j})\right]$$

Combining this with Eq. (8-44) and simplifying,

$$\eta_{overall} = 1 - \frac{\sum_j \left\{\left[1-\eta(D_{p,j})\right]c(D_{p,j})_{in}\right\}}{c_{in}} = 1 - \frac{\sum_j c(D_{p,j})_{in}}{c_{in}} + \frac{\sum_j \left[\eta(D_{p,j})c(D_{p,j})_{in}\right]}{c_{in}}$$

$$= 1 - 1 + \sum_j \left[\eta(D_{p,j})g(D_{p,j})_{in}\right]$$

Finally,

$$\boxed{\eta_{overall} = \sum_j \left[\eta(D_{p,j})g(D_{p,j})_{in}\right]} \tag{8-45}$$

Thus the overall collection efficiency is equal to the inlet mass fraction for a certain class (interval of particle sizes) times the fractional efficiency for that class, summed over all the classes. At the outlet of the air cleaner, the mass fraction differs from that at the inlet, not only in magnitude, but also in *distribution*. It can be shown that

$$\boxed{g(D_{p,j})_{out} = \frac{1-\eta(D_{p,j})}{1-\eta_{overall}}g(D_{p,j})_{in}} \tag{8-46}$$

Since grade efficiency increases from 0 to 1 with increasing particle size, large particles are removed, while the tiniest particles are not. The outlet mass distribution thus shifts to the left, towards smaller particles.

To compare collectors whose efficiencies are very close to 100%, it is useful to use the term penetration. ***Penetration*** (P) is defined as $(1 - \eta)$. Thus to compare two collectors whose efficiencies are 99.5% and 99.8%, emphasis can be gained by saying the penetration of one collector is 0.5% while the penetration of the other is 0.2%.

Example 8.3 - Overall Removal Efficiency of a Polydisperse Aerosol

Given: It has been proposed to remove particles from the air described in Example 8.2 with a particle collection device whose fractional (grade) efficiency is given below.

$D_{p,j}$ (μm)	$\eta(D_{p,j})$
5.5	0.42
7.5	0.50
11	0.60
15.5	0.68
22	0.72

$D_{p,j}$ (μm)	$\eta(D_{p,j})$
32	0.80
42.5	0.83
62.5	0.93
88	0.98

The overall dust concentration is measured, and found to be 49.2 mg/m^3.

To do: Determine if this device will be able to bring the workplace air into compliance with the OSHA standard, which specifies that the maximum allowable concentration for nonrespirable nuisance dust is 15 mg/m^3.

Solution: The minimum overall removal efficiency is

$$\eta_{min} = 1 - \frac{c_{max\ allowable}}{c_{in}} = 1 - \frac{15}{49.2} = 0.695$$

or 69.5%. Using the inlet mass fraction $g(D_{p,J}) = m_J/m_t$ given in Table E8.2, the overall removal efficiency is computed using Eq. (8-45),

$$\eta_{overall} = \sum_J \left[\eta(D_{p,J}) g(D_{p,J})_{in}\right] = (0.42)(0) + (0.50)(0) + (0.60)(0.0002) + ... + (0.98)(0.1567) = 0.869$$

or 86.9%. Consequently the collector is capable of satisfying the OSHA standard.

Discussion: It remains to be shown that all the air in workplace can be cleaned with this efficiency, that pockets of unusually high dust concentration do not exist, etc.

8.3.1 Air Cleaners in Series

Consider a collection system consisting of several (m) air cleaners arranged in series as shown in Figure 8.8. The overall collection efficiency is

$$\eta_{overall} = 1 - \frac{c_{out}}{c_{in}} = 1 - \frac{c_1}{c_{in}} \frac{c_2}{c_1} \frac{c_3}{c_2} ... \frac{c_{m-1}}{c_{m-2}} \frac{c_{out}}{c_{m-1}}$$

where the efficiency of each collector can be expressed as

$$\eta_1 = 1 - \frac{c_1}{c_{in}} \qquad \eta_2 = 1 - \frac{c_2}{c_1} \quad ... \quad \eta_m = 1 - \frac{c_{out}}{c_{m-1}}$$

which can be rearranged as

$$\frac{c_1}{c_{in}} = 1 - \eta_1 \qquad \frac{c_2}{c_1} = 1 - \eta_2 \quad ... \quad \frac{c_{out}}{c_{m-1}} = 1 - \eta_m$$

from which

$$\boxed{\eta_{overall} = 1 - (1-\eta_1)(1-\eta_2)...(1-\eta_m) = 1 - \prod_{J=1}^{m}(1-\eta_J)} \qquad (8\text{-}47)$$

Thus the overall collection efficiency is not the arithmetic average of the separate collection efficiencies, but unity minus the product of the penetration ($P_J = 1 - \eta_J$) of each air cleaner in series,

$$\boxed{\eta_{overall} = 1 - (P_1 P_2 ... P_m) = 1 - \prod_{J=1}^{m} P_J} \qquad (8\text{-}48)$$

Readers must keep in mind that although the overall efficiency increases when a second air cleaner is added in series downstream of an existing cleaner, the efficiency (η_2) of the downstream cleaner *decreases* compared to its value if it were the first cleaner in the series. This is because the particle distribution downstream of the first cleaner shifts toward smaller particles, the largest particles having been removed. The air entering the downstream cleaner now contains disproportionately more small particles, which are more difficult to remove (as seen in Figure 8.7), thus lowering η_2. More rigorous analysis of air cleaners in series requires Eq. (8-46) to calculate the mass fraction leaving the first air cleaner and entering the second, and then Eq. (8-45) to calculate the overall efficiency of the second air cleaner. Only then is Eq. (8-47) strictly valid. This type of analysis is well-suited for spreadsheets.

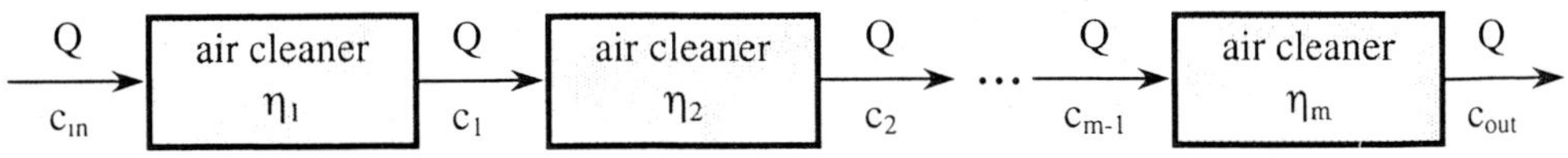

Figure 8.8 A group of m air cleaners in series, with the same volumetric flow rate through each cleaner.

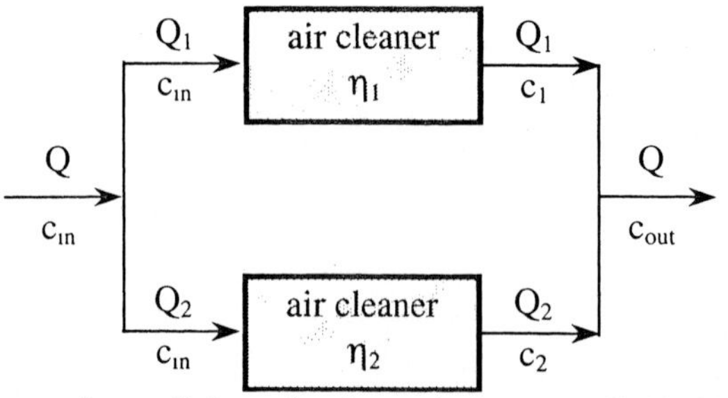

Figure 8.9 Two air cleaners in parallel; total volumetric flow rate divided between the air cleaners.

8.3.2 Air Cleaners in Parallel

Consider two air cleaners arranged in parallel as shown in Figure 8.9. The efficiencies and the volumetric flow rates through each air cleaner are not, in general, the same although the concentration entering each air cleaner is the same. The contaminant mass flow rate leaving each air cleaner is

$$c_1Q_1 = (1-\eta_1)c_{in}Q_1 \qquad c_2Q_2 = (1-\eta_2)c_{in}Q_2$$

The overall collection efficiency of the parallel configuration is then

$$\eta_{overall} = 1 - \frac{c_1Q_1 + c_2Q_2}{c_{in}Q} = 1 - \frac{(1-\eta_1)c_{in}Q_1 + (1-\eta_2)c_{in}Q_2}{c_{in}Q} = 1 - \left[(1-\eta_1)\frac{Q_1}{Q} + (1-\eta_2)\frac{Q_2}{Q}\right]$$

which can be re-written as

$$\boxed{\eta_{overall} = 1\text{-}\left[(1-\eta_1)f_1 + (1-\eta_2)f_2\right]} \qquad (8\text{-}49)$$

where the ***volumetric flow rate fractions*** (f_1 and f_2) are defined as

$$\boxed{f_1 = \frac{Q_1}{Q} \qquad f_2 = \frac{Q_2}{Q}} \qquad (8\text{-}50)$$

Thus the overall collection efficiency is not the arithmetic average of each air cleaner's efficiency, but depends on the manner in which the volumetric flow rates are split as well as the magnitudes of η_1 and η_2.

Example 8.4 - The Most Economical Retrofit of a Particle Collection System

Given: A coal-fired electric utility boiler produces an exhaust gas stream of 30,000 ACFM at 400. K. Fly ash particles are removed with a cyclone particle collector that has an overall collection efficiency of 75%. New state environmental regulations require that 95% of the particles must be removed. Three options are being considered:

(a) replace the existing 75% collector with a new 95% collector
(b) add one or more new collectors in series with the existing collector
(c) devise a parallel configuration in which a fraction of the flow passes through the existing collector and the remainder passes through a new collector

The capital cost for new collectors is as follows:

η (%)	capital cost ($/SCFM)
75	30
80	40
85	46
90	52

η (%)	capital cost ($/SCFM)
95	58
99	70
99.5	80

To do: Recommend the option with the lowest capital cost.

Solution: The standard volumetric flow rate is

$$Q_{STP} = Q_{actual}\frac{T_{STP}}{T} = \left(30,000\frac{ft^3}{min}\right)\frac{298.15\text{ K}}{400.\text{ K}} = 22,360\text{ SCFM}$$

Option A – *Replace the existing collector with a new 95% collector*: The cost is easily determined as follows:

$$\text{cost} = \frac{\$58.}{\text{SCFM}}(22,360\text{ SCFM}) = \$1,300,000$$

Option B – *Series configuration*: The overall collection efficiency is given by Eq. (8-47). For only two collectors in series,

$$\eta_{overall} = 1-(1-\eta_1)(1-\eta_2)$$

from which the new collector must have an efficiency (η_2) equal to

$$\eta_2 = 1-\frac{1-\eta_{overall}}{1-\eta_1} = 1-\frac{1-0.95}{1-0.75} = 0.80$$

The cost of an 80% collector placed in series with the existing collector is

$$\text{cost} = \frac{\$40.}{\text{SCFM}}(22,360\text{ SCFM}) = \$894,000$$

Option C – *Parallel configuration*: The overall efficiency for a parallel configuration is given by Eq. (8-49), with $\eta_1 = 0.75$ and η_2 equal to one of the above efficiencies. To achieve an overall efficiency of 95%, there is a unique flow spilt for each configuration, such that $f_1 + f_2 = 1$. Solving for f_2,

$$f_2 = \frac{\eta_{overall}-\eta_1}{\eta_2-\eta_1} = \frac{0.95-0.75}{\eta_2-0.75} = \frac{0.20}{\eta_2-0.75}$$

The following is a summary of the cost for each of the above collectors arranged in a parallel configuration:

η_2 (%)	f_2	Q_2 (SCFM)	cost
75	∞	(physically impossible)	-
80	4	(physically meaningless)	-
85	2	(physically meaningless)	-
90	1.33	(physically meaningless)	-
95	1.0	22,360	$1,300,000
99	0.833	18,600	$1,300,000
99.5	0.816	18,200	$1,460,000

The costs have been rounded to three significant digits. Note that a predicted value of f_2 greater than unity is not possible. Also note that the case with $\eta_2 = 95\%$ is the same as Option A.

In summary, it is clear that Option B (the series configuration) has the lowest capital cost.

Discussion: Option B is best. However, space is required for the second collector in series with the first. If there is no room for a second collector to follow the first, Option A is recommended. The parallel configuration (Option C) is not recommended since it does not save money. Furthermore, since such a large percentage of the flow would have to be rerouted to the new collector, the second collector would still be large, and considerable space would need to be available. Two collectors in series (Option B) would actually perform less effectively than indicated here, since the particle distribution exiting the first collector would shift to a higher percentage of *small* particles, as discussed above. Thus the collection efficiency of the second collector would be reduced.

8.4 Equations of Particle Motion

To design particle collection devices and to predict their performance, engineers must be able to predict the trajectories of particles passing through the devices. The objective of this section is to establish equations that predict the trajectories of particles. As an aerosol particle moves through air, it disturbs the air, changing the velocity and pressure fields of the air flow. In addition, particles collide with each other and are influenced by each others' wakes. Exact analysis of particles moving in air is therefore nearly impossible, even for simple air flows, since the equations for the air flow and for each particle's trajectory are coupled. Fortunately, some simplifications are possible, which make the problem more tractable. Two major assumptions are made here:

1. Particles move independently of each other.
2. Particles do not influence the flow field of the carrier gas.

The first assumption is valid if two conditions are met: (a) particle collisions are infrequent and inconsequential, and (b) particles are not significantly affected when they pass through each others' wakes. A useful rule of thumb that quantifies these conditions for a monodisperse aerosol is that the average distance between particles is at least 10 times the particle diameter. Assuming that 8 particles are located at the corners of a cube of dimension L, one finds that $L/D_p > 10$ when

$$\boxed{\frac{4\pi\rho_p}{3c} = \frac{8}{\left(c_{number}\right)D_p^{\ 3}} < 1000} \tag{8-51}$$

where c is the mass concentration of particles, ρ_p is the particle density, and $c_{number} = n_t/V$ is the ***particle number concentration***, defined as the total number of particles per unit volume of gas. Table 8.3 illustrates these upper limits and indicates that the particle number concentration has to be exceedingly large for the particles to influence each other. For water droplets (ρ_p = 1,000 kg/m^3), application of Eq. (8-51) leads to a particle mass concentration c = 4.2 kg/m^3, corresponding to the upper limit of 1,000 in Eq. (8-51). For most problems in indoor air pollution, particle concentrations of water and particulate pollutants are *hundreds* of times smaller than 4.2 kg/m^3. Consequently the first assumption above is clearly valid, and one can calculate the trajectory of one individual particle at a time.

The second assumption is more difficult to quantify. Large objects moving through air generate air flows due to displacement effects and the effects of aerodynamic wakes. For example, a pedestrian walking along the side of a road feels the "wind" created by a passing vehicle. The same effect can be experienced when a person walks past another person in otherwise quiescent air. Automobiles and people, however, are very large objects; particles usually associated with indoor air pollution are many orders of magnitude smaller. As shown below, most particles of interest to indoor

Table 8.3 Particle number concentrations beyond which particles influence each other.

D_p (μm)	c_{number} (particles/m^3)
1	8×10^{15}
10	8×10^{12}
100	8×10^{9}

air pollution move at extremely small speeds relative to the surrounding air, and therefore do not significantly alter the air flow field. The second assumption is critical to the analysis that follows. The approach is to first solve for the flow field of the carrier gas independently of the particles, and then to compute the particle trajectories through this "frozen" carrier gas flow field. Mathematically, the assumption is equivalent to an ***uncoupling*** of the equations.

The flow field of the carrier gas can be expressed analytically if one is so fortunate as to have a system of simple geometry, or it may be established experimentally and the data stored numerically. If an analytical expression for the velocity field exists, one can compute the particle trajectories explicitly. For most industrial applications only experimentally measured velocity data are available and a computer is needed to compute the particle trajectories.

An equation of motion for a particle can be derived from Newton's second law. Namely, particle mass times acceleration of the particle is equal to the vector sum of all the forces acting on the particle. Only two forces are considered here – gravitational force (net weight) and aerodynamic drag. Newton's second law is therefore

$$\boxed{m_{\text{particle}}\vec{a} = \sum \vec{F} = \vec{F}_{\text{gravity}} + \vec{F}_{\text{drag}}} \tag{8-52}$$

Other forces, such as magnetic forces, etc., are ignored here.

8.4.1 Gravitational Force

The net gravitational force is equal to the weight of the particle minus the buoyancy force on the particle, where the buoyancy force is equal to the weight of the air displaced by the particle. For a spherical particle,

$$\boxed{\vec{F}_{\text{gravity}} = m_{\text{particle}}\vec{g} - m_{\text{air}}\vec{g} = \pi\frac{D_p^{\,3}}{6}\rho_p\vec{g} - \pi\frac{D_p^{\,3}}{6}\rho\vec{g} = \pi\frac{D_p^{\,3}}{6}\left(\rho_p - \rho\right)\vec{g}} \tag{8-53}$$

where $\vec{g}$ is the acceleration of gravity, ρ is the density of the air, and ρ_p is the density of the particle. The density of a particle is typically of order 1,000 times greater than the density of air ($\rho_p >> \rho$). Thus the force of buoyancy on a particle is often neglected compared to its weight,

$$\boxed{\vec{F}_{\text{gravity}} = \pi\frac{D_p^{\,3}}{6}\left(\rho_p - \rho\right)\vec{g} \approx \pi\frac{D_p^{\,3}}{6}\rho_p\vec{g}} \tag{8-54}$$

The approximation of Eq. (8-54) is *not* made in the material which follows, so that the equations are more general.

8.4.2 Aerodynamic Drag

Consider a particle moving at some arbitrary velocity $\vec{v}$ through air, at a location in which the air is moving at some other arbitrary velocity $\vec{U}$ along a streamline of the air flow. As the particle moves relative to the air, as in Figure 8.10, the air produces an ***aerodynamic drag force*** on the particle. As indicated in the sketch, the aerodynamic drag force ($\vec{F}_{\text{drag}}$) acts in the direction of $-\vec{v}_r$, i.e. in the direction opposite to the velocity of the particle relative to the air. The net force on the particle is the vector sum of the gravitational force and the drag force, as also sketched in Figure 8.10. The particle in Figure 8.10 veers to the right and decelerates since the net force in this example is somewhat opposite to the particle's direction and to the right. For a sphere of diameter D_p, $\vec{F}_{\text{drag}}$ can be written as

$$\boxed{\vec{F}_{\text{drag}} = -c_D\frac{\rho}{2}\frac{\pi D_p^{\,2}}{4}\left(\vec{v} - \vec{U}\right)\left|\vec{v} - \vec{U}\right| = -c_D\frac{\rho}{2}\frac{\pi D_p^{\,2}}{4}\vec{v}_r\left|\vec{v}_r\right|} \tag{8-55}$$

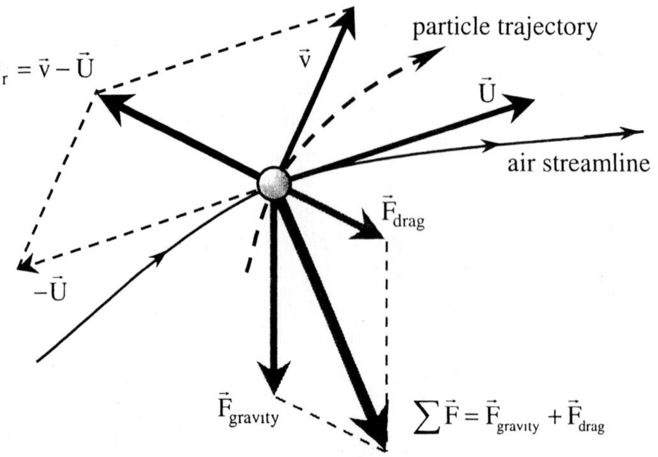

Figure 8.10 Particle moving through air, which is also moving; $\vec{U}$ is the air velocity and $\vec{v}$ is the particle velocity; gravitational force, drag force, and net force on the particle are shown.

where

- $\vec{U}$ = carrier fluid (air) velocity
- $\vec{v}$ = particle velocity
- $\vec{v}_r$ = ***relative velocity*** (particle velocity relative to the air),

$$\boxed{\vec{v}_r = \vec{v} - \vec{U} = \hat{i}(v_x - U_x) + \hat{j}(v_y - U_y) + \hat{k}(v_z - U_z) = \hat{i}v_{rx} + \hat{j}v_{ry} + \hat{k}v_{rz}} \tag{8-56}$$

- $|\vec{v}_r| = |\vec{v} - \vec{U}| = v_r$ = ***magnitude of the relative velocity***, also called the ***relative speed***,

$$\boxed{|\vec{v} - \vec{U}| = \sqrt{(v_x - U_x)^2 + (v_y - U_y)^2 + (v_z - U_z)^2} = \sqrt{v_{rx}^2 + v_{ry}^2 + v_{rz}^2}} \tag{8-57}$$

- c_D = ***drag coefficient*** for a sphere, based on projected frontal area

Figure 8.11 shows the drag coefficient for a sphere as a function of Reynolds number. Similar curves

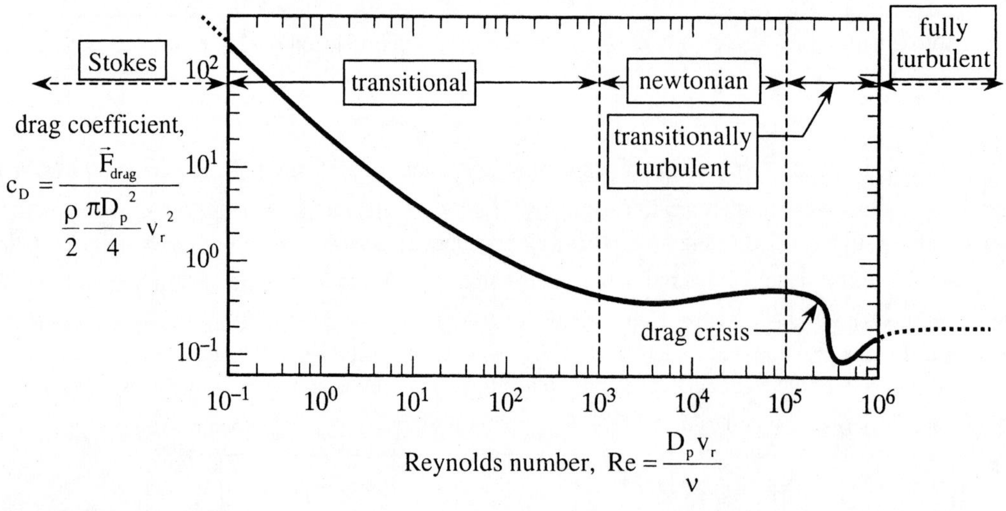

Figure 8.11 Drag coefficient of a sphere versus Re (redrawn from Incropera and DeWitt, 1981).

exist for other geometric shapes. The drag coefficient includes both drag caused by an unsymmetrical pressure distribution (***form drag***) and drag caused by shear stresses acting on the sphere's surface (***viscous drag***). The marked reduction in drag on a sphere at a Reynolds number around 200,000 is caused by transition from laminar to turbulent boundary layer flow separation (delayed flow separation) on the downstream side of the sphere; this sudden decrease in drag is sometimes called the ***drag crisis***. It is common practice to divide the flow into five flow regimes, as labeled in Figure 8.11:

- ***Stokes flow regime***: $\mathrm{Re} < 10^{-1}$ $c_D \approx 24/\mathrm{Re}$
- ***Transitional flow regime***: $10^{-1} < \mathrm{Re} < 10^{3}$ $c_D = \text{variable}$
- ***Newtonian flow regime***: $10^{3} < \mathrm{Re} < 10^{5}$ $c_D \approx 0.4$
- ***Transitionally turbulent flow regime***: $10^{5} < \mathrm{Re} < 10^{6}$ $c_D = \text{variable}$
- ***Fully turbulent flow regime***: $10^{6} < \mathrm{Re}$ $c_D \approx 0.2$

Reynolds number is defined in terms of the relative velocity and the particle diameter,

$$\boxed{\mathrm{Re} = \frac{\rho D_p \left|\vec{v} - \vec{U}\right|}{\mu} = \frac{\rho D_p \left|\vec{v}_r\right|}{\mu} = \frac{D_p \left|\vec{v}_r\right|}{\nu} = \frac{D_p v_r}{\nu}} \qquad (8\text{-}58)$$

where μ is the ***dynamic viscosity*** and $\nu = \mu/\rho$ is the ***kinematic viscosity***. The relationship between sphere drag coefficient and Reynolds number for $\mathrm{Re} < 10^5$ can be expressed by the following empirical equation with an accuracy of around 10%:

$$\boxed{c_D = 0.4 + \frac{24}{\mathrm{Re}} + \frac{6}{\left(1 + \sqrt{\mathrm{Re}}\right)}} \qquad (8\text{-}59)$$

If the Reynolds number is very small, the Stokes flow equation is adequate, and much simpler to employ:

- <u>Re < 0.1</u>: $$\boxed{c_D = \frac{24}{\mathrm{Re}}} \qquad (8\text{-}60)$$

If Re is larger than 0.1, but the range of Reynolds numbers in an application is narrow, the following empirical expressions are suggested by Willeke and Baron (1993):

- <u>0.1 < Re < 5</u>: $$\boxed{c_D = \frac{24}{\mathrm{Re}}\left(1 + 0.0916\,\mathrm{Re}\right)} \qquad (8\text{-}61)$$

- <u>5 < Re < 1000</u>: $$\boxed{c_D = \frac{24}{\mathrm{Re}}\left(1 + 0.158\,\mathrm{Re}^{2/3}\right)} \qquad (8\text{-}62)$$

It must be emphasized that the speed used in the Reynolds number is based on the *relative* speed (v_r) between the particle and the fluid, as defined by Eq. (8-56). For particles moving in motionless air, the relative velocity *is* the particle velocity. For particles traveling through a moving fluid, great care must be taken to evaluate the relative velocity.

Either the dynamic viscosity (μ) or the kinematic viscosity (ν) of the carrier gas must be known in order to calculate Re in Eq. (8-58). For most gases, μ is a strong function of temperature, but a very weak function of pressure. For air, the dynamic viscosity can be expressed empirically as a function of temperature (T), in units of kg/(m s). A well-known, highly accurate equation for μ as a function of T is ***Sutherland's law***,

$$\boxed{\mu = \mu_0 \left(\frac{T}{T_0}\right)^{1.5} \left(\frac{T_0 + S}{T + S}\right)} \qquad (8\text{-}63)$$

where $\mu_0 = 1.71 \times 10^{-5}$ kg/(m s), $T_0 = 273.15$ K, S = 110.4 K, and T must be in units of K. Other (less accurate) equations for μ can be obtained by performing a least-squares polynomial fit of experimental data; e.g. the following third-order polynomial fit (from Appendix A.11):

$$\boxed{\mu = \left(1.3554\times10^{-6} + 0.6738\times10^{-7}T - 3.808\times10^{-11}T^2 + 1.183\times10^{-14}T^3\right)\frac{\text{kg}}{\text{m s}}} \tag{8-64}$$

where T must again be in units of K. The kinematic viscosity (ν) can be calculated from either of these by definition, $\nu = \mu/\rho$. Shown in Figure 8.12 are μ and ν for air as functions of temperature at standard pressure. The values calculated by Eq. (8-64) overlap those calculated by Eq. (8-63) except at large values of T, where the polynomial curve fit deviates from Sutherland's law.

Air consists of molecules that travel at very high speed. The average distance traveled by air molecules between collisions with each other is called the ***mean free path*** (λ). Mean free path can be expressed in terms of viscosity by the following equation (Jenning, 1988; Flagan and Seinfeld, 1988):

$$\boxed{\lambda = \frac{\dfrac{\mu}{0.499}\sqrt{\dfrac{\pi}{8}}}{\sqrt{\rho P}}} \tag{8-65}$$

For air at STP, $\mu_{STP} = 1.830 \times 10^{-5}$ kg/(m s), $\rho_{STP} = 1.184$ kg/m^3, $P_{STP} = 101{,}325$ Pa, and $\lambda_{STP} = 0.06635$ μm. Assuming air to be a ideal gas (often called a perfect gas),

$$\boxed{P = \rho\frac{R_u}{M}T = \rho RT} \tag{8-66}$$

where R_u is the universal gas constant, and R is the specific gas constant, as defined in Chapter 1. The mean free path at temperatures and pressures other than STP can be expressed as

$$\boxed{\lambda(T,P) = \lambda_{STP}\left(\frac{\mu}{\mu_{STP}}\right)\left(\frac{P_{STP}}{P}\right)\sqrt{\frac{T}{T_{STP}}}} \tag{8-67}$$

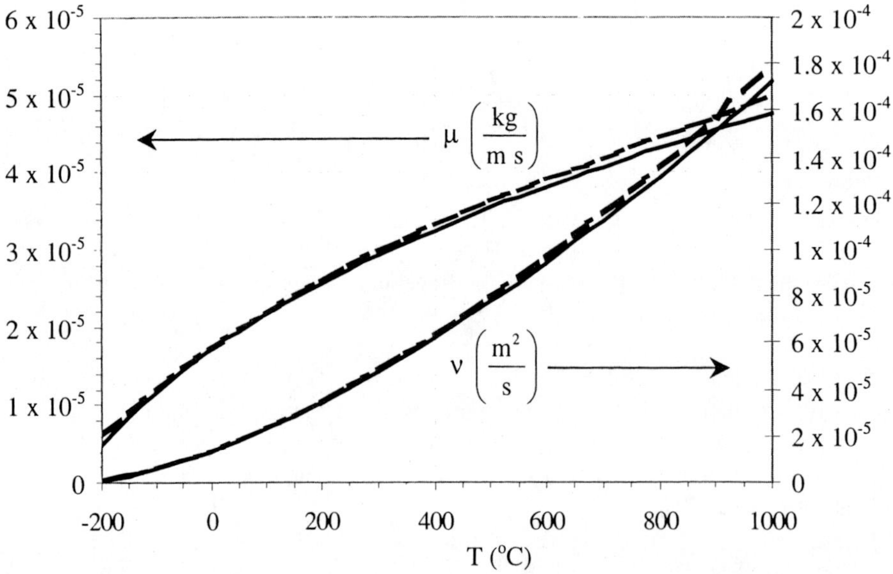

Figure 8.12 Dynamic viscosity (μ) and kinematic viscosity (ν) of air as functions of temperature at standard atmospheric pressure; solid line from Eq. (8-63), dashed line from Eq. (8-64).

A fluid is considered a continuum fluid if the moving particle is considerably larger than the mean free path of the fluid; the particle moves through the fluid as if the fluid were a continuous medium. On the other hand, if the particle is comparable in size (or smaller) than the mean free path, the particle is affected by collisions with individual molecules, and at times slips between molecules. Such motion is called ***free molecular flow*** or more generally, ***slip flow***.

The parameter used to distinguish the continuum regime from the free molecular regime is the Knudsen number (Kn), defined as the ratio of the mean free path to a characteristic length of the particle. For spherical particles,

$$\boxed{\mathrm{Kn} = \frac{\lambda}{D_p}} \tag{8-68}$$

Three regimes can be classified according to the value of the Knudsen number as follows:

- $\mathrm{Kn} < 0.1$ ***continuum regime***
- $0.1 < \mathrm{Kn} < 10$ ***transitional flow regime***
- $\mathrm{Kn} > 10$ ***free molecular flow regime***

Note that since particle diameter is in the denominator of Eq. (8-68), smaller particles have larger Knudsen numbers. Note also that the "transitional flow regime" defined here is not the same as the transitional flow regime defined previously for drag on a sphere.

Engineers are familiar with the equations of continuum fluid mechanics, but many contaminant particles, e.g. smoke, fume, or fine dust, are in the transitional or free molecular flow regimes. When dealing with these particles, one must modify Eq. (8-55). It is expedient and accurate to insert a parameter called the ***Cunningham slip factor*** (C), also called the ***Cunningham correction factor*** or simply the ***slip factor***, into the denominator of the equation for aerodynamic drag. The slip factor can be expressed by the following (Jenning, 1988):

$$\boxed{C = 1 + \mathrm{Kn}\left[2.514 + 0.80\exp\left(-\frac{0.55}{\mathrm{Kn}}\right)\right]} \tag{8-69}$$

For spherical particles, Eq. (8-55) thus becomes

$$\boxed{\vec{F}_{drag} = -c_D \frac{\rho}{2C}\frac{\pi D_p^{\,2}}{4}\left(\vec{v} - \vec{U}\right)\left|\vec{v} - \vec{U}\right| = -c_D \frac{\rho}{2C}\frac{\pi D_p^{\,2}}{4}\vec{v}_r\left|\vec{v}_r\right|} \tag{8-70}$$

The relationship between the slip factor and the diameter of spherical particles in air at STP is shown in Table 8.4.

Table 8.4 Knudsen number and Cunningham correction factor for spherical particles in air at STP.

D_p (μm)	Kn	C
0.001	66.37	220.5
0.002	33.19	110.5
0.005	13.27	44.56
0.01	6.637	22.57
0.02	3.319	11.59
0.05	1.327	5.039
0.1	0.6637	2.900
0.2	0.3319	1.885

D_p (μm)	Kn	C
0.5	0.1327	1.335
1	0.0664	1.167
2	0.0332	1.083
5	0.0133	1.033
10	0.0066	1.017
20	0.0033	1.008
50	0.0013	1.003
100	0.0007	1.002

It is clear that for large particles, $D_p >> \lambda$, the slip factor (C) is essentially unity and can be omitted from Eq. (8-70). The slip factor is particularly significant for submicron particles. It should be noted that the slip factor is independent of particle velocity, but depends only on particle size and the physical properties of the fluid.

8.4.3 Effect of Particle Shape on Aerodynamic Drag

Particles vary in geometry, e.g. perfect spheres such as condensed vapors, cylindrical or flat filaments such as cotton fibers or asbestos in which the ratio of length to width is large, platelets such as silica or mica, feathery agglomerates such as soot, and irregularly shaped fragments such as coal dust, foundry sand, or metal grinding particles. If particles are not spheres the drag may be quite different than for spheres of the same mass. To accommodate nonspherical particles, Fuchs (1964) suggested introducing a unitless ***dynamic shape factor*** (X, Greek letter chi), defined as the ratio of the drag force of the nonspherical particle and the drag force of a sphere having the same volume and velocity. Thus the drag force of Eq. (8-70) is further modified for nonspherical particles as

$$\vec{F}_{drag} = -Xc_D \frac{\rho}{2C} \frac{\pi D_{e,p}^2}{4} \left(\vec{v} - \vec{U}\right)\left|\vec{v} - \vec{U}\right| = -Xc_D \frac{\rho}{2C} \frac{\pi D_{e,p}^2}{4} \vec{v}_r \left|\vec{v}_r\right| \tag{8-71}$$

where $D_{e,p}$ is the ***equivalent volume diameter*** defined in terms of the actual particle volume (V_p) as

$$D_{e,p} = \left(\frac{6V_p}{\pi}\right)^{\frac{1}{3}} \tag{8-72}$$

and the drag coefficient is expressed in terms of the Reynolds number based on this equivalent volume diameter. The Cunningham correction factor should also be computed on the basis of the equivalent volume diameter. The equivalent volume diameter is the diameter an actual particle would have if it were a sphere. For flow beyond the Stokes flow regime, one may use Eq. (8-71) to describe the drag in lieu of a better expression, but readers are urged to consult other sources (Fuchs, 1964; Davies, 1966; Strauss, 1966; Hidy and Brock, 1970; Hinds, 1982; Cheng et al., 1988; Lee and Leith, 1989) for more details. Table 8.5 shows values of the dynamic shape factor (X) for a variety of common shapes. Dynamic shape factors for particles that have a length, width, and height of comparable value are close to unity, and may be omitted for purposes of indoor air pollution analysis. For these cases, $D_{e,p}$ may be replaced by either the length, width, or height. Leith (1987) suggests that the dynamic shape factor (X) can be estimated as follows:

$$X = \left(0.33 + 0.67\frac{D_{s,p}}{D_{p,p}}\right)\frac{D_{e,p}}{D_{p,p}} \tag{8-73}$$

where $D_{p,p}$ and $D_{s,p}$ are called the ***equivalent projected area diameter*** and ***equivalent surface area diameter*** respectively,

- $D_{p,p}$ = diameter of a sphere with the same projected area as the actual particle, where projected area is the cross-sectional area of the particle normal to the direction of flow
- $D_{s,p}$ = diameter of a sphere with the same surface area as the actual particle

Sauter mean diameter is defined as the total volume of all particles in an aerosol divided by the total surface area of all the particles. From this definition, Sauter diameter can be thought of as a mean volume-to-surface diameter.

Spherical particles whose density is equal to 1000 kg/m^3, the density of water, are called ***unit density spheres***. Such particles have a ***specific gravity*** (SG) equal to 1.0, where SG is defined as the ratio of particle density (ρ_p) to water density (ρ_{water}),

Table 8.5 Dynamic shape factors, averaged over all orientations unless otherwise noted (abstracted from Fuchs, 1964 and Strauss, 1966).

shape		X
sphere		1.00
cube		1.08
cylinder (L/D = 4):		
	axis horizontal	1.32
	axis vertical	1.07
ellipsoid, across polar axis, with ratio of major to minor diameters = 4		1.20
parallelepiped with square base, with various values of height to base:		
	0.25	1.15
	0.50	1.07
	2.00	1.16
	3.00	1.22
	4.00	1.31
clusters of spheres		
	chain of 2	1.12
	chain of 3	1.27
	3 compact	1.15
	chain of 4	1.32
	4 compact	1.17

$$SG = \frac{\rho_p}{\rho_{water}} \tag{8-74}$$

For the most part, the density of a particle is the density of the compound of which it is composed. In the event the particle contains voids, is a loose feathery agglomerate, or is a composite material, the density is more difficult to define. Details of how to cope with these circumstances can be found in Hinds (1982).

8.4.4 Predicting the Trajectory of a Spherical Particle

For a spherical particle of diameter D_p, the net gravitational force is given by Eq. (8-54), and the drag force is given by Eq. (8-70). Substitution of these expressions into Eq. (8-52) yields the equation of motion of a single spherical particle in air,

$$\pi\frac{D_p^{\ 3}}{6}\rho_p\frac{d\vec{v}}{dt} = \pi\frac{D_p^{\ 3}}{6}\vec{g}\left(\rho_p - \rho\right) - c_D\frac{\rho}{2C}\pi\frac{D_p^{\ 2}}{4}\left(\vec{v}-\vec{U}\right)\left|\vec{v}-\vec{U}\right| \tag{8-75}$$

Consider the general motion of a spherical particle traveling in the x-y plane through a moving two- dimensional gas stream in which air velocity $\vec{U}$ varies, and in which the gravity vector is in the negative y direction. Let the particle have initial velocity $\vec{v}(0)$. Figure 8.10 depicts such motion. Equation (8-75) reduces to

$$\frac{d\vec{v}}{dt} = -\hat{j}g\frac{\rho_p - \rho}{\rho_p} - \frac{3\rho c_D}{4C\rho_p D_p}\left(\vec{v}-\vec{U}\right)\left|\vec{v}-\vec{U}\right| \tag{8-76}$$

where

$$\left(\vec{v}-\vec{U}\right)=\hat{i}\left(v_x-U_x\right)+\hat{j}\left(v_y-U_y\right)=\hat{i}v_{rx}+\hat{j}v_{ry} \tag{8-77}$$

and

$$v_r=\left|\vec{v}-\vec{U}\right|=\sqrt{{v_{rx}}^2+{v_{ry}}^2} \tag{8-78}$$

The above reduces to a pair of coupled differential equations,

$$\frac{dv_x}{dt}=-\frac{3c_D\rho}{4C\rho_pD_p}v_{rx}\sqrt{{v_{rx}}^2+{v_{ry}}^2} \tag{8-79}$$

and

$$\frac{dv_y}{dt}=-g\frac{\rho_p-\rho}{\rho_p}-\frac{3c_D\rho}{4C\rho_pD_p}v_{ry}\sqrt{{v_{rx}}^2+{v_{ry}}^2} \tag{8-80}$$

where drag coefficient (c_D) is a function of Reynolds number, as in Eqs. (8-59) through (8-62). The Reynolds number of Eq. (8-58) reduces to

$$\text{Re}=\frac{\rho D_p\sqrt{{v_{rx}}^2+{v_{ry}}^2}}{\mu}=\frac{\rho D_pv_r}{\mu}=\frac{D_pv_r}{\nu} \tag{8-81}$$

8.4.5 Particle Trajectory in the Stokes Flow Regime

If the particle's motion is entirely within the Stokes flow regime, Eq. (8-60) applies; replacing c_D by 24/Re *decouples* Eqs. (8-79) and (8-80), resulting in a simplified set of uncoupled equations,

$$\frac{dv_x}{dt}=-\frac{\left(v_x-U_x\right)}{\tau_pC} \tag{8-82}$$

and

$$\frac{dv_y}{dt}=-g\frac{\rho_p-\rho}{\rho_p}-\frac{\left(v_y-U_y\right)}{\tau_pC} \tag{8-83}$$

where τ_p is called the ***particle relaxation time*** since it possesses the units of time, and because it is customary to use this name when it appears in first-order differential equations like Eq. (8-82) or (8-83). τ_p is defined as

$$\tau_p=\frac{\rho_pD_p^{\,2}}{18\mu} \tag{8-84}$$

Thus if U_x and U_y are constant or known functions of x and y, Eqs. (8-82) and (8-83) can be solved to predict v_x and v_y as functions of time and known initial conditions, $v_x(0)$ and $v_y(0)$.

To illustrate this kind of Stokes flow, consider the case in which a particle enters an air stream that is moving to the right at constant speed U. (The components of the air velocity are $U_x = U$ and $U_y = 0$.) The particle's initial velocity components are $v_x(0)$ and $v_y(0)$, which are arbitrary. Shown in Figure 8.13 is the particle at some arbitrary time, along with the forces acting on it. Equations (8-82) and (8-83) become simple first-order ordinary differential equations with constant coefficients. It is left as an exercise for the reader to rewrite these equations in standard form, i.e. in the form of Eq. (1-59), for which analytical solutions are possible. The solutions are

$$v_x(t)=v_x(0)\exp\left(-\frac{t}{\tau_pC}\right)+U\left[1-\exp\left(-\frac{t}{\tau_pC}\right)\right] \tag{8-85}$$

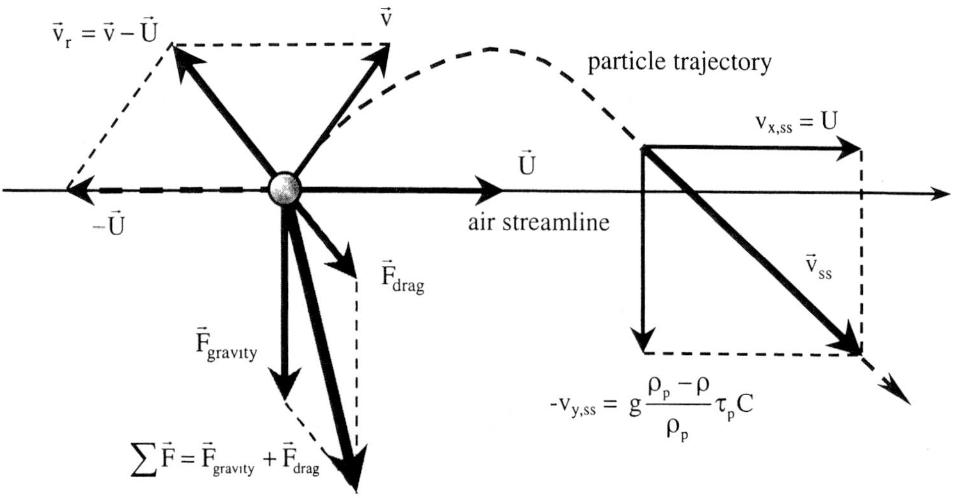

Figure 8.13 Particle moving through air, which is moving horizontally; forces on the particle and the particle trajectory are shown.

and

$$\boxed{v_y(t) = v_y(0)\exp\left(-\frac{t}{\tau_p C}\right) - Cg\frac{\rho_p - \rho}{\rho_p}\tau_p\left[1 - \exp\left(-\frac{t}{\tau_p C}\right)\right]} \tag{8-86}$$

After some time, $t >> \tau_p$, the transients can be neglected, and the particle maintains a steady-state velocity, $\vec{v}_{ss}$, as also shown in Figure 8.13. Equations (8-85) and (8-86) reduce to

$$\boxed{\vec{v}_{ss} = \vec{v}_{\text{steady state}} = U \cdot \hat{i} + \left(-g\frac{\rho_p - \rho}{\rho_p}\tau_p C\right) \cdot \hat{j}} \tag{8-87}$$

The net force on the particle in steady-state conditions is zero as shown in Figure 8.14.

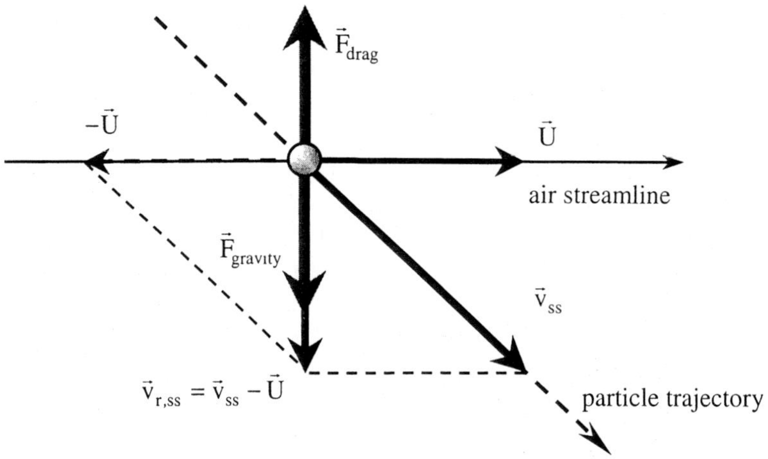

Figure 8.14 Particle moving through air, which is moving horizontally; the forces acting on the particle and the particle trajectory are shown for steady-state conditions.

As shown, the aerodynamic drag force is vertical, exactly balancing the gravitational force. By Newton's law, the particle experiences no further acceleration – it moves in a straight diagonal line at constant speed as sketched. This analysis justifies the commonplace assumption that small particles move horizontally at a speed equal to the carrier gas, and simultaneously drift downward relative to the gas at a speed equal to the ***terminal settling velocity*** (v_t), sometimes simply called the ***terminal velocity*** or the ***fall velocity***. In the case being examined here,

$$\boxed{v_t = -v_{y,ss} = Cg\frac{\rho_p - \rho}{\rho_p}\tau_p} \tag{8-88}$$

One must keep in mind that this is true only for spherical particles in Stokes flow, and is only an approximation for other flow regimes. This behavior is often assumed as a general proposition, which may be adequate providing the accuracy requirements (or lack thereof) of the analysis allow it.

If the Reynolds numbers are unknown and one suspects that the flow is apt to be beyond the Stokes flow regime, differential equations (8-79) and (8-80) remain coupled and numerical methods are required to compute the particle velocity and trajectory. For brevity, only two-dimensional motion is described. To solve these equations numerically, they are first rewritten in standard form for first-order ordinary differential equations,

$$\boxed{\frac{dv_x}{dt} = B_x - A_x v_x \qquad \text{and} \qquad \frac{dv_y}{dt} = B_y - A_y v_y} \tag{8-89}$$

where, since $v_{rx} = v_x - U_x$ and $v_{ry} = v_y - U_y$,

$$\boxed{A_x = A_y = \left(\frac{3c_D\rho}{4C\rho_p D_p}\right)\sqrt{\left(v_x - U_x\right)^2 + \left(v_y - U_y\right)^2}} \tag{8-90}$$

$$\boxed{B_x = A_x U_x} \tag{8-91}$$

and

$$\boxed{B_y = A_y U_y - g\frac{\rho_p - \rho}{\rho_p}} \tag{8-92}$$

In the general case, air velocity components (U_x and U_y) change with location (x and y) as the particle moves. Thus, two additional equations need to be solved to predict the trajectory, namely

$$\boxed{v_x = \frac{dx}{dt} \qquad \text{and} \qquad v_y = \frac{dy}{dt}} \tag{8-93}$$

Four coupled ordinary differential equations, Eqs. (8-89) and Eqs. (8-93), must be solved simultaneously. This is accomplished by numerical means; the ***Runge-Kutta method*** (Appendix A.12) is recommended.

Example 8.5 - Trajectory of Particles in a Boundary Layer

Given: In a production line, a high-speed press punches holes in a strip of metal ribbon that is formed into electrical connectors for the automotive industry. The speed of the operation requires the metal to be bathed in cutting oil (ρ = 891. kg/m^3) to cool and lubricate the punch. Insufficient local ventilation is provided and drops of cutting oil are ejected upward into the air with an initial velocity of around 30. m/s at an angle approximately 150° from the positive x-axis. The injection point is x = 0.0, y = 1.0 cm. The range of particle diameters is estimated to be 50 μm < D_p < 200 μm. Room air passes over the horizontal surface as a boundary layer in the positive x-direction with horizontal velocity component approximated by

$$U_x = U_\infty \sin\left(\frac{y}{\delta}\right)$$

where U_∞ = 5.0 m/s and δ is the boundary layer thickness, which is equal to 10. cm. Above the boundary layer thickness ($y > \delta$), $U_x = U_\infty$ = constant. The air has negligible vertical velocity component. The workers claim that the large drops are carried far downwind while small particles are not. The plant engineer claims that that they are wrong – the large particles should settle close to where they are injected because of their weight, and no particles will be transported beyond about one meter from the point of injection.

To do: Use a numerical technique to predict the velocity and trajectory of these oil particles traveling in the given air stream.

Solution: The authors used Mathcad to perform Runge-Kutta marching. Trajectories for the two extreme particle diameters (D_p = 50 and 200 μm) are shown in Figure E8.5. The 50 μm particles settle at around 0.74 m downstream of the injection site, while the 200 μm particles settle much further downstream, at about 1.8 m.

Discussion: The results show that the workers are right and the plant engineer is wrong. Unlike trajectories in air moving uniformly in the x-direction, the calculations show that large particles travel the farthest. The reason for this behavior is that large particles possess more inertia, and rise to greater values of y before their upward motion is damped by aerodynamic drag and gravity. At these large values of y, the larger particles encounter higher air velocities that sweep them farther downwind than the smaller particles. As seen in Figure E8.5, the 50 μm particles never rise above the boundary layer, which is 0.10 m high.

8.5 Freely Falling Particles in Quiescent Media

Consider the motion of a spherical particle in quiescent air ($\vec{U} = 0$) as shown in Figure 8.15. Since the only motion is downward, the instantaneous downward ***settling velocity***, $v_s(t)$, is defined as

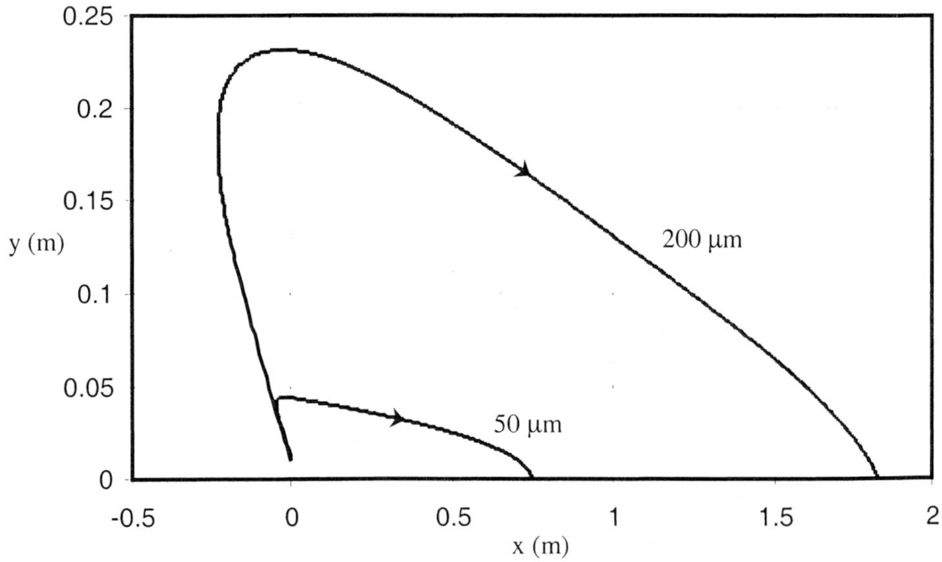

Figure E8.5 Particle trajectories for 50 and 200 μm diameter oil drops injected into a boundary layer, flow from left to right.

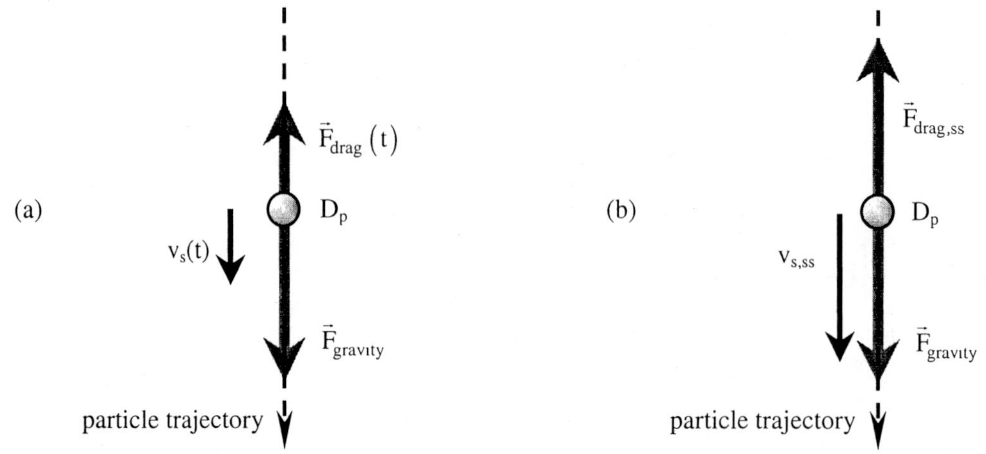

Figure 8.15 Particle settling in quiescent air; the particle trajectory is straight down, and the forces acting on the particle are shown for (a) arbitrary and (b) steady-state conditions.

$$\boxed{v_s(t) = -v_y(t)} \tag{8-94}$$

at any time t. As the particle begins to fall, as in Figure 8.15a, the downward gravitational force is greater than the upward aerodynamic drag force; thus the particle accelerates downward. Eventually, the settling velocity increases to a point where the aerodynamic drag force balances the gravitational force, and the particle settles at constant speed (no net force, no net acceleration) as in Figure 8.15b. This steady-state settling speed, $v_{s,ss}$, is equal to the terminal velocity, v_t.

8.5.1 Small Particles (Stokes Flow Regime)

To illustrate a number of important concepts while keeping the algebra simple, consider only the motion of *small* particles in which the Reynolds number is always much less than unity. For particles in the Stokes flow regime, Eq. (8-60) yields

$$\boxed{c_D = \frac{24}{Re} = \frac{24\mu}{\rho D_p v_s}} \tag{8-95}$$

Eq. (8-83), for the vertical component of velocity, is the only differential equation which needs to be solved in this case. After setting $\vec{U} = 0$, substitution of Eq. (8-94) into Eq. (8-83) yields

$$\boxed{\frac{dv_s}{dt} = g\frac{\rho_p - \rho}{\rho_p} - \frac{v_s}{\tau_p C}} \tag{8-96}$$

where τ_p is the particle relaxation time, as defined previously in Eq. (8-84). Eq. (8-96) is a first-order ordinary differential equation in standard form with constant coefficients, as discussed in previous chapters; an analytical solution is available. If the particle starts from rest, the downward velocity is

$$\boxed{v_s(t) = Cg\frac{\rho_p - \rho}{\rho_p}\tau_p\left[1 - \exp\left(-\frac{t}{\tau_p C}\right)\right]} \tag{8-97}$$

As with any first-order system, the settling velocity (v_s) increases rapidly with time, achieving 63.2% of its steady-state value ($v_{s,ss} = v_t$) after an elapsed time $t = \tau_p C$; it achieves 99.9% of its final value in t $\approx 7\tau_p C$. When $t >> \tau_p C$, steady-state conditions occur, equivalent to setting the left hand side of Eq. (8-96) to zero. Under these conditions the downward velocity is constant and equal to the terminal velocity (v_t), defined previously by Eq. (8-88). The above results pertain only to spherical particles falling in quiescent fluid in which the Reynolds number is small, specifically,

$$\text{Re} = \frac{\rho D_p v_t}{\mu} = \frac{\rho D_p C g \frac{\rho_p - \rho}{\rho_p} \tau_p}{\mu} = \frac{\rho D_p C g \frac{\rho_p - \rho}{\rho_p} \frac{\rho_p D_p^2}{18\mu}}{\mu} = \frac{\rho(\rho_p - \rho) D_p^3 g C}{18\mu^2} < 0.1 \quad (8\text{-}98)$$

Table 8.6 shows values of C, τ_p, v_t, and Re for unit density spheres (SG = 1) for Stokes flow in air at STP. Terminal velocity (v_t) is shown in Figure 8.16 for spherical particles of various sizes and three specific gravities for Stokes flow in air at STP. The final table entry for D_p = 50 μm is somewhat

Table 8.6 Gravimetric settling of unit density spheres in quiescent air at STP.

D_p (μm)	C	τ_p (s)	v_t (m/s)	Re
0.01	22.57	3.03×10^{-10}	6.71×10^{-8}	4.34×10^{-11}
0.02	11.59	1.21×10^{-9}	1.38×10^{-7}	1.78×10^{-10}
0.05	5.039	7.59×10^{-9}	3.75×10^{-7}	1.21×10^{-9}
0.1	2.900	3.03×10^{-8}	8.62×10^{-7}	5.58 x 10-⁹
0.2	1.885	1.21×10^{-7}	2.24×10^{-6}	2.90×10^{-8}
0.5	1.335	7.59×10^{-7}	9.93×10^{-6}	3.21×10^{-7}
1	1.167	3.03×10^{-6}	3.47×10^{-5}	2.24×10^{-6}
2	1.083	1.21×10^{-5}	1.29×10^{-4}	1.67×10^{-5}
5	1.033	7.59×10^{-5}	7.68×10^{-4}	2.48×10^{-4}
10	1.017	3.03×10^{-4}	3.02×10^{-3}	1.96×10^{-3}
20	1.008	1.21×10^{-3}	1.20×10^{-2}	1.55×10^{-2}
50	1.003	7.59×10^{-3}	7.46×10^{-2}	2.41×10^{-1}

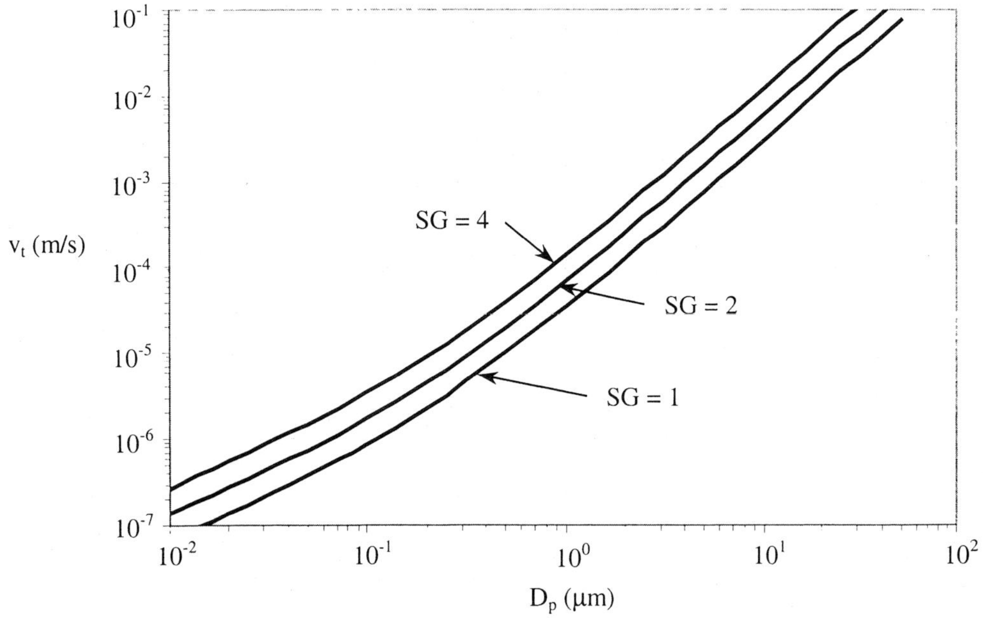

Figure 8.16 Terminal velocity of particles of various diameters (log scale) and three specific gravities settling in quiescent air at STP; Stokes flow regime is assumed.

questionable since the Reynolds number is greater than 0.1, although some authors employ the Stokes flow assumption up to Re ≈ 1. It should be noted that the time constants ($\tau_p C$) in Table 8.6 are very small (on the order of milliseconds even for particles as large as 50 μm). In indoor air pollution problems, characteristic time scales are of the order of seconds or minutes; consequently the transient portion of vertical fall is of negligible importance for such small particles. From a practical point of view the settling velocity for these small particles (relative to air) can always be set equal to the terminal velocity,

$$\boxed{v_s \approx v_t} \tag{8-99}$$

From the data in Table 8.6, it is obvious that *very small* particles, such as microorganisms suspended in air (Kowalski et al., 1999), have negligibly small settling velocities and tend to stay suspended. For such cases, settling can be ignored with no significant loss of accuracy.

8.5.2 Intermediate Particles (Transitional Flow Regime)

Suppose the particle is large enough that the Reynolds number is between 0.1 and 1,000, in the transitional flow regime. The drag coefficient (c_D) is a more complicated function of Reynolds number, as in Eqs. (8-59), (8-61), or (8-62). The steady-state vertical settling velocity, i.e. the terminal velocity, can be found by setting the left hand side of Eq. (8-80) to zero and manipulating the remaining variables, resulting in

$$\boxed{v_t = \sqrt{\frac{4}{3}\frac{(\rho_p - \rho)}{\rho}\frac{D_p g C}{c_D}}} \tag{8-100}$$

where relative velocity in Eq. (8-80) has been set to $-v_t$ since the particle trajectory is straight down. In Eq. (8-100), v_t is a function of c_D, which is a function of Reynolds number, which is in turn a function of v_t, namely

$$\boxed{Re = \frac{\rho D_p v_t}{\mu}} \tag{8-101}$$

Hence, calculation of the terminal velocity of particles in this transitional flow regime requires solution of three simultaneous equations, Eq. (8-100), Eq. (8-101), and one of the equations for c_D as a function of Re, Eqs. (8-59), (8-61), or (8-62). Numerical solutions are usually required. The authors have written a Mathcad program that calculates the terminal velocity (v_t) of a spherical particle of any diameter and density, falling in quiescent air at a given temperature and pressure. Equation (8-59) was assumed for c_D. Figure 8.17 shows a plot of v_t as a function of particle diameter for unit density spheres falling in air at STP. For comparison, the Stokes flow approximation is shown on the same plot. As noted previously, the Stokes flow approximation begins to break down for particles larger than approximately 50 μm, in which Re > 0.1. For large particles, the "exactly" predicted value of v_t is much smaller than that predicted with Stokes assumption.

Example 8.6 - Terminal Velocity of a Particle from a Volcano

Given: A volcano has erupted, spewing stones, steam, and ash several thousand feet into the atmosphere. Consider a solid particle of diameter 200 μm, falling in air which is at –50. °C and 55. kPa. The density of the particle is 2200 kg/m^3 (SG = 2.2).

To do: Estimate the terminal velocity of this particle at this altitude.

Solution: The density of the air is calculated from the ideal gas law,

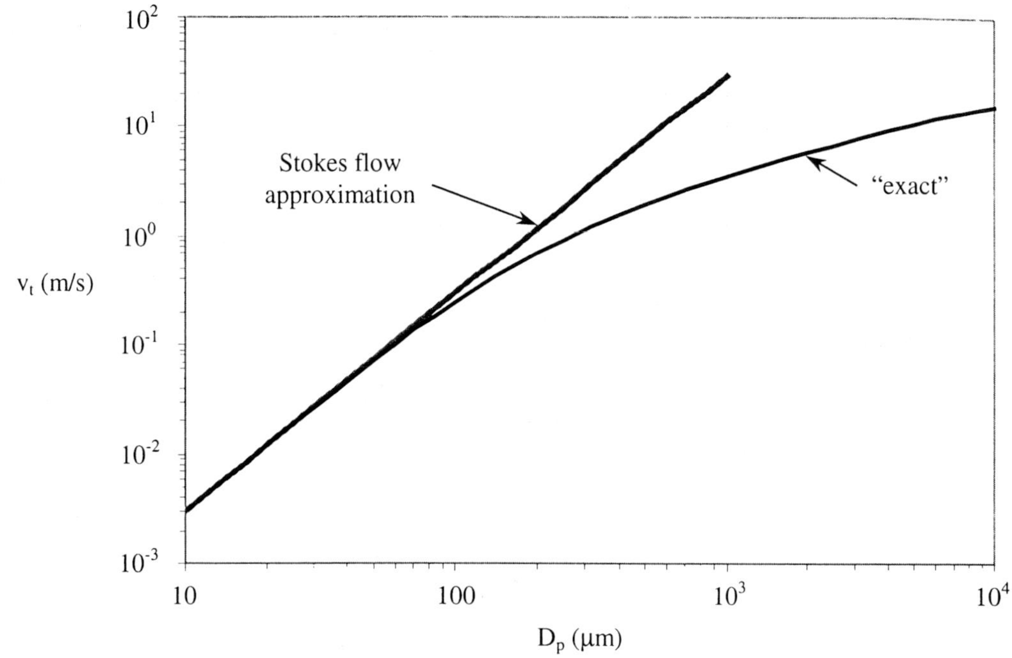

Figure 8.17 Terminal velocity of unit density spheres of various diameters (log scale) in quiescent air at STP; the Stokes approximate solution is shown for comparison.

$$\rho = \frac{PM_{air}}{R_u T} = \frac{55{,}000\frac{N}{m^2}\left(28.97\frac{kg}{kmol}\right)}{8.3143\frac{kJ}{kmol\ K}(223.15\ K)}\left(\frac{kJ}{1000\ N\ m}\right) = 0.8588\frac{kg}{m^3}$$

The viscosity of the air is calculated from Sutherland's law, Eq. (8-63); the result is μ = 1.452 x 10^{-5} kg/(m s). The mean free path of air molecules at these conditions is calculated from Eq. (8-65), which results in λ = 8.390 x 10^{-8} m (0.08390 μm). For this value of λ, the Knudsen number is Kn = 4.195 x 10^{-4} from Eq. (8-68), and from Eq. (8-69) the Cunningham slip factor is 1.001. The authors used Mathcad to solve the simultaneous equation set, Eq. (8-59) for c_D, Eq. (8-100) for v_t, and Eq. (8-101) for Re. The results are:

$$Re = 18.1$$

$$c_D = 2.87$$

and

$$v_t = 1.53\frac{m}{s}$$

The Mathcad program that solves these equations is available on the book's web site.

Discussion: This terminal velocity is valid only at the given altitude; v_t changes as the particle falls to lower altitudes.

8.5.3 Larger Particles (Newtonian Flow Regime):

For Reynolds numbers between 1,000 and 100,000, in the newtonian flow regime, the Cunningham slip factor (C) is close enough to unity to be omitted (C = 1). Additionally, the drag coefficient (c_D) is nearly constant: $c_D = 0.40$. The terminal velocity can then be found analytically from Eq. (8-80),

$$\boxed{v_t = \sqrt{\frac{4}{3}\frac{(\rho_p - \rho)}{\rho}\frac{D_p g}{0.40}} \quad \text{for } 10^3 < \text{Re} < 10^5} \qquad (8\text{-}102)$$

An example using Eq. (8-102) is provided below.

Example 8.7 - Falling Bullets

Given: Great concern is exercised in the US about instructing individuals to use firearms properly. On the other hand, TV, movies, newspapers, and magazines often show individuals firing pistols and rifles into the air, with reckless abandon. Rarely, if ever, does anyone seem to care about harm that can be done when the bullets subsequently fall to earth.

To do: Estimate the terminal velocity of a 9.0 mm bullet of density $\rho_p = 8{,}000\ \text{kg/m}^3$. For simplicity, assume the bullet is spherical.

Solution: For a start, assume the bullet falls in the newtonian flow regime ($10^3 < \text{Re} < 10^5$) so that c_D is constant (0.4); after the terminal velocity (v_t) is computed, verify that the Reynolds number lies in this range. Using Eq. (8-102),

$$v_t = \sqrt{\frac{4}{3}\frac{(\rho_p - \rho)}{\rho}\frac{D_p g}{0.40}} = \sqrt{\frac{4}{3}\frac{(8000 - 0.859)\frac{\text{kg}}{\text{m}^3}}{1.2\frac{\text{kg}}{\text{m}^3}}\frac{(0.0090\ \text{m})9.81\frac{\text{m}}{\text{s}^2}}{0.40}} = 44.3\frac{\text{m}}{\text{s}} \cong 99.\frac{\text{miles}}{\text{hr}}$$

The final Reynolds number is

$$\text{Re} = \frac{D_p v_t}{\nu} = \frac{(0.0090\ \text{m})\left(44.3\frac{\text{m}}{\text{s}}\right)}{1.51\times 10^{-5}\ \frac{\text{m}^2}{\text{s}}} = 2.6\times 10^4$$

which validates the assumption that the bullet falls in the newtonian flow regime. A 9.0 mm bullet thus settles with a terminal velocity of nearly 100 miles per hour!

Discussion: One could also use the Mathcad program from Example 8.5 to calculate the *trajectory* of the falling bullet through still air. Regardless of the initial velocity of the bullet, a steady-state terminal velocity of close to 100 mph is eventually reached. Ouch; beware of falling bullets!

8.5.4 Aerodynamic Diameter

Aerodynamic diameter ($D_{p,aero}$) is the diameter assigned to a particle of unknown density and shape that possesses the same terminal velocity as a perfect sphere of density 1,000 kg/m^3 (SG = 1.0), e.g. a sphere of water. If the unknown particle settling in air has an observed terminal velocity $v_{t,observed}$ within the Stokes flow regime,

$$\boxed{D_{p,aero} = \sqrt{\frac{18 v_{t,observed}\,\mu}{(\rho_w - \rho) g \cdot C(D_{p,aero})}}} \qquad (8\text{-}103)$$

where $C(D_{p,aero})$ is the Cunningham correction factor based on the aerodynamic diameter, and ρ_w is the density of water. If the unknown particle happens to be a sphere of diameter (D_p), then by substitution of Eqs. (8-84) and (8-88),

$$D_{aero} = D_p \sqrt{\frac{C(D_p)\,(\rho_p - \rho)}{C(D_a)\,(\rho_w - \rho)}} \tag{8-104}$$

where $C(D_p)$ and $C(D_{p,aero})$ are the Cunningham correction factors based on the actual and aerodynamic diameters respectively, and ρ_p is the actual particle density. For particles larger than about 1 μm, the Cunningham slip factors approach unity, and Eq. (8-104) simplifies to

$$D_{p\ aero} = D_p \sqrt{\frac{(\rho_p - \rho)}{(\rho_w - \rho)}} \tag{8-105}$$

8.5.5 Particles Settling in Quiescent Water

For particles in a liquid medium, settling is called ***sedimentation***. Sedimentation is described by the same model used for settling in a gaseous medium. Note that since the gravitational force, Eq. (8-53), includes the effect of buoyancy, the above equations are valid even for particles lighter than water, e.g. wood, bubbles, etc. The signs take care of themselves (v_t and ρ_p - ρ are negative for particles lighter than water).

8.6 Horizontally Moving Particles In Quiescent Air

To illustrate the dominating effect of viscosity in reducing the relative motion between a particle and the carrier gas, consider the horizontal velocity $v_x(t)$ of a sphere in quiescent air where the initial Reynolds number is less than 0.1. Since subsequent Reynolds numbers are even smaller than the initial value, the particle's motion is entirely within the Stokes flow regime. The differential equation for the horizontal velocity component is obtained from Eq. (8-82). After setting the horizontal component of the air velocity to zero,

$$\frac{dv_x}{dt} = -\frac{v_x}{\tau_p C}$$

If the particle's initial velocity is $v_x(0)$, the horizontal velocity component at any subsequent time is found by integration of the above, resulting in

$$v_x(t) = v_x(0)\exp\left(-\frac{t}{\tau_p C}\right) \tag{8-106}$$

The particle's horizontal displacement can be found by integrating its horizontal velocity component; the result is

$$\int_0^{x(t)} dx = \int_0^t v_x dt = \tau_p C v_x(0)\left[1 - \exp\left(-\frac{t}{\tau_p C}\right)\right] \tag{8-107}$$

The maximum horizontal displacement is called the ***penetration distance*** (ℓ), also called the ***pulvation distance*** or ***stopping distance***, and is found by setting t >> $\tau_p C$, which yields

$$\ell = \tau_p C v_x(0) \tag{8-108}$$

Table 8.7 is a compilation of penetration distances for water drops of several sizes possessing an initial horizontal velocity of 1.0 m/s. From Table 8.7, it can be seen that penetration distances are small, indicating that viscosity damps relative motion very quickly. Thus at times of significance to indoor air pollution, small particles traveling in air moving at velocity $\vec{U}(t) = \hat{i}U_x(t) + \hat{j}U_y(t)$ may be assumed to possess the following instantaneous velocity components:

Table 8.7 Penetration distance of unit density spheres in quiescent air at STP for particles with an initial velocity of $v_x(0) = 1.00$ m/s.

D_p (μm)	penetration distance (m)
0.01	6.86×10^{-9}
0.10	8.82×10^{-8}
1.0	3.55×10^{-6}
10.0	3.09×10^{-4}

- horizontal direction: horizontal particle velocity component equal to the horizontal carrier gas velocity component,

$$\boxed{v_x(t) \approx U_x(t)} \tag{8-109}$$

- vertical direction: vertical particle velocity component equal to the vertical carrier gas velocity component minus the terminal velocity,

$$\boxed{v_y(t) \approx U_y(t) - v_t} \tag{8-110}$$

8.7 Gravimetric Settling in a Room

If one is interested in estimating the particle concentration in a room as a function of time, a crude model of the sedimentation process can be used. The model assumes that the particles have only the two velocity components of Eqs. (8-109) and (8-110). The assumptions ignore transient behavior and take the particles to be in equilibrium with the carrier gas everywhere inside the room. It is often difficult to predict or measure small gas velocities in a room. Thus it is useful to consider upper and lower limits to the particle concentration, knowing that reality lies somewhere in between. The lower limit is defined for the case of no mixing (called the ***laminar settling model***), and the upper limit is defined as one for which mixing is a maximum (called the ***well-mixed settling model*** or ***turbulent settling model***):

(a) *Laminar settling model*: All particles of the same size fall at a uniform speed equal to their terminal velocity, and there is no mixing mechanism to redistribute particles as they fall; they settle to the floor and remain there.

(b) *Well-mixed settling model*: All particles of the same size fall at a uniform speed equal to their terminal velocity, and the ones that hit the floor remain there. However, there is an idealized mixing mechanism that completely redistributes the remaining particles so that even though the concentration decreases with time, it is always uniform within the volume.

Consider a room of volume V, height H, and horizontal cross-sectional area A as shown in Figure 8.18, which illustrates both models.

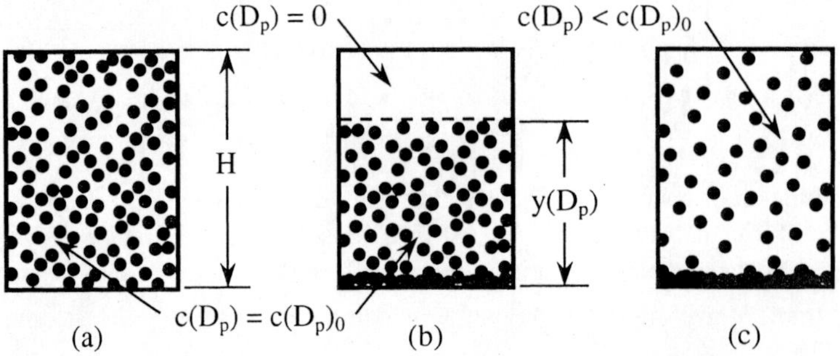

Figure 8.18 Gravimetric settling of a monodisperse aerosol in quiescent room air: (a) initial condition for both cases, t = 0, (b) laminar settling model for t > 0, (c) well-mixed model for t > 0.

Let $c(D_p)_0$ be the initial mass concentration of particles of diameter D_p in the room (Figure 8.18a). In the laminar model all particles of the same size fall uniformly at terminal velocity v_t such that those near the bottom settle to the floor. At any subsequent time there are *no* particles of that size above a certain height $y(D_p)$, and the concentration remains at $c(D_p)_0$ below this height (Figure 8.18b). The same argument applies to particles of other sizes, except the values of $y(D_p)$ are different because they settle at different velocities. For the well-mixed model, on the other hand, some particles settle, but those that remain are mixed throughout the room volume such that $c(D_p)$ decreases with time (Figure 8.18c).

8.7.1 Laminar Settling Model

Suppose the laminar settling model is valid. The rate of change of mass of particles of size D_p suspended in the room air is equal to the rate of deposition of particles onto the floor of the room, as expressed by the following differential equation:

$$\frac{d\left(Ay(D_p)c(D_p)_0\right)}{dt} = Ac(D_p)_0 \frac{dy(D_p)}{dt} = -v_t Ac(D_p)_0$$

or

$$\frac{dy(D_p)}{dt} = -v_t$$

Integrating,

$$\int_0^{y(D_p)} dy(D_p) = -\int_0^t v_t dt$$

which yields

$$\boxed{y(D_p) = H - v_t t} \tag{8-111}$$

Let the average mass concentration of particles be denoted by $\bar{c}(D_p)$. Thus,

$$\boxed{\bar{c}(D_p) = \frac{Ay(D_p)c(D_p)_0}{V} = \frac{A\left(H - v_t t\right)c(D_p)_0}{V} = c(D_p)_0\left(1 - \frac{v_t t}{H}\right)} \tag{8-112}$$

The average concentration of these particles decreases linearly with time until a time (t_c) called the ***critical time*** elapses, where

$$\boxed{t_c = \frac{H}{v_t}} \tag{8-113}$$

and all particles of the size whose terminal velocity is v_t have settled to the floor.

8.7.2 Well-Mixed Settling Model

On the other hand, suppose the well-mixed settling model is valid. As particles of a particular size fall to the floor with velocity v_t, an idealized mixer instantaneously redistributes the remaining particles throughout the room. The rate of change of mass of particles of this size suspended in the room air is equal to the rate of deposition onto the floor,

$$\frac{d\left(c(D_p)V\right)}{dt} = V\frac{dc(D_p)}{dt} = -v_t Ac(D_p)$$

which can be integrated to yield

$$\boxed{\frac{c(D_p)}{c(D_p)_0} = \frac{\bar{c}(D_p)}{c(D_p)_0} = \exp\left(-\frac{v_t t}{H}\right)} \tag{8-114}$$

Since the well-mixed model presumes that the concentration is the same throughout the enclosure at any instant of time, the term $c(D_p)$ in Eq. (8-114) is also equal to the average concentration, $\overline{c}(D_p)$.

The well-mixed model predicts that the average mass concentration decreases exponentially, while the laminar model predicts that it decreases linearly. In the well-mixed model $\overline{c}(D_p)/c(D_p)_0 = 0.368$ at $t = t_c$, while in the laminar model it is zero. In the well-mixed model, the average mass concentration does not decrease to 0.001 (0.1%) of its initial value until nearly seven of these time constants, i.e. until $t \approx 7t_c$.

The laminar model leads one to believe that the dust will be removed too quickly because it ignores unavoidable thermal currents, drafts, diffusion, etc. that redistribute particles. The laminar model also predicts an infinite concentration gradient at an interface (see Figure 8.18b) that cannot exist in nature. The well-mixed model overestimates the time to clean the air because it exaggerates the mixing mechanisms. For small, inhalable particles, it is however the more realistic model to use. Certainly, it is the more *conservative* model. In either case, it has been assumed in the analyses above that any particle that hits the floor stays there. In reality, some of the particles can be re-entrained into the room by air currents; models of this re-entrainment process are beyond the scope of this text. An additional source of error is the fact that some particles are deposited or adsorbed on other surfaces in the room, including the side walls, as discussed in Chapter 5.

8.8 Gravimetric Settling in Ducts

Gravimetric settling in ducts can also be analyzed using the concepts of a laminar settling model and a well-mixed settling model. Figure 8.19 illustrates the laminar and well-mixed models for flow in a horizontal duct of rectangular cross section (A = WH).

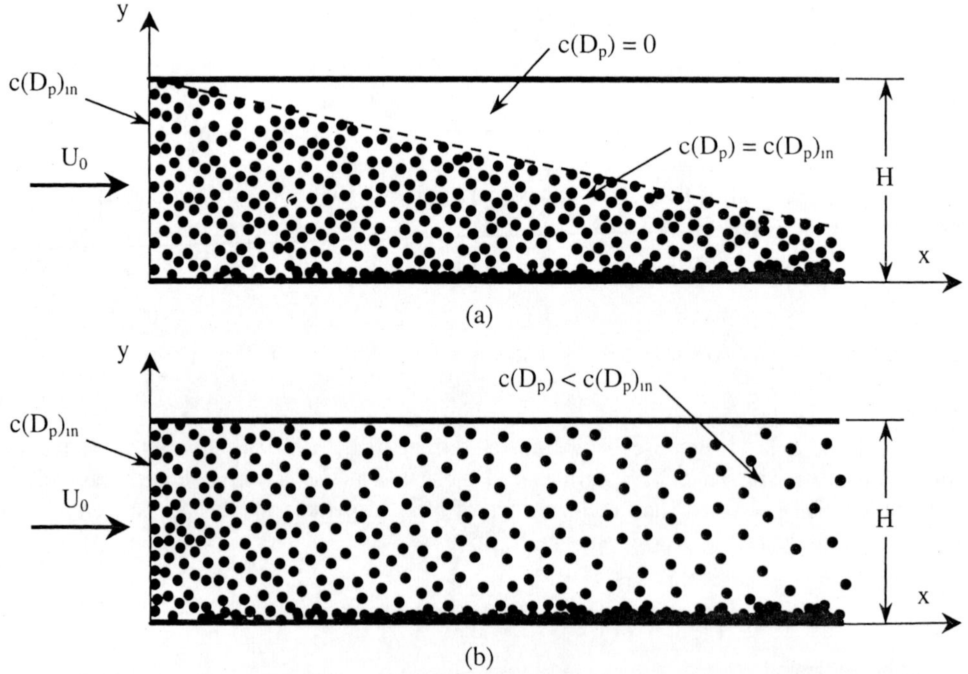

Figure 8.19 Gravimetric settling of a monodisperse aerosol in a horizontal duct with uniform air flow: (a) laminar conditions, (b) well-mixed conditions.

It is assumed that the horizontal velocity of the gas everywhere in the duct is equal to the average duct velocity (U_0), i.e. there is ***plug flow*** in the gas phase. At the duct inlet, the mass concentration of particles of diameter D_p is $c(D_p)_{in}$.

8.8.1 Laminar Settling Model

In the laminar settling model, all particles of the same size fall at their terminal velocity (v_t) and move with a horizontal velocity equal to that of the carrier gas, $v_x = U_0$. Thus at a downstream distance x, the *uppermost* particles have fallen a distance (H - y),

$$H - y = v_t t = v_t \frac{x}{U_0}$$

The average mass concentration of particles of a certain size $\bar{c}(D_p)$ at a distance x from the inlet can be written as

$$\bar{c}(D_p) = c(D_p)_{in} \frac{y}{H}$$

Combining the above two equations yields

$$\boxed{\frac{\bar{c}(D_p)}{c(D_p)_{in}} = 1 - \frac{x}{H}\frac{v_t}{U_0}} \tag{8-115}$$

Since particles settle to the floor of the duct, the duct can be thought of as a simple particle collector. The grade efficiency of the duct, $\eta(D_p)$, for particles of size D_p is defined as

$$\boxed{\eta(D_p) = 1 - \frac{\bar{c}(D_p)}{c(D_p)_{in}} = \frac{x}{H}\frac{v_t}{U_0}} \tag{8-116}$$

Equation (8-116) also applies to gravimetric settling in fully established flow between parallel plates. See Flagan and Seinfeld (1988) for the derivation. At a ***critical distance*** ($x = L_c$) downstream in the duct, the collection efficiency is 100% and the duct contains *no* particles of the size defined by the terminal velocity. The critical distance is defined by

$$\boxed{L_c = H\frac{U_0}{v_t}} \tag{8-117}$$

and Eq. (8-116) becomes

$$\boxed{\eta(D_p) = \frac{x}{H}\frac{v_t}{U_0} = \frac{x}{H}\frac{v_t H}{L_c v_t} = \frac{x}{L_c}} \tag{8-118}$$

Flagen and Seinfeld (1988) also show that the laminar collection efficiency given by Eq. (8-116) is valid for *all* laminar duct flows, irrespective of the velocity profile, so long as the profile does not change in the flow-wise direction, i.e. so long as the flow is ***fully established***, or ***fully developed***, which means that $\partial(U_x)/\partial x = 0$.

8.8.2 Well-Mixed Settling Model

Following the assumptions of the well-mixed model used in the previous section, it can be assumed that the mass concentration $c(D_p)$ of particles of a particular diameter D_p is uniform in a volume element of the duct (Adx). The difference in the mass concentration of particles entering and leaving the elemental volume is equal to the rate of deposition within the volume. Thus,

$$c(D_p)AU_0 = \left[c(D_p) + dc(D_p)\right]AU_0 + c(D_p)v_t W dx$$

which can be rearranged and integrated to obtain

$$\int_{c(D_p)_{in}}^{c(D_p)} \frac{dc(D_p)}{c(D_p)} = -\int_0^x \frac{v_t}{U_0 H} dx$$

which simplifies to

$$\boxed{\frac{c(D_p)}{c(D_p)_{in}} = \exp\left(-\frac{v_t}{U_0}\frac{x}{H}\right)} \tag{8-119}$$

Using the above definition of grade efficiency, and realizing that for the well-mixed model $\bar{c}(D_p) = c(D_p)$,

$$\boxed{\eta(D_p) = 1 - \frac{\bar{c}(D_p)}{c(D_p)_{in}} = 1 - \exp\left(-\frac{v_t}{U_0}\frac{x}{H}\right) = 1 - \exp\left(-\frac{v_t}{U_0}\frac{x}{H}\frac{W}{W}\right) = 1 - \exp\left(-\frac{v_t A_s}{Q}\right)} \tag{8-120}$$

where

- A_s = area of lower collecting surface, $A_s = xW$
- Q = volumetric flow rate, $Q = U_0HW$

The grade efficiency can also be written in terms of the critical length, as defined by Eq. (8-117),

$$\boxed{\eta(D_p) = 1 - \exp\left(-\frac{x}{L_c}\right)} \tag{8-121}$$

Comparison of the collection efficiencies for laminar and well-mixed settling models shows differences similar to those concluded for settling in rooms:

(a) The laminar settling model overestimates deposition because it ignores turbulence and diffusion that mix and redistribute particles.
(b) The well-mixed model exaggerates mixing but nevertheless provides a more accurate and conservative design estimate.
(c) At the critical downstream distance, L_c, the well-mixed settling model predicts a collection efficiency of 63.2%, while the laminar settling model predicts 100%.
(d) At a downstream distance of approximately $7L_c$, the well-mixed settling model predicts a collection efficiency of 99.9%.

It is common practice in industrial ventilation to design for ***duct transport velocities*** of 3,500 to 4,500 FPM (approximately 18 to 23 m/s) to minimize gravimetric settling (see Table 6-6). The above expressions can be used to examine the settling one can expect at these velocities. Table 8.8 is a compilation of the duct lengths (expressed nondimensionally as x/H) at which 1% and 10% of unit density spheres settle.

Table 8.8 Gravimetric settling of unit density spheres in ducts; duct velocity = 22.86 m/s (4500 FPM), air at STP.

		efficiency = 1%		**efficiency = 10%**	
D_p (μm)	**v_t (m/s)**	**$(x/H)_{lam}$**	**$(x/H)_{turb}$**	**$(x/H)_{lam}$**	**$(x/H)_{turb}$**
10	3.1×10^{-3}	74.6	74.6	745.8	785.8
50	7.3×10^{-2}	3.12	3.12	31.2	32.9
100	2.5×10^{-1}	0.92	0.92	9.2	9.7
200	0.7	0.33	0.33	3.3	3.4
600	2.5	0.09	0.09	0.9	1.0
1,000	4.0	0.06	0.06	0.6	0.6

It can be seen that 3,500-4,500 FPM is not an exaggerated recommendation. For particles less than 100 micrometers, settling is minimal in duct lengths up to 30H; however deposition is serious for larger particles. Thus it is a wise practice to provide a particle collector (cyclone or gravity drop out chamber) whenever a ventilation system is connected to a source that generates particles, particularly if it contains large particles. If this practice is not followed, a ventilation duct may unknowingly become a particle collector, which after a period of time reduces the duct cross-sectional area and alters the pressure drop and volumetric flow rate for which the system of ducts was designed. The volumetric flow rate in the duct decreases and, in extreme cases, the flow may cease.

If the flow in the rectangular duct is fully established, it can be shown that the expression for the collection efficiency, Eq. (8-121), is also valid. It can also be shown that if the duct cross-sectional area is circular, the expression for the collection efficiency for the well-mixed model is the same but duct diameter (D) should replace duct height (H) in Eqs. (8-119) and (8-120). Anand and McFarland (1989) analyzed particle deposition for turbulent flow in transport lines of circular cross-sectional area at arbitrary angles of inclination. They found that there is an optimum inside diameter at which deposition is a minimum, and that this optimum diameter is independent of tube length but depends on particle diameter, flow rate, and angle of inclination.

8.8.3 Applications

Gravimetric settling is not considered to be an efficient air pollution control method. Although particles carried in a duct settle, this is not a desired situation since the settled dust has to be removed periodically because it adds unwanted weight (vertical load) to ducts not designed to suffer such loading. Nonetheless duct systems are often equipped with ***drop out boxes*** at key junctions to allow large particles to be collected prior to entry of the process gas stream into conventional particle removal systems. In addition, drop out boxes with access doors provide the opportunity for workers to enter duct systems to perform routine maintenance and to remove settled dust. The pressure drop and fan costs associated with drop out boxes are generally negligible.

One application of gravimetric settling is in the field of industrial hygiene where one may only be interested in knowing the concentration of *respirable* dust to which workers are exposed. Devices using gravimetric settling are called ***elutriators***; these devices allow nonrespirable dust to settle, but allow respirable dust to pass through and be captured or measured.

Example 8.8 - Gravimetric Settling in a Duct with a Converging Cross Section

Given: Consider flow of a polydisperse dust-laden gas stream with volumetric rate Q and total inlet dust mass concentration c_{in} passing through a duct of width W with a horizontal floor and a top that is inclined downward at angle θ to the horizontal, as sketched in Figure E8.8. The duct height varies with axial distance from the inlet as follows:

$$H(x) = H_0 - x \tan\theta$$

To do: Predict the grade efficiency of particles of diameter D_p using the laminar settling model and the well-mixed settling model. Compare $\eta(D_p)$ from the two models and discuss the results.

Solution: Only one particle size is considered at a time. For conciseness, functions of particle diameter D_p are written without the functional form. E.g. $c(D_p)_{in}$ is written as c_{in}, $v_t(D_p)$ is written as v_t, etc.

Laminar settling model: It is assumed that the gas velocity has no vertical component ($U_y = 0$) and that U_x does not vary in the vertical direction, but increases in the x-direction owing to the reduction in duct height. Volumetric flow rate (Q) is constant through the duct. An expression can be written for $U(x) = U_x$ as a function of distance x from the entrance of the duct:

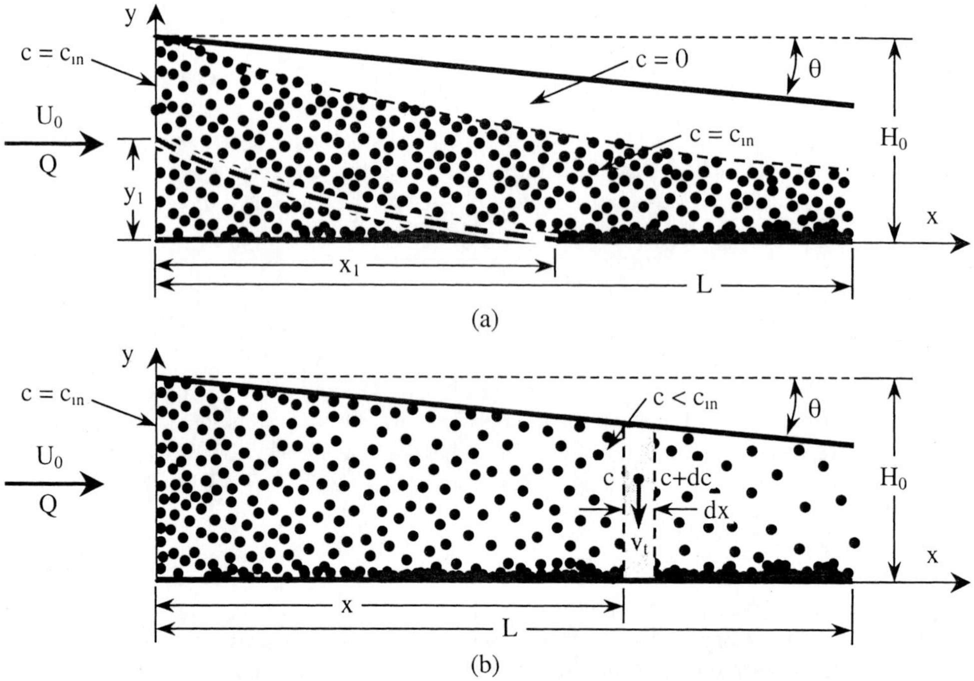

Figure E8.8 Gravimetric settling of particles of diameter D_p in a converging duct of width W and length L; (a) laminar settling model, (b) well-mixed settling model.

$$U(x) = \frac{Q}{WH(x)} = \frac{Q}{W\left(H_0 - x\tan\theta\right)}$$

Consider a particle of diameter D_p at height y_l at the inlet. Its trajectory is tracked until it settles to the duct floor at downstream distance x_l. As sketched in Figure E8.8a, the trajectory is *not* linear because although terminal settling velocity v_t is constant, the particle's x-component of velocity increases as it moves downstream. The rate at which particles settle to the floor over the distance $0 \le x \le x_l$ is

$$\text{settling rate} = y_l c_{in} W U_0$$

since all particles entering at $y < y_l$ settle before $x = x_l$. Over distance x_l, the grade efficiency is

$$\eta\left(D_p\right) = \frac{y_l c_{in} W U_0}{H_0 c_{in} W U_0} = \frac{y_l}{H_0}$$

The particle falls through distance y_l in time t_l, where $y_l = v_t\, t_l$. Thus $\eta\left(D_p\right) = \frac{v_t t_l}{H_0}$ where t_l is

$$t_l = \int_0^{t_l} dt = \int_0^{x_l} \frac{dx}{U(x)} = \int_0^{x_l} \frac{W(H_0 - x\tan\theta)dx}{Q} = \frac{Wx_l\left(H_0 - \frac{x_l}{2}\tan\theta\right)}{Q}$$

Setting x_l as total duct length L, the grade efficiency for particles of diameter D_p can be expressed as

$$\eta\left(D_p\right) = \frac{Wv_t L\left(H_0 - \frac{L}{2}\tan\theta\right)}{H_0 Q} = \frac{Wv_t L\left(H_0 - \frac{L}{2}\tan\theta\right)}{H_0\left(U_0 H_0 W\right)} = \frac{v_t}{U_0}\frac{L}{H_0}\left(1 - \frac{L\tan\theta}{2H_0}\right)$$

These results contain a term that accounts for the sloping top of the duct, thus satisfying an engineer's intuition. In the limit, as angle θ approaches zero, the above expression also approaches Eq. (8-116).

Well-mixed settling model: The following mass balance can be written for a differential element of length dx located at distance x downstream of the inlet, as sketched in Figure E8.8b:

$$Qc = Q(c + dc) + v_t cW dx$$

Separation of variables yields

$$\frac{dc}{c} = -\left(\frac{v_t W}{Q}\right) dx$$

At any downstream distance x, the volumetric flow rate of the air is equal to Q = U(x)WH(x). Even though U(x) and H(x) vary with x, the volumetric flow rate Q is constant. Thus, since the quantity in parentheses in the above differential equation is constant, the equation can be integrated analytically over the length of the duct (from x = 0, $c = c_{in}$ to x = L, $c = c_L$). The result is

$$\frac{c_L}{c_{in}} = \exp\left(-\frac{v_t LW}{Q}\right)$$

Noting that at the inlet, $Q = U_0 H_0 W$, the grade efficiency for particles of diameter D_p thus becomes

$$\eta(D_p) = 1 - \frac{c_L}{c_{in}} = 1 - \exp\left(-\frac{v_t LW}{Q}\right) = 1 - \exp\left(-\frac{v_t}{U_0}\frac{L}{H_0}\right)$$

Note that this is the same expression that would have been obtained if the duct were of constant height or if the duct *diverged* rather than converged. It is clear that the well-mixed model fails to account for the geometry of the duct, which is contrary to an engineer's intuition. However, settling velocity v_t is constant regardless of the x-component of velocity, and this is the key to understanding these results.

Discussion: The laminar settling model results depend on angle of inclination θ, while the well-mixed model results do not. The "real" solution is most likely somewhere between these two extremes.

8.9 Clouds

Consider an ensemble of particles suspended in air, henceforth called a ***cloud***. Clouds can be produced by explosions, smoke generators, insecticide mists, or discharges from chimneys, automobiles, etc. A common example is exhaled tobacco smoke. It is assumed that the cloud has discernible boundaries and that it rises or falls relative to the surrounding air with a detectable velocity. It is desirable to predict the velocity at which the cloud settles. In the analysis that follows it is assumed that the cloud contains monodisperse particles. It is observed that clouds with a small particle number concentration (c_{number} = number of particles per unit volume) settle with a velocity similar to the terminal velocity of the individual particles (v_t). It is also observed that when either the particle number concentration is large or the cloud is large (or both), the cloud settles with a much larger velocity called the ***cloud settling velocity*** (v_c).

From the concepts of ideal flow in Chapter 7, it can be shown that the velocity field produced by one particle does not significantly influence the flow field produced by another particle if the distance between particles exceeds about ten particle diameters. Of course, particle motion is dictated by viscous effects, and one must broaden the analysis to account for the influence of the wakes produced by particles. If the wake of one particle influences the motion of another particle, one can anticipate unique dynamics of the bulk aerosol. The equations that govern the flow of an aerosol with a large particle number concentration are the topic in multiphase flow. The subject is complex because the equations describing the motion of the aerosol and the carrier gas are coupled.

The behavior of concentrated aerosols is called ***colligative behavior*** (Phalen et al., 1994). Colligative behavior can occur for concentrated aerosols within a confined space or in an unbounded environment. In an unbounded environment, the concentrated aerosol is called a cloud. In lieu of such a detailed analysis Fuchs (1964) and Hinds (1982) describe a simple way to predict the settling velocity of a cloud. Consider the clouds shown in Figure 8.20. It is assumed that the clouds satisfy the following:

- the cloud contains monodisperse spherical particles of diameter (D_p) and density (ρ_p)
- the particle number concentration (c_{number}) is uniform within the cloud
- the cloud itself is spherical with cloud diameter D_c as sketched in Figure 8.20.
- the temperature of the particles is equal to that of the air inside and outside of the cloud
- the surrounding air is motionless
- when the particle number concentration is sufficiently large it causes the cloud to settle as an entity with speed (v_c) greater than the individual particle terminal speed (v_t)

If the cloud settles with speed v_t, then ambient air passes through the envelope and the drag on the cloud is the sum of the drag on each individual particle. On the other hand, if the air and particles within the cloud settle at speed $v_c > v_t$, then with respect to the cloud, ambient air passes around the cloud as if its outer boundary were an impervious spherical envelope. Under these conditions, aerodynamic drag on the cloud is the drag produced by air passing around the spherical envelope. The settling speed v_c of a solid sphere of diameter D_c whose mass is equal to the mass of the particles within the cloud is found by equating the weight of such a sphere to the drag force on the sphere:

$$W_{sphere} = m_{sphere} g = c_{number} \frac{\pi D_c^3}{6} \frac{\pi D_p^3}{6} (\rho_p - \rho) g = drag_{sphere} = \rho \frac{v_c^2}{2} c_D \frac{\pi D_c^2}{4}$$

where c_D is the drag coefficient for the sphere of diameter D_c. Solving the above for v_c,

$$v_c = \sqrt{\frac{2\pi g}{9 c_D} \frac{\rho_p - \rho}{\rho} n_c D_c D_p^3} \qquad (8\text{-}122)$$

There may be occasions when processes occur within the cloud that change D_p, c_{number}, or D_c, and engineers want to anticipate whether the cloud's motion might change from one regime to the other. For these situations it is useful to define the following dimensionless parameter:

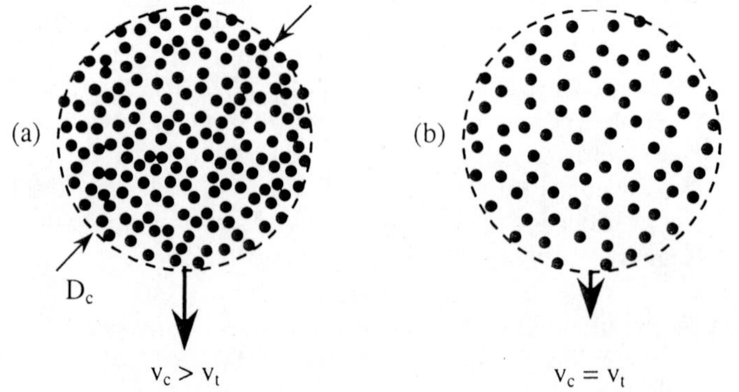

Figure 8.20 Clouds with large and small number concentrations of monodisperse particles; (a) large particle concentration, cloud settling velocity larger than individual particle settling velocity, $v_c > v_t$; (b) small particle concentration, cloud settling velocity equal to individual particle settling velocity, $v_c = v_t$.

$$\boxed{\beta = \frac{v_c}{v_t}}$$ (8-123)

If it is assumed that individual particles are governed by Stokes flow, Eq. (8-88) can be substituted for v_t in Eq. (8-123). After substitution of Eq. (8-122) into Eq. (8-123), β becomes

$$\boxed{\beta = \frac{\sqrt{\dfrac{2\pi g}{9c_D}\dfrac{\rho_p - \rho}{\rho} n_c D_c D_p^3}}{\tau_p g \dfrac{\rho_p - \rho}{\rho_p} C} = \frac{6\mu}{C}\sqrt{\frac{2\pi n_c}{g c_D}\frac{D_c}{D_p}\frac{1}{\rho(\rho_p - \rho)}}}$$ (8-124)

where C is the Cunningham slip factor and τ_p is the particle relaxation time, as defined previously in Eq. (8-84). Thus,

- if $\beta > 1$ the cloud settles with velocity v_c as given by Eq. (8-122)
- if $\beta \leq 1$ the cloud settles with velocity v_t as given by Eq. (8-88)

When the cloud is large enough that the Reynolds number of the spherical cloud exceeds 1,000, Figure 8.11 shows that one can assume $c_D = 0.4$. It is useful to use the product of the number concentration and cloud diameter ($c_{number}D_c$) as a variable. Assuming that the cloud particles are unit density spheres (ρ_p = 1,000 kg/m^3), Table 8.9 shows the maximum value of ($c_{number}D_c$) at which the cloud at STP settles with a velocity equal to v_t.

To acquire a frame of reference that relates particle number concentration (c_{number}) and mass concentration (c), consider a typical atmospheric cloud, which has a particle mass concentration of 1 g/m^3 (10^{-3} kg/m^3). For a cloud containing a monodisperse aerosol of unit density spheres of diameter D_p = 1 μm, the relationship between number concentration and mass concentration is shown in Table 8.10. Figure 8.21 shows cloud settling velocity (v_c) as a function of particle size (D_p) for different values of ($c_{number}D_c$).

As a result of these calculations it can be seen that small, highly dense aerosol clouds may settle with a velocity considerably larger than one would expect based on the size of the individual particles. On the other hand, large clouds with modest particle number concentrations may also settle faster than one expects based on the size of particles.

Table 8.9 Clouds of unit density spheres with settling velocity equal to particle terminal velocity.

D_p (μm)	v_t (m/s)	maximum $c_{number}D_c$ (particles/m^2)
0.1	8.5×10^{-7}	4.98×10^4
1.0	3.6×10^{-4}	9.94×10^6
10.0	3.6×10^{-3}	6.33×10^5

Table 8.10 Comparison of number concentration and mass concentration for a cloud composed of unit density spheres.

D_p (μm)	c (kg/m^3)	c_{number} (particles/m^3)
1.0	0.1	2×10^{14}
1.0	5.23×10^{-2}	1×10^{14}
1.0	5.23×10^{-4}	1×10^{12}
1.0	5.23×10^{-6}	1×10^{10}

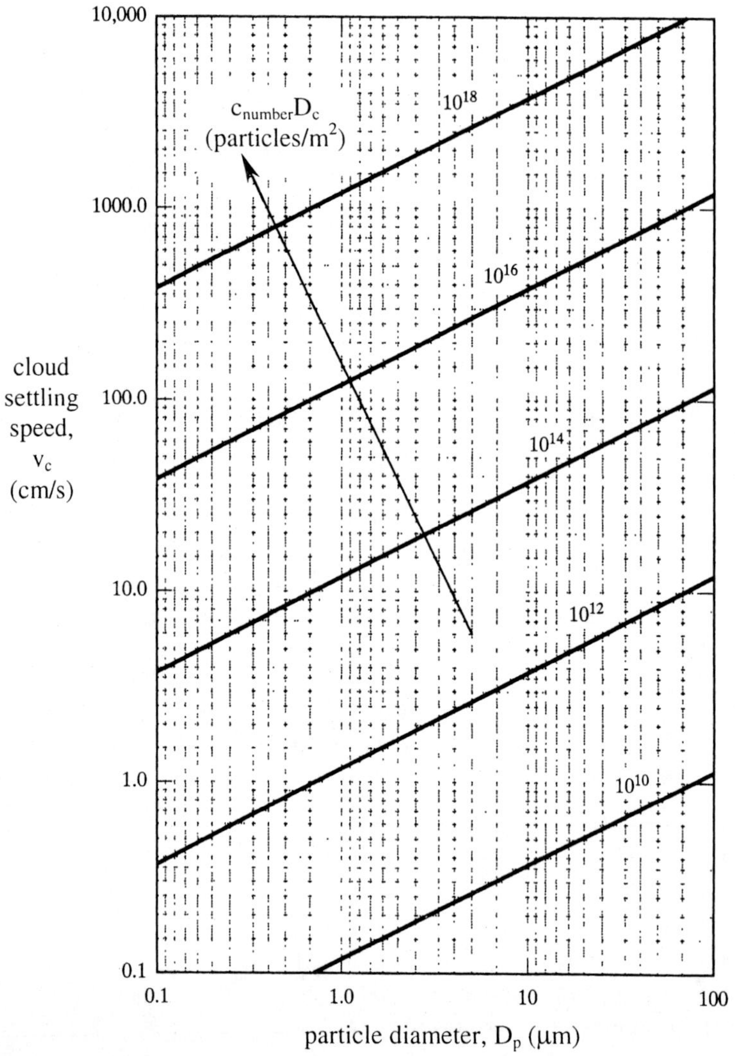

Figure 8.21 Cloud settling velocity as a function of particle size and concentration (adapted from Heinsohn & Kabel, 1999).

The analysis is a useful approximation, but users must be aware of several physical processes affecting cloud behavior that the model ignores. From everyday observations, (watch smokers blow smoke rings) users should recall that clouds behave as an entity early in their lifetime, but with the passage of time, clouds tend to be governed by the motion of individual particles as particles coagulate and diffuse into the air, and the cloud breaks up.

1. The boundaries of the cloud expand and the particle number concentration decreases near the boundaries.
2. The assumption that the air is motionless (relative to the cloud) inside the cloud's envelope ignores that fact that there is circulation of air even within small clouds.
3. As a cloud of particles falls, it loses its spherical shape and the drag coefficient changes. The cloud may even divide into two or more fragments.
4. The air in a room is never motionless; there are air currents whose velocity exceeds v_t and may even exceed v_c.

5. Particles coagulate, thus increasing D_p and decreasing c_{number} and D_c. Coagulation decreases β and may bring about a change of flow regimes.
6. Alternatively, D_p may decrease due to evaporation, thus reducing the settling velocity. Evaporation increases β and may bring about a change of flow regimes.

8.10 Stokes Number

Stokes number is an important dimensionless parameter in particle dynamics just as Reynolds number is important in fluid dynamics. Rather than using dimensional analysis (the Buckingham Pi theorem) to show the uniqueness of the Stokes number, it is easier to transform the equation for particle motion into a dimensionless equation, and allow the Stokes number to emerge through the derivation. This process is called ***inspectional analysis*** as contrasted with dimensional analysis. To begin inspectional analysis, it is necessary to define a characteristic velocity U_0 and a characteristic length L. The characteristic time is equal to L/U_0. The characteristic velocity and length are two terms that an engineer can easily identify, which are relevant to the problem; numerical values can readily be defined. Typical values might be the velocity of an aerosol entering a device and the width or diameter of the inlet. The choice of U_0 and L is generally dictated by the phenomenon being modeled.

Consider a spherical particle traveling with velocity $\vec{v}$ through a carrier gas that has velocity $\vec{U}$ (see Figure 8.10). The equation of motion for the particle is given by Eq. (8-75). For simplicity, ***gravity is neglected in this analysis***; gravity acts independently on the particles, based on the orientation of the flow. The Reynolds number was defined previously in terms of the relative velocity in Eq. (8-58). Substituting this Reynolds number into Eq. (8-75) and simplifying,

$$\frac{d\vec{v}}{dt} = -\frac{3}{4}\frac{c_D \mu \, Re}{C D_p^2 \rho_p}\left(\vec{v} - \vec{U}\right) \tag{8-125}$$

Eq. (8-125) can be reduced to a dimensionless equation by multiplying and dividing all velocities and lengths by the known characteristic velocity U_0 and length L, respectively. In addition, a characteristic time scale is formed from L and U_0, namely (L/U_0), and t is multiplied and divided by this time scale:

$$\frac{d\left(\frac{\vec{v}}{U_0}\right)U_0}{d\left(\frac{tU_0}{L}\right)\frac{L}{U_0}} = -\frac{3}{4}\frac{c_D \mu \, Re}{C D_p^2 \rho_p}\left[\frac{\vec{v}}{U_0} - \frac{\vec{U}}{U_0}\right]U_0$$

which can be re-written as

$$\frac{d\vec{v}^*}{dt^*} = -\left[\frac{18 \mu L}{D_p^2 \rho_p U_0}\right]\frac{c_D \, Re}{24 C}\left(\vec{v}^* - \vec{U}^*\right) \tag{8-126}$$

where the dimensionless particle and carrier gas velocities are defined as

$$\vec{v}^* = \frac{\vec{v}}{U_0} \tag{8-127}$$

and

$$\vec{U}^* = \frac{\vec{U}}{U_0} \tag{8-128}$$

and the dimensionless time is defined as

$$t^* = \frac{tL}{U_0} \tag{8-129}$$

A dimensionless parameter called the ***Stokes number*** (Stk), may be defined from the group of variables in square brackets in Eq. (8-126) as

$$\text{Stk} = \frac{D_p^2 \rho_p U_0}{18\mu L} = \frac{D_p^2 \rho_p}{18\mu}\frac{U_0}{L} = \frac{\tau_p U_0}{L} \tag{8-130}$$

where τ_p is the relaxation time, as defined previously in Eq. (8-84). Stokes number is proportional to the ratio of the penetration distance $\ell = C\tau_p U_0$ from Eq. (8-108) to the characteristic distance (L). Equation (8-126) can now be wrntten in terms of dimensionless parameters as

$$\frac{d\vec{v}^*}{dt^*} = -\left[\frac{c_D \, \text{Re}}{24(\text{Stk})C}\right]\left(\vec{v}^* - \vec{U}^*\right) \tag{8-131}$$

The term in square brackets in Eq. (8-131) does not have a special name, but can be thought of as a scaling parameter. Since the differential equation is dimensionless, solutions to widely different problems are the same providing the initial values of the dimensionless variables and the scaling parameter are the same. For example, provided that the scaling parameters of the particles are the same, the dynamics of dust entering a dust sampler is similar to the dynamics of rain entering a jet engine during take-off. The dynamics of dust generated by trucks traveling on unpaved roads is similar to the movement of welding fume from a workbench. In general, if two different physical phenomena (A and B) are such that

$$\left[\frac{c_D \, \text{Re}}{24(\text{Stk})C}\right]_A = \left[\frac{c_D \, \text{Re}}{24(\text{Stk})C}\right]_B$$

the motion of the particles written in dimensionless terms is equivalent. If the effect of the slip coefficient (C) is neglected, and the flow in both situations A and B is within the Stokes flow regime ($c_D = 24/\text{Re}$), the fluid dynamics is scaled by the Stokes numbers alone (Stk_A, Stk_B).

8.11 Inertial Deposition in Curved Ducts

It is important to understand the deposition of particles in a curved duct (bend) since it occurs in many situations of practical interest. In the field of health, inhaled particles pass through a number of bifurcations (divisions) as the air flow divides again and again in passing through the bronchial tree, as discussed in Chapter 2. This is important since the highest incidence of carcinogenic tumors occurs in the upper generations of the bronchial system. In terms of particle sampling, industrial dust control, and pneumatic transfer of powders, the impaction of particles in bends through inertial deposition has important practical consequences. In the analysis that follows, ***gravitational settling is neglected***; it acts independently, based on the orientation of the flow. The reader should consult Balashazy et al. (1990) for an analysis including gravitational settling.

As an aerosol flows through a curved duct, the inertia of a particle causes it to move radially outward, as in Figure 8.22. The velocity of the carrier gas can be obtained from a solution of the Navier-Stokes equations. For purposes of illustration, it is assumed that the cross-sectional area is rectangular (width W) and that the tangential and radial velocity components are given by

$$U_\theta r^n = \text{constant} = C_n \qquad U_r = 0 \tag{8-132}$$

where n lies between (or is equal to) -1 and 1. When n = 1 the flow is irrotational (vorticity = curl $\vec{U}$ = 0), and when n = -1 the flow is in solid body rotation (vorticity = constant = twice the angular

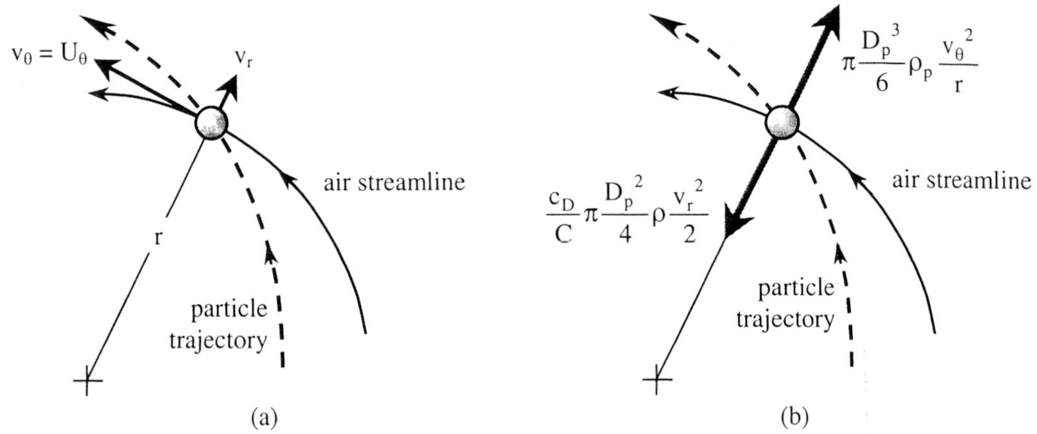

Figure 8.22 Quasi-static equilibrium of a particle of diameter D_p and density ρ_p in curvilinear flow; (a) particle velocity components, and (b) forces acting on the particle.

velocity). When n is between these two extremes, the flow is rotational, but with a smaller value of vorticity than that for the case of solid body rotation. For lack of a better phrase, flow fields that can be described by Eq. (8-132) where $-1 < n < 1$, are called ***n-degree rotational***. Table 8.11 is a summary of these cases. Figure 8.23 shows the velocity profiles for rotational and irrotational flow through a bend in a duct. The volumetric flow rate through a curved duct of constant rectangular cross section and constant radius of curvature is

$$\boxed{Q = \int_{r_1}^{r_2} W U_\theta dr} \qquad (8\text{-}133)$$

where W is the depth perpendicular to the r-θ plane, and r_1 and r_2 are the radii of the inner and outer wall respectively. Upon substitution of Eq. (8-132) into (8-133), the constant can be evaluated and the tangential velocity can be expressed in terms of Q for the above three classes of flow:

Table 8.11 Exponent n and constant C_n for three types of circular flows.

description	**n**	**C_n, Eq. (8-132)**
solid body rotation	$n = -1$	$C_n = U_\theta/r$
n-degree rotational	$-1 < n < 1$	$C_n = U_\theta r^n$
irrotational	$n = 1$	$C_n = rU_\theta$

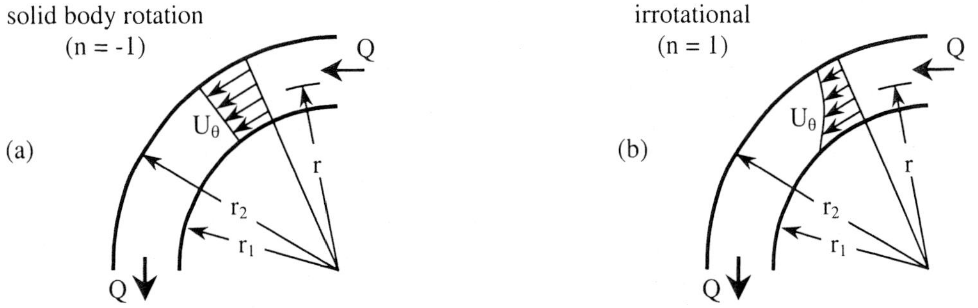

Figure 8.23 Velocity profiles in a curved duct for (a) solid body rotational flow and (b) irrotational flow.

- *Solid body rotation* (n = -1):

$$C_n = \frac{2Q}{W\left(r_2^2 - r_1^2\right)} \qquad U_\theta = \frac{2Qr}{W\left(r_2^2 - r_1^2\right)} \tag{8-134}$$

- *n-degree rotational flow* (-1 < n < 1):

$$C_n = \frac{(1-n)Q}{W\left(r_2^{1-n} - r_1^{1-n}\right)} \qquad U_\theta = \frac{(1-n)Q}{W\left(r_2^{1-n} - r_1^{1-n}\right)r^n} \tag{8-135}$$

- *Irrotational flow* (n = 1):

$$C_n = \frac{Q}{W \ln\left(\frac{r_2}{r_1}\right)} \qquad U_\theta = \frac{Q}{rW \ln\left(\frac{r_2}{r_1}\right)} \tag{8-136}$$

Rather than solving the full set of equations describing the motion of particles in a moving flow field, a useful approximation is available by making the following assumptions:

(a) quasi-static equilibrium
(b) Stokes flow
(c) negligible gravimetric settling

Consistent with the assumption of quasi-static equilibrium is the assumption that the tangential velocity of the particle and the gas are equal. In the tangential direction, there is no relative motion between the particle and the carrier gas. In the radial direction, it is useful to use the imaginary concept of ***centrifugal force***. It is assumed that the centrifugal force of the particle is equal to the viscous drag. Since the gas velocity has no radial component ($U_r = 0$), the particle's relative velocity is its absolute velocity in the radial direction (v_r), as seen in Figure 8.22. Thus, the particle velocity has the following tangential and radial components:

tangential direction (θ):

$$v_\theta = U_\theta = \frac{C_n}{r^n} \tag{8-137}$$

radial direction (r):

$$\rho_p \frac{\pi D_p^3}{6} \frac{U_\theta^2}{r} = \frac{c_D}{C} \frac{\pi D_p^2}{4} \frac{\rho v_r^2}{2}$$

from which

$$v_r = U_\theta \sqrt{\frac{\rho_p}{\rho} \frac{4C}{3c_D} \frac{D_p}{r}} \tag{8-138}$$

It is assumed that the particle's Reynolds number, based on the relative velocity, is small such that Stokes flow can be assumed,

$$c_D = \frac{24}{Re} \qquad Re = \frac{\rho D_p v_r}{\mu} = \frac{D_p v_r}{\nu} \tag{8-139}$$

Eq. (8-138) for the particle's radial velocity component then simplifies to

$$v_r = C \frac{U_\theta^2}{r} \tau_p \tag{8-140}$$

where τ_p is given by Eq. (8-84). Equation (8-140) is the same as Eq. (8-88) except that gravitational acceleration is replaced by centrifugal acceleration. Deposition of particles on the outer wall of the bend can therefore be modeled in a fashion similar to gravimetric deposition in horizontal ducts, through use of either the laminar (no mixing) settling model or the turbulent (well-mixed) settling model. The particles for which inertial separation is important are usually sufficiently large that the Cunningham slip factor (C) is close to unity, but for completeness C is included in the analysis which follows.

8.11.1 Laminar Settling Model

Figure 8.24 depicts inertial deposition for the case of a monodisperse aerosol entering a 90° bend. As was the case for laminar settling in ducts, the particle concentration remains uniform ($c = c_{in}$) over that portion of the duct containing contaminant particles (the shaded region in Figure 8.24). However, outside of that portion, c is identically zero (unshaded region), since particles migrate across air streamlines, impact the wall, and stick to the outer wall of the bend.

It is useful to find the angle θ_{impact} (in *radians*) at which a particle entering the bend at radius r_{in} impacts the outer wall. Such an expression can be found by forming a ratio of the two particle velocity components, keeping in mind that $v_\theta = U_\theta$, and making use of Eq. (8-132) for U_θ and Eq. (8-140) for v_r, The result is

$$\frac{v_\theta}{v_r} = \frac{r\dfrac{d\theta}{dt}}{\dfrac{dr}{dt}} = r\frac{d\theta}{dr} = \frac{U_\theta}{v_r} = \frac{r}{\tau_p C U_\theta} = \frac{r^{1+n}}{\tau_p C C_n}$$

which simplifies to

$$\boxed{d\theta = \frac{r^n}{C_n \tau_p C} dr} \tag{8-141}$$

where C_n is given by Eq. (8-134), (8-135), or (8-136) depending on the value of n, and must be evaluated in terms of other flow parameters. To calculate θ_{impact} for a given value of r_{in} (or vice-versa), Eq. (8-141) must be integrated from $\theta = 0$ to θ_{impact} and from $r = r_{in}$ to r_2, where the particle impacts the outer wall.

$$\int_0^{\theta_{impact}} d\theta = \int_{r_{in}}^{r_2} \frac{r^n dr}{C_n \tau_p C}$$

The above integration can be performed analytically for circular flow as follows:

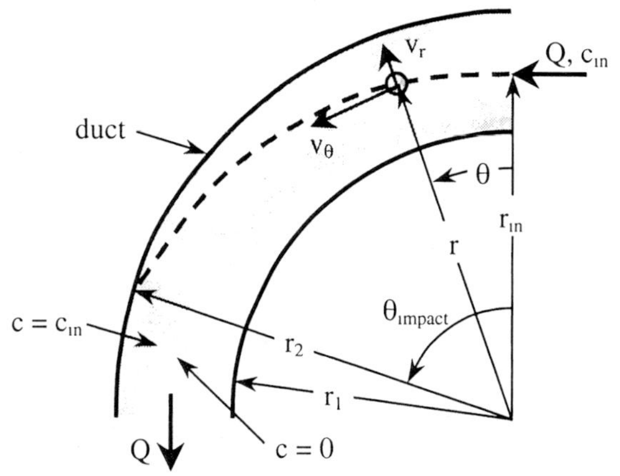

Figure 8.24 Particle trajectory (dashed line) and mass concentration in a curved duct of rectangular cross section for the laminar flow model; particle enters at $r = r_{in}$, $\theta = 0$, and impacts the outer wall at $r = r_2$, $\theta = \theta_{impact}$; particle shown at arbitrary time.

- *Irrotational (n=1) flow and n-degree rotational flow* (n ≠ -1):

$$\theta_{impact} = \frac{r_2^{\,n+1} - r_{in}^{\,n+1}}{(n+1)C_n\tau_p C} \qquad r_{in} = \left[r_2^{\,n+1} - (n+1)C_n\tau_p C\theta_{impact}\right]^{\frac{1}{1+n}} \tag{8-142}$$

- *Rotational flow* (n = -1):

$$\theta_{impact} = -\frac{1}{C_n\tau_p C}\ln\left(\frac{r_{in}}{r_2}\right) \qquad r_{in} = r_2\exp\left(-C_n\tau_p C\theta_{impact}\right) \tag{8-143}$$

The grade efficiency (fractional efficiency), $\eta(D_p)$, for particles of diameter D_p can be calculated as a function of angle θ from the inlet of the duct bend. From Figure 8.24, it is seen that all particles injected at the bend inlet at an initial radius greater than r_{in} impact the outer wall and are collected. All particles with initial radius less than r_{in} remain in the flow. Thus, at $\theta = \theta_{impact}$, $\eta(D_p)$ is simply the fraction of particles with initial radius greater than r_{in}, and

$$\eta(D_p)_{\theta=\theta_{impact}} = \frac{r_2 - r_{in}}{r_2 - r_1} = \frac{1-\frac{r_{in}}{r_2}}{1-\frac{r_1}{r_2}} \tag{8-144}$$

Furthermore, any arbitrary angle θ from the inlet of the duct bend can be thought of as θ_{impact} for some value of initial injection radius r_{in}, and $\eta(D_p)$ can be written for any angle θ by substituting Eq. (8-142) or (8-143) into Eq. (8-144), but dropping the subscript "impact" on θ. For example, for irrotational flow (n = 1), this procedure results in

$$\eta(D_p)_{n=1} = \frac{1-\sqrt{1-\dfrac{2\theta Q\tau_p C}{W r_2^{\,2}\ln\left(\frac{r_2}{r_1}\right)}}}{1-\frac{r_1}{r_2}} \tag{8-145}$$

where Eq. (8-136) has also been applied for C_n. Similarly, for solid body rotation (n = -1), it can be shown that

$$\eta(D_p)_{n=-1} = \frac{1-\exp\left(\dfrac{-2Q\theta\tau_p C}{W\left(r_2^{\,2}-r_1^{\,2}\right)}\right)}{1-\frac{r_1}{r_2}} \tag{8-146}$$

Equations (8-145) and (8-146) can be rearranged in terms of an average Stokes number (Stk_{avg}) defined in terms of an average velocity (U_{avg}), namely

$$Stk_{avg} = \frac{\tau_p U_{avg}}{r_2} = \frac{\tau_p Q}{r_2 W(r_2 - r_1)} \tag{8-147}$$

where

$$U_{avg} = \frac{Q}{W(r_2 - r_1)} \tag{8-148}$$

For irrotational flow (n = 1), the grade efficiency becomes

$$\boxed{\eta\left(D_p\right)_{n=1} = \frac{1-\sqrt{1-2Stk_{avg}C\theta\dfrac{\left(r_2-r_1\right)}{r_2\ln\left(\dfrac{r_2}{r_1}\right)}}}{1-\dfrac{r_1}{r_2}}} \tag{8-149}$$

and for solid body rotation (n = -1),

$$\boxed{\eta\left(D_p\right)_{n=-1} = \frac{1-\exp\left(-2Stk_{avg}C\theta\dfrac{r_2}{r_1+r_2}\right)}{1-\dfrac{r_1}{r_2}}} \tag{8-150}$$

A similar development can be performed for n-rotational flows. The efficiency can be shown to be

$$\boxed{\eta\left(D_p\right)_{n} = \frac{1-\left[1-\dfrac{\left(1-n^2\right)\theta r_2\left(r_2-r_1\right)Stk_{avg}C}{r_2^{1+n}\left(r_2^{1-n}-r_1^{1-n}\right)}\right]^{\frac{1}{1+n}}}{1-\dfrac{r_1}{r_2}}} \tag{8-151}$$

A useful parameter to define is the ***critical angle*** (θ_c) at which *all* particles of diameter D_p entering the bend impact on the outer wall (and are assumed to be collected), i.e. $\eta(D_p) = 1$. As an example, consider irrotational flow (n = 1). The critical angle θ_c (in *radians*) can be calculated either by setting $r_{in} = r_1$ in Eq. (8-142), with C_n from Eq. (8-136), or by setting $\eta(D_p)$ equal to 1 in Eq. (8-149). The critical angle for irrotational flow is thus

$$\boxed{\theta_c\left(D_p\right) = \left(r_2+r_1\right)\frac{\ln\left(\dfrac{r_2}{r_1}\right)}{2r_2Stk_{avg}C}} \tag{8-152}$$

θ_c is a function of particle diameter D_p; the smaller the particle, the bigger θ_c. Theoretically, then, the laminar settling model predicts that as long as the duct bend sweeps an angle greater than or equal to θ_c, the air exiting the bend will be free of all particles of diameter greater than or equal to D_p.

8.11.2 Well-Mixed Model

The well-mixed model assumes that within the flow field there exists a mixing mechanism that redistributes particles remaining in the duct after those along the outer radius impact the outer wall. For this reason, the particle concentration varies with angle θ but not with radius. To analyze the flow, consider an element of the flow as shown in Figure 8.25. A mass balance is constructed for the element:

$$cQ = \left(c+dc\right)Q + cv_r\left(r_2\right)Wr_2d\theta \tag{8-153}$$

which after rearrangement and integration yields

$$\boxed{\int_{c_0}^{c_\theta}\frac{dc}{c} = -\int_0^{\theta}\frac{v_r\left(r_2\right)Wr_2}{Q}d\theta} \tag{8-154}$$

where $v_r(r_2)$ is the particle's radial velocity, Eq. (8-140), evaluated at the outer radius (r_2). The collection efficiency (grade efficiency or fractional efficiency) at angle θ for particles of diameter D_p is defined as

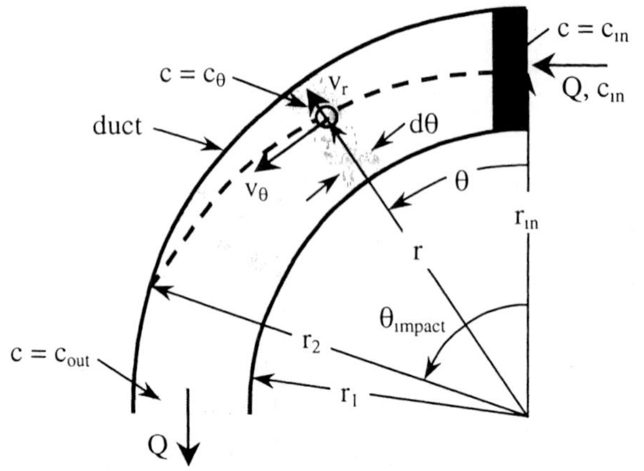

Figure 8.25 Particle trajectory (dashed line) and mass concentration in a curved duct of rectangular cross section for the well-mixed model; particle enters at $r = r_{in}$, $\theta = 0$, and impacts the outer wall at $r = r_2$, $\theta = \theta_{impact}$; particle shown at arbitrary time at location (r,θ). Four control volumes are shown with degree of shading indicating how mass concentration decreases with θ.

$$\boxed{\eta(D_p) = 1 - \frac{c_\theta(D_p)}{c_{in}(D_p)}} \tag{8-155}$$

The right hand side of Eq. (8-155) can be found by integrating Eq. (8-154), replacing $v_r(r_2)$ by Eq. (8-140), simplifying, and introducing the Stokes number based on average velocity from Eq. (8-147). The grade efficiency can then be written as

$$\boxed{\eta(D_p) = 1 - \exp\left[-(\text{Stk}_{avg})K\theta C\right]} \tag{8-156}$$

where K is a constant which depends on the value of n, i.e. on the category of flow:

$$\boxed{K(\text{irrotational}) = \frac{r_2 - r_1}{r_2\left[\ln\left(\frac{r_2}{r_1}\right)\right]^2}} \tag{8-157}$$

$$\boxed{K(\text{solid body rotation}) = \frac{4r_2^3}{(r_2 + r_1)(r_2^2 - r_1^2)}} \tag{8-158}$$

$$\boxed{K(\text{n-degree rotational}) = \frac{(1-n)^2 (r_2 - r_1) r_2^{(1-2n)}}{\left(r_2^{1-n} - r_1^{1-n}\right)^2}} \tag{8-159}$$

The assumption of quasi-static equilibrium is valid for a large class of flows when one recalls the analysis of Section 8.4. The time required for small particles to achieve equilibrium is very small and only several times longer than $\tau_p C$. Thus it is reasonable to assume viscosity acts rapidly on particles flowing in a curved duct. The assumption of Stokes flow is also not restrictive since it is only the particle's radial velocity component that produces the relative velocity. The radial velocity is small since it depends on the relaxation time (τ_p). The restrictive assumption in the above analysis is that the gas velocity is given by Eq. (8-132). The velocities entering the bend depend on conditions upstream of the inlet. If air enters the bend from quiescent room air, the inlet velocity profile will be nearly flat.

If a long rectangular duct precedes the bend, the velocity profile entering the bend is fully established, or well on its way to becoming so. In either case, the velocity profile in the bend is not given by Eq. (8-132). The velocity profile is also dictated by the pressure distribution in the bend. The radial and tangential pressure gradients for irrotational and rotational flow are quite different. A thorough analysis of particle motion depends on solving the equations of motion of the air in the bend.

Example 8.9 - Particle Classifier

Given: A company processes agricultural materials, grains, corn, rice, etc. One of the processes is a milling operation. Significant fugitive dust is produced. Enclosures and exhaust air (Q, in CFM) are needed to capture the dust. A classifier is needed to separate no less than 50% of the particles larger than 100 μm ($D_p > 100$ μm) which are returned for reprocessing. The smaller particles are removed by filters (a baghouse – see Chapter 9). Your supervisor suggests constructing a simple device consisting of a 180-degree elbow of rectangular cross section containing louvers on the outside surface, as in Figure E8.9a. The volumetric flow rate of air in the elbow is Q. Centrifugal force sends large particles in the radial direction; the particles pass through the louvers and are drawn off by a slip stream and removed by other means.

To do: Compute the grade efficiency curves (similar to Figure 8.7) that will enable operators to select the proper volumetric flow rate Q to achieve a certain removal efficiency (η). Assume that the gas flow is irrotational and well mixed. Plot the results for a classifier whose dimensions are:

$r_1 = 0.30$ m $\qquad r_2 = 0.70$ m $\qquad W = 0.40$ m $\qquad s = 0.070$ cm

and which separates unit density particles ($\rho_p = 1{,}000$ kg/m^3) traveling in an air stream at 300 K.

Solution: From Eq. (8-157),

$$K = \frac{r_2 - r_1}{r_2\left[\ln\left(\frac{r_2}{r_1}\right)\right]^2} = \frac{0.70\text{ m} - 0.30\text{ m}}{(0.70\text{ m})\left[\ln\left(\frac{0.70\text{ m}}{0.30\text{ m}}\right)\right]^2} = 0.796$$

From Eq. (8-147),

$$\text{Stk}_{\text{avg}} = \frac{\tau_p Q}{r_2 W (r_2 - r_1)} = \frac{\tau_p Q}{(0.70\text{ m})(0.40\text{ m})(0.70\text{ m} - 0.30\text{ m})} = \frac{\tau_p Q}{0.112\text{ m}^3}$$

and from Eq. (8-156), for $\theta = \pi$ (180-degrees), the fractional efficiency is

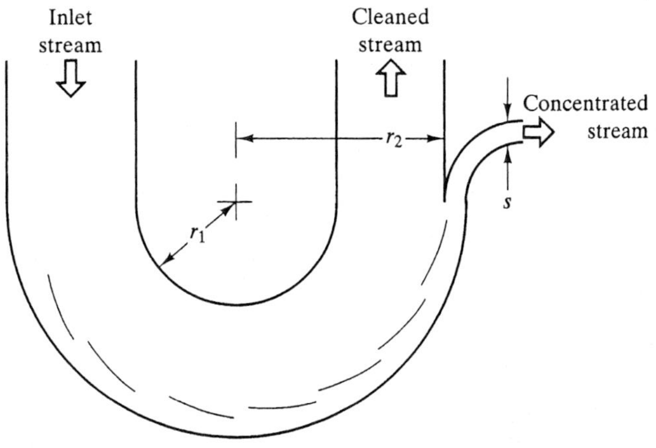

Figure E8.9a Centrifugal particle classifier (from Heinsohn & Kabel, 1999).

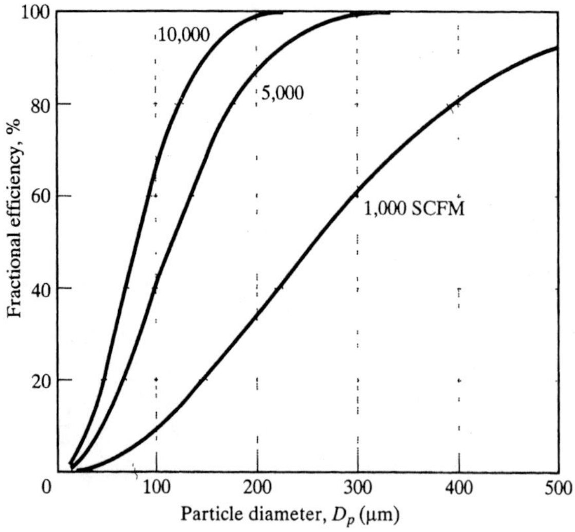

Figure E8.9b Fractional efficiency of a centrifugal particle classifier for three volumetric flow rates; classifier dimensions: $r_1 = 0.3$ m, $r_2 = 0.7$ m, W = 0.4 m, s = 0.07 cm (from Heinsohn & Kabel, 1999).

$$\eta(D_p) = 1 - \exp\left[-\left(\mathrm{Stk}_{avg}\right)K\theta C\right] = 1 - \exp\left(-\frac{\tau_p Q}{0.112\ \mathrm{m}^3}0.796\pi(1.00)\right) = 1 - \exp\left(-22.3\tau_p Q\right)$$

where the Cunningham slip factor (C) is assumed to equal unity for such large particles, as was shown in Table 8.4. Figure E8.9b shows the fractional efficiency at three volumetric flow rates.

Discussion: Clearly the lowest volumetric flow rate classifies particles poorly. At the three higher flow rates, the device removes 200 and 300 μm particles efficiently. Use of the well-mixed model is a reasonable selection since Reynolds numbers in the elbow are surely large enough to establish turbulent flow. The assumption of irrotational flow, however, needs to be justified. The next level of sophistication is to use ***computational fluid dynamics*** (CFD) computer programs to predict the trajectories of particles in the three-dimensional velocity field of the 180-degree elbow. CFD is discussed in Chapter 10.

8.12 - Closure

If the number concentration of a contaminant gas is low (valid for most problems in indoor air pollution), the motion of one particle is not affected by the motion or wake of another particle. This fortunate state of affairs accommodates the wonderful simplification of modeling the motion of each particle separately. The motion of a particle is governed by Newton's second law in which the aerodynamic drag force can be expressed as a function of a conventional drag coefficient (modified by the Cunningham slip factor for very small particles) and the square of the relative velocity. Computers facilitate calculations which are repeated for each particle (or small range of particles) in an aerosol. Further simplification occurs for small particles in the Stokes flow regime in which the drag coefficient can be expressed in simple terms. The number, area, and mass distributions of aerosols in indoor air pollution can often be described as log normal and the log-normal size distribution function can be used in conjunction with the equations of particle motion to describe the motion of an entire aerosol. Gravitational and inertial forces exerted on particles can be included in the equations of motion, which allows engineers to model the performance of mechanical particle collectors.

Chapter 8 - Problems

1. A puff of smoke from a wood-burning stove has an initial diameter of $D_c(0) = 5.0$ cm, and contains a unit density aerosol with diameter $D_p = 0.10$ μm. Initially the particle number concentration, $c_{number}(0)$, is 10^{15} particles/m^3. It is observed that the diameter of the smoke puff increases steadily at the rate of 1 cm/min. Assuming that no smoke particles escape, plot the cloud velocity versus time.

2. High-speed stamping machines used to manufacture small electrical devices require a fluid to be sprayed (as a mist) on the dies for lubrication and cooling. Unfortunately an aerosol is formed around the machine that irritates the surface of the nasal cavity of the operator. From time to time the aerosol is analyzed and found to be log normal. The results of a recent analysis are shown below. The total particle concentration is 10,000 particles per cm^3 and the density of the fluid is 1.2 gm/cm^3. Estimate the aerosol mass concentration (in units of mg/m^3) and compute the median particle size based on number and mass.

range of D_p (μm)	fraction of particles in range
0 - 2	0
2 - 8	0.0359
8 - 14	0.1999
14 - 20	0.2642

range of D_p (μm)	fraction of particles in range
20 - 30	0.2940
30 - 45	0.1556
45 - 60	0.0365
60 - 80	0.0139

3. An aerosol produced during a bagging operation is log normal with a mass median diameter of 10 μm and a geometric standard deviation of 2.5. What is the median particle size based on number? If the specific gravity of the dust is 2.5 and the number concentration is 100,000 particles/m^3, what is the mass concentration (in units of mg/m^3)?

4. A thin layer of finely ground corn starch is placed on the surface of newly printed pages leaving a four-roll, color offset printing press. The purpose is to keep the pages from sticking together. Unfortunately some of the corn starch becomes airborne and settles on the rolls, ink reservoir, fountain (wetting) solution, etc. Ultimately it is transferred to the rolls and produces what printers humorously call "hickeys" on the printed page. A sample of the air above the press is passed through a filter, and analyzed under the microscope. The number of particles (by number) within size intervals is as follows:

range of D_p (μm)	number of particles in range
2 - 3	10
3 - 4	0
4 - 5	0
5 - 6	30
6 - 7	100
7 - 8	210
8 - 9	450
9 - 10	910

range of D_p (μm)	number of particles in range
10 - 11	850
11 - 12	800
12 - 13	440
13 - 14	230
14 - 15	90
15 - 16	30
16 - 17	20
17 - 20	0

Is the size distribution log normal? What is the median size with respect to number and with respect to mass?

5. Paint particles from the exhaust of a spray booth in an auto body repair shop are sampled. The mass of particles per unit volume of air is found to be 2 g/m^3 at STP. The particle size distribution is found to be:

range of D_p (μm)	number of particles in range
2 - 3	1
3 - 4	1
4 - 5	1
5 - 6	3
6 - 7	10
7 - 8	21
8 - 9	45
9 - 10	91

range of D_p (μm)	number of particles in range
10 - 11	85
11 - 12	80
12 - 13	44
13 - 14	23
14 - 15	9
15 - 16	3
16 - 17	2
17 - 20	0

(a) Draw a histogram of the size distribution.
(b) Compute the arithmetic mean particle diameter and standard deviation.
(c) Compute the geometric standard deviation and geometric mean particle diameter.
(d) Plot the actual cumulative size distribution (based on diameter) on log probability paper. Is the distribution nearly log normal?
(e) For comparison, using the data in part (c), and on the same plot as that of part (d), plot the cumulative size distribution *assuming* a log-normal size distribution.
(f) Compute the mass median diameter.
(g) If the particle density is 1,800 kg/m^3, how many particles are there per cubic meter of air?

6. Particles are generated by brick curing in a baking oven. The exhaust from the oven is discharged to the atmosphere. A sample of the exhaust gas is taken and the particle number concentration is found to be 20,000 particles/cm^3. A light scattering instrument records the following size distribution:

range of D_p (μm)	number of particles in range
0.30 - 0.40	6
0.40 - 0.53	15
0.53 - 0.71	41
0.71 - 0.94	88
0.94 - 1.28	150
1.28 - 1.68	180
1.68 - 2.33	205

range of D_p (μm)	number of particles in range
2.23 - 3.00	156
3.00 - 4.00	91
4.00 - 5.30	44
5.30 - 7.10	18
7.10 - 9.40	5
9.40 - 12.80	1
12.80 – 15.00	0

(a) Test to see if the size distribution is log normal by plotting the data on log probability paper.
(b) Find the arithmetic mean diameter based on number, the geometric mean diameter based on number, the median particle diameter based on number, and the median particle diameter based on mass.
(c) What is the mass concentration of fine particles, for which $D_p < 2.5$ μm?
(d) Find the particle mass concentration (in units of mg/m^3) in the process gas stream. Is the stream in compliance with the new source performance standards, which are 0.050 g/(dry ft^3) at STP?
(e) A particle removal system with the following fractional efficiency (based on mass) is suggested. What is the overall particle removal efficiency and what is the exit mass concentration?

$D_p(\mu m)$	$\eta(D_p)$	$D_p(\mu m)$	$\eta(D_p)$
0	0	6.0	0.68
0.5	0.10	7.0	0.76
1.0	0.19	8.0	0.82
2.0	0.32	9.0	0.87
3.0	0.42	10.0	0.92
4.0	0.52	12.0	0.97
5.0	0.60	14.0	0.99

7. The size distribution shown in Table E8.2 was obtained by drawing workplace air (at 5.0 °C, 101. kPa) through a filter for 100 seconds at a volumetric flow rate of 0.15 liter/min. and then counting and sizing the particles on special filter paper. A total of 240 particles were collected and counted. If the particle density is 1,800 kg/m^3, compute the particle mass concentration (in units of mg/m^3) in the sampled air. The company management wishes to install a cyclone air cleaner to bring the workplace in compliance with the inert dust OSHA PEL of 15 mg/m^3. The grade efficiency of the collection device is shown below. What is the particle mass concentration (in units of mg/m^3) leaving the device? Will it be in compliance with the OSHA PEL?

$D_p(\mu m)$	$\eta(D_p)$ (%)	$D_p(\mu m)$	$\eta(D_p)$ (%)
0	0	50	86
10	60	60	90
20	72	70	93
30	78	80	98
40	82	100	99.9

8. An aerosol has the size distribution (based on number) shown in Table 8.2. If the mass concentration of the entire aerosol is 100 mg/m^3, what is the mass concentration between 5 μm < D_p < 20 μm?

9. The aerosol described in Example 8.2 flows through a horizontal duct that is 23.3 m long and has a 1.0 m by 1.0 m square cross-sectional area. The particle concentration entering the duct is 100 mg/m^3. After 2,000 hours, what is the mass of particles that have settled on the floor of the duct?

10. A baghouse contains 72 identical bags (N_b = 72). The total volumetric flow rate is 7,200 SCFM. The volumetric flow rate into each bag is the same. The particle collection efficiency of each bag is η_b = 99%. Over time the cell plate corrodes and allows 10% of the inlet flow (720 SCFM and particles) to bypass the bags and mix with the air leaving the bags.

(a) What is the new overall efficiency of the baghouse?

(b) If k bags break allowing all the dusty air to pass to the clean side of the baghouse, show that the rate of change of the overall collection efficiency (η) with respect to the number of broken bags (k) is given by $\frac{d\eta}{dk} = -\frac{\eta_b}{N_b}$.

11. A duct carries 25. °C, 101. kPa air with suspended water drops of 100 μm diameter. The volumetric flow rate is 1.0 m^3/s. The duct makes a 90-degree turn upward, and some of the drops impact the outer elbow wall and are removed from the air stream. The dimensions of the duct are r_1 = 0.50 m, r_2 = 1.0 m and width W = 1.0 m. The viscosity of air (at 25. °C) is $\mu = 1.8 \times 10^{-5}$ kg/(m s). Assuming well-mixed irrotational flow, what percentage of the water drops will be removed?

12. A widely used experimental technique to study the velocity, temperature, and concentration in a moving gas stream uses a laser to excite small particles injected into the stream. By choosing unique particles and lasers, velocities can be studied (laser Doppler velocimetry) or temperatures and concentrations can be determined by analyzing the emission spectra. It has been suggested that large particles can be used but others believe that gravimetric settling will result in bogus information. You have been asked to study particle motion in a horizontal round duct for the following conditions:

- duct diameter (D) = 0.10 m
- volumetric flow rate (Q) = 0.070 m^3/min
- initial particle velocity is in the axial direction, with $v_x(0)$ = 10. times U_{avg}
- T = 25. °C, P = 1.0 atm
- particle density (ρ_p) = 1,200 kg/m^3

(a) If the air velocity is uniform and constant and $U_r = 0$, $U_\phi = 0$, plot a graph of L/D versus D_p for D_p = 10, 100, and 1000 μm, where L is the horizontal distance a particle travels before it encounters the duct wall. Assume the particles are injected initially at the axis of the duct.

(b) Repeat Part (a) if the flow is fully established, i.e. the axial air velocity is $U_x(r)/U_{avg} = 2\,[1-(r/R)^2]$, $U_{avg} = Q/A$

(c) A polydisperse aerosol is injected along the axis at a mass flow rate of 10. mg/min, $v_x(0)$ = 10. times U_{avg}. The mass median diameter is 15. μm and the geometric standard deviation is 1.2. If the air flow velocity is fully established as in Part (b) above, plot the mass fraction of the deposited particles as a function of L/D.

13. A remote sensor monitors an aerosol in a foundry using electric arc furnaces and determines that

- the particle size distribution is log normal
- the geometric mean diameter based on number is 20 μm
- the geometric standard deviation is 1.8
- the overall number concentration is 1,500 particles per cm^3

On the basis of mass (ρ_p = 8,000 kg/m^3) what percentage of the particles are greater than 40 μm in diameter?

14. The mass concentration of particles in a quarry is 25. mg/m^3. The density of the particles is 1,200 kg/m^3. The size distribution is log normal. The geometric standard deviation is 2.0 and the mass median diameter is 10. μm. What is the concentration of particles (in units of mg/m^3) between 5 and 25 μm?

15. Contaminant collectors 1 and 2 are arranged in series, and remove smoke and fume from a stream of air (at volumetric flow rate Q_t) taken from the workplace. The efficiencies of both collectors are the same ($\eta_1 = \eta_2 = 60\%$). What is the overall efficiency of the pair of units? The company wishes to increase the overall efficiency to 95% and plans to buy an additional collector (A) with an efficiency ($\eta_A \geq 95\%$). Which configuration below requires the smallest amount of air to be diverted through the new unit (Q_A), and what is the minimum value of Q_A/Q_t?

(a) Units A and 1 in parallel followed by unit 2.
(b) Unit 1 followed by units 2 and A arranged in parallel.
(c) Unit A in parallel with units 1 and 2 arranged in series.

16. Two pollution control devices were installed in series many years ago. The collection efficiency of each device is 75.% and is independent of volumetric flow rate. Company management wishes to increase the overall collection efficiency to 95.% by purchasing a new device with a collection efficiency of 95.%. The new device is to be installed in parallel with the two existing devices in series. What fraction of the flow should be diverted through the new device? Repeat for the case in which the new device is placed in parallel with the first existing device, with the combination in series with the second existing device.

17. Compute the terminal velocity of the following "particles" in air at STP:

(a) baseball (D_p = 9.0 cm, ρ_p = 950. kg/m^3)
(b) ping-pong ball (D_p = 3.0 cm, overall density = 2.0 kg/m^3)
(c) agricultural dust (D_p = 5.0 μm, ρ_p = 840. kg/m^3)

18. Consider particles of various sizes and densities settling in quiescent air.

(a) Calculate how long it will take the following particles to achieve 34% of their steady-state terminal velocity if they have an initial downward velocity of 100. m/s in quiescent air at -10. °C and 0.90 atm:

- D_p = 1.0 μm, ρ_p = 980 kg/m^3
- D_p = 10. μm, ρ_p = 980 kg/m^3
- D_p = 100 μm, ρ_p = 980 kg/m^3
- D_p = 25. μm, ρ_p = 3,000 kg/m^3

(b) If the particles in Part (a) have an initial velocity of 100. m/s in the horizontal direction in quiescent air at -10. °C and 0.90 atm, what is the displacement and elapsed time before their horizontal components of velocity have decreased to 0.0010 m/s?

19. Compute the penetration distance of a unit density sphere, D_p = 500. μm, injected horizontally into still air at STP with a velocity of 100. m/s.

20. A 1.0 mm sphere (density 8,000 kg/m^3) is injected into still air (1.0 atm, 25. °C) with an initial velocity of 50. m/s inclined at 60. degrees to the horizontal. How far will it travel in the horizontal direction? What maximum height will it achieve?

21. A spherical particle (D_p = 10. μm, 1,800 kg/m^3) is injected into still air at STP in the horizontal direction with a velocity of 5.0 m/s. After 10 seconds, find the x and y displacement from the injection point.

22. A 6.0 mm water drop is given an initial downward velocity of 30. m/s into an air stream that is traveling uniformly to the right at 5.0 m/s. What is the steady-state vertical component of the particle velocity? How long does take for the drop to decelerate to a downward velocity of 15. m/s?

23. An aerosol consists of glass spheres (ρ_p = 2,724 kg/m^3) falling in an upward moving stream of hot air. The air temperature and pressure are 1,000 K and 0.80 atm. For particles of the following diameters: 1.0, 10.0 and 100.0 μm,

(a) What is the gas viscosity (kg/m s)?
(b) What is the Cunningham correction factor (C)?
(c) What are the aerodynamic diameters of the particles?
(d) What gas velocity will levitate the particles?

24. A water drop 20. μm in diameter is traveling in an irrotational flow field where the irrotational constant C ($C = rU_\theta$) is equal to 10. m^2/s. If the tangential velocity of the particle is equal to the tangential air velocity, what is the radial velocity of the particle at a radius of 1.0 m?

25. Consider the gravimetric settling of particles on the floor of a long horizontal duct of square cross section through which an aerosol flows at a uniform velocity U_0. Assume that settling is governed by the well-mixed model. Define R as the local rate of settling at a distance x from the inlet divided by the total inlet mass flow rate/duct cross-sectional area. Show that a graph of R versus (v_t/U_0) for constant values of (x/H) have maximum values and that the maximum values of R occur when $(v_t/U_0) = H/X$.

26. An aerosol of mass concentration (c_{in}) enters a horizontal duct of width (W) and height (H). The volumetric flow rate is Q. What is the total rate of deposition (D, in units of kg/s) over the entire length (L) of the duct?

27. An aerosol of uniform mass concentration (c_{in}) enters a horizontal duct of width (W) and height (H). The volumetric flow rate is Q. Show that the rate of change of deposition (D, in units of kg/s) with respect to the length of duct (L) is given by the following equation:

$$\frac{dD}{dL} = v_t c_{in} W \exp\left(-\frac{LWv_t}{Q}\right)$$

28. Air from an enclosure containing a foundry shake-out process is to be ducted to a baghouse to remove particles of molding sand. The dust mass concentration (c_0) entering the 0.10 m by 0.10 m duct is 10. g/m^3. The horizontal duct is 50. m long.

(a) Foundry dust settles to the bottom of the duct and collects in a layer in which the bulk density (ρ_b) is 5,000 kg/m^3. Assuming that the volumetric flow rate (Q) is constant and that settling proceeds according to the well-mixed model, derive an expression for the thickness (h) of the deposited dust layer as functions of time (t), distance downstream of the inlet (x), volumetric flow rate (Q), dust terminal velocity (v_t), and duct width (W).

(b) Assuming that the fan maintains a constant volumetric flow rate ($Q = 0.10\ m^3/s$), how long will it take to block one half of the cross-sectional area? Where will this occur?

(c) Assuming that the fan maintains a constant volumetric flow rate ($Q = 0.10\ m^3/s$), draw a graph of the friction head loss (h_f) versus time, where

$$h_f = f\frac{L}{W}\frac{U^2}{2g} = f\frac{L}{W}\frac{Q^2}{2gA^2}$$

The term (f) is the friction factor, which for a first approximation can be given by its laminar value, 64/Re, where Re is the Reynolds number

$$Re = \frac{\rho WU}{\mu} = \frac{4\rho Q}{\pi \mu W}$$

(d) Repeat Part (c) above but assume that the friction factor has to be taken from the Moody chart which can be approximated by the empirical equation

$$\frac{1}{\sqrt{f}} = -2.0\log_{10}\left[\frac{\left(\frac{\varepsilon}{W}\right)}{3.7} + \frac{2.51}{Re\sqrt{f}}\right]$$

where ε is the duct roughness height (0.20 mm).

(e) Repeat Part (d) above but take into account that as dust fills the duct, the volumetric flow rate decreases in accordance with the fan operating curve, $h_f(m) = h_p(m)$, i.e.

$$\int_0^L \frac{f}{W}\frac{U^2}{2g}dx = 75 - 52Q^2$$

where Q has units of m^3/s. Draw a graph of volumetric flow rate versus time.

29. A settling chamber is to be constructed in the form of N horizontal circular trays. The inner radius is R_1 and the outer radius is R_2. Air enters the center of the apparatus, divides equally between the N trays and flows radially outward. The distance separating the trays is (h), the overall height of the unit is H.

(a) Assuming the well-mixed model, show that the fractional efficiency can be expressed as

$$\eta(D_p) = 1 - \exp\left[-\frac{g\rho_p D_p^2}{18Q\mu}\frac{H}{h}\left(R_2^2 - R_1^2\right)\right]$$

(b) Assuming that the flow is reversed and travels radially inward, compute the fractional gravimetric collection efficiency.

30. A horizontal duct of length (L = 10. m), width (W = 1.0 m), and height (H = 0.50 m) is used as a gravimetric settling device. Three volumetric flow rates are considered: Q = 1.0, 1.25, and 1.5 m^3/s.

(a) Using the well-mixed model, draw curves of fractional efficiency $\eta(D_p)$ versus particle diameter D_p for unit density spheres and for the three volumetric flow rates. Note that the fractional efficiency is defined as

$$\eta(D_p) = 1 - \frac{c(D_p)_{out}}{c(D_p)_{in}}$$

(b) An aerosol of unit density spheres enters the duct at an overall concentration of 100 mg/m^3 and volumetric flow rate of 1.25 m^3/s. The entering aerosol has a log-normal size distribution in which the mass median diameter is 30. μm, and the geometric standard deviation is 1.3. What is the overall mass concentration (in units of mg/m^3) leaving the duct and what is the mass median particle size (in μm)?

31. Air carrying dust particles flows between two horizontal parallel plates separated by distance H. The velocity profile is given by

$$\frac{U}{U_0} = 1 - \left(\frac{2z - H}{H}\right)^2$$

where z is measured from the bottom plate and U_0 is the velocity along the mid-plane. At x = 0 the particle mass concentration at all points between the plates is equal to c_0 and the particle velocity in the flow-wise direction is equal to the local gas velocity. Assume that gravimetric settling can be described as laminar and that at any downstream location the particle velocity in the flow-wise direction is equal to the local air velocity. Write a general expression for the collection efficiency and show that all the particles will be removed within a critical downstream distance x_c given by

$$x = H\frac{U_0}{v_t}$$

32. The flow through a 90-degree elbow is irrotational and well mixed. Show that at the exit the mass concentration (c) of particles of a particular diameter D_p varies with particle size as follows:

$$\frac{dc}{dD_p} = -c_0\beta D_p \exp\left(-\beta\frac{D_p^{\ 2}}{2}\right) \quad \text{where} \quad \beta = \frac{\pi Q\rho_p}{18\mu W\left[R_2 \ln\left(\frac{R_2}{R_1}\right)\right]^2}$$

33. [*Design Problem*] A duct (0.30 m high by 0.60 m wide) is placed on the roof of a building for make-up air to a newly renovated optics laboratory. Unfortunately the installers ignored an unpaved access road directly beneath the side of the building where the duct inlet is located. Vehicles using the access road produce fugitive dust that enters the make-up air duct. The dust drawn into the duct has a concentration of 10 mg/m^3, a density 1,500 kg/m^3, and a log-normal size distribution (mass median diameter of 20. μm and geometric standard deviation of 2.0). The duct is 50. meters long and has a 90-degree elbow at the inlet (r_1 = 1 meters). What is the concentration and size distribution of particles entering the laboratory if the average velocity in the duct is 10 m/s?

34. [*Design Problem*] Compute and plot the trajectories of coal dust particles traveling over an infinitely long mound of coal having a semicircular cross-sectional area of radius (a). Assume that the velocity field around the mound can be described by the stream function

$$\psi = U_0 r\left[1 - \left(\frac{a}{r}\right)^2\right]\sin\theta$$

where the radius (r) is measured from the center of the mound, and the positive x-axis corresponds to a value of θ equal to zero. Air with a uniform velocity (U_o) approaches the mound in the negative x-direction. Begin the analysis with particles at points along a vertical line x = 1.5a, y = h, where the particle velocity $v_x = -1.5U_0$. Some coal dust particles will impact the mound and others will pass over it to settle to the ground downwind of the mound. Analyze the particle trajectories and determine where each particle impacts either the mound or the ground. Carry out the computations for particles originating at values of h from 0 to 1.5a.

(a) Plot the trajectory of a particle whose diameter is 50. μm.
(b) Plot a graph of (h/a) versus the normalized displacement (x/a) at which the particle impacts a solid surface. Repeat the calculation for several particle sizes. Assume the following physical conditions: ρ_p = 1,000 kg/m^3, U_0 = 1.0 m/s, D_p = 20., 30., 50. and 100. μm, a = 10. m.
(c) Repeat Part (b) but omit the mound. Discuss the significance of the mound in enhancing the collection of particles.
(d) A polydisperse particle stream having a log-normal size distribution (mass median diameter 30. μm, geometric standard deviation = 1.2) enters the air stream at x = 1.5a, y = a with an initial velocity v_p = -1.5 m/s in the negative x-direction. Plot the mass fraction versus y at the point x = -1.5a.

35. For each case below, calculate the terminal settling velocity of the particle, and indicate whether the flow is in the Stokes flow regime, transitional flow regime, or newtonian flow regime. The carrier fluid is quiescent air in all cases. Hint: The Mathcad program on the book's web site may be used for convenience.

(a) D_p = 10 μm, ρ_p = 1000 kg/m^3, T = 20 °C, P = 101,300 Pa
(b) D_p = 400 μm, ρ_p = 1000 kg/m^3, T = 20 °C, P = 101,300 Pa
(c) D_p = 1.0 μm, ρ_p = 6200 kg/m^3, T = 0 °C, P = 50,000 Pa
(d) D_p = 3000 μm, ρ_p = 6200 kg/m^3, T = 0 °C, P = 50,000 Pa

36. A spherical particle of density 6900 kg/m^3 and diameter 120. μm is thrown from a grinding wheel at a speed of 35. m/s at an angle 25° above the horizon. The air is quiescent and at STP conditions. The initial particle position is x = 0, and y = 0.34 m above the ground (ground is at y = 0 m).

(a) How far will the particle travel horizontally before it hits the ground? Hint: The Mathcad program on the book's web site may be used for convenience.
(b) What is the maximum height of the particle's trajectory?
(c) Attach a plot of the particle's trajectory.

37. A polydisperse aerosol enters a particle collector which has grade efficiency $\eta(D_{p,J})$ defined for each class j of a grouped data sample. The inlet mass fraction $g(D_{p,J})_{in}$ and fractional mass concentration $c(D_{p,J})_{in}$ for each class of the grouped data set are also known, as illustrated in Figure P8.37 for one class, j. The volumetric flow rate through the device is Q, and the overall collection efficiency of the particle collector $\eta_{overall}$ can be calculated. Prove that

$$g\left(D_{p,J}\right)_{out} = g\left(D_{p,J}\right)_{in} \frac{1-\eta\left(D_{p,J}\right)}{1-\eta_{overall}}$$

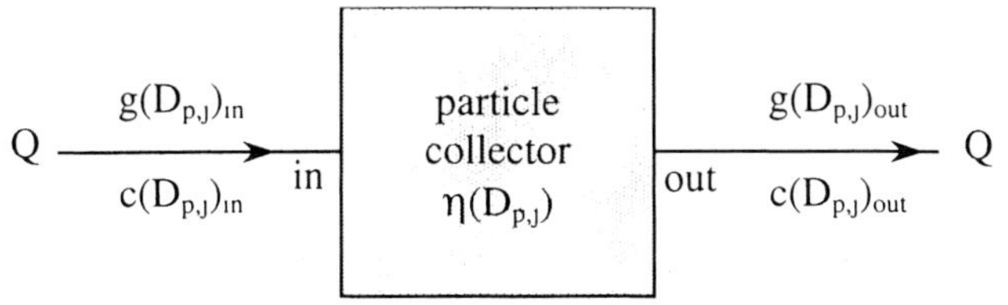

Figure P8.37 Schematic diagram of the particle collector of Problem 8.37.

Removing Particles from a Gas Stream

In this chapter you will learn:

- to describe the physical phenomena underlying particle removal systems
- to model the performance of particle removal systems

The primary concern of this book is indoor air quality, which is determined by the design of general or (preferably) local ventilation systems. Of additional concern is what happens to contaminant particles removed from the indoor environment; otherwise, indoor air quality is improved at the expense of *outdoor* air quality. Of particular concern in the present chapter is how contaminant particles removed from indoor air are subsequently removed from the gas stream prior to discharge to the atmosphere. It is therefore important for indoor air quality engineers to be familiar with various types of particle removal devices.

Many devices that remove particles from a gas stream rely on the fact that solid or liquid particles have densities approximately one thousand times larger than that of the carrier gas. Thus particles experience gravitational and inertial forces larger than the forces experienced by the carrier gas, as discussed in Chapter 8. Such devices are often called ***inertial separation collectors***. Secondly, electric charge can be placed on solid and liquid particles such that an external electric field can exert forces on particles but not on the carrier gas. Particle removal systems capitalize on these forces to separate particles from the gas stream in which they are immersed.

9.1 Cyclone Collectors

The equations derived in Chapter 8 can be used to estimate the particle collection efficiency of a large class of particle collection devices called ***cyclone collectors***, or simply ***cyclones***. Cyclones have no moving parts, can remove both solid and liquid particles, and may have a modest pressure drop. Traditionally, cyclones are used to remove coarse particles; as a rule cyclones do not remove very small particles although they can be designed to do so.

9.1.1 Straight-Through Cyclone

Shown in Figure 9.1 is a schematic diagram of a ***straight-through cyclone collector*** of working length L and radius r_2. The helical turning vanes (at angle α) impart rotation to the air such that

$$\boxed{\tan \alpha = \frac{U_\theta}{U_x}} \qquad (9\text{-}1)$$

Flow through the cyclone can be conceived as a wrapped-around bend of total angular displacement, θ_t (in radians), where

$$\boxed{\theta_t = \frac{U_\theta(r_2)}{r_2}\frac{L}{U_x} = \frac{L}{r_2}\tan \alpha} \qquad (9\text{-}2)$$

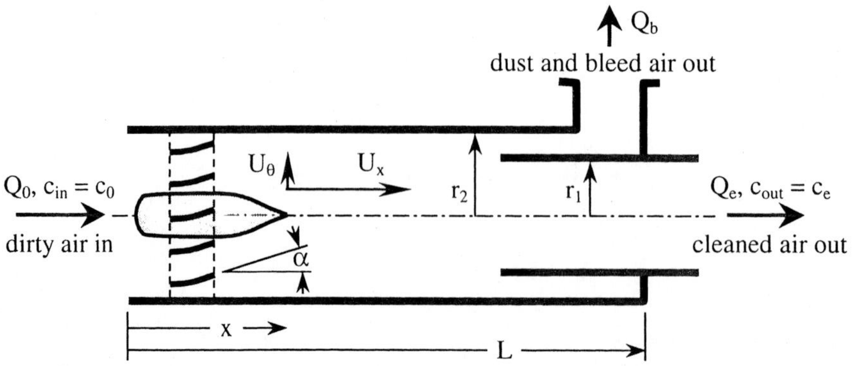

Figure 9.1 Straight-through cyclone collector with helical turning blades; bleed air is required to remove particles.

The ***number of turns*** (N_e) is equal to ($\theta_t/2\pi$). The width of the bend (W) can be scaled as follows:

$$W = \frac{L}{N_e} = 2\pi r_2 \frac{U_x}{U_\theta(r_2)} = \frac{2\pi r_2}{\tan\alpha} \tag{9-3}$$

Air flowing through cyclones is turbulent, and neither irrotational nor of constant vorticity. A realistic estimate is difficult to make but Strauss (1966) assumed the flow to be n-degree rotational with n equal to 0.5 for a reverse-flow cyclone. The assumption that n = 0.5 can also be used for straight-through cyclones, and the grade efficiency for straight-through cyclones can be obtained from Eqs. (8-156) and (8-159),

$$\eta(D_p) = 1 - \exp\left[-\frac{L(\tan\alpha)^2 \tau_p Q}{8\pi r_2^3\left(r_2 + r_1 - \sqrt{r_1 r_2}\right)}\right] \tag{9-4}$$

where r_1 is the inner radius of the outlet annulus, as defined in Figure 9.1.

9.1.2 Reverse-Flow Cyclone

Shown in Figure 9.2 is a schematic diagram of a ***reverse-flow cyclone collector*** of major diameter D_2 and working length L_1. A reverse-flow cyclone can also be conceived as a wrapped-around bend; the width (W) of the bend can be approximated as the width of the inlet duct. The grade efficiency (for particles of diameter D_p) can be expressed by Eq. (8-156) with K given by Eq. (8-159), where r_2 is replaced by the radius of the exit duct ($D_e/2$), and with a suitable value of exponent n (such as 0.5).

9.1.3 Particle Removal and Bleed Air

Since cyclones have no moving parts and are subject to wear only by abrasion, they are inherently reliable. Care must be taken, however, to remove collected particles so that they do not become resuspended in the outlet flow. In straight-through cyclones it is necessary to remove some of the air (***bleed air***) in order to remove the particles that migrate radially outward. If used as a concentrator, bleed air transports particles downstream to the next step in the process. If used as an air-cleaning device, bleed air reduces the amount of clean air available for use. In reverse-flow cyclones, bleed air may not be required. For example, if a reverse-flow cyclone is mounted with a conical hopper beneath the outlet, it is not necessary to bleed air to remove particles. Particles migrating radially outward strike the cyclone walls and fall downward into the hopper where they can be removed at a later time as a batch process, or they can be removed continually by hopper valves. In either scenario, only a minimal amount of bleed air is required.

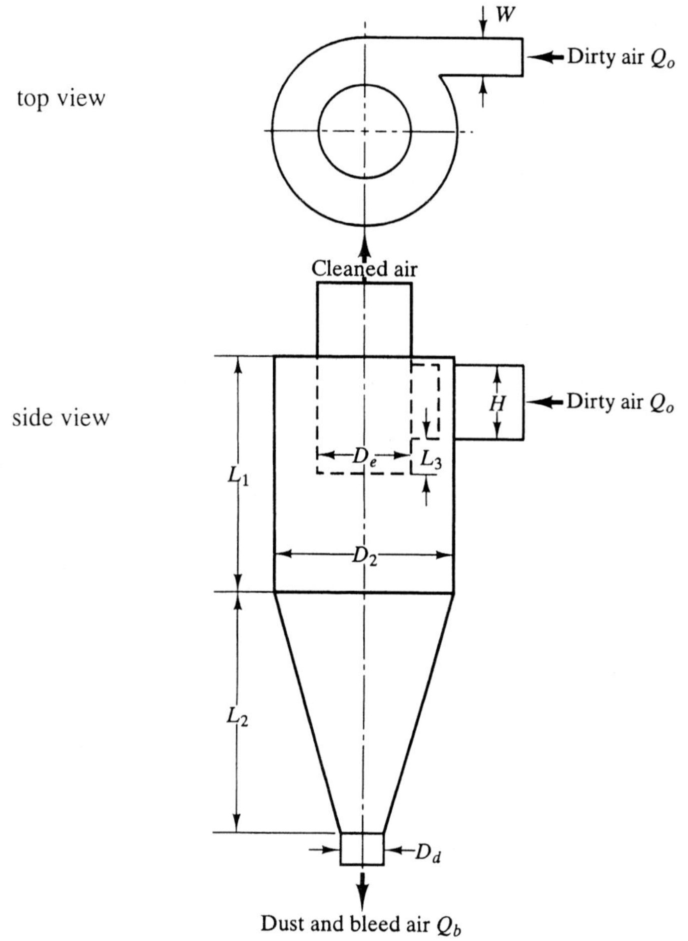

Figure 9.2 Lapple standard reverse-flow cyclone collector with tangential entry, and with characteristic dimensions given by Eqs. (9-5) and (9-6).

9.1.4 Lapple Standard Reverse-Flow Cyclone

The cyclone shown in Figure 9.2 is called a ***Lapple standard reverse-flow cyclone***. The configuration is called standard because large and small collectors are geometrically similar; the dimensions of the collector are scaled to the outer diameter D_2. Over the years, many researchers have proposed different scale values, yet they are all remarkably alike (Herrick and Davis, 2000). The choice of the scale factors and the grade efficiency for these configurations has received a great deal of attention (see for example Fuchs, 1964; Strauss, 1966; Danielson, 1973; Crawford, 1976; Licht, 1980; Boysan et al., 1982; Cooper and Alley, 1984; Moore and McFarland, 1993 and 1996). The grade efficiency of these configurations is expressed in forms similar to those in Chapter 8. Calvert and Englund (1984) and Herrick and Davis (2000) summarize the scale values and performance equations for several of the standard designs.

For simplicity, the Lapple standard reverse-flow cyclone is used in this text as a representative example of the configurations presented by Calvert and Englund (1984). The scale factors of the Lapple cyclone are:

$$L_1 = L_2 = 2D_2 \qquad W = D_d = \frac{D_2}{4} \qquad L_3 = \frac{D_2}{8} \tag{9-5}$$

$$H = D_e = \frac{D_2}{2} \qquad N_e = \frac{L_1 + \frac{L_2}{2}}{H} \tag{9-6}$$

where D_2 is the major diameter of the cyclone as defined in Figure 9.2. Note that substitution of the first and third equations above into the fourth equation yields $N_e = 6$ for any Lapple standard cyclone, regardless of size. In other words, the air swirls around approximately six times inside the cyclone. An attraction for Lapple standard cyclones is that their performance can be described by a single normalized ***grade (fractional) efficiency curve***, as shown in Figure 9.3.

There are many empirical expressions for the grade efficiency of cyclones in which $\eta(D_p)$ is expressed as function of volumetric flow rate and the dimensions of the cyclone. While some expressions have been shown to be more accurate, none are as straightforward or as universal as the curve shown in Figure 9.3. An algebraic curve fit for the plot in Figure 9.3 was generated by Theodore and DePaola (1980),

$$\eta(D_p) = \frac{1}{1 + \left(\frac{D_p}{D_{p,cut}}\right)^{-2}} \tag{9-7}$$

It is recommended that Eq. (9-7) be used instead of reading data from Figure 9.3, since the curve fit is excellent. The dimensionless abscissa is the actual particle diameter (D_p) divided by the ***cut diameter*** ($D_{p,cut}$), defined as the diameter of a particle that can be removed with a collection efficiency of 50%. The value of the cut diameter is given by

$$D_{p,cut} = \sqrt{\frac{9\mu H W^2}{2\pi N_e Q(\rho_p - \rho)}} \approx \sqrt{\frac{9\mu H W^2}{2\pi N_e Q \rho_p}} \tag{9-8}$$

where the approximation is valid when the particle density is much greater than the air density.

The pressure drop across a cyclone is proportional to the square of the volumetric flow rate. Calvert and Englund (1984) suggest an equation for the pressure drop (δP) of a Lapple standard cyclone; their equation reduces to

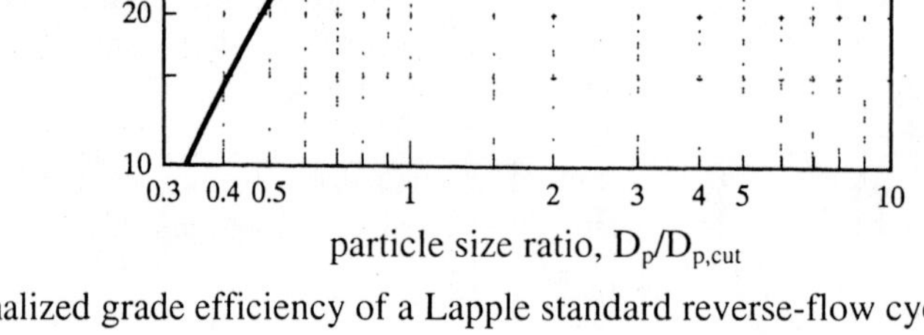

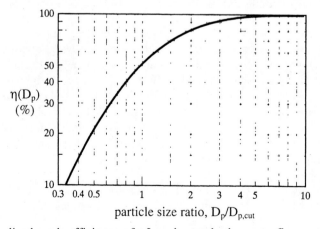

Figure 9.3 Normalized grade efficiency of a Lapple standard reverse-flow cyclone.

$$\boxed{\delta P = 40.96\ \rho \left(\frac{Q}{WH} \right)^2} \tag{9-9}$$

where the terms and the required units are:

- δP = pressure drop through the cyclone (in units of cm H_2O)
- ρ = actual gas density (in units of g/cm^3)
- Q = actual volumetric flow rate (in units of m^3/s)
- W, H = inlet dimensions (in units of m)

An acceptable pressure drop through a Lapple standard cyclone is generally less than 20 cm of water.

9.1.5 Applications

Cyclones are incorporated in the inlets of military gas turbine engines for helicopters and in the outlets of coal-fired boilers, and find use as inlet air cleaners for diesel engines of construction vehicles, etc. Cyclones can be arranged in parallel within a collector so that large volumetric flow rates can be cleaned. Cyclones are also used in the powder industry in which the goal is to increase the concentration of particles in an air stream; cyclones used for this purpose are called ***concentrators***. Cyclones can be constructed of inexpensive plastic, or of exotic materials to withstand very high temperatures or highly corrosive gases. They can remove solid as well as liquid particles.

Effective removal of a particle in a cyclone requires maximization of the particle's radial velocity, v_r, which is proportional to U_θ^2/r. To achieve high removal efficiency, engineers must either maximize the gas volumetric flow rate (Q) through the cyclone or reduce the diameter (D_2) of the cyclone. Since the energy loss associated with pressure drop varies with the cube of Q, maximizing the flow rate can be quite costly. Reducing the cyclone diameter while coping with a large volumetric flow rate requires using many small cyclones arranged in parallel; this is often the most economical course of action.

Cyclone collectors have been used for decades in series with other collectors of higher efficiency. Placed upstream of high-efficiency filters or electrostatic precipitators, cyclones reduce the particle mass concentration by removing large particles, and allow the high-efficiency collector to remove smaller particles. In other words, cyclones are used principally as the *first* stage of a ***multistage collection system***; a higher efficiency collector is used as the second stage. Without the cyclone removing the bulk of the suspended solids, the second stage would quickly become clogged or overloaded and unable to remove the small particles.

With the advent of ***computational fluid dynamics*** (CFD) modeling programs, engineers can now accurately predict the velocity and pressure fields of air flow inside cyclones, as shown for example in Figure 9.4 (Boysan et al., 1982). Once the velocity field is known, the techniques of Chapter 8 can be employed to predict particle trajectories. Since CFD methods improve steadily with time, there is reason to be optimistic that the performance of cyclones to capture smaller particles will also improve, and that novel next-generation centrifugal collectors will appear with high removal efficiencies. Modern commercially available CFD codes have built-in particle tracking algorithms that enable users to conveniently inject and track particles of specified density, diameter, and initial location at a specified initial velocity. Several examples of CFD solutions of relevance to indoor air quality, some of which involve this kind of particle tracking, are presented in Chapter 10.

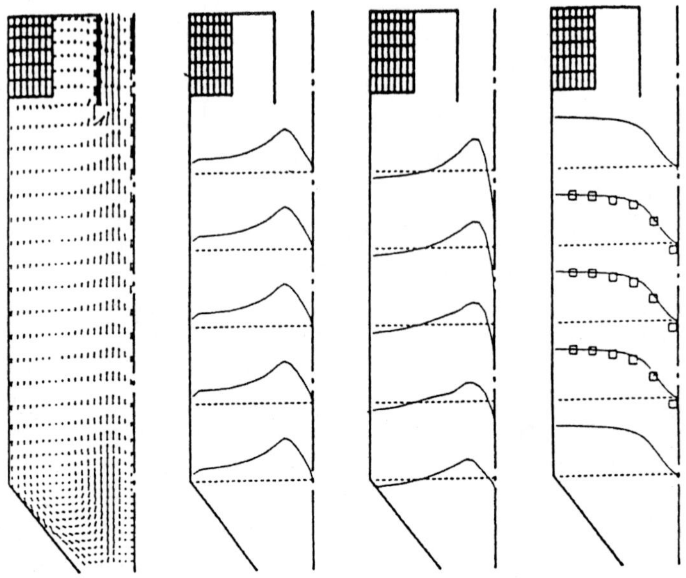

Figure 9.4 Predicted velocity and pressure distribution in a reverse-flow cyclone; (a) velocity vectors, (b) tangential velocity, (c) axial velocity, (d) pressure distribution with experimental data (from Boysan et al., 1982).

Example 9.1 - Fractional Efficiency of a Reverse-Flow Cyclone
Given: A Lapple standard reverse-flow cyclone with D_2 = 2.0 cm is used to capture unit density (ρ_p = 1,000 kg/m^3) spherical particles.

To do: Compute and plot the fractional efficiency of this cyclone as a function of volumetric flow rate at STP, which varies from 1.0 x 10^{-7} to 1.0 x 10^{-6} m^3/s. Also find the range of cut diameters ($D_{p,cut}$) for this range of volumetric flow rates.

Solution: The viscosity of air at STP is μ = 1.83 x 10^{-5} kg/(m s). Equations (9-5), (9-6), and (9-8) are used to find the cut diameter,

$$L_1 = L_2 = 2D_2 = 2(0.020 \text{ m}) = 0.040 \text{ m} \qquad W = L_3 = D_d = \frac{D_2}{4} = \frac{0.020 \text{ m}}{4} = 0.0050 \text{ m}$$

$$H = D_e = \frac{D_2}{2} = \frac{0.020 \text{ m}}{2} = 0.010 \text{ m} \qquad N_e = \frac{L_1 + \frac{L_2}{2}}{H} = \frac{0.040 \text{ m} + \frac{0.040 \text{ m}}{2}}{0.010 \text{ m}} = 6.0$$

$$D_{p,cut} = \sqrt{\frac{9\mu H W^2}{2\pi N_e Q \rho_p}} = \sqrt{\frac{9\left(1.83\times10^{-5}\frac{\text{kg}}{\text{m}\cdot\text{s}}\right)(0.010 \text{ m})(0.0050 \text{ m})^2}{2\pi(6.0)Q\left(1000\frac{\text{kg}}{\text{m}^3}\right)}} = \frac{3.30\times10^{-8}}{\sqrt{Q}}\sqrt{\frac{\text{m}^5}{\text{s}}}$$

where for Q in units of m^3/s, the units of $D_{p,cut}$ are meters, as seen in the above expression. For the range of Q given in the problem statement, $D_{p,cut}$ ranges from 105. μm at Q = 1.0 x 10^{-7} m^3/s to 33.0 μm at Q = 1.0 x 10^{-6} m^3/s. The fractional efficiency is calculated from the curve fit given by Eq. (9-7). The authors used Excel to generate plots of the fractional efficiency curves for this cyclone, which are shown in Figure E9.1 for four values of volumetric flow rate.

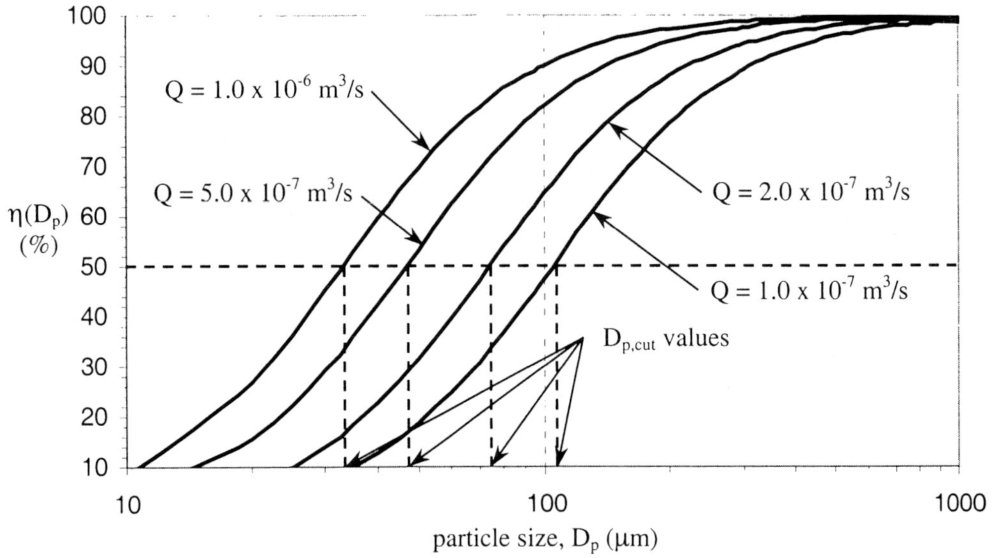

Figure E9.1 Fractional efficiency of a reverse-flow cyclone with D_2 = 2.0 cm for particles of density 1,000 kg/m^3 in air at STP for four volumetric flow rates.

Discussion: Fractional (grade) efficiency is strongly dependent on volumetric flow rate, shifting to smaller $D_{p,cut}$ with increasing Q. The predictions are accurate to at most two significant digits.

9.2 Other Inertial Separation Collectors and Sampling Issues

9.2.1 Mist Eliminators

Another important use of inertial separation is the removal of liquid droplets from a process gas stream. Collectors that achieve this function are called ***mist eliminators*** or ***demisters***. These devices contain a labyrinth of passages producing curvilinear flow that causes liquid drops to acquire a radial velocity. The drops impact a solid surface, coalesce, and form a liquid film that is removed by other means. Mist eliminators are used in wet scrubbers because the scrubbing fluid contains undesirable material that should not be emitted to the atmosphere. There is a large variety of mist eliminators (Figures 9.5 and 9.6), but all have the common feature of causing the liquid droplets to impact a solid surface, thus removing them from the gas stream.

9.2.2 Cascade Impactors

To size and select particle collection equipment, it is necessary that engineers describe the size distribution statistically using the techniques described in Chapter 8. Ultimately however, one needs to classify particles on the basis of mass, since compliance with air pollution regulations is based on the mass emitted by a process. An experimental device to measure the size distribution based on mass is the ***cascade impactor*** whose operation is depicted in Figure 9.7a. There are N ***stages***, each of which collects particles of a certain range of diameters, referred to as a ***class*** in Chapter 8. A commercial version of this device, with N = 8, is the Anderson 1 ACFM non-viable air sampler shown in Figure 9.7b. This sampler is called ***non-viable*** because particles are collected onto filter paper that is mounted on the top of each impaction plate; no effort is made to distinguish between living and nonliving particles. Another example of a non-viable cascade impactor is provided in Figure 3.3. A cascade impactor is said to be ***viable*** when an agar plate or glass petri dish is used in place of filter paper so that viable particles, such as bacteria, can be collected on each plate, and then studied later under a microscope.

Figure 9.5 DEMISTER® mist eliminators (used with the permission of Koch-Otto York Separations Technology Co., Inc.).

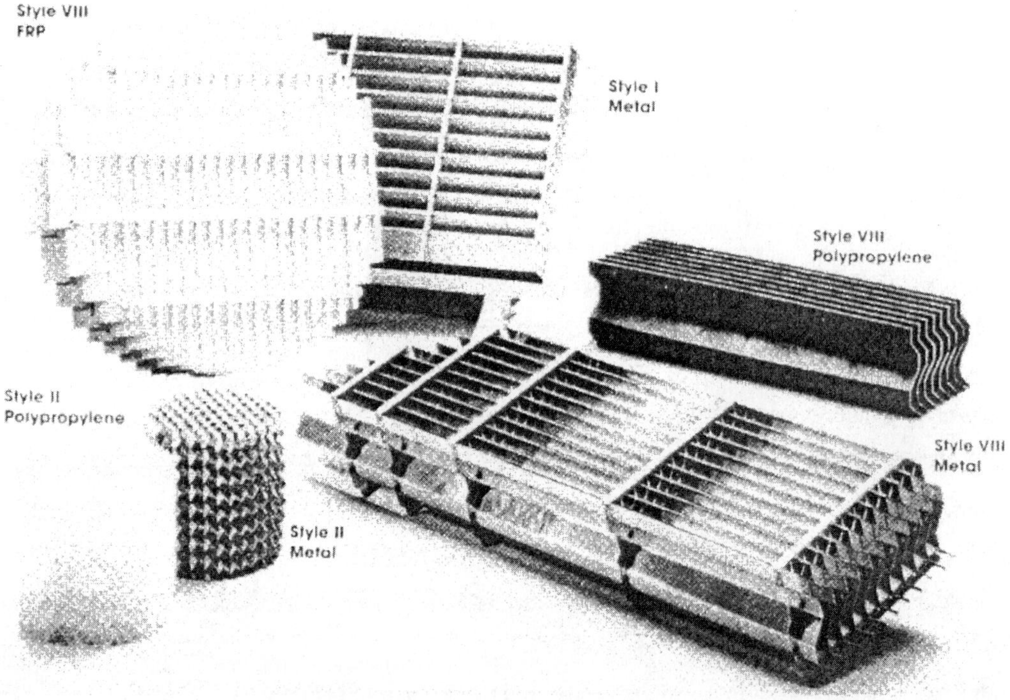

Figure 9.6 FLEXICHEVRON® mist eliminators (used with the permission of Koch-Otto York Separations Technology Co., Inc.).

The Anderson impactor shown in Figure 9.7b contains eight aluminum stages held together by three spring-clamps and O-ring gaskets. Each stage contains a large number of precision-drilled orifices of the same diameter. When an aerosol is drawn into the impactor, each orifice in a stage produces an air jet directing the aerosol toward its collecting surface, which is called an ***impaction plate***. The orifice diameter and the distance to the impaction plate become

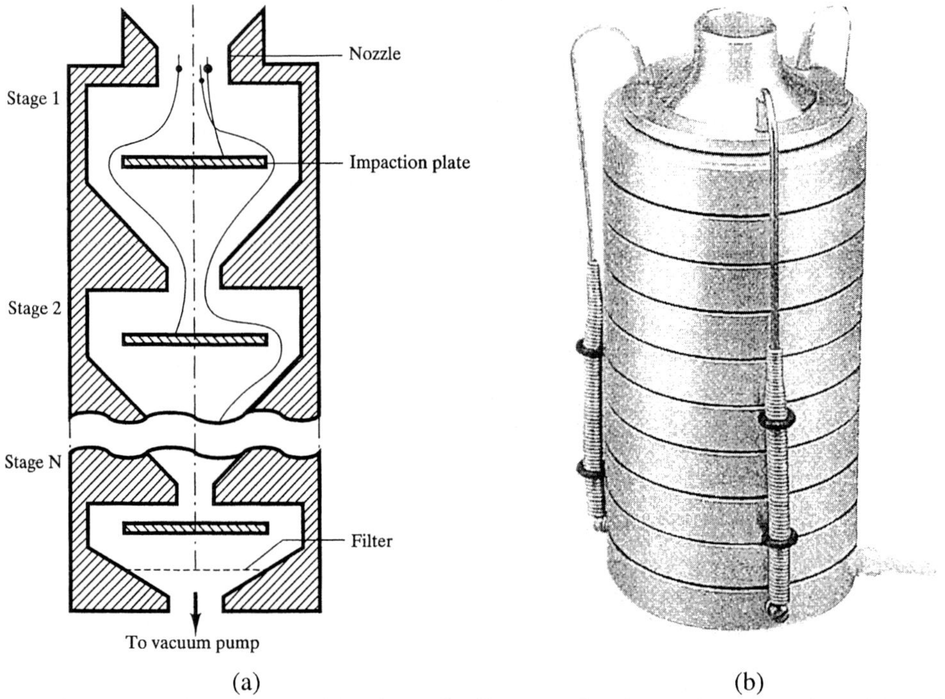

Figure 9.7 Cascade impactor: (a) schematic diagram, showing trajectories of particles of three different diameters (adapted from Willeke and Baron, 1993); (b) Andersen eight-stage, non-viable, 1 ACFM ambient air sampler (from Andersen Instruments Inc.).

progressively smaller in each succeeding stage. The aerodynamic dimensions and density of the particle determine whether the particle impacts the impaction plate of a particular stage. The range of particle sizes collected on each stage depends on the velocity of the air jet of that stage and the cutoff particle size of the previous stage. Any particle not collected on the first stage follows the air stream around the periphery of the plate to the next stage where it either impacts the collecting surface of that stage or passes on to the succeeding stage, and so on until the jet velocity is sufficient for impaction. The last stage contains a filter that removes most of the remaining particles. As seen in Figure 9.8, a well designed impactor has very little overlap in diameter range between stages, and each stage has an extremely steep grade efficiency curve through the majority of its corresponding particle size range.

If the air to be sampled contains particles larger than 10 μm, the impactor inlet should be replaced by a ***preseparator*** (also called a ***preimpactor***), that is designed to collect particles larger than about 10 μm to prevent them from entering subsequent stages of the sampler. The preseparator for the impactor of Figure 9.7b is an impaction chamber with a single 0.530 inch-diameter inlet orifice and three outlet tubes ending 1.00 inch above the impaction surface. An example particle collection efficiency plot for the cascade impactor of Figure 9.7b, but with a preimpactor, is shown in Figure 9.8.

In field measurements, the impactor is run at volumetric flow rate Q for some time t. After the impactor is taken apart, the mass of particles collected on each stage is determined with an analytical balance and a grouped data table is constructed, similar to those in Chapter 8, with the discriminating parameter being the mass of particles in each stage (class). The mass fraction and cumulative mass fraction for each class are computed, and the data yield the mass distribution

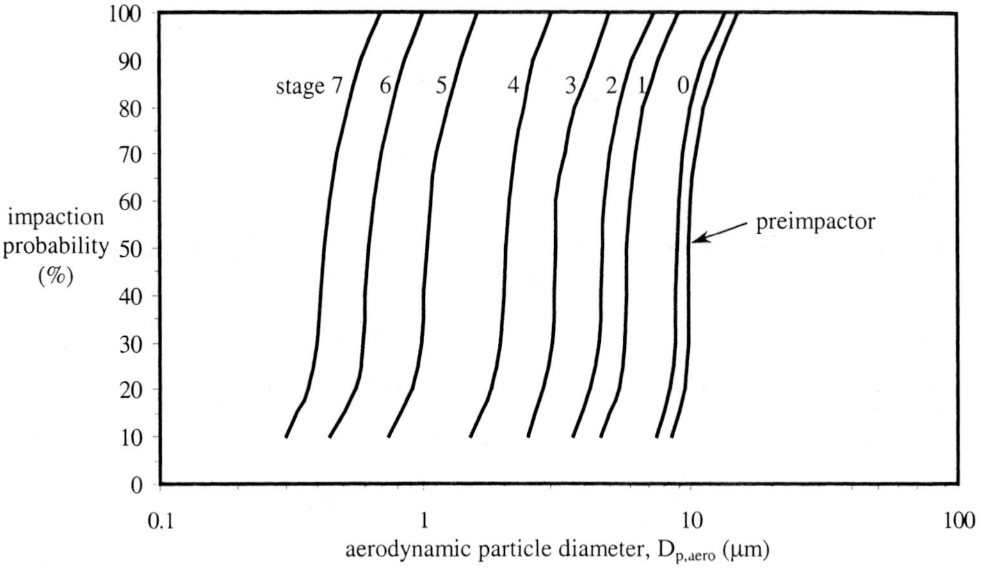

Figure 9.8 Particle collection efficiency for each stage of an Andersen eight-stage, 1 ACFM ambient air sampler with preimpactor (redrawn from Andersen Instruments, Inc.).

that can be combined with the collector fractional (grade) efficiency to compute the overall collection efficiency via Eq. (8-45). The data can also be plotted on log probability graph paper to test for log normality. The mass concentration of the entire aerosol is equal to the total mass collected on all the stages divided by the total volume of sampled gas, V = Qt.

9.2.3 Sampling Issues

The cascade impactor is a rugged instrument that can be used to measure the size distribution of ambient aerosols or aerosols in process gas streams with various temperatures, pressures, and velocity profiles. The accuracy of the measurement requires obtaining an aerosol sample that is the same as what exists in the process gas stream. Obtaining a representative sample is ensured if

(a) the velocity of the aerosol entering the sampling probe is equal to the velocity of the process gas stream
(b) the gas velocity is perpendicular to the plane of the sampling probe inlet

Criterion (a) is called ***isokinetic sampling*** and (b) is called ***isoaxial sampling***. Both are illustrated in Figure 9.9, in which air streamlines are depicted as solid lines and particle trajectories are depicted as dashed lines. With isokinetic, isoaxial sampling (Figure 9.9a), particles enter straight into the probe as they follow the streamlines of the gas, and U in the probe = U_0. With subisokinetic, isoaxial sampling (Figure 9.9b), the volumetric flow rate is too low, $U < U_0$, and some particles cross gas streamlines, entering the probe when they should not. In this case, the measured particle concentration is *greater* than the actual particle concentration. With superisokinetic, isoaxial sampling (Figure 9.9c), the volumetric flow rate is too high, $U > U_0$, and some particles again cross gas streamlines. In this case some particles which *should* enter the probe do not; the measured particle concentration is *less* than the actual particle concentration. Finally, a non-isoaxial misaligned sampling probe, even if isokinetic (Figure 9.9d) introduces additional errors. Some particles may cross streamlines and enter the probe even though the air streamlines do not, while other particles may miss the probe even though the air streamlines

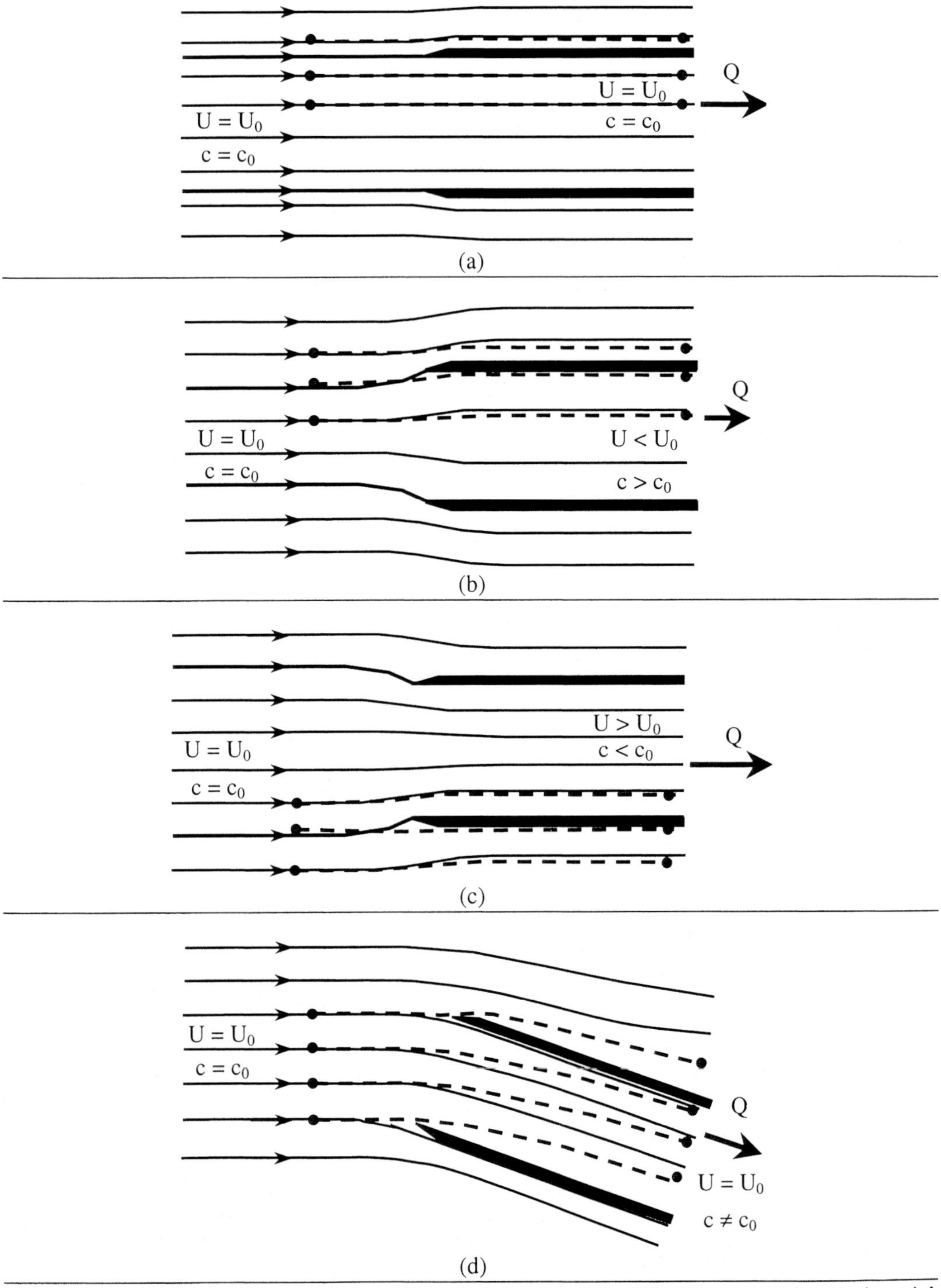

Figure 9.9 Particle trajectories (dashed) and streamlines (solid) for (a) isokinetic, isoaxial sampling, (b) subisokinetic, isoaxial sampling, (c) superisokinetic, isoaxial sampling, (d) isokinetic, but non-isoaxial sampling (misaligned sampling probe); dividing streamlines in (a), (b), and (c) are bold.

around them enter the probe. Particles may impact the inside of the probe and be collected there, never making it to the sampling instrument. In addition, large misalignments can lead to flow separation which introduces further error.

Figure 9.10 shows the sampling train used to obtain an aerosol sample from a process gas stream following EPA Method 5 (CFR 40, Part 60). The probe withdraws a gas sample at predetermined points in the process gas stream. The particles are collected on a filter contained in a heated region, and the condensable gases (including water vapor) are collected in impingers immersed in an ice bath. Prior to sampling, engineers measure the velocity profile in the duct. Once the velocity profile is known, the vacuum pump is used to control the sampling volumetric flow rate to ensure isokinetic conditions. ***Aspiration efficiency*** (η_a), also called ***aspiration coefficient***, is defined as the ratio of the measured particle mass concentration in the sample (c) to that in the process gas stream (c_0),

$$\boxed{\eta_a = \frac{c}{c_0}} \tag{9-10}$$

Thus

- $\eta_a > 1$ implies a particle concentration in excess of what exists in the process gas stream
- $\eta_a < 1$ implies a particle concentration less than what exists in the process gas stream

Figure 9.11 illustrates the consequences of failing to sample at isokinetic conditions (Brockman, 1993). In the final analysis, failure to sample under both isokinetic *and* isoaxial conditions can cause sources that are in compliance to appear to be out of compliance, or can cause sources that are out of compliance to appear to be in compliance.

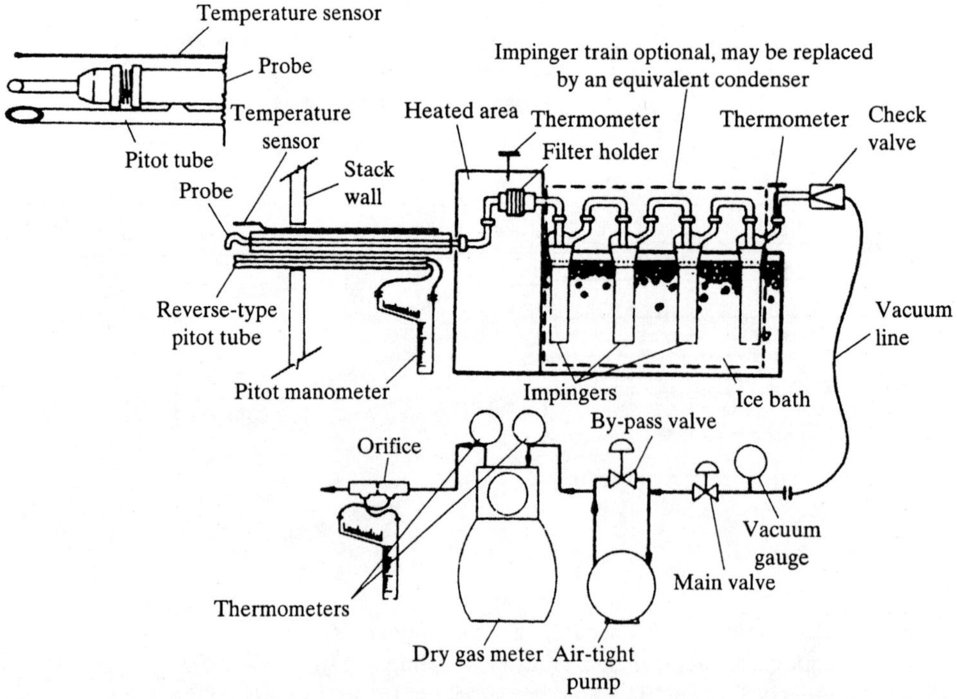

Figure 9.10 Components of an EPA Method 5 sampling train (adapted from EPA document CFR 40, Part 60, App 5, Method 5, p 625, 1 July 1989).

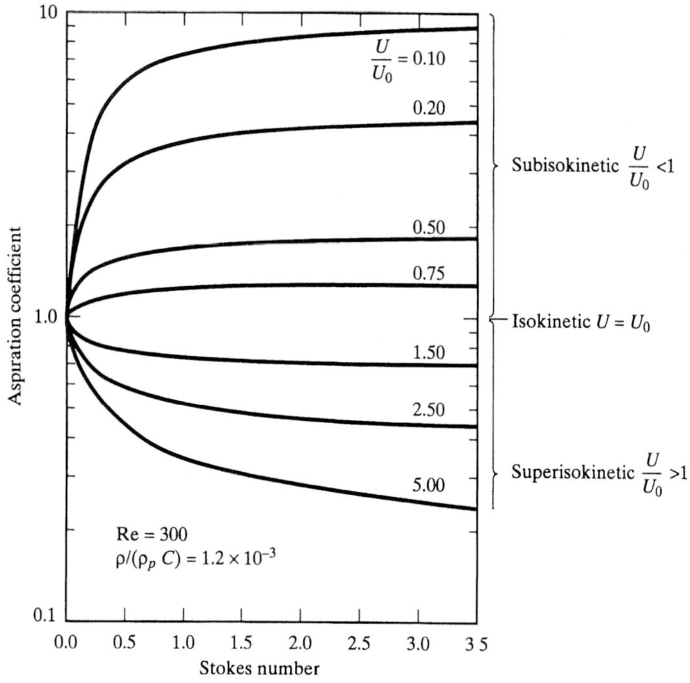

Figure 9.11 Aspiration coefficient ($\eta_a = c/c_0$) versus Stokes number, Stk, based on the diameter of the sampling probe for an axially aligned sampling probe at several different velocity ratios (U/U_0) (from Heinsohn & Kabel, 1999).

9.3 Impaction between Moving Particles

Impaction is the general name given to the collision of a particle with a collecting surface as the air carrying the particle passes around the collecting surface. ***Interception*** accounts for the fact that even though the particle's center of gravity may not collide with a collecting surface, collision still occurs if the distance between the path of the center of gravity and surface is less than the particle's radius. ***Diffusion*** accounts for the fact that in addition to inertial effects, particles exhibit Brownian movement, and very small particles may migrate toward a collecting surface and be removed from the gas stream. A spherical particle's diffusion coefficient can be shown to be inversely proportional to its diameter such that above about 1 μm, diffusion is negligible. All three processes (impaction, interception, and diffusion) occur simultaneously, and for convenience are hereafter subsumed under the phrase *impaction*. The reader should consult standard texts on particle dynamics (Fuchs, 1964; Davies, 1966; Hidy and Brock, 1970; Crawford, 1976; Licht, 1980; Hinds, 1982; Flagan and Seinfeld, 1988; Heinsohn, 1991; Willeke and Baron, 1993) to obtain a full understanding of impaction, interception, and diffusion.

Impaction occurs between small particles and large bodies so long as there is relative motion between the two; e.g. rain drops falling through quiescent dusty air, high speed aerosol particles passing through slower moving water droplets in a venturi scrubber, etc. For generality, consider impaction between a small particle (of diameter D_p) possessing the velocity of the carrier gas, and a large collecting body (of diameter D_c) moving through the gas stream at a different velocity, as illustrated in Figure 9.12. It is assumed that both particles are spheres. Throughout this section the phrase ***small particle*** or ***contaminant particle*** refers to the aerosol particle (of diameter D_p) that is to be removed by the larger ***collecting particle*** or ***collecting drop*** (of diameter D_c), composed of scrubbing liquid.

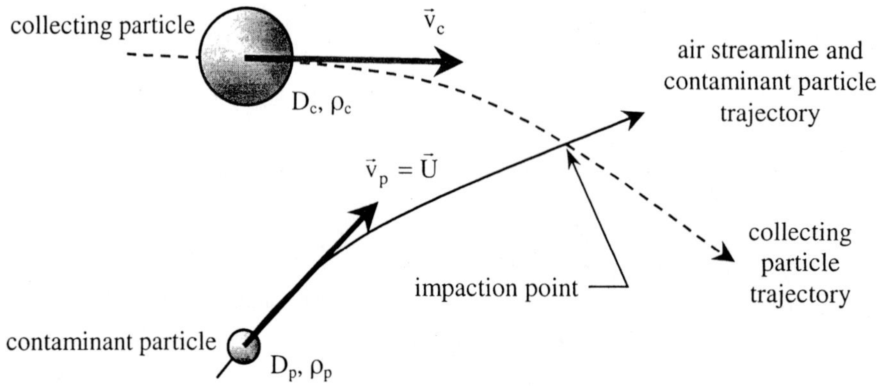

Figure 9.12 Impaction between a contaminant particle and a (larger) collecting particle.

It is assumed that if a contaminant particle impacts a collecting particle, it sticks to the collecting particle and is removed from the gas stream. The consequent growth in size and mass of the collecting particle is neglected. The effectiveness of the removal process is expressed as the ***single drop collection efficiency*** (η_d) (sometimes called the more ambiguous ***single particle collection efficiency***). η_d is defined as the rate (by mass) with which contaminant particles are removed by impacting with a single collecting particle divided by the mass flow rate of particles in a stream tube of cross-sectional area equal to that of the collecting particle. For a spherical collecting particle,

$$\boxed{\eta_d = \frac{\text{mass removal rate}}{c_0 \left|\vec{U}_0 - \vec{v}_c\right| \frac{\pi D_c^{\ 2}}{4}}} \tag{9-11}$$

where $\left|\vec{U}_0 - \vec{v}_c\right|$ is the magnitude of the velocity of the gas stream carrying the particles relative to that of the collector, and c_0 is the concentration of contaminant particles in the gas stream approaching the collector. If the collecting particle is at rest, the magnitude of the relative velocity is equal to the gas speed, U_0, as illustrated in Figure 9.13. Figure 9.14 shows the limiting upstream radius (r_1), within which contaminant particles are removed from the air stream.

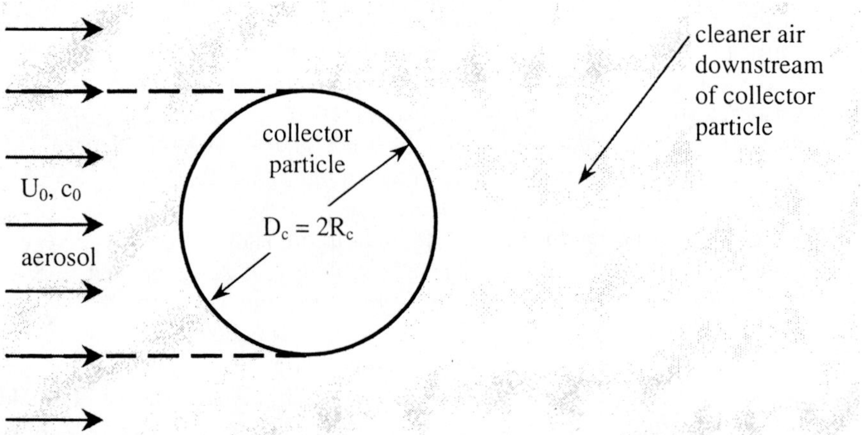

Figure 9.13 Aerosol particles collected by a larger collector particle; the stream tube used to define the single drop collection efficiency is bounded by the dashed lines.

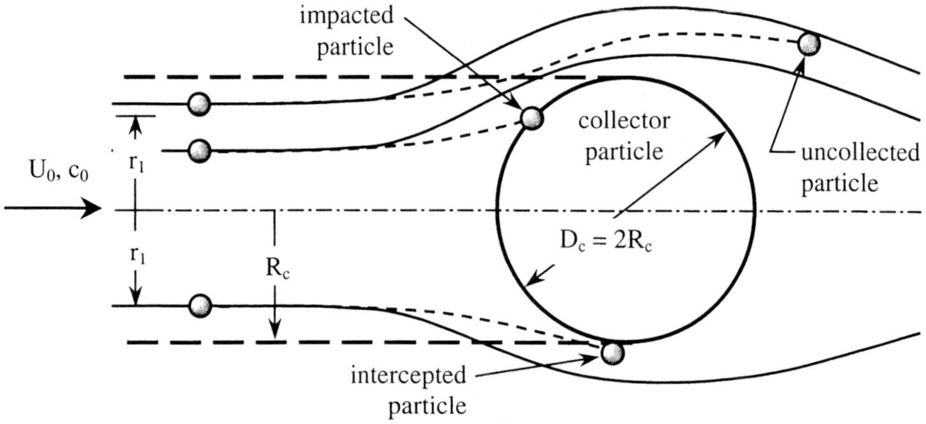

Figure 9.14 Illustration of sphere-on-sphere particle collection; solid lines are air streamlines, dashed lines are particle pathlines. Particles within radius r_l upstream of the collecting particle impact the collecting particle, particles outside of radius r_l pass by uncollected, and particles at radius r_l intercept the collecting particle.

Contaminant particles initially *within* the stream tube defined by radius r_l collide with (*impact*) the collector, while particles initially *outside* this stream tube miss the collector altogether, even for cases in which $r_l < R_c$, as shown in Figure 9.14. For the limiting case of particles initially *at* radius r_l, their surfaces just barely collide with (*intercept*) the collector, as also sketched. Not shown in Figure 9.14 is the removal of very small particles that diffuse to the surface of the collector. The single drop, sphere-on-sphere removal efficiency (η_d) is defined as the fraction of particles in the stream tube defined by the collector that are removed by the processes of impaction, interception, and diffusion:

$$\boxed{\eta_d = \frac{c_0 \pi r_l^2 U_0}{c_0 \pi R_c^2 U_0} = \left(\frac{r_l}{R_c}\right)^2} \tag{9-12}$$

Similar expressions can be defined for impaction of spheres on cylinders or for any pair of impacting bodies.

9.3.1 Single Drop Collection Efficiency

The single drop collection efficiency (η_d) of spheres impacting on spheres may be written as a function of Stokes number (Stk), as defined in Chapter 8:

$$\boxed{\text{Stk} = \frac{\tau_p |\vec{v}_r|}{D_c}} \tag{9-13}$$

where τ_p is the relaxation time constant for the particle, also defined in Chapter 8 as

$$\boxed{\tau_p = \frac{\rho_p D_p^2}{18\mu}} \tag{9-14}$$

and $\vec{v}_r$ is the velocity of the particle relative to the collector,

$$\boxed{\vec{v}_r = \vec{v}_p - \vec{v}_c} \tag{9-15}$$

where $\vec{v}_p$ is the velocity of the particle and $\vec{v}_c$ is the velocity of the collector. If gravimetric settling is neglected, the particle travels at a velocity equal to the gas velocity, $\vec{v}_p = \vec{U}_0$. A graph of the single drop collection efficiency for spheres impacting on spheres shows that η_d is zero at

Stokes number Stk = 0.083, and asymptotically approaches unity for large values of Stk. Calvert and Englund (1984) recommend that for flows in which the Stokes number exceeds 0.2, the single drop collection efficiency of spheres impacting on spheres can be approximated by

$$\eta_d = \left(\frac{Stk}{Stk + 0.7} \right)^2 \qquad \text{for } Stk > 0.2 \tag{9-16}$$

Example 9.2 – Single Drop Collection Efficiency of Rain Drops falling through Dusty Air
Given: Consider rain drops with a diameter range of 5 μm ≤ D_c ≤ 500 μm, falling through air containing dust particles with a diameter range of 5 μm ≤ D_p ≤ 50 μm.

To do: Compute and plot the single drop collection efficiency (η_d) as a function of D_c and D_p.

Solution: Equation (9-16) predicts the single drop collection efficiency (η_d) as a function of Stokes number. Since the dust particle is motionless, the relative velocity is equal to the terminal settling velocity of the rain drop (v_c) which in turn varies with its diameter (D_c), as discussed in Chapter 8. The particle relaxation time (τ_p) is a function of dust particle diameter (D_p), as given by Eq. (9-14). To compute the efficiency, the drop's terminal settling velocity (v_c) is calculated for a particular value of D_c, and then the efficiency for various values of D_p is calculated. The process is repeated for various combinations of D_c and D_p, recognizing that the drag coefficient c_D is a function of Reynolds number, which complicates the calculations. The authors used Mathcad to accomplish this task. Figure E9.2 shows the single drop collection efficiency of a raindrop falling through dusty air.

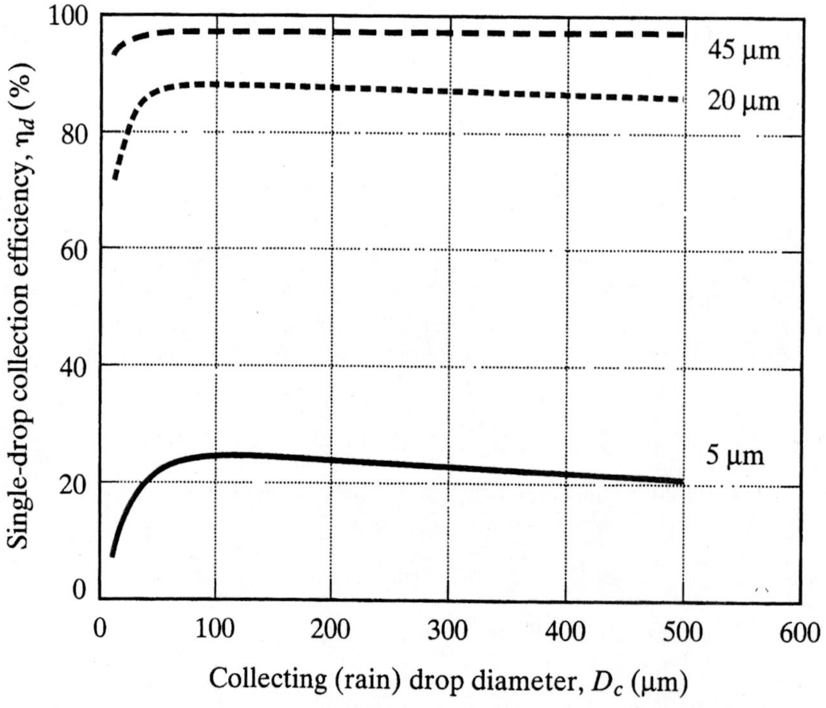

Figure E9.2 Ability of rain to remove airborne dust. Single drop collection efficiency of various size rain drops (μm) falling freely through still air containing dust particles, D_p = 5, 20, and 45 μm, ρ_p = 1,000 kg/m^3 (from Heinsohn & Kabel, 1999).

Discussion: Readers may be surprised to see the efficiency reach a maximum and then fall for large drop diameters. The explanation turns on how settling velocity varies with drop diameter. For small drop diameters, the Reynolds number is small, Stokes flow exists, and the settling velocity increases with the square of drop diameter. Thus Stokes number increases linearly with D_c and the efficiency rises with D_c. For large values of D_c, the Reynolds number is large and the drag coefficient approaches a constant (0.4). Thus, settling velocity increases with the *square root* of D_c, and Stokes number is inversely proportional to the square root of D_c. As D_c increases, Stokes number therefore falls and the efficiency falls as well. The results shown in Figure E9.2 should be checked to ensure that Stokes number (Stk) is greater than 0.2 for each case.

9.3.2 Spray Chambers

To illustrate impaction as a method for control of small particles, consider the impaction between a falling stream of (collecting) water drops of diameter D_c and an ascending stream of small contaminant particles of diameter D_p. In nature this occurs when falling rain drops impact dust suspended in the air. In this case the process is called ***scavenging*** or ***washout***. Two industrial processes in which scavenging is used are:

- spray chamber used to remove small particles of cutting fluid generated by a high-speed punch
- water spray used to control dust generated by a conveyor that discharges material to a stockpile

The overall effectiveness of a ***counterflow spray chamber***, as in Figure 9.15, can be modeled using an approach suggested by Calvert (1972) and Crawford (1976). The following assumptions are made:

- The height of the control region is L. The diameter of the collecting drops (D_c) and the absolute velocity ($\vec{v}_c$) of the falling drops are constants.
- The diameter of the contaminant particles (D_p) is constant and the number concentration of contaminant particles (c_{number}), varies only with height (z).
- The number concentration of collecting drops ($c_{number,c}$) is uniform; the number of encounters between particles and collecting drops (n) is uniform.
- The volumetric flow rates of air (Q_a) and spray liquid (Q_s) are constants.
- Gravimetric settling of the small contaminant particles is negligible; it is assumed that the velocity of these particles ($\vec{v}_p$) is equal to the air velocity ($\vec{U}_a$) which is constant,

$$\boxed{\vec{v}_p = \vec{U}_a} \tag{9-17}$$

- The velocity of impaction is the velocity of the particle relative to that of the collector,

$$\boxed{\vec{v}_{impaction} = \vec{v}_p - \vec{v}_c = \vec{U}_a - \vec{v}_c} \tag{9-18}$$

Because the air and particles flow upward and the drops fall downward, the impaction speed is the sum of the air speed and the drop speed, which in turn is equal to the magnitude of the settling velocity ($v_{t,c}$) of the collector particle in quiescent air:

$$\boxed{v_{t,c} = U_a + v_c} \tag{9-19}$$

The number of collecting drops per unit volume of air ($c_{number,c}$) can be calculated from the mass flow rate of the scrubbing liquid ($\dot{m}_c$), which is a parameter controlled by the engineer; $\dot{m}_c$ can be expressed by

$$\boxed{\dot{m}_c = \bar{\rho}_c v_c A} \tag{9-20}$$

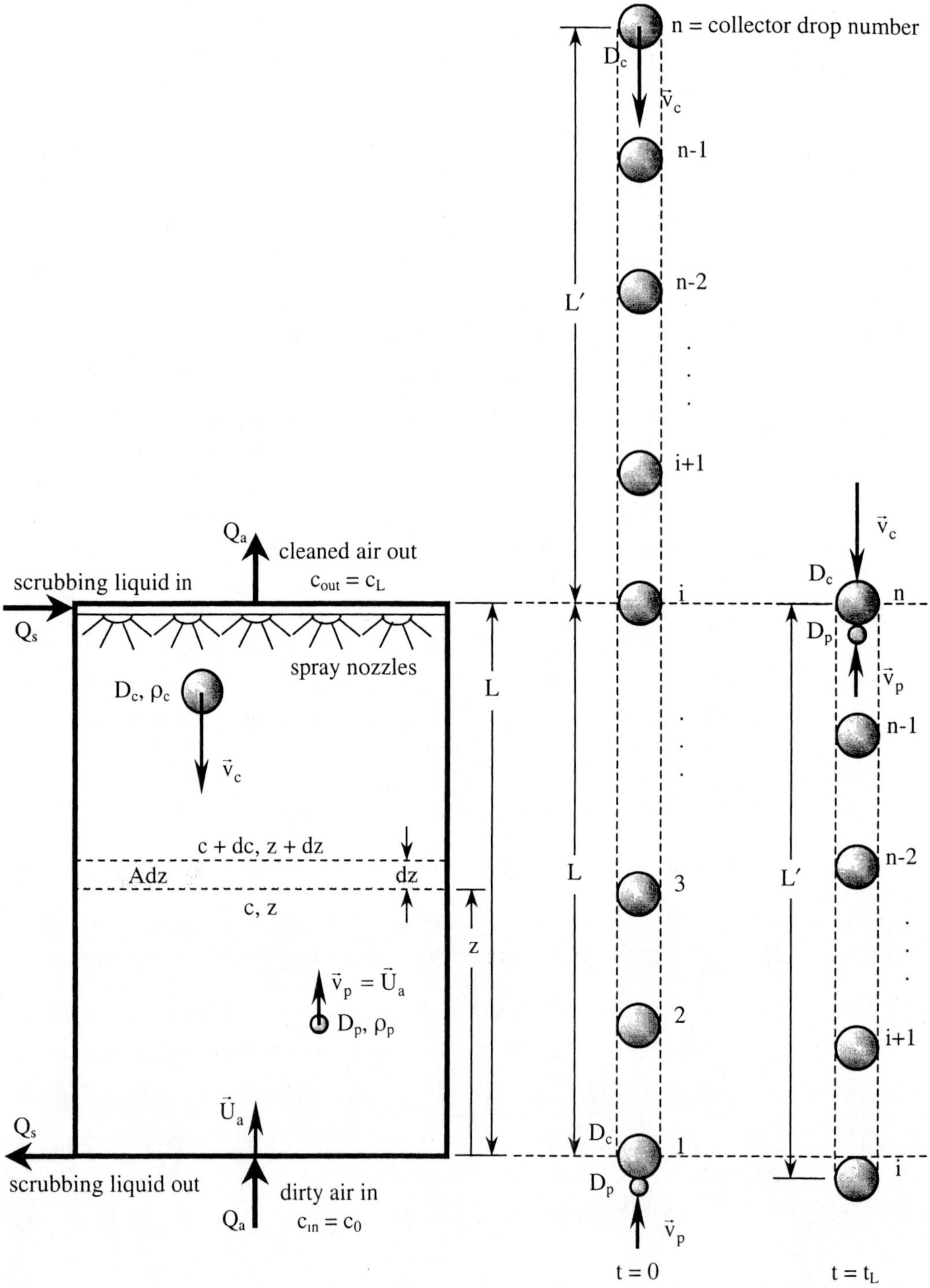

Figure 9.15 Schematic diagram of a vertical counterflow spray chamber illustrating encounters between upward traveling contaminant particles and falling collector drops; at t = 0, contaminant particles at the bottom of the chamber encounter drop 1, and at $t = t_L$, contaminant particles at the top of the chamber encounter drop n.

where $\overline{\rho}_c$ is the bulk density, defined as the mass of collector drops per volume of duct, v_c is the speed of the downward falling collecting drops, and A is the cross-sectional area of the spray chamber. The bulk density can be constructed from the following:

$$\overline{\rho}_c = \frac{\text{mass of drops}}{\text{volume of duct}} = \left(\frac{\text{mass of drop}}{\text{volume of drop}}\right)(\text{volume of drop})\left(\frac{\text{number of drops}}{\text{volume of duct}}\right)$$

which can be written as

$$\boxed{\overline{\rho}_c = \rho_c \frac{\pi D_c^3}{6} c_{\text{number,c}}} \tag{9-21}$$

The volumetric flow rate of the scrubbing or collecting liquid (Q_s) is equal to the mass flow rate of the collecting liquid ($\dot{m}_c$) divided by the density of the collecting liquid (ρ_c). Using Eqs. (9-20) and (9-21),

$$Q_s = \frac{\dot{m}_c}{\rho_c} = \frac{\overline{\rho}_c v_c A}{\rho_c} = \frac{\pi D_c^3}{6} c_{\text{number,c}} v_c A$$

Solving for the number of collecting drops per unit volume of air,

$$\boxed{c_{\text{number,c}} = \frac{6Q_s}{\pi D_c^3 v_c A}} \tag{9-22}$$

Multiplying and dividing by the volumetric flow rate of air (Q_a), where $Q_a = U_a A$,

$$\boxed{c_{\text{number,c}} = \frac{Q_s}{Q_a} \frac{6}{\pi D_c^3} \frac{U_a}{v_c}} \tag{9-23}$$

Alternatively, upon substitution of Eq. (9-19), Eq. (9-23) can be written as

$$\boxed{c_{\text{number,c}} = \frac{Q_s}{Q_a} \frac{6}{\pi D_c^3} \frac{U_a}{v_{t,c} - U_a}} \tag{9-24}$$

Since the only spatial variable is the height, the collecting drops can be imagined to travel one behind the other in a column of height (L + L′). Impaction occurs as small particles rise and encounter drops. The single drop collection efficiency (η_d) can be calculated from Eq. (9-16), and is constant since the velocities are constant. The parameter "n" in Figure 9.15 is the number of times small particles encounter collector drops as the contaminant particles travel upward through spray chamber height L. The overall removal of small particles can be imagined as the flow of an aerosol through a stream tube containing n collectors in series, each with a single drop collection efficiency (for particles of diameter D_p). Thus the grade efficiency for the column for particles of diameter D_p can be constructed as follows:

$$\eta(D_p) = 1 - \frac{c_{\text{out}}(D_p)}{c_{\text{in}}(D_p)} = 1 - \left(\frac{c_{\text{out}}}{c_{n-1}}\right)\left(\frac{c_{n-1}}{c_{n-2}}\right)\cdots\left(\frac{c_3}{c_2}\right)\left(\frac{c_2}{c_1}\right)\left(\frac{c_1}{c_{\text{in}}}\right)$$

Note that for brevity the notation (D_p) has been dropped from the concentrations on the right hand side of the above equation. Each of the ratios (c_2/c_1), etc., can be expressed in terms of the single drop collection efficiency, η_d,

$$\eta_d = 1 - \frac{c_i}{c_{i-1}} \qquad \frac{c_i}{c_{i-1}} = 1 - \eta_d$$

Combination of the above two equations yields

$$\boxed{\eta(D_p) = 1 - (1 - \eta_d)^n} \tag{9-25}$$

The total number of encounters can be deduced using Figure 9.15. It must be kept in mind that as the contaminant particles travel upward, collecting drops travel downward such that at the instant a particle has risen through height L, drop number n is being encountered. The total number of encounters is thus

$$n = c_{\text{number,c}} \frac{\pi D_c^{\ 2}}{4}(L + L')$$

which can be combined with Eq. (9-23) to produce the following expression for the total number of encounters:

$$n = \frac{Q_s}{Q_a} \frac{6}{\pi D_c^{\ 3}} \frac{U_a}{v_c} \frac{\pi D_c^{\ 2}}{4}(L + L') = \frac{3}{2} \frac{Q_s}{Q_a} \frac{U_a}{v_c} \frac{1}{D_c}(L + L')$$

Let t_L represent the time it takes for the small particles to rise through height L, which is also the time it takes for drop number n to fall through height L′,

$$\boxed{t_L = \frac{L}{U_a} = \frac{L'}{v_c}} \tag{9-26}$$

Thus

$$n = \frac{3}{2} \frac{Q_s}{Q_a} \frac{U_a}{v_c} \frac{1}{D_c}(t_L U_a + t_L v_c) = \frac{3}{2} \frac{Q_s}{Q_a} \frac{t_L U_a}{D_c} \frac{U_a + v_c}{v_c}$$

or, using Eq. (9-26) and (9-19),

$$\boxed{n = \frac{3}{2} \frac{Q_s}{Q_a} \frac{v_{t,c}}{v_c} \frac{L}{D_c}} \tag{9-27}$$

Substitution of Eq. (9-27) into Eq. (9-25) completes the analysis. Since the collecting drops fall by gravity alone, the relative velocity between falling collector drops and rising dust particles is small, which in turn produces a small single drop collection efficiency. The overall collection efficiency of spray chambers is thus inherently low unless L is large.

An alternative model of a vertical (counter-flow) spray chamber can be formulated by considering an infinitesimal well-mixed volume element at height z, as also sketched in Figure 9.15. It is assumed that the particle concentration varies only in the vertical direction (z-direction) and that there are no variations in the transverse direction. Consider a differential volume Adz. Conservation of mass for particles can be written as

$$AU_a c = AU_a(c + dc) + \eta_d c(v_c + U_a) c_{\text{number,c}} \frac{\pi D_c^{\ 2}}{4} Adz$$

where $c_{\text{number,c}}$ is given by Eq. (9-23). After substitution into the above,

$$\boxed{\frac{dc}{c} = -\frac{3}{2}\eta_d \frac{v_c + U_a}{v_c} \frac{Q_s}{Q_a} \frac{dz}{D_c}} \tag{9-28}$$

where Q_s and Q_a are the volumetric flow rates of the scrubbing liquid and the air, respectively. The ratio Q_s/Q_a is called by a number of names, such as the ***liquid to gas ratio*** and the ***reflux ratio***, and is given the symbol R,

$$\boxed{R = \frac{Q_s}{Q_a}} \tag{9-29}$$

The settling velocity of the collecting drop is sufficiently large so as to overcome the upcoming gas velocity. Thus the absolute velocity of the collecting drop (v_c) plus the carrier gas velocity (U_a) is equal to the settling velocity ($v_{t,c}$) of the collecting drop, as was given previously by Eq. (9-19). Integrating Eq. (9-28) over the height of the spray tower (L), one obtains the following expression for the grade efficiency of a vertical, counter-flow spray chamber:

$$\boxed{\eta(D_p) = 1 - \frac{c_{out}(D_p)}{c_{in}(D_p)} = 1 - \exp\left[-\frac{3}{2}\eta_d \frac{v_{t,c}}{v_c}\frac{Q_s}{Q_a}\frac{L}{D_c}\right]} \tag{9-30}$$

Eq. (9-25) combined with Eq. (9-27) is not equivalent to Eq. (9-30), although they both predict the grade efficiency of a vertical spray chamber. Differences between the two expressions can be resolved if the exponential term in Eq. (9-30) and the power term in Eq. (9-25) are both expanded in Taylor series. As the number of collecting drops (n) encountered over the height of the tower (L) becomes large, it can be shown that both expressions yield similar numerical results. If the model introduced at the beginning of this section is repeated, but the encounters between contaminant particles and collecting droplets are described by a Poisson distribution, an alternative form of Eq. (9-25) would emerge in a form that is compatible with Eq. (9-30).

Example 9.3 - Fractional Efficiency of a Counter-Flow Spray Chamber

***Given*:** Airborne dust generated by a crusher in a limestone quarry is captured and ducted to a counterflow spray chamber to be removed. Assume the dust particles are unit density spheres of diameter D_p. The air stream travels upward with velocity (U_a) equal to 1.00 ft/sec (0.3048 m/s). The scrubbing liquid is water, and the diameter (D_c) of the collecting particles is 1,000 μm (1.0 mm). The spray tower height (L) is 5.0 m, and its diameter (D) is 1.0 m. The volumetric flow rate of the air is thus

$$Q_a = U_a \frac{\pi D^2}{4} = 0.3048 \frac{m}{s} \frac{\pi (1.0\ m)^2}{4} = 0.239 \frac{m^3}{s}$$

***To do*:** Compute and plot the fractional efficiency (grade efficiency) of the counterflow spray chamber for several values of reflux ratio.

***Solution*:** For the collecting drops, which are 1,000 μm in diameter, the terminal settling velocity ($v_{t,c}$) can be calculated using the technique described in Chapter 8. The result is $v_{t,c}$ = 3.55 m/s (downward) for air at STP. This is the speed with which a dust particle and a collecting particle impact on one another. Since the gas velocity is 0.3048 m/s (upward), the velocity of the collecting drops (v_c) is 3.55 - 0.3048 = 3.245 m/s downward. The single drop collection efficiency (η_d) is given by Eq. (9-16), where the Stokes number is given by Eq. (9-13) using the terminal settling velocity,

$$Stk = \frac{\tau_p |\vec{v}_r|}{D_c} = \frac{\tau_p v_{t,c}}{D_c}$$

The grade efficiency of the spray chamber is given by Eq. (9-30),

$$\eta(D_p) = 1 - \exp\left[-\frac{3}{2}\frac{v_{t,c}}{v_c}\frac{L}{D_c}\eta_d \frac{Q_s}{Q_a}\right] = 1 - \exp[-K\eta_d R]$$

where R is the reflux ratio defined by Eq. (9-29), and K is a nondimensional constant formed from the remaining terms in the exponent,

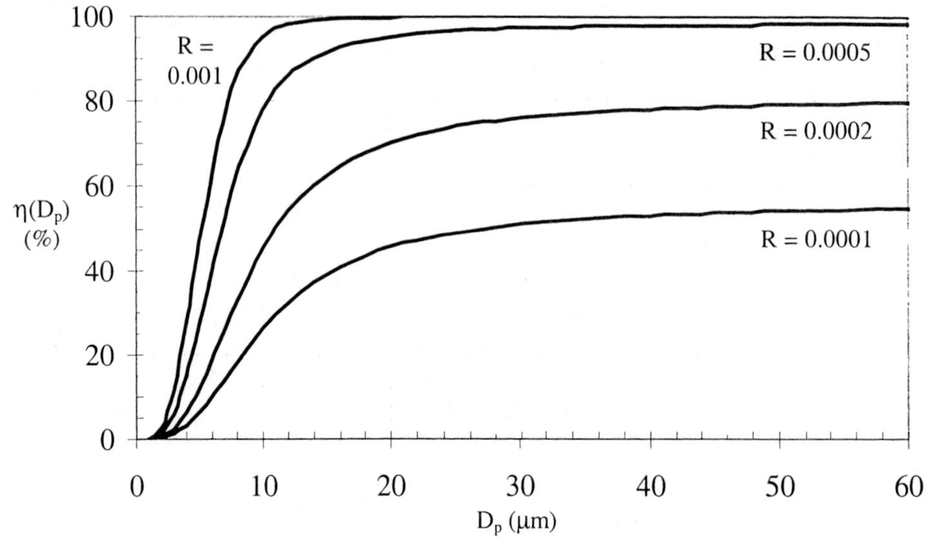

Figure E9.3 Grade efficiency of a counterflow gravity spray chamber at four values of reflux ratio, and for the parameters and dimensions of Example 9.3.

$$K = \frac{3}{2}\frac{v_{t,c}}{v_c}\frac{L}{D_c}$$

Thus, $\eta(D_p)$ can be calculated for selected values of R, and for a range of particle diameter, D_p. The authors used Excel to perform the calculations for four values of R. The results are shown in Figure E9.3.

Discussion: At large values of D_p, the grade efficiency asymptotes to a maximum value that depends on reflux ratio R. This maximum value of $\eta(D_p)$ increases with R as shown in Table E9.3, and can be calculated by setting $\eta_d = 1$ in Eq. (9-30).

Vertical spray towers require considerable height to maximize efficiency. In industrial buildings there is often insufficient vertical distance to accommodate a tall vertical spray chamber. Under these conditions, engineers often install a ***cross-flow (transverse) spray chamber*** instead, in which the gas and particles flow in the *horizontal* direction and the scrubbing liquid is sprayed downward, transverse to the carrier gas flow. By modeling the device in a manner similar to the above, and assuming that well-mixed conditions exist in a unit volume Adx, where A is the chamber cross-sectional area and x is in the direction of flow, the grade efficiency of the device can be expressed by an equation similar to Eq. (9-30).

Table E9.3 Maximum (asymptotic) grade efficiency as a function of reflux ratio for a counterflow gravity spray chamber.

$R = Q_s/Q_a$	maximum $\eta(D_p)$ (%)
0.0001	55.98
0.0002	80.62
0.0005	98.35
0.0010	99.97

9.3.3 Transverse Packed Bed Scrubbers

An alternative design is the ***transverse packed bed scrubber*** shown schematically in Figure 9.16. The aerosol flows horizontally with velocity U_a and encounters a packed bed of inert solid material with ***bed porosity*** ε. A scrubbing liquid of volumetric flow rate Q_s flows over and around the packing and removes particles that impact the packing. The scrubbing liquid drains out the bottom and is treated in a secondary process to remove the captured particles. The purpose of the packing is to establish curved passageways so that inertial separation of contaminant particles can occur. The curved surfaces create centrifugal forces that enable particles to acquire radial velocities (v_r) and strike the surface of the wetted packing, as illustrated in Figure 9.16b. A second function of the packing is to maximize a_p, the total bed packing surface area per bed volume.

$$\boxed{a_p = \frac{A_{\text{packing,total}}}{V_{\text{bed}}}} \tag{9-31}$$

For this reason, packing elements are generally not spherical, and can be highly irregular in shape, with a variety of designs such as those shown in Figure 9.17. There are two characteristic dimensions associated with packing. D_{packing} is the ***average overall packing diameter*** of the packing elements as illustrated in Figure 9.16a. D_c is the ***characteristic diameter*** of the curved passageways through which the air and particles navigate, as illustrated in Figure 9.16b. For simple packing shapes like rings and saddles (Figure 9.17a through d), D_c is approximately equal to D_{packing}. However, for more complex packing elements (Figure 9.17e and f), D_c can be significantly smaller than D_{packing}. Packing material may be plastic, ceramic, or metallic, and D_{packing} ranges from less than 1/4 inch to several inches, depending on the application.

To model the performance of a transverse packed bed scrubber, it is assumed that the aerosol enters the bed with uniform velocity (U_a) and particle mass concentration ($c_{in} = c_0$). It is also assumed that as the aerosol passes through the labyrinth of open passages in the bed, particles acquire radial velocities given by Eq. (8-140),

$$\boxed{v_r = C\frac{U_\theta^2}{r_c}\tau_p} \tag{9-32}$$

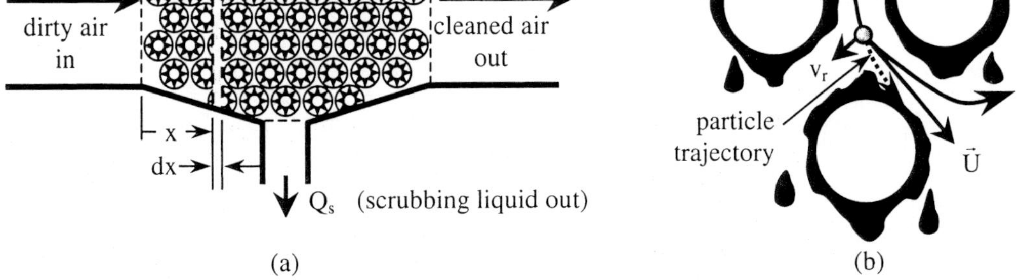

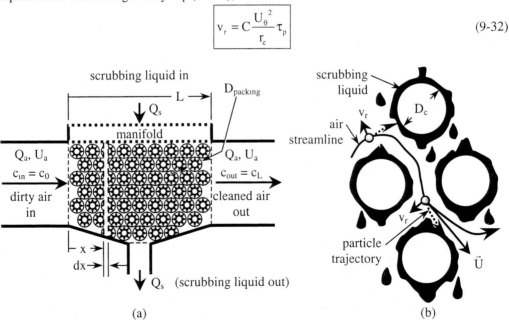

Figure 9.16 Transverse packed bed scrubber: (a) overall schematic diagram; (b) magnified view of curvilinear air flow through packed bed, generating radial particle velocities (v_r) that cause particles to impact wetted packing material of characteristic diameter D_c.

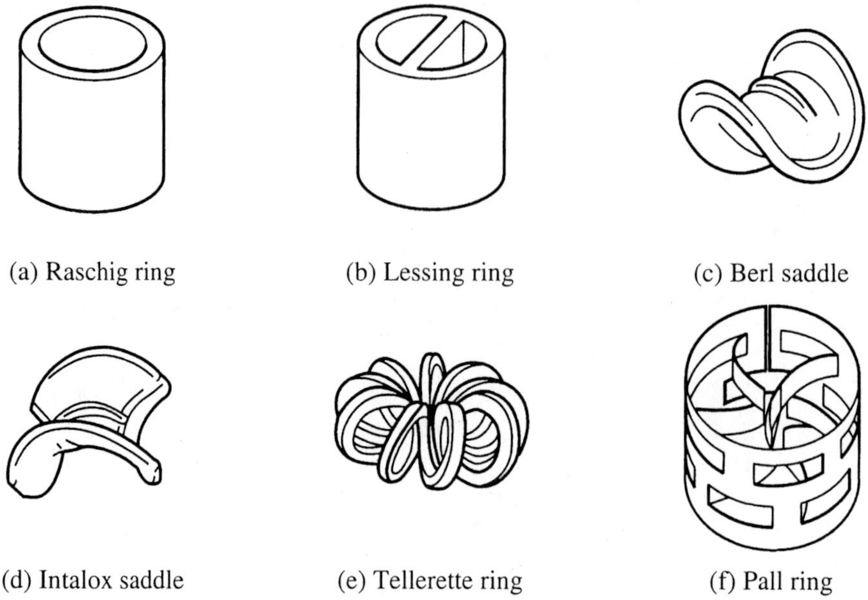

Figure 9.17 Packing elements of various sizes, shapes, and materials, depending on application.

where r_c is equal to the characteristic radius of the packing element ($r_c = D_c/2$), C is the Cunningham slip factor, τ_p is the particle relaxation time defined in Chapter 8, and U_θ is the local tangential component of gas velocity inside the bed. For contaminant particles several micrometers in diameter and larger, $C \cong 1$. For a tightly packed bed, the air is continually turning around the packing material as sketched in Figure 9.16b; U_θ is thus approximately equal to the magnitude of the air velocity inside the bed, $U_\theta \approx |\vec{U}|$. Since the packing blocks the gas flow, the gas velocity inside the bed is larger than the approach velocity (U_a),

$$\boxed{U_\theta = \frac{U_a}{\varepsilon}} \tag{9-33}$$

where by definition ε is less than 1. Consider an element of the packing as sketched in Figure 9.16a (of volume $A_{bed}dx$), where A_{bed} is the cross-sectional area of the packed bed. An equation for conservation of mass of contaminant particles can be written as

$$Q_a c = Q_a (c + dc) + v_r a_p c A_{bed} dx$$

which upon substitution of Eqs. (9-32) and (9-33) and recognition that $Q_a = U_a A_{bed}$ yields

$$\frac{dc}{c} = -\frac{v_r a_p A_{bed} dx}{Q_a} = -2C\frac{\tau_p}{D_c}\left(\frac{U_a}{\varepsilon}\right)^2 a_p \frac{A_{bed} dx}{U_a A_{bed}}$$

where Q_a is the volumetric flow rate of air through the packed bed scrubber. The above equation reduces to

$$\boxed{\frac{dc}{c} = -2C\frac{\tau_p a_p U_a}{D_c \varepsilon^2} dx} \tag{9-34}$$

Integrating over the length of the packed bed (L), the removal efficiency for particles of diameter D_p can be written as

$$\eta(D_p) = 1 - \frac{c_L}{c_0} = 1 - \exp\left[-2C \frac{\tau_p a_p U_a}{D_c \varepsilon^2} L \right]$$

or

$$\eta(D_p) = 1 - \exp\left[-2C(\text{Stk}) \frac{a_p}{\varepsilon^2} L \right] \tag{9-35}$$

where Stokes number (Stk) is defined as

$$\text{Stk} = \frac{\tau_p U_a}{D_c} = \frac{\tau_p Q_a}{D_c A_{bed}} \tag{9-36}$$

The volumetric flow rate of the scrubbing liquid (Q_s) does not enter the calculation unless it is of such magnitude that globs of liquid fall off the packing and reduce the porosity (ε) of the bed.

The pressure drop across a packed bed can be estimated using a number of empirical equations for flow through porous beds. Reynolds number is defined in terms of the average overall packing diameter ($D_{packing}$) of an element of packing and the approaching air speed (U_a),

$$\text{Re} = \frac{\rho U_a D_{packing}}{\mu} = \frac{\rho Q_a D_{packing}}{A_{bed} \mu} \tag{9-37}$$

If Re is less than about 10, and the volumetric flow rate of liquid, Q_s, does not produce suspended drops of liquid, Crawford (1976) reports that the bed pressure drop can be estimated from the following:

$$\delta P = 200 \frac{\mu Q_a L f_f^{\,2}}{A_{bed} D_{packing}^{\,2} \Phi^2 (1 - f_f)^3} \tag{9-38}$$

The parameter f_f is called the ***packing density***, or alternatively the ***solids fraction*** or the ***solidity***, and is defined as the bed bulk density (ρ_{bulk}) divided by the density of the solid packing material ($\rho_{packing}$),

$$f_f = \frac{\rho_{bulk}}{\rho_{packing}} \tag{9-39}$$

and Φ is the ***packing area ratio***, defined as the ratio of the projected frontal area of a sphere of diameter $D_{packing}$ to the total surface area of one element of packing material ($A_{packing}$).

$$\Phi = \frac{\pi D_{packing}^{\,2}}{4 A_{packing}} \tag{9-40}$$

The units of δP are dictated by the units used in the variables in Eq. (9-38), but are typically converted to an equivalent column height of water, e.g. cm or inches of H_2O.

Example 9.4 - Particle Removal Efficiency of a Transverse Packed Bed Scrubber

Given: A 1.0-foot long (L = 0.305 m) packed bed is composed of 1.0-inch ceramic Raschig rings ($D_{packing} = D_c = 0.0254$ m). For such packing, $a_p = 58.$ ft^{-1} and $\varepsilon = 0.73$. The aerosol particles to be removed are unit density spheres ($\rho_p = 1{,}000$ kg/m^3) of particle diameter (D_p) between 1 and 50 μm. Three gas velocities (U_a) are to be analyzed: 0.50 ft/s (0.152 m/s), 0.75 ft/s (0.229 m/s), and 1.0 ft/s (0.305 m/s).

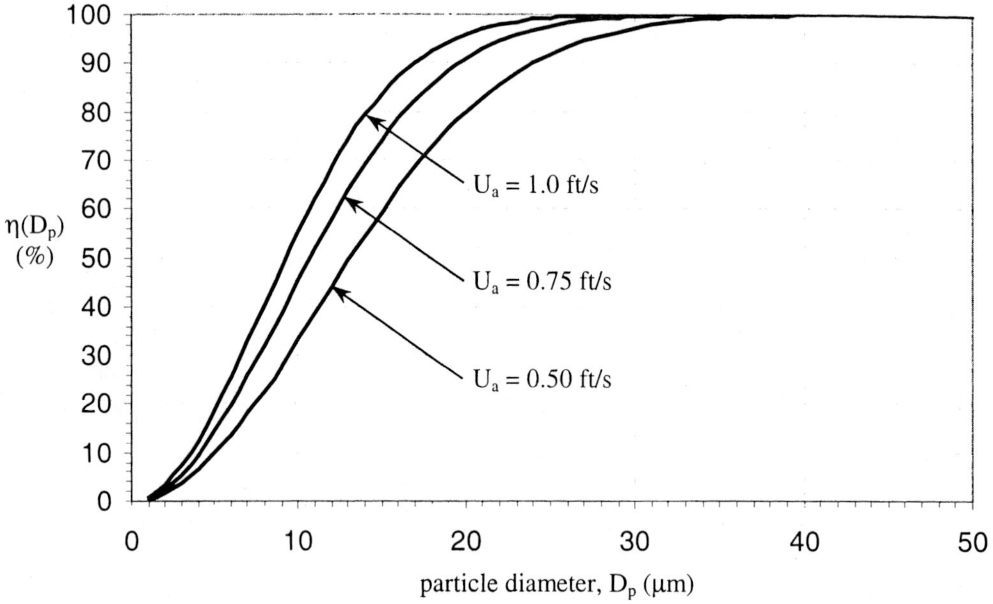

Figure E9.4 Grade efficiency of a transverse packed bed scrubber that uses water to clean unit density spherical contaminant particles; three air velocities are modeled.

To do: Compute and plot grade efficiency curves of this packed bed scrubber for the three given air velocities.

Solution: Stokes number Stk is found from Eq. (9-36), and grade efficiency $\eta(D_p)$ for aerosol particles of diameter D_p is found from Eq. (9-35). The authors used Excel to generate a plot of $\eta(D_p)$ versus D_p for the three given air velocities U_a. The results are shown in Figure E9.4.

Discussion: For these relatively large contaminant particles, Cunningham correction factor is close to unity, ranging from 1.167 at D_p = 1.0 μm to 1.003 at D_p = 50. μm. Although included in the above calculations, an assumption that C = 1 would not make a huge difference in the results.

The transverse packed bed scrubber is ideally suited for removing droplets of liquids that are miscible in water, or solid particles that can be easily separated from water. The scrubber is poorly suited for removing sticky solid particles or liquids that are apt to form a sticky gum in the presence of water. The packing can tolerate a certain amount of surface coating, but the scrubber should not be used if the coating build-up reduces the bed porosity.

Since packing increases the surface area per scrubber volume, it can improve particle removal efficiency in a significant manner. For example and comparison, the space requirements (volume of the device) and water requirements (Q_s) of the spray tower and the packed bed are examined below. In both cases, the scrubber is required to remove particles of 30 μm diameter with an efficiency of 80%, i.e. η(30 μm) = 0.80.

- *Spray chamber of Example 9.3*: With Q_a = 0.239 m^3/s and a required reflux ratio, R = Q_s/Q_a = 3.0 x 10^{-4}, the water requirements are

$$Q_s = RQ_a = (3.0\times10^{-4})0.239\frac{m^3}{s}\left(3.281\frac{ft}{m}\right)^3\left(\frac{60s}{min}\right)\left(\frac{1\ gal}{0.13368\ ft^3}\right) = 1.14\ GPM$$

and the volume requirements are

$$\frac{V_{tower}}{Q_a} = \frac{AL}{AU_a} = \frac{L}{U_a} = \frac{5.0\ m}{0.305\frac{m}{s}}\left(\frac{min}{60\ s}\right) = 0.273\ min = 0.273\frac{ft^3}{ACFM}$$

- *Transverse scrubber*: With the required gas velocity = 0.152 m/s, the water requirements are small, simply enough to bathe the packing and remove the deposited particles. Q_s could easily be less than one GPM. The volume requirements are

$$\frac{V_{bed}}{Q_a} = \frac{A_{bed}L}{A_{bed}U_a} = \frac{L}{U_a} = \frac{1.0\ ft}{1.52\frac{m}{s}}\left(\frac{m}{3.281\ ft}\right)\left(\frac{min}{60\ s}\right) = 0.00334\ min = 0.00334\frac{ft^3}{ACFM}$$

The transverse packed bed scrubber is more than 80 times smaller per ACFM than the spray tower, and yet offers the opportunity to use less water.

9.3.4 Venturi Scrubbers

A second example in which impaction is used to remove particles from a gas stream is the ***venturi scrubber***. In Figure 9.18 is a sketch of a venturi scrubber along with a downstream cyclone separator used to remove the collecting (scrubbing) liquid drops after they leave the scrubber.

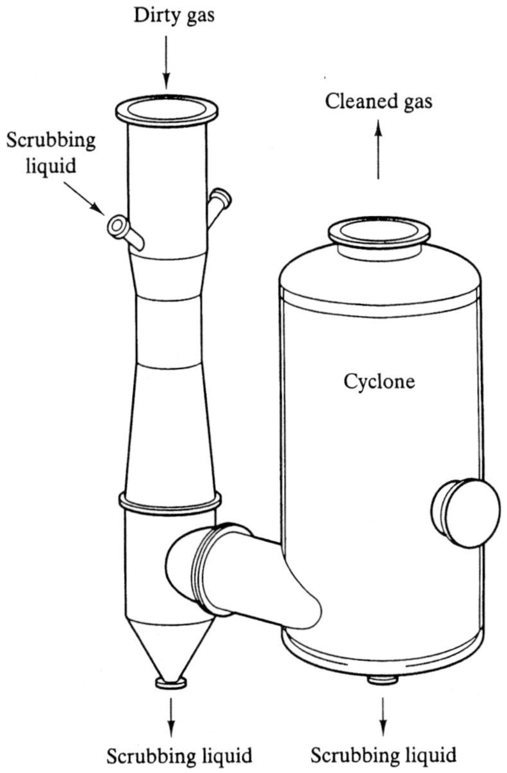

Figure 9.18 Venturi scrubber with cyclone collector to remove drops of scrubbing liquid (redrawn from Heinsohn & Kabel, 1999).

Some venturi scrubbers have a constant throat cross-sectional area, but in other cases it is desirable to vary the throat cross-section. Figure 9.19 shows three ways to control the throat cross-sectional area, namely with a horizontal sliding gate, with a moveable plug, and with a rotatable hinge. Figure 9.20 depicts the throat of a venturi scrubber that produces a large relative velocity and high overall collection efficiency.

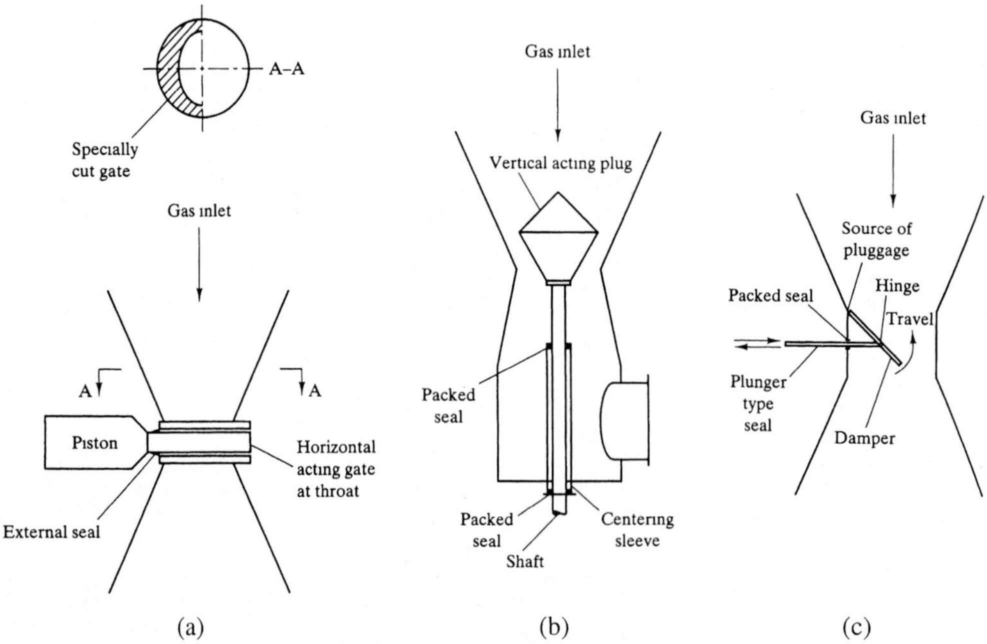

Figure 9.19 Methods to control the throat area of a venturi scrubber: (a) gate, (b) plug, and (c) hinge (printed with the permission of D. R. Technology, Inc.).

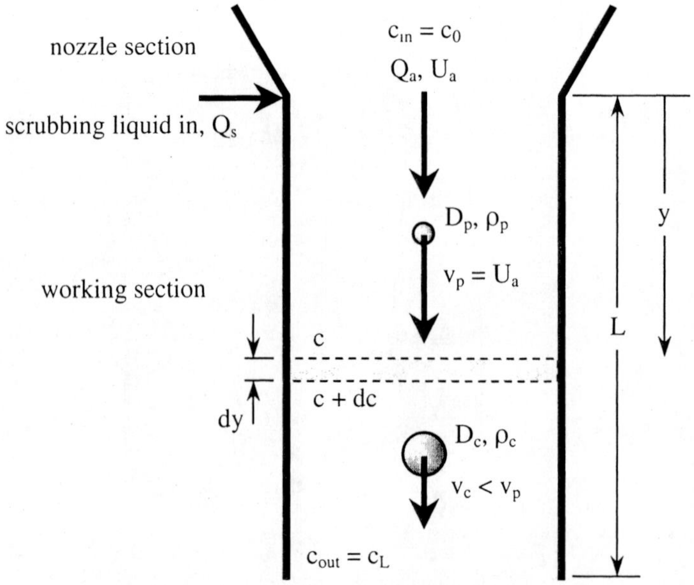

Figure 9.20 Schematic diagram of a venturi scrubber throat section, showing one contaminant particle and one collecting particle.

The operation of a venturi scrubber is based on the following:

- Air containing the small contaminant particles is brought to high velocity by passing it through a constricted area (throat) nozzle. Velocities of 50 to 100 m/s are common.
- Drops of the collecting liquid are injected into the aerosol at the inlet of the throat section. The collecting liquid drops may be produced by spray nozzles, or may be generated as the high-velocity gas shears small drops from the scrubbing liquid that flows down the sides of the venturi. The drops have low velocity as they enter the gas stream but accelerate as they travel downstream and approach (if not achieve) the gas velocity.
- The high-speed small contaminant particles impact the slower moving collecting drops and are removed from the gas stream. The single drop collection efficiency decreases as the collecting drops accelerate down the throat and the relative velocity between drop and collector decreases.
- The collecting drops containing the impacted small particles are removed from the gas stream by a cyclone or some other conventional particle removal system.

The collection efficiency of a venturi scrubber can be modeled using equations developed earlier if the following assumptions are made:

- The diameter of the collecting drops (D_c) and the diameter of the dust particles (D_p) are constant.
- Steady-state, well-mixed conditions exist at any value of y (see Figure 9.20 for definition of y; y is downward in the direction of flow).
- Only spatial variations in the direction of flow (y) are considered; variations transverse to this direction are assumed to be zero.
- The velocity of the small particles ($\vec{v}_p$) equals the air velocity ($\vec{U}_a$) and is constant throughout the throat.
- Evaporation of the scrubbing liquid is neglected.
- The pressure and temperature of the gas in the throat section are constants.

Writing a contaminant mass balance for the elemental volume shown in Figure 9.20,

$$AU_a c = AU_a\left(c + dc\right) + v_r c \frac{\pi D_c^2}{4} \eta_d c_{number,c} A dy$$

which after substitution of Eq. (9-23) reduces to

$$\boxed{\frac{dc}{c} = -\frac{3}{2}\eta_d \frac{v_r}{v_c}\frac{Q_s}{Q_a}\frac{dy}{D_c}} \tag{9-41}$$

where v_r is the magnitude of the relative velocity, and η_d is the single drop collection efficiency from Eq. (9-16). However, for an *accelerating drop*, v_c and hence also v_r change with time; η_d is therefore not constant. The acceleration of the collecting drop is given by Eq. (8-76). The vertical component of collector drop velocity (v_c) increases from $v_c(0)$ at the throat inlet and may ultimately reach the gas velocity at some point downstream of the inlet. At any point downstream of the inlet, $v_c(y)$ can be found by solving the equation of motion for a falling sphere as given in Chapter 8. Because the distance y is now measured in the *downward* direction, the equation of motion becomes

$$\frac{dv_c}{dt} = g - \frac{3}{4}c_D \frac{\rho}{\rho_c}\frac{1}{D_c}\left(v_c - U_a\right)\left|v_c - U_a\right| = g - \frac{3}{4}c_D \frac{\rho}{\rho_c}\frac{1}{D_c} v_r\left|v_r\right|$$

where $v_r = v_c - U_a$, and the Cunningham correction factor is ignored due to the relatively large size of the drops. The drag coefficient can be obtained from equations given in Chapter 8, namely

$$c_D = 0.4 + \frac{24}{Re} + \frac{6}{1+\sqrt{Re}} \qquad \text{where} \qquad Re = \frac{\rho D_c v_r}{\mu}$$

The above equations can be combined into a first-order ordinary differential equation in standard form,

$$\boxed{\frac{dv_c}{dt} = B - Av_c} \tag{9-42}$$

where

$$\boxed{A = \frac{3}{4} c_D \frac{\rho v_r}{\rho_c D_c} \qquad B = g - AU_a} \tag{9-43}$$

It must be emphasized that only collector drops accelerate. Thus the equations of motion for only the collecting drop need to be written and solved. The small contaminant particles that one wishes to remove are assumed to travel at a velocity equal to the gas velocity. Solution of Eq. (9-42) is best handled by the Runge-Kutta technique, as in previous examples.

Example 9.5 - Performance of a Venturi Scrubber

Given: An aqueous cutting fluid is used in a manufacturing process. The cutting fluid is toxic, and the airborne droplets can damage other equipment on the production floor. A venturi scrubber is used to prevent small drops of the cutting fluid from escaping into the air inside a manufacturing plant. (The venturi scrubber captures the cutting fluid, and returns the liquid to a sump.) The following properties are known:

- throat length, L = 1.0 m
- gas velocity in the throat, U_a = 60. m/s
- collecting (scrubber) drop diameter, D_c = 400 μm
- contaminant particle density, ρ_p = 1,500 kg/m^3
- three reflux ratios are available: $R = Q_s/Q_a = 1.0 \times 10^{-4}$, 3.0×10^{-4}, and 6.0×10^{-4}

To do: Model the performance of the venturi scrubber; compute and plot the following two operational characteristics:

(a) particle mass concentration as a function of location in the throat, c(y)
(b) fractional (grade) efficiency curves for the three given reflux ratios

Solution:

(a) The Runge-Kutta algorithm is employed to solve Eq. (9-42), the differential equation describing the instantaneous velocity of the collecting particle, v_c. The program begins at the throat (y = 0) with an initial velocity $v_c(0) = 0.01U_a$, and marches in time; the total distance traveled by the collecting drop (y) is monitored. When the total distance traveled by the collecting drop equals the length of the throat, (y = L) the program is terminated. The authors used Mathcad to perform the calculations; the results are shown in Figures E9.5a and E9.5b.

Discussion, Part (a): Examination of the graphical results, Figure E9.5a, shows the importance of the diameter of the collecting particles. For a given reflux ratio, $R = Q_s/Q_a$, the larger the drop diameter, the fewer the number of drops per unit volume of air. Thus it is not surprising that the grade efficiency at a given contaminant particle diameter D_p is inversely proportional to collector drop diameter, D_c, for a given reflux ratio. If the computation is repeated for several reflux ratios, it is seen that as the reflux ratio increases, both the number concentration of drops and the grade efficiency increase, as seen in Table E9.5.

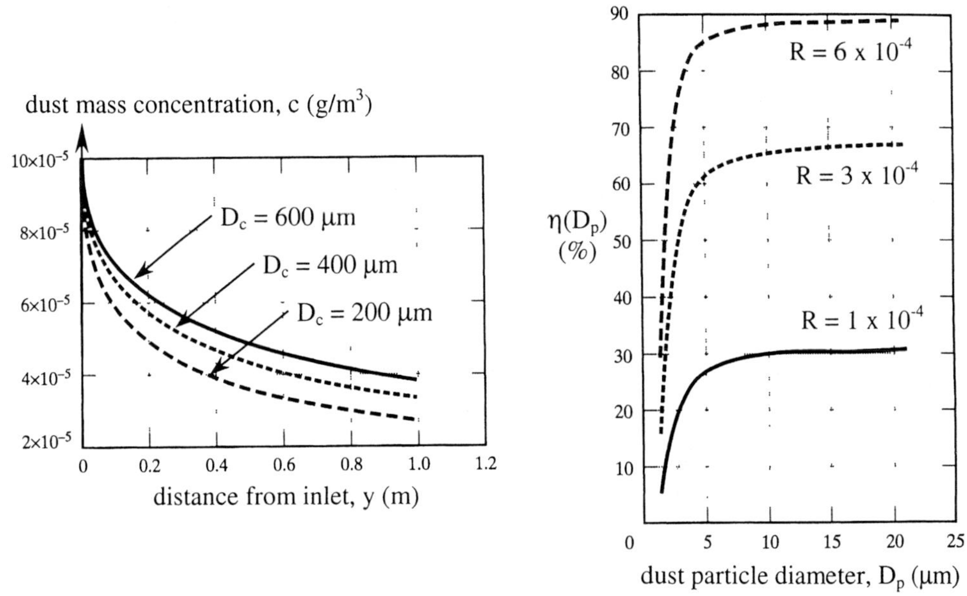

(a) (b)

Figure E9.5 Performance of a venturi scrubber: (a) Dust mass concentration versus distance downstream of the venturi inlet for three collecting water drop diameters, D_c = 200, 400 and 600 μm, for dust diameter D_p = 10. μm and density ρ_p = 1,500 kg/m^3, and for venturi inlet conditions, $c_{in} = c_0$ = 0.10 g/m^3, U_a = 60. m/s. (b) Grade efficiency curves for three reflux ratios, $R = Q_s/Q_a$ = 1.0 x 10^{-4}, 3.0 x 10^{-4}, and 6.0 x 10^{-4}.

Table E9.5 Grade efficiency for 10 μm particles as a function of collecting drop diameter and reflux ratio.

D_c (μm)	$\eta(D_p = 10\ \mu m)$				
	R = 1 x 10^{-4}	R = 1.5 x 10^{-4}	R = 3 x 10^{-4}	R = 6 x 10^{-4}	R = 1 x 10^{-3}
200	0.54	0.69	0.90	0.99	1.00
400	0.47	0.50	0.75	0.94	0.99
600	0.30	0.42	0.67	0.89	0.975

The graphical results show clearly that dust removal occurs most effectively in the top portion of the throat where the relative velocity between dust and water drop is the largest. In fact, for the data shown in the results, more than 50% of the dust is removed in the first 20% of the throat. Thus lengthening the throat achieves little improvement in dust removal.

(b) To predict and plot the fractional (grade) efficiency, $\eta(D_p)$, of a venturi scrubber in ways that can accommodate changes in the operating variables, the authors modified the Mathcad program to compute and store the concentration at the throat exit for a particular particle diameter (D_p), and to repeat the computation for the next particle of somewhat larger size ($D_p + \Delta D_p$). Figure E9.5b shows the grade efficiency of the venturi scrubber for three values of reflux ratio.

Discussion, Part (b): The results show clearly that for large enough reflux ratios, the venturi scrubber is a high-efficiency device for all particles larger than 10 μm, since the grade efficiency achieves its maximum value at about this size range. The engineers should experiment with different values of reflux ratio (R) and throat length (L). They will find that the maximum efficiency improves with increases in both R and L.

In the analysis of both the spray chamber and the venturi scrubber it was assumed that all of the spray drops were the same size (D_c). Considering the ways sprays are formed, it is highly unlikely that this will actually be achieved in practice. The assumption that the gas volumetric flow rate and velocity are constant is also weak since the injected spray drops evaporate, lower the gas temperature, and increase the total mass flow rate. It was also assumed that the particles to be collected were of constant diameter (D_p); this is also highly unlikely. Modeling a system in which the dust particles and/or spray drops have their own size distributions is tedious but not conceptually difficult. Chapter 8 should be recalled for guidance.

A parameter often used to characterize the spray is the ***Sauter diameter***, which is equal to the total volume of the spray drops in a volume of the carrier gas divided by the total surface area of these drops. Since gas absorption depends on absorber surface area, the magnitude of the Sauter diameter for a spray is a measure of the absorption rate of the spray. For a monodisperse spray of spherical drops, readers can show that the Sauter diameter is equal to the drop diameter divided by six.

9.3.5 Contact Scrubbers

Contact scrubber is the generic name of a device that captures dust from a gas stream by the impaction of dust and drops of scrubbing liquid that are produced by various types of geometric configurations. An example of a contact scrubber is shown in Figure 9.21. Impaction in a ***disk and donut scrubber*** occurs in the annular space between the disk and donut. Atomization occurs as the scrubbing liquid flows over the disk and encounters high-velocity air passing through the annular opening between the disk and donut. The dust continues to impact the drops of scrubbing liquid in the space below the disk. The clearance between the disk and donut can be varied if it is desirable to adjust the device to varying gas volumetric flow rates. One drawback of the disk and donut design is that the pressure drop across the disk and donut cannot be recovered as it can in the diverging section of a venturi scrubber. Thus the overall pressure drop/ACFM would be larger than in the venturi scrubber.

An additional type of contact scrubber that has been used for many years and is successful in removing large particles is typified by the one shown in Figure 9.22. As the aerosol passes upward through a thin layer of scrubbing liquid it produces a "frothy-foam" of air and water drops in which centrifugal forces cause the particles to impact liquid drops in the foam and in so doing are removed from the air. There are no moving parts in this type of contact scrubber, aside from the fan that blows the aerosol through the device. The pressure drop is small and is proportional to the thickness of scrubbing liquid through which the aerosol passes. A mist eliminator must be placed downstream of the device to capture ***carry-over*** liquid particles. The height of the scrubbing liquid in the device must be maintained at a prescribed level, and the liquid must be cleaned regularly to remove solids that collect in the scrubbing liquid. These solids lessen the ability of the contact scrubber to capture particles but most importantly, small solid particles contained in the carryover clog the mist eliminator. If for any reason the original aerosol contains drops of oil, the task of cleaning the scrubbing liquid in both types of collectors becomes more difficult owing to balls of dirty sticky oil that accumulate in the scrubbing liquid. Sticky particles that accumulate in the mist eliminator can clog it completely over a period of time.

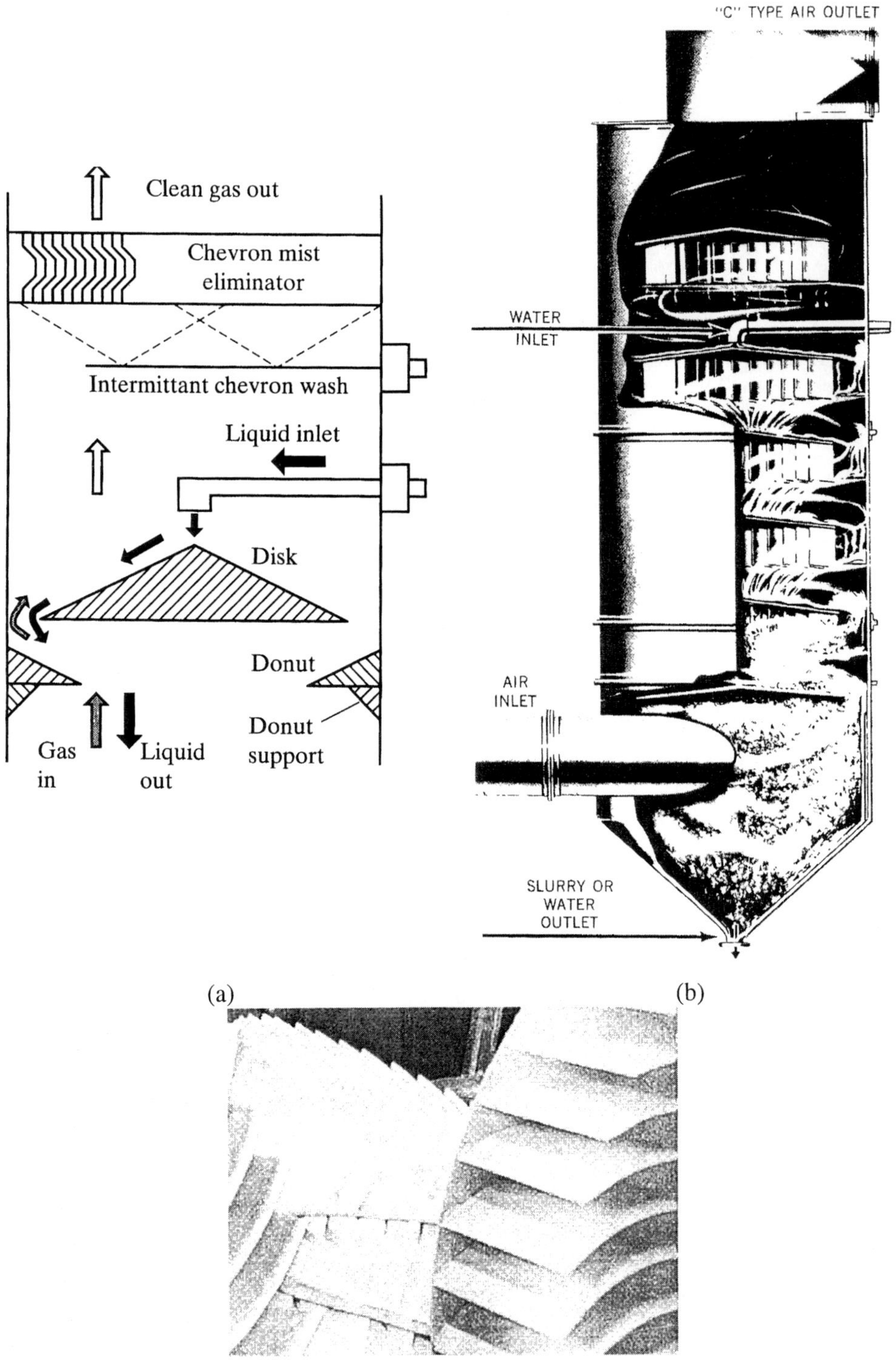

Figure 9.21 Non-clogging impact contact scrubbers that contain no moving parts: (a) one stage of a disk and donut design, (b) flow diagram, and (c) vane stages of the CMI-Schneible MULTI-WASH scrubber (courtesy of CMI-Schneible Co.).

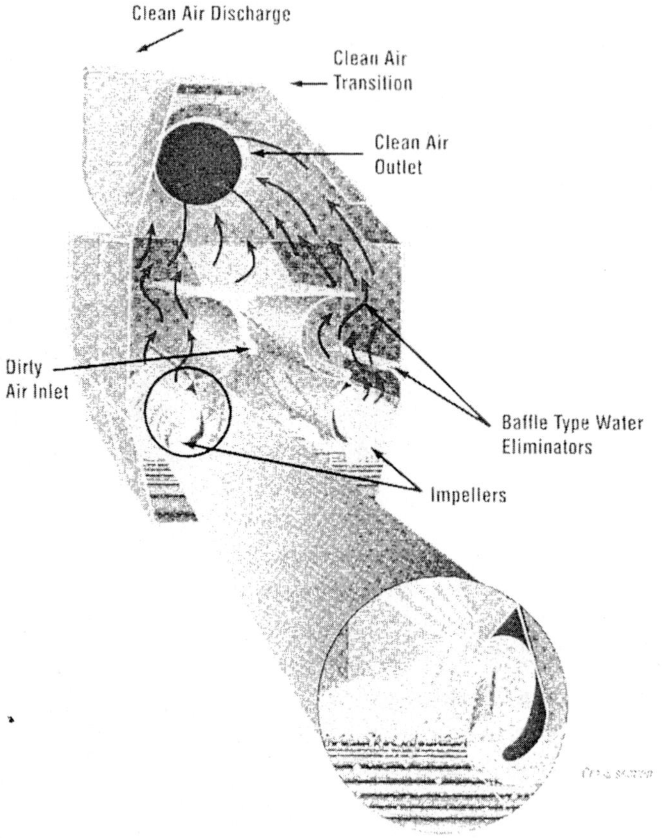

Figure 9.22 Contact scrubber which utilizes frothy foam (with permission from Torit Products, Donaldson Co.).

9.3.6 General Application of Wet Scrubbers

Spray chambers, packed bed scrubbers, venturi scrubbers, and contact scrubbers all use a scrubbing liquid (usually water), and can therefore collectively be called ***wet scrubbers***. Wet scrubbers can solve air pollution problems but they can also create problems in wastewater treatment since the scrubbing liquid must be treated before it is returned to the scrubber. While solid particles can be removed by wet scrubbers, the need to remove these particles before returning the water to the scrubber presents serious engineering problems by itself. Fabric filters or electrostatic precipitators are better suited for the removal of dry particle matter.

Wet scrubbers can be designed and fabricated of appropriate materials to process industrial gas streams that are very hot, caustic, or acidic. Wet scrubbers are primarily used to remove gases and vapors that are absorbed (soluble, miscible, etc.) in the scrubbing liquid. Pollutants that are not soluble, for example solid particles, oily sticky droplets, fumes, etc., are apt to clog parts of the scrubber and necessitate significant maintenance. Nozzles used to produce liquid collecting drops are prone to clogging or erosion. While the volumetric flow rate of the scrubbing liquid can be monitored, it is difficult to know which nozzles are clogged or if the designed droplet size distribution is maintained.

Wet scrubbers that are used to clean hot gas streams produce conspicuous steam plumes that may cause unwanted public attention, and that may also require a larger than usual supply of water. Vaporizing a great deal of the scrubbing water increases the mass and volumetric flow rate of the scrubber which is a factor that has been ignored in the present analysis. If the volumetric flow rate of the process gas stream varies over a wide range, the collection efficiency of a wet scrubber may also vary since collection efficiency is sensitive to gas volumetric flow rate. The overall pressure drop across a scrubber is apt to be large owing to the high velocities often present. (The diffuser section downstream of the throat of a venturi scrubber is designed to recover some of this pressure drop.) There are also pressure drops across the inlet header and outlet mist eliminator. The manufacturer's literature is needed to compute these pressure drops.

9.4 - Filtration

Filtration is the name given to the removal of particles as they pass through some permeable material such as paper, felt, or woven cloth, or through a bed of collectors. See Billings and Wilder (1970) for a more detailed discussion of filtration. Filtration is the most common method of collecting particles, and applications are found in general HVAC air cleaners, specialized laboratory air cleaners (e.g. clean rooms), and in health care equipment and facilities. Filtration units are often part of large air pollution control systems such as are those used to collect particles discharged by electric utility boilers, kilns, etc.; volumetric flow rates can reach hundreds of thousands of CFM. Filtration units may also be small uncomplicated units that capture particles generated by individual machines inside a plant, e.g. downdraft grinding benches, grinding wheels, wood sanding machines, etc. It is often more economical to capture particles by a small filter mounted to a machine rather than to install ducts connecting the machine to a large central air cleaning system for the entire plant.

9.4.1 Baghouses

Baghouse is the name given to a large filtration system containing many fabric filter bags arranged in modules operating in parallel. There are three principal types of baghouse, each using a different method to remove the collected dust from filters:

- shaker fabric baghouse
- reverse-flow baghouse
- pulse-jet baghouse

Shaker Fabric Baghouse: Figure 9.23 shows a schematic diagram of one module of a ***shaker fabric baghouse***. Individual cylindrical filters in the baghouse are approximately 10 inches in diameter and 10-15 ft long, and are held taught by hangers attached to the top of the closed bag. The bags are packed closely with only a few inches separating one from another. Dozens of bags are contained in a module, and a baghouse consists of many modules arranged in parallel to receive the dust laden process gas stream. The bottom of each bag is clamped to a ***tube sheet***, which has openings for each bag as illustrated in Figure 9.23. Dust laden air enters the bottom of the module, passes through the tube sheet, and into the inside of the filter bags. As the dirty air travels upward, clean air can pass through the bags, but most of the particles are trapped inside and deposited on the inside surfaces of the bags. A ***dust cake*** therefore accumulates on the inside of each bag. Once formed, the dust cake functions as a filter. In a sense, the woven fabric is merely the substrate to hold the dust cake – it is the dust cake that performs the bulk of the collection. Cleaned air passes through the outside of the bag into a plenum that receives cleaned air from other modules. Cleaned air finally leaves the module through an outlet duct. Periodically an entire module is taken off-line; it is isolated by damper valves, and its individual filters are cleaned by shaking them at the top. The dust cake on the inside of the bag fractures and falls below to the hopper. The module is allowed to rest for a few minutes until the dust settles, and the cleaned module is placed back on-line. Note that the dust cake is assumed to be ***friable***, which means that it easily crumbles to powder; otherwise it may not dislodge from the bag.

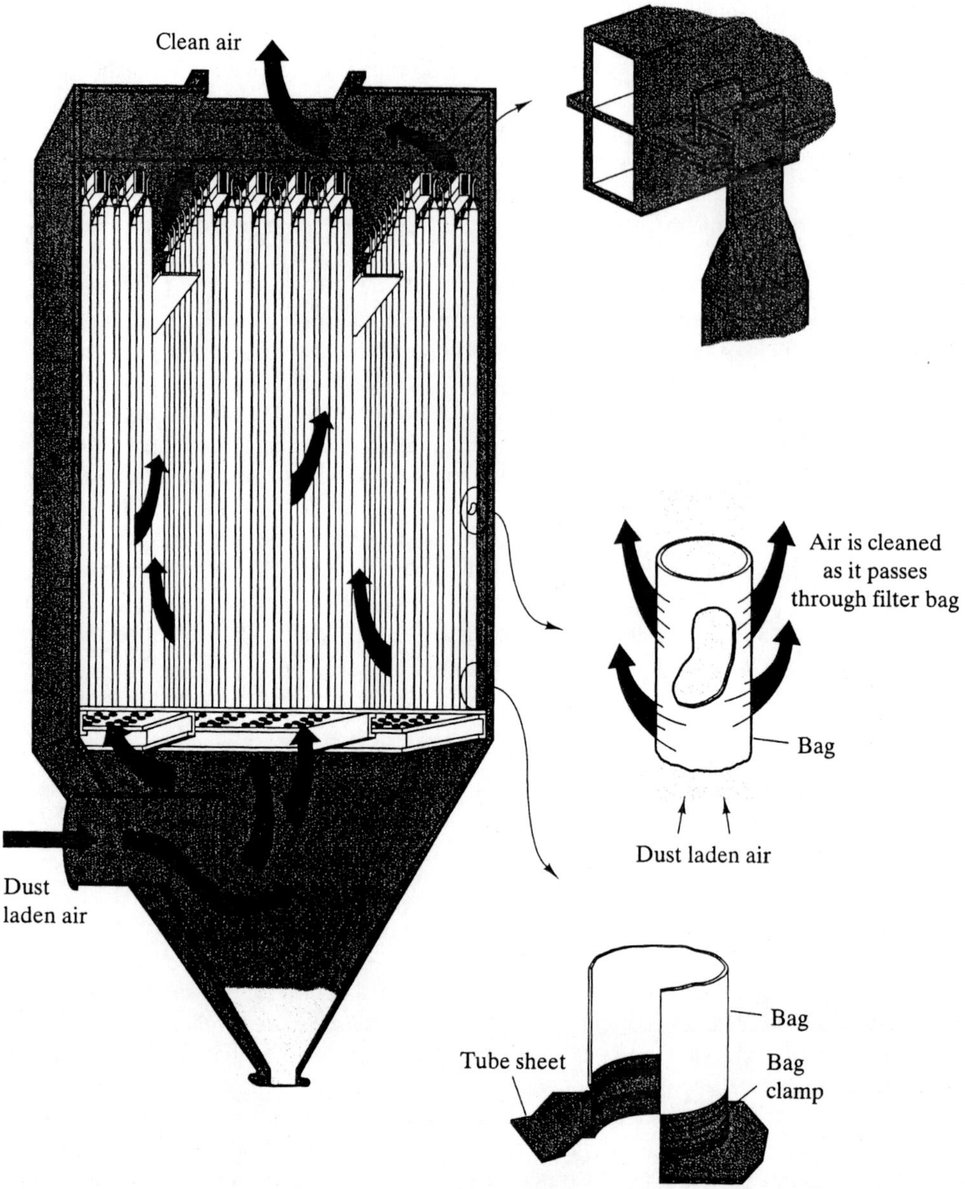

Figure 9.23 One module of a shaker fabric baghouse; bag closed with shaker at the top, open and sealed at the bottom; there are 128 bags in this module.

<u>Reverse-Flow Baghouse</u>: Figure 9.24 shows a schematic diagram of one module of a ***reverse-flow baghouse***. The fundamental features of a reverse-flow baghouse are the same as those of a shaker baghouse, except instead of shaking the bags to remove the dust cake, reverse air flow is used. A pulse of cleaned air is forced into the module from the top. This reverse-flow air compresses portions of each filter bag, as illustrated in Figure 9.24, whereupon the dust cake is dislodged and falls to the hopper below. Meritt and Vann Bush (1997) report that low-frequency sonic horns are installed in approximately 80% of the reverse-flow baghouses used for coal-fired utility boilers (excluding those using dry flue gas desulfurization). Sonic energy augments the fabric movement caused by reverse flow and dislodges additional dust cake material from the fabric.

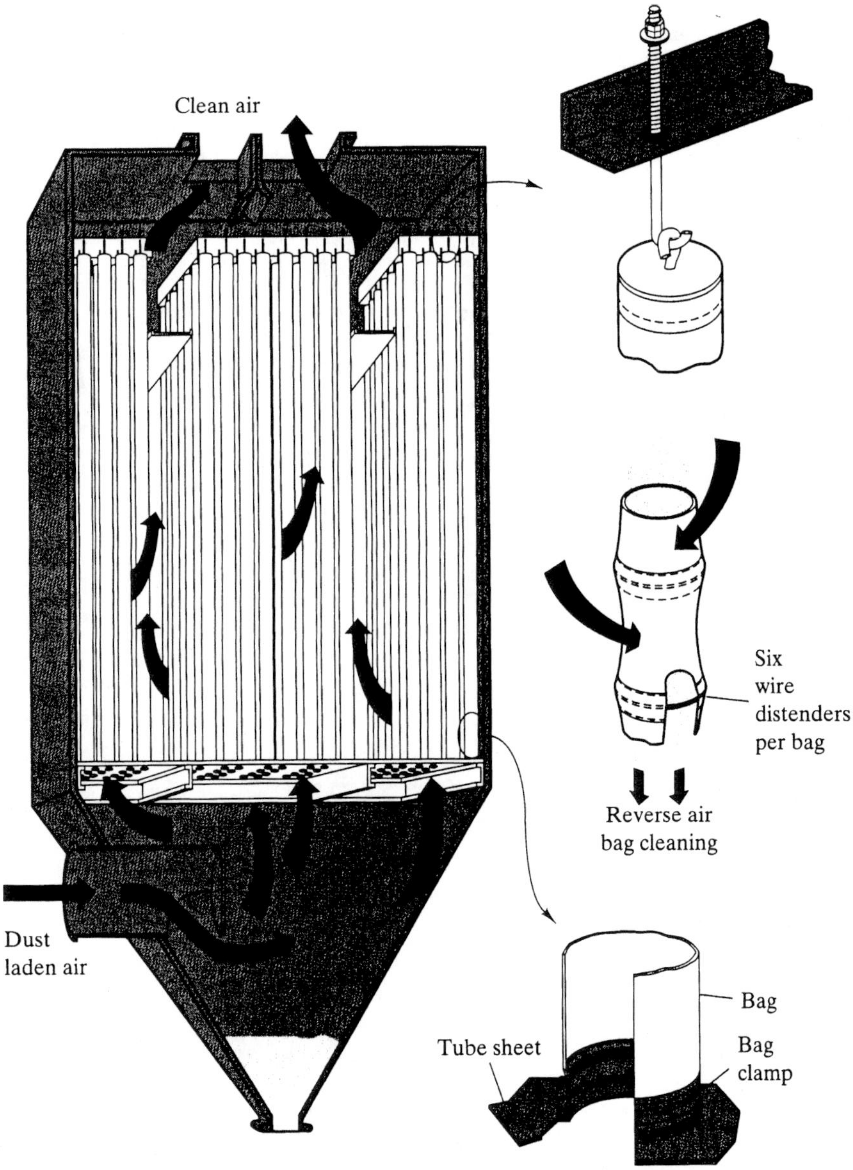

Figure 9.24 One module of a reverse-flow baghouse; bag closed and hung from the top, open and sealed at the bottom; there are 112 bags in this module.

Pulse-Jet Baghouse: The fundamental difference between a shaker or reverse-flow baghouse and a pulse-jet baghouse is that dust laden air passes radially *inward* into the bag in a pulse-jet baghouse, but flows radially *outward* through the bag in shaker and reverse-flow baghouses. Thus dust accumulates on the *outside* surface of the pulse-jet bag and on the *inside* surface of shaker and reverse-flow bags. A schematic diagram of a ***pulse-jet baghouse*** is shown in Figure 9.25. A pulse-jet baghouse contains cylindrical fiber filters surrounding cylindrical wire cages that maintain the cylindrical shape and allow air to flow radially inward through the filters. The filter bags are closed at the bottom but open at the top, where they are clamped to a tube sheet. Dust laden air enters the plenum surrounding the bags; cleaned air passes inward through the bag and

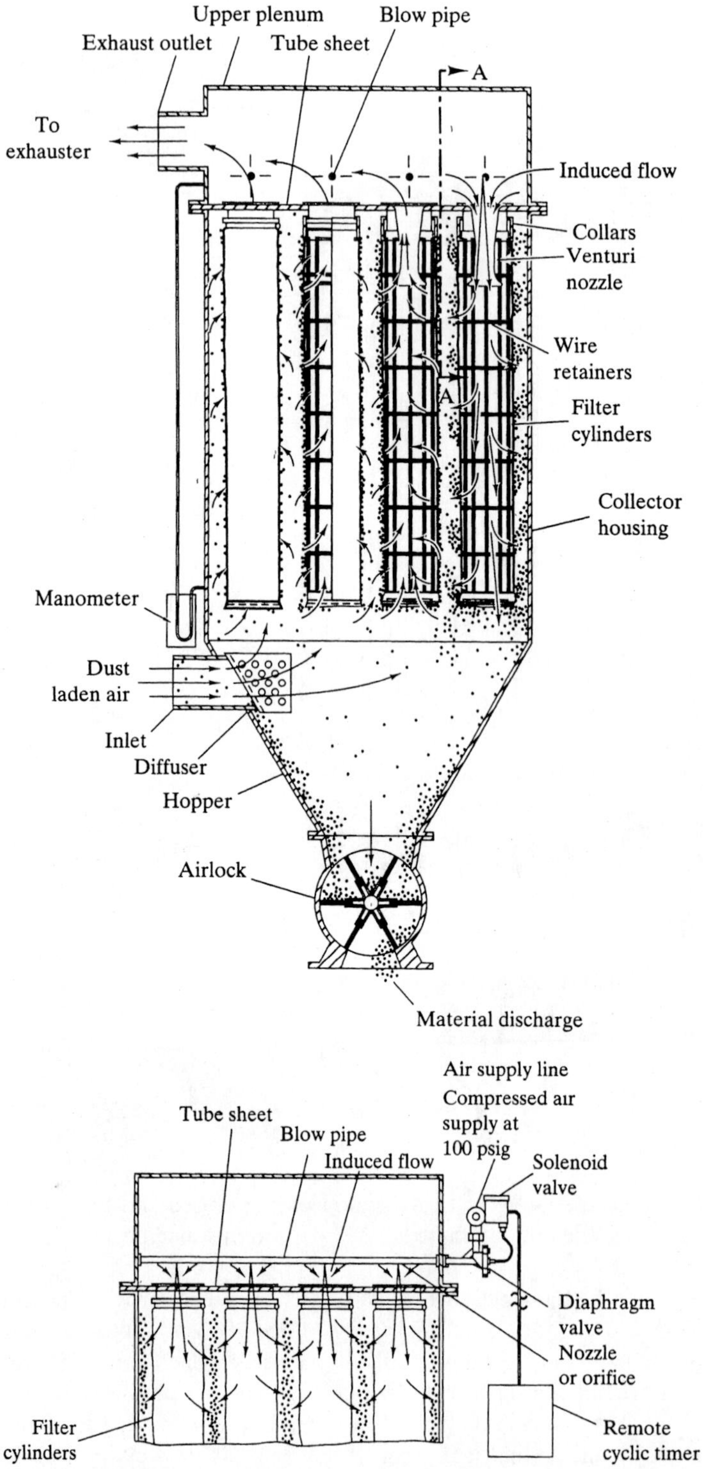

Figure 9.25 Pulse-jet baghouse; cutaway view at top shows internal wire cage; bag open and sealed at the top, closed at the bottom; there are 16 bags in this module.

exits the top of the bag, while dust builds up on the outside surface of the bag. Periodically a high-pressure pulse of air passes through nozzles at the top of each filter bag. The air pulse travels downward, flexes the bag outward, and dislodges dust collected on the outside of the filter; the dust then falls into the hopper below. Following cleaning, the high-pressure air jet ceases, dust-laden gas again tries to enter the bag, and dust is once again collected on its exterior surface.

Two major features differentiate pulse-jet filters from reverse-flow and shaker filters:

(a) Reverse-flow and shaker filters are cleaned every one-half to several hours. Pulse-jet filters are cleaned every few minutes. Thus the weight of the collected dust (dust cake) can become large for reverse-flow and shaker filters but is very small for pulse-jet filters. The dust cake from a reverse-flow baghouse weighed 40-60 lbf/bag, as measured by Carr and Smith (1984), who also report that only about 10% of the dust is removed per cleaning cycle, implying that much of the dust is trapped within the filter fibers, as illustrated in Figure 9.26.

(b) Pulse-jet filters are composed of ***felt*** material whereas shaker and reverse-flow filters are composed of woven fabric. Woven fabrics are constructed of yarn woven to provide the tensile strength needed by the fabric to withstand the large dead weight of the dust cake and flexure properties needed to withstand the cleaning process. Felts consist of fibers that adhere to one another because of adhesive applied to the surface of each fiber. Felts do not have the tensile strength of woven fabrics, so felt filters surround a cylindrical wire structure to maintain its cylindrical shape. Figure 9.27 shows the fundamental differences between woven fabric and felt.

Filter collectors may be woven fibers as seen in Figure 9.27a and 9.27b, or a bed of tightly packed matted fibers (felt), as in Figure 9.27c. Alternatively, the bed may consist of a layer of individual collected particles attached to the filter fabric through which the aerosol passes, as in Figure 9.26. Particles are removed as they impact on the collectors. Industrial filters often combine both phenomena since the dust cake on the upstream (dirty) side of a filter acts as a filter bed while the filter fabric itself removes additional particles and supports the dust cake.

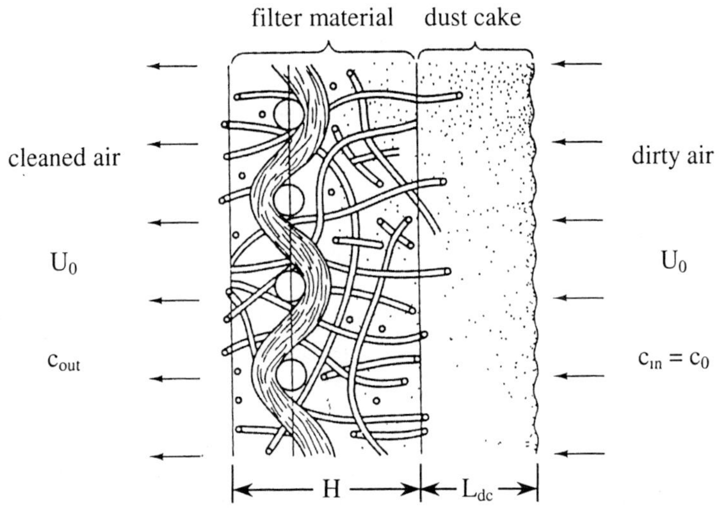

Figure 9.26 Schematic diagram of filter material of thickness H and dust cake of thickness L_{dc}, which varies with time.

(a)

(b)

(c)

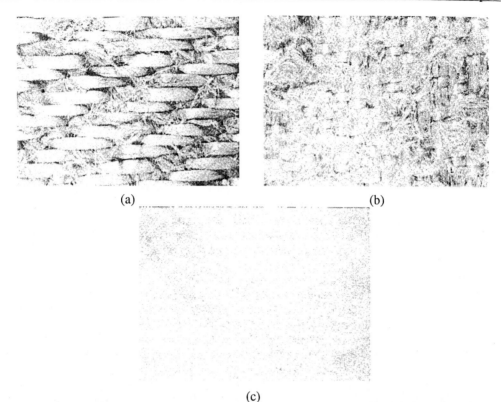

Figure 9.27 Typical filter material; (a) woven fiberglass with a Teflon® B finish that has become the industry standard for boiler applications, (b) highly texturized woven fiberglass used for very fine particles, (c) felted material suitable for low-acid applications and for high concentrations of abrasive particles.

9.4.2 Small Indoor Filtration Units

Examples of smaller filtration units used inside a plant and dedicated to individual machines are seen in Figures 9.28 and 9.29. The unit in Figure 9.28 is called a ***cartridge pulse-jet downflow dust collector***. The principle of operation is the same as that of a pulse-jet baghouse, except that instead of felt, the unit uses unique cartridge filters that remove a variety of dry particles, even those at elevated gas temperatures. During normal operation, contaminated air enters the top of the unit, and dust collects on the outsides of the cartridge filters as cleaned air is sucked out through the inside of the cartridge. During the cleaning cycle, a jet of clean air is pulsed in the opposite direction through some of the cartridges, dislodging the dust, most of which falls into a hopper at the bottom of the unit. The remaining cartridges operate as usual during the pulse, and are themselves cleaned at a later phase of the cycle. Elaborate hopper valves to remove dust from the hoppers are not needed for these small units. Rather, the captured dust is collected in drums and removed when it is necessary to do so. The unit requires an external fan to draw the aerosol into the machine and an external high-pressure air supply to operate the pulse jets.

A small indoor shaker filter is seen in Figure 9.29. Dirty air enters the inlet near the bottom of the unit and is drawn through a one piece cloth ***envelope filter*** by a centrifugal fan on top of the unit. The filter assembly is shaken periodically to remove collected dust. In some less sophisticated units the automatic shaker control can be eliminated and the bags shaken manually by a lever mounted on the outside of the unit. The systems in Figures 9.28 and 9.29 should not be used to collect sticky or wet particles because they cannot be dislodged from the filter media.

Figure 9.28 Cartridge pulse-jet downflow dust collector (with permission of Torit Products, Donaldson Company).

9.4.3 Fundamental Concepts of Filtering

Collection of small contaminant particles of diameter D_p onto larger filter fibers of diameter D_f is provided by impaction, interception, and diffusion. The interaction is largely the same as that described previously for small particles interacting with liquid collecting drops (see Figure 9.14), except that the collecting elements are now cylinders of diameter D_f rather than spheres of diameter D_c. However, because the aerosol velocity is low, *diffusion* is of more importance in filtration than it is in scrubbing. The ***single fiber collection efficiency*** for spheres impacting on cylinders (η_f) is reported by Calvert and Englund (1984) as the following:

$$\boxed{\eta_f = \left(\frac{\text{Stk}}{\text{Stk} + 0.85}\right)^{2\,2}} \tag{9-44}$$

where Stokes number (Stk) is defined by Eq. (9-13), but with D_f in place of D_c. Note that η_f is a function of particle diameter D_p, but for brevity is not written as $\eta_f(D_p)$ in the equations which follow. When the Stokes number is about 0.05 or less, the single fiber impaction collection efficiency is negligible, but diffusion becomes important. Figure 9.30 illustrates the mechanism of diffusion. Two particles are shown; although both are initially at the same distance from the collecting fiber's centerline, one is collected and one is not, due to the random (Brownian) motion caused by diffusion.

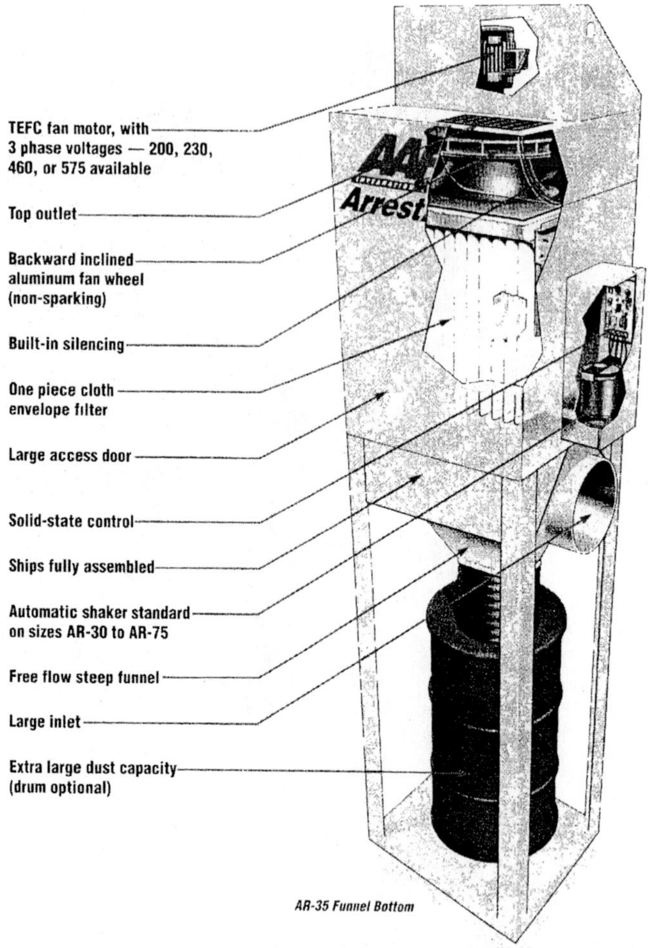

Figure 9.29 Small shaker filter suitable for an individual machine in the workplace (with permission of Torit Products, Donaldson Company).

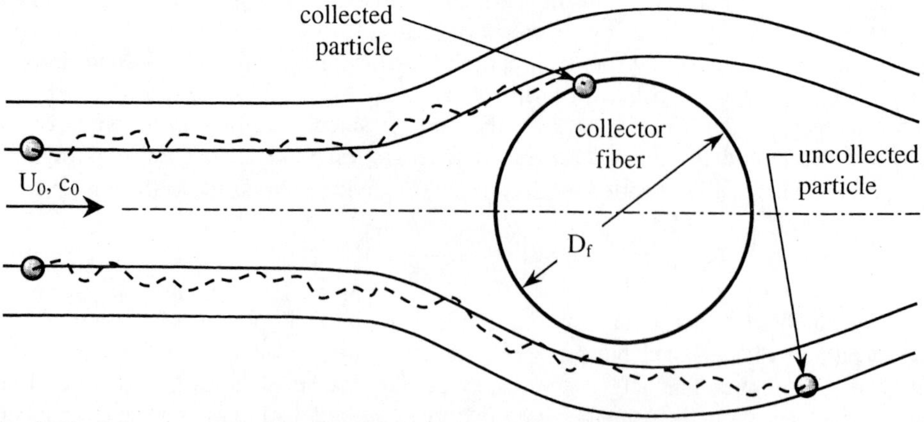

Figure 9.30 Sphere-on-cylinder single filter fiber particle collection due to diffusion; the top particle impacts the collecting fiber, while the bottom particle does not; solid lines are air streamlines, and dashed lines are particle trajectories.

Hinds (1982) suggests that the following equation for total single fiber collection efficiency can be used to include impaction (η_I), interception (η_R), diffusion (η_D), and an enhanced collection due to a combination of diffusion and interception (η_{DR}) for large Peclet numbers,

$$\boxed{\eta_f = \eta_R + \eta_I + \eta_D + \eta_{DR}} \tag{9-45}$$

Interception efficiency is given by

$$\boxed{\eta_R = \frac{2\left(R_{pf}+1\right)\ln\left(R_{pf}+1\right)-\left(R_{pf}+1\right)+\dfrac{1}{\left(R_{pf}+1\right)}}{2Ku}} \tag{9-46}$$

where R_{pf} is the ratio of particle diameter to filter fiber diameter,

$$\boxed{R_{pf} = \frac{D_p}{D_f}} \tag{9-47}$$

and Ku is the ***Kuwabara hydrodynamic factor*** that accounts for alteration of the flow field as air flows around fibers in close proximity to one another,

$$\boxed{Ku = -\frac{1}{2}\ln f_f - 0.75 + f_f - \frac{f_f^{\,2}}{4}} \tag{9-48}$$

where f_f is the ***fiber solids fraction***, defined as the ratio of the volume of solids to overall volume of the filter, discussed in more detail below. The impaction efficiency is given by

$$\boxed{\eta_I = \frac{J\ Stk}{2(Ku)^2}} \tag{9-49}$$

where J is a dimensionless empirical function of R_{pf} and f_f (Hinds, 1982). For $R_{pf} < 0.4$,

$$\boxed{J = \left(29.6 - 28.0 f_f^{\,0.62}\right) R_{pf}^{\,2} - 27.5 R_{pf}^{\,2.8}} \tag{9-50}$$

The diffusion efficiency is

$$\boxed{\eta_D = \frac{2}{(Pe)^{0.67}}} \tag{9-51}$$

where the ***Peclet number*** (Pe) is defined as

$$\boxed{Pe = \frac{D_f U_0}{D}} \tag{9-52}$$

and D is the ***particle diffusion coefficient***,

$$\boxed{D = \frac{kTC}{3\pi D_p \mu}} \tag{9-53}$$

k is Boltzmann's constant = 1.38×10^{-23} J/(molecule K), and C is the Cunningham correction factor. The combined diffusion and interception efficiency is given by the following if Pe is greater than about a hundred:

$$\boxed{\eta_{DR} = \frac{1.24 R_{pf}^{\,0.67}}{\sqrt{Ku \cdot Pe}}} \tag{9-54}$$

To estimate the grade efficiency of a filter, consider one-dimensional flow of an aerosol through a filter as shown in Figure 9.31, where H is the total thickness of the filter, and A_f is the ***face area*** (frontal area) of the filter. The velocity inside the filter (U) is larger than the velocity approaching the filter (U_0) owing to the blockage produced by fibers within the filter.

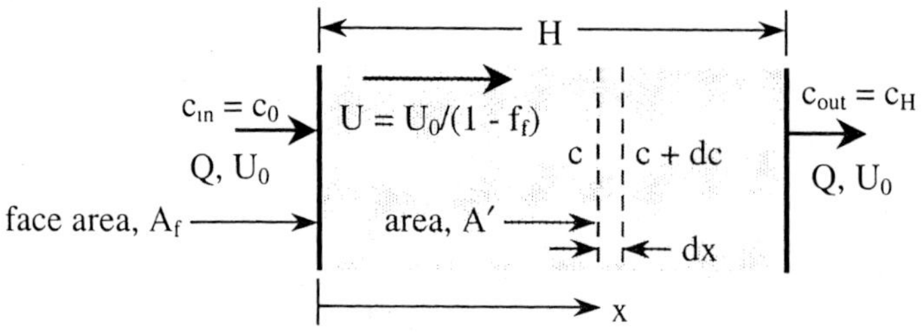

Figure 9.31 Schematic diagram illustrating an analytical model of flow through a filter.

To model the overall collection efficiency, it is necessary to define the following parameters:

- fiber solids fraction (f_f)
- length of fiber per unit volume of filter (L_f)

The ***porosity*** (ε) of the filter is defined as the fraction of the overall filter volume that is open. The porosity is also called the ***voidage*** or ***void fraction***. The ***fiber packing density*** (f_f), also called the ***fiber solids fraction***, is the fraction of the overall filter volume that is composed of solids (fibers). f_f is related to the porosity by

$$\boxed{f_f = 1 - \text{porosity} = 1 - \varepsilon} \tag{9-55}$$

By similar reasoning, f_f is the ratio of the bulk density of the filter to the density of the fiber material, as per Eq. (9-39). Similarly, f_f is equal to the ***blockage***, the fraction of the cross-sectional area of the filter that is composed of solid matter,

$$\boxed{f_f = 1 - \frac{A'}{A_f}} \tag{9-56}$$

where A′ is the cross-sectional area through which air actually moves inside the filter. The speed of the aerosol inside the filter (U) is related to the approach speed (U_0) by

$$\boxed{U = \frac{U_0 A_f}{A'} = \frac{U_0}{1 - f_f} = \frac{U_0}{\varepsilon}} \tag{9-57}$$

The ***length of fiber per unit volume of filter*** (L_f) is defined as follows. For simplicity, imagine that the filter is composed of a single fiber of length L_{fiber} and of diameter D_f convoluted like a long spaghetti noodle into filter volume V defined by the outer boundaries of the filter. The fiber solids fraction (f_f) can then be written as the ratio of fiber volume to total filter volume,

$$f_f = \frac{\frac{\pi}{4} D_f^{\,2} L_{fiber}}{V}$$

In terms of fiber solids fraction and fiber diameter, L_f can thus be written as

$$\boxed{L_f = \frac{L_{fiber}}{V} = \frac{4 f_f}{\pi D_f^{\,2}}} \tag{9-58}$$

Consider air carrying contaminant particles as it passes through the filter. A mass balance for the contaminant in the elemental filter volume shown in Figure 9.31 can be written as

$$cUA' = (c + dc)UA' + cUD_f L_f \eta_f A_f dx$$

Using Eqs. (9-56) and (9-58), the above reduces to

$$\frac{dc}{c} = -\frac{D_f \eta_f L_f}{1-f_f} dx = -\frac{4}{\pi} \eta_f \frac{f_f}{1-f_f} \frac{1}{D_f} dx$$

which can be integrated (from c_{in} to c_{out} on the left, and from x = 0 to H on the right) to obtain the grade efficiency, $\eta(D_p)$, of the filter for particles of diameter D_p,

$$\boxed{\eta(D_p) = 1 - \frac{c_{out}(D_p)}{c_{in}(D_p)} = 1 - \exp\left[-\frac{4}{\pi} \eta_f \frac{f_f}{1-f_f} \frac{H}{D_f}\right]} \quad (9\text{-}59)$$

Upon examination, each term in Eq. (9-59) is of order unity except (H/D_f) which can be considerably larger than unity. For this reason the grade efficiency should be virtually 100% for all particle sizes. However, an exception may occur when the single fiber collection efficiency (η_f) is very small, as occurs for a narrow band of particle diameters in the vicinity of 0.1 micrometers, and may lead to a trough in the grade efficiency curve. In this range, collection by both diffusion and impaction may be small. The location and depth of the trough vary with face velocity, U_0, and with filter fiber diameter, D_f, as illustrated in Figures 9.32 and 9.33 respectively (Hinds, 1982). In Figure 9.32, grade efficiency plots are shown for an example filter with $f_f = 0.05$ and H = 1.0 mm, and for two values of U_0. Note the trough in grade efficiency for particles in the range $D_p \approx 0.02$ to 1 μm. The trough shifts to smaller particles and increases in depth as face velocity increases. As filter thickness increases, however, H/D_f becomes large and the trough in the grade efficiency curve is less of a concern. Thicker filters have higher pressure drops, however, so some optimization may be necessary for a particular filtration application.

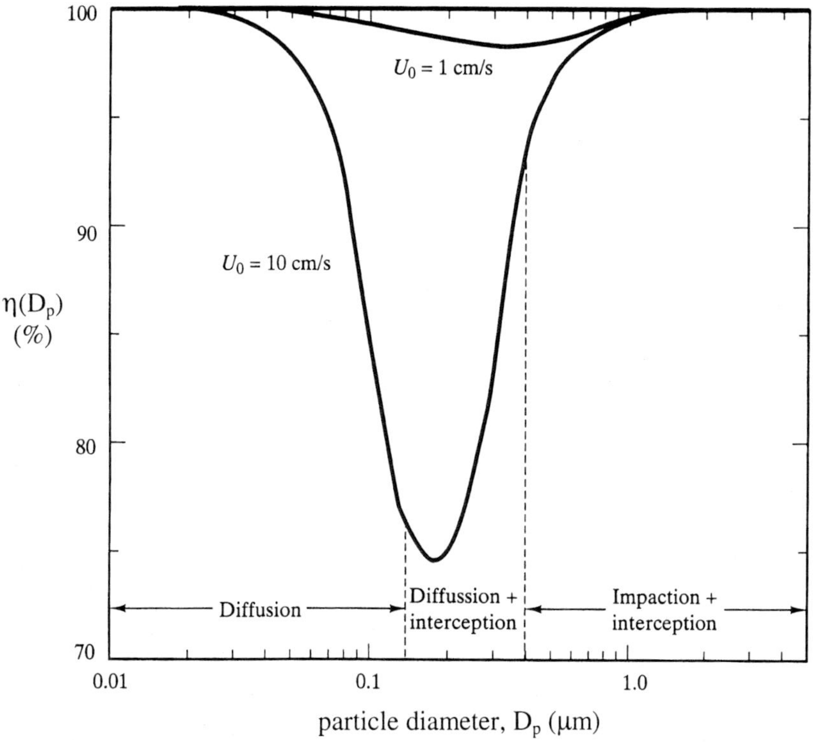

Figure 9.32 Filter grade efficiency for two face velocities; filter thickness H = 1.0 mm, solids fraction $f_f = 0.05$, single fiber diameter $D_f = 2$ μm (adapted from Hinds, 1982).

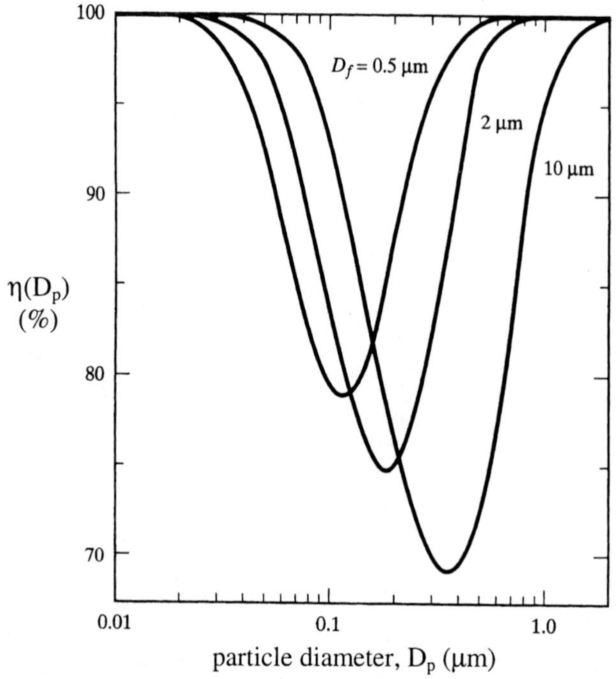

Figure 9.33 Filter grade efficiency for three fiber diameters (D_f); face velocity U_0 = 1.0 cm/s, solids fraction f_f = 0.05, filter thickness (H) adjusted to produce the same pressure drop for all three cases (adapted from Hinds, 1982).

Figure 9.33 compares the grade efficiency curve for filters with three different filter fiber diameters, for the case with f_f = 0.05 and U_0 = 1.0 cm/s. Filter thickness (H) was adjusted as necessary to produce the same pressure drop for all three cases. The trough shifts to larger particles and increases in depth as fiber diameter increases. For many applications it is important to be able to predict conditions producing the trough of low filter grade efficiency as seen in Figures 9.32 and 9.33. Particles associated with this trough are said to *penetrate* the filter. From Eq. (9-59), it is clear that for fixed values of the solids fraction f_f, fiber diameter D_f, and filter thickness H, the most penetrating particles produce the lowest filter grade efficiency.

Equation (9-59) assumes that the filter material is itself the primary agent collecting the particles. This is correct for felt fibers (as used in pulse-jet baghouse filters), but not true for fabric filters (as used in reverse-flow or shaker baghouse filters) except during the early stages of collection. When the collected dust on the filter accumulates into a dust cake and becomes the primary medium removing the particles, the above derivation needs to be repeated using physical properties of the dust cake rather than those of the fabric to arrive at an expression similar to Eq. (9-59). The thicker the dust cake, the less the penetration, but pressure drop and tension in the bag increase as the dust cake grows in thickness (and weight); this is the reason why dust cakes must be periodically removed from the filter bag.

The ***most penetrating particle diameter*** ($D_{p,min}$) can be found by differentiating Eq. (9-59) with respect to D_p and setting the derivative equal to zero. Since the particles are small, particle removal by inertia and gravimetric settling is unimportant and the single fiber collection efficiency is dominated by interception and diffusion. For these conditions, Lee and Liu (1980) suggest that the single fiber collection efficiency can be expressed by

$$\eta_f = 2.6(Pe)^{-0.67}\left(\frac{1-f_f}{Ku}\right)^{0.33} + \frac{R_{pf}^{\;2}\left(\frac{1-f_f}{Ku}\right)}{1+R_{pf}} \qquad (9\text{-}60)$$

where R_{pf} and Ku were defined previously in Eqs. (9-47) and (9-48). Differentiating Eq. (9-59) is complicated by the fact that the diffusion coefficient in the Peclet number is also dependent on particle size, D_p. For particles in which the Knudsen number ($K_n = \lambda/D_p$) lies between 0.075 and 1.3, Lee and Liu (1980) show that

$$D_{p,min} = 0.885\frac{kT\sqrt{\lambda}}{\mu}\frac{Ku}{1-f_f}\left(\frac{D_f^{\;2}}{U}\right)^{2/9} \qquad (9\text{-}61)$$

where k is Boltzmann's constant. Examining Eq. (9-61), the larger the filtration velocity (U), or the smaller the solids fraction f_f, or the smaller the single fiber diameter D_f, the smaller the size of the most penetrating particle.

The ***minimum single fiber collection efficiency*** ($\eta_{f,min}$) associated with the most penetrating particle can be found by substituting Eq. (9-61) into Eq. (9-44) and simplifying. The result is

$$\eta_{f,min} = 2.35\left[\left(\frac{1-f_f}{Ku}\right)^5 \frac{J^4}{U^4 D_f^{\;10}}\right]^{1/9} \qquad (9\text{-}62)$$

where J is defined for $R_{pf} < 0.4$ by Eq. (9-50). From this, one concludes that the minimum efficiency decreases with decreasing values of the solids fraction f_f, or increasing values of fiber diameter D_f or filtration velocity U.

One of the attractive features of filters is the relative independence of the grade efficiency to volumetric flow rate (provided that the filter thickness is large enough so that the troughs in Figures 9.32 and 9.33 are insignificant as discussed above). Thus filters are well suited for batch processes where the volumetric flow rate is not constant. To compare baghouses such as those shown in Figures 9.23 to 9.25, it is customary to refer to the volumetric flow rate in terms of the air-to-cloth ratio. ***Air-to-cloth ratio*** is defined as the ratio of total volumetric flow rate of cleaned air (Q_a) to total filter face area (A_f), and has dimensions of velocity,

$$\text{air-to-cloth ratio} = \frac{Q_a}{A_f} \qquad (9\text{-}63)$$

Superficial velocity (U_0) is defined as the velocity of the carrier gas approaching the filter face. Consequently the air-to-cloth ratio and the superficial velocity are the same when this velocity is normal to the filter face.

If the process gas stream (at flow rate Q_a) enters the filter normal to its surface (of area A_f), the air-to-cloth ratio is equal to the average air velocity entering the filter (U_0). With cyclones, scrubbers, and other inertial separation collectors, the overall collection efficiency is strongly dependent on volumetric flow rate (Q_a). Thus for industrial batch processes in which the volumetric flow rate is subject to change, inertial separation collectors have an inherent disadvantage. With filters, the overall collection efficiency ($\eta_{overall}$) is virtually 100% for all volumetric flow rates except for the unique conditions producing the minimum single fiber collection efficiency ($\eta_{f,min}$).

The pressure drop across a filter increases with time, air-to-cloth ratio, and incoming particle mass concentration ($c_{in} = c_0$). It is assumed that the dust cake seen in Figure 9.26 is of ***dust cake thickness*** L_{dc}, which varies with time. After a period of time t, this thickness can be expressed as

$$L_{dc}(t) = \frac{w(t)}{\rho_p f_{f,dc}} \tag{9-64}$$

where

- $f_{f,dc}$ is the solids fraction of the dust cake
- ρ_p is the density of the particulate matter
- w(t) is the mass of dust per unit area of filter after a period of time t; w(t) is called the ***dust cake loading***

Dust cake loading can be written as

$$w(t) = c_0 U_0 t \eta_{overall} \approx c_0 U_0 t \tag{9-65}$$

because the overall collection efficiency is virtually 100%. Combining Eqs. (9-64) and (9-65), the thickness of the dust cake becomes

$$L_{dc}(t) = \frac{c_0 U_0 t}{\rho_p f_{f,dc}} \tag{9-66}$$

The overall pressure drop ($\delta P_{overall}$) across the filter and dust cake can be written as

$$\delta P_{overall} = \delta P_{filter\ material} + \delta P_{dust\ cake} = \delta P_f + \delta P_{dc} \tag{9-67}$$

where δP_f and δP_{dc} are the pressure drops across the filter material and the dust cake respectively. Since the velocities are low, these pressure drops can be expressed by ***Darcy's law*** which is applicable to flow through porous media,

$$\left(\frac{dP}{dx}\right)_f = -C_f \mu U_0 \quad \text{and} \quad \left(\frac{dP}{dx}\right)_{dc} = -C_{dc} \mu U_0 \tag{9-68}$$

where C_f and C_{dc} are constants related to the porosity of the filter material and dust cake respectively. The overall pressure drop is thus

$$\delta P_{overall} = C_f \mu U_0 H + C_{dc} \mu U_0 L_{dc}(t) = C_f \mu U_0 H + C_{dc} \mu U_0 \frac{c_0 U_0 t}{\rho_p f_{f,dc}} = (C_f \mu H) U_0 + \left(\frac{C_{dc} \mu}{\rho_p f_{f,dc}}\right) c_0 U_0^2 t$$

Defining K_f and K_{dc} by the groupings in parentheses in the above equation, the overall pressure drop becomes

$$\delta P_{overall} = K_f U_0 + K_{dc} c_0 U_0^2 t \tag{9-69}$$

The constant K_f is called the ***residual drag*** of the filter, and the constant K_{dc} is called the ***dust cake specific resistance***. Values of K_f can be obtained from filter manufacturers or by experiment. In the absence of these values, Calvert and Englund (1984) suggest using K_f = 350. N min/m^3. The range of K_{dc} values is wide and engineers should conduct experiments with dust cakes obtained from their particular process to determine accurate values of K_{dc}. Table 9.1 summarizes typical values of air-to-cloth ratios for felt (pulse-jet baghouse) and woven fabric (shaker and reverse-flow baghouse) filters, and K_{dc} for woven fabric filters. Table 9.2 lists the maximum operating temperature of several fabrics commonly used for filtering.

Table 9.1 Air-to-cloth ratio for felt and woven fabric filters and dust cake specific resistance for woven fabric filters (abstracted from Calvert and Englund, 1984).

particle type	air-to-cloth ratio for felt (m/min)	air-to-cloth ratio for woven fabric (m/min)	K_{dc} for woven cloth (s^{-1})
alumina	2.44	0.58	1.98×10^3
asphalt		0.76-2.23	1.70×10^4
calcium sulfate		2.28	4.02×10^3
carbon black	1.52	0.34-0.49	$2.2\text{-}5.61 \times 10^5$
cement	2.44	0.46-0.64	$1.2\text{-}7.01 \times 10^5$
coal	2.44	0.76	
cocoa chocolate	3.65	0.85	
copper		0.18-0.82	$1.5\text{-}6.5 \times 10^5$
cosmetics	3.04	0.92	
dolomite		1.00	6.72×10^6
electric furnace		0.46-1.22	$0.45\text{-}7.14 \times 10^6$
flour	3.66	0.76	4.3×10^4
fly ash	1.52	0.58-1.80	$0.70\text{-}1.51 \times 10^5$
foundry dust		0.64	$6 \times 10^3\text{-}1.2 \times 10^6$
gypsum	3.05	0.76	$0.63\text{-}1.9 \times 10^5$
iron oxide	2.13	0.43-1.00	$3 \times 10^4\text{-}7.14 \times 10^6$
lead oxide	1.89	0.30	5.7×10^5
lime kiln	3.05	0.7	9×10^4
limestone	2.44	0.82	
milk powder			4.5×10^4
oats	4.27	1.6	1.5×10^4
pigments	2.13	0.61	$2.28\text{-}2.88 \times 10^4$
soap	1.52	0.69	$1.62\text{-}3.12 \times 10^4$
tobacco	3.65	1.07	3.6×10^5
wood sawdust	3.66	1.07	
zinc		.55-0.92	$0.7\text{-}5.01 \times 10^5$
zinc oxide	1.52	0.18-0.36	$1.84\text{-}4.0 \times 10^5$

It must be remembered that the pressure drop for a new, unused filter is somewhat less than that of a filter that has undergone several cleaning cycles, due to the build-up non-removable dust within the filter material (see Figure 9.26). In reverse-flow and shaker fabric baghouses that use woven fabric filters, the largest pressure drop is through the dust cake. In pulse-jet filter units that use felt material, the dust cake is thin and the pressure drop through the filter is of more importance. There are occasions when Eq. (9-69) is shortened to

$$\boxed{\delta P_{\text{overall}} = S U_0} \tag{9-70}$$

Table 9.2 Maximum operating temperature of common filter fabrics (abstracted from Calvert and Englund, 1984).

material	T_{max}
cotton	180 °F
dacron	275 °F
fiberglass	500 °F

material	T_{max}
nomex	375 °F
nylon	200 °F
orlon	260 °F

material	T_{max}
polypropylene	200 °F
teflon	450 °F
wool	200 °F

where parameter S is called the ***filter drag***. It can be seen that the filter drag is a property of not only the filter and the particles to be collected, but also the thickness of the dust cake which in turn varies linearly with time and dust concentration. For an entire baghouse in which the separate modules are taken off-line and cleaned in a prescribed fashion, the overall pressure drop of the baghouse may vary linearly with air-to-cloth ratio (U_0) and it may be possible to measure the filter drag for the entire baghouse.

As sketched in Figure 9.34, the pressure drop increases linearly with time, after some short time of adjustment. The slope of δP with respect to t varies with the square of velocity (air-to-cloth ratio). Thus in baghouses consisting of several modular units on a staggered cleaning schedule, engineers must decide what maximum pressure drop they are willing to tolerate, and must adjust the cleaning cycle to clean each unit after the consequent predetermined period of time.

9.4.4 High-Temperature Filtration

Many industrial processes generate hot waste gases containing particles. Conventional fabric filters require the gas to be cooled prior to filtration. Cooling can be accomplished by ***evaporative cooling*** (spraying water in the gas stream) if the waste gases are very dry. On the other hand, if the water vapor content in the gas stream is large, evaporative cooling could produce a humid gas stream poorly suited for filtration. In this case, ambient air (possibly dried), may have to be used to cool the gas stream. The choice of cooling method is dictated by the maximum water content that can be tolerated in the gas stream prior to filtration. If cooled by spraying water, the final volumetric flow rate may be smaller than the original volumetric flow rate. If cooled with ambient air, the final volumetric flow rate may be larger than the original volumetric flow rate. A larger volumetric flow rate requires a larger baghouse, accompanied by higher capital cost and higher operating costs.

Example 9.6 - Cooling a Hot and Dry Gas Stream

Given: A 10,000 ACFM dry (0% water vapor) discharge gas stream containing small particles exits a metallurgical kiln at T_1 = 600. °C and P_1 = 101. kPa. To capture the particles it is necessary to reduce the gas temperature to T_3 = 200. °C and P_3 = 101. kPa before passing the gas into a baghouse. To keep the bags from ***blinding*** (being totally blocked by particles), it has been found necessary to maintain the ***humidity ratio***, ω_3 = (mass of water vapor) / (mass of dry air), below 0.65. The gas stream enters the cooler at state (1), is cooled by material entering at state (2), and leaves the cooler to enter the baghouse at state (3). Three cooling methods are to be considered:

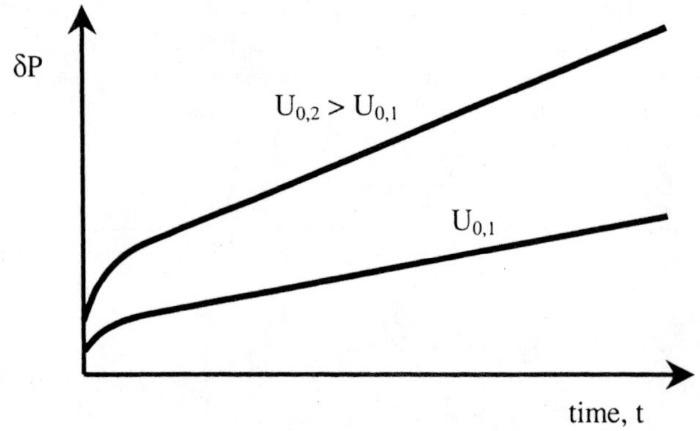

Figure 9.34 Pressure drop across a filter versus time for an aerosol with two different face velocities ($U_{0,1}$ and $U_{0,2}$) and constant inlet particle concentration (c_0).

- Method 1: cool with dried air (0% water vapor) at $T_2 = 25.$ °C
- Method 2: cool with a water spray at $T_2 = 25.$ °C
- Method 3: cool with ambient air at $T_2 = 25.$ °C in which the relative humidity, Φ, is 50.% and $\omega_2 = 0.010$

Humidity ratio (ω) is related to the partial pressure of water vapor ($P_{\text{water vapor}}$) and that of dry air ($P_{\text{dry air}}$). Using relationships developed for gas mixtures in Chapter 1,

$$\omega = \frac{m_{\text{water vapor}}}{m_{\text{dry air}}} = \frac{M_{\text{water vapor}} n_{\text{water vapor}}}{M_{\text{dry air}} n_{\text{dry air}}} = \frac{M_{\text{water}}}{M_{\text{dry air}}} \frac{n_{\text{water vapor}}}{n} \frac{n}{n_{\text{dry air}}} = \frac{M_{\text{water}}}{M_{\text{dry air}}} \frac{P_{\text{water vapor}}}{P} \frac{P}{P_{\text{dry air}}}$$

where n is the total number of mols in the gas mixture and P is its total pressure. Since P is the sum of partial pressures $P_{\text{dry air}}$ and $P_{\text{water vapor}}$, the above reduces to

$$\omega = \frac{M_{\text{water}}}{M_{\text{dry air}}} \frac{P_{\text{water vapor}}}{P - P_{\text{water vapor}}} = \frac{18.02}{28.97} \frac{P_{\text{water vapor}}}{P - P_{\text{water vapor}}} = 0.662 \frac{P_{\text{water vapor}}}{P - P_{\text{water vapor}}}$$

Thus a maximum value of $\omega = 0.65$ corresponds to a water vapor partial pressure of $P_{\text{water}} = 51.6$ kPa. For purposes of calculation one can round this to a maximum value of $P_{\text{water}} \approx 50$ kPa.

To do: Analyze the three cooling methods.

Solution: The data are summarized below:

$\rho_1 = P_1/(RT_1) = 0.4031 \text{ kg/m}^3$ $\qquad Q_1 = Q_{\text{dry air},1} = 10{,}000$ ACFM (283.39 actual m^3/min)

$\rho_3 = P_3/RT_3 = 0.7440 \text{ kg/m}^3$ $\qquad \dot{m}_1 = \dot{m}_{\text{dry air},1} = Q_1\rho_1 = 114.23$ kg/min

water vapor specific enthalpy: $\hat{h}_{w,3}$ (200. °C, 50. kPa) = 2877.7 kJ/kg

Method 1 (dilution with dried air) - Cool with dry air $T_2 = 25.$ °C. Defining the cooler as an adiabatic open system (a control volume) and applying conservation of mass,

$$\dot{m}_3 = \dot{m}_1 + \dot{m}_2 = \dot{m}_{\text{dry air},1} + \dot{m}_{\text{dry air},2}$$

and conservation of energy,

$$0 = \sum\left(\dot{m}\hat{h}\right)_{\text{out}} - \sum\left(\dot{m}\hat{h}\right)_{\text{in}} = \dot{m}_3\hat{h}_3 - \dot{m}_{\text{dry air},1}\hat{h}_{\text{dry air},1} - \dot{m}_{\text{dry air},2}\hat{h}_{\text{dry air},2}$$

$$= \left(\dot{m}_{\text{dry air},1} + \dot{m}_{\text{dry air},2}\right)\hat{h}_{\text{dry air},3} - \dot{m}_{\text{dry air},1}\hat{h}_{\text{dry air},1} - \dot{m}_{\text{dry air},2}\hat{h}_{\text{dry air},2}$$

$$= \dot{m}_{\text{dry air},1}\left(\hat{h}_{\text{dry air},3} - \hat{h}_{\text{dry air},1}\right) + \dot{m}_{\text{dry air},2}\left(\hat{h}_{\text{dry air},3} - \hat{h}_{\text{dry air},2}\right)$$

Since air behaves as an ideal gas, each specific enthalpy difference can be replaced by the specific heat at constant pressure (c_P) times the temperature difference,

$$0 = \dot{m}_{\text{dry air},1} c_P \left(T_3 - T_1\right) + \dot{m}_{\text{dry air},2} c_P \left(T_3 - T_2\right)$$

c_P can be removed from the above equation since it appears in both terms on the right. Solving for the mass flow rate at state 2,

$$\dot{m}_{\text{dry air},2} = -\frac{\dot{m}_{\text{dry air},1}\left(T_3 - T_1\right)}{\left(T_3 - T_2\right)}$$

Plugging in the values,

$$\dot{m}_{\text{dry air},2} = -\frac{\dot{m}_{\text{dry air},1}\left(T_3 - T_1\right)}{\left(T_3 - T_2\right)} = -114.23\frac{\text{kg}}{\text{min}}\frac{(200. - 600.)\text{K}}{(200. - 25.0)\text{K}} = 261.\frac{\text{kg}}{\text{min}}$$

The volumetric flow rate exiting the cooler can now be calculated.

$$Q_3 = \frac{\dot{m}_3}{\rho_3} = \frac{\dot{m}_{\text{dry air},1} + \dot{m}_{\text{dry air},2}}{\rho_3} = \frac{114.2\frac{\text{kg}}{\text{min}} + 261.\frac{\text{kg}}{\text{min}}}{0.744\frac{\text{kg}}{\text{m}^3}} = 504.\frac{\text{m}^3}{\text{min}}$$

This flow rate is about 78 percent larger than the input flow rate, Q_1.

Q_3 (Method 1) = 504. m^3/min

Method 2 (evaporative cooling) - Cool with a water spray at T_2 = 25. °C in which $\hat{h}_{w,2} = \hat{h}_f$ (25. °C) = 104.9 kJ/kg. Apply conservation of mass and energy, as previously. Although the air can be treated as an ideal gas, the water vapor cannot. Conservation of mass becomes

$$\dot{m}_3 = \dot{m}_1 + \dot{m}_2 = \dot{m}_{\text{dry air},1} + \dot{m}_{\text{water},2}$$

and conservation of energy becomes

$$0 = \sum\left(\dot{m}\hat{h}\right)_{\text{out}} - \sum\left(\dot{m}\hat{h}\right)_{\text{in}} = \dot{m}_3\hat{h}_3 - \dot{m}_{\text{dry air},1}\hat{h}_{\text{dry air},1} - \dot{m}_{\text{water},2}\hat{h}_{\text{water},2}$$

$$= \left(\dot{m}_{\text{dry air},1}\hat{h}_{\text{dry air},3} + \dot{m}_{\text{water},2}\hat{h}_{\text{water},3}\right) - \dot{m}_{\text{dry air},1}\hat{h}_{\text{dry air},1} - \dot{m}_{\text{water},2}\hat{h}_{\text{water},2}$$

Solving for the mass flow rate of cooling water,

$$\dot{m}_{\text{water},2} = \frac{\dot{m}_{\text{dry air},1}\left(\hat{h}_{\text{dry air},3} - \hat{h}_{\text{dry air},1}\right)}{\left(\hat{h}_{\text{water},2} - \hat{h}_{\text{water},3}\right)} = \frac{\dot{m}_{\text{dry air},1}c_P\left(T_3 - T_1\right)}{\left(\hat{h}_{\text{water},2} - \hat{h}_{\text{water},3}\right)}$$

$$= \frac{114.2\frac{\text{kg}}{\text{min}}1.004\frac{\text{kJ}}{\text{kg}\cdot\text{K}}(200. - 600.)\,\text{K}}{(104.9 - 2877.7)\frac{\text{kJ}}{\text{kg}}} = 16.54\frac{\text{kg}}{\text{min}}$$

The volumetric flow rate exiting the cooler is

$$Q_3 = \frac{\dot{m}_3}{\rho_3} = \frac{\dot{m}_{\text{dry air},1} + \dot{m}_{\text{water},2}}{\rho_3} = \frac{114.2\frac{\text{kg}}{\text{min}} + 16.54\frac{\text{kg}}{\text{min}}}{0.744\frac{\text{kg}}{\text{m}^3}} = 176.\frac{\text{m}^3}{\text{min}}$$

This flow rate is about 38 percent *smaller* than the input flow rate, Q_1.

Q_3 (Method 2) = 176. m^3/min

Method 3 (dilution with moist ambient air) - Cool with ambient air T_2 = 25. °C containing water vapor at 50% relative humidity, ω_2 = 0.01, and $\hat{h}_{w,2} \cong \hat{h}_g$ (25. °C) = 2547.2 kJ/kg. Applying conservation of mass,

$$\dot{m}_3 = \dot{m}_{\text{dry air},3} + \dot{m}_{\text{water},3} = \dot{m}_1 + \dot{m}_2 = \dot{m}_{\text{dry air},1} + \dot{m}_{\text{dry air},2} + \dot{m}_{\text{water},2}$$

Using the definition of humidity ratio (ω),

$$\dot{m}_{\text{water},2} = \omega_2\dot{m}_{\text{dry air},2}$$

and the above becomes

$$\dot{m}_{\text{dry air},3} + \dot{m}_{\text{water},3} = \dot{m}_{\text{dry air},1} + \dot{m}_{\text{dry air},2}\left(1 + \omega_2\right)$$

Applying conservation of energy,

$$0=\sum\left(\dot{m}\hat{h}\right)_{out}-\sum\left(\dot{m}\hat{h}\right)_{in}$$

$$=\dot{m}_{dry\ air,3}\hat{h}_{dry\ air,3}+\dot{m}_{water,3}\hat{h}_{water,3}-\dot{m}_{dry\ air,1}\hat{h}_{dry\ air,1}-\dot{m}_{dry\ air,2}\hat{h}_{dry\ air,2}-\dot{m}_{water,2}\hat{h}_{water,2}$$

$$=\dot{m}_{dry\ air,1}\left(\hat{h}_{dry\ air,3}-\hat{h}_{dry\ air,1}\right)+\dot{m}_{dry\ air,2}\left(\hat{h}_{dry\ air,3}-\hat{h}_{dry\ air,2}\right)+\dot{m}_{water,2}\left(\hat{h}_{water,3}-\hat{h}_{water,2}\right)$$

$$=\dot{m}_{dry\ air,1}\left(\hat{h}_{dry\ air,3}-\hat{h}_{dry\ air,1}\right)+\dot{m}_{dry\ air,2}\left(\hat{h}_{dry\ air,3}-\hat{h}_{dry\ air,2}\right)+\omega_2\dot{m}_{dry\ air,2}\left(\hat{h}_{water,3}-\hat{h}_{water,2}\right)$$

Solving for the mass flow rate of cooling water,

$$\dot{m}_{dry\ air,2}=-\frac{\dot{m}_{dry\ air,1}\left(\hat{h}_{dry\ air,3}-\hat{h}_{dry\ air,1}\right)}{\left(\hat{h}_{dry\ air,3}-\hat{h}_{dry\ air,2}\right)+\omega_2\left(\hat{h}_{water,3}-\hat{h}_{water,2}\right)}=-\frac{\dot{m}_{dry\ air,1}c_P\left(T_3-T_1\right)}{c_P\left(T_3-T_2\right)+\omega_2\left(\hat{h}_{water,3}-\hat{h}_{water,2}\right)}$$

$$=-\frac{114.2\frac{kg}{min}1.004\frac{kJ}{kg\cdot K}\left(200.-600.\right)K}{1.004\frac{kJ}{kg\cdot K}\left(200.-25.\right)K+0.010\left(2877.7-2547.2\right)\frac{kJ}{kg}}=256.2\frac{kg}{min}$$

The volumetric flow rate exiting the cooler is

$$Q_3=\frac{\dot{m}_3}{\rho_3}=\frac{\dot{m}_{dry\ air,1}+\dot{m}_{dry\ air,2}}{\rho_3}=\frac{114.2\frac{kg}{min}+256.2\frac{kg}{min}}{0.744\frac{kg}{m^3}}=498.\frac{m^3}{min}$$

This flow rate is about 76 percent *larger* than the input flow rate, Q_1.

Q_3 (Method 3) = 498. m^3/min

Discussion: It is clear that a water spray reduces both the temperature and the volumetric flow rate of gas entering the baghouse. If cooling is attempted by using either bone dry air at 25. °C or ambient air at 25. °C, 50% relative humidity, the required volumetric flow rate of cooling gas entering the baghouse must be considerably larger than that of the original hot gas stream.

If the particles to be removed are large and can be removed by conventional fabric filters, cooling alone by one of the methods described in Example 9.6 can be used. If however the original aerosol consists of submicron particles and the gas stream is hot, inexpensive fiber filter material cannot be used and the task of achieving regulatory compliance is doubly difficult.

9.4.5 Applications

Fabric filters are one of the most commonly used particle collectors in industry. In terms of total volume of air cleaned, they clean nearly as much air as do electrostatic precipitators. The attraction of filtration is based on its versatility, variable size, and the fact that the collection efficiency is virtually independent of volumetric flow rate. No other particle collector has a collection efficiency that is independent of the volumetric flow rate. Thus for batch processes or processes in which the volumetric flow rate is variable, fabric filters have an inherent advantage. The principal limitations to fabric filters are:

- process gas temperature cannot exceed 500 °F unless exotic filter material is used
- condensation of vapors cannot be tolerated; the dew point must be considerably below the actual gas temperature
- precautions must be taken to prevent fire and explosion when collecting combustible materials

- the collected dust must be friable, since non-disposable filters cannot be cleaned if they collect particles that adhere strongly to the bags or to other particles

Within these limitations it is obvious that filters are applicable to countless industrial operations in which the volumetric flow rate (Q) may be huge or quite small. For very hot and dry process gas streams, it may be possible to spray water into the gas to decrease its temperature and still retain a low dew point temperature.

To clean a large volumetric flow rate, it is advantageous to place a large number of bags in parallel within a modular unit and arrange the modular units themselves in parallel. In so doing each modular unit can be cleaned in a staggered fashion. Figure 9.35 shows a reverse-flow or shaker baghouse consisting of five modules arranged in parallel. The pressure drop across each module is essentially the same since they are in parallel, but the volumetric flow rate through any one module decreases with time as its dust cake grows in thickness. After a selected amount of time, one module is taken off-line and its bags are cleaned. Some time (several seconds to minutes) is allowed for the dust to settle, and then the module is put back into operation. An important design parameter engineers must select is the ***cleaning cycle time*** (t_c) for each module. Two items are to be optimized: the overall pressure drop is not to exceed a certain selected value, and the physical deterioration of the bags is to be minimized. Figure 9.36 shows the volumetric flow rate in each of the five modules of Figure 9.35 as functions of time.

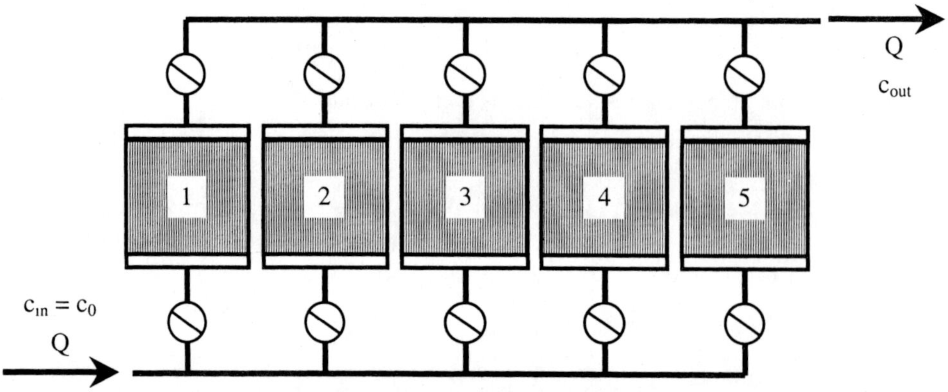

Figure 9.35 Schematic diagram of a baghouse containing five modules in parallel.

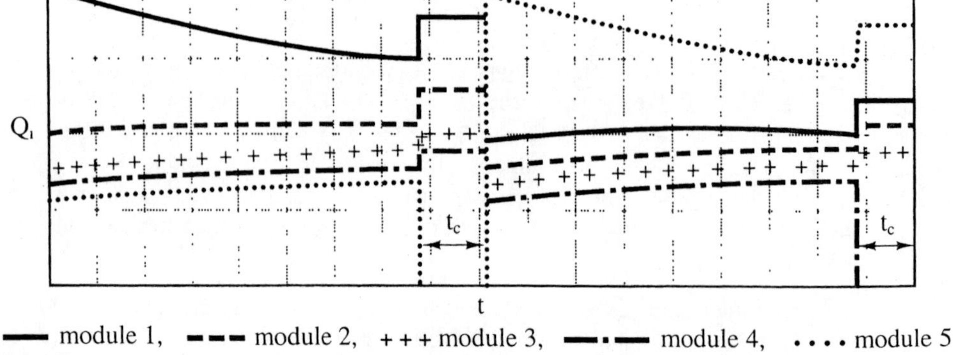

Figure 9.36 Volumetric flow rate (Q_i) in each module (i) versus time during two cleaning cycles (adapted from Heinsohn & Kabel, 1999).

As the thickness of the dust cake in one module grows, its volumetric flow rate decreases until such time as the dust cake is removed, whereupon the volumetric flow rate achieves its maximum value as the module is put back on-line. By staggering the times the modules are taken off-line, the total volumetric flow rate, which is the sum of the volumetric flow rates through each module, can be maintained at an acceptable overall pressure drop. A method to estimate the flow distribution and pressure drop in a multichamber baghouse has been proposed by Pontius and Smith (1988). Pulse-jet baghouses do not need to be organized in modular units; since all bags are cleaned more or less on a continuous basis and no appreciable dust cake accumulates, the pressure drop and volumetric flow rate across each module do not vary significantly with time.

Example 9.7 - Control of Particles from a Lime Kiln using a Fabric Filter

***Given*:** Exhaust gas containing lime dust particles leaves a lime kiln at 325. °F (162.8 °C) with a volumetric flow rate of 50,000 SCFM (23.61 m^3/s). The mass concentration (c_0) of dust is measured to be 10. g/m^3 at STP. It is desired to remove 99.% of the particles using a reverse-flow cleaning cycle of 1 per hour. Nomex bags, 10. inches in diameter and 15. ft long ($A_{per\ bag}$ = bag surface area per bag = 39.09 ft^2/bag), will be used.

***To do*:** Estimate the number of bags required and the overall pressure drop across the bags.

***Solution*:** From Table 9.1, a typical air-to-cloth ratio (U_0) of 0.70 m/min (0.01167 m/s) can be used and the dust cake constants can be assumed to be K_f = 350. N min/m^3 and $K_{dc} = 9.0 \times 10^4\ s^{-1}$. The actual volumetric flow rate is

$$Q(\text{ACFM}) = Q(\text{SCFM})\frac{T}{T_{STP}} = 50,000\frac{ft^3}{min}\frac{(162.8+273.15)\,K}{298.15\ K} = 73,100\frac{ft^3}{min}$$

which in SI units is 34.5 m^3/s. The total required bag surface area is

$$A_{bags} = \frac{Q_{actual}}{U_0} = \frac{34.5\frac{m^3}{s}}{0.70\frac{m}{min}}\left(\frac{60\ s}{min}\right) = 2960\ m^2 = 31,800\ ft^2$$

The minimum required total number of bags is thus

$$N_{bags} = \frac{A_{bags}}{A_{per\ bag}} = \frac{31,800\ ft^2}{39.09\frac{ft^2}{bag}} = 814\ \text{bags}$$

If the cleaning cycle is 1 per hour (3600 s), the maximum overall pressure drop across the bags is calculated from Eq. (9-69),

$$\delta P_{overall} = K_f U_0 + K_{dc} c_0 U_0^{\,2} t = 350\frac{N\cdot min}{m^3}0.70\frac{m}{min}\left(\frac{406.9\ \text{inches water}}{101,300\ \frac{N}{m^2}}\right)$$

$$+9.0\times10^4\frac{1}{s}\left(10.\frac{g}{m^3}\right)\left(0.01167\frac{m}{s}\right)^2(3600\ s)\left(\frac{kg}{1000\ g}\right)\left(\frac{s^2N}{kg\cdot m}\right)\left(\frac{406.9\ \text{inches water}}{101,300\ \frac{N}{m^2}}\right)$$

$$= 2.76\ \text{inches water}$$

Discussion: If the configuration of Figure 9.24 is used (112 bags per module), a minimum of 8 modules is required (for a total of 896 bags – the extra 82 bags further distribute the load). The fan should be sized to overcome a pressure drop of approximately 3 inches across the bags plus the pressure drop across the system of ducts and damper valves, etc. associated with the baghouse.

9.4.6 Wet Filtration

Collection of *liquid* contaminant particles is subject to the same physical principles as collection of solid particles (described in earlier sections). For a clean liquid of low viscosity, small liquid particles (mist) coalesce into larger particles (drops) and fall by gravity. For example, during splash filling (discussed in Chapter 4) small drops of liquid are generated and leave with the displaced air unless steps are taken to remove them. Figure 9.37 illustrates ways mesh and chevron mist eliminators can be placed inside a drum to remove such liquid particles. The pressure drop across a wet filter may be high because the porosity of a wetted filter is generally lower than for same filter used to collect dry particles.

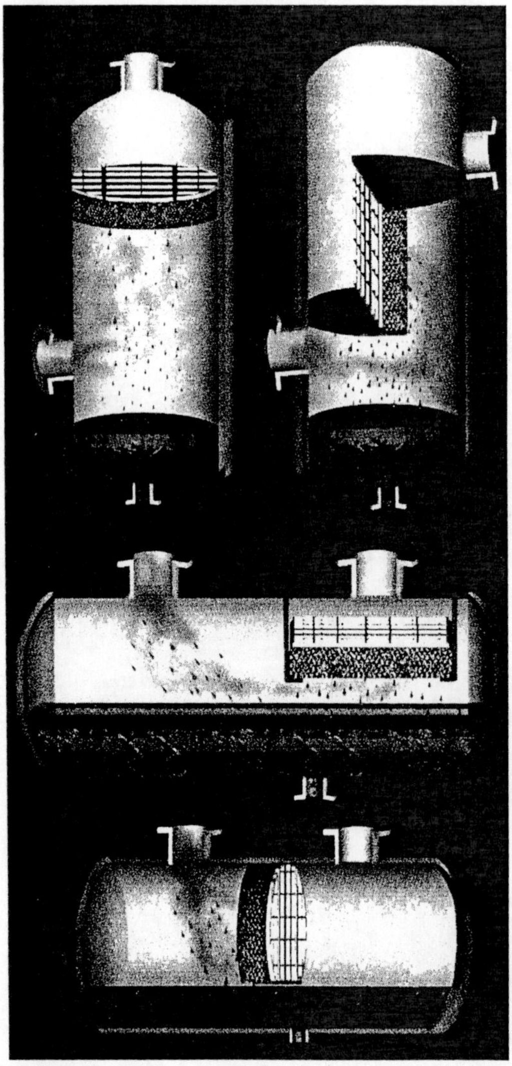

Figure 9.37 TEX-MESH, knockout drum mist eliminators (with permission from Amistco Separation Products, Inc.).

Engineers must be careful in the use of wet filters for air pollution applications. If the waste gas stream contains both liquid and solid particles (as is often the case in a waste gas stream) the collected material may be difficult to remove, the filter may become blocked, and the pressure drop may become excessive. Secondly if the liquid's viscosity is high, it may be difficult for the liquid mist to coalesce to form drops that fall by gravity. For example, before returning the crankcase emissions from large diesel engines to the engine's air inlet, liquid and solid particles have to be separated from the gas stream. If these particles are collected on mist eliminators, they are difficult to remove. The viscosity of crankcase oil is not particularly small and crankcase emissions also contain particles of carbon produced by "blow-by gas". Consequently crankcase emissions contain sticky particles that are difficult to remove from the wet filter. Lastly, if a process is shut down at the end of each day, or over a weekend, the temperature of the collection equipment decreases. Filters are apt to become clogged and remain so even when the temperature is raised and the process is resumed. In short, before wet filtration should be adopted engineers must convince themselves that the liquid is clean and has a low viscosity, and that the collected material does not clog the filter.

9.4.7 Oil Mist Collectors

Many machines produce a mist of petroleum-based or synthetic coolants, e.g. automatic screw machines, CNC machining centers, grinders, etc. In addition, some industrial processes produce mists, e.g. parts washing, vacuum pumps, food processing, etc. that enter the workplace environment unless steps are take to prevent it. Enclosing the process poses no problem but the task then becomes how to capture the mist and recycle it. While filters can capture mist, the fan operating the system tends to accumulate liquid, and inevitably puddles of liquid accumulate on the floor. A device that successfully captures mist is seen in Figure 9.38. The device is mounted on the machine generating the mist and captures mist particles from the process. Mist-laden air is drawn into the machine by its fan. The mist is drawn through the rotating perforated drum where it is trapped and coalesces until it grows to droplet size. The droplets are thrown free of the perforated drum to the inner wall of the casing where they flow down the wall through a circumferential slot into a collection chamber. Clean air then returns to the workplace, and the fluid can then drain back to the unit for reuse or disposal.

9.5 Electrostatic Precipitators

Another method to clean unwanted aerosol particles from the air without using a scrubbing liquid is ***electrostatic precipitation***, which uses electrical charge to force particles to drift towards collecting plates, onto which the particles impact and are collected. Such a device is called an ***electrostatic precipitator*** (***ESP***).

9.5.1 Classification of ESPs

Electrostatic precipitators can be classified as single-stage or two-stage ESPs, depending on the number of stages involved in ionization and collection. ***Single-stage ESPs*** use a single set of electrodes to produce the corona, paired with a collecting electric field to remove particles. The key features of a large single-stage ESP are:

- Flow straighteners and/or baffle plates are placed upstream of the charging and collecting sections to distribute the flow uniformly.
- Vertically aligned corona wires are spaced several inches apart, mounted midway between vertically aligned collecting plates that are nearly one foot apart; dead weights keep the corona wires under tension and vertical alignment.
- Corona wires are maintained at a *negative* voltage of several thousand volts by electrical connections at the top of the ESP; the ionizer is therefore referred to as a ***negative***

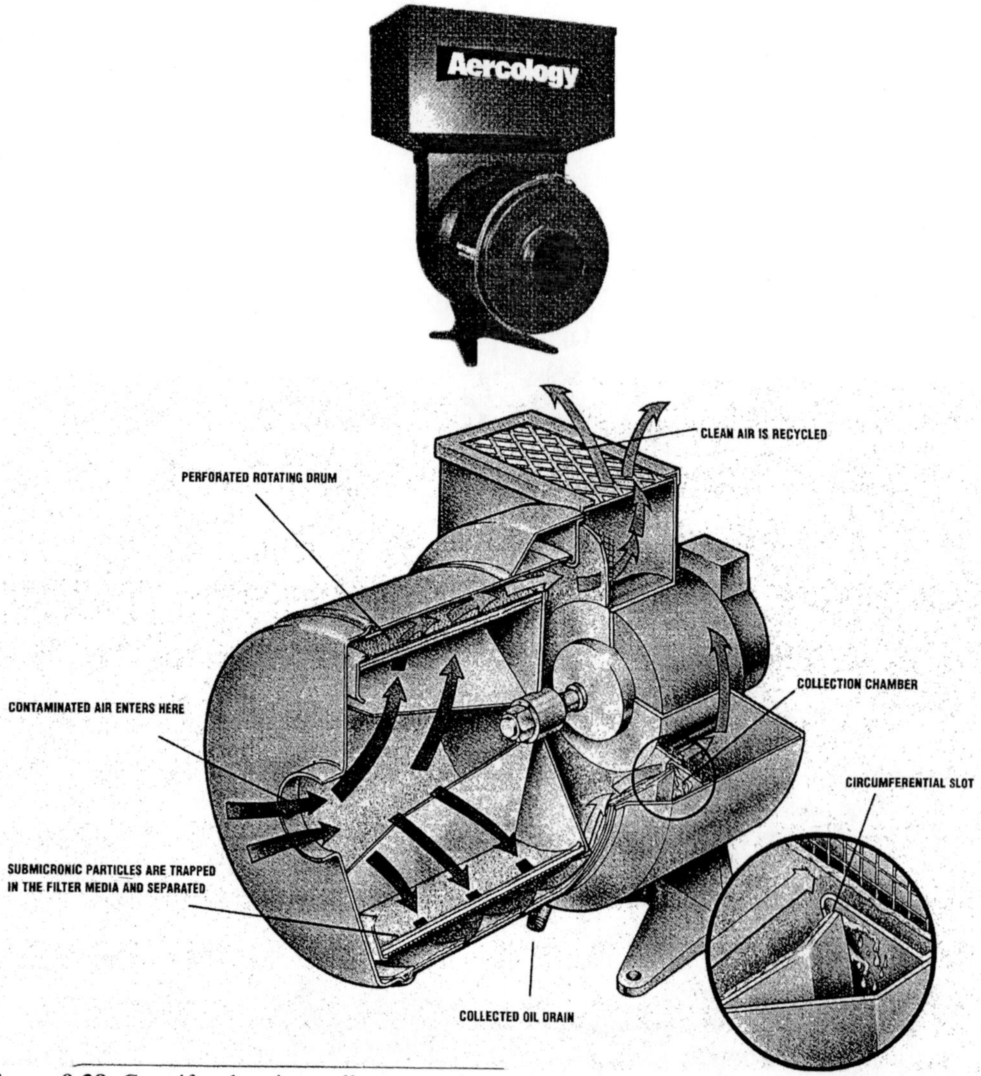

Figure 9.38 Centrifugal mist collector and separator; photograph and schematic diagram explaining operation (with permission of Torit Products, Donaldson Company).

ionizer. A ***corona***, which provides electrons for transfer to the airborne dust particles, is generated around each wire. Unfortunately, ozone (O_3) is also produced in the corona.

- The collecting plates are grounded and thus provide the positively charged electrode that attracts the negatively charged dust particles; each collecting plate can be several feet wide and ten to twenty feet high in large ESPs.
- Mechanical ***rappers*** on the top of the ESP are hammer-like devices which strike the collection plate periodically (at a ***rapping frequency***) to dislodge the collected dust; the gas velocity between the plates is low such that dislodged clumps of dust fall nearly vertically to dust hoppers below and are not resuspended into the gas stream.
- A modular design divides the ESP into several parallel legs, each leg consisting of several modules in series. Modules in parallel operate at the same voltage and rapping frequency, but modules in series may operate at different voltages or rapping frequencies.

A common configuration of ESP, mostly for outdoor use, is the ***single-stage plate-wire electrostatic precipitator***. These large ESPs are commonly used to remove particles from process gas streams having large steady volumetric flow rates, such as electric utility boilers. Batch processes and processes in which the gas properties vary considerably with time do not lend themselves to ESPs because the collection efficiency of an ESP is strongly dependent on small changes in these properties. Figure 9.39 is a top view schematic diagram of a large single-stage, plate-wire ESP.

Singe-stage ESPs are primarily large outdoor devices; smaller ESPs used to clean *indoor* air are typically ***two-stage ESPs*** that have one set of electrodes to charge the particles (the ***ionizer*** stage) and a second set of electrodes to collect the particles (the ***collector*** stage). Two-stage ESPs offer users the opportunity to independently optimize particle charging and particle collecting. Figure 9.40 shows a schematic diagram of a typical two-stage ESP. In the case illustrated here, the particles are *positively* charged as they pass through the ionizer stage, and collect on *negatively* charged (or grounded) collector plates in the collector stage. This type of ESP is called a ***positive ionizer***; the voltage is of opposite sign compared to that of the negative ionizing single-stage ESP described above. Although some two-stage ESPs are of the negative ionizer design, positive ionizers are preferred for indoor ESPs because positive ionization produces much less ozone (O_3) than does negative ionization. Two-stage indoor ESPs are designed to remove dust, fumes, and pollen from the workplace, homes, offices, and public places such as restaurants. Indoor ESPs often include a filter, called a ***pre-filter***, to remove particles upstream of the ESP, and an activated charcoal adsorber downstream of the ESP, called an ***after-filter***, to remove volatile organic compounds and odors. Such devices are referred to as ***electronic air filters*** (or ***electronic air cleaners***); they are manufactured as stand-alone devices to be placed in the home, or to be placed just below the ceiling in a public area, as in Figure 9.41. They can also be incorporated into the central duct work for (forced air) heating and air conditioning systems. In indoor air cleaning applications, pollens and interior airborne dust concentrations are small, and the collecting plates retain the collected dust. Cleaning is achieved by dismantling the device and periodically cleaning the plates with detergent solutions.

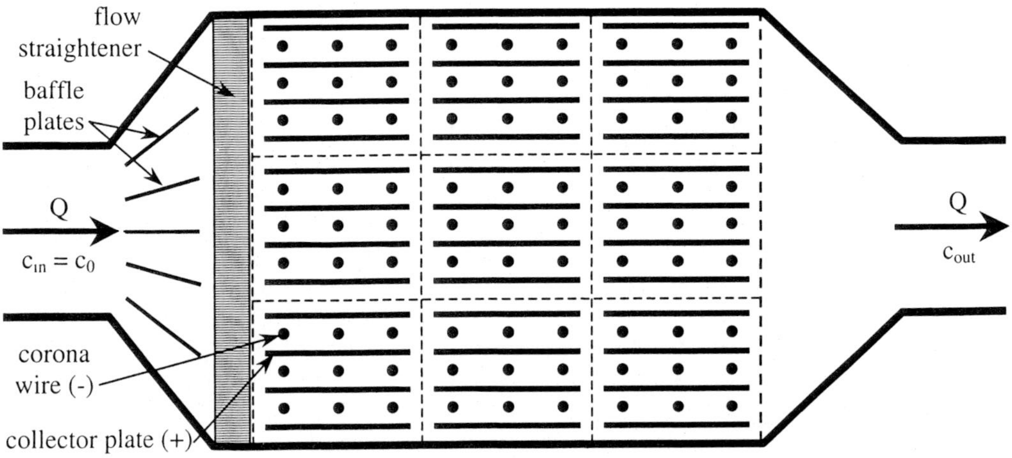

Figure 9.39 Top view of a negative ionization, single-stage, plate-wire ESP, with three parallel legs, each of which has three modules in series; circles represent the negatively charged corona wires, lines represent the positively charged collector plates.

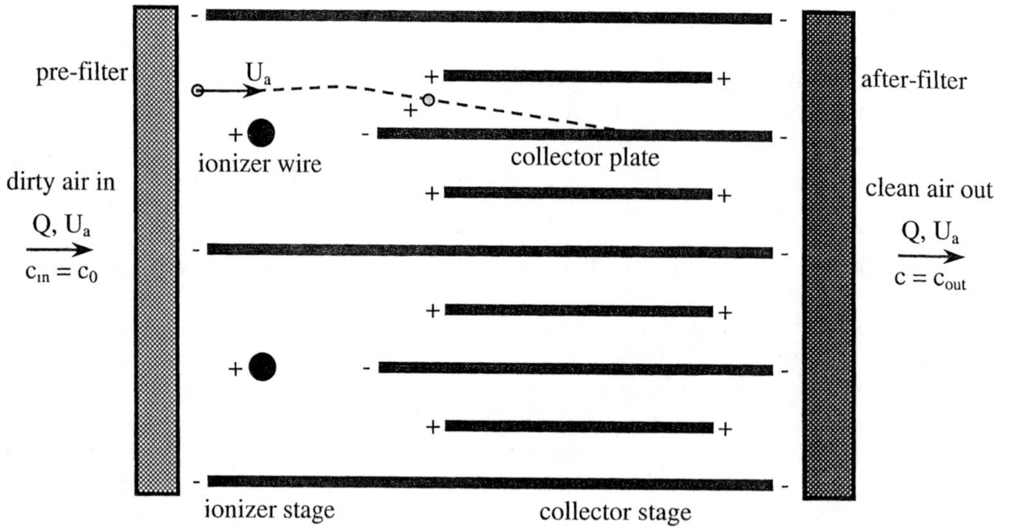

Figure 9.40 Schematic diagram of a positive ionization, two-stage, plate-wire ESP; dashed line indicates a particle trajectory.

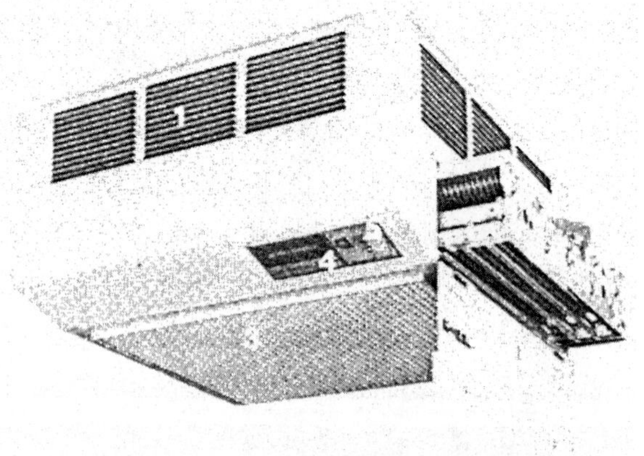

Figure 9.41 Smokemaster ceiling-mounted two-stage electrostatic precipitator that removes smoke, fume and small particles from public places; 1 – discharge louvers, 2 – housing, 3 – prefilters and grille, 4 – indicator lamp, 5 – speed control, 6 – ESP cells, 7 – access door (with permission of Air Quality Engineering).

Figure 9.41 illustrates features of a typical electronic air cleaner. The *Smokemaster Electronic Air Cleaner* is a two-stage ESP that is hung from the ceiling. Room air enters the cleaner from the bottom, passes through charged collecting plates where it is cleaned, and is then blown outward along the ceiling. An internal axial flow fan is used to circulate air since the pressure drop across the device is very small. The inlet and exhaust streams assist in circulating room air and improve ventilation effectiveness (see Chapter 5). The collecting plates are removed and washed with detergent on a routine basis to remove the collected particles. An upstream filter and downstream activated charcoal adsorber bed can be incorporated to improve the air quality of the room served by the device, but become two additional components that need to be maintained during routine servicing. These devices are sometimes used in the industrial workplace to capture welding fume, or fume and mists from machine tools. While capable of removing such contaminants, engineers are advised to capture such contaminants at the point where they are generated using local ventilation techniques, as discussed in earlier chapters, rather than allowing the contaminants to enter the workplace atmosphere, employing a general ventilation strategy. In any event, the maintenance to clean fume, mist, etc. from the collecting plates is more difficult than merely washing office dust from the collecting plates.

Other indoor electronic air cleaners are similar to the above, but are designed to be placed in a room, and often look like a piece of furniture. The device possesses the same components described above but in addition, HEPA or ULPA filters can be placed at the exit to remove particles that cause an allergic response such as pollens, plant spores, animal hair, or skin flakes. The device is well suited for individuals suffering from asthma or vexing seasonal allergies. The fan must be carefully designed so as not to produce annoying noise, but maintenance of the components is easier to perform compared to the device in Figure 9.41.

9.5.2 Fundamental Concepts

To illustrate the essential features that govern the collection of charged particles in a two-stage ESP, consider a contaminant particle that has been positively charged in the ionizer stage of the ESP, and then passes between two parallel collecting plates of the collector stage. The particle's trajectory is shown in Figure 9.42. The plates are separated by distance H and are each of length L in the flow-wise (x) direction. An external power supply establishes a uniform electrical field (E = constant) between the plates. The upper plate in the diagram is positively charged (the anode), and the lower plate in the diagram is either grounded or negatively charged (the cathode). Gravimetric settling is neglected. Dust particles of diameter D_p possessing a positive charge q_p enter the region with a speed equal to the gas speed U_a,

$$v_p = U_a$$

Charged particles experience a transverse electrical force of magnitude F_{el} perpendicular to the flow direction,

$$\boxed{F_{el} = q_p E} \tag{9-71}$$

which produces a transverse velocity called the ***drift velocity*** (w). For simplicity, a state of quasi-equilibrium is assumed in which the transverse electrical force equals the opposing drag force,

$$F_{el} = F_{drag}$$

Substitution of Eq. (9-71) and the magnitude of the drag force yields

$$\boxed{q_p E = c_D \frac{\pi D_p^{\,2}}{4C} \frac{\rho w^2}{2}} \tag{9-72}$$

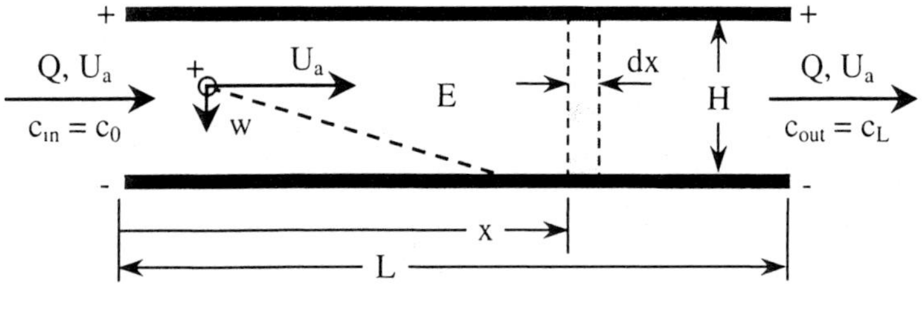

Figure 9.42 Top view of the collector stage of a positive ionization, parallel-plate electrostatic precipitator; contaminant particle moves with velocity components U_a and w, follows the dashed trajectory line, impacts the collecting plate, and adheres.

where c_D is the drag coefficient and C is the Cunningham correction factor, discussed previously in Chapter 8. Assuming Stokes flow,

$$c_D = \frac{24}{Re} = \frac{24\mu}{\rho D_p w}$$

where the relative velocity is the drift velocity (w). Solving for the drift velocity,

$$\boxed{w = \frac{q_p E_\infty C}{3\pi\mu D_p}} \qquad (9\text{-}73)$$

To predict the grade efficiency, $\eta(D_p)$, for particles of diameter D_p, consider an element between the plates of length dx, as shown in Figure 9.42. Let the depth of the element (into the page in Figure 9.42) be denoted by W. The air within the element is assumed to be well mixed, and a mass balance can be written for the element as follows:

$$Qc - wcWdx - Q\left(c + \frac{dc}{dx}dx\right) = 0$$

which can be simplified and re-written as

$$\frac{dc}{c} = -\frac{wW}{Q}dx$$

and then integrated over the length of the duct (x = 0 to L and c = c_0 to c_L),

$$\int_{c_0}^{c_L} \frac{dc}{c} = -\int_0^L \frac{wW}{Q}dx$$

The integration yields

$$\boxed{\frac{c_L}{c_0} = \exp\left(-\frac{wA_s}{Q}\right)} \qquad (9\text{-}74)$$

where A_s is the total area of the collecting electrode (the area of one of the collecting plates),

$$\boxed{A_s = WL} \qquad (9\text{-}75)$$

The grade efficiency is thus

$$\boxed{\eta(D_p) = 1 - \frac{c_{out}(D_p)}{c_{in}(D_p)} = 1 - \frac{c_L(D_p)}{c_0(D_p)} = 1 - \exp\left(-\frac{wA_s}{Q}\right)} \qquad (9\text{-}76)$$

This expression is called the ***Deutsch equation***. The reader must realize that the term Q is the *actual* volumetric flow rate, evaluated at the local temperature and pressure inside the ESP,

$$\boxed{Q = U_a A_{cross\text{-}section} = U_a HW} \qquad (9\text{-}77)$$

where U_a is the actual gas velocity and $A_{cross\text{-}section}$ is the cross-sectional area of the passage between plates. When the Deutsch equation is placed alongside the equation for well-mixed settling in a horizontal duct (Chapter 8), it is seen that the expressions are similar except that the drift velocity (w) has replaced the gravimetric settling velocity, or terminal velocity (v_t). For a more thorough description of ESPs, readers should consult White (1963) or Oglesby and Nichols (1970).

Even though Eq. (9-76) was derived for the simplified geometry of Figure 9.42, it is nevertheless used for ESPs of complicated geometry because of its simplicity and clear statement about the actual volumetric flow rate (Q) and area of the collecting plate (A_s). To account for the complex design of actual ESPs, the drift velocity is replaced by an empirical parameter derived from experiment, called the ***precipitation parameter*** or ***migration velocity*** (with units of velocity). The symbol (w) is nevertheless retained. Likewise, area A_s is replaced by the total surface area of *all* the collecting plates. Table 9.3 lists typical values of the precipitation parameter for a variety of particles removed by ESPs.

Equation Eq. (9-76) enables engineers to understand how collection efficiency depends on collecting plate area (A_s), temperature, mass flow rate of the process gas stream, and electrical properties of the collected dust. The relationship between the precipitation parameter and other operating parameters is not obvious. To appreciate this relationship, consider the physical phenomena that dictate the value of the drift velocity, Eq. (9-73):

(a) how particles acquire charge
(b) how space charge and the electrical resistance of the collected dust affect the electric field strength

9.5.3 Particle Charging

Two fundamental equations from the subject of electricity and magnetism, particularly the field of electrostatics, are

$$\boxed{\vec{\nabla}\cdot\vec{E} = \frac{q_v}{\varepsilon_0}} \tag{9-78}$$

and

$$\boxed{\vec{E} = -\vec{\nabla}V} \tag{9-79}$$

where $\vec{E}$ is the electric field strength vector, ε_0 is the permitivity of free space, q_v is called the space charge density (to be defined later), and V is the voltage. Equation (9-73) shows that one of the parameters affecting drift velocity is the charge on a dust particle (q_p).

To understand the process by which uncharged dust particles acquire negative charge, consider a two-stage, negative ionization, plate-wire electrostatic precipitator, a schematic of which is shown in Figure 9.43. The ionizer stage, consisting of negatively charged corona wires (the cathode) and positively charged or grounded plates (the anode), is shown on the left side of the figure. High-energy electrons (e^-) are generated in the ***active corona region***, of radius r_c, which is called the ***corona radius***. Within the corona ($r < r_c$) these energized electrons acquire very large velocities. When they strike neutral gas molecules, lower energy electrons called

Table 9.3 Typical precipitation parameters or migration velocities, w (in units of cm/s) for plate-wire electrostatic precipitators (abstracted from Vatavuk, EPA Report 1990).

process (ESP temperature)	**design efficiency (%)**			
	95.0	**99.0**	**99.5**	**99.9**
bituminous coal fly ash (300 °F)	12.6	10.1	9.3	8.2
sub-bituminous coal fly ash in a tangentially-fired boiler (300 °F)	17.0	11.8	10.3	8.8
cement kiln (600 °F)	1.5	1.5	1.8	1.8
glass plant (500 °F)	1.6	1.6	1.5	1.5
incinerator fly ash (250 °F)	15.3	11.4	10.6	9.4
iron/steel sinter plant dust with mechanical precollector (300 °F)	6.8	6.2	6.6	6.3
lime dust	1.5	1.5	1.5	1.5

thermal electrons (e^-_{th}) are dislodged from the molecules, which become positively charged gas ions. These ions move toward the wire where they become neutral. The original and newly formed electrons are free to ionize other gas molecules or travel outward toward the plate. At radius r_c the magnitude of the electric field strength decreases to $E = E_c$ such the electrons are no longer sufficiently energetic to dissociate or ionize neutral molecules, but now merely attach themselves to gas molecules. The process that generates the thermal electrons is called an ***avalanche***; these thermal electrons then travel outwards toward the collecting plate, which has opposite charge. As they travel outward, they encounter neutral molecules (not shown) and transfer their charge in the ***passive region***. Negatively charged gas ions (not shown) encounter dust particles, and transfer their charge to dust particles in the passive region. As the dust particles move through the passive region, they acquire negative charge and begin to drift towards the ionizer plate, as shown in Figure 9.43. However, when the fully charged dust particles travel to the collector stage, they acquire drift velocity, w, in the opposite direction. Finally, the dust particles strike a collecting plate, adhere to it, and transfer their charge to the external circuit. Two distinct regions can be defined, the ***corona region*** and the ***passive region***. These are discussed in detail below.

<u>Corona region</u>: A corona is a thin annular region close to the negatively charged wire, $r < r_c$, in which the magnitude of the electric field strength is very large, $E > E_c$. E_c is the electric field strength capable of ionizing gas molecules. Crawford (1976) states that the electric field strength at the outer radius (r_c) of the corona (E_c, in units of volts per meter) can be expressed as

$$\boxed{E_c = 3.0\times10^6 f\left[\frac{T_0}{T}\frac{P}{P_0} + 0.030\sqrt{\frac{T_0}{T}\frac{P}{P_0}\frac{1}{r_c}}\right]} \tag{9-80}$$

with

$$\boxed{r_c = r_w + 0.030\sqrt{r_w}} \tag{9-81}$$

where r_w is the radius of the corona wire (in meters), and r_c is also in meters. Parameter f is equal to 1 for a smooth corona wire and is greater than 1 if the corona wire is rough. From Eq. (9-79) it is seen that the voltage gradient is equal to the electric field strength; thus small diameter wires charged to a large voltage produce large voltage gradients close to their surfaces.

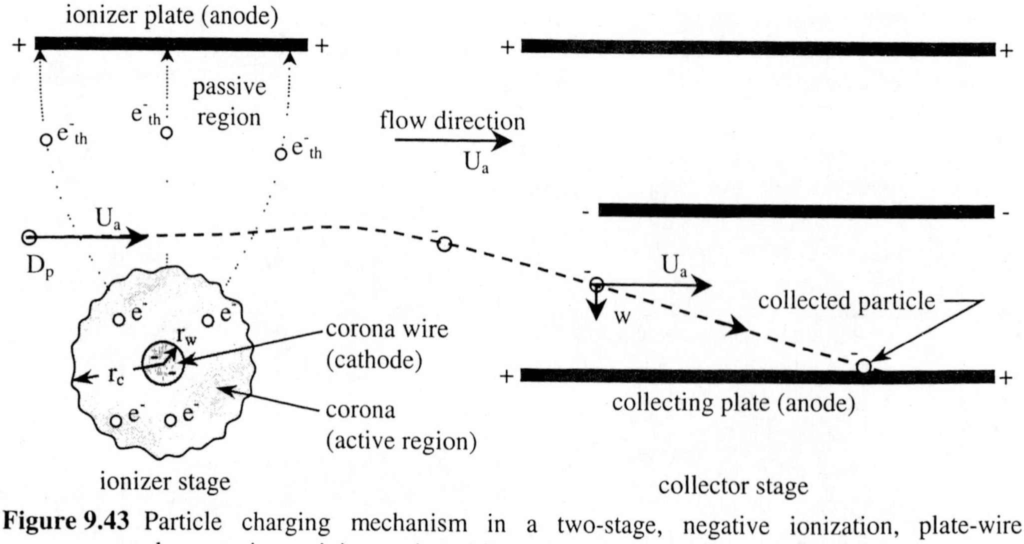

Figure 9.43 Particle charging mechanism in a two-stage, negative ionization, plate-wire electrostatic precipitator; dotted lines represent thermal electron trajectories, and the dashed line represents a dust particle trajectory.

Passive region: In the passive region, in which $r > r_c$ and $E < E_c$, electrons attach themselves to gas molecules through a process called ***electron transfer***. The number of gas molecules is much larger than the number of dust particles; thus electrons are more likely to attach themselves to gas molecules, producing negative ions, than to attach themselves to dust particles. It must also be recalled from Chapter 8 that a dust particle is many orders of magnitude larger in diameter than a gas molecule. Negatively charged gas ions travel outward toward the positively charged ionizer plate. Along the way they encounter dust particles and transfer their negative charge to the dust particles through a process called ***charge transfer***. Through countless encounters with gas ions, dust particles acquire negative charge. The charge acquired by dust particles has an upper limit called the ***saturation charge*** (q_{ps}) because as they become charged they produce their own electric field that ultimately repels gas ions traveling toward them. As the charged dust particles continue traveling downstream towards the collector stage, they are attracted to an oppositely charged collecting plate, eventually migrating toward the plate at a velocity equal to the drift velocity (w).

Readers must note that two-stage, *positive* ionization, plate-wire ESPs operate in a similar manner to the two-stage negative ionization, plate-wire ESP described above, except that all electrical charges are reversed in sign. In positive ionization ESPs, electrons are stripped *from* gas molecules, which intern strip electrons from dust particles as they pass through the ionizer stage; the dust particles therefore acquire *positive* charge.

Particles acquire charge by two distinct physical processes: ***field charging*** (also called ***bombardment charging***) and ***diffusion charging***. These are discussed in detail below.

Field charging: Imagine a dust particle subjected to a stream of "bombarding" ions. Let q_{pf} be defined as the ***field charge*** (in coulombs, C). The rate at which charge is transferred to the dust particle by field charging is given by

$$\boxed{\frac{dq_{pf}}{dt} = \frac{Nq_e}{4\varepsilon_0}\kappa q_{pf,s}\left(1 - \frac{q_{pf}}{q_{pf,s}}\right)^2} \tag{9-82}$$

where $q_{pf,s}$ is called the ***saturation field charge*** (also in coulombs, C),

$$\boxed{q_{pf,s} = \frac{3\kappa}{\kappa + 2}\pi\varepsilon_0 D_p{}^2 E} \tag{9-83}$$

and

- N = number of gas ions per unit volume
- q_e = charge (in coulombs) per ion = charge on one electron = -1.6029×10^{-19} C
- ε_0 = permitivity of free space = 8.85×10^{-12} $C^2/(N\ m^2)$
- κ = dielectric constant of the dust particle
- E = magnitude of the electric field strength driving the stream of ions
- D_p = diameter of the dust particle

The time it takes to acquire charge is given by

$$\boxed{t = \frac{\dfrac{4\varepsilon_0}{\kappa N q_e}\dfrac{q_{pf}}{q_{pf,s}}}{1 - \dfrac{q_{pf}}{q_{pf,s}}}} \tag{9-84}$$

For electric fields encountered in ESPs, the parameter $4\varepsilon_0/(\kappa N q_e)$ in Eq. (9-84) is of order 0.01 to 0.1 seconds. By contrast, the typical time (t) in which dust particles reside inside an ESP is of the order of several seconds. Thus for design purposes, one can assume that if field charging is the dominant charging mechanism, dust particles always posses their saturation charge ($q_{pf,s}$).

Diffusion charging: As a continuum, gas ions possess a velocity dictated both by their charge (Nq_e) and by the magnitude of the local electric field strength (E). Individual gas molecules also possess velocities given by a Gaussian distribution dictated by the gas temperature, and there is a small number of ions possessing sufficiently large velocities as to be unaffected by the repelling electric field on a charged dust particle. Charging that occurs by the process of molecular diffusion is called diffusion charging. The rate at which a particle acquires charge by diffusion is given by an ordinary differential equation,

$$\frac{dq_{pd}}{dt} = Nq_e\sqrt{\frac{kT\pi}{2m}}D_p^{\,2}\exp\left[-\frac{q_e q_{pd}}{2\pi\varepsilon_0 kTD_p}\right]$$

where q_{pd} is defined as the ***diffusion charge***. The above differential equation can be integrated to yield an expression for the diffusion charge on the particle,

$$\boxed{q_{pd} = \frac{2\pi\varepsilon_0 kTD_p}{q_e}\ln\left[1+\frac{ND_p q_e^{\,2} t}{2\varepsilon_0\sqrt{2m\pi kT}}\right]} \tag{9-85}$$

where m is the mass of the ion and k is Boltzmann's constant. It is seen that charge acquired by the process of diffusion charging has no upper limit, whereas charge acquired by the process of field charging has an upper limit which is the saturation charge ($q_{pf,s}$). Thus at any time, the total charge on a dust particle (q_p) can be thought of as the sum of Eqs. (9-83) and (9-85),

$$\boxed{q_p = q_{pf,s} + q_{pd}} \tag{9-86}$$

Since field charging depends on the square of particle diameter whereas diffusion charging depends on particle diameter to the first power, diffusion charging is important for small particles ($D_p < 2$ μm), while field charging is more important for larger particles ($D_p > 2$ μm). It should be noted that both particle charging mechanisms depend on the ion density (Nq_e).

9.5.4 Electric Field Strength

Space charge density, q_v, is the total charge per unit volume of gas and is composed of charges on ions and charges on particles,

$$\boxed{q_v = Nq_e + n_p q_p} \tag{9-87}$$

where N and n_p are the number of ions and number of particles, respectively, per unit volume of gas. q_v has dimensions of charge per volume (with units C/m^3). Up to this point the magnitude (E) of the electric field strength driving ions toward the dust particles has been treated as a function of location between the electrodes, but independent of the space charge density between the electrodes. This is a crude assumption since $\vec{E}$ is related to q_v by Eq. (9-78). For computational simplicity, consider an axisymmetric geometry, namely a single-stage, negative ionization, wire-cylinder ESP configuration, as sketched in Figure 9.44. It is assumed that the space charge density (q_v) is constant. Since the electric field strength varies only in the radial direction, magnitude E is equal to the radial component of the electric field strength vector, $E = E_r$. Thus, from Eq. (9-78),

$$\frac{d(rE)}{dr} = \frac{d(rE_r)}{dr} = \frac{q_v}{\varepsilon_0}$$

Integration yields

$$\boxed{E = \frac{q_v r}{2\varepsilon_0} + \frac{C_1}{r}} \tag{9-88}$$

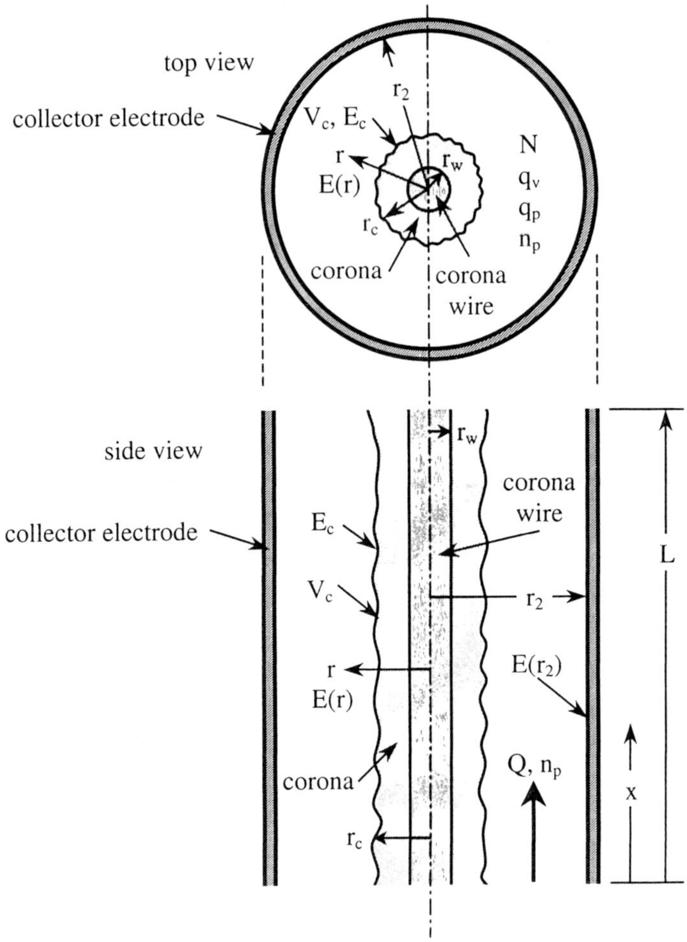

Figure 9.44 Schematic diagram of a single-stage, negative ionization, wire-cylinder electrostatic precipitator with constant space charge density.

From Eq. (9-79), the component of electric field strength in the radial (r) direction is also equal to the voltage gradient in that direction,

$$E = E_r = -\frac{dV}{dr}$$

A second integration yields

$$V = \frac{q_v r^2}{4\varepsilon_0} - C_1 \ln r + C_2 \tag{9-89}$$

The constant C_1 is evaluated at the edge of the corona, where $r = r_c$ and $E = E_c$, and the constant C_2 is evaluated at the outer electrode, where $r = r_2$. The voltage on the outer electrode, $V(r_2)$, is zero since it is grounded. E thus becomes

$$E = \frac{q_v r}{2\varepsilon_0} + \frac{V_c - q_v\left(r_2^2 - r_c^2\right)4\varepsilon_0}{r \ln\left(\frac{r_2}{r_c}\right)} \tag{9-90}$$

where V_c is the magnitude of the voltage at the edge of the corona,

$$V_c = \frac{q_v}{4\varepsilon_0}\left[r_c^2 - r_w^2 - 2r_w^2 \ln\left(\frac{r_c}{r_w}\right)\right] + r_w E_c \ln\left(\frac{r_c}{r_w}\right) \approx \frac{q_v r_c^2}{4\varepsilon_0} + r_w E_c \ln\left(\frac{r_c}{r_w}\right) \tag{9-91}$$

Figure 9.45 shows how E and V vary with radius and q_v.

When the space charge density is negligible, E decreases with radius, but for non-negligible values of q_v, E *increases* with radius. Thus it is seen that *space charge suppresses the electric field.* Space charge decreases E near the corona wire where one wants it to be large and increases E at the outer electrode where one wants it to be low. Large values of E at the outer electrode produce energetic electron activity near the collecting electrode where one does not want energetic activity. Indeed large E near the collecting electrode is apt to produce a ***back corona*** or ionization within the collected dust particles (dust cake), which further suppresses the electric field.

9.5.5 Dust Cake Resistivity and Back Corona

Consider the layer of dust adhering to the collecting plate, hereafter called the dust cake. In terms of the equivalent electrical circuit, as shown in Figure 9.46, the electrical resistance of the dust cake is in series with the resistance of the corona and the passive region between the electrodes. The ***dust cake resistance*** (R_{dc}) can be expressed as

$$R_{dc} = \frac{\rho_{dc} L_{dc}(t)}{A_s} \tag{9-92}$$

where A_s is the total surface area of the collecting electrode (a constant), $L_{dc}(t)$ is the thickness of the dust cake which increases with time, and ρ_{dc} is the ***dust cake resistivity***. The voltage across the dust cake (V_{dc}) can also be expressed as

$$V_{dc} = IR_{dc} \tag{9-93}$$

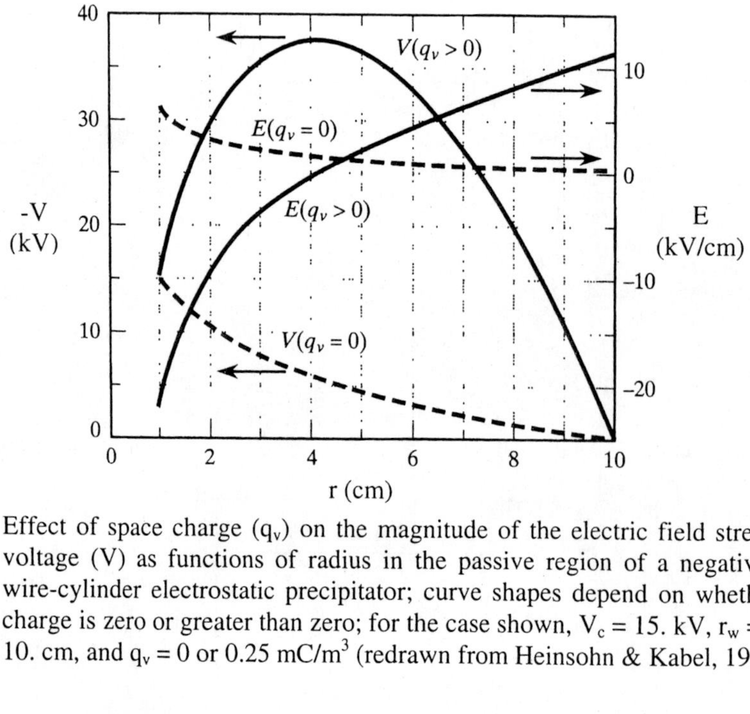

Figure 9.45 Effect of space charge (q_v) on the magnitude of the electric field strength (E) and voltage (V) as functions of radius in the passive region of a negative ionization, wire-cylinder electrostatic precipitator; curve shapes depend on whether the space charge is zero or greater than zero; for the case shown, V_c = 15. kV, r_w = 1.0 cm, r_c = 10. cm, and q_v = 0 or 0.25 mC/m^3 (redrawn from Heinsohn & Kabel, 1999).

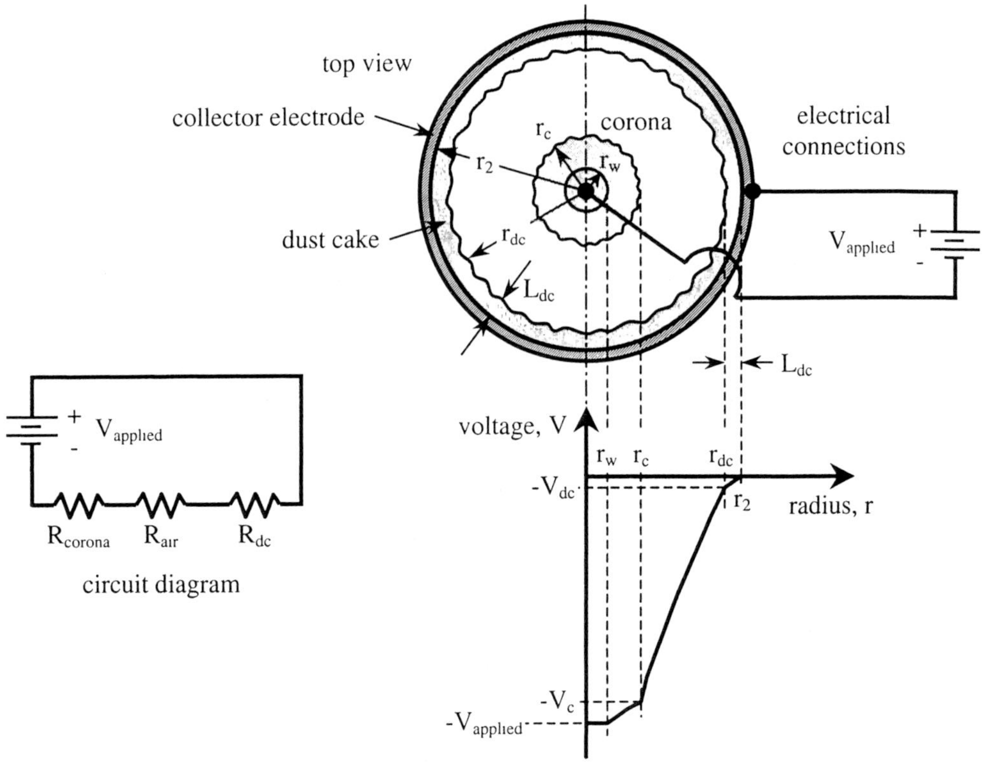

Figure 9.46 Schematic diagram, circuit diagram, and voltage variation with radius for a single-stage, negative ionization, wire-cylinder electrostatic precipitator; changes in voltage are affected by resistance in the corona, air, and dust cake.

where I is the current passing between the corona and collecting electrode and R_{dc} is the electrical resistance of the dust cake. The thickness of the dust cake is a function of the rapping frequency. As dust cake thickness, $L_{dc}(t)$, increases with time, so does dust cake resistance, R_{dc}. As R_{dc} increases, the voltage drop across the dust cake (V_{dc}) also increases, and may reach a value such that a corona forms on the deposited dust particles and a concentrated discharge (spark) occurs within the dust cake. Any discharge suppresses the voltage difference between the corona wire and the outer edge of the dust cake, which is the voltage difference that produces the desired corona. Suppressing the corona has the disastrous effect of reducing the generation of ions upon which the entire ESP operates. ESP operators may not be aware of such an occurrence because as far as they are concerned the circuit current (I) and overall applied voltage drop ($V_{applied}$) across the entire ESP may not have changed. For these reasons, proper ESP operation requires one to monitor the dust cake resistivity (ρ_{dc}) and to optimize the rapping frequency. If the rapping frequency is too low, a back corona forms. If rapping is too frequent, small particles may be reintroduced into the gas stream and produce a visible "puffing".

Dust cake resistance is a quantity that varies over a very wide range, i.e. four to five orders of magnitude, depending on such variables as dust cake porosity, temperature, concentration of water vapor, and presence in the process gas stream of ionizable gas species such as NH_3 and SO_2. Figure 9.47 shows a typical variation of dust cake resistivity with temperature and concentration of water vapor and ionizing species. At temperatures below 300 °F, electrical charge tends to travel over a surface film on particles as the charge finds its way to the collecting plate. Furthermore, the higher the concentration of water vapor and ionizing species, the easier it

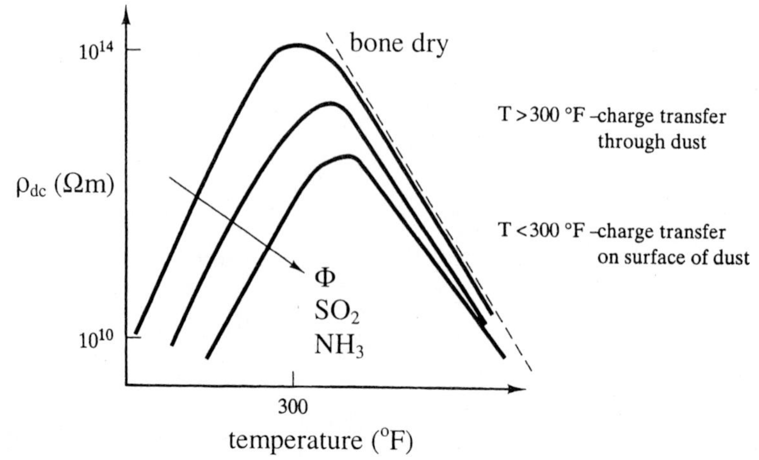

Figure 9.47 Dependence of dust cake resistivity on temperature, moisture, SO_2 concentration, and NH_3 concentration; direction of increasing moisture (relative humidity, Φ), SO_2 concentration, and/or NH_3 concentration indicated by arrow.

is for the charge to travel through the conducting surface film. Above 300 °F, electrical charge tends to pass through the dust particles rather than on their surface and the resistivity becomes independent of water vapor. In general, the higher the temperature the lower the resistivity of the dust particle. Power plant operators are faced with an ironic choice in selecting the sulfur content of the coal they purchase. High-sulfur coal produces SO_2 which reduces the dust cake resistivity and enhances the ESP particle collection efficiency. Countering this advantage however, is the fact that larger SO_2 concentrations in the exhaust gas complicate the task of satisfying SO_2 emission standards.

To produce a low resistance dust cake, designers would like to operate ESPs at either low or high gas temperatures. Low gas temperatures present a problem because a plume's buoyancy is adversely affected. Increasing the temperature of the exhaust stream merely wastes energy and worsens the overall thermodynamic efficiency of a process. An attractive idea is to design a process where the ESP is located upstream of heat-exchangers used to extract energy prior to discharging exhaust gases. It may be possible to design such "***hot side ESPs***" into new facilities, but it is generally not possible to do so when retrofitting an existing process where large pieces of equipment cannot be moved easily. Some engineers have found that hot side ESPs have failed to fulfill their promise.

9.5.6 Applications

Principal applications for single-stage ESPs are the removal of particles from large steady flows found in coal-fired electric utility boilers, furnaces to make molten glass, lime and cement kilns, etc. ESPs can be designed to operate at high temperatures and collect liquid as well as solid particles. It is important that the collected particles be removed by rapping. Since collection efficiency depends strongly on actual volumetric flow rate, changes in operating temperature or changes in the mass flow rate of the process gas stream can have significant effects on the collection efficiency. Serious deterioration in ESP performance may also occur because rapping produces "puffs" or changes in gas composition and alters the dust resistivity (ρ_{dc}). The pressure drop across an ESP is modest, and the manufacturer's literature should be consulted for specific data.

Example 9.8 - Control of Particles from a Lime Kiln using an ESP

Given: The kiln described previously in Example 9.7. The volumetric flow rate of the exhaust gas is 50,000 SCFM, and a collection efficiency of 99.% is sought. In the previous example problem, the actual volumetric flow rate was computed to be 73,100 ACFM.

To do: Estimate the total collecting surface area of an ESP that can be used to capture lime dust from the kiln.

Solution: From Table 9.3, the precipitation parameter (w) associated with lime dust is found to be 1.5 cm/s (about 0.0492 ft/s). The magnitude of the area of the collecting plates can be estimated from the Deutsch equation, Eq. (9-76),

$$\eta_{\text{overall}} = 1 - \exp\left(-\frac{wA_s}{Q}\right)$$

where Q refers to the actual volumetric flow rate. Solving for the plate surface area,

$$A_s = -\frac{Q}{w}\ln\left(1-\eta_{\text{overall}}\right)$$

which yields

$$A_s = -\frac{73{,}100\,\dfrac{\text{ft}^3}{\text{min}}}{0.0492\,\dfrac{\text{ft}}{\text{s}}}\left(\frac{\text{min}}{60\ \text{s}}\right)\ln\left(1-0.99\right) \cong 110{,}000\ \text{ft}^2$$

The value is reported to two significant digits because of the uncertainty in the value of w. Note that although the Deutsch equation was derived for a single passageway between two parallel plates, and A_s was defined as the surface area of one collecting plate, the equation is also valid for the case of *many* collecting plates, as in Figure 9.39. Here, A_s represents the total surface area of *all* the collecting plates combined.

Discussion: While not asked for in the problem, readers should realize that a dust concentration of 10 g/m^3 is huge, but is nevertheless what can be expected in the exhaust gases from a lime kiln. For this reason, it would be wise to consider using a cyclone collector upstream of the ESP to remove the large particles that are certainly expected to be present. Without removing the large particles, upstream modules of the ESP would be overloaded and performance would suffer.

9.6 Engineering Design - Selecting and Sizing Particle Collectors

The discussion that follows reflects the point of view of engineers acting as consultants to industry or engineers employed by companies that wish to purchase a particle collector. The perspective of engineers employed by companies that manufacture and market particle collectors is somewhat different.

The first step in complying with state and federal particle emission standards is to select materials entering a process and/or to alter the process to minimize generation of particles in the first place. Once this approach has been exhausted, engineers must be able to recommend the most economical particle collection system. An economic analysis can be used to estimate the total initial cost (TIC) and total annual cost (TAC). Prior to an economic analysis, decisions must be made regarding which generic class of collector to use and what the dimensions and operating conditions should be for that collector. The purpose of this section is to outline methods for the selection and sizing of particle collection systems whose performance can be predicted by the equations in this chapter and in Chapter 8. The following are constraints (boundary conditions) within which engineers operate, along with typical units used for each variable:

- Physical conditions at the inlet to the control device:
  - total actual volumetric flow rate (Q_t, ACFM or actual m^3/s)
  - temperature (T, K), pressure (P, kPa), relative humidity (Φ, %), and chemical composition (mass fraction f_J or mol fraction y_J, unitless) of the process gas stream
  - particle mass concentration (c_{inlet}, mg/m^3) and size distribution
  - physical conditions within and around the plant that affect the size and location of the control device
- Physical conditions at the outlet of the control device:
  - state and federal compliance standards, expressed in units of $Mg/10^6$ BTU, mg/m^3, etc., which dictate the particle removal efficiency

If the particles are sticky or the process gas stream contains large molecular weight hydrocarbon vapors that may condense and make the particles sticky, none of the particle removal systems discussed here will perform well. Under these conditions, a direct-flame thermal oxidation process with a regenerative heat exchanger may be the only way to remove combustible particles and hydrocarbon vapors. If the particles are noncombustible but sticky, the removal task is particularly difficult. Care must also be taken in the design of the hoppers, and in the methods to empty and transfer dust from the hoppers while preserving the positive or negative pressure in the hoppers. Hoppers clog from time to time, and provision must be made for rappers, air cannons, and access ports to remove blockages.

Inertial separation collectors (standard cyclones): Knowing the required grade efficiency, $\eta(D_p)$, of particles to be removed, one can use Figure 9.3 to determine the ratio $D_p/D_{p,cut}$:

- If the cyclone dimensions are known, the volumetric flow rate (Q_J) that an individual cyclone must accommodate to achieve the desired efficiency is computed. If Q_J is less than total flow rate Q_t, a number (n) of cyclones is arranged in *parallel* where $n = Q_t/Q_J$.
- For cyclones in *series*, the volumetric flow rate through an individual cyclone is constant and equal to the total ($Q_J = Q_t$). Various cyclone outer diameters ($D_{2,J}$) are selected, and their efficiencies are computed. A configuration of cyclones in series is devised such that the overall efficiency ($\eta_{overall}$) is equal to or better than the desired efficiency.

The pressure drop for a cyclone, or a configuration of cyclones, can be predicted. One must be sure that the material used to construct the cyclones can withstand the temperature, pressure, and corrosive nature of the process gas stream.

Wet scrubbers: The location of the scrubber affects the dimensions and type of scrubber, and its ancillary water treatment equipment. The type of scrubber best suited to the process must be selected, e.g. counterflow scrubber with or without packing, transverse scrubber with or without packing, cyclone scrubber, or venturi scrubber. The volumetric flow rate of the scrubbing liquid (Q_s) and the scrubbing drop diameter (D_c) or the characteristic diameter of packing material (also D_c) must be selected in order to achieve the desired efficiency. The relationships for grade or fractional efficiency, $\eta(D_p)$, are given in this chapter.

Fabric filters (baghouse): Vendors can provide information about the best fabric to be used, depending on the dust to be collected and the physical properties of the process gas stream. Selecting the type of collector (reverse-flow, shaker, pulse-jet, etc.), and air-to-cloth ratio defines the total bag surface area. Appropriate values of the air-to-cloth ratio for the dust to be collected can be found in Table 9.1 and similar tables provided by vendors. Selecting the cleaning cycle and overall pressure drop is a matter of trading-off the initial cost of the baghouse against the annual operating maintenance and (electric power) fan costs. An oversized baghouse has a large initial capital cost but can have minimal annual operational costs for fans and maintenance. An undersized baghouse with its lower initial capital cost may lead to higher annual costs. If combustible dusts are collected, fire sprinklers inside the baghouse may be necessary to extinguish fires. Blow-out ports should be provided to minimize the effects of an explosion.

Electrostatic precipitators (ESPs): The first step is to select an appropriate precipitation parameter (w) from Table 9.3 or similar table provided by ESP vendors to characterize the particles to be removed. Using equations given earlier in this chapter, the area of the collecting plates needed to achieve the desired removal efficiency is computed. Manufacturers of ESPs should be asked to visit the plant and perform experiments using a slip stream of the actual process gas stream to determine a unique value of the precipitation parameter (w) for the plant. The actual design of the ESP components (corona wires, rappers, electric rectifiers, friability monitors, etc.) and calculation of dust cake resistance and operating variables are based on methods more sophisticated than presented in this chapter.

9.7 Hoppers

Nearly all the particle collection systems discussed in this chapter have hoppers at their base. ***Hoppers*** are containers, generally cylindrical, with a cone at the base used to temporarily store collected particles prior to their disposal. The included angle of the cone is typically 45 to 60 degrees. If the exit opening is circular, the material is removed by a rotary valve that may also accommodate a hopper under positive or negative pressure. If the exit opening is a slot, the material is removed by a screw feeder (auger). Hoppers are typically beset with problems in which particles adhere to the hopper walls and themselves, and cannot be removed at the desired rate. Feed bins delivering feed stock to a process have the same problems, although only hoppers are discussed in this section. Hopper blockage is characterized by the following:

- ***arching*** or ***bridging*** - Material near the exit compacts and does not flow in spite of the considerable mass of material above it. Arching is illustrated in Figure 9.48a.
- ***ratholing*** - Material adheres to the walls of the hopper, reducing the hopper's capacity. The flow rate of material is also reduced as it must pass through a narrow central passage. Ratholing is illustrated in Figure 9.48b.
- discharge occurs in spurts and may even be accompanied by forces that damage the hopper and rotary valve or screw feeder.

These problems are functions of the size and size distribution of particles, moisture, level of material in the hopper, adhesive properties of the dust, and how it compacts as a function of time.

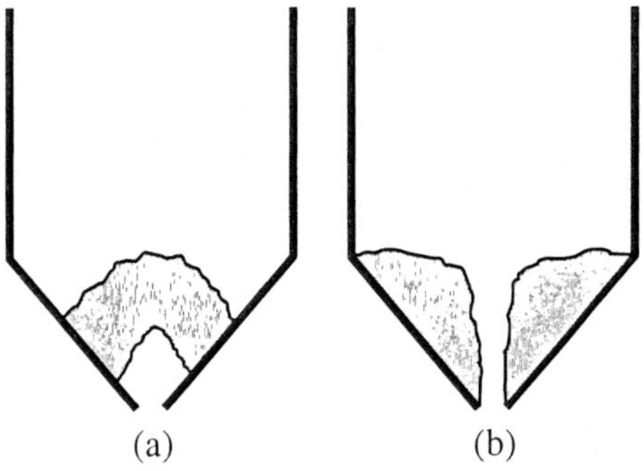

Figure 9.48 Two types of hopper blockage: (a) arching, (b) ratholing.

Complicating the matter further is the fact that there occurs some side-to-side and top-to-bottom segregation of particle sizes when hoppers are filled. Falling particles spread outward and strike the edges of the hopper. Fine particles concentrate in the center of the hopper and coarse particles accumulate at the outside of the hopper. Side-to-side segregation is important when fine particles constitute less than 10% of the material. The second segregation occurs from top-to-bottom when fine particles accumulate in a layer on top of the larger particles. Falling heavy particles penetrate and fluidize this layer and ultimately settle to the bottom. Top-to-bottom segregation is important when small particles constitute more than 20% of the mixture.

9.6.1 Blockage Remedies

The tale-tale signs of blockage are hoppers severely dented by workers hammering on their sides, or hoppers with "poke holes" – access ports which enable workers to dislodge material manually. Babysitting blocked hoppers increases downtime, slows production schedules, and increases labor costs. The least expensive hopper design is a 45-degree conical hopper with a single rotary valve. If blockages prove costly, the first step is to increase the angle to 60 degrees. If blockage persists in a hopper with a single circular discharge and rotary valve, a common remedy is to change the discharge to a long slot and screw feeder (auger). A slot and a standard-pitch screw feeder causes material at the back of the hopper to flow, but it is dragged under non-flowing material towards the front end of the screw. This dragging action compacts cohesive particles and produces a permanent arch; it also increases the torque on the screw.

Hoppers can be fitted with "air cannons" and mechanical vibrators to dislodge material. These devices increase costs and generate considerable noise, often in excess of OSHA standards. Such steps should be viewed as remedial and may be avoided if more time and attention is given to the design of the hopper and the schedules used to fill and empty the hopper. Shown in Figure 9.49 is a novel design called the ***Diamondback Hopper***, which incorporates an arch-breaking concept to minimize (if not eliminate) the problems discussed above. The unique feature about the Diamondback Hopper is that hopper cross section changes from circular to oval, and then from oval to circular on one side of the hopper as seen in Figure 9.49. When the flow converges in one dimension only, the angle at which the solids flow is much flatter than when solids flow in two dimensions. Secondly, when the flow channel expands to the vertical bin walls, ratholes cannot develop. The entire hopper can have such a design, or an existing cone can be retrofitted where problems have arisen in the past. The design can accommodate a rotary discharge valve for a circular discharge or a screw feed for a slot discharge.

9.8 Closure

As long as the number of particles per unit volume of gas is small, particles travel independently of one another, and their motion can be predicted by equations derived in Chapter 8. Such predictions are crucial to modeling the performance of particle collection systems. In this chapter, the grade (fractional) efficiencies of cyclones, cascade impactors, spray chambers, wet scrubbers, venturi scrubbers, filters, and electrostatic precipitators are derived. Some guidance is given so that engineers can use the analyses in this chapter as aids in selecting particle collection device(s) for an APCS, or to design new particle collection devices. Lastly, some of the problems encountered in hoppers are discussed, along with remedies.

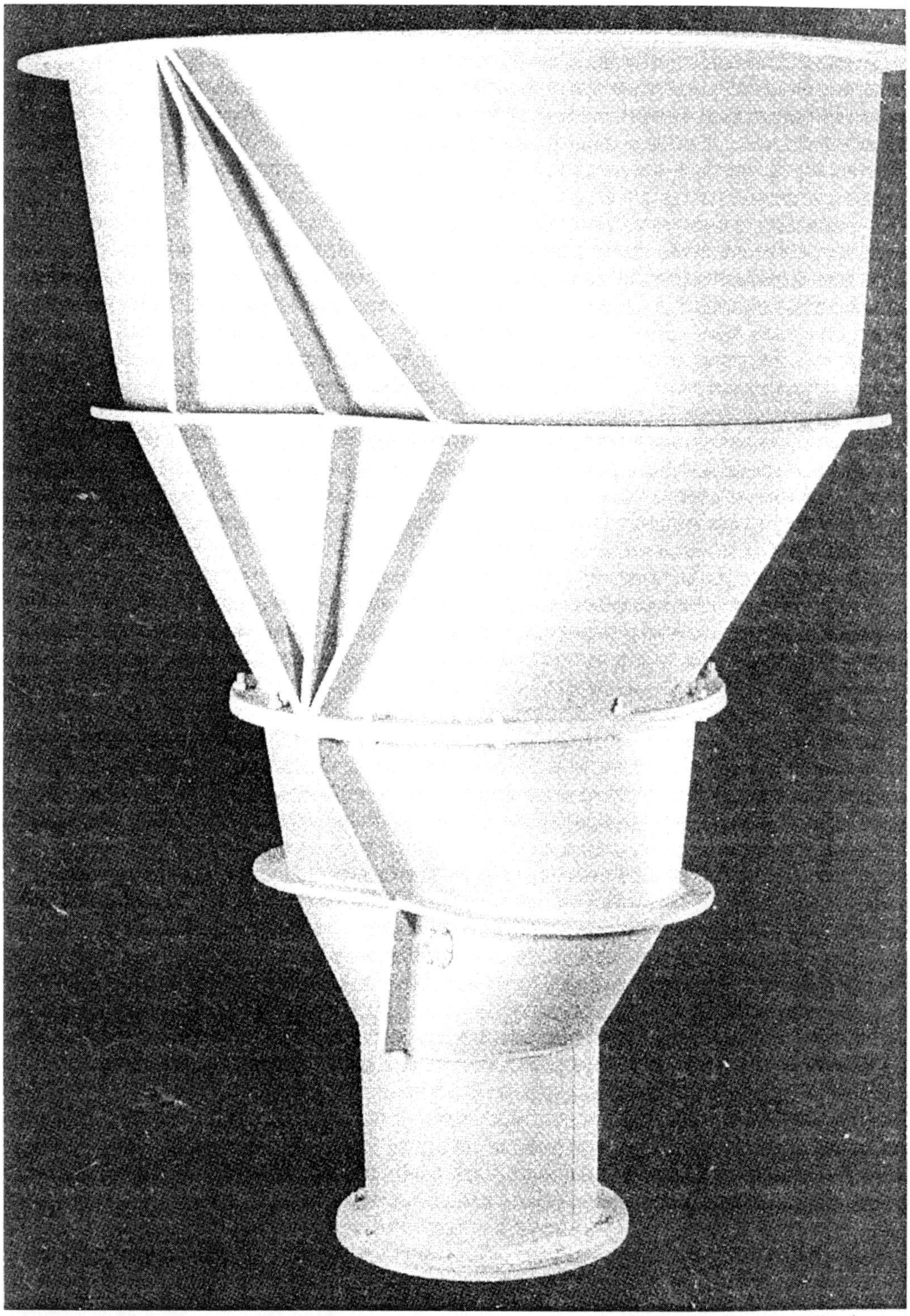

Figure 9.49 The Diamondback Hopper, invented by Jerry R. Johnson, Ph.D. (from Diamondback Technology Inc.).

Chapter 9 - Problems

1. Diesel engines used in construction vehicles are equipped with an air cleaner consisting of a number of standard Lapple reverse-flow cyclones arranged in parallel, followed by a cartridge paper filter. The total volumetric flow rate of air is 10. SCFM. It is necessary to remove all particles (ρ_p = 1,500 kg/m^3) larger than 10. microns with an efficiency no less than 80%. Each reverse-flow cyclone has an overall diameter of D_2 = 2.0 cm. How many cyclones are needed?

2. Consider raindrops of diameter D_c falling through air containing suspended dust particles of diameter D_p. Show that the change in the single drop collection efficiency (η_d) varies with D_p in the following manner:

$$\frac{d\eta_d}{dD_p} = 2.8\frac{\eta_d}{D_p\left(0.70+\text{Stk}\right)}$$

where Stk is the Stokes number.

3. An aerosol (D_p = 3.0 μm, ρ_p = 1,700 kg/m^3) enters the standard cyclone described in Example 9.1 at a velocity of 15. m/s and a temperature of 350. K. If the inlet mass concentration is 100. mg/m^3, what exit concentration can be expected?

4. Hot exhaust (assume air) from a fluidized bed combustor contains large particles of diameter D_p = 50. μm, and density ρ_p = 2,500 kg/m^3. The gas temperature and pressure are 1,000 °C and 1,000 kPa respectively. What is the particle settling velocity in such a gas? It has been suggested that 80.% of these particles can be removed by a bank of reverse-flow cyclones. Assuming collection occurs at the above temperature and pressure, how many cyclones of the dimensions below should be installed in parallel? The total gas mass flow rate is 1.0 kg/s.

- outer diameter = 10. cm
- exit diameter = 5.0 cm
- height and width of inlet = 5.0 cm and 2.5 cm

5. Predict the fractional efficiency curve for an air cleaning unit consisting of a straight-through cyclone and a reverse-flow cyclone. Air and particles from the bleed flow (Q_b) of the straight-through cyclone are sent to the reverse-flow cyclone. The flow from the reverse-flow cyclone is combined with the flow exiting the straight-through cyclone. Assume that a fan draws a total volumetric flow rate (Q_0) into the unit and produces a bleed flow of 10.% (Q_b/Q_0 = 0.10). Assume also that the flow through both devices is n-degree rotational with n = 0.50. Assume that the particles are unit-density spheres and that the air is at STP. The following dimensions of the cyclones apply:

straight-through cyclone	**reverse-flow cyclone**
D_2 = 5.0 cm	D_2 = 10. cm
D_1 = 4.5 cm	D_1 = 5.0 cm
L = 50. cm	$L_1 = L_2$ = 20. cm
vane angle = $\pi/4$	W = 2.5 cm, H = 5.0 cm

Draw the fractional efficiency curve for Q_0 = 0.050, 0.075, and 0.10 m^3/s.

6. Derive an expression for the gravimetric settling efficiency of a horizontal duct of rectangular cross section in which the base is horizontal but the roof slopes upward at an angle θ in the flow-wise direction. The duct width (W) is constant. Flow enters the duct at a Reynolds number of

100,000 and the particles are large such that the drag coefficient is constant. Show the collection efficiency as a function of air properties and some (but not necessarily all) of the following:

volumetric flow rate (Q)	flow-wise distance (x)
angle (θ)	local average velocity [U(x)]
duct width (W)	initial height (H_0)
dust particle diameter (D_p)	particle density (ρ_p)

7. A spray chamber uses water to capture small particles produced in grinding marble. Draw the fractional efficiency curves for $R = Q_s/Q_a = 0.0010$ and 0.00020 if the spray chamber has the following properties:

$D_c = 1.0$ mm	$L = 5.0$ m
$U_a = 3.0$ m/s	$c_0 = 1{,}000$ mg/m^3
$\rho_p = 4{,}000$ kg/m^3	$\rho_c = 1{,}000$ kg/m^3
$D_p = 2.0, 5.0\ 10., 20., 40.\ \mu$m	

8. Show that single drop collection efficiency (η_d) varies with relative speed (v_r) and Stokes number (Stk) as follows:

$$\frac{d\eta_d}{dv_r} = \frac{Stk}{v_r}\frac{1.4(Stk)}{(Stk+0.7)^3}$$

9. Spherical water drops (D_c = 1.0 mm) are sprayed on dust generated when a conveyor discharges coal (ρ = 1,500 kg/m^3) to a stockpile. Assume the air and coal dust velocities are negligible. The coal dust concentration is 100 mg/m^3. Estimate the single drop collection efficiency assuming the drops to fall at their terminal settling velocity. If there are 1,000 drops/m^3 of air, what is the initial rate of removal of coal dust, in mg/(m^{3}s)?

10. Consider a transverse flow scrubber through which an aerosol passes in the horizontal direction. Assume that there is only one spatial variable (in the direction of flow) and that the following are known:

- duct dimensions: height H, length L, width W
- scrubbing liquid volumetric flow rate: Q_s that produces a uniform stream of drops of diameter D_c
- uniform air velocity: $U_a = Q_a/(HW)$
- contaminant aerosol particles: particle diameter D_p and density ρ_p, inlet and outlet mass concentrations $c_{in} = c(0)$ and $c_{out} = c(L)$
- scrubber drop velocity: $\vec{v}_c = U_a\vec{i} - v_{t,c}\vec{j}$, where $v_{t,c}$ is the constant terminal settling speed, and $v_{t,c}$ is much smaller than U_a

Assume that the single drop collection efficiency (η_d) between drop and aerosol particle is known and constant. Show that the grade efficiency is equal to

$$\eta(D_p) = 1 - \exp\left[-\frac{\frac{3}{2}\eta_d\frac{Q_s}{Q_a}\frac{H}{D_c}v_{t,c}L}{v_{t,c}L + U_aH}\right]$$

11. You are employed by a firm that makes bituminous concrete (roadway asphalt). The process produces a sticky aerosol of mass concentration c_0 = 2.0 g/m^3 at STP. It has been decided to design a venturi scrubber to bring the process into compliance with state air pollution regulations. You have been asked to conduct a preliminary analysis of the overall collection process and to produce a series of graphs showing how the grade efficiency varies with the following operating and process parameters:

- collecting drop diameter (D_c)
- throat length (L_t)
- reflux ratio, which is the ratio of the volumetric flow rate of the scrubbing liquid to that of the gas ($R = Q_s/Q_a$)
- particle density, ρ_p = 1,500 kg/m^3
- throat diameter, D_{throat} = 0.60 m
- throat gas velocity, U_a = 15. m/s
- gas temperature and pressure, T = 25. C, P = 1.0 atm
- scrubbing liquid is water and enters the throat with negligible velocity
- once particles impact water drops, they are removed, and liquid particles do not impact one another

(a) Develop an analytical model that predicts the grade efficiency by impaction. State all assumptions carefully and fully. Use a numerical technique to integrate the differential equations so that the grade efficiency can be computed for a venturi throat of arbitrary length and for aerosol particles of any diameter (D_p). Compute the velocity of collecting drops of various diameters (D_c = 0.25, 1.0, 2.0 and 4.0 mm) versus throat length.
(b) Using Eq. (9-42), Eq. (9-43) ,and Eq. (9-16), compute the single drop collection efficiency as a function of throat length for D_c = 1.0 and 2.0 mm, and for D_p = 2.0 μm.
(c) Plot the collection efficiency at D_p = 2.0 μm as a function of throat length (L_t) for L_t = 0.10 to 10. m, assuming D_c = 1.0 mm and R = 1/9,000.
(d) Plot the collection efficiency at D_p = 2.0 μm as a function of throat length assuming D_c = 1.0 mm and R = 1/3,000, 1/9,000, and 1/15,000.
(e) Plot the collection efficiency at D_p = 2.0 μm as a function of collecting drop diameter (D_c) for D_c = 0.25, 1.0, 2.0, 3.0, and 4.0 mm, assuming L_t = 3.0 m and R = 1/9,000.

12. Fog particles of diameter D_p in air pass through a window screen of wire diameter D_f with an approach velocity U_0 normal to the screen. Some of the particles are collected by the screen. Calculate the collection efficiency of each of the following fog aerosol particles.

	D_p (μm)	D_f (μm)	U_0 (m/s)
(a)	20	1,000	0.5
(b)	20	2,000	1.0
(c)	10	2,000	4.0
(d)	10	2,000	2.0

13. Consider a filter of thickness H consisting of spherical particles of diameter D_c. The filter bed has the following properties:

- U_0 = approach velocity $U_0 = Q_a/A_f$, where Q_a is the volumetric flow rate through a bed of frontal area A_f
- η_d = single drop collection efficiency, defined as the efficiency of each bed particle D_c to collect contaminant particles of diameter D_p
- f_f = solids fraction, defined as the fraction of bed composed of solid material, the solids fraction is equal to (1 - ε) where ε is the porosity of the bed

(a) Show that the grade efficiency in terms of the above parameters is

$$\eta\left(D_p\right) = 1 - \exp\left[-\frac{3}{2}\eta_d \frac{H}{D_c}\frac{f_f}{1-f_f}\right]$$

(b) Derive an expression for the rate of change of grade efficiency with respect to face velocity (U_0), if the single drop collection efficiency is governed by Eq. (9-16).

14. You wish to install a baghouse to capture particles removed from the air in a foundry. The baghouse manufacturer claims that the filter resistance (K_1) is 10,000 (N s)/m^3 and estimates that the dust cake resistance (K_2) is 80,000 sec^{-1}. The dust concentration entering the unit is 5.0 gm/m^3. The recommended air-to-cloth ratio is 1.0 m/min. If the baghouse is to operate at a maximum pressure drop of 500 N/m^2, how often should the bags be cleaned?

15. A horizontal duct of square cross section (W by W) and length (L) carries an aerosol at volumetric flow rate (Q). The inlet dust mass concentration is $c_{in} = c(0)$. The duct is mounted only at the inlet end and has no other supports. Dust settles to the floor of the duct. After a while a sufficient amount of dust accumulates to produce a critical bending moment M_c that fractures the duct mounting. Write an expression to predict the time (t_c) when the mounting fails.

16. A jewelry manufacturer asks you to design an inertial separation collector followed by a filter to capture small particles of silver generated by cutting tools. The capture efficiency of the collector should be at least 80.%. The inertial separation collector consists of N individual Lapple standard reverse-flow cyclones (Figure 9.2) arranged in parallel. How many cyclone collectors are needed? The following characterize the particles and the gas stream:

particles	**gas stream (air)**	**individual cyclones**
D_p = 25. μm	Q(actual) = 0.0010 m^3/s	D_2 = 4.0 cm
ρ_p = 10,500 kg/m^3	P = 100. kPa, T = 300. K	N_e = 6.0

17. Consider a cylindrical (toroidal) filter for an off-road vehicle. Air containing particles at an inlet mass concentration $c_{in} = c_0$ is drawn inward in the radial direction through the filter of height (H), outer diameter (D_2) and inner diameter (D_1). The volumetric flow rate of air is (Q). The filter is composed of fibers of diameter (D_f) and solids fraction (f_f). The single fiber collection efficiency is η_f. Show that the grade efficiency (fractional efficiency) of the filter can be expressed as

$$\eta(D_p) = 1 - \exp\left[-2\frac{\eta_f}{\pi}\frac{f_f}{1-f_f}\frac{D_2 - D_1}{D_f}\right]$$

18. An electrostatic precipitator with a collecting area of 5,000 ft^2 is designed to remove 95.% of the fly ash particles from a 250. °F exhaust gas stream from a municipal incinerator. Your supervisor suggests that the same design can be used to remove 95.% of the particles from a 590. K, 101. kPa, 150. kg/min gas stream from a cement kiln.

(a) What capture efficiency do you believe the original design will achieve when used for the cement kiln?

(b) If your supervisor's expectation is to be achieved, what should be the area of the collecting surface?

19. An electrostatic precipitator is used to remove particles from the exhaust of a pulverized coal furnace in an electric power generating station. The collection efficiency is 95.%. Management decides to increase the collection efficiency to 98.% and to increase the exhaust gas temperature from 350 to 380 K to increase plume buoyancy. The exhaust gas mass flow rate remains the same. Estimate the change in area of the collecting electrodes needed to accomplish these tasks.

20. The dust concentration in the air of a foundry increases during the day. The workplace is of height H and floor area A. Circulating fans within the room ensure that the dust mass concentration varies only with time and not location,

$$c(x,y,z,t) = c(t)$$

At the end of the workday the mass concentration is c(0), dust is no longer produced, but the fan remains on. Assuming settling is governed by the well-mixed model, write an expression that predicts the concentration at any instant of time after the workday ends,

$$c(t) = \text{function of } (c(0), D_p, \rho_p, \mu, H, A)$$

and the total mass of dust that settles on the floor $m_{settled}(t)$ over an elapsed time,

$$m_{settled}(t) = \text{function of } (c(0), D_p, \rho_p, \mu, H, A)$$

10

Application of CFD to Indoor Air Quality

In this chapter you will learn:

- fundamentals of computational fluid dynamics (CFD)
- capabilities and limitations of CFD as applied to indoor air quality engineering

If the air and contaminants in a room are well mixed, the differential equations that describe their behavior reduce to ordinary differential equations, enabling straightforward integration. One can then predict important global features of general ventilation (such as mass concentration in the room) as functions of time; this type of analysis is presented in Chapter 5. If contaminants are to be withdrawn locally with some type of hood, and if the geometrical arrangement of the source and hood inlet is similar to a well-tested standard arrangement, the empirically based design techniques and design plates of Chapter 6 can be applied. If a flow field can be assumed to be largely irrotational, the potential flow techniques of Chapter 7 can be applied in order to solve for the velocity field. Once the velocity field is known, the techniques of Chapter 8 can be applied to predict trajectories of contaminant particles. However, for the general case in which the air and contaminants in a room are *not* well mixed, and/or the source and hood inlet are *not* similar to any standard arrangement, and/or viscous rotational forces and perhaps turbulence cannot be ignored, more rigorous analysis is necessary. Such is the domain of ***computational fluid dynamics*** (CFD).

Some of the pioneering applications of CFD for indoor air quality engineering (worldwide) include those of Nielsen et al. (1978) who calculated the velocity fields in ventilated rooms of various geometries; Zarouri et al. (1983) who computed trajectories and concentrations of particles in a grinding booth; Yang and Lloyd (1985) who calculated velocity, temperature, and concentration fields of turbulent buoyant flows resulting from fires in aircraft cabins; Dellagi et al. (1986) and Cornu et al. (1986) who calculated velocity and concentration fields inside paint spray booths for automobiles; and Ye an Pui (1990) who analyzed particle deposition in a tube containing an abrupt contraction. A useful summary of some of the early work on flow patterns and diffusion of contaminants in rooms can be found in Kurabuchi and Kusuda (1987) and in Murakami and Kato (1989).

10.1 Fundamentals of CFD

This chapter does not present a comprehensive analysis of grid generation techniques, CFD algorithms, or numerical stability; that is not its intent. Rather, a brief introduction to CFD is presented; it is assumed that readers have access to an in-house or commercially available working CFD code. All examples presented here have been obtained with the commercial computational fluid dynamics code, ***FLUENT***, run in double precision on a personal computer. Other CFD codes would yield similar, but not identical results.

10.1.1 Equations of Motion

The equations of motion for laminar flow of a viscous, incompressible, Newtonian fluid are conservation of mass (the ***continuity equation***),

$$\boxed{\vec{\nabla}\cdot\vec{U}=0} \tag{10-1}$$

and conservation of linear momentum (the ***Navier-Stokes equation***),

$$\boxed{\frac{\partial\vec{U}}{\partial t}+\left(\vec{U}\cdot\vec{\nabla}\right)\vec{U}=-\frac{1}{\rho}\vec{\nabla}P+\vec{g}+\nu\nabla^2\vec{U}} \tag{10-2}$$

where P is the pressure, $\vec{U}$ is the velocity vector of the fluid, ρ is its density, and ν is its kinematic viscosity ($\nu = \mu/\rho$); both ρ and ν are assumed to be constant for incompressible flow. Eqs. (10-1) and (10-2) are called ***transport equations***; the former representing transport of mass and the latter representing transport of linear momentum throughout the computational domain. Note that Eq. (10-1) is a scalar equation, while Eq. (10-2) is a vector equation. For three-dimensional flow in cartesian coordinates, there are thus four coupled transport equations for four unknowns, U, V, W, and P. If the flow were compressible, Eqs. (10-1) and (10-2) would need to be modified appropriately. However, for indoor air flows, the air is typically at low enough Mach number that it behaves as a nearly incompressible fluid. For a steady-state flow field, the first term in Eq. (10-2) is zero.

10.1.2 Solution Procedure

To solve Eqs. (10-1) and (10-2) numerically for the simple case of steady flow, the following steps are performed, some of which are examined in more detail later:

1. A ***computational domain*** is chosen and a ***grid*** is generated; the domain is divided into many small elements called ***cells***. For 2-D domains, the cells are areas, while for 3-D domains the cells are volumes.
2. ***Boundary conditions*** are defined on each ***edge*** of the computational domain (2-D flows) or on each ***face*** of the domain (3-D flows).
3. Fluid properties are specified (temperature and pressure, viscosity, etc.)
4. Numerical parameters (relaxation factors, etc.) and solution algorithms are selected; these are specific to each CFD code, and will not be discussed here.
5. Values for all flow field variables are specified for each cell; these are ***initial guesses***, which are known to be incorrect, but are necessary as a starting point.
6. Beginning with the initial guesses, discretized forms of Eqs. (10-1) and (10-2) are solved iteratively, usually at the center of each cell. If one were to put all the terms of Eq. (10-2) on one side of the equation, the solution would be "exact" when the sum of these terms, defined as the ***residual***, is zero for every cell in the domain. In a CFD solution, however, the sum is *never* identically zero, but (hopefully) decreases with progressive iterations. A residual can be thought of as a measure of how much the solution to a given transport equation deviates from exact, and one monitors the residual of each transport equation to help determine when the solution has converged. Sometimes hundreds or even thousands of iterations are required to converge on a final solution, and the residuals decrease by several orders of magnitude. This technique of iterating towards a steady-state solution is referred to in numerical literature as ***relaxation***.
7. Once the solution has converged, flow field variables such as velocity and pressure are plotted and analyzed graphically. Users can also define and analyze additional functions, called ***user defined functions***, which are formed by algebraic combinations of flow field variables. Most commercial CFD codes have built in ***post-processors***, designed for quick graphical analysis of the flow field. There are also stand-alone post-processor software

packages available for this purpose. Since the graphics output is often displayed in vivid colors, CFD has earned the nickname "colorful fluid dynamics".

8. ***Global properties*** of the flow field, such as pressure drop, and ***integral properties***, such as forces (lift and drag) and moments acting on a surface, are calculated from the converged solution. With most CFD codes, this can in fact be done "on the fly" as the iterations proceed. In many cases, it is wise to monitor these quantities along with the residuals; when a solution has converged, the global and integral properties should settle down to constant values.

For *unsteady* flow, a physical time step is specified, some initial conditions are assigned, and an iteration loop is carried out to solve the transport equations to simulate changes in the flow field over this small span of time. Since the changes are small, a relatively small number of iterations (on the order of tens or hundreds) is required between each time step. Upon convergence of this "inner loop", the code marches to the next time step. If a flow has a steady-state solution, it is sometimes easier to find by marching in time – after enough time has past, the flow field variables settle down to their steady-state values. Some CFD codes take advantage of this fact by internally specifying a pseudo-time step (***artificial time***), and marching towards a steady-state solution. In such cases, the pseudo-time step can even be different for different cells in the computational domain, and can be tuned to decrease convergence time. Other "tricks" are often used to speed up computational time, such as ***multigridding***, in which the flow field variables are updated first on a coarse grid so that global features of the flow are quickly established; that solution is then interpolated to finer and finer grids, the final grid being the one specified by the user. In commercial CFD codes, several layers of multigridding may occur "behind the scenes" during the iteration process, without user input (or awareness). Readers can learn more about computational algorithms and other numerical techniques that improve convergence by reading books devoted to computational methods, such as Tannehill, Anderson, and Pletcher (1997).

10.1.3 Additional Equations

If contaminant particles are to be injected into the computational domain, the techniques of Chapter 8 can be applied to solve for individual particle trajectories. It is assumed that the particles themselves do not influence the flow field; hence, particle trajectories can be calculated *after* the flow solution is obtained (post-processing). However, if *gaseous* contaminants are released into the computational domain, Eqs. (10-1) and (10-2) must be modified so as to apply to a gas mixture, and these equations must be supplemented by an additional equation of mass conservation for each contaminant. For laminar flow, Eq. (4-22) applies, and is repeated here:

$$\boxed{\frac{\partial c_j}{\partial t} + \vec{\nabla}\cdot\left(\vec{U}c_j\right) = D_{ja}\nabla^2 c_j} \tag{10-3}$$

where c_j is the mass concentration of species j, D_{ja} is the diffusion coefficient of species j into the gas mixture (which is predominantly air), and $\vec{U}$ is the ***bulk velocity*** of the gas mixture. Alternatively, Eq. (10-3) can be written in terms of mass fraction (f_j) for each species j. By definition,

$$f_j = \frac{m_j}{m} = \frac{m_j}{V}\frac{V}{m} = \frac{c_j}{\bar{\rho}}$$

where $\bar{\rho}$ is the ***bulk density*** of the gas mixture; this density is not constant throughout the computational domain since gases of various densities are mixed non-uniformly in the domain. Thus, Eq. (10-3) becomes

$$\frac{\partial\left(\bar{\rho}f_j\right)}{\partial t}+\vec{\nabla}\cdot\left(\bar{\rho}\vec{U}f_j\right)=D_{ja}\nabla^2\left(\bar{\rho}f_j\right) \tag{10-4}$$

If heat transfer is important in the problem, another transport equation, the energy equation, must also be solved. If temperature differences lead to significant changes in density, an equation of state (such as the ideal gas law) is used. If buoyancy is important, the effect of temperature on density is reflected in the gravity term in Eq. (10-2). If the flow is turbulent, additional transport equations to model the enhanced mixing and diffusion of turbulence must be solved as well. There are many ***turbulence models*** in use today; the most popular ones are the k-ε model, the k-ω model, and the q-ω model. These so-called "two-equation" turbulence models add two more transport equations, which must be solved simultaneously with the equations of mass, linear momentum, energy, and species. Furthermore, additional terms appear in the momentum, energy, and species equations due to the turbulence. Detailed analysis of turbulence models is beyond the scope of this text; readers are referred to Wilcox (1998).

10.1.4 Grid Generation and Resolution

The first step (and arguably the most important step) in a CFD solution is generation of a grid (also called a ***mesh***), which defines the nodes or cells at which flow variables (velocity, pressure, etc.) are calculated throughout the computational domain. Modern commercial CFD codes come with their own grid generator, and third-party grid generation programs are also available. Most CFD codes can run with either structured or unstructured grids. A ***structured grid*** consists of planar cells with four edges (2-D) or volumetric cells with six faces (3-D). Although the cells can be distorted from rectangular, each cell is numbered according to indices (i,j,k) for coordinates x, y, and z. An illustration of a two-dimensional structured grid is shown in Figure 10.1a. To construct this grid, 9 ***nodes*** are marked on the top and bottom edges; these nodes correspond to 8 ***intervals*** along these edges. Similarly, 5 nodes are marked on the left and right edges, corresponding to 4 intervals along these edges. The intervals correspond to i = 1 through 8 and j = 1 through 4, and are numbered and marked in Figure 10.1a. An internal grid is then generated by connecting nodes one-for-one across the domain such that rows (j = constant) and columns (i = constant) are clearly defined, even though the cells themselves are distorted (not necessarily rectangular). The grids used in this chapter are generated with FLUENT's grid generation package, ***GAMBIT***. In a two-dimensional structured grid, each cell is uniquely specified by an index pair (i,j). For example, the shaded cell in Figure 10.1a is at (i = 4, j = 3). Readers should note that some CFD codes number *nodes* rather than intervals. An ***unstructured grid*** consists of cells of various shapes, but typically triangles (2-D) and tetrahedrons (3-D) are used. An unstructured grid for the same domain can be generated, using the *same* interval distribution on the edges; this grid is shown in Figure 10.1b. Unlike the structured grid, one cannot uniquely identify cells in the unstructured grid by indices i and j; instead, cells are numbered in some other fashion internally in the CFD code.

For complex geometries, especially those with highly curved surfaces, an unstructured grid is usually much easier for the user to generate. However, there are advantages to structured grids. In general, a CFD solution will converge more rapidly, and often more accurately, with a structured grid. In addition, less cells are generated with a structured grid than with an unstructured grid. In Figure 10.1, for example, the structured grid has 8 x 4 = 32 cells, while the unstructured grid has 70 cells, even though the identical node distribution is applied at the edges. In boundary layers, where flow variables change rapidly normal to the wall and highly resolved grids are required close to the wall, structured grids enable much finer resolution than do unstructured grids for the same number of cells. This can be seen by comparing the grids of

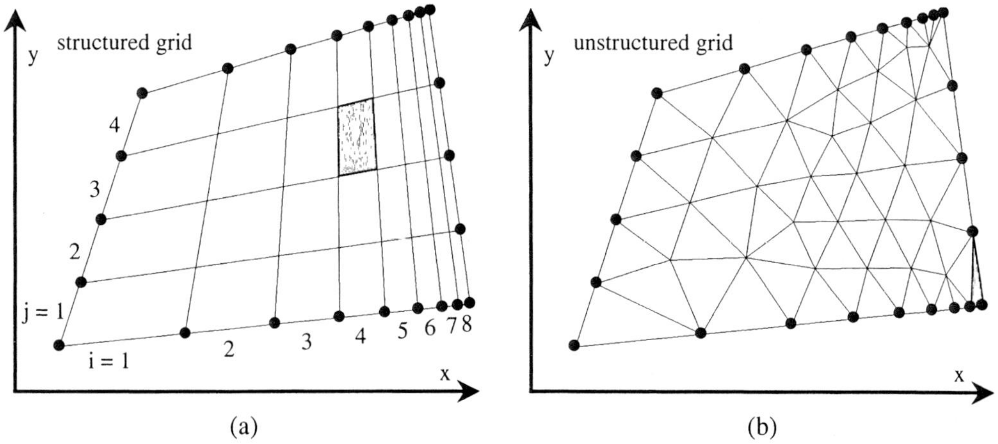

Figure 10.1 Sample two-dimensional grid with 9 nodes (represented by the dots) and 8 intervals on top and bottom edges, and 5 nodes and 4 intervals on left and right edges: (a) structured grid with i as the index for coordinate x and j as the index for coordinate y; the shaded cell is at (i = 4, j = 3); (b) unstructured grid generated from the same node distribution on the edges; the shaded cell is highly skewed.

Figure 10.1 near the far right edge. The cells of the structured grid are thin and tightly packed near the right edge, while those of the unstructured grid are not. As a general rule, therefore, *one should use structured grids whenever possible*. Finally, a ***hybrid grid*** is one that combines structured and unstructured grids. For example, one can mate a structured grid close to a wall with an unstructured grid outside of the region of influence of the boundary layer. In any case, users must also be careful that individual cells are not highly skewed, as this can lead to convergence difficulties and inaccuracies in the numerical solution. The shaded cell in Figure 10.1b is an example of a cell with moderately high ***skewness***.

Generation of a good grid is often tedious and time-consuming; engineers who use CFD on a regular basis will agree that grid generation often takes more time than does the CFD solution itself. However, time spent generating a good grid is time well spent, since the CFD results will be more reliable and will converge more rapidly. A good grid is critical to an accurate CFD solution; a poorly resolved grid may even lead to an *incorrect* solution. It is important, therefore, for users of CFD to test if their solution is ***grid resolved***. The standard method to test for adequate grid resolution is to increase the resolution (by a factor of 2 in all directions if feasible) and repeat the simulation. If the results do not change appreciably, the original grid is deemed adequate. If, on the other hand, there are significant differences between the two solutions, the original grid is likely of inadequate resolution. In such a case, an even finer grid should be tried until the grid is adequately resolved. This method of testing for grid resolution is time consuming and unfortunately not always feasible, especially for large engineering problems in which the solution pushes computer resources to their limits. In a two-dimensional simulation, if one doubles the number of intervals on each edge the number of cells increases by a factor of $2^2 = 4$; the required computation time for the CFD solution also increases by approximately a factor of 4. For three-dimensional flows, doubling the number of intervals in each direction increases the cell count by a factor of $2^3 = 8$; readers can see how grid resolution studies can easily get beyond the range of a computer's memory capacity and/or CPU availability.

10.1.5 Boundary Conditions

While the equations of motion, the computational domain, and even the grid may be the same for two CFD problems, the imposed boundary conditions are the factor that leads to a unique CFD solution. There are several types of boundary conditions available; the most relevant ones are listed and briefly described here. The names are those used by FLUENT; other CFD codes may use somewhat different terminology. In the descriptions below, the words "face" or "plane" are used, implying three-dimensional flow. For two-dimensional flows, the word "edge" or "line" should be substituted for "face" or "plane".

- ***Axis***: For axisymmetric flows, this boundary condition is applied to an edge that represents the axis of symmetry. Fluid can flow *parallel* to the axis, but cannot flow *through* the axis. In the case of swirling axisymmetric flows, fluid can flow *tangentially* in a circular path around the axis of symmetry. Swirling axisymmetric flows are sometimes called ***rotationally symmetric***.
- ***Fan***: A pressure rise is specified across a plane to which is assigned the fan boundary condition. This boundary condition is similar to "interior" except for the forced pressure rise. The CFD code does not solve the detailed, unsteady flow field through individual fan blades, but simply models the plane as an infinitesimally thin fan that increases the pressure across the plane. If the pressure rise is specified as zero, this boundary condition behaves the same as "interior".
- ***Interior***: Flow crosses through the face without any user-forced changes, just as it would cross from one interior cell to another. This boundary condition is necessary for situations in which the computational domain is divided into separate zones or blocks, and enables communication between zones. The authors have found this boundary condition useful for post-processing as well, since a pre-defined face is present in the flow field, on whose surface one can plot velocity vectors, pressure contours, etc.
- ***Periodic***: Flow field variables along one face of a periodic boundary are numerically linked to a second face of identical shape. Thus, flow leaving (crossing) the first periodic boundary can be imagined as entering (crossing) the second periodic boundary with identical properties (velocity, pressure, temperature, etc.). This boundary condition is useful for flows with repetitive geometries, such as flow through an array of heat exchanger tubes.
- ***Pressure inlet***: Fluid flows into the computational domain. Users specify the pressure along the inlet face (for example, flow coming into the computational domain from a pressurized tank of known pressure or from the far field where the ambient pressure is known). Flow properties such as temperature, mass fractions, and turbulence properties must also be specified at a pressure inlet. Velocity is *not* specified at a pressure inlet, as this would lead to mathematical over-specification. Rather, velocity at a pressure inlet adjusts itself to match the rest of the flow field.
- ***Pressure outlet***: Fluid flows out of the computational domain. Users specify the pressure along the outlet face; in many cases this is atmospheric pressure (at the outlet of a duct open to ambient air for example). Flow properties such as temperature, mass fractions, and turbulence properties are also specified at a pressure outlet, but are not used unless the solution demands reverse flow across the outlet. *Reverse flow at a pressure outlet is usually an indication that the computational domain is not large enough, and should be avoided.* As discussed above for a pressure inlet, velocity is *not* specified at a pressure outlet, but adjusts itself to match the rest of the flow field.
- ***Symmetry***: Flow field variables are mirror-imaged across a symmetry plane, fluid can flow *parallel* to the plane, but fluid cannot flow *through* the plane. Mathematically, *gradients* of flow field variables are zero across a symmetry plane. For physical flows with one or more symmetry planes, this boundary condition enables one to model only a *portion* of the physical flow domain, thereby conserving computer resources.
- ***Velocity inlet***: Fluid flows into the computational domain. Users specify the velocity and composition (mass fractions) of the incoming flow along the inlet face. If energy and/or

turbulence equations are being solved, the temperature and/or turbulence properties of the incoming flow need to be specified as well. Pressure is *not* specified at a velocity inlet, as this would lead to mathematical over-specification. Rather, pressure at a velocity inlet adjusts itself to match the rest of the flow field.

- ***Wall***: Fluid cannot pass through a wall, thus the normal component of velocity at a wall is zero. In addition, because of the no slip condition, the tangential component of velocity at a stationary wall is also zero. This latter condition is relaxed for the case of high Knudsen number flows (free molecular flow regime), as discussed in Chapter 8. If the energy equation is being solved, either wall temperature or wall heat flux must also be specified (but not both). If turbulence equations are being solved, wall roughness must be specified, and users must choose among various kinds of turbulence wall treatments (wall functions, etc.) which are beyond the scope of the present text (see Wilcox, 1998). Moving walls and walls with specified shear stresses can also be simulated in most CFD codes, as can walls which absorb or produce gaseous contaminants, and walls which collect or reflect aerosol particles that impact them.

10.2 Flow around a Circular Cylinder

The best way to learn computational fluid dynamics is through examples and practice. Readers are encouraged to experiment with various grids, boundary conditions, numerical parameters, etc. in order to get a feel for CFD. Before tackling a complicated problem, it is best to solve simpler problems, especially ones for which analytical or empirical solutions are known (for comparison and verification). The authors chose flow around a circular cylinder as the first example in order to illustrate many of the capabilities and limitations of CFD. Following that is a collection of example problems more directly relevant to indoor air quality engineering.

10.2.1 Laminar Flow Grid Resolution Tests

Consider flow around a circular cylinder of diameter D = 20.0 cm. The two-dimensional computational domain used for this simulation is sketched in Figure 10.2. Only the upper half of the flow field is solved due to symmetry along the bottom edge of the computational domain; a symmetry boundary condition is specified along this edge to ensure that no flow crosses the plane of symmetry. With the symmetry boundary condition imposed, the required computational domain size is reduced by a factor of two. A stationary wall boundary condition is applied at the cylinder surface. The left half of the far field outer edge of the domain has a velocity inlet boundary condition, on which is specified the velocity components $U = U_\infty$ and $V = 0$. A pressure outlet boundary condition is specified along the right half of the outer edge of the domain.

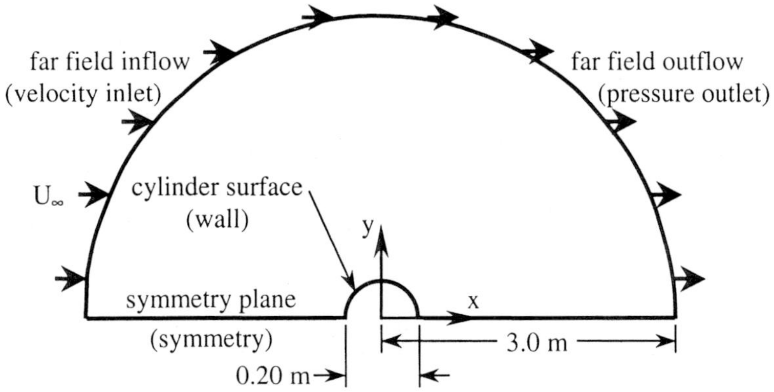

Figure 10.2 Computational domain used to simulate two-dimensional flow of air over a circular cylinder (not to scale); applied boundary conditions are in parentheses.

Three two-dimensional structured grids are generated for comparison: coarse (30 radial intervals x 60 intervals along the cylinder surface = 1,800 cells), medium (60 x 120 = 7,200 cells), and fine (120 x 240 = 28,800 cells), as seen in Figure 10.3. Note that only a small portion of the grid is shown here; the full grid extends fifteen cylinder diameters outward from the origin, and the cells get progressively larger further away from the cylinder.

(a)
(b)
(c)

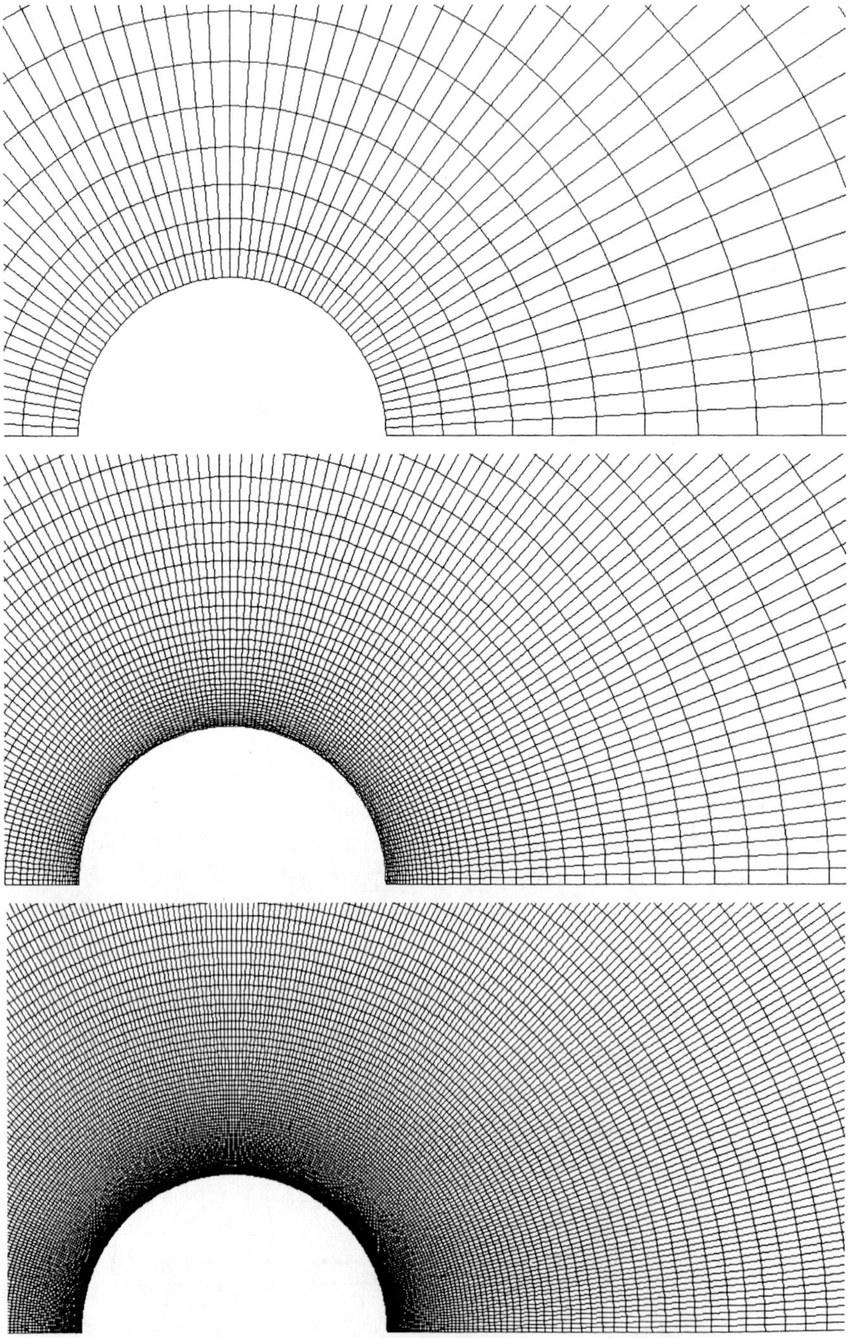

Figure 10.3 Structured 2-D grids around a circular cylinder: (a) coarse (30 x 60), (b) medium (60 x 120), and (c) fine (120 x 240) grid; bottom edge is a line of symmetry.

Consider streaming flow of air at STP and at velocity U_∞ = 1.2 m/s from left to right around this circular cylinder. The Reynolds number of the flow, based on cylinder diameter (D = 20.0 cm), is approximately 1500. Experiments at this Reynolds number reveal that the boundary layer is laminar, and separates several degrees upstream of the top of the cylinder. The wake, however, is turbulent; such a mixture of laminar and turbulent flow is particularly difficult for CFD codes. The measured drag coefficient at this Reynolds number is $c_D \approx 0.90$ (Tritton, 1977). FLUENT solutions are obtained for each of the above three grids, assuming steady laminar flow. All three cases converge without problems, but the results do not necessarily agree with physical intuition or with empirical data. Streamlines are shown in Figure 10.4 for the three grid resolutions and for a fourth case (turbulent flow, to be discussed shortly). In all cases, the image is mirrored about the symmetry line so that even though only the top half of the flow field is solved, the full flow field is visualized. For the coarse resolution case (Figure 10.4a), the boundary layer separates well past the top of the cylinder, and c_D is only 0.25. Clearly, the boundary layer is not well-enough resolved to yield the proper boundary layer separation point, and the drag is more than a factor of three times smaller than it should be. For the medium resolution case (Figure 10.4b), the flow field is significantly different. The boundary layer separates upstream of the top of the cylinder, which is more in line with the experimental results, and c_D has increased to about 0.36 – closer, but still less than half of the experimental value. Most disturbing are the huge

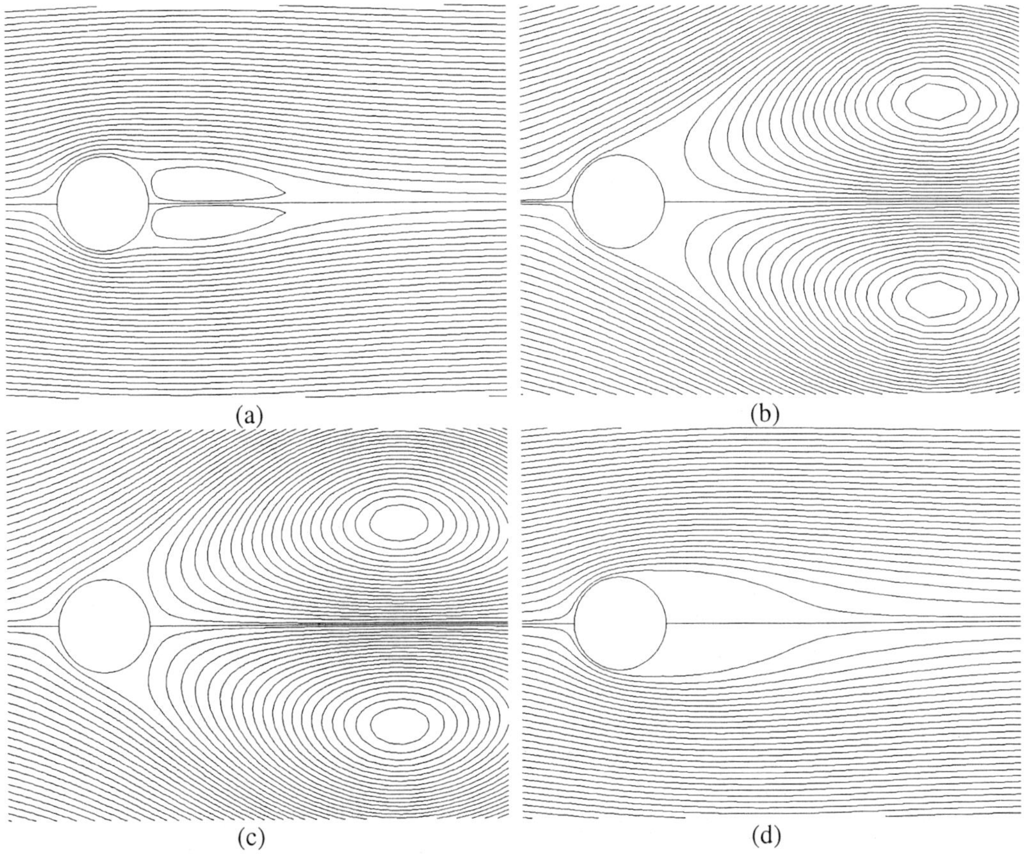

Figure 10.4 Streamlines produced by steady-state CFD solutions for flow over a circular cylinder at Re = 1500: (a) laminar flow, coarse grid (30 x 60), (b) laminar flow, medium grid (60 x 120), (c) laminar flow, fine grid (120 x 240), and (d) turbulent (realizable k-ε) flow, medium grid (60 x 120).

recirculating eddies in the cylinder's wake, which are much larger than expected. Does refining the grid even further improve the numerical results? Figure 10.4c shows streamlines for the fine resolution case. The results look qualitatively similar to those of the medium resolution câse, but the drag coefficient is even worse ($c_D = 0.28$).

Why do the simulations shown in Figure 10.4a, b, and c have such poor agreement with experiment? The answer is two-fold:

(1) Although the boundary layer is indeed laminar, the wake of the cylinder is actually *turbulent* at this Reynolds number. The solutions shown in Figure 10.4a through c are laminar, and therefore erroneous.

(2) All four cases in Figure 10.4 are for the upper half plane only, the solutions are steady-state, and symmetry is enforced. In reality, flow over a circular is unsteady and highly non-symmetric; vortices are shed alternately from the top and the bottom of the cylinder, forming the well-known ***Karman vortex street*** (Tritton, 1977; Cimbala et al., 1988).

10.2.2 Turbulent Flow and Unsteady Flow

To address the first of the above two points, a fourth case is run, this time for turbulent flow (realizable k-ε turbulence model) and with the medium resolution grid; the streamlines are shown in Figure 10.4d. Although the boundary layer separates a bit too late (almost directly at the top of the cylinder), the predicted drag coefficient is $c_D = 0.99$, much closer to the experimental value. The huge recirculating wake eddies are replaced by smaller, more physically realistic eddies (due to enhanced mixing and turbulent diffusion in the flow).

To address the second point, a fifth simulation is run with a full grid (top and bottom, both halves equivalent to the medium grid case described above). The symmetry condition is removed, and the simulation is laminar but *unsteady*. Instantaneous streamlines at fifteen time steps are shown in Figure 10.5, representing approximately one full cycle of the formation of alternating vortices (the Karman vortex street). The results appear more physically realistic, at least qualitatively. Unfortunately, the time-averaged drag coefficient ($c_{D,avg}$) is overpredicted at a value of about 1.7. The instantaneous value of c_D is plotted in Figure 10.6 as a function of time, with labels corresponding to each time frame of Figure 10.5. When a new vortex is beginning to shed from the cylinder, the drag is near its lowest value, e.g. at times (e) and (l). When the vortex is fully formed and energetic, but still close to the cylinder, the drag is near its highest value, e.g. at times (a), (i), and (o). This may be explained by the low pressure in the cylinder wake near its surface due to the presence of the vortex. In the present calculations, there is an asymmetry between the top and bottom vortex shedding processes, the cause of which is unknown. The main reasons for discrepancies between these unsteady simulations and experiments are: inadequate grid resolution in the downstream wake, laminar flow solutions in a wake that should be turbulent, and forced two-dimensionality in a flow that is inherently three-dimensional (all turbulent flows are three-dimensional, although they may be two-dimensional in the mean).

10.2.3 Further Improvements to the Simulation

Are there steps that can be taken to further improve this CFD simulation? Yes, but only with significant increase of computational recourses. The authors ran a FLUENT simulation of this same flow, but as an unsteady *turbulent* flow field. Unfortunately, the solution asymptotes to the turbulent steady-state solution of Figure 10.4d, implying that turbulent dissipation terms in the turbulence model are strong enough to damp out any unsteadiness. The next step would be ***large eddy simulation*** (LES) in which large unsteady features of the turbulent eddies are resolved, while small-scale dissipative turbulent eddies are modeled. Beyond LES, there is ***direct numerical simulation*** (DNS), in which turbulent eddies of *all* scales are resolved. Both LES and DNS require extremely fine, fully three-dimensional grids, large computers, and a huge amount of CPU time (Cimbala, 1995). Neither LES nor DNS is attempted here.

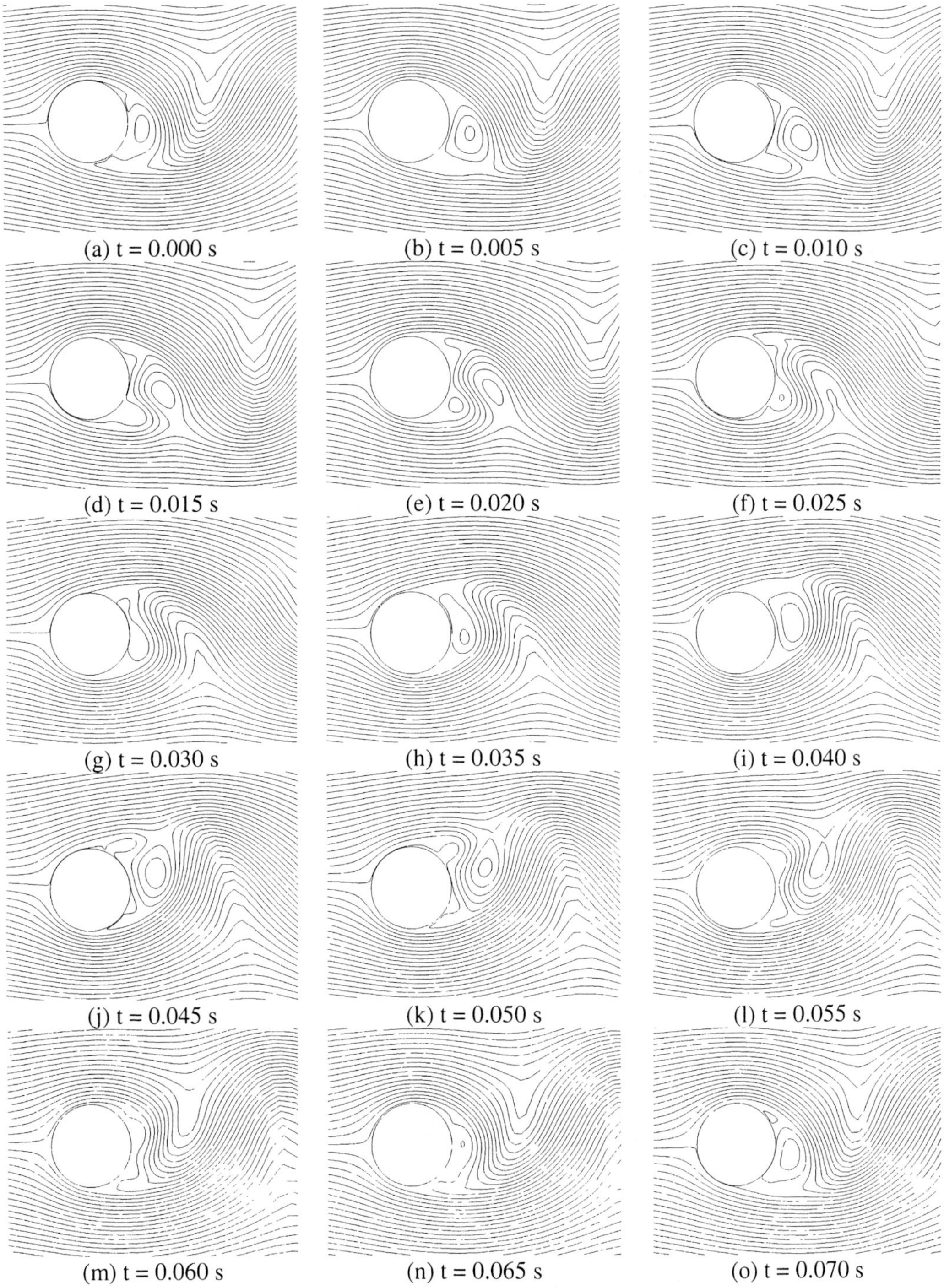

Figure 10.5 Instantaneous streamlines produced by unsteady laminar CFD simulation of flow over a circular cylinder at Re = 1500; time t is relative to some arbitrary time, and $\Delta t = 0.005$ s between successive images.

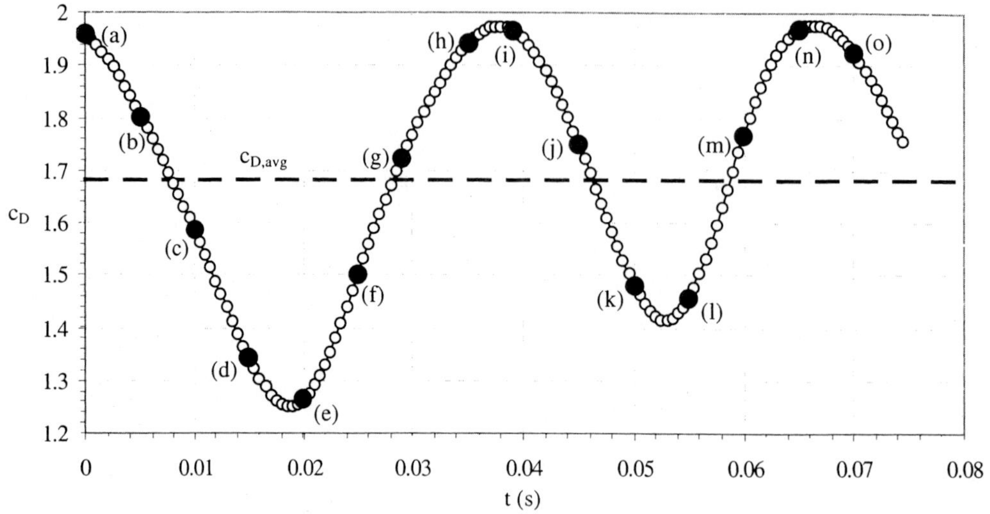

Figure 10.6 Instantaneous drag coefficient calculated from unsteady laminar CFD simulation of flow over a circular cylinder at Re = 1500; filled circles labeled (a) through (o) correspond to the instantaneous streamlines shown in Figure 10.5.

10.2.4 Conclusions from the Cylinder Calculations

This exercise with "simple" flow over a circular cylinder has demonstrated some of the capabilities of computational fluid dynamics, but has also revealed several aspects of CFD about which one must be cautious. Poor grid resolution can lead to incorrect solutions, particularly with respect to boundary layer separation. Laminar solutions can differ significantly from turbulent solutions, and it is not always clear which to choose. For the case at hand, for example, the boundary layer remains laminar, but the wake is turbulent. As a result, the laminar solution does a better job at predicting boundary layer separation, but the turbulent solution does a better job at predicting the wake and the drag coefficient. Forced numerical flow symmetry is not always wise, even for cases in which the physical geometry is entirely symmetric, because *symmetric geometry does not guarantee symmetric flow*. Forced steady flow can yield incorrect results when the flow is inherently unstable and/or oscillatory. Likewise, forced two-dimensionality can yield incorrect results when the flow is inherently three-dimensional. Turbulence models enable solution of turbulent flow fields of interest to engineers, but do not always predict correct results since no turbulence model is universally applicable to all flows. In theory, direct numerical simulation should always be able to predict correct results, but unfortunately computers need to improve by several orders of magnitude before DNS (or even LES) becomes feasible for practical engineering analyses, especially at high Reynolds numbers.

10.3 Modeling of Air Flows with Gaseous Contaminants

In the field of indoor air quality engineering, simulation of the flow of air and gaseous contaminants in rooms, hoods, etc. is inherently difficult. Modeling of air and gaseous contaminants involves simultaneous solution of the equations of motion, Eqs. (10-1) and (10-2), for the bulk flow *plus* an equation of conservation of mass, Eq. (10-3), for each species. The large sizes associated with rooms and air pollution control systems lead to high Reynolds number, turbulent flows which require solution of additional transport equations. The flow fields are usually three-dimensional, often unsteady, and may involve large temperature gradients. Gaseous contaminants complicate the situation, since chemical reactions may occur that change the properties of the contaminants as they move and diffuse in the air. These difficulties perhaps explain why computational fluid dynamics has yet to make a significant contribution to this field. Nevertheless, as computers continue to improve in both speed and value, CFD will become more and more useful as a tool, first for researchers and ultimately for engineers designing and analyzing flow fields associated with indoor air quality.

The following examples, along with those in Section 10.4, are directly applicable to indoor air quality engineering. Although these flows are two-dimensional (or axisymmetric), steady, and simplistic, they are good examples of the potential usefulness of CFD. The solutions are certainly more realistic and detailed than those of the much simpler well-mixed and one-dimensional analytical models of previous chapters. Extension to fully three-dimensional simulations, while time-consuming and computer intensive, is not conceptually difficult.

10.3.1 Suction of Air into a Round Opening

Empirically obtained velocity isopleths (curves of constant velocity magnitude) for flow drawn into a plain circular opening and into a flanged circular opening are presented in Chapter 6. Since these flows are axisymmetric, they are easily simulated with CFD for comparison. A structured grid is generated for the case of a 0.20 m diameter round inlet, into which air at STP is drawn. In the simulation, a flange of radial length 0.10 m can be turned "on" (wall boundary condition) or "off" (internal boundary condition) to simulate a flanged or unflanged (plain) circular inlet, respectively. A schematic diagram of the computational domain is shown in Figure 10.7. A smooth-walled pipe (of diameter 0.20 m) extends 1.0 m to the left of the inlet face. A pressure outlet boundary condition is specified at the outlet of the pipe, so that air is drawn into the inlet, through the pipe, and out the pipe outlet. The computational domain extends 10.0 m (50 inlet diameters) in all directions radially from the opening, and a pressure inlet boundary condition is specified at this far field boundary. The pressure in the far field is assumed to be atmospheric (zero gage pressure), with a turbulence intensity of 0.1% and a turbulent length scale of 0.1 m (very quiet ambient conditions). When the gage pressure at the pipe outlet is set to –40. Pa, the average face velocity across the plain inlet is about 6.3 m/s. The reader may ask why the pipe is necessary – why can't the pressure outlet boundary condition be applied directly to the face of the inlet? The answer is two-fold. First, without the pipe extending some distance from the face, reverse flow problems are encountered at the face. Second, it is more physically realistic to apply constant suction pressure at some downstream location in the pipe rather than at the face itself.

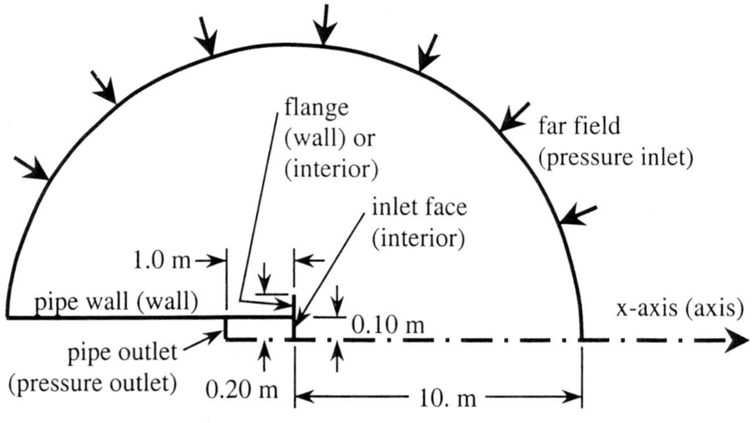

Figure 10.7 Axisymmetric computational domain used to simulate suction of air into a plain or flanged round inlet (not to scale); applied boundary conditions are in parentheses.

Results from CFD calculations using the computational domain sketched in Figure 10.7 are shown in Figure 10.8 for both the plain inlet and the flanged inlet. In both cases, steady-state turbulent flow is specified, using the realizable k-ε turbulence model with FLUENT's enhanced wall treatment option. Streamlines for the plain round inlet converge upon the face from all directions; a small recirculating eddy forms just inside the pipe, downstream of the inlet face, as the air cannot negotiate the turn around the sharp pipe edge. For the flanged case, the air is seen to flow around the flange; while streamlines in the vicinity of the flange are vastly different, streamlines further away from the flange and upstream of the face do not appear to be significantly different from those of the plain case. In fact, as one zooms out, the streamlines for both the plain and flanged round inlet look nearly identical, and asymptote to rays from the far field to the origin (see Figure 10.9). This is not surprising since from "far away" the opening looks like a point sink, regardless of whether a flange is present or not. The isopleths of Figures 10.8b and 10.8d compare favorably with those shown in Figures 6.9 and 6.10.

Finally, in Figure 10.10 is shown the decay of normalized speed along the axis of symmetry from the CFD results, compared with those obtained from the empirical curve fits given in Chapter 6. The agreement between CFD and experiment is reasonable, but not perfect. Even more striking is the surprisingly minor difference between on-axis normalized speed for the plain inlet compared to the flanged inlet. While the reach of a flanged inlet extends somewhat further *radially*, the flange has apparently little effect on the *axial* reach of the inlet.

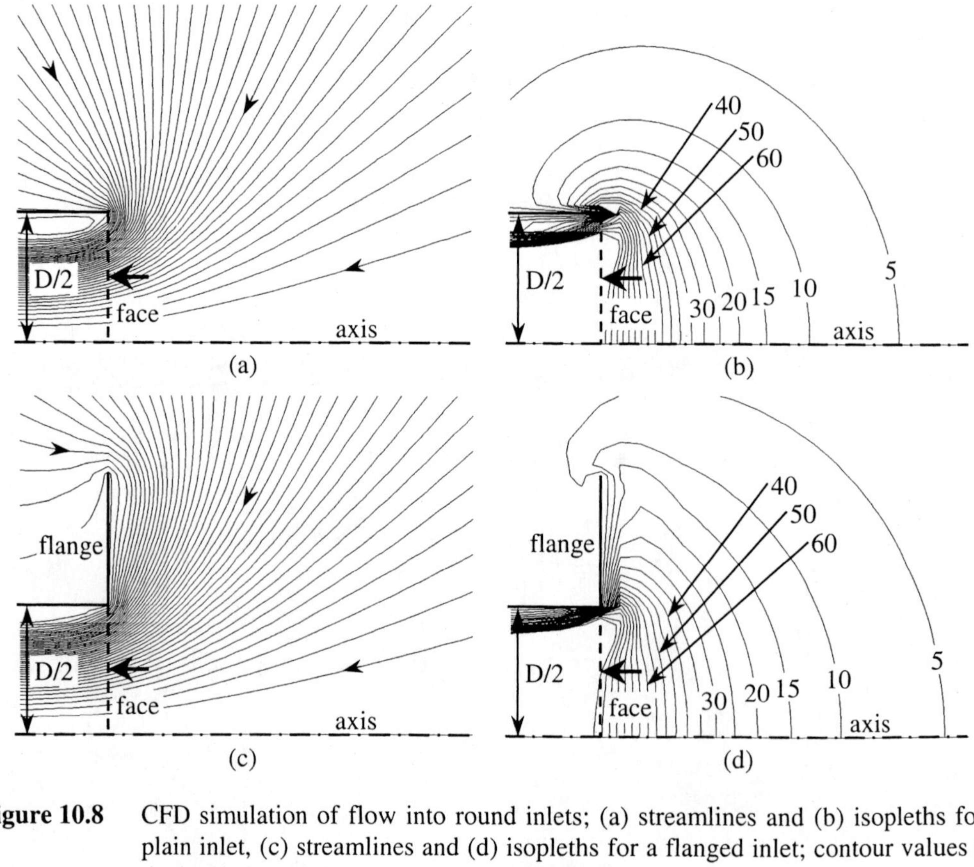

Figure 10.8 CFD simulation of flow into round inlets; (a) streamlines and (b) isopleths for a plain inlet, (c) streamlines and (d) isopleths for a flanged inlet; contour values for the isopleths are shown as a percentage of average speed through the face.

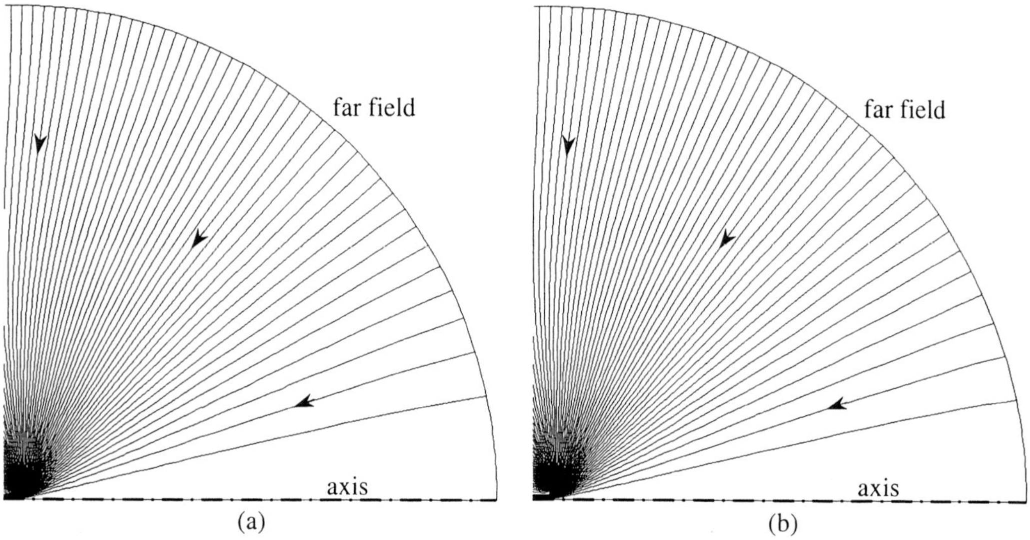

Figure 10.9 Streamlines calculated from the CFD simulation (front half only), viewed from "far away": (a) plain round inlet, (b) flanged round inlet.

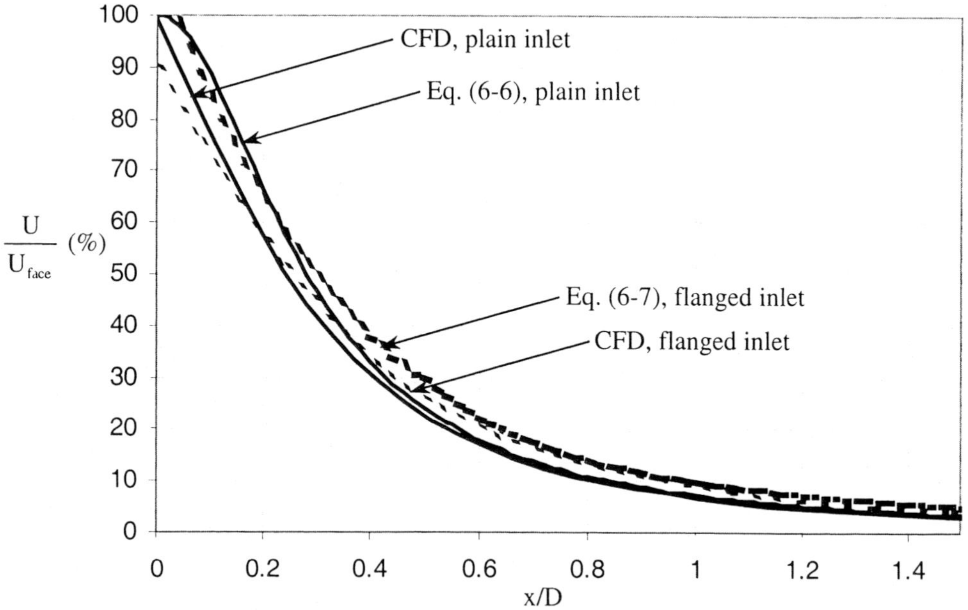

Figure 10.10 Decay of normalized air speed along the axis; solid lines for plain round inlet, dashed lines for flanged round inlet.

10.3.2 Mixing of Gaseous Contaminants in a Ventilated Room

As discussed in Chapter 5, there are two main strategies of general ventilation: dilution ventilation and displacement ventilation. For the former, it is assumed that gaseous contaminants are well mixed in the room. CFD is used here to calculate and compare contaminant concentrations in a room ventilated by dilution ventilation and in the same room ventilated by displacement ventilation. In addition, CFD is used to examine how appropriate is the well-mixed assumption for the case of dilution ventilation with and without a ceiling fan. For simplicity, a two-dimensional room is modeled; an overall view of the computational domain is shown in Figure 10.11. A uniform structured grid is used, with square cells of dimension 0.020 m everywhere for simplicity. The room is 4.0 m long and 2.0 m high with a source of carbon monoxide (CO) sitting on the floor in the center of the room. The source is modeled numerically as a 0.20 m wide velocity inlet boundary, through which an air/CO mixture is injected into the room at very low velocity. The mass fraction of CO in this gas mixture is specified so as to control the mass flow rate of CO into the room. In two-dimensional approximations like this, it is convenient to imagine that the room is of unit depth (1.0 m) in the direction perpendicular to the computational plane, so that calculations of volumetric flow rates, mass concentrations, etc. have familiar (three-dimensional) units. The source is thus imagined to be a slot, 0.20 m wide and 1.0 m deep (into the page in Figure 10.11). Exhaust air from the room leaves through a 0.20 m wide opening in the right portion of the ceiling. Supply air can come from any combination of three 0.20 m wide openings: upper left, lower left, and lower right. When only the upper left supply is activated, dilution ventilation is simulated. When this supply is turned off, but both lower supplies are activated, displacement ventilation is simulated. For simplicity, air supply diffusers are not modeled; a more realistic configuration would have the supply air fanning out in all directions rather than entering the room as a single jet. In all cases, steady-state turbulent flow is specified, using the realizable k-ε turbulence model with FLUENT's enhanced wall treatment option. The turbulence intensity at the air supply is set to 50% for the dilution ventilation case, and to 1% for the displacement ventilation case, since displacement ventilation systems are designed so as to stir room air as little as possible. At all velocity inlets in these example problems, the turbulent length scale is specified as the characteristic dimension of the duct (here 0.20 m).

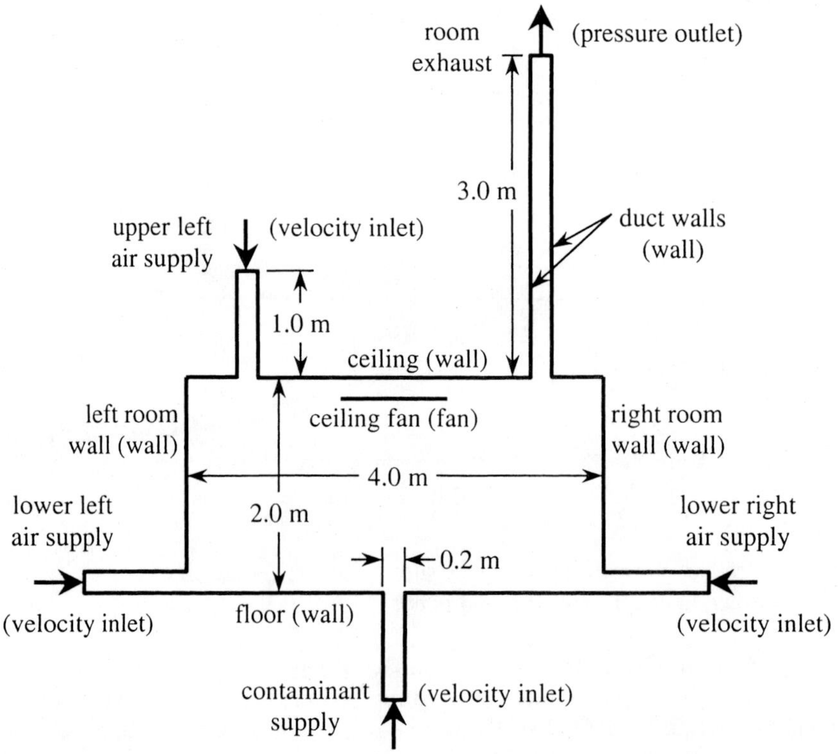

Figure 10.11 Computational domain used to simulate ventilation in a 4.0 x 2.0 m two-dimensional room; applied boundary conditions are in parentheses.

As seen in Figure 10.11, the supply ducts extend 1.0 m beyond the boundaries of the room, and the exhaust duct extends 3.0 m above the ceiling. The velocity inlet and pressure outlet boundary conditions are specified at the outer edges of the ducts, as shown in the figure. An internal boundary condition is specified at each junction where the duct meets the room. The extended ducts enable the inflowing air and contaminants to mix somewhat before entering the room, and are necessary to avoid reverse flow at the pressure outlets. In addition, boundary layers form on the side walls of the ducts, which is more physically realistic than if the velocity inlets were to be placed directly on the room walls and ceiling. The total room volume is 4.0 x 2.0 x 1.0 m = 8.0 m^3, but when the ducts are included, the total volume becomes 9.4 m^3. A ceiling fan in the center of the room is modeled by applying the fan boundary condition described in Section 10.1.5. The fan can blow air up or down depending on the sign of the pressure rise specified across the fan. When the fan is off ($\delta P_{fan} = 0$), the fan boundary condition degenerates computationally to that of an interior boundary condition, as if it were not there at all.

The mass flow rate of CO into the room (source, S_{CO}) is specified as S_{CO} = 412. mg/hr. Rather than injecting pure CO, a mixture of air and CO is supplied from the bottom of the domain at a speed of 0.0010 m/s and with a CO mass fraction of 4.675 x 10^{-4} in order to generate a diluted source of contaminant into the room. In both the dilution and displacement ventilation cases, fresh air is supplied from the air supply ducts at a total volumetric flow rate of 20.2 m^3/hr. In addition, 0.72 m^3/hr of air is added to the room through the contaminant supply, for a total fresh air supply of Q_{supply} = 20.9 m^3/hr. This is equivalent to N = 2.22 room air changes per hour (based on the total domain volume of 9.4 m^3). Assuming well-mixed conditions in the room, as in Chapter 5, the steady-state well-mixed mass concentration of carbon monoxide in the room can be calculated,

$$c_{ss,CO} = \frac{S_{CO}}{Q_{supply}} = \frac{412.\dfrac{mg}{hr}}{20.2\dfrac{m^3}{hr}} = 20.4\frac{mg}{m^3}$$

Or, in terms of mol fraction in units of parts per million, the steady-state well-mixed level of CO in the room can be calculated from Eq. (1-30),

$$PPM_{CO} = c_{ss,CO}\frac{24.5}{M_{CO}} = 20.4\frac{24.5}{28.0} = 17.9 \text{ PPM}$$

Readers should recall from Chapter 5 that with the well-mixed approximation, one cannot distinguish between dilution and displacement ventilation. In other words, the well-mixed assumption predicts that the carbon monoxide level is 17.9 PPM *everywhere* in the room regardless of the locations of the source or of the supply and exhaust ducts. This value is less than half of the OSHA PEL (50 PPM) and is used as a base for comparison with CFD predictions.

CFD results for the case of dilution ventilation are shown in Figure 10.12; Figure 10.12a shows flow streamlines, and Figure 10.12b shows CO levels in the room. Contours of constant CO mol fraction above the OSHA PEL of 50 PPM are not shown in Figure 10.12b. (CO levels in the white region near the source are *greater than* 50 PPM.) The streamlines show a large counterclockwise circulating eddy in the room, with smaller clockwise circulating eddies in three corners of the room. Contaminant gas from the floor is swept up and to the right, creating much higher concentrations of CO on the right half of the room compared to the left half, which receives fresh air from the air supply. Turbulent diffusion spreads contaminant around the room, however, such that nowhere in the room is the air totally devoid of CO. The CO level in the center of the room, for example, is around 5 PPM. Of particular interest is the CO level exiting the room through the exhaust duct. The average mol fraction of CO at the exhaust plane is approximately

17. PPM, very close to the value predicted by the well-mixed approximation. It is left as an exercise to show that this *must* be the case, regardless of the locations of the air supplies, source, or exhaust. One must conclude that with the ceiling fan turned off, the well-mixed assumption is not very realistic for this ventilation configuration. Readers must keep in mind, however, that the simulation is steady-state and two-dimensional, and there are no blockages (furniture, machinery, people, etc.) in the room. None of these assumptions is valid for any real room; more accurate CFD predictions are possible by relaxing these assumptions, but at the expense of a much more complex grid and vastly increased computer time and memory requirements. As a side note, the fundamental differences between a blowing jet and a suction inlet, as discussed in Chapter 7, are clearly seen in Figure 10.12a. The streamlines near the air supply (blowing jet) are nearly straight and normal to the supply face, while those near the exhaust (suction inlet) converge on the exhaust face nearly evenly from all directions.

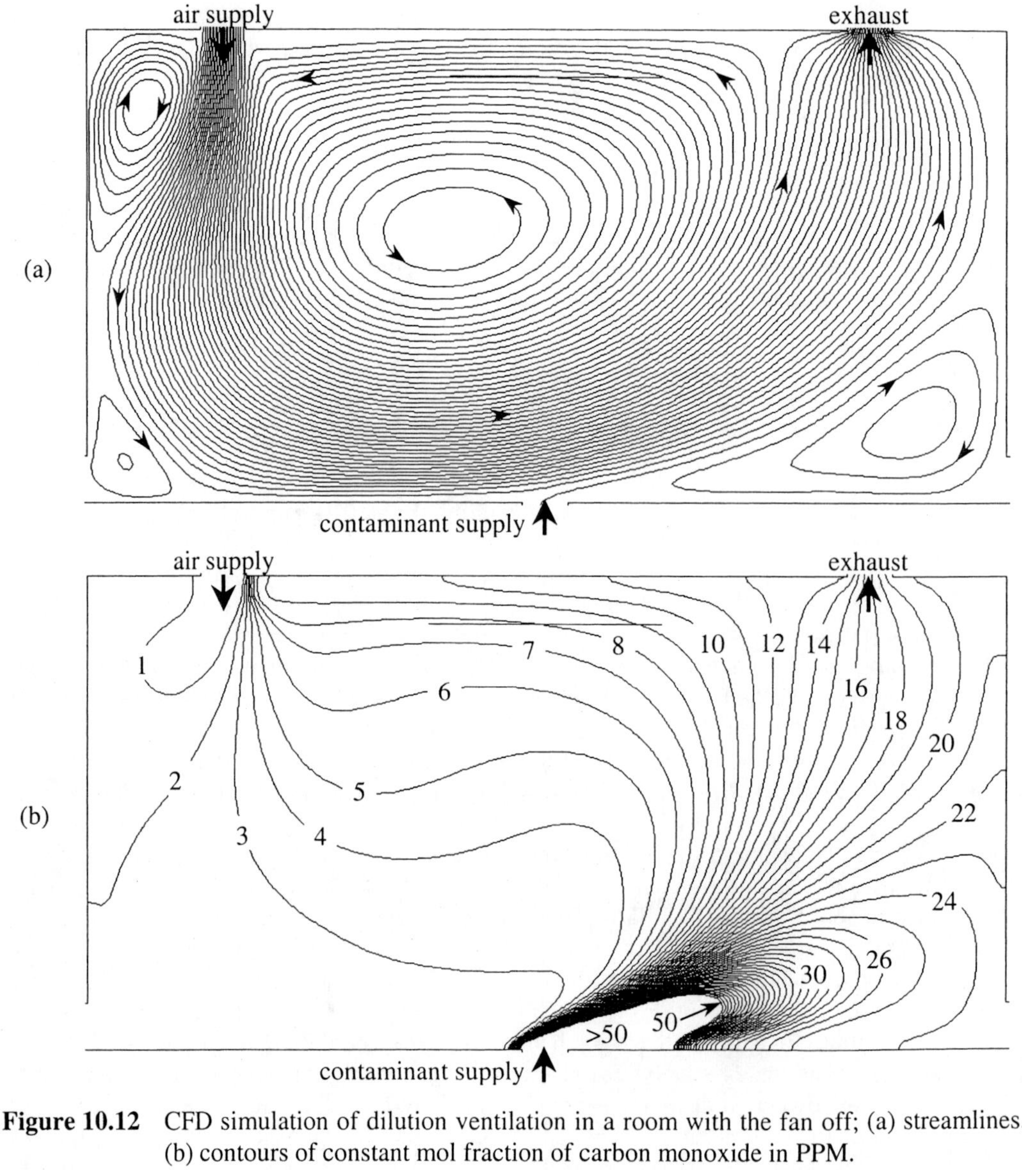

Figure 10.12 CFD simulation of dilution ventilation in a room with the fan off; (a) streamlines, (b) contours of constant mol fraction of carbon monoxide in PPM.

Next, consider an identical case except that the ceiling fan is turned *on* with a pressure rise of 5.0 Pa in the downward direction. The streamlines and CO levels are shown in Figure 10.13. The maximum air velocity occurs in the center of the room, and is about 3.8 m/s in magnitude, which is more than a factor of 100 times larger than the speed of the supply air, which is 0.028 m/s. For this reason, the streamlines are dominated by air currents produced by the fan. Two large circulating eddies are established in the room; the one on the left rotates clockwise while the one on the right rotates counterclockwise. The fan thoroughly mixes air and CO in the room, as clearly seen by the contour plots of Figure 10.13b. The CO concentration varies from about 16.4 to 16.7 PPM throughout the entire room except very close to the air supply or the contaminant source. The air exhausting from the room has an average CO level of about 16.7 PPM, again very close to the steady-state well-mixed approximation, as it must be. Hence, with the ceiling fan turned on, the well-mixed assumption of Chapter 5 is quite acceptable!

(a)

(b)

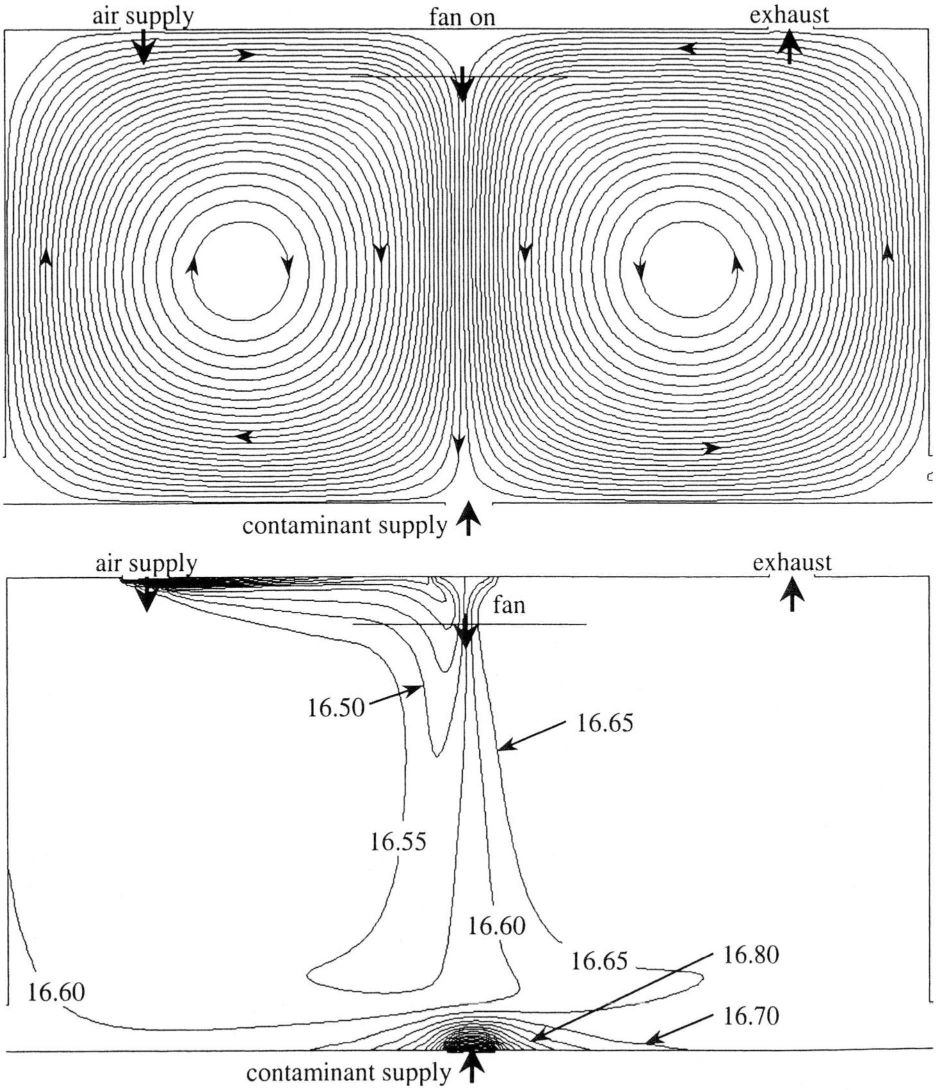

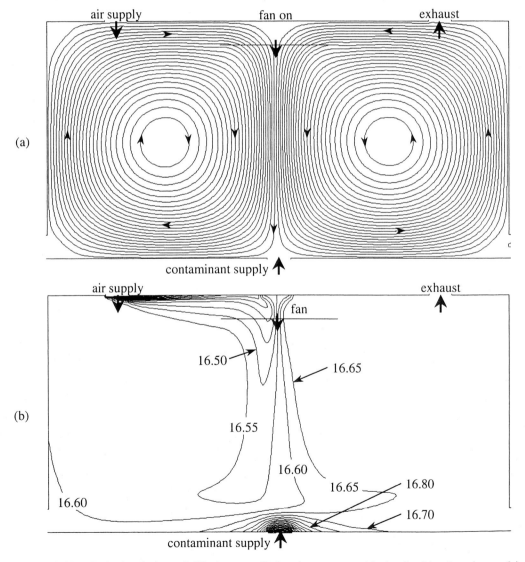

Figure 10.13 CFD simulation of dilution ventilation in a room with the fan blowing down; (a) streamlines, (b) contours of constant mol fraction of carbon monoxide in PPM.

Finally, results of a third CFD simulation are shown in Figure 10.14 – the case of *displacement* ventilation with the ceiling fan turned off. For this case, the upper left air supply is turned off (numerically the velocity inlet boundary condition is changed to a wall boundary condition), and the lower left and lower right air supplies are turned on. Since the combined face area of these two floor-level air supplies is twice that of the upper air supply, the inlet air velocity is reduced by a factor of two to 0.014 m/s for each air supply. Hence the volumetric flow rate of fresh air is the same as that of the other two cases. The turbulence intensity of the supply air is also reduced for this displacement ventilation simulation as discussed previously. The streamlines show that the air travels parallel to the floor for some distance from both sides of the room, and then slowly moves upward towards the exhaust opening, dragging contaminant with it. Two large counter-rotating eddies are formed; in this case, the eddy on the left is counterclockwise, while the one on the right is clockwise (opposite of the case with the fan blowing air downward). Since

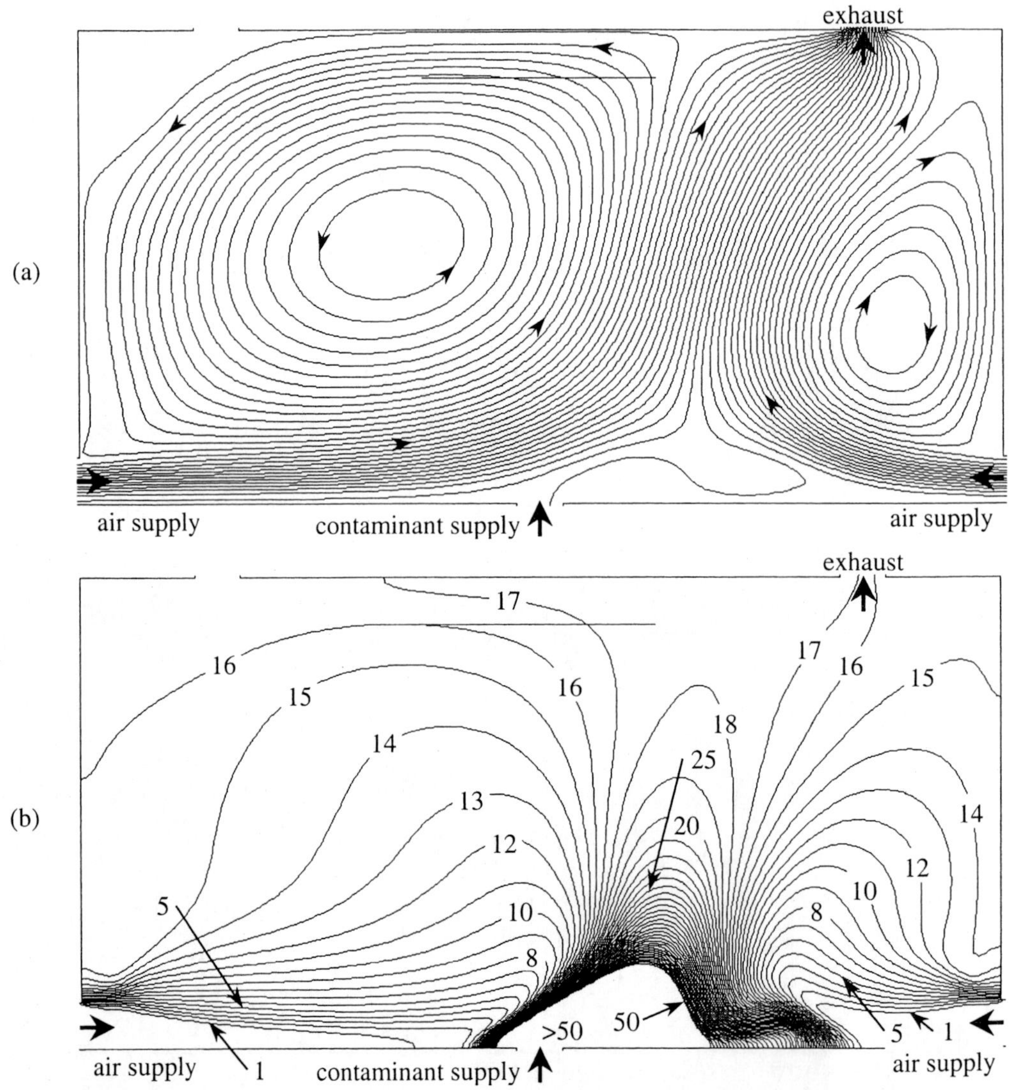

Figure 10.14 CFD simulation of displacement ventilation in a room with the fan off; (a) streamlines, (b) contours of constant mol fraction of carbon monoxide in PPM.

the exhaust opening is located on the right side of the room, the streamlines here are not symmetric; a better displacement ventilation design might have the exhaust opening in the center of the ceiling instead. The mol fraction of CO leaving the room through the exhaust duct is again around 17 PPM as seen in Figure 10.14b. An overall stratification of CO is apparent in the room; CO levels are much higher near the ceiling than near the floor, except of course near the source, where the CO levels are very high. As a side note, in displacement ventilation systems this same stratification effect occurs with *temperature*, and the effect is even enhanced by gravity (buoyancy). Chilled air entering the room at floor level pushes or "displaces" warmer air towards the top of the room, much like fresh air in the present simulation pushes contaminated air towards the top of the room. The reader can easily see why displacement ventilation may be a wise choice for very tall rooms like hotel lobbies and atriums. More sophisticated three-dimensional simulations of a room with displacement ventilation have been reported by Cimbala et al. (1998).

Which is better, dilution ventilation or displacement ventilation? The answer depends on many factors, such as the locations of people in the room and whether or not the fan is turned on. In the present simulations, the lowest CO level of the three cases is encountered by a person standing on the left side of the room with dilution ventilation and the ceiling fan off. This clean air is at the expense of his colleague, however, who may be working on the right side of the room, breathing air with a much higher CO concentration. The stratification imposed by displacement ventilation provides cleaner air to *both* sides of the room, but only near the floor; CO levels are much higher but relatively uniform near the ceiling. This situation is not desirable for rooms with low ceilings as in the present example, but may be preferable for rooms with high ceilings. When the ceiling fan is turned on, the room air is very well mixed, and the CO is more equitably distributed – the level is around 17 PPM *everywhere* in the room, even directly above the source. Not shown here is the case in which the fan is running in the opposite direction, nor the case of displacement ventilation with the fan on; both of these cases yield well-mixed air, with CO levels similar to those shown in Figure 10.13b, i.e. around 17 PPM everywhere in the room. The bottom line of course is that none of these general ventilation systems should be used to remove carbon monoxide (or other contaminants) from this room. Rather, a *local* ventilation system (hood) is a much wiser choice.

Example 10.1 – Comparison of two Ventilation Configurations in an Engine Laboratory
Given: A 4.0 m long by 2.5 m high by 7.0 m wide laboratory is being constructed to test the performance of small lawnmower engines. Seven workbenches, each approximately 1.0 m wide and each with a hood, are to be placed side by side along the left wall as sketched in Figure E10.1a. (Note that "width" here refers to the distance into the page in Figure E10.1a; a two-dimensional slice through the center of the laboratory is shown.) A fan is located at the top of each hood, and the duct work consists of a sudden contraction, two sharp 90° bends, and a straight duct, 12.2 m in length which exhausts into the atmosphere. The lab is to be ventilated by a dilution ventilation system with fresh air supplied and exhausted through the ceiling; both ceiling openings span nearly the entire width of the room. The openings in the ceiling have already been cut, one in the center of the room and one on the right side of the room as shown in the diagram, but it has yet to be decided which of the two will serve as the fresh air supply and which will serve as the room exhaust. In either case, fresh air is to be supplied at a constant rate of four to five room air changes per hour. CO enters the work space at a rate of approximately 2600 mg/hr from each running engine (one engine per work station). The chief engineer is particularly concerned about the scenario in which a worker forgets to turn on the hood exhaust fan, and carbon monoxide escapes into the room, where it is inhaled by the worker. The approximate location of the worker's mouth is indicated in Figure E10.1a. The architect thinks it would be better to have the fresh air supply located in the center of the room, close to the workers so that

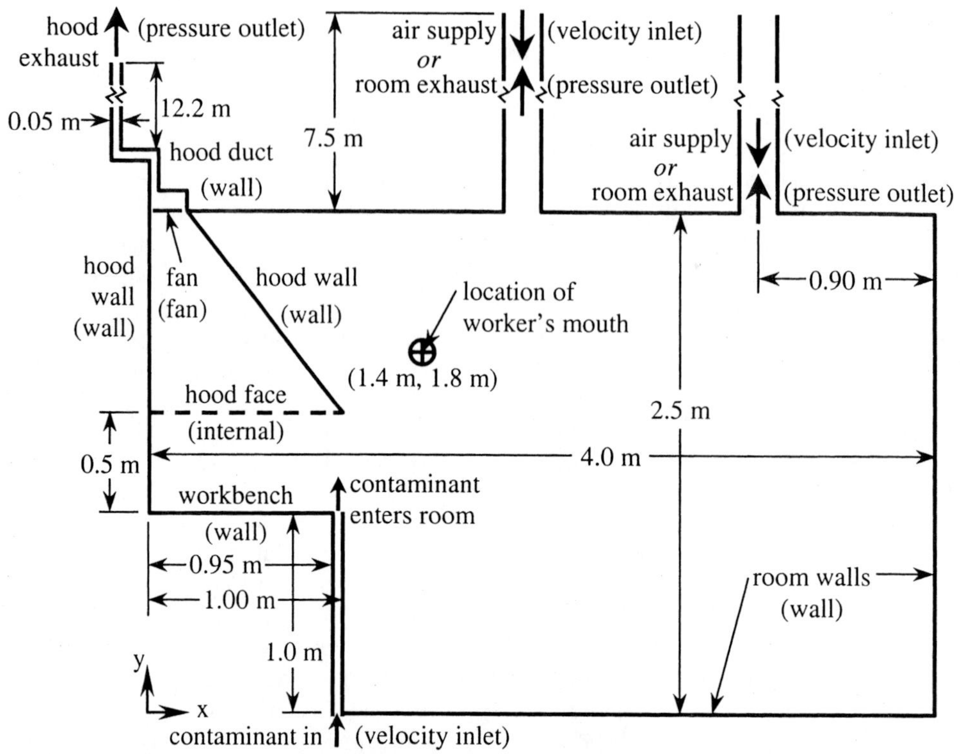

Figure E10.1a Two-dimensional computational domain used to simulate dilution ventilation in a 4.0 x 2.5 x 7.0 m room with local ventilation through hoods on the left side of the room; applied boundary conditions are in parentheses.

they stand nearly directly below the supply. The chief engineer disagrees, claiming that it would be better to have the room *exhaust* closer to the workers, and the air supply on the right side of the room. You are consulted to settle the dispute, and decide to use CFD to simulate how the air and CO mix in the room for the two cases.

To do: Simulate both ventilation options under conditions in which the hood fans are not turned on, and discuss which option would be the wiser choice.

Solution: Since the geometry is nearly two-dimensional, a 2-D simulation is performed using the geometry shown in Figure E10.1a as the computational domain. One can imagine a one-meter wide slice of the room, containing one worker and one hood. The two ceiling ducts extend 7.5 m above the ceiling to avoid reverse flow in the computations. Fresh air is supplied through the velocity inlet of the air supply duct at a speed of 0.060 m/s with a turbulence intensity and length scale of 10% and 0.20 m respectively. The volumetric flow rate of air into the room is thus 43.2 m^3/hr. To simulate suction provided by the HVAC system, the gage pressure is specified as -2.3×10^{-3} Pa at the outlet of the room exhaust duct. The pressure at the outlet of the hood exhaust duct is set to zero gage pressure (atmospheric pressure). The worst-case scenario is modeled, in which carbon monoxide enters the room right at the edge of the workbench, and the hood exhaust fan is off. In the simulation, CO is introduced through a velocity inlet into a 0.050 m duct from below the floor of the room, as sketched. The bulk speed of the air and CO entering this duct is set to 0.0060 m/s with a turbulence intensity of 20% and a turbulence length scale of 0.050 m. The mass fraction of CO entering the contaminant supply duct is specified as 2.0×10^{-3} at the velocity inlet.

Readers can verify that the mass flow rate of CO into the computational domain (the source) is thus S = 2600 mg/hr. The total amount of fresh air entering the room is 44.3 m^3/hr, which is the sum of the volumetric flow rate of fresh air supplied through the ceiling and that supplied through the contaminant supply duct. As a first approximation, the well-mixed assumption of Chapter 5 is invoked to determine a base level steady-state value of CO mass concentration in the room,

$$c_{ss,CO} = \frac{S_{CO}}{Q_{supply}} = \frac{2600\,\frac{mg}{hr}}{44.3\,\frac{m^3}{hr}} = 58.7\,\frac{mg}{m^3}$$

In terms of mol fraction, the steady-state well-mixed level of CO in the room is 51. PPM, just above the OSHA PEL of 50 PPM for CO.

The two ventilation options are simulated using FLUENT, assuming steady-state turbulent flow with the realizable k-ε turbulence model and enhanced wall treatment. The hood exhaust fan is off in both cases. Streamlines and CO mol fraction contours are shown in Figure E10.1b for the case in which fresh air is supplied through the duct on the right and room air is exhausted through the duct in the middle of the ceiling. Contours of constant CO mol fraction above 500 PPM are not shown in the figure. A large clockwise eddy forms in the middle of the room and draws CO from the workbench towards the exhaust. The CO level near the worker's breathing zone is around 30 to 40 PPM, below the OSHA PEL. Higher concentrations of CO are seen in the small counterclockwise eddy above the hood, where CO levels are around 80 to 90 PPM. A counterclockwise eddy is also seen inside the hood itself, where the CO level is much higher, around 250 to 260 PPM. Finally, the large counterclockwise eddy near the right wall of the room contains very clean air with CO levels less than 5 PPM.

CFD results are shown in Figure E10.1c for the opposite ventilation option, in which the air supply is through the middle duct and the room exhaust is through the duct at the right. A large counterclockwise eddy is seen on the right side of the room, and a smaller clockwise eddy forms on the left, drawing carbon monoxide from the workbench and mixing it with room air. The CO level near the worker's breathing zone is around 130 to 140 PPM, above the OSHA PEL. The overall CO concentrations in both the room and the hood are in fact much higher than those for the first case, although the average CO mol fraction leaving the room through the room exhaust duct is about 45 PPM, about the same as that of the first case. Note that in both cases a small amount of CO escapes through the hood exhaust duct even though the hood fan is off.

The first case, the one recommended by the chief engineer, is clearly the better option. Apparently, placing the exhaust near the worker's head allows more fresh air to flow into his or her breathing area from below. This may seem counterintuitive, especially since the worker's mouth is on nearly a direct line between the CO source and the room exhaust. It would be difficult if not impossible to predict this result without CFD.

Discussion: The two-D approximation is realistic only for the case in which all seven engines are running and all seven workers forget to turn on their hood exhaust fans (a highly unlikely scenario). If only one or two of the seven hood fans are off, the carbon monoxide would diffuse in the direction perpendicular to the page in Figure E10.1a; CO levels in the room would then be much smaller than those calculated here. A full three-dimensional simulation is necessary for such detailed analysis. Both the streamlines and CO concentration levels would also change significantly if one were to include obstructions caused by the workers standing next to the workbench and/or furniture, equipment, etc. in the rest of the room.

(i)

(ii)

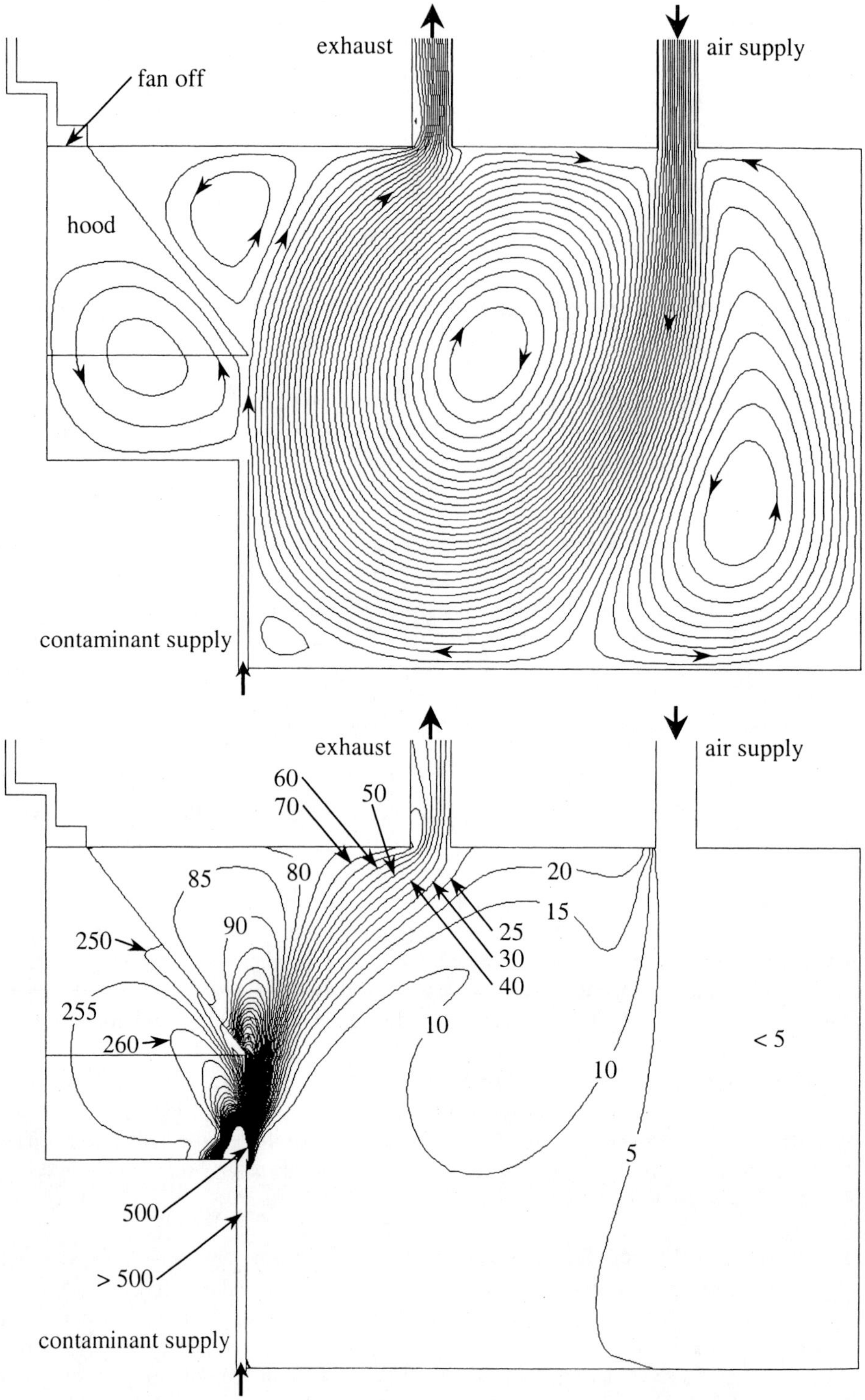

Figure E10.1b CFD simulation of flow in a laboratory with the hood exhaust fan turned off; fresh air supplied through the supply duct on the right side of the ceiling; (i) streamlines, (ii) contours of constant mol fraction of carbon monoxide in PPM.

(i)

(ii)

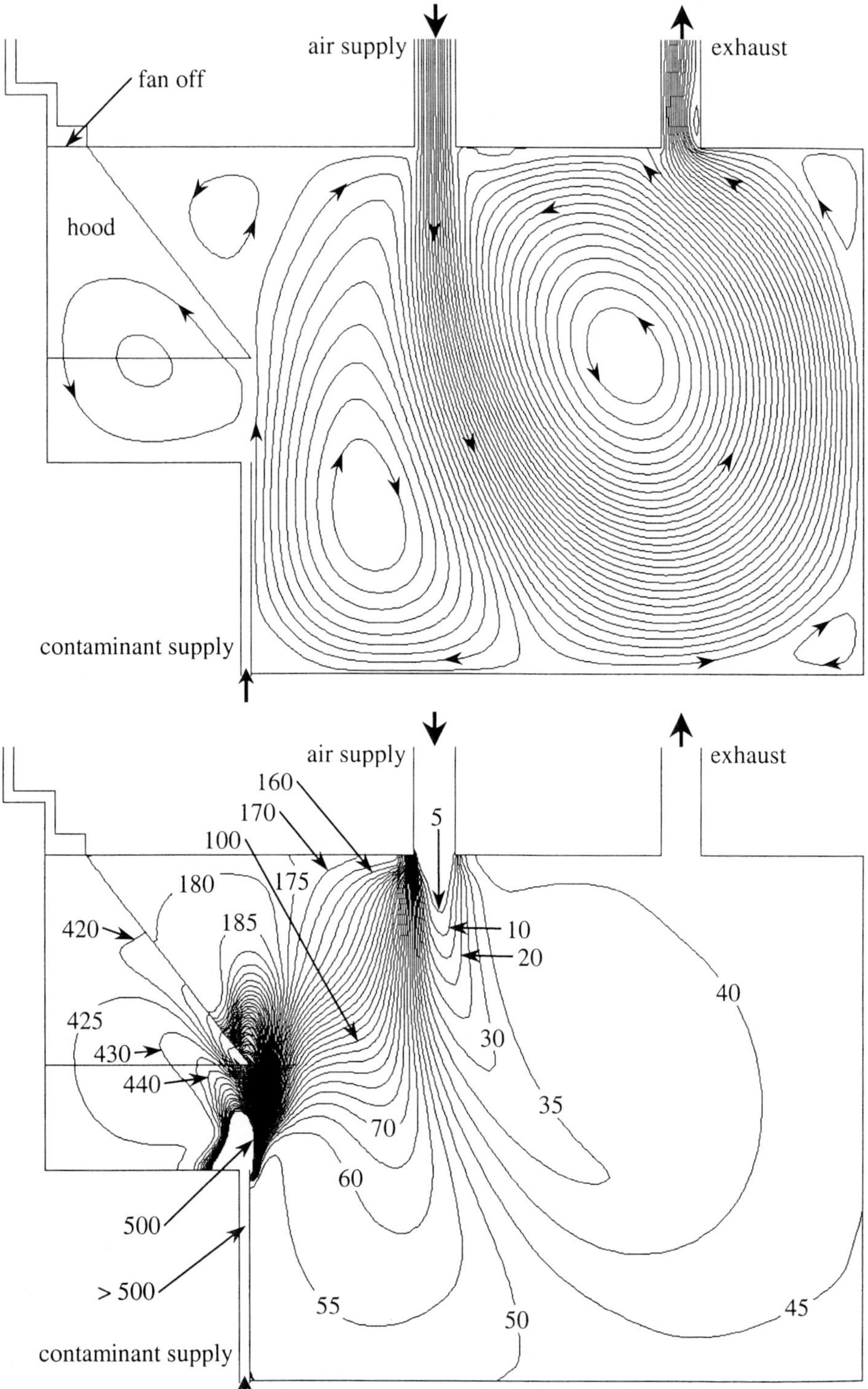

Figure E10.1c CFD simulation of flow in a laboratory with the hood exhaust fan turned off; fresh air supplied through the supply duct in the middle of the ceiling; (i) streamlines, (ii) contours of constant mol fraction of carbon monoxide in PPM.

Example 10.1 clearly shows the usefulness of computational fluid dynamics as a design tool for indoor air quality engineering. While the simulation presented here is two-dimensional, the fundamental concepts of grid generation, specification of boundary conditions, iteration procedure, post-processing, etc., are the same for three-dimensional flow fields. As computer speed and memory continue to improve, fully three-dimensional simulations of room ventilation will become more commonplace in the HVAC industry. In the following example, flow and contaminant concentration in this same room is modeled, but with the hood exhaust fan operating at various values of pressure rise across the fan.

Example 10.2 – Withdrawal of Carbon Monoxide through a Hood in a Ventilated Room
Given: Consider the same engine testing laboratory as described in Example 10.1 and as sketched in Figure E10.1a. Despite the clear evidence presented in that example, the architect overrules the chief engineer, and installs the fresh air supply in the center of the ceiling and the room exhaust on the right side of the ceiling (the configuration of Figure E10.1c). Now the chief engineer needs to select an appropriate fan for the hood exhaust.

To do: Use CFD to simulate the flow of air and CO in the room and through the hood exhaust duct for various values of fan pressure rise, δP_{fan}.

Solution: The two-dimensional approximation is again employed for simplicity and to decrease the required computational time. The computational domain is the same as that shown in Figure E10.1a. All parameters are in fact identical to the second case of Example 10.1 (fresh air supply rate, CO supply rate, pressures specified at the room exhaust and hood exhaust pressure outlets, turbulence intensities, etc.). The only change from the previous example occurs in the hood exhaust fan boundary condition. Namely, several values of δP_{fan} are selected, and FLUENT is run to convergence for each value of δP_{fan}.

Streamlines and CO mol fraction contours are shown in Figure E10.2a for the case in which δP_{fan} = 50.0 mPa. Contours of constant CO mol fraction above 500 PPM are not shown in the figure, as previously. A large counterclockwise eddy forms on the right side of the room, and a smaller clockwise eddy appears to the left of the air supply jet. The overall features of the streamlines are not much different than those shown in Figure E10.1c, except near the workbench and hood. Whereas the streamlines flow nearly vertically past the workbench when the fan is off, there is a clear split in the flow streamlines at the edge of the hood for the case with the fan on. Some of the room air enters the hood, drawing contaminant with it into the hood. The CO level near the worker's breathing zone is around 40 PPM, significantly below that of Figure E10.1c, and just below the OSHA PEL. The air is fairly clean on the entire right half of the laboratory, with values less than 15 PPM. Higher concentrations of CO are again seen in the small counterclockwise eddy above the hood, where CO levels are around 55 to 60 PPM. A counterclockwise eddy is also seen inside the hood itself, where the CO level is much higher, around 230 to 250 PPM, but about half of the values shown in Figure E10.1c. Finally, several streamlines clearly pass through the hood exhaust fan and up the hood exhaust duct; the CO level in the air going up the hood exhaust duct is between 210 and 215 PPM. Notice that since much of the carbon monoxide is drawn through the hood, the CO level exiting the room through the ceiling exhaust has decreased to a value between 10 and 15 PPM.

CFD results are shown in Figure E10.2b for the case in which δP_{fan} = 200. mPa. The large counterclockwise eddy fills up nearly the entire room, and a much smaller clockwise rotating eddy fills in the space above the hood. The streamlines coming from the air supply are noticeably tilted to the left, and many of them enter directly into the hood. At this larger fan

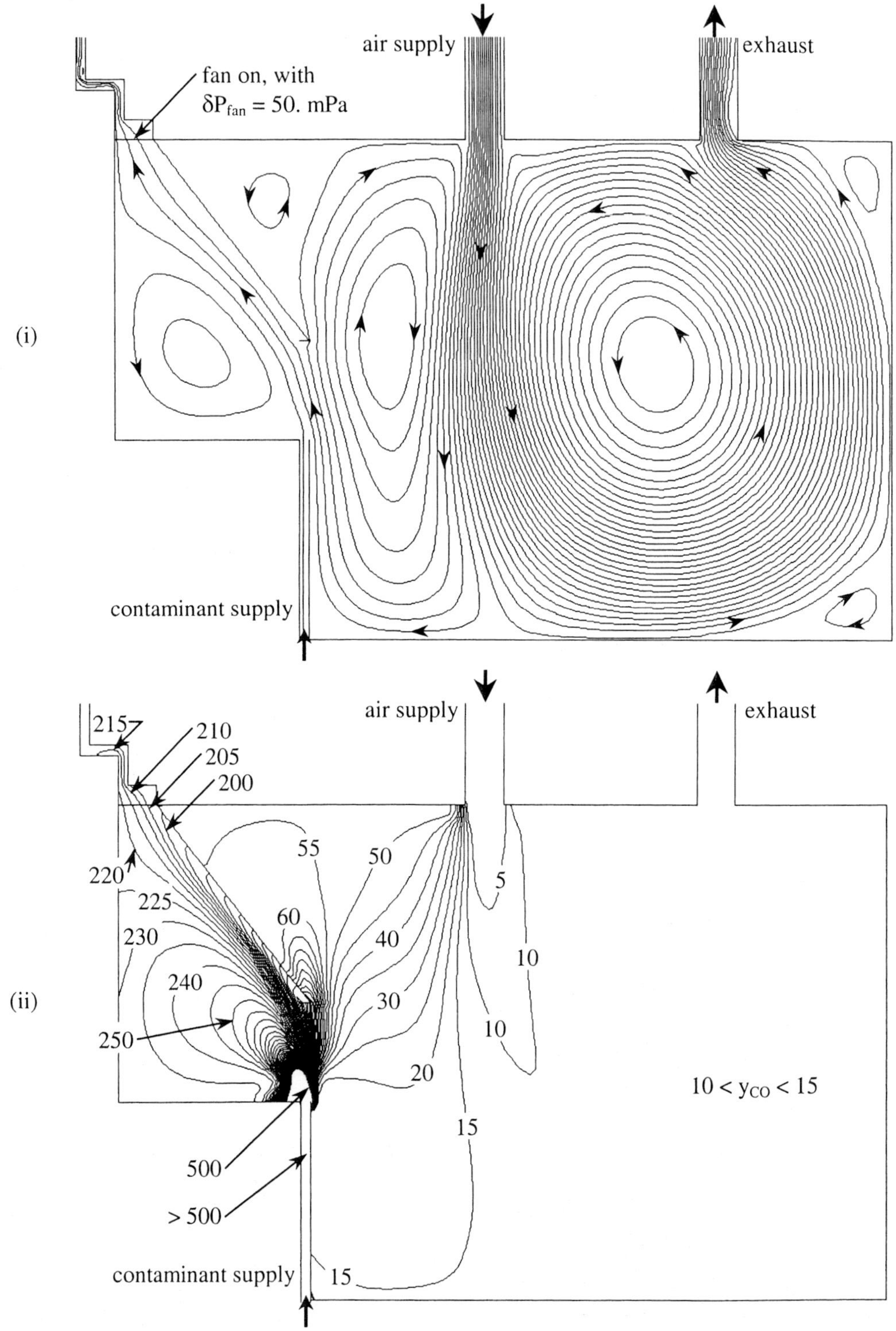

Figure E10.2a CFD simulation of flow in a laboratory with the hood fan running at δP_{fan} = 50.0 mPa; fresh air supplied from the center of the ceiling; (i) streamlines, (ii) contours of constant mol fraction of carbon monoxide in PPM.

(i)

(ii)

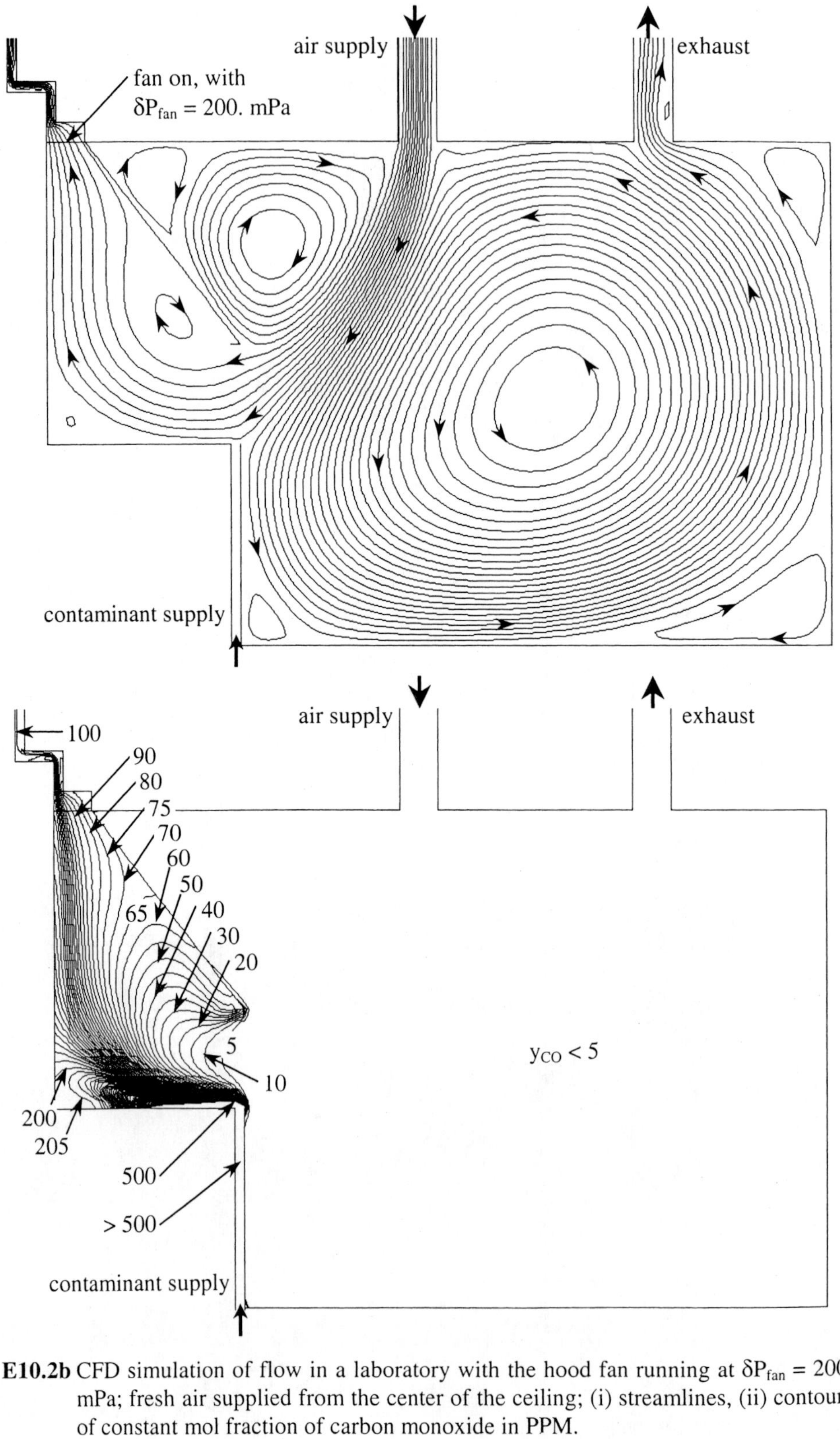

Figure E10.2b CFD simulation of flow in a laboratory with the hood fan running at δP_{fan} = 200. mPa; fresh air supplied from the center of the ceiling; (i) streamlines, (ii) contours of constant mol fraction of carbon monoxide in PPM.

pressure rise, air crosses the vertical plane from left to right at every location between the edge of the workbench and the edge of the hood above it. Nearly all of the carbon monoxide is drawn into the hood and up the hood exhaust as it should; this value of fan pressure rise is thus considered adequate for proper operation of the hood as a local ventilation system. Note that since air drawn into the hood mixes with carbon monoxide, the CO level in the air going up the hood exhaust duct is around 100 PPM, less than half of that calculated for the case with δP_{fan} = 50. mPa. The CO level in the entire room is less than 5 PPM (which is the smallest value of constant CO mol fraction plotted in the contour plot). Finer resolution contour plots (not shown) reveal that the CO level near the worker's breathing zone is in fact less than 1 PPM.

Finally, Figure E10.2c shows how CO mol fraction at the worker's breathing zone location decreases with increasing δP_{fan}. There appears to be a discontinuity or "bump" in Figure E10.2c around δP_{fan} = 50 mPa. The FLUENT solution had much more difficulty converging at this value of δP_{fan} as well. The authors conjecture that under these conditions the physical flow would tend to oscillate at the edge of the hood where the streamlines split into those entering the hood and those remaining in the room. In other words, the solution here is forced to be steady whereas the physical flow wants to be *unsteady*, leading to the convergence difficulties and the discontinuity in Figure E10.2c. At even higher values of δP_{fan}, namely around 1000 mPa, the strong suction provided by the hood exhaust fan leads to reverse flow from the pressure outlet at the top of the room exhaust duct. In other words, the hood fan is too strong for the room ventilation system.

Discussion: The required fan pressure rise is low, only about 200 mPa, which is much less than one inch of water column, because of the two-dimensional assumption. Although the hood exhaust duct is only 0.050 m from wall to wall, it is 1.0 m wide (into the page) in these simulations. An actual duct would most likely have a much smaller aspect ratio, or would be round in cross-section, either of which would significantly increase the velocity and pressure drop through the hood exhaust duct, and would thereby demand a more realistic value of δP_{fan}.

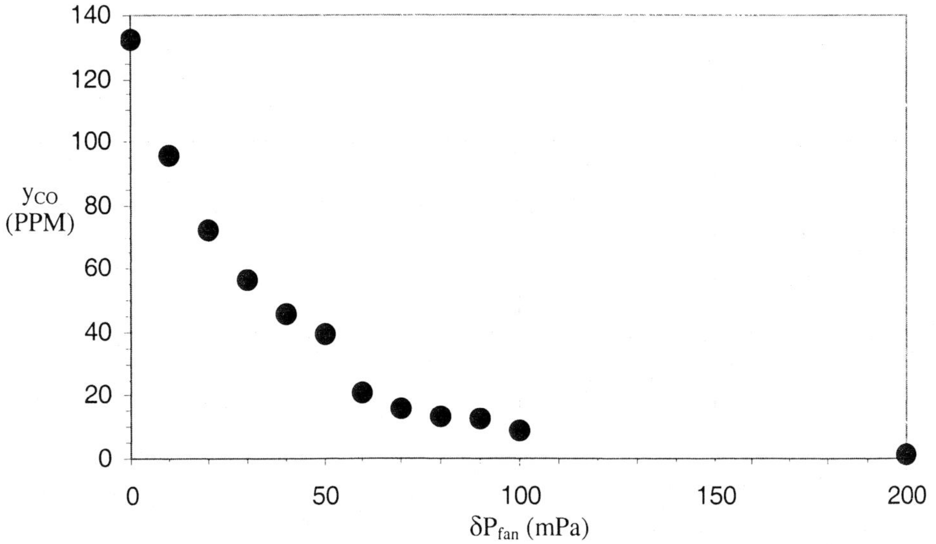

Figure E10.2c Carbon monoxide mol fraction near the worker's breathing zone as a function of hood exhaust fan pressure rise; fresh air supplied from the center of the ceiling.

10.4 Modeling of Aerosol Particle Trajectories

In the preceding section, it was shown that modeling of air and gaseous contaminants requires simultaneous solution of the equations of motion (mass and linear momentum) for the bulk flow *plus* an equation of conservation of mass for each gaseous species. Modeling of the trajectories of aerosol particles is somewhat simpler conceptually and is less computer intensive because a two-step solution procedure can be employed. In the first step, equations of motion for the air flow are solved, assuming that contaminant particles in the air do not significantly affect the bulk air flow field. In the second step, the trajectory of a particle of specified size and density is calculated based on its initial location, initial velocity, and the flow field calculated in the first step. The air flow field is said to be "frozen" with respect to the particle's trajectory; in other words, motion of the particle does not change the air flow, and it is not necessary to return to the first step. The equations for particle trajectories are first-order ordinary differential equations – much easier to solve than the partial differential transport equations for air and gaseous contaminant flow. Once an air flow is calculated for a given computational domain, the trajectories of particles of various sizes, densities, and initial locations and velocities can be calculated very quickly using Runge-Kutta time marching techniques as discussed in previous chapters. Modern commercially available CFD codes, such as FLUENT, have convenient built-in particle trajectory post-processing capabilities that enable dozens of particle trajectories to be calculated and displayed onto the computer screen in a matter of seconds. Some examples of particle trajectory calculations are described here. Although simplistic, these examples illustrate the capability of CFD to model the performance of more complex air pollution control devices that are designed to remove dust or contaminant particles from an air stream.

10.4.1 Particle Trajectories in a Curved Duct

Due to their larger size, density, and inertia compared to air molecules, aerosol particles do not necessarily follow air streamlines, and in fact can actually *cross* curved streamlines and be separated from the air – a process known as inertial separation, as discussed in Chapter 9. As a simple illustration of inertial separation, flow through a 90° bend in a two-dimensional duct is solved numerically, using a 30 x 100 structured grid. The computational domain with all the specified boundary conditions is sketched in Figure 10.15.

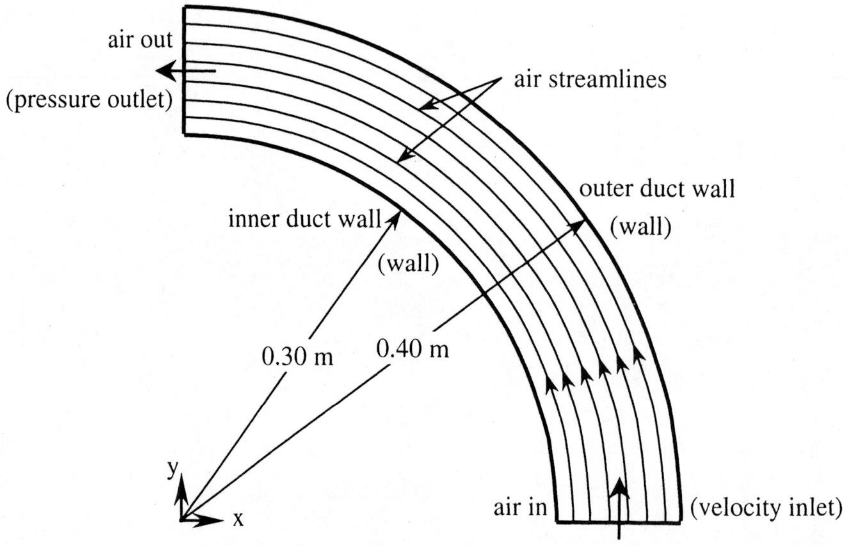

Figure 10.15 Computational domain used to simulate flow of air and aerosol particles through a 90° two-dimensional duct; applied boundary conditions are in parentheses.

The inner and outer radii of the duct are 0.30 m and 0.40 m respectively. The air speed at the inlet is 5.0 m/s, and gravity is turned off. The air flow is solved first; the streamlines are simply circular arcs running parallel to the duct walls, as also shown in Figure 10.15. The trajectories of unit density spherical particles are then calculated, based on this "frozen" air velocity field. Particles of four different diameters are tested. In all cases, eight particles of diameter D_p are injected at points along the inlet with an initial velocity equal to that of the air flow. The resulting trajectories are shown in Figure 10.16.

The smallest injected particles (D_p = 1 μm, Figure 10.16a) follow the air streamlines with negligible deviation; these particles are not massive enough to have significant inertial effects. The 10-micron particles (Figure 10.16b) behave similarly, although careful comparison of the two cases reveals that the 10-micron trajectories deviate slightly outward from the air streamlines. Particles much larger than 10 μm have more significant inertial effects. For example, consider the trajectories of 50-micron particles, as shown in Figure 10.16c. Of the eight 50 μm particles released at the duct inlet, four impact the outer wall of the duct bend; the other four escape through the duct outlet, but move radially outward past the centerline, crossing air streamlines

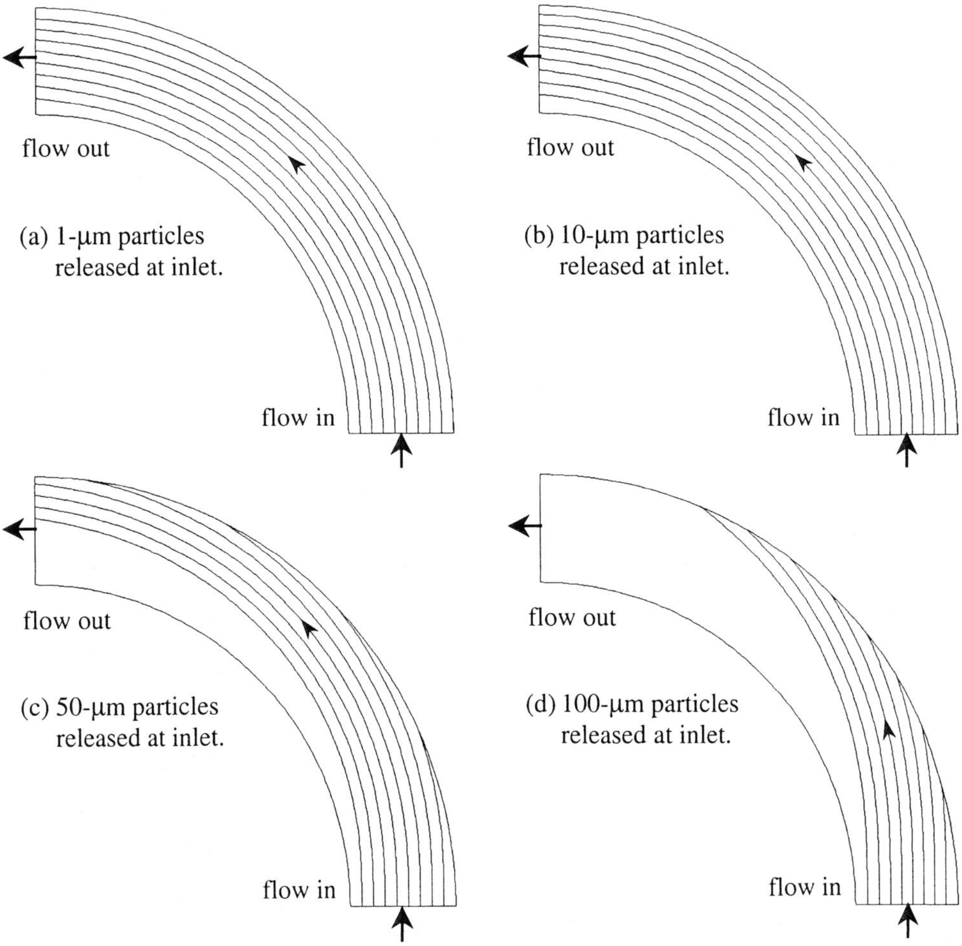

Figure 10.16 CFD simulation of particle trajectories in a 90° duct bend; particles are unit density spheres, injected at points along the inlet with initial velocity equal to that of the air; particle diameter D_p = (a) 1, (b) 10, (c) 50, and (d) 100 μm.

along the way. FLUENT has several options regarding particles that impact a wall, the two most common options being reflection and entrapment. In these simulations, the entrapment option is selected; particles that impact a wall are assumed to adhere to the wall. Finally, Figure 10.16d shows trajectories of 100-micron particles. All eight particles impact (and adhere) to the outer wall of the duct since they cannot negotiate the sharp curve.

Example 10.3 – Aerosol Particle Trajectories near a Flanged Round Opening
Given: The circular pipe with flanged round opening described in Section 10.3.1 is mounted with its face parallel to the horizon (pipe axis aligned vertically). It is used as a snorkel fume extractor, similar to the one shown in Figure 1.26. Unit density spherical particles of various diameters (D_p) are generated at a point located at x = 0.10 m (axial distance one pipe radius from the face), and y = 0.10 m (radial distance one pipe radius from the axis), as illustrated in Figure E10.3a. The particles are injected into the flow in all directions with a maximum initial speed of about 14. m/s.

To do: Plot the trajectories of unit density spherical particles of various diameters, and determine the maximum particle diameter that can successfully be sucked into the snorkel inlet.

Solution: Only the worst-case scenario is considered, i.e. when the aerosol particles have an initial velocity in a direction *opposite* to the air flow. From Figure 10.8c, it is seen that at the specified initial particle location, the direction of the air flow is approximately 45° from the pipe axis. It is therefore assumed that the initial particle velocity is approximately 14. m/s opposite to this direction, as indicated in Figure E10.3a. Rounding off on the conservative side, particle velocity $\vec{v}$ is set to

$$\vec{v} = \left(10.0\vec{i} + 10.0\vec{j}\right)\frac{m}{s}$$

The air flow field is already known from the CFD simulation described in Section 10.3.1. This flow field is assumed to be "frozen", which means that particle motion does not affect air flow. Particles of the following diameters are injected: D_p = 10, 25, 50, 60, 70, 80, 90, 100, and 200 μm. Their calculated trajectories are shown in Figure E10.3b. While each particle starts at the same location and moves at the same initial velocity, their trajectories are vastly different. All the particle trajectories curve towards the axis due to aerodynamic drag in that direction, but the

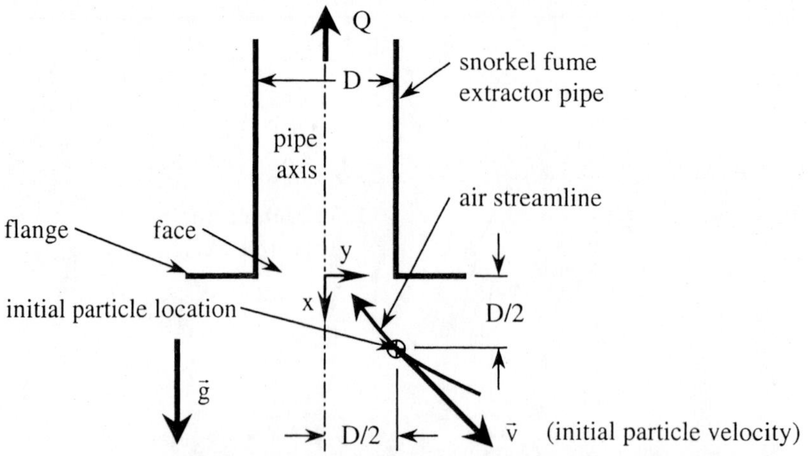

Figure E10.3a Schematic diagram illustrating the initial location and velocity of aerosol particles injected into an air flow field generated by a flanged round inlet; the air streamline that passes through the initial particle location is also shown.

larger particles are less affected by the surrounding air flow because of their larger inertia (and weight – gravity is in the x-direction here). It is hard to distinguish the trajectories of 10 and 25 μm particles; after quickly turning around, they follow the air streamlines almost exactly, and get sucked into the opening. Particles of diameters 50, 60, 70, and 80 μm are flung progressively further out from the starting point, but they too eventually turn around and get captured by the snorkel. It is clear from the figure that particles of diameter 90 μm and larger do *not* get sucked into the opening. Instead, these heavy particles eventually fall straight down due to gravity.

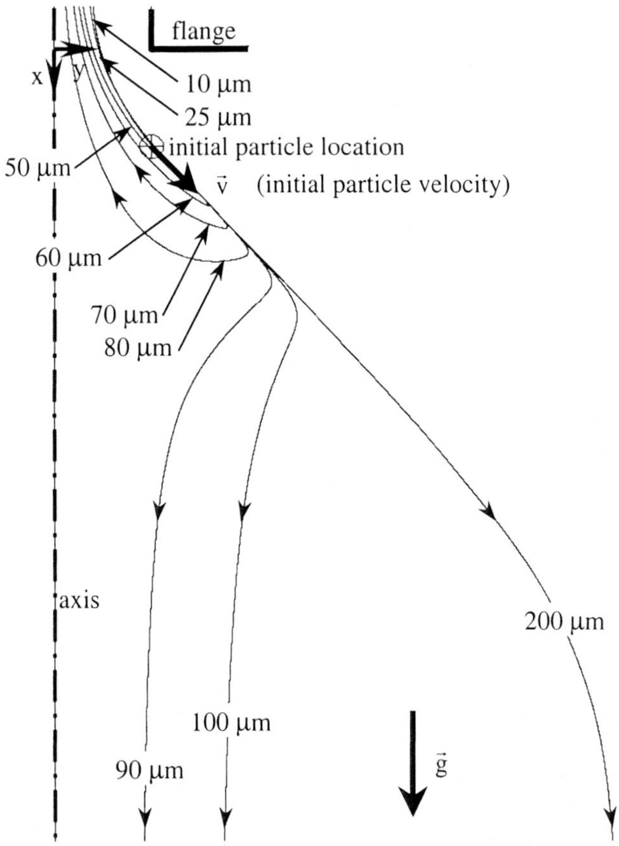

Figure E10.3b CFD simulation of particle trajectories near a flanged round inlet; unit density spherical particles of diameter D_p = 10, 25, 50, 60, 70, 80, 90, 100, and 200 μm are injected at the indicated point with initial velocity equal to 14. m/s.

Discussion: The worst-case scenario is modeled, in which the particles are injected in a direction opposite that of the air. If particles were to be injected at an angle other than the one simulated, they would have a much better chance of being captured. One can conclude with some confidence, therefore, that all unit density spherical particles of diameter less than about 80 or 90 microns will be captured by the snorkel. If heavier particles, such as welding fume, were to be captured, new trajectories would need to be calculated. Since gravity acts *against* the suction flow of air in the present configuration, the cut-off diameter for liquid metal particles would potentially be much smaller than the 80 or 90 microns calculated here. Finally, workers and machinery near the particle injection location would certainly interfere with the air flow field. If the geometry were known, a more complex three-dimensional grid could be generated to account for such disturbances.

10.4.2 Simulation of Flow through a Cascade Impactor

Flow of air and particles through a cascade impactor presents a more complicated example of inertial separation. A crude three-stage, axisymmetric cascade impactor is considered here for illustrative purposes. The impactor is 30.0 mm long, with inlet and outlet diameters of 4.0 mm. The first stage orifice is 2.0 mm in diameter. The orifice diameter of the second stage is half of that (1.0 mm), and that of the third stage is half again (0.50 mm). The collector plate of each stage is progressively closer to the outlet of its respective orifice as well, with distances of 2.5, 1.5, and 0.5 mm respectively for the first, second, and third stage of the cascade. Air at STP is drawn through the device with an inlet speed of 0.10 m/s. The flow field is solved using FLUENT, assuming steady laminar flow. Air streamlines are shown in Figure 10.17a. Only one plane of the flow is solved (the x-y plane in FLUENT, corresponding to the r-θ plane defined for axisymmetric flows in Chapter 7); the graphical results have been mirrored about the axis of symmetry (the vertical x-axis in Figure 10.17) for display purposes. Except for some recirculation eddies near the sharp upper corners, the air flows smoothly from the inlet (at the top of the figure) through all three stages of the cascade impactor.

Trajectories of aerosol particles entering the cascade impactor are then calculated, based on the "frozen" velocity field. Particles of five different diameters are tested: D_p = 1, 2, 5, 10, and 15 μm. In all cases, unit density spherical particles of diameter D_p are injected at ten points along a horizontal line just below the inlet with initial particle velocity equal to that of the air flow. The resulting trajectories are shown in Figure 10.17b through f. Note that because of the mirror imaging, twenty rather than ten particle trajectories are plotted for each case. The wall boundary conditions are specified such that particles impacting the collector plates are entrapped (adhere), while those striking the outer walls of the cascade impactor reflect (bounce) off the wall. In these calculations, however, none of the particles ever hits the outer wall of the cascade impactor.

In Figure 10.17b, particles of diameter 1 micron are injected. All ten of these tiny particles escape the cascade impactor, and their trajectories approximately follow the air streamlines except near the region of the third stage collector plate, from which they narrowly escape impaction. For the case of 2-micron unit density spherical particles (Figure 10.17c), all ten of the injected particles flow past the first two collector plates, but impact (and adhere to) the third collector plate. All but one of the ten injected 5-micron particles impact the second stage collector (Figure 10.17d). Careful tracing of this particle's trajectory verifies that it is the particle that originates closest to the outer radius of the inlet, as expected. All of the injected 10-micron particles impact the second stage collector plate (Figure 10.17e), and all of the 15-micron particles impact the first stage collector plate (Figure 10.17f). Even though this is a crude three-stage cascade impactor, its ability to sharply sort particles by size is remarkable. Commercially available units can have ten or more stages, and are finely tuned for even better particle classification (see for example Figure 9.8). CFD is shown here to be a useful tool for the design of cascade impactors and other inertial separation devices.

10.5 Closure

Although neither as ubiquitous as spreadsheets like Excel nor as easy to use as mathematical solvers like Mathcad, computational fluid dynamics codes are continually improving and are becoming more commonplace. Once the realm of specialized scientists who wrote their own codes, commercial CFD codes with numerous features and user-friendly interfaces can now be obtained at reasonable cost, and are available to engineers of all disciplines. As shown in this chapter, however, a poor grid, improper choice of laminar versus turbulent flow, inappropriate boundary conditions, and/or any of a number of other miscues can lead to CFD solutions which are physically incorrect, even though the colorful graphical output always looks

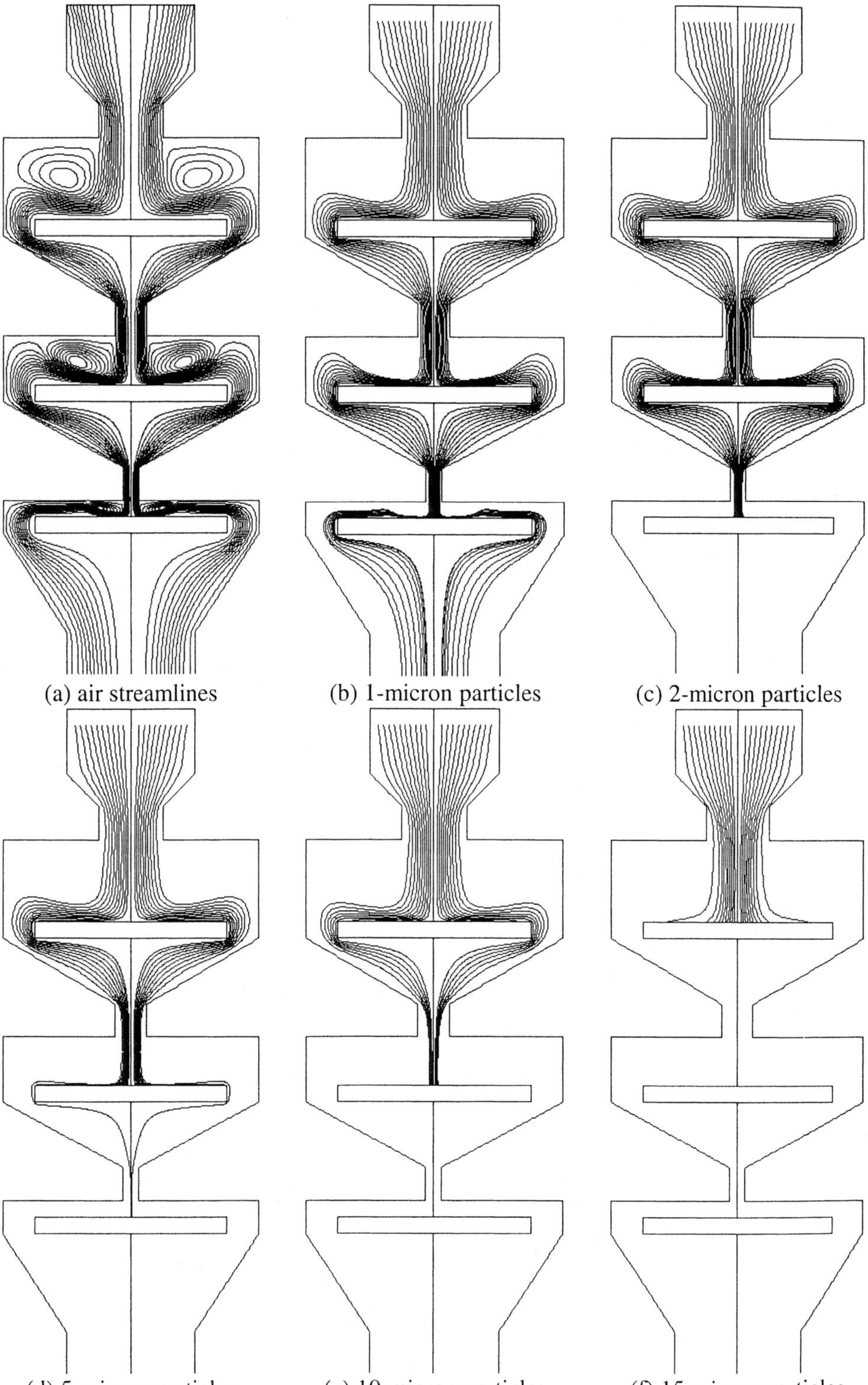

Figure 10.17 CFD simulation of flow in a 3-stage cascade impactor; (a) streamlines, (b)-(f): trajectories of unit density particles of diameter D_p = (b) 1, (c) 2, (d) 5, (e) 10, and (f) 15 μm.

pretty. Therefore it is imperative that CFD users be well grounded in the fundamentals of fluid mechanics in order to avoid erroneous answers from a CFD simulation. Bearing this caution in mind, CFD has enormous potential for application to indoor air quality engineering, especially when one considers the seemingly unbounded increases in computer speed and memory that are projected. In this final chapter, we have attempted to show both the capabilities and the limitations of CFD as applied to indoor air quality engineering. Examples have been presented for two categories of contaminants – gaseous and particulate. While the examples in this chapter are simple and two-dimensional, the methodology learned here is applicable to the much more complicated geometries and flow fields one might encounter in the field of indoor air quality engineering.

Chapter 10 - Problems

1. Consider room ventilation in which there is a source of gaseous contaminant somewhere in the room, a fresh air supply, and an exhaust. If steady-state conditions are achieved, prove that the average mol fraction of contaminant exiting the room at the exhaust must equal the steady-state well-mixed value, regardless of where the source, air supply, or exhaust are located, and regardless of whether the air in the room is well mixed or not.

2. Create a rectangular two-dimensional computational domain as sketched in Figure P10.2 to study the development of a boundary layer along a flat plate. Generate a uniform structured grid in the domain. Use 20 intervals on the sides and 80 intervals on the top and bottom such that the size of each rectangular cell is 0.05 x 0.05 m. Inject air at STP into the domain at a velocity of 20.0 m/s. Run a steady-state laminar CFD simulation of this case to convergence.

(a) Zoom in to visualize the boundary layer growing on the wall. Calculate the total drag force on the wall. Is this grid fine enough to adequately define the boundary layer?

(b) Double the grid resolution and re-do the CFD simulation to convergence. Calculate the total drag force on the wall. Is the solution grid resolved?

(c) Rearrange the nodes of Part (b) such that nodes near the wall are very close together, while those at the top are at least two orders of magnitude further apart. Create a grid and run the CFD simulation. Calculate the total drag force on the wall. Compare the results of this "boundary layer grid" with those of Parts (a) and (b) and comment.

3. Repeat Problem 10.2, but for steady-state *turbulent* flow. The realizable k-ε turbulence model with enhanced wall treatment is recommended. Specify the turbulence intensity at 10% and the turbulent length scale at 0.1 m. Does the turbulent solution converge more or less rapidly than the laminar solution? Explain. Compare the drag force on the wall for the two cases, laminar and turbulent. Calculate the Reynolds number based on flat plate length. The boundary layer on a smooth flat plate becomes fully turbulent at a Reynolds number around 3×10^6; which solution is more physical – laminar or turbulent? Discuss.

4. Create a rectangular two-dimensional computational domain to represent a room of dimensions 4.0 x 2.0 m (and of unit depth perpendicular to this plane). Somewhere along the top edge that defines the ceiling, create a velocity inlet that spans 0.20 m along this edge. Likewise, generate a pressure outlet of the same size somewhere else along the ceiling as sketched in Figure P10.4. Generate a uniform structured grid in the domain. Use 40 intervals on the side walls and 80 intervals on the floor and ceiling such that the size of each rectangular cell is 0.05 x 0.05 m. Blow air at STP into the room at a velocity such that N = 5.0 room air changes per hour. Run a steady-state laminar CFD simulation of this case to convergence.

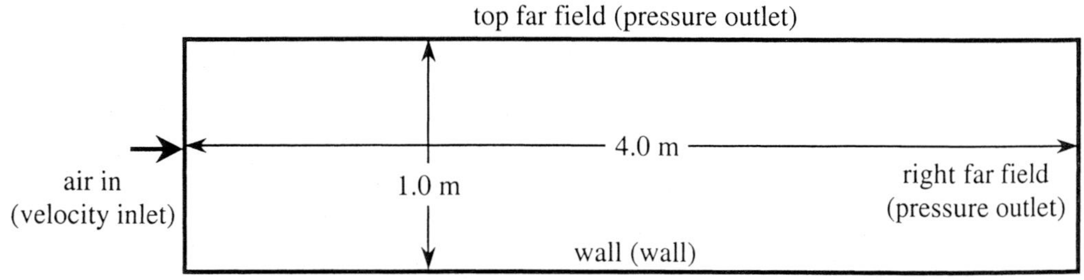

Figure P10.2 Computational domain for simulation of a flat plate boundary layer; boundary conditions are given in parentheses.

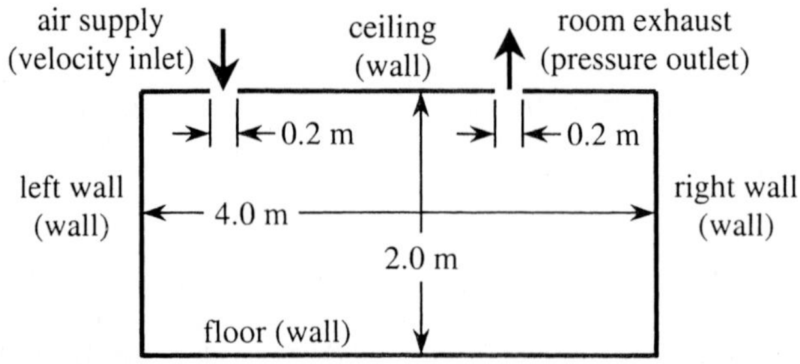

Figure P10.4 Computational domain for simulation of ventilation in a two-dimensional room without air supply or exhaust ducts; boundary conditions are given in parentheses.

(a) Is there reverse flow across any cells along the pressure outlet? Explain why an exhaust duct extending out of the room might make the simulation more physically meaningful.
(b) Repeat for steady-state *turbulent* flow. Use the realizable k-ε turbulence model with enhanced wall treatment. Set the turbulence intensity to 10% at the inlet, with a turbulent length scale of 0.2 m. Compare the streamlines – is there much difference between the laminar and turbulent streamlines?

5. Consider the same room as in the above problem. This time, generate a duct, 0.20 m across and 3.0 m in length, that intersects the room somewhere along the ceiling, as sketched in Figure P10.5. The boundary condition at the top end of the duct should be a velocity inlet. The boundary condition at the junction between the duct and the room should be specified as interior. Likewise, generate another duct of the same dimensions somewhere else along the ceiling; the boundary condition at the top end of this duct should be a pressure outlet. Create a structured grid such that the size of each cell is 0.05 x 0.05 m. Blow air at STP into the room at a velocity such that N = 5.0 room air changes per hour. Run a steady-state turbulent CFD simulation of this case to convergence. Use the realizable k-ε turbulence model with enhanced wall treatment. Set the turbulence intensity to 10% at the velocity inlet, with a turbulent length scale of 0.2 m.

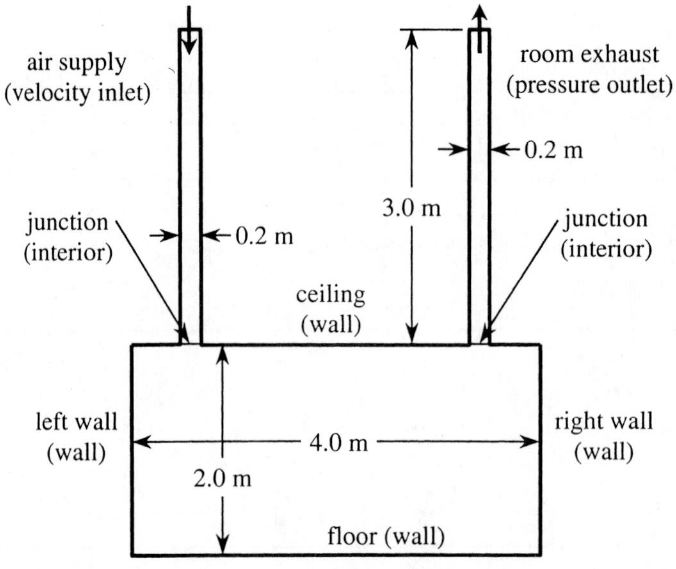

Figure P10.5 Computational domain for simulation of ventilation in a two-dimensional room with duct extensions; boundary conditions are given in parentheses.

(a) Do you get reverse flow across some of the cells in the pressure outlet? Explain why an exhaust duct extending out of the room makes this simulation more physically meaningful than that of the previous problem.
(b) Double the grid resolution in both directions and repeat the CFD run. Is your solution grid resolved? Why or why not?
(c) Create an *unstructured* (triangular) grid with the same number of boundary intervals as the original grid used in Part (a). Is the number of cells less than, the same as, or greater than that for the structured grid case? Run the CFD code to convergence with this grid. Is less, the same, or more CPU time required compared to the structured grid case? Are the results nearly the same? Why or why not?

6. Repeat the calculations of Example 10.3, except with a plain round opening instead of a flanged round opening. This problem requires two steps. First, run a steady-state turbulent CFD simulation to convergence. Use the realizable k-ε turbulence model with enhanced wall treatment. Set the turbulence intensity to 0.1% and the turbulent length scale to 0.1 m (to simulate very quiet ambient conditions at the far-field pressure inlet). Second, calculate the trajectories of particles of unit density (in FLUENT, choose particles of liquid water). Which of the two openings (plain or flanged) is able to suck in larger particles? Inject particles as follows:

(a) x = 0.10 m, y = 0.10 m, U = 10. m/s, V = 10. m/s (same as in Example 10.3)
(b) x = 0.10 m, y = 0.10 m, U = -10. m/s, V = -10. m/s
(c) x = 0.20 m, y = 0.01 m, U = -5.0 m/s, V = 5.0. m/s

7. Repeat the above problem, except use particles of density 7,800 kg/m^3 to simulate steel particles ejected from a grinder. Estimate the particle diameter below which ejected particles will get sucked into the inlet.

8. Simulate a two-dimensional Aaberg inlet, as discussed in Chapter 6. Use air at STP and dimensions and velocities that correspond to those of Example 6.2. Compare two cases: blowing jet on and blowing jet off. For the case with the blowing jet off, simply set the jet velocity at the velocity inlet to zero. Alternatively, change the velocity inlet boundary condition to a wall boundary condition. In either case, the Aaberg inlet degenerates to a flanged inlet. For both cases draw isopleths as a percentage of average speed across the face of the opening. Does the Aaberg inlet significantly increase the reach as claimed?

9. Generate a crude two-dimensional model of one of the last generations of branching bronchiole in the lung (see Chapter 2). The computational domain should look something like that sketched in Figure P10.9. Note that because of symmetry, only the upper half needs to be modeled.

(a) Using information given in Chapter 2, estimate the average air speed into such a bronchiole passage during inhalation for an average adult male at rest.
(b) Generate an appropriate grid and run a CFD simulation to convergence assuming laminar steady-state flow, using the air velocity estimated in Part (a). Plot streamlines of the flow field and calculate the pressure drop from inlet to outlet.
(c) Inject particles at several points along the inlet plane and plot their trajectories. Use unit density spherical particles of various diameters, and assume their initial velocity is the same as that of the air entering the passage. At approximately what cut-off diameter do particles get trapped in the bronchiole passage due to inertial separation? (Particles smaller than this diameter continue on downstream to the next branch, perhaps into alveoli.)

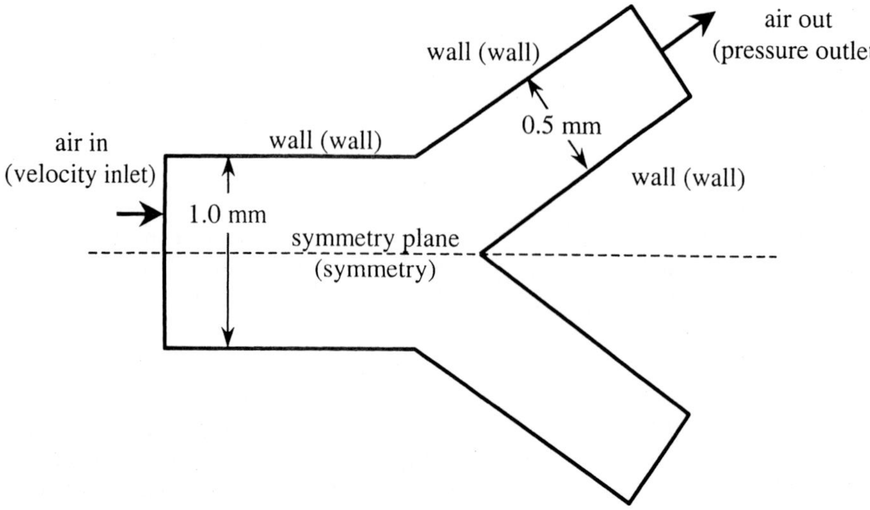

Figure P10.9 Computational domain for simulation of flow through branching bronchiole; boundary conditions are given in parentheses.

10. Consider a crude simulation of evaporation of benzene from an open barrel (see Chapter 4). Set up the computational domain sketched in Figure P10.10. The boundary condition at the bottom of the barrel is a wall representing the liquid/vapor interface. The mass fraction of benzene at this boundary is set to 1 (100%) to simulate evaporation of liquid into vapor. Pure air at 0.20 m/s enters the domain through the velocity inlet. The gage pressure in the far field is zero.

(a) Generate a grid (a structured grid is best, but a hybrid grid is also acceptable here). The grid should be structured and fine inside the barrel. The grid should be fine near the top of the barrel, but can stretch out towards the far field.

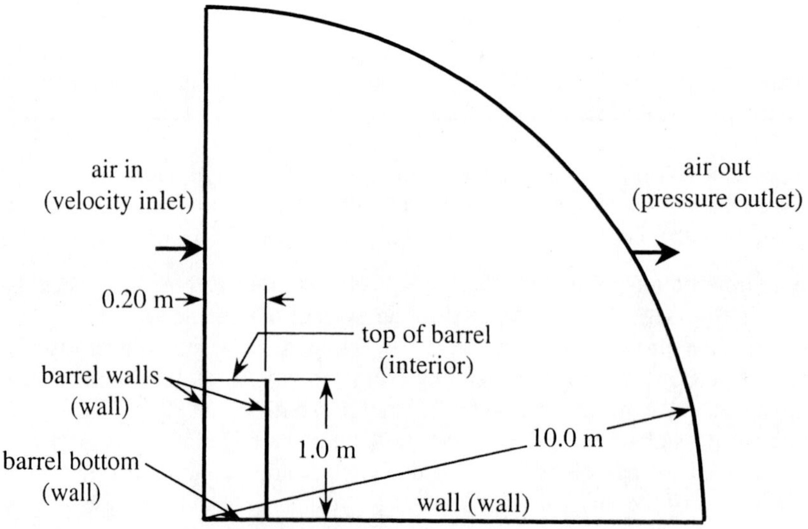

Figure P10.10 Computational domain for simulation of evaporation of benzene from an open barrel (not to scale); boundary conditions are given in parentheses.

(b) Solve for the air flow field without any benzene. Assume steady-state laminar flow. Refine the grid as necessary to obtain a good solution. Plot the air streamlines.
(c) Set the benzene mass fraction to 1 (that is, 100%) at the barrel bottom and run the CFD simulation to convergence. Plot contours of benzene mass fraction throughout the domain. Is the mass concentration of benzene greater than the OSHA PEL anywhere outside the barrel?

11. Construct a computational domain consisting of a two-dimensional duct of height 0.20 m with a sharp 90° bend. The length of duct upstream of the bend should be about five duct heights, while that downstream of the bend should be at least twenty duct heights. Construct a second computational domain consisting of the same duct, but with a gentle round 90° bend such that the total duct length is identical to that of the first duct. Apply U = 10. m/s at the velocity inlet to the duct. The outlet of the duct in both cases should be a pressure outlet at zero gage pressure. Generate a grid and run a steady-state turbulent flow CFD solution to convergence for both cases. In post-processing, calculate the average pressure at the velocity inlet for both cases and compare. Which one has the higher pressure drop? Explain.

12. Design an axisymmetric cascade impactor which is able to separate and collect anthrax spores. Run a CFD simulation of laminar air flow through the impactor, and then calculate and plot trajectories of particles of diameter less than, equal to, and greater than that of anthrax spores (see Chapter 2).

13. Consider the two-dimensional room of Figure 10.11. Instead of carbon monoxide contaminant in the room, however, the HVAC problem of room heating will be modeled. For simplicity, specify the temperature on all room walls to be 15.0 °C, and specify approximately two room air changes per hour through the ventilation system. Adjust the temperature of the fresh air at the velocity inlet(s) until the average temperature in the room is a comfortable 21. °C.

(a) Run a steady-state turbulent CFD simulation to convergence for the case of dilution ventilation (supply and exhaust through the ceiling) with the ceiling fan off. Plot air streamlines and temperature contours.
(b) Repeat for the case of dilution ventilation with the ceiling fan on (use δP_{fan} = 2.0 Pa).
(c) Repeat for the case of displacement ventilation with the ceiling fan off.
(d) Compare and discuss the streamlines and temperature distribution in the room.

14. Repeat the above problem, but for the case of room cooling. Set the wall temperature to 31.0 °C, and adjust the temperature of the fresh air at the velocity inlet(s) until the average temperature in the room is a comfortable 21. °C. Which ventilation/fan combination would be most comfortable for workers in the room?

15. Simulate the three split-flow ventilation paint booth configurations of Example 5.12. For simplicity use a two-dimensional approximation; namely imagine that the booths are one unit width (1.0 m) into the page of Figure E5.12. Use the dimensions and flow rates specified in the example problem, and place the contaminant supply along the bottom (floor) of the booth. Assume turbulent flow (the realizable k-ε turbulence model is recommended). Hint: With a little forethought, all three booths can be simulated using the same grid; wall and fan boundary conditions are easily converted to internal boundary conditions to simulate differences between the three booths. For Booth C, a duct needs to be added to the computational domain from the right side of the booth, over the top of the booth, and down to the left side of the booth. Use a duct that is 0.2 m across, and insert a fan boundary condition somewhere along this duct. The fan pressure rise will need to be adjusted so as to mix the recirculated air with the make-up air supply.

If the pressure rise is too low, make-up air may flow into the recirculation duct. If the pressure is too high, reverse flow will be encountered at the supply air inlet boundary condition. Run the CFD simulation to convergence for all three cases. Compare the mass concentrations in the booths and discuss. How valid are the approximations made in Example 5.12?

16. Simulate a vertical laminar flow clean room (see Figure 5.11a). For simplicity use two-dimensional flow. Assume that the middle half of the ceiling is the fresh air supply, and assume that the entire floor is the exhaust opening. Plot streamlines in the room.

17. Simulate a vertical horizontal flow clean room (see Figure 5.11b). For simplicity use two-dimensional flow. Assume that the entire left wall of the room is the fresh air supply, and assume that the lower half of the right wall is the exhaust opening. Plot streamlines in the room. Can you think of a way to avoid the large recirculating eddy that develops in the upper right corner of the room?

Appendices

Appendix A.1 Comparison of OSHA permissible exposure limit (PEL) for common industrial materials, dusts, and fumes in 1989 and 1997[a,s] (abstracted from US NIOSH, Dept. of Health and Human Services, 1989, 1997).

Industrial materials:

substance (CAS number)	1989 PEL[t] (PPM)	1997 PEL[t] (PPM)
acetaldehyde (75-07-0)	100	200
acetic acid (64-19-7)	10	10
acetic anhydride (108-24-1)	5 (ceiling)	5
acetone (67-64-1)	750	1000
acrolein (107-02-8)	0.1	0.1
allyl alcohol (107-18-6)	2[d]	2
ammonia (7664-41-7)	35 (STEL)	50
aniline (62-53-3)	2[d]	5[d]
benzene (71-43-2)	10 (ceiling)	1, STEL=5
boron trifluoride (7637-07-2)	1 (ceiling)	1 (ceiling)
n-butyl acetate (123-86-4)	150	150
n-butyl alcohol (71-36-3)	50 (ceiling)	100
n-butyl mercaptan (109-79-5)	0.5	10
n-butylamine, butylamine (109-73-9)	5 (ceiling)[d]	5 (ceiling)[d]
carbon disulfide (75-15-0)	4 (ceiling)	20
carbon monoxide (630-08-0)	35	50, ceiling=200
carbon tetrachloride (56-23-5)	2 (ceiling)	10, ceiling=25
chlorine (7782-50-5)	0.5	1 (ceiling)
chlorobenzene (108-90-7)	75	75
chloroform (67-66-3)	2	50 (ceiling)
cyclohexane (110-82-7)	300	300
decaborane (17702-41-9)	0.05[d]	0.05[d]
diazomethane (334-88-3)	0.2	0.2
diborane (19287-45-7)	0.1	0.1
dichloroethyl ether (111-44-4)	5[d]	15[d] (ceiling)
diethylamine (109-89-7)	10	25
diisopropylamine (108-18-9)	5[d]	5[d]
dimethylamine (124-40-3)	10	10
dioxane (123-91-1)	25[d]	100[d]
diphenyl (92-52-4)	0.2	0.2
ethyl acetate (141-78-6)	400	400
ethyl acrylate (140-88-5)	5[d]	25[d]
ethyl alcohol (64-17-5)	1000	1000
ethyl benzene (100-41-4)	100	100
ethyl chloride (75-00-3)	1000	1000
ethyl formate (109-94-4)	100	100
ethyl mercaptan (75-08-1)	0.5	10 (ceiling)
ethylamine (75-04-7)	10	10
ethylene dibromide (106-93-4)	#	20, ceiling=30
ethylene dichloride (107-06-2)	1	50, ceiling=1000
ethylene oxide (75-21-8)	1	1
fluorine (7782-41-4)	0.1	0.1
formaldehyde (50-00-0)	3 (ceiling)	0.75, STEL=2
formic acid (64-18-6)	5	5
furfural (98-01-1)	2[d]	5[d]

Appendix A.1 (continued)

substance (CAS number)	1989 PEL[t] (PPM)	1997 PEL[t] (PPM)
n-heptane (142-82-5)	400	500
hexachloroethane (67-72-1)	1[d]	1[d]
n-hexane (110-54-3)	50	500
hydrazine (302-01-2)	0.1[d]	1[d]
hydrogen bromide (10035-10-6)	3 (ceiling)	3
hydrogen chloride (7647-01-0)	5 (ceiling)	5 (ceiling)
hydrogen cyanide (74-90-8)	4.7 (STEL)[d]	10[d]
hydrogen fluoride (7664-39-3)	3	3
hydrogen peroxide (7722-84-1)	1	1
hydrogen sulfide (7783-06-4)	10	20 (ceiling)
iodine (7553-56-2)	0.1 (ceiling)	0.1 (ceiling)
isobutyl acetate (110-19-0)	150	150
isobutyl alcohol (78-83-1)	50	100
isopropyl acetate (108-21-4)	250	250
isopropyl alcohol (67-63-0)	400	400
isopropyl ether (108-21-3)	500	500
mercury, vapor (7439-97-6)	0.05 mg/m^3	0.1 mg/m^3 (ceiling)
methacrylic acid (79-41-4)	20	0, NIOSH=20[d]
methyl acetate (79-20-9)	200	200
methyl acrylate (96-33-3)	10[d]	10[d]
methyl alcohol (67-56-1)	200	200
methyl bromide (74-83-9)	5[d]	5[d] (ceiling)
methyl chloride (74-87-3)	100	100, ceiling=200
methyl ethyl ketone, 2-butanone (78-93-3)	200	200
methyl formate (107-31-3)	100	100
methyl iodide ((74-88-4)	2[d]	5[d]
methyl isocyanate (624-83-9)	0.02[d]	0.02[d]
methyl mercaptan (74-93-1)	0.5	10 (ceiling)
methylamine (74-89-5)	10	10
methylene chloride (75-09-2)	#	25, STEL=125
naphthalene (91-20-3)	10	10
nickel carbonyl (13463-39-3)	0.001	0.001
nitric acid (7697-37-2)	2	2
nitric oxide (10102-43-9)	25	25
nitrobenzene (98-95-3)	1[d]	1[d]
nitroethane (79-24-3)	100	100
nitrogen dioxide (10102-44-0)	1 (STEL)	5 (ceiling)
nitromethane (75-52-5)	100	100
o-nitrotoluene (88-72-2)	2[d]	5[d]
octane (111-65-9)	300	500
ozone (10028-15-6)	0.1	0.1
pentaborane (19624-22-7)	0.005	0.005
perchloroethylene, tetrachloroethylene (127-18-4)	25	100, ceiling=200
phenol (108-95-2)	5[d]	5[d]
phosgene (75-44-5)	0.1	0.1
phosphoric acid (7664-38-2)	1 mg/m^3	1 mg/m^3
picric acid (88-89-1)	0.1[d] mg/m^3	0.1[d] mg/m^3
n-propyl acetate (109-50-4)	200	200

Appendix A.1 (continued)

substance (CAS number)	1989 PEL[t] (PPM)	1997 PEL[t] (PPM)
n-propyl alcohol (71-23-8)	200	200
propylene oxide (75-56-9)	20	100
pyridine (110-86-1)	5	5
styrene, monomer (100-42-5)	50	100, ceiling=200
sulfur dioxide (7446-09-5)	2	5
sulfuric acid (7664-93-9)	1 mg/m^3	1 mg/m^3
toluene (108-88-3)	100	200, ceiling=300
trichloroethylene (79-01-6)	50	100, ceiling=200
1, 2, 3-trichloropropane (96-18-4)	10	50
turpentine (8006-64-2)	100	100
o-xylene (95-47-6)	100	100

Industrial dusts and fumes:

substance (CAS number)	1989 PEL (mg/m^3)	1997 PEL (mg/m^3)
chromium metal (7440-47-3)	1	1
coal dust, respirable fraction		
SiO_2 <5% by mass	2	2.4/(%SiO_2 + 2)
SiO_2 >5% by mass	0.1	10/(%SiO_2 + 2)
cobalt metal dust and fume (7440-48-4)	0.05	0.1
copper		
dust (7440-50-8)	1	1
fume (7440-50-8)	0.1	0.1
grain dust (oat, wheat, barley)	10	10
graphite respirable dust (7782-42-5)	2.5	2.5 (NIOSH)
gypsum, (7778-18-9)		
respirable dust	5	5
total dust	15	15
iron oxide, total dust (1309-37-1)	10	10
kaolin		
respirable dust	5	5
total dust	10	15
limestone, total respirable dust (1317-67-3)	15	15
magnesium oxide, total dust (1309-48-4)	10	15
manganese fume (7439-96-5)	1	5 (ceiling)
marble (1317-65-3)		
respirable dust	5	5
total dust	15	15
molybdenum, insoluble compounds (7439-98-7)		
respirable dust	5	-
total dust	10	15
nickel, metal dust (7440-02-0)	-	1
insoluble compounds	1	-
soluble compounds	0.1	-
oil mist, mineral	5	5
particulates, not otherwise regulated		
respirable dust	5	5
total dust	15	15

Appendix A.1 (continued)

substance (CAS number)	1989 PEL (mg/m^3)	1997 PEL (mg/m^3)
platinum (7440-06-4)		
metal	1	none
soluble salts	0.002	0.002
Portland cement, (65997-15-1)		
respirable dust	5	5
total dust	10	15
rhodium (7440-16-6)		
insoluble compounds	0 .1	0 .1
soluble compounds	0 .001	0 .001
selenium (7782-49-2)	0.2	0.2
silica		
amorphous, crystalline silica < 1% (68855-54-9)	6	-
cristobalite silica, respirable quartz (14464-46-1)	0.05	5/(%SiO_2 by mass+2)
cristobalite silica, total quartz (14464-46-1)	-	15/(%SiO_2 by mass+2)
crystalline silica, respirable quartz (14808-60-7)	0.1	10/(%SiO_2 by mass+2)
crystalline silica, total quartz (14808-60-7)	-	30/(%SiO_2 by mass+2)
crystalline tripoli, respirable (1317-95-9)	0.1	-
fused (60676-86-0)	0.1	-
tridymite silica, respirable quartz (15468-32-3)	0.05	5/(%SiO_2 by mass+2)
tridymite silica, total quartz (15468-32-3)	-	15/(%SiO_2 by mass+2)
silicon, (7440-21-3)		
respirable dust	5	5
total dust	10	15
silver, metal (7440-22-4)	0.01	0.01
tin		
organic compounds (7440-31-5)	0.1	0.1
oxide (7440-31-5)	2	-
tungsten, soluble compounds (7440-33-1)	1	-
vanadium, respirable dust (1314-62-1)	0.05	0.5 (ceiling)
vegetable oil mist		
respirable dust	5	5
total dust	15	15
wood dust		
general	5	-
respirable dust	-	5
total dust	-	15
western red cedar	2.5	-
zinc oxide fume, (1314-13-2)		
respirable dust	5	5
total dust	10	15
zirconium compounds (7440-67-7)	5	5

[s] see US Dept. of Labor (1989, 1997) for ceiling and maximum transitory values
[#] in process of rulemaking at the time
[a] see Appendix A.20 for PEL of additional materials; note that PEL and TLV are reviewed on a regular basis and readers should always use currently approved values
[d] denotes additional entry through the skin
[t] TWA (time weighted average) PEL is assumed unless indicated otherwise

Appendix A.2 Emission factors for particles from uncontrolled metallurgical processes (abstracted from US EPA, 1972; Bond and Straub, 1972; and US NTIS, 1979, 1980a, 1980b, 1981, and 1986).

process	EF (kg particles/Mg raw material)
aluminum production	
aluminum hydroxide	100.00
bauxite grinding	3.00
crucible furnace	0.95
prebake cell	47.00
reverberatory furnace	2.60
secondary aluminum	
smelting	
crucible furnace	0.95
reverberatory furnace	2.15
sweating furnace	7.20
gray iron fugitive emissions	
cleaning, finishing	8.50
cooling	5.00
core making, baking	0.60
inoculation	1.5-2.5
magnesium treatment	2.50
pouring	2.50
sand handling, preparation, mulling	20.00
scrap and charge handling	0.30
shakeout	16.00
gray iron furnaces	
cupolas (average)	8.55
less than 48 in ID	6.45
48-60 in ID	9.75
greater than 60 in ID	9.45
electric arc	5.00
electric induction	0.75
reverberatory	1.00

Appendix A.2 (continued)

process	EF (kg particles/Mg raw material)
iron and steel mills	
blast furnaces	
cast house (kg/Mg hot metal)	
tap hole and trough	0.15
slips (kg/slip)	39.5
electric arc furnaces (kg/Mg steel)	
charging, tapping, slagging	0.7
melting and refining (carbon steel)	19.0
melting, refining, charging, tapping, slagging	
alloy steel	5.65
carbon steel	25.0
machine scarfing (kg/Mg steel)	0.05
miscellaneous (boilers, soaking pits, slab reheat furnaces, blast furnace gas, coke oven gas) (kg/ 1000 mJ)	0.015
open hearth furnaces (kg/Mg steel)	
melting and refining	10.55
sintering (kg/Mg finished sinter)	
basic oxygen furnaces (kg/Mg steel)	
BOF monitor (all sources)	0.25
charging, at source	0.3
discharging (breaker and hot screens)	3.40
hot metal transfer, at source	0.095
tapping, at source	0.46
top blown melting and refining	14.25
windbox leaving grate	5.56
teeming (kg/Mg steel)	
leaded steel at source	0.405
unleaded steel at source	0.035
lead products	
cable covering	0.30
type metal production	0.35
melting of red brass (< 7% zinc)	
electric furnaces	1.50
crucible or pot furnaces	1.65
reverberatory furnaces	8.40
rotary furnace	10.60
melting of bronze	
crucible furnace	1.90
rotary furnace	15.30

Appendix A.2 (continued)

process	EF (kg particles/Mg raw material)
primary lead smelting	
blast furnace	180.50
dross reverberatory furnace	10.00
fugitive emissions	
conveyor loading, car charging (sinter)	0.25
dross kettle	0.24
ladle operation	0.46
lead casting	0.44
ore mixing and pelletizing	1.13
reverberatory furnace leakage	1.50
silver retort building	0.90
sinter crushing, screening, discharge	0.75
sinter machine leakage	0.34
sinter transfer to dump	0.10
slag cooling	0.24
zinc fuming furnace vents	2.30
ore crushing	1.00
sintering (updraft)	106.50
secondary copper smelting	
cupola	
insulated copper wire	115
scrap copper and brass	35
reverberatory	
brass and bronze	18
copper	2.55
rotary, brass and bronze	150
crucible, pot for brass and bronze	10.50
electric arc	
copper	2.50
brass and bronze	5.50
electric induction	
copper	3.50
brass and bronze	10.00
secondary lead	
fugitive emissions	
casting	0.44
smelting	1.4-7.9
sweating	0.8-1.8
smelting	
blast (cupola)	96.50
reverberatory	73.50
sweating	16-35

Appendix A.2 (continued)

process	EF (kg particles/Mg raw material)
secondary zinc smelting	
crushing, screening	0.5-3.8
electric resistance sweating	<5
galvanizing	2.50
kcttle sweating	
general metallic scrap	5.50
residual scrap	12.50
muffle distillation, oxidation	10-20
muffle sweating	5.4-16.0
retort and muffle distillation	
casting	0.1-0.2
muffle distillation	22.50
pouring	0.2-0.4
retort distillation, oxidation	10-20
retort reduction	23.50
reverberatory	
general metallic scrap	6.50
residual scrap	16.0
rotary sweating	5.5-12.5
sodium carbonate leaching	
calcining	44.50
crushing, screening	0.5-3.8
fugitive emissions	
casting	0.0075
crucible melting furnace	0.0025
crushing, screening	2.13
electric induction melting	0.0025
electric resistance sweating	0.25
kettle (pot) melting	0.0025
kettle (pot) sweating	0.28
muffle sweating	0.54
retort and muffle distillation	1.18
reverberatory melting furnace	0.0025
reverberatory sweating	0.63
rotary sweating	0.45

Appendix A.3 Emission factors for volatile hydrocarbons from uncontrolled sources (abstracted from US EPA, 1972 and 1980; Bond and Straub, 1972; Hoogheem et al., 1979; US NIOSH, 1981; and US NTIS, 1979, 1980a, 1980b, 1981, and 1986).

source	emission factor, EF
degreasers	
all	1000 kg/Mg
cold cleaner (0.39 kg/hr m2)	430 kg/Mg
conveyorized	850 kg/Mg
fabric scouring	500 kg/Mg
open top vapor (0.73 kg/hr m^2)	775 kg/Mg
nonmethane hydrocarbon vapor emissions	
compressor seals	
hydrocarbon service	0.98 lbm/hr
hydrogen service	0.10 lbm/hr
drains (all)	0.070 lbm/hr
heavy liquid streams	0.029 lbm/hr
light liquid/two-phase streams	0.085 lbm/hr
flanges (all)	0.00058 lbm/hr
gas-vapor	0.0005 lbm/hr
heavy liquid streams	0.0007 lbm/hr
light liquid/two-phase streams	0.0005 lbm/hr
pump seals	
heavy liquid streams	0.045 lbm/hr
light liquid streams	0.26 lbm/hr
relief valves (all)	0.19 lbm/hr
gas-vapor streams	0.36 lbm/hr
heavy liquid streams	0.019 lbm/hr
light liquid/two-phase streams	0.013 lbm/hr
valves	
gas-vapor streams	0.047 lbm/hr
heavy liquid streams	0.0007 lbm/hr
light liquid/ two-phase streams	0.023 lbm/hr
rotogravure printing	840 kg/Mg
dryer exhaust	130 kg/Mg
fugitive	30 kg/Mg
printed product	

Appendix A.3 (continued)

source	emission factor, EF
surface coating	
enamel	420 kg/Mg
interior printed panels	
basecoat	
conventional paint	2.4 kg/ 100 m^2
ultraviolet coating	0.24 kg/100 m^2
water borne	0.2 kg/ 100 m^2
ink	
conventional paint	0.3 kg/ 100 m^2
ultraviolet coating	0.1 kg/ 100 m^2
water borne	0.1 kg/ 100 m^2
topcoat	
conventional paint	1.8 kg/ 100 m^2
ultraviolet coating	Negligible
water borne	0.4 kg/ 100 m^2
filler	
conventional paint	3.0 kg/ 100 m^2
ultraviolet coating	Negligible
water borne	0.3 kg/100 m^2
sealer	
conventional paint	0.5 kg/ 100 m^2
ultraviolet coating	0
water borne	0.2 kg/ 100 m^2
lacquer	770 kg/Mg
paint	560 kg/Mg
plywood veneer dryers	
Douglas fir, heartwood	6.7 kg/10,000 m^2
Douglas fir, sapwood	
steam fired	2.3 kg/ 10,000 m^2
gas fired	38.6 kg/10,000 m^2
larch	1.0 kg/ 10,000 m^2
southern pine	15.1 kg/10,000 m^2
primer (zinc chromate)	660 kg/Mg
varnish and shellac	500 kg/Mg

Appendix A.3 (continued)

source	emission factor, EF
transfer of hydrocarbons by tank cars and trucks	
gasoline in transit (fully loaded)	0.001-0.009 kg/m^3
splash loading-balance service	
crude oil	0.6 kg/m^3
gasoline	1.0 kg/m^3
JP-4	0.3 kg/m^3
splash loading-normal service	
crude oil	0.8 kg/m^3
gasoline	1.4 kg/m^3
JP-4	0.5 kg/m^3
kerosene	0.005 kg/m^3
no. 2	0.004 kg/m^3
no. 6	0.00004 kg/m^3
submerged loading-balance service	
crude oil	0.6 kg/m^3
gasoline	1.0 kg/m^3
JP-4	0.3 kg/m^3
submerged loading-normal service	
crude oil	0.4 kg/m^3
gasoline	0.6 kg/m^3
JP-4	0.18 kg/m^3
kerosene	0.002 kg/m^3
no. 2	0.001 kg/m^3
no. 6	0.00001 kg/m^3

Solvent emission rate relative to evaporation of carbon tetrachloride.

hydrocarbon (HC)	(HC evaporation rate / CCl_4 evaporation rate)
acetone	0.91
benzene	0.49
sec-butanol	0.094
n-butanol	0.035
carbon tetrachloride	1.00
cyclohexane	0.02
ethers (petroleum)	1.00
hexane	1.13
methyl ethyl ketone	0.45
methylene chloride	1.47
mineral spirits	0.0063
naphtha, coal tar	0.015-0.12
naphtha, safety (standard)	0.015-0.12
perchloroethylene	0.27
toluene	0.12
1,1,1-trichloroethane	1.39
trichloroethylene	0.69
trichlorotrifluroethane	2.80
o-xylene	0.055

Appendix A.4 Emission factors for uncontrolled mineral processes (abstracted from US EPA, 1972; and US NTIS, 1979, 1980a, 1980b, 1981, and 1986).

process	EF (kg/Mg raw material)
asphalt concrete	
fugitive particles	
cold and dried aggregate elevator	0.10
screening hot aggregate	0.013
unloading aggregate to bins	0.05
hot mix dryer drum (particles)	2.45
stack gas	
aldehydes	
1-butanal	0.0012
formaldehyde	0.000075
3-methylbutanal	0.008
2-methylpropanal	0.00065
carbon monoxide	0.019
particles	0.137
polycyclic organic compounds	0.000013
sulfur dioxide	0.146 times % sulfur
volatile organic hydrocarbons	0.1
brick	
curing and firing (fluorides)	
gas, oil, or coal-fired kiln	0.4
raw material handling (particles)	
drying	35
grinding	38
storage	17
ceramic clay (particles)	
drying	35
grinding	38
storage	17
clay and fly-ash sintering (particles)	
clay mixed with coke	20
fly ash	55
natural clay	6
concrete batching (particles)	
loading dry-batch truck	0.02
loading mix truck	0.01
loading mixer with raw materials	0.01
transfer of cement to silos	0.12
transfer of sand and aggregate to bins	0.02

Appendix A.4 (continued)

process	EF (kg/Mg raw material)
glass fiber (particles)	
forming-wool	
flame attenuation	1
rotary spun	29
glass furnace-wool	
electric	0.25
gas-recuperative	13-15
gas-regenerative	11
gas-unit melter	4.5
glass furnace-textile	
recuperative	1
regenerative	8
unit melter	3
mixing and weighing	0.3
oven curing and cooling-textile	0.6
oven curing-wool	
flame attenuation	3
rotary spun	4.5
rotary spun	
cooling-wool	0.65
storage bins	0.1
unloading and conveying	1.5
glass frit smelters	
rotary furnace	
fluorides	2.5
particles	8
glass manufacture (soda-lime)	
particles	1
gypsum (particles)	
calciner	45
conveying	0.35
primary grinder	0.5
raw material dryer	20
lime (particles)	
calcining	
crushing, primary	15.5
rotary kiln	100
vertical kiln	4
mineral wool (particles)	
blow furnace	8.5
cooler	1
cupola	11
curing oven	2
reverberatory furnace	2.5

Appendix A.4 (continued)

process	EF (kg/Mg raw material)
phosphate rock processing (particles)	
calcining	7.7
drying	2.9
grinding	0.8
open storage piles	20
transfer and storage	1
Portland cement (particles)	
dry process	
dryers, grinders	48
kilns	123
wet process	
dryers, grinders	16
kilns	100
stone quarrying and processing (particles)	
crushing	
fines mill	3
primary	0.25
recrushing and screening	2.5
screening, conveying, and handling	1
secondary crushing and screening	0.75
storage pile losses	5
tertiary crushing and screening	3

Appendix A.5 Emission factors for uncontrolled chemical processes (abstracted from US EPA, 1972; and US NTIS, 1979, 1980a, 1980b, 1981, and 1986).

process	EF (kg/Mg raw material)
ammonium sulfate	
fluidized bed dryers	
particles	109
volatile organic hydrocarbons	0.11
rotary dryers	
particles	23
volatile organic hydrocarbons	0.74
carbon black	
oil furnace	
CO boiler and incinerator	
carbon monoxide	0.88
hydrocarbons	0.99
hydrogen sulfide	0.11
particles	1.04
sulfur oxides	17.5
flare	
carbon monoxide	122
hydrocarbons	1.85
hydrogen sulfide	1
particles	1.35
sulfur oxides	25
main process vent	
carbon monoxide	1.4
hydrocarbons	50
hydrogen sulfide	30
particles	3.27
charcoal (without chemical recovery plant)	
acetic acid	116
carbon monoxide	160
crude methanol	76
hydrocarbons (as methane)	50
particles	200
paint and varnish manufacture	
paint	
particles	1 (kg/Mg pigment)
undefined hydrocarbons	15 (kg/Mg pigment)
varnish (undefined hydrocarbons)	
acrylic	10 (kg/Mg pigment)
alkyd	80 (kg/Mg pigment)
bodying oil	20 (kg/Mg pigment)
oleoresinous	75 (kg/Mg pigment)

Appendix A.5 (continued)

process	EF (kg/Mg raw material)
plastics manufacturing	
polypropylene	
particles	1.5
propylene gas	0.35
polyvinyl chloride	
particles	17.5
vinyl chloride gas	8.5
printing ink (condensed organics)	
vehicle cooking	
alkyds	80
general	60
oils	20
oleoresinous	75
synthetic fibers	
dacron, oil mist	3.5
nylon	
hydrocarbons	3.5
oil mist	7.5
viscose rayon	
carbon disulfide	27.5
hydrogen sulfide	3
synthetic rubber	
alkanes	
dethylheptane	0.5
pentane	1
alkenes	
butadiene	20
butylene	1.5
methyl propene	7.5
pentadiene	0.5
carbonyls	
acrolein	1.5
acrylonitrile	8.5
ethanenitrile	0.5

Appendix A.6 Emission factors for uncontrolled food processes (abstracted from US EPA, 1972; and US NTIS, 1979, 1980a, 1980b, 1981, and 1986).

process	EF
coffee roasting	
direct-fired	
aldehydes	0.1 kg/Mg
organic acids	0.45 kg/Mg
particles	3.8 kg/Mg
indirect-fired	
aldehydes	0.1 kg/Mg
organic acids	0.45 kg/Mg
particles	2.1 kg/Mg
instant coffee spray dryer, particles	0.7 kg/Mg
stoner and cooler, particles	0.7 kg/Mg
cotton gin, particles	
cleaner	0.45 kg/bale
miscellaneous	13.6 kg/bale
stick and burr machine	1.36 kg/bale
unloading fan	2.27 kg/bale
feed and grain mills and elevators (particles)	
grain processing	
barley flour milling	1.5 kg/Mg
barley or wheat cleaner	0.1 kg/Mg
corn meal	2.5 kg/Mg
milo cleaner	0.2 kg/Mg
soybean	305 kg/Mg
terminal elevators	
drying	3 kg/Mg
screening and cleaning	2.5 kg/Mg
shipping or receiving	0.5 kg/Mg
transferring, conveying, etc.	1 kg/Mg
fermentation, beer and whiskey	
hydrocarbons (whiskey)	0.024 kg/Mg
particles	
drying spent grains	2.5 kg/Mg
grain handling	1.5 kg/Mg
meat smoking	
aldehydes (HCOH)	0.04 kg/Mg
carbon monoxide	0.3 kg/Mg
hydrocarbons (methane)	0.035 kg/Mg
organic acids (acetic)	0.1 kg/Mg
particles	0.15 kg/Mg

Appendix A.7 Emission factors for indoor processes and activities.

Cigarettes (abstracted from Committee on Indoor Pollution, 1981 and Lofroth et al., 1989):

general

	mainstream (inhaled)	sidestream (smoldering)
duration (sec)	20	550
particles (no./cigarette)	1.05×10^{12}	3.5×10^{12}
tobacco burned (mg)	347	441

emissions (mg/cigarette)

	mainstream (inhaled)	sidestream (smoldering)
acrolein	0.084	0.825
CO	18.3	86.3
HCN	0.24	0.16
NH_3	0.16	7.4
NOx	0.014	0.051
particles		
filtered cigarette		
nicotine	0.46	1.27
tar	10.2	34.5
unfiltered cigarette		
nicotine	0.92	1.69
tar	20.8	44.1

Gas ranges and kerosene space heaters (μg/kcal) (abstracted from Meyer, 1983):

compound	gas range (2500 kcal/hr)	space heater (2800 kcal/hr)
CH_20	7.1	
CO	890	632
CO_2	209,000	200,000
NO	31	76
NO_2	85	46
SO_2	0.8	

Formaldehyde (abstracted from Meyer, 1983):

material	emission factor, EF [mg/(m^2 day)]
100% cotton drapery fabric	0.2-0.7
fiberglass ceiling panel	2.8
fiberglass insulation	0.45
foam backed carpet	0.12
hardwood paneling (UF)	1-34
latex-backed fabric	0.19
nylon upholstery fabric	0.018
OF-foam insulation	1-50
paper cups and plates	0.33-0.7
particleboard (std, UF)	2-34
plywood (UF-bonded)	1-34

Appendix A.7 (continued)

Common household aerosols in 1970 (abstracted from Meyer, 1983):

material	emission factor, EF (g/month)
air fresheners	28-56
deodorant spray	112-140
disinfectant sprays	112
dust sprays	28-56
furniture polish	56
hairspray	84-112
oven cleaners	84
shaving foam	84-112

Small stoves (abstracted from Butcher and Ellenbecker, 1982):

material	emission factor, EF (g/kg fuel)
anthracite coal	
carbon monoxide	21
particles	0.5
bituminous coal	
carbon monoxide	116
particles	10.4
wood	
carbon monoxide	100
particles	1.6-6.4

Appendix A.8 Temperature (°C) versus vapor pressure for industrial volatile liquids (abstracted from Perry and Chilton, 1973).

substance	1 mm	5 mm	10 mm	20 mm	40 mm	60 mm	100 mm	200 mm	400 mm	760 mm
acetaldehyde	-81.5	-65.1	-56.8	-47.8	-37.8	-31.4	-22.6	-10.0	4.9	20.2
acetic acid	-17.2	6.3	17.5	29.9	43.0	51.7	63.0	80.0	99.0	118.1
acetic anhydride	1.7	24.8	36.0	48.3	62.1	70.8	82.2	100.0	119.8	139.6
acetone	-59.4	-40.5	-31.1	-20.8	-9.4	-2.0	7.7	22.7	39.5	56.5
acrolein	-64.5	-46.0	-36.7	-26.3	-15.0	-7.5	2.5	17.5	34.5	52.5
allyl alcohol	-20.0	0.2	10.5	21.7	33.4	40.3	50.0	64.5	80.2	96.6
aniline	34.8	57.9	69.4	82.0	96.7	106.0	119.9	140.1	161.9	184.4
benzene	-36.7	-19.6	-11.5	-2.6	7.6	15.4	26.1	42.2	60.6	80.1
carbon disulfide	-73.8	54.3	-44.7	-34.3	-22.5	-15.3	-5.1	10.4	28.0	46.5
carbon tetrachloride	-50.0	-30.0	-19.6	-8.2	4.3	12.3	23.0	38.3	57.8	76.7
chlorobenzene	-13.0	10.6	22.2	35.3	49.7	58.3	70.7	89.4	110.0	132.2
chloroform	-58.0	-39.1	-29.7	-19.0	-7.1	0.5	10.4	25.9	47.7	61.3
cyclohexane	-45.3	-25.4	-15.9	-5.0	6.7	14.7	25.5	42.0	60.8	80.7
dimethylamine	-87.7	-72.2	-64.6	-56.0	-46.7	-40.7	-32.6	-20.4	-7.1	7.4
diphenyl	70.6	101.8	117.0	134.2	152.5	165.2	180.7	204.2	229.2	254.9
ethyl acetate	-43.4	-23.5	-13.5	-3.0	9.1	16.6	27.0	42.0	59.3	77.1
ethyl acrylate	-29.5	-8.7	2.0	13.0	26.0	33.5	44.5	61.5	80.0	99.5
ethyl alcohol	-31.3	-12.0	-2.3	8.0	19.0	26.0	34.9	48.4	63.5	78.4
ethyl benzene	-9.8	13.9	25.9	38.6	52.8	61.8	74.1	92.7	113.8	136.2
ethyl chloride	-89.8	-73.9	-65.8	-56.8	-47.0	-40.6	-32.0	-18.6	-3.9	12.3
ethyl formate	-60.5	-42.2	-33.0	-22.7	-11.5	-4.3	5.4	20.0	37.1	54.3
ethyl mercaptan	-76.7	-59.1	-50.2	-40.7	-29.8	-22.4	-13.0	1.5	17.7	35.0
ethylamine	-82.3	-66.4	-58.3	-48.6	-39.8	-33.4	-25.1	-12.3	2.0	16.6
ethylene dibromide	-27.0	4.7	18.6	32.7	48.0	57.9	70.4	89.8	110.1	131.5
ethylene dichloride	-44.5	-24.0	-13.6	-2.6	10.0	18.1	29.4	45.7	64.0	82.4
formaldehyde	-	-	-88.0	-79.6	-70.6	-65.0	-57.3	-46.0	-33.0	-19.5
formic acid	-20.0	-5.0	2.1	10.3	24.0	32.4	43.8	61.4	80.3	100.6
furfural	18.5	42.6	54.8	67.8	82.1	91.5	103.4	121.8	141.8	161.8
n-heptane	-34.0	-12.7	-2.1	9.5	22.3	30.6	41.8	58.7	78.0	98.4
hexachloroethane	32.7	49.8	73.5	87.6	102.3	112.0	124.2	143.1	163.8	185.6
n-hexane	-53.9	-34.5	-25.0	-14.1	-2.3	5.4	15.8	31.6	49.6	68.7

Appendix A.8 (continued)

substance	1 mm	5 mm	10 mm	20 mm	40 mm	60 mm	100 mm	200 mm	400 mm	760 mm
isobutyl acetate	-21.2	1.4	12.8	25.5	39.2	48.0	59.7	77.6	97.5	118.0
isobutyl alcohol	-9.0	11.6	21.7	32.4	44.1	51.7	61.5	75.9	91.4	108.0
isopropyl acetate	-38.3	-17.4	-7.2	4.2	17.0	25.1	35.7	51.7	69.8	89.0
isopropyl alcohol	-26.1	-7.0	2.4	12.7	23.8	30.5	39.5	53.0	67.8	82.5
methacrylic acid	25.5	48.5	60.0	72.7	86.4	95.3	106.6	123.9	142.5	161.0
methyl acetate	-57.2	-38.6	-29.3	-19.1	-7.9	-0.5	9.4	24.0	40.0	57.8
methyl acrylate	-43.7	-23.6	-13.5	-2.7	9.2	17.3	28.0	43.9	61.0	80.2
methyl alcohol	-44.0	-25.3	-16.2	-6.0	5.0	12.1	21.2	34.8	49.9	64.7
methyl chloride	-	-99.5	-92.4	-94.8	-76.0	-70.4	-63.0	-51.2	-38.0	-24.0
methyl formate	-74.2	-57.0	-48.6	-39.2	-28.7	-21.9	-12.9	0.8	16.0	32.0
methyl iodide	-	-55.0	-45.8	-35.6	-24.2	-16.9	-7.0	8.0	25.3	42.4
methylamine	-95.8	-81.3	-73.8	-65.9	-56.9	-51.3	-43.7	-32.4	-19.7	-6.3
methylene chloride	-70.0	-52.1	-43.3	-33.4	-22.3	-15.7	-6.3	8.0	24.1	40.7
naphthalene	52.6	74.2	85.8	101.7	119.3	130.2	145.5	167.7	193.2	217.9
nitrobenzene	44.4	71.6	84.9	99.3	115.4	125.8	139.9	161.2	185.8	210.6
nitroethane	-21.0	1.5	12.5	24.8	38.0	46.5	57.8	74.8	94.0	114.0
nitromethane	-29.0	-7.9	2.8	14.1	27.5	35.5	46.6	63.5	82.0	101.2
o-nitrotoluene	50.0	79.1	93.8	109.6	126.3	137.6	151.5	173.7	197.7	222.3
octane	-14.0	8.3	19.2	31.5	45.1	53.8	65.7	83.6	104.0	125.6
perchloroethylene	-20.6	2.4	13.8	26.3	40.1	49.2	61.3	79.8	100.0	120.8
phenol	40.1	62.5	73.8	86.0	100.1	108.4	121.4	139.0	160.0	181.9
n-propyl acetate	-26.7	-5.4	5.0	16.0	28.8	37.0	47.8	64.0	82.0	101.8
n-propyl alcohol	-15.0	5.0	14.7	25.3	36.4	43.5	52.8	66.8	82.0	97.8
propylene oxide	-75.0	-57.8	-49.0	-39.3	-28.4	-21.3	-12.0	2.1	17.8	34.5
pyridine	-18.9	2.5	13.2	24.8	38.0	46.8	57.8	75.0	95.6	115.4
styrene, monomer	-7.0	18.0	30.8	44.6	59.8	69.5	82.0	101.3	122.5	145.2
toluene	-26.7	-4.4	6.4	18.4	31.8	40.3	51.9	69.5	89.5	110.6
trichloroethylene	-43.8	-22.8	-12.4	-1.0	11.9	20.0	31.4	48.0	67.0	86.7
1, 2, 3-trichloropropane	9.0	33.7	46.0	59.3	74.0	83.6	96.1	115.6	137.0	158.0
o-xylene	-3.8	20.2	32.1	45.1	59.5	68.8	81.3	100.2	121.7	144.4

Appendix A.9 Henry's law constant and diffusion coefficients of contaminants in air and water for T = 0 to 25 °C (abstracted from Crawford, 1976 except where noted: [R]Reid et al., 1977; [P]Perry and Chilton, 1973; [S]Machay et al., 1981; [V]Vargaftik, 1975; [M]Mackay and Yeun, 1983).

substance	Henry's law constant, H′ $(10^7\ N/m^2)^*$	air diffusion coefficient, $D_{j,air}$ $(10^{-5}\ m^2/s)^*$	water diffusion coefficient, $D_{j,water}$ $(10^{-9}\ m^2/s)^*$
acetic acid		1.06	1.19
acetone		0.83	1.16
acetonitrile			1.26
acetylene	13.5	1.7	2.0
ammonia	0.03	2.2	2.0
aniline		0.75	0.92
benzene	3.05[S]	0.77	1.02
benzoic acid			1.00
benzyl alcohol			0.82
biphenyl	0.03[S]		
bromine	0.747	1.0	1.3
n-butane		0.96[V]	0.89
n-butanol		0.89	0.77
carbon dioxide	16.5	1.5	2.0
carbon monoxide	587.0	2.0	2.0
carbon disulfide		0.89[P]	
carbon tetrachloride	11.1[5]	0.62	0.82
carbonyl sulfide	26.3	1.3	1.5
chloroform	2.66[L]	0.87	0.92
chlorine	6.82	1.2	1.5
chlorobenzene	2.0[5]	0.62	0.86[M]
cyclohexane	18.0[5]	0.86[P]	
dibromochloropropane	0.021	0.69	0.72
diethylamine		0.88	0.97
ethane	281[5]	1.5	1.4
ethyl alcohol		1.02	0.84
ethyl acetate		0.72	1.00
ethylbenzene	4.44[5]	0.66[P]	0.81[R]
ethylene	116.0	1.6	1.5
ethylene dibromide	85.66	0.81	0.89
ethylene dichloride	0.61[C]		
ethyl formate		0.84[P]	
ethylene glycol			1.16
formaldehyde			
formic acid		1.31[P]	0.69[P]
furfural			1.04[R]
glycerol			0.82
glycine			1.06

Appendix A.9 (continued)

substance	Henry's law constant, H′ (10^7 N/m^2)*	air diffusion coefficient, $D_{j,air}$ (10^{-5} m^2/s)*	water diffusion coefficient, $D_{j,water}$ (10^{-9} m^2/s)*
heptane		0.71[R]	
hexane	944[S]	0.8[R]	
hydrogen cyanide	0.064	1.5	1.8
hydrogen sulfide	5.52	1.7	1.6
isobutyl acetate	0.61[C]		
isopropyl alcohol		1.07[V]	0.87[R]
methane	374	2.2	1.8
methyl alcohol		1.33	0.84
methyl acetate		0.84[P]	
methyl chloroform	0.346	0.78	0.81
methyl chloride	13.3[L]	1.3	1.5
methylene chloride	1.67		
methyl formate		0.87[P]	
naphthalene	0.043[S]	0.51[P]	
nitric oxide	291.0	2.0	2.4
nitrous oxide	22.7	1.5	1.8
nitrobenzene		0.86[V]	
octane	1667[S]		
oxalic acid			1.53
ozone	46.4		2.0
perchloroethylene	2.42	0.74	0.76
phosgene		0.80	
phosphine	398[S]	1.6	
propane		0.88	0.97
propylene	57.3		1.1
n-propyl acetate		0.67[P]	
propyl alcohol		0.85[P]	1.1[P]
pyridine			0.58
sulfur dioxide	0.485	1.3	1.7
toluene	3.72[S]	0.71	0.844[M]
trichloroethylene	0.922	0.78	0.81
urethane			1.06
o-xylene	2.78[S]		

* multiply by exponent shown; e.g. for acetylene, $H' = 13.5 \times 10^7$ N/m^2, $D_{j,air} = 1.7 \times 10^{-5}$ m^2/s, $D_{j,water}$ = 2.0×10^{-9} m^2/s

C critical tables; see Crawford (1976) for details

Appendix A.10 Critical temperatures and critical pressures for common toxicants (abstracted from CRC Press, 1975 and ACGIH Ventilation Manual, 1988).

name	formula	T_c (°C)	P_c (atm)
acetaldehyde	C_4H_4O	187.8	54.7
acetic acid	$C_2H_4O_2$	321.6	57.1
acetone	C_3H_6O	235.5	47
acetonitrile	C_2H_3N	274.7	47.7
aniline	C_6H_7N	425.6	52.3
benzene	C_6H_6	288.9	48.6
benzyl chloride	C_6H_5Cl	359.2	44.6
boron trifluoride	BF_3	-12.3	49.2
carbon disulfide	CS_2	279	78
carbon tetrachloride	CCl_4	283.4	45.6
diethylamine	$C_4H_{11}N$	223.3	36.6
dimethylamine	C_2H_7N	164.6	52.4
ethylene oxide	C_2H_5O	195.8	71
hydrogen chloride	HCl	51.4	82.1
hydrogen cyanide	HCN	183.5	48.9
hydrogen sulfide	H_2S	100.4	88.9
methyl alcohol	CH_40	240	78.5
methylamine	CH_5N	156.9	40.2
methylene chloride	CH_2Cl_2	237	60
methyl mercaptan	CH_4S	196.8	71.4
naphthalene	$C_{10}H_8$	474.8	40.6
nitric oxide	NO	-93	64
ozone	O_3	-5.2	67
phenol	C_6H_6O	421.1	60.5
propylene oxide	C_3H_6O	209	48.6
styrene	C_8H_8	374.4	39.4
triethylamine	$C_6H_{15}N$	258.9	30
toluene	C_7H_8	320.8	41.6

Appendix A.11 Thermophysical properties of air (abstracted from Schmidt et al., 1984).

T (K)	ρ (kg/m^3)	C_p (kJ/kg K)	μ (N s/m^2)	ν (m^2/s)	κ (W/m K)	α (m^2/s)	Pr
200	1.7458	1.007	132.5 x 10^{-7}	7.590 x 10^{-6}	18.1 x 10^{-3}	10.3 x 10^{-6}	0.737
250	1.3947	1.006	159.6 x 10^{-7}	11.44 x 10^{-6}	22.3 x 10^{-3}	15.9 x 10^{-6}	0.720
300	1.1614	1.007	184.6 x 10^{-7}	15.89 x 10^{-6}	26.3 x 10^{-3}	22.5 x 10^{-6}	0.707
350	0.9950	1.009	208.2 x 10^{-7}	20.92 x 10^{-6}	30.0 x 10^{-3}	29.9 x 10^{-6}	0.700
400	0.8711	1.014	230.1 x 10^{-7}	26.41 x 10^{-6}	33.8 x 10^{-3}	38.3 x 10^{-6}	0.690
450	0.7740	1.021	250.7 x 10^{-7}	32.39 x 10^{-6}	37.3 x 10^{-3}	47.2 x 10^{-6}	0.686
500	0.6964	1.030	270.1 x 10^{-7}	38.79 x 10^{-6}	40.7 x 10^{-3}	56.7 x 10^{-6}	0.684
550	0.6329	1.040	288.4 x 10^{-7}	45.57 x 10^{-6}	43.9 x 10^{-3}	66.7 x 10^{-6}	0.683
600	0.5804	1.051	305.8 x 10^{-7}	52.69 x 10^{-6}	46.9 x 10^{-3}	76.9 x 10^{-6}	0.685
650	0.5356	1.063	322.5 x 10^{-7}	60.21 x 10^{-6}	49.7 x 10^{-3}	87.3 x 10^{-6}	0.690
700	0.4975	1.075	338.8 x 10^{-7}	68.10 x 10^{-6}	52.4 x 10^{-3}	98.0 x 10^{-6}	0.695
750	0.4643	1.087	354.6 x 10^{-7}	76.37 x 10^{-6}	54.9 x 10^{-3}	109.0 x 10^{-6}	0.702
800	0.4354	1.099	369.8 x 10^{-7}	84.93 x 10^{-6}	57.3 x 10^{-3}	120.0 x 10^{-6}	0.709
850	0.4097	1.110	384.3 x 10^{-7}	93.80 x 10^{-6}	59.6 x 10^{-3}	131.0 x 10^{-6}	0.716
900	0.3868	1.121	398.1 x 10^{-7}	102.9 x 10^{-6}	62.0 x 10^{-3}	143.0 x 10^{-6}	0.720
950	0.3666	1.131	411.3 x 10^{-7}	112.2 x 10^{-6}	64.3 x 10^{-3}	155.0 x 10^{-6}	0.723
1000	0.3482	1.141	424.4 x 10^{-7}	121.9 x 10^{-6}	66.7 x 10^{-3}	168.0 x 10^{-6}	0.726

Formulas for interpolation; T in absolute temperature (K), and σ = standard deviation:

$\rho = \frac{348.59}{T} \quad (\sigma = 9\times10^{-4})$; $f(T) = A + BT + CT^2 + DT^3$, where coefficients are given below:

f(T)	A	B	C	D	σ
c_p	1.0507	-3.645 x 10^{-4}	8.388 x 10^{-7}	-3.848 x 10^{-10}	4 x 10^{-4}
μ x 10^7	13.554	0.6738	-3.808 x 10^{-4}	1.183 x 10^{-7}	0.4192
κ x 10^3	-2.450	0.1130	-6.287 x 10^{-5}	1.891 x 10^{-8}	0.1198
α x 10^6	-11.064	7.04 x 10^{-2}	1.528 x 10^{-4}	-4.476 x 10^{-8}	0.4417
Pr	0.8650	-8.488 x 10^{-4}	-1.234 x 10^{-6}	-5.232 x 10^{-10}	1.623 x 10^{-3}

Sutherland's law for air viscosity:

$$\mu = \mu_0 \left(\frac{T}{T_0}\right)^{1.5} \left(\frac{T_0 + S}{T + S}\right)$$

where

- μ_0 = 1.71 x 10^{-5} kg/(m s)
- T_0 = 273.15 K
- S = 110.4 K
- T must be in units of K.

Appendix A.12 Runge-Kutta method for solving ordinary differential equations.

Consider a first-order ordinary differential equation (ODE) for y as a function of t,

$$\boxed{\frac{dy}{dt} = B - Ay} \qquad \text{(A-1)}$$

Assume that the starting or initial condition $y(t_{start})$ at some time $t = t_{start}$ is known (t_{start} is often but not necessarily zero). If coefficients A and B are constants, Eq. (A-1) can be solved analytically, as discussed in Chapter 1. However, if A and/or B are functions of t and/or of y, an analytical solution may be difficult, if not impossible to find. In such cases, the ***Runge-Kutta marching technique*** is useful for obtaining an *approximate* numerical solution of Eq. (A-1). Subroutines to perform Runge-Kutta marching are built into modern mathematical programs such as Mathcad; nevertheless, readers should be familiar with how the method works. While leaving out much of the details, this Appendix provides enough information about the algorithm so that readers can write a computer program to perform Runge-Kutta marching. Readers are encouraged to learn more about this technique by studying Press et al. (1986) or other books on numerical methods.

The most simple-minded algorithm for solving Eq. (A-1) numerically is the explicit Euler method, where one marches in time steps of duration Δt, incrementing y based on its slope with respect to time. Letting subscript n denote the time step at hand, one can march to the next time step, n+1, as follows:

$$\boxed{y_{n+1} = y_n + \Delta t\left(\frac{dy}{dt}\right)_n} \qquad \text{(A-2)}$$

where the slope in Eq. (A-2) is evaluated from the right hand side of Eq. (A-1) at time step n,

$$\boxed{\left(\frac{dy}{dt}\right)_n = B_n - A_n y_n} \qquad \text{(A-3)}$$

Once the value of y_{n+1} is known, Eq. (A-2) is re-solved at the new time step; this process is continued from the starting time, t_{start}, till some ending time, t_{end}. While this technique is simple to program on a computer, it is inherently unstable, only first-order accurate, and requires very small Δt in order to achieve reasonable results. The main problem here is that the slope is evaluated only at time step n, and is assumed to be constant throughout the time interval Δt. In reality, as one marches towards time step n+1, the slope does *not* remain constant. Mathematicians and engineers have developed clever algorithms to modify the slope such that information is used from one or more values of t between t_n and t_{n+1}. These algorithms include the implicit Euler method and various kinds of predictor-corrector techniques, which can be formulated to first-, second-, or higher-order accuracy.

The Runge-Kutta technique is fourth-order accurate, and can be thought of as a kind of predictor-corrector technique in that the final value of y_{n+1} at $t = t_{n+1}$ is calculated as

$$\boxed{y_{n+1} = y_n + \Delta y_{final}} \qquad \text{(A-4)}$$

where increment Δy_{final} is a ***weighted average*** of four "***trial increments***," namely Δy_1, Δy_2, Δy_3, and Δy_4, evaluated from slopes calculated at $t = t_n$, $t_{n+1/2}$, $t_{n+1/2}$, and t_{n+1}, respectively, as indicated in Figure A.12.1,

$$\boxed{\Delta y_{final} = \frac{\Delta y_1}{6} + \frac{\Delta y_2}{3} + \frac{\Delta y_3}{3} + \frac{\Delta y_4}{6}} \qquad \text{(A-5)}$$

The first trial increment, Δy_1, is evaluated as in the explicit Euler method, using the slope evaluated at point 1, at $t = t_n$,

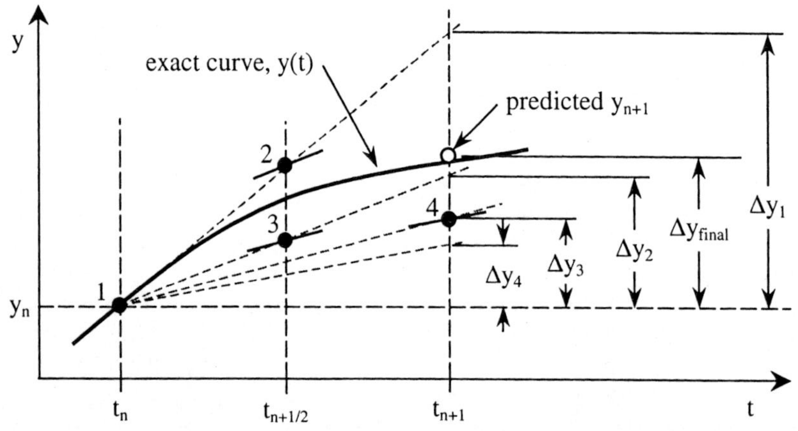

n+1

Figure A.12.1 Illustration of the four points where trial increments Δy_1, Δy_2, Δy_3, and Δy_4 are calculated for the fourth-order Runge-Kutta marching technique; point 1 is the original point at $t = t_n$, and points 2, 3, and 4 are trial points.

$$\boxed{\Delta y_1 = \Delta t\left(\frac{dy}{dt}\right)_{t=t_n, y=y_n}} \tag{A-6}$$

This slope is used to predict a trial midpoint (point 2) at $t = t_{n+1/2}$ (halfway to the next time step),

$$y_2 = y_n + \frac{\Delta y_1}{2}$$

The second slope is calculated at point 2, which is used to predict the second trial increment (Δy_3) and the second trial midpoint (point 3),

$$\boxed{\Delta y_2 = \Delta t\left(\frac{dy}{dt}\right)_{t=t_n+\frac{\Delta t}{2}, y=y_2}} \tag{A-7}$$

$$y_3 = y_n + \frac{\Delta y_2}{2}$$

The third slope is calculated at point 3, which is used to predict the third trial increment (Δy_4) and the third trial point (point 4), this time at $t = t_{n+1}$:

$$\boxed{\Delta y_3 = \Delta t\left(\frac{dy}{dt}\right)_{t=t_n+\frac{\Delta t}{2}, y=y_3}} \tag{A-8}$$

$$y_4 = y_n + \Delta y_3$$

The fourth slope is calculated at trial point 4, as predicted by the third slope.

$$\boxed{\Delta y_4 = \Delta t\left(\frac{dy}{dt}\right)_{t=t_n+\Delta t, y=y_4}} \tag{A-9}$$

Finally, a weighted average of all four slopes is used to predict y_{n+1}, using Eqs. (A-4) and (A-5). As seen in Eq. (A-5), increments Δy_2 and Δy_3 have twice the importance (weight) as the other two.

The technique can be applied to more than one differential equation simultaneously. In addition, an nth-order ODE can be solved by the Runge-Kutta method by splitting it into n first-order ODEs.

Appendix A.13 Newton-Raphson method for solving simultaneous equations.

The ***Newton-Raphson*** (N-R) method is a numerical technique useful for solving implicit equations. For illustrative purposes, suppose ϕ and ψ are implicit functions of x and y,

$$\phi = -\exp(\phi + x)\cos(\psi + y) \qquad \psi = -\exp(\phi + x)\sin(\psi + y) \tag{A-10}$$

Equations (A-10) are implicit in that they cannot be solved analytically for ϕ and ψ at specified values of x and y. In other words, one cannot write explicit equations for $\phi(x_1,y_1)$ or $\psi(x_1,y_1)$; fortunately, values for $\phi(x_1,y_1)$ and $\psi(x_1,y_1)$ can be obtained numerically by the N-R method. The first step is to re-write Eqs. (A-10) by defining two functions, F and G, which are identically zero,

$$F = \phi + \exp(\phi + x)\cos(\psi + y) \qquad G = \psi + \exp(\phi + x)\sin(\psi + y) \tag{A-11}$$

These functions are then differentiated with respect to ϕ and ψ,

$$\begin{aligned} \frac{\partial F}{\partial \phi} &= 1 + \exp(\phi + x)\cos(\psi + y) & \frac{\partial G}{\partial \phi} &= \exp(\phi + x)\sin(\psi + y) \\ \frac{\partial F}{\partial \psi} &= -\exp(\phi + x)\sin(\psi + y) & \frac{\partial G}{\partial \psi} &= 1 + \exp(\phi + x)\cos(\psi + y) \end{aligned} \tag{A-12}$$

Note that

$$\frac{\partial F}{\partial \phi} = \frac{\partial G}{\partial \psi} \qquad \frac{\partial F}{\partial \psi} = -\frac{\partial G}{\partial \phi} \tag{A-13}$$

Also, since F and G are continuous functions of ϕ and ψ, the chain rule can be applied,

$$dF = \frac{\partial F}{\partial \phi}d\phi + \frac{\partial F}{\partial \psi}d\psi \qquad dG = \frac{\partial G}{\partial \phi}d\phi + \frac{\partial G}{\partial \psi}d\psi \tag{A-14}$$

Eqs. (A-14) can be written in first-order finite difference form as

$$F_{i+1} - F_i = \frac{\partial F}{\partial \phi}(\phi_{i+1} - \phi_i) + \frac{\partial F}{\partial \psi}(\psi_{i+1} - \psi_i) \qquad G_{i+1} - G_i = \frac{\partial G}{\partial \phi}(\phi_{i+1} - \phi_i) + \frac{\partial G}{\partial \psi}(\psi_{i+1} - \psi_i) \tag{A-15}$$

The first step in the N-R method is to guess values of $\phi(x_1,y_1)$ and $\psi(x_1,y_1)$; the corresponding (incorrect) values of F and G are denoted as F_i and G_i. In Eqs. (A-15), F_{i+1} and G_{i+1} are defined as the *correct* values of $\phi(x_1,y_1)$ and $\psi(x_1,y_1)$; F_{i+1} and G_{i+1} are zero by definition. Equations (A-15) can be solved simultaneously for the quantities $(\phi_{i+1}-\phi_i)$ and $(\psi_{i+1}-\psi_i)$,

$$(\phi_{i+1} - \phi_i) = \frac{G_i \dfrac{\partial F}{\partial \psi} - F_i \dfrac{\partial G}{\partial \psi}}{J} \qquad (\psi_{i+1} - \psi_i) = \frac{F_i \dfrac{\partial G}{\partial \phi} - G_i \dfrac{\partial F}{\partial \phi}}{J} \tag{A-16}$$

where J is the ***Jacobian***,

$$J = \frac{\partial F}{\partial \phi}\frac{\partial G}{\partial \psi} - \frac{\partial F}{\partial \psi}\frac{\partial G}{\partial \phi} \tag{A-17}$$

The Newton-Raphson algorithm is summarized by the following steps:

1. Guess values of $\phi(x_1,y_1)$ and $\psi(x_1,y_1)$; call these ϕ_i and ψ_i respectively.
2. Calculate F and G from Eq. (A-11); call these F_i and G_i.
3. Calculate the partial derivatives of F and G from Eq. (A-12).
4. Calculate $(\phi_{i+1}-\phi_i)$ and $(\psi_{i+1}-\psi_i)$ using Eq. (A-16); solve for ϕ_{i+1} and ψ_{i+1}.
5. Repeat Steps 2 through 4 using ϕ_{i+1} and ψ_{i+1} in place of ϕ_i and ψ_i until convergence.

Appendix A.14 Newton's method for solving a single implicit equation.

At particular values of x and y, namely x_1 and y_1, suppose one needs to calculate the value ϕ which satisfies an implicit analytical function of ϕ, x, and y. Newton's method is an uncomplicated iterative way to find ϕ numerically. For illustrative purposes, consider the following function:

$$1 = \left[\frac{\frac{x_1}{c}}{\cosh\frac{\phi}{k}}\right]^2 + \left[\frac{\frac{y_1}{c}}{\sinh\frac{\phi}{k}}\right]^2 \tag{A-18}$$

where k is a constant. Newton's method begins by defining a function $G(\phi,x,y)$ which is identically zero at the correct value of ϕ. From Eq. (A-18),

$$G(\phi, x_1, y_1) = \left[\frac{\frac{x_1}{c}}{\cosh\frac{\phi}{k}}\right]^2 + \left[\frac{\frac{y_1}{c}}{\sinh\frac{\phi}{k}}\right]^2 - 1 \tag{A-19}$$

The partial derivative of G with respect to ϕ can also be calculated at (ϕ,x_1,y_1). Here,

$$\frac{\partial G}{\partial \phi}(\phi, x_1, y_1) = -\frac{\frac{2}{k}\left(\frac{x_1}{c}\right)^2 \sinh\frac{\phi}{k}}{\left(\cosh\frac{\phi}{k}\right)^3} - \frac{\frac{2}{k}\left(\frac{y_1}{c}\right)^2 \cosh\frac{\phi}{k}}{\left(\sinh\frac{\phi}{k}\right)^3} \tag{A-20}$$

The algorithm proceeds according to the following steps:

1. Guess a value of ϕ at known values of x_1 and y_1; call this ϕ_i.
2. Calculate G at (ϕ_i,x_1,y_1) from Eq. (A-19); call this G_i.
3. Calculate the partial derivative of G from Eq. (A-20); call this $\left(\frac{\partial G}{\partial \phi}\right)_i$
4. Calculate a new value of ϕ (call it ϕ_{i+1}) at (ϕ_i,x_1,y_1) from a first-order approximation,

$$\phi_{i+1} = \phi_i - \frac{G_i}{\left(\frac{\partial G}{\partial \phi}\right)_i} \tag{A-21}$$

5. Repeat Steps 2 through 4 using ϕ_{i+1} in place of ϕ_i until convergence.

Convergence is defined, of course, when G_i approaches zero to within some desired tolerance.

Appendix A.15 Iteration methods in the BASIC programming language.

Bisection Method

```
1   Rem: Basic iteration program to find the remaining
    root of f (x, y)= c if either x or y is known
2   Rem: Assume f (x, y) is a monotonically increasing
    function in y. If f (x, y) is a monotonically
    decreasing function in y, reverse the instructions
    in line 60 to read..... yl = yg and in line 70 to
    read, . . . . yh = yg
3   Rem: input known values of c and xl 10 yl = 0.001
11  yh = 100
15  Rem:  The  function  f(x,y)  must  be  studied  to
    determine whether it is monotonically increasing or
    decreasing and how to choose the appropriate values
    of yh and yl
20  yg = (yl+yh)/2
30  fg = f(xl,yg)
40  r = fg/c
50  if abs(r-1) <= 0.001 then goto 90
60  if abs(r) > 1 then yh = yg
70  if abs(r) < 1 then yl = yg
80  goto 20
90  print xl,yg
```

Newton's Method

```
1   Rem: Assume y = f(x) is a known differentiable
    function and you want to find the value of x at
    which Y = C
2   Rem: Expand f(x) as a Taylor Series about x0, i.e.
    x0 = xguess. Keep the first two terms and solve for
    x x = x0+(C-f(x0))/f'(x0)
3   Rem: Define error (ER) as x-x0. Thus ER = (C-
    f(x0))/ f' (x0)
30  Rem: Input C, xguess and the functions f (x), f'
    (x)
40  x = xguess
45  C = C
50  A = f (x)
55  B = f' (x)
60  ER = (C-B) /A
65  if abs (ER) <= 0.001 then goto 80
70  x = x+ER
75  goto 50
80  Print x
```

Appendix A.16 Overview of ASHRAE Standard 62-2001.

Table A.16.1 Outdoor requirements for ventilation (extracted from Table 2 of ASHRAE Standard 62-2001, 2001).

type of room	estimated max. occupancy (person/1000 ft^2)	ventilation requirement (CFM/person)	ventilation requirement (CFM/ft^2)
auditorium	150	15	
auto repair room			1.5
bar, cocktail lounge	100	30	
beauty shop	25	25	
cafeteria, fast food	100	20	
classroom, education	50	15	
clothier, furniture			0.3
commercial dry cleaner	30	30	
commercial laundry	10	25	
conference room	50	20	
corridor			0.05
duplicating, printing			0.5
elevator			1.0
gymnasium playing floor	30	20	
hospital, operating room	20	30	
hospital, patient room	10	25	
laboratory, education	30	20	
meat processing	10	15	
office space	7	20	
photography darkroom	10		0.5
prison cell	20	20	
shipping and receiving	10		1.5
smoking lounge	70	60	
spectator area	150	15	
supermarket	8	15	
transportation vehicle	150	15	
warehouse	5		0.5

Appendix A.16 (continued)
Overview of Draft Addendum n to ASHRAE Standard 62-2001.

The required outdoor airflow rate for a *zone* is the sum of people and area components,

$$\boxed{v_z = R_P\,P_z + R_B\,A_z \text{ (CFM/person)}}$$

where

- R_P = minimum people ventilation requirement (CFM/person)
- P_z = design number of people in a zone
- R_B = minimum area ventilation requirement (CFM/ft^2)
- A_z = zone floor area (ft^2)

The minimum outdoor air intake flow rate for a *system* serving multiple zones is

$$\boxed{V_O = V_{OU}/E}$$

where

- V_{OU} = uncorrected outdoor air requirement for the air handler,

$$\boxed{V_{OU} = D\,\Sigma_{\text{All Zones}}(R_P\,P_z) + \Sigma_{\text{All Zones}}(R_B\,A_z)}$$

and

- D = occupant diversity factor for the system, $0 < D \le 1$ (dimensionless)
- E = system ventilation efficiency, a function of maximum zone outside air fraction, that is determined by zone air flow rates and air distribution system characteristics

Table A.16.2 Minimum ventilation rates in breathing zone; values assume no smoking and an air density of 0.075 lbm dry air per cubic foot (extracted from Table 6.1 of Draft Addendum n to ASHRAE Standard 62-2001).

			default values	
occupancy category	**R_P = people OA rate (SCFM/person)**	**R_A = people OA rate (SCFM/ft^2)**	**occupant density (people/1000 ft^2)**	**combined OA rate (SCFM/person)**
classrooms (age 9 plus)	10	0.12	35	13.49
health club	20	0.06	10	26
office building	5	0.06	5	17
retail sales	7.5	0.12	15	15.5

Example: Determine the ventilation requirement for a 1000 ft^2 office for which the actual occupancy has not been determined:

R_P = 5 SCFM/person, P_D = 5 people/1000 ft^2, R_A = 0.06 SCFM/ft^2, A_z = 1000 ft^2.

$P_z = A_z/P_D$ = (1000 ft^2)/(5 people/1000 ft^2) = 5 people

v_z = (5 SCFM/person)(5 people) + (0.06 SCFM/ft^2)(1000 ft^2) = 85. SCFM.

Discussion: Note that, since this example was based on the default occupant density, the ventilation rate is 85./5 = 17. cfm/person, which agrees with the default rate shown in the table.

Appendix A.17 Saturation pressure (vapor pressure) of pure water as a function of temperature (adapted from Perry and Chilton, 1973 and Cengel, 1997).

T (°C)	P_{sat} (kPa)	T (°F)	P_{sat} (psi)
-106.67	3.4122×10^{-7}	-160	4.9490×10^{-8}
-101.11	1.1169×10^{-6}	-150	1.6200×10^{-7}
-95.56	3.3977×10^{-6}	-140	4.9280×10^{-7}
-90.00	9.6733×10^{-6}	-130	1.4030×10^{-6}
-84.44	2.5904×10^{-5}	-120	3.7570×10^{-6}
-78.89	6.5617×10^{-5}	-110	9.5170×10^{-6}
-73.33	1.5796×10^{-4}	-100	2.2910×10^{-5}
-67.78	3.6266×10^{-4}	-90	5.2600×10^{-5}
-62.22	7.9772×10^{-4}	-80	1.1570×10^{-4}
-56.67	1.6844×10^{-3}	-70	2.4430×10^{-4}
-51.11	3.4281×10^{-3}	-60	4.9720×10^{-4}
-45.56	6.7403×10^{-3}	-50	9.7760×10^{-4}
-42.78	9.3355×10^{-3}	-45	1.3540×10^{-3}
-40.00	1.2831×10^{-2}	-40	1.8610×10^{-3}
-37.22	1.7513×10^{-2}	-35	2.5400×10^{-3}
-34.44	2.3718×10^{-2}	-30	3.4400×10^{-3}
-31.67	3.1902×10^{-2}	-25	4.6270×10^{-3}
-28.89	4.2616×10^{-2}	-20	6.1810×10^{-3}
-26.11	5.6564×10^{-2}	-15	8.2040×10^{-3}
-23.33	7.4601×10^{-2}	-10	1.0820×10^{-2}
-20.56	9.7836×10^{-2}	-5	1.4190×10^{-2}
-17.78	1.2748×10^{-1}	0	1.8490×10^{-2}
-15.00	1.6520×10^{-1}	5	2.3960×10^{-2}
-12.22	2.1284×10^{-1}	10	3.0870×10^{-2}
-9.44	2.7282×10^{-1}	15	3.9570×10^{-2}
-8.89	2.8655×10^{-1}	16	4.1560×10^{-2}
-7.78	3.1585×10^{-1}	18	4.5810×10^{-2}
-6.67	3.4784×10^{-1}	20	5.0450×10^{-2}
-5.56	3.8280×10^{-1}	22	5.5520×10^{-2}

T (°C)	P_{sat} (kPa)	T (°F)	P_{sat} (psi)
-4.44	4.2092×10^{-1}	24	6.1050×10^{-2}
-3.33	4.6250×10^{-1}	26	6.7080×10^{-2}
-2.22	5.0780×10^{-1}	28	7.3650×10^{-2}
-1.11	5.5709×10^{-1}	30	8.0800×10^{-2}
-0.56	5.8336×10^{-1}	31	8.4610×10^{-2}
0.00	6.1074×10^{-1}	32	8.8580×10^{-2}
0.01	6.1130×10^{-1}	32.018	8.8662×10^{-2}
5	8.7210×10^{-1}	41	1.2649×10^{-1}
10	1.2276×10^{0}	50	1.7805×10^{-1}
15	1.7051×10^{0}	59	2.4730×10^{-1}
20	2.3390×10^{0}	68	3.3924×10^{-1}
25	3.1690×10^{0}	77	4.5963×10^{-1}
30	4.2460×10^{0}	86	6.1583×10^{-1}
35	5.6280×10^{0}	95	8.1628×10^{-1}
40	7.3840×10^{0}	104	1.0710×10^{0}
45	9.5930×10^{0}	113	1.3914×10^{0}
50	1.2349×10^{1}	122	1.7911×10^{0}
55	1.5758×10^{1}	131	2.2855×10^{0}
60	1.9940×10^{1}	140	2.8921×10^{0}
65	2.5030×10^{1}	149	3.6303×10^{0}
70	3.1190×10^{1}	158	4.5237×10^{0}
75	3.8580×10^{1}	167	5.5956×10^{0}
80	4.7390×10^{1}	176	6.8734×10^{0}
85	5.7830×10^{1}	185	8.3876×10^{0}
90	7.0140×10^{1}	194	1.0173×10^{1}
95	8.4550×10^{1}	203	1.2263×10^{1}
100	1.0133×10^{2}	212	1.4697×10^{1}
105	1.2082×10^{2}	221	1.7524×10^{1}
110	1.4327×10^{2}	230	2.0780×10^{1}

Appendix A.17 (continued)

T (°C)	P_{sat} (kPa)	T (°F)	P_{sat} (psi)
115	1.6906×10^{2}	239	2.4520×10^{1}
120	1.9853×10^{2}	248	2.8794×10^{1}
125	2.3210×10^{2}	257	3.3663×10^{1}
130	2.7010×10^{2}	266	3.9175×10^{1}
135	3.1300×10^{2}	275	4.5397×10^{1}
140	3.6130×10^{2}	284	5.2402×10^{1}
145	4.1540×10^{2}	293	6.0249×10^{1}
150	4.7580×10^{2}	302	6.9009×10^{1}
155	5.4310×10^{2}	311	7.8770×10^{1}
160	6.1780×10^{2}	320	8.9605×10^{1}
165	7.0050×10^{2}	329	1.0160×10^{2}
170	7.9170×10^{2}	338	1.1483×10^{2}
175	8.9200×10^{2}	347	1.2937×10^{2}
180	1.0021×10^{3}	356	1.4534×10^{2}
185	1.1227×10^{3}	365	1.6283×10^{2}
190	1.2544×10^{3}	374	1.8194×10^{2}
195	1.3978×10^{3}	383	2.0273×10^{2}
200	1.5538×10^{3}	392	2.2536×10^{2}
205	1.7230×10^{3}	401	2.4990×10^{2}
210	1.9062×10^{3}	410	2.7647×10^{2}
215	2.1040×10^{3}	419	3.0516×10^{2}
220	2.3180×10^{3}	428	3.3620×10^{2}
225	2.5480×10^{3}	437	3.6956×10^{2}
230	2.7950×10^{3}	446	4.0538×10^{2}

T (°C)	P_{sat} (kPa)	T (°F)	P_{sat} (psi)
235	3.0600×10^{3}	455	4.4382×10^{2}
240	3.3440×10^{3}	464	4.8501×10^{2}
245	3.6480×10^{3}	473	5.2910×10^{2}
250	3.9730×10^{3}	482	5.7624×10^{2}
255	4.3190×10^{3}	491	6.2642×10^{2}
260	4.6880×10^{3}	500	6.7994×10^{2}
265	5.0810×10^{3}	509	7.3694×10^{2}
270	5.4990×10^{3}	518	7.9757×10^{2}
275	5.9420×10^{3}	527	8.6182×10^{2}
280	6.4120×10^{3}	536	9.2999×10^{2}
285	6.9090×10^{3}	545	1.0021×10^{3}
290	7.4360×10^{3}	554	1.0785×10^{3}
295	7.9930×10^{3}	563	1.1593×10^{3}
300	8.5810×10^{3}	572	1.2446×10^{3}
305	9.2020×10^{3}	581	1.3346×10^{3}
310	9.8560×10^{3}	590	1.4295×10^{3}
315	1.0547×10^{4}	599	1.5297×10^{3}
320	1.1274×10^{4}	608	1.6352×10^{3}
330	1.2845×10^{4}	626	1.8630×10^{3}
340	1.4586×10^{4}	644	2.1155×10^{3}
350	1.6513×10^{4}	662	2.3950×10^{3}
360	1.8651×10^{4}	680	2.7051×10^{3}
370	2.1030×10^{4}	698	3.0502×10^{3}
374.14	2.2090×10^{4}	705.452	3.2039×10^{3}

Appendix A.18 Common physical constants and conversions.

density

1 kg/m^3 = 1000 g/cm^3 = 10^6 mg/m^3 = 0.0624 lbm/ft^3 = 0.001941 slug/ft^3

energy

1 joule = 1 kg m^2/s^2
1 BTU = 778.16 ft lbf = 1.055 x 10'° ergs = 252 cal
1 cal = 4.186 joules
1 erg = 1 g cm^2/s^2

force

1 newton = 1 kg m/s^2 = 10^5 dynes
1 dyne = 1 g cm/s^2
1 lbf = 4.448 x 10^5 dynes = 4.448 newtons

length

1 m = 100 cm = 1000 mm = 3.280 ft = 39.37 cm
1 μm = 10^{-6} m = 10^4 Angstroms
1 mile = 5280 ft = 1609.344 m

mass

1 kg = 1000 g = 2.2046 lbm = 6.8521 x 10^{-2} slugs
1 slug = 1 lbf s^2/ft = 32.174 lbm
1 ton = 2000 lbm
1 metric ton = 1000 kg
1 ounce (troy) = 3.110347 x 10^{-2} kg

power

1 watt = 1 joule/s = 1 kg m^2/s^3
1 HP = 550 ft IN /s = 746 watts = 2545 BTU/hr
1 MW = 10^3 kW = 10^6 W

pressure

1 atm = 14.696 lbf/in^2 = 760 torr = 101.325 kPa = 33.92 inches of water column
1 kPa = 1000 Pa = 1000 newton/m^2 = 1 kJ/m^3
1 mm Hg = 1 torr = 0.01934 lbf/in^2
1 bar = 10^6 dynes /cm^2 = 14.504 lbf/in^2

temperature

T(R) = 1.8 T(K)
T(K) = T(C) + 273.15
T(R) = T(F) + 459.67
T(C) = 5[T(F)-32]/9

time

1 hr = 60 min = 3600 s
1 ms = 10^{-3} s
1 μs = 10^{-6} s = 1000 ns

universal gas constant

R_u = 8.314 J/(gmol K) = 8.314 kJ /(kmol K) = 0.082 (L atm)/(gmol K) = 1.987 BTU/(lbmol R) = 1.987 cal/(gmol K) = 1545.33 (ft lbf) / (lbmol R)

volume

1 gal (liquid) = 0.13368 ft^3 = 3.785 liters
1 liter = 10^{-3} m^3 = 1000.028 cm^3
1 barrel (petroleum) = 42 gal
1 fluid ounce = 2.957352 x 10^{-5} m^3

Appendix A.19 International atomic weights (adapted from the IUPAC Commission on Atomic Weights and Isotopic Abundances web site, http://www.chem.qmw.ac.uk/iupac/AtWt/, 2001).

name	symbol	atomic no.	atomic weight
actinium	Ac	89	227
aluminum	Al	13	26.981538
americium	Am	95	243
antimony	Sb	51	121.76
argon	Ar	18	39.948
arsenic	As	33	74.9216
astatine	At	85	210
barium	Ba	56	137.327
berkelium	Bk	97	247
beryllium	Be	4	9.012182
bismuth	Bi	83	208.98038
bohrium	Bh	107	262
boron	B	5	10.811
bromine	Br	35	79.904
cadmium	Cd	48	112.411
cesium	Cs	55	132.90545
calcium	Ca	20	40.078
californium	Cf	98	251
carbon	C	6	12.0107
cerium	Ce	58	140.116
chlorine	Cl	17	35.453
chromium	Cr	24	51.9961
cobalt	Co	27	58.9332
copper	Cu	29	63.546
curium	Cm	96	247
dubnium	Db	105	262
dysprosium	Dy	66	162.5
einsteinium	Es	99	252

name	symbol	atomic no.	atomic weight
erbium	Er	68	167.259
europium	Eu	63	151.964
fermium	Fm	100	257
fluorine	F	9	18.998403
francium	Fr	87	223
gadolinium	Gd	64	157.25
gallium	Ga	31	69.723
germanium	Ge	32	72.64
gold	Au	79	196.96655
hafnium	Hf	72	178.49
hassium	Hs	108	277
helium	He	2	4.002602
holmium	Ho	67	164.93032
hydrogen	H	1	1.00794
indium	In	49	114.818
iodine	I	53	126.90447
iridium	Ir	77	192.217
iron	Fe	26	55.845
krypton	Kr	36	83.8
lanthanum	La	57	138.9055
lawrencium	Lr	103	262
lead	Pb	82	207.2
lithium	Li	3	6.941
lutetium	Lu	71	174.967
magnesium	Mg	12	24.305
manganese	Mn	25	54.938049
meitnerium	Mt	109	268
mendelevium	Md	101	258

Appendix A.19 (continued)

name	symbol	atomic no.	atomic weight
mercury	Hg	80	200.59
molybdenum	Mo	42	95.94
neodymium	Nd	60	144.24
neon	Ne	10	20.1797
neptunium	Np	93	237
nickel	Ni	28	58.6934
niobium	Nb	41	92.90638
nitrogen	N	7	14.0067
nobelium	No	102	259
osmium	Os	76	190.23
oxygen	O	8	15.9994
palladium	Pd	46	106.42
phosphorus	P	15	30.973761
platinum	Pt	78	195.078
plutonium	Pu	94	244
polonium	Po	84	209
potassium	K	19	39.0983
praseodymium	Pr	59	140.90765
promethium	Pm	61	145
protactinium	Pa	91	231.03588
radium	Ra	88	226
radon	Rn	86	222
rhenium	Re	75	186.207
rhodium	Rh	45	102.9055
rubidium	Rb	37	85.4678
ruthenium	Ru	44	101.07
rutherfordium	Rf	104	261
samarium	Sm	62	150.36
scandium	Sc	21	44.95591

name	symbol	atomic no.	atomic weight
seaborgium	Sg	106	266
selenium	Se	34	78.96
silicon	Si	14	28.0855
silver	Ag	47	107.8682
sodium	Na	11	22.98977
strontium	Sr	38	87.62
sulfur	S	16	32.065
tantalum	Ta	73	180.9479
technetium	Tc	43	98
tellurium	Te	52	127.6
terbium	Tb	65	158.92534
thallium	Tl	81	204.3833
thorium	Th	90	232.0381
thulium	Tm	69	168.93421
tin	Sn	50	118.71
titanium	Ti	22	47.867
tungsten	W	74	183.84
ununbium	Uub	112	285
ununnilium	Uun	110	281
ununquadium	Uuq	114	289
unununium	Uuu	111	272
uranium	U	92	238.02891
vanadium	V	23	50.9415
xenon	Xe	54	131.293
ytterbium	Yb	70	173.04
yttrium	Y	39	88.90585
zinc	Zn	30	65.39
zirconium	Zr	40	91.224

Appendix A.20 Properties of common industrial materials (odor recognition thresholds to two significant digits from data in Ruth, 1986; data from Hellman and Small, 1974 used where unavailable in Ruth; OSHA PEL data from US NIOSH, 1997).

chemical name (synonyms, common names)	M (g/mol)	CAS	odor threshold (PPM) low value	odor threshold (PPM) high value	odor description	OSHA PEL (PPM)
acetaldehyde (ethanal, ethyl aldehyde)	44.1	75-07-0	0.00011	2.3	green, sweet, fruity	200
acetic acid (ethanoic acid, vinegar)	60.1	64-19-7	1.0	100	sour, vinegar-like	10
acetic anhydride (acetic acid anhydride)	102.1	108-24-1	0.13	0.35	pungent, vinegar-like	5
acetone (dimethyl ketone, 2-propanone)	58.1	67-64-1	20.	680	minty chemical, sweet	1000
acetonitrile (cyanomethane, methyl cyanide)	41.1	75-05-8	42.	42.	ether-like	40
acrolein (acrylic aldehyde, propenal)	56.1	107-02-8	0.023	16.	burnt, sweet	0.1
acrylic acid (acroleic acid, ethylenecarboxylic acid)	72.1	79-10-7	0.096	1.1	rancid, sweet	2
acrylonitrile (cyanoethylene, VCN, vinyl cyanide)	53.1	107-13-1	3.7	36.	onion-garlic, pungent	2
aldrin (HHDN, octalene)	364.9	309-00-2	0.017	0.027	mild chemical odor	0.25[d]
allyl alcohol (propenol, vinyl carbinol)	58.1	107-18-6	0.82	2.1	mustard, pungent	2
allyl chloride (3-chloropropene)	76.5	107-05-1	0.45	24.	green, garlic, oniony	1
allyl glycidyl ether (AGE, glycidyl allyl ether)	114.2	106-92-3	9.4	9.4	sweet	10
ammonia (aqueous ammonia)	17.0	7664-41-7	0.038	57.	pungent, irritating	50
n-amyl acetate (amyl acetic ester)	130.2	628-63-7	0.0050	7.0	fruity, banana, pear	100
aniline (aminobenzene, aniline oil)	93.1	62-53-3	0.000053	92.	aromatic	5[d]
arsine (arsenic hydride, hydrogen arsenide	78.0	7784-42-1	0.26	0.63	garlic-like	0.05
benzene (benzol, phenyl hydride)	78.1	71-43-2	1.4	85.	sweet, solventy	1
benzyl chloride (chloromethylbenzene)	126.6	100-44-7	0.045	0.30	solventy	1
boron trifluoride (boron fluoride, trifluoroborane)	67.8	7637-07-2	1.6	1.6	pugnent, suffocating	1[c]
bromine (molecular bromine)	158.8	7726-95-6	0.051	3.8	bleachy, penetrating	0.1
n-butyl acetate (butyl acetate, butyl ethanoate)	116.2	123-86-4	7.0	20.	fruity	150
n-butyl alcohol (n-butanol, butyl alcohol)	74.1	71-36-3	0.12	50.	sweet	100
sec-butyl alcohol (2-butanol, butylene hydrate)	74.1	78-92-2	43.	43.	strong, pleasant	150
tert-butyl alcohol (trimethyl carbinol)	74.1	75-65-0	72.	72.	camphor-like	100
n-butylamine (1-aminobutane, butylamine)	73.2	109-73-9	0.080	2.0	fishy, ammonia-like	5[c,d]
n-butyl mercaptan (butanethiol)	90.2	109-79-5	0.00043	0.00090	repulsive	10
carbon disulfide (carbon bisulfide)	76.1	75-15-0	0.0078	7.4	sweet, disagreeable	20
carbon monoxide (carbon oxide, flue gas)	28.0	630-08-0			odorless	50
carbon tetrachloride (Freon® 10, Halon® 104)	153.8	56-23-5	9.6	240	sweet, pungent	10

Appendix A.20 (continued)

chemical name (synonyms, common names)	M (g/mol)	CAS	odor threshold (PPM) low value	odor threshold (PPM) high value	odor description	OSHA PEL (PPM)
chlorine (molecular chlorine)	70.9	7782-50-5	0.010	5.2	bleachy, pungent	1[c]
chlorine dioxide (chlorine oxide)	67.5	10049-04-4	0.11	0.11	sharp, pungent	0.1
chloroacetaldehyde (2-chloroethanal)	78.5	107-20-0	0.94	0.94	sharp, irritating	1[c]
chlorobenzene (chlorobenzol, phenyl chloride)	112.6	108-90-7	0.21	61.	sweet, almond-like	75
chlorobromomethane (CB, CBM, Halon® 1011)	129.4	74-97-5	400	400	sweet	200
chloroform (trichloromethane)	119.4	67-66-3	51.	210	pleasant odor	50
chloropicrin (nitrochloroform)	164.4	76-06-2	0.81	1.1	sharp, penetrating	0.1
p-cresol (4-cresol, 4-methyl phenol)	108.2	106-44-5	0.00027	5.0	sweet, creosote tar	5[d]
crotonaldehyde (2-butenal, propylene aldehyde)	70.1	4170-30-3	0.037	1.0	pungent, suffocating	2
cumene (cumol, isopropyl benzene)	120.2	98-82-8	0.0080	1.3	sharp, aromatic	50[d]
cyclohexane (benzene hexahydride)	84.2	110-82-7	0.42	0.42	sweet, aromatic	300
cyclohexanol (anol, cyclohexyl alcohol, hexalin)	100.2	108-93-0	98.	98.	camphor-like	50
cyclohexanone (anone, cyclohexyl ketone)	98.2	108-94-1	0.12	100	sweet, peppermint-like	50
decaborane (decaboron tetradecahydride)	122.2	17702-41-9	0.072	0.072	intense, bitter, chocolate	0.05[d]
diacetone alcohol (diacetone)	116.2	123-42-2	0.28	100	sweet, faintly minty	50
diazomethane (azimethylene, diazirine)	42.1	334-88-3			musty	0.2
diborane (boroethane, boron hydride)	27.7	19287-45-7	1.8	3.5	repulsively sweet	0.1
para-dichlorobenzene (p-DCB, dichlorocide)	147.0	106-46-7	15.	30.	mothballs	75
1,1-dichloroethane (ethylidene chloride)	99.0	75-34-3	110	200	chloroform-like	100
dichloroethyl ether (2,2'-dichloroethyl ether)	143.0	111-44-4	15.	370	chlorinated solvent-like	15[c,d]
1,2-dichloroethylene (acetylene dichloride)	97.0	540-59-0	0.085	500	acrid, ethereal	200
diethylamine (diethamine, N-ethylethanamine)	73.1	109-89-7	0.020	38.	fishy, ammonia-like	25
diisobutyl ketone (DIBK, valerone)	142.3	108-83-8	0.11	0.32	sweet, ester	50
diisopropylamine (DIPA)	101.2	108-18-9	0.13	0.82	fishy, amine	20[d]
dimethylamine (N-methylmethanamine)	45.1	124-40-3	0.021	30.	fishy, ammonia-like	10
dioxane (1,4-dioxane)	88.1	123-91-1	0.0030	170	ether-like	100[d]
diphenyl (biphenyl, phenyl benzene)	154.2	92-52-4	0.0010	0.048	pleasant	0.2
epichlorohydrin (1-chloro-2,3-epoxypropane)	92.5	106-89-8	13.	21.	chloroform-like	5
ethanolamine (ethylolamine, 2-aminoethanol)	61.1	141-43-5	2.1	4.3	ammonia	3
ethyl acetate (acetic ester, acetic ether)	88.1	141-78-6	0.0055	180	fruity, pleasant	400

Appendix A.20 (continued)

chemical name (synonyms, common names)	M (g/mol)	CAS	odor threshold (PPM) low value	odor threshold (PPM) high value	odor description	OSHA PEL (PPM)
ethyl acrylate (ethyl propenoate)	100.1	140-88-5	0.00020	7.8	earthy, acrid, hot plastic	25
ethyl alcohol (alcohol, ethanol, grain alcohol)	46.1	64-17-5	0.18	5,100	ethereal, vinous	1000
ethylamine (aminoethane, monoethylamine)	45.1	75-04-7	0.26	220	sharp, ammonia-like	10
ethyl benzene (ethylbenzol, phenylethane)	106.2	100-41-4	2.0	200	aromatic	100
ethyl bromide (bromoethane, monobromoethane)	109.0	74-96-4	200	200	ether-like	200
ethyl chloride (chloroethane, hydrochloric ether)	64.5	75-00-3			ether-like	1000
ethyl ether (diethyl ether, ethyl oxide, ether)	74.1	60-29-7	0.33	0.99	sweet, ether-like	400
ethyl formate (ethyl methanoate)	74.1	109-94-4			fruity, irritating	100
ethyl mercaptan (ethanethiol, mercaptoethane)	62.1	75-08-1	0.000013	0.036	garlic	10[c]
ethyl silicate (tetraethoxysilane, tetraethyl silicate)	208.3	78-10-4	85.	85.	alcohol-like, sharp	100
ethylene dibromide (ethylene bromide)	187.9	106-93-4	10.	10.	mild, sweet	20
ethylene dichloride (1,2-dichloroethane)	99.0	107-06-2	5.9	110	sweet	50
ethylene glycol (antifreeze, glycol, glycol alcohol)	62.1	107-21-1	25.	25.	sweet	none
ethylene oxide (dimethylene oxide, oxirane)	44.1	75-08-1	290	780	sweet, olefinic	1
fluorine (fluorine-19)	38.0	7782-41-4	3.9	3.9	pungent, irritating	0.1
formaldehyde (methanal, methyl aldehyde)	30.0	50-00-0	1.2	60.	hay-like, pungent	0.75
formic acid (hydrogen carboxylic acid)	46.0	64-18-6	0.024	20.	pungent, penetrating	5
furfural (fural, furfuraldehyde, 2-furfuraldehyde)	96.1	98-01-1	0.0061	5.1	almonds	5[d]
heptane (n-heptane, normal heptane)	100.2	142-82-5	49.	310	gasoline-like	500
hydrazine (diamine, hydrazine base)	32.1	302-01-2	2.3	3.1	fishy, ammonia-like	1[d]
hydrogen bromide (hydrobromic acid)	80.9	10035-10-6	2.0	2.0	sharp irritating	3
hydrogen chloride (hydrochloric acid)	36.5	7647-01-0	4.7	33.	irritating, pungent	5[c]
hydrogen cyanide (hydrocyanic acid)	27.0	74-90-8	0.82	4.5	bitter, almond-like	10[d]
hydrogen fluoride (hydrofluoric acid)	20.0	7664-39-3	0.041	0.16	strong, irritating	3
hydrogen peroxide (hydroperoxide, peroxide)	34.0	7722-84-1			slightly sharp	1
hydrogen selenide (selenium hydride)	81.0	7783-07-5	0.00048	3.6	decayed horseradish	0.05
hydrogen sulfide (hydrosulfuric acid, sewer gas)	34.1	7783-06-4	0.00050	0.010	rotten eggs	20
iodine (iodine crystals, molecular iodine)	253.8	7553-56-2	0.87	0.87	sharp	0.1[c]
isoamyl alcohol (amyl alcohol, isopentyl alcohol)	88.2	123-51-3	0.12	7.0	alcohol	100
isobutyl acetate (2-methylpropyl acetate)	116.2	110-19-0	0.35	0.50	fruity, floral	150

Appendix A.20 (continued)

chemical name (synonyms, common names)	M (g/mol)	CAS	odor threshold (PPM) low value	odor threshold (PPM) high value	odor description	OSHA PEL (PPM)
isobutyl alcohol (isobutanol, 2-methyl-1-propanol)	74.1	78-83-1	0.00089	0.043	sweet, musty	100
isopropyl acetate (isobutyl ester of acetic acid)	102.2	108-21-4	0.046	360	fruity	250
isopropyl alcohol (rubbing alcohol, isopropanol)	60.1	67-63-0	3.2	200	pleasant	400
isopropylamine (2-propylamine)	59.1	75-31-0	0.21	200	pungent, ammonia-like	5
isopropyl ether (diisopropyl ether)	102.2	108-21-3	0.017	300	sweet, sharp, ether	500
kerosene (fuel oil No. 1, coal oil)	170.	8008-20-6	0.080	0.080	fuel smell	
lead (plumbum)	207.2	7439-92-1			oderless particles	0.0059
maleic anhydride (toxilic anhydride)	98.1	108-31-6	0.46	0.49	acrid	0.25
mercury vapor (colloidal mercury, quicksilver)	200.6	7439-97-6			oderless	0.01[c]
methane (methyl hydride)	16.0	74-82-8			oderless	
methyl acetate (methyl ethanoate)	74.1	72-20-9	200	300	fragrant, fruity	200
methyl acrylate (methyl propenoate)	86.1	96-33-3	20.	20.	sharp, sweet, fruity	10[d]
methyl alcohol (methanol, wood alcohol)	32.1	67-56-1	10.	20,000	sweet	200
methyl amyl alcohol (methyl isobutyl carbinol)	102.2	108-11-2	0.33	0.52	sweet, alcohol	25
methyl bromide (bromomethane)	95.0	74-83-9	21.	1,000	sweetish	5[c,d]
methyl Cellosolve® (EGME, 2-methoxyethanol)	76.1	109-86-4	0.093	93.	mild, non-residual	25
methyl chloride (chloromethane)	50.5	74-87-3	10.	10.	sweet, etheral	100
methyl chloroform (1,1,1-trichloroethane)	133.4	71-55-6	100	700	chloroform-like	350
methyl ethyl ketone (MEK, 2-butanone)	72.1	78-93-3	0.25	50.	sweet, acetone-like	200
methyl formate (methyl methanoate)	60.1	107-31-3	200	2,800	pleasant	100
methyl iodide (iodomethane, monoiodomethane)	141.9	74-88-4			pungent, ether-like	5[d]
methyl isocyanate (MIC)	57.1	624-83-9			sharply pungent	0.02[d]
methyl mercaptan (mercaptomethane)	48.1	74-93-1	0.000020	0.042	sulfidy, pungent	10[c]
methyl methacrylate (methacrylate monomer)	100.1	80-62-6	0.050	0.34	arid, fruity, sulfidy	100
alpha-methyl stryene (AMS, isopropenyl benzene)	118.2	98-83-9	0.052	200	sweet, sharp	100[c]
methylamine (aminomethane)	31.1	74-89-5	0.020	9.5	fishy, pungent	10
methylcyclohexane (cyclohexylmethane)	98.2	108-87-2	500	500	faint, benzene-like	500
methylcyclohexanol (hexahydrocresol)	114.2	25639-42-3	500	500	weak, like cocunut oil	100
methylene chloride (dichloromethane)	84.9	75-09-2	160	620	sweet, chloroform-like	25
morpholine (diethylene imidoxide)	87.1	110-91-8	0.0098	0.14	fishy, amine	20

Appendix A.20 (continued)

chemical name (synonyms, common names)	M (g/mol)	CAS	odor threshold (PPM) low value	odor threshold (PPM) high value	odor description	OSHA PEL (PPM)
naphthalene (tar camphor, white tar, mothball)	128.2	91-20-3	0.29	24.	mothball, tar-like	10
nickel carbonyl (nickel tetracarbonyl)	170.7	13463-39-3	0.030	3.0	musty	0.001
nitric acid (hydrogen nitrate, engravers acid)	63.0	7697-37-2	0.29	0.97	acrid, suffocating	2
nitric oxide (nitrogen monoxide)	30.0	10102-43-9	0.29	0.98		25
nitrobenzene (nitrobenzol, oil of mirbane)	123.1	98-95-3	0.0047	1.9	shoe polish, pungent	1[d]
nitroethane (nitroetan)	75.1	79-24-3	200	200	mild, fruity	100
nitrogen dioxide (nitrogen peroxide)	46.0	10102-44-0	1.1	5.3	sweetish, acrid	5[c]
nitromethane (nitrocarbol)	61.0	75-52-5	100	100	mild, fruity	100
1-nitropropane (nitropropane, 1-NP)	89.1	108-03-2	300	300	mild, fruity	25
o-nitrotoluene (o-methylnitrobenzene)	137.1	88-72-2			weak aromatic	5[d]
octane (n-octane, normal-octane)	114.2	111-65-9	160	260	gasoline-like	500
oxygen difluoride (fluorine monoxide)	54.0	7783-41-7	0.091	0.45	foul	0.05
ozone (triatomic oxygen)	48.0	10028-15-6	0.00051	0.52	clover-like, pungent	0.1
pentaborane (pentaboron nonahydride)	63.1	19624-22-7	0.97	0.97	strong, pungent	0.005
pentane (n-pentane, normal-pentane)	72.2	109-66-0	2.2	1,000	gasoline-like	1000
perchloroethylene (PERC, perchlorethylene)	165.8	127-18-4	4.6	69.	chlorinated solvent	100
perchloryl fluoride (chlorine oxyfluoride)	102.5	7616-94-6	11.	11.	sweet	3
phenol (carbolic acid, hydroxybenzene)	94.1	108-95-2	0.047	5.8	medicinal, sweet	5[d]
phenyl ether (diphenyl ether)	170.2	101-84-8	0.0010	0.10	disagreeable	1
phosgene (carbon oxychloride, carbonyl chloride)	98.9	75-44-5	0.50	0.99	musty hay-like, green corn	0.1
phosphine (hydrogen phosphide)	34.0	7803-51-2	0.020	2.6	oniony, mustard, fishy	0.3
phosphoric acid (orthophosphoric acid)	98.0	7664-38-2			oderless	0.25
n-propyl acetate (propylacetate)	102.2	109-60-4	0.050	25.	mild fruity	200
n-propyl alcohol (n-propanol, propyl alcohol)	60.1	71-23-8	0.031	200	sweet, alcohol	200
n-propyl nitrate (propyl ester of nitric acid)	105.1	627-13-4	49.	49.	ether-like	25
propylene dichloride (1,2-dichloropropane)	113.0	78-87-5	0.25	130	sweet	75
propylene oxide (1,2-epoxy propane)	58.1	75-56-9	10.	210	sweet, alcoholic	100
pyridine (azabenzene, azine)	79.1	110-86-1	0.0028	4.6	burnt, sickening	5
quinone (1,4-benzoquinone)	108.1	106-51-4	0.091	0.091	acrid	0.1
refrigerant-12 (dichlorodifluoromethane, Freon®)	120.9	75-71-8			ether-like	1000

Appendix A.20 (continued)

chemical name (synonyms, common names)	M (g/mol)	CAS	odor threshold (PPM) low value	odor threshold (PPM) high value	odor description	OSHA PEL (PPM)
styrene (ethenyl benzene, styrol, vinyl benzene)	104.2	100-42-5	0.048	200	solventy, rubber-like	100
sulfur dioxide (sulfurous acid anhydride)	64.1	7446-09-5	0.45	4.8	irritating, pungent	5
sulfur monochloride (sulfur chloride)	135.0	10025-67-9			nauseating	1
sulfuric acid (battery acid, hydrogen sulfate)	98.1	7664-93-9	0.25	0.25	pungent, irritating	0.25
1,1,2,2-tetrachloroethane (acetylene tetrachloride)	167.9	79-34-5	3.1	5.1	sickly sweet	5
tetrahydrofuran (diethylene oxide, THF)	72.1	109-99-9	2.5	60.	ether-like	200
toluene (methyl benzene, methyl benzol, toluol)	92.1	108-88-3	2.1	70.	rubbery, mothballs	200
toluene-2,4-diisocyanate (TDI, 2,4-TDI)	174.2	584-84-9	0.45	2.4	sweet, fruity, acrid	0.02[c]
trichloroethylene (TCE, trichloroethene)	131.4	79-01-6	0.21	400	chloroform-like	100
1,2,3-trichloropropane (trichlorohydrin)	147.4	96-18-4			strong, acrid	50
triethylamine (TEA)	101.2	121-44-8	0.087	0.27	fishy, amine	25
vinyl acetate (ethenyl acetate, VAC)	86.1	108-05-4	0.10	0.47	sour, sharp	4 (NIOSH)
vinyl toluene (methylstyrene, tolyethylene)	118.2	25013-15-4	50.	50.	strong, disagreeable	100
o-xylene (1,2-dimethylbenzene, o-xylol)	106.2	95-47-6	0.080	40.	sweet	100
xylidene (aminodimethylbenzene, aminoxylene)	121.2	1300-73-8	0.0049	0.0049	weak, amine-like	5

[c] ceiling value
[d] denotes additional entry through the skin

References

"The Air Pollution Consultant; Guidance on HAP Early Reduction Compliance Extensions", (author unknown), Vol. 1, Issue 2, pp 2.1-2.13, Nov/Dec 1991.

Abt, E, Suh, H H, Catalano, P and Koutrakis, P, "Relative Contribution of Outdoor and Indoor Particle Sources to Indoor Concentrations", Environmental Science and Technology, Vol. 34, No. 17, pp 3579-3587, 2000.

ACGIH (American Conference of Governmental and Industrial Hygienists), "Air Sampling Instruments", 4th ed., Cincinnati, OH, 1972.

ACGIH (American Conference of Governmental and Industrial Hygienists), "Threshold Limit Values for Chemical Substances and Physical Agents in the Workroom Environment with Intended Changes for 1977", PO Box 1937, Cincinnati, Ohio, 45201, 1977.

ACGIH (American Conference of Governmental and Industrial Hygienists), "Industrial Ventilation, A Manual of Recommended Practice", Committee on Industrial Ventilation, PO Box 16153 Lansing MI, 48901, 20th ed. 1988; most recent is 24th ed. 2001.

ACGIH (American Conference of Governmental and Industrial Hygienists), "Threshold Limit Values for Chemical Substances and Biological Exposure Indices from 1989-1990", 6550 Glenway Ave. Bldg D-7, Cincinnati, Ohio 45211-9438, 1990.

AIChE (American Institute of Chemical Engineers), Center for Chemical Process Safety, "Safety, Health, and Loss Prevention in Chemical Processes" Problems for Undergraduate Engineering Curricula and the Instructor's Guide, AIChE, 345 East 47th Street, New York, NY 10017, 1990.

Alden, J L and Kane, J M, "Design of Industrial Ventilation Systems", Industrial Press Inc, 5th ed., NY, 1982.

Alenandesson, R and Hedenstierna, G, "Pulmonary Function in Wood Workers Exposed to Formaldehyde: a Prospective Study", Archives Environmental Health, Vol. 44, B No. 1, pp 5-11, 1989.

American Welding Society, 550 NW, LeJeuene Road, Miami FL, 33126, 1983.

Ames, B N, Magaw, R, and Gold, L S, "Ranking Possible Carcinogenic Hazards", Science, Vol. 236, pp 271-280, 17 April 1987.

Amoore, J E and Hautala, E, "Odor as an Aid to Chemical Safety: Odor Thresholds Compared with Threshold Limit Values and Volatilities for 214 Industrial Chemicals in Air and Water Pollution" Journal of Applied Toxicology, Vol. 3, No. 6, pp 272-283, 1983.

Anand, N K and McFarland, A R, "Particle Deposition in Aerosol Sampling Lines Caused by Turbulent Diffusion and Gravimetric Settling", American Industrial Hygiene Association Journal, Vol. 50, No. 6, pp 307-312, 1989.

Anastas M Y, "General Methods for Computation of Flow into Local Exhaust Hoods", "Ventilation '91", edited by R T Hughes, H D Goodfellow, and G S Rajhans, ACGIH, Cincinnati, OH, p 429, 1993.

Anastas, M Y and Hughes, R T, "Center-line Velocity Models for Flanged Local Exhaust Openings", Applied Industrial Hygiene, Vol. 3, No. 12, pp 342-347, 1988.

ASHRAE (American Society of Heating, Refrigerating and Air Conditioning Engineers, Inc.), "Ventilation for Acceptable Indoor Air Quality, ASHRAE Standard 62-1989, 1791 Tullie Circle, NE, Atlanta GA, 30329, 1989.

ASHRAE (American Society of Heating, Refrigerating and Air Conditioning Engineers, Inc.), "ASHRAE Handbook & Products Directory, 1980 Systems", Chapter 22, Local Exhaust Systems, Atlanta, GA, pp 22.1-22.14, 1984.

ASHRAE (American Society of Heating, Refrigerating and Air Conditioning Engineers, Inc.), "ASHRAE Fundamentals Handbook" ASHRAE, 1791 Tullie Circle, NE, Atlanta, GA, 30329, 2001; editions every four years: 1985, 1989, 1993, 1997...

ASHRAE (American Society of Heating, Refrigerating and Air Conditioning Engineers, Inc.), "ASHRAE HVAC Applications Handbook" ASHRAE, 1791 Tullie Circle, NE, Atlanta, GA, 30329, 1999; editions every four years: 1983, 1987, 1991, 1995...

ASHRAE (American Society of Heating, Refrigerating and Air-Conditioning Engineers), "ASHRAE Handbook & Products Directory, 1999 Systems, Chapter 28, "Ventilation of the Industrial Environment", and Chapter 29, "Industrial Local Exhaust Systems", ASHRAE, 1791 Tullie Circle NE, Atlanta, GA, 1999.

ASHRAE (American Society of Heating, Refrigerating and Air-Conditioning Engineers), "ASHRAE Journal HVAC & R Software Directory", ASHRAE, 1791 Tullie Circle NE, Atlanta, GA 30329, 1990.

Ayer, J and Hyde, C, "VOC Emission Reduction Study at the Hill Air Force Base Building 515 Painting Facility", Report AD-A242-109, NTIS, 1990.

Baker, A J, Kelso, R M, Noronha, W P, and Woods, J B, "On the Maturing of CFD in Design of Room Air Ventilation Systems", Proceedings of Building Systems: Room Air Contaminant Distribution, Ed. L L Christianson, ASHRAE, pp 149-152, 1989.

Balashazy, I and Hofmann, W, "Particle Deposition Onto Indoor Residential Surfaces", Environmental Science and Technology, Vol. 24, No. 6, pp 745-772, 1993.

Balashazy, I, Martonen, T B, and Hofmann, W, "Simultaneous Sedimentation and Impaction of Aerosols in Two-Dimensional Channel Bends", Aerosol Science and Technology, Vol. 13, pp 20-34, 1990.

Bare, J C, "Indoor Air Pollution Source Database", Journal of the Air Pollution Control Association, Vol. 38, No. 5, pp 670-671, 1988.

Bastress, E K, Niedzwecki, J M, and Nugent, A E, "Ventilation Requirements for Grinding, Buffing and Polishing Operations", NTIS, PB-277332, 1974.

Baturin, V V, "Fundamentals of Industrial Ventilation", Pergamon Press, New York, 1972.

Bejam, A, "Heat Transfer", Wiley & Sons, NY, 1993.

Belding, H S and Hatch, T F, "Index for Evaluating Heat Stress in Terms of Resulting and Physiological Strains", Heating, Piping and Air Conditioning, pp 207-239, 1955.

Bendelius, A G and Hettinger, J C, "Environmental Control within the Sydney Harbor Tunnel", "Ventilation '91", edited by R T Hughes, H D Goodfellow, and G S Rajhans, ACGIH, Cincinnati, OH, pp 517-526, 1993.

Bender, M, "Fume Hoods, Open Canopy Type - their Ability to Capture Pollutants in Various Environments", American Hygiene Association Journal, Vol. 40, pp 118-127, 1979.

Benner, C L, Bayona, J M, Caka, F M, Hongmao, T, Lewis, L, Crawford, J, Lamb, J D, Lee, M L, Lewis, E A, Hansen, L D, and Eatough, D J, "Chemical Composition of Environmental Tobacco Smoke; 2. Particulate-Phase Compounds" Journal of Environmental Science and Technology, Vol. 23, No. 6, pp 688-699, 1989.

Billings, C E and Vanderslice, S F, "Methods for Control of Indoor Air Quality", Environment International, Vol. 8, pp 497-504, 1982.

Billings, C E and Wilder, J, "Handbook of Fabric Filter Technology", Vol. I and II, US Government Printing Office, PB200648, PB-200649, 1970.

Bird, R B, Stewart, W E, and Lightfoot, E N, "Transport Phenomena", Wiley, New York, 1960.

Bond, R and Staub, C P, Eds., Prober, R (Coordinating Ed), "Handbook of Environmental Control" Vol. I, Air Pollution, CRC Press, Cleveland OH, 1972.

Bossard, F C, LeFever, J J, LeFever, J B, and Stout, K S, "A Manual of Mine Ventilation Design Practices", Floyd C Bossard and Associates, Inc, Box 3837, Butte, Montana, 59702, 1983.

Boubel, R W, Fox, D L, Turner, B, and Stern, A C, "Fundamentals of Air Pollution", Academic Press, San Diego, CA, 1994.

Boysan, F, Ayers, W H, and Swithenbank, J A, "A Fundamental Mathematical Modeling Approach to Cyclone Design", Trans. Inst. Chemical Engineering, Vol. 60, pp 222-230, 1982.

Braconnier, R, "Bibliographic Review of Velocity Fields in the Vicinity of Local Exhaust Hood Openings", American Industrial Hygiene Association Journal, Vol. 49, No. 4, pp 185-198, 1988; Errata, Vol. 49, pp 475-478, 1988.

Braconnier, R, Regnier, R, and Bonthoux F, "An Experimental and Numerical Study of the Capture of Pollutants Over a Surface-Treating Tank Equipped with a Suction Slot", "Ventilation '91", edited by R T Hughes, H D Goodfellow, and G S Rajhans, ACGIH, Cincinnati, OH, pp 96-105, 1993.

Breum, N O and Orhede E, "Dilution Versus Displacement Ventilation - Environmental Conditions in a Garment Plant", Am. Ind. Hyg. Assoc. J., Vol. 55, No. 2, pp 140-148, 1994.

Breum, N, "Air Exchange Efficiency of Displacement Ventilation in a Printing Plant", Annals of Occupational Hygiene, Vol. 32, No. 4, pp 481-488, 1988.

Brockman, J, "Sampling and Transport of Aerosols", Chapter 6, pp 77-108, "Aerosol Measurement", edited by K. Willeke and P Baron, Van Nostrand Reinhold, New York, 1993.

Brown, S, "The Product Liability Handbook: Prevention, Risk, Consequences and Forensics of Product Failure", New York, Van Nostrand Reinhold, 1991.

Bruno, B and Heinsohn, R J, "Using the Sequential Box Model to Predict Transient Solvent Concentrations Arising from Applying a Surface Coating Inside a Confined Space", Transactions of American Society of Heating, Refrigerating and Air-Conditioning Engineers, Vol. 98, Part 1, pp 73-81, 1992.

Budavari, S (editor), "(The) Merck Index", Merck Research Laboratories, Division of Merck and Co. Inc., Whitehouse Station, NJ, 1996.

Bunicore, A J and Davis, W T, Eds., "Air Pollution Engineering Manual", 3rd ed., Air Waste Management Association, Pittsburgh PA, 1992.

Burgess, W A, Ellenbecker, M J, and Treitman, R T, "Ventilation for Control of the Work Environment", John Wiley and Sons, New York, 1989.

Burton, D J, "Industrial Ventilation - a Self Study Companion to the ACGIH Ventilation Manual", 2nd ed., ACGIH, 6500 Glenway Ave., Bldg D-5, Cincinnati, OH, 45211, 1984.

Burton, D J, "Practical Applications of an Aaberg Hood", Occupational Health and Safety, Vol. 69, No. 11, pp 37-39, Nov 2000.

Busnaina, A A, "Industrial Applications of Computational Fluid Dynamics", "Computers in Engineering", V A Tipnis and E M Patton editors, Vol. III, ASME, 345 East 47th St, NY, NY, 10017, 1988.

Butcher, S S and Ellenbecker, M J, "Particulate Emissions for Small Wood Stoves and Coal Stoves", Journal of the Air Pollution Control Association, Vol. 32, No. 4, pp 380-384, 1982.

Cadle, R D, "Particle Size", Reinhold Press, New York, 1965.

Cain, W S, Leaderer, B P, Isseroff, R, Berglund, L G, Huey, R J, Lipsitt, E D, and Perlman, D, "Ventilation Requirements in Buildings - I. Control of Occupancy Odor and Tobacco Smoke Odor", Atmospheric Environment, Vol. 17, No. 6, pp 1183-1197, 1983.

Calvert, S and Englund H M, Eds., "Handbook of Air Pollution Technology", Wiley-Interscience, New York, 1984.

Calvert, S, "Scrubber Handbook", US Government Printing Office, PB-213016, 1972.

Canon Communications, Inc., Microcontamination, Vol. 5, No. 6, pp 21, June 1987.

Cano-Ruiz, J A, Kong, D, Balas, R B, and Nazaroff, W W, "Removal of Reactive Gases at Indoor Surfaces: Combining Mass Transport and Surface Kinetics", Atmospheric Environment, Vol. 27A, No. 13, pp 2039-2050, 1993.

Carr, R C and Smith, W B, "Fabric Filter Technology for Utility Coal-Fired Power Plants", J. of the Air Pollution Control Assoc., Vol. 31, No. 4, pp 399-412, 1984.

Cena, K and Clark, J A, editors, "Bioengineering, Thermal Physiology, and Comfort", Elsevier Scientific, Amsterdam, 1981.

Cengel, Y A, "Introduction to Thermodynamics and Heat Transfer", McGraw-Hill, New York, pp 120-127, 1997.

Cesta, T, "Capture of Pollutants from a Buoyant Point Source Using a Lateral Exhaust Hood with and without Assistance from an Air Curtain", Ventilation '88, Vincent, J H, editor, Pergamon Press, NY, pp 63-79, 1989.

Chang, H K, "Mechanisms of Gas Transport during Ventilation by High-Frequency Oscillation", Journal of Applied Physiology, Vol. 56, No. 3, pp 553-563, 1984.

Chang, T Y and Rudy, S J, "Roadway Tunnel Air Quality Models", Environmental Science and Technology, Vol. 24, No. 5, pp 672-676, 1990.

Cheng, Y S, Yeh, H C, and Allen, M D, "Dynamic Shape Factor of a Plate-Like Particle", Aerosol Science and Technology, Vol. 8, pp 109-123, 1988.

Cheong, K W and Riffat, S B, Performance Testing of HVAC Using Tracer Gas Techniques", "Ventilation '91", edited by R T Hughes, H D Goodfellow, and G S Rajhans, ACGIH, Cincinnati, OH, pp 473-479, 1993.

Choi, S J, All, R D, Overcash, M R, and Lim, P K, "Capillary-Flow Mechanism for Fugitive Emissions of Volatile Organics from Valves and Flanges: Model Development, Experimental Evidence, and Implications", Environmental Science and Technology, Vol. 26, No. 3, pp 478-484, 1992.

Churchill, R V, Brown, J W, and Verhey, R F, "Complex Variables and Applications", ed. 3, McGraw-Hill, New York, 1976.

Cimbala, J M, "Direct Numerical Simulations and Modeling of a Spatially Evolving Turbulent Wake", 10th Symp. on Turbulent Shear Flows, The Pennsylvania State University, University Park, PA, pp 6.25-6.30, Aug., 1995.

Cimbala, J M, Garanich, J S, Settles, G S, and Miller, J D, "Combined Schlieren Imaging and Numerical Analysis of Displacement Ventilation", Bulletin of the Am. Physical Soc., Div. of Fluid Dynamics, Vol. 43, No. 9, p 2016, 1998.

Cimbala, J M, Nagib, H M, and Roshko, A, "Large Structure in the Far Wakes of Two-Dimensional Bluff Bodies", J. Fluid Mech., Vol. 190, pp 265-298, 1988.

Clark, R P and Cox, R N, "The Generation of Aerosols from the Human Body", Article 95 in "Airborne Transmission and Airborne Infection, Eds., J F P Hers and K C Winkler, Oosthoek Publishing Co., Utrecht, the Netherlands, pp 413-426, 1973.

Clark, R P, "Skin Scales Among Airborne Particles", Journal of Hygiene (Cambridge), Vol. 72, pp 47-51, 1974.

Colucci, A V and Strieter, R P, "Dose Considerations in the Sulfur Dioxide Exposed Exercising Asthmatic", Environmental Health Perspectives, Vol. 52, pp 221-232, 1983.

Cometto-Muniz, J E and Cain, W S, "Efficacy of Volatile Organic Compounds in Evoking Nasal Pungency and Odor", Archives of Environmental Health, Vol. 48, No. 5, pp 309-314, 1993.

Commission of the European Communities, Directorate-General for Science, Research and Development, "Guidelines for Ventilation Requirements in Buildings", Report No. 11, Joint Research Center -Environment Institute, Luxembourg Office for Publications, Brussels, Luxembourg, 1992.

Committee On Indoor Pollution, "Indoor Pollution", National Academy Press, Washington DC, 1981.

Conroy, L M, Ellenbecker, M J, and Flynn, M R, "Prediction and Measurement of Velocity into Flanged Slot Hoods", American Industrial Hygiene Association Journal, Vol. 49, No. 5, pp 226-234, 1988.

Constance, J D, "Controlling In-Plant Airborne Contaminants", Marcel Dekker Inc, New York, 1983.

Cookson, M O C M and Moffatt, M, "Asthma: An Epidemic in the Absence of Infection", Science, Vol. 275, pp 41-42, 1997.

Cooper, C D and Alley, F C, "Air Pollution Control: a Design Approach", PWS Engineering Publishers, Boston, MA, 1986.

Cooper, D W and Horowitz, M, "Exposures from Indoor Powder Releases: Models and Experiments", American Industrial Hygiene Association Journal, Vol. 47, No. 4, pp 214-218, April 1986.

Cornu, J C, "A Method for Measuring the Capture Efficiency of Fume-Extracting Welding Guns", "Ventilation '91", edited by R T Hughes, H D Goodfellow, and G S Rajhans, ACGIH, Cincinnati, OH, pp 185-189, 1993.

Cornu, J C, Leleu, J, Gerber, J M, Vincent, R, Voirin, D, and Aubertin, G, "Determination of Ventilation Criteria for Closed Paint Booths - Validation on 14 Booths in Operation", "Ventilation '85", edited by H D Goodfellow, Elsevier Science Publishers, Amsterdam, pp 193-204, 1986.

Corsi, R L, "Indoor Air Quality: a Time for Recognition", Environmental Management, Air Waste and Management Association, pp 10-145, September 2000.

Cowherd, C Jr, Grelinger, M A, Englehart, P J, Kent, R F and Wong, K F, "An Apparatus and Methodology for Predicting the Dustiness of Materials", American Industrial Hygiene Association Journal, Vol. 50, No. 3, pp 123-130, 1989.

Cox, C S, and Wathes, C M (editors), "Bioaerosols Handbook", Lewis Publishers, Boca Raton, 1995.

Cox, G V and Strickland, G D, "Risk is Normal to Life Itself", American Industrial Hygiene Association Journal, Vol. 49, pp 223-227, 1988.

Crabb, C, "Antimicrobials Meet Resistance", Chemical Engineering, pp 59-62, September 2000.

Cralley, L J and Cralley, L V (editors), "Patty's Industrial Hygiene and Toxicology", Volume I - "General Principles", Volume II - "Toxicology", and Volume III - "Theory and Rationale of Industrial Hygiene Practice", John Wiley and Sons, New York, 1979.

Crawford, M, "Air Pollution Control Theory", McGraw Hill, New York, 1976.

Crawford, W A, "On the Health Effects of Environmental Tobacco Smoke", Archives of Environmental Health, Vol. 43, No. 1, pp 34-37, 1988.

Crowl, D A and Louvar, J F, "Chemical Process Safety: Fundamentals with Applications", Prentice Hall, Englewood Cliffs, NJ, 1990.

Crump, J G and Seinfeld, J H, "Turbulent Deposition and Gravitational Sedimentation of an Aerosol in a Vessel of Arbitrary Shape", Journal of Aerosol Science, Vol. 12, No. 5, pp 405-415, 1981.

Danielson, J A, "Air Pollution Engineering Manual", 2nd ed., EPA Publication AP-40, US Government Printing Office, EPA Office of Air and Water Programs, Office of Air Quality Planning and Standards, Research Triangle Park, NC, May 1973.

Davies, C N, Ed., "Aerosol Science", Academic Press, New York, 1966.

Davies, C N, "Cigarette Smoke: Generation and Properties of the Aerosol", J Aerosol Science, Vol. 19, No. 4, pp 463-469, 1988.

de Nevers, N, "Air Pollution Control Engineering", Ed. 2, McGraw-Hill, Boston, 2000.

de Nevers, N, "Air pollution Control", McGraw Hill, New York, 1995.

DeGasperi, J, "A Fan for all Seasons", Mechanical Engineering, Vol. 121, No. 12, pp 58-60, December 1999.

Dellagi, F, Dumaine, J Y, and Aubertin, G, "Numerical Simulation of Air Flows - Application to the Ventilation of a Paint-Booth", "Ventilation '85", edited by H D Goodfellow, Elsevier Science Publishers, Amsterdam, pp 391-403, 1986.

DeMarini, D M and Lewtas, J, "Mutagenicity and Carcinogenicity of Complex Combustion Emissions: Emerging Molecular Data to Improve Risk Assessment", Toxicological and Environmental Chemistry, Vol. 49, pp 157-166, 1995.

DeMarini, D M, Shelton, M L, and Bell, D A, "Mutation Spectra of Chemical Fractions of a Complex Mixture: Role of Nitroarenes in the Mutagenic Specificity of Municipal Waste Incinerator Emissions", "Mutation Research Fundamental and Molecular Mechanisms of Mutagenesis, edited by M Ashby, J Gentile, K Sankaranarayanan and B Glickman, Elsevier, The Netherlands, Vol. 349, pp 1-20, 1996.

Deming, W E, "Out of the Crisis", MIT Press, 1986.

Dennis, R and Bubenick, D V, "Fugitive Emissions Control for Solid Materials Handling Operations", Journal of the Air Pollution Control Association, Vol. 33, No. 12, pp 1156-1161, 1983.

Dewees, D N, "Does the Danger from Asbestos in Buildings Warrant the Cost of Taking it out?", American Scientist, Vol. 75, pp 285-288, May-June 1987.

Dixon, J R, "Design Engineering: Inventiveness, Analysis, and Decision Making", McGraw Hill, New York, 1966.

Dodge, T M, "Ventilating the English Channel Tunnel", ASHRAE Journal, pp 70-74, October 1993.

Dols, W S, "A Tool for Modeling Airflow & Contaminant Transport", ASHRAE Journal, Vol. 43, No. 3, pp 35-42, Feb. 2001.

Donovan, R P, Locke, B R, and Ensor, D S, "Measuring Particle Emissions from Clean room Equipment", Microcontamination, Vol. 5, No. 10, pp 36-39, 60-63, 10 Oct 1987.

Dravnieks, A, Schmidtsdorff, W and Meilgaard, M, "Odor Thresholds by Forced-Choice Dynamic Triangle Olfactometry: Reproducibility and Methods of Calculation", Journal of the Air Pollution Control Association, Vol. 36, No. 8, pp 900-905, August 1986.

Drivas, P J, "Calculation of Evaporative Emissions from Multicomponent Liquid Spills", Environmental Sciences and Technology, Vol. 16, pp 726-728, 1982.

Dunn, J E and Tichenor, B A, "Compensating for Sink Effects in Emissions Test Chambers by Mathematical Modeling", Atmospheric Environment, Vol. 22, No. 5, pp 885-894, 1988.

Eatough, D J, Benner, C L, Bayona, J M, Richards, G, Lamb, J D, Lee, M L, Lewis, E A, and Hansen L D, "Chemical Composition of Environmental Tobacco Smoke; 1. Gas-Phase Acids and Bases", Journal of Environmental Science and Technology, Vol. 23, No. 6, pp 679-687, 1989.

Eisner, A D, Martonen, R T B, "Simulation of Heat and Mass Transfer Processes in a Surrogate Bronchial System Developed for Hygroscopic Aerosol Studies", Aerosol Science and Technology, Vol. 11, pp 39--57, 1989.

Eklund, B, "Practical Guidance for Flux Chamber Measurements of Fugitive Volatile Organic Emission Rates", Journal of Air Waste Management Association, Vol. 42, pp 1583-1591, 1992.

Ellul, J, "Propaganda", Vintage Books, New York, 1965.

Elsom, D M, "Atmospheric Pollution: a Global Problem", 2nd ed., Blackwell, Oxford, 1992.

Emmerich, S J and Persily, A K, "Literature Review on CO2-based Demand-Controlled Ventilation", ASHRAE Trans., Vol.103, Part 2, Paper no. 4075, pp 229-243, 1997.

Engelmann, R L, Pendergrass, W R, White, J R, and Hall, M E, "The Effectiveness of Stationary Automobiles as Shelters in Accidental Releases of Toxic Materials", Atmospheric Environment, Vol. 26A, No. 17, pp 3119-3125, 1992.

Engle, L, "Intraregional Gas Mixing and Distribution", Chapter 7, "Gas Mixing in the Lung", Engle, L A and Paiva, M, Eds., Marcel Dekker, New York, pp 287-358, 1985.

Envirex Co., "An Evaluation of Occupational Health Hazard Control Technology for the Foundry Industry," NIOSH Pub. 79-114, 1978.

Erkey, C, Madras, G, Orejuela, and Akgerman, A, "Supercritical Carbon Dioxide Extraction of Organics from Soil", Environ. Sci. Technology., Vol. 27, No. 6, pp 1225-1231, 1993.

Esmen, N A and Weyel, D A, "Aerodynamics of Multiple Orifice Hoods", Ventilation '85, edited by H D Goodfellow, Elsevier Science Publishers, Amsterdam, pp 735-741, 1986.

Esmen, N A and Weyel, D A, "Air Flow Generated by Flanged Suction Hoods with Adjacent Planes", Ventilation '88, Vincent J H, editor, Pergamon Press, NY, pp 47-54, 1989.

Esmen, N A, "Characterization of Contaminant Concentrations in Enclosed Spaces", Environmental Science and Technology, Vol. 12, No. 3, pp 337-339, March 1978.

Esmen, N A, Weyel, D A, and McGuigan, F P, "Aerodynamic Properties of Exhaust Hoods", American Industrial Hygiene Association Journal, Vol. 47, No. 8, pp 448-454, 1986.

Fanger, P O, "The Olf and Decipol", ASHRAE Journal, Vol. 40, No. 10, pp 35-38, Oct 1988.

Finlayson-Pitts, B J and Pitts, J N, Jr., "Atmospheric Chemistry: Fundamentals and Experimental Techniques", Wiley Interscience, New York, 1986.

Flagan, R C and Seinfeld, J H, "Fundamentals of Air Pollution Engineering", Prentice Hall, Englewood Cliffs, New Jersey, 1988.

Fleischer, R L, "Basement Ventilation Needed to Lower Indoor Radon to Acceptable Levels", Journal of the American Air Pollution Control Association, Vol. 38, No. 7, pp 914-916, 1988.

Fletcher, B and Johnson, A E, "The Capture Efficiency of Local Exhaust Ventilation Hoods and the Role of Capture Velocity", Ventilation '85, edited by H D Goodfellow, Elsevier Science Publishers, Amsterdam, pp 369-379, 1986.

Flynn, M R and Ellenbecker, M J, "Capture Efficiency of Flanged Circular Local Exhaust Hoods", Ann. Occup. Hygiene, Vol. 30, No. 4, pp 497-513, 1986.

Flynn, M R and Ellenbecker, M J, "Empirical Validation of Theoretical Velocity Fields into Flanged Circular Hoods", American Industrial Hygiene Association Journal, Vol. 48, No. 4, pp 380-389, 1987.

Flynn, M R and Ellenbecker, M J, "The Potential Flow Solution for Air Flow into a Flanged Circular Hood", American Industrial Hygiene Association Journal, Vol. 46, No. 6, pp 318-322, 1985.

Flynn, M R and Miller, C T, "Comparison of Models for Flow through Flanged and Plain Circular Hoods", Annals Occupational Hygiene, Vol. 32, No. 2, pp 373-384, 1988.

Flynn, M R and Miller, C T, "The Boundary Integral Equation Method (BIEM) for Modeling Local Exhaust Hood Flow Fields", American Industrial Hygiene Association Journal, Vol. 50, No. 5, pp 281-288, 1989.

Flynn, M R and Taeheung, K, "On the Use of Computational Fluid Dynamics for the Simulation of Contaminant Control System Performance", "Ventilation '91", edited by R T Hughes, H D Goodfellow, and G S Rajhans, ACGIH, Cincinnati, OH, pp 441-447, 1993.

Fogler, H S, "Elements of Chemical Reaction Engineering", 2nd ed., Prentice Hall, New Jersey, 1992.

Folinsbee, L J, McDonnell, W F, and Horstman, D H, "Pulmonary Function and Symptom Responses After 6.6-Hour Exposure to 0.12 PPM Ozone with Moderate Exercise" Journal of the Air Pollution Control Association, Vol. 38, No. 1, pp 28-35, 1988.

Fontaine, J R, Braconnier, R, Rapp, R, and Serieys, J C, "EOL: a computational Fluid Dynamics Software Designed to Solve Air Quality Problems", "Ventilation '91", edited by R T Hughes, H D Goodfellow, and G S Rajhans, ACGIH, Cincinnati, OH, pp 449-460, 1993.

Fontaine, J R, Gardin, P, Soumoy, V, and Aubertin, G, "Criteria for the Evaluation of General Ventilation Systems: Numerical and Physical Simulations", Ventilation '88, Vincent J H, editor, Pergamon Press, NY, pp 143-153, 1989.

Formica, P N, "Controlled and Uncontrolled Emission Rates and Applicable Limitations for Eighty Processes", EPA-340/1-78-004, April 1978.

Fox, R W and McDonald, A T, "Introduction to Fluid Mechanics", 3rd ed., John Wiley and Sons, New York, NY, 1985.

Fredrickson, R A, "Ventilation System Design for Controlling Otto Fuel II-Contaminated Air", "Ventilation '91", edited by R T Hughes, H D Goodfellow, and G S Rajhans, ACGIH, Cincinnati, OH, pp 73-80, 1993.

Friedlander, S K, "Smoke, Dust and Haze", Wiley Interscience, New York, 1977.

Fuchs, N A, "The Mechanics of Aerosols", Pergamon Press, New York, 1964.

Gale, G E, Torre-Bueno, R, Moon, R E, Saltzman, H A, and Wagner, P T, Journal of Applied Physiology, Vol. 58, pp 978-988, 1985.

Gallily, I, Schiby, D, Cohen, A H, Hollander, W, Schless, D, and Stober, W, "On the Inertial Separation of Nonspherical Aerosol Particles from Laminar Flows. I. The Cylindrical Case", Aerosol Science and Technology, Vol. 5, pp 267-286, 1986.

Ganong, W F "Review of Medical Physiology", 18th ed. Appleton and Lange, Stamford CT, 1997.

Garrison, R P and Byers, D H, "Noise Characteristics of Circular Nozzles for High Velocity/Low Volume Exhaust Ventilation", American Industrial Hygiene Association Journal, Vol. 41, No. 10, pp 713-720, 1980a.

Garrison, R P and Byers, D H, "Static Pressure, Velocity, and Noise Characteristics of Rectangular Nozzles for High Velocity/Low Volume Exhaust Ventilation", American Industrial Hygiene Association Journal, Vol. 41, No. 12, pp 855-863, 1980b.

Garrison, R P and Erig, M, "Velocity Characteristics of Local Exhaust Inlets Facing an External Boundary Surface" American Industrial Hygiene Association Journal, Vol. 49, No. 4, pp 176-184, 1988.

Geary, D F, "New Guidelines on Legionella", Journal of the American Society of Heating, Air Conditioning and Refrigerating Engineering, pp 44-49, September 2000.

Georgopoulos, P G, Walia, A, Roy, A, and Lioy, P J, "Integrated Exposure and Dose Modeling and Analysis System. 1. Formulation and Testing of Microenvironmental and Pharmacokinetic Components", Environmental Science and Technology, Vol. 31, No. 1 pp 17-27, 1997.

Giardino, N J and Andelman, J B, "Characterization of the Emissions of Trichloroethylene, Chloroform, and 1,2-Dibromo-3-Chloropropane in a Full-Size Experimental Shower", Journal of Exposure Analytical Environmental Epidemiology, Vol. 6, No. 4, p 413, 1996.

Giardino, N J, Esmen, N A, and Andelman, J B, "Modeling Volatilization of Trichloroethylene from a Domestic Shower Spray: The Role of Drop-Size Distribution", Environmental Science and Technology, Vol. 26, No. 8, pp 1602-1606, 1992.

Girman, J R and Hodgson, A T, "Source Characterization and Personal Exposure to Methylene Chloride from Consumer Products", Paper No. 86-52.7, Proceedings 79th Annual Meeting of the Air Pollution Control Association, 22-27 June 1986.

Godish, T, "Air Quality", 3rd Ed., Lewis Publishers, Boca Raton, FL, pp 464, 1998.

Godish, T, "Sick Buildings, Definition, Diagnosis and Mitigation", CRC Press, Lewis Publishers, Boca Raton, 1995.

Gold, L S, Stone, T H, Stern, B R, Manley, N B, and Ames, B, "Rodent Carcinogens: Setting Priorities" , Science, Vol. 258, pp 261-265, 1992.

Goodfellow, H D and Smith, J, "Industrial Ventilation - a Review and Update", American Industrial Hygiene Association Journal, Vol. 43, pp 175-184, March 1982.

Goodfellow, H D, "Advanced Design of Ventilation Systems for Contaminant Control", Elsevier Science Publishers, Amsterdam, the Netherlands, 1986.

Goodfellow, H D, "Air Pollution Engineering Manual", 2nd ed., edited by W T Davis, Air Waste Management Association, Wiley-Interscience Publication, New York, pp 135-143, 2000.

Gough, M, "Zero Risk or Acceptable Risk", Am. Ind. Hyg. Assoc. J., Vol. 52, pp A-556 to A-560, 1991.

Gradon, L and Orlicki, D, "Deposition of Inhaled Aerosol Particles in a Generation of the Tracheobronchial Tree", Journal of Aerosol Science, Vol. 21, No. 1, pp 3-19, 1990.

Gradon, L and Yu, C P, "Diffusional Particle Deposition in the Human and Mouth", Aerosol Science and Technology, Vol. 11, 213-220, 1989.

Grant, E L and Ireson, W G, "Principles of Engineering Economy", 5th ed., Ronald Press, NY, 1970.

Guerin, M R, Higgins, C E, and Jenkins, R A, "Measuring Environmental Emissions from Tobacco Combustion: Sidestream Cigarette Smoke Literature Review", Atmospheric Environment, Vol. 21, No. 2, pp 291-297, 1987.

Guyton, A C, "Textbook of Medical Physiology", 7th ed., W B Saunders Co, Philadelphia, PA, 1986.

Haberlin, G M and Heinsohn, R J, "Predicting Solvent Concentrations from Coating the Inside of Bulk Storage Tanks, AIHA Journal, Vol. 54, No. 1, pp 1-9, 1993; see also, Haberlin G M, "Calculation of Contaminant Concentrations while Coating the Interior of a Bulk Storage Fuel Tank", MS Thesis, Mechanical Engineering Department, Penn State Univ., 1990; see also AD-A224930/9/WEP, NTIS, 1993.

Hagapain, J H and Bastress, K E, "Recommended Industrial Ventilation Guidelines", A D Little Co., Cambridge MA, NIOSH Technical Report, Contract No. CDC-99-74-33, January 1976.

Hahn R S, and Lindsay, R P, "Principles of Grinding", Machinery, New York, Part I July 1971 pp 55-62, Part II August 1971 pp 33-39, Part III September 1971 pp 33-39, Part IV October 1971 pp 57-67, Part V November 1971 pp 48-53, 1971.

Halitsky, J, "A Jet Plume Model for Short Stacks", Journal of the Air Pollution Control Association, Vol. 39, No. 6, pp 856-858, 1989.

Hall, R E and DeAngelis, D G, "EPA's Research Program for Controlling Residential Wood Combustion Emissions", Journal of the Air Pollution Control Association, Vol. 30, No. 8, pp 862-867, 1980.

Hamilton, R G, Chapman, M D, Platts-mills, T A E, and Adkinson, N F, "House Dust Aeroallergen Measurements in Clinical Practice: a Guide to Allergen-Free Home and Work Environments", Immunology and Allergy Practice, Vol. 14, No. 3, pp 9-25, 1992.

Hampl, V, Johnston, O E, and Murdock, Jr. D L, "Application of an Air Curtain-Exhaust System at a Milling Process", American Industrial Hygiene Association Journal, Vol. 49, No. 4, pp 167-175, 1988.

Hanna, L M and Scherer, P W, "A Theoretical Model of Localized Heat and Water Vapor Transport in the Human Respiratory Tract", Transactions of the ASME, Journal Biomechanics Engineering, Vol. 108, pp 19-27, 1986a.

Hanna, L M and Scherer, P W, "Regional Control of Local Airway Heat and Water Vapor Losses", Journal of Applied Physiology, Vol. 61, No. 2, pp 624-632, 1986b.

Hanna, L M and Scherer, P W, "Measurement of Local Mass Transfer Coefficients in a Cast Model of the Human Upper Respiratory Tract", Transactions of the ASME, Journal of Biomechanics Engineering, Vol. 108, pp 12-18, 1986c.

Hanna, S R, Briggs, G A, and Hosker, R P Jr., "Handbook of Atmospheric Diffusion", DOE/TIC-11223 (DE82002045), US Dept. of Energy, NTIS, 1982.

Haselton, F R and Scherer, P W, "Bronchial Bifurcations and Respiratory Mass Transport", Science, Vol. 208, pp 69-71, 1980.

Hatch, T and Choate, S P, "Statistical Description of the Size Properties of Non-Uniform Particulate Substances", J. Franklin Inst., Vol. 207, p. 369, 1929.

Hatch, T F and Gross, P, "Pulmonary Deposition and Retention of Inhaled Aerosols", Academic Press, New York, 1964.

Hattis, D, Wasson J M, Page G S, Stern B and Franklin C A, "Acid Particles and the Tracheobronchial Region of the Respiratory System - An Irritation-Signaling Model for Possible Health Effects', Journal of the Air Pollution Control Association, Vol. 37, No. 9, pp 1060-1066, 1987.

Hawthorne, A R and Matthews, T G, "Models for Estimating Organic Emissions from Building Materials: Formaldehyde Example", Atmospheric Environment, Vol. 21, No. 2, pp 419-424, 1987.

Hayashi, T, Howell, R H, Shibata, M, and Tsuji, K, "Industrial Ventilation and Air Conditioning", CRC Press, Boca Raton, Florida, 1985.

Hayes, F C and Stoecker, W F, "Design Data for Air Curtains" Paper 2121, Trans ASHRAE, Part II, pp 168-180, 1969b.

Hayes, F C and Stoecker, W F, "Heat Transfer Characteristics of the Air Curtain", Paper 2120, Trans ASHRAE, Part II, Vol. 75, pp 153-167, 1969a.

Hays, S M, Gobbell, R V, and Ganick, N R, "Indoor Air Quality: Solutions and Strategies", McGraw-Hill, NY, pp320, 1994.

Heinsohn, R J and Choi, M S, "Advanced Design Methods in Industrial Ventilation", Ventilation '85, edited by H D Goodfellow, Elsevier Science Publishers, Amsterdam, pp 81-109, 1986.

Heinsohn, R J and Kabel, R L, "Sources and Control of Air Pollution", Prentice Hall, New Jersey, 1999.

Heinsohn, R J, "Industrial Ventilation: Engineering Principles", Wiley-Interscience, New York, NY, 1991.

Heinsohn, R J, "Macro-Modeling Using Sequential Box Models", Proceedings, "Buildings Systems: Room Air and Air Contaminant Distribution", University of Illinois, Dec 5-8, 1989, edited by L L Christianson, ASHRAE, pp 200-205, 1989.

Heinsohn, R J, "Predicting Solvent Vapor Exposure to Workers Applying Coatings Inside Confined Spaces", "Ventilation '91", 3rd International Symposium on Ventilation for Contaminant Control", edited by R T Hughes, H D Goodfellow, and G S Rajhans, ACGIH, 6500 Glenway Ave, Bldg D7, Cincinnati, OH, pp 547-552, 1993.

Heinsohn, R J, "Predicting Transient Concentrations at Arbitrary Points in an Undivided Space", Proceedings of the "IAQ 91 Healthy Buildings Conference", ASHRAE, 1791 Tullie Circle, NE, Atlanta GA, 30329, pp 192-196, 1991.

Heinsohn, R J, Johnson, D, and Davis, J W, "Grinding Booth for Large Castings", American Industrial Hygiene Association Journal, Vol. 48, No. 8, pp 587-595, 1982.

Heinsohn, R J, O'Donnell W and Tao J, "Automobile Passenger Compartment Ventilation" Paper Number 3669, ASHRAE annual meeting 1993, also in Transactions of the American Society of Heating, Refrigerating and Air-Conditioning Engineers, 1993.

Heinsohn, R J, Yu, S T, Merkle, C L, and Settles, G, Viscous Turbulent Flow in Push Pull Ventilation Systems", "Ventilation '85, Edited by H. D. Goodfellow, Elsevier Science, Amsterdam, pp 529-566, 1986.

Hellman, T M and Small, F H, "Characterization of the Odor Properties of 101 Petrochemicals Using Sensory Methods", Journal of the Air Pollution Control Association, Vol. 24, No. 10, pp 979-982, October 1974.

Hemeon, W C L, "Plant and Process Ventilation", Industrial Press, New York, 1963.

Henson, D A and Fox, J A , "Transient Flows in Tunnel Complexes of the Type Proposed for the Channel Tunnel", Proceedings Institution of Mechanical Engineers, Vol. 188, pp 153-161, 1974.

Henson, D A and Fox, J A, "Application to the Channel Tunnel of a Method of Calculating the Transient Flows in Complex Tunnel Systems", Proceedings of the Institution of Mechanical Engineering, Vol. 188, pp 162-167, 1974.

Henson, D A and Lowndes, J F L, "Design of a Ventilation System for an English Channel Tunnel", Journal of the American Society of Heating Refrigerating and Air Conditioning Engineering", pp 23-28, February 1978.

Herrick, T J and Davis, W T, "Cyclones and Inertial Separators", Chapter 3, "Air Pollution Engineering Manual", ed. W. T. Davis, Air & Waste Management Association, Wiley-Interscience & John Wiley Sons, New York, pp 66-73, 2000.

Herrigel, E, "Zen in the Art of Archery", Pantheon Books Inc, New York, 1953.

Hertzberg, M, "Autoignition Temperatures for Coal Particles Dispersed in Air", Fuel, Vol. 70, pp 1115-1124, Oct 1991.

Hertzberg, M, Cashdollar, K L, and Zlochower, I A, "Flammability Limit Measurements for Dusts and Gases: Ignition Energy Requirements and Pressure Dependencies", 21st Symposium (International) on Combustion, The Combustion Institute, Pittsburgh PA, pp 303-313, 1986.

Hertzberg, M, Cashdollar, K L, Ng, D L, and Conti, R S, "Domains of Flammability and Thermal Ignitability for Pulverized Coals and Other Dusts: Particle Size Dependencies and Microscopic Residue Analysis", 19th Symposium (International) on Combustion, The Combustion Institute, Pittsburgh PA, pp 1169-1180, 1982.

Hidy, G M and Brock, J R, "The Dynamics of Aerocolloidal Systems", Pergamon Press, New York, 1970.

Hinds, W C, "Aerosol Technology", Wiley Interscience, New York 1982.

Hirschfelder, J O, Curtiss, C F, and Bird, R B, "Molecular Theory of Gases and Liquids", John Wiley and Sons, New York, 1954.

Holmberg, S, Li, Y, and Fuchs, N, "Ventilation by Vertical and Horizontal Displacement: A Numerical Comparison of Air Change Efficiency and Thermal Comfort Parameters", "Ventilation '91", edited by R T Hughes, H D Goodfellow, and G S Rajhans, ACGIH, Cincinnati, OH, pp 367-374, 1993.

Holmen, B A, Niemeir, D A and Meng, Y, "Time-Analysis of Above-Road Particulate Matter at the Caldecott Tunnel Exit", Journal of the Air & Waste Management Association, Vol. 51, No. 4, pp 601-615, 2001.

Hosker, R P Jr., "Methods of Estimating Wake Flow and Effluent Dispersion Near Simple Block-Like Buildings", Report NUREG/CR-2521, ERL-ARL-108, National Oceanic and Atmospheric Administration, Washington, DC, 1982.

Howell, R H and Sauer, H J, "Bibliography On Available Computer Programs in the General Area of Heating, Refrigerating, Air Conditioning and Ventilating," 2nd ed., (prepared by the University of Missouri at Rolla for ASHRAE, sponsored by DOE, July 1980).

Hunt, G R and Ingham, D B, "Effects of Exhaust Inlet Size on the Fluid Flow Patterns Created by an Aaberg Exhaust Hood", J. Aerosol Science, Vol. 23, Supp. 1, pp S575-S578, 1992.

Incropera, F P and DeWitt, D P, "Fundamentals of Heat Transfer", 3rd ed., John Wiley & Sons, New York, 1990.

Ingebrethsen, B J, Heavner, D L, Angel, A L, Corner, J M, Steichen, T J, and Green, C R, "A Comparative Study of Environmental Tobacco Smoke Particulate Mass Measurements in an Environmental Chamber", Journal of the Air Pollution Control Association, Vol. 38, No. 4, pp 413-417, 1988.

Ishizu, Y, "General Equation for the Estimation of Indoor Pollution", Environmental Science and Technology, Vol. 14, No. 10, pp 1254-1257, Oct 1980.

Jenkins, B M, Jones, A D, Turn, S Q, and Williams, R B, "Emission Factors for Polycyclic Aromatic Hydrocarbons from Biomass Burning", Environmental Science and Technology, Vol. 30, No. 8, pp 2462-2469, 1996.

Jenning, S G, "The Mean Free Path", Journal of Aerosol Science, Vol. 19, No. 2, pp 159-166, 1988.

Johanson, G, "Modeling of Respiratory Exchange of Polar Solvents", Annals Occupational Hygiene, Vol. 35, No. 3, pp 323-339, 1991.

Jones, D L, Burklin, C E, Seaman, J C, Jones, J W, and Corsi, R L, "Models to Estimate Volatile Organic Hazardous Air Pollutant Emissions from Municipal Sewer Systems", Journal of the Air & Waste Management Association, Vol. 46, pp 657-666, 1996.

Kashdan, E R, Coy, D W, Spivey, J J, and Cesta, T, "Technical Manual: Hood System Capture of Process Fugitive Particulate Emissions" EPA/600/7-86/016, PB86-190444, National Technical Information Service, Springfield VA, 22161, April 1986a.

Kashdan, E R, Coy, D W, Spivey, J J, Cesta, T, and Harmon, D L, "Highlights from Technical Manual On Hood System Capture of Process Fugitive Particulate Emissions", Ventilation '85, edited by H D Goodfellow, Elsevier Science Publishers, Amsterdam, pp 497-520, 1986b.

Kasper, G, Nida, T and Yang, M, "Measurements of Viscous Drag on Cylinders and Chains of Spheres with Aspect Ratios Between 2 and 50", Journal of Aerosol Science, Vol. 16, No. 6, pp 535-556, 1985.

Kelley, T J, Smith, D L, and Satola J, "Emission Rates of Formaldehyde from Materials and Consumer Products in California Homes", Environmental Science and Technology, Vol. 33, No. 1, pp 81-88, 1999.

Kirchhoff, R H, "Potential Flows", Marcel Dekker Inc, NY, 1985.

Klaassen, C D, Amdur, M O, and Doull, J (editors), "Casarett and Doull's Toxicology", ed. 3, Macmillan Publishing Company, New York, 1986.

Kleeman, M J, Schauer, J J, and Cass, G R, "Size and Composition Distribution of Fine Particulate Matter Emitted from Wood Burning, Meat Charbroiling, and Cigarettes", Environmental Science and Technology, Vol. 33, pp 3516-3523, 1999.

Kleinman, M T, "Sulfur Dioxide and Exercise; Relationships Between Response and Absorption in Upper Airways" Journal of Air Pollution Control Association, Vol. 34, No. 1, pp 32-37, 1984.

Klein, M K, "The Air Flow Characteristics on an Annular Exhaust Hood", Applied Industrial Hygiene Association, Vol. 3, No. 4, pp 105-109, 1988.

Kowalski, W, Bahnfleth, W, and Whittam, T, "Filtration of Airborne Microorganisms: Modeling and Prediction", ASHRAE Transactions Vol. 105, No. 2, pp 4-17, 1999.

Krarti, M and Ayari, A, "Ventilation for Enclosed Parking Garages", ASHRAE Journal, Vol. 43, No. 2, February pp 52-57, 2001.

Kreibel D and Smith, T J, "A Nonlinear Pharmacologic Model of the Acute Effects of Ozone on the Human Lungs", Environmental Research, Vol. 51, pp 120-146, 1990.

Kreith, F, "Principles of Heat Transfer", 3rd ed., Harper & Row, New York, 1973.

Kulmula, I, "Experimental Validation of Potential and Turbulent Flow Models for a Two-Dimensional jet-Enhanced Exhaust Hood", American Industrial Hygiene Association, Vol. 61, pp183-191, March/April 2000.

Kumar, A, Vatcha, N S, and Schmelzle, J, "Estimate Emissions from Atmospheric Releases of Hazardous Substances", Environmental Engineering World, Vol. 2, No. 6, pp 20-23, November-December 1996.

Kundu, P K, "Fluid Mechanics", Academic Press, San Diego, CA, 1990.

Kunkel, B A, "A Comparison of Evaporative Source Strength Models for Toxic Chemical Spills" Report AFGL-TR-0307, Air Force Geophysics Laboratory, Air Force Systems Command, USAF, 50 pages, 16 Nov 1983.

Kurabuchi, T and Kusuda, T, "Numerical Prediction for Indoor Air Movement", ASHRAE Journal, Vol. No. 12, pp 26-30, December 1987.

Kurabuchi, T, Sakamoto, Y, and Kaizuka, M, "Numerical Predictions of Indoor Airflows by Means of the K-E Turbulence Model", Proceedings, "Building Systems: Room Air and Air Contaminant Distribution", edited by L L Christianson, ASHRAE, pp 57-67, 1989.

Lave, L B, "Health and Safety Risk Analyses: Information for Better Decisions", Science Vol. 236, pp 291-295, 17 April 1987.

Leaderer, B P, "Air Pollutant Emissions from Kerosene Space Heaters", Science, Vol. 218, pp 1113-5, 10 Dec 1982.

Leaderer, B P, Cain, W S, Isseroff, R, and Berglund, L G, "Ventilation Requirements in Buildings - II Particulate Matter and Carbon Monoxide from Cigarette Smoking" , Atmospheric Environment, Vol. 18, No. 1, pp 99-106, 1984.

Leaderer, B P, Zagraniski, R T, Berwick, M, and Stolwijk, J A J, "Assessment of Exposure to Indoor Air Contaminants from Combustion Sources: Methodology and Application", American Journal of Epidemiology, Vol. 24, No. 2, pp 275-289, 1986.

Lee, C T and Leith, D, "Drag Force on Agglomerated Spheres in Creeping Flow", Journal of Aerosol Science, Vol. 20, No. 5, pp 503-513, 1989.

Lee, K W and Liu, B Y H, "On the Minimum Efficiency and the Most Penetrating Particle Size for Fibrous Filters", Journal of the Air Pollution Control Association, Vol. 30, No. 4, pp 377-381, 1980.

Lehr, J H, "Rational Readings on Environmental Concerns", Van Nostrand Reinhold, New York, 1992.

Leith, D, " Drag on Nonspherical Objects", Aerosol Science and Technology, Vol. 6, pp 153-161, 1987.

Lemaire, T and Luscuere, P, "Investigating Computer Modeling of Clean room Air Flow Patterns", Microcontamination, Vol. 9, No. 8, pp 19-26, August 1991.

Leonardos, G, Kendall, D, and Barnard, N, "Odor Threshold Determinations of 53 Odorant Chemicals", Journal of the Air Pollution Control Association, Vol. 19, No. 2, pp 91-95, January 1969.

Leung, H W and Paustenbach, D J, "Application of Pharmacokinetics to Derive Biological Exposure Indexes from Threshold Limit Values", American Industrial Hygiene Association Journal, Vol. 49, No. 9, pp 445-450, 1988.

Levitzky, M G, "Pulmonary Physiology", 2nd ed., McGraw Hill, New York, 1986.

Lewis, B and von Elbe G, "Combustion, Flames and Explosions of Gases", Academic press, New York, 1951.

Licht, W, "Air Pollution Control Engineering", Marcel Dekker Inc., New York, 1980.

Lipmann, M, "Health Effects of Ozone, a Critical Review", Journal of the Air Pollution Control Association, Vol. 39, No. 5, pp 672-695, 1989.

Lipmann, M, "Health Effects of Tropospheric Ozone", Environmental Science and Technology, Vol. 25, No. 12, pp 1954-1961, 1991.

Lipmann, M, "Role of Science Advisory Groups in Establishing Standards for Ambient Air Pollutants", Aerosol Science and Technology, Vol. 6, pp 93-114, 1987.

Little, J C, "Applying the Two-Resistance Theory to Contaminant Volatilization in Showers", Environmental Science and Technology, Vol. 26, No. 7, pp 1341-1349, 1992.

Lodge, J P Jr. (editor) "Methods of Air Sampling and Analysis", 3rd ed., Lewis Publishers, New York, 1988.

Lofroth, G, Burton, R M, Forehand, L, Hammond, S K, Seila, R L, Zweidlnger, R B, and Lewtas, J, "Characterization of Environmental Tobacco Smoke", Environmental Science and Technology, Vol. 23, No. 5, pp 610-614, 1989.

Longo, M L, Bisagno, A M, Zasadzinski, J A N, Bruni, R, and Waring, A J, "A Function of Lung Surfactant Protein SP-B", Science, Vol. 261, pp 453-456, 1993.

Mackay, D and Paterson, S, "Model Describing the Rates of Transfer Processes of Organic Chemicals Between Atmosphere and Water", Environmental Science and Technology, Vol. 20, No. 8, pp 810-816, 1986.

Mackay, D and Shiu, W Y, "A Critical Review of Henry's Law Constants for Chemicals of Environmental Interest", Journal of Physical and Chemical Reference Data, Vol. 10, No. 4, pp 1175-1191, 1981.

Mackay, D and Yeun, A T K, "Mass Coefficient Correlations for Volatilization of Organic Solutes from Water", Environmental Science and Technology, Vol. 17, No. 4, pp 211-217, 1983.

Mackay, D, Bobra, A, Chan, D W, and Shiu, W Y, "Vapor Pressure Correlations for Low-Volatility Environmental Chemicals", Environmental Science and Technology, Vol. 16, No. 10, pp 645-649, 1982.

Maroni, M, Seifert, B, and Lindvall, T, "Indoor Air Quality", Elsevier Science, Amsterdam, The Netherlands, 1995.

Marple, V A , Rubow, K L, and Olson, B A, "Inertial, Gravitational, Centrifugal and Thermal Collection Techniques" Chapter 11, pp 206-228, "Aerosol Measurement", edited by K. Willeke and P. A. Baron, Van Nostrand Reinhold, New York, 1993.

Martonen, T B, "Deposition Patterns of Cigarette Smoke in Human Airways", American Industrial Hygiene Journal, Vol. 53, pp 6-18, 1992.

Martonen, T B, Zang, Z, and Yang, Y, "Interspecies Modeling of Inhaled Particle Deposition Patterns", Journal of Aerosol Science, Vol. 23, No. 4, pp 389-406, 1992.

Martonen, T B, Zhang, Z, and Lessmann, R C, "Fluid Dynamics of Human Larynx and Upper Tracheobronchial Airways", Aerosol Science and Technology, Vol. 19, pp 133-156, 1993.

Matthews, T G, Hawthorne, A R, and Thompson, C V, "Formaldehyde Sorption and Desorption Characteristics of Gypsum Wallboard", Environmental Science and Technology, Vol. 21, No. 7, pp 629-634, 1987.

Matthews, T G, Wilson, D L, Thompson, A J, Mason, M A, Bailey, S N, and Nelms, L H, "Interlaboratory Comparison of Formaldehyde Emissions from Particleboard Underlayment in Small-Scale Environmental Chambers", Journal of the Air Pollution Control Association, Vol. 37, No. 11, pp 1320-1326, 1987.

McCabe, W L, Smith, J C, and Harriott, P, "Unit Operations of Chemical Engineering", 5th ed., McGraw Hill, New York, 1993.

McCartney, M L, "Sensitivity Analysis Applied to Coburn-Foster-Kane Models of Carboxyhemoglobin Formation", American Industrial Hygiene Association Journal, Vol. 51, No. 3, pp 169-177, 1990.

McDermott, H J, "Handbook of Ventilation for Contaminant Control", Ann Arbor Science Publishers Inc., Ann Arbor, Michigan, 1976.

McKay, E A, "Tunneling to New York", Invention and Technology, pp 23-29, Fall 1998.

McKone, T E and Knezovich J P, "The Transfer of Trichloroethylene (TCE) from a Shower to Indoor Air: Experimental Measurements and their Applications", Journal of Air Waste Management, Vol. 41, pp 832-837, 1991.

McKone, T E, "Human Exposure to Volatile Organic Compounds from Common Household Tap Water: The Indoor Inhalation Pathway", Environmental Science and Technology, Vol. 21, No. 12, pp 1194-1201, 1987.

Menkes, H, Cohen, B, Permutt, S, Beatty, T, and Shelhamer, J, "Characterization and Interpretation of Forced Expiration", Annals of Biomedical Engineering, Vol. 9, pp 501-511, 1981.

Meritt, R L and Vann Bush, P, "Status and Future of Baghouses in the Utility Industry", Journal of the Air & Waste Management Association, Vol. 47, pp 704-709, 1997.

Meyer, B and Hermanns, K, "Reducing Indoor Air Formaldehyde Concentrations", Air Pollution Control Association Journal, Vol. 35, pp 816-821, 1985.

Meyer, B, "Indoor Air Quality", Addison-Wesley Publishing Co, Reading, MA, 1983.

Meyer, W C, "Avoid Legionellosis Lawsuits over Cooling Towers", Chemical Engineering, pp 113-117, Sept 2000.

Miller, C A, Ryan, J V, and Lombardo, T, "Characterization of Air Toxics from an Oil-Fired Firetube Boiler", Journal of the Air Waste Management Association, Vol. 46, pp 742-748, 1996.

Miller, F J, Overton, J H, Jaskot, R H and Menzel, D B, "A Model of the Regional Uptake of Gaseous Pollutants in the Lung, 1. The Sensitivity of the Uptake of Ozone in the Human Lung to Lower Respiratory Track Secretions and to Exercise,", Journal of Toxicological Environmental Health, Vol. 79, pp 11-27, 1985.

Milne-Thomson, L M, "Theoretical Hydrodynamics", MacMillan Co, New York, 1950.

Moffat, "Handbook of Indoor Air Quality Management", Prentice Hall, NY, pp 640, 1997.

Molburg, J C and Rubin, E S, "Air Pollution Costs for Coal-to-Electricity Systems", J. Air Pollution Control Assoc., Vol. 33, No. 5, pp 523-530, 1983.

Moore, M E and McFarland, A R, "Performance Modeling of Single-Inlet Aerosol Sampling Cyclones", Environmental Science and Technology, Vol. 27, No. 9, pp 1842-1848, 1993.

Moore, M E, and McFarland, A R, "Design Methodology for Multiple Inlet Cyclones", Environmental Science and Technology, Vol. 30, pp 271-276, 1996.

Morton, B R, "Forced plumes", Journal of Fluid Mechanics, 5:151-163, 1959.

Morton, B R, Taylor, G I, and Turner, J S, "Turbulent Gravitational Convection from Maintained and Transient Sources", Proceedings of the Royal Society, A 234:1-23, 1956.

Mossman, B T, Bignon, J, Corn, M, Seaton, A, and Gee J B L, "Asbestos: Scientific Developments and Implications for Public Policy" Science, Vol. 247, 294-301, 1990.

Moya, J, Howard-Reed, C, and Corsi, R L, "Volatilization of Chemicals from Tap Water to Indoor Air from Contaminated Water Used for Showering", Environmental Science and Technology, Vol. 33, pp 2321-2327, 1999.

Muller, W J, Hess, G D, and Scherer, P W, "A Model of Cigarette Smoke Deposition" American Industrial Hygiene Association Journal, Vol. 51, No. 5, pp 245-256, 1990.

Mumma, S A, "Overview of integrating dedicated outdoor air systems with parallel terminal systems", ASHRAE Trans., Vol.107, Part 1, Paper no. AT-01-7-1, pp 545-552, 2001.

Murakami, S and Kato, S, "Current Status of Numerical and Experimental Methods for Analyzing Flow Field and Diffusion Field in a Room", Proceedings, "Building Systems: Room Air and Air Contaminant Distribution" edited L L Christianson, ASHRAE, pp 39-56, 1989.

Murray, J F, "The Normal Lung," Ed. 2, W B Saunders Company, Philadelphia, PA, 1986.

Mycok, J C, McKenna, J D and Theodore, A J, "Handbook of Air Pollution Control Engineering and Technology", Lewis Publishers, Boca Raton, Florida 1995.

Nadel, E R, "Physiological Adaptations to Aerobic Training", American Scientist, Vol. 73, pp 334-342, 1985.

Nagy, G Z, "The Odor Impact Model", Air Waste Management Assoc. J., Vol. 41, No. 10, pp 1360-1362, 1991.

National Research Council, Committee On Airliner Cabin Air Quality, "The Airliner Cabin Environment", National Academy Press, Washington DC, 1986.

National Safety Council, "Injury Facts", 2000 Edition, Washington DC, 2000.

Nazaroff, W W and Cass, G R, "Mathematical Modeling of Indoor Aerosol Dynamics", Environmental Science and Technology, Vol. 23, No. 2, pp 157-166, 1989.

Nazaroff, W W and Teichman, K, "Indoor Radon", Environmental Science and Technology, Vol. 24, No. 6, pp 774-782, 1990.

Neveril, R B, Price, J U and Engdahl, K L, "Capital and Operating Costs of Selected Air Pollution Control Systems", J of Air Pollution Control Assoc, Part I, Vol. 28, No. 8, pp 829-836, Aug 1978; Vol. 28, Part II, No. 9, pp 963-968, Sept 1978; Vol. 28, Part III, No. 10, pp 1069-1072, Oct 1978; Vol. 28, Part IV, No. 11, pp 1171-1172, Nov 1978; Vol. 28,Part V, No. 12, pp 1253-1256, Dec 1978.

Nichols, A and Zeckhauser, R, "The Dangers of Caution: Conservatism in Assessment and the Mismanagement of Risk", Harvard University, #E-85-11, November 1985.

Nielsen, P V, "Numerical Prediction of Air Distribution in Rooms - Status and Potentials", Proceedings, "Building Systems: Room Air and Air Contaminant Distribution", ASHRAE, edited by L L Christianson, pp 31-38, 1989.

Nielsen, P V, Restivo, A, and Whitelaw, J H, "The Velocity Characteristics of Ventilated Rooms", Journal of Fluids Engineering, Transactions of the ASME, Vol. 100, pp 291-298, Sep. 1978.

Niemela, R, Toppila, E, and Rolin, I, "Characterization of Supply Air Distribution in Large Industrial Premises by the Tracer Gas Technique" Ventilation '85, Elsevier Science Publishers, H D Goodfellow editor, Amsterdam, pp 797-805, 1986.

Nirmalakhandan, N N and Speece, R E, "QSAR Model for Predicting Henry's Constant", Environmental Science and Technology, Vol. 22, No. 11, pp 1349-1357, 1988.

Nixon, W and Egan, M J, "Modeling Study of Regional Deposition of Inhaled Aerosols with Special Reference to Effects of Ventilation Asymmetry", Journal of Aerosol Science, Vol. 18, No. 5, pp 563-579, 1987.

O'Brien, D M and Hurley, D E, "An Evaluation of Engineering Control Technology for Spray Painting", DHHS(NIOSH) Publication No. 81-121, 1981.

Occupational Health and Safety, Vol. 58, No. 3, March 1989.

Oglesby, S and Nichols, G B, "A Manual of Electrostatic Precipitator Technology, Parts I and II", PB 196380 and 196381, National Technical Information Service, 1970.

Okrent, D, "The Safety Goals of the Nuclear Regulatory Commission", Science Vol. 236, pp 296-300, 17 April 1987.

Olander, L, "Industrial Ventilation Literature", "Ventilation '91", edited by R T Hughes, H D Goodfellow, and G S Rajhans, ACGIH, Cincinnati, OH, pp 17-26, 1993.

Oladakar III, G B and Conrad, F C, "Estimation of Effect of Environmental Tobacco Smoke On Air Quality within Passenger Cabins of Commercial Aircraft", Environmental Science and Technology, Vol. 21, No. 10, pp 994-999, 1987.

Orwell, G, " Politics and the English Language" from the Collected Essays in Vol. 4, Journalism and Letters of George Orwell, Angus, A and Angus, I (editors), Harcourt, Brace and Jovanovich, New York, pp 127-139, 1968.

Otis, A B, "Mechanical Factors in Distribution of Pulmonary Ventilation", Journal of Applied Physiology, Vol. 8, pp 427-443, 1956.

Ott, W R, "A Physical Explanation of the Lognormality of Pollutant Concentrations", Journal of Air Waste Management Association, Vol. 40, pp 1378-1383, 1990.

Paiva, M, "Theoretical Studies of Gas Mixing in the Lung", Chapter 6, "Gas Mixing in the Lung", Engle L A and Paiva M, Eds., Marcel Dekker, New York, pp 221-286, 1985.

Panton, R L, "Incompressible Flow", ed. 2, Wiley-Interscience, New York, 1996.

Paustenbach, D and Langner, R, "Corporate Occupational Exposure Limits: The Current State of Affairs", American Industrial Hygiene Association Journal, Vol. 47, No. 12, pp 809-818, 1986.

Pedersen, L G and Nielsen, P V, "Exhaust System Reinforced by Flow", "Ventilation '91", edited by R T Hughes, H D Goodfellow, and G S Rajhans, ACGIH, Cincinnati, OH, pp 203-208, 1993.

Perera, M D A E S, Walker, R R, and Oglesby, O D, "Infiltration Measurements in Naturally Ventilated, Large Multicelled Buildings", Ventilation '85, Elsevier Science Publishers, H D Goodfellow editor, Amsterdam, pp 807-827, 1986.

Permutt, S and Menkes, H A, "Spirometry, Analysis of Forced Expiration within the Time Domain", contained in the "Lung in the Transition Between Health and Disease", Macklen, P T and Permutt, S (editors), Marcel Dekker, Inc, New York, pp 113-152, 1979.

Perra, F P and Ahmed, A K, "Respiratory Particles", Ballinger Publishing Company, Cambridge MA, 1979.

Perry, R H and Chilton, C H, "Chemical Engineers' Handbook", 5th ed., McGraw Hill, New York, 1973.

Perry, R H, Green, D W, and Maloney, J G, "Perry's Chemical Engineers Handbook", 6th ed., McGraw Hill, New York, 1984.

Petersen, R. L., "Stack Heights, Air Intake Locations, and Indoor Air Quality", "Ventilation '91", edited by R T Hughes, H D Goodfellow, and G S Rajhans, ACGIH, Cincinnati, OH, pp 333-341, 1993.

Phalen, R F, Oldam, M J, Mannix, R C, and Schum, G M, "Cigarette Smoke Deposition in the Tracheobronchial Tree: Evidence for Colligative Effects", Aerosol Science and Technology, Vol. 20, pp 215-226, 1994.

Pickrell, J A, Griffis, L C, Mokler, B V, Kanapilly, G M, and Hobbs, C H, "Formaldehyde Release from Selected Consumer Products: Influence of Chamber Loading, Multiple Products, Relative Humidity and Temperature", Environmental Science and Technology, Vol. 18, pp 682-686, 1984.

Pipes, L A and Harvill, L R, "Applied Mathematics for Engineers and Physicists", 3rd edition, McGraw Hill, NY, 1970.

Pontius, D H and Smith, W B, "Method for Computing Flow Distribution and Pressure Drop in Multichamber Baghouses", Journal of the Air Pollution Control Association, Vol. 38, No. 1, pp 39-45, 1988.

Press, W H, Flannery, B P, Teukolsky, S A, and Vettering, W T, "Numerical Recipes – The Art of Scientific Computing", Cambridge University Press, New York, 1986.

Prugh, R W, "Plant Safety" in Kirk-Othmer Encyclopedia of Chemical Technology, ed. 3, 18:60, 1982.

Pruppacher, H R and Klett, J D, "Microphysics of Clouds and Precipitation" D Reidel Publishing CO, Dordrecht, Holland, 1978.

Reid, R C, Prausnitz, J M, and Poling, B E, "The Properties of Gases and Liquids", 4th ed., McGraw Hill, New York, 1987.

Reiss, R, Ryan, P B, and Koutrakis, P, "Modeling Ozone Deposition Onto Indoor Residential Surfaces", Environmental Science and Technology, Vol. 28, pp 504-513, 1994.

Repace, J L and Lowery, A H, "Indoor Air Pollution, Tobacco Smoke, and Public Health", Science, Vol. 208, pp 464-472, May 1980.

Repace, J L and Lowery, A H, "Tobacco Smoke, Ventilation, and Indoor Air Quality", Transactions of the American Society of Heating, Refrigerating and Air-Conditioning Engineers, 88 Part 1, Paper HO-82-6 No. 2, pp 895-914, 1982.

Ricci, P F and Molton, L S, "Regulating Cancer Risks", Environmental Science and Technology, Vol. 19, No. 6, pp 473-479, 1985.

Ricci, P F and Molton, L S, "Risk and Benefit in Environmental Law", Science, Vol. 214, 4, pp 1096-1100, Dec 1981.

Rich, G, "A Primer on Risk Calculations", Pollution Engineering, Vol. 22, No. 5, pp 94-99, 1990.

Rothenberg, S J and Swift, D L, "Aerosol Deposition in the Human Lung at Variable Tidal Volumes: Calculation of Fractional Deposition, Journal of Aerosol Science, Vol. 3, pp 215-226, 1984.

Rothman, S "Physiology and Biochemistry of the Skin", Chicago, 1954.

Rothstein, M A, "West's Handbook Series: Occupational Safety and Health Law", 2nd ed., West Publishing Co, St. Paul, MN, 1983.

Russell, M and Gruber, M, "Risk Assessment in Environmental Policy-Making", Science Vol. 236, pp 286-290, 17 April 1987.

Ruth, J H, "Odor Thresholds and Irritation Levels of Several Chemical Substances: A Review", Journal of American Industrial Hygiene Association, Vol. 47, pp A-142 to A-151, March 1986.

Ryan, P B, Spengler, J D and Letz, R, "The Effects of Kerosene Heaters On Indoor Pollutant Concentrations: A Monitoring and Modeling Study", Atmospheric Environment, Vol. 17, No. 7, pp 1339-1345, 1983.

Ryan, P B, Spengler, J D, and Halfpenny, P F, "Sequential Box Models for Indoor Air Quality: Application to Airliner Clean Cabin Air Quality", Paper 86-7.4, 79th Annual Meeting of the Air Pollution Control Association, 22-27 June 1986, also Atmospheric Environment, Vol. 22, No. 6, pp 1031-1038, 1988.

Sandberg, M, "The Use of Moments for Assessing Air Quality in Ventilated Rooms", American Society of Heating, Refrigerating and Air-Conditioning Engineers, Building and Environment, Vol. 18, No. 4, pp 9-25, 1983.

Sargent, E V and Kirk, G D, "Establishing Airborne Exposure Control Limits in the Pharmaceutical Industry", American Industrial Hygiene Association Journal, Vol. 49, No. 6, pp 309-313, 1988.

Sax, N I, "Dangerous Properties of Industrial Materials" ed. 5, Van Nostrand Reinhold, New York, 1979.

Schaper, M, "Development of a Database for Sensory Irritants and its Use in Establishing Occupational Exposure Limits", American Industrial Hygiene Association Journal, Vol. 54, No. 9, pp 488-544, 1993.

Schmidt, F W, Henderson, R E, and Wolgemuth, C H, "Introduction to Thermal Sciences", Ed. 2, John Wiley & Sons, 1993.

Schulman, L L and Scire, J S, "Building Downwash Screening Modeling for the Downwind Recirculation Cavity", Journal of Air and Waste Management Association, Vol. 43, Aug, pp 1122-1127, 1993.

Schwartz, S E and Andreae, M O, "Uncertainty in Climate Change Caused by Aerosols", Science, Vol. 272, pp 1121-1122, 24 May 1996.

Sciloa, V., "The Practical Application of Reduced Flow Push-Pull Plating Tank Exhaust Systems", "Ventilation '91", edited by R T Hughes, H D Goodfellow, and G S Rajhans, ACGIH, Cincinnati, OH, pp 439-48, 1993.

Scruton, R, "The End of Courage", National Review, Vol. 4, No. 35, pp 52-53, May 31, 1999.

Seinfeld, J H and Pandis, S N, "Atmospheric Chemistry and Physics", Wiley Interscience, New York, 1998.

Seinfeld, J H, "Atmospheric Chemistry and Physics of Air Pollution", Wiley Interscience Publications, New York, 1986.

Shapiro, A H, "The Dynamics and Thermodynamics of Compressible Flow", Vol. I, The Ronald Press Co, New York, 1953.

Shen, T T, "Estimation of Organic Compound Emissions from Waste Lagoons", Journal of the Air Pollution Control Association, Vol. 32, No. 1, pp 80-82, 1982.

Shepherd, J L, Corsi, R L, and Kemp, J, "Chloroform in Indoor Air and Wastewater: The Role of Residential Washing Machines", Journal of the Air & Waste Management Association, Vol. 46, pp 631-642, 1996.

Sherwood, T K, Pigford, R L, and Wilke, C R, "Mass Transfer", McGraw-Hill Book Co, New York, 1975.

Shusterman, D, "Critical Review: The Health Significance of Environmental Odor Pollution", Archives of Environmental Health, Vol. 47, No. 1, pp 76-91, January/February 1992.

Silberstein, S, Grot, R A, Ishiguro, K, and Mulligan, J L, "Validation of Models for Predicting Formaldehyde Concentrations in Residences due to Pressed-Wood Products", Journal of the American Air Pollution Association, Vol. 38, No. 11, pp 1403-1411, 1988.

Silverthorne, D S, "Human Physiology, an Integrated Approach", Prentice Hall, Upper Saddle River, NJ, 1998.

Siu, R G H, "The Tao of Science", MIT Press, Cambridge, Mass, 1957.

Skaret, E and Mathisen, H M, "Ventilation Efficiency - A Guide to Efficient Ventilation", Trans ASHRAE, Part 2B, pp 480-495, 1983.

Skaret, E, "Ventilation by Displacement - Characterization and Design Implications", Ventilation '85, Elsevier Science Publishers, H D Goodfellow editor, Amsterdam, pp 827-842, 1986.

Slonim, N B and Hamilton, L H, "Respiratory Physiology", 5th ed., C V Mosby Co, St Louis, MO, 1987.

Slovic, P, "Perception of Risk", Science Vol. 236, pp 280-285, 17 April 1987.

Slutsky, A S, Drazen, J M, Ingram, R H, Kamm, R D, Shapiro, A H, Fredberg, J J, Loring, S H, and Lehr, J, "Effective Pulmonary Ventilation with Small-Volume Oscillations at High Frequency", Science, Vol. 209, pp 609-610, 1980.

Slutsky, A S, Kamm, R D and Drazen, J M, "Alveolar Ventilation at High Frequencies Using Tidal Volumes Smaller than the Anatomical Dead Space", Chapter 4, "Gas Mixing in the Lung", Engle L A and Paiva M, Eds., Marcel Dekker, New York, pp 137-176, 1985.

Smith, J M and Van Ness, H C, "Introduction to Chemical Engineering Thermodynamics", 3rd ed. , McGraw Hill, New York, 1975.

Spengler, J D, Samet, J, and McCarthy, J F, "Indoor Air Quality Handbook", McGraw-Hill, pp 1488, 2000.

Stahl, W H, Ed., "Compilation of Odor and Taste Threshold Value Data", American Society of Testing and Materials, ASTM DS 48A, 1978.

Starr, C, "Social Benefit Versus Technological Risk", Science, Vol. 165, pp 1232-1238, 19 Sept 1969.

Staub-Reinhault, "Odor Threshold Values", Luft, Vol. 32, No. 10, pp 28-31, 1972.

Steering Committee on Identification of Toxic and Potentially Toxic Chemicals for Consideration by The National Toxicology Program, Board on Toxicology and Environmental Health Hazards, Commission on Life Sciences, National Research Council (NRC), "Toxicity Testing, Strategies to Determine Needs and Priorities", National Academy Press, Washington DC, 1984.

Sterling, T D, Dimich, H, and Kobayashi, D, "Indoor Byproduct Levels of Tobacco Smoke: A Critical Review of the Literature", J. Air Pollution Control Association, Vol. 32, No. 3, pp 250-259, March 1982.

Stiver, W and Mackay, D, "Evaporation Rate of Spills of Hydrocarbons and Petroleum Mixtures", Environmental Science and Technology, Vol. 18, No. 11, pp 834-840, 1984.

Stiver, W, Shiu, W Y, and Mackay, D, "Evaporation Times and Rates of Specific Hydrocarbons in Oil Spills", Environmental Sciences and Technology, Vol. 23, No. 1, pp 101-105, 1989.

Stock, T H, "Formaldehyde Concentrations Inside Conventional Housing", Journal of the American Air Pollution Control Association, Vol. 37, No. 8, pp 913-918, 1987.

Strauss, W, "Industrial Gas Cleaning", Pergamon Press, New York, 1966.

Streeter, V L, "Fluid Dynamics", McGraw Hill, New York, 1948.

Streeter, V L, Wylie, E B, and Bedford, K W, "Fluid Mechanics", ed. 9, WCB/McGraw-Hill, New York, 1998.

Striebig, B A, Private Communications, May 2002.

Talty, J T, "Industrial Hygiene Engineering, Recognition, Measurement, Evaluation and Control", 2nd ed., Noyes Publications, Noyes Data Corporation, Park Ridge, NJ 07656, 1988.

Tannehill, J C, Anderson, D A, and Pletcher, R H, "Computational Fluid Mechanics and Heat Transfer", Ed. 2, Taylor and Francis, Washington, DC, 1997.

Taylor, G I, "Dispersion of Soluble Matter in Solvent Flowing Slowly through a Tube", Proc. Roy. Soc. (London), Ser. A: Vol. 219, p. 186, 1953.

Taylor, G I, "The Dispersion of Matter in Turbulent Flow through a Pipe", Proc. Roy. Soc. (London), Ser. A.: Vol. 220, p. 440, 1954.

Tencer, G M, "Electronic Communication for Hygienists: Bulletin Board Systems", American Industrial Hygiene Journal, Vol. 55, No3, pp 257-260, 1994.

Theodore, L and Bunicore, A J, "Air Pollution Control Equipment", Springer-Verlag, , New York, 1994.

Theodore, L and DePaola, V, "Predicting Cyclone Efficiency", Journal of the Air Pollution Control Association, Vol. 30, No. 10, 1980.

Thurow, L C, "A Weakness in Process Technology", Science, Vol. 238, pp 1659-1663, 18 December 1987.

Tichenor, B A and Mason, M A, "Organic Emissions from Consumer Products and Building Materials in the Indoor Environment", Journal of Air Pollution Control Association, Vol. 38, No. 3, pp 264-268, 1988.

Tichenor, B A, Sparks, L E, and Jackson, M D, "Emission of Perchlorothylene from Dry Cleaned Fabrics", Atmospheric Environment, Vol. 24A, No. 5 pp 1219-1229, 1990.

Tilton, B E, "Health Effects of Tropospheric Ozone", Environmental Science and Technology, Vol. 23, No. 3, pp 257-263, 1989.

Tim Suden, K D, Flynn, M R, and Goodman, R, "Computer Simulation in the Design of Local Exhaust Hoods for Shielded Metal Arc Welding", American Industrial Hygiene Association Journal, Vol. 51, No. 3, pp 115-126, March 1990.

Topmiller, J L and Hampl, V, "Recent Results from Research on Wood Dust Emission Control", "Ventilation '91", edited by R T Hughes, H D Goodfellow, and G S Rajhans, ACGIH, Cincinnati, OH, pp 229-234, 1993.

Traynor, G W, Allen, J R, Apte, M G, Girman, J R, and Hollowell, C D, "Pollutant Emissions from Portable Kerosene-Fired Space Heaters", Environmental Science and Technology, Vol. 17, No. 6, pp 369-371, 1983.

Traynor, G W, Anthon, D W, and Hollowell, C D, "Technique for Determining Pollutant Emissions from Gas-Fired Range", Atmospheric Environment, Vol. 16, No. 12, pp 2979-2987, 1982.

Traynor, G W, Apte, M G, Carruthers, A R, Dillworth, J F, Grimsrud, D T and Gundel, L A, "Indoor Air Pollution due to Emissions from Wood-Burning Stoves", Environmental Science and Technology, Vol. 21, No. 7, pp 691-697, 1987.

Traynor, G W, Apte, M G, Carruthers, A R, Dillworth, J F, Prill, R J, Grimsrud, D T, and Turk, B H, "The Effects of Infiltration and Insulation On the Source Strengths and Indoor Air Pollution from Combustion Space Heating Appliances", Journal of the Air Pollution Control Association, Vol. 38, No. 8, pp 1011-1015, 1988.

Traynor, G W, Girman, J R, Apte, M G, Dillworth, J F, and White, P D, "Indoor Air Pollution due to Emissions from Unvented Gas-Fired Space Heaters", Air Pollution Control Association Journal, Vol. 35, No. 3, pp 231-237, 1988.

Treybal, R E, "Mass-Transfer Operations", 3rd ed., McGraw-Hill Co, New York, 1975.

Tritton, D J, "Physical Fluid Dynamics", Van Nostrand Reinhold, New York, 1977.

Turner, D B, "Workbook of Atmospheric Dispersion Estimates", Lewis Publishers, Boca Raton, 1994.

Turner, J S, "Buoyancy Effects in Fluids", Cambridge University Press, 1973.

Ultman, J S, "Exercise and Regional Dosimetry: Factors Governing Gas Transport in the Lower Airways", Susceptibility to Inhaled Pollutants, ASTM STP 1024, M J Utell and R Frank eds., American Society for Testing and Materials, Philadelphia, pp 111-126, 1989.

Ultman, J S, "Gas Transport in the Conducting Airways", Chapter 3, "Gas Mixing in the Lung", Engle L A and Paiva M, Eds., Marcel Dekker, New York, pp 63-136, 1985.

Ultman, J S, "Transport and Uptake of Inhaled Gases", Air Pollution, the Automobile, and Public Health", A Y Watson, R R Bates and D Kennedy, Eds., Health Effects Institute, National Academy Press, Washington, DC, pp 323-366, 1988.

US Department of Labor, Air Contaminants - Permissible Exposure Limits, Title 29 Code of Federal Regulations, Part 1910.1000, 1989.

US Department of Labor, Air Contaminants - Permissible Exposure Limits, Title 29 Code of Federal Regulations, Part 1910.1000, updated values as of 1997.

US EPA Air Programs Publication AP-42 "Compilation of Air Pollutant Emission Factors, Volume I: Stationary Point and Area Sources", 1980.

US EPA Publication, "Volatile Organic Compound (VOC) Species Manual", 2nd ed. EPA 450/4-80-015, July 1980.

US EPA Publication, "A Manual for the Preparation of Engineering Assessments", Chemical Engineering Branch, Economics and Technology Division, Office of Toxic Substances, EPA, Washington DC, 20460, 1986.

US EPA Publication, "User's Guide, Emission Control Technologies and Emission Factors for Unpaved Road Fugitive Emissions", EPA/625/5-87/022, September 1987.

US EPA Publication, "Reference Guide to Odor Thresholds for Hazardous Air Pollutants Listed in the Clean Air Act Amendments of 1990", EPA/600/R-92/047, 1992.

US NIOSH, Department of Health Education and Welfare (now HHS), Public Health Service, Center for Disease Control, National Institute of Occupational Safety and Health, "The Industrial Environment - It's Evaluation and Control", US Government Printing Office, Washington DC 20402, 1973.

US NIOSH, Department of Health Education and Welfare (now HHS), Public Health Service, Center for Disease Control, "Occupational Diseases, a Guide to their Recognition", Pub No. 77-181, US Government Printing Office, revised 1977.

US NIOSH, Department of Health and Human Services (formerly HEW), Public Health Service, Centers for Disease Control, National Institute for Occupational Safety and Health, DHHS 81-118, 1981.

US NIOSH, Department of Health and Human Services (formerly HEW), Public Health Service, Centers for Disease Control, National Institute for Occupational Safety and Health, "Registry of Toxic Effects of Chemical Substances", Vols. I, II, and III edited by Tatken, R L and Lewis, R J, 1981-82.

US NIOSH, Department of Health and Human Services (formerly HEW), Public Health Service, Centers for Disease Control, National Institute for Occupational Safety and Health, "NIOSH Pocket Guide to Chemical Hazards", US Government Printing Office, 1985.

US NIOSH, Department of Health and Human Services (formerly HEW), Public Health Service, Centers for Disease Control, National Institute for Occupational Safety and Health, "NIOSH Pocket Guide to Chemical Hazards", US Government Printing Office, 1997.

US NTIS, "Compilation of Air Pollutant Emission Factors", (Including Supplements 1-7), 3rd ed., Supplement 9, NTIS PB81-244097, July 1979.

US NTIS, "Compilation of Air Pollutant Emission Factors", (Including Supplements 1-7), 3rd ed., Supplement 10, NTIS PB80-199045, February 1980a.

US NTIS, "Compilation of Air Pollutant Emission Factors", (Including Supplements 1-7), 3rd ed., Supplement 11, NTIS PB81-178014, October 1980b.

US NTIS, "Compilation of Air Pollutant Emission Factors", (Including Supplements 1-7), 3rd ed., Supplement 12, NTIS PB82-101213, April 1981.

US NTIS, "Compilation of Air Pollutant Emission Factors", (Supplement A), NTIS PB87-150959, Oct 1986.

US Office of Technology Assessment, "Preventing Illness and Injury in the Workplace", 1985.

US Office of the Federal Register, Code of Federal Regulations (CFR), Title 29, Parts 1900 to 1910, (commonly called "OSHA General Industry Standards") Revised as of 1 July 1988.

US Office of the Federal Register, Code of Federal Regulations (CFR) 40 "Protection of Environment", Part 60, Superintendent of Documents, US Government Printing Office, pp 195-1013, 1989.

Valenti, M, "News and Notes" in Mechanical Engineering, Vol. 120, No. 2, p. 10, February, 1999.

Van, N Q and Howell, R H, "Influence of Initial Turbulent Intensity On the Development of Plane Air Curtain Jets", ASHRAE Transactions, Vol. 85, Part 1, pp 208-228, 1976.

Vargaftik, N B, "Tables On the Thermophysical Properties of Liquids and Gases", 2nd ed., Hemisphere Publishing Corp, Washington, DC, 1975.

Vatavuk, W M, "OAQPS Control Cost Manual", 4th ed., EPA 450/3-90-006, Office of Air Quality Planning and Standards, Research triangle Park, NC, 1990.

Vedal, S, "Ambient Particles and Health: Lines that Divide", Journal of the Air & Waste Management Association, Vol. 47, pp 551-581, 1997.

Verschueren, K, "Handbook of Environmental Data on Organic Chemicals" 2nd ed., Van Nostrand Reinhold Co, New York, 1983.

Viscusi, W K, "Fatal Tradeoffs: Public and Private Responsibilities for Risk", Oxford University Press, New York, 1992.

Visser, G T, "A Wind-Tunnel Study of the Dust Emissions from the Continuous Dumping of Coal", Atmospheric Environment, Vol. 26A, No. 8, pp 1453-1460, 1992.

Wadden, R A and Scheff, P A, "Engineering Design for the Control of Workplace Hazards", McGraw-Hill Book Company, New York, 1987.

Wadden, R A and Scheff, P A, "Indoor Air Pollution", Wiley-Interscience Publication, John Wiley & Sons, New York, 1983.

Wark, K, Warner, C F, and Davis, W T, "Air Pollution: Its Origin and Control", Addison Wesley, Reading, MA, 1998.

Waters, M A, Selvin, S, and Pappaport, S M, "A Measure of Goodness-of-Fit for the Lognormal Model Applied to Occupational Exposures", American Industrial Hygiene Association Journal, Vol. 52, No. 11, pp 493-502, 1991.

Weibel, E, "Morphology of the Lung", Academic Press, New York, 1963.

Weinstein, A S, "Products Liability and the Reasonably Safe Product: a Guide for Management, Design, and Marketing", New York, John Wiley and Sons, 1978.

Weschler, C J, Shields, H C, and Rainer, D, "Concentrations of Volatile Organic Compounds at a Building with Health and Comfort Complaints", American Industrial Hygiene Association Journal, Vol. 51, No. 5, pp 261-268, 1990.

West, J B, "Respiratory Physiology - the Essentials", Williams and Wilkins Co, Baltimore, MD, 1974.

Whitby, K T, Anderson, G R, and Rubow, K L, "Dynamic Method for Evaluating Room-Size Air Purifiers", Transactions of American Society of Heating and Air Conditioning Engineers, paper No. 2771, Part 2A, pp 172-184, 1983.

Whitby, K T, Charlson, R E, Wilson, W E, and Stevens, R K, "The Size of Particle Matter in Air", Science, Vol. 183, pp 1098-1100, 1974.

White, F M, "Fluid Mechanics", ed. 4, McGraw-Hill, New York, 1999.

White, H T, "Industrial Electrostatic Precipitation", Addison-Wesley, 1963.

White, M C, Infante, P F, and Chu, K C, "A Quantitative Estimate of Leukemia Mortality Associated with Occupational Exposure to Benzene", Journal for the Society of Risk Analysis, Vol. 2, No3, pp 195-204, 1982.

Wilcox, D C, "Turbulence Modeling for CFD", DCW Industries, Inc., La Canada, CA, 1998.

Willeke, K and Baron, P A (editors), "Aerosol Measurement", Van Nostrand and Reinhold, New York, 1993.

Williamson, S J, "Fundamentals of Air Pollution", Addison and Wesley, Reading MA, 1973.

Wilson, R and Crouch, E A C, "Risk Assessment and Comparisons: An Introduction", Science Vol. 236, pp 267-270, 17 April 1987.

Woodson, T T, "Introduction to Engineering Design", McGraw Hill, New York, 1966.

Xu, G B and Yu, C P, "Deposition of Diesel Exhaust Particles in Mammalian Lungs", Journal of Aerosol Science and Technology, Vol. 7, pp 117-123, 1987.

Xu, M, Nematollahi, M, Sextro, R G, Gadgil, A J, and Nazaroff, W W, "Deposition of Tobacco Smoke Particles in a Low Ventilation Room", Aerosol Science and Technology, Vol. 20, pp 194-206, 1994.

Yang, K T and Lloyd, J R, "Turbulent Buoyant Flow in Vented Simple and Complex Enclosures", contained in "Natural Convection, Fundamentals and Applications", edited by Kakac, S, Aung, W, and Viskanta R, Hemisphere Publishing Corp., New York, pp 303-329, 1985.

Yaws, C L, Hopper, J R, Wang, X, and Rathinsamy, A K, "Calculating Solubility & Henry's Law Constants for Gases in Water", Chemical Engineering, McGraw Hill, New York, Vol. 106, No. 6, pp 102-105, June 1999.

Yaws, C, Yang, H C, and Pan, X, "Henry's Law Constants for 362 Organic Compounds in Water", Chemical Engineering, Vol. 98, No. 11, pp 179-185, 1991.

Ye, Y and Pui, D Y H, "Particle Deposition in a Tube with an Abrupt Contraction", J. of Aerosol Science, Vol. 21, No. 1, pp 29-40, 1990.

Yousefi, V and Annegarn, H J, "Aerodynamic Aspects of Exhaust Ventilation", "Ventilation '91", edited by R T Hughes, H D Goodfellow, and G S Rajhans, ACGIH, Cincinnati, OH, pp 413-425, 1993.

Yu, C P and Xu, G B, "Predicted Deposition of Diesel Particles in Young Humans", Journal of Aerosol Science, Vol. 18, No. 4, pp 419-429, 1987.

Yu, J-W and Neretnieks, I, "Single-Component and Multicomponent Adsorption Equilibria on Activated Carbon of Methylcyclohexane, Toluene, and Isobutyl Methyl Ketone", Industrial Engineering Chemistry Research, Vol. 29, No. 2, pp 220-231, 1990.

Zarouri, M D, Heinsohn, R J, and Merkle, C L, "Computer-Aided Design of a Grinding Booth for Large Castings", Paper No. 2767, ASHRAE Transactions, Part 2A, pp 95-118, 1983.

Index

B

C

D

F

G

H

M

N

O

Q

R

S

U

V

W